Applied Linear Algebra

Peter J. Olver
University of Minnesota

Chehrzad Shakiban
University of St. Thomas

Upper Saddle River, NJ 07458

Library of Congress Cataloging-in-Publication Data
Olver, Peter J.
Applied linear algebra / Peter J. Olver, Chehrzad Shakiban.
1st ed. / Peter J. Olver
p. cm.
Includes bibliographical references and index,
ISBN 0-13-147382-4
1. Algebra, Linear 1. Title
CIP data available

Executive Acquisitions Editor: *George Lobell*
Editor-in-Chief: *Sally Yagan*
Production Editor: *Debbie Ryan*
Senior Managing Editor: *Linda Mihatov Behrens*
Assistant Managing Editor: *Bayani Mendoza de Leon*
Executive Managing Editor: *Kathleen Schiaparelli*
Assistant Manufacturing Manager/Buyer: *Michael Bell*
Marketing Manager: *Halee Dinsey*
Marketing Assistant: *Joon Moon*
Interior & Cover Designer: *Kristine Carney*
Art Director: *Jonathan Boylan*
Director of Creative Services: *Paul Belfanti*
Editorial Assistant: *Jennifer Urban*
Cover Image: *Lara Swimmer/Esto. All rights reserved*

© 2006 Pearson Education, Inc.
Pearson Prentice Hall
Pearson Education, Inc.
Upper Saddle River, NJ 07458

All rights reserved. No part of this book may
be reproduced, in any form or by any means,
without permission in writing from the publisher.

Pearson Prentice Hall™ is a trademark of Pearson Education, Inc.

Printed in the United States of America

10 9 8 7 6 5 4 3 2 1

ISBN 0-13-147382-4

Pearson Education Ltd., *London*
Pearson Education Australia Pty, Limited, *Sydney*
Pearson Education Singapore, Pte. Ltd.
Pearson Education North Asia Ltd., *Hong Kong*
Pearson Education Canada, Ltd., *Toronto*
Pearson Educacion de Mexico, S.A. de C.V.
Pearson Education – Japan, *Tokyo*
Pearson Education Malaysia, Pte. Ltd.

To our children, Parizad, Sheehan, and Noreen.
You are the light of our life.

Contents

Preface

Applied mathematics rests on two central pillars: calculus and linear algebra. While calculus has its roots in the universal laws of Newtonian physics, linear algebra arises from a much more mundane issue: the need to solve simple systems of linear algebraic equations. Despite its humble origins, linear algebra ends up playing a comparably profound role in both applied and theoretical mathematics, as well as in all of science and engineering, computer science, probability and statistics, economics, numerical analysis, mathematical biology, and many other disciplines. Nowadays, a proper grounding in both calculus and linear algebra is an essential prerequisite for a successful career in science, engineering, and mathematics.

Since Newton, and, to an even greater extent after Einstein, modern science has been confronted with the inherent nonlinearity of the macroscopic universe. But most of our insight and progress is based on linear approximations. Moreover, at the atomic level, quantum mechanics remains an inherently linear theory. (The complete reconciliation of linear quantum theory with the nonlinear relativistic universe remains the holy grail of modern physics.) Only with the advent of large scale computers have we been able to begin to investigate the full complexity of natural phenomena. But computers rely on numerical algorithms, and these in turn require manipulating and solving systems of algebraic equations. Now, rather than just a handful of equations, we may be confronted by gigantic systems containing thousands (or even millions) of unknowns. Without the discipline of linear algebra to formulate systematic, efficient solution algorithms, as well as the consequent insight into how to proceed when the numerical solution is insufficiently accurate, we would be unable to make progress in the linear regime, let alone make sense of the truly nonlinear physical universe.

Linear algebra can thus be viewed as the mathematical apparatus needed to solve potentially huge linear systems, to understand their underlying structure, and to apply what is learned in other contexts. The term "linear" is the key, and, in fact, refers not just to linear algebraic equations, but also linear differential equations, both ordinary and partial, linear boundary value problems, linear integral equations, linear iterative systems, linear control systems, and so on. It is a profound truth that, while outwardly different, all linear systems are remarkably similar at their core. Basic mathematical principles such as linear superposition, the interplay between homogeneous and inhomogeneous systems, the Fredholm alternative characterizing solvability, orthogonality, positive definiteness and minimization principles, and the role of eigenvalues in dynamics, to name but a few, reoccur in many ostensibly unrelated contexts.

In the late nineteenth and early twentieth centuries, mathematicians came to the realization that all of these disparate techniques could be subsumed in the edifice now known as linear algebra. Understanding, and, more importantly, exploiting the apparent similarities between, say, algebraic equations and differential equations, requires us to become more sophisticated, that is, more abstract, in our mode of thinking. The abstraction process distills the essence of the problem away from all its distracting particularities, and, seen in this light, all linear systems rest on a common mathematical framework. Don't be afraid! Abstraction is not new in your mathematical education. In elementary algebra, you already learned to deal with variables, which are the abstraction of numbers. Later, the abstract concept of a function formalized particular relations between variables, say distance and time, or energy, mass and velocity. In linear algebra, the abstraction is raised to yet a further level, in that one views apparently different types of objects (vectors, matrices, functions, ...) and systems (algebraic, differential, integral, ...) in a common conceptual framework. (And this is by no means the end of the mathematical abstraction process; modern category theory, [29], abstractly unites different conceptual frameworks.)

In applied mathematics, we do not introduce abstraction for its intrinsic beauty. Our ultimate purpose is to develop effective methods and algorithms for applications in science, engineering, computing, statistics, etc. For us, abstraction is driven by the need to understand, and is only justified if it aids in the solution to real world problems. Whereas to the beginning student the initial concepts may seem designed merely to bewilder and confuse, one must reserve judgment until genuine applications appear. Patience and perseverance are vital. Once we have acquired some familiarity with basic linear algebra, significant, interesting applications will be readily forthcoming. In this text, we encounter graph theory, mechanical structures, electrical circuits, thermodynamics, quantum mechanics, the geometry underlying computer graphics, animation, and games, signal and image processing, interpolation and approximation of functions and data, statistical analysis, dynamical systems modeled by linear differential equations, vibrations, resonance and damping, probability and stochastic processes, splines and modern font design, bars, beams and other continua, and numerical solution techniques for linear algebraic equations and boundary value problems, to name a few. Further applications of the material you learn here will appear throughout your mathematical and scientific career.

This textbook has two inter-related pedagogical goals. The first is to explain basic techniques that are used in modern, real world problems of the type you will encounter in engineering, science, numerical analysis, and mathematics itself. But we have not written a mere mathematical cookbook—a collection of linear algebraic recipes and algorithms. We believe that is important for the applied mathematician, as well as the scientist and engineer, to not just learn mathematical techniques and how to apply them in a variety of settings, but, even more importantly, to understand why they work and how they are derived from first principles. In our approach, applications go hand-in-hand with theory, each reinforcing and explaining the other. To this end, we try to lead the reader through the reasoning that leads to the important results. We do not shy away from stating theorems and writing out proofs, particularly when they lead to insight into the methods and their range of applicability. We hope to spark that eureka moment, when you realize "Yes, of course! I could have come up with that if I'd only sat down and thought it out." Most concepts in linear algebra are not all that difficult at their core, and, by grasping their essence, you will not only know how to apply them in routine contexts, you will understand what is required to adapt to unusual or recalcitrant problems. And, the further you go on in your studies or work, the more you realize that very few real world problems fit neatly into the idealized framework outlined in a textbook. So it is (applied) mathematical reasoning and not mere linear algebraic technique that is the core and raison d'etre of this text!

Applied mathematics can be broadly divided into three mutually reinforcing components. The first is modeling—how one derives the governing equations from physical principles. The second is solution techniques and algorithms—methods for solving the model equations. The third, perhaps least appreciated but in many ways most important, are the frameworks that incorporate disparate analytical methods into a few broad themes. The key paradigms of applied linear algebra to be covered in this text include

- Gaussian Elimination and factorization of matrices,
- linearity and linear superposition,
- span, linear independence, and basis,
- inner products and norms,
- compatibility of linear systems via the Fredholm alternative,
- positive definiteness and minimization principles,
- orthonormality and the Gram–Schmidt process,
- least squares solutions, interpolation, and approximation
- eigenvalues and eigenvectors/eigenfunctions,
- symmetry, self-adjointness, and orthonormal bases,
- singular values and condition number,
- vibrations, quasi-periodicity, damping and resonance,
- iteration, including Markov processes and numerical solution methods,
- finite element approximations by restriction to finite-dimensional subspaces.

As you will see, these are all interconnected parts of a very general applied mathematical structure of remarkable power and practicality. Understanding such broad themes of applied mathematics is our overarching objective.

Linear algebra is of importance in its own right, but also motivates solution techniques for linear differential equations. Traditionally, initial value problems for linear ordinary differential equations form an integral part of the course, inspiring and applying eigenvalue methods. But it is our contention that linear boundary value problems are where one really starts to appreciate the power of abstract linear algebra techniques. For this reason, we have included as a final chapter the case of one-dimensional boundary value problems described by linear ordinary differential equations. Students who finish this material will, we hope, be motivated to continue their studies in applied mathematics. The next topic is the dynamics of continuous media, modeled by linear partial differential equations, leading immediately, by direct analogy with the matrix eigenvalue solution method, to Fourier series. And then on to equilibrium and dynamics in higher dimensions, and, finally, nonlinear methods. But this must all be left for another course and another textbook.

Indeed, this book began life as a part of a much larger work, [47], whose goal is to similarly cover the full range of modern applied mathematics, both linear and nonlinear, at an advanced undergraduate level. Our inspirational source was and continues to be the visionary texts of Gil Strang, [59, 60]. Based on our students' reactions, our goal has been to present a more linearly ordered and less ambitious development of the subject, while retaining the excitement and interconnectedness of theory and applications that is evident in Strang's works. Stay tuned for updates on the impending completion of our applied mathematics book.

COURSES AND PREREQUISITES

This book is designed for three possible audiences:

- A beginning, in depth course in the fundamentals of linear algebra and its applications for highly motivated and mathematically mature students.
- A second undergraduate course in linear algebra, with an emphasis on those methods and topics which are important in applications.
- A beginning graduate level course in linear mathematics for students in engineering, physical science, computer science, numerical methods, statistics, and even mathematical biology, finance, and economics, as well as master's students in applied mathematics.

Although most students will have already encountered some basic linear algebra —matrices, vectors, linear systems, etc.—the text makes no such assumptions. Indeed, the first chapter starts at the very beginning by introducing algebraic linear systems, matrices and vectors, followed by very basic Gaussian elimination. We do assume that the reader has taken a standard two year calculus sequence. Very basic one-variable calculus—derivatives and integrals—will be used without comment. Multivariable calculus will only appear fleetingly and in an inessential way. The ability to handle scalar, linear constant coefficient ordinary differential equations is also assumed, although we do briefly review the basic techniques in Chapter 7. Proofs by induction will be used on occasion. But the most essential prerequisite is a certain degree of mathematical maturity and willingness to handle the increased level of abstraction that lies at the heart of contemporary applied mathematics.

SURVEY OF TOPICS

Besides the basics of matrices, vectors and Gaussian elimination, the first chapter includes some of the less familiar topics from linear systems theory, particularly the (permuted) LU decomposition. We also discuss some of the practical numerical issues underlying the solution algorithms, including the computational efficiency of Gaussian elimination coupled with Back Substitution as contrasted with methods based on the inverse matrix, as well as using pivoting to mitigate the effects of round-off error. Because our goal is learning practical algorithms used in real applications, matrix inverse and determinants do not receive undue emphasis—indeed, the most efficient way to compute a determinant is by Gaussian elimination, which remains *the* algorithm throughout the first portion of the text.

Chapter 2 is the heart of linear algebra, and a successful course rests on the students' ability to assimilate the absolutely essential concepts of vector space, subspace, span, linear independence, basis, and dimension. While these ideas may have been encountered in an introductory ordinary differential equation course, it is rare, in our experience, that students at this level are at all comfortable with them. The underlying mathematics is not particularly difficult, but enabling the students to adequately grasp the level of abstraction remains the most challenging part of the course. To this end, we have included a wide range of illustrative examples; of particular note is the connection between functions and their sample vectors. Students should start by making sure they understand how a concept applies to vectors in Euclidean space $\mathbb{R}^n$ before pressing on to function space and other situations. While one could design a course that, at least until the final chapter, completely avoids infinite-dimensional function spaces, we are convinced that, at this level, they should be integrated into the subject right from the start. Indeed, modern linear analysis and applied mathematics, including Fourier methods, boundary value problems, partial differential equations, numerical solution techniques, signal processing, control theory, modern physics, quantum chemistry, and many, many others, all essentially rely on basic vector space notions, and so learning to deal with the full range of examples is the secret to future success. In the final two sections, we concentrate on the fundamental subspaces associated with a matrix—kernel or null space, range or column space, corange or row space and cokernel or left null space. (We prefer the former terms since they naturally carry over to the function space context.) The role of these spaces in the characterization of solutions to linear systems, e.g., the basic superposition principles, is emphasized.

Chapter 3 introduces norms and inner products. Again, we develop both the finite-dimensional and function space cases in tandem. The classification of inner products on Euclidean space is used to motivate the class of positive definite matrices, which reappear throughout the text. Gram matrices, constructed out of inner products of elements of inner product spaces, are a particularly fruitful source of positive definite and semi-definite matrices. Tests for positive definiteness rely on Gaussian elimination and the connections between the $L D L^T$ factorization of symmetric matrices and the process of completing the square in a quadratic form. We have deferred treating complex vector spaces until this chapter, since only the definition of an inner product is not an evident adaptation of the real version.

Chapter 4 is devoted to solving the most basic multivariable minimization problem: a quadratic function of several variables. The solution is reduced, by a purely algebraic computation, to a linear system, and then solved in practice by Gaussian elimination. Applications include finding the closest element of a subspace to a given point, which is reinterpreted as the least squares solution to an incompatible linear system. Polynomial (and more general) interpolation and least squares approximation of data and functions are handled in a common framework.

Chapter 5 exploits the numerous advantages of orthogonality in a wide range of applied problems. Use of adapted orthogonal bases creates a dramatic speed-up in basic computational algorithms throughout mathematics and its applications, such as the least squares minimization problem whose solution corresponds to orthogonal projection onto a subspace. The orthogonality of the fundamental matrix subspaces leads to a linear algebraic version of the Fredholm alternative for compatibility of linear systems. We develop several versions of the basic Gram–Schmidt process for converting an arbitrary basis into an orthogonal basis, and use it to construct orthogonal polynomials and functions. When implemented on bases of $\mathbb{R}^n$, the algorithm becomes the $Q R$ factorization of a nonsingular matrix. The final section uses (complex) orthogonal bases to develop the discrete Fourier representation of a sampled signal, culminating in the Fast Fourier Transform, with applications to denoising and data compression.

Chapter 6 is devoted to certain striking applications of the preceding developments. We introduce a general mathematical framework that incorporates a wide range of equilibrium problems in mechanics and electrical circuits. We start with simple mass-spring chains, followed by electrical circuits, and finish by analyzing the equilibrium configurations and instabilities of general structures. These applications serve as an excellent illustration of the powerful linear algebra techniques that have been developed in the preceding chapters. In the final chapter, the function space version of this framework will play a key role in the analysis of boundary value problems arising in continuum mechanics.

Chapter 7 turns to the general foundations of linear algebra, and includes applications in geometry and graphics. We will see that matrices are a particular instance of the general concept of a linear function, which also includes linear differential operators, linear integral operators, and so on. Basic facts about linear systems, e.g., linear superposition, that were already established in the algebraic context are shown to be of very general applicability, including linear ordinary differential equations and boundary value problems. Linear functions on Euclidean space represent basic geometrical transformations, and so play a key role in modern computer graphics, design, animation, and gaming, and the study of symmetry.

Chapters 8–10 deal with eigenvalues and their applications in linear dynamical systems, both continuous and discrete. After motivating the definition of eigenvalue and eigenvector by the need to solve linear systems of ordinary differential equations, the remainder of Chapter 8 explores the basic theory and a range of applications, including eigenvector bases, diagonalization, singular values, the Schur decomposition, and the Jordan canonical form. Practical computational schemes for eigenvalues and eigenvectors are postponed until Chapter 10.

Chapter 9 applies eigenvalue methods to solve and study linear systems of ordinary differential equations. After completing the development of basic solution techniques begun in the preceding chapter, the focus shifts to understanding the qualitative properties of solutions and particularly the role of eigenvalues in the stability of equilibria. The two-dimensional case is discussed in full detail, culminating in a complete classification of the possible phase portraits and stability properties. Matrix exponentials are introduced as an alternative means for solving first order homogeneous systems, and applied to the inhomogeneous version, and in geometry and computer graphics. Our focus then shifts to second order linear systems, which model dynamical motions and vibrations in mechanical structures and electrical circuits. In the absence of frictional damping and instabilities, solutions are quasiperiodic combinations of normal modes. The effects of damping, and of periodic forcing and its potentially catastrophic role in resonance, are discussed in detail.

Chapter 10 employs eigenvalues to analyze discrete dynamics, as governed by linear iterative systems or powers of matrices. The formulation of stability properties are based on the spectral radius, the Gerschgorin circle theorem and matrix norms. Applications to Markov chains arising in probabilistic and stochastic processes are developed in some detail. As practical alternatives to Gaussian elimination, we discuss basic iterative solution methods for linear systems, including the Jacobi, Gauss–Seidel and SOR schemes, and, finally, the semi-direct method of conjugate gradients. The chapter concludes with basic iterative methods for computing eigenvalues and eigenvectors: the power method and the $Q\,R$ algorithm, including a new proof of its convergence.

The final Chapter 11 is not as traditional for a linear algebra text. Its ostensible purpose is to solve simple boundary value problems arising in the equilibrium mechanics of one-dimensional continuous media, including stretching of bars and bending of beams. More importantly, it provides the bridge between the finite-dimensional matrix systems modeling the equilibrium of discrete structures and the infinite-dimensional function space methods arising in continuum mechanics—elasticity, fluid mechanics, electromagnetism, thermomechanics, and so on. The same abstract equilibrium framework reappears in a more powerful, infinite-dimensional guise. A crucial leap in mathematical sophistication changes the finite-dimensional operation of transposing a matrix into the adjoint of a linear differential operator, and minimizing a quadratic function into minimization of a quadratic functional. Finite-dimensional linear algebraic structures and solution methods serve as our guide in the more advanced infinite-dimensional universe. They directly inspire the calculus of generalized functions, including the delta function and Green's functions, as a powerful new solution paradigm. The chapter concludes with a presentation of the basics of the finite element numerical solution method for positive definite boundary value problems, which makes crucial use of the associated minimization principle. In this way, the chapter serves to both summarize and reinforce the earlier material, and as a launching pad into the rapidly expanding universe that is modern applied mathematics.

COURSE OUTLINES

The text includes far more material than can be comfortably covered in a single semester. A full year's course in linear algebra would be able to do it justice. If you do not have this luxury, several possible semester and quarter courses can be extracted from the wealth of material and applications.

First, let us describe a few thematic threads that can be extracted from the text

- The core: 1.1–4, 1.5 (omit Gauss–Jordan), 1.6, 1.8–9, 2.1–5, 3.1–2, 3.4, 3.5 (first subsection), 4.1–2, 5.1, 5.2 (first subsection), 5.3 (omit Householder's Method), 5.5 (first subsection), 5.6, 8.1–3, 8.4 (first subsection).
- Least squares, interpolation, and singular values: 4.3–4, 5.4–5, 8.5, 11.4 (splines).
- Linear systems of differential equations: 8.6 (just the Jordan canonical form—optional), 9.1–4.
- Structures and circuits: 2.6, 6.1–3, 9.1–2, 9.5–6.
- Numerical methods: 1.7, 3.3, 5.3, then selected topics from 8.5, 8.6 (the Schur decomposition), 10.1–3, 10.5–6, 11.6 (requires earlier material in Chapter 11).
- Markov processes: 10.1–2, 10.3 (the Gerschgorin Theorem), 10.4.
- Discrete Fourier methods: 3.6, 5.7.
- Geometry and graphics: 5.3 (first subsection), 5.5 (first subsection), 7.1–3, 9.4.
- Boundary value problems: 11.1–5.

The core constitutes the essential material to appear in any course. Elementary topics (e.g. matrices, vectors, determinants) can be omitted or covered rapidly if the class has already seen them.

For a first course, we recommend covering the core, and, if time permits, one other thread, our preference being the material on structures and circuits. One option for saving time is to concentrate on finite-dimensional vector spaces, leaving the function space material for a later course.

For a second course in linear algebra, the students are presumably already familiar with basic matrix methods such as Gaussian Elimination, determinants, inverses, eigenvalues and, perhaps, first order systems of ordinary differential equations. Thus, most of chapter 1, except perhaps the matrix factorizations, should be reviewed at a rapid pace. On the other hand, the more abstract fundamentals, including vector spaces, span, linear independence and basis, are still not fully mastered, and one should expect to spend a significant fraction of the early part of the course covering these essential topics in detail. While the dot product should be familiar, general inner products will also need to be developed in detail, although more general norms can be omitted if time is an issue. Beyond the core material, there should be time for most of the structures and circuits thread, particularly if the students do not need to see the elementary material on linear systems of differential equations, and the least squares thread also, although one may wish to substitute one of the other threads depending on the audience and interest of the instructor.

Similar considerations hold for a beginning graduate level course for scientists and engineers. Here, the emphasis should be on applications required by the students, and function space methods should be firmly built into the class. As before, vector spaces, bases, etc. are still essential, and the key to success is the students' assimilation of the material in Chapter 2.

COMMENTS ON INDIVIDUAL CHAPTERS

Chapter 1: Presuming the students have already seen matrices, vectors and Gaussian elimination, most of this material will be review and should be covered at a fairly rapid pace. The LU decomposition and the emphasis on Forward and Back Substitution, while de-emphasizing matrix inverses, might be new. Sections 1.7 and (presuming students have familiarity with the basics of determinants from an earlier course) 1.9 can be safely omitted.

Chapter 2: The crux of the course. A key decision is whether to include infinite-dimensional vector spaces throughout the course, as is done in the text, or to do an abbreviated version that only covers finite-dimensional spaces, or, even more restrictively, only $\mathbb{R}^n$. Our recommendation is to include functions spaces from the start, but this can interfere with getting to some of the more advanced matrix algorithms and results in a semester course. The last section, on graph theory, can be skipped unless you plan on covering the second and third sections of Chapter 6.

Chapter 3: Inner products are essential, but, under time constraints, one can omit Section 3.3 on more general norms, as they only start to play an essential role in Chapter 10. Section 3.6, on complex vector spaces, is mainly required in Section 5.7 on the discrete Fourier transform.

Chapter 4: The solution of quadratic minimization problems are a key component of the course. Least squares, interpolation, and approximation of data are more optional, depending on the audience, although most would argue that this course is where they should be learned.

Chapter 5: The basics of orthogonality, as covered in Sections 5.1–3, is an essential part of the students' knowledge. Which of the other sections of this chapter to include depends on the taste of the instructor. Our own suggestion is that, if a choice needs to be made, the final section, which is used in the applications in Chapters 6 and 11, should be of higher priority.

Chapter 6 provides a welcome relief from the theory for the more applied students in the class, and is one of our favorite parts to teach. While it may be omitted in an abbreviated course, the material is particularly appealing for a class with engineering students.

Chapter 7: This material is the most abstract of the book, and could be entirely omitted if time is an issue. If you do have some time, our suggestion is to cover Sections 7.1–2 to introduce linear transformations and some applications in geometry.

Chapter 8: Eigenvalues are absolutely essential. The motivational material on systems of differential equations in Section 8.1 could be postponed until the following chapter, when solution methods are begun in earnest. Sections 8.2 and 8.3 are the heart of the matter. Of the remaining three sections, the material on symmetric matrices should have the highest priority, while singular values are of increasing importance in contemporary applications.

Chapter 9 studies linear ordinary differential equations, and justifies the material on eigenvalues. In a semester course, time is now becoming an issue, and perhaps only the first 2 or 3 sections can be covered in any depth. Sections 9.4 and, to a lesser extent, 9.5 rely on the material in Section 6.1, at least.

Chapter 10: If time permits, the first couple of sections are well worth covering. For a numerically oriented class, Section 10.5 would be a priority, whereas 10.4 contains appealing probabilistic and stochastic applications.

Chapter 11: The final chapter is the launching pad into linear and nonlinear analysis. See the earlier discussion for how to motivate this material, and where it is pointing towards. Probably only for an extended course, but this is one of our favorite chapters, as it finally reveals the true range of linear algebra, reinforcing and justifying all of the finite-dimensional constructions that appeared before.

EXERCISES AND SOFTWARE

Exercises appear at the end of almost every subsection, and come in a wide variety of flavors. Each exercise set starts with some straightforward computational problems to test and reinforce the new techniques and ideas. Ability to solve these basic problems is a minimal requirement for successfully learning the material. More advanced and more theoretical exercises appear later on in the set. Some are routine, but others are challenging computational problems, computer-based exercises and projects, details of proofs that were not given in the text, additional practical and theoretical results of interest, further developments in the subject, etc. some will challenge even the most advanced student. When time is an issue, don't be afraid to only assign a couple of parts of a multi-part exercise. We have found the True/False exercises to be a particularly useful indicator of a student's level of understanding. Emphasize to the students that a full answer is not merely a T or F, but must include a detailed explanation of the reason, e.g., a proof or a counterexample, a reference to a result in the text, etc.

As a guide, some of the exercises are marked with special signs:

◇ indicates an exercise that is used at some point in the text, or is important for further development of the subject.

♡ indicates a project—usually an exercise with multiple interdependent parts.

♠ indicates an exercise that requires (or at least strongly recommends) use of a computer. The student could either be asked to write their own computer code in, say, MATLAB , MAPLE , or MATHEMATICA , or make use of pre-existing software packages.

♣ = ♠ + ♡ indicates a computer project.

The book's dedicated web site

`http://www.math.umn.edu/~olver/ala.html`

will contain a list of known errors, commentary, feedback, and resources, as well as a number of illustrative MATLAB programs that we've used when teaching the course. A goal for a subsequent edition is to more directly incorporate the software in the text.

CONVENTIONS AND NOTATIONS

Equations are numbered consecutively within chapters, so that, for example, (3.12) refers to the 12th equation in Chapter 3.

Theorems, Lemmas, Propositions, Definitions, and Examples are also numbered consecutively within each chapter, using a common index. Thus, in Chapter 1, Lemma 1.2 follows Definition 1.1, and precedes Theorem 1.3 and Example 1.4. We find this numbering system to be the most helpful for navigating through the book.

References (books, papers, etc.) are listed alphabetically at the end of the text, and are referred to by number. Thus, [47] is the 47th listed reference, which happens to be our forthcoming applied mathematics text.

■ indicates the end of a proof.

● indicates the end of an example.

$\mathbb{R}, \mathbb{C}, \mathbb{Z}, \mathbb{Q}$ denote, respectively, the real numbers, the complex numbers, the integers, and the rational numbers.

Modular arithmetic: $j = k \bmod n$ is the unique number $0 \le j < n$ such that $k = j + m\,n$ for some integer m. If k, n are both positive integers, then j is the remainder upon dividing k by n.

$\displaystyle\sum_{i=1}^{n} a_i = a_1 + a_2 + \cdots + a_n$ and $\displaystyle\prod_{i=1}^{n} a_i = a_1 a_2 \cdots a_n$ are standard notations for the sum and product of the quantities $a_1, \ldots, a_n$.

We consistently use bold face lower letters, e.g., $\mathbf{v}$, $\mathbf{x}$, $\mathbf{a}$, to denote vectors (almost always column vectors), whose entries are indicated by subscripts v_1, x_i, etc. Matrices are denoted by ordinary capital letters, e.g., A, C, K, M—but not all such letters refer to matrices; for instance, V sometimes refers to a vector space. The entries of a matrix, say A, are indicated by the corresponding subscripted lower case letters, a_{ij}.

We use $S = \{\, f \mid C \,\}$ to denote a set, where f is a formula for the members of the set and C is a list of conditions. For example, $\{\, x \mid 0 \le x \le 1 \,\}$ means the closed unit interval from 0 to 1, also denoted $[\,0, 1\,]$, while $\{\, a\,x^2 + b\,x + c \mid a, b, c \in \mathbb{R} \,\}$ is the set of real quadratic polynomials, and $\{0\}$ is the set consisting only of the number 0. We use $x \in S$ to indicate that x is an element of the set S, while $y \notin S$ indicates that y is not an element.

The subset sign $A \subset B$ includes the possibility that the sets might be equal, although for emphasis we sometimes write $A \subseteq B$. On the other hand, $A \subsetneq B$ specifically implies that the two sets are not equal. We can also write $A \subset B$ as $B \supset A$. We use $B \setminus A = \{\, x \mid x \in B, x \notin A \,\}$ to denote the set theoretic difference, meaning all elements of B that do not belong to the subset A.

An arrow $\to$ is used in two senses: first, to indicate convergence of a sequence: $x_n \to x^\star$ as $n \to \infty$; second, to indicate a function, so $f\colon X \to Y$ means that f defines a function from the domain set X to the image or target set Y.

We use $\equiv$ to emphasize when two functions agree everywhere, so $f(x) \equiv 1$ means that f is the constant function, equal to 1 at all values of x.

Angles are always measured in radians (although occasionally degrees will be mentioned in descriptive sentences). All trigonometric functions are evaluated on radians. (Make sure your calculator is locked in radian mode!)

We decided to use $\operatorname{ph} z$ for the phase of a complex number, also known as its "argument". We prefer this to the more common $\arg z$, because arg is also used (but not in this text) to refer to the argument of a function $f(z)$, while "phase" is completely unambiguous.

We always use $\log x$ for the natural (base e) logarithm (never the ugly modern version $\ln x$) while $\log_a x = \log x / \log a$ is used for logarithms with base a.

We will employ a variety of standard notations for derivatives. In the case of ordinary derivatives, the most basic is the Leibnizian notation $\dfrac{du}{dx}$ for the derivative of u with respect to x. An alternative is the Newtonian notation u'. Higher order derivatives are similar, with u'' denoting $\dfrac{d^2u}{dx^2}$, while $u^{(n)}$ denotes the nth order derivative $\dfrac{d^n u}{dx^n}$. If the function depends on time, t, instead of space, x, then we use dots, $\dot{u}$, $\ddot{u}$, instead of primes. We use the full Leibniz notation

$$\frac{\partial u}{\partial x}, \quad \frac{\partial u}{\partial t}, \quad \frac{\partial^2 u}{\partial x^2}, \quad \frac{\partial^2 u}{\partial x \, \partial t},$$

for partial derivatives of functions of several variables. Unless specifically indicated, all functions are assumed to be sufficiently smooth that any indicated derivatives exist and mixed partial derivatives are equal, cf. [2].

Definite integrals are denoted by $\int_0^1 f(x)\,dx$, while $\int f(x)\,dx$ is the corresponding indefinite integral or anti-derivative.

A full notation index can be found at the end of the book.

HISTORICAL BACKGROUND

Mathematics is both a historical and a social activity, and many of the algorithms, theorems and formulas are named after famous (and, on occasion, not-so-famous) mathematicians and scientists—usually, but not necessarily, the one(s) who first came up with the idea. In the text, we do indicate first names, approximate dates, and geographic locations of most of the named contributors. Readers who are interested in additional historical details, complete biographies, and, when available, portraits or photos, are urged to consult the wonderful University of St. Andrews web site:

`http://www-history.mcs.st-andrews.ac.uk/history/index.html`

SOME FINAL REMARKS

To the student: You are about to learn modern applied linear algebra. We hope you enjoy the experience and profit from it in your future studies and career. Please send us your comments. Did you find our explanations helpful or confusing? Were enough examples included in the text? Were the exercises of sufficient variety and appropriate level to enable you to learn the material? Do you have suggestions for improvements?

To the instructor: Thank you for adopting our text! We hope you enjoy teaching from it as much as we enjoyed writing it. Whatever your experience, we want to hear from you. Let us know which parts you liked and which you didn't. Which sections worked and which were less successful. Which parts your students enjoyed, which parts they struggled with, and which parts they disliked. How can we improve it?

Like every author, we sincerely hope that we have eliminated all errors in the text. But, from experience, we know that no matter how many times you proofread, mistakes still manage to squeeze through (or, worse, be generated during the editing process). So, particularly for this first edition, we ask your indulgence to overlook the few (we hope) that remain. Even better, email us with your questions, typos, mathematical errors, comments, suggestions, and so on.

ACKNOWLEDGMENTS

First, let us express our profound gratitude to Gil Strang for his continued encouragement from the very beginning of this undertaking. Readers familiar with his groundbreaking texts and remarkable insight can readily find his influence throughout our book. We would like to thank Mikhail Shvartsman for helping us with the arduous task of writing out the solutions to the exercises. We thank Pavel Belik, Tim Garoni, Markus Keel, Don Kahn, Cristina Santa Marta, Nilima Nigam, Greg Pierce, Fadil Santosa, Wayne Schmaedeke, Jackie Shen, Peter Shook, Thomas Scofield, and Richard Varga, as well as our classes and students, particularly Taiala Carvalho, Colleen Duffy, and Ryan Lloyd, and last, but certainly not least, our father/father-in-law Frank W.J. Olver and son Sheehan Olver, for proofreading, corrections, remarks, and many useful suggestions. We also greatly appreciated the helpful feedback from the reviewers of the book: Augustin Banyaga, Penn State University, Robert Cramer, University of Colorado, James H. Curry, University of Colorado, Jerome Dancis, University of Maryland, Bruno Harris, Brown University, Norman L. Johnson, University of Iowa, Cerry M. Klein, University of Missouri, Doron Lubinsky, Georgia Tech, Juan Manfredi, University of Pittsburgh, Fabio Augusto Milner, Purdue University, Tzuong–Tsieng Moh, Purdue University, Paul S. Muhly, University of Iowa, Juan Carlos Álvarez Paiva, Brooklyn Polytechnic University, John F. Rossi, Georgia Tech, Brian Shader, University of Wyoming, Shagi–Di Shih, University of Wyoming, Tamas Wiandt, Rochester Institute of Technology, and two anonymous reviewers, for their insightful comments on earlier drafts of this material. We are particularly indebted to Dennis Kletzing for his marvelous job typesetting the final text, which required coming to grips with our highly non-standard plain TeX macro system.

And of course, it goes without saying that we owe an immense debt to George Lobell at Prentice–Hall for encouraging us to pursue this project, and for consistently supporting our efforts throughout its writing, and revising, and further revising, and . . . but the end is finally here!

Peter J. Olver
University of Minnesota
olver@math.umn.edu

Cheri Shakiban
University of St. Thomas
c9shakiban@stthomas.edu

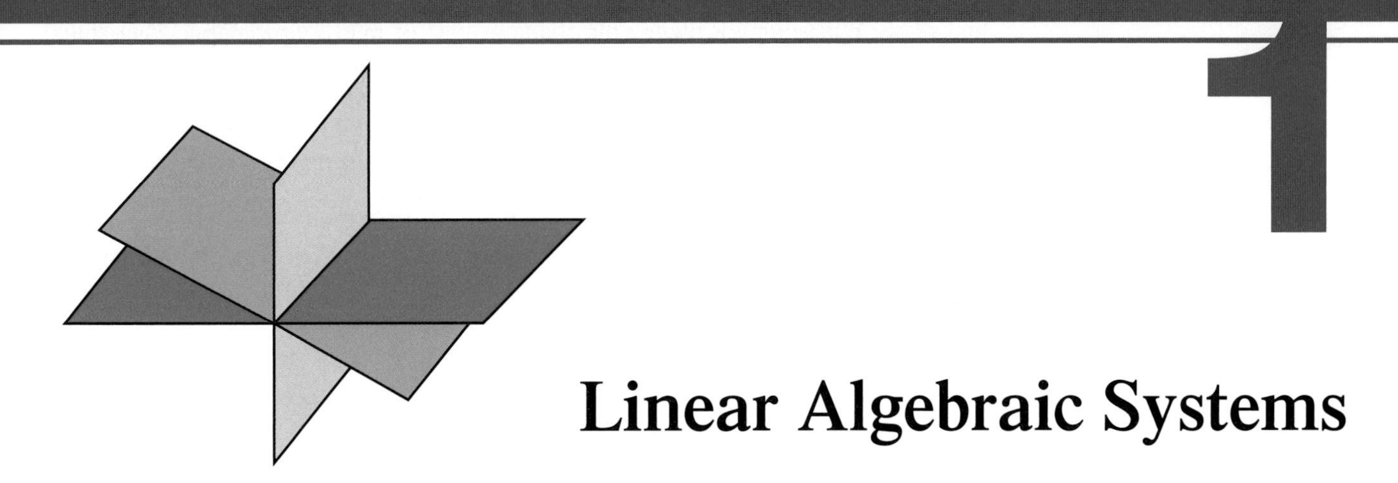

Linear Algebraic Systems

Linear algebra is the core of modern applied mathematics. Its humble origins are to be found in the need to solve "elementary" systems of linear algebraic equations. But its ultimate scope is vast, impinging on all of mathematics, both pure and applied, as well as numerical analysis, statistics, physics, engineering, mathematical biology, financial mathematics, and every other discipline in which mathematical methods are required. A thorough grounding in the methods and theory of linear algebra is an essential prerequisite for understanding and harnessing the power of modern mathematics in applications.

In the first chapter, our focus will be on the most basic method for solving linear algebraic systems, known as *Gaussian Elimination* in honor of one of the all-time mathematical greats—the early nineteenth century German mathematician Carl Friedrich Gauss. As the father of linear algebra, his name will occur repeatedly throughout this text. Gaussian Elimination is quite elementary, but remains one of *the* most important algorithms in applied (as well as theoretical) mathematics. Our initial focus will be on the most important class of systems: those involving the same number of equations as unknowns—although we will eventually develop techniques for handling completely general linear systems. While the former typically have a unique solution, general linear systems may have either no solutions or infinitely many solutions. Since physical models require existence and uniqueness of their solution, the systems arising in applications often (but not always) involve the same number of equations as unknowns. Nevertheless, the ability to confidently handle all types of linear systems is a basic prerequisite for further progress in the subject. In contemporary applications, particularly those arising in numerical solutions of differential equations, in signal and image processing, and elsewhere, the governing linear systems can be huge, sometimes involving millions of equations in millions of unknowns, challenging even the most powerful supercomputer. So, a systematic and careful development of solution techniques is essential. Section 1.7 discusses some of the practical issues and limitations in computer implementations of the Gaussian Elimination method for large systems arising in applications.

Modern linear algebra relies on the basic concepts of scalar, vector, and matrix, and so we must quickly review the fundamentals of matrix arithmetic. Gaussian Elimination can be profitably reinterpreted as a certain matrix factorization, known as the (permuted) $L U$ decomposition, which provides valuable insight into the solution algorithms. Matrix inverses and determinants are also discussed in brief, primarily for their theoretical properties. As we shall see, formulas relying on the inverse or the determinant are extremely inefficient, and so, except in low-dimensional or highly structured environments, are to be avoided in almost all practical computations. In the theater of applied linear algebra, Gaussian Elimination and matrix factorization are the stars, while inverses and determinants are relegated to the supporting cast.

1.1 SOLUTION OF LINEAR SYSTEMS

Gaussian Elimination is a simple, systematic algorithm to solve systems of linear equations. It is the workhorse of linear algebra, and, as such, of absolutely fundamental importance in applied mathematics. In this section, we review the method in the most important case, in which there are the same number of equations as unknowns. The general situation will be deferred until Section 1.8.

To illustrate, consider an elementary system of three linear equations

$$\begin{aligned} x + 2y + z &= 2, \\ 2x + 6y + z &= 7, \\ x + y + 4z &= 3, \end{aligned} \tag{1.1}$$

in three unknowns x, y, z. Linearity* refers to the fact that the unknowns only appear to the first power, and there are no product terms like $x\,y$ or $x\,y\,z$. The basic solution method is to systematically employ the following fundamental operation:

> *Linear System Operation* #1: Add a multiple of one equation to another equation.

Before continuing, you might try to convince yourself that this operation doesn't change the solutions to the system. Our goal is to judiciously apply the operation and so be led to a much simpler linear system that is easy to solve, and, moreover has the same solutions as the original. Any linear system that is derived from the original system by successive application of such operations will be called an *equivalent system*. By the preceding remark, *equivalent linear systems have the same solutions*.

The systematic feature is that we successively eliminate the variables in our equations in order of appearance. We begin by eliminating the first variable, x, from the second equation. To this end, we subtract twice the first equation from the second, leading to

$$\begin{aligned} x + 2y + z &= 2, \\ 2y - z &= 3, \\ x + y + 4z &= 3. \end{aligned} \tag{1.2}$$

Next, we eliminate x from the third equation by subtracting the first equation from

*The "official" definition of linearity will be deferred until Chapter 7.

it:

$$\begin{aligned} x + 2y + z &= 2, \\ 2y - z &= 3, \\ -y + 3z &= 1. \end{aligned} \tag{1.3}$$

The equivalent system (1.3) is already simpler than the original (1.1). Notice that the second and third equations do not involve x (by design) and so constitute a system of two linear equations for two unknowns. Moreover, once we have solved this subsystem for y and z, we can substitute the answer into the first equation, and we need only solve a single linear equation for x.

We continue on in this fashion, the next phase being the elimination of the second variable, y, from the third equation by adding $\frac{1}{2}$ the second equation to it. The result is

$$\begin{aligned} x + 2y + z &= 2, \\ 2y - z &= 3, \\ \tfrac{5}{2} z &= \tfrac{5}{2}, \end{aligned} \tag{1.4}$$

which is the simple system we are after. It is in what is called *triangular form*, which means that, while the first equation involves all three variables, the second equation only involves the second and third variables, and the last equation only involves the last variable.

Any triangular system can be straightforwardly solved by the method of *Back Substitution*. As the name suggests, we work backwards, solving the last equation first, which requires that $z = 1$. We substitute this result back into the penultimate equation, which becomes $2y - 1 = 3$, with solution $y = 2$. We finally substitute these two values for y and z into the first equation, which becomes $x + 5 = 2$, and so the solution to the triangular system (1.4) is

$$x = -3, \qquad y = 2, \qquad z = 1. \tag{1.5}$$

Moreover, since we only used our basic linear system operation to pass from (1.1) to the triangular system (1.4), this is also the solution to the original system of linear equations, as you can check. We note that the system (1.1) has a unique—meaning one and only one—solution, namely (1.5).

And that, barring a few minor complications that can crop up from time to time, is all that there is to the method of Gaussian Elimination! It is extraordinarily simple, but its importance cannot be overemphasized. Before exploring the relevant issues, it will help to reformulate our method in a more convenient matrix notation.

EXERCISES 1.1

1.1.1. Solve the following systems of linear equations by reducing to triangular form and then using Back Substitution.

(a) $\begin{aligned} x - y &= 7 \\ x + 2y &= 3 \end{aligned}$

(b) $\begin{aligned} 6u + v &= 5 \\ 3u - 2v &= 5 \end{aligned}$

(c) $\begin{aligned} p + q - r &= 0 \\ 2p - q + 3r &= 3 \\ -p - q &= 6 \end{aligned}$

(d) $\begin{aligned} 2u - v + 2w &= 2 \\ -u - v + 3w &= 1 \\ 3u - 2w &= 1 \end{aligned}$

(e) $\begin{aligned} 5x_1 + 3x_2 - x_3 &= 9 \\ 3x_1 + 2x_2 - x_3 &= 5 \\ x_1 + x_2 + x_3 &= -1 \end{aligned}$

(f) $\begin{aligned} x + z - 2w &= -3 \\ 2x - y + 2z - w &= -5 \\ -6y - 4z + 2w &= 2 \\ x + 3y + 2z - w &= 1 \end{aligned}$

(g) $$\begin{aligned} 3x_1 + x_2 &= 1 \\ x_1 + 3x_2 + x_3 &= 1 \\ x_2 + 3x_3 + x_4 &= 1 \\ x_3 + 3x_4 &= 1 \end{aligned}$$

1.1.2. How should the coefficients a, b, and c be chosen so that the system $ax+by+cz=3$, $ax-y+cz=1$, $x+by-cz=2$, has the solution $x=1$, $y=2$ and $z=-1$?

♡ **1.1.3.** The system $2x=-6$, $-4x+3y=3$, $x+4y-z=7$, is in *lower triangular form*.

(a) Formulate a method of *Forward Substitution* to solve it.

(b) What happens if you reduce the system to the (upper) triangular form using the algorithm in this section?

(c) Devise an algorithm that uses our linear system operation to reduce a system to lower triangular form and then solve it by Forward Substitution.

(d) Check your algorithm by applying it to one or two of the systems in Exercise 1.1.1. Are you able to solve them in all cases?

1.2 MATRICES AND VECTORS

A *matrix* is a rectangular array of numbers. Thus,

$$\begin{pmatrix} 1 & 0 & 3 \\ -2 & 4 & 1 \end{pmatrix}, \quad \begin{pmatrix} \pi & 0 \\ e & \frac{1}{2} \\ -1 & .83 \\ \sqrt{5} & -\frac{4}{7} \end{pmatrix}, \quad (.2 \quad -1.6 \quad .32), \quad \begin{pmatrix} 0 \\ 0 \end{pmatrix}, \quad \begin{pmatrix} 1 & 3 \\ -2 & 5 \end{pmatrix},$$

are all examples of matrices. We use the notation

$$A = \begin{pmatrix} a_{11} & a_{12} & \dots & a_{1n} \\ a_{21} & a_{22} & \dots & a_{2n} \\ \vdots & \vdots & \ddots & \vdots \\ a_{m1} & a_{m2} & \dots & a_{mn} \end{pmatrix} \tag{1.6}$$

for a general matrix of size $m \times n$ (read "m by n"), where m denotes the number of *rows* in A and n denotes the number of *columns*. Thus, the preceding examples of matrices have respective sizes 2×3, 4×2, 1×3, 2×1, and 2×2. A matrix is *square* if $m=n$, i.e., it has the same number of rows as columns. A *column vector* is a $m\times 1$ matrix, while a *row vector* is a $1\times n$ matrix. As we shall see, column vectors are by far the more important of the two, and the term "vector" without qualification will always mean "column vector". A 1×1 matrix, which has but a single entry, is both a row and a column vector.

The number that lies in the ith row and the jth column of A is called the (i,j) *entry* of A, and is denoted by a_{ij}. The row index always appears first and the column index second.* Two matrices are equal, $A=B$, if and only if they have the same size, and *all* their entries are the same: $a_{ij}=b_{ij}$ for $i=1,\dots,m$ and $j=1,\dots,n$.

A general linear system of m equations in n unknowns will take the form

$$\begin{aligned} a_{11}x_1 + a_{12}x_2 + \cdots + a_{1n}x_n &= b_1, \\ a_{21}x_1 + a_{22}x_2 + \cdots + a_{2n}x_n &= b_2, \\ \vdots \qquad\qquad \vdots \qquad\qquad &\;\;\vdots \\ a_{m1}x_1 + a_{m2}x_2 + \cdots + a_{mn}x_n &= b_m. \end{aligned} \tag{1.7}$$

*In tensor analysis, [1], a sub- and super-script notation is adopted, with a^i_j denoting the (i,j) entry of the matrix A. This has certain advantages, but, to avoid possible confusion with powers, we shall stick with the simpler subscript notation throughout this text.

As such, it is composed of three basic ingredients: the $m \times n$ *coefficient matrix* A, with entries a_{ij} as in (1.6), the column vector

$$\mathbf{x} = \begin{pmatrix} x_1 \\ x_2 \\ \vdots \\ x_n \end{pmatrix}$$

containing the *unknowns*, and the column vector

$$\mathbf{b} = \begin{pmatrix} b_1 \\ b_2 \\ \vdots \\ b_m \end{pmatrix}$$

containing *right hand sides*. In our previous example,

$$\begin{aligned} x + 2y + z &= 2, \\ 2x + 6y + z &= 7, \\ x + y + 4z &= 3, \end{aligned}$$

the coefficient matrix

$$A = \begin{pmatrix} 1 & 2 & 1 \\ 2 & 6 & 1 \\ 1 & 1 & 4 \end{pmatrix}$$

can be filled in, entry by entry, from the coefficients of the variables appearing in the equations; the vector $\mathbf{x} = \begin{pmatrix} x \\ y \\ z \end{pmatrix}$ lists the variables, while the entries of $\mathbf{b} = \begin{pmatrix} 2 \\ 7 \\ 3 \end{pmatrix}$ are the right hand sides of the equations.

REMARK: We will consistently use bold face lower case letters to denote vectors, and ordinary capital letters to denote general matrices.

EXERCISES 1.2

1.2.1. Let $A = \begin{pmatrix} -2 & 0 & 1 & 3 \\ -1 & 2 & 7 & -5 \\ 6 & -6 & -3 & 4 \end{pmatrix}$.

(a) What is the size of A?

(b) What is its $(2, 3)$ entry?

(c) $(3, 1)$ entry?

(d) 1st row?

(e) 2nd column?

1.2.2. Write down examples of

(a) a 3×3 matrix (b) a 2×3 matrix

(c) a matrix with 3 rows and 4 columns

(d) a row vector with 4 entries

(e) a column vector with 3 entries

(f) a matrix that is both a row vector and a column vector.

1.2.3. For which values of x, y, z, w are the matrices $\begin{pmatrix} x + y & x - z \\ y + w & x + 2w \end{pmatrix}$ and $\begin{pmatrix} 1 & 0 \\ 2 & 1 \end{pmatrix}$ equal?

1.2.4. For each of the systems in Exercise 1.1.1, write down the coefficient matrix A and the vectors $\mathbf{x}$ and $\mathbf{b}$.

1.2.5. Write out and solve the linear systems corresponding to the indicated matrix, vector of unknowns and right hand side.

(a) $A = \begin{pmatrix} 1 & -1 \\ 2 & 3 \end{pmatrix}$, $\mathbf{x} = \begin{pmatrix} x \\ y \end{pmatrix}$, $\mathbf{b} = \begin{pmatrix} -1 \\ -3 \end{pmatrix}$

(b) $A = \begin{pmatrix} 1 & 0 & 1 \\ 1 & 1 & 0 \\ 0 & 1 & 1 \end{pmatrix}, \mathbf{x} = \begin{pmatrix} u \\ v \\ w \end{pmatrix}, \mathbf{b} = \begin{pmatrix} -1 \\ -1 \\ 2 \end{pmatrix}$

(c) $A = \begin{pmatrix} 3 & 0 & -1 \\ -2 & -1 & 0 \\ 1 & 1 & -3 \end{pmatrix}, \mathbf{x} = \begin{pmatrix} x_1 \\ x_2 \\ x_3 \end{pmatrix},$ $\mathbf{b} = \begin{pmatrix} 1 \\ 0 \\ 1 \end{pmatrix}$

(d) $A = \begin{pmatrix} 1 & 1 & -1 & -1 \\ -1 & 0 & 1 & 2 \\ 1 & -1 & 1 & 0 \\ 0 & 2 & -1 & 1 \end{pmatrix}, \mathbf{x} = \begin{pmatrix} x \\ y \\ z \\ w \end{pmatrix},$ $\mathbf{b} = \begin{pmatrix} 0 \\ 4 \\ 1 \\ 5 \end{pmatrix}$

Matrix Arithmetic

Matrix arithmetic involves three basic operations: *matrix addition*, *scalar multiplication*, and *matrix multiplication*. First we define *addition* of matrices. You are only allowed to add two matrices of the *same size*, and matrix addition is performed entry by entry. For example,

$$\begin{pmatrix} 1 & 2 \\ -1 & 0 \end{pmatrix} + \begin{pmatrix} 3 & -5 \\ 2 & 1 \end{pmatrix} = \begin{pmatrix} 4 & -3 \\ 1 & 1 \end{pmatrix}.$$

Therefore, if A and B are $m \times n$ matrices, their sum $C = A + B$ is the $m \times n$ matrix whose entries are given by $c_{ij} = a_{ij} + b_{ij}$ for $i = 1, \ldots, m$ and $j = 1, \ldots, n$. When defined, matrix addition is commutative, $A + B = B + A$, and associative, $A + (B + C) = (A + B) + C$, just like ordinary addition.

A *scalar* is a fancy name for an ordinary number—the term merely distinguishes it from a vector or a matrix. For the time being, we will restrict our attention to real scalars and matrices with real entries, but eventually complex scalars and complex matrices must be dealt with. We will consistently identify a scalar $c \in \mathbb{R}$ with the 1×1 matrix (c) in which it is the sole entry, and so will omit the redundant parentheses in the latter case. *Scalar multiplication* takes a scalar c and an $m \times n$ matrix A and computes the $m \times n$ matrix $B = c\,A$ by multiplying each entry of A by c. For example,

$$3\begin{pmatrix} 1 & 2 \\ -1 & 0 \end{pmatrix} = \begin{pmatrix} 3 & 6 \\ -3 & 0 \end{pmatrix}.$$

In general, $b_{ij} = c\,a_{ij}$ for $i = 1, \ldots, m$ and $j = 1, \ldots, n$. Basic properties of scalar multiplication are summarized at the end of this section.

Finally, we define *matrix multiplication*. First, the product between a row vector $\mathbf{a}$ and a column vector $\mathbf{x}$ having the *same* number of entries is the *scalar* or 1×1 matrix defined by the following rule:

$$\mathbf{a}\,\mathbf{x} = (\,a_1\; a_2\; \ldots\; a_n\,) \begin{pmatrix} x_1 \\ x_2 \\ \vdots \\ x_n \end{pmatrix} = a_1 x_1 + a_2 x_2 + \cdots + a_n x_n = \sum_{k=1}^{n} a_k x_k. \qquad (1.8)$$

More generally, if A is an $m \times n$ matrix and B is an $n \times p$ matrix, so that the number of *columns* in A equals the number of *rows* in B, then the matrix product $C = A\,B$ is defined as the $m \times p$ matrix whose (i, j) entry equals the vector product of the ith row of A and the jth column of B. Therefore,

$$c_{ij} = \sum_{k=1}^{n} a_{ik}\, b_{kj}. \qquad (1.9)$$

Note that our restriction on the sizes of A and B guarantees that the relevant row and column vectors will have the same number of entries, and so their product is defined.

For example, the product of the coefficient matrix A and vector of unknowns $\mathbf{x}$ for our original system (1.1) is given by

$$A\,\mathbf{x} = \begin{pmatrix} 1 & 2 & 1 \\ 2 & 6 & 1 \\ 1 & 1 & 4 \end{pmatrix} \begin{pmatrix} x \\ y \\ z \end{pmatrix} = \begin{pmatrix} x + 2y + z \\ 2x + 6y + z \\ x + y + 4z \end{pmatrix}.$$

The result is a column vector whose entries reproduce the left hand sides of the original linear system! As a result, we can rewrite the system

$$A\,\mathbf{x} = \mathbf{b} \tag{1.10}$$

as an equality between two column vectors. This result is general; a linear system (1.7) consisting of m equations in n unknowns can be written in the matrix form (1.10) where A is the $m \times n$ coefficient matrix (1.6), $\mathbf{x}$ is the $n \times 1$ column vector of unknowns, and $\mathbf{b}$ is the $m \times 1$ column vector containing the right hand sides. This is one of the principal reasons for the non-evident definition of matrix multiplication. Component-wise multiplication of matrix entries turns out to be almost completely useless in applications.

Now, the bad news. Matrix multiplication is *not* commutative—that is, BA is not necessarily equal to $A\,B$. For example, BA may not be defined even when $A\,B$ is. Even if both are defined, they may be different sized matrices. For example the product $s = \mathbf{r}\,\mathbf{c}$ of a row vector $\mathbf{r}$, a $1 \times n$ matrix, and a column vector $\mathbf{c}$, an $n \times 1$ matrix with the same number of entries, is a 1×1 matrix or scalar, whereas the reversed product $C = \mathbf{c}\,\mathbf{r}$ is an $n \times n$ matrix. For instance,

$$\begin{pmatrix} 1 & 2 \end{pmatrix} \begin{pmatrix} 3 \\ 0 \end{pmatrix} = 3, \qquad \text{whereas} \qquad \begin{pmatrix} 3 \\ 0 \end{pmatrix} \begin{pmatrix} 1 & 2 \end{pmatrix} = \begin{pmatrix} 3 & 6 \\ 0 & 0 \end{pmatrix}.$$

In computing the latter product, don't forget that we multiply the *rows* of the first matrix by the *columns* of the second. Moreover, even if the matrix products $A\,B$ and $B\,A$ have the same size, which requires both A and B to be square matrices, we may still have $A\,B \neq B\,A$. For example,

$$\begin{pmatrix} 1 & 2 \\ 3 & 4 \end{pmatrix} \begin{pmatrix} 0 & 1 \\ -1 & 2 \end{pmatrix} = \begin{pmatrix} -2 & 5 \\ -4 & 11 \end{pmatrix} \neq \begin{pmatrix} 3 & 4 \\ 5 & 6 \end{pmatrix} = \begin{pmatrix} 0 & 1 \\ -1 & 2 \end{pmatrix} \begin{pmatrix} 1 & 2 \\ 3 & 4 \end{pmatrix}.$$

On the other hand, matrix multiplication is associative, so $A\,(B\,C) = (A\,B)\,C$ whenever A has size $m \times n$, B has size $n \times p$, and C has size $p \times q$; the result is a matrix of size $m \times q$. The proof of associativity is a tedious computation based on the definition of matrix multiplication that, for brevity, we omit.† Consequently, the one difference between matrix algebra and ordinary algebra is that you need to be careful not to change the order of multiplicative factors without proper justification.

Since matrix multiplication acts by multiplying rows by columns, one can compute the columns in a matrix product $A\,B$ by multiplying the matrix A and the individual columns of B. For example, the two columns of the matrix product

$$\begin{pmatrix} 1 & -1 & 2 \\ 2 & 0 & -2 \end{pmatrix} \begin{pmatrix} 3 & 4 \\ 0 & 2 \\ -1 & 1 \end{pmatrix} = \begin{pmatrix} 1 & 4 \\ 8 & 6 \end{pmatrix}$$

†A much simpler, but more abstract proof can be found in Exercise 7.1.44.

are obtained by multiplying the first matrix with the individual columns of the second:

$$\begin{pmatrix} 1 & -1 & 2 \\ 2 & 0 & -2 \end{pmatrix} \begin{pmatrix} 3 \\ 0 \\ -1 \end{pmatrix} = \begin{pmatrix} 1 \\ 8 \end{pmatrix}, \qquad \begin{pmatrix} 1 & -1 & 2 \\ 2 & 0 & -2 \end{pmatrix} \begin{pmatrix} 4 \\ 2 \\ 1 \end{pmatrix} = \begin{pmatrix} 4 \\ 6 \end{pmatrix}.$$

In general, if we use $\mathbf{b}_k$ to denote the kth column of B, then

$$A\,B = A\,(\mathbf{b}_1 \quad \mathbf{b}_2 \quad \ldots \quad \mathbf{b}_p) = (A\,\mathbf{b}_1 \quad A\,\mathbf{b}_2 \quad \ldots \quad A\,\mathbf{b}_p), \tag{1.11}$$

indicating that the kth column of their matrix product is $A\,\mathbf{b}_k$.

There are two special matrices. The first is the *zero matrix*, all of whose entries are 0. We use $\mathrm{O}_{m \times n}$ to denote the $m \times n$ zero matrix, often written as just O if the size is clear from the context. The zero matrix is the additive unit, so $A + \mathrm{O} = A = \mathrm{O} + A$ when O has the same size as A. In particular, we will use a bold face $\mathbf{0}$ to denote a column vector with all zero entries.

The role of the multiplicative unit is played by the square *identity matrix*

$$\mathrm{I} = \mathrm{I}_n = \begin{pmatrix} 1 & 0 & 0 & \cdots & 0 & 0 \\ 0 & 1 & 0 & \cdots & 0 & 0 \\ 0 & 0 & 1 & \cdots & 0 & 0 \\ \vdots & \vdots & \vdots & \ddots & \vdots & \vdots \\ 0 & 0 & 0 & \cdots & 1 & 0 \\ 0 & 0 & 0 & \cdots & 0 & 1 \end{pmatrix}$$

of size $n \times n$. The entries along the *main diagonal* (which runs from top left to bottom right) are equal to 1, while the *off-diagonal* entries are all 0. As you can check, if A is any $m \times n$ matrix, then $\mathrm{I}_m A = A = A\,\mathrm{I}_n$. We will sometimes write the preceding equation as just $\mathrm{I}\,A = A = A\,\mathrm{I}$, since each matrix product is well-defined for exactly one size of identity matrix.

The identity matrix is a particular example of a *diagonal matrix*. In general, a square matrix A is diagonal if all its off-diagonal entries are zero: $a_{ij} = 0$ for all $i \neq j$. We will sometimes write $D = \operatorname{diag}(c_1, \ldots, c_n)$ for the $n \times n$ diagonal matrix with diagonal entries $d_{ii} = c_i$. Thus, $\operatorname{diag}(1, 3, 0)$ refers to the diagonal matrix

$$\begin{pmatrix} 1 & 0 & 0 \\ 0 & 3 & 0 \\ 0 & 0 & 0 \end{pmatrix},$$

while the $n \times n$ identity matrix can be written as $\mathrm{I}_n = \operatorname{diag}(1, 1, \ldots, 1)$.

Let us conclude this section by summarizing the basic properties of matrix arithmetic. In the accompanying table, A, B, C are matrices; c, d are scalars; O is a zero matrix; and I is an identity matrix. All matrices are assumed to have the correct sizes so that the indicated operations are defined.

Basic Matrix Arithmetic

Matrix Addition:	Commutativity	$A + B = B + A$
	Associativity	$(A + B) + C = A + (B + C)$
	Zero Matrix	$A + \mathrm{O} = A = \mathrm{O} + A$
	Inverse	$A + (-A) = \mathrm{O}, \quad -A = (-1)A$
Scalar Multiplication:	Associativity	$c(dA) = (cd)A$
	Distributivity	$c(A + B) = (cA) + (cB)$
		$(c + d)A = (cA) + (dA)$
	Unit	$1A = A$
	Zero	$0A = \mathrm{O}$
Matrix Multiplication:	Associativity	$(AB)C = A(BC)$
	Distributivity	$A(B + C) = AB + AC$
		$(A + B)C = AC + BC$
	Identity Matrix	$A\mathrm{I} = A = \mathrm{I}A$
	Zero Matrix	$A\mathrm{O} = \mathrm{O}, \quad \mathrm{O}A = \mathrm{O}$

EXERCISES 1.2

1.2.6. (a) Write down the 4×4 identity and zero matrices.
(b) Write down their sum and their product. Does the order of multiplication matter?

1.2.7. Consider the matrices

$$A = \begin{pmatrix} 1 & -1 & 3 \\ -1 & 4 & -2 \\ 3 & 0 & 6 \end{pmatrix}, \quad B = \begin{pmatrix} -6 & 0 & 3 \\ 4 & 2 & -1 \end{pmatrix}, \quad C = \begin{pmatrix} 2 & 3 \\ -3 & -4 \\ 1 & 2 \end{pmatrix}.$$

Compute the indicated combinations where possible.

(a) $3A - B$ (b) AB (c) BA
(d) $(A + B)C$ (e) $A + BC$
(f) $A + 2CB$ (g) $BCB - \mathrm{I}$
(h) $A^2 - 3A + \mathrm{I}$ (i) $(B - \mathrm{I})(C + \mathrm{I})$

1.2.8. Which of the following pairs of matrices commute under matrix multiplication?

(a) $\begin{pmatrix} 1 & 2 \\ -2 & 1 \end{pmatrix}, \begin{pmatrix} 2 & 3 \\ 5 & 0 \end{pmatrix}$

(b) $\begin{pmatrix} 3 & -1 \\ 0 & 2 \\ 1 & 4 \end{pmatrix}, \begin{pmatrix} 4 & 2 & -2 \\ 5 & 2 & 4 \end{pmatrix}$

(c) $\begin{pmatrix} 3 & 0 & -1 \\ -2 & -1 & 2 \\ 2 & 0 & 0 \end{pmatrix}, \begin{pmatrix} 2 & 0 & -1 \\ 1 & 1 & -1 \\ 2 & 0 & -1 \end{pmatrix}$

1.2.9. List the diagonal entries of

$$A = \begin{pmatrix} 1 & 2 & 3 & 4 \\ 5 & 6 & 7 & 8 \\ 9 & 10 & 11 & 12 \\ 13 & 14 & 15 & 16 \end{pmatrix}.$$

1.2.10. Write out the following diagonal matrices:
(a) $\operatorname{diag}(1, 0, -1)$ (b) $\operatorname{diag}(2, -2, 3, -3)$

1.2.11. *True or False*:
(a) The sum of two diagonal matrices of the same size is a diagonal matrix.
(b) The product is also diagonal.

♡ **1.2.12.** (a) Show that if $D = \begin{pmatrix} a & 0 \\ 0 & b \end{pmatrix}$ is a 2×2 diagonal matrix with $a \neq b$, then the only matrices that commute (under matrix multiplication) with D are other 2×2 diagonal matrices.
(b) What if $a = b$?
(c) Find all matrices that commute with

$$D = \begin{pmatrix} a & 0 & 0 \\ 0 & b & 0 \\ 0 & 0 & c \end{pmatrix},$$

where a, b, c are all different.

(d) Answer the same question for the case when $a \neq b = c$.

(e) Prove that a matrix A commutes with an $n \times n$ diagonal matrix D with all *distinct* diagonal entries if and only if A is a diagonal matrix.

1.2.13. Show that the matrix products $A\,B$ and $B\,A$ have the same size if and only if A and B are square matrices of the same size.

1.2.14. Find all matrices B that commute (under matrix multiplication) with $A = \begin{pmatrix} 1 & 2 \\ 0 & 1 \end{pmatrix}$.

1.2.15. (a) Show that, if A and B are commuting square matrices, then $(A+B)^2 = A^2 + 2AB + B^2$.

(b) Find a pair of 2×2 matrices A and B such that $(A+B)^2 \neq A^2 + 2AB + B^2$.

1.2.16. Show that if the matrices A and B commute, then they necessarily are both square and the same size.

1.2.17. Let A be an $m \times n$ matrix. What are the permissible sizes for the zero matrices appearing in the identities $A\,\mathrm{O} = \mathrm{O}$, $\mathrm{O}\,A = \mathrm{O}$?

1.2.18. Let A be an $m \times n$ matrix and let c be a scalar. Show that if $c\,A = \mathrm{O}$, then either $c = 0$ or $A = \mathrm{O}$.

1.2.19. *True or false*: If $A\,B = \mathrm{O}$ then either $A = \mathrm{O}$ or $B = \mathrm{O}$.

1.2.20. *True or false*: If A, B are square matrices of the same size, then $A^2 - B^2 = (A+B)(A-B)$.

1.2.21. Prove that $A\mathbf{v} = \mathbf{0}$ for every vector $\mathbf{v}$ (with the appropriate number of entries) if and only if $A = \mathrm{O}$ is the zero matrix. *Hint*: If you are stuck, first try to find a proof when A is a small matrix, e.g., of size 2×2.

1.2.22. (a) Under what conditions is the square A^2 of a matrix defined?

(b) Show that A and A^2 commute.

(c) How many matrix multiplications are needed to compute A^n?

1.2.23. Find a nonzero matrix $A \neq \mathrm{O}$ such that $A^2 = \mathrm{O}$.

◇ **1.2.24.** Let A have a row all of whose entries are zero.

(a) Explain why the product AB also has a zero row.

(b) Find an example where BA does not have a zero row.

1.2.25. (a) Find all solutions $X = \begin{pmatrix} x & y \\ z & w \end{pmatrix}$ to the matrix equation $A\,X = \mathrm{I}$ when $A = \begin{pmatrix} 2 & 1 \\ 3 & 1 \end{pmatrix}$.

(b) Find all solutions to $X\,A = \mathrm{I}$. Are they the same?

1.2.26. (a) Find all solutions $X = \begin{pmatrix} x & y \\ z & w \end{pmatrix}$ to the matrix equation $A\,X = B$ when $A = \begin{pmatrix} 0 & 1 \\ -1 & 3 \end{pmatrix}$ and $B = \begin{pmatrix} 1 & 2 \\ -1 & 1 \end{pmatrix}$.

(b) Find all solutions to $X\,A = B$. Are they the same?

1.2.27. (a) Find all solutions $X = \begin{pmatrix} x & y \\ z & w \end{pmatrix}$ to the matrix equation $A\,X = X\,B$ when $A = \begin{pmatrix} 1 & 2 \\ -1 & 0 \end{pmatrix}$ and $B = \begin{pmatrix} 0 & 1 \\ 3 & 0 \end{pmatrix}$.

(b) Can you find a pair of nonzero matrices $A \neq B$ such that the matrix equation $A\,X = X\,B$ has a nonzero solution $X \neq \mathrm{O}$?

1.2.28. Let A be a matrix and c a scalar. Find all solutions to the matrix equation $c\,A = \mathrm{I}$.

◇ **1.2.29.** Let $\mathbf{z}$ be the $n \times 1$ column vector all of whose entries are equal to 1.

(a) Show that if A is an $m \times n$ matrix, the ith entry of the product $\mathbf{v} = A\mathbf{z}$ is the ith *row sum* of A, meaning the sum of all the entries in its ith row.

(b) Let W denote the $n \times n$ matrix whose diagonal entries are equal to $\dfrac{1-n}{n}$ and whose off-diagonal entries are all equal to $\dfrac{1}{n}$. Prove that the row sums of $B = AW$ are all zero.

(c) Check both results when

$$A = \begin{pmatrix} 1 & 2 & -1 \\ 2 & 1 & 3 \\ -4 & 5 & -1 \end{pmatrix}.$$

Remark: If the columns of A represent experimental data values, then the entries of $\frac{1}{n}A\mathbf{z}$ represent the means or averages of the data values, while $B = AW$ corresponds to data that has been normalized to have mean 0.

◇ **1.2.30.** Prove that matrix multiplication is associative: $A(BC) = (AB)C$ when defined.

♡ **1.2.31.** The *commutator* of two matrices A, B, is defined to be the matrix

$$C = [\,A, B\,] = A\,B - B\,A. \tag{1.12}$$

(a) Explain why $[\,A, B\,]$ is defined if and only if A and B are square matrices of the same size.

(b) Show that A and B commute under matrix multiplication if and only if $[\,A, B\,] = \mathrm{O}$.

(c) Compute the commutator of the following matrices:

(i) $\begin{pmatrix} 1 & 0 \\ 1 & -1 \end{pmatrix}, \begin{pmatrix} 2 & 1 \\ -2 & 0 \end{pmatrix}$;

(ii) $\begin{pmatrix} -1 & 3 \\ 3 & -1 \end{pmatrix}, \begin{pmatrix} 1 & 7 \\ 7 & 1 \end{pmatrix}$;

(iii) $\begin{pmatrix} 0 & -1 & 0 \\ 1 & 0 & 0 \\ 0 & 0 & 1 \end{pmatrix}, \begin{pmatrix} 1 & 0 & 0 \\ 0 & 0 & -1 \\ 0 & 1 & 0 \end{pmatrix}$;

(d) Prove that the commutator is

(i) *Bilinear*:

$$[\,cA + dB, C\,] = c[\,A, B\,] + d[\,B, C\,]$$
$$[\,A, cB + dC\,] = c[\,A, B\,] + d[\,A, C\,]$$

for any scalars c, d;

(ii) *Skew-symmetric*: $[\,A, B\,] = -[\,B, A\,]$;

(iii) satisfies the the *Jacobi identity*:

$$[\,[\,A, B\,], C\,] + [\,[\,C, A\,], B\,] + [\,[\,B, C\,], A\,] = \mathrm{O},$$

for any square matrices A, B, C of the same size.

Remark: The commutator plays a very important role in geometry, symmetry, and quantum mechanics. See Section 9.4 as well as [40, 46, 70] for further developments.

◇ **1.2.32.** The *trace* of a $n \times n$ matrix $A \in \mathcal{M}_{n\times n}$ is defined to be the sum of its diagonal entries: $\operatorname{tr} A = a_{11} + a_{22} + \cdots + a_{nn}$.

(a) Compute the trace of

(i) $\begin{pmatrix} 1 & -1 \\ 2 & 3 \end{pmatrix}$ (ii) $\begin{pmatrix} 1 & 3 & 2 \\ -1 & 0 & 1 \\ -4 & 3 & -1 \end{pmatrix}$

(b) Prove that $\operatorname{tr}(A + B) = \operatorname{tr} A + \operatorname{tr} B$.

(c) Prove that $\operatorname{tr}(A\,B) = \operatorname{tr}(B\,A)$.

(d) Prove that the commutator matrix $C = A\,B - B\,A$ has zero trace: $\operatorname{tr} C = 0$.

(e) Is part (c) valid if A has size $m \times n$ and B has size $n \times m$?

◇ **1.2.33.** Justify the alternative formula for multiplying a matrix A and a column vector $\mathbf{x}$:

$$A\,\mathbf{x} = x_1\,\mathbf{c}_1 + x_2\,\mathbf{c}_2 + \cdots + x_n\,\mathbf{c}_n, \qquad (1.13)$$

where $\mathbf{c}_1, \ldots, \mathbf{c}_n$ are the columns of A and $x_1, \ldots, x_n$ the entries of $\mathbf{x}$.

♡ **1.2.34.** The basic definition of matrix multiplication $A\,B$ tells us to multiply rows of A times columns of B. Remarkably, if you suitably interpret the operation, you can also compute $A\,B$ by multiplying columns of A times rows of B! Suppose A is an $m \times n$ matrix with columns $\mathbf{c}_1, \ldots, \mathbf{c}_n$. Suppose B is an $n \times p$ matrix with rows $\mathbf{r}_1, \ldots, \mathbf{r}_n$. Then we claim that

$$A\,B = \mathbf{c}_1\,\mathbf{r}_1 + \mathbf{c}_2\,\mathbf{r}_2 + \cdots + \mathbf{c}_n\,\mathbf{r}_n, \qquad (1.14)$$

where each summand is a matrix of size $m \times p$.

(a) Verify the particular case

$$\begin{pmatrix} 1 & 2 \\ 3 & 4 \end{pmatrix}\begin{pmatrix} 0 & -1 \\ 2 & 3 \end{pmatrix} = \begin{pmatrix} 1 \\ 3 \end{pmatrix}\begin{pmatrix} 0 & -1 \end{pmatrix} + \begin{pmatrix} 2 \\ 4 \end{pmatrix}\begin{pmatrix} 2 & 3 \end{pmatrix} = \begin{pmatrix} 0 & -1 \\ 0 & -3 \end{pmatrix} + \begin{pmatrix} 4 & 6 \\ 8 & 12 \end{pmatrix} = \begin{pmatrix} 4 & 5 \\ 8 & 9 \end{pmatrix}$$

does agree with the usual method for computing the matrix product.

(b) Use this method to compute the matrix products

(i) $\begin{pmatrix} -2 & 1 \\ 3 & 2 \end{pmatrix}\begin{pmatrix} 1 & -2 \\ 1 & 0 \end{pmatrix}$

(ii) $\begin{pmatrix} 1 & -2 & 0 \\ -3 & -1 & 2 \end{pmatrix}\begin{pmatrix} 2 & 5 \\ -3 & 0 \\ 1 & -1 \end{pmatrix}$

(iii) $\begin{pmatrix} 3 & -1 & 1 \\ -1 & 2 & 1 \\ 1 & 1 & -5 \end{pmatrix}\begin{pmatrix} 2 & 3 & 0 \\ 3 & -1 & 4 \\ 0 & 4 & 1 \end{pmatrix}$

and verify that you get the same answer as the traditional method.

(c) Explain why (1.13) is a special case of (1.14).

(d) Prove that (1.14) gives the correct formula for the matrix product.

♡ **1.2.35.** *Matrix Polynomials*. Let $p(x) = c_n\,x^n + c_{n-1}\,x^{n-1} + \cdots + c_1\,x + c_0$ be a polynomial function. If A is a square matrix, we define the corresponding *matrix polynomial* $p(A) = c_n\,A^n + c_{n-1}\,A^{n-1} + \cdots + c_1\,A + c_0\,\mathrm{I}$; the constant term becomes a scalar multiple of the identity matrix. For instance, if $p(x) = x^2 - 2x + 3$, then $p(A) = A^2 - 2A + 3\,\mathrm{I}$.

(a) Write out the matrix polynomials $p(A)$, $q(A)$ when $p(x) = x^3 - 3x + 2$, $q(x) = 2x^2 + 1$.

(b) Evaluate $p(A)$ and $q(A)$ when

$$A = \begin{pmatrix} 1 & 2 \\ -1 & -1 \end{pmatrix}.$$

(c) Show that the matrix product $p(A)\,q(A)$ is the matrix polynomial corresponding to the product polynomial $r(x) = p(x)\,q(x)$.

(d) *True or false*: If $B = p(A)$ and $C = q(A)$ then $B\,C = C\,B$. Check your answer in the particular case of part (b).

♡ **1.2.36.** The matrix S is said to be a *square root* of the matrix A if $S^2 = A$.

(a) Show that $S = \begin{pmatrix} 1 & 1 \\ 3 & -1 \end{pmatrix}$ is a square root of the matrix $A = \begin{pmatrix} 4 & 0 \\ 0 & 4 \end{pmatrix}$. Can you find another square root of A?

(b) Explain why only square matrices can have a square root.

(c) Find all real square roots of the 2×2 identity matrix $\mathrm{I} = \begin{pmatrix} 1 & 0 \\ 0 & 1 \end{pmatrix}$.

(d) Does

$$-\mathrm{I} = \begin{pmatrix} -1 & 0 \\ 0 & -1 \end{pmatrix}$$

have a real square root?

♡ **1.2.37.** A *block matrix* has the form $M = \begin{pmatrix} A & B \\ C & D \end{pmatrix}$ in which A, B, C, D are matrices with respective sizes $i \times k, i \times l, j \times k, j \times l$.

(a) What is the size of M?

(b) Write out the block matrix M when

$$A = \begin{pmatrix} 1 \\ 3 \end{pmatrix}, \qquad B = \begin{pmatrix} 1 & -1 \\ 0 & 1 \end{pmatrix},$$

$$C = \begin{pmatrix} 1 \\ -2 \\ 1 \end{pmatrix}, \qquad D = \begin{pmatrix} 1 & 3 \\ 2 & 0 \\ 1 & -1 \end{pmatrix}.$$

(c) Show that if $N = \begin{pmatrix} P & Q \\ R & S \end{pmatrix}$ is a block matrix whose blocks have the same size as those of M, then

$$M + N = \begin{pmatrix} A + P & B + Q \\ C + R & D + S \end{pmatrix},$$

i.e., matrix addition can be done in blocks.

(d) Show that if $P = \begin{pmatrix} X & Y \\ Z & W \end{pmatrix}$ has blocks of a compatible size, the matrix product is

$$M P = \begin{pmatrix} AX + BZ & AY + BW \\ CX + DZ & CY + DW \end{pmatrix}.$$

Explain what "compatible" means.

(e) Write down a compatible block matrix P for the matrix M in part (b). Then validate the block matrix product identity for your chosen matrices.

1.3 GAUSSIAN ELIMINATION—REGULAR CASE

With the basic matrix arithmetic operations in hand, let us now return to our primary task. The goal is to develop a systematic method for solving linear systems of equations. While we could continue to work directly with the equations, matrices provide a convenient alternative that begins by merely shortening the amount of writing, but ultimately leads to profound insight into the structure of linear systems and their solutions.

We begin by replacing the system (1.7) by its matrix constituents. It is convenient to ignore the vector of unknowns, and form the *augmented matrix*

$$M = \left(A \mid \mathbf{b} \right) = \left(\begin{array}{cccc|c} a_{11} & a_{12} & \dots & a_{1n} & b_1 \\ a_{21} & a_{22} & \dots & a_{2n} & b_2 \\ \vdots & \vdots & \ddots & \vdots & \vdots \\ a_{m1} & a_{m2} & \dots & a_{mn} & b_m \end{array} \right) \tag{1.15}$$

which is an $m \times (n + 1)$ matrix obtained by tacking the right hand side vector onto the original coefficient matrix. The extra vertical line is included just to remind us that the last column of this matrix plays a special role. For example, the augmented matrix for the system (1.1), i.e.,

$$\begin{aligned} x + 2y + z &= 2,\\ 2x + 6y + z &= 7,\\ x + y + 4z &= 3, \end{aligned} \qquad \text{is} \qquad M = \left(\begin{array}{ccc|c} 1 & 2 & 1 & 2\\ 2 & 6 & 1 & 7\\ 1 & 1 & 4 & 3 \end{array}\right). \tag{1.16}$$

Note that one can immediately recover the equations in the original linear system from the augmented matrix. Since operations on equations also affect their right hand sides, keeping track of everything is most easily done through the augmented matrix.

For the time being, we will concentrate our efforts on linear systems that have the same number, n, of equations as unknowns. The associated coefficient matrix A is square, of size $n \times n$. The corresponding augmented matrix $M = \left(A \mid \mathbf{b}\right)$ then has size $n \times (n+1)$.

The matrix operation that assumes the role of Linear System Operation #1 is:

> *Elementary Row Operation* #1: Add a scalar multiple of one row of the augmented matrix to another row.

For example, if we add -2 times the first row of the augmented matrix (1.16) to the second row, the result is the row vector

$$-2\,(1 \quad 2 \quad 1 \quad 2) + (2 \quad 6 \quad 1 \quad 7) = (0 \quad 2 \quad -1 \quad 3).$$

The result can be recognized as the second row of the modified augmented matrix

$$\left(\begin{array}{ccc|c} 1 & 2 & 1 & 2\\ 0 & 2 & -1 & 3\\ 1 & 1 & 4 & 3 \end{array}\right) \tag{1.17}$$

that corresponds to the first equivalent system (1.2). When elementary row operation #1 is performed, it is critical that the result replaces the row being added to—*not* the row being multiplied by the scalar. Notice that the elimination of a variable in an equation—in this case, the first variable in the second equation—amounts to making its entry in the coefficient matrix equal to zero.

We shall call the $(1, 1)$ entry of the coefficient matrix the *first pivot*. The precise definition of pivot will become clear as we continue; the one key requirement is that a pivot be *nonzero*. Eliminating the first variable x from the second and third equations amounts to making all the matrix entries in the column below the pivot equal to zero. We have already done this with the $(2, 1)$ entry in (1.17). To make the $(3, 1)$ entry equal to zero, we subtract (that is, add -1 times) the first row from the last row. The resulting augmented matrix is

$$\left(\begin{array}{ccc|c} 1 & 2 & 1 & 2\\ 0 & 2 & -1 & 3\\ 0 & -1 & 3 & 1 \end{array}\right),$$

which corresponds to the system (1.3). The *second pivot* is the $(2, 2)$ entry of this matrix, which is 2, and is the coefficient of the second variable in the second equation. Again, the pivot must be nonzero. We use the elementary row operation of adding $\frac{1}{2}$ of the second row to the third row to make the entry below the second pivot equal to 0; the result is the augmented matrix

$$N = \left(\begin{array}{ccc|c} 1 & 2 & 1 & 2\\ 0 & 2 & -1 & 3\\ 0 & 0 & \frac{5}{2} & \frac{5}{2} \end{array}\right)$$

that corresponds to the triangular system (1.4). We write the final augmented matrix as

$$N = \bigl(U \mid \mathbf{c}\bigr), \qquad \text{where} \quad U = \begin{pmatrix} 1 & 2 & 1 \\ 0 & 2 & -1 \\ 0 & 0 & \frac{5}{2} \end{pmatrix}, \qquad \mathbf{c} = \begin{pmatrix} 2 \\ 3 \\ \frac{5}{2} \end{pmatrix}.$$

The corresponding linear system has vector form

$$U\mathbf{x} = \mathbf{c}. \tag{1.18}$$

Its coefficient matrix U is *upper triangular*, which means that all its entries below the main diagonal are zero: $u_{ij} = 0$ whenever $i > j$. The three nonzero entries on its diagonal, $1, 2, \frac{5}{2}$, including the last one in the $(3, 3)$ slot, are the three pivots. Once the system has been reduced to triangular form (1.18), we can easily solve it by Back Substitution, as before.

The preceding algorithm for solving a linear system of n equations in n unknowns is known as *regular Gaussian Elimination*. A square matrix A will be called *regular** if the algorithm successfully reduces it to upper triangular form U with all non-zero pivots on the diagonal. In other words, for regular matrices, as the algorithm proceeds, each successive pivot appearing on the diagonal must be nonzero; otherwise, the matrix is not regular. We then use the pivot row to make all the entries lying in the column below the pivot equal to zero through elementary row operations. The solution is found by applying Back Substitution to the resulting triangular system.

Let us state this algorithm in the form of a program, written in a general "pseudocode" that can be easily translated into any specific language, e.g., C++ , FORTRAN , JAVA , MAPLE , MATHEMATICA or MATLAB . By convention, the same letter $M = (m_{ij})$ will be used to denote the current augmented matrix at each stage in the computation, keeping in mind that its entries will change as the algorithm progresses. We initialize $M = \bigl(A \mid \mathbf{b}\bigr)$. The final output of the program, assuming A is regular, is the augmented matrix $M = \bigl(U \mid \mathbf{c}\bigr)$, where U is the upper triangular matrix whose diagonal entries are the pivots, while $\mathbf{c}$ is the resulting vector of right hand sides in the triangular system $U\mathbf{x} = \mathbf{c}$.

Gaussian Elimination—Regular Case

```
start
    for j = 1 to n
        if m_jj = 0, stop; print "A is not regular"
        else for i = j + 1 to n
            set l_ij = m_ij / m_jj
            add -l_ij times row j of M to row i of M
        next i
    next j
end
```

*Strangely, there is no commonly accepted term to describe these kinds of matrices. For lack of a better alternative, we propose to use the adjective "regular" in the sequel.

EXERCISES 1.3

1.3.1. Write out the augmented matrix for the following linear systems. Then solve the system by first applying elementary row operations of type #1 to place the augmented matrix in upper triangular form, followed by Back Substitution.

(a) $\begin{aligned} x_1 + 7x_2 &= 4 \\ -2x_1 - 9x_2 &= 2 \end{aligned}$ (b) $\begin{aligned} 3z - 5w &= -1 \\ 2z + w &= 8 \end{aligned}$

(c) $\begin{aligned} x - 2y + z &= 0 \\ 2y - 8z &= 8 \\ -4x + 5y + 9z &= -9 \end{aligned}$

(d) $\begin{aligned} p + 4q - 2r &= 1 \\ -2p - 3r &= -7 \\ 3p - 2q + 2r &= -1 \end{aligned}$

(e) $\begin{aligned} x_1 - 2x_3 &= -1 \\ x_2 - x_4 &= 2 \\ -3x_2 + 2x_3 &= 0 \\ -4x_1 + 7x_4 &= -5 \end{aligned}$

(f) $\begin{aligned} -x + 3y - z + w &= -2 \\ x - y + 3z - w &= 0 \\ y - z + 4w &= 7 \\ 4x - y + z &= 5 \end{aligned}$

1.3.2. For each of the following augmented matrices write out the corresponding linear system of equations. Solve the system by applying Gaussian Elimination to the augmented matrix.

(a) $\left(\begin{array}{cc|c} 3 & 2 & -4 \\ -4 & -3 & -1 \end{array}\right)$

(b) $\left(\begin{array}{ccc|c} 1 & 2 & 0 & -3 \\ -1 & 2 & 1 & -6 \\ -2 & 0 & -3 & 1 \end{array}\right)$

(c) $\left(\begin{array}{ccc|c} 3 & -1 & 2 & -3 \\ 0 & -2 & -5 & -1 \\ 6 & -2 & 1 & -3 \end{array}\right)$

(d) $\left(\begin{array}{cccc|c} 2 & -1 & 0 & 0 & 0 \\ -1 & 2 & -1 & 0 & 1 \\ 0 & -1 & 2 & -1 & 1 \\ 0 & 0 & -1 & 2 & 0 \end{array}\right)$

1.3.3. Solve the following linear systems by Gaussian Elimination.

(a) $\begin{pmatrix} 1 & -1 \\ 1 & 2 \end{pmatrix}\begin{pmatrix} x \\ y \end{pmatrix} = \begin{pmatrix} 7 \\ 3 \end{pmatrix}$

(b) $\begin{pmatrix} 6 & 1 \\ 3 & -2 \end{pmatrix}\begin{pmatrix} u \\ v \end{pmatrix} = \begin{pmatrix} 5 \\ 5 \end{pmatrix}$

(c) $\begin{pmatrix} 2 & 1 & 2 \\ -1 & 3 & 3 \\ 4 & -3 & 0 \end{pmatrix}\begin{pmatrix} u \\ v \\ w \end{pmatrix} = \begin{pmatrix} 3 \\ -2 \\ 7 \end{pmatrix}$

(d) $\begin{pmatrix} 5 & 3 & -1 \\ 3 & 2 & -1 \\ 1 & 1 & 2 \end{pmatrix}\begin{pmatrix} x_1 \\ x_2 \\ x_3 \end{pmatrix} = \begin{pmatrix} 9 \\ 5 \\ -1 \end{pmatrix}$

(e) $\begin{pmatrix} 1 & 1 & -1 \\ 2 & -1 & 3 \\ -1 & -1 & 3 \end{pmatrix}\begin{pmatrix} p \\ q \\ r \end{pmatrix} = \begin{pmatrix} 0 \\ 3 \\ 5 \end{pmatrix}$

(f) $\begin{pmatrix} -1 & 1 & 1 & 0 \\ 2 & -1 & 0 & 1 \\ 1 & 0 & 2 & 3 \\ 0 & 1 & -1 & -2 \end{pmatrix}\begin{pmatrix} a \\ b \\ c \\ d \end{pmatrix} = \begin{pmatrix} 1 \\ 0 \\ 1 \\ 0 \end{pmatrix}$

(g) $\begin{pmatrix} 2 & -3 & 1 & 1 \\ 1 & -1 & -2 & -1 \\ 3 & -2 & 1 & 2 \\ 1 & 3 & 2 & 1 \end{pmatrix}\begin{pmatrix} x \\ y \\ z \\ w \end{pmatrix} = \begin{pmatrix} -1 \\ 0 \\ 5 \\ 3 \end{pmatrix}$

1.3.4. Find the equation of the parabola $y = ax^2 + bx + c$ that goes through the points $(1, 6)$, $(2, 4)$, $(3, 0)$.

1.3.5. Which of the following matrices are regular?

(a) $\begin{pmatrix} 2 & 1 \\ 1 & 4 \end{pmatrix}$ (b) $\begin{pmatrix} 0 & -1 \\ 3 & -2 \end{pmatrix}$

(c) $\begin{pmatrix} 3 & -2 & 1 \\ -1 & 4 & -3 \\ 3 & -2 & 5 \end{pmatrix}$

(d) $\begin{pmatrix} 1 & -2 & 3 \\ -2 & 4 & -1 \\ 3 & -1 & 2 \end{pmatrix}$

(e) $\begin{pmatrix} 1 & 3 & -3 & 0 \\ -1 & 0 & -1 & 2 \\ 3 & 3 & -6 & 1 \\ 2 & 3 & -3 & 5 \end{pmatrix}$

1.3.6. The techniques that are developed for solving linear systems are also applicable to systems with complex coefficients, whose solutions may also be complex. Use Gaussian Elimination to solve the following complex linear systems.

(a) $\begin{aligned} -\mathrm{i}\,x_1 + (1 + \mathrm{i})x_2 &= -1 \\ (1 - \mathrm{i})x_1 + x_2 &= -3\,\mathrm{i} \end{aligned}$

(b) $\begin{aligned} \mathrm{i}\,x + (1 - \mathrm{i})z &= 2\,\mathrm{i} \\ 2\,\mathrm{i}\,y + (1 + \mathrm{i})z &= 2 \\ -x + 2\,\mathrm{i}\,y + \mathrm{i}\,z &= 1 - 2\,\mathrm{i} \end{aligned}$

(c) $\begin{aligned} (1 - \mathrm{i})x + 2y &= \mathrm{i} \\ -\mathrm{i}\,x + (1 + \mathrm{i})y &= -1 \end{aligned}$

(d) $$\begin{aligned}(1+\mathrm{i})x + \mathrm{i}\,y + (2+2\mathrm{i})z &= 0\\ (1-\mathrm{i})x + 2y + \mathrm{i}\,z &= 0\\ (3-3\mathrm{i})x + \mathrm{i}\,y + (3-11\mathrm{i})z &= 6\end{aligned}$$

1.3.7. (a) Write down an example of a system of 5 linear equations in 5 unknowns with regular diagonal coefficient matrix.

(b) Solve your system.

(c) Explain why solving a system whose coefficient matrix is diagonal is very easy.

◇ **1.3.8.** A linear system is called *homogeneous* if all the right hand sides are zero, and so takes the matrix form $A\mathbf{x} = \mathbf{0}$. Explain why the solution to a homogeneous system with regular coefficient matrix is $\mathbf{x} = \mathbf{0}$.

♠ **1.3.9.** (a) Write a pseudocode program for Back Substitution. The input will consist of an upper triangular matrix U with nonzero diagonal entries and a vector $\mathbf{c}$. The output should be the solution $\mathbf{x}$ to the triangular system $U\mathbf{x} = \mathbf{c}$.

(b) Implement your code on a computer, and test it on some representative systems.

1.3.10. Under what conditions do two 2×2 upper triangular matrices commute?

1.3.11. A matrix is called *lower triangular* if all entries above the diagonal are zero. Show that a matrix is both lower and upper triangular if and only if it is a diagonal matrix.

◇ **1.3.12.** A square matrix is called *strictly lower* (*upper*) *triangular* if all entries on or above (below) the main diagonal are 0.

(a) Prove that any square matrix can be uniquely written as a sum $A = L+D+U$, with L strictly lower triangular, D diagonal, and U strictly upper triangular.

(b) Decompose $A = \begin{pmatrix} 3 & 1 & -1 \\ 1 & -4 & 2 \\ -2 & 0 & 5 \end{pmatrix}$ in this manner.

◇ **1.3.13.** A square matrix A is called *nilpotent* if $A^k = \mathrm{O}$ for some $k \geq 1$.

(a) Show that $A = \begin{pmatrix} 0 & 1 & 2 \\ 0 & 0 & 1 \\ 0 & 0 & 0 \end{pmatrix}$ is nilpotent.

(b) Show that any strictly upper triangular matrix is nilpotent.

(c) Find a nilpotent matrix which is neither lower nor upper triangular.

Elementary Matrices

A key observation is that elementary row operations can, in fact, be realized by matrix multiplication. To this end, we introduce the first type of "elementary matrix". (Later we will meet two other types of elementary matrix, corresponding to two other kinds of elementary row operation.)

Definition 1.1 The *elementary matrix* E associated with an elementary row operation for m–rowed matrices is the matrix obtained by applying the row operation to the $m \times m$ identity matrix I_m.

For example, applying the elementary row operation that adds -2 times the first row to the second row of the 3×3 identity matrix $\mathrm{I} = \begin{pmatrix} 1 & 0 & 0 \\ 0 & 1 & 0 \\ 0 & 0 & 1 \end{pmatrix}$ results in the corresponding elementary matrix $E_1 = \begin{pmatrix} 1 & 0 & 0 \\ -2 & 1 & 0 \\ 0 & 0 & 1 \end{pmatrix}$. We claim that, if A is *any* 3–rowed matrix, then multiplying $E_1 A$ has the same effect as the given elementary row operation. For example,

$$\begin{pmatrix} 1 & 0 & 0 \\ -2 & 1 & 0 \\ 0 & 0 & 1 \end{pmatrix}\begin{pmatrix} 1 & 2 & 1 \\ 2 & 6 & 1 \\ 1 & 1 & 4 \end{pmatrix} = \begin{pmatrix} 1 & 2 & 1 \\ 0 & 2 & -1 \\ 1 & 1 & 4 \end{pmatrix},$$

which you may recognize as the first elementary row operation we used to solve our illustrative example. If we set

$$E_1 = \begin{pmatrix} 1 & 0 & 0 \\ -2 & 1 & 0 \\ 0 & 0 & 1 \end{pmatrix}, \qquad E_2 = \begin{pmatrix} 1 & 0 & 0 \\ 0 & 1 & 0 \\ -1 & 0 & 1 \end{pmatrix}, \qquad E_3 = \begin{pmatrix} 1 & 0 & 0 \\ 0 & 1 & 0 \\ 0 & \frac{1}{2} & 1 \end{pmatrix}, \tag{1.19}$$

then multiplication by E_1 will subtract twice the first row from the second row, multiplication by E_2 will subtract the first row from the third row, and multiplication by E_3 will add $\frac{1}{2}$ the second row to the third row—precisely the row operations used to place our original system in triangular form. Therefore, performing them in the correct order (and using the associativity of matrix multiplication), we conclude that when

$$A = \begin{pmatrix} 1 & 2 & 1 \\ 2 & 6 & 1 \\ 1 & 1 & 4 \end{pmatrix}, \qquad \text{then} \quad E_3 E_2 E_1 A = U = \begin{pmatrix} 1 & 2 & 1 \\ 0 & 2 & -1 \\ 0 & 0 & \frac{5}{2} \end{pmatrix}. \tag{1.20}$$

The reader is urged to check this by directly multiplying the indicated matrices.

In general, then, an $m \times m$ *elementary matrix E of the first type* will have all 1's on the diagonal, one nonzero entry c in some off-diagonal position (i, j), with $i \neq j$, and all other entries equal to zero. If A is any $m \times n$ matrix, then the matrix product $E A$ is equal to the matrix obtained from A by the elementary row operation adding c times row j to row i. (Note that the order of i and j is reversed.)

To undo the operation of adding c times row j to row i, we must perform the inverse row operation that subtracts c (or, equivalently, adds $-c$) times row j from row i. The corresponding *inverse elementary matrix* again has 1's along the diagonal and $-c$ in the (i, j) slot. Let us denote the inverses of the particular elementary matrices (1.19) by L_i, so that, according to our general rule,

$$L_1 = \begin{pmatrix} 1 & 0 & 0 \\ 2 & 1 & 0 \\ 0 & 0 & 1 \end{pmatrix}, \qquad L_2 = \begin{pmatrix} 1 & 0 & 0 \\ 0 & 1 & 0 \\ 1 & 0 & 1 \end{pmatrix}, \qquad L_3 = \begin{pmatrix} 1 & 0 & 0 \\ 0 & 1 & 0 \\ 0 & -\frac{1}{2} & 1 \end{pmatrix}. \tag{1.21}$$

Note that the products

$$L_1 E_1 = L_2 E_2 = L_3 E_3 = \mathrm{I} \tag{1.22}$$

yield the 3×3 identity matrix, reflecting the fact that the matrices represent mutually inverse row operations. (A more thorough discussion of matrix inverses will be postponed until Section 1.5.)

The product of the latter three elementary matrices (1.21) is equal to

$$L = L_1 L_2 L_3 = \begin{pmatrix} 1 & 0 & 0 \\ 2 & 1 & 0 \\ 1 & -\frac{1}{2} & 1 \end{pmatrix}. \tag{1.23}$$

The matrix L is called a *special lower triangular* matrix, where "lower triangular" means that all the entries above the main diagonal are 0, while "special" indicates that all the entries on the diagonal are equal to 1. Observe that the entries of L below the diagonal are the same as the corresponding nonzero entries in the L_i. This is a general fact that holds when the lower triangular elementary matrices are multiplied in the correct order. More generally, the following elementary consequence of the laws of matrix multiplication will be used extensively.

Lemma 1.2 If L and $\widehat{L}$ are lower triangular matrices of the same size, so is their product $L \widehat{L}$. If they are both special lower triangular, so is their product. Similarly, if $U, \widehat{U}$ are (special) upper triangular matrices, so is their product $U \widehat{U}$.

The LU Factorization

We have almost arrived at our first important result. Let us compute the product of the matrices L and U in (1.20), (1.23). Using associativity of matrix multiplication, equations (1.22), and the basic property of the identity matrix I, we conclude that

$$\begin{aligned} LU &= (L_1L_2L_3)(E_3E_2E_1A) = L_1L_2(L_3E_3)E_2E_1A = L_1L_2\,\mathrm{I}\,E_2E_1A \\ &= L_1(L_2E_2)E_1A = L_1\,\mathrm{I}\,E_1A = (L_1E_1)A = \mathrm{I}\,A = A. \end{aligned}$$

In other words, we have *factored* the coefficient matrix $A = L\,U$ into a product of a special lower triangular matrix L and an upper triangular matrix U with the nonzero pivots on its main diagonal. By similar reasoning, the same holds true for almost all square matrices.

THEOREM 1.3 A matrix A is regular if and only if it can be factored

$$A = L\,U, \tag{1.24}$$

where L is a special lower triangular, having all 1's on the diagonal, and U is upper triangular with nonzero diagonal entries, which are the pivots of A. The nonzero off-diagonal entries l_{ij} for $i > j$ appearing in L prescribe the elementary row operations that bring A into upper triangular form; namely, one subtracts l_{ij} times row j from row i at the appropriate step of the Gaussian Elimination process.

In practice, to find the $L\,U$ factorization of a square matrix A, one applies the regular Gaussian Elimination algorithm to reduce A to its upper triangular form U. The entries of L can be filled in during the course of the calculation with the negatives of the multiples used in the elementary row operations. If the algorithm fails to be completed, which happens whenever zero appears in any diagonal pivot position, then the original matrix is *not* regular, and does *not* have an $L\,U$ factorization.

EXAMPLE 1.4 Let us compute the $L\,U$ factorization of the matrix

$$A = \begin{pmatrix} 2 & 1 & 1 \\ 4 & 5 & 2 \\ 2 & -2 & 0 \end{pmatrix}.$$

Applying the Gaussian Elimination algorithm, we begin by adding -2 times the first row to the second row, and then adding -1 times the first row to the third. The result is the matrix $\begin{pmatrix} 2 & 1 & 1 \\ 0 & 3 & 0 \\ 0 & -3 & -1 \end{pmatrix}$. The next step adds the second row to the third row, leading to the upper triangular matrix

$$U = \begin{pmatrix} 2 & 1 & 1 \\ 0 & 3 & 0 \\ 0 & 0 & -1 \end{pmatrix},$$

whose diagonal entries are the pivots. The corresponding lower triangular matrix is

$$L = \begin{pmatrix} 1 & 0 & 0 \\ 2 & 1 & 0 \\ 1 & -1 & 1 \end{pmatrix};$$

its entries lying below the main diagonal are the *negatives* of the multiples we used during the elimination procedure. For instance, the (2, 1) entry indicates that we added -2 times the first row to the second row, and so on. The reader might wish to verify the resulting factorization

$$\begin{pmatrix} 2 & 1 & 1 \\ 4 & 5 & 2 \\ 2 & -2 & 0 \end{pmatrix} = A = LU = \begin{pmatrix} 1 & 0 & 0 \\ 2 & 1 & 0 \\ 1 & -1 & 1 \end{pmatrix} \begin{pmatrix} 2 & 1 & 1 \\ 0 & 3 & 0 \\ 0 & 0 & -1 \end{pmatrix}. \qquad \bullet$$

EXERCISES 1.3

1.3.14. What elementary row operations do the following matrices represent? What size matrices do they apply to?

(a) $\begin{pmatrix} 1 & -2 \\ 0 & 1 \end{pmatrix}$ (b) $\begin{pmatrix} 1 & 0 \\ 7 & 1 \end{pmatrix}$

(c) $\begin{pmatrix} 1 & 0 & 0 \\ 0 & 1 & -5 \\ 0 & 0 & 1 \end{pmatrix}$ (d) $\begin{pmatrix} 1 & 0 & 0 \\ 0 & 1 & 0 \\ \frac{1}{2} & 0 & 1 \end{pmatrix}$

(e) $\begin{pmatrix} 1 & 0 & 0 & 0 \\ 0 & 1 & 0 & -3 \\ 0 & 0 & 1 & 0 \\ 0 & 0 & 0 & 1 \end{pmatrix}$

1.3.15. Write down the elementary matrix corresponding to the following row operations on 4×4 matrices:

(a) Add the third row to the fourth row.

(b) Subtract the fourth row from the third row.

(c) Add 3 times the last row to the first row.

(d) Subtract twice the second row from the fourth row.

1.3.16. Compute the product $L_3 L_2 L_1$ of the elementary matrices (1.21). Compare your answer with (1.23).

1.3.17. Determine the product $E_3 E_2 E_1$ of the elementary matrices in (1.19). Is this the same as the product $E_1 E_2 E_3$? Which is easier to predict?

1.3.18. (a) Explain, using their interpretation as elementary row operations, why elementary matrices do not generally commute: $E\widetilde{E} \neq \widetilde{E}E$.

(b) Which pairs of the elementary matrices listed in (1.19) commute?

(c) Can you formulate a general rule that tells in advance whether two given elementary matrices commute?

1.3.19. Determine which of the following 3×3 matrices is

(i) upper triangular,

(ii) special upper triangular,

(iii) lower triangular, and/or

(iv) special lower triangular:

(a) $\begin{pmatrix} 1 & 2 & 0 \\ 0 & 3 & 2 \\ 0 & 0 & -2 \end{pmatrix}$ (b) $\begin{pmatrix} 1 & 0 & 0 \\ 0 & 1 & 0 \\ 0 & 0 & 1 \end{pmatrix}$

(c) $\begin{pmatrix} 1 & 0 & 0 \\ 2 & 0 & 0 \\ 0 & 3 & 3 \end{pmatrix}$ (d) $\begin{pmatrix} 1 & 0 & 0 \\ 0 & 1 & 0 \\ 1 & -4 & 1 \end{pmatrix}$

(e) $\begin{pmatrix} 0 & 0 & 0 \\ 0 & 3 & 1 \\ 0 & 1 & 0 \end{pmatrix}$

1.3.20. Write down the explicit requirements on its entries a_{ij} that a square matrix A be

(a) diagonal

(b) upper triangular

(c) special upper triangular

(d) lower triangular

(e) special lower triangular

◇ **1.3.21.** (a) Explain why the product of two lower triangular matrices is lower triangular.

(b) What can you say concerning the diagonal entries of the product of two lower triangular matrices?

(c) Explain why the product of two special lower triangular matrices is also special lower triangular.

1.3.22. Find the LU factorization of the following matrices:

(a) $\begin{pmatrix} 1 & 3 \\ -1 & 0 \end{pmatrix}$ (b) $\begin{pmatrix} 1 & 3 \\ 3 & 1 \end{pmatrix}$

(c) $\begin{pmatrix} -1 & 1 & -1 \\ 1 & 1 & 1 \\ -1 & 1 & 2 \end{pmatrix}$ (d) $\begin{pmatrix} 2 & 0 & 3 \\ 1 & 3 & 1 \\ 0 & 1 & 1 \end{pmatrix}$

(e) $\begin{pmatrix} -1 & 0 & 0 \\ 2 & -3 & 0 \\ 1 & 3 & 2 \end{pmatrix}$ (f) $\begin{pmatrix} 1 & 0 & -1 \\ 2 & 3 & 2 \\ -3 & 1 & 0 \end{pmatrix}$

(g) $\begin{pmatrix} 1 & 0 & -1 & 0 \\ 0 & 2 & -1 & -1 \\ -1 & 3 & 0 & 2 \\ 0 & -1 & 2 & 1 \end{pmatrix}$

(h) $\begin{pmatrix} 1 & 1 & -2 & 3 \\ -1 & 2 & 3 & 0 \\ -2 & 1 & 1 & -2 \\ 3 & 0 & 1 & 5 \end{pmatrix}$

(i) $\begin{pmatrix} 2 & 1 & 3 & 1 \\ 1 & 4 & 0 & 1 \\ 3 & 0 & 2 & 2 \\ 1 & 1 & 2 & 2 \end{pmatrix}$

1.3.23. Given the factorization

$$A = \begin{pmatrix} 2 & -1 & 0 \\ -6 & 4 & -1 \\ 4 & -6 & 7 \end{pmatrix} = \begin{pmatrix} 1 & 0 & 0 \\ -3 & 1 & 0 \\ 2 & -4 & 1 \end{pmatrix} \begin{pmatrix} 2 & -1 & 0 \\ 0 & 1 & -1 \\ 0 & 0 & 3 \end{pmatrix},$$

explain, without computing, which elementary row operations are used to reduce A to upper triangular form. Be careful to state which order they should be applied. Then check the correctness of your answer by performing the elimination.

1.3.24. (a) Write down a 4×4 special lower triangular matrix whose entries below the diagonal are different, nonzero numbers.

(b) Explain which elementary row operation each entry corresponds to.

(c) Indicate the order in which the elementary row operations should be performed by labeling the entries $1, 2, 3, \ldots$.

◇ **1.3.25.** Let $t_1, t_2, \ldots$ be distinct real numbers. Find the LU factorization of the following *Vandermonde matrices*:

(a) $\begin{pmatrix} 1 & 1 \\ t_1 & t_2 \end{pmatrix}$ (b) $\begin{pmatrix} 1 & 1 & 1 \\ t_1 & t_2 & t_3 \\ t_1^2 & t_2^2 & t_3^2 \end{pmatrix}$

(c) $\begin{pmatrix} 1 & 1 & 1 & 1 \\ t_1 & t_2 & t_3 & t_4 \\ t_1^2 & t_2^2 & t_3^2 & t_4^2 \\ t_1^3 & t_2^3 & t_3^3 & t_4^3 \end{pmatrix}$

Can you spot a pattern? Test your conjecture with the 5×5 Vandermonde matrix.

1.3.26. *True or false*: If A has a zero entry on its main diagonal, it is not regular.

1.3.27. In general, how many elementary row operations does one need to perform in order to reduce a regular $n \times n$ matrix to upper triangular form?

1.3.28. Prove that if A is a regular 2×2 matrix, then its LU factorization is unique. In other words, if $A = LU = \widehat{L}\,\widehat{U}$ where $L, \widehat{L}$ are special lower triangular and $U, \widehat{U}$ are upper triangular, then $L = \widehat{L}$ and $U = \widehat{U}$. (The general case appears in Proposition 1.30.)

◇ **1.3.29.** Prove directly that the matrix $A = \begin{pmatrix} 0 & 1 \\ 1 & 0 \end{pmatrix}$ does not have an LU factorization.

◇ **1.3.30.** Suppose A is regular.

(a) Show that the matrix obtained by multiplying each column of A by the sign of its pivot is also regular and, moreover, has all positive pivots.

(b) Show that the matrix obtained by multiplying each row of A by the sign of its pivot is also regular and has all positive pivots.

(c) Check these results in the particular case

$$A = \begin{pmatrix} -2 & 2 & 1 \\ 1 & 0 & 1 \\ 4 & 2 & 3 \end{pmatrix}.$$

Forward and Back Substitution

Once we know the LU factorization of a regular matrix A, we are able to solve any associated linear system $A\mathbf{x} = \mathbf{b}$ in two easy stages:

(1) First, solve the lower triangular system

$$L\mathbf{c} = \mathbf{b} \qquad (1.25)$$

for the vector $\mathbf{c}$ by *Forward Substitution*. This is the same as Back Substitution, except one solves the equations for the variables in the direct order—from first to last. Explicitly,

$$c_1 = b_1, \qquad c_i = b_i - \sum_{j=1}^{i-1} l_{ij}\,c_j, \qquad \text{for} \quad i = 2, 3, \ldots, n, \qquad (1.26)$$

noting that the previously computed values of $c_1, \ldots, c_{i-1}$ are used to determine c_i.

(2) Second, solve the resulting upper triangular system

$$U\,\mathbf{x} = \mathbf{c} \tag{1.27}$$

by *Back Substitution*. The values of the unknowns

$$x_n = \frac{c_n}{u_{nn}}, \qquad x_i = \frac{1}{u_{ii}}\left(c_i - \sum_{j=i+1}^{n} u_{ij}\,x_j \right), \qquad \text{for} \quad i = n-1, \ldots, 2, 1, \tag{1.28}$$

are successively computed, but now in reverse order. It is worth pointing out that the requirement that each pivot $u_{ii} \neq 0$ is essential here, as otherwise we would not be able to solve for the corresponding variable x_i.

Note that the combined algorithm does indeed solve the original system, since if

$$U\,\mathbf{x} = \mathbf{c} \quad \text{and} \quad L\,\mathbf{c} = \mathbf{b}, \quad \text{then} \quad A\,\mathbf{x} = L\,U\,\mathbf{x} = L\,\mathbf{c} = \mathbf{b}.$$

EXAMPLE 1.5 With the $L\,U$ decomposition

$$\begin{pmatrix} 2 & 1 & 1 \\ 4 & 5 & 2 \\ 2 & -2 & 0 \end{pmatrix} = \begin{pmatrix} 1 & 0 & 0 \\ 2 & 1 & 0 \\ 1 & -1 & 1 \end{pmatrix} \begin{pmatrix} 2 & 1 & 1 \\ 0 & 3 & 0 \\ 0 & 0 & -1 \end{pmatrix}$$

found in Example 1.4, we can readily solve any linear system with the given coefficient matrix by Forward and Back Substitution. For instance, to find the solution to

$$\begin{pmatrix} 2 & 1 & 1 \\ 4 & 5 & 2 \\ 2 & -2 & 0 \end{pmatrix} \begin{pmatrix} x \\ y \\ z \end{pmatrix} = \begin{pmatrix} 1 \\ 2 \\ 2 \end{pmatrix},$$

we first solve the lower triangular system

$$\begin{pmatrix} 1 & 0 & 0 \\ 2 & 1 & 0 \\ 1 & -1 & 1 \end{pmatrix} \begin{pmatrix} a \\ b \\ c \end{pmatrix} = \begin{pmatrix} 1 \\ 2 \\ 2 \end{pmatrix}, \qquad \text{or, explicitly,} \qquad \begin{aligned} a \qquad\quad &= 1, \\ 2a + b \qquad &= 2, \\ a - b + c &= 2. \end{aligned}$$

The first equation says $a = 1$; substituting into the second, we find $b = 0$; the final equation yields $c = 1$. We then use Back Substitution to solve the upper triangular system

$$\begin{pmatrix} 2 & 1 & 1 \\ 0 & 3 & 0 \\ 0 & 0 & -1 \end{pmatrix} \begin{pmatrix} x \\ y \\ z \end{pmatrix} = \begin{pmatrix} a \\ b \\ c \end{pmatrix} = \begin{pmatrix} 1 \\ 0 \\ 1 \end{pmatrix}, \qquad \text{which is} \qquad \begin{aligned} 2x + y + z &= 1, \\ 3y \qquad &= 0, \\ -z &= 1. \end{aligned}$$

We find $z = -1$, then $y = 0$, and then $x = 1$, which is indeed the solution. ●

Thus, once we have found the $L\,U$ factorization of the coefficient matrix A, the Forward and Back Substitution processes quickly produce the solution to any system $A\,\mathbf{x} = \mathbf{b}$. Moreover, they can be straightforwardly programmed on a computer. In practice, to solve a system from scratch, it is a matter of taste whether you work directly with the augmented matrix, or first determine the $L\,U$ factorization of the coefficient matrix, and then apply Forward and Back Substitution to compute the solution.

EXERCISES 1.3

1.3.31. Given the LU factorizations you calculated in Exercise 1.3.22, solve the associated linear systems $A\mathbf{x} = \mathbf{b}$ where $\mathbf{b}$ is the column vector with all entries equal to 1.

1.3.32. In each of the following problems, find the $A = LU$ factorization of the coefficient matrix, and then use Forward and Back Substitution to solve the corresponding linear systems $A\mathbf{x} = \mathbf{b}_j$ for each of the indicated right hand sides:

(a) $A = \begin{pmatrix} -1 & 3 \\ 3 & 2 \end{pmatrix}$,
$\mathbf{b}_1 = \begin{pmatrix} 1 \\ -1 \end{pmatrix}$, $\mathbf{b}_2 = \begin{pmatrix} 2 \\ 5 \end{pmatrix}$, $\mathbf{b}_3 = \begin{pmatrix} 0 \\ 3 \end{pmatrix}$

(b) $A = \begin{pmatrix} -1 & 1 & -1 \\ 1 & 1 & 1 \\ -1 & 1 & 2 \end{pmatrix}$,
$\mathbf{b}_1 = \begin{pmatrix} 1 \\ -1 \\ 1 \end{pmatrix}$, $\mathbf{b}_2 = \begin{pmatrix} -3 \\ 0 \\ 2 \end{pmatrix}$

(c) $A = \begin{pmatrix} 9 & -2 & -1 \\ -6 & 1 & 1 \\ 2 & -1 & 0 \end{pmatrix}$,
$\mathbf{b}_1 = \begin{pmatrix} 2 \\ -1 \\ 0 \end{pmatrix}$, $\mathbf{b}_2 = \begin{pmatrix} 1 \\ 2 \\ 5 \end{pmatrix}$

(d) $A = \begin{pmatrix} 2.0 & .3 & .4 \\ .3 & 4.0 & .5 \\ .4 & .5 & 6.0 \end{pmatrix}$,
$\mathbf{b}_1 = \begin{pmatrix} 1 \\ 0 \\ 0 \end{pmatrix}$, $\mathbf{b}_2 = \begin{pmatrix} 0 \\ 1 \\ 0 \end{pmatrix}$, $\mathbf{b}_3 = \begin{pmatrix} 0 \\ 0 \\ 1 \end{pmatrix}$

(e) $A = \begin{pmatrix} 1 & 0 & -1 & 0 \\ 0 & 2 & 3 & -1 \\ -1 & 3 & 2 & 2 \\ 0 & -1 & 2 & 1 \end{pmatrix}$,
$\mathbf{b}_1 = \begin{pmatrix} 1 \\ 0 \\ -1 \\ 1 \end{pmatrix}$, $\mathbf{b}_2 = \begin{pmatrix} 0 \\ -1 \\ 0 \\ 1 \end{pmatrix}$

(f) $A = \begin{pmatrix} 1 & -2 & 0 & 2 \\ 4 & 1 & -1 & -1 \\ -8 & -1 & 2 & 1 \\ -4 & -1 & 1 & 2 \end{pmatrix}$,
$\mathbf{b}_1 = \begin{pmatrix} 1 \\ 0 \\ 0 \\ 0 \end{pmatrix}$, $\mathbf{b}_2 = \begin{pmatrix} 3 \\ 0 \\ -1 \\ 2 \end{pmatrix}$, $\mathbf{b}_3 = \begin{pmatrix} 2 \\ 3 \\ -2 \\ 1 \end{pmatrix}$

1.4 PIVOTING AND PERMUTATIONS

The method of Gaussian Elimination presented so far applies only to regular matrices. But not every square matrix is regular; a simple class of examples is matrices whose upper left, i.e., (1, 1), entry is zero, and so cannot serve as the first pivot. More generally, the algorithm cannot proceed whenever a zero entry appears in the current pivot position on the diagonal. What then to do? The answer requires revisiting the source of the method.

Consider, as a specific example, the linear system

$$\begin{aligned} 2y + z &= 2, \\ 2x + 6y + z &= 7, \\ x + y + 4z &= 3. \end{aligned} \tag{1.29}$$

The augmented coefficient matrix is

$$\left(\begin{array}{ccc|c} 0 & 2 & 1 & 2 \\ 2 & 6 & 1 & 7 \\ 1 & 1 & 4 & 3 \end{array}\right).$$

In this case, the (1, 1) entry is 0, and so is not a legitimate pivot. The problem, of course, is that the first variable x does not appear in the first equation, and so we cannot use it to eliminate x in the other two equations. But this "problem" is actually a bonus—we already have an equation with only two variables in it, and so we only need to eliminate x from one of the other two equations. To be systematic, we rewrite the system in a different order,

$$\begin{aligned} 2x + 6y + z &= 7, \\ 2y + z &= 2, \\ x + y + 4z &= 3, \end{aligned}$$

by interchanging the first two equations. In other words, we employ

Linear System Operation #2: Interchange two equations.

Clearly, this operation does not change the solution and so produces an equivalent linear system. In our case, the augmented coefficient matrix

$$\left(\begin{array}{ccc|c} 2 & 6 & 1 & 7 \\ 0 & 2 & 1 & 2 \\ 1 & 1 & 4 & 3 \end{array}\right),$$

can be obtained from the original by performing the second type of row operation:

Elementary Row Operation #2: Interchange two rows of the matrix.

The new nonzero upper left entry, 2, can now serve as the first pivot, and we may continue to apply elementary row operations of Type #1 to reduce our matrix to upper triangular form. For this particular example, we eliminate the remaining nonzero entry in the first column by subtracting $\frac{1}{2}$ the first row from the last:

$$\left(\begin{array}{ccc|c} 2 & 6 & 1 & 7 \\ 0 & 2 & 1 & 2 \\ 0 & -2 & \frac{7}{2} & -\frac{1}{2} \end{array}\right).$$

The (2, 2) entry serves as the next pivot. To eliminate the nonzero entry below it, we add the second to the third row:

$$\left(\begin{array}{ccc|c} 2 & 6 & 1 & 7 \\ 0 & 2 & 1 & 2 \\ 0 & 0 & \frac{9}{2} & \frac{3}{2} \end{array}\right).$$

We have now placed the system in upper triangular form, with the three pivots 2, 2, and $\frac{9}{2}$ along the diagonal. Back Substitution produces the solution $x = \frac{5}{6}$, $y = \frac{5}{6}$, $z = \frac{1}{3}$.

The row interchange that is required when a zero shows up in the diagonal pivot position is known as *pivoting*. Later, in Section 1.7, we will discuss practical reasons for pivoting even when a diagonal entry is nonzero. Let us distinguish the class of matrices that can be reduced to upper triangular form by Gaussian Elimination with pivoting. These matrices will prove to be of fundamental importance throughout linear algebra.

Definition 1.6 A square matrix is called *nonsingular* if it can be reduced to upper triangular form with all non-zero elements on the diagonal—the pivots—by elementary row operations of Types 1 and 2.

In contrast, a *singular* square matrix cannot be reduced to such upper triangular form by such row operations, because at some stage in the elimination procedure the diagonal entry and all the entries below it are zero. Every regular matrix is nonsingular, but, as we just saw, not every nonsingular matrix is regular. Uniqueness of solutions is the key defining characteristic of nonsingularity.

THEOREM 1.7 A linear system $A\mathbf{x} = \mathbf{b}$ has a unique solution for *every* choice of right hand side $\mathbf{b}$ if and only if its coefficient matrix A is square and nonsingular.

We are able to prove the "if" part of this theorem, since nonsingularity implies reduction to an equivalent upper triangular form that has the same solutions as the original system. The unique solution to the system is then found by Back Substitution. The "only if" part will be proved in Section 1.8.

The revised version of the Gaussian Elimination algorithm, valid for all nonsingular coefficient matrices, is implemented by the accompanying pseudocode program. The starting point is the augmented matrix $M = \left(A \mid \mathbf{b} \right)$ representing the linear system $A\mathbf{x} = \mathbf{b}$. After successful termination of the program, the result is an augmented matrix in upper triangular form $M = \left(U \mid \mathbf{c} \right)$ representing the equivalent linear system $U\mathbf{x} = \mathbf{c}$. One then uses Back Substitution to determine the solution $\mathbf{x}$ to the linear system.

Gaussian Elimination– Nonsingular Case

```
start
    for j = 1 to n
        if m_kj = 0 for all k ≥ j, stop; print "A is singular"
        if m_jj = 0 but m_kj ≠ 0 for some k > j, switch rows k and j
        for i = j + 1 to n
            set l_ij = m_ij / m_jj
            add − l_ij times row j to row i of M
        next i
    next j
end
```

EXERCISES 1.4

1.4.1. Determine whether the following matrices are singular or nonsingular:

(a) $\begin{pmatrix} 0 & 1 \\ 1 & 2 \end{pmatrix}$ (b) $\begin{pmatrix} -1 & 2 \\ 4 & -8 \end{pmatrix}$

(c) $\begin{pmatrix} 0 & 1 & 2 \\ -1 & 1 & 3 \\ 2 & -2 & 0 \end{pmatrix}$ (d) $\begin{pmatrix} 1 & 1 & 3 \\ 2 & 2 & 2 \\ 3 & -1 & 1 \end{pmatrix}$

(e) $\begin{pmatrix} 1 & 2 & 3 \\ 4 & 5 & 6 \\ 7 & 8 & 9 \end{pmatrix}$

(f) $\begin{pmatrix} -1 & 1 & 0 & -3 \\ 2 & -2 & 4 & 0 \\ 1 & -2 & 2 & -1 \\ 0 & 1 & 0 & 1 \end{pmatrix}$

(g) $\begin{pmatrix} 0 & -1 & 0 & 1 \\ 1 & 0 & -1 & 0 \\ 0 & 2 & 0 & -2 \\ 2 & 0 & 2 & 0 \end{pmatrix}$

(h) $\begin{pmatrix} 1 & -2 & 0 & 2 \\ 4 & 1 & -1 & -1 \\ -8 & -1 & 2 & 1 \\ -4 & -1 & 1 & 2 \end{pmatrix}$

1.4.2. Classify the following matrices as

(i) regular (ii) nonsingular, and/or

(iii) singular:

(a) $\begin{pmatrix} 2 & 1 \\ 1 & 4 \end{pmatrix}$ (b) $\begin{pmatrix} 3 & -2 & 1 \\ -1 & 4 & 4 \\ 2 & 2 & 5 \end{pmatrix}$

(c) $\begin{pmatrix} 1 & -2 & 3 \\ -2 & 4 & -1 \\ 3 & -1 & 2 \end{pmatrix}$

(d) $\begin{pmatrix} 1 & 3 & -3 & 0 \\ -1 & 0 & -1 & 2 \\ 3 & -2 & 6 & 1 \\ 2 & -1 & 3 & 5 \end{pmatrix}$

1.4.3. Solve the following systems of equations by Gaussian Elimination:

(a) $\begin{aligned} x_1 - 2x_2 + 2x_3 &= 15 \\ x_1 - 2x_2 + x_3 &= 10 \\ 2x_1 - x_2 - 2x_3 &= -10 \end{aligned}$

(b) $\begin{aligned} 2x_1 - x_2 &= 1 \\ -4x_1 + 2x_2 - 3x_3 &= -8 \\ x_1 - 3x_2 + x_3 &= 5 \end{aligned}$

(c) $\begin{aligned} x_2 - x_3 &= 4 \\ -2x_1 - 5x_2 &= 2 \\ x_1 + x_3 &= -8 \end{aligned}$

(d) $\begin{aligned} x - y + z - w &= 0 \\ -2x + 2y - z + w &= 2 \\ -4x + 4y + 3z &= 5 \\ x - 3y + w &= 4 \end{aligned}$

(e) $\begin{aligned} -3x_2 + 2x_3 &= 0 \\ x_3 - x_4 &= 2 \\ x_1 - 2x_3 &= -1 \\ -4x_1 + 7x_4 &= -5 \end{aligned}$

1.4.4. Find the equation $z = ax + by + c$ for the plane passing through the three points $\mathbf{p}_1 = (0, 2, -1)$, $\mathbf{p}_2 = (-2, 4, 3)$, $\mathbf{p}_3 = (2, -1, -3)$.

1.4.5. Show that a 2×2 matrix $A = \begin{pmatrix} a & b \\ c & d \end{pmatrix}$ is

(a) nonsingular if and only if $ad - bc \neq 0$,

(b) regular if and only if $ad - bc \neq 0$ and $a \neq 0$.

1.4.6. *True or false*: A singular matrix cannot be regular.

◇ **1.4.7.** Explain why the solution to the homogeneous system $A\mathbf{x} = \mathbf{0}$ with nonsingular coefficient matrix is $\mathbf{x} = \mathbf{0}$.

1.4.8. Write out the details of the proof of the "if" part of Theorem 1.7: if A is nonsingular, then the linear system $A\mathbf{x} = \mathbf{b}$ has a unique solution for every $\mathbf{b}$.

Permutation Matrices

As with the first type of elementary row operation, row interchanges can be accomplished by multiplication by a second type of elementary matrix, which is found by applying the row operation to the identity matrix of the appropriate size. For instance, interchanging rows 1 and 2 of the 3×3 identity matrix produces the elementary interchange matrix

$$P = \begin{pmatrix} 0 & 1 & 0 \\ 1 & 0 & 0 \\ 0 & 0 & 1 \end{pmatrix}.$$

The result PA of multiplying any 3–rowed matrix A on the left by P is the same as interchanging the first two rows of A. For instance,

$$\begin{pmatrix} 0 & 1 & 0 \\ 1 & 0 & 0 \\ 0 & 0 & 1 \end{pmatrix} \begin{pmatrix} 1 & 2 & 3 \\ 4 & 5 & 6 \\ 7 & 8 & 9 \end{pmatrix} = \begin{pmatrix} 4 & 5 & 6 \\ 1 & 2 & 3 \\ 7 & 8 & 9 \end{pmatrix}.$$

Multiple row interchanges are accomplished by combining such elementary interchange matrices. Each such combination of row interchanges corresponds to a unique permutation matrix.

Definition 1.8 A *permutation matrix* is a matrix obtained from the identity matrix by any combination of row interchanges.

In particular, applying a row interchange to a permutation matrix produces another permutation matrix. The following result is easily established.

Lemma 1.9 A matrix P is a permutation matrix if and only if each row of P contains all 0 entries except for a single 1, and, in addition, each column of P also contains all 0 entries except for a single 1.

In general, if a permutation matrix P has a 1 in position (i, j), then the effect of multiplication by P is to move the jth row of A into the ith row of the product $P A$.

EXAMPLE 1.10 There are six different 3×3 permutation matrices, namely

$$\begin{pmatrix} 1 & 0 & 0 \\ 0 & 1 & 0 \\ 0 & 0 & 1 \end{pmatrix}, \quad \begin{pmatrix} 0 & 1 & 0 \\ 0 & 0 & 1 \\ 1 & 0 & 0 \end{pmatrix}, \quad \begin{pmatrix} 0 & 0 & 1 \\ 1 & 0 & 0 \\ 0 & 1 & 0 \end{pmatrix}, \\ \begin{pmatrix} 0 & 1 & 0 \\ 1 & 0 & 0 \\ 0 & 0 & 1 \end{pmatrix}, \quad \begin{pmatrix} 0 & 0 & 1 \\ 0 & 1 & 0 \\ 1 & 0 & 0 \end{pmatrix}, \quad \begin{pmatrix} 1 & 0 & 0 \\ 0 & 0 & 1 \\ 0 & 1 & 0 \end{pmatrix}. \tag{1.30}$$

These have the following effects: if A is a matrix with row vectors $\mathbf{r}_1, \mathbf{r}_2, \mathbf{r}_3$, then multiplication on the left by each of the six permutation matrices produces, respectively,

$$\begin{pmatrix} \mathbf{r}_1 \\ \mathbf{r}_2 \\ \mathbf{r}_3 \end{pmatrix}, \quad \begin{pmatrix} \mathbf{r}_2 \\ \mathbf{r}_3 \\ \mathbf{r}_1 \end{pmatrix}, \quad \begin{pmatrix} \mathbf{r}_3 \\ \mathbf{r}_1 \\ \mathbf{r}_2 \end{pmatrix}, \quad \begin{pmatrix} \mathbf{r}_2 \\ \mathbf{r}_1 \\ \mathbf{r}_3 \end{pmatrix}, \quad \begin{pmatrix} \mathbf{r}_3 \\ \mathbf{r}_2 \\ \mathbf{r}_1 \end{pmatrix}, \quad \begin{pmatrix} \mathbf{r}_1 \\ \mathbf{r}_3 \\ \mathbf{r}_2 \end{pmatrix}.$$

Thus, the first permutation matrix, which is the identity, does nothing. The fourth, fifth and sixth represent row interchanges. The second and third are non-elementary permutations, and can be realized by a pair of successive row interchanges. ●

An elementary combinatorial argument proves that there are a total of

$$n! = n\,(n-1)\,(n-2) \cdots 3 \cdot 2 \cdot 1 \tag{1.31}$$

different permutation matrices of size $n \times n$. Moreover, the product $P = P_1 P_2$ of any two permutation matrices is also a permutation matrix. An important point is that multiplication of permutation matrices is *noncommutative*—the order in which one permutes makes a difference. Switching the first and second rows, and then switching the second and third rows *does not* have the same effect as first switching the second and third rows and then switching the first and second rows!

EXERCISES 1.4

1.4.9. Write down the elementary 4×4 permutation matrix

(a) P_1 that permutes the second and fourth rows, and

(b) P_2 that permutes the first and fourth rows.

(c) Do P_1 and P_2 commute?

(d) Explain what the matrix products $P_1 P_2$ and $P_2 P_1$ do to a 4×4 matrix.

1.4.10. Write down the permutation matrix P such that

(a) $P \begin{pmatrix} u \\ v \\ w \end{pmatrix} = \begin{pmatrix} v \\ w \\ u \end{pmatrix}$

(b) $P \begin{pmatrix} a \\ b \\ c \\ d \end{pmatrix} = \begin{pmatrix} d \\ c \\ a \\ b \end{pmatrix}$

(c) $P\begin{pmatrix} a \\ b \\ c \\ d \end{pmatrix} = \begin{pmatrix} b \\ a \\ d \\ c \end{pmatrix}$

(d) $P\begin{pmatrix} x_1 \\ x_2 \\ x_3 \\ x_4 \\ x_5 \end{pmatrix} = \begin{pmatrix} x_4 \\ x_1 \\ x_3 \\ x_2 \\ x_5 \end{pmatrix}$

1.4.11. Construct a multiplication table that shows all possible products of the 3×3 permutation matrices (1.30). List all pairs that commute.

1.4.12. Write down all 4×4 permutation matrices that

(a) fix the third row of a 4×4 matrix A;

(b) take the third row to the fourth row;

(c) interchange the second and third rows.

1.4.13. *True or false*:

(a) Every elementary permutation matrix satisfies $P^2 = \mathrm{I}$.

(b) Every permutation matrix satisfies $P^2 = \mathrm{I}$.

(c) A matrix that satisfies $P^2 = \mathrm{I}$ is necessarily a permutation matrix.

1.4.14. Let P and Q be permutation matrices and $\mathbf{v}$ a fixed vector.

(a) Under what conditions does the equation $P\,\mathbf{v} = Q\,\mathbf{v}$ imply that $P = Q$?

(b) Answer the same question when $P\,A = Q\,A$, where A is a fixed matrix.

1.4.15. Let P be the 3×3 permutation matrix such that the product $P A$ permutes the first and third rows of the 3×3 matrix A.

(a) Write down P.

(b) *True or false*: The product $A P$ is obtained by permuting the first and third columns of A.

(c) Does the same conclusion hold for any permutation matrix: the effect of $P A$ on the rows of a square matrix A is the same as the effect of $A P$ on the columns of A?

♡ **1.4.16.** A common notation for a permutation π of the integers $\{1, \ldots, m\}$ is as a $2 \times m$ matrix

$$\begin{pmatrix} 1 & 2 & 3 & \cdots & m \\ \pi(1) & \pi(2) & \pi(3) & \cdots & \pi(m) \end{pmatrix},$$

indicating that π takes i to $\pi(i)$.

(a) Show that such a permutation corresponds to the permutation matrix with 1's in positions $(\pi(j), j)$ for $j = 1, \ldots, m$.

(b) Write down the permutation matrices corresponding to the following permutations:

(i) $\begin{pmatrix} 1 & 2 & 3 \\ 2 & 1 & 3 \end{pmatrix}$

(ii) $\begin{pmatrix} 1 & 2 & 3 & 4 \\ 4 & 2 & 3 & 1 \end{pmatrix}$

(iii) $\begin{pmatrix} 1 & 2 & 3 & 4 \\ 1 & 4 & 2 & 3 \end{pmatrix}$

(iv) $\begin{pmatrix} 1 & 2 & 3 & 4 & 5 \\ 5 & 4 & 3 & 2 & 1 \end{pmatrix}$

Which are elementary matrices?

(c) Write down, using the preceding notation, the permutations corresponding to the following permutation matrices:

(i) $\begin{pmatrix} 0 & 0 & 1 \\ 1 & 0 & 0 \\ 0 & 1 & 0 \end{pmatrix}$

(ii) $\begin{pmatrix} 0 & 0 & 1 & 0 \\ 0 & 0 & 0 & 1 \\ 1 & 0 & 0 & 0 \\ 0 & 1 & 0 & 0 \end{pmatrix}$

(iii) $\begin{pmatrix} 0 & 1 & 0 & 0 \\ 0 & 0 & 1 & 0 \\ 0 & 0 & 0 & 1 \\ 1 & 0 & 0 & 0 \end{pmatrix}$

(iv) $\begin{pmatrix} 0 & 0 & 0 & 1 & 0 \\ 1 & 0 & 0 & 0 & 0 \\ 0 & 0 & 1 & 0 & 0 \\ 0 & 0 & 0 & 0 & 1 \\ 0 & 1 & 0 & 0 & 0 \end{pmatrix}$

♡ **1.4.17.** Justify the statement that there are $n\,!$ different $n \times n$ permutation matrices.

1.4.18. Consider the following combination of elementary row operations of type #1:

(i) Add row i to row j.

(ii) Subtract row j from row i.

(iii) Add row i to row j again.

Prove that the net effect is to interchange -1 times row i with row j. Thus, we can *almost* produce an elementary row operation of type #2 by a combination of elementary row operations of type #1. Lest you be tempted to try, Exercise 1.9.18 proves that one *cannot* produce a bona fide row interchange by a combination of elementary row operations of type #1.

The Permuted *LU* Factorization

As we now know, any nonsingular matrix A can be reduced to upper triangular form by elementary row operations of types #1 and #2. The row interchanges merely reorder the equations. If one performs all of the required row interchanges in advance, then the elimination algorithm can proceed without requiring any further pivoting. Thus, the matrix obtained by permuting the rows of A in the prescribed manner is regular. In other words, if A is a nonsingular matrix, then there is a permutation matrix P such that the product PA is regular, and hence admits an LU factorization. As a result, we deduce the general *permuted LU factorization*

$$PA = LU, \tag{1.32}$$

where P is a permutation matrix, L is special lower triangular, and U is upper triangular with the pivots on the diagonal. For instance, in the preceding example, we permuted the first and second rows, and hence equation (1.32) has the explicit form

$$\begin{pmatrix} 0 & 1 & 0 \\ 1 & 0 & 0 \\ 0 & 0 & 1 \end{pmatrix} \begin{pmatrix} 0 & 2 & 1 \\ 2 & 6 & 1 \\ 1 & 1 & 4 \end{pmatrix} = \begin{pmatrix} 1 & 0 & 0 \\ 0 & 1 & 0 \\ \frac{1}{2} & -1 & 1 \end{pmatrix} \begin{pmatrix} 2 & 6 & 1 \\ 0 & 2 & 1 \\ 0 & 0 & \frac{9}{2} \end{pmatrix}.$$

We have now established the following generalization of Theorem 1.3.

THEOREM 1.11 Let A be an $n \times n$ matrix. Then the following conditions are equivalent:

(i) A is nonsingular.

(ii) A has n nonzero pivots.

(iii) A admits a permuted LU factorization: $PA = LU$.

A practical method to construct a permuted LU factorization of a given matrix A would proceed as follows. First set up $P = L = \mathrm{I}$ as $n \times n$ identity matrices. The matrix P will keep track of the permutations performed during the Gaussian Elimination process, while the entries of L below the diagonal are gradually replaced by the negatives of the multiples used in the corresponding row operations of type #1. Each time two rows of A are interchanged, the same two rows of P will be interchanged. Moreover, any pair of entries that both lie *below* the diagonal in these same two rows of L must also be interchanged, while entries lying on and above its diagonal need to stay in their place. At a successful conclusion to the procedure, A will have been converted into the upper triangular matrix U, while L and P will assume their final form. Here is an illustrative example.

EXAMPLE 1.12 Our goal is to produce a permuted LU factorization of the matrix

$$A = \begin{pmatrix} 1 & 2 & -1 & 0 \\ 2 & 4 & -2 & -1 \\ -3 & -5 & 6 & 1 \\ -1 & 2 & 8 & -2 \end{pmatrix}.$$

To begin the procedure, we apply row operations of type #1 to eliminate the entries below the first pivot. The updated matrices* are

$$A = \begin{pmatrix} 1 & 2 & -1 & 0 \\ 0 & 0 & 0 & -1 \\ 0 & 1 & 3 & 1 \\ 0 & 4 & 7 & -2 \end{pmatrix}, \qquad L = \begin{pmatrix} 1 & 0 & 0 & 0 \\ 2 & 1 & 0 & 0 \\ -3 & 0 & 1 & 0 \\ -1 & 0 & 0 & 1 \end{pmatrix}, \qquad P = \begin{pmatrix} 1 & 0 & 0 & 0 \\ 0 & 1 & 0 & 0 \\ 0 & 0 & 1 & 0 \\ 0 & 0 & 0 & 1 \end{pmatrix},$$

*Here, we are adopting computer programming conventions, where updates of a matrix are all given the same name.

where L keeps track of the row operations, and we initialize P to be the identity matrix. The $(2,2)$ entry of the new A is zero, and so we interchange its second and third rows, leading to

$$A = \begin{pmatrix} 1 & 2 & -1 & 0 \\ 0 & 1 & 3 & 1 \\ 0 & 0 & 0 & -1 \\ 0 & 4 & 7 & -2 \end{pmatrix}, \qquad L = \begin{pmatrix} 1 & 0 & 0 & 0 \\ -3 & 1 & 0 & 0 \\ 2 & 0 & 1 & 0 \\ -1 & 0 & 0 & 1 \end{pmatrix}, \qquad P = \begin{pmatrix} 1 & 0 & 0 & 0 \\ 0 & 0 & 1 & 0 \\ 0 & 1 & 0 & 0 \\ 0 & 0 & 0 & 1 \end{pmatrix}.$$

We interchanged the same two rows of P, while in L we only interchanged the already computed entries in its second and third rows that lie in its first column below the diagonal. We then eliminate the nonzero entry lying below the $(2,2)$ pivot, leading to

$$A = \begin{pmatrix} 1 & 2 & -1 & 0 \\ 0 & 1 & 3 & 1 \\ 0 & 0 & 0 & -1 \\ 0 & 0 & -5 & -6 \end{pmatrix}, \qquad L = \begin{pmatrix} 1 & 0 & 0 & 0 \\ -3 & 1 & 0 & 0 \\ 2 & 0 & 1 & 0 \\ -1 & 4 & 0 & 1 \end{pmatrix}, \qquad P = \begin{pmatrix} 1 & 0 & 0 & 0 \\ 0 & 0 & 1 & 0 \\ 0 & 1 & 0 & 0 \\ 0 & 0 & 0 & 1 \end{pmatrix}.$$

A final row interchange places the matrix in upper triangular form:

$$U = \begin{pmatrix} 1 & 2 & -1 & 0 \\ 0 & 1 & 3 & 1 \\ 0 & 0 & -5 & -6 \\ 0 & 0 & 0 & -1 \end{pmatrix}, \qquad L = \begin{pmatrix} 1 & 0 & 0 & 0 \\ -3 & 1 & 0 & 0 \\ -1 & 4 & 1 & 0 \\ 2 & 0 & 0 & 1 \end{pmatrix}, \qquad P = \begin{pmatrix} 1 & 0 & 0 & 0 \\ 0 & 0 & 1 & 0 \\ 0 & 0 & 0 & 1 \\ 0 & 1 & 0 & 0 \end{pmatrix}.$$

Again, we performed the same row interchange on P, while only interchanging the third and fourth row entries of L that lie below the diagonal. You can verify that

$$\begin{aligned} PA &= \begin{pmatrix} 1 & 2 & -1 & 0 \\ -3 & -5 & 6 & 1 \\ -1 & 2 & 8 & -2 \\ 2 & 4 & -2 & -1 \end{pmatrix} \\ &= \begin{pmatrix} 1 & 0 & 0 & 0 \\ -3 & 1 & 0 & 0 \\ -1 & 4 & 1 & 0 \\ 2 & 0 & 0 & 1 \end{pmatrix} \begin{pmatrix} 1 & 2 & -1 & 0 \\ 0 & 1 & 3 & 1 \\ 0 & 0 & -5 & -6 \\ 0 & 0 & 0 & -1 \end{pmatrix} = LU, \end{aligned} \tag{1.33}$$

as promised. Thus, by rearranging the equations in the order first, third, fourth, second, as prescribed by P, we obtain an equivalent linear system with regular coefficient matrix PA. ●

Once the permuted LU factorization is established, the solution to the original system $A\mathbf{x} = \mathbf{b}$ is obtained by applying the same Forward and Back Substitution algorithm presented above. Explicitly, we first multiply the system $A\mathbf{x} = \mathbf{b}$ by the permutation matrix, leading to

$$PA\mathbf{x} = P\mathbf{b} = \widehat{\mathbf{b}}, \tag{1.34}$$

whose right hand side $\widehat{\mathbf{b}}$ has been obtained by permuting the entries of $\mathbf{b}$ in the same fashion as the rows of A. We then solve the two triangular systems

$$L\mathbf{c} = \widehat{\mathbf{b}} \quad \text{and} \quad U\mathbf{x} = \mathbf{c} \tag{1.35}$$

by, respectively, Forward and Back Substitution.

EXAMPLE 1.12 *Continued*

Suppose we wish to solve the linear system

$$\begin{pmatrix} 1 & 2 & -1 & 0 \\ 2 & 4 & -2 & -1 \\ -3 & -5 & 6 & 1 \\ -1 & 2 & 8 & -2 \end{pmatrix} \begin{pmatrix} x \\ y \\ z \\ w \end{pmatrix} = \begin{pmatrix} 1 \\ -1 \\ 3 \\ 0 \end{pmatrix}.$$

In view of the $PA = LU$ factorization established in (1.33), we need only solve the two auxiliary lower and upper triangular systems (1.35). The lower triangular system is

$$\begin{pmatrix} 1 & 0 & 0 & 0 \\ -3 & 1 & 0 & 0 \\ -1 & 4 & 1 & 0 \\ 2 & 0 & 0 & 1 \end{pmatrix} \begin{pmatrix} a \\ b \\ c \\ d \end{pmatrix} = \begin{pmatrix} 1 \\ 3 \\ 0 \\ -1 \end{pmatrix};$$

whose right hand side was obtained by applying the permutation matrix P to the right hand side of the original system. Its solution, namely $a = 1, b = 6, c = -23$, $d = -3$, is obtained through Forward Substitution. The resulting upper triangular system is

$$\begin{pmatrix} 1 & 2 & -1 & 0 \\ 0 & 1 & 3 & 1 \\ 0 & 0 & -5 & -6 \\ 0 & 0 & 0 & -1 \end{pmatrix} \begin{pmatrix} x \\ y \\ z \\ w \end{pmatrix} = \begin{pmatrix} 1 \\ 6 \\ -23 \\ -3 \end{pmatrix}.$$

Its solution, $w = 3$, $z = 1$, $y = 0$, $x = 2$, which is also the solution to the original system, is easily obtained by Back Substitution. ●

EXERCISES 1.4

1.4.19. For each of the listed matrices A and vectors $\mathbf{b}$, find a permuted LU factorization of the matrix, and use your factorization to solve the system $A\mathbf{x} = \mathbf{b}$.

(a) $\begin{pmatrix} 0 & 1 \\ 2 & -1 \end{pmatrix}, \begin{pmatrix} 3 \\ 2 \end{pmatrix}$

(b) $\begin{pmatrix} 0 & 0 & -4 \\ 1 & 2 & 3 \\ 0 & 1 & 7 \end{pmatrix}, \begin{pmatrix} 1 \\ 2 \\ -1 \end{pmatrix}$

(c) $\begin{pmatrix} 0 & 1 & -3 \\ 0 & 2 & 3 \\ 1 & 0 & 2 \end{pmatrix}, \begin{pmatrix} 1 \\ 2 \\ -1 \end{pmatrix}$

(d) $\begin{pmatrix} 1 & 2 & -1 & 0 \\ 3 & 6 & 2 & -1 \\ 1 & 1 & -7 & 2 \\ 1 & -1 & 2 & 1 \end{pmatrix}, \begin{pmatrix} 1 \\ 0 \\ 0 \\ 3 \end{pmatrix}$

(e) $\begin{pmatrix} 0 & 1 & 0 & 0 \\ 2 & 3 & 1 & 0 \\ 1 & 4 & -1 & 2 \\ 7 & -1 & 2 & 3 \end{pmatrix}, \begin{pmatrix} -1 \\ -4 \\ 0 \\ 5 \end{pmatrix}$

(f) $\begin{pmatrix} 0 & 0 & 2 & 3 & 4 \\ 0 & 1 & -7 & 2 & 3 \\ 1 & 4 & 1 & 1 & 1 \\ 0 & 0 & 1 & 0 & 2 \\ 0 & 0 & 1 & 7 & 3 \end{pmatrix}, \begin{pmatrix} -3 \\ -2 \\ 0 \\ 0 \\ -7 \end{pmatrix}$

1.4.20. For each of the following linear systems find a permuted LU factorization of the coefficient matrix and then use it to solve the system by Forward and Back Substitution.

(a) $$\begin{aligned} 4x_1 - 4x_2 + 2x_3 &= 1 \\ -3x_1 + 3x_2 + x_3 &= 3 \\ -3x_1 + x_2 - 2x_3 &= -5 \end{aligned}$$

(b) $$\begin{aligned} y - z + w &= 0 \\ y + z &= 1 \\ x - y + z - 3w &= 2 \\ x + 2y - z + w &= 4 \end{aligned}$$

(c) $$\begin{aligned} x - y + 2z + w &= 0 \\ -x + y - 3z &= 1 \\ x - y + 4z - 3w &= 2 \\ x + 2y - z + w &= 4 \end{aligned}$$

◇ **1.4.21.** (a) Explain why

$$\begin{pmatrix} 0 & 1 & 0 \\ 1 & 0 & 0 \\ 0 & 0 & 1 \end{pmatrix}\begin{pmatrix} 0 & 1 & 3 \\ 2 & -1 & 1 \\ 2 & -2 & 0 \end{pmatrix} = \begin{pmatrix} 1 & 0 & 0 \\ 0 & 1 & 0 \\ 1 & -1 & 1 \end{pmatrix}\begin{pmatrix} 2 & -1 & 1 \\ 0 & 1 & 3 \\ 0 & 0 & 2 \end{pmatrix},$$

$$\begin{pmatrix} 0 & 0 & 1 \\ 0 & 1 & 0 \\ 1 & 0 & 0 \end{pmatrix}\begin{pmatrix} 0 & 1 & 3 \\ 2 & -1 & 1 \\ 2 & -2 & 0 \end{pmatrix} = \begin{pmatrix} 1 & 0 & 0 \\ 1 & 1 & 0 \\ 0 & 1 & 1 \end{pmatrix}\begin{pmatrix} 2 & -2 & 0 \\ 0 & 1 & 1 \\ 0 & 0 & 2 \end{pmatrix},$$

and

$$\begin{pmatrix} 0 & 0 & 1 \\ 1 & 0 & 0 \\ 0 & 1 & 0 \end{pmatrix}\begin{pmatrix} 0 & 1 & 3 \\ 2 & -1 & 1 \\ 2 & -2 & 0 \end{pmatrix} = \begin{pmatrix} 1 & 0 & 0 \\ 0 & 1 & 0 \\ 1 & 1 & 1 \end{pmatrix}\begin{pmatrix} 2 & -2 & 0 \\ 0 & 1 & 3 \\ 0 & 0 & -2 \end{pmatrix}$$

are all legitimate permuted LU factorizations of the same matrix. List the elementary row operations that are being used in each case.

(b) Use each of the factorizations to solve the linear system

$$\begin{pmatrix} 0 & 1 & 3 \\ 2 & -1 & 1 \\ 2 & -2 & 0 \end{pmatrix}\begin{pmatrix} x \\ y \\ z \end{pmatrix} = \begin{pmatrix} -5 \\ -1 \\ 0 \end{pmatrix}.$$

Do you always obtain the same result? Explain why or why not.

1.4.22. (a) Find three different permuted LU factorizations of the matrix

$$A = \begin{pmatrix} 0 & 1 & 2 \\ 1 & 0 & -1 \\ 1 & 1 & 3 \end{pmatrix}.$$

(b) How many different permuted LU factorizations does A have?

1.4.23. What is the maximal number of permuted LU factorizations a regular 3×3 matrix can have? Give an example of such a matrix.

1.4.24. *True or false*: The pivots of a nonsingular matrix are uniquely defined.

♠ **1.4.25.** (a) Write a pseudocode program implementing the algorithm for finding the permuted LU factorization of a matrix.

(b) Program your algorithm and test it on the examples in Exercise 1.4.19.

1.5 MATRIX INVERSES

The inverse of a matrix is analogous to the reciprocal $a^{-1} = 1/a$ of a scalar. We already encountered the inverses of matrices corresponding to elementary row operations. In this section, we will study inverses of general square matrices. We begin with the formal definition.

Definition 1.13 Let A be a square matrix of size $n \times n$. An $n \times n$ matrix X is called the *inverse* of A if it satisfies

$$XA = \mathrm{I} = AX, \tag{1.36}$$

where $\mathrm{I} = \mathrm{I}_n$ is the $n \times n$ identity matrix. The inverse is commonly denoted by $X = A^{-1}$.

REMARK: Noncommutativity of matrix multiplication requires that we impose both conditions in (1.36) in order to properly define an inverse to the matrix A. The first condition, $XA = \mathrm{I}$, says that X is a *left inverse*, while the second, $AX = \mathrm{I}$, requires that X also be a *right inverse*. Rectangular matrices might have either a left inverse or a right inverse, but, as we shall see, *only* square matrices have both, and so only square matrices can have full-fledged inverses. However, not every square matrix has an inverse. Indeed, not every scalar has an inverse: $0^{-1} = 1/0$ is not defined since the equation $0x = 1$ has no solution.

EXAMPLE 1.14 Since

$$\begin{pmatrix} 1 & 2 & -1 \\ -3 & 1 & 2 \\ -2 & 2 & 1 \end{pmatrix} \begin{pmatrix} 3 & 4 & -5 \\ 1 & 1 & -1 \\ 4 & 6 & -7 \end{pmatrix} = \begin{pmatrix} 1 & 0 & 0 \\ 0 & 1 & 0 \\ 0 & 0 & 1 \end{pmatrix} = \begin{pmatrix} 3 & 4 & -5 \\ 1 & 1 & -1 \\ 4 & 6 & -7 \end{pmatrix} \begin{pmatrix} 1 & 2 & -1 \\ -3 & 1 & 2 \\ -2 & 2 & 1 \end{pmatrix},$$

we conclude that when

$$A = \begin{pmatrix} 1 & 2 & -1 \\ -3 & 1 & 2 \\ -2 & 2 & 1 \end{pmatrix},$$

then

$$A^{-1} = \begin{pmatrix} 3 & 4 & -5 \\ 1 & 1 & -1 \\ 4 & 6 & -7 \end{pmatrix}.$$

Observe that there is no obvious way to anticipate the entries of A^{-1} from the entries of A. ●

EXAMPLE 1.15 Let us compute the inverse $X = \begin{pmatrix} x & y \\ z & w \end{pmatrix}$, when it exists, of a general 2×2 matrix $A = \begin{pmatrix} a & b \\ c & d \end{pmatrix}$. The right inverse condition

$$A\,X = \begin{pmatrix} a\,x + b\,z & a\,y + b\,w \\ c\,x + d\,z & c\,y + d\,w \end{pmatrix} = \begin{pmatrix} 1 & 0 \\ 0 & 1 \end{pmatrix} = \mathrm{I}$$

holds if and only if x, y, z, w satisfy the linear system

$$a\,x + b\,z = 1, \qquad a\,y + b\,w = 0,$$
$$c\,x + d\,z = 0, \qquad c\,y + d\,w = 1.$$

Solving by Gaussian Elimination (or directly), we find

$$x = \frac{d}{a\,d - b\,c}, \qquad y = -\frac{b}{a\,d - b\,c},$$
$$z = -\frac{c}{a\,d - b\,c}, \qquad w = \frac{a}{a\,d - b\,c},$$

provided the common denominator $a\,d - b\,c \neq 0$ does not vanish. Therefore, the matrix

$$X = \frac{1}{a\,d - b\,c} \begin{pmatrix} d & -b \\ -c & a \end{pmatrix}$$

forms a right inverse to A. However, a short computation shows that it also defines a left inverse:

$$X\,A = \begin{pmatrix} x\,a + y\,c & x\,b + y\,d \\ z\,a + w\,c & z\,b + w\,d \end{pmatrix} = \begin{pmatrix} 1 & 0 \\ 0 & 1 \end{pmatrix} = \mathrm{I},$$

and hence $X = A^{-1}$ is the inverse to A.

The denominator appearing in the preceding formulae has a special name; it is called the *determinant* of the 2×2 matrix A, and denoted

$$\det \begin{pmatrix} a & b \\ c & d \end{pmatrix} = a\,d - b\,c. \tag{1.37}$$

Thus, the determinant of a 2×2 matrix is the product of the diagonal entries minus the product of the off-diagonal entries. (Determinants of larger square matrices will be discussed in Section 1.9.) Thus, the 2×2 matrix A is invertible, with

$$A^{-1} = \frac{1}{a\,d - b\,c} \begin{pmatrix} d & -b \\ -c & a \end{pmatrix}, \tag{1.38}$$

if and only if $\det A \neq 0$. For example, if

$$A = \begin{pmatrix} 1 & 3 \\ -2 & -4 \end{pmatrix},$$

then $\det A = 2 \neq 0$. We conclude that A has an inverse, which, by (1.38), is

$$A^{-1} = \frac{1}{2}\begin{pmatrix} -4 & -3 \\ 2 & 1 \end{pmatrix} = \begin{pmatrix} -2 & -\frac{3}{2} \\ 1 & \frac{1}{2} \end{pmatrix}.$$

EXAMPLE 1.16 We already learned how to find the inverse of an elementary matrix of type #1: we just negate the one nonzero off-diagonal entry. For example, if

$$E = \begin{pmatrix} 1 & 0 & 0 \\ 0 & 1 & 0 \\ 2 & 0 & 1 \end{pmatrix}, \qquad \text{then} \quad E^{-1} = \begin{pmatrix} 1 & 0 & 0 \\ 0 & 1 & 0 \\ -2 & 0 & 1 \end{pmatrix}.$$

This is because the inverse of the elementary row operation that adds twice the first row to the third row is the operation of subtracting twice the first row from the third row.

EXAMPLE 1.17 Let

$$P = \begin{pmatrix} 0 & 1 & 0 \\ 1 & 0 & 0 \\ 0 & 0 & 1 \end{pmatrix}$$

denote the elementary matrix that has the effect of interchanging rows 1 and 2 of a 3–rowed matrix. Then $P^2 = \mathrm{I}$, since performing the interchange twice returns us to where we began. This implies that $P^{-1} = P$ is its own inverse. Indeed, the same result holds for all elementary permutation matrices that correspond to row operations of type #2. However, it is not true for more general permutation matrices.

The following fundamental result will be established later in this chapter.

THEOREM 1.18 A square matrix A has an inverse if and only if it is nonsingular.

Consequently, an $n \times n$ matrix will have an inverse if and only if it can be reduced to upper triangular form, with n nonzero pivots on the diagonal, by a combination of elementary row operations. Indeed, "invertible" is often used as a synonym for "nonsingular". All other matrices are singular and do not have an inverse as defined above. Before attempting to prove Theorem 1.18, we need to first become familiar with some elementary properties of matrix inverses.

Lemma 1.19 The inverse of a square matrix, if it exists, is unique.

Proof Suppose both X and Y satisfy (1.36), so $X A = \mathrm{I} = A X$ and $Y A = \mathrm{I} = A Y$. Then, by associativity, $X = X\mathrm{I} = X(A Y) = (XA)Y = \mathrm{I}Y = Y$, and hence $X = Y$. ■

Inverting a matrix twice brings us back to where we started.

Lemma 1.20 If A is invertible, then A^{-1} is also invertible and $(A^{-1})^{-1} = A$.

Proof The matrix inverse equations $A^{-1} A = \mathrm{I} = A A^{-1}$ are sufficient to prove that A is the inverse of A^{-1}. ■

Lemma 1.21 If A and B are invertible matrices of the same size, then their product, $A B$, is invertible, and

$$(A B)^{-1} = B^{-1}A^{-1}. \tag{1.39}$$

Note that the order of the factors is reversed under inversion.

Proof Let $X = B^{-1}A^{-1}$. Then, by associativity,

$$X (A B) = B^{-1}A^{-1}A B = B^{-1}B = \mathrm{I},$$
$$(A B) X = A B B^{-1}A^{-1} = A A^{-1} = \mathrm{I}.$$

Thus X is both a left and a right inverse for the product matrix $A B$ and the result follows. ■

EXAMPLE 1.22 One verifies, directly, that the inverse of

$$A = \begin{pmatrix} 1 & 2 \\ 0 & 1 \end{pmatrix} \quad \text{is} \quad A^{-1} = \begin{pmatrix} 1 & -2 \\ 0 & 1 \end{pmatrix},$$

while the inverse of

$$B = \begin{pmatrix} 0 & 1 \\ -1 & 0 \end{pmatrix} \quad \text{is} \quad B^{-1} = \begin{pmatrix} 0 & -1 \\ 1 & 0 \end{pmatrix}.$$

Therefore, the inverse of their product

$$C = A B = \begin{pmatrix} 1 & 2 \\ 0 & 1 \end{pmatrix}\begin{pmatrix} 0 & 1 \\ -1 & 0 \end{pmatrix} = \begin{pmatrix} -2 & 1 \\ -1 & 0 \end{pmatrix}$$

is given by

$$C^{-1} = B^{-1}A^{-1} = \begin{pmatrix} 0 & -1 \\ 1 & 0 \end{pmatrix}\begin{pmatrix} 1 & -2 \\ 0 & 1 \end{pmatrix} = \begin{pmatrix} 0 & -1 \\ 1 & -2 \end{pmatrix}.$$ ●

We can straightforwardly generalize the preceding result. The inverse of a k-fold product of invertible matrices is the product of their inverses, *in the reverse order*:

$$(A_1 A_2 \cdots A_{k-1} A_k)^{-1} = A_k^{-1} A_{k-1}^{-1} \cdots A_2^{-1} A_1^{-1}. \tag{1.40}$$

Warning: In general, $(A + B)^{-1} \neq A^{-1} + B^{-1}$. This equation is not even true for scalars (1×1 matrices)!

EXERCISES 1.5

1.5.1. Verify by direct multiplication that the following matrices are inverses, i.e., both conditions in (1.36) hold:

(a) $A = \begin{pmatrix} 2 & 3 \\ -1 & -1 \end{pmatrix}$, $A^{-1} = \begin{pmatrix} -1 & -3 \\ 1 & 2 \end{pmatrix}$

(b) $A = \begin{pmatrix} 2 & 1 & 1 \\ 3 & 2 & 1 \\ 2 & 1 & 2 \end{pmatrix}$,

$$A^{-1} = \begin{pmatrix} 3 & -1 & -1 \\ -4 & 2 & 1 \\ -1 & 0 & 1 \end{pmatrix}$$

(c) $A = \begin{pmatrix} -1 & 3 & 2 \\ 2 & 2 & -1 \\ -2 & 1 & 3 \end{pmatrix}$,

$$A^{-1} = \begin{pmatrix} -1 & 1 & 1 \\ \frac{4}{7} & -\frac{1}{7} & -\frac{3}{7} \\ -\frac{6}{7} & \frac{5}{7} & \frac{8}{7} \end{pmatrix}$$

1.5.2. Let

$$A = \begin{pmatrix} 1 & 2 & 0 \\ 0 & 1 & 3 \\ 1 & -1 & -8 \end{pmatrix}.$$

Find the right inverse of A by setting up and solving the linear system $A X = \mathrm{I}$. Verify that the resulting matrix X is also a left inverse.

1.5.3. Write down the inverse of each of the following elementary matrices:

(a) $\begin{pmatrix} 0 & 1 \\ 1 & 0 \end{pmatrix}$ (b) $\begin{pmatrix} 1 & 0 \\ 5 & 1 \end{pmatrix}$

(c) $\begin{pmatrix} 1 & -2 \\ 0 & 1 \end{pmatrix}$ (d) $\begin{pmatrix} 1 & 0 & 0 \\ 0 & 1 & -3 \\ 0 & 0 & 1 \end{pmatrix}$

(e) $\begin{pmatrix} 1 & 0 & 0 & 0 \\ 0 & 1 & 0 & 0 \\ 0 & 6 & 1 & 0 \\ 0 & 0 & 0 & 1 \end{pmatrix}$

(f) $\begin{pmatrix} 0 & 0 & 0 & 1 \\ 0 & 1 & 0 & 0 \\ 0 & 0 & 1 & 0 \\ 1 & 0 & 0 & 0 \end{pmatrix}$

1.5.4. Show that the inverse of

$$L = \begin{pmatrix} 1 & 0 & 0 \\ a & 1 & 0 \\ b & 0 & 1 \end{pmatrix} \quad \text{is} \quad L^{-1} = \begin{pmatrix} 1 & 0 & 0 \\ -a & 1 & 0 \\ -b & 0 & 1 \end{pmatrix}.$$

However, the inverse of

$$M = \begin{pmatrix} 1 & 0 & 0 \\ a & 1 & 0 \\ b & c & 1 \end{pmatrix} \quad \text{is \textit{not}} \quad \begin{pmatrix} 1 & 0 & 0 \\ -a & 1 & 0 \\ -b & -c & 1 \end{pmatrix}.$$

What is M^{-1}?

1.5.5. Explain why a matrix with a row of all zeros does not have an inverse.

1.5.6. (a) Write down the inverse of the matrices

$$A = \begin{pmatrix} 1 & 1 \\ 2 & 1 \end{pmatrix} \quad \text{and} \quad B = \begin{pmatrix} 1 & -1 \\ 1 & 2 \end{pmatrix}.$$

(b) Write down the product matrix $C = AB$ and its inverse C^{-1} using the inverse product formula.

1.5.7. (a) Find the inverse of the *rotation matrix*

$$R_\theta = \begin{pmatrix} \cos\theta & -\sin\theta \\ \sin\theta & \cos\theta \end{pmatrix},$$

where θ is a fixed angle.

(b) Use your result to solve the system

$$x = a\cos\theta - b\sin\theta,$$
$$y = a\sin\theta + b\cos\theta,$$

for a and b in terms of x and y.

(c) Prove that, for any $a \in \mathbb{R}$ and $0 < \theta < \pi$, the matrix $R_\theta - a\,\mathrm{I}$ has an inverse.

1.5.8. (a) Write down the inverses of each of the 3×3 permutation matrices (1.30).

(b) Which ones are their own inverses, $P^{-1} = P$?

(c) Can you find a non-elementary permutation matrix P which is its own inverse: $P^{-1} = P$?

1.5.9. Find the inverse of the following permutation matrices

(a) $\begin{pmatrix} 0 & 0 & 0 & 1 \\ 0 & 0 & 1 & 0 \\ 0 & 1 & 0 & 0 \\ 1 & 0 & 0 & 0 \end{pmatrix}$ (b) $\begin{pmatrix} 0 & 1 & 0 & 0 \\ 0 & 0 & 1 & 0 \\ 0 & 0 & 0 & 1 \\ 1 & 0 & 0 & 0 \end{pmatrix}$

(c) $\begin{pmatrix} 1 & 0 & 0 & 0 \\ 0 & 0 & 0 & 1 \\ 0 & 1 & 0 & 0 \\ 0 & 0 & 1 & 0 \end{pmatrix}$

(d) $\begin{pmatrix} 1 & 0 & 0 & 0 & 0 \\ 0 & 0 & 1 & 0 & 0 \\ 0 & 0 & 0 & 0 & 1 \\ 0 & 1 & 0 & 0 & 0 \\ 0 & 0 & 0 & 1 & 0 \end{pmatrix}$

1.5.10. Explain how to write down the inverse permutation using the notation of Exercise 1.4.16. Apply your method to the examples in Exercise 1.5.9, and check the result by verifying that it produces the inverse permutation matrix.

1.5.11. Show that

$$A = \begin{pmatrix} 0 & a & 0 & 0 & 0 \\ b & 0 & c & 0 & 0 \\ 0 & d & 0 & e & 0 \\ 0 & 0 & f & 0 & g \\ 0 & 0 & 0 & h & 0 \end{pmatrix}$$

is not invertible for any value of the entries.

1.5.12. Find all real 2×2 matrices that are their own inverses: $A^{-1} = A$.

1.5.13. Show that if a square matrix A satisfies $A^2 - 3A + I = \mathrm{O}$, then $A^{-1} = 3I - A$.

1.5.14. Prove that if $c \neq 0$ is any nonzero scalar and A is an invertible matrix, then the scalar product matrix cA is invertible, and $(cA)^{-1} = \dfrac{1}{c} A^{-1}$.

1.5.15. Show that if A is a nonsingular matrix, so is any power A^n.

1.5.16. Prove that a diagonal matrix $D = \operatorname{diag}(d_1, \ldots, d_n)$ is invertible if and only if all its diagonal entries are nonzero, in which case $D^{-1} = \operatorname{diag}(1/d_1, \ldots, 1/d_n)$.

1.5.17. Prove that if U is a nonsingular upper triangular matrix, then the diagonal entries of U^{-1} are the reciprocals of the diagonal entries of U.

◇ **1.5.18.** Two matrices A and B are said to be *similar*, written $A \sim B$, if there exists an invertible matrix S such that $B = S^{-1} A S$. Prove:

(a) $A \sim A$

(b) If $A \sim B$, then $B \sim A$.

(c) If $A \sim B$ and $B \sim C$, then $A \sim C$.

♡ **1.5.19.** (a) A block matrix $D = \begin{pmatrix} A & \mathrm{O} \\ \mathrm{O} & B \end{pmatrix}$ is called *block diagonal* if A and B are square matrices, not necessarily of the same size, while the O's are zero matrices of the appropriate sizes. Prove that D has an inverse if and only if both A and B do, and $D^{-1} = \begin{pmatrix} A^{-1} & \mathrm{O} \\ \mathrm{O} & B^{-1} \end{pmatrix}$.

(b) Find the inverse of

$$\begin{pmatrix} 1 & 2 & 0 \\ 2 & 1 & 0 \\ 0 & 0 & 3 \end{pmatrix} \quad \text{and} \quad \begin{pmatrix} 1 & -1 & 0 & 0 \\ 2 & -1 & 0 & 0 \\ 0 & 0 & 1 & 3 \\ 0 & 0 & 2 & 5 \end{pmatrix}$$

by using this method.

1.5.20. (a) Show that $B = \begin{pmatrix} 1 & 1 & 0 \\ -1 & -1 & 1 \end{pmatrix}$ is a left inverse of $A = \begin{pmatrix} 1 & -1 \\ 0 & 1 \\ 1 & 1 \end{pmatrix}$.

(b) Show that A does not have a right inverse.

(c) Can you find any other left inverses of A?

1.5.21. Prove that the rectangular matrix

$$A = \begin{pmatrix} 1 & 2 & -1 \\ 1 & 2 & 0 \end{pmatrix}$$

has a right inverse, but no left inverse.

1.5.22. (a) Are there any nonzero real scalars that satisfy $(a + b)^{-1} = a^{-1} + b^{-1}$?

(b) Are there any nonsingular real 2×2 matrices that satisfy $(A + B)^{-1} = A^{-1} + B^{-1}$?

Gauss–Jordan Elimination

The principal algorithm used to compute the inverse of a nonsingular matrix is known as *Gauss–Jordan Elimination*, in honor of Gauss and Wilhelm Jordan, a nineteenth century German engineer. A key fact is that we only need to solve the right inverse equation

$$AX = \mathrm{I} \tag{1.41}$$

in order to compute $X = A^{-1}$. The left inverse equation in (1.36), namely $XA = \mathrm{I}$, will then follow as an automatic consequence. In other words, for square matrices, a right inverse is automatically a left inverse, and conversely! A proof will appear below.

The reader may well ask, then, why use both left and right inverse conditions in the original definition? There are several good reasons. First of all, a non-square matrix may satisfy one of the two conditions—having either a left inverse or a right inverse—but can never satisfy both. Moreover, even when we restrict our attention to square matrices, starting with only one of the conditions makes the logical development of the subject considerably more difficult, and not really worth the extra effort. Once we have established the basic properties of the inverse of a square matrix, we can then safely discard the superfluous left inverse condition. Finally, when we generalize the notion of an inverse to linear operators in Chapter 7, then, unlike square matrices, we *cannot* dispense with either of the conditions.

Let us write out the individual columns of the right inverse equation (1.41). The jth column of the $n \times n$ identity matrix I is the vector $\mathbf{e}_j$ that has a single 1 in the jth slot and 0's elsewhere, so

$$\mathbf{e}_1 = \begin{pmatrix} 1 \\ 0 \\ 0 \\ \vdots \\ 0 \\ 0 \end{pmatrix}, \quad \mathbf{e}_2 = \begin{pmatrix} 0 \\ 1 \\ 0 \\ \vdots \\ 0 \\ 0 \end{pmatrix}, \quad \cdots \quad \mathbf{e}_n = \begin{pmatrix} 0 \\ 0 \\ 0 \\ \vdots \\ 0 \\ 1 \end{pmatrix}. \tag{1.42}$$

According to (1.11), the jth column of the matrix product $A X$ is equal to $A\,\mathbf{x}_j$, where $\mathbf{x}_j$ denotes the jth column of the inverse matrix X. Therefore, the single matrix equation (1.41) is equivalent to n linear systems

$$A\mathbf{x}_1 = \mathbf{e}_1, \quad A\mathbf{x}_2 = \mathbf{e}_2, \quad \ldots \quad A\mathbf{x}_n = \mathbf{e}_n, \tag{1.43}$$

all having the same coefficient matrix. As such, to solve them we should form the n augmented matrices $M_1 = \left(A \mid \mathbf{e}_1\right), \ldots, M_n = \left(A \mid \mathbf{e}_n\right)$, and then apply our Gaussian Elimination algorithm to each. But this would be a waste of effort. Since the coefficient matrix is the same, we will end up performing *identical* row operations on each augmented matrix. Clearly, it will be more efficient to combine them into one large augmented matrix $M = \left(A \mid \mathbf{e}_1 \cdots \mathbf{e}_n\right) = \left(A \mid \mathrm{I}\right)$, of size $n \times (2n)$, in which the right hand sides $\mathbf{e}_1, \ldots, \mathbf{e}_n$ of our systems are placed into n different columns, which we then recognize as reassembling the columns of an $n \times n$ identity matrix. We may then simultaneously apply our elementary row operations to reduce, if possible, the large augmented matrix so that its first n columns are in upper triangular form.

EXAMPLE 1.23 For example, to find the inverse of the matrix

$$A = \begin{pmatrix} 0 & 2 & 1 \\ 2 & 6 & 1 \\ 1 & 1 & 4 \end{pmatrix},$$

we form the large augmented matrix

$$\left(\begin{array}{ccc|ccc} 0 & 2 & 1 & 1 & 0 & 0 \\ 2 & 6 & 1 & 0 & 1 & 0 \\ 1 & 1 & 4 & 0 & 0 & 1 \end{array}\right).$$

Applying the same sequence of elementary row operations as in Section 1.4, we first interchange the rows

$$\left(\begin{array}{ccc|ccc} 2 & 6 & 1 & 0 & 1 & 0 \\ 0 & 2 & 1 & 1 & 0 & 0 \\ 1 & 1 & 4 & 0 & 0 & 1 \end{array}\right),$$

and then eliminate the nonzero entries below the first pivot,

$$\left(\begin{array}{ccc|ccc} 2 & 6 & 1 & 0 & 1 & 0 \\ 0 & 2 & 1 & 1 & 0 & 0 \\ 0 & -2 & \frac{7}{2} & 0 & -\frac{1}{2} & 1 \end{array}\right).$$

Next we eliminate the entry below the second pivot:

$$\left(\begin{array}{ccc|ccc} 2 & 6 & 1 & 0 & 1 & 0 \\ 0 & 2 & 1 & 1 & 0 & 0 \\ 0 & 0 & \frac{9}{2} & 1 & -\frac{1}{2} & 1 \end{array}\right).$$

At this stage, we have reduced our augmented matrix to the form $\left(U \mid C \right)$ where U is upper triangular. This is equivalent to reducing the original n linear systems $A\mathbf{x}_i = \mathbf{e}_i$ to n upper triangular systems $U\mathbf{x}_i = \mathbf{c}_i$. We can therefore perform n back substitutions to produce the solutions $\mathbf{x}_i$, which would form the individual columns of the inverse matrix $X = \left(\mathbf{x}_1 \;\; \cdots \;\; \mathbf{x}_n \right)$. In the more common version of the Gauss–Jordan scheme, one instead continues to employ elementary row operations to fully reduce the augmented matrix. The goal is to produce an augmented matrix $\left(\mathrm{I} \mid X \right)$ in which the left hand $n \times n$ matrix has become the identity, while the right hand matrix is the desired solution $X = A^{-1}$. Indeed, $\left(\mathrm{I} \mid X \right)$ represents the n trivial linear systems $\mathrm{I}\mathbf{x} = \mathbf{x}_i$ whose solutions $\mathbf{x} = \mathbf{x}_i$ are the columns of the inverse matrix X.

Now, the identity matrix has 0's below the diagonal, just like U. It also has 1's along the diagonal, whereas U has the pivots (which are all nonzero) along the diagonal. Thus, the next phase in the reduction process is to make all the diagonal entries of U equal to 1. To proceed, we need to introduce the last, and least, of our linear systems operations.

Linear System Operation #3: Multiply an equation by a nonzero constant.

This operation clearly does not affect the solution, and so yields an equivalent linear system. The corresponding elementary row operation is:

Elementary Row Operation #3: Multiply a row of the matrix by a nonzero scalar.

Dividing the rows of the upper triangular augmented matrix $\left(U \mid C \right)$ by the diagonal pivots of U will produce a matrix of the form $\left(V \mid B \right)$ where V is *special upper triangular*, meaning it has all 1's along the diagonal. In our particular example, the result of these three elementary row operations of Type #3 is

$$\left(\begin{array}{ccc|ccc} 1 & 3 & \frac{1}{2} & 0 & \frac{1}{2} & 0 \\ 0 & 1 & \frac{1}{2} & \frac{1}{2} & 0 & 0 \\ 0 & 0 & 1 & \frac{2}{9} & -\frac{1}{9} & \frac{2}{9} \end{array}\right),$$

where we multiplied the first and second rows by $\frac{1}{2}$ and the third row by $\frac{2}{9}$.

We are now over half-way towards our goal. We need only make the entries above the diagonal of the left hand matrix equal to zero. This can be done by elementary row operations of Type #1, but now we work backwards. First, we eliminate the nonzero entries in the third column lying above the $(3, 3)$ entry by subtracting one half the third row from the second and also from the first:

$$\left(\begin{array}{ccc|ccc} 1 & 3 & 0 & -\frac{1}{9} & \frac{5}{9} & -\frac{1}{9} \\ 0 & 1 & 0 & \frac{7}{18} & \frac{1}{18} & -\frac{1}{9} \\ 0 & 0 & 1 & \frac{2}{9} & -\frac{1}{9} & \frac{2}{9} \end{array}\right).$$

Finally, we subtract 3 times the second row from the first to eliminate the remaining nonzero off-diagonal entry, thereby completing the Gauss–Jordan procedure:

$$\left(\begin{array}{ccc|ccc} 1 & 0 & 0 & -\frac{23}{18} & \frac{7}{18} & \frac{2}{9} \\ 0 & 1 & 0 & \frac{7}{18} & \frac{1}{18} & -\frac{1}{9} \\ 0 & 0 & 1 & \frac{2}{9} & -\frac{1}{9} & \frac{2}{9} \end{array}\right).$$

The left hand matrix is the identity, and therefore the final right hand matrix is our desired inverse:

$$A^{-1} = \begin{pmatrix} -\frac{23}{18} & \frac{7}{18} & \frac{2}{9} \\ \frac{7}{18} & \frac{1}{18} & -\frac{1}{9} \\ \frac{2}{9} & -\frac{1}{9} & \frac{2}{9} \end{pmatrix}. \tag{1.44}$$

The reader may wish to verify that the final result does satisfy both inverse conditions $A A^{-1} = \mathrm{I} = A^{-1} A$. ●

We are now able to complete the proofs of the basic results on inverse matrices. First, we need to determine the elementary matrix corresponding to an elementary row operation of type #3. Again, this is obtained by performing the row operation in question on the identity matrix. Thus, the elementary matrix that multiplies row i by the nonzero scalar c is the diagonal matrix having c in the ith diagonal position, and 1's elsewhere along the diagonal. The inverse elementary matrix is the diagonal matrix with $1/c$ in the ith diagonal position and 1's elsewhere on the main diagonal; it corresponds to the inverse operation that divides row i by c. For example, the elementary matrix that multiplies the second row of a 3–rowed matrix by 5 is

$$E = \begin{pmatrix} 1 & 0 & 0 \\ 0 & 5 & 0 \\ 0 & 0 & 1 \end{pmatrix},$$

and has inverse

$$E^{-1} = \begin{pmatrix} 1 & 0 & 0 \\ 0 & \frac{1}{5} & 0 \\ 0 & 0 & 1 \end{pmatrix}.$$

In summary:

Lemma 1.24 Every elementary matrix is nonsingular, and its inverse is also an elementary matrix of the same type.

The Gauss–Jordan method tells us how to reduce any nonsingular square matrix A to the identity matrix by a sequence of elementary row operations. Let $E_1, E_2, \ldots, E_N$ be the corresponding elementary matrices. The elimination procedure that reduces A to I amounts to multiplying A by a succession of elementary matrices:

$$E_N E_{N-1} \cdots E_2 E_1 A = \mathrm{I}. \tag{1.45}$$

We claim that the product matrix

$$X = E_N E_{N-1} \cdots E_2 E_1 \tag{1.46}$$

is the inverse of A. Indeed, formula (1.45) says that $X A = \mathrm{I}$, and so X is a left inverse. Furthermore, each elementary matrix has an inverse, and so by (1.40), X itself is invertible, with

$$X^{-1} = E_1^{-1} E_2^{-1} \cdots E_{N-1}^{-1} E_N^{-1}. \tag{1.47}$$

Therefore, multiplying formula (1.45), namely $X A = \mathrm{I}$, on the left by X^{-1} leads to $A = X^{-1}$. Lemma 1.20 implies $X = A^{-1}$, as claimed, completing the proof of Theorem 1.18. Finally, equating $A = X^{-1}$ to the product (1.47), and invoking Lemma 1.24, we have established the following result.

Proposition 1.25 Every nonsingular matrix A can be written as the product of elementary matrices.

EXAMPLE 1.26 The 2×2 matrix

$$A = \begin{pmatrix} 0 & -1 \\ 1 & 3 \end{pmatrix}$$

is converted into the identity matrix by first interchanging its rows, $\begin{pmatrix} 1 & 3 \\ 0 & -1 \end{pmatrix}$, then scaling the second row by -1, $\begin{pmatrix} 1 & 3 \\ 0 & 1 \end{pmatrix}$, and, finally, subtracting 3 times the second row from the first to obtain

$$\begin{pmatrix} 1 & 0 \\ 0 & 1 \end{pmatrix} = \mathrm{I}.$$

The corresponding elementary matrices are

$$E_1 = \begin{pmatrix} 0 & 1 \\ 1 & 0 \end{pmatrix}, \qquad E_2 = \begin{pmatrix} 1 & 0 \\ 0 & -1 \end{pmatrix}, \qquad E_3 = \begin{pmatrix} 1 & -3 \\ 0 & 1 \end{pmatrix}.$$

Therefore, by (1.46),

$$A^{-1} = E_3 E_2 E_1 = \begin{pmatrix} 1 & -3 \\ 0 & 1 \end{pmatrix} \begin{pmatrix} 1 & 0 \\ 0 & -1 \end{pmatrix} \begin{pmatrix} 0 & 1 \\ 1 & 0 \end{pmatrix} = \begin{pmatrix} 3 & 1 \\ -1 & 0 \end{pmatrix},$$

while

$$A = E_1^{-1} E_2^{-1} E_3^{-1} = \begin{pmatrix} 0 & 1 \\ 1 & 0 \end{pmatrix} \begin{pmatrix} 1 & 0 \\ 0 & -1 \end{pmatrix} \begin{pmatrix} 1 & 3 \\ 0 & 1 \end{pmatrix} = \begin{pmatrix} 0 & -1 \\ 1 & 3 \end{pmatrix}.$$

●

As an application, let us prove that the inverse of a nonsingular triangular matrix is also triangular. Specifically:

Proposition 1.27 If L is a lower triangular matrix with all nonzero entries on the main diagonal, then L is nonsingular and its inverse L^{-1} is also lower triangular. In particular, if L is special lower triangular, so is L^{-1}. A similar result holds for upper triangular matrices.

Proof It suffices to note that if L has all nonzero diagonal entries, one can reduce L to the identity by elementary row operations of Types #1 and #3, whose associated elementary matrices are all lower triangular. Lemma 1.2 implies that the product (1.46) is then also lower triangular. If L is special, then all the pivots are equal to 1. Thus, no elementary row operations of Type #3 are required, and so L can be reduced to the identity matrix by elementary row operations of Type #1 alone. Therefore, its inverse is a product of special lower triangular matrices, and hence is itself special lower triangular. A similar argument applies in the upper triangular cases. ■

EXERCISES 1.5

1.5.23. (a) Write down the elementary matrix that multiplies the third row of a 4×4 matrix by 7.
(b) Write down its inverse.

1.5.24. Find the inverse of each of the following matrices, if possible, by applying the Gauss–Jordan Method.

(a) $\begin{pmatrix} 1 & -2 \\ 3 & -3 \end{pmatrix}$ (b) $\begin{pmatrix} 1 & 3 \\ 3 & 1 \end{pmatrix}$

(c) $\begin{pmatrix} \frac{3}{5} & -\frac{4}{5} \\ \frac{4}{5} & \frac{3}{5} \end{pmatrix}$ (d) $\begin{pmatrix} 1 & 2 & 3 \\ 4 & 5 & 6 \\ 7 & 8 & 9 \end{pmatrix}$

(e) $\begin{pmatrix} 1 & 0 & -2 \\ 3 & -1 & 0 \\ -2 & 1 & -3 \end{pmatrix}$

(f) $\begin{pmatrix} 1 & 2 & 3 \\ 3 & 5 & 5 \\ 2 & 1 & 2 \end{pmatrix}$

(g) $\begin{pmatrix} 2 & 1 & 2 \\ 4 & 2 & 3 \\ 0 & -1 & 1 \end{pmatrix}$

(h) $\begin{pmatrix} 2 & 1 & 0 & 1 \\ 0 & 0 & 1 & 3 \\ 1 & 0 & 0 & -1 \\ 0 & 0 & -2 & -5 \end{pmatrix}$

(i) $\begin{pmatrix} 1 & -2 & 1 & 1 \\ 2 & -3 & 3 & 0 \\ 3 & -7 & 2 & 4 \\ 0 & 2 & 1 & 1 \end{pmatrix}$

1.5.25. Write each of the matrices in Exercise 1.5.24 as a product of elementary matrices.

1.5.26. Express

$$A = \begin{pmatrix} \frac{\sqrt{3}}{2} & -\frac{1}{2} \\ \frac{1}{2} & \frac{\sqrt{3}}{2} \end{pmatrix}$$

as a product of elementary matrices.

1.5.27. Use the Gauss–Jordan Method to find the inverse of the following complex matrices:

(a) $\begin{pmatrix} \mathrm{i} & 1 \\ 1 & \mathrm{i} \end{pmatrix}$ (b) $\begin{pmatrix} 1 & 1-\mathrm{i} \\ 1+\mathrm{i} & 1 \end{pmatrix}$

(c) $\begin{pmatrix} 0 & 1 & -\mathrm{i} \\ \mathrm{i} & 0 & -1 \\ -1 & \mathrm{i} & 1 \end{pmatrix}$

(d) $\begin{pmatrix} 1 & 0 & \mathrm{i} \\ \mathrm{i} & -1 & 1+\mathrm{i} \\ -3\,\mathrm{i} & 1-\mathrm{i} & 1+\mathrm{i} \end{pmatrix}$

1.5.28. Can two nonsingular linear systems have the same solution and yet not be equivalent?

♡ **1.5.29.** (a) Suppose $\tilde{A}$ is obtained from A by applying an elementary row operation. Let $C = AB$, where B is any matrix of the appropriate size. Explain why $\tilde{C} = \tilde{A}B$ can be obtained by applying the same elementary row operation to C.

(b) Illustrate by adding -2 times the first row to the third row of

$$A = \begin{pmatrix} 1 & 2 & -1 \\ 2 & -3 & 2 \\ 0 & 1 & -4 \end{pmatrix}$$

and then multiplying the result on the right by

$$B = \begin{pmatrix} 1 & -2 \\ 3 & 0 \\ -1 & 1 \end{pmatrix}.$$

Check that the resulting matrix is the same as first multiplying AB and then applying the same row operation to the product matrix.

Solving Linear Systems with the Inverse

The primary motivation for introducing the matrix inverse is that it provides a compact formula for the solution to any linear system with an invertible coefficient matrix.

THEOREM 1.28 If A is nonsingular, then $\mathbf{x} = A^{-1}\mathbf{b}$ is the unique solution to the linear system $A\mathbf{x} = \mathbf{b}$.

Proof We merely multiply the system by A^{-1}, which yields $\mathbf{x} = A^{-1}A\mathbf{x} = A^{-1}\mathbf{b}$. Moreover, $A\mathbf{x} = AA^{-1}\mathbf{b} = \mathbf{b}$, proving that $\mathbf{x} = A^{-1}\mathbf{b}$ is indeed the solution. ■

For example, let us return to the linear system (1.29). Since we computed the inverse of its coefficient matrix in (1.44), a "direct" way to solve the system is to multiply the right hand side by the inverse matrix:

$$\begin{pmatrix} x \\ y \\ z \end{pmatrix} = \begin{pmatrix} -\frac{23}{18} & \frac{7}{18} & \frac{2}{9} \\ \frac{7}{18} & \frac{1}{18} & -\frac{1}{9} \\ \frac{2}{9} & -\frac{1}{9} & \frac{2}{9} \end{pmatrix} \begin{pmatrix} 2 \\ 7 \\ 3 \end{pmatrix} = \begin{pmatrix} \frac{5}{6} \\ \frac{5}{6} \\ \frac{1}{3} \end{pmatrix},$$

reproducing our earlier solution.

However, while æsthetically appealing, the solution method based on the inverse matrix is hopelessly inefficient as compared to direct Gaussian Elimination, and, despite what you may have learned, *should not be used in practical computations*. (A complete justification of this dictum will be provided in Section 1.7.) On the other hand, the inverse does play a useful role in theoretical developments, as well as providing insight into the design of practical algorithms. But the principal message of applied linear algebra is that LU decomposition and Gaussian Elimination are fundamental; matrix inverses are to be avoided in all but the most elementary computations.

REMARK: The reader may have learned a version of the Gauss–Jordan algorithm for solving a single linear system that replaces the Back Substitution step by a complete reduction of the coefficient matrix to the identity. In other words, to solve $A\mathbf{x} = \mathbf{b}$, we start with the augmented matrix $M = \left(A \mid \mathbf{b} \right)$ and use all three types of elementary row operations to produce (assuming nonsingularity) the fully reduced form $\left(\mathrm{I} \mid \mathbf{d} \right)$, representing the trivially soluble, equivalent system $\mathbf{x} = \mathbf{d}$, which is the solution to the original system. However, Back Substitution is more efficient, and remains the method of choice in practical computations.

EXERCISES 1.5

1.5.30. Solve the following systems of linear equations by computing the inverses of their coefficient matrices.

(a) $\begin{aligned} x + 2y &= 1 \\ x - 2y &= -2 \end{aligned}$ (b) $\begin{aligned} 3u - 2v &= 2 \\ u + 5v &= 12 \end{aligned}$

(c) $\begin{aligned} x - y + 3z &= 3 \\ x - 2y + 3z &= -2 \\ x - 2y + z &= 2 \end{aligned}$ (d) $\begin{aligned} y + 5z &= 3 \\ x - y + 3z &= -1 \\ -2x + 3y &= 5 \end{aligned}$

(e) $\begin{aligned} x + 4y - z &= 3 \\ 2x + 7y - 2z &= 5 \\ -x - 5y + 2z &= -7 \end{aligned}$

(f) $\begin{aligned} x + y &= 4 \\ 2x + 3y - w &= 11 \\ -y - z + w &= -7 \\ z - w &= 6. \end{aligned}$

(g) $\begin{aligned} x - 2y + z + 2u &= -2 \\ x - y + z - u &= 3 \\ 2x - y + z + u &= 3 \\ -x + 3y - 2z - u &= 2 \end{aligned}$

1.5.31. For each of the nonsingular matrices in Exercise 1.5.24, use your computed inverse to solve the associated linear system $A\mathbf{x} = \mathbf{b}$, where $\mathbf{b}$ is the column vector of the appropriate size that has all 1's as its entries.

The *LDV* Factorization

The second phase of the Gauss–Jordan process leads to a slightly more detailed version of the LU factorization. Let D denote the diagonal matrix having the same diagonal entries as U; in other words, D contains the pivots on its diagonal and zeros everywhere else. Let V be the special upper triangular matrix obtained from U by dividing each row by its pivot, so that V has all 1's on the diagonal. We already encountered V during the course of the Gauss–Jordan procedure. It is easily seen that $U = DV$, which implies the following result.

THEOREM 1.29 A matrix A is regular if and only if it admits a factorization

$$A = L\,D\,V, \tag{1.48}$$

where L is a special lower triangular matrix, D is a diagonal matrix having the nonzero pivots on the diagonal, and V is a special upper triangular matrix.

For the matrix appearing in Example 1.4, we have $U = D\,V$, where

$$U = \begin{pmatrix} 2 & 1 & 1 \\ 0 & 3 & 0 \\ 0 & 0 & -1 \end{pmatrix}, \qquad D = \begin{pmatrix} 2 & 0 & 0 \\ 0 & 3 & 0 \\ 0 & 0 & -1 \end{pmatrix}, \qquad V = \begin{pmatrix} 1 & \frac{1}{2} & \frac{1}{2} \\ 0 & 1 & 0 \\ 0 & 0 & 1 \end{pmatrix}.$$

This leads to the factorization

$$A = \begin{pmatrix} 2 & 1 & 1 \\ 4 & 5 & 2 \\ 2 & -2 & 0 \end{pmatrix} = \begin{pmatrix} 1 & 0 & 0 \\ 2 & 1 & 0 \\ 1 & -1 & 1 \end{pmatrix} \begin{pmatrix} 2 & 0 & 0 \\ 0 & 3 & 0 \\ 0 & 0 & -1 \end{pmatrix} \begin{pmatrix} 1 & \frac{1}{2} & \frac{1}{2} \\ 0 & 1 & 0 \\ 0 & 0 & 1 \end{pmatrix} = LDV.$$

Proposition 1.30 If $A = L\,U$ is regular, then the factors L and U are uniquely determined. The same holds for the $A = L\,D\,V$ factorization.

Proof Suppose $L\,U = \widetilde{L}\,\widetilde{U}$. Since the diagonal entries of all four matrices are non-zero, Proposition 1.27 implies that they are invertible. Therefore,

$$\tilde{L}^{-1}L = \tilde{L}^{-1}LUU^{-1} = \tilde{L}^{-1}\tilde{L}\tilde{U}U^{-1} = \tilde{U}U^{-1}. \tag{1.49}$$

The left hand side of the matrix equation (1.49) is the product of two special lower triangular matrices, and so, by Lemma 1.2, is itself special lower triangular. The right hand side is the product of two upper triangular matrices, and hence is upper triangular. But the only way a special lower triangular matrix could equal an upper triangular matrix is if they both equal the diagonal identity matrix. Therefore, $\widetilde{L}^{-1}L = \mathrm{I} = \widetilde{U}\,U^{-1}$, and so $\widetilde{L} = L$ and $\widetilde{U} = U$, proving the first result. The $L\,D\,V$ version is an immediate consequence. ■

As you may have guessed, the more general cases requiring one or more row interchanges lead to a permuted $L\,D\,V$ factorization in the following form.

THEOREM 1.31 A matrix A is nonsingular if and only if there is a permutation matrix P such that

$$P\,A = L\,D\,V, \tag{1.50}$$

where L, D, V are, respectively, special lower triangular, diagonal, and special upper triangular matrices.

Uniqueness does not hold for the more general permuted factorizations (1.32), (1.50), since there may be several permutation matrices that place a matrix in regular form; an explicit example can be found in Exercise 1.4.21. Moreover, unlike regular elimination, the pivots, i.e., the diagonal entries of U, are no longer uniquely defined, but depend on the particular combination of row interchanges employed during the course of the computation.

EXERCISES 1.5

1.5.32. Produce the $L\,D\,V$ or a permuted $L\,D\,V$ factorization of the following matrices:

(a) $\begin{pmatrix} 1 & 2 \\ -3 & 1 \end{pmatrix}$ (b) $\begin{pmatrix} 0 & 4 \\ -7 & 2 \end{pmatrix}$

(c) $\begin{pmatrix} 2 & 1 & 2 \\ 2 & 4 & -1 \\ 0 & -2 & 1 \end{pmatrix}$

(d) $\begin{pmatrix} 1 & 1 & 5 \\ 1 & 1 & -2 \\ 2 & -1 & 3 \end{pmatrix}$

(e) $\begin{pmatrix} 2 & -3 & 2 \\ 1 & -1 & 1 \\ 1 & -1 & 2 \end{pmatrix}$

(f) $\begin{pmatrix} 1 & -1 & 1 & 2 \\ 1 & -4 & 1 & 5 \\ 1 & 2 & -1 & -1 \\ 3 & 1 & 1 & 6 \end{pmatrix}$

(g) $\begin{pmatrix} 1 & 0 & 2 & -3 \\ 2 & -2 & 0 & 1 \\ 1 & -2 & -2 & -1 \\ 0 & 1 & 1 & 2 \end{pmatrix}$

1.5.33. Using the LDV factorization for the matrices you found in parts (a)–(g) of Exercise 1.5.32, solve the corresponding linear systems $A\mathbf{x} = \mathbf{b}$, for the indicated vector $\mathbf{b}$.

(a) $\begin{pmatrix} 1 \\ 2 \end{pmatrix}$ (b) $\begin{pmatrix} -1 \\ -2 \end{pmatrix}$ (c) $\begin{pmatrix} 1 \\ -3 \\ 2 \end{pmatrix}$

(d) $\begin{pmatrix} -1 \\ 4 \\ -1 \end{pmatrix}$ (e) $\begin{pmatrix} -1 \\ -2 \\ 5 \end{pmatrix}$ (f) $\begin{pmatrix} 2 \\ -9 \\ 3 \\ 4 \end{pmatrix}$

(g) $\begin{pmatrix} 6 \\ -4 \\ 0 \\ -3 \end{pmatrix}$

1.6 TRANSPOSES AND SYMMETRIC MATRICES

Another basic operation on matrices is to interchange their rows and columns. If A is an $m \times n$ matrix, then its *transpose*, denoted A^T, is the $n \times m$ matrix whose (i, j) entry equals the (j, i) entry of A; thus

$$B = A^T \quad \text{means that} \quad b_{ij} = a_{ji}.$$

For example, if

$$A = \begin{pmatrix} 1 & 2 & 3 \\ 4 & 5 & 6 \end{pmatrix}, \quad \text{then} \quad A^T = \begin{pmatrix} 1 & 4 \\ 2 & 5 \\ 3 & 6 \end{pmatrix}.$$

Observe that the rows of A become the columns of A^T and vice versa. In particular, the transpose of a row vector is a column vector, while the transpose of a column vector is a row vector; if $\mathbf{v} = \begin{pmatrix} 1 \\ 2 \\ 3 \end{pmatrix}$, then $\mathbf{v}^T = (\,1 \quad 2 \quad 3\,)$. The transpose of a scalar, considered as a 1×1 matrix, is itself: $c^T = c$.

REMARK: Most vectors appearing in applied mathematics are column vectors. To conserve vertical space in this text, we will often use the transpose notation, e.g., $\mathbf{v} = (v_1, v_2, v_3)^T$, as a compact way of writing column vectors.

In the square case, transposition can be viewed as "reflecting" the matrix entries across the main diagonal. For example,

$$\begin{pmatrix} 1 & 2 & -1 \\ 3 & 0 & 5 \\ -2 & -4 & 8 \end{pmatrix}^T = \begin{pmatrix} 1 & 3 & -2 \\ 2 & 0 & -4 \\ -1 & 5 & 8 \end{pmatrix}.$$

In particular, the transpose of a lower triangular matrix is upper triangular and vice-versa.

Transposing twice returns you to where you started:

$$(A^T)^T = A. \tag{1.51}$$

Unlike inversion, transposition *is* compatible with matrix addition and scalar multiplication:

$$(A + B)^T = A^T + B^T, \quad (c\,A)^T = c\,A^T. \tag{1.52}$$

Transposition is also compatible with matrix multiplication, but with a twist. Like the inverse, the transpose *reverses* the order of multiplication:

$$(A\,B)^T = B^T A^T. \tag{1.53}$$

Indeed, if A has size $m \times n$ and B has size $n \times p$, so they can be multiplied, then A^T has size $n \times m$ and B^T has size $p \times n$, and so, in general, one has no choice but to multiply $B^T A^T$ in that order. Formula (1.53) is a straightforward consequence of the basic laws of matrix multiplication. An important special case is the product between a row vector $\mathbf{v}^T$ and a column vector $\mathbf{w}$ with the same number of entries. In this case,

$$\mathbf{v}^T \mathbf{w} = (\mathbf{v}^T \mathbf{w})^T = \mathbf{w}^T \mathbf{v}, \tag{1.54}$$

because their product is a scalar and so, as noted above, equals its own transpose.

Lemma 1.32 If A is a nonsingular matrix, so is A^T, and its inverse is denoted

$$A^{-T} = (A^T)^{-1} = (A^{-1})^T. \tag{1.55}$$

Thus, transposing a matrix and then inverting yields the same result as first inverting and then transposing.

Proof Let $X = (A^{-1})^T$. Then, according to (1.53),

$$X\,A^T = (A^{-1})^T A^T = (A\,A^{-1})^T = \mathrm{I}^T = \mathrm{I}.$$

The proof that $A^T X = \mathrm{I}$ is similar, and so we conclude that $X = (A^T)^{-1}$. ■

EXERCISES 1.6

1.6.1. Write down the transpose of the following matrices:

(a) $\begin{pmatrix} 1 \\ 5 \end{pmatrix}$ (b) $\begin{pmatrix} 1 & 1 \\ 0 & 2 \end{pmatrix}$

(c) $\begin{pmatrix} 1 & 2 \\ 2 & 1 \end{pmatrix}$ (d) $\begin{pmatrix} 1 & 2 & -1 \\ 2 & 0 & 2 \end{pmatrix}$

(e) $\begin{pmatrix} 1 & 2 & -3 \end{pmatrix}$ (f) $\begin{pmatrix} 1 & 2 \\ 3 & 4 \\ 5 & 6 \end{pmatrix}$

(g) $\begin{pmatrix} 1 & 2 & -1 \\ 0 & 3 & 2 \\ 1 & 1 & 5 \end{pmatrix}$.

1.6.2. Let

$$A = \begin{pmatrix} 3 & -1 & -1 \\ 1 & 2 & 1 \end{pmatrix}, \quad B = \begin{pmatrix} -1 & 2 \\ 2 & 0 \\ -3 & 4 \end{pmatrix}.$$

Compute A^T and B^T. Then compute $(A\,B)^T$ and $(B\,A)^T$ without first computing $A\,B$ or $B\,A$.

1.6.3. Show that $(A\,B)^T = A^T B^T$ if and only if A and B are square commuting matrices.

◇ **1.6.4.** Prove formula (1.53).

1.6.5. Find a formula for the transposed product $(A\,B\,C)^T$ in terms of A^T, B^T and C^T.

1.6.6. *True or false*: Every square matrix A commutes with its transpose A^T.

◇ **1.6.7.** A square matrix is called *normal* if it commutes with its transpose: $A^T A = A A^T$. Find all normal 2×2 matrices.

1.6.8. (a) Prove that the inverse transpose operation (1.55) respects matrix multiplication: $(A B)^{-T} = A^{-T} B^{-T}$.

(b) Verify this identity for

$$A = \begin{pmatrix} 1 & -1 \\ 1 & 0 \end{pmatrix}, \qquad B = \begin{pmatrix} 2 & 1 \\ 1 & 1 \end{pmatrix}.$$

1.6.9. Prove that if A is an invertible matrix, then $A A^T$ and $A^T A$ are also invertible.

1.6.10. If $\mathbf{v}, \mathbf{w}$ are column vectors with the same number of entries, does $\mathbf{v}\,\mathbf{w}^T = \mathbf{w}\,\mathbf{v}^T$?

1.6.11. Is there a matrix analogue of formula (1.54), namely $A^T B = B^T A$?

◇ **1.6.12.** (a) Let A be an $m \times n$ matrix. Let $\mathbf{e}_j$ denote the $1 \times n$ column vector with a single 1 in the jth entry, as in (1.42). Explain why the product $A\mathbf{e}_j$ equals the jth column of A.

(b) Similarly, let $\widehat{\mathbf{e}}_i$ be the $1 \times m$ column vector with a single 1 in the ith entry. Explain why the triple product $\widehat{\mathbf{e}}_i^T A\mathbf{e}_j = a_{ij}$ equals the (i, j) entry of the matrix A.

◇ **1.6.13.** Let A and B be $m \times n$ matrices.

(a) Suppose that $\mathbf{v}^T A\,\mathbf{w} = \mathbf{v}^T B\,\mathbf{w}$ for all vectors $\mathbf{v}, \mathbf{w}$. Prove that $A = B$.

(b) Give an example of two matrices such that $\mathbf{v}^T A\,\mathbf{v} = \mathbf{v}^T B\,\mathbf{v}$ for all vectors $\mathbf{v}$, but $A \neq B$.

◇ **1.6.14.** (a) Explain why the inverse of a permutation matrix equals its transpose: $P^{-1} = P^T$.

(b) If $A^{-1} = A^T$, is A necessarily a permutation matrix?

◇ **1.6.15.** Let A be a square matrix and P a permutation matrix of the same size.

(a) Explain why the product $A\,P^T$ has the effect of applying the permutation defined by P to the columns of A.

(b) Explain the effect of multiplying $P\,A\,P^T$. *Hint*: Try this on some 3×3 examples first.

♡ **1.6.16.** Let $\mathbf{v}, \mathbf{w}$ be $n \times 1$ column vectors.

(a) Prove that in most cases the inverse of the $n \times n$ matrix $A = \mathrm{I} - \mathbf{v}\,\mathbf{w}^T$ has the form $A^{-1} = \mathrm{I} - c\,\mathbf{v}\,\mathbf{w}^T$ for some scalar c. Find all $\mathbf{v}, \mathbf{w}$ for which such a result is valid.

(b) Illustrate the method when

$$\mathbf{v} = \begin{pmatrix} 1 \\ 3 \end{pmatrix} \quad \text{and} \quad \mathbf{w} = \begin{pmatrix} -1 \\ 2 \end{pmatrix}.$$

(c) What happens when the method fails?

Factorization of Symmetric Matrices

A particularly important class of square matrices is those that are unchanged by the transpose operation.

Definition 1.33 A square matrix is called *symmetric* if it equals its own transpose: $A = A^T$.

Thus, A is symmetric if and only if its entries satisfy $a_{ji} = a_{ij}$ for all i, j. In other words, entries lying in "mirror image" positions relative to the main diagonal must be equal. For example, the most general symmetric 3×3 matrix has the form

$$A = \begin{pmatrix} a & b & c \\ b & d & e \\ c & e & f \end{pmatrix}.$$

Note that any diagonal matrix, including the identity, is symmetric. A lower or upper triangular matrix is symmetric if and only if it is, in fact, a diagonal matrix.

The $L\,D\,V$ factorization of a nonsingular matrix takes a particularly simple form if the matrix also happens to be symmetric. This result will form the foundation of some significant later developments.

THEOREM 1.34 A symmetric matrix A is regular if and only if it can be factored as

$$A = L\,D\,L^T, \tag{1.56}$$

where L is a special lower triangular matrix and D is a diagonal matrix with nonzero diagonal entries.

Proof We already know, according to Theorem 1.29, that we can factor

$$A = L\,D\,V. \tag{1.57}$$

We take the transpose of both sides of this equation:

$$A^T = (L\,D\,V)^T = V^T D^T L^T = V^T D\,L^T, \tag{1.58}$$

since diagonal matrices are automatically symmetric: $D^T = D$. Note that V^T is special lower triangular, and L^T is special upper triangular. Therefore (1.58) is the $L\,D\,V$ factorization of A^T.

In particular, if A is symmetric, then

$$L\,D\,V = A = A^T = V^T D\,L^T.$$

Uniqueness of the $L\,D\,V$ factorization implies that

$$L = V^T \quad \text{and} \quad V = L^T$$

(which are two versions of the same equation). Replacing V by L^T in (1.57) establishes the factorization (1.56). ■

REMARK: If $A = L\,D\,L^T$, then A is necessarily symmetric. Indeed,

$$A^T = (L\,D\,L^T)^T = (L^T)^T D^T L^T = L\,D\,L^T = A.$$

However, not every symmetric matrix has an $L\,D\,L^T$ factorization. A simple example is the irregular but nonsingular 2×2 matrix $\begin{pmatrix} 0 & 1 \\ 1 & 0 \end{pmatrix}$.

EXAMPLE 1.35 The problem is to find the $L\,D\,L^T$ factorization of the particular symmetric matrix

$$A = \begin{pmatrix} 1 & 2 & 1 \\ 2 & 6 & 1 \\ 1 & 1 & 4 \end{pmatrix}.$$

This requires performing the usual Gaussian Elimination algorithm. Subtracting twice the first row from the second and also the first row from the third produces the matrix

$$\begin{pmatrix} 1 & 2 & 1 \\ 0 & 2 & -1 \\ 0 & -1 & 3 \end{pmatrix}.$$

We then add one half of the second row of the latter matrix to its third row, resulting in the upper triangular form

$$U = \begin{pmatrix} 1 & 2 & 1 \\ 0 & 2 & -1 \\ 0 & 0 & \frac{5}{2} \end{pmatrix} = \begin{pmatrix} 1 & 0 & 0 \\ 0 & 2 & 0 \\ 0 & 0 & \frac{5}{2} \end{pmatrix} \begin{pmatrix} 1 & 2 & 1 \\ 0 & 1 & -\frac{1}{2} \\ 0 & 0 & 1 \end{pmatrix} = D\,V,$$

which we further factor by dividing each row of U by its pivot. On the other hand, the special lower triangular matrix associated with the preceding row operations is

$$L = \begin{pmatrix} 1 & 0 & 0 \\ 2 & 1 & 0 \\ 1 & -\frac{1}{2} & 1 \end{pmatrix},$$

which, as guaranteed by Theorem 1.34, is the transpose of $V = L^T$. Therefore, the desired $A = L\,U = L\,D\,L^T$ factorizations of this particular symmetric matrix are

$$\begin{pmatrix} 1 & 2 & 1 \\ 2 & 6 & 1 \\ 1 & 1 & 4 \end{pmatrix} = \begin{pmatrix} 1 & 0 & 0 \\ 2 & 1 & 0 \\ 1 & -\frac{1}{2} & 1 \end{pmatrix} \begin{pmatrix} 1 & 2 & 1 \\ 0 & 2 & -1 \\ 0 & 0 & \frac{5}{2} \end{pmatrix}$$
$$= \begin{pmatrix} 1 & 0 & 0 \\ 2 & 1 & 0 \\ 1 & -\frac{1}{2} & 1 \end{pmatrix} \begin{pmatrix} 1 & 0 & 0 \\ 0 & 2 & 0 \\ 0 & 0 & \frac{5}{2} \end{pmatrix} \begin{pmatrix} 1 & 2 & 1 \\ 0 & 1 & -\frac{1}{2} \\ 0 & 0 & 1 \end{pmatrix}. \qquad \bullet$$

EXAMPLE 1.36 Let us look at a general 2×2 symmetric matrix $A = \begin{pmatrix} a & b \\ b & c \end{pmatrix}$. Regularity requires that the first pivot be $a \neq 0$. A single row operation will place A in upper triangular form

$$U = \begin{pmatrix} a & c \\ 0 & \dfrac{a\,c - b^2}{a} \end{pmatrix}.$$

The associated lower triangular matrix is $L = \begin{pmatrix} 1 & 0 \\ \dfrac{b}{a} & 1 \end{pmatrix}$. Thus, $A = L\,U$. Finally,

$$D = \begin{pmatrix} a & 0 \\ 0 & \dfrac{ac - b^2}{a} \end{pmatrix}$$

is just the diagonal part of U, and we find $U = D\,L^T$, so that the $L\,D\,L^T$ factorization is explicitly given by

$$\begin{pmatrix} a & b \\ b & c \end{pmatrix} = \begin{pmatrix} 1 & 0 \\ \dfrac{b}{a} & 1 \end{pmatrix} \begin{pmatrix} a & 0 \\ 0 & \dfrac{ac - b^2}{a} \end{pmatrix} \begin{pmatrix} 1 & \dfrac{b}{a} \\ 0 & 1 \end{pmatrix}. \tag{1.59}$$

$\bullet$

EXERCISES 1.6

1.6.17. Find all values of a, b, and c for which the following matrices are symmetric:

(a) $\begin{pmatrix} 3 & a \\ 2a - 1 & a - 2 \end{pmatrix}$ (b) $\begin{pmatrix} 1 & a & 2 \\ -1 & b & c \\ b & 3 & 0 \end{pmatrix}$

(c) $\begin{pmatrix} 3 & a + 2b - 2c & -4 \\ 6 & 7 & b - c \\ -a + b + c & 4 & b + 3c \end{pmatrix}$

1.6.18. List all symmetric

(a) 3×3 permutation matrices,

(b) 4×4 permutation matrices.

1.6.19. *True or false*: If A is symmetric, then A^2 is symmetric.

◇ **1.6.20.** *True or false*: If A is a nonsingular symmetric matrix, then A^{-1} is also symmetric.

◇ **1.6.21.** *True or false*: If A and B are symmetric $n \times n$ matrices, so is $A\,B$.

1.6.22. (a) Show that every diagonal matrix is symmetric.

(b) Show that an upper (lower) triangular matrix is symmetric if and only if it is diagonal.

1.6.23. Let A be a symmetric matrix.

(a) Show that A^n is symmetric for any nonnegative integer n.

(b) Show that $2A^2 - 3A + \mathrm{I}$ is symmetric.

(c) Show that any matrix polynomial $p(A)$ of A, cf. Exercise 1.2.35, is a symmetric matrix.

1.6.24. Show that if A is any matrix, then $K = A^T A$ and $L = A A^T$ are both well-defined, symmetric matrices.

1.6.25. Find the $L D L^T$ factorization of the following symmetric matrices:

(a) $\begin{pmatrix} 1 & 1 \\ 1 & 4 \end{pmatrix}$ (b) $\begin{pmatrix} -2 & 3 \\ 3 & -1 \end{pmatrix}$

(c) $\begin{pmatrix} 1 & -1 & -1 \\ -1 & 3 & 2 \\ -1 & 2 & 0 \end{pmatrix}$

(d) $\begin{pmatrix} 1 & -1 & 0 & 3 \\ -1 & 2 & 2 & 0 \\ 0 & 2 & -1 & 0 \\ 3 & 0 & 0 & 1 \end{pmatrix}$

1.6.26. Find the $L D L^T$ factorization of the matrices

$$M_2 = \begin{pmatrix} 2 & 1 \\ 1 & 2 \end{pmatrix}, \quad M_3 = \begin{pmatrix} 2 & 1 & 0 \\ 1 & 2 & 1 \\ 0 & 1 & 2 \end{pmatrix},$$

and

$$M_4 = \begin{pmatrix} 2 & 1 & 0 & 0 \\ 1 & 2 & 1 & 0 \\ 0 & 1 & 2 & 1 \\ 0 & 0 & 1 & 2 \end{pmatrix}.$$

◇ **1.6.27.** Prove that the 3×3 matrix $A = \begin{pmatrix} 1 & 2 & 1 \\ 2 & 4 & -1 \\ 1 & -1 & 3 \end{pmatrix}$ cannot be factored as $A = L D L^T$.

◇ **1.6.28.** Suppose $A = L U$ is a regular matrix. Write down the $L U$ factorization of A^T. Prove that A^T is also regular, and its pivots are the *same* as the pivots of A.

♡ **1.6.29.** *Skew–Symmetric Matrices*: An $n \times n$ matrix J is called *skew-symmetric* if $J^T = -J$.

(a) Show that every diagonal entry of a skew-symmetric matrix is zero.

(b) Write down an example of a nonsingular skew-symmetric matrix.

(c) Can you find a regular skew-symmetric matrix?

(d) Show that if J is a nonsingular skew-symmetric matrix, then J^{-1} is also skew-symmetric. Verify this fact for the matrix you wrote down in part (b).

(e) Show that if J and K are skew-symmetric, then so are J^T, $J + K$, and $J - K$. What about $J K$?

(f) Prove that if J is a skew-symmetric matrix, then $\mathbf{v}^T J \mathbf{v} = 0$ for *any* vector $\mathbf{v}$.

1.6.30. (a) Prove that every square matrix can be expressed as the sum, $A = S + J$, of a symmetric matrix $S = S^T$ and a skew-symmetric matrix $J = -J^T$.

(b) Write $\begin{pmatrix} 1 & 2 \\ 3 & 4 \end{pmatrix}$ and $\begin{pmatrix} 1 & 2 & 3 \\ 4 & 5 & 6 \\ 7 & 8 & 9 \end{pmatrix}$ as the sum of symmetric and skew-symmetric matrices.

1.7 PRACTICAL LINEAR ALGEBRA

For pedagogical and practical reasons, the examples and exercises we have chosen to illustrate the algorithms are all based on relatively small matrices. When dealing with matrices of moderate size, the differences between the various approaches to solving linear systems (Gauss, Gauss–Jordan, matrix inverse, etc.) are relatively unimportant, particularly if one has a decent computer or even hand calculator to do the tedious parts. However, real-world applied mathematics deals with much larger linear systems, and the design of efficient algorithms is a must. For example, numerical solution schemes for ordinary differential equations will typically lead to matrices with thousands of entries, while numerical schemes for partial differential equations arising in fluid and solid mechanics, weather prediction, image and video processing, quantum mechanics, molecular dynamics, chemical processes, etc., will often require dealing with matrices with more than a million entries. It is not hard for such systems to tax even the most sophisticated supercomputer. Thus, it is essential that we understand the computational details of competing methods in order to compare their efficiency, and thereby gain some experience with the issues underlying the design of high performance numerical algorithms.

The most basic question is: how many arithmetic operations* are required to complete an algorithm? The number will directly influence the time spent running the algorithm on a computer. We shall keep track of additions and multiplications separately, since the latter typically take longer to process. But we shall not distinguish between addition and subtraction, nor between multiplication and division, as these typically rely on the same floating point algorithm. We shall also assume that the matrices and vectors we deal with are *generic*, with few, if any, zero entries. Modifications of the basic algorithms for *sparse matrices*, meaning those that have lots of zero entries, are an important topic of research, since these include many of the large matrices that appear in applications to differential equations. We refer the interested reader to more advanced treatments of numerical linear algebra, such as [18, 32, 51, 67], for further developments.

First, when multiplying an $n \times n$ matrix A and an $n \times 1$ column vector $\mathbf{b}$, each entry of the product $A\mathbf{b}$ requires n multiplications of the form $a_{ij}\, b_j$ and $n-1$ additions to sum the resulting products. Since there are n entries, this means a total of n^2 multiplications and $n(n-1) = n^2 - n$ additions. Thus, for a matrix of size $n = 100$, one needs about $10,000$ distinct multiplications and a similar number of additions. If $n = 1{,}000{,}000 = 10^6$, then $n^2 = 10^{12}$, which is phenomenally large, and the total time required to perform the computation becomes a significant issue†.

Let us next look at the (regular) Gaussian Elimination algorithm, referring back to our pseudocode program for the notational details. First, we count how many arithmetic operations are based on the jth pivot m_{jj}. For each of the $n-j$ rows lying below it, we must perform one division to compute the factor $l_{ij} = m_{ij}/m_{jj}$ used in the elementary row operation. The entries in the column below the pivot will be set to zero automatically, and so we need only compute the updated entries lying strictly below and to the right of the pivot. There are $(n-j)^2$ such entries in the coefficient matrix and an additional $n-j$ entries in the last column of the augmented matrix. Let us concentrate on the former for the moment. For each of these, we replace m_{ik} by $m_{ik} - l_{ij}\, m_{jk}$, and so must perform one multiplication and one addition. For the jth pivot, there are a total of $(n-j)(n-j+1)$ multiplications—including the initial $n-j$ divisions needed to produce the l_{ij}—and $(n-j)^2$ additions needed to update the coefficient matrix. Therefore, to reduce a regular $n \times n$ matrix to upper triangular form requires a total‡ of

$$
\begin{aligned}
\sum_{j=1}^{n} (n-j)(n-j+1) &= \frac{n^3 - n}{3} \qquad \text{multiplications and} \\
\sum_{j=1}^{n} (n-j)^2 &= \frac{2n^3 - 3n^2 + n}{6} \qquad \text{additions.}
\end{aligned}
\tag{1.60}
$$

Thus, when n is large, both involve approximately $\frac{1}{3} n^3$ operations.

We should also be keeping track of the number of operations on the right hand side of the system. No pivots appear there, and so there are

$$\sum_{j=1}^{n} (n-j) = \frac{n^2 - n}{2} \tag{1.61}$$

multiplications and the same number of additions required to produce the right hand side in the resulting triangular system $U\mathbf{x} = \mathbf{c}$. For large n, this count is consid-

*For simplicity, we will only count basic arithmetic operations. But it is worth noting that other issues, such as the number of I/O operations, may also play a role in estimating the computational complexity of a numerical algorithm.

†See Exercise 1.7.8 for more sophisticated computational algorithms that speed up matrix multiplication.

‡In Exercise 1.7.4, the reader is asked to prove these summation formulae by induction.

erably smaller than the coefficient matrix totals (1.60). We note that the Forward Substitution equations (1.26) require precisely the same number of arithmetic operations to solve $L\mathbf{c} = \mathbf{b}$ for the right hand side of the upper triangular system. Indeed, the jth equation

$$c_j = b_j - \sum_{k=1}^{j-1} l_{jk} c_k$$

requires $j - 1$ multiplications and the same number of additions, giving a total of

$$\sum_{j=1}^{n} j = \frac{n^2 - n}{2}$$

operations of each type. Therefore, to reduce a linear system to upper triangular form, it makes no difference in computational efficiency whether one works directly with the augmented matrix or employs Forward Substitution after the LU factorization of the coefficient matrix has been established.

The Back Substitution phase of the algorithm can be similarly analyzed. To find the value of

$$x_j = \frac{1}{u_{jj}} \left(c_j - \sum_{k=j+1}^{n} u_{jk} x_k \right)$$

once we have computed $x_{j+1}, \ldots, x_n$, requires $n - j + 1$ multiplications/divisions and $n - j$ additions. Therefore, the Back Substitution phase of the algorithm requires

$$\begin{aligned} \sum_{j=1}^{n} (n - j + 1) &= \frac{n^2 + n}{2} \qquad \text{multiplications, along with} \\ \sum_{j=1}^{n} (n - j) &= \frac{n^2 - n}{2} \qquad \text{additions.} \end{aligned} \tag{1.62}$$

For n large, both of these are approximately equal to $\frac{1}{2} n^2$. Comparing the counts, we conclude that the bulk of the computational effort goes into the reduction of the coefficient matrix to upper triangular form.

Combining the two counts (1.61–62), we discover that, once we have computed the $A = LU$ decomposition of the coefficient matrix, the Forward and Back Substitution process requires n^2 multiplications and $n^2 - n$ additions to solve a linear system $A\mathbf{x} = \mathbf{b}$. This is exactly the *same* as the number of multiplications and additions needed to compute the product $A^{-1}\mathbf{b}$. Thus, even if we happen to know the inverse of A, it is still *just as efficient* to use Forward and Back Substitution to compute the solution!

On the other hand, the computation of A^{-1} is decidedly more inefficient. There are two possible strategies. First, we can solve the n linear systems (1.43), namely

$$A\mathbf{x} = \mathbf{e}_i, \qquad i = 1, \ldots, n, \tag{1.63}$$

for the individual columns of A^{-1}. This requires first computing the LU decomposition, which uses about $\frac{1}{3} n^3$ multiplications and a similar number of additions, followed by applying Forward and Back Substitution to each of the systems, using $n \cdot n^2 = n^3$ multiplications and $n\,(n^2 - n) \approx n^3$ additions, for a grand total of about $\frac{4}{3} n^3$ operations of each type in order to compute A^{-1}. Gauss–Jordan Elimination fares no better (in fact, slightly worse), also requiring about the same number, $\frac{4}{3} n^3$, of each type of arithmetic operation. Both algorithms can be made more efficient by exploiting the fact that there are lots of zeros on the right hand sides of the systems (1.63). Designing the algorithm to avoid adding or subtracting a preordained 0, or multiplying or dividing by a preordained ± 1, reduces the total number of operations required to compute A^{-1} to exactly n^3 multiplications and $n\,(n-1)^2 \approx n^3$ additions. (Details are relegated to the exercises.) And don't forget we still need to multiply $A^{-1}\mathbf{b}$ to solve the original system. As a result, solving a linear system with the inverse matrix requires approximately *three* times as many arithmetic operations, and so would take three times as long to complete, as the more elementary Gaussian Elimination and Back Substitution algorithm. This justifies our earlier contention that matrix inversion is inefficient, and, except in very special situations, should never be used for solving linear systems in practice.

EXERCISES 1.7

1.7.1. Solve the following linear systems by

(i) Gaussian Elimination with Back Substitution.;

(ii) the Gauss–Jordan algorithm to convert the augmented matrix to the fully reduced form $\left(\,\mathrm{I} \mid \mathbf{x}\,\right)$ with solution $\mathbf{x}$;

(iii) computing the inverse of the coefficient matrix, and then multiplying it with the right hand side. Keep track of the number of arithmetic operations you need to perform to complete each computation, and discuss their relative efficiency.

(a) $$\begin{aligned} x - 2\,y &= 4 \\ 3\,x + y &= -7 \end{aligned}$$

(b) $$\begin{aligned} 2\,x - 4\,y + 6\,z &= 6 \\ 3\,x - 3\,y + 4\,z &= -1 \\ -4\,x + 3\,y - 4\,z &= 5 \end{aligned}$$

(c) $$\begin{aligned} x - 3\,y \qquad\;\; &= 1 \\ 3\,x - 7\,y + 5\,z &= -1 \\ -2\,x + 6\,y - 5\,z &= 0. \end{aligned}$$

1.7.2. (a) Let A be an $n \times n$ matrix. Which is faster to compute, A^2 or A^{-1}? Justify your answer.

(b) What about A^3 versus A^{-1}?

(c) How many operations are needed to compute A^k? *Hint*: When $k > 3$, you can get away with less than $k - 1$ matrix multiplications!

1.7.3. Which is faster: Back Substitution or multiplying a matrix times a vector? How much faster?

◇ **1.7.4.** Use induction to prove the summation formulae (1.60), (1.61) and (1.62).

♡ **1.7.5.** Let A be a general $n \times n$ matrix. Determine the exact number of arithmetic operations needed to compute A^{-1} using

(a) Gaussian Elimination to factor $P\,A = L\,U$ and then Forward and Back Substitution to solve the n linear systems (1.63);

(b) the Gauss–Jordan method. Make sure your totals do not count adding or subtracting a known 0, or multiplying or dividing by a known ± 1.

1.7.6. Count the number of arithmetic operations needed to solve a system the "old-fashioned" way, by using elementary row operations of all three types, in the same order as the Gauss–Jordan scheme, to fully reduce the augmented matrix $M = \left(\,A \mid \mathbf{b}\,\right)$ to the form $\left(\,\mathrm{I} \mid \mathbf{d}\,\right)$, with $\mathbf{x} = \mathbf{d}$ being the solution.

1.7.7. An alternative solution strategy, also called *Gauss–Jordan* in some texts, is, once a pivot is in position, to use elementary row operations of type #1 to eliminate all entries both above and below it, thereby reducing the augmented matrix to diagonal form $\left(\,D \mid \mathbf{c}\,\right)$ where $D = \operatorname{diag}(d_1, \ldots, d_n)$ is a diagonal matrix containing the pivots. The solutions $x_i = c_i / d_i$ are then obtained by simple division. Is this strategy

(a) more efficient,

(b) less efficient, or

(c) the same

as Gaussian Elimination with Back Substitution? Justify your answer with an exact operations count.

♡ **1.7.8.** Here, we describe a remarkable algorithm for matrix multiplication discovered by Strassen, [62]. Let $A = \begin{pmatrix} A_1 & A_2 \\ A_3 & A_4 \end{pmatrix}$, $B = \begin{pmatrix} B_1 & B_2 \\ B_3 & B_4 \end{pmatrix}$, and $C = \begin{pmatrix} C_1 & C_2 \\ C_3 & C_4 \end{pmatrix} = A\,B$ be block matrices of size $n = 2m$, where all blocks are of size $m \times m$.

(a) Let

$$\begin{aligned} D_1 &= (A_1 + A_4)\,(B_1 + B_4), \\ D_2 &= (A_1 - A_3)\,(B_1 + B_2), \\ D_3 &= (A_2 - A_4)\,(B_3 + B_4), \\ D_4 &= (A_1 + A_2)\,B_4, \\ D_5 &= (A_3 + A_4)\,B_1, \\ D_6 &= A_4(B_1 - B_3), \\ D_7 &= A_1(B_2 - B_4). \end{aligned}$$

Show that

$$\begin{aligned} C_1 &= D_1 + D_3 - D_4 - D_6, \\ C_2 &= D_4 + D_7, \\ C_3 &= D_5 - D_6, \\ C_4 &= D_1 - D_2 - D_5 + D_7. \end{aligned}$$

(b) How many arithmetic operations are required when A and B are 2×2 matrices? How does this compare with the usual method of multiplying 2×2 matrices?

(c) In the general case, suppose we use standard matrix multiplication for the matrix products in $D_1, \ldots, D_7$. Prove that Strassen's method is faster than the direct algorithm for computing AB by a factor of $\approx \frac{7}{8}$.

(d) When A and B have size $n \times n$ with $n = 2^r$, we can recursively apply Strassen's method to multiply the $2^{r-1} \times 2^{r-1}$ blocks A_i, B_i. Prove that the resulting algorithm requires a total of $7^r = n^{\log_2 7} = n^{2.80735}$ multiplications and

$$6(7^{r-1} - 4^{r-1}) \le 7^r = n^{\log_2 7}$$

additions/subtractions, versus n^3 multiplications and $n^3 - n^2 \approx n^3$ additions for the ordinary matrix multiplication algorithm. How much faster is Strassen's method when $n = 2^{10}$? 2^{25}? 2^{100}?

(e) How might you proceed if the size of the matrices does not happen to be a power of 2? Further developments of these ideas can be found in [11, 32].

Tridiagonal Matrices

Of course, in special cases, the actual arithmetic operation count might be considerably reduced, particularly if A is a sparse matrix with many zero entries. A number of specialized techniques have been designed to handle sparse linear systems. A particularly important class are the *tridiagonal matrices*

$$A = \begin{pmatrix} q_1 & r_1 & & & & \\ p_1 & q_2 & r_2 & & & \\ & p_2 & q_3 & r_3 & & \\ & & \ddots & \ddots & \ddots & \\ & & & p_{n-2} & q_{n-1} & r_{n-1} \\ & & & & p_{n-1} & q_n \end{pmatrix} \tag{1.64}$$

with all entries zero except for those on the main diagonal, namely $a_{ii} = q_i$, the *subdiagonal*, meaning the $n-1$ entries $a_{i+1,i} = p_i$ immediately below the main diagonal, and the *superdiagonal*, meaning the entries $a_{i,i+1} = r_i$ immediately above the main diagonal. (Blank entries indicate a 0.) Such matrices arise in the numerical solution of ordinary differential equations and the spline fitting of curves for interpolation and computer graphics. If $A = LU$ is regular, it turns out that the factors are lower and upper *bidiagonal matrices*, of the form

$$L = \begin{pmatrix} 1 & & & & & \\ l_1 & 1 & & & & \\ & l_2 & 1 & & & \\ & & \ddots & \ddots & & \\ & & & l_{n-2} & 1 & \\ & & & & l_{n-1} & 1 \end{pmatrix}, \qquad U = \begin{pmatrix} d_1 & u_1 & & & & \\ & d_2 & u_2 & & & \\ & & d_3 & u_3 & & \\ & & & \ddots & \ddots & \\ & & & & d_{n-1} & u_{n-1} \\ & & & & & d_n \end{pmatrix}. \tag{1.65}$$

Multiplying out $L\,U$ and equating the result to A leads to the equations

$$\begin{aligned}
d_1 &= q_1, & u_1 &= r_1, & l_1 d_1 &= p_1,\\
l_1 u_1 + d_2 &= q_2, & u_2 &= r_2, & l_2 d_2 &= p_2,\\
&\ \vdots & &\ \vdots & &\ \vdots\\
l_{j-1} u_{j-1} + d_j &= q_j, & u_j &= r_j, & l_j d_j &= p_j,\\
&\ \vdots & &\ \vdots & &\ \vdots\\
l_{n-2} u_{n-2} + d_{n-1} &= q_{n-1}, & u_{n-1} &= r_{n-1}, & l_{n-1} d_{n-1} &= p_{n-1},\\
l_{n-1} u_{n-1} + d_n &= q_n. & & & &
\end{aligned} \tag{1.66}$$

These elementary algebraic equations can be successively solved for the entries of L and U in the following order: $d_1, u_1, l_1, d_2, u_2, l_2, d_3, u_3 \ldots$. The original matrix A is regular provided none of the diagonal entries $d_1, d_2, \ldots$ are zero, which allows the recursive procedure to successfully proceed to termination.

Once the $L\,U$ factors are in place, we can apply Forward and Back Substitution to solve the tridiagonal linear system $A\,\mathbf{x} = \mathbf{b}$. We first solve the lower triangular system $L\,\mathbf{c} = \mathbf{b}$ by Forward Substitution, which leads to the recursive equations

$$c_1 = b_1, \quad c_2 = b_2 - l_1 c_1, \quad \ldots \quad c_n = b_n - l_{n-1} c_{n-1}. \tag{1.67}$$

We then solve the upper triangular system $U\,\mathbf{x} = \mathbf{c}$ by Back Substitution, again recursively:

$$x_n = \frac{c_n}{d_n}, \quad x_{n-1} = \frac{c_{n-1} - u_{n-1} x_n}{d_{n-1}}, \quad \ldots \quad x_1 = \frac{c_1 - u_1 x_2}{d_1}. \tag{1.68}$$

As you can check, there are a total of $5n - 4$ multiplications/divisions and $3n - 3$ additions/subtractions required to solve a general tridiagonal system of n linear equations—a striking improvement over the general case.

EXAMPLE 1.37 Consider the $n \times n$ tridiagonal matrix

$$A = \begin{pmatrix} 4 & 1 & & & & & \\ 1 & 4 & 1 & & & & \\ & 1 & 4 & 1 & & & \\ & & 1 & 4 & 1 & & \\ & & & \ddots & \ddots & \ddots & \\ & & & & 1 & 4 & 1 \\ & & & & & 1 & 4 \end{pmatrix}$$

in which the diagonal entries are all $q_i = 4$, while the entries immediately above and below the main diagonal are all $p_i = r_i = 1$. According to (1.66), the tridiagonal factorization (1.65) has $u_1 = u_2 = \cdots = u_{n-1} = 1$, while

$$d_1 = 4, \quad l_j = 1/d_j, \quad d_{j+1} = 4 - l_j, \qquad j = 1, 2, \ldots, n-1.$$

The computed values are

j	1	2	3	4	5	6	7
d_j	4.0	3.75	3.733333	3.732143	3.732057	3.732051	3.732051
l_j	.25	.266666	.267857	.267942	.267948	.267949	.267949

These converge rapidly to

$$d_j \to 2 + \sqrt{3} = 3.732051\ldots, \qquad l_j \to 2 - \sqrt{3} = .267949\ldots,$$

which makes the factorization for large n almost trivial. The numbers $2 \pm \sqrt{3}$ are the roots of the quadratic equation $x^2 - 4x + 1 = 0$, and are characterized as the fixed points of the nonlinear iterative system $d_{j+1} = 4 - 1/d_j$. ●

EXERCISES 1.7

1.7.9. For each of the following tridiagonal systems find the LU factorization of the coefficient matrix, and then solve the system.

(a) $\begin{pmatrix} 1 & 2 & 0 \\ -1 & -1 & 1 \\ 0 & -2 & 3 \end{pmatrix} \mathbf{x} = \begin{pmatrix} 4 \\ -1 \\ -6 \end{pmatrix}$

(b) $\begin{pmatrix} 1 & -1 & 0 & 0 \\ -1 & 2 & 1 & 0 \\ 0 & -1 & 4 & 1 \\ 0 & 0 & -5 & 6 \end{pmatrix} \mathbf{x} = \begin{pmatrix} 1 \\ 0 \\ 6 \\ 7 \end{pmatrix}$

(c) $\begin{pmatrix} 1 & 2 & 0 & 0 \\ -1 & -3 & 0 & 0 \\ 0 & -1 & 4 & -1 \\ 0 & 0 & -1 & -1 \end{pmatrix} \mathbf{x} = \begin{pmatrix} 0 \\ -2 \\ -3 \\ 1 \end{pmatrix}$

1.7.10. (a) Find the LU factorization of the $n \times n$ tridiagonal matrix A_n with all 2's along the diagonal and all -1's along the sub- and super-diagonals for $n = 3, 4$ and 5.

(b) Use your factorizations to solve the system $A_n \mathbf{x} = \mathbf{b}$, where $\mathbf{b} = (1, 1, 1, \ldots, 1)^T$.

(c) Do the entries in the factors approach a limit as n gets larger and larger?

♠ **1.7.11.** Answer Exercise 1.7.10 if the super-diagonal entries of A_n are changed to $+1$.

1.7.12. *True or false*:

(a) The product of two tridiagonal matrices is tridiagonal.

(b) The inverse of a tridiagonal matrix is tridiagonal.

♠ **1.7.13.** Find the LU factorizations of

$$\begin{pmatrix} 4 & 1 & 1 \\ 1 & 4 & 1 \\ 1 & 1 & 4 \end{pmatrix}, \quad \begin{pmatrix} 4 & 1 & 0 & 1 \\ 1 & 4 & 1 & 0 \\ 0 & 1 & 4 & 1 \\ 1 & 0 & 1 & 4 \end{pmatrix},$$

$$\begin{pmatrix} 4 & 1 & 0 & 0 & 1 \\ 1 & 4 & 1 & 0 & 0 \\ 0 & 1 & 4 & 1 & 0 \\ 0 & 0 & 1 & 4 & 1 \\ 1 & 0 & 0 & 1 & 4 \end{pmatrix}.$$

Do you see a pattern? Try the 6×6 version. The following exercise should now be clear.

♡ **1.7.14.** A *tricirculant matrix*

$$C = \begin{pmatrix} q_1 & r_1 & & & & & p_1 \\ p_2 & q_2 & r_2 & & & & \\ & p_3 & q_3 & r_3 & & & \\ & & \ddots & \ddots & \ddots & & \\ & & & & p_{n-1} & q_{n-1} & r_{n-1} \\ r_n & & & & & p_n & q_n \end{pmatrix}$$

is tridiagonal except for its $(1, n)$ and $(n, 1)$ entries. Tricirculant matrices arise in the numerical solution of periodic boundary value problems and in spline interpolation.

(a) Prove that if $C = LU$ is regular, its factors have the form

$$\begin{pmatrix} 1 & & & & & & \\ l_1 & 1 & & & & & \\ & l_2 & 1 & & & & \\ & & l_3 & 1 & & & \\ & & & \ddots & \ddots & & \\ & & & & l_{n-2} & 1 & \\ m_1 & m_2 & m_3 & \ldots & m_{n-2} & l_{n-1} & 1 \end{pmatrix},$$

$$\begin{pmatrix} d_1 & u_1 & & & & v_1 \\ & d_2 & u_2 & & & v_2 \\ & & d_3 & u_3 & & v_3 \\ & & & \ddots & \ddots & \vdots \\ & & & d_{n-2} & u_{n-2} & v_{n-2} \\ & & & & d_{n-1} & u_{n-1} \\ & & & & & d_n \end{pmatrix}.$$

(b) Compute the LU factorization of the $n \times n$ tricirculant matrix

$$C_n = \begin{pmatrix} 1 & -1 & & & & -1 \\ -1 & 2 & -1 & & & \\ & -1 & 3 & -1 & & \\ & & \ddots & \ddots & \ddots & \\ & & & -1 & n-1 & -1 \\ -1 & & & & -1 & 1 \end{pmatrix}$$

for $n = 3, 5$ and 6. What goes wrong when $n = 4$?

♡ **1.7.15.** A matrix A is said to have *band width* k if all entries that are more than k slots away from the main diagonal are zero: $a_{ij} = 0$ whenever $|i - j| > k$.

(a) Show that a tridiagonal matrix has band width 1.

(b) Write down an example of a 6×6 matrix of band width 2 and one of band width 3.

(c) Prove that the L and U factors of a regular banded matrix have the same band width.

(d) Find the LU factorization of the matrices you wrote down in part (b).

(e) Use the factorization to solve the system $A\mathbf{x} = \mathbf{b}$, where $\mathbf{b}$ is the column vector with all entries equal to 1.

(f) How many arithmetic operations are needed to solve $A\mathbf{x} = \mathbf{b}$ if A is banded?

(g) Prove or give a counterexample: the inverse of a banded matrix is banded.

Pivoting Strategies

Let us now investigate the practical side of pivoting. As we know, in the irregular situations when a zero shows up in a diagonal pivot position, a row interchange is required to proceed with the elimination algorithm. But even when a nonzero pivot element is in place, there may be good numerical reasons for exchanging rows in order to install a more desirable element in the pivot position. Here is a simple example:

$$.01x + 1.6y = 32.1, \qquad x + .6y = 22. \tag{1.69}$$

The exact solution to the system is easily found:

$$x = 10, \qquad y = 20.$$

Suppose we are working with a very primitive calculator that only retains 3 digits of accuracy. (Of course, this is not a very realistic situation, but the example could be suitably modified to produce similar difficulties no matter how many digits of accuracy our computer retains.) The augmented matrix is

$$\left(\begin{array}{cc|c} .01 & 1.6 & 32.1 \\ 1 & .6 & 22 \end{array}\right).$$

Choosing the $(1, 1)$ entry as our pivot, and subtracting 100 times the first row from the second produces the upper triangular form

$$\left(\begin{array}{cc|c} .01 & 1.6 & 32.1 \\ 0 & -159.4 & -3188 \end{array}\right).$$

Since our calculator has only three–place accuracy, it will round the entries in the second row, producing the augmented coefficient matrix

$$\left(\begin{array}{cc|c} .01 & 1.6 & 32.1 \\ 0 & -159.0 & -3190 \end{array}\right).$$

The solution by Back Substitution gives

$$y = 3190/159 = 20.0628\ldots \simeq 20.1,$$

and then

$$x = 100(32.1 - 1.6y) = 100(32.1 - 32.16) \simeq 100(32.1 - 32.2) = -10.$$

The relatively small error in y has produced a very large error in x—not even its sign is correct!

The problem is that the first pivot, .01, is much smaller than the other element, 1, that appears in the column below it. Interchanging the two rows before performing the row operation would resolve the difficulty—even with such an inaccurate calculator! After the interchange, we have

$$\left(\begin{array}{cc|c} 1 & .6 & 22 \\ .01 & 1.6 & 32.1 \end{array}\right),$$

which results in the rounded-off upper triangular form

$$\left(\begin{array}{cc|c} 1 & .6 & 22 \\ 0 & 1.594 & 31.88 \end{array}\right) \simeq \left(\begin{array}{cc|c} 1 & .6 & 22 \\ 0 & 1.59 & 31.9 \end{array}\right).$$

The solution by Back Substitution now gives a respectable answer:

$$y = 31.9/1.59 = 20.0628\ldots \simeq 20.1,$$
$$x = 22 - .6y = 22 - 12.06 \simeq 22 - 12.1 = 9.9.$$

The general strategy, known as *Partial Pivoting*, says that at each stage, we should use the largest (in absolute value) legitimate (i.e., in the pivot column on or below the diagonal) element as the pivot, even if the diagonal element is nonzero. Partial pivoting can help suppress the undesirable effects of round-off errors during the computation.

In a computer implementation of pivoting, there is no need to waste processor time physically exchanging the row entries in memory. Rather, one introduces a separate array of pointers that serve to indicate which original row is currently in which permuted position. More concretely, one initializes n row pointers $\rho(1) = 1, \ldots, \rho(n) = n$. Interchanging row i and row j of the coefficient or augmented matrix is then accomplished by merely interchanging $\rho(i)$ and $\rho(j)$. Thus, to access a matrix element that is currently in row i of the augmented matrix, one merely retrieves the element that is in row $\rho(i)$ in the computer's memory. An explicit implementation of this strategy is provided in the accompanying pseudocode program.

Gaussian Elimination With Partial Pivoting

start
- **for** $i = 1$ **to** n
  - **set** $\rho(i) = i$
- **next** i
- **for** $j = 1$ **to** n
  - **if** $m_{\rho(i),j} = 0$ **for all** $i \geq j$**, stop; print** "A is singular"
  - **choose** $i > j$ **such that** $m_{\rho(i),j}$ **is maximal**
  - **interchange** $\rho(i) \longleftrightarrow \rho(j)$
  - **for** $i = j+1$ **to** n
    - **set** $l_{\rho(i)j} = m_{\rho(i)j}/m_{\rho(j)j}$
    - **for** $k = j+1$ **to** $n+1$
      - **set** $m_{\rho(i)k} = m_{\rho(i)k} - l_{\rho(i)j}m_{\rho(j)k}$
    - **next** k
  - **next** i
- **next** j

end

Partial pivoting will solve most problems, although there can still be difficulties. For instance, it does not accurately solve the system

$$10\,x + 1600\,y = 3210, \qquad x + .6\,y = 22,$$

obtained by multiplying the first equation in (1.69) by 1000. The tip-off is that, while the entries in the column containing the pivot are smaller, those in its row are much larger. The solution to this difficulty is *Full Pivoting*, in which one also performs column interchanges—preferably with a column pointer—to move the largest legitimate element into the pivot position. In practice, a column interchange amounts to reordering the variables in the system, which, as long as one keeps proper track of the order, also doesn't change the solutions. Thus, switching the order of x, y leads to the augmented matrix

$$\left(\begin{array}{cc|c} 1600 & 10 & 3210 \\ .6 & 1 & 22 \end{array}\right)$$

in which the first column now refers to y and the second to x. Now Gaussian Elimination will produce a reasonably accurate solution to the system.

Finally, there are some matrices that are hard to handle even with sophisticated pivoting strategies. Such *ill-conditioned* matrices are typically characterized by being "almost" singular. A famous example of an ill-conditioned matrix is the $n \times n$ *Hilbert matrix*

$$H_n = \begin{pmatrix} 1 & \frac{1}{2} & \frac{1}{3} & \frac{1}{4} & \cdots & \frac{1}{n} \\ \frac{1}{2} & \frac{1}{3} & \frac{1}{4} & \frac{1}{5} & \cdots & \frac{1}{n+1} \\ \frac{1}{3} & \frac{1}{4} & \frac{1}{5} & \frac{1}{6} & \cdots & \frac{1}{n+2} \\ \frac{1}{4} & \frac{1}{5} & \frac{1}{6} & \frac{1}{7} & \cdots & \frac{1}{n+3} \\ \vdots & \vdots & \vdots & \vdots & \ddots & \vdots \\ \frac{1}{n} & \frac{1}{n+1} & \frac{1}{n+2} & \frac{1}{n+3} & \cdots & \frac{1}{2n-1} \end{pmatrix}. \tag{1.70}$$

Later, in Proposition 3.34, we will prove that H_n is nonsingular for all n. However, the solution of a linear system whose coefficient matrix is a Hilbert matrix H_n, even for moderately large n, is a very challenging problem, even using high precision computer arithmetic.[§] This is because the larger n is, the closer H_n is, in a sense, to being singular. A full discussion of the so-called condition number of a matrix can be found in Sections 8.5 and 10.3.

The reader is urged to try the following computer experiment. Fix a moderately large value of n, say 20. Choose a column vector $\mathbf{x}$ with n entries chosen at random. Compute $\mathbf{b} = H_n \mathbf{x}$ directly. Then try to solve the system $H_n \mathbf{x} = \mathbf{b}$ by Gaussian Elimination, and compare the result with the original vector $\mathbf{x}$. If you obtain an accurate solution with $n = 20$, try $n = 50$ or 100. This will give you a good indicator of the degree of arithmetic precision used by your computer hardware, and the accuracy of the numerical solution algorithm(s) in your software.

[§]In computer algebra systems such as MAPLE or MATHEMATICA, one can use exact rational arithmetic to perform the computations. Then the important issues are time and computational efficiency. Incidentally, there is an explicit formula for the inverse of a Hilbert matrix; see Exercise 1.7.25.

EXERCISES 1.7

1.7.16. (a) Find the exact solution to the linear system

$$\begin{pmatrix} .1 & 2.7 \\ 1.0 & .5 \end{pmatrix}\begin{pmatrix} x \\ y \end{pmatrix} = \begin{pmatrix} 10. \\ -6.0 \end{pmatrix}.$$

(b) Solve the system using Gaussian Elimination with 2 digit rounding.

(c) Solve the system using Partial Pivoting and 2 digit rounding.

(d) Compare your answers and discuss.

1.7.17. (a) Find the exact solution to the linear system $x - 5y - z = 1,\ \frac{1}{6}x - \frac{5}{6}y + z = 0,\ 2x - y = 3$.

(b) Solve the system using Gaussian Elimination with 4 digit rounding.

(c) Solve the system using Partial Pivoting and 4 digit rounding. Compare your answers.

1.7.18. Answer Exercise 1.7.17 for the system $x + 4y - 3z = -3,\ 25x + 97y - 35z = 39,\ 35x - 22y + 33z = -15$.

1.7.19. Employ 2 digit arithmetic with rounding to compute an approximate solution of the linear system $0.2x + 2y - 3z = 6,\ 5x + 43y + 27z = 58,\ 3x + 23y - 42z = -87$, using the following methods:

(a) Regular Gaussian Elimination with Back Substitution;

(b) Gaussian Elimination with Partial Pivoting;

(c) Gaussian Elimination with Full Pivoting.

(d) Compare your answers and discuss their accuracy.

1.7.20. Solve the following systems by hand, using pointers instead of physically interchanging the rows:

(a) $\begin{pmatrix} 0 & 1 & -2 \\ 1 & -1 & 1 \\ 3 & 1 & 0 \end{pmatrix}\begin{pmatrix} x \\ y \\ z \end{pmatrix} = \begin{pmatrix} 1 \\ 2 \\ 1 \end{pmatrix}$

(b) $\begin{pmatrix} 0 & -1 & 0 & -1 \\ 0 & 0 & -2 & 1 \\ 1 & 0 & 2 & 0 \\ -1 & 0 & 0 & 3 \end{pmatrix}\begin{pmatrix} x \\ y \\ z \\ w \end{pmatrix} = \begin{pmatrix} 1 \\ 0 \\ 0 \\ 1 \end{pmatrix}$

(c) $\begin{pmatrix} 3 & -1 & 2 & -1 \\ 6 & -2 & 4 & 3 \\ 3 & 1 & 0 & -2 \\ -1 & 3 & -2 & 0 \end{pmatrix}\begin{pmatrix} x \\ y \\ z \\ w \end{pmatrix} = \begin{pmatrix} 1 \\ 2 \\ 1 \\ 1 \end{pmatrix}$

(d) $\begin{pmatrix} 0 & -1 & 5 & -1 \\ 1 & -2 & 0 & 1 \\ 2 & -3 & -3 & -1 \\ 2 & 0 & 1 & -1 \end{pmatrix}\begin{pmatrix} x \\ y \\ z \\ w \end{pmatrix} = \begin{pmatrix} 1 \\ -2 \\ 3 \\ 0 \end{pmatrix}$

1.7.21. Solve the following systems using Partial Pivoting and pointers:

(a) $\begin{pmatrix} 1 & 5 \\ 2 & -3 \end{pmatrix}\begin{pmatrix} x \\ y \end{pmatrix} = \begin{pmatrix} 3 \\ -2 \end{pmatrix}$

(b) $\begin{pmatrix} 1 & 2 & -1 \\ 4 & -2 & 1 \\ 3 & 5 & -1 \end{pmatrix}\begin{pmatrix} x \\ y \\ z \end{pmatrix} = \begin{pmatrix} 1 \\ 3 \\ 1 \end{pmatrix}$

(c) $\begin{pmatrix} 1 & -3 & 6 & -1 \\ 2 & -5 & 0 & 1 \\ -1 & -6 & 4 & -2 \\ 3 & 0 & 2 & -1 \end{pmatrix}\begin{pmatrix} x \\ y \\ z \\ w \end{pmatrix} = \begin{pmatrix} 1 \\ -2 \\ 0 \\ 1 \end{pmatrix}$

(d) $\begin{pmatrix} .01 & 4 & 2 \\ 2 & -802 & 3 \\ 7 & .03 & 250 \end{pmatrix}\begin{pmatrix} x \\ y \\ z \end{pmatrix} = \begin{pmatrix} 1 \\ 2 \\ 122 \end{pmatrix}$

1.7.22. Use Full Pivoting with pointers to solve the systems in Exercise 1.7.21.

♠ **1.7.23.** (a) Write out a pseudo-code algorithm, using both row and column pointers, for Gaussian Elimination with Full Pivoting.

(b) Implement your code on a computer, and try it on the systems in Exercise 1.7.21.

♠ **1.7.24.** Implement the computer experiment with Hilbert matrices outlined in the last paragraph of the section.

♠ **1.7.25.** Let H_n be the $n \times n$ Hilbert matrix (1.70), and $K_n = H_n^{-1}$ its inverse. It can be proved, [32, p. 513], that the (i, j) entry of K_n is

$$(-1)^{i+j}(i + j - 1) \cdot \binom{n+i-1}{n-j}\binom{n+j-1}{n-i}\binom{i+j-2}{i-1}^2,$$

where

$$\binom{n}{k} = \frac{n!}{k!\,(n-k)!}$$

is the standard binomial coefficient. (*Warning*: Proving this formula is a nontrivial combinatorial challenge.)

(a) Write down the inverse of the Hilbert matrices H_3, H_4, H_5 by either using the formula or using the Gauss–Jordan Method with exact rational arithmetic. Check your results by multiplying the matrix by its inverse.

(b) Recompute the inverses on your computer using floating point arithmetic and compare with the exact answers.

(c) Try using floating point arithmetic to find K_{10} and K_{20}. Test the answer by multiplying the Hilbert matrix by its computed inverse.

1.8 GENERAL LINEAR SYSTEMS

So far, we have only treated linear systems involving the same number of equations as unknowns, and then only those with nonsingular coefficient matrices. These are precisely the systems that always have a unique solution. We now turn to the problem of solving a general linear system of m equations in n unknowns. The cases not treated as yet are non-square systems, with $m \neq n$, as well as square systems with singular coefficient matrices. The basic idea underlying the Gaussian Elimination Algorithm for nonsingular systems can be straightforwardly adapted to these cases, too. One systematically applies the same two types of elementary row operation to reduce the coefficient matrix to a simplified form that generalizes the upper triangular form we aimed for in the nonsingular situation.

Definition 1.38 An $m \times n$ matrix is said to be in *row echelon form* if it has the following "staircase" structure:

$$U = \begin{pmatrix} \circledast & * & \cdots & * & * & \cdots & * & * & \cdots & \cdots & * & * & * & \cdots & * \\ 0 & 0 & \cdots & 0 & \circledast & \cdots & * & * & \cdots & \cdots & * & * & * & \cdots & * \\ 0 & 0 & \cdots & 0 & 0 & \cdots & 0 & \circledast & \cdots & \cdots & * & * & * & \cdots & * \\ \vdots & \vdots & \ddots & \vdots & \vdots & \ddots & \vdots & \vdots & & \ddots & & \vdots & \vdots & \ddots & \vdots \\ 0 & 0 & \cdots & 0 & 0 & \cdots & 0 & 0 & \cdots & \cdots & 0 & \circledast & * & \cdots & * \\ 0 & 0 & \cdots & 0 & 0 & \cdots & 0 & 0 & \cdots & \cdots & 0 & 0 & 0 & \cdots & 0 \\ \vdots & \vdots & \ddots & \vdots & \vdots & \ddots & \vdots & \vdots & \ddots & \ddots & \vdots & \vdots & \vdots & \ddots & \vdots \\ 0 & 0 & \cdots & 0 & 0 & \cdots & 0 & 0 & \cdots & \cdots & 0 & 0 & 0 & \cdots & 0 \end{pmatrix}$$

The entries indicated by $\circledast$ are the *pivots*, and must be nonzero. The first r rows of U each contain exactly one pivot, but not all columns are required to include a pivot entry. The entries below the "staircase", indicated by the solid line, are all zero, while the non-pivot entries above the staircase, indicated by stars, can be anything. The last $m - r$ rows are identically zero, and do not contain any pivots. There may, in exceptional situations, be one or more initial all zero columns. Here is an explicit example of a matrix in row echelon form:

$$\begin{pmatrix} 3 & 1 & 0 & 2 & 5 & -1 \\ 0 & -1 & -2 & 1 & 8 & 0 \\ 0 & 0 & 0 & 0 & 2 & -4 \\ 0 & 0 & 0 & 0 & 0 & 0 \end{pmatrix}$$

The three pivots are the first nonzero entries in the three nonzero rows, namely, 3, -1, and 2.

Proposition 1.39 Any matrix can be reduced to row echelon form by a sequence of elementary row operations of Types #1 and #2.

In matrix language, Proposition 1.39 implies that if A is any $m \times n$ matrix, then there exists an $m \times m$ permutation matrix P and an $m \times m$ special lower triangular matrix L such that

$$PA = LU, \tag{1.71}$$

where U is in row echelon form. As with a square matrix, the entries of L below the diagonal correspond to the row operations of type #1, while P keeps track of row interchanges. The factorization (1.71) is not unique.

A constructive proof of this result is based on the general Gaussian Elimination algorithm, which proceeds as follows. Starting on the left of the matrix, one searches for the first column that is not identically zero. Any of the nonzero entries in that column may serve as the pivot. Partial pivoting indicates that it is probably best to choose the largest one, although this is not essential for the algorithm to proceed. One places the chosen pivot in the first row of the matrix via a row interchange, if necessary. The entries below the pivot are made equal to zero by the appropriate elementary row operations of Type #1. One then proceeds iteratively, performing the same reduction algorithm on the submatrix consisting of all entries strictly to the right and below the pivot. The algorithm terminates when either there is a nonzero pivot in the last row, or all of the rows lying below the last pivot are identically zero, and so no more pivots can be found.

EXAMPLE 1.40 The easiest way to learn the general Gaussian Elimination algorithm is to follow through an illustrative example. Consider the linear system

$$\begin{aligned} x + 3y + 2z - u \qquad &= a, \\ 2x + 6y + z + 4u + 3v &= b, \\ -x - 3y - 3z + 3u + v &= c, \\ 3x + 9y + 8z - 7u + 2v &= d, \end{aligned} \tag{1.72}$$

of 4 equations in 5 unknowns, where a, b, c, d are given numbers.* The coefficient matrix is

$$A = \begin{pmatrix} 1 & 3 & 2 & -1 & 0 \\ 2 & 6 & 1 & 4 & 3 \\ -1 & -3 & -3 & 3 & 1 \\ 3 & 9 & 8 & -7 & 2 \end{pmatrix}. \tag{1.73}$$

To solve the system, we introduce the augmented matrix

$$\left(\begin{array}{rrrrr|c} 1 & 3 & 2 & -1 & 0 & a \\ 2 & 6 & 1 & 4 & 3 & b \\ -1 & -3 & -3 & 3 & 1 & c \\ 3 & 9 & 8 & -7 & 2 & d \end{array}\right),$$

obtained by appending the right hand side of the system. The upper left entry is nonzero, and so can serve as the first pivot. We eliminate the entries below it by elementary row operations, resulting in

$$\left(\begin{array}{rrrrr|l} 1 & 3 & 2 & -1 & 0 & a \\ 0 & 0 & -3 & 6 & 3 & b - 2a \\ 0 & 0 & -1 & 2 & 1 & c + a \\ 0 & 0 & 2 & -4 & 2 & d - 3a \end{array}\right).$$

Now, the second column contains no suitable nonzero entry to serve as the second pivot. (The top entry already lies in a row containing a pivot, and so cannot be used.)

*It will be convenient to work with the right hand side in general form, although the reader may prefer, at least initially, to assign numerical values to a, b, c, d.

Therefore, we move on to the third column, choosing the $(2,3)$ entry, -3, as our second pivot. Again, we eliminate the entries below it, leading to

$$\left(\begin{array}{ccccc|c} 1 & 3 & 2 & -1 & 0 & a \\ 0 & 0 & -3 & 6 & 3 & b-2a \\ 0 & 0 & 0 & 0 & 0 & c-\frac{1}{3}b+\frac{5}{3}a \\ 0 & 0 & 0 & 0 & 4 & d+\frac{2}{3}b-\frac{13}{3}a \end{array}\right).$$

The fourth column has no pivot candidates, and so the final pivot is the 4 in the fifth column. We interchange the last two rows in order to place the coefficient matrix in row echelon form:

$$\left(\begin{array}{ccccc|c} 1 & 3 & 2 & -1 & 0 & a \\ 0 & 0 & -3 & 6 & 3 & b-2a \\ 0 & 0 & 0 & 0 & 4 & d+\frac{2}{3}b-\frac{13}{3}a \\ 0 & 0 & 0 & 0 & 0 & c-\frac{1}{3}b+\frac{5}{3}a \end{array}\right). \tag{1.74}$$

There are three pivots, $1, -3$, and 4, sitting in positions $(1,1)$, $(2,3)$, and $(3,5)$. Note the staircase form, with the pivots on the steps and everything below the staircase being zero. Recalling the row operations used to construct the solution (and keeping in mind that the row interchange that appears at the end also affects the entries of L), we find the factorization (1.71) takes the explicit form

$$\begin{pmatrix} 1 & 0 & 0 & 0 \\ 0 & 1 & 0 & 0 \\ 0 & 0 & 0 & 1 \\ 0 & 0 & 1 & 0 \end{pmatrix}\begin{pmatrix} 1 & 3 & 2 & -1 & 0 \\ 2 & 6 & 1 & 4 & 3 \\ -1 & -3 & -3 & 3 & 1 \\ 3 & 9 & 8 & -7 & 2 \end{pmatrix}$$

$$= \begin{pmatrix} 1 & 0 & 0 & 0 \\ 2 & 1 & 0 & 0 \\ 3 & -\frac{2}{3} & 1 & 0 \\ -1 & \frac{1}{3} & 0 & 1 \end{pmatrix}\begin{pmatrix} 1 & 3 & 2 & -1 & 0 \\ 0 & 0 & -3 & 6 & 3 \\ 0 & 0 & 0 & 0 & 4 \\ 0 & 0 & 0 & 0 & 0 \end{pmatrix}.$$

We shall return to find the solution to our linear system after a brief theoretical interlude. ●

Warning: In the augmented matrix, pivots can *never* appear in the last column, representing the right hand side of the system. Thus, even if $c - \frac{1}{3}b + \frac{5}{3}a \neq 0$, that entry does not qualify as a pivot.

We now introduce the most important numerical quantity associated with a matrix.

Definition 1.41 The *rank* of a matrix A is the number of pivots.

For instance, the rank of the matrix (1.73) equals 3, since its reduced row echelon form, i.e., the first five columns of (1.74), has three pivots. Since there is at most one pivot per row and one pivot per column, the rank of an $m \times n$ matrix is bounded by both m and n, and so

$$0 \leq r = \operatorname{rank} A \leq \min\{m, n\}. \tag{1.75}$$

The only $m \times n$ matrix of rank 0 is the zero matrix O—which is the only matrix without any pivots.

Proposition 1.42 A square matrix of size $n \times n$ is nonsingular if and only if its rank is equal to n.

Indeed, the only way an $n \times n$ matrix can end up having n pivots is if its reduced row echelon form is upper triangular with nonzero diagonal entries. But a matrix that reduces to such triangular form is, by definition, nonsingular.

Interestingly, the rank of a matrix *does not depend* on which elementary row operations are performed along the way to row echelon form. Indeed, performing a different sequence of row operations—say using Partial Pivoting versus no pivoting—can produce a completely different reduced form. The remarkable fact, though, is that all such row echelon forms end up having exactly the same number of pivots, and this number is the rank of the matrix. A formal proof of this fact will appear in Chapter 2.

Once the coefficient matrix has been reduced to row echelon form $\left(U \mid \mathbf{c}\right)$, the solution to the equivalent linear system $U\mathbf{x} = \mathbf{c}$ proceeds as follows. The first step is to see if there are any equations that do not have a solution. Suppose one of the rows in the echelon form U is identically zero, but the corresponding entry in the last column $\mathbf{c}$ of the augmented matrix is nonzero. What linear equation would this represent? Well, the coefficients of all the variables are zero, and so the equation is of the form

$$0 = c_i, \tag{1.76}$$

where i is the row's index. If $c_i \neq 0$, then the equation cannot be satisfied—it is *inconsistent*. The reduced system does not have a solution. Since the reduced system was obtained by elementary row operations, the original linear system is *incompatible*, meaning it also has no solutions. *Note*: It only takes one inconsistency to render the entire system incompatible. On the other hand, if $c_i = 0$, so the entire row in the augmented matrix is zero, then (1.76) is merely $0 = 0$, and is trivially satisfied. Such all zero rows do not affect the solubility of the system.

In our example, the last row in the echelon form (1.74) is all zero, and hence the last entry in the final column must also vanish in order that the system be compatible. Therefore, the linear system (1.72) will have a solution if and only if the right hand sides a, b, c, d satisfy the linear constraint

$$\tfrac{5}{3}a - \tfrac{1}{3}b + c = 0. \tag{1.77}$$

In general, if the system is incompatible, there is nothing else to do. Otherwise, every all zero row in the row echelon form of the coefficient matrix also has a zero entry in the last column of the augmented matrix; the system is *compatible* and admits one or more solutions. (If there are no all zero rows in the coefficient matrix, meaning that every row contains a pivot, then the system is automatically compatible.) To find the solution(s), we split the variables in the system into two classes.

Definition 1.43 In a linear system $U\mathbf{x} = \mathbf{c}$ in row echelon form, the variables corresponding to columns containing a pivot are called *basic variables*, while the variables corresponding to the columns without a pivot are called *free variables*.

The solution to the system then proceeds by an adaptation of the Back Substitution procedure. Working in reverse order, each nonzero equation is solved for the basic variable associated with its pivot. The result is substituted into the preceding equations before they in turn are solved. The solution then specifies all the basic variables as certain combinations of the remaining free variables. As their name indicates, the free variables, if any, are allowed to take on any values whatsoever, and so serve to parametrize the general solution to the system.

EXAMPLE 1.44 Let us illustrate the solution procedure with our particular system (1.72). The values $a = 0$, $b = 3$, $c = 1$, $d = 1$, satisfy the consistency constraint (1.77), and the corresponding reduced augmented matrix (1.74) is

$$\left(\begin{array}{ccccc|c} 1 & 3 & 2 & -1 & 0 & 0 \\ 0 & 0 & -3 & 6 & 3 & 3 \\ 0 & 0 & 0 & 0 & 4 & 3 \\ 0 & 0 & 0 & 0 & 0 & 0 \end{array}\right).$$

The pivots are found in columns 1, 3, 5, and so the corresponding variables, x, z, v, are basic; the other variables, y, u, are free. Our task is to solve the reduced system

$$\begin{aligned} x + 3y + 2z - u \qquad &= 0, \\ -3z + 6u + 3v &= 3, \\ 4v &= 3, \\ 0 &= 0, \end{aligned}$$

for the basic variables x, z, v in terms of the free variables y, u. As before, this is done in the reverse order, successively substituting the resulting values in the preceding equation. The result is the general solution

$$v = \tfrac{3}{4}, \qquad z = -1 + 2u + v = -\tfrac{1}{4} + 2u, \qquad x = -3y - 2z + u = \tfrac{1}{2} - 3y - 3u.$$

The free variables y, u remain completely arbitrary; any assigned values will produce a solution to the original system. For instance, if $y = -1, u = \pi$, then $x = \frac{7}{2} - 3\pi$, $z = -\frac{1}{4} + 2\pi$, $v = \frac{3}{4}$. But keep in mind that this is merely one of an infinite number of valid solutions. ●

In general, if the $m \times n$ coefficient matrix of a system of m linear equations in n unknowns has rank r, there are $m - r$ all zero rows in the row echelon form, and these $m - r$ equations must have zero right hand side in order that the system be compatible and have a solution. Moreover, there are a total of r basic variables and $n - r$ free variables, and so the general solution depends upon $n - r$ parameters.

Summarizing the preceding discussion, we have learned that there are only three possible outcomes for the solution to a system of linear equations.

THEOREM 1.45 A system $A\mathbf{x} = \mathbf{b}$ of m linear equations in n unknowns has either

(i) exactly one solution,

(ii) infinitely many solutions, or

(iii) no solution.

Case (iii) occurs if the system is incompatible, producing a zero row in the echelon form that has a nonzero right hand side. Case (ii) occurs if the system is compatible and there are one or more free variables, and so the rank of the coefficient matrix is strictly less than the number of columns: $r < n$. Case (i) occurs for nonsingular square coefficient matrices, and, more generally, for compatible systems for which $r = n$, implying there are no free variables. Since $r \leq m$, this case can only arise if the coefficient matrix has at least as many rows as columns, i.e., the linear system has at least as many equations as unknowns. A linear system can *never* have a finite number—other than 0 or 1—of solutions. As a consequence, any linear system that admits two or more solutions automatically has infinitely many!

Warning: This property requires linearity, and is *not* valid for nonlinear systems. For instance, the real quadratic equation $x^2 + x - 2 = 0$ has exactly two real solutions: $x = 1$ or $x = -2$.

EXAMPLE 1.46 Consider the linear system

$$y + 4z = a, \qquad 3x - y + 2z = b, \qquad x + y + 6z = c,$$

consisting of three equations in three unknowns. The augmented coefficient matrix is

$$\left(\begin{array}{ccc|c} 0 & 1 & 4 & a \\ 3 & -1 & 2 & b \\ 1 & 1 & 6 & c \end{array}\right).$$

Interchanging the first two rows, and then eliminating the elements below the first pivot leads to

$$\left(\begin{array}{ccc|c} 3 & -1 & 2 & b \\ 0 & 1 & 4 & a \\ 0 & \frac{4}{3} & \frac{16}{3} & c - \frac{1}{3}b \end{array}\right).$$

The second pivot is in the $(2, 2)$ position, but after eliminating the entry below it, we find the row echelon form to be

$$\left(\begin{array}{ccc|c} 3 & -1 & 2 & b \\ 0 & 1 & 4 & a \\ 0 & 0 & 0 & c - \frac{1}{3}b - \frac{4}{3}a \end{array}\right).$$

Since there is a row of all zeros, the original coefficient matrix is singular, and its rank is only 2.

The consistency condition follows from this last row in the reduced echelon form, which requires

$$\tfrac{4}{3}a + \tfrac{1}{3}b - c = 0.$$

If this is not satisfied, the system has no solutions; otherwise it has infinitely many. The free variable is z, since there is no pivot in the third column. The general solution is

$$y = a - 4z, \qquad x = \tfrac{1}{3}b + \tfrac{1}{3}y - \tfrac{2}{3}z = \tfrac{1}{3}a + \tfrac{1}{3}b - 2z,$$

where z is arbitrary. ●

Geometrically, Theorem 1.45 is telling us about the possible configurations of linear subsets (lines, planes, etc.) of an n-dimensional space. For example, a single linear equation $ax + by + cz = d$, with $(a, b, c) \neq \mathbf{0}$, defines a plane P in three-dimensional space. The solutions to a system of three linear equations in three unknowns belong to all three planes; that is, they lie in their *intersection* $P_1 \cap P_2 \cap P_3$. Generically, three planes intersect in a single common point; this is case (i) of the theorem, and occurs if and only if the coefficient matrix is nonsingular. The case of infinitely many solutions occurs when the three planes intersect on a common line, or, even more degenerately, when they all coincide. On the other hand, parallel planes, or planes intersecting in parallel lines, have no common point of intersection, and this occurs when the system is incompatible and has no solutions. There are no other possibilities: the total number of points in the intersection is either 0, 1, or ∞. Some sample geometric configurations appear in Figure 1.1.

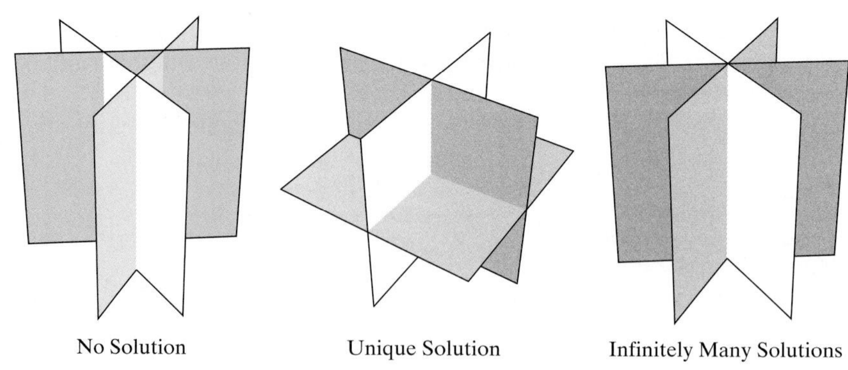

Figure 1.1 Intersecting planes.

EXERCISES 1.8

1.8.1. Which of the following systems has

(i) a unique solution?

(ii) infinitely many solutions?

(iii) no solution?

In each case, find all solutions:

(a) $\begin{aligned} x - 2y &= 1 \\ 3x + 2y &= -3 \end{aligned}$ (b) $\begin{aligned} 2x + y + 3z &= 1 \\ x + 4y - 2z &= -3 \end{aligned}$

(c) $\begin{aligned} x + y - 2z &= -3 \\ 2x - y + 3z &= 7 \\ x - 2y + 5z &= 1 \end{aligned}$

(d) $\begin{aligned} x - 2y + z &= 6 \\ 2x + y - 3z &= -3 \\ x - 3y + 3z &= 10 \end{aligned}$

(e) $\begin{aligned} x - 2y + 2z - w &= 3 \\ 3x + y + 6z + 11w &= 16 \\ 2x - y + 4z + w &= 9 \end{aligned}$

(f) $\begin{aligned} 3x - 2y + z &= 4 \\ x + 3y - 4z &= -3 \\ 2x - 3y + 5z &= 7 \\ x - 8y + 9z &= 10 \end{aligned}$

(g) $\begin{aligned} x + 2y + 17z - 5w &= 50 \\ 9x - 16y + 10z - 8w &= 24 \\ 2x - 5y - 4z &= -13 \\ 6x - 12y + z - 4w &= -1 \end{aligned}$

1.8.2. Determine if the following systems are compatible and, if so, find the general solution:

(a) $\begin{aligned} 6x_1 + 3x_2 &= 12 \\ 4x_1 + 2x_2 &= 9 \end{aligned}$ (b) $\begin{aligned} 8x_1 + 12x_2 &= 16 \\ 6x_1 + 9x_2 &= 13 \end{aligned}$

(c) $\begin{aligned} x_1 + 2x_2 &= 1 \\ 2x_1 + 5x_2 &= 2 \\ 3x_1 + 6x_2 &= 3 \end{aligned}$

(d) $\begin{aligned} 2x_1 - 6x_2 + 4x_3 &= 2 \\ -x_1 + 3x_2 - 2x_3 &= -1 \end{aligned}$

(e) $\begin{aligned} 2x_1 + 2x_2 + 3x_3 &= 1 \\ x_2 + 2x_3 &= 3 \\ 4x_1 + 5x_2 + 7x_3 &= 15 \end{aligned}$

(f) $\begin{aligned} x_1 + x_2 + x_3 + 9x_4 &= 8 \\ x_2 + 2x_3 + 8x_4 &= 7 \\ -3x_1 + x_3 - 7x_4 &= 9 \end{aligned}$

(g) $\begin{aligned} x_1 + 2x_2 + 3x_3 + 4x_4 &= 1 \\ 2x_1 + 4x_2 + 6x_3 + 5x_4 &= 0 \\ 3x_1 + 4x_2 + x_3 + x_4 &= 0, \\ 4x_1 + 6x_2 + 4x_3 - x_4 &= 0 \end{aligned}$

1.8.3. Graph the following planes and determine whether they have a common intersection: $x + y + z = 1$, $x + y = 1$, $x + z = 1$.

1.8.4. Let

$$A = \left(\begin{array}{ccc|c} a & 0 & b & 2 \\ a & 2 & a & b \\ b & 2 & a & a \end{array}\right)$$

be the augmented matrix for a linear system. For which values of a and b does the system have

(i) a unique solution?

(ii) infinitely many solutions?

(iii) no solution?

1.8.5. For which values of b and c does the system $x_1 + x_2 + bx_3 = 1$, $bx_1 + 3x_2 - x_3 = -2$, $3x_1 + 4x_2 + x_3 = c$, have

(a) no solution?

(b) exactly one solution?

(c) infinitely many solutions?

1.8.6. Determine the general (complex) solution to the following systems:

(a) $2x + (1+\mathrm{i})y - 2\mathrm{i}z = 2\mathrm{i}$
$(1-\mathrm{i})x + y - 2\mathrm{i}z = 0$

(b) $x + 2\mathrm{i}y + (2-4\mathrm{i})z = 5+5\mathrm{i}$
$(-1+\mathrm{i})x + 2y + (4+2\mathrm{i})z = 0$
$(1-\mathrm{i})x + (1+4\mathrm{i})y - 5\mathrm{i}z = 10+5\mathrm{i}$

(c) $x_1 + \mathrm{i}x_2 + x_3 = 1+4\mathrm{i}$
$-x_1 + x_2 - \mathrm{i}x_3 = -1$
$\mathrm{i}x_1 - x_2 - x_3 = -1-2\mathrm{i}$

(d) $(2+\mathrm{i})x + \mathrm{i}y + (2+2\mathrm{i})z + (1+12\mathrm{i})w = 0$
$(1-\mathrm{i})x + y + (2-\mathrm{i})z + (8+2\mathrm{i})w = 0$
$(3+2\mathrm{i})x + \mathrm{i}y + (3+3\mathrm{i})z + 19\mathrm{i}w = 0$

1.8.7. Determine the rank of the following matrices:

(a) $\begin{pmatrix} 1 & 1 \\ 1 & -2 \end{pmatrix}$ (b) $\begin{pmatrix} 2 & 1 & 3 \\ -2 & -1 & -3 \end{pmatrix}$

(c) $\begin{pmatrix} 1 & -1 & 1 \\ 1 & -1 & 2 \\ -1 & 1 & 0 \end{pmatrix}$ (d) $\begin{pmatrix} 2 & -1 & 0 \\ 2 & -1 & 1 \\ 1 & 1 & -1 \end{pmatrix}$

(e) $\begin{pmatrix} 3 \\ 0 \\ -2 \end{pmatrix}$ (f) $\begin{pmatrix} 0 & -1 & 2 & 5 \end{pmatrix}$

(g) $\begin{pmatrix} 0 & -3 \\ 4 & -1 \\ 1 & 2 \\ -1 & -5 \end{pmatrix}$

(h) $\begin{pmatrix} 1 & -1 & 2 & 1 \\ 2 & 1 & -1 & 0 \\ 1 & 2 & -3 & -1 \\ 4 & -1 & 3 & 2 \\ 0 & 3 & -5 & -2 \end{pmatrix}$

(i) $\begin{pmatrix} 0 & 0 & 0 & 3 & 1 \\ 1 & 2 & -3 & 1 & -2 \\ 2 & 4 & -2 & 1 & -2 \end{pmatrix}$

1.8.8. Write out a $PA = LU$ factorization for each of the matrices in Exercise 1.8.7.

1.8.9. Construct a system of three linear equations in three unknowns that has

(a) one and only one solution;

(b) more than one solution;

(c) no solution.

1.8.10. Find a coefficient matrix A such that the associated linear system $A\mathbf{x} = \mathbf{b}$ has

(a) infinitely many solutions for every $\mathbf{b}$,

(b) 0 or ∞ solutions, depending on $\mathbf{b}$,

(c) 0 or 1 solution depending on $\mathbf{b}$,

(d) exactly 1 solution for all $\mathbf{b}$.

1.8.11. Give an example of a *nonlinear* system of two equations in two unknowns that has

(a) no solution,

(b) exactly two solutions,

(c) exactly three solutions,

(d) infinitely many solutions.

1.8.12. What does it mean if a linear system has a coefficient matrix with a column of all 0's?

1.8.13. *True or false*: One can find an $m \times n$ matrix of rank r for any $0 \le r \le \min\{m, n\}$.

1.8.14. *True or false*: Every $m \times n$ matrix has

(a) exactly m pivots;

(b) at least one pivot.

♡ **1.8.15.** (a) Prove that the product $A = \mathbf{v}\,\mathbf{w}^T$ of a nonzero $m \times 1$ column vector $\mathbf{v}$ by a nonzero $1 \times n$ row vector $\mathbf{w}^T$ is an $m \times n$ matrix of rank $r = 1$.

(b) Compute the following rank one products:

(i) $\begin{pmatrix} 1 \\ 3 \end{pmatrix}\begin{pmatrix} -1 & 2 \end{pmatrix}$,

(ii) $\begin{pmatrix} 4 \\ 0 \\ -2 \end{pmatrix}\begin{pmatrix} -2 & 1 \end{pmatrix}$,

(iii) $\begin{pmatrix} 2 \\ -3 \end{pmatrix}\begin{pmatrix} 1 & 3 & -1 \end{pmatrix}$.

(c) Prove that every rank one matrix can be written in the form $A = \mathbf{v}\mathbf{w}^T$.

1.8.16. Find the rank of the matrix

$$\begin{pmatrix} a & a\,r & \ldots & a\,r^{n-1} \\ a\,r^n & a\,r^{n+1} & \ldots & a\,r^{2n-1} \\ \vdots & \vdots & \ddots & \vdots \\ a\,r^{(n-1)n} & a\,r^{(n-1)n+1} & \ldots & a\,r^{n^2-1} \end{pmatrix}$$

when $a, r \ne 0$.

1.8.17. Find the rank of the $n \times n$ matrix

$$\begin{pmatrix} 1 & 2 & 3 & \ldots & n \\ n+1 & n+2 & n+3 & \ldots & 2n \\ 2n+1 & 2n+2 & 2n+3 & \ldots & 3n \\ \vdots & \vdots & \vdots & \ddots & \vdots \\ n^2-n+1 & n^2-n+2 & \ldots & \ldots & n^2 \end{pmatrix}.$$

1.8.18. Find two matrices A, B such that $\operatorname{rank} AB \ne \operatorname{rank} BA$.

◇ **1.8.19.** Let A be an $m \times n$ matrix of rank r.

(a) Suppose $C = \begin{pmatrix} A & B \end{pmatrix}$ is an $m \times k$ matrix, $k > n$, whose first n columns are the same as the columns of A. Prove that rank $C \geq$ rank A. Give an example with rank $C =$ rank A; with rank $C >$ rank A.

(b) Let $E = \begin{pmatrix} A \\ D \end{pmatrix}$ be a $j \times n$ matrix, $j > m$, whose first m rows are the same as those of A. Prove that rank $E \geq$ rank A. Give an example with rank $E =$ rank A; with rank $E >$ rank A.

◇ **1.8.20.** Let A be a singular square matrix. Prove that there exist elementary matrices $E_1, \ldots, E_N$ such that $A = E_1 E_2 \cdots E_N Z$, where Z is a matrix with at least one all zero row.

◇ **1.8.21.** (a) If A is an $m \times n$ matrix and $M = \begin{pmatrix} A \mid \mathbf{b} \end{pmatrix}$ the augmented matrix for the linear system $A\mathbf{x} = \mathbf{b}$. Show that either

(i) rank $A =$ rank M, or

(ii) rank $A =$ rank $M - 1$.

(b) Prove that the system is compatible if and only if case (i) holds.

Homogeneous Systems

A linear system with all 0's on the right hand side is called *homogeneous*. In matrix notation, a homogeneous system takes the form

$$A\mathbf{x} = \mathbf{0}, \tag{1.78}$$

where the zero vector $\mathbf{0}$ indicates that every entry on the right hand side is zero. Homogeneous systems are always compatible, since $\mathbf{x} = \mathbf{0}$ is a solution, known as the *trivial solution*. If the homogeneous system has a nontrivial solution $\mathbf{x} \neq \mathbf{0}$, then Theorem 1.45 assures us that it must have infinitely many solutions. This will occur if and only if the reduced system has one or more free variables.

THEOREM 1.47 A homogeneous linear system $A\mathbf{x} = \mathbf{0}$ of m equations in n unknowns has a nontrivial solution $\mathbf{x} \neq \mathbf{0}$ if and only if the rank of A is $r < n$. If $m < n$, the system *always* has a nontrivial solution. If $m = n$, the system has a nontrivial solution if and only if A is singular.

Thus, homogeneous systems with fewer equations than unknowns always have infinitely many solutions. Indeed, the coefficient matrix of such a system has more columns than rows, and so at least one column cannot contain a pivot, meaning that there is at least one free variable in the general solution formula.

EXAMPLE 1.48 Consider the homogeneous linear system

$$2x_1 + x_2 + 5x_4 = 0, \quad 4x_1 + 2x_2 - x_3 + 8x_4 = 0, \quad -2x_1 - x_2 + 3x_3 - 4x_4 = 0,$$

with coefficient matrix

$$A = \begin{pmatrix} 2 & 1 & 0 & 5 \\ 4 & 2 & -1 & 8 \\ -2 & -1 & 3 & -4 \end{pmatrix}.$$

Since there are only three equations in four unknowns, we already know that the system has infinitely many solutions, including the trivial solution $x_1 = x_2 = x_3 = x_4 = 0$.

When solving a homogeneous system, the final column of the augmented matrix consists of all zeros. As such, it will never be altered by row operations, and so it is a waste of effort to carry it along during the process. We therefore perform the Gaussian Elimination algorithm directly on the coefficient matrix A. Working with the $(1,1)$ entry as the first pivot, we first obtain

$$\begin{pmatrix} 2 & 1 & 0 & 5 \\ 0 & 0 & -1 & -2 \\ 0 & 0 & 3 & 1 \end{pmatrix}.$$

The (2, 3) entry is the second pivot, and we apply one final row operation to place the matrix in row echelon form

$$\begin{pmatrix} 2 & 1 & 0 & 5 \\ 0 & 0 & -1 & -2 \\ 0 & 0 & 0 & -5 \end{pmatrix}.$$

This corresponds to the reduced homogeneous system

$$2x_1 + x_2 + 5x_4 = 0, \qquad -x_3 - 2x_4 = 0, \qquad -5x_4 = 0.$$

Since there are three pivots in the final row echelon form, the rank of the coefficient matrix A is 3. There is one free variable, namely x_2. Using Back Substitution, we easily obtain the general solution

$$x_1 = -\tfrac{1}{2}t, \qquad x_2 = t, \qquad x_3 = x_4 = 0,$$

which depends upon a single free parameter $t = x_2$. ●

EXAMPLE 1.49 Consider the homogeneous linear system

$$2x - y + 3z = 0, \quad -4x + 2y - 6z = 0, \quad 2x - y + z = 0, \quad 6x - 3y + 3z = 0,$$

with coefficient matrix

$$A = \begin{pmatrix} 2 & -1 & 3 \\ -4 & 2 & -6 \\ 2 & -1 & 1 \\ 6 & -3 & 3 \end{pmatrix}.$$

The system admits the trivial solution $x = y = z = 0$, but in this case we need to complete the elimination algorithm before we know for sure whether or not there are other solutions. After the first stage in the reduction process, the coefficient matrix becomes

$$\begin{pmatrix} 2 & -1 & 3 \\ 0 & 0 & 0 \\ 0 & 0 & -2 \\ 0 & 0 & -6 \end{pmatrix}.$$

To continue, we need to interchange the second and third rows to place a nonzero entry in the final pivot position; after that, the reduction to the row echelon form

$$\begin{pmatrix} 2 & -1 & 3 \\ 0 & 0 & -2 \\ 0 & 0 & 0 \\ 0 & 0 & 0 \end{pmatrix}$$

is immediate. Thus, the system reduces to the equations

$$2x - y + 3z = 0, \qquad -2z = 0, \qquad 0 = 0, \qquad 0 = 0.$$

The third and fourth equations are trivially compatible, as they must be in the homogeneous case. The rank of the coefficient matrix is equal to two, which is less than the number of columns, and so, even though the system has more equations than unknowns, it has infinitely many solutions. These can be written in terms of the free variable y; the general solution is $x = \frac{1}{2}y$, $z = 0$, where y is arbitrary. ●

EXERCISES 1.8

1.8.22. Solve the following homogeneous linear systems.

(a) $\begin{aligned} x + y - 2z &= 0 \\ -x + 4y - 3z &= 0 \end{aligned}$

(b) $\begin{aligned} 2x + 3y - z &= 0 \\ -4x + 3y - 5z &= 0 \\ x - 3y + 3z &= 0 \end{aligned}$

(c) $\begin{aligned} -x + y - 4z &= 0 \\ -2x + 2y - 6z &= 0 \\ x + 3y + 3z &= 0 \end{aligned}$

(d) $\begin{aligned} x + 2y - 2z + w &= 0 \\ -3x + z - 2w &= 0 \end{aligned}$

(e) $\begin{aligned} -x + 3y - 2z + w &= 0 \\ -2x + 5y + z - 2w &= 0 \\ 3x - 8y + z - 4w &= 0 \end{aligned}$

(f) $\begin{aligned} -y + z &= 0 \\ 2x - 3w &= 0 \\ x + y - 2w &= 0 \\ y - 3z + w &= 0 \end{aligned}$

1.8.23. Find all solutions to the homogeneous system $A\mathbf{x} = \mathbf{0}$ for the coefficient matrix

(a) $\begin{pmatrix} 3 & -1 \\ -9 & 3 \end{pmatrix}$ (b) $\begin{pmatrix} 2 & -1 & 4 \\ 3 & 1 & 2 \end{pmatrix}$

(c) $\begin{pmatrix} 1 & -2 & 3 & -3 \\ 2 & 1 & 4 & 0 \end{pmatrix}$ (d) $\begin{pmatrix} 1 & 2 & 3 \\ 4 & 5 & 6 \\ 7 & 8 & 9 \end{pmatrix}$

(e) $\begin{pmatrix} 0 & 2 & -1 \\ -2 & 0 & 3 \\ 1 & 3 & 0 \end{pmatrix}$

(f) $\begin{pmatrix} 1 & 2 & 0 \\ -1 & -3 & 2 \\ 4 & 7 & 2 \\ -1 & 1 & 6 \end{pmatrix}$

(g) $\begin{pmatrix} 1 & -2 & 3 & 1 \\ 1 & -1 & 0 & -1 \\ 2 & -1 & -3 & 3 \\ 1 & 0 & -3 & -3 \end{pmatrix}$

(h) $\begin{pmatrix} 0 & 0 & 3 & -3 \\ 1 & -1 & 0 & 3 \\ 2 & -2 & 1 & 5 \\ -1 & 1 & 1 & -4 \end{pmatrix}$

1.8.24. Let U be an upper triangular matrix. Show that the homogeneous system $U\mathbf{x} = \mathbf{0}$ admits a nontrivial solution if and only if U has at least one 0 on its diagonal.

1.8.25. Find the solution to the homogeneous system $2x_1 + x_2 - 2x_3 = 0$, $2x_1 - x_2 - 2x_3 = 0$. Then solve the inhomogeneous version where the right hand sides are changed to a, b, respectively. What do you observe?

1.8.26. Answer Exercise 1.8.25 for the system $2x_1 + x_2 + x_3 - x_4 = 0$, $2x_1 - 2x_2 - x_3 + 3x_4 = 0$.

1.8.27. Find all values of k for which the following homogeneous systems of linear equations have a nontrivial solution:

(a) $\begin{aligned} x + ky &= 0 \\ kx + 4y &= 0 \end{aligned}$

(b) $\begin{aligned} x_1 + kx_2 + 4x_3 &= 0 \\ kx_1 + x_2 + 2x_3 &= 0 \\ 2x_1 + kx_2 + 8x_3 &= 0 \end{aligned}$

(c) $\begin{aligned} x + ky + 2z &= 0 \\ 3x - ky - 2z &= 0 \\ (k+1)x - 2y - 4z &= 0 \\ kx + 3y + 6z &= 0 \end{aligned}$

1.9 DETERMINANTS

You may be surprised that, so far, we have not mentioned determinants—a topic that typically assumes a central role in basic linear algebra. Determinants can be useful in low dimensional and highly structured problems, and have many fascinating properties. But, like matrix inverses, they are almost completely irrelevant when it comes to large scale applications and practical computations. Indeed, for most matrices, the best way to compute a determinant is (surprise) Gaussian Elimination! Consequently, from a computational standpoint, the determinant adds no new information concerning the linear system and its solutions. However, for completeness and in preparation for certain later developments (particularly computing eigenvalues of small matrices), you should be familiar with a few of the basic facts and properties of determinants. In this final section, we shall provide a very brief introduction to the basics.

The determinant of a square matrix* A is a scalar, written $\det A$, that can distinguish between singular and nonsingular matrices. We already encountered, (1.37), the determinant of a 2×2 matrix:†

$$\det \begin{pmatrix} a & b \\ c & d \end{pmatrix} = ad - bc.$$

The key fact is that the determinant is nonzero if and only if the matrix has an inverse. Our goal is to find an analogous quantity for general square matrices.

There are many different ways to define determinants. The difficulty is that the actual formula is very unwieldy—see (1.85) below—and not well motivated. We prefer an axiomatic approach that explains how our elementary row operations affect the determinant.

THEOREM 1.50 The *determinant* of a square matrix A is the uniquely defined scalar quantity that obeys the following axioms:

(i) Adding a multiple of one row to another does not change the determinant.

(ii) Interchanging two rows changes the sign of the determinant.

(iii) Multiplying a row by any scalar (including zero) multiplies the determinant by the same scalar.

(iv) The determinant of an upper triangular matrix is equal to the product of its diagonal entries: $\det U = u_{11} u_{22} \cdots u_{nn}$.

Checking that all four of these axioms hold in the 2×2 case is left as an elementary exercise. Suppose, in particular, we multiply a row of the matrix A by the zero scalar. The resulting matrix has a row of all zeros, and, by property (iii), has zero determinant. Since any matrix with a zero row can be obtained in this fashion, we conclude:

Lemma 1.51 Any matrix with one or more all zero rows has zero determinant.

Using these properties, one is able to compute the determinant of any square matrix by Gaussian Elimination, which is, in fact, the fastest and most practical computational method in all but the simplest situations.

THEOREM 1.52 If $A = L\,U$ is a regular matrix, then

$$\det A = \det U = u_{11} u_{22} \cdots u_{nn} \tag{1.79}$$

equals the product of the pivots. More generally, if A is nonsingular, and requires k row interchanges to arrive at its permuted factorization $P A = L\,U$, then

$$\det A = \det P \ \det U = (-1)^k \, u_{11} u_{22} \cdots u_{nn}. \tag{1.80}$$

Finally, A is singular if and only if

$$\det A = 0. \tag{1.81}$$

---

*Non-square matrices do not have determinants.

†Some authors use vertical lines to indicate the determinant: $\begin{vmatrix} a & b \\ c & d \end{vmatrix} = \det \begin{pmatrix} a & b \\ c & d \end{pmatrix}$.

Proof In the regular case, we only need elementary row operations of type #1 to reduce A to upper triangular form U, and axiom (i) says these do not change the determinant. Therefore, $\det A = \det U$, in accordance with (1.79). The nonsingular case follows similarly. By axiom (ii), each row interchange changes the sign of the determinant, and so $\det A$ equals $\det U$ if there have been an even number of interchanges, but equals $-\det U$ if there have been an odd number. For the same reason, the determinant of the permutation matrix P equals $+1$ if there have been an even number of row interchanges, and -1 for an odd number. Finally, if A is singular, then we can reduce it to a matrix with at least one row of zeros by elementary row operations of types #1 and #2. Lemma 1.51 implies that the resulting matrix has zero determinant, and so $\det A = 0$, also. ■

EXAMPLE 1.53 Let us compute the determinant of the 4×4 matrix

$$A = \begin{pmatrix} 1 & 0 & -1 & 2 \\ 2 & 1 & -3 & 4 \\ 0 & 2 & -2 & 3 \\ 1 & 1 & -4 & -2 \end{pmatrix}.$$

We perform our usual Gaussian Elimination algorithm, successively leading to the matrices

$$A \longmapsto \begin{pmatrix} 1 & 0 & -1 & 2 \\ 0 & 1 & -1 & 0 \\ 0 & 2 & -2 & 3 \\ 0 & 1 & -3 & -4 \end{pmatrix} \longmapsto \begin{pmatrix} 1 & 0 & -1 & 2 \\ 0 & 1 & -1 & 0 \\ 0 & 0 & 0 & 3 \\ 0 & 0 & -2 & -4 \end{pmatrix}$$

$$\longmapsto \begin{pmatrix} 1 & 0 & -1 & 2 \\ 0 & 1 & -1 & 0 \\ 0 & 0 & -2 & -4 \\ 0 & 0 & 0 & 3 \end{pmatrix},$$

where we used a single row interchange to obtain the final upper triangular form. Owing to the row interchange, the determinant of the original matrix is -1 times the product of the pivots:

$$\det A = -1 \cdot (1 \cdot 1 \cdot (-2) \cdot 3) = 6.$$

In particular, this tells us that A is nonsingular. But, of course, this was already evident, since we successfully reduced the matrix to upper triangular form with 4 nonzero pivots. ●

There are a variety of other approaches to evaluating determinants. However, except for very small (2×2 or 3×3) matrices or other special situations, the most efficient algorithm for computing the determinant of a matrix is to apply Gaussian Elimination, with pivoting if necessary, and then invoke the relevant formula from Theorem 1.52. In particular, the determinantal criterion (1.81) for singular matrices, while of theoretical interest, is unnecessary in practice, since we will have already detected whether the matrix is singular during the course of the elimination procedure by observing that it has less than the full number of pivots.

Let us finish by stating a few of the basic properties of determinants. Proofs are outlined in the exercises.

Proposition 1.54 The determinant of the product of two square matrices of the same size is the product of their determinants:

$$\det(A\,B) = \det A \;\det B. \tag{1.82}$$

Therefore, even though matrix multiplication is not commutative, and so $A B \neq B A$ in general, both matrix products have the same determinant:

$$\det(A B) = \det A \ \det B = \det B \ \det A = \det(B A),$$

because ordinary (scalar) multiplication *is* commutative. In particular, setting $B = A^{-1}$ and using axiom (iv), we find that the determinant of the inverse matrix is the reciprocal of the matrix's determinant.

Corollary 1.55 If A is a nonsingular matrix, then

$$\det A^{-1} = \frac{1}{\det A}. \tag{1.83}$$

We also note that the determinant is unaffected by the transpose operation.

Proposition 1.56 Transposing a matrix does not change its determinant:

$$\det A^T = \det A. \tag{1.84}$$

Finally, for later reference, we end with the general formula for the determinant of an $n \times n$ matrix A with entries a_{ij}:

$$\det A = \sum_{\pi} (\operatorname{sign} \pi)\, a_{\pi(1),1}\, a_{\pi(2),2} \cdots a_{\pi(n),n}. \tag{1.85}$$

The sum is over all possible permutations π of the rows of A. The *sign* of the permutation, written $\operatorname{sign} \pi$, equals the determinant of the corresponding permutation matrix P, so $\operatorname{sign} \pi = \det P = +1$ if the permutation is composed of an even number of row interchanges and -1 if composed of an odd number. For example, the six terms in the well-known formula

$$\det \begin{pmatrix} a_{11} & a_{12} & a_{13} \\ a_{21} & a_{22} & a_{23} \\ a_{31} & a_{32} & a_{33} \end{pmatrix} = \begin{matrix} a_{11} a_{22} a_{33} + a_{31} a_{12} a_{23} + a_{21} a_{32} a_{13} \\ -a_{11} a_{32} a_{23} - a_{21} a_{12} a_{33} - a_{31} a_{22} a_{13} \end{matrix} \tag{1.86}$$

for a 3×3 determinant correspond to the six possible permutations (1.30) of a 3 rowed matrix.

A proof that the formula (1.85) satisfies the defining properties of the determinant listed in Theorem 1.50 is not hard. The reader might wish to try out the 3×3 case to be convinced that it works. The explicit formula proves that the determinant function is well-defined, and formally completes the proof of Theorem 1.50.

However, the explicit determinant formula (1.85) is not used in practice. It contains $n!$ terms, which, as soon as n is of moderate size, renders it completely useless for practical computations. For instance, the determinant of a 10×10 matrix contains $10! = 3{,}628{,}800$ terms, while a 100×100 determinant would require summing 9.3326×10^{157} terms, each of which is a product of 100 matrix entries! The most efficient way to compute determinants is still our mainstay—Gaussian Elimination, coupled with the fact that the determinant is $\pm$ the product of the pivots! On this note, we conclude our brief introduction to the subject.

EXERCISES 1.9

1.9.1. Use Gaussian Elimination to find the determinant of the following matrices:

(a) $\begin{pmatrix} 2 & -1 \\ -4 & 3 \end{pmatrix}$ (b) $\begin{pmatrix} 0 & 1 & -2 \\ -1 & 0 & 3 \\ 2 & -3 & 0 \end{pmatrix}$

(c) $\begin{pmatrix} 1 & 2 & 3 \\ 2 & 5 & 8 \\ 3 & 8 & 10 \end{pmatrix}$ (d) $\begin{pmatrix} 0 & 1 & -1 \\ -2 & 1 & 3 \\ 2 & 7 & -8 \end{pmatrix}$

(e) $\begin{pmatrix} 5 & -1 & 0 & 2 \\ 0 & 3 & -1 & 5 \\ 0 & 0 & -4 & 2 \\ 0 & 0 & 0 & 3 \end{pmatrix}$

(f) $\begin{pmatrix} 1 & -2 & 1 & 4 \\ 2 & -4 & 0 & 0 \\ 3 & -4 & 2 & 5 \\ 0 & 2 & -4 & -9 \end{pmatrix}$

(g) $\begin{pmatrix} 1 & -2 & 1 & 4 & -5 \\ 1 & 1 & -2 & 3 & -3 \\ 2 & -1 & -1 & 2 & 2 \\ 5 & -1 & 0 & 5 & 5 \\ 2 & 2 & 0 & 4 & -1 \end{pmatrix}$

1.9.2. Verify the determinant product formula (1.82) when

$$A = \begin{pmatrix} 1 & -1 & 3 \\ 2 & -1 & 1 \\ 4 & -2 & 0 \end{pmatrix}, \quad B = \begin{pmatrix} 0 & 1 & -1 \\ 1 & -3 & -2 \\ 2 & 0 & 1 \end{pmatrix}.$$

1.9.3. (a) Give an example of a non-diagonal 2×2 matrix for which $A^2 = \mathrm{I}$.

(b) If $A^2 = \mathrm{I}$, show that $\det A = \pm 1$.

1.9.4. If $A^2 = A$, what can you say about $\det A$?

1.9.5. *True or false*: If true, explain why. If false, give an explicit counterexample.

(a) If $\det A \neq 0$ then A^{-1} exists.

(b) $\det(2A) = 2 \det A$.

(c) $\det(A + B) = \det A + \det B$.

(d) $\det A^{-T} = \dfrac{1}{\det A}$.

(e) $\det(A B^{-1}) = \dfrac{\det A}{\det B}$.

(f) $\det[\,(A+B)(A-B)\,] = \det(A^2 - B^2)$.

(g) If A is an $n \times n$ matrix with $\det A = 0$, then $\operatorname{rank} A < n$.

(h) If $\det A = 1$ and $A B = \mathrm{O}$, then $B = \mathrm{O}$.

1.9.6. For what values of a, b, c is the matrix $\begin{pmatrix} 0 & a & -b \\ -a & 0 & c \\ b & -c & 0 \end{pmatrix}$ invertible?

1.9.7. Prove that similar matrices have the same determinant: $\det A = \det B$ whenever $B = S^{-1} A S$.

1.9.8. Prove that if A is a $n \times n$ matrix and c is a scalar, then $\det(c\,A) = c^n \det A$.

1.9.9. Prove that the determinant of a lower triangular matrix is the product of its diagonal entries.

1.9.10. (a) Show that if A has size $n \times n$, then $\det(-A) = (-1)^n \det A$.

(b) Prove that, for n odd, any $n \times n$ skew-symmetric matrix $A = -A^T$ is singular.

(c) Find a nonsingular skew-symmetric matrix.

◇ **1.9.11.** Prove directly that the 2×2 determinant formula (1.37) satisfies the four determinant axioms listed in Theorem 1.50.

◇ **1.9.12.** In this exercise, we prove the determinantal product formula (1.82).

(a) Prove that if E is any elementary matrix (of the appropriate size), then

$$\det(E\,B) = \det E \ \det B.$$

(b) Use induction to prove that if

$$A = E_1 E_2 \cdots E_N$$

is a product of elementary matrices, then $\det(A\,B) = \det A \ \det B$. Explain why this proves the product formula whenever A is a nonsingular matrix.

(c) Prove that if Z is a matrix with a zero row, then $Z\,B$ also has a zero row, and so

$$\det(Z\,B) = 0 = \det Z \ \det B.$$

(d) Use Exercise 1.8.20 to complete the proof of the product formula.

1.9.13. Prove (1.83).

◇ **1.9.14.** Prove (1.84). *Hint*: Use Exercise 1.6.28 in the regular case. Then extend to the nonsingular case. Finally, explain why the result also holds for singular matrices.

1.9.15. Write out the formula for a 4×4 determinant. It should contain $24 = 4!$ terms.

◇ **1.9.16.** Show that (1.85) satisfies all four determinant axioms, and hence is the correct formula for a determinant.

◇ **1.9.17.** Prove that axiom (iv) in Theorem 1.50 can be proved as a consequence of the first three axioms and the property $\det \mathrm{I} = 1$.

◇ **1.9.18.** Prove that one cannot produce an elementary row operation of type #2 by a combination of elementary row operations of type #1.

♡ **1.9.19.** Show that

(a) if
$$A = \begin{pmatrix} a & b \\ c & d \end{pmatrix}$$
is regular, then its pivots are
$$a \quad \text{and} \quad \frac{\det A}{a};$$

(b) if
$$A = \begin{pmatrix} a & b & e \\ c & d & f \\ g & h & j \end{pmatrix}$$
is regular, then its pivots are
$$a, \quad \frac{ad - bc}{a}, \quad \text{and} \quad \frac{\det A}{ad - bc}.$$

(c) Can you generalize this observation to regular $n \times n$ matrices?

♡ **1.9.20.** In this exercise, we justify the use of "elementary column operations" to compute determinants. Prove that

(a) adding a scalar multiple of one column to another does not change the determinant;

(b) multiplying a column by a scalar multiplies the determinant by the same scalar;

(c) interchanging two columns changes the sign of the determinant.

(d) Explain how to use elementary column operations to reduce a matrix to lower triangular form and thereby compute its determinant.

(e) Use this method to compute
$$\det \begin{pmatrix} 0 & 1 & 2 \\ -1 & 3 & 5 \\ 2 & -3 & 1 \end{pmatrix}.$$

◇ **1.9.21.** Find the determinant of the Vandermonde matrices listed in Exercise 1.3.25. Can you guess the general $n \times n$ formula?

♡ **1.9.22.** *Cramer's Rule.*

(a) Show that the nonsingular system $ax + by = p$, $cx + dy = q$ has the solution given by the determinantal ratios
$$x = \frac{1}{\Delta} \det \begin{pmatrix} p & b \\ q & d \end{pmatrix}, \qquad y = \frac{1}{\Delta} \det \begin{pmatrix} a & p \\ c & q \end{pmatrix}, \tag{1.87}$$
where
$$\Delta = \det \begin{pmatrix} a & b \\ c & d \end{pmatrix}.$$

(b) Use Cramer's Rule (1.87) to solve the systems

(i) $x + 3y = 13$, $4x + 2y = 0$ (ii) $x - 2y = 4$, $3x + 6y = -2$

(c) Prove that the solution to
$$ax + by + cz = p, \quad dx + ey + fz = q, \quad gx + hy + jz = r,$$
with
$$\Delta = \det \begin{pmatrix} a & b & c \\ d & e & f \\ g & h & j \end{pmatrix} \neq 0$$
is
$$x = \frac{1}{\Delta} \det \begin{pmatrix} p & b & c \\ q & e & f \\ r & h & j \end{pmatrix}, \quad y = \frac{1}{\Delta} \det \begin{pmatrix} a & p & c \\ d & q & f \\ g & r & j \end{pmatrix}, \quad z = \frac{1}{\Delta} \det \begin{pmatrix} a & b & p \\ d & e & q \\ g & h & r \end{pmatrix}. \tag{1.88}$$

(d) Use Cramer's Rule (1.88) to solve

(i) $x + 4y = 3$, $4x + 2y + z = 2$, $-x + y - z = 0$,

(ii) $3x + 2y - z = 1$, $x - 3y + 2z = 2$, $2x - y + z = 3$.

(e) Can you see the pattern that will generalize to n equations in n unknowns?

Remark: Although elegant, Cramer's rule is not a very practical solution method.

◇ **1.9.23.** (a) Show that if
$$D = \begin{pmatrix} A & O \\ O & B \end{pmatrix}$$
is a block diagonal matrix, where A and B are square matrices, then $\det D = \det A \det B$.

(b) Prove that the same holds for a block upper triangular matrix
$$\det \begin{pmatrix} A & C \\ O & B \end{pmatrix} = \det A \det B.$$

(c) Use this method to compute the determinant of the following matrices:

(i) $\begin{pmatrix} 3 & 2 & -2 \\ 0 & 4 & -5 \\ 0 & 3 & 7 \end{pmatrix}$

(ii) $\begin{pmatrix} 1 & 2 & -2 & 5 \\ -3 & 1 & 0 & -5 \\ 0 & 0 & 1 & 3 \\ 0 & 0 & 2 & -2 \end{pmatrix}$

(iii) $\begin{pmatrix} 1 & 2 & 0 & 4 \\ -3 & 1 & 4 & -1 \\ 0 & 3 & 1 & 8 \\ 0 & 0 & 0 & -3 \end{pmatrix}$

(iv) $\begin{pmatrix} 5 & -1 & 0 & 0 \\ 2 & 5 & 0 & 0 \\ 2 & 4 & 4 & -2 \\ 3 & -2 & 9 & -5 \end{pmatrix}$

CHAPTER

2

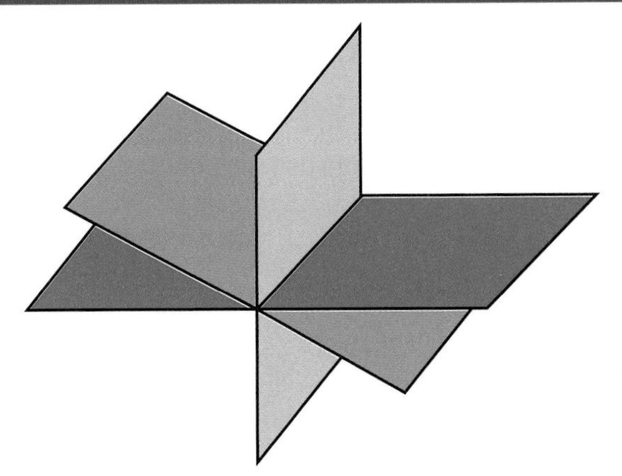

Vector Spaces and Bases

Vector spaces and their ancillary structures provide the common language of linear algebra, and, as such, are an essential prerequisite for understanding contemporary applied mathematics. The key concepts of vector space, subspace, linear independence, span, and basis will appear, not only in linear systems of algebraic equations and the geometry of n-dimensional Euclidean space, but also in the analysis of linear differential equations, linear boundary value problems, Fourier analysis, signal processing, numerical methods, and many, many other fields. Therefore, to master the methods of modern linear algebra and its applications, the first order of business is to acquire a firm working knowledge of fundamental vector space constructions.

One of the great triumphs of mathematics was the recognition that many seemingly unrelated constructions are, in fact, different manifestations of the same underlying mathematical structure. The abstract notion of a vector space serves to unify spaces of ordinary vectors, spaces of functions, such as polynomials, exponentials, trigonometric functions, etc., as well as spaces of matrices, spaces of linear operators, and so on, all in a common conceptual framework. Moreover, proofs that might look rather complicated in any particular context often turn out to be relatively transparent when recast in the more inclusive vector space language. The price that one pays for the increased level of abstraction is that, while the underlying mathematics is not all that difficult, students typically take a long time to assimilate the underlying concepts. In our opinion, the best way to approach the subject is to think in terms of concrete examples. First, make sure you understand what is being said in the case of ordinary Euclidean space. Once this is grasped, the next important case to consider is an elementary function space, e.g., the space of continuous scalar functions. With the two most important cases firmly in hand, the leap to the general abstract formulation should not be too painful. Patience is essential; ultimately, the only way to truly understand an abstract concept like a vector space is by working with it in concrete applications! And always keep in mind that the effort expended here will be amply rewarded later on.

Following an introduction to vector spaces and subspaces, we concentrate on the notions of span and linear independence, first in the context of ordinary vectors, but then in more generality, with an emphasis on function spaces. These are then combined into the all-important definition of a basis of a vector space, leading to a linear algebraic characterization of its dimension. Here is where the distinction between finite-dimensional and infinite-dimensional vector spaces first becomes apparent, although the full ramifications of this dichotomy will take time to unfold. We will then study the four fundamental subspaces associated with a matrix—its range, kernel, corange, and cokernel — and explain how they help us understand the structure and the solutions of linear algebraic systems. Of particular significance is the linear superposition principle that enables us to combine solutions to linear systems. Superposition is the hallmark of linearity, and will apply not only to linear algebraic equations, but also to linear ordinary differential equations, linear boundary value problems, linear partial differential equations, linear integral equations, etc. The final section in this chapter develops some interesting applications in graph theory that serve to illustrate the fundamental matrix subspaces; these results will be used in our later study of electrical circuits.

2.1 REAL VECTOR SPACES

A vector space is the abstract reformulation of the quintessential properties of n-dimensional* *Euclidean space* $\mathbb{R}^n$, which is defined as the set of all real (column) vectors with n entries. The basic laws of vector addition and scalar multiplication in $\mathbb{R}^n$ serve as the motivation for the general definition.

Definition 2.1 A *vector space* is a set V equipped with two operations:

(i) *Addition*: adding any pair of vectors $\mathbf{v}, \mathbf{w} \in V$ gives another vector $\mathbf{v} + \mathbf{w} \in V$;

(ii) *Scalar Multiplication*: multiplying a vector $\mathbf{v} \in V$ by a scalar $c \in \mathbb{R}$ produces a vector $c\,\mathbf{v} \in V$;

subject to the following axioms, for all $\mathbf{u}, \mathbf{v}, \mathbf{w} \in V$ and all scalars $c, d \in \mathbb{R}$:

(a) *Commutativity of Addition*: $\mathbf{v} + \mathbf{w} = \mathbf{w} + \mathbf{v}$.

(b) *Associativity of Addition*: $\mathbf{u} + (\mathbf{v} + \mathbf{w}) = (\mathbf{u} + \mathbf{v}) + \mathbf{w}$.

(c) *Additive Identity*: There is a zero element $\mathbf{0} \in V$ satisfying $\mathbf{v} + \mathbf{0} = \mathbf{v} = \mathbf{0} + \mathbf{v}$.

(d) *Additive Inverse*: For each $\mathbf{v} \in V$ there is an element $-\mathbf{v} \in V$ such that $\mathbf{v} + (-\mathbf{v}) = \mathbf{0} = (-\mathbf{v}) + \mathbf{v}$.

(e) *Distributivity*: $(c + d)\,\mathbf{v} = (c\,\mathbf{v}) + (d\,\mathbf{v})$, and $c\,(\mathbf{v} + \mathbf{w}) = (c\,\mathbf{v}) + (c\,\mathbf{w})$.

(f) *Associativity of Scalar Multiplication*: $c\,(d\,\mathbf{v}) = (c\,d)\,\mathbf{v}$.

(g) *Unit for Scalar Multiplication*: the scalar $1 \in \mathbb{R}$ satisfies $1\,\mathbf{v} = \mathbf{v}$.

REMARK: For most of this chapter we will deal with real vector spaces, in which the scalars are ordinary real numbers $c \in \mathbb{R}$. Complex vector spaces, where complex scalars are allowed, will be introduced in Section 3.6. Vector spaces over other fields are studied in abstract algebra, [30].

In the beginning, we will refer to the individual elements of a vector space as "vectors", even though, as we shall see, they might also be functions, or matrices, or even more general objects. Unless dealing with certain specific examples such as a space of functions or matrices, we will use bold face, lower case Latin letters $\mathbf{v}$,

*The precise definition of dimension will appear later, in Theorem 2.29.

$\mathbf{w}, \ldots$ to denote the elements of our vector space. We will usually use a bold face $\mathbf{0}$ to denote the unique† zero element of our vector space, while ordinary 0 denotes the real number zero.

The following identities are elementary consequences of the vector space axioms:

(h) $0\,\mathbf{v} = \mathbf{0}$;

(i) $(-1)\,\mathbf{v} = -\,\mathbf{v}$;

(j) $c\,\mathbf{0} = \mathbf{0}$.

(k) If $c\,\mathbf{v} = \mathbf{0}$, then either $c = 0$ or $\mathbf{v} = \mathbf{0}$.

Let us, for example, prove (h). Let $\mathbf{z} = 0\,\mathbf{v}$. Then, by the distributive property,

$$\mathbf{z} + \mathbf{z} = 0\,\mathbf{v} + 0\,\mathbf{v} = (0 + 0)\,\mathbf{v} = 0\,\mathbf{v} = \mathbf{z}.$$

Adding $-\,\mathbf{z}$ to both sides of this equation, and making use of axioms (b), (d), and then (c), we conclude that $\mathbf{z} = \mathbf{0}$, which completes the proof. Verification of the other three properties is left as an exercise for the reader.

EXAMPLE 2.2 As noted above, the prototypical example of a real vector space is the space $\mathbb{R}^n$ consisting of column vectors or n-tuples of real numbers $\mathbf{v} = \left(v_1, v_2, \ldots, v_n \right)^T$. Vector addition and scalar multiplication are defined in the usual manner:

$$\mathbf{v} + \mathbf{v} = \begin{pmatrix} v_1 + w_1 \\ v_2 + w_2 \\ \vdots \\ v_n + w_n \end{pmatrix}, \qquad c\mathbf{v} = \begin{pmatrix} c v_1 \\ c v_2 \\ \vdots \\ c v_n \end{pmatrix}, \qquad \text{whenever} \quad \mathbf{v} = \begin{pmatrix} v_1 \\ v_2 \\ \vdots \\ v_n \end{pmatrix}, \mathbf{w} = \begin{pmatrix} w_1 \\ w_2 \\ \vdots \\ w_n \end{pmatrix}.$$

The zero vector is $\mathbf{0} = \left(0, \ldots, 0 \right)^T$. The fact that vectors in $\mathbb{R}^n$ satisfy all of the vector space axioms is an immediate consequence of the laws of vector addition and scalar multiplication. ●

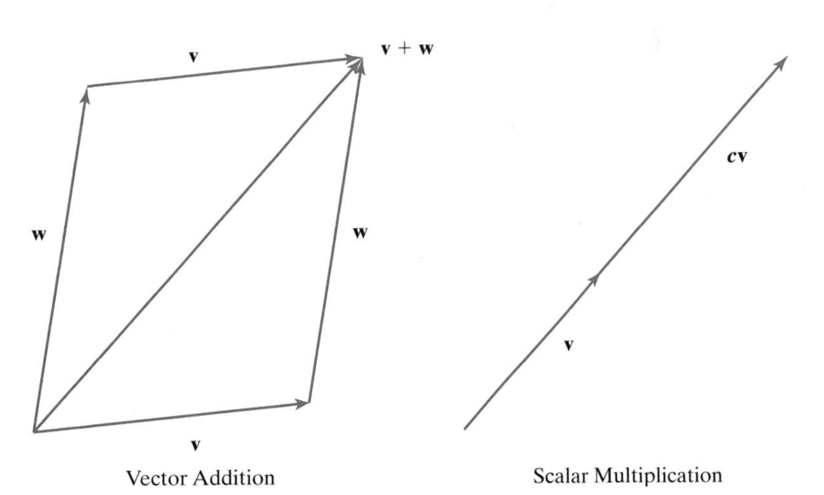

Figure 2.1 Vector space operations in $\mathbb{R}^n$.

EXAMPLE 2.3 Let $\mathcal{M}_{m \times n}$ denote the space of all real matrices of size $m \times n$. Then $\mathcal{M}_{m \times n}$ forms a vector space under the laws of matrix addition and scalar multiplication. The zero element is the zero matrix O. (We are ignoring matrix multiplication, which

†See Exercise 2.1.12.

is *not* a vector space operation.) Again, the vector space axioms are immediate consequences of the basic laws of matrix arithmetic. The preceding example of the vector space $\mathbb{R}^n = \mathcal{M}_{n\times 1}$ is a particular case when the matrices have only one column. ●

EXAMPLE 2.4 Consider the space

$$\mathcal{P}^{(n)} = \left\{ \, p(x) = a_n x^n + a_{n-1} x^{n-1} + \cdots + a_1 x + a_0 \, \right\} \tag{2.1}$$

consisting of all real polynomials of degree $\leq n$. Addition of polynomials is defined in the usual manner; for example,

$$(x^2 - 3x) + (2x^2 - 5x + 4) = 3x^2 - 8x + 4.$$

Note that the sum $p(x) + q(x)$ of two polynomials of degree $\leq n$ also has degree $\leq n$. The zero element of $\mathcal{P}^{(n)}$ is the zero polynomial. We can multiply polynomials by scalars — real constants—in the usual fashion; for example if $p(x) = x^2 - 2x$, then $3\,p(x) = 3x^2 - 6x$. The proof that $\mathcal{P}^{(n)}$ satisfies the vector space axioms is an easy consequence of the basic laws of polynomial algebra. ●

Warning: It is not true that the sum of two polynomials of degree n also has degree n; for example

$$(x^2 + 1) + (-x^2 + x) = x + 1$$

has degree 1 even though the two summands have degree 2. This means that the set of polynomials of degree $= n$ is *not* a vector space.

Warning: You might be tempted to identify a scalar with a constant polynomial, but one should really regard these as two completely different objects — one is a *number*, while the other is a *constant function*. To add to the confusion, one typically uses the same notation for these two objects; for instance, 0 could either mean the real number 0 or the constant function taking the value 0 everywhere, which is the zero element, $\mathbf{0}$, of this vector space. The reader needs to exercise due care when interpreting each occurrence.

For much of analysis, including differential equations, Fourier theory, numerical methods, etc., the most important vector spaces consist of functions that have certain prescribed properties. The simplest such example is the following.

EXAMPLE 2.5 Let $I \subset \mathbb{R}$ be an interval.[‡] Consider the *function space* $\mathcal{F} = \mathcal{F}(I)$ whose elements are all real-valued functions $f(x)$ defined for all $x \in I$. The claim is that the function space $\mathcal{F}$ has the structure of a vector space. Addition of functions in $\mathcal{F}$ is defined in the usual manner: $(f+g)(x) = f(x)+g(x)$ for all $x \in I$. Multiplication by scalars $c \in \mathbb{R}$ is the same as multiplication by constants, $(cf)(x) = c f(x)$. The zero element is the constant function that is identically 0 for all $x \in I$. The proof of the vector space axioms is straightforward. Observe that we are ignoring all additional operations that affect functions such as multiplication, division, inversion, composition, etc.; these are irrelevant as far as the vector space structure of $\mathcal{F}$ goes. ●

[‡]An *interval* is a subset $I \subset \mathbb{R}$ that contains all real numbers between $a < b$, and can be either
- *closed*, meaning that it includes its endpoints: $I = [a,b] = \{\, x \mid a \leq x \leq b \,\}$;
- *open*, which does not include either endpoint: $I = (a,b) = \{\, x \mid a < x < b \,\}$; or
- *half open*, which includes one but not the other endpoint, so $I = [a,b)$ or $(a,b]$. An open endpoint is allowed to be infinite; in particular, $(-\infty,\infty) = \mathbb{R}$ is another way of writing the entire real line.

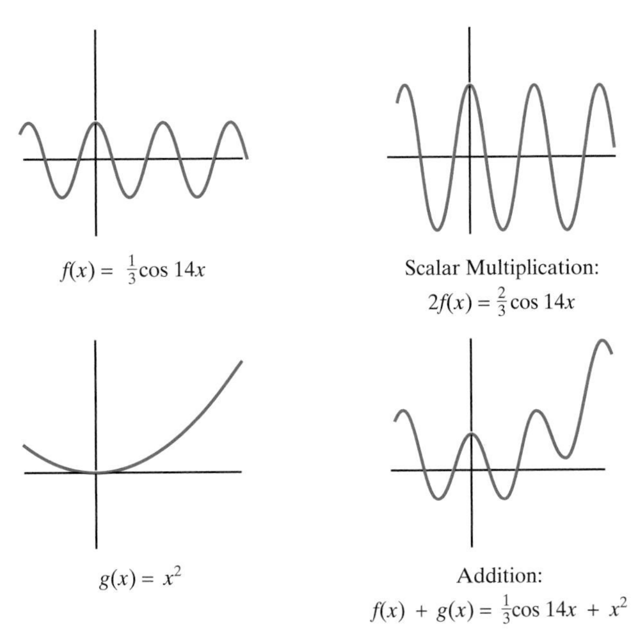

Figure 2.2 Vector space operations in function space.

EXAMPLE 2.6 The preceding examples are all, in fact, special cases of an even more general construction. A clue is to note that the last example of a function space does not make any use of the fact that the domain of the functions is a real interval. Indeed, the same construction produces a function space $\mathcal{F}(I)$ corresponding to *any* subset $I \subset \mathbb{R}$.

Even more generally, let S be *any* set. Let $\mathcal{F} = \mathcal{F}(S)$ denote the space of all real-valued functions $f \colon S \to \mathbb{R}$. Then we claim that V is a vector space under the operations of function addition and scalar multiplication. More precisely, given functions f and g, we define their sum to be the function $h = f + g$ such that $h(x) = f(x) + g(x)$ for all $x \in S$. Similarly, given a function f and a real scalar $c \in \mathbb{R}$, we define the scalar multiple $g = c\, f$ to be the function such that $g(x) = c f(x)$ for all $x \in S$. The proof of the vector space axioms is straightforward, and the reader should be able to fill in the necessary details.

In particular, if $S \subset \mathbb{R}$ is an interval, then $\mathcal{F}(S)$ coincides with the space of scalar functions described in the preceding example. If $S \subset \mathbb{R}^n$ is a subset of Euclidean space, then the elements of $\mathcal{F}(S)$ are real-valued functions $f(x_1, \ldots, x_n)$ depending upon the n variables corresponding to the coordinates of points $\mathbf{x} = (x_1, \ldots, x_n) \in S$ in the domain. In this fashion, the set of real-valued functions defined on any domain in $\mathbb{R}^n$ forms a vector space.

Another useful example is to let $S = \{x_1, \ldots, x_n\} \subset \mathbb{R}$ be a finite set of real numbers. A real-valued function $f \colon S \to \mathbb{R}$ is defined by its values $f(x_1)$, $f(x_2)$, $\ldots$, $f(x_n)$ at the specified points. In applications, these objects serve to indicate the *sample values* of a scalar function $f(x) \in \mathcal{F}(\mathbb{R})$ taken at the *sample points* $x_1, \ldots, x_n$. For example, if $f(x) = x^2$ and the sample points are $x_1 = 0, x_2 = 1, x_3 = 2, x_4 = 3$, then the corresponding sample values are $f(x_1) = 0$, $f(x_2) = 1$, $f(x_3) = 4$, $f(x_4) = 9$. When measuring a physical quantity, such as temperature, velocity, pressure, etc., one typically records only a finite set of sample values. The intermediate, non-recorded values between the sample points are then reconstructed through some form of interpolation—a topic that we shall visit in depth in Chapters 4 and 5.

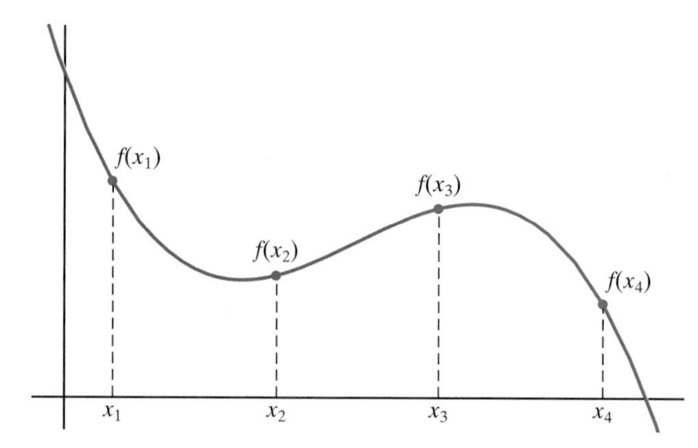

Figure 2.3 Sampled function.

Interestingly, the sample values $f(x_i)$ can be identified with the entries f_i of a vector

$$\mathbf{f} = \left(f_1, f_2, \ldots, f_n \right)^T = \left(f(x_1), f(x_2), \ldots, f(x_n) \right)^T \in \mathbb{R}^n,$$

known as the *sample vector*. Every sampled function $f : S \to \mathbb{R}$ corresponds to a unique vector $\mathbf{f} \in \mathbb{R}^n$ and vice versa. (But keep in mind that different scalar functions $f(x) \in \mathcal{F}(\mathbb{R})$ may have the same sample values.) Addition of sample functions corresponds to addition of their sample vectors, as does scalar multiplication. Thus, *the vector space of sample functions* $\mathcal{F}(S) = \mathcal{F}\{x_1, \ldots, x_n\}$ *is the same as the vector space* $\mathbb{R}^n$! The identification of sampled functions as vectors is of fundamental importance in modern signal processing, as we will see in Section 5.7. ●

EXAMPLE 2.7 The preceding construction admits yet a further generalization. We continue to let S be an arbitrary set. Let V be a vector space. The claim is that the space $\mathcal{F}(S, V)$ consisting of all V–valued functions $\mathbf{f}: S \to V$ is a vector space. In other words, we replace the particular vector space $\mathbb{R}$ in the preceding example by a general vector space V, and the same conclusion holds. The operations of function addition and scalar multiplication are defined in the evident manner: $(\mathbf{f} + \mathbf{g})(x) = \mathbf{f}(x) + \mathbf{g}(x)$ and $(c\,\mathbf{f})(x) = c\,\mathbf{f}(x)$ for $x \in S$, where we are using the vector addition and scalar multiplication operations on V to induce corresponding operations on V–valued functions. The proof that $\mathcal{F}(S, V)$ satisfies all of the vector space axioms proceeds as before.

The most important example is when $S \subset \mathbb{R}^n$ is a domain in Euclidean space and $V = \mathbb{R}^m$ is itself a Euclidean space. In this case, the elements of $\mathcal{F}(S, \mathbb{R}^m)$ consist of vector-valued functions $\mathbf{f}: S \to \mathbb{R}^m$, so that

$$\mathbf{f}(\mathbf{x}) = (f_1(x_1, \ldots, x_n), \ldots, f_m(x_1, \ldots, x_n))^T$$

is a column vector consisting of m functions of n variables, all defined on a common domain S. The general construction implies that addition and scalar multiplication of vector-valued functions is done componentwise; for example

$$2\begin{pmatrix} x^2 \\ e^x - 4 \end{pmatrix} - \begin{pmatrix} \cos x \\ x \end{pmatrix} = \begin{pmatrix} 2x^2 - \cos x \\ 2e^x - x - 8 \end{pmatrix}.$$

Of particular importance are the vector fields arising in physics, including gravitational force fields, electromagnetic fields, fluid velocity fields, and many others. ●

EXERCISES 2.1

2.1.1. Show that the set of complex numbers $x + \mathrm{i}\, y$ forms a real vector space under the operations of addition $(x + \mathrm{i}\, y) + (u + \mathrm{i}\, v) = (x + u) + \mathrm{i}\,(y + v)$ and scalar multiplication $c\,(x + \mathrm{i}\, y) = c\,x + \mathrm{i}\, c\, y$. (But complex multiplication is *not* a real vector space operation.)

2.1.2. Show that the positive quadrant

$$Q = \{\, (x, y) \mid x, y > 0 \,\} \subset \mathbb{R}^2$$

forms a vector space if we define addition by $(x_1, y_1) + (x_2, y_2) = (x_1 x_2, y_1 y_2)$ and scalar multiplication by $c\,(x, y) = (x^c, y^c)$.

◇ **2.1.3.** Let S be any set. Carefully justify the validity of all the vector space axioms for the space $\mathcal{F}(S)$ consisting of all real-valued functions $f\colon S \to \mathbb{R}$.

2.1.4. Let $S = \{0, 1, 2, 3\}$.

(a) Find the sample vectors corresponding to the functions $1, \cos \pi x, \cos 2\pi x, \cos 3\pi x$.

(b) Is a function uniquely determined by its sample values?

2.1.5. Find two different functions $f(x)$ and $g(x)$ that have the *same* sample vectors $\mathbf{f}, \mathbf{g}$ at the sample points $x_1 = 0, x_2 = 1, x_3 = -1$.

2.1.6. (a) Let $x_1 = 0, x_2 = 1$. Find the unique linear function $f(x) = a\,x + b$ that has the sample vector $\mathbf{f} = (3, -1)^T$.

(b) Let $x_1 = 0, x_2 = 1, x_3 = -1$. Find the unique quadratic function $f(x) = a\,x^2 + b\,x + c$ with sample vector $\mathbf{f} = (1, -2, 0)^T$.

2.1.7. Let $\mathcal{F}(\mathbb{R}^2, \mathbb{R}^2)$ denote the vector space consisting of all functions $\mathbf{f}\colon \mathbb{R}^2 \to \mathbb{R}^2$.

(a) Which of the following functions $\mathbf{f}(x, y)$ are elements?

(i) $x^2 + y^2$, (ii) $\begin{pmatrix} x - y \\ x\,y \end{pmatrix}$,

(iii) $\begin{pmatrix} e^x \\ \cos y \end{pmatrix}$, (iv) $\begin{pmatrix} 1 \\ 3 \end{pmatrix}$,

(v) $\begin{pmatrix} x & y \\ -y & x \end{pmatrix}$, (vi) $\begin{pmatrix} x \\ y \\ x + y \end{pmatrix}$.

(b) Sum all of the elements of $\mathcal{F}(\mathbb{R}^2, \mathbb{R}^2)$ you identified in part (a). Then multiply your sum by the scalar -5.

(c) Carefully describe the zero element of the vector space $\mathcal{F}(\mathbb{R}^2, \mathbb{R}^2)$.

◇ **2.1.8.** A *planar vector field* is a function which assigns a vector $\mathbf{v}(x, y) = \begin{pmatrix} v_1(x, y) \\ v_2(x, y) \end{pmatrix}$ to each point $\begin{pmatrix} x \\ y \end{pmatrix} \in \mathbb{R}^2$. Explain why the set of all planar vector fields forms a vector space.

♡ **2.1.9.** Let $h, k > 0$ be fixed. Let

$$S = \{\, (i\,h, j\,k) \mid 1 \le i \le m, 1 \le j \le n \,\}$$

be points in a rectangular planar grid. Show that the function space $\mathcal{F}(S)$ can be identified with the vector space of $m \times n$ matrices $\mathcal{M}_{m\times n}$.

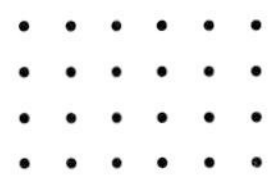

2.1.10. The space $\mathbb{R}^\infty$ is defined as the set of all infinite real sequences $\mathbf{a} = (a_1, a_2, a_3, \ldots)$ where $a_i \in \mathbb{R}$. Define addition and scalar multiplication in such a way as to make $\mathbb{R}^\infty$ into a vector space. Explain why all the vector space axioms are valid.

2.1.11. Prove the basic vector space properties (i), (j), (k) following Definition 2.1.

◇ **2.1.12.** Prove that a vector space has only one zero element $\mathbf{0}$.

◇ **2.1.13.** Suppose that V and W are vector spaces. The *Cartesian product space*, denoted by $V \times W$ is defined as the set of all ordered pairs $(\mathbf{v}, \mathbf{w})$ where $\mathbf{v} \in V, \mathbf{w} \in W$, with vector addition $(\mathbf{v}, \mathbf{w}) + (\widehat{\mathbf{v}}, \widehat{\mathbf{w}}) = (\mathbf{v} + \widehat{\mathbf{v}}, \mathbf{w} + \widehat{\mathbf{w}})$ and scalar multiplication $c\,(\mathbf{v}, \mathbf{w}) = (c\,\mathbf{v}, c\,\mathbf{w})$.

(a) Prove that $V \times W$ is a vector space.

(b) Explain why $\mathbb{R} \times \mathbb{R}$ is the same as $\mathbb{R}^2$.

(c) More generally, explain why $\mathbb{R}^m \times \mathbb{R}^n$ is the same as $\mathbb{R}^{m+n}$.

2.1.14. Use Exercise 2.1.13 to show that the space of pairs $(f(x), a)$ where f is a continuous scalar function and a is a real number forms a vector space. What is the zero element? Be precise! Write out the laws of vector addition and scalar multiplication.

2.2 SUBSPACES

In the preceding section, we were introduced to the most basic vector spaces that arise in this text. Almost all of the vector spaces used in applications appear as subsets of these prototypical examples.

Definition 2.8 A *subspace* of a vector space V is a subset $W \subset V$ which is a vector space in its own right—under the same operations of vector addition and scalar multiplication and the same zero element.

In particular, a subspace W *must* contain the zero element of V. Proving that a given subset of a vector space forms a subspace is particularly easy: we only need check its *closure* under addition and scalar multiplication.

Proposition 2.9 A nonempty subset $W \subset V$ of a vector space is a subspace if and only if

(a) for every $\mathbf{v}, \mathbf{w} \in W$, the sum $\mathbf{v} + \mathbf{w} \in W$, and

(b) for every $\mathbf{v} \in W$ and every $c \in \mathbb{R}$, the scalar product $c\,\mathbf{v} \in W$.

Proof The proof is immediate. For example, let us check commutativity. The subspace elements $\mathbf{v}, \mathbf{w} \in W$ can be regarded as elements of V, in which case $\mathbf{v} + \mathbf{w} = \mathbf{w} + \mathbf{v}$ because V is a vector space. But the closure condition implies that the common sum also belongs to W, and so the commutativity axiom also holds for elements of W. Establishing the validity of the other axioms is equally easy. ■

It will sometimes be convenient to combine the two closure conditions. Thus, to prove $W \subset V$ is a subspace it suffices to check that $c\,\mathbf{v} + d\,\mathbf{w} \in W$ for every $\mathbf{v}, \mathbf{w} \in W$ and $c, d \in \mathbb{R}$.

EXAMPLE 2.10 Let us list some examples of subspaces of the three-dimensional Euclidean space $\mathbb{R}^3$.

(a) The trivial subspace $W = \{\mathbf{0}\}$. Demonstrating closure is easy: since there is only one element $\mathbf{0}$ in W, we just need to check that $\mathbf{0} + \mathbf{0} = \mathbf{0} \in W$ and $c\,\mathbf{0} = \mathbf{0} \in W$ for any scalar c.

(b) The entire space $W = \mathbb{R}^3$. Here closure is immediate because $\mathbb{R}^3$ is a vector space in its own right.

(c) The set of all vectors of the form $(x, y, 0)^T$, i.e., the $x\,y$ coordinate plane. To prove closure, we check that all sums

$$(x, y, 0)^T + (\widehat{x}, \widehat{y}, 0)^T = (x + \widehat{x}, y + \widehat{y}, 0)^T$$

and scalar multiples

$$c\,(x, y, 0)^T = (c\,x, c\,y, 0)^T$$

of vectors in the $x\,y$–plane remain in the plane.

(d) The set of solutions $(x, y, z)^T$ to the homogeneous linear equation

$$3x + 2y - z = 0. \tag{2.2}$$

Indeed, if $\mathbf{x} = (x, y, z)^T$ is a solution, then so is any scalar multiple $c\,\mathbf{x} = (c\,x, c\,y, c\,z)^T$ since

$$3(c\,x) + 2(c\,y) - (c\,z) = c\,(3x + 2y - z) = 0.$$

Moreover, if $\widehat{\mathbf{x}} = (\widehat{x}, \widehat{y}, \widehat{z})^T$ is a second solution, the sum

$$\mathbf{x} + \widehat{\mathbf{x}} = (x + \widehat{x}, y + \widehat{y}, z + \widehat{z})^T$$

is also a solution, since

$$3(x + \widehat{x}) + 2(y + \widehat{y}) - (z + \widehat{z}) = (3x + 2y - z) + (3\widehat{x} + 2\widehat{y} - \widehat{z}) = 0.$$

The solution space is, in fact, the two-dimensional plane passing through the origin with normal vector $(3, 2, -1)^T$.

(e) The set of all vectors lying in the plane spanned by the vectors $\mathbf{v}_1 = (2, -3, 0)^T$ and $\mathbf{v}_2 = (1, 0, 3)^T$. In other words, we consider all vectors of the form

$$\mathbf{v} = a\,\mathbf{v}_1 + b\,\mathbf{v}_2 = a\begin{pmatrix} 2 \\ -3 \\ 0 \end{pmatrix} + b\begin{pmatrix} 1 \\ 0 \\ 3 \end{pmatrix} = \begin{pmatrix} 2a+b \\ -3a \\ 3b \end{pmatrix},$$

where $a, b \in \mathbb{R}$ are arbitrary scalars. If $\mathbf{v} = a\,\mathbf{v}_1 + b\,\mathbf{v}_2$ and $\mathbf{w} = \widehat{a}\,\mathbf{v}_1 + \widehat{b}\,\mathbf{v}_2$ are any two vectors in the span, so is

$$\begin{aligned} c\,\mathbf{v} + d\,\mathbf{w} &= c\,(a\,\mathbf{v}_1 + b\,\mathbf{v}_2) + d\,(\widehat{a}\,\mathbf{v}_1 + \widehat{b}\,\mathbf{v}_2) \\ &= (a\,c + \widehat{a}\,d)\mathbf{v}_1 + (b\,c + \widehat{b}\,d)\mathbf{v}_2 = \tilde{a}\,\mathbf{v}_1 + \tilde{b}\,\mathbf{v}_2, \end{aligned}$$

where $\tilde{a} = a\,c + \widehat{a}\,d$, $\tilde{b} = b\,c + \widehat{b}\,d$. This demonstrates that the span is a subspace of $\mathbb{R}^3$. The reader might already have noticed that this subspace is the same plane defined by the equation (2.2). ●

EXAMPLE 2.11 The following subsets of $\mathbb{R}^3$ are *not* subspaces.

(a) The set P of all vectors of the form $(x, y, 1)^T$, i.e., the plane parallel to the xy coordinate plane passing through $(0, 0, 1)^T$. Indeed, $(0, 0, 0) \notin P$, which is the most basic requirement for a subspace. In fact, neither of the closure axioms hold for this subset.

(b) The positive orthant $\mathcal{O}^+ = \{x \geq 0, y \geq 0, z \geq 0\}$. Although $\mathbf{0} \in \mathcal{O}^+$, and the sum of two vectors in $\mathcal{O}^+$ also belongs to $\mathcal{O}^+$, multiplying by negative scalars takes us outside the orthant, violating closure under scalar multiplication.

(c) The unit sphere $S_1 = \{\, x^2 + y^2 + z^2 = 1 \,\}$. Again, $\mathbf{0} \notin S_1$. More generally, curved surfaces, such as the paraboloid $P = \{\, z = x^2 + y^2 \,\}$, are not subspaces. Although $\mathbf{0} \in P$, most scalar multiples of elements of P do not belong to P. For example, $(1, 1, 2)^T \in P$, but $2(1, 1, 2)^T = (2, 2, 4)^T \notin P$.

In fact, there are only four fundamentally different types of subspaces $W \subset \mathbb{R}^3$ of three-dimensional Euclidean space:

(i) the entire three-dimensional space $W = \mathbb{R}^3$,

(ii) a plane passing through the origin,

(iii) a line passing through the origin,

(iv) a point—the trivial subspace $W = \{\mathbf{0}\}$.

We can establish this observation by the following argument. If $W = \{\mathbf{0}\}$ contains only the zero vector, then we are in case (iv). Otherwise, $W \subset \mathbb{R}^3$ contains a nonzero vector $\mathbf{0} \neq \mathbf{v}_1 \in W$. But since W must contain all scalar multiples $c\mathbf{v}_1$ of this element, it includes the entire line in the direction of $\mathbf{v}_1$. If W contains another vector $\mathbf{v}_2$ that does not lie in the line through $\mathbf{v}_1$, then it must contain the entire plane $\{c\,\mathbf{v}_1 + d\,\mathbf{v}_2\}$ spanned by $\mathbf{v}_1, \mathbf{v}_2$. Finally, if there is a third vector $\mathbf{v}_3$ not contained in this plane, then we claim that $W = \mathbb{R}^3$. This final fact will be an immediate consequence of general results in this chapter, although the interested reader might try to prove it directly before proceeding. ●

EXAMPLE 2.12 Let $I \subset \mathbb{R}$ be an interval, and let $\mathcal{F}(I)$ be the space of real-valued functions $f : I \to \mathbb{R}$. Let us look at some of the most important examples of subspaces of $\mathcal{F}(I)$. In each case, we need only verify the closure conditions to verify that the given subset is indeed a subspace.

(a) The space $\mathcal{P}^{(n)}$ of polynomials of degree $\leq n$, which we already encountered.

(b) The space $\mathcal{P}^{(\infty)} = \bigcup_{n\geq 0} \mathcal{P}^{(n)}$ consisting of all polynomials. Closure means that the sum of any two polynomials is a polynomial, as is any scalar (constant) multiple of a polynomial.

(c) The space $C^0(I)$ of all continuous functions. Closure of this subspace relies on knowing that if $f(x)$ and $g(x)$ are continuous, then both $f(x) + g(x)$ and $cf(x)$, for any $c \in \mathbb{R}$, are also continuous—two basic results from calculus, [2, 58].

(d) More restrictively, we can consider the subspace $C^n(I)$ consisting of all functions $f(x)$ that have n continuous derivatives $f'(x), f''(x), \ldots, f^{(n)}(x)$ on* I. Again, we need to know that if $f(x)$ and $g(x)$ have n continuous derivatives, so does $c\,f(x) + d\,g(x)$ for any $c, d \in \mathbb{R}$.

(e) The space $C^\infty(I) = \bigcap_{n\geq 0} C^n(I)$ of infinitely differentiable or *smooth* functions is also a subspace. This can be proved directly, or follows from the general fact that the intersection of subspaces is a subspace, cf. Exercise 2.2.23.

(f) The space $\mathcal{A}(I)$ of analytic functions on the interval I. Recall that a function $f(x)$ is called *analytic* at a point a if it is smooth, and, moreover, its Taylor series

$$f(a) + f'(a)(x-a) + \tfrac{1}{2} f''(a)(x-a)^2 + \cdots = \sum_{n=0}^{\infty} \frac{f^{(n)}(a)}{n!}(x-a)^n \qquad (2.3)$$

converges to $f(x)$ for all x sufficiently close to a. (The series is not required to converge on the entire interval I.) Not every smooth function is analytic, and so $\mathcal{A}(I) \subsetneq C^\infty(I)$. An explicit example of a smooth but non-analytic function can be found in Exercise 2.2.28.

(g) The set of all *mean zero* functions. The *mean* or *average* of an integrable function defined on a closed interval $I = [a, b]$ is the real number

$$\overline{f} = \frac{1}{b-a} \int_a^b f(x)\,dx. \qquad (2.4)$$

In particular, f has *mean zero* if and only if $\displaystyle\int_a^b f(x)\,dx = 0$. Since $\overline{f+g} = \overline{f} + \overline{g}$, and $\overline{cf} = c\overline{f}$, sums and scalar multiples of mean zero functions also have mean zero, proving closure.

(h) Fix $x_0 \in I$. The set of all functions $f(x)$ that vanish at the point, $f(x_0) = 0$, is a subspace. Indeed, if $f(x_0) = 0$ and $g(x_0) = 0$, then clearly $(cf + dg)(x_0) = cf(x_0) + dg(x_0) = 0$, proving closure. This example can evidently be generalized to functions that vanish at several points, or even on an entire subset $S \subset I$.

(i) The set of all solutions $u = f(x)$ to the homogeneous linear differential equation

$$u'' + 2u' - 3u = 0.$$

Indeed, if $f(x)$ and $g(x)$ are solutions, so is $f(x) + g(x)$ and $c\,f(x)$ for any $c \in \mathbb{R}$. Note that we do *not* need to actually solve the equation to verify these

*We use one-sided derivatives at any endpoint belonging to the interval.

claims! They follow directly from linearity:

$$(f+g)''+2(f+g)'-3(f+g) = (f''+2f'-3f)+(g''+2g'-3g) = 0,$$
$$(cf)''+2(cf)'-3(cf) = c(f''+2f'-3f) = 0. \qquad \bullet$$

Note: In the last three examples, 0 is essential for the indicated set of functions to be a subspace. The set of functions such that $f(x_0) = 1$, say, is not a subspace. The set of functions with a fixed nonzero mean, say $\overline{f} = 3$, is also not a subspace. Nor is the set of solutions to an inhomogeneous ordinary differential equation, say $u'' + 2u' - 3u = x - 3$. None of these subsets contain the zero function, nor do they satisfy the closure conditions.

EXERCISES 2.2

2.2.1. (a) Prove that the set of all vectors $(x, y, z)^T$ such that $x - y + 4z = 0$ forms a subspace of $\mathbb{R}^3$.
(b) Explain why the set of all vectors that satisfy $x - y + 4z = 1$ does not form a subspace.

2.2.2. Which of the following are subspaces of $\mathbb{R}^3$? Justify your answers!
(a) The set of all vectors $(x, y, z)^T$ satisfying $x + y + z + 1 = 0$.
(b) The set of vectors of the form $(t, -t, 0)^T$ for $t \in \mathbb{R}$.
(c) The set of vectors of the form $(r - s, r + 2s, -s)^T$ for $r, s \in \mathbb{R}$.
(d) The set of vectors whose first component equals 0.
(e) The set of vectors whose last component equals 1.
(f) The set of all vectors $(x, y, z)^T$ with $x \geq y \geq z$.
(g) The set of all solutions to the equation $z = x - y$.
(h) The set of all solutions to $z = x\,y$.
(i) The set of all solutions to $x^2 + y^2 + z^2 = 0$.
(j) The set of all solutions to the system $x\,y = y\,z = x\,z$.

2.2.3. Graph the following subsets of $\mathbb{R}^3$ and use this to explain which are subspaces:
(a) The line $(t, -t, 3t)^T$ for $t \in \mathbb{R}$.
(b) The helix $(\cos t, \sin t, t)^T$.
(c) The surface $x - 2y + 3z = 0$.
(d) The unit ball $x^2 + y^2 + z^2 < 1$.
(e) The cylinder $(y + 2)^2 + (z - 1)^2 = 5$.
(f) The intersection of the cylinders $(x - 1)^2 + y^2 = 1$ and $(x + 1)^2 + y^2 = 1$.

2.2.4. Show that if $W \subset \mathbb{R}^3$ is a subspace containing the vectors $(1, 2, -1)^T$, $(2, 0, 1)^T$, $(0, -1, 3)^T$, then $W = \mathbb{R}^3$.

2.2.5. *True or false*: An interval is a vector space.

2.2.6. (a) Can you construct an example of a subset $S \subset \mathbb{R}^2$ with the property that $c\,\mathbf{v} \in S$ for any $c \in \mathbb{R}$, $\mathbf{v} \in S$, and yet S is not a subspace?
(b) What about an example where $\mathbf{v} + \mathbf{w} \in S$ for every $\mathbf{v}, \mathbf{w} \in S$, and yet S is not a subspace?

2.2.7. Determine which of the following sets of vectors $\mathbf{x} = \left(x_1, \ldots, x_n\right)^T$ are subspaces of $\mathbb{R}^n$:
(a) all equal entries $x_1 = \cdots = x_n$;
(b) all positive entries: $x_i \geq 0$;
(c) first and last entries equal to zero: $x_1 = x_n = 0$;
(d) entries add up to zero: $x_1 + \cdots + x_n = 0$;
(e) first and last entries differ by one: $x_1 - x_n = 1$.

2.2.8. Prove that the set of all solutions $\mathbf{x}$ of the linear system $A\,\mathbf{x} = \mathbf{b}$ forms a subspace if and only if the system is homogeneous.

2.2.9. A square matrix is called *strictly lower triangular* if all entries on or above the main diagonal are 0. Prove that the space of strictly lower triangular matrices forms a subspace of the vector space of all $n \times n$ matrices.

◇ **2.2.10.** The *trace* of an $n \times n$ matrix $A \in \mathcal{M}_{n\times n}$ is defined to be the sum of its diagonal entries: $\operatorname{tr} A = a_{11} + a_{22} + \cdots + a_{nn}$. Prove that the set of trace zero matrices, $\operatorname{tr} A = 0$, is a subspace of $\mathcal{M}_{n\times n}$.

2.2.11. (a) Is the set of $n \times n$ matrices with $\det A = 1$ a subspace of $\mathcal{M}_{n\times n}$?
(b) What about the matrices with $\det A = 0$?

2.2.12. Which of the following are subspaces of the vector space of $n \times n$ matrices $\mathcal{M}_{n\times n}$? The set of all
(a) regular matrices;

(b) nonsingular matrices;

(c) singular matrices;

(d) lower triangular matrices;

(e) special lower triangular matrices;

(f) diagonal matrices;

(g) symmetric matrices;

(h) skew-symmetric matrices.

2.2.13. Which of the following are vector spaces? Justify your answer!

(a) The set of all row vectors of the form $(a, 3a)$.

(b) The set of all vectors of the form $(a, a+1)$.

(c) The set of all continuous functions for which $f(-1) = 0$.

(d) The set of all periodic functions of period 1, i.e., $f(x+1) = f(x)$.

(e) The set of all non-negative functions: $f(x) \geq 0$.

(f) The set of all even polynomials: $p(x) = p(-x)$.

(g) The set of all polynomials $p(x)$ that have $x - 1$ as a factor.

(h) The set of all quadratic forms $q(x, y) = ax^2 + bxy + cy^2$.

2.2.14. Let $V = C^0(\mathbb{R})$ be the vector space consisting of all continuous functions $f: \mathbb{R} \to \mathbb{R}$. Explain why the set of all functions such that $f(1) = 0$ forms a subspace, but the set of functions such that $f(0) = 1$ does not. For which values of a, b does the set of functions such that $f(a) = b$ form a subspace?

2.2.15. Determine which of the following conditions describe subspaces of the vector space C^1 consisting of all continuously differentiable scalar functions $f(x)$.

(a) $f(2) = f(3)$ (b) $f'(2) = f(3)$

(c) $f'(x) + f(x) = 0$ (d) $f(2-x) = f(x)$

(e) $f(x+2) = f(x) + 2$ (f) $f(-x) = e^x f(x)$

(g) $f(x) = a + b|x|$ for some $a, b \in \mathbb{R}$,

2.2.16. Let $V = C([a, b])$ be the vector space consisting of all functions $f(t)$ which are defined and continuous on the interval $0 \leq t \leq 1$. Which of the following conditions define subspaces of V? Explain your answer.

(a) $f(0) = 0$ (b) $f(0) = 2f(1)$

(c) $f(0)f(1) = 1$

(d) $f(0) = 0$ or $f(1) = 0$

(e) $f(1-t) = -tf(t)$

(f) $f(1-t) = 1 - f(t)$

(g) $f\left(\frac{1}{2}\right) = \int_0^1 f(t)\, dt$

(h) $\int_0^1 (t-1) f(t)\, dt = 0$

(i) $\int_0^t f(s) \sin s\, ds = \sin t$.

2.2.17. Prove that the set of solutions to the second order ordinary differential equation $u'' = xu$ forms a vector space.

2.2.18. Show that the set of solutions to $u'' = x + u$ does not form a vector space.

2.2.19. (a) Prove that $C^1([a, b], \mathbb{R}^2)$, which is the space of continuously differentiable parametrized plane curves $\mathbf{f}: [a, b] \to \mathbb{R}^2$, forms a vector space.

(b) Is the subset consisting of all curves that go through the origin a subspace?

2.2.20. A *planar vector field* $\mathbf{v}(x, y) = (u(x, y), v(x, y))^T$ is called *irrotational* if it has zero divergence:

$$\nabla \cdot \mathbf{v} = \frac{\partial u}{\partial x} + \frac{\partial v}{\partial y} \equiv 0.$$

Prove that the set of all irrotational vector fields forms a subspace of the space of all vector fields.

2.2.21. Let $C \subset \mathbb{R}^\infty$ denote the set of all convergent sequences of real numbers, where $\mathbb{R}^\infty$ was defined in Exercise 2.2.21. Is C a subspace?

2.2.22. Show that if W and Z are subspaces of V, then

(a) their intersection $W \cap Z$ is a subspace of V,

(b) their sum $W + Z = \{\, \mathbf{v} + \mathbf{z} \mid \mathbf{w} \in W, \mathbf{z} \in Z \,\}$ is also a subspace, but

(c) their union $W \cup Z$ is not a subspace of V, unless $W \subset Z$ or $Z \subset W$.

◇ **2.2.23.** Let V be a vector space. Prove that the intersection $\bigcap W_i$ of any collection (finite or infinite) of subspaces $W_i \subset V$ is a subspace.

♡ **2.2.24.** Let $W \subset V$ be a subspace. A subspace $Z \subset V$ is called a *complementary subspace* to W if

(i) $W \cap Z = \{\mathbf{0}\}$, and

(ii) $W + Z = V$, i.e., every $\mathbf{v} \in V$ can be written as $\mathbf{v} = \mathbf{w} + \mathbf{z}$ for $\mathbf{w} \in W$ and $\mathbf{z} \in Z$.

(a) Show that the x and y axes are complementary subspaces of $\mathbb{R}^2$.

(b) Show that the lines $x = y$ and $x = 3y$ are complementary subspaces of $\mathbb{R}^2$.

(c) Show that the line $(a, 2a, 3a)^T$ and the plane $x + 2y + 3z = 0$ are complementary subspaces of $\mathbb{R}^3$.

(d) Prove that if $\mathbf{v} = \mathbf{w} + \mathbf{z}$ then the summands $\mathbf{w} \in W$ and $\mathbf{z} \in Z$ are uniquely determined.

2.2.25. (a) Show that $V_0 = \{\, (\mathbf{v}, \mathbf{0}) \mid \mathbf{v} \in V \,\}$ and $W_0 = \{\, (\mathbf{0}, \mathbf{w}) \mid \mathbf{w} \in W \,\}$ are complementary subspaces, as in Exercise 2.2.24, of the Cartesian product space $V \times W$, as defined in Exercise 2.1.13.

(b) Prove that the *diagonal* $D = \{(\mathbf{v}, \mathbf{v})\}$ and the *anti-diagonal* $A = \{(\mathbf{v}, -\mathbf{v})\}$ are complementary subspaces of $V \times V$.

2.2.26. (a) Show that the set of even functions, $f(-x) = f(x)$ is a subspace of the vector space of all functions $\mathcal{F}(\mathbb{R})$.

(b) Show that the set of odd functions $g(-x) = -g(x)$ forms a complementary subspace, as defined in Exercise 2.2.24.

(c) Explain why every function can be uniquely written as the sum of an even function and an odd function.

2.2.27. Show that the subspace of skew-symmetric $n \times n$ matrices forms a complementary subspace to the space of symmetric $n \times n$ matrices. Explain why this implies that every square matrix can be uniquely written as the sum of a symmetric and a skew-symmetric matrix.

2.2.28. Define $f(x) = \begin{cases} e^{-1/x}, & x > 0, \\ 0, & x \le 0. \end{cases}$

(a) Prove that all derivatives of f vanish at the origin: $f^{(n)}(0) = 0$ for $n = 0, 1, 2, \ldots$.

(b) Prove that $f(x)$ is not analytic by showing that its Taylor series at $a = 0$ does not converge to $f(x)$ when $x > 0$.

2.2.29. Let $f(x) = \dfrac{1}{1+x^2}$.

(a) Find the Taylor series of f at $a = 0$.

(b) Prove that the Taylor series converges for $|x| < 1$, but diverges for $|x| \ge 1$.

(c) Prove that $f(x)$ is analytic at $x = 0$.

♡ **2.2.30.** Let V be a vector space. A subset of the form $A = \{\mathbf{w} + \mathbf{a} \mid \mathbf{w} \in W\}$ where $W \subset V$ is a subspace and $\mathbf{a} \in V$ is a fixed vector is known as an *affine subspace* of V.

(a) Show that the affine subspace $A \subset V$ is a genuine subspace if and only if $\mathbf{a} \in W$.

(b) Draw the affine subspaces $A \subset \mathbb{R}^2$ when

(i) W is the x-axis and $\mathbf{a} = (2, 1)^T$

(ii) W is the line $y = \frac{3}{2}x$ and $\mathbf{a} = (1, 1)^T$,

(iii) W is the line $\{(t, -t)^T \mid t \in \mathbb{R}\}$, and $\mathbf{a} = (2, -2)^T$.

(c) Prove that every affine subspace $A \subset \mathbb{R}^2$ is either a point, a line, or all of $\mathbb{R}^2$.

(d) Show that the plane $x - 2y + 3z = 1$ is an affine subspace of $\mathbb{R}^3$.

(e) Show that the set of all polynomials such that $p(0) = 1$ is an affine subspace of $\mathcal{P}^{(n)}$.

2.3 SPAN AND LINEAR INDEPENDENCE

The definition of the span of a collection of elements of a vector space generalizes, in a natural fashion, the geometric notion of two vectors spanning a plane in $\mathbb{R}^3$. As such, it forms the first of two universal methods for constructing subspaces of vector spaces.

Definition 2.13 Let $\mathbf{v}_1, \ldots, \mathbf{v}_k$ belong to a vector space V. A sum of the form

$$c_1\mathbf{v}_1 + c_2\mathbf{v}_2 + \cdots + c_k\mathbf{v}_k = \sum_{i=1}^{k} c_i\mathbf{v}_i, \tag{2.5}$$

where the coefficients $c_1, c_2, \ldots, c_k$ are any scalars, is known as a *linear combination* of the elements $\mathbf{v}_1, \ldots, \mathbf{v}_k$. Their *span* is the subset $W = \operatorname{span}\{\mathbf{v}_1, \ldots, \mathbf{v}_k\} \subset V$ consisting of all possible linear combinations.

For instance, $3\mathbf{v}_1 + \mathbf{v}_2 - 2\mathbf{v}_3$, $8\mathbf{v}_1 - \frac{1}{3}\mathbf{v}_3 = 8\mathbf{v}_1 + 0\mathbf{v}_2 - \frac{1}{3}\mathbf{v}_3$, $\mathbf{v}_2 = 0\mathbf{v}_1 + 1\mathbf{v}_2 + 0\mathbf{v}_3$, and $\mathbf{0} = 0\mathbf{v}_1 + 0\mathbf{v}_2 + 0\mathbf{v}_3$ are four different linear combinations of the three vector space elements $\mathbf{v}_1, \mathbf{v}_2, \mathbf{v}_3 \in V$. The key observation is that the span always forms a subspace.

Proposition 2.14 The span $W = \operatorname{span}\{\mathbf{v}_1, \ldots, \mathbf{v}_k\}$ of any finite collection of vector space elements forms a subspace of the underlying vector space.

Proof We need to show that if

$$\mathbf{v} = c_1\mathbf{v}_1 + \cdots + c_k\mathbf{v}_k \quad \text{and} \quad \widehat{\mathbf{v}} = \widehat{c}_1\mathbf{v}_1 + \cdots + \widehat{c}_k\mathbf{v}_k$$

are any two linear combinations, then their sum

$$\mathbf{v} + \widehat{\mathbf{v}} = (c_1 + \widehat{c}_1)\mathbf{v}_1 + \cdots + (c_k + \widehat{c}_k)\mathbf{v}_k$$

is also a linear combination, as is any scalar multiple

$$a\mathbf{v} = (ac_1)\mathbf{v}_1 + \cdots + (ac_k)\mathbf{v}_k.$$

Both facts are immediately evident from the formulas. ■

EXAMPLE 2.15 Examples of subspaces spanned by vectors in $\mathbb{R}^3$:

(a) If $\mathbf{v}_1 \neq \mathbf{0}$ is any non-zero vector in $\mathbb{R}^3$, then its span is the line $\{ c\,\mathbf{v}_1 \mid c \in \mathbb{R} \}$ consisting of all vectors *parallel* to $\mathbf{v}_1$. If $\mathbf{v}_1 = \mathbf{0}$, then its span just contains the origin.

(b) If $\mathbf{v}_1$ and $\mathbf{v}_2$ are any two vectors in $\mathbb{R}^3$, then their span is the set of all vectors of the form $c_1\mathbf{v}_1 + c_2\mathbf{v}_2$. Typically, such a span forms a plane passing through the origin. However, if $\mathbf{v}_1$ and $\mathbf{v}_2$ are parallel, then their span is just a line. The most degenerate case is when $\mathbf{v}_1 = \mathbf{v}_2 = \mathbf{0}$, where the span is just a point—the origin.

(c) If we are given three non-coplanar vectors $\mathbf{v}_1, \mathbf{v}_2, \mathbf{v}_3$, then their span is all of $\mathbb{R}^3$, as we shall prove below. However, if they all lie in a plane, then their span is the plane—unless they are all parallel, in which case their span is a line—or, in the completely degenerate situation $\mathbf{v}_1 = \mathbf{v}_2 = \mathbf{v}_3 = \mathbf{0}$, a single point.

Thus, any subspace of $\mathbb{R}^3$ can be realized as the span of some set of vectors. One can consider subspaces spanned by four or more vectors in $\mathbb{R}^3$, but these continue to be limited to being either a point (the origin), a line, a plane, or the entire three-dimensional space. ●

Figure 2.4 Plane and line spanned by two vectors.

A crucial question is to determine when a given vector belongs to the span of a prescribed collection.

EXAMPLE 2.16 Let $W \subset \mathbb{R}^3$ be the plane spanned by the vectors $\mathbf{v}_1 = (1, -2, 1)^T$ and $\mathbf{v}_2 = (2, -3, 1)^T$. Question: Is the vector $\mathbf{v} = (0, 1, -1)^T$ an element of W? To answer, we need to see whether we can find scalars c_1, c_2 such that

$$\mathbf{v} = c_1\,\mathbf{v}_1 + c_2\,\mathbf{v}_2;$$

that is,

$$\begin{pmatrix} 0 \\ 1 \\ -1 \end{pmatrix} = c_1 \begin{pmatrix} 1 \\ -2 \\ 1 \end{pmatrix} + c_2 \begin{pmatrix} 2 \\ -3 \\ -1 \end{pmatrix} = \begin{pmatrix} c_1 + 2c_2 \\ -2c_1 - 3c_2 \\ c_1 - c_2 \end{pmatrix}.$$

Thus, c_1, c_2 must satisfy the linear algebraic system

$$c_1 + 2c_2 = 0, \qquad -2c_1 - 3c_2 = 1, \qquad c_1 - c_2 = -1.$$

Applying Gaussian Elimination, we find the solution $c_1 = -2, c_2 = 1$, and so $\mathbf{v} = -2\mathbf{v}_1 + \mathbf{v}_2$ does belong to the span. On the other hand, $\widetilde{\mathbf{v}} = (1, 0, 0)^T$ does not belong to W. Indeed, there are no scalars c_1, c_2 such that $\widetilde{\mathbf{v}} = c_1\mathbf{v}_1 + c_2\mathbf{v}_2$, because the corresponding linear system is incompatible. ●

Warning: It is entirely possible for different sets of vectors to span the *same* subspace. For instance, $\mathbf{e}_1 = (1, 0, 0)^T$ and $\mathbf{e}_2 = (0, 1, 0)^T$ span the $x\,y$ plane in $\mathbb{R}^3$, as do the three coplanar vectors $\mathbf{v}_1 = (1, -1, 0)^T$, $\mathbf{v}_2 = (-1, 2, 0)^T$, $\mathbf{v}_3 = (2, 1, 0)^T$.

EXAMPLE 2.17

Let $V = \mathcal{F}(\mathbb{R})$ denote the space of all scalar functions $f(x)$.

(a) The span of the three monomials $f_1(x) = 1$, $f_2(x) = x$, and $f_3(x) = x^2$ is the set of all functions of the form

$$f(x) = c_1 f_1(x) + c_2 f_2(x) + c_3 f_3(x) = c_1 + c_2\,x + c_3\,x^2,$$

where c_1, c_2, c_3 are arbitrary scalars (constants). In other words,

$$\operatorname{span}\{1, x, x^2\} = \mathcal{P}^{(2)}$$

is the subspace of all quadratic (degree ≤ 2) polynomials. In a similar fashion, the space $\mathcal{P}^{(n)}$ of polynomials of degree $\le n$ is spanned by the monomials $1, x, x^2, \ldots, x^n$.

(b) The next example plays a key role in many applications. Let $\omega \in \mathbb{R}$ be fixed. Consider the two basic trigonometric functions $f_1(x) = \cos\omega x$, $f_2(x) = \sin\omega x$ of frequency ω, and hence period $2\pi/\omega$. Their span consists of all functions of the form

$$f(x) = c_1 f_1(x) + c_2 f_2(x) = c_1 \cos\omega x + c_2 \sin\omega x. \tag{2.6}$$

For example, the function $\cos(\omega x + 2)$ lies in the span because, by the addition formula for the cosine,

$$\cos(\omega x + 2) = (\cos 2)\cos\omega x - (\sin 2)\sin\omega x$$

is a linear combination of $\cos\omega x$ and $\sin\omega x$.

We can express a general function in the span in the alternative *phase-amplitude form*

$$f(x) = c_1\cos\omega x + c_2\sin\omega x = r\cos(\omega x - \delta), \tag{2.7}$$

in which $r \ge 0$ is known as the *amplitude* and $0 \le \delta < 2\pi$ the *phase shift*. Indeed, expanding the right hand side, we find

$$r\cos(\omega x - \delta) = r\cos\delta\,\cos\omega x + r\sin\delta\,\sin\omega x,$$

and hence

$$c_1 = r\cos\delta, \qquad c_2 = r\sin\delta.$$

Thus, (r, δ) are the polar coordinates of the point $\mathbf{c} = (c_1, c_2) \in \mathbb{R}^2$ prescribed by the coefficients. Thus, any combination of $\sin\omega x$ and $\cos\omega x$ can be rewritten as a single cosine, with a phase shift. Figure 2.5 shows the particular function $3\cos(2x - 1)$ which has amplitude $r = 3$, frequency $\omega = 2$, and phase shift $\delta = 1$. The first peak appears at $x = \delta/\omega = \frac{1}{2}$.

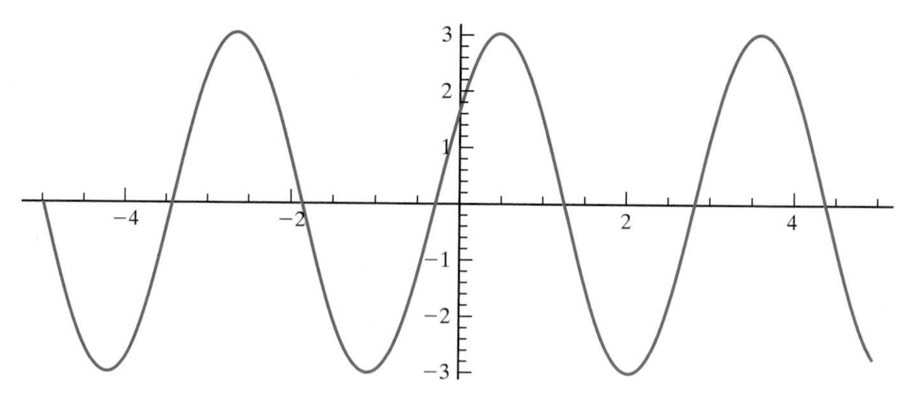

Figure 2.5 Graph of $3\cos(2x-1)$.

(c) The space $\mathcal{T}^{(2)}$ of *quadratic trigonometric polynomials* is spanned by the functions

$$1, \quad \cos x, \quad \sin x, \quad \cos^2 x, \quad \cos x \, \sin x, \quad \sin^2 x.$$

Its general element is a linear combination

$$q(x) = c_0 + c_1 \cos x + c_2 \sin x + c_3 \cos^2 x + c_4 \cos x \, \sin x + c_5 \sin^2 x, \tag{2.8}$$

where $c_0, \ldots, c_5$ are arbitrary constants. A more useful spanning set for the same subspace consists of the trigonometric functions

$$1, \quad \cos x, \quad \sin x, \quad \cos 2x, \quad \sin 2x. \tag{2.9}$$

Indeed, by the double angle formulas, both

$$\cos 2x = \cos^2 x - \sin^2 x, \qquad \sin 2x = 2 \sin x \cos x,$$

have the form of a quadratic trigonometric polynomial (2.8), and hence both belong to $\mathcal{T}^{(2)}$. On the other hand, we can write

$$\cos^2 x = \tfrac{1}{2}\cos 2x + \tfrac{1}{2}, \qquad \cos x \, \sin x = \tfrac{1}{2}\sin 2x, \qquad \sin^2 x = -\tfrac{1}{2}\cos 2x + \tfrac{1}{2},$$

in terms of the functions (2.9). Therefore, the original linear combination (2.8) can be written in the alternative form

$$\begin{aligned} q(x) &= \left(c_0 + \tfrac{1}{2}c_3 + \tfrac{1}{2}c_5\right) + c_1 \cos x + c_2 \sin x \\ &\qquad + \left(\tfrac{1}{2}c_3 - \tfrac{1}{2}c_5\right)\cos 2x + \tfrac{1}{2}c_4 \sin 2x \\ &= \widehat{c}_0 + \widehat{c}_1 \cos x + \widehat{c}_2 \sin x + \widehat{c}_3 \cos 2x + \widehat{c}_4 \sin 2x, \end{aligned} \tag{2.10}$$

and so the functions (2.9) do indeed span $\mathcal{T}^{(2)}$. It is worth noting that we first characterized $\mathcal{T}^{(2)}$ as the span of 6 functions, whereas the second characterization only required 5 functions. It turns out that 5 is the minimal number of functions needed to span $\mathcal{T}^{(2)}$, but the proof of this fact will be deferred until Chapter 5.

(d) The homogeneous linear ordinary differential equation

$$u'' + 2u' - 3u = 0 \tag{2.11}$$

considered in part (i) of Example 2.12 has two solutions: $f_1(x) = e^x$ and $f_2(x) = e^{-3x}$. (Now may be a good time for you to review the basic techniques for solving linear, constant coefficient ordinary differential equations, cf. [6].) Its general solution is, in fact, a linear combination

$$u = c_1 f_1(x) + c_2 f_2(x) = c_1 e^x + c_2 e^{-3x},$$

where c_1, c_2 are arbitrary scalars. Thus, the vector space of solutions to (2.11) is described as the span of these two basic solutions. The fact that there are no other solutions is not obvious, but relies on the basic uniqueness theorem for ordinary differential equations; further details can be found in Theorem 7.34. ●

REMARK: One can also define the span of an infinite collection of elements of a vector space. To avoid convergence issues, one should only consider finite linear combinations (2.5). For example, the span of the monomials $1, x, x^2, x^3, \ldots$ is the subspace $\mathcal{P}^{(\infty)}$ of all polynomials—*not* the space of analytic functions or convergent Taylor series. Similarly, the span of the functions $1, \cos x, \sin x, \cos 2x, \sin 2x, \cos 3x, \sin 3x, \ldots$ is the space of all *trigonometric polynomials*, of fundamental importance in the theory of Fourier series, [16].

EXERCISES 2.3

2.3.1. Show that $\begin{pmatrix} -1 \\ 2 \\ 3 \end{pmatrix}$ belongs to the subspace of $\mathbb{R}^3$ spanned by $\begin{pmatrix} 2 \\ -1 \\ 2 \end{pmatrix}, \begin{pmatrix} 5 \\ -4 \\ 1 \end{pmatrix}$ by writing it as a linear combination of the spanning vectors.

2.3.2. Show that $\begin{pmatrix} -3 \\ 7 \\ 6 \\ 1 \end{pmatrix}$ is in the subspace of $\mathbb{R}^4$ spanned by $\begin{pmatrix} 1 \\ -3 \\ -2 \\ 0 \end{pmatrix}, \begin{pmatrix} -2 \\ 6 \\ 3 \\ 4 \end{pmatrix}$ and $\begin{pmatrix} -2 \\ 4 \\ 6 \\ -7 \end{pmatrix}$.

2.3.3. (a) Determine whether $\begin{pmatrix} 1 \\ -2 \\ -3 \end{pmatrix}$ is in the span of $\begin{pmatrix} 1 \\ 1 \\ 0 \end{pmatrix}$ and $\begin{pmatrix} 0 \\ 1 \\ 1 \end{pmatrix}$.

(b) Is $\begin{pmatrix} 1 \\ -2 \\ -1 \end{pmatrix}$ in the span of $\begin{pmatrix} 1 \\ 2 \\ 2 \end{pmatrix}, \begin{pmatrix} 1 \\ -2 \\ 0 \end{pmatrix}$ and $\begin{pmatrix} 0 \\ 3 \\ 4 \end{pmatrix}$?

(c) Is $\begin{pmatrix} 3 \\ 0 \\ -1 \\ -2 \end{pmatrix}$ in the span of $\begin{pmatrix} 1 \\ 2 \\ 0 \\ 1 \end{pmatrix}, \begin{pmatrix} 0 \\ -1 \\ 3 \\ 0 \end{pmatrix}$ and $\begin{pmatrix} 2 \\ 0 \\ 1 \\ -1 \end{pmatrix}$?

2.3.4. Which of the following sets of vectors span all of $\mathbb{R}^2$?

(a) $\begin{pmatrix} 1 \\ -1 \end{pmatrix}$ (b) $\begin{pmatrix} 2 \\ -1 \end{pmatrix}, \begin{pmatrix} 1 \\ 3 \end{pmatrix}$

(c) $\begin{pmatrix} 2 \\ -1 \end{pmatrix}, \begin{pmatrix} -1 \\ 2 \end{pmatrix}$ (d) $\begin{pmatrix} 6 \\ -9 \end{pmatrix}, \begin{pmatrix} -4 \\ 6 \end{pmatrix}$

(e) $\begin{pmatrix} 1 \\ -1 \end{pmatrix}, \begin{pmatrix} 2 \\ -1 \end{pmatrix}, \begin{pmatrix} 3 \\ -1 \end{pmatrix}$

(f) $\begin{pmatrix} 0 \\ 0 \end{pmatrix}, \begin{pmatrix} 1 \\ -1 \end{pmatrix}, \begin{pmatrix} 2 \\ -2 \end{pmatrix}$

2.3.5. (a) Graph the subspace of $\mathbb{R}^3$ spanned by the vector $\mathbf{v}_1 = (3, 0, 1)^T$.

(b) Graph the subspace spanned by the vectors $\mathbf{v}_1 = (3, -2, -1)^T$, $\mathbf{v}_2 = (-2, 0, -1)^T$.

(c) Graph the span of $\mathbf{v}_1 = (1, 0, -1)^T$, $\mathbf{v}_2 = (0, -1, 1)^T$, $\mathbf{v}_3 = (1, -1, 0)^T$.

2.3.6. Let U be the subspace of $\mathbb{R}^3$ spanned by $\mathbf{u}_1 = (1, 2, 3)^T$, $\mathbf{u}_2 = (2, -1, 0)^T$. Let V be the subspace spanned by $\mathbf{v}_1 = (5, 0, 3)^T$, $\mathbf{v}_2 = (3, 1, 3)^T$. Is V a subspace of U? Are U and V the same?

2.3.7. (a) Let S be the subspace of $\mathcal{M}_{2\times 2}$ consisting of all symmetric 2×2 matrices. Show that S is spanned by the matrices $\begin{pmatrix} 1 & 0 \\ 0 & 0 \end{pmatrix}, \begin{pmatrix} 0 & 0 \\ 0 & 1 \end{pmatrix}$ and $\begin{pmatrix} 0 & 1 \\ 1 & 0 \end{pmatrix}$.

(b) Find a spanning set of the space of symmetric 3×3 matrices.

2.3.8. (a) Determine whether the polynomials $x^2 + 1$, $x^2 - 1$, $x^2 + x + 1$ span $\mathcal{P}^{(2)}$.

(b) Do $x^3 - 1$, $x^2 + 1$, $x - 1$, 1 span $\mathcal{P}^{(3)}$?

(c) What about x^3, $x^2 + 1$, $x^2 - x$, $x + 1$?

2.3.9. Which of the following functions lies in the subspace spanned by the functions 1, x, $\sin x$ and $\sin^2 x$?

(a) $3 - 5x$ (b) $x^2 + \sin^2 x$

(c) $\sin x - 2\cos x$ (d) $\cos^2 x$

(e) $x \sin x$ (f) e^x

2.3.10. Write the following trigonometric functions in phase-amplitude form:

(a) $\sin 3x$, (b) $\cos x - \sin x$

(c) $3\cos 2x + 4\sin 2x$ (d) $\cos x \sin x$

2.3.11. (a) Prove that the set of solutions to the homogeneous ordinary differential equation $u'' - 4u' + 3u = 0$ forms a vector space.

(b) Write the solution space as the span of a finite number of functions.

(c) What is the minimal number of functions needed to span the solution space?

2.3.12. Explain why the functions $1, \cos x, \sin x$ span the solution space to the third order ordinary differential equation $u''' + u' = 0$.

2.3.13. Find a finite set of real functions that spans the solution space to the following homogeneous ordinary differential equations:

(a) $u' - 2u = 0$ (b) $u'' + 4u = 0$

(c) $u'' - 3u' = 0$ (d) $u'' + 4u' + 3u = 0$

(e) $u'' + u' + u = 0$ (f) $u''' - 5u'' = 0$

(g) $u^{(4)} + u = 0$

2.3.14. Consider the boundary value problem $u'' + 4u = 0$, $0 \le x \le \pi$, $u(0) = 0$, $u(\pi) = 0$.

(a) Prove, without solving, that the set of solutions forms a vector space.

(b) Write this space as the span of one or more functions. *Hint*: First solve the differential equation; then find out which solutions satisfy the boundary conditions.

2.3.15. Which of the following functions lie in the span of the vector-valued functions

$$\mathbf{f}_1(x) = \begin{pmatrix} 1 \\ x \end{pmatrix}, \quad \mathbf{f}_2(x) = \begin{pmatrix} x \\ 1 \end{pmatrix}, \quad \mathbf{f}_3(x) = \begin{pmatrix} x \\ 2x \end{pmatrix}?$$

(a) $\begin{pmatrix} 2 \\ 1 \end{pmatrix}$ (b) $\begin{pmatrix} 1-2x \\ 1-x \end{pmatrix}$

(c) $\begin{pmatrix} 1-2x \\ -1-x \end{pmatrix}$ (d) $\begin{pmatrix} 1+x^2 \\ 1-x^2 \end{pmatrix}$

(e) $\begin{pmatrix} 2-x \\ 0 \end{pmatrix}$

2.3.16. *True or false*: The zero vector belongs to the span of any collection of vectors.

2.3.17. Prove or give a counter-example: if $\mathbf{z}$ is a linear combination of $\mathbf{u}, \mathbf{v}, \mathbf{w}$, then $\mathbf{w}$ is a linear combination of $\mathbf{u}, \mathbf{v}, \mathbf{z}$.

◇ **2.3.18.** Suppose $\mathbf{v}_1, \ldots, \mathbf{v}_m$ span V. Let $\mathbf{v}_{m+1}, \ldots, \mathbf{v}_n \in V$ be any other elements. Prove that the combined collection $\mathbf{v}_1, \ldots, \mathbf{v}_n$ also spans V.

◇ **2.3.19.** (a) Show that if $\mathbf{v}$ is a linear combination of $\mathbf{v}_1, \ldots, \mathbf{v}_m$, and each $\mathbf{v}_j$ is a linear combination of $\mathbf{w}_1, \ldots, \mathbf{w}_n$, then $\mathbf{v}$ is a linear combination of $\mathbf{w}_1, \ldots, \mathbf{w}_n$.

(b) Suppose $\mathbf{v}_1, \ldots, \mathbf{v}_m$ span V. Let $\mathbf{w}_1, \ldots, \mathbf{w}_n \in V$ be any other elements. Suppose that each $\mathbf{v}_i$ can be written as a linear combination of $\mathbf{w}_1, \ldots, \mathbf{w}_n$. Prove that $\mathbf{w}_1, \ldots, \mathbf{w}_n$ also span V.

2.3.20. The span of an infinite collection $\mathbf{v}_1, \mathbf{v}_2, \mathbf{v}_3, \ldots \in V$ of vector space elements is defined as the set of all *finite* linear combinations $\sum_{i=1}^{n} c_i \mathbf{v}_i$, where $n < \infty$ is finite but arbitrary.

(a) Prove that the span defines a subspace of the vector space V.

(b) What is the span of the monomials $1, x, x^2, x^3, \ldots$?

Linear Independence and Dependence

Most of the time, all of the vectors used to form a span are essential. For example, we cannot use fewer than two vectors to span a plane in $\mathbb{R}^3$ since the span of a single vector is at most a line. However, in degenerate situations, some of the spanning elements may be redundant. For instance, if the two vectors are parallel, then their span is a line, but only one of the vectors is really needed to prescribe the line. Similarly, the subspace spanned by the polynomials $p_1(x) = x - 2$, $p_2(x) = 3x + 4$, $p_3(x) = -x + 1$, is the vector space $\mathcal{P}^{(1)}$ consisting of all linear polynomials. But only two of the polynomials are really required to span $\mathcal{P}^{(1)}$. (The reason will become clear soon, but you may wish to see if you can demonstrate this on your own.) The elimination of such superfluous spanning elements is encapsulated in the following important definition.

Definition 2.18 The vectors $\mathbf{v}_1, \ldots, \mathbf{v}_k \in V$ are called *linearly dependent* if there exist scalars $c_1, \ldots, c_k$, *not all zero*, such that

$$c_1 \mathbf{v}_1 + \cdots + c_k \mathbf{v}_k = \mathbf{0}. \tag{2.12}$$

Vectors that are not linearly dependent are called *linearly independent.*

The restriction that not all the c_i's are zero is essential: if $c_1 = \cdots = c_k = 0$, then the linear combination (2.12) is automatically zero. To check linear independence, one needs to show that the *only* linear combination that produces the zero vector (2.12) is this trivial one. In other words, $c_1 = \cdots = c_k = 0$ is the *one and only* solution to the vector equation (2.12).

EXAMPLE 2.19 Some examples of linear independence and dependence:

(a) The vectors

$$\mathbf{v}_1 = \begin{pmatrix} 1 \\ 2 \\ -1 \end{pmatrix}, \qquad \mathbf{v}_2 = \begin{pmatrix} 0 \\ 3 \\ 1 \end{pmatrix}, \qquad \mathbf{v}_3 = \begin{pmatrix} -1 \\ 4 \\ 3 \end{pmatrix},$$

are linearly dependent, because

$$\mathbf{v}_1 - 2\mathbf{v}_2 + \mathbf{v}_3 = \mathbf{0}.$$

On the other hand, the first two vectors $\mathbf{v}_1, \mathbf{v}_2$ are linearly independent. To see this, suppose that

$$c_1 \mathbf{v}_1 + c_2 \mathbf{v}_2 = \begin{pmatrix} c_1 \\ 2c_1 + 3c_2 \\ -c_1 + c_2 \end{pmatrix} = \begin{pmatrix} 0 \\ 0 \\ 0 \end{pmatrix}.$$

For this to happen, c_1, c_2 must satisfy the homogeneous linear system

$$c_1 = 0, \qquad 2c_1 + 3c_2 = 0, \qquad -c_1 + c_2 = 0,$$

which, as you can check, has only the trivial solution $c_1 = c_2 = 0$.

(b) In general, any collection $\mathbf{v}_1, \ldots, \mathbf{v}_k$ that includes the zero vector, say $\mathbf{v}_1 = \mathbf{0}$, is automatically linearly dependent, since $1\,\mathbf{0} + 0\,\mathbf{v}_2 + \cdots + 0\,\mathbf{v}_k = \mathbf{0}$ is a nontrivial linear combination that adds up to $\mathbf{0}$.

(c) Two vectors $\mathbf{v}, \mathbf{w} \in V$ are linearly dependent if and only if they are *parallel*, meaning that one is a scalar multiple of the other. Indeed, if $\mathbf{v} = a\,\mathbf{w}$, then $\mathbf{v} - a\,\mathbf{w} = \mathbf{0}$ is a nontrivial linear combination summing to zero. Vice versa, if $c\,\mathbf{v} + d\,\mathbf{w} = \mathbf{0}$ and $c \neq 0$, then $\mathbf{v} = -(d/c)\mathbf{w}$, while if $c = 0$ but $d \neq 0$, then $\mathbf{w} = \mathbf{0}$.

(d) The polynomials

$$p_1(x) = x - 2, \qquad p_2(x) = x^2 - 5x + 4, \qquad p_3(x) = 3x^2 - 4x, \qquad p_4(x) = x^2 - 1$$

are linearly dependent, since

$$p_1(x) + p_2(x) - p_3(x) + 2\,p_4(x) \equiv 0$$

is a nontrivial linear combination that vanishes identically. On the other hand, the first three polynomials,

$$p_1(x) = x - 2, \quad p_2(x) = x^2 - 5x + 4, \quad p_3(x) = 3x^2 - 4x,$$

are linearly independent. Indeed, if the linear combination

$$\begin{aligned} c_1 p_1(x) &+ c_2 p_2(x) + c_3 p_3(x) \\ &= (c_2 + 3c_3)x^2 + (c_1 - 5c_2 - 4c_3)x - 2c_1 + 4c_2 \equiv 0 \end{aligned}$$

is the zero polynomial, then its coefficients must vanish, and hence c_1, c_2, c_3 are required to solve the homogeneous linear system

$$c_2 + 3c_3 = 0, \quad c_1 - 5c_2 - 4c_3 = 0, \quad -2c_1 + 4c_2 = 0.$$

But this has only the trivial solution $c_1 = c_2 = c_3 = 0$, and so linear independence follows. ●

REMARK: In the last example, we are using the basic fact that a polynomial is identically zero,

$$p(x) = a_0 + a_1 x + a_2 x^2 + \cdots + a_n x^n \equiv 0 \quad \text{for all} \quad x,$$

if and only if its coefficients all vanish: $a_0 = a_1 = \cdots = a_n = 0$. This is equivalent to the "obvious" fact that the basic monomial functions $1, x, x^2, \ldots, x^n$ are linearly independent. Exercise 2.3.36 asks for a *bona fide* proof.

EXAMPLE 2.20 The set of quadratic trigonometric functions

$$1, \quad \cos x, \quad \sin x, \quad \cos^2 x, \quad \cos x \, \sin x, \quad \sin^2 x,$$

that were used to define the vector space $\mathcal{T}^{(2)}$ of quadratic trigonometric polynomials, are, in fact, linearly dependent. This is a consequence of the basic trigonometric identity

$$\cos^2 x + \sin^2 x \equiv 1,$$

which can be rewritten as a nontrivial linear combination

$$1 + 0 \cos x + 0 \sin x + (-1) \cos^2 x + 0 \cos x \, \sin x + (-1) \sin^2 x \equiv 0$$

that equals the zero function. On the other hand, the alternative spanning set

$$1, \quad \cos x, \quad \sin x, \quad \cos 2x, \quad \sin 2x$$

is linearly independent, since the only identically zero linear combination,

$$c_0 + c_1 \cos x + c_2 \sin x + c_3 \cos 2x + c_4 \sin 2x \equiv 0,$$

turns out to be the trivial one $c_0 = \cdots = c_4 = 0$. However, the latter fact is not as obvious, and requires a bit of work to prove directly; see Exercise 2.3.37. An easier proof, based on orthogonality, will appear in Chapter 5. ●

Let us now focus our attention on the linear independence or dependence of a set of vectors $\mathbf{v}_1, \ldots, \mathbf{v}_k \in \mathbb{R}^n$ in Euclidean space. We begin by forming the $n \times k$ matrix $A = (\mathbf{v}_1 \ \ldots \ \mathbf{v}_k)$ whose *columns* are the given vectors. (The fact that we use column vectors is essential here.) Our analysis is based on the very useful formula

$$A\mathbf{c} = c_1 \mathbf{v}_1 + \cdots + c_k \mathbf{v}_k, \quad \text{where} \quad \mathbf{c} = \begin{pmatrix} c_1 \\ c_2 \\ \vdots \\ c_k \end{pmatrix}, \tag{2.13}$$

that expresses any linear combination in terms of matrix multiplication. For example,

$$\begin{pmatrix} 1 & 3 & 0 \\ -1 & 2 & 1 \\ 4 & -1 & -2 \end{pmatrix} \begin{pmatrix} c_1 \\ c_2 \\ c_3 \end{pmatrix} = \begin{pmatrix} c_1 + 3c_2 \\ -c_1 + 2c_2 + c_3 \\ 4c_1 - c_2 - 2c_3 \end{pmatrix}$$
$$= c_1 \begin{pmatrix} 1 \\ -1 \\ 4 \end{pmatrix} + c_2 \begin{pmatrix} 3 \\ 2 \\ -1 \end{pmatrix} + c_3 \begin{pmatrix} 0 \\ 1 \\ -2 \end{pmatrix}.$$

Formula (2.13) follows directly from the rules of matrix multiplication; see also Exercise 1.2.34(c). It enables us to reformulate the notions of linear independence and span of vectors in $\mathbb{R}^n$ in terms of linear algebraic systems. The key result is the following:

THEOREM 2.21 Let $\mathbf{v}_1, \ldots, \mathbf{v}_k \in \mathbb{R}^n$ and let $A = (\,\mathbf{v}_1 \ \ldots \ \mathbf{v}_k\,)$ be the corresponding $n \times k$ matrix.

(a) The vectors $\mathbf{v}_1, \ldots, \mathbf{v}_k \in \mathbb{R}^n$ are linearly dependent if and only if there is a non-zero solution $\mathbf{c} \neq \mathbf{0}$ to the homogeneous linear system $A\mathbf{c} = \mathbf{0}$.

(b) The vectors are linearly independent if and only if the only solution to the homogeneous system $A\mathbf{c} = \mathbf{0}$ is the trivial one, $\mathbf{c} = \mathbf{0}$.

(c) A vector $\mathbf{b}$ lies in the span of $\mathbf{v}_1, \ldots, \mathbf{v}_k$ if and only if the linear system $A\mathbf{c} = \mathbf{b}$ is compatible, i.e., has at least one solution.

Proof We prove the first statement, leaving the other two as exercises for the reader. The condition that $\mathbf{v}_1, \ldots, \mathbf{v}_k$ be linearly dependent is that there is a nonzero vector

$$\mathbf{c} = \left(c_1, \ldots, c_k\right)^T \neq \mathbf{0} \quad \text{such that} \quad A\,\mathbf{c} = c_1\,\mathbf{v}_1 + \cdots + c_k\,\mathbf{v}_k = \mathbf{0}.$$

Therefore, linear dependence requires the existence of a nontrivial solution to the homogeneous linear system $A\mathbf{c} = \mathbf{0}$. ■

EXAMPLE 2.22 Let us determine whether the vectors

$$\mathbf{v}_1 = \begin{pmatrix} 1 \\ 2 \\ -1 \end{pmatrix}, \quad \mathbf{v}_2 = \begin{pmatrix} 3 \\ 0 \\ 4 \end{pmatrix}, \quad \mathbf{v}_3 = \begin{pmatrix} 1 \\ -4 \\ 6 \end{pmatrix}, \quad \mathbf{v}_4 = \begin{pmatrix} 4 \\ 2 \\ 3 \end{pmatrix} \tag{2.14}$$

are linearly independent or linearly dependent. We combine them as column vectors into a single matrix

$$A = \begin{pmatrix} 1 & 3 & 1 & 4 \\ 2 & 0 & -4 & 2 \\ -1 & 4 & 6 & 3 \end{pmatrix}.$$

According to Theorem 2.21, we need to figure out whether there are any nontrivial solutions to the homogeneous equation $A\mathbf{c} = \mathbf{0}$; this can be done by reducing A to row echelon form

$$U = \begin{pmatrix} 1 & 3 & 1 & 4 \\ 0 & -6 & -6 & -6 \\ 0 & 0 & 0 & 0 \end{pmatrix}. \tag{2.15}$$

The general solution to the homogeneous system $A\mathbf{c} = \mathbf{0}$ is

$$\mathbf{c} = (2c_3 - c_4, -c_3 - c_4, c_3, c_4)^T,$$

where c_3, c_4—the free variables—are arbitrary. Any nonzero choice of c_3, c_4 will produce a nontrivial linear combination

$$(2c_3 - c_4)\mathbf{v}_1 + (-c_3 - c_4)\mathbf{v}_2 + c_3\mathbf{v}_3 + c_4\mathbf{v}_4 = \mathbf{0}$$

that adds up to the zero vector. We conclude that the vectors (2.14) are linearly dependent. ●

In fact, in this particular case, we didn't even need to complete the row reduction if we only need to check linear (in)dependence. According to Theorem 1.47, any coefficient matrix with more columns than rows automatically has a nontrivial solution to the associated homogeneous system. This implies:

Lemma 2.23 Any collection of $k > n$ vectors in $\mathbb{R}^n$ is linearly dependent.

Warning: The converse to this lemma is *not* true. For example, $\mathbf{v}_1 = (1, 2, 3)^T$ and $\mathbf{v}_2 = (-2, -4, -6)^T$ are two linearly dependent vectors in $\mathbb{R}^3$ since $2\mathbf{v}_1 + \mathbf{v}_2 = \mathbf{0}$. For a collection of n or fewer vectors in $\mathbb{R}^n$, one does need to analyze the homogeneous linear system.

Lemma 2.23 is a particular case of the following general characterization of linearly independent vectors.

Proposition 2.24 A set of k vectors in $\mathbb{R}^n$ is linearly independent if and only if the corresponding $n \times k$ matrix A has rank k. In particular, this requires $k \leq n$.

Or, to state the result another way, the vectors are linearly independent if and only if the homogeneous linear system $A\mathbf{c} = \mathbf{0}$ has no free variables. Proposition 2.24 is an immediate corollary of Theorems 2.21 and 1.47.

EXAMPLE 2.22 ***Continued***

Let us now see which vectors $\mathbf{b} \in \mathbb{R}^3$ lie in the span of the vectors (2.14). According to Theorem 2.21, this will be the case if and only if the linear system $A\mathbf{c} = \mathbf{b}$ has a solution. Since the resulting row echelon form (2.15) has a row of all zeros, there will be a compatibility condition on the entries of $\mathbf{b}$, and hence not every vector lies in the span. To find the precise condition, we augment the coefficient matrix, and apply the same row operations, leading to the reduced augmented matrix

$$\left(\begin{array}{cccc|c} 1 & 3 & 1 & 4 & b_1 \\ 0 & -6 & -6 & -6 & b_2 - 2b_1 \\ 0 & 0 & 0 & 0 & b_3 + \frac{7}{6}b_2 - \frac{4}{3}b_1 \end{array}\right).$$

Therefore, $\mathbf{b} = (b_1, b_2, b_3)^T$ lies in the span if and only if $-\frac{4}{3}b_1 + \frac{7}{6}b_2 + b_3 = 0$. Thus, these four vectors only span a plane in $\mathbb{R}^3$. ●

The same method demonstrates that a collection of vectors will span all of $\mathbb{R}^n$ if and only if the row echelon form of the associated matrix contains no all zero rows, or, equivalently, the rank is equal to n, the number of rows in the matrix.

Proposition 2.25 A collection of k vectors spans $\mathbb{R}^n$ if and only if their $n \times k$ matrix has rank n. In particular, this requires $k \geq n$.

Warning: Not every collection of n or more vectors in $\mathbb{R}^n$ will span all of $\mathbb{R}^n$. A counterexample was already provided by the vectors (2.14).

EXERCISES 2.3

2.3.21. Determine whether the given vectors are linearly independent or linearly dependent:

(a) $\begin{pmatrix}1\\2\end{pmatrix}, \begin{pmatrix}2\\1\end{pmatrix}$ (b) $\begin{pmatrix}1\\3\end{pmatrix}, \begin{pmatrix}-2\\-6\end{pmatrix}$

(c) $\begin{pmatrix}2\\1\end{pmatrix}, \begin{pmatrix}-1\\3\end{pmatrix}, \begin{pmatrix}5\\2\end{pmatrix}$

(d) $\begin{pmatrix}1\\3\\-2\end{pmatrix}, \begin{pmatrix}0\\2\\-1\end{pmatrix}$

(e) $\begin{pmatrix}0\\1\\1\end{pmatrix}, \begin{pmatrix}1\\-1\\0\end{pmatrix}, \begin{pmatrix}3\\-1\\2\end{pmatrix}$

(f) $\begin{pmatrix}2\\1\\3\end{pmatrix}, \begin{pmatrix}1\\-2\\1\end{pmatrix}, \begin{pmatrix}2\\-3\\0\end{pmatrix}, \begin{pmatrix}0\\-1\\4\end{pmatrix}$

(g) $\begin{pmatrix}4\\2\\0\\-6\end{pmatrix}, \begin{pmatrix}-6\\-3\\0\\9\end{pmatrix}$

(h) $\begin{pmatrix}2\\1\\-1\\3\end{pmatrix}, \begin{pmatrix}-1\\3\\1\\0\end{pmatrix}, \begin{pmatrix}5\\1\\2\\-3\end{pmatrix}$

(i) $\begin{pmatrix}1\\0\\1\\0\end{pmatrix}, \begin{pmatrix}1\\0\\0\\1\end{pmatrix}, \begin{pmatrix}2\\2\\1\\0\end{pmatrix}, \begin{pmatrix}1\\2\\3\\4\end{pmatrix}$

2.3.22. (a) Show that the vectors $\begin{pmatrix}1\\0\\2\\1\end{pmatrix}, \begin{pmatrix}-2\\3\\-1\\1\end{pmatrix}, \begin{pmatrix}2\\-2\\1\\-1\end{pmatrix}$ are linearly independent.

(b) Which of the following vectors are in their span?

(i) $\begin{pmatrix}1\\1\\2\\1\end{pmatrix}$ (ii) $\begin{pmatrix}1\\0\\0\\0\end{pmatrix}$

(iii) $\begin{pmatrix}0\\1\\0\\0\end{pmatrix}$ (iv) $\begin{pmatrix}0\\0\\0\\0\end{pmatrix}$

(c) Suppose $\mathbf{b} = (a, b, c, d)^T$ lies in their span. What conditions must a, b, c, d satisfy?

2.3.23. (a) Show that the vectors

$$\begin{pmatrix}1\\1\\1\\0\end{pmatrix}, \begin{pmatrix}1\\1\\-1\\0\end{pmatrix}, \begin{pmatrix}1\\-1\\0\\1\end{pmatrix}, \begin{pmatrix}1\\-1\\0\\-1\end{pmatrix}$$

are linearly independent.

(b) Show that they also span $\mathbb{R}^4$.

(c) Write $(1, 0, 0, 1)^T$ as a linear combination of them.

2.3.24. Determine whether the given row vectors are linearly independent or linearly dependent:

(a) $(2, 1), (-1, 3), (5, 2)$

(b) $(1, 2, -1), (2, 4, -2)$

(c) $(1, 2, 3), (1, 4, 8), (1, 5, 7)$

(d) $(1, 1, 0), (1, 0, 3), (2, 2, 1), (1, 3, 4)$

(e) $(1, 2, 0, 3), (-3, -1, 2, -2), (3, -4, -4, 5)$

(f) $(2, 1, -1, 3), (-1, 3, 1, 0), (5, 1, 2, -3)$

2.3.25. *True or false*: The six 3×3 permutation matrices (1.30) are linearly independent.

2.3.26. *True or false*: A set of vectors is linearly dependent if the zero vector belongs to their span.

2.3.27. Does a single vector ever define a linearly dependent set?

2.3.28. Let $\mathbf{x}$ and $\mathbf{y}$ be linearly independent elements of a vector space V. Show that $\mathbf{u} = a\mathbf{x} + b\mathbf{y}$, and $\mathbf{v} = c\mathbf{x} + d\mathbf{y}$ are linearly independent if and only if $ad - bc \neq 0$. Is the entire collection $\mathbf{x}, \mathbf{y}, \mathbf{u}, \mathbf{v}$ linearly independent?

2.3.29. Prove or give a counterexample to the following statement: If $\mathbf{v}_1, \ldots, \mathbf{v}_k$ are elements of a vector space V and do not span V, then $\mathbf{v}_1, \ldots, \mathbf{v}_k$ are linearly independent.

◇ **2.3.30.** Prove parts (b) and (c) of Theorem 2.21.

◇ **2.3.31.** (a) Prove that if $\mathbf{v}_1, \ldots, \mathbf{v}_m$ are linearly independent, then any subset, e.g., $\mathbf{v}_1, \ldots, \mathbf{v}_k$ with $k < m$, is also linearly independent.

(b) Does the same hold true for linearly dependent vectors? Prove or give a counterexample.

2.3.32. (a) Determine whether the polynomials $f_1(x) = x^2 - 3$, $f_2(x) = 2 - x$, $f_3(x) = (x - 1)^2$ are linearly independent or linearly dependent.

(b) Do they span the vector space of all quadratic polynomials?

2.3.33. Determine whether the given functions are linearly independent or linearly dependent:

(a) $2 - x^2, 3x, x^2 + x - 2$

(b) $3x - 1, x(2x + 1), x(x - 1)$
(c) e^x, e^{x+1}
(d) $\sin x, \sin(x + 1)$
(e) e^x, e^{x+1}, e^{x+2}
(f) $\sin x, \sin(x + 1), \sin(x + 2)$
(g) $e^x, x e^x, x^2 e^x$
(h) e^x, e^{2x}, e^{3x}
(i) $x+y, x-y+1, x+3y+2$—these are functions of two variables.

2.3.34. Show that the functions $f(x) = x$ and $g(x) = |x|$ are linearly independent when considered as functions on all of $\mathbb{R}$, but are linearly dependent when considered as functions defined only on $\mathbb{R}^+ = \{x > 0\}$.

♡ **2.3.35.** (a) Prove that the polynomials $p_i(x) = \sum_{j=0}^{n} a_{ij} x^j$, $i = 1, \ldots, k$, are linearly independent if and only if the $k \times (n + 1)$ matrix A whose entries are their coefficients a_{ij}, $1 \le i \le k$, $0 \le j \le n$, has rank k.
(b) Formulate a similar matrix condition for testing whether another polynomial $q(x)$ lies in their span.
(c) Use (a) to determine whether $p_1(x) = x^3 - 1$, $p_2(x) = x^3 - 2x + 4$, $p_3(x) = x^4 - 4x$, $p_4(x) = x^2 + 1$, $p_5(x) = -x^4 + 4x^3 + 2x + 1$ are linearly independent or linearly dependent.
(d) Does the polynomial $q(x) = x^3$ lie in their span? If so produce a linear combination that adds up to $q(x)$.

2.3.36. The Fundamental Theorem of Algebra, [22], states that a non-zero polynomial of degree n has at most n distinct real roots $p(x) = 0$. Use this fact to prove linear independence of the monomial functions $1, x, x^2, \ldots, x^n$. *Remark*: An elementary proof of this fact can be found in Exercise 4.4.28.

♡ **2.3.37.** Let $x_1, x_2, \ldots, x_n$ be a distinct set of sample points.
(a) Prove that the functions $f_1(x), \ldots, f_k(x)$ are linearly independent if their sample vectors $\mathbf{f}_1, \ldots, \mathbf{f}_k$ are linearly independent vectors in $\mathbb{R}^n$.
(b) Give an example of linearly independent functions that have linearly dependent sample vectors.
(c) Use this method to prove that the functions 1, $\cos x$, $\sin x$, $\cos 2x$, $\sin 2x$, are linearly independent. *Hint*: You will need at least 5 sample points.

2.3.38. Suppose $\mathbf{f}_1(t), \ldots, \mathbf{f}_k(t)$ are vector-valued functions from $\mathbb{R}$ to $\mathbb{R}^n$.
(a) Prove that if $\mathbf{f}_1(t_0), \ldots, \mathbf{f}_k(t_0)$ are linearly independent vectors in $\mathbb{R}^n$ at one fixed t_0, then $\mathbf{f}_1(t), \ldots, \mathbf{f}_k(t)$ are linearly independent functions.
(b) Show that

$$\mathbf{f}_1(t) = \begin{pmatrix} 1 \\ t \end{pmatrix} \quad \text{and} \quad \mathbf{f}_2(t) = \begin{pmatrix} 2t - 1 \\ 2t^2 - t \end{pmatrix}$$

are linearly independent functions, even though at each fixed t_0, the vectors $\mathbf{f}_1(t_0), \mathbf{f}_2(t_0)$ are linearly dependent. Therefore, the converse to the result in part (a) is not valid.

♡ **2.3.39.** The *Wronskian* of a pair of differentiable functions $f(x), g(x)$ is the scalar function

$$W[f(x), g(x)] = \det \begin{pmatrix} f(x) & g(x) \\ f'(x) & g'(x) \end{pmatrix} = f(x)g'(x) - f'(x)\,g(x). \tag{2.16}$$

(a) Prove that if f, g are linearly dependent, then $W[f(x), g(x)] \equiv 0$.
(b) Prove that if $W[f(x), g(x) \not\equiv 0$, then f, g are linearly independent.
(c) Let $f(x) = x^3$, $g(x) = |x|^3$. Prove that $f, g \in \mathrm{C}^2$ are twice continuously differentiable and linearly independent, but $W[f(x), g(x)] \equiv 0$. Thus, the Wronskian is *not* a fool-proof test for linear independence.
Remark: It can be proved, [6], that if f, g both satisfy a second order linear ordinary differential equation, then f, g are linearly dependent if and only if $W[f(x), g(x)] \equiv 0$.

2.4 BASES AND DIMENSION

In order to span a vector space or subspace, we must use a sufficient number of distinct elements. On the other hand, including too many elements in the spanning set will violate linear independence, and cause redundancies. The optimal spanning sets are those that are also linearly independent. By combining the properties of span and linear independence, we arrive at the all-important concept of a "basis".

Definition 2.26 A *basis* of a vector space V is a finite collection of elements $\mathbf{v}_1, \ldots, \mathbf{v}_n \in V$ that

(a) spans V, and

(b) is linearly independent.

Bases are absolutely fundamental in all areas of linear algebra and linear analysis, including matrix algebra, Euclidean geometry, statistical analysis, solutions to linear differential equations, both ordinary and partial, linear boundary value problems, Fourier analysis, signal and image processing, data compression, control systems, and so on.

EXAMPLE 2.27 The *standard basis* of $\mathbb{R}^n$ consists of the n vectors

$$\mathbf{e}_1 = \begin{pmatrix} 1 \\ 0 \\ 0 \\ \vdots \\ 0 \\ 0 \end{pmatrix}, \quad \mathbf{e}_2 = \begin{pmatrix} 0 \\ 1 \\ 0 \\ \vdots \\ 0 \\ 0 \end{pmatrix}, \quad \ldots \quad \mathbf{e}_n = \begin{pmatrix} 0 \\ 0 \\ 0 \\ \vdots \\ 0 \\ 1 \end{pmatrix}, \tag{2.17}$$

so that $\mathbf{e}_i$ is the vector with 1 in the ith slot and 0's elsewhere. We already encountered these vectors (1.42)—they are the columns of the $n \times n$ identity matrix. They clearly span $\mathbb{R}^n$, since we can write any vector

$$\mathbf{x} = \begin{pmatrix} x_1 \\ x_2 \\ \vdots \\ x_n \end{pmatrix} = x_1 \mathbf{e}_1 + x_2 \mathbf{e}_2 + \cdots + x_n \mathbf{e}_n \tag{2.18}$$

as a linear combination, whose coefficients are its entries. Moreover, the only linear combination that yields the zero vector $\mathbf{x} = \mathbf{0}$ is the trivial one $x_1 = \cdots = x_n = 0$, which shows that $\mathbf{e}_1, \ldots, \mathbf{e}_n$ are linearly independent. ●

In the three-dimensional case $\mathbb{R}^3$, a common physical notation for the standard basis is

$$\mathbf{i} = \mathbf{e}_1 = \begin{pmatrix} 1 \\ 0 \\ 0 \end{pmatrix}, \quad \mathbf{j} = \mathbf{e}_2 = \begin{pmatrix} 0 \\ 1 \\ 0 \end{pmatrix}, \quad \mathbf{k} = \mathbf{e}_3 = \begin{pmatrix} 0 \\ 0 \\ 1 \end{pmatrix}. \tag{2.19}$$

This is but one of many possible bases for $\mathbb{R}^3$. Indeed, any three non-coplanar vectors can be used to form a basis. This is a consequence of the following general characterization of bases in Euclidean space.

THEOREM 2.28 Every basis of $\mathbb{R}^n$ consists of exactly n vectors. Furthermore, a set of n vectors $\mathbf{v}_1, \ldots, \mathbf{v}_n \in \mathbb{R}^n$ is a basis if and only if the $n \times n$ matrix $A = (\mathbf{v}_1 \ \ldots \ \mathbf{v}_n)$ is nonsingular: $\operatorname{rank} A = n$.

Proof This is a direct consequence of Theorem 2.21. Linear independence requires that the only solution to the homogeneous system $A\mathbf{c} = \mathbf{0}$ is the trivial one $\mathbf{c} = \mathbf{0}$. On the other hand, a vector $\mathbf{b} \in \mathbb{R}^n$ will lie in the span of $\mathbf{v}_1, \ldots, \mathbf{v}_n$ if and only if the linear system $A\mathbf{c} = \mathbf{b}$ has a solution. For $\mathbf{v}_1, \ldots, \mathbf{v}_n$ to span all of $\mathbb{R}^n$, this must hold for all possible right hand sides $\mathbf{b}$. Theorem 1.7 tells us that both results require that A be nonsingular, i.e., have maximal rank n. ■

Thus, every basis of n-dimensional Euclidean space $\mathbb{R}^n$ contains the same number of vectors, namely n. This is a general fact, and motivates a linear algebra characterization of dimension.

THEOREM 2.29 Suppose the vector space V has a basis $\mathbf{v}_1, \ldots, \mathbf{v}_n$. Then every other basis of V has the same number of elements in it. This number is called the *dimension* of V, and written $\dim V = n$.

The proof of Theorem 2.29 rests on the following lemma.

Lemma 2.30 Suppose $\mathbf{v}_1, \ldots, \mathbf{v}_n$ span a vector space V. Then every set of $k > n$ elements $\mathbf{w}_1, \ldots, \mathbf{w}_k \in V$ is linearly dependent.

Proof Let us write each element

$$\mathbf{w}_j = \sum_{i=1}^{n} a_{ij}\,\mathbf{v}_i, \qquad j = 1, \ldots, k,$$

as a linear combination of the spanning set. Then

$$c_1\,\mathbf{w}_1 + \cdots + c_k\,\mathbf{w}_k = \sum_{i=1}^{n} \sum_{j=1}^{k} a_{ij}\,c_j\mathbf{v}_i.$$

This linear combination will be zero whenever $\mathbf{c} = (c_1, c_2, \ldots, c_k)^T$ solves the homogeneous linear system

$$\sum_{j=1}^{k} a_{ij}\,c_j = 0, \qquad i = 1, \ldots, n,$$

consisting of n equations in $k > n$ unknowns. Theorem 1.47 guarantees that every homogeneous system with more unknowns than equations always has a non-trivial solution $\mathbf{c} \neq \mathbf{0}$, and this immediately implies that $\mathbf{w}_1, \ldots, \mathbf{w}_k$ are linearly dependent. ■

Proof of Theorem 2.29 Suppose we have two bases containing a different number of elements. By definition, the smaller basis spans the vector space. But then Lemma 2.30 tell us that the elements in the larger purported basis must be linearly dependent. This contradicts our assumption that the latter is a basis. ■

As a direct consequence, we can now give a precise meaning to the optimality of bases.

THEOREM 2.31 Suppose V is an n-dimensional vector space. Then

(a) Every set of more than n elements of V is linearly dependent.

(b) No set of less than n elements spans V.

(c) A set of n elements forms a basis if and only if it spans V.

(d) A set of n elements forms a basis if and only if it is linearly independent.

In other words, once we know the dimension of a vector space, to check that a collection having the correct number of elements forms a basis, we only need establish one of the two defining properties: span or linear independence. Thus, n elements that span an n-dimensional vector space are automatically linearly independent and hence form a basis; vice versa, n linearly independent elements of n-dimensional vector space automatically span the space and so form a basis.

EXAMPLE 2.32 The standard basis of the space $\mathcal{P}^{(n)}$ of polynomials of degree $\leq n$ is given by the $n+1$ monomials $1, x, x^2, \ldots, x^n$. We conclude that the vector space $\mathcal{P}^{(n)}$ has dimension $n+1$. Any other basis of $\mathcal{P}^{(n)}$ must contain precisely $n+1$ polynomials. But, not every collection of $n+1$ polynomials in $\mathcal{P}^{(n)}$ is a basis—they must be linearly independent. We conclude that no set of n or fewer polynomials can span $\mathcal{P}^{(n)}$, while any collection of $n+2$ or more polynomials of degree $\leq n$ is automatically linearly dependent. ●

By definition, every vector space of dimension $1 \leq n < \infty$ has a basis. If a vector space V has no basis, it is either the trivial vector space $V = \{\mathbf{0}\}$, which by convention has dimension 0, or its dimension is infinite. An infinite-dimensional vector space contains an infinite collection of linearly independent elements, and hence no (finite) basis. Examples of infinite-dimensional vector spaces include most spaces of functions, such as the spaces of continuous, differentiable, or mean zero functions, as well as the space of *all* polynomials, and the space of solutions to a linear homogeneous partial differential equation. (On the other hand, the solution space for a homogeneous linear ordinary differential equation turns out to be a finite-dimensional vector space.) There is a well-developed concept of a "complete basis" of certain infinite-dimensional function spaces, [52, 54], but this requires more delicate analytical considerations that lie beyond our present abilities. Thus, in this book, the term "basis" *always* means a finite collection of vectors in a finite-dimensional vector space.

Proposition 2.33 *If $\mathbf{v}_1, \ldots, \mathbf{v}_n$ span the vector space V, then $\dim V \leq n$.*

Thus, every (nonzero) vector space, spanned by a finite number of elements is necessarily finite-dimensional, and so admits a basis. Indeed, one can find the basis by successively looking at the spanning vectors, and retaining those that cannot be expressed as linear combinations of their predecessors in the list. Therefore, $m = \dim V$ is the maximal number of linearly independent vectors in the set $\mathbf{v}_1, \ldots, \mathbf{v}_n$. The details of the proof are left to the reader; see Exercise 2.4.22.

Lemma 2.34 *The elements $\mathbf{v}_1, \ldots, \mathbf{v}_n$ form a basis of V if and only if every $\mathbf{x} \in V$ can be written uniquely as a linear combination of the basis elements:*

$$\mathbf{x} = c_1 \mathbf{v}_1 + \cdots + c_n \mathbf{v}_n = \sum_{i=1}^{n} c_i \mathbf{v}_i. \tag{2.20}$$

Proof The condition that the basis span V implies that every $\mathbf{x} \in V$ can be written as some linear combination of the basis elements. Suppose we can write an element

$$\mathbf{x} = c_1 \mathbf{v}_1 + \cdots + c_n \mathbf{v}_n = \widehat{c}_1 \mathbf{v}_1 + \cdots + \widehat{c}_n \mathbf{v}_n$$

as two different combinations. Subtracting one from the other, we find

$$(c_1 - \widehat{c}_1) \mathbf{v}_1 + \cdots + (c_n - \widehat{c}_n) \mathbf{v}_n = \mathbf{0}.$$

The basis elements are linearly independent if and only if all $c_i - \widehat{c}_i = 0$, meaning that the linear combinations are one and the same. ■

The coefficients $(c_1, \ldots, c_n)$ in (2.20) are called the *coordinates* of the vector $\mathbf{x}$ with respect to the given basis. For the standard basis (2.17) of $\mathbb{R}^n$, the coordinates of a vector $\mathbf{x} = (x_1, \ldots, x_n)^T$ are its entries—i.e., its usual Cartesian coordinates, cf. (2.18).

EXAMPLE 2.35 *A Wavelet Basis*. The vectors

$$\mathbf{v}_1 = \begin{pmatrix} 1 \\ 1 \\ 1 \\ 1 \end{pmatrix}, \qquad \mathbf{v}_2 = \begin{pmatrix} 1 \\ 1 \\ -1 \\ -1 \end{pmatrix}, \qquad \mathbf{v}_3 = \begin{pmatrix} 1 \\ -1 \\ 0 \\ 0 \end{pmatrix}, \qquad \mathbf{v}_4 = \begin{pmatrix} 0 \\ 0 \\ 1 \\ -1 \end{pmatrix} \tag{2.21}$$

form a basis of $\mathbb{R}^4$. This is verified by performing Gaussian Elimination on the corresponding 4×4 matrix

$$A = \begin{pmatrix} 1 & 1 & 1 & 0 \\ 1 & 1 & -1 & 0 \\ 1 & -1 & 0 & 1 \\ 1 & -1 & 0 & -1 \end{pmatrix}$$

to check that it is nonsingular. This is a very simple example of a *wavelet basis*. Wavelets play an increasingly central role in modern signal and digital image processing, [16, 66].

How do we find the coordinates of a vector, say $\mathbf{x} = \left(4, -2, 1, 5\right)^T$, relative to the wavelet basis? We need to find the coefficients c_1, c_2, c_3, c_4 so that

$$\mathbf{x} = c_1 \mathbf{v}_1 + c_2 \mathbf{v}_2 + c_3 \mathbf{v}_3 + c_4 \mathbf{v}_4.$$

We use (2.13) to rewrite this equation in matrix form $\mathbf{x} = A\,\mathbf{c}$, where

$$\mathbf{c} = (c_1, c_2, c_3, c_4)^T\,.$$

Solving the resulting linear system by Gaussian Elimination produces $c_1 = 2, c_2 = -1, c_3 = 3, c_4 = -2$, which are the coordinates of

$$\mathbf{x} = \begin{pmatrix} 4 \\ -2 \\ 1 \\ 5 \end{pmatrix} = 2\mathbf{v}_1 - \mathbf{v}_2 + 3\mathbf{v}_3 - 2\mathbf{v}_4 = 2\begin{pmatrix} 1 \\ 1 \\ 1 \\ 1 \end{pmatrix} - \begin{pmatrix} 1 \\ 1 \\ -1 \\ -1 \end{pmatrix} + 3\begin{pmatrix} 1 \\ -1 \\ 0 \\ 0 \end{pmatrix} - 2\begin{pmatrix} 0 \\ 0 \\ 1 \\ -1 \end{pmatrix}$$

in the wavelet basis. ●

In general, to find the coordinates of a vector $\mathbf{x}$ with respect to a new basis of $\mathbb{R}^n$ requires the solution of a linear system of equations, namely

$$A\,\mathbf{c} = \mathbf{x} \quad \text{for} \quad \mathbf{c} = A^{-1}\mathbf{x}. \tag{2.22}$$

The columns of $A = (\,\mathbf{v}_1\ \mathbf{v}_2 \ldots \mathbf{v}_n\,)$ are the basis vectors, $\mathbf{x} = (x_1, x_2, \ldots, x_n)^T$ are the Cartesian coordinates of $\mathbf{x}$, with respect to the standard basis $\mathbf{e}_1, \ldots, \mathbf{e}_n$, while $\mathbf{c} = (c_1, c_2, \ldots, c_n)^T$ contains its coordinates with respect to the new basis $\mathbf{v}_1, \ldots, \mathbf{v}_n$. In practice, one finds the coordinates by Gaussian Elimination, *not* matrix inversion.

Why would one want to change bases? The answer is *simplification* and *speed*—many computations and formulas become much easier, and hence faster, to perform in a basis that is adapted to the problem at hand. In signal processing, wavelet bases are particularly appropriate for denoising, compression, and efficient storage of signals, including audio, still images, videos, medical and geophysical images, and so on. These processes would be quite time-consuming—if not impossible in complicated situations like video and three-dimensional image processing—to accomplish in the standard basis. Additional examples will appear throughout the text.

EXERCISES 2.4

2.4.1. Determine which of the following sets of vectors are bases of $\mathbb{R}^2$:

(a) $\begin{pmatrix} 1 \\ -3 \end{pmatrix}, \begin{pmatrix} -2 \\ 5 \end{pmatrix}$ (b) $\begin{pmatrix} 1 \\ -1 \end{pmatrix}, \begin{pmatrix} -1 \\ 1 \end{pmatrix}$

(c) $\begin{pmatrix} 1 \\ 2 \end{pmatrix}, \begin{pmatrix} 2 \\ 1 \end{pmatrix}$ (d) $\begin{pmatrix} 3 \\ 5 \end{pmatrix}, \begin{pmatrix} 0 \\ 0 \end{pmatrix}$

(e) $\begin{pmatrix} 2 \\ 0 \end{pmatrix}, \begin{pmatrix} -1 \\ 2 \end{pmatrix}, \begin{pmatrix} 0 \\ -1 \end{pmatrix}$

2.4.2. Determine which of the following are bases of $\mathbb{R}^3$:

(a) $\begin{pmatrix} 2 \\ 1 \\ 5 \end{pmatrix}, \begin{pmatrix} 1 \\ 5 \\ 2 \end{pmatrix}$

(b) $\begin{pmatrix} 0 \\ 1 \\ -5 \end{pmatrix}, \begin{pmatrix} -1 \\ 3 \\ 0 \end{pmatrix}, \begin{pmatrix} 1 \\ 3 \\ 0 \end{pmatrix}$

(c) $\begin{pmatrix} 0 \\ 4 \\ -1 \end{pmatrix}, \begin{pmatrix} -1 \\ 0 \\ 1 \end{pmatrix}, \begin{pmatrix} 1 \\ -8 \\ 1 \end{pmatrix}$

(d) $\begin{pmatrix} 2 \\ 0 \\ -2 \end{pmatrix}, \begin{pmatrix} -1 \\ 2 \\ -1 \end{pmatrix}, \begin{pmatrix} 0 \\ -1 \\ 0 \end{pmatrix}, \begin{pmatrix} -1 \\ 2 \\ 1 \end{pmatrix}$

2.4.3. Find a basis for

(a) the plane given by the equation $z - 2y = 0$ in $\mathbb{R}^3$;

(b) the plane given by the equation $4x + 3y - z = 0$ in $\mathbb{R}^3$;

(c) the hyperplane $x + 2y + z - w = 0$ in $\mathbb{R}^4$.

2.4.4. Let

$$\mathbf{v}_1 = \begin{pmatrix} 1 \\ 0 \\ 2 \end{pmatrix}, \quad \mathbf{v}_2 = \begin{pmatrix} 3 \\ -1 \\ 1 \end{pmatrix}, \quad \mathbf{v}_3 = \begin{pmatrix} 2 \\ -1 \\ -1 \end{pmatrix}, \quad \mathbf{v}_4 = \begin{pmatrix} 4 \\ -1 \\ 3 \end{pmatrix}.$$

(a) Do $\mathbf{v}_1, \mathbf{v}_2, \mathbf{v}_3, \mathbf{v}_4$ span $\mathbb{R}^3$? Why or why not?

(b) Are $\mathbf{v}_1, \mathbf{v}_2, \mathbf{v}_3, \mathbf{v}_4$ linearly independent? Why or why not?

(c) Do $\mathbf{v}_1, \mathbf{v}_2, \mathbf{v}_3, \mathbf{v}_4$ form a basis for $\mathbb{R}^3$? Why or why not? If not, is it possible to choose some subset which is a basis?

(d) What is the dimension of the span of $\mathbf{v}_1, \mathbf{v}_2, \mathbf{v}_3, \mathbf{v}_4$? Justify your answer.

2.4.5. Answer Exercise 2.4.4 when

$$\mathbf{v}_1 = \begin{pmatrix} 1 \\ -1 \\ 2 \end{pmatrix}, \quad \mathbf{v}_2 = \begin{pmatrix} 2 \\ -2 \\ 5 \end{pmatrix}, \quad \mathbf{v}_3 = \begin{pmatrix} 0 \\ -2 \\ 1 \end{pmatrix}, \quad \mathbf{v}_4 = \begin{pmatrix} 1 \\ 3 \\ -1 \end{pmatrix}.$$

2.4.6. (a) Show that

$$\begin{pmatrix} 4 \\ 0 \\ 1 \end{pmatrix}, \begin{pmatrix} 2 \\ 1 \\ 0 \end{pmatrix} \quad \text{and} \quad \begin{pmatrix} 2 \\ -1 \\ 1 \end{pmatrix}, \begin{pmatrix} 0 \\ 2 \\ -1 \end{pmatrix}$$

are two different bases for the plane $x - 2y - 4z = 0$.

(b) Show how to write both elements of the second basis as linear combinations of the first.

(c) Can you find a third basis?

♡ **2.4.7.** A basis $\mathbf{v}_1, \ldots, \mathbf{v}_n$ of $\mathbb{R}^n$ is called *right handed* if the $n \times n$ matrix $A = \begin{pmatrix} \mathbf{v}_1 & \cdots & \mathbf{v}_n \end{pmatrix}$ whose columns are the basis vectors has positive determinant: $\det A > 0$. If $\det A < 0$, the basis is called *left-handed*.

(a) Which of the following form right handed bases of $\mathbb{R}^3$?

(i) $\begin{pmatrix} 1 \\ 0 \\ 1 \end{pmatrix}, \begin{pmatrix} -1 \\ 1 \\ 1 \end{pmatrix}, \begin{pmatrix} -1 \\ 1 \\ 0 \end{pmatrix}$

(ii) $\begin{pmatrix} 2 \\ 1 \\ 1 \end{pmatrix}, \begin{pmatrix} 1 \\ 2 \\ 1 \end{pmatrix}, \begin{pmatrix} 1 \\ 1 \\ 2 \end{pmatrix}$

(iii) $\begin{pmatrix} -1 \\ 2 \\ 3 \end{pmatrix}, \begin{pmatrix} 1 \\ -2 \\ -2 \end{pmatrix}, \begin{pmatrix} 1 \\ -2 \\ 2 \end{pmatrix}$

(iv) $\begin{pmatrix} 3 \\ 2 \\ 1 \end{pmatrix}, \begin{pmatrix} 1 \\ 2 \\ 3 \end{pmatrix}, \begin{pmatrix} 2 \\ 1 \\ 3 \end{pmatrix}$

(b) Show that if $\mathbf{v}_1, \mathbf{v}_2, \mathbf{v}_3$ is a left handed basis of $\mathbb{R}^3$, then $\mathbf{v}_2, \mathbf{v}_1, \mathbf{v}_3$ and $-\mathbf{v}_1, \mathbf{v}_2, \mathbf{v}_3$ are both right handed bases.

(c) What sort of basis has $\det A = 0$?

2.4.8. Find a basis for and the dimension of the following subspaces:

(a) The space of solutions to the linear system $A\mathbf{x} = \mathbf{0}$, where $A = \begin{pmatrix} 1 & 2 & -1 & 1 \\ 3 & 0 & 2 & -1 \end{pmatrix}$.

(b) The set of all quadratic polynomials $p(x) = a x^2 + b x + c$ that satisfy $p(1) = 0$.

(c) The space of all solutions to the homogeneous ordinary differential equation $u''' - u'' + 4u' - 4u = 0$.

2.4.9. Find a basis for and the dimension of the span of

(a) $\begin{pmatrix} 3 \\ 1 \\ -1 \end{pmatrix}, \begin{pmatrix} -6 \\ -2 \\ 2 \end{pmatrix}$

(b) $\begin{pmatrix} 2 \\ 0 \\ 1 \end{pmatrix}, \begin{pmatrix} 0 \\ -1 \\ 3 \end{pmatrix}, \begin{pmatrix} 2 \\ 1 \\ -2 \end{pmatrix}$

(c) $\begin{pmatrix} 1 \\ 0 \\ -1 \\ 2 \end{pmatrix}, \begin{pmatrix} 0 \\ 1 \\ 1 \\ 3 \end{pmatrix}, \begin{pmatrix} 2 \\ -1 \\ -3 \\ 1 \end{pmatrix}, \begin{pmatrix} 1 \\ -2 \\ 1 \\ 1 \end{pmatrix}$

2.4.10. (a) Prove that $1 + t^2, t + t^2, 1 + 2t + t^2$ is a basis for the space of quadratic polynomials $\mathcal{P}^{(2)}$.

(b) Find the coordinates of $p(t) = 1 + 4t + 7t^2$ in this basis.

2.4.11. (a) Show that $1, (1-t), (1-t)^2, (1-t)^3$ is a basis for $\mathcal{P}^{(3)}$.

(b) Write $p(t) = 1 + t^3$ in terms of the basis elements.

2.4.12. Let $\mathcal{P}^{(4)}$ denote the vector space consisting of all polynomials $p(x)$ of degree ≤ 4.

(a) Are $p_1(x) = x^3 - 3x + 1$, $p_2(x) = x^4 - 6x + 3$, $p_3(x) = x^4 - 2x^3 + 1$, linearly independent elements of $\mathcal{P}^{(4)}$?

(b) What is the dimension of the subspace of $\mathcal{P}^{(4)}$ spanned by p_1, p_2, p_3?

2.4.13. Let $S = \left\{0, \frac{1}{4}, \frac{1}{2}, \frac{3}{4}\right\}$.

(a) Show that the sample vectors corresponding to the functions 1, $\cos \pi x$, $\cos 2\pi x$ and $\cos 3\pi x$ form a basis for the vector space of all sample functions on S.

(b) Write the sampled version of the function $f(x) = x$ in terms of this basis.

2.4.14. (a) Prove that the vector space of all 2×2 matrices is a four-dimensional vector space by exhibiting a basis.

(b) Generalize your result and prove that the vector space $\mathcal{M}_{m \times n}$ consisting of all $m \times n$ matrices has dimension mn.

2.4.15. Determine all values of the scalar k for which the following four matrices form a basis for $\mathcal{M}_{2\times 2}$:

$$A_1 = \begin{pmatrix} 1 & -1 \\ 0 & 0 \end{pmatrix}, \quad A_2 = \begin{pmatrix} k & -3 \\ 1 & 0 \end{pmatrix},$$

$$A_3 = \begin{pmatrix} 1 & 0 \\ -k & 2 \end{pmatrix}, \quad A_4 = \begin{pmatrix} 0 & k \\ -1 & -2 \end{pmatrix}.$$

2.4.16. Prove that the space of diagonal $n \times n$ matrices forms an n-dimensional vector space.

2.4.17. (a) Find a basis for and the dimension of the space of upper triangular 2×2 matrices.

(b) Can you generalize your result to upper triangular $n \times n$ matrices?

2.4.18. (a) What is the dimension of the vector space of 2×2 symmetric matrices?

(b) Of skew-symmetric matrices?

(c) Generalize to 3×3 case.

(d) What about $n \times n$ matrices?

♡ **2.4.19.** A matrix is said to be a *semi-magic square* if its row sums and column sums (i.e., the sum of entries in an individual row or column) all add up to the same number. An example is

$$\begin{pmatrix} 8 & 1 & 6 \\ 3 & 5 & 7 \\ 4 & 9 & 2 \end{pmatrix},$$

whose row and column sums are all equal to 15.

(a) Explain why the set of all semi-magic squares forms a subspace.

(b) Prove that the 3×3 permutation matrices (1.30) span the space of semi-magic squares. What is its dimension?

(c) A *magic square* also has the diagonal and *anti-diagonal* (running from top right to bottom left) also add up to the common row and column sum; the preceding 3×3 example is magic. Does the set of 3×3 magic squares form a vector space? If so, what is its dimension?

(d) Write down a formula for all 3×3 magic squares.

◇ **2.4.20.** Show, by example, how the uniqueness result in Lemma 2.34 fails if one has a linearly dependent set of vectors.

◇ **2.4.21.** Suppose that $\mathbf{v}_1, \ldots, \mathbf{v}_n$ form a basis for $\mathbb{R}^n$. Let A be a nonsingular matrix. Prove that $A\mathbf{v}_1, \ldots, A\mathbf{v}_n$ also form a basis for $\mathbb{R}^n$. What is this basis if you start with the standard basis: $\mathbf{v}_i = \mathbf{e}_i$?

◇ **2.4.22.** Show that if $\mathbf{v}_1, \ldots, \mathbf{v}_n$ span $V \neq \{\mathbf{0}\}$, then one can choose a subset $\mathbf{v}_{i_1}, \ldots, \mathbf{v}_{i_m}$ that forms a basis of V. Thus, $\dim V = m \leq n$. Under what conditions is $\dim V = n$?

◇ **2.4.23.** Prove that if $\mathbf{v}_1, \ldots, \mathbf{v}_n$ are a basis of V, then any subset, e.g., $\mathbf{v}_{i_1}, \ldots, \mathbf{v}_{i_k}$, is linearly independent.

◇ **2.4.24.** (a) Prove that if $\mathbf{v}_1, \ldots, \mathbf{v}_m$ forms a basis for $V \subsetneq \mathbb{R}^n$, then $m < n$.

(b) Under the hypotheses of part (b), prove that there exist vectors $\mathbf{v}_{m+1}, \ldots, \mathbf{v}_n \in \mathbb{R}^n \setminus V$ such that the complete collection $\mathbf{v}_1, \ldots, \mathbf{v}_n$ forms a basis for $\mathbb{R}^n$.

(c) Illustrate by constructing bases of $\mathbb{R}^3$ that include

(i) the basis $\left(1, 1, \frac{1}{2}\right)^T$ of the line $x = y = 2z$;

(ii) the basis $\left(1, 0, -1\right)^T, \left(0, 1, -2\right)^T$ of the plane $x + 2y + z = 0$.

◇ **2.4.25.** Let $W \subset V$ be a subspace.

(a) Prove that $\dim W \leq \dim V$.

(b) Prove that if $\dim W = \dim V = n < \infty$, then $W = V$. Equivalently, if $W \subsetneq V$ is a proper subspace of a finite-dimensional vector space, then $\dim W < \dim V$.

(c) Give an example where the result is false if $\dim V = \infty$.

◇ **2.4.26.** Let $W, Z \subset V$ be complementary subspaces in a vector space V, as in Exercise 2.2.24.

(a) Prove that if $\{\mathbf{w}_1, \ldots, \mathbf{w}_j\}$ form a basis for W and $\{\mathbf{z}_1, \ldots, \mathbf{z}_k\}$ a basis for Z, then $\{\mathbf{w}_1, \ldots, \mathbf{w}_j, \mathbf{z}_1, \ldots, \mathbf{z}_k\}$ form a basis for V.

(b) Prove that $\dim W + \dim Z = \dim V$.

◇ **2.4.27.** Let $f_1(x), \ldots, f_n(x)$ be scalar functions. Suppose that *every* set of sample points $x_1, \ldots, x_m \in \mathbb{R}$, for all finite $m \geq 1$, leads to linearly dependent sample vectors $\mathbf{f}_1, \ldots, \mathbf{f}_n \in \mathbb{R}^m$. Prove that $f_1(x), \ldots, f_n(x)$ are linearly dependent functions. *Hint*: Given sample points $x_1, \ldots, x_m$, let $V_{x_1,\ldots,x_m} \subset \mathbb{R}^n$ be the subspace consisting of all vectors $\mathbf{c} = \left(c_1, \ldots, c_m\right)^T$ such that $c_1 \mathbf{f}_1 + \cdots + c_n \mathbf{f}_n = \mathbf{0}$. First, show that one can select sample points $x_1, x_2, x_3, \ldots$ such that $\mathbb{R}^n \supsetneq V_{x_1} \supsetneq V_{x_1,x_2} \supsetneq \cdots$. Then, apply Exercise 2.4.25 to conclude that $V_{x_1,\ldots,x_n} = \{\mathbf{0}\}$.

2.5 THE FUNDAMENTAL MATRIX SUBSPACES

Let us now return to the general study of linear systems of equations, which we write in our usual matrix form

$$A\,\mathbf{x} = \mathbf{b}. \tag{2.23}$$

As before, A is an $m \times n$ matrix, where m is the number of equations, so $\mathbf{b} \in \mathbb{R}^m$, and n is the number of unknowns, i.e., the entries of $\mathbf{x} \in \mathbb{R}^n$. We already know how to solve the system, at least when the coefficient matrix is not too large: just apply a variant of Gaussian Elimination. Our goal now is to understand the solution and thereby prepare ourselves for more sophisticated problems and solution techniques.

Kernel and Range

There are four important vector subspaces associated with any matrix. The first two are defined as follows.

Definition 2.36 The *range* of an $m \times n$ matrix A is the subspace $\operatorname{rng} A \subset \mathbb{R}^m$ spanned by its columns. The *kernel* of A is the subspace $\ker A \subset \mathbb{R}^n$ consisting of all vectors which are annihilated by A, so

$$\ker A = \left\{ \mathbf{z} \in \mathbb{R}^n \;\middle|\; A\,\mathbf{z} = \mathbf{0} \right\} \subset \mathbb{R}^n. \tag{2.24}$$

The range is also known as the *column space* or the *image* of the matrix. By definition, a vector $\mathbf{b} \in \mathbb{R}^m$ belongs to $\operatorname{rng} A$ if and only if it can be written as a linear combination,

$$\mathbf{b} = x_1 \mathbf{v}_1 + \cdots + x_n \mathbf{v}_n,$$

of the columns of $A = (\,\mathbf{v}_1\ \mathbf{v}_2 \ldots \mathbf{v}_n\,)$. By our basic matrix multiplication formula (2.13), the right hand side of this equation equals the product $A\,\mathbf{x}$ of the matrix A with the column vector $\mathbf{x} = (x_1, x_2, \ldots, x_n)^T$, and hence $\mathbf{b} = A\,\mathbf{x}$ for some $\mathbf{x} \in \mathbb{R}^n$. Thus,

$$\operatorname{rng} A = \left\{ A\,\mathbf{x} \;\middle|\; \mathbf{x} \in \mathbb{R}^n \right\} \subset \mathbb{R}^m, \tag{2.25}$$

and so *a vector* $\mathbf{b}$ *lies in the range of* A *if and only if the linear system* $A\,\mathbf{x} = \mathbf{b}$ *has a solution.* The compatibility conditions for linear systems can thereby be reinterpreted as the requirements for a vector to lie in the range of the coefficient matrix.

A common alternative name for the kernel is the *null space*. The kernel or null space of A is the set of solutions $\mathbf{z}$ to the homogeneous system $A\,\mathbf{z} = \mathbf{0}$. The proof that $\ker A$ is a subspace requires us to verify the usual closure conditions: suppose that $\mathbf{z}, \mathbf{w} \in \ker A$, so that $A\,\mathbf{z} = \mathbf{0} = A\,\mathbf{w}$. Then, for any scalars c, d,

$$A(c\,\mathbf{z} + d\,\mathbf{w}) = c\,A\,\mathbf{z} + d\,A\,\mathbf{w} = \mathbf{0},$$

which implies that $c\,\mathbf{z} + d\,\mathbf{w} \in \ker A$. Closure of $\ker A$ can be re-expressed as the following important *superposition principle* for solutions to a homogeneous system of linear equations.

THEOREM 2.37 If $\mathbf{z}_1, \ldots, \mathbf{z}_k$ are individual solutions to the *same* homogeneous linear system $A\,\mathbf{z} = \mathbf{0}$, then so is any linear combination $c_1\,\mathbf{z}_1 + \cdots + c_k\,\mathbf{z}_k$.

Warning: The set of solutions to an inhomogeneous linear system $A\,\mathbf{x} = \mathbf{b}$ with $\mathbf{b} \neq \mathbf{0}$ is *not* a subspace. Linear combinations of solutions are not, in general, solutions to the same inhomogeneous system.

Superposition is the reason why linear systems are so much easier to solve, since one only needs to find relatively few solutions in order to construct the general solution as a linear combination. In Chapter 7 we shall see that superposition applies to completely general linear systems, including linear differential equations, both ordinary and partial, linear boundary value problems, linear integral equations, linear control systems, etc.

EXAMPLE 2.38 Let us compute the kernel of the matrix

$$A = \begin{pmatrix} 1 & -2 & 0 & 3 \\ 2 & -3 & -1 & -4 \\ 3 & -5 & -1 & -1 \end{pmatrix}.$$

Our task is to solve the homogeneous system $A\mathbf{x} = \mathbf{0}$, we only need perform the elementary row operations on A itself. The resulting row echelon form

$$U = \begin{pmatrix} 1 & -2 & 0 & 3 \\ 0 & 1 & -1 & -10 \\ 0 & 0 & 0 & 0 \end{pmatrix}$$

corresponds to the equations $x - 2y + 3w = 0$, $\quad y - z - 10w = 0$. The free variables are z, w, and the general solution is

$$\mathbf{x} = \begin{pmatrix} x \\ y \\ z \\ w \end{pmatrix} = \begin{pmatrix} 2z + 17w \\ z + 10w \\ z \\ w \end{pmatrix} = z \begin{pmatrix} 2 \\ 1 \\ 1 \\ 0 \end{pmatrix} + w \begin{pmatrix} 17 \\ 10 \\ 0 \\ 1 \end{pmatrix}.$$

The result describes the most general vector in $\ker A$, which is thus the two-dimensional subspace of $\mathbb{R}^4$ spanned by the linearly independent vectors

$$(2, 1, 1, 0)^T, \quad (17, 10, 0, 1)^T.$$

This example is indicative of a general method for finding a basis for $\ker A$, to be developed in more detail below. ●

Once we know the kernel of the coefficient matrix A, i.e., the space of solutions to the homogeneous system $A\,\mathbf{z} = \mathbf{0}$, we are able to completely characterize the solutions to the inhomogeneous linear system (2.23).

THEOREM 2.39 The linear system $A\mathbf{x} = \mathbf{b}$ has a solution $\mathbf{x}^\star$ if and only if $\mathbf{b}$ lies in the range of A. If this occurs, then $\mathbf{x}$ is a solution to the linear system if and only if

$$\mathbf{x} = \mathbf{x}^\star + \mathbf{z}, \tag{2.26}$$

where $\mathbf{z} \in \ker A$ is an element of the kernel of the coefficient matrix.

Proof We already demonstrated the first part of the theorem. If $A\mathbf{x} = \mathbf{b} = A\mathbf{x}^\star$ are any two solutions, then their difference $\mathbf{z} = \mathbf{x} - \mathbf{x}^\star$ satisfies

$$A\mathbf{z} = A(\mathbf{x} - \mathbf{x}^\star) = A\mathbf{x} - A\,\mathbf{x}^\star = \mathbf{b} - \mathbf{b} = \mathbf{0},$$

and hence $\mathbf{z}$ is in the kernel of A. Therefore, $\mathbf{x}$ and $\mathbf{x}^\star$ are related by formula (2.26), which proves the second part of the theorem. ■

Therefore, to construct the most general solution to an inhomogeneous system, we need only know one *particular solution* $\mathbf{x}^\star$, along with the general solution $\mathbf{z} \in \ker A$ to the corresponding homogeneous system. This construction should remind the reader of the method for solving inhomogeneous linear ordinary differential equations. Indeed, both linear algebraic systems and linear ordinary differential equations are but two particular instances in the general theory of linear systems, to be developed in Chapter 7.

EXAMPLE 2.40 Consider the system $A\mathbf{x} = \mathbf{b}$, where

$$A = \begin{pmatrix} 1 & 0 & -1 \\ 0 & 1 & -2 \\ 1 & -2 & 3 \end{pmatrix}, \qquad \mathbf{x} = \begin{pmatrix} x_1 \\ x_2 \\ x_3 \end{pmatrix}, \qquad \mathbf{b} = \begin{pmatrix} b_1 \\ b_2 \\ b_3 \end{pmatrix},$$

where the right hand side of the system will remain unspecified for the moment. Applying our usual Gaussian Elimination procedure to the augmented matrix

$$\left(\begin{array}{ccc|c} 1 & 0 & -1 & b_1 \\ 0 & 1 & -2 & b_2 \\ 1 & -2 & 3 & b_3 \end{array}\right)$$

leads to the row echelon form

$$\left(\begin{array}{ccc|c} 1 & 0 & -1 & b_1 \\ 0 & 1 & -2 & b_2 \\ 0 & 0 & 0 & b_3 + 2b_2 - b_1 \end{array}\right).$$

Therefore, the system has a solution if and only if the compatibility condition

$$-b_1 + 2b_2 + b_3 = 0 \tag{2.27}$$

holds. This equation serves to characterize the vectors $\mathbf{b}$ that belong to the range of the matrix A, which is therefore a plane in $\mathbb{R}^3$.

To characterize the kernel of A, we take $\mathbf{b} = \mathbf{0}$, and solve the homogeneous system $A\,\mathbf{z} = \mathbf{0}$. The row echelon form corresponds to the reduced system

$$z_1 - z_3 = 0, \qquad z_2 - 2\,z_3 = 0.$$

The free variable is z_3, and the equations are solved to give

$$z_1 = c, \qquad z_2 = 2\,c, \qquad z_3 = c,$$

where c is an arbitrary scalar. The general solution to the homogeneous system is $\mathbf{z} = (c, 2\,c, c)^T = c\,(1, 2, 1)^T$, and so the kernel is the line in the direction of the vector $(1, 2, 1)^T$.

If we take $\mathbf{b} = (3, 1, 1)^T$—which satisfies (2.27) and hence lies in the range of A—then the general solution to the inhomogeneous system $A\mathbf{x} = \mathbf{b}$ is

$$x_1 = 3 + c, \quad x_2 = 1 + 2c, \quad x_3 = c,$$

where c is arbitrary. We can write the solution in the form (2.26), namely

$$\mathbf{x} = \begin{pmatrix} 3+c \\ 1+2c \\ c \end{pmatrix} = \begin{pmatrix} 3 \\ 1 \\ 0 \end{pmatrix} + c \begin{pmatrix} 1 \\ 2 \\ 1 \end{pmatrix} = \mathbf{x}^\star + \mathbf{z},$$

where, as in (2.26), $\mathbf{x}^\star = (3, 1, 0)^T$ plays the role of the particular solution, and $\mathbf{z} = c\,(1, 2, 1)^T$ is the general element of the kernel. ●

We can characterize the situations when the linear system has a unique solution in any of the following equivalent ways.

Proposition 2.41 If A is an $m \times n$ matrix, then the following conditions are equivalent:

(i) $\ker A = \{\mathbf{0}\}$, i.e., the homogeneous system $A\mathbf{x} = \mathbf{0}$ has the unique solution $\mathbf{x} = \mathbf{0}$.

(ii) $\operatorname{rank} A = n$.

(iii) The linear system $A\mathbf{x} = \mathbf{b}$ has no free variables.

(iv) The system $A\mathbf{x} = \mathbf{b}$ has a unique solution for each $\mathbf{b} \in \operatorname{rng} A$.

Thus, while existence of a solution may depend upon the particularities of the right hand side $\mathbf{b}$, uniqueness is universal: if for any one $\mathbf{b}$, e.g., $\mathbf{b} = \mathbf{0}$, the system admits a unique solution, then all $\mathbf{b} \in \operatorname{rng} A$ also admit unique solutions. Specializing even further to square matrices, we can now characterize invertible matrices by looking either at their kernel or their range.

Proposition 2.42 If A is a square $n \times n$ matrix, then the following four conditions are equivalent:

(i) A is nonsingular;

(ii) $\operatorname{rank} A = n$;

(iii) $\ker A = \{\mathbf{0}\}$;

(iv) $\operatorname{rng} A = \mathbb{R}^n$.

EXERCISES 2.5

2.5.1. Characterize the range and kernel of the following matrices:

(a) $\begin{pmatrix} 8 & -4 \\ -6 & 3 \end{pmatrix}$ (b) $\begin{pmatrix} 1 & -1 & 2 \\ -2 & 2 & -4 \end{pmatrix}$

(c) $\begin{pmatrix} 1 & 2 & 3 \\ -2 & 4 & 1 \\ 4 & 0 & 5 \end{pmatrix}$

(d) $\begin{pmatrix} 1 & -1 & 0 & 1 \\ -1 & 0 & 1 & -1 \\ 1 & -2 & 1 & 1 \\ 1 & 2 & -3 & 1 \end{pmatrix}$

2.5.2. For each of the following matrices, write the kernel as the span of a finite number of vectors. Is the kernel a point, line, plane or all of $\mathbb{R}^3$?

(a) $\begin{pmatrix} 2 & -1 & 5 \end{pmatrix}$ (b) $\begin{pmatrix} 1 & 2 & -1 \\ 3 & -2 & 0 \end{pmatrix}$

(c) $\begin{pmatrix} 2 & 6 & -4 \\ -1 & -3 & 2 \end{pmatrix}$ (d) $\begin{pmatrix} 1 & 2 & 5 \\ 0 & 4 & 8 \\ 1 & -6 & -11 \end{pmatrix}$

(e) $\begin{pmatrix} 2 & -1 & 1 \\ -1 & 1 & -2 \\ 3 & -1 & 1 \end{pmatrix}$ (f) $\begin{pmatrix} 1 & -2 & 3 \\ -3 & 6 & -9 \\ -2 & 4 & -6 \\ 3 & 0 & -1 \end{pmatrix}$

2.5.3. (a) Find the kernel and range of the coefficient matrix for the system $x - 3y + 2z = a$, $2x - 6y + 2w = b$, $z - 3w = c$.

(b) Write down compatibility conditions on a, b, c for a solution to exist.

2.5.4. Suppose $\mathbf{x}^\star = \begin{pmatrix} 1 \\ 2 \\ 3 \end{pmatrix}$ is a particular solution to the equation

$$\begin{pmatrix} 1 & -1 & 0 \\ -1 & 0 & 1 \\ 0 & 1 & -1 \end{pmatrix} \mathbf{x} = \mathbf{b}.$$

(a) What is $\mathbf{b}$?

(b) Find the general solution.

2.5.5. Write the general solution to the following linear systems in the form (2.26). Clearly identify the particular solution $\mathbf{x}^\star$ and the element $\mathbf{z}$ of the kernel.

(a) $x - y + 3z = 1$

(b) $\begin{pmatrix} 1 & -2 & 0 \\ 2 & 3 & 1 \end{pmatrix} \begin{pmatrix} x \\ y \\ z \end{pmatrix} = \begin{pmatrix} 3 \\ -1 \end{pmatrix}$

(c) $\begin{pmatrix} 1 & -1 & 0 \\ 2 & 0 & -4 \\ 2 & -1 & -2 \end{pmatrix} \begin{pmatrix} x \\ y \\ z \end{pmatrix} = \begin{pmatrix} -1 \\ -6 \\ -4 \end{pmatrix}$

(d) $\begin{pmatrix} 2 & -1 & 1 \\ 4 & -1 & 2 \\ 0 & 1 & 3 \end{pmatrix} \begin{pmatrix} x \\ y \\ z \end{pmatrix} = \begin{pmatrix} 0 \\ 1 \\ -1 \end{pmatrix}$

(e) $\begin{pmatrix} 1 & -2 \\ 2 & -4 \\ -3 & 6 \\ -1 & 2 \end{pmatrix} \begin{pmatrix} u \\ v \end{pmatrix} = \begin{pmatrix} -1 \\ -2 \\ 3 \\ 1 \end{pmatrix}$

(f) $\begin{pmatrix} 1 & -3 & 2 & 0 \\ -1 & 5 & 1 & 1 \\ 2 & -8 & 1 & -1 \end{pmatrix} \begin{pmatrix} p \\ q \\ r \\ s \end{pmatrix} = \begin{pmatrix} 4 \\ -3 \\ 7 \end{pmatrix}$

(g) $\begin{pmatrix} 0 & -1 & 2 & -1 \\ 1 & -3 & 0 & 1 \\ -2 & 5 & 2 & -3 \\ 1 & 1 & -8 & 5 \end{pmatrix} \begin{pmatrix} x \\ y \\ z \\ w \end{pmatrix} = \begin{pmatrix} -2 \\ -3 \\ 4 \\ 5 \end{pmatrix}$

2.5.6. Prove that the average of all the entries in each row of A is 0 if and only if $(1, 1, \ldots, 1)^T \in \ker A$.

2.5.7. Given $a, r \neq 0$, characterize the kernel and the range of the matrix

$$\begin{pmatrix} a & a\,r & \ldots & a\,r^{n-1} \\ a\,r^n & a\,r^{n+1} & \ldots & a\,r^{2n-1} \\ \vdots & \vdots & \ddots & \vdots \\ a\,r^{(n-1)n} & a\,r^{(n-1)n+1} & \ldots & a\,r^{n^2-1} \end{pmatrix}.$$

Hint: See Exercise 1.8.16.

◇ **2.5.8.** A *projection matrix* is a square matrix P that satisfies $P^2 = P$.

(a) Prove that $\mathbf{w} \in \operatorname{rng} P$ if and only if $P\,\mathbf{w} = \mathbf{w}$.

(b) Show that $\operatorname{rng} P$ and $\ker P$ are complementary subspaces, as defined in Exercise 2.2.24, so every $\mathbf{v} \in \mathbb{R}^n$ can be uniquely written as $\mathbf{v} = \mathbf{w} + \mathbf{z}$ where $\mathbf{w} \in \operatorname{rng} P$, $\mathbf{z} \in \ker P$.

2.5.9. *True or false*: If A is a square matrix, $\ker A \cap \operatorname{rng} A = \{\mathbf{0}\}$.

◇ **2.5.10.** Let A be an $m \times n$ matrix. Suppose that

$$C = \begin{pmatrix} A \\ B \end{pmatrix}$$

is a $(m + k) \times n$ matrix whose first m rows are the same as A. Prove that $\ker C \subseteq \ker A$. Thus, appending more rows cannot increase the size of a matrix's kernel. Give an example where $\ker C \neq \ker A$.

◇ **2.5.11.** Let A be an $m \times n$ matrix. Suppose that $C = (\,A \;\; B\,)$ is an $m \times (n + k)$ matrix whose first n columns are the same as A. Prove that $\operatorname{rng} C \supseteq \operatorname{rng} A$. Thus, appending more columns cannot decrease the size of a matrix's range. Give an example where $\operatorname{rng} C \neq \operatorname{rng} A$.

The Superposition Principle

The principle of superposition lies at the heart of linearity. For homogeneous systems, superposition allows one to generate new solutions by combining known solutions. For inhomogeneous systems, superposition combines the solutions corresponding to different inhomogeneities.

Suppose we know particular solutions $\mathbf{x}_1^\star$ and $\mathbf{x}_2^\star$ to two inhomogeneous linear systems

$$A\mathbf{x} = \mathbf{b}_1, \qquad A\mathbf{x} = \mathbf{b}_2,$$

that have the *same* coefficient matrix A. Consider the system

$$A\,\mathbf{x} = c_1\,\mathbf{b}_1 + c_2\,\mathbf{b}_2,$$

whose right hand side is a linear combination or *superposition* of the previous two. Then a particular solution to the combined system is given by the *same* superposition of the previous solutions:

$$\mathbf{x}^\star = c_1\,\mathbf{x}_1^\star + c_2\,\mathbf{x}_2^\star.$$

The proof is easy:

$$A\,\mathbf{x}^\star = A(c_1\,\mathbf{x}_1^\star + c_2\,\mathbf{x}_2^\star) = c_1\,A\,\mathbf{x}_1^\star + c_2\,A\,\mathbf{x}_2^\star = c_1\,\mathbf{b}_1 + c_2\,\mathbf{b}_2.$$

In physical applications, the inhomogeneities $\mathbf{b}_1, \mathbf{b}_2$ typically represent external forces, and the solutions $\mathbf{x}_1^\star, \mathbf{x}_2^\star$ represent the respective responses of the physical apparatus. The linear superposition principle says that if we know how the system responds to the individual forces, we immediately know its response to any combination thereof. The precise details of the system are irrelevant—all that is required is its linearity.

EXAMPLE 2.43 For example, the system

$$\begin{pmatrix} 4 & 1 \\ 1 & 4 \end{pmatrix}\begin{pmatrix} x_1 \\ x_2 \end{pmatrix} = \begin{pmatrix} f_1 \\ f_2 \end{pmatrix}$$

models the mechanical response of a pair of masses connected by springs, subject to external forcing. The solution $\mathbf{x} = (x_1, x_2)^T$ represents the displacements of the masses, while the entries of the right hand side $\mathbf{f} = (f_1, f_2)^T$ are the applied forces. (Details can be found in Chapter 6.) We can directly determine the response of the system $\mathbf{x}_1^\star = \left(\frac{4}{15}, -\frac{1}{15}\right)^T$ to a unit force $\mathbf{e}_1 = (1, 0)^T$ on the first mass, and the response $\mathbf{x}_2^\star = \left(-\frac{1}{15}, \frac{4}{15}\right)^T$ to a unit force $\mathbf{e}_2 = (0, 1)^T$ on the second mass. Superposition gives the response of the system to a general force, since we can write

$$\mathbf{f} = \begin{pmatrix} f_1 \\ f_2 \end{pmatrix} = f_1\,\mathbf{e}_1 + f_2\,\mathbf{e}_2 = f_1\begin{pmatrix} 1 \\ 0 \end{pmatrix} + f_2\begin{pmatrix} 0 \\ 1 \end{pmatrix},$$

and hence

$$\mathbf{x} = f_1\,\mathbf{x}_1^\star + f_2\,\mathbf{x}_2^\star = f_1\begin{pmatrix} \frac{4}{15} \\ -\frac{1}{15} \end{pmatrix} + f_2\begin{pmatrix} -\frac{1}{15} \\ \frac{4}{15} \end{pmatrix} = \begin{pmatrix} \frac{4}{15}f_1 - \frac{1}{15}f_2 \\ -\frac{1}{15}f_1 + \frac{4}{15}f_2 \end{pmatrix}. \qquad \bullet$$

The preceding construction is easily extended to several inhomogeneities, and the result is the general *Superposition Principle* for inhomogeneous linear systems.

THEOREM 2.44 Suppose that we know particular solutions $\mathbf{x}_1^\star, \ldots, \mathbf{x}_k^\star$ to each of the inhomogeneous linear systems

$$A\,\mathbf{x} = \mathbf{b}_1, \qquad A\,\mathbf{x} = \mathbf{b}_2, \qquad \ldots \qquad A\,\mathbf{x} = \mathbf{b}_k, \tag{2.28}$$

all with the same coefficient matrix, and where $\mathbf{b}_1, \ldots, \mathbf{b}_k \in \operatorname{rng} A$. Then, for any choice of scalars $c_1, \ldots, c_k$, a particular solution to the combined system

$$A\,\mathbf{x} = c_1\mathbf{b}_1 + \cdots + c_k\mathbf{b}_k \tag{2.29}$$

is the corresponding superposition

$$\mathbf{x}^\star = c_1\,\mathbf{x}_1^\star + \cdots + c_k\,\mathbf{x}_k^\star \tag{2.30}$$

of individual solutions. The general solution to (2.29) is

$$\mathbf{x} = \mathbf{x}^\star + \mathbf{z} = c_1\,\mathbf{x}_1^\star + \cdots + c_k\,\mathbf{x}_k^\star + \mathbf{z}, \tag{2.31}$$

where $\mathbf{z} \in \ker A$ is the general solution to the homogeneous system $A\,\mathbf{z} = \mathbf{0}$.

For instance, if we know particular solutions $\mathbf{x}_1^\star, \ldots, \mathbf{x}_m^\star$ to

$$A\mathbf{x} = \mathbf{e}_i, \qquad \text{for each} \quad i = 1, \ldots, m, \tag{2.32}$$

where $\mathbf{e}_1, \ldots, \mathbf{e}_m$ are the standard basis vectors of $\mathbb{R}^m$, then we can reconstruct a particular solution $\mathbf{x}^\star$ to the general linear system $A\mathbf{x} = \mathbf{b}$ by first writing

$$\mathbf{b} = b_1\mathbf{e}_1 + \cdots + b_m\mathbf{e}_m$$

as a linear combination of the basis vectors, and then using superposition to form

$$\mathbf{x}^\star = b_1\mathbf{x}_1^\star + \cdots + b_m\mathbf{x}_m^\star. \tag{2.33}$$

However, for linear algebraic systems, the practical value of this insight is rather limited. Indeed, in the case when A is square and nonsingular, the superposition formula (2.33) is merely a reformulation of the method of computing the inverse of the matrix. Indeed, the vectors $\mathbf{x}_1^\star, \ldots, \mathbf{x}_m^\star$ that satisfy (2.32) are just the columns of A^{-1} (why?), while (2.33) is precisely the solution formula $\mathbf{x}^\star = A^{-1}\mathbf{b}$ that we abandoned in practical computations, in favor of the more efficient Gaussian Elimination process. Nevertheless, this idea turns out to have important implications in more general situations, such as linear differential equations and boundary value problems.

EXERCISES 2.5

2.5.12. Find the solution $\mathbf{x}_1^\star$ to the system

$$\begin{pmatrix} 1 & 2 \\ -3 & -4 \end{pmatrix}\begin{pmatrix} x \\ y \end{pmatrix} = \begin{pmatrix} 1 \\ 0 \end{pmatrix},$$

and the solution $\mathbf{x}_2^\star$ to

$$\begin{pmatrix} 1 & 2 \\ -3 & -4 \end{pmatrix}\begin{pmatrix} x \\ y \end{pmatrix} = \begin{pmatrix} 0 \\ 1 \end{pmatrix}.$$

Express the solution to

$$\begin{pmatrix} 1 & 2 \\ -3 & -4 \end{pmatrix}\begin{pmatrix} x \\ y \end{pmatrix} = \begin{pmatrix} 1 \\ 4 \end{pmatrix}$$

as a linear combination of $\mathbf{x}_1^\star$ and $\mathbf{x}_2^\star$.

2.5.13. Let $A = \begin{pmatrix} 1 & 2 & -1 \\ 2 & 5 & -1 \\ 1 & 3 & 2 \end{pmatrix}$. Given that

$$\mathbf{x}_1^\star = \begin{pmatrix} 5 \\ -1 \\ 2 \end{pmatrix} \quad \text{solves} \quad A\mathbf{x} = \mathbf{b}_1 = \begin{pmatrix} 1 \\ 3 \\ 6 \end{pmatrix}$$

and

$$\mathbf{x}_2^\star = \begin{pmatrix} -11 \\ 5 \\ -1 \end{pmatrix} \quad \text{solves} \quad A\mathbf{x} = \mathbf{b}_2 = \begin{pmatrix} 0 \\ 4 \\ 2 \end{pmatrix},$$

find a solution to $A\mathbf{x} = 2\mathbf{b}_1 + \mathbf{b}_2 = \begin{pmatrix} 2 \\ 10 \\ 14 \end{pmatrix}$.

2.5.14. (a) Show that

$$\mathbf{x}_1^\star = \begin{pmatrix} 1 \\ 1 \\ 0 \end{pmatrix} \quad \text{and} \quad \mathbf{x}_2^\star = \begin{pmatrix} -3 \\ 3 \\ -2 \end{pmatrix}$$

are particular solutions to the system

$$\begin{pmatrix} 2 & -1 & -5 \\ 1 & -4 & -6 \\ 3 & 2 & -4 \end{pmatrix}\mathbf{x} = \begin{pmatrix} 1 \\ -3 \\ 5 \end{pmatrix}.$$

(b) Find the general solution.

2.5.15. A physical apparatus moves 2 meters under a force of 4 Newtons. Assuming linearity, how far will it move under a force of 10 Newtons?

2.5.16. Applying a unit external force in the horizontal direction moves a mass 3 units to the right, while applying a unit force in the vertical direction moves it up 2 units. Assuming linearity, where will the mass move under the applied force $\mathbf{f} = (2, -3)^T$?

2.5.17. Suppose $\mathbf{x}_1^\star$ and $\mathbf{x}_2^\star$ are both solutions to $A\mathbf{x} = \mathbf{b}$. List all linear combinations of $\mathbf{x}_1^\star$ and $\mathbf{x}_2^\star$ that solve the system.

2.5.18. *True or false*: If $\mathbf{x}_1^\star$ solves $A\mathbf{x} = \mathbf{c}$, and $\mathbf{x}_2^\star$ solves $B\mathbf{x} = \mathbf{d}$, then $\mathbf{x}^\star = \mathbf{x}_1^\star + \mathbf{x}_2^\star$ solves $(A+B)\mathbf{x} = \mathbf{c}+\mathbf{d}$.

◇ **2.5.19.** Under what conditions on the coefficient matrix A will the systems in (2.32) all have a solution?

◇ **2.5.20.** Let A be a nonsingular $m \times m$ matrix.

(a) Explain in detail why the solutions $\mathbf{x}_1^\star, \ldots, \mathbf{x}_m^\star$ to the systems (2.32) are the columns of the matrix inverse A^{-1}.

(b) Illustrate your argument in the case

$$A = \begin{pmatrix} 0 & 1 & 2 \\ -1 & 1 & 3 \\ 1 & 0 & 1 \end{pmatrix}.$$

Adjoint Systems, Cokernel, and Corange

A linear system of m equations in n unknowns is based on an $m \times n$ coefficient matrix A. The transposed matrix A^T will be of size $n \times m$, and forms the coefficient matrix of an associated linear system, consisting of n equations in m unknowns.

Definition 2.45 The *adjoint** to a linear system $A\mathbf{x} = \mathbf{b}$ of m equations in n unknowns is the linear system

$$A^T \mathbf{y} = \mathbf{f} \tag{2.34}$$

consisting of n equations in m unknowns $\mathbf{y} \in \mathbb{R}^m$ with right hand side $\mathbf{f} \in \mathbb{R}^n$.

EXAMPLE 2.46 Consider the linear system

$$\begin{aligned} x_1 - 3x_2 - 7x_3 + 9x_4 &= b_1, \\ x_2 + 5x_3 - 3x_4 &= b_2, \\ x_1 - 2x_2 - 2x_3 + 6x_4 &= b_3, \end{aligned} \tag{2.35}$$

of three equations in four unknowns. Its coefficient matrix

$$A = \begin{pmatrix} 1 & -3 & -7 & 9 \\ 0 & 1 & 5 & -3 \\ 1 & -2 & -2 & 6 \end{pmatrix} \quad \text{has transpose} \quad A^T = \begin{pmatrix} 1 & 0 & 1 \\ -3 & 1 & -2 \\ -7 & 5 & -2 \\ 9 & -3 & 6 \end{pmatrix}.$$

Thus, the adjoint system to (2.35) is the following system of four equations in three unknowns:

$$\begin{aligned} y_1 + \qquad\quad y_3 &= f_1, \\ -3y_1 + y_2 - 2y_3 &= f_2, \\ -7y_1 + 5y_2 - 2y_3 &= f_3, \\ 9y_1 - 3y_2 + 6y_3 &= f_4. \end{aligned} \tag{2.36}$$

●

On the surface, there appears to be no direct connection between the solutions to a linear system and its adjoint. Nevertheless, as we shall soon see (and then in even greater depth in Sections 5.6 and 8.5) the two are linked in a number of remarkable, but subtle ways. As a first step in this direction, we use the adjoint system to define the remaining two fundamental subspaces associated with a coefficient matrix A.

**Warning*: Some texts misuse the term "adjoint" to describe the *adjugate* or *cofactor matrix*, [60]. The constructions are completely unrelated, and the adjugate will play no role in this book.

Definition 2.47

The *corange* of an $m \times n$ matrix A is the range of its transpose,

$$\operatorname{corng} A = \operatorname{rng} A^T = \{ A^T \mathbf{y} \mid \mathbf{y} \in \mathbb{R}^m \} \subset \mathbb{R}^n. \tag{2.37}$$

The *cokernel* or *left null space* of A is the kernel of its transpose,

$$\operatorname{coker} A = \ker A^T = \{ \mathbf{w} \in \mathbb{R}^m \mid A^T \mathbf{w} = \mathbf{0} \} \subset \mathbb{R}^m, \tag{2.38}$$

that is, the set of solutions to the homogeneous adjoint system.

The corange coincides with the subspace of $\mathbb{R}^n$ spanned by the rows* of A, and is sometimes referred to as the *row space*. As a direct consequence of Theorem 2.39, the adjoint system $A^T \mathbf{y} = \mathbf{f}$ has a solution if and only if $\mathbf{f} \in \operatorname{rng} A^T = \operatorname{corng} A$.

EXAMPLE 2.48

To solve the linear system (2.35) just presented, we perform Gaussian Elimination on the augmented matrix

$$\left(\begin{array}{cccc|c} 1 & -3 & -7 & 9 & b_1 \\ 0 & 1 & 5 & -3 & b_2 \\ 1 & -2 & -2 & 6 & b_3 \end{array} \right),$$

reducing it to the row echelon form

$$\left(\begin{array}{cccc|c} 1 & -3 & -7 & 9 & b_1 \\ 0 & 1 & 5 & -3 & b_2 \\ 0 & 0 & 0 & 0 & b_3 - b_2 - b_1 \end{array} \right).$$

Thus, the system has a solution if and only if

$$-b_1 - b_2 + b_3 = 0,$$

which is the condition required for $\mathbf{b} \in \operatorname{rng} A$. For such vectors, the general solution is

$$\mathbf{x} = \begin{pmatrix} b_1 + 3b_2 - 8x_3 \\ b_2 - 5x_3 + 3x_4 \\ x_3 \\ x_4 \end{pmatrix} = \begin{pmatrix} b_1 + 3b_2 \\ b_2 \\ 0 \\ 0 \end{pmatrix} + x_3 \begin{pmatrix} -8 \\ -5 \\ 1 \\ 0 \end{pmatrix} + x_4 \begin{pmatrix} 0 \\ 3 \\ 0 \\ 1 \end{pmatrix}.$$

In the second expression, the first vector represents a particular solution, while the two remaining terms constitute the general element of $\ker A$.

The solution to the adjoint system (2.36) is also obtained by Gaussian Elimination, starting with its augmented matrix

$$\left(\begin{array}{ccc|c} 1 & 0 & 1 & f_1 \\ -3 & 1 & -2 & f_2 \\ -7 & 5 & -2 & f_3 \\ 9 & -3 & 6 & f_4 \end{array} \right).$$

The resulting row echelon form is

$$\left(\begin{array}{ccc|c} 1 & 0 & 1 & f_1 \\ 0 & 1 & 1 & f_2 + 3f_1 \\ 0 & 0 & 0 & f_3 - 5f_2 - 8f_1 \\ 0 & 0 & 0 & f_4 + 3f_2 \end{array} \right).$$

*Or, more precisely, the column vectors obtained by transposing the rows.

Thus, there are two consistency constraints required for a solution to the adjoint system:

$$-8 f_1 - 5 f_2 + f_3 = 0, \qquad 3 f_2 + f_4 = 0.$$

These are the conditions required for the right hand side to belong to the corange: $\mathbf{f} \in \operatorname{rng} A^T = \operatorname{corng} A$. If satisfied, the adjoint system has the general solution depending on the single free variable y_3:

$$\mathbf{y} = \begin{pmatrix} f_1 - y_3 \\ 3 f_1 + f_2 - y_3 \\ y_3 \end{pmatrix} = \begin{pmatrix} f_1 \\ 3 f_1 + f_2 \\ 0 \end{pmatrix} + y_3 \begin{pmatrix} -1 \\ -1 \\ 1 \end{pmatrix}.$$

In the latter formula, the first term represents a particular solution, while the second is the general element of the cokernel $\ker A^T = \operatorname{coker} A$. ●

The Fundamental Theorem of Linear Algebra

The four fundamental subspaces associated with an $m \times n$ matrix A, then, are its range, corange, kernel, and cokernel. The range and cokernel are subspaces of $\mathbb{R}^m$, while the kernel and corange are subspaces of $\mathbb{R}^n$. The *Fundamental Theorem of Linear Algebra*[†] states that their dimensions are fixed by the rank (and size) of the matrix.

THEOREM 2.49 Let A be an $m \times n$ matrix of rank r. Then

$$\begin{aligned} \dim \operatorname{corng} A = \dim \operatorname{rng} A &= \operatorname{rank} A = \operatorname{rank} A^T = r, \\ \dim \ker A = n - r, & \qquad \dim \operatorname{coker} A = m - r. \end{aligned} \tag{2.39}$$

Thus, the rank of a matrix, i.e., the number of pivots, indicates the number of linearly independent columns, which, remarkably, is always the same as the number of linearly independent rows. A matrix and its transpose are guaranteed to have the same rank, i.e., the same number of pivots, despite the fact that their row echelon forms are quite different, and are almost never transposes of each other. Theorem 2.49 also establishes our earlier contention that the rank of a matrix is an *intrinsic* quantity, since it equals the common dimension of its range and corange, and so does not depend on which specific elementary row operations are employed during the reduction process, nor on the final row echelon form.

Let us turn to the proof of the Fundamental Theorem 2.49. Since the dimension of a subspace is prescribed by the number of vectors in any basis, we need to relate bases of the fundamental subspaces to the rank of the matrix. Before trying to digest the general argument, it is better to first understand how to construct the required bases in a particular example. Consider the matrix

$$A = \begin{pmatrix} 2 & -1 & 1 & 2 \\ -8 & 4 & -6 & -4 \\ 4 & -2 & 3 & 2 \end{pmatrix}. \tag{2.40}$$

Its row echelon form

$$U = \begin{pmatrix} 2 & -1 & 1 & 2 \\ 0 & 0 & -2 & 4 \\ 0 & 0 & 0 & 0 \end{pmatrix}$$

is obtained in the usual manner. There are two pivots, and thus the rank of A is $r = 2$.

[†]Not to be confused with the Fundamental Theorem of Algebra, which states that every (nonconstant) polynomial has a complex root; see [22].

Kernel: The general solution to the homogeneous system $A\mathbf{x} = \mathbf{0}$ can be expressed as a linear combination of $n - r$ linearly independent vectors, whose coefficients are the free variables for the system corresponding to the $n - r$ columns without pivots. In fact, these vectors form a basis for the kernel, which thus has dimension $n - r$.

In our example, the pivots are in columns 1 and 3, and so the free variables are x_2, x_4. Applying Back Substitution to the reduced homogeneous system $U\mathbf{x} = \mathbf{0}$, we find the general solution

$$\mathbf{x} = \begin{pmatrix} \frac{1}{2}x_2 - 2x_4 \\ x_2 \\ 2x_4 \\ x_4 \end{pmatrix} = x_2 \begin{pmatrix} \frac{1}{2} \\ 1 \\ 0 \\ 0 \end{pmatrix} + x_4 \begin{pmatrix} -2 \\ 0 \\ 2 \\ 1 \end{pmatrix} \tag{2.41}$$

written as a linear combination of the vectors

$$\mathbf{z}_1 = \left(\tfrac{1}{2}, 1, 0, 0\right)^T, \quad \mathbf{z}_2 = \left(-2, 0, 2, 1\right)^T.$$

We claim that $\mathbf{z}_1$, $\mathbf{z}_2$ form a basis of $\ker A$. By construction, they span the kernel, and linear independence follows easily since the only way in which the linear combination (2.41) could vanish is if both free variables vanish: $x_2 = x_4 = 0$.

Corange: The corange is the subspace of $\mathbb{R}^n$ spanned by the rows[‡] of A. As we prove below, applying an elementary row operation to a matrix does not alter its corange. Since the row echelon form U is obtained from A by a sequence of elementary row operations, we conclude that $\operatorname{corng} A = \operatorname{corng} U$. Moreover, the row echelon structure implies that the r nonzero rows of U are necessarily linearly independent, and hence form a basis of both $\operatorname{corng} U$ and $\operatorname{corng} A$, which therefore have dimension $r = \operatorname{rank} A$. In our example, then, a basis for $\operatorname{corng} A$ consists of the vectors

$$\mathbf{s}_1 = \left(2, -1, 1, 2\right)^T, \quad \mathbf{s}_2 = \left(0, 0, -2, 4\right)^T,$$

coming from the nonzero rows of U. The reader can easily check their linear independence, as well as the fact that every row of A lies in their span.

Range: There are two methods for computing a basis of the range or column space. The first proves that it has dimension equal to the rank. This has the important, and remarkable consequence that the space spanned by the rows of a matrix and the space spanned by its columns always have the same dimension, even though they are usually different subspaces of different vector spaces.

Now, the row echelon structure implies that the columns of U which contain the pivots form a basis for its range, i.e., $\operatorname{rng} U$. In our example, these are its first and third columns, and you can check that they are linearly independent and span the full column space. But the range of A is *not* the same as the range of U, and so, unlike the corange, we cannot directly use a basis for $\operatorname{rng} U$ as a basis for $\operatorname{rng} A$. However, the linear dependencies among the columns of A and U *are* the same, and this implies that the r columns of A that end up containing the pivots will form a basis for $\operatorname{rng} A$. In our example (2.40), the pivots lie in the first and third columns of U, and hence the first and third columns of A, namely

$$\mathbf{v}_1 = \begin{pmatrix} 2 \\ -8 \\ 4 \end{pmatrix}, \quad \mathbf{v}_3 = \begin{pmatrix} 1 \\ -6 \\ 3 \end{pmatrix},$$

[‡]Or, more correctly, the transposes of the rows, since the elements of $\mathbb{R}^n$ are supposed to be column vectors.

form a basis for $\operatorname{rng} A$. This means that every column of A can be written uniquely as a linear combination of its first and third columns. Again, skeptics may wish to check this.

An alternative method to find a basis for the range is to recall that $\operatorname{rng} A = \operatorname{corng} A^T$, and hence we can employ the previous algorithm to compute $\operatorname{corng} A^T$. In our example, applying Gaussian Elimination to

$$A^T = \begin{pmatrix} 2 & -8 & 4 \\ -1 & 4 & -2 \\ 1 & -6 & 3 \\ 2 & -4 & 2 \end{pmatrix}$$

leads to the row echelon form

$$\widehat{U} = \begin{pmatrix} 2 & -8 & 4 \\ 0 & -2 & 1 \\ 0 & 0 & 0 \\ 0 & 0 & 0 \end{pmatrix}. \tag{2.42}$$

Note that the row echelon form of A^T is *not* the transpose of the row echelon form of A. However, they do have the same number of pivots since, as we now know, both A and A^T have the same rank, namely 2. Solving the homogeneous system $\widehat{U}\mathbf{y} = \mathbf{0}$, we conclude that

$$\mathbf{y}_1 = \begin{pmatrix} 2 \\ -8 \\ 4 \end{pmatrix}, \qquad \mathbf{y}_2 = \begin{pmatrix} 0 \\ -2 \\ 1 \end{pmatrix},$$

forms an alternative basis for $\operatorname{rng} A$.

Cokernel: Finally, to determine a basis for the cokernel, we apply the algorithm for finding a basis for $\ker A^T = \operatorname{coker} A$. Since the ranks of A and A^T coincide, there are now $m - r$ free variables, which is the same as the dimension of $\ker A^T$. In our particular example, using the reduced form (2.42), the only free variable is y_3, and the general solution to the homogeneous adjoint system $A^T\mathbf{y} = \mathbf{0}$ is

$$\mathbf{y} = \begin{pmatrix} 0 \\ \frac{1}{2}y_3 \\ y_3 \end{pmatrix} = y_3 \begin{pmatrix} 0 \\ \frac{1}{2} \\ 1 \end{pmatrix}.$$

We conclude that $\operatorname{coker} A$ is one-dimensional, with basis $\left(0, \frac{1}{2}, 1\right)^T$.

Summarizing, given an $m \times n$ matrix A with row echelon form U, to find a basis for

- $\operatorname{rng} A$: choose the r columns of A where the pivots appear in U;
- $\ker A$: write the general solution to $A\mathbf{x} = \mathbf{0}$ as a linear combination of the $n - r$ basis vectors whose coefficients are the free variables;
- $\operatorname{corng} A$: choose the r nonzero rows of U;
- $\operatorname{coker} A$: write the general solution to the adjoint system $A^T\mathbf{y} = \mathbf{0}$ as a linear combination of the $m - r$ basis vectors whose coefficients are the free variables.

Let us conclude this section by justifying these constructions for general matrices, and thereby complete the proof of the Fundamental Theorem 2.49.

Kernel: If A has rank r, then the general element of the kernel, i.e., solution to the homogeneous system $A\mathbf{x} = \mathbf{0}$, can be written as a linear combination of $n - r$ vectors whose coefficients are the free variables. Moreover, the only combination that yields the zero solution $\mathbf{x} = \mathbf{0}$ is when all the free variables are zero, since any nonzero value for a free variable, say $x_i \neq 0$, gives a solution $\mathbf{x} \neq \mathbf{0}$ whose ith entry (at least) is nonzero. Thus, the only linear combination of the $n - r$ kernel basis vectors summing up to $\mathbf{0}$ is the trivial one, which implies their linear independence.

Corange: We need to prove that elementary row operations do not change the corange. To see this for row operations of the first type, suppose, for instance, that $\widehat{A}$ is obtained by adding a times the first row of A to the second row. If $\mathbf{r}_1, \mathbf{r}_2, \mathbf{r}_3, \ldots, \mathbf{r}_m$ are the rows of A, then the rows of $\widehat{A}$ are $\mathbf{r}_1, \widehat{\mathbf{r}}_2 = \mathbf{r}_2 + a\,\mathbf{r}_1, \mathbf{r}_3, \ldots, \mathbf{r}_m$. If

$$\mathbf{v} = c_1\,\mathbf{r}_1 + c_2\,\mathbf{r}_2 + c_3\,\mathbf{r}_3 + \cdots + c_m\,\mathbf{r}_m$$

is any vector belonging to $\operatorname{corng} A$, then

$$\mathbf{v} = \widehat{c}_1\,\mathbf{r}_1 + c_2\,\widehat{\mathbf{r}}_2 + c_3\,\mathbf{r}_3 + \cdots + c_m\,\mathbf{r}_m, \qquad \text{where} \quad \widehat{c}_1 = c_1 - a\,c_2,$$

is also a linear combination of the rows of the new matrix, and hence lies in $\operatorname{corng} \widehat{A}$. The converse is also valid—$\mathbf{v} \in \operatorname{corng} \widehat{A}$ implies $\mathbf{v} \in \operatorname{corng} A$—and we conclude that elementary row operations of type #1 do not change $\operatorname{corng} A$. The proofs for the other two types of elementary row operations are even easier, and are left to the reader.

The basis for $\operatorname{corng} A$ will be the first r nonzero pivot rows $\mathbf{s}_1, \ldots, \mathbf{s}_r$ of U. Since the other rows, if any, are all $\mathbf{0}$, the pivot rows clearly span $\operatorname{corng} U = \operatorname{corng} A$. To prove their linear independence, suppose

$$c_1\mathbf{s}_1 + \cdots + c_r\mathbf{s}_r = \mathbf{0}. \tag{2.43}$$

Let $u_{1k} \neq 0$ be the first pivot. Since all entries of U lying below the pivot are zero, the kth entry of (2.43) is $c_1\,u_{1k} = 0$, which implies that $c_1 = 0$. Next, suppose $u_{2l} \neq 0$ is the second pivot. Again, using the row echelon structure of U, the lth entry of (2.43) is found to be $c_1\,u_{1l} + c_2\,u_{2l} = 0$, and so $c_2 = 0$ since we already know $c_1 = 0$. Continuing in this manner, we deduce that only the trivial linear combination $c_1 = \cdots = c_r = 0$ will satisfy (2.43), proving linear independence. Thus, $\mathbf{s}_1, \ldots, \mathbf{s}_r$ form a basis for $\operatorname{corng} U = \operatorname{corng} A$, which therefore has dimension $r = \operatorname{rank} A$.

Range: In general, a vector $\mathbf{b} \in \operatorname{rng} A$ if and only if it can be written as a linear combination of the columns: $\mathbf{b} = A\,\mathbf{x}$. But, as we know, the general solution to the linear system $A\,\mathbf{x} = \mathbf{b}$ is expressed in terms of the free and basic variables; in particular, we are allowed to set all the free variables to zero, and so end up writing $\mathbf{b}$ in terms of the basic variables alone. This effectively expresses $\mathbf{b}$ as a linear combination of the pivot columns of A only, which proves that they span $\operatorname{rng} A$. To prove their linear independence, suppose some linear combination of the pivot columns adds up to $\mathbf{0}$. Interpreting the coefficients as basic variables, this would correspond to a vector $\mathbf{x}$, all of whose free variables are zero, satisfying $A\,\mathbf{x} = \mathbf{0}$. But our solution to this homogeneous system expresses the basic variables as combinations of the free variables which, if the latter are all zero, are also zero when the right hand sides all vanish. This shows that, under these assumptions, $\mathbf{x} = \mathbf{0}$, and hence the pivot columns are linearly independent.

Cokernel: By the preceding arguments, $\operatorname{rank} A = \operatorname{rank} A^T = r$, and hence the general element of $\operatorname{coker} A = \ker A^T$ can be written as a linear combination of $m - r$ basis vectors whose coefficients are the free variables in the homogeneous adjoint system $A^T\mathbf{y} = \mathbf{0}$. Linear independence of the basis elements follows as in the case of the kernel.

EXERCISES 2.5

2.5.21. For each of the following matrices find bases for the

(i) range,

(ii) corange,

(iii) kernel, and

(iv) cokernel.

(a) $\begin{pmatrix} 1 & -3 \\ 2 & -6 \end{pmatrix}$

(b) $\begin{pmatrix} 0 & 0 & -8 \\ 1 & 2 & -1 \\ 2 & 4 & 6 \end{pmatrix}$

(c) $\begin{pmatrix} 1 & 1 & 2 & 1 \\ 1 & 0 & -1 & 3 \\ 2 & 3 & 7 & 0 \end{pmatrix}$

(d) $\begin{pmatrix} 1 & -3 & 2 & 2 & 1 \\ 0 & 3 & -6 & 0 & -2 \\ 2 & -3 & -2 & 4 & 0 \\ 3 & -3 & -6 & 6 & 3 \\ 1 & 0 & -4 & 2 & 3 \end{pmatrix}$

2.5.22. Find a set of columns of the matrix

$$\begin{pmatrix} -1 & 2 & 0 & -3 & 5 \\ 2 & -4 & 1 & 1 & -4 \\ -3 & 6 & 2 & 0 & 8 \end{pmatrix}$$

that form a basis for its range. Then express each column as a linear combination of the basis columns.

2.5.23. Find bases for the range and corange of

$$\begin{pmatrix} 1 & -3 & 0 \\ 2 & -6 & 4 \\ -3 & 9 & 1 \end{pmatrix}.$$

Make sure they have the same number of elements. Then write each row and column as a linear combination of the appropriate basis vectors.

2.5.24. For each of the following matrices A:

(a) Determine the rank and the dimensions of the four fundamental subspaces.

(b) Find bases for both the kernel and cokernel.

(c) Find explicit conditions on vectors $\mathbf{b}$ which guarantee that the system $A\mathbf{x} = \mathbf{b}$ has a solution.

(d) Write down a specific *nonzero* vector $\mathbf{b}$ that satisfies your conditions, and then find all possible solutions $\mathbf{x}$.

(i) $\begin{pmatrix} 1 & 2 \\ -2 & -4 \end{pmatrix}$ (ii) $\begin{pmatrix} 3 & -1 & -2 \\ -6 & 2 & 4 \end{pmatrix}$

(iii) $\begin{pmatrix} 1 & 5 \\ -2 & 3 \\ 2 & 7 \end{pmatrix}$ (iv) $\begin{pmatrix} 2 & -5 & -1 \\ 1 & -6 & -4 \\ 3 & -4 & 2 \end{pmatrix}$

(v) $\begin{pmatrix} 2 & 5 & 7 \\ 6 & 13 & 19 \\ 3 & 8 & 11 \\ 1 & 2 & 3 \end{pmatrix}$

(vi) $\begin{pmatrix} 1 & 2 & 3 & 4 \\ 3 & 2 & 4 & 1 \\ 1 & -2 & 2 & 7 \\ 3 & 6 & 5 & -2 \end{pmatrix}$

(vii) $\begin{pmatrix} 2 & 4 & 0 & -6 & 0 \\ 1 & 2 & 3 & 15 & 0 \\ 3 & 6 & -1 & 15 & 5 \\ -3 & -6 & 2 & 21 & -6 \end{pmatrix}$

2.5.25. Find the dimension of and a basis for the subspace spanned by the following sets of vectors. *Hint*: First identify the subspace with the range of a certain matrix.

(a) $\begin{pmatrix} 1 \\ 2 \\ -1 \end{pmatrix}, \begin{pmatrix} 2 \\ 2 \\ 0 \end{pmatrix}$

(b) $\begin{pmatrix} 1 \\ 1 \\ -1 \end{pmatrix}, \begin{pmatrix} 2 \\ 2 \\ -2 \end{pmatrix}, \begin{pmatrix} -3 \\ -3 \\ 3 \end{pmatrix}$

(c) $\begin{pmatrix} 1 \\ 0 \\ 1 \\ 0 \end{pmatrix}, \begin{pmatrix} 1 \\ 0 \\ 0 \\ 1 \end{pmatrix}, \begin{pmatrix} 2 \\ 2 \\ 1 \\ 0 \end{pmatrix}, \begin{pmatrix} 1 \\ 2 \\ 3 \\ -3 \end{pmatrix}$

(d) $\begin{pmatrix} 1 \\ 0 \\ -3 \\ 2 \end{pmatrix}, \begin{pmatrix} 0 \\ 1 \\ 2 \\ -3 \end{pmatrix}, \begin{pmatrix} -3 \\ -4 \\ 1 \\ 6 \end{pmatrix}, \begin{pmatrix} 1 \\ -3 \\ -8 \\ 7 \end{pmatrix}, \begin{pmatrix} 2 \\ 1 \\ -6 \\ 9 \end{pmatrix}$

(e) $\begin{pmatrix} 1 \\ 1 \\ -1 \\ 1 \\ 1 \end{pmatrix}, \begin{pmatrix} 2 \\ -1 \\ 2 \\ 2 \\ 1 \end{pmatrix}, \begin{pmatrix} 3 \\ 0 \\ 1 \\ 3 \\ 2 \end{pmatrix}, \begin{pmatrix} 0 \\ -3 \\ 4 \\ 0 \\ -1 \end{pmatrix}, \begin{pmatrix} 1 \\ 3 \\ -1 \\ 2 \\ 1 \end{pmatrix}, \begin{pmatrix} 1 \\ 0 \\ 3 \\ 2 \\ 0 \end{pmatrix}$

2.5.26. Show that the set of all vectors of the form $\mathbf{v} = (a - 3b, a + 2c + 4d, b + 3c - d, c - d)^T$, where a, b, c, d are real numbers, forms a subspace of $\mathbb{R}^4$, and find its dimension.

2.5.27. Find a basis of the solution space of the following homogeneous linear systems.

(a) $x_1 - 2x_3 = 0,$
$x_2 + x_4 = 0.$

(b) $2x_1 + x_2 - 3x_3 + x_4 = 0,$
$2x_1 - x_2 - x_3 - x_4 = 0.$

(c) $x_1 - x_2 - 2x_3 + 4x_4 = 0,$
$2x_1 + x_2 - x_4 = 0,$
$-2x_1 + 2x_3 - 2x_4 = 0.$

2.5.28. Find bases for the range of

$$\begin{pmatrix} 1 & 2 & -1 \\ 0 & 3 & -3 \\ 2 & -4 & 6 \\ 1 & 5 & -4 \end{pmatrix}$$

using both of the indicated methods. Demonstrate that they are indeed both bases for the same subspace by showing how to write each basis in terms of the other.

2.5.29. Show that

$$\mathbf{v}_1 = \begin{pmatrix} 1 \\ 2 \\ 0 \\ -1 \end{pmatrix}, \quad \mathbf{v}_2 = \begin{pmatrix} -3 \\ 1 \\ 1 \\ -1 \end{pmatrix}, \quad \mathbf{v}_3 = \begin{pmatrix} 2 \\ 0 \\ -4 \\ 3 \end{pmatrix}$$

and

$$\mathbf{w}_1 = \begin{pmatrix} 3 \\ 2 \\ -4 \\ 2 \end{pmatrix}, \quad \mathbf{w}_2 = \begin{pmatrix} 2 \\ 3 \\ -7 \\ 4 \end{pmatrix}, \quad \mathbf{w}_3 = \begin{pmatrix} 0 \\ 3 \\ -3 \\ 1 \end{pmatrix}$$

are two bases for the same three-dimensional subspace $V \subset \mathbb{R}^4$.

2.5.30. (a) Prove that if A is a symmetric matrix, then $\ker A = \operatorname{coker} A$ and $\operatorname{rng} A = \operatorname{corng} A$.

(b) Use this observation to produce bases for the four fundamental subspaces associated with $A = \begin{pmatrix} 1 & 2 & 0 \\ 2 & 6 & 2 \\ 0 & 2 & 2 \end{pmatrix}$.

(c) Is the converse to part (a) true?

2.5.31. Let A be a 4×4 matrix and let U be its row echelon form.

(a) Suppose columns 1, 2, 4 of U form a basis for its range. Do columns 1, 2, 4 of A form a basis for its range? If so, explain why; if not, construct a counterexample.

(b) Suppose rows 1, 2, 3 of U form a basis for its corange. Do rows 1, 2, 3 of A form a basis for its corange? If so, explain why; if not, construct a counterexample.

(c) Suppose you find a basis for $\ker U$. Is it also a basis for $\ker A$?

(d) Suppose you find a basis for $\operatorname{coker} U$. Is it also a basis for $\operatorname{coker} A$?

2.5.32. (a) Write down a matrix of rank r whose first r rows do *not* form a basis for its row space.

(b) Can you find an example that can be reduced to row echelon form without any row interchanges?

2.5.33. Can you devise a nonzero matrix whose row echelon form is the same as the row echelon form of its transpose?

◇ **2.5.34.** Explain why the elementary row operations of types #2 and #3 do not change the corange of a matrix.

2.5.35. Let A be an $m \times n$ matrix. Prove that $\operatorname{rng} A = \mathbb{R}^m$ if and only if $\operatorname{rank} A = m$.

2.5.36. Prove or give a counterexample: If U is the row echelon form of A, then $\operatorname{rng} U = \operatorname{rng} A$.

◇ **2.5.37.** (a) Devise an alternative method for finding a basis of the corange of a matrix. *Hint*: Look at the two methods for finding a basis for the range.

(b) Use your method to find a basis for the corange of

$$\begin{pmatrix} 1 & 3 & -5 & 2 \\ 2 & -1 & 1 & -4 \\ 4 & 5 & -9 & 2 \end{pmatrix}.$$

Is it the same basis as found by the method in the text?

◇ **2.5.38.** Prove that $\ker A \subseteq \ker A^2$. More generally, prove $\ker A \subseteq \ker B A$ for any (compatible) matrix B.

◇ **2.5.39.** Prove that $\operatorname{rng} A \supseteq \operatorname{rng} A^2$. More generally, prove $\operatorname{rng} A \supseteq \operatorname{rng}(A B)$ for any (compatible) matrix B.

2.5.40. Suppose A is an $m \times n$ matrix, and B and C are nonsingular matrices of sizes $m \times m$ and $n \times n$, respectively. Prove that $\operatorname{rank} A = \operatorname{rank} B A = \operatorname{rank} A C = \operatorname{rank} B A C$.

◇ **2.5.41.** Suppose A is a nonsingular $n \times n$ matrix.

(a) Prove that any $n \times (n+k)$ matrix of the form $\begin{pmatrix} A & B \end{pmatrix}$, where B has size $n \times k$, has rank n.

(b) Prove that any $(n+k) \times n$ matrix of the form $\begin{pmatrix} A \\ C \end{pmatrix}$, where C has size $k \times n$, has rank n.

2.5.42. *True or false*: If $\ker A = \ker B$, then $\operatorname{rank} A = \operatorname{rank} B$.

◇ **2.5.43.** Let A be an $m \times n$ matrix of rank r. Suppose $\mathbf{v}_1, \ldots, \mathbf{v}_n$ are a basis for $\mathbb{R}^n$ such that $\mathbf{v}_{r+1}, \ldots, \mathbf{v}_n$ form a basis for $\ker A$. Prove that $\mathbf{w}_1 = A\mathbf{v}_1, \ldots, \mathbf{w}_r = A\mathbf{v}_r$ form a basis for $\operatorname{rng} A$.

◇ **2.5.44.** (a) Suppose A, B are $m \times n$ matrices such that $\ker A = \ker B$. Prove that there is a nonsingular $m \times m$ matrix M such that $MA = B$. *Hint*: Use Exercise 2.5.43.

(b) Use this to conclude that if $A\mathbf{x} = \mathbf{b}$ and $B\mathbf{x} = \mathbf{c}$ have the same solutions then they are equivalent linear systems, i.e., one can be obtained from the other by a sequence of elementary row operations.

◇ **2.5.45.** Let A be an $m \times n$ matrix and let V be a subspace of $\mathbb{R}^n$.

(a) Show that $W = AV = \{ A\mathbf{v} \mid \mathbf{v} \in V \}$ forms a subspace of $\operatorname{rng} A$.

(b) If $\dim V = k$, show that $\dim W \le \min\{k, r\}$, where $r = \operatorname{rank} A$. *Hint*: Use Exercise 2.4.25.

◇ **2.5.46.** (a) Show that an $m \times n$ matrix has a left inverse if and only if it has rank n. *Hint*: Use Exercise 2.5.45.

(b) Show that it has a right inverse if and only if it has rank m.

(c) Conclude that only nonsingular square matrices have both left and right inverses.

2.6 GRAPHS AND INCIDENCE MATRICES

We now present an intriguing application of linear algebra to graph theory. A *graph* consists of a finite number of points, called *vertices*, and finitely many lines or curves connecting them, called *edges*. Each edge connects exactly two vertices, which, for later convenience, are assumed to always be distinct, so that no edge forms a *loop* that connects a vertex to itself. However, we do permit two vertices to be connected by multiple edges. Some examples of graphs appear in Figure 2.6; the vertices are the black dots and the edges are the lines connecting them.

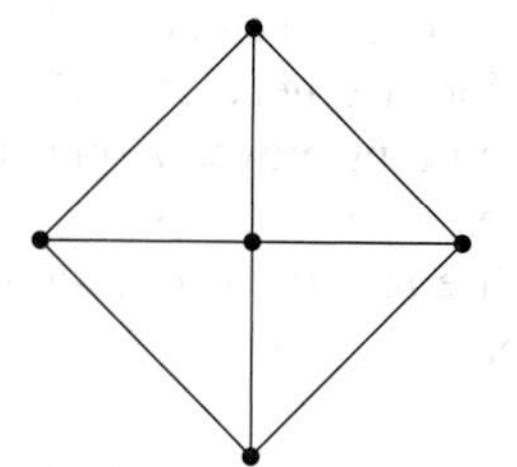
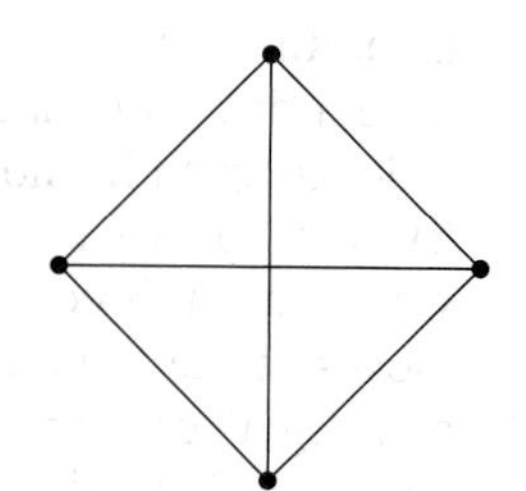
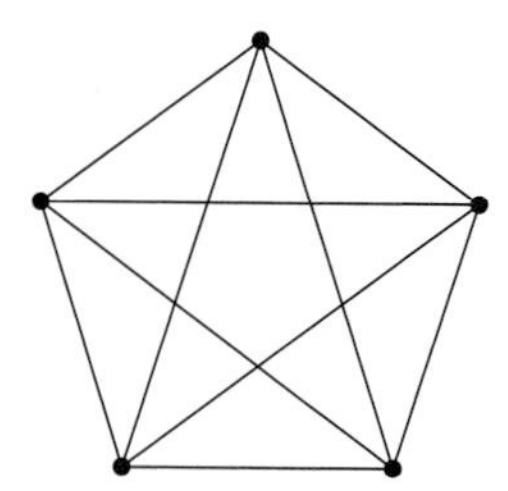

Figure 2.6 Three different graphs.

Graphs arise in a multitude of applications. A particular case that will be considered in depth is electrical networks, where the edges represent wires, and the vertices represent the nodes where the wires are connected. Another example is the framework for a building—the edges represent the beams, and the vertices the joints where the beams are connected. In each case, the graph encodes the topology—meaning interconnectedness—of the system, but not its geometry—lengths of edges, angles, etc.

In a planar representation of the graph, the edges are allowed to cross over each other at non-nodal points without meeting—think of a network where the (insulated) wires lie on top of each other, but do not interconnect. Thus, the first graph in Figure 2.6 has 5 vertices and 8 edges; the second has 4 vertices and 6 edges—the two central edges do not meet; the final graph has 5 vertices and 10 edges.

Two graphs are considered to be the same if there is a one-to-one correspondence between their edges and their vertices, so that matched edges connect matched vertices. In an electrical network, moving the nodes and wires around without cutting or rejoining will have no effect on the underlying graph. Consequently, there are many ways to draw a given graph; three representations of one and the same graph appear in Figure 2.7.

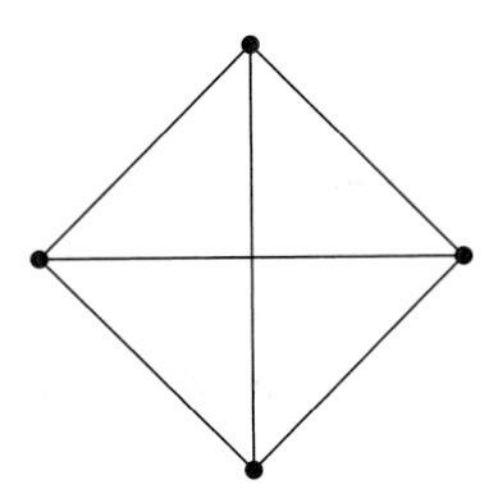
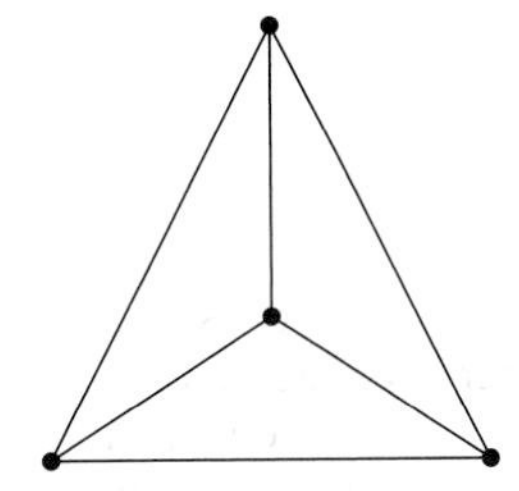
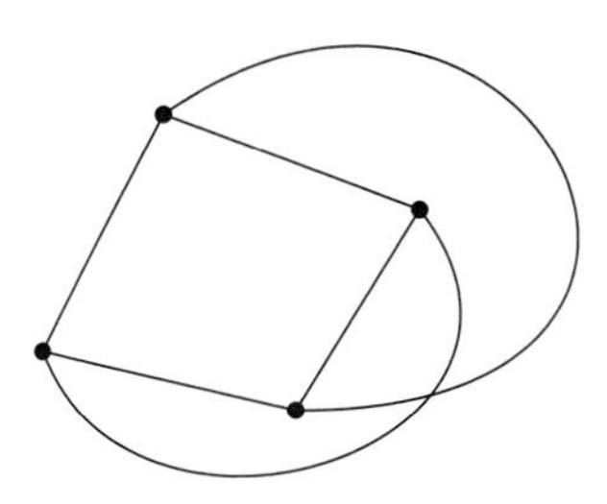

Figure 2.7 Three versions of the same graph.

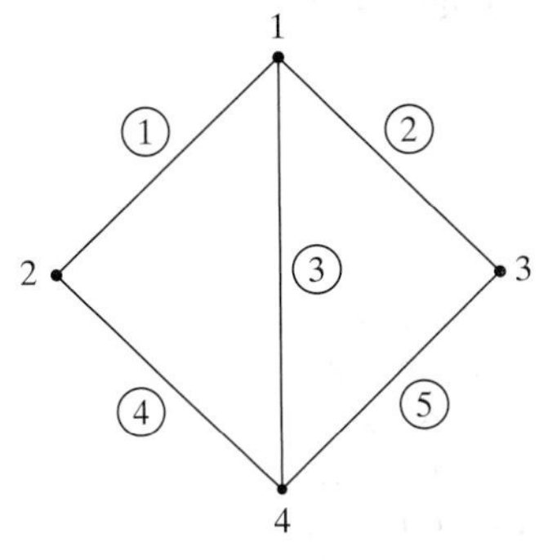

Figure 2.8 A simple graph.

A *path* in a graph is an ordered list of distinct edges $e_1, \ldots, e_k$ connecting (not necessarily distinct) vertices $v_1, \ldots, v_{k+1}$ so that edge e_i connects vertex v_i to v_{i+1}. For instance, in the graph in Figure 2.8, one path starts at vertex 1, then goes in order along the edges labeled as 1, 4, 3, 2, successively passing through the vertices 1, 2, 4, 1, 3. Observe that while an edge cannot be repeated in a path, a vertex may be. A graph is *connected* if you can get from any vertex to any other vertex by a path, which is the most important case for applications. We note that every graph can be decomposed into a disconnected collection of connected subgraphs.

A *circuit* is a path that ends up where it began, i.e., $v_{k+1} = v_1$. For example, the circuit consisting of edges 1, 4, 5, 2 starts at vertex 1, then goes to vertices 2, 4, 3 in order, and finally returns to vertex 1. In a closed circuit, the choice of starting vertex is not important, and we identify circuits that go around the edges in the same order. Thus, for example, the edges 4, 5, 2, 1 represent the same circuit as above.

In electrical circuits, one is interested in measuring currents and voltage drops along the wires in the network represented by the graph. Both of these quantities have a direction, and therefore we need to specify an orientation on each edge in order to quantify how the current moves along the wire. The orientation will be fixed by specifying the vertex the edge "starts" at, and the vertex it "ends" at. Once we assign a direction to an edge, a current along that wire will be positive if it moves in the same direction, i.e., goes from the starting vertex to the ending one, and negative if it moves in the opposite direction. The direction of the edge does *not* dictate the direction of the current—it just fixes what directions positive and negative values of current represent. A graph with directed edges is known as a *directed graph* or *digraph* for short. The edge directions are represented by arrows; examples of digraphs can be seen in Figure 2.9.

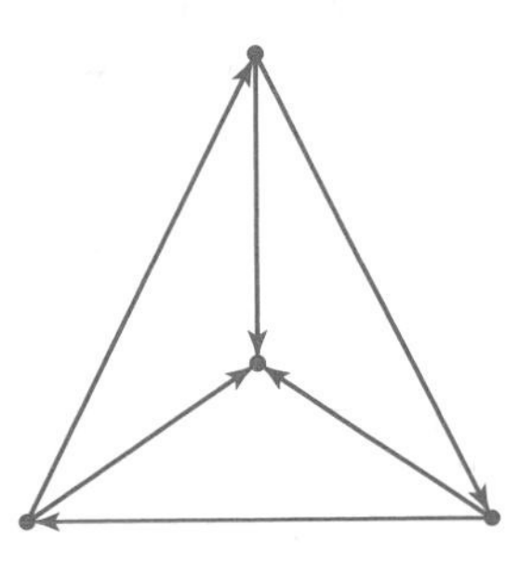
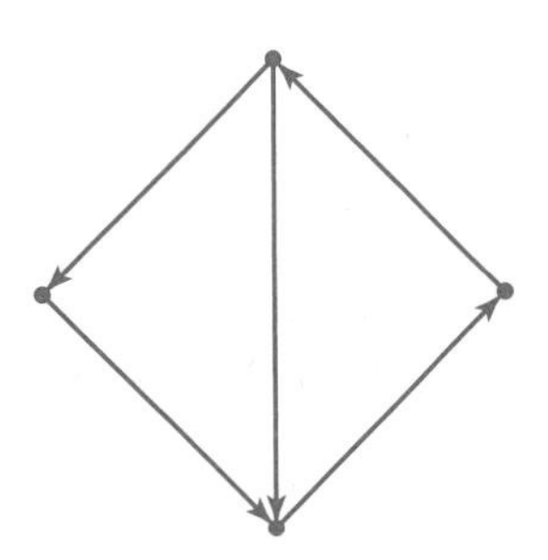
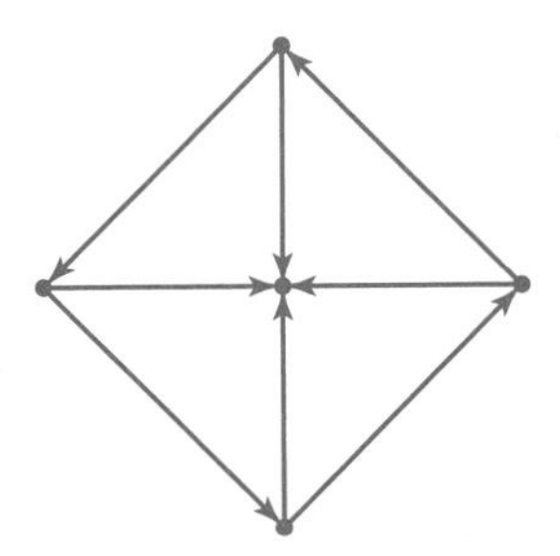

Figure 2.9 Digraphs.

Consider a digraph D consisting of n vertices connected by m edges. The *incidence matrix* associated with D is an $m \times n$ matrix A whose rows are indexed by the edges and whose columns are indexed by the vertices. If edge k starts at vertex i and ends at vertex j, then row k of the incidence matrix will have $+1$ in its (k, i) entry and -1 in its (k, j) entry; all other entries in the row are zero. Thus, our

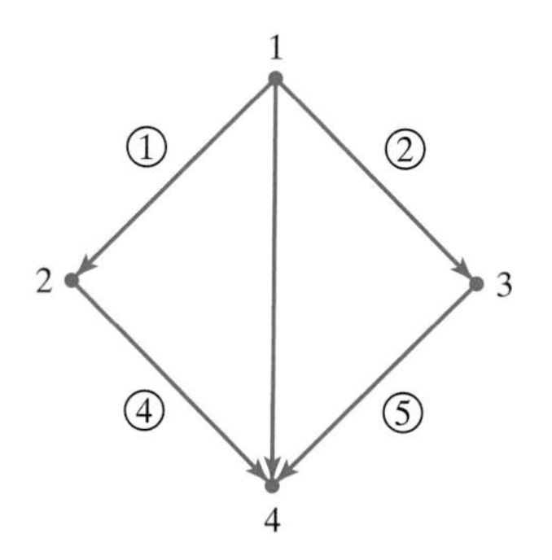

Figure 2.10 A simple digraph.

convention is that $+1$ represents the outgoing vertex at which the edge starts, and -1 the incoming vertex at which it ends.

A simple example is the digraph in Figure 2.10, which consists of five edges joined at four different vertices. Its 5×4 incidence matrix is

$$A = \begin{pmatrix} 1 & -1 & 0 & 0 \\ 1 & 0 & -1 & 0 \\ 1 & 0 & 0 & -1 \\ 0 & 1 & 0 & -1 \\ 0 & 0 & 1 & -1 \end{pmatrix}. \tag{2.44}$$

Thus the first row of A tells us that the first edge starts at vertex 1 and ends at vertex 2. Similarly, row 2 says that the second edge goes from vertex 1 to vertex 3. Clearly, one can completely reconstruct any digraph from its incidence matrix.

EXAMPLE 2.50 The matrix

$$A = \begin{pmatrix} 1 & -1 & 0 & 0 & 0 \\ -1 & 0 & 1 & 0 & 0 \\ 0 & -1 & 1 & 0 & 0 \\ 0 & 1 & 0 & -1 & 0 \\ 0 & 0 & -1 & 1 & 0 \\ 0 & 0 & 1 & 0 & -1 \\ 0 & 0 & 0 & 1 & -1 \end{pmatrix}. \tag{2.45}$$

qualifies as an incidence matrix because each row contains a single $+1$, a single -1, and the other entries are 0. Let us construct the digraph corresponding to A. Since A has five columns, there are five vertices in the digraph, which we label by the numbers $1, 2, 3, 4, 5$. Since it has seven rows, there are 7 edges. The first row has its $+1$ in column 1 and its -1 in column 2, and so the first edge goes from vertex 1 to vertex 2. Similarly, the second edge corresponds to the second row of A and so goes from vertex 3 to vertex 1. The third row of A indicates an edge from vertex 3 to vertex 2; and so on. In this manner, we construct the digraph drawn in Figure 2.11. ●

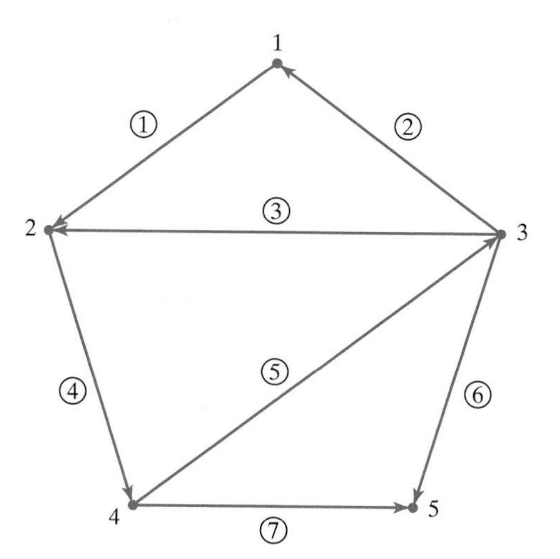

Figure 2.11 Another digraph.

The incidence matrix serves to encode important geometric and quantitative information of the digraph it represents. In particular, its kernel and cokernel have topological significance. For example, the kernel of the incidence matrix (2.45) is spanned by the single vector $\mathbf{z} = \left(1, 1, 1, 1, 1\right)^T$, and represents the fact that the sum of the entries in any given row of A is zero. This observation holds in general for connected digraphs.

Proposition 2.51 If A is the incidence matrix for a connected digraph, then $\ker A$ is one-dimensional, with basis $\mathbf{z} = \left(1, 1, \ldots, 1\right)^T$.

Proof If edge k connects vertices i and j, then the kth equation in $A\mathbf{z} = \mathbf{0}$ is $z_i = z_j$. The same equality holds, by a simple induction, if the vertices i and j are connected by a path. Therefore, if D is connected, all the entries of $\mathbf{z}$ are equal, and the result follows. ■

Applying the Fundamental Theorem 2.49, we immediately deduce:

Corollary 2.52 If A is the incidence matrix for a connected digraph with n vertices, then $\operatorname{rank} A = n - 1$.

Next, let us look at the cokernel of an incidence matrix. Consider the particular example (2.44) corresponding to the digraph in Figure 2.10. We need to compute the kernel of the transposed incidence matrix

$$A^T = \begin{pmatrix} 1 & 1 & 1 & 0 & 0 \\ -1 & 0 & 0 & 1 & 0 \\ 0 & -1 & 0 & 0 & 1 \\ 0 & 0 & -1 & -1 & -1 \end{pmatrix}. \tag{2.46}$$

Solving the homogeneous system $A^T\mathbf{y} = \mathbf{0}$ by Gaussian Elimination, we discover that $\operatorname{coker} A = \ker A^T$ is spanned by the two vectors

$$\mathbf{y}_1 = \left(1, 0, -1, 1, 0\right)^T, \quad \mathbf{y}_2 = \left(0, 1, -1, 0, 1\right)^T.$$

Each of these vectors represents a *circuit* in the digraph. Keep in mind that their entries are indexed by the edges, so a nonzero entry indicates the direction to traverse the corresponding edge. For example, $\mathbf{y}_1$ corresponds to the circuit that starts out along edge 1, then goes along edge 4 and finishes by going along edge 3 in the reverse direction, which is indicated by the minus sign in its third entry. Similarly, $\mathbf{y}_2$ represents the circuit consisting of edge 2, followed by edge 5, and then edge 3, backwards. The fact that $\mathbf{y}_1$ and $\mathbf{y}_2$ are linearly independent vectors says that the two circuits are "independent".

The general element of $\operatorname{coker} A$ is a linear combination $c_1\mathbf{y}_1 + c_2\mathbf{y}_2$. Certain values of the constants lead to other types of circuits; for example $-\mathbf{y}_1$ represents the same circuit as $\mathbf{y}_1$, but traversed in the opposite direction. Another example is

$$\mathbf{y}_1 - \mathbf{y}_2 = \left(1, -1, 0, 1, -1\right)^T,$$

which represents the square circuit going around the outside of the digraph along edges 1, 4, 5, 2, the fifth and second edges taken in the reverse direction. We can view this circuit as a combination of the two triangular circuits; when we add them together the middle edge 3 is traversed once in each direction, which effectively "cancels" its contribution. (A similar cancellation occurs in the calculus of line integrals, [2, 58].) Other combinations represent "virtual" circuits; for instance, one can interpret $2\mathbf{y}_1 - \frac{1}{2}\mathbf{y}_2$ as two times around the first triangular circuit plus one half of the other triangular circuit, taken in the reverse direction—whatever that might mean.

Let us summarize the preceding discussion.

THEOREM 2.53 Each circuit in a digraph D is represented by a vector in the cokernel of its incidence matrix, whose entries are $+1$ if the edge is traversed in the correct direction, -1 if in the opposite direction, and 0 if the edge is not in the circuit. The dimension of the cokernel of A equals the number of independent circuits in D.

REMARK: A full proof that the cokernel of the incidence matrix of a general digraph has a basis consisting entirely of independent circuits requires a more in depth analysis of the properties of graphs than we can provide in this abbreviated treatment. Full details can be found in [5, §II.3].

The preceding two theorems have an important and remarkable consequence. Suppose D is a connected digraph with m edges and n vertices and A its $m \times n$ incidence matrix. Corollary 2.52 implies that A has rank $r = n-1 = n - \dim\ker A$. On the other hand, Theorem 2.53 tells us that $l = \dim\operatorname{coker} A$ equals the number of independent circuits in D. The Fundamental Theorem 2.49 says that $r = m - l$. Equating these two formulas for the rank, we find $r = n-1 = m-l$, or $n+l = m+1$. This celebrated result is known as *Euler's formula* for graphs, first discovered by the extraordinarily prolific eighteenth century Swiss mathematician Leonhard Euler*.

THEOREM 2.54 If G is a connected graph, then

$$\#\,\text{vertices} + \#\,\text{independent circuits} = \#\,\text{edges} + 1. \tag{2.47}$$

REMARK: If the graph is *planar*, meaning that it can be graphed in the plane without any edges crossing over each other, then the number of independent circuits is equal to the number of "holes", i.e., the number of distinct polygonal regions bounded by the edges of the graph. For example, the pentagonal digraph in Figure 2.11 bounds three triangles, and so has three independent circuits.

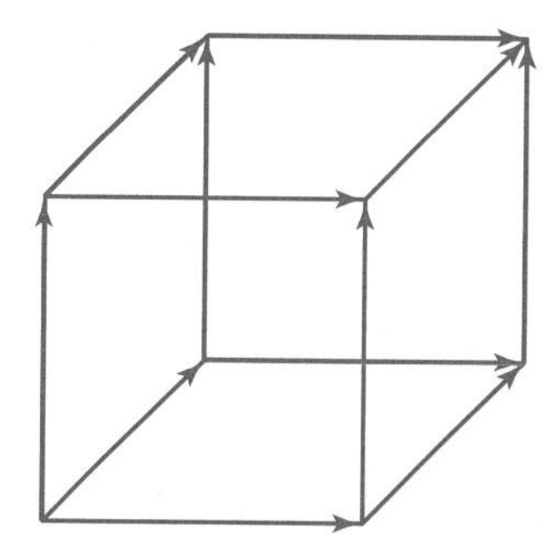

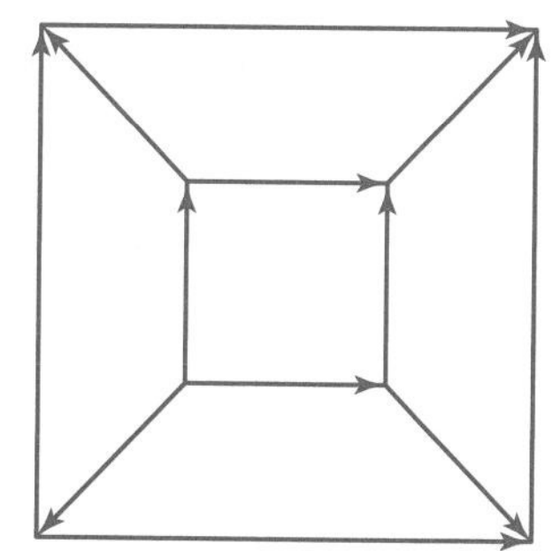

Figure 2.12 A cubical digraph.

EXAMPLE 2.55 Consider the graph corresponding to the edges of a cube, as illustrated in Figure 2.12, where the second figure represents the same graph squashed down onto a plane. The graph has 8 vertices and 12 edges. Euler's formula (3.84) tells us that there are 5 independent circuits. These correspond to the interior square and four

*Pronounced "Oiler". Euler spent most of his career in Russia.

trapezoids in the planar version of the digraph, and hence to circuits around 5 of the 6 faces of the cube. The "missing" face does indeed define a circuit, but it can be represented as the sum of the other five circuits, and so is not independent. In Exercise 2.6.6, the reader is asked to write out the incidence matrix for the cubical digraph and explicitly identify the basis of its kernel with the circuits. ●

Further development of the many remarkable connections between graph theory and linear algebra would, unfortunately, take us too far afield. The interested reader is encouraged to consult a more specialized text in graph theory, e.g., [5].

EXERCISES 2.6

2.6.1. Draw the digraph represented by the following incidence matrices:

(a) $\begin{pmatrix} -1 & 0 & 1 & 0 \\ 1 & 0 & 0 & -1 \\ 0 & -1 & 1 & 0 \\ 0 & 1 & 0 & -1 \end{pmatrix}$

(b) $\begin{pmatrix} 1 & 0 & -1 & 0 \\ 0 & 1 & 0 & -1 \\ -1 & 1 & 0 & 0 \\ 0 & 0 & 1 & -1 \end{pmatrix}$

(c) $\begin{pmatrix} 0 & 1 & 0 & 0 & -1 \\ -1 & 0 & 1 & 0 & 0 \\ 0 & 0 & 0 & -1 & 1 \\ 0 & -1 & 1 & 0 & 0 \end{pmatrix}$

(d) $\begin{pmatrix} -1 & 0 & 1 & 0 & 0 \\ 0 & -1 & 0 & 1 & 0 \\ 1 & -1 & 0 & 0 & 0 \\ 0 & 0 & 0 & -1 & 1 \\ 0 & 0 & -1 & 0 & 1 \end{pmatrix}$

(e) $\begin{pmatrix} 0 & 1 & -1 & 0 & 0 & 0 & 0 \\ -1 & 0 & 0 & 1 & 0 & 0 & 0 \\ 0 & 0 & 0 & -1 & 1 & 0 & 0 \\ 0 & -1 & 0 & 0 & 0 & 0 & 1 \\ 0 & 0 & -1 & 0 & 0 & 1 & 0 \\ 0 & 0 & 0 & 0 & 0 & 1 & -1 \end{pmatrix}$

2.6.2. (a) Draw the graph corresponding to the 6×7 incidence matrix whose nonzero (i, j) entries equal 1 if $j = i$ and -1 if $j = i + 1$, for $i = 1$ to 6.

(b) Find a basis for its kernel and cokernel.

(c) How many circuits are in the digraph?

2.6.3. Write out the incidence matrix of the following digraphs.

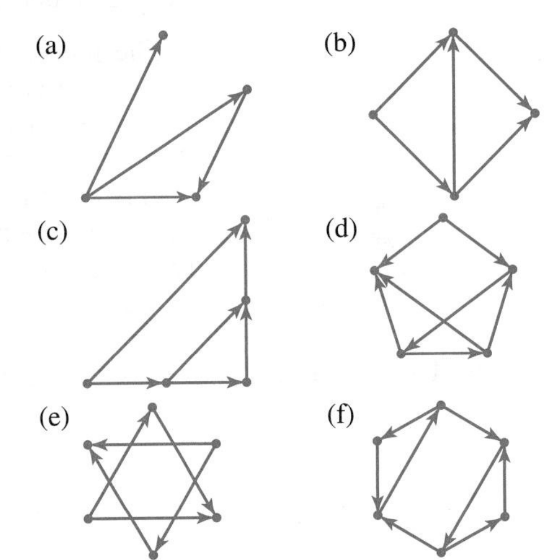

2.6.4. For each of the digraphs in Exercise 2.6.3, see if you can predict a collection of independent circuits. Verify your prediction by constructing a suitable basis of the cokernel of the incidence matrix and identifying each basis vector with a circuit.

♡ **2.6.5.** (a) Write down the incidence matrix A for the indicated digraph.

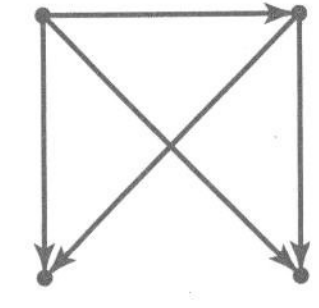

(b) What is the rank of A?

(c) Determine the dimensions of its four fundamental subspaces.

(d) Find a basis for its kernel and cokernel.

(e) Determine explicit conditions on vectors $\mathbf{b}$ which guarantee that the system $A\mathbf{x} = \mathbf{b}$ has a solution.

(f) Write down a specific nonzero vector $\mathbf{b}$ that satisfies your conditions, and then find all possible solutions.

◇ **2.6.6.** (a) Write out the incidence matrix for the cubical digraph and identify the basis of its cokernel with the circuits.

(b) Find three circuits which do not correspond to any of your basis elements, and express them as a linear combination of the basis circuit vectors.

♡ **2.6.7.** Write out the incidence matrix for the other Platonic solids:

(a) tetrahedron (b) octahedron

(c) dodecahedron (d) icosahedron

(You will need to choose an orientation for the edges.) Show that, in each case, the number of independent circuits equals the number of faces minus 1.

♡ **2.6.8.** A connected graph is called a *tree* if it has no circuits.

(a) Find the incidence matrix for each of the following directed trees:

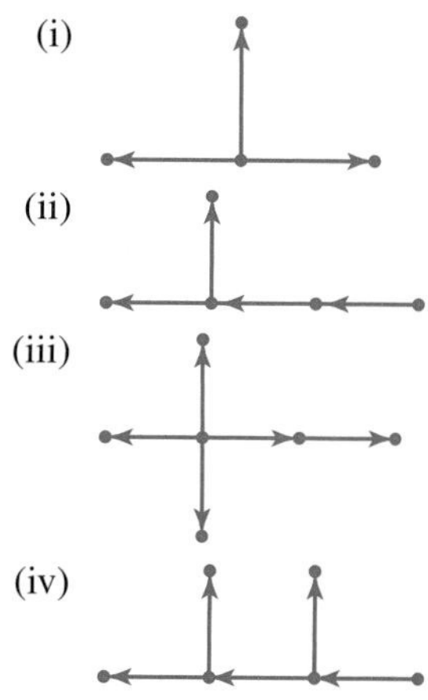

(b) Draw all distinct trees with 4 vertices. Assign a direction to the edges, and write down the corresponding incidence matrices.

(c) Prove that a connected graph on n vertices is a tree if and only if it has precisely $n-1$ edges.

♡ **2.6.9.** A *complete graph* K_n on n vertices has one edge joining every distinct pair of vertices.

(a) Draw K_3, K_4 and K_5.

(b) Choose an orientation for each edge and write out the resulting incidence matrix of each digraph.

(c) How many edges does K_n have?

(d) How many independent circuits?

♡ **2.6.10.** The *complete bipartite digraph* $K_{m,n}$ is based on two disjoint sets of, respectively, m and n vertices. Each vertex in the first set is connected to each vertex in the second set by a single edge.

(a) Draw $K_{2,3}$, $K_{2,4}$, and $K_{3,3}$.

(b) Write the incidence matrix of each digraph.

(c) How many edges does $K_{m,n}$ have?

(d) How many independent circuits?

♡ **2.6.11.** (a) Construct the incidence matrix A for the disconnected digraph D in the figure.

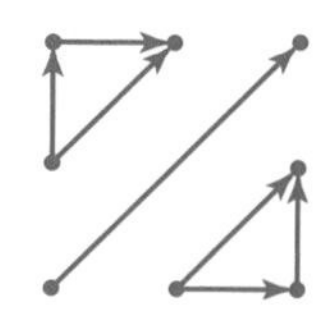

(b) Verify that $\dim \ker A = 3$, which is the same as the number of connected components, meaning the maximal connected subgraphs in D.

(c) Can you assign an interpretation to your basis for $\ker A$?

(d) Try proving the general statement that $\dim \ker A$ equals the number of connected components in the digraph D.

◇ **2.6.12.** Prove that a graph with n nodes and n edges must have at least one circuit.

2.6.13. How does altering the direction of the edges of a digraph affect its incidence matrix? The cokernel of its incidence matrix? Can you realize this operation by matrix multiplication?

♡ **2.6.14.** (a) Explain why two digraphs are equivalent under relabeling of vertices and edges if and only if their incidence matrices satisfy $PAQ = B$, where P, Q are permutation matrices.

(b) Decide which of the following incidence matrices produce the equivalent digraphs:

(i) $\begin{pmatrix} 1 & 0 & -1 & 0 \\ 0 & 1 & 0 & -1 \\ -1 & 1 & 0 & 0 \\ 0 & 0 & 1 & -1 \end{pmatrix}$

(ii) $\begin{pmatrix} 0 & -1 & 1 & 0 \\ -1 & 0 & 1 & 0 \\ 1 & 0 & 0 & -1 \\ 0 & -1 & 0 & 1 \end{pmatrix}$

(iii) $\begin{pmatrix} 1 & 0 & 0 & -1 \\ 0 & 1 & 0 & -1 \\ 1 & 0 & -1 & 0 \\ 0 & 0 & -1 & 1 \end{pmatrix}$

(iv) $\begin{pmatrix} 1 & -1 & 0 & 0 \\ 1 & 0 & -1 & 0 \\ 0 & -1 & 0 & 1 \\ 0 & 0 & -1 & 1 \end{pmatrix}$

(v) $\begin{pmatrix} 1 & 0 & 0 & -1 \\ 0 & 0 & -1 & 1 \\ 0 & -1 & 1 & 0 \\ 1 & -1 & 0 & 0 \end{pmatrix}$

(vi) $\begin{pmatrix} 1 & -1 & 0 & 0 \\ 0 & -1 & 1 & 0 \\ 0 & -1 & 0 & 1 \\ -1 & 0 & 1 & 0 \end{pmatrix}$

(c) How are the cokernels of equivalent incidence matrices related?

2.6.15. *True or false*: If A and B are incidence matrices of the same size and $\operatorname{coker} A = \operatorname{coker} B$, then the corresponding digraphs are equivalent.

2.6.16. (a) Explain why the incidence matrix for a disconnected graph can be written in block diagonal matrix form $A = \begin{pmatrix} B & O \\ O & C \end{pmatrix}$ under an appropriate labeling of the vertices.

(b) Show how to label the vertices of the digraph in Exercise 2.6.3e so that its incidence matrix is in block form.

CHAPTER 3

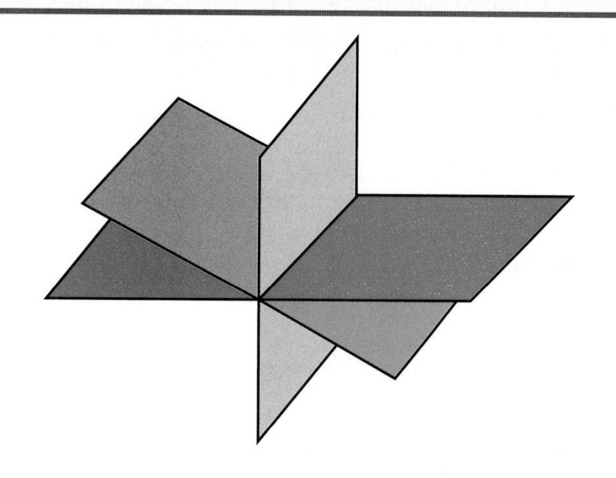

Inner Products and Norms

The geometry of Euclidean space is founded on the familiar properties of length and angle. The abstract concept of a norm on a vector space formalizes the geometrical notion of the length of a vector. In Euclidean geometry, the angle between two vectors is specified by their dot product, which is itself formalized by the abstract concept of an inner product. Inner products and norms lie at the heart of linear (and nonlinear) analysis, both in both finite-dimensional vector spaces and infinite-dimensional function spaces. It is impossible to overemphasize their importance for both theoretical developments, practical applications, and in the design of numerical solution algorithms. We begin this chapter with a discussion of the basic properties of inner products, illustrated by some of the most important examples.

Mathematical analysis is founded on inequalities. The most fundamental is the Cauchy–Schwarz inequality, which is valid in any inner product space. The more familiar triangle inequality for the associated norm is then derived as a simple consequence. Not every norm comes from an inner product, and, in such cases, the triangle inequality becomes part of the general definition. Both inequalities retain their validity in both finite-dimensional and infinite-dimensional vector spaces. Indeed, their abstract formulation exposes the key ideas for the proof, avoiding all distracting particularities appearing in the explicit formulas.

The characterization of general inner products on Euclidean space will lead us to the noteworthy class of positive definite matrices. The most common are the Gram matrices, whose entries are the inner products between selected elements of an inner product space. Positive definite matrices appear in a wide variety of applications, including minimization, least squares, mechanical systems, electrical circuits, and the differential equations describing dynamical processes. (Later, we will generalize the notion of positive definiteness to more general linear operators, governing the boundary value problems of continuous media.) The test for positive definiteness relies on Gaussian Elimination, and we can reinterpret the resulting matrix factorization as the process of completing the square for the associated quadratic form.

So far, we have confined our attention to real vector spaces. Complex numbers, vectors and functions also arise in numerous applications, and so, in the final section, we formally introduce complex vector spaces. Most of the theory proceeds in direct analogy with the real version, but the notions of inner product and norm on complex vector spaces require some thought. Applications of complex vector spaces and their inner products are of particular significance in Fourier analysis and signal processing, and are absolutely essential in modern quantum mechanics.

3.1 INNER PRODUCTS

The most basic example of an inner product is the familiar *dot product*

$$\langle \mathbf{v}, \mathbf{w} \rangle = \mathbf{v} \cdot \mathbf{w} = v_1 w_1 + v_2 w_2 + \cdots + v_n w_n = \sum_{i=1}^{n} v_i w_i, \tag{3.1}$$

between (column) vectors $\mathbf{v} = (v_1, v_2, \ldots, v_n)^T$, $\mathbf{w} = (w_1, w_2, \ldots, w_n)^T$, lying in the Euclidean space $\mathbb{R}^n$. A key observation is that the dot product (3.1) is equal to the matrix product

$$\mathbf{v} \cdot \mathbf{w} = \mathbf{v}^T \mathbf{w} = \begin{pmatrix} v_1 & v_2 & \ldots & v_n \end{pmatrix} \begin{pmatrix} w_1 \\ w_2 \\ \vdots \\ w_n \end{pmatrix} \tag{3.2}$$

between the row vector $\mathbf{v}^T$ and the column vector $\mathbf{w}$.

The dot product is the cornerstone of Euclidean geometry. The key fact is that the dot product of a vector with itself,

$$\mathbf{v} \cdot \mathbf{v} = v_1^2 + v_2^2 + \cdots + v_n^2,$$

is the sum of the squares of its entries, and hence, by the classical Pythagorean Theorem, equals the square of its length; see Figure 3.1. Consequently, the *Euclidean norm* or *length* of a vector is found by taking the square root:

$$\| \mathbf{v} \| = \sqrt{\mathbf{v} \cdot \mathbf{v}} = \sqrt{v_1^2 + v_2^2 + \cdots + v_n^2} \,. \tag{3.3}$$

Note that every nonzero vector $\mathbf{v} \neq \mathbf{0}$ has positive Euclidean norm, $\| \mathbf{v} \| > 0$, while only the zero vector has zero norm: $\| \mathbf{v} \| = 0$ if and only if $\mathbf{v} = \mathbf{0}$. The elementary properties of dot product and Euclidean norm serve to inspire the abstract definition of more general inner products.

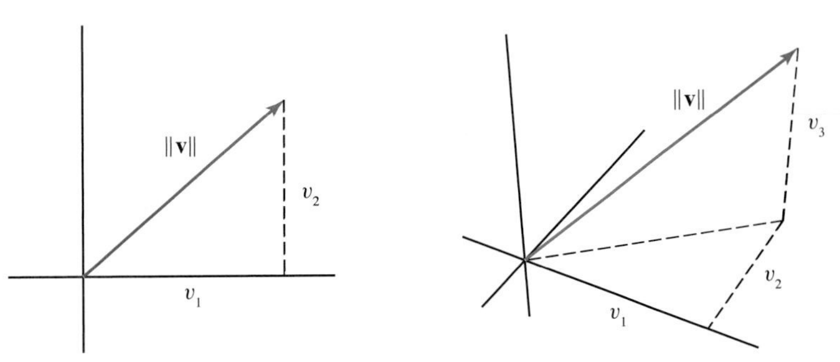

Figure 3.1 The Euclidean norm in $\mathbb{R}^2$ and $\mathbb{R}^3$.

Definition 3.1 An *inner product* on the real vector space V is a pairing that takes two vectors $\mathbf{v}, \mathbf{w} \in V$ and produces a real number $\langle \mathbf{v}, \mathbf{w} \rangle \in \mathbb{R}$. The inner product is required to satisfy the following three axioms for all $\mathbf{u}, \mathbf{v}, \mathbf{w} \in V$, and scalars $c, d \in \mathbb{R}$.

(i) *Bilinearity*:

$$\begin{aligned} \langle c\mathbf{u} + d\mathbf{v}, \mathbf{w} \rangle &= c\langle \mathbf{u}, \mathbf{w} \rangle + d\langle \mathbf{v}, \mathbf{w} \rangle, \\ \langle \mathbf{u}, c\mathbf{v} + d\mathbf{w} \rangle &= c\langle \mathbf{u}, \mathbf{v} \rangle + d\langle \mathbf{u}, \mathbf{w} \rangle. \end{aligned} \tag{3.4}$$

(ii) *Symmetry*:

$$\langle \mathbf{v}, \mathbf{w} \rangle = \langle \mathbf{w}, \mathbf{v} \rangle. \tag{3.5}$$

(iii) *Positivity*:

$$\langle \mathbf{v}, \mathbf{v} \rangle > 0 \quad \text{whenever} \quad \mathbf{v} \neq \mathbf{0}, \quad \text{while} \quad \langle \mathbf{0}, \mathbf{0} \rangle = 0. \tag{3.6}$$

A vector space equipped with an inner product is called an *inner product space*. As we shall see, a vector space can admit many different inner products. Verification of the inner product axioms for the Euclidean dot product is straightforward, and left as an exercise for the reader.

Given an inner product, the associated *norm* of a vector $\mathbf{v} \in V$ is defined as the positive square root of the inner product of the vector with itself:

$$\| \mathbf{v} \| = \sqrt{\langle \mathbf{v}, \mathbf{v} \rangle}\,. \tag{3.7}$$

The positivity axiom implies that $\| \mathbf{v} \| \geq 0$ is real and non-negative, and equals 0 if and only if $\mathbf{v} = \mathbf{0}$ is the zero vector.

EXAMPLE 3.2 While certainly the most common inner product on $\mathbb{R}^2$, the dot product

$$\mathbf{v} \cdot \mathbf{w} = v_1 w_1 + v_2 w_2$$

is by no means the only possibility. A simple example is provided by the *weighted inner product*

$$\langle \mathbf{v}, \mathbf{w} \rangle = 2 v_1 w_1 + 5 v_2 w_2, \qquad \mathbf{v} = \begin{pmatrix} v_1 \\ v_2 \end{pmatrix}, \quad \mathbf{w} = \begin{pmatrix} w_1 \\ w_2 \end{pmatrix}. \tag{3.8}$$

Let us verify that this formula does indeed define an inner product. The symmetry axiom (3.5) is immediate. Moreover,

$$\begin{aligned} \langle c\,\mathbf{u} + d\,\mathbf{v}, \mathbf{w} \rangle &= 2 (c u_1 + d v_1) w_1 + 5 (c u_2 + d v_2) w_2 \\ &= c\,(2 u_1 w_1 + 5 u_2 w_2) + d\,(2 v_1 w_1 + 5 v_2 w_2) \\ &= c\,\langle \mathbf{u}, \mathbf{w} \rangle + d\,\langle \mathbf{v}, \mathbf{w} \rangle, \end{aligned}$$

which verifies the first bilinearity condition; the second follows by a very similar computation. (Or, one can use the symmetry axiom to deduce the second bilinearity identity from the first; see Exercise 3.1.9.) Moreover, $\langle \mathbf{0}, \mathbf{0} \rangle = 0$, while

$$\langle \mathbf{v}, \mathbf{v} \rangle = 2 v_1^2 + 5 v_2^2 > 0 \quad \text{whenever} \quad \mathbf{v} \neq \mathbf{0},$$

since at least one of the summands is strictly positive. This establishes (3.8) as a legitimate inner product on $\mathbb{R}^2$. The associated *weighted norm* $\| \mathbf{v} \| = \sqrt{2 v_1^2 + 5 v_2^2}$ defines an alternative, "non-Pythagorean" notion of length of vectors and distance between points in the plane.

A less evident example of an inner product on $\mathbb{R}^2$ is provided by the expression

$$\langle \mathbf{v}, \mathbf{w} \rangle = v_1 w_1 - v_1 w_2 - v_2 w_1 + 4 v_2 w_2. \tag{3.9}$$

Bilinearity is verified in the same manner as before, and symmetry is immediate. Positivity is ensured by noticing that

$$\langle \mathbf{v}, \mathbf{v} \rangle = v_1^2 - 2 v_1 v_2 + 4 v_2^2 = (v_1 - v_2)^2 + 3 v_2^2 \geq 0$$

is always non-negative, and, moreover, is equal to zero if and only if $v_1 - v_2 = 0$, $v_2 = 0$, i.e., only when $\mathbf{v} = \mathbf{0}$. We conclude that (3.9) defines yet another inner product on $\mathbb{R}^2$, with associated norm

$$\| \mathbf{v} \| = \sqrt{\langle \mathbf{v}, \mathbf{v} \rangle} = \sqrt{v_1^2 - 2 v_1 v_2 + 4 v_2^2}\,. \qquad \bullet$$

The second example (3.8) is a particular case of a general class of inner products.

EXAMPLE 3.3 Let $c_1, \ldots, c_n > 0$ be a set of *positive* numbers. The corresponding *weighted inner product* and *weighted norm* on $\mathbb{R}^n$ are defined by

$$\langle \mathbf{v}, \mathbf{w} \rangle = \sum_{i=1}^{n} c_i v_i w_i, \qquad \| \mathbf{v} \| = \sqrt{\langle \mathbf{v}, \mathbf{v} \rangle} = \sqrt{\sum_{i=1}^{n} c_i v_i^2}\,. \tag{3.10}$$

The numbers c_i are the *weights*. Observe that the larger the weight c_i, the more the ith coordinate of $\mathbf{v}$ contributes to the norm. Weighted norms are particularly relevant in statistics and data fitting, [17], where one wants to emphasize certain quantities and de-emphasize others; this is done by assigning appropriate weights to the different components of the data vector $\mathbf{v}$. Section 4.3 on least squares approximation methods will contain further details. $\bullet$

EXERCISES 3.1

3.1.1. Prove that the formula $\langle \mathbf{v}, \mathbf{w} \rangle = v_1 w_1 - v_1 w_2 - v_2 w_1 + b\, v_2 w_2$ defines an inner product on $\mathbb{R}^2$ if and only if $b > 1$.

3.1.2. Which of the following formulas for $\langle \mathbf{v}, \mathbf{w} \rangle$ define inner products on $\mathbb{R}^2$?

(a) $2 v_1 w_1 + 3 v_2 w_2$ (b) $v_1 w_2 + v_2 w_1$

(c) $(v_1 + v_2)(w_1 + w_2)$

(d) $v_1^2 w_1^2 + v_2^2 w_2^2$

(e) $\sqrt{v_1^2 + v_2^2}\,\sqrt{w_1^2 + w_2^2}$

(f) $2 v_1 w_1 + (v_1 - v_2)(w_1 - w_2)$

(g) $4 v_1 w_1 - 2 v_1 w_2 - 2 v_2 w_1 + 4 v_2 w_2$

3.1.3. Show that $\langle \mathbf{v}, \mathbf{w} \rangle = v_1 w_1 + v_1 w_2 + v_2 w_1 + v_2 w_2$ does *not* define an inner product on $\mathbb{R}^2$.

3.1.4. Prove that each of the following formulas for $\langle \mathbf{v}, \mathbf{w} \rangle$ defines an inner product on $\mathbb{R}^3$. Verify all the inner product axioms in careful detail:

(a) $v_1 w_1 + 2 v_2 w_2 + 3 v_3 w_3$

(b) $4 v_1 w_1 + 2 v_1 w_2 + 2 v_2 w_1 + 4 v_2 w_2 + v_3 w_3$

(c) $2 v_1 w_1 - 2 v_1 w_2 - 2 v_2 w_1 + 3 v_2 w_2 - v_2 w_3 - v_3 w_2 + 2 v_3 w_3$

3.1.5. The *unit circle* for an inner product on $\mathbb{R}^2$ is defined as the set of all vectors of unit length: $\| \mathbf{v} \| = 1$. Graph the unit circles for

(a) the Euclidean inner product,

(b) the weighted inner product (3.8),

(c) the non-standard inner product (3.9).

(d) Prove that cases (b), (c) are, in fact, both ellipses.

◇ **3.1.6.** (a) Explain why the formula for the Euclidean norm in $\mathbb{R}^2$ follows from the Pythagorean Theorem.

(b) How do you use the Pythagorean Theorem to justify the formula for the Euclidean norm in $\mathbb{R}^3$? *Hint*: Look at Figure 3.1.

◇ **3.1.7.** Prove that the norm on an inner product space satisfies $\| c\, \mathbf{v} \| = |c|\, \| \mathbf{v} \|$ for any scalar c and vector $\mathbf{v}$.

3.1.8. Prove that $\langle a\, \mathbf{v} + b\, \mathbf{w}, c\, \mathbf{v} + d\, \mathbf{w} \rangle = a c\, \| \mathbf{v} \|^2 + (a d + b c) \langle \mathbf{v}, \mathbf{w} \rangle + b d\, \| \mathbf{w} \|^2$.

◇ **3.1.9.** Prove that the second bilinearity formula (3.4) is a consequence of the first and the other two inner product axioms.

◇ **3.1.10.** Let V be an inner product space.

(a) Prove that $\langle \mathbf{x}, \mathbf{v} \rangle = 0$ for all $\mathbf{v} \in V$ if and only if $\mathbf{x} = \mathbf{0}$.

(b) Prove that $\langle \mathbf{x}, \mathbf{v} \rangle = \langle \mathbf{y}, \mathbf{v} \rangle$ for all $\mathbf{v} \in V$ if and only if $\mathbf{x} = \mathbf{y}$.

(c) Let $\mathbf{v}_1, \ldots, \mathbf{v}_n$ be a basis for V. Prove that $\langle \mathbf{x}, \mathbf{v}_i \rangle = \langle \mathbf{y}, \mathbf{v}_i \rangle$, $i = 1, \ldots, n$, if and only if $\mathbf{x} = \mathbf{y}$.

◇ **3.1.11.** (a) Prove the identity

$$\langle \mathbf{u}, \mathbf{v} \rangle = \tfrac{1}{4}\left(\| \mathbf{u} + \mathbf{v} \|^2 - \| \mathbf{u} - \mathbf{v} \|^2 \right), \qquad (3.11)$$

which allows one to reconstruct an inner product from its norm.

(b) Use (3.11) to find the inner product on $\mathbb{R}^2$ corresponding to the norm

$$\| \mathbf{v} \| = \sqrt{v_1^2 - 3\, v_1 v_2 + 5\, v_2^2}\,.$$

3.1.12. (a) Show that, for all vectors $\mathbf{x}$ and $\mathbf{y}$ in an inner product space,

$$\| \mathbf{x} + \mathbf{y} \|^2 + \| \mathbf{x} - \mathbf{y} \|^2 = 2 \left(\| \mathbf{x} \|^2 + \| \mathbf{y} \|^2 \right).$$

(b) Interpret this result pictorially for vectors in $\mathbb{R}^2$ under the Euclidean norm.

3.1.13. Suppose $\mathbf{u}, \mathbf{v}$ satisfy $\| \mathbf{u} \| = 3$, $\| \mathbf{u} + \mathbf{v} \| = 4$, and $\| \mathbf{u} - \mathbf{v} \| = 6$. What must $\| \mathbf{v} \|$ equal? Does your answer depend upon which norm is being used?

3.1.14. Let A be any $n \times n$ matrix. Prove that the dot product identity $\mathbf{v} \cdot (A\,\mathbf{w}) = (A^T \mathbf{v}) \cdot \mathbf{w}$ is valid for any vectors $\mathbf{v}, \mathbf{w} \in \mathbb{R}^n$.

◇ **3.1.15.** Prove that $A = A^T$ is a symmetric $n \times n$ matrix if and only if $(A\,\mathbf{v}) \cdot \mathbf{w} = \mathbf{v} \cdot (A\,\mathbf{w})$ for all $\mathbf{v}, \mathbf{w} \in \mathbb{R}^n$.

3.1.16. Suppose $\langle \mathbf{v}, \mathbf{w} \rangle$ defines an inner product on a vector space V. Explain why it also defines an inner product on any subspace $W \subset V$.

3.1.17. Prove that if $\langle \mathbf{v}, \mathbf{w} \rangle$ and $\langle\!\langle \mathbf{v}, \mathbf{w} \rangle\!\rangle$ are two different inner products on the same vector space V, then their sum $\langle\!\langle\!\langle \mathbf{v}, \mathbf{w} \rangle\!\rangle\!\rangle = \langle \mathbf{v}, \mathbf{w} \rangle + \langle\!\langle \mathbf{v}, \mathbf{w} \rangle\!\rangle$ defines an inner product on V.

◇ **3.1.18.** Let V and W be inner product spaces with respective inner products $\langle \mathbf{v}, \widetilde{\mathbf{v}} \rangle$ and $\langle\!\langle \mathbf{w}, \widetilde{\mathbf{w}} \rangle\!\rangle$. Show that $\langle\!\langle\!\langle (\mathbf{v}, \mathbf{w}), (\widetilde{\mathbf{v}}, \widetilde{\mathbf{w}}) \rangle\!\rangle\!\rangle = \langle \mathbf{v}, \widetilde{\mathbf{v}} \rangle + \langle\!\langle \mathbf{w}, \widetilde{\mathbf{w}} \rangle\!\rangle$ for $\mathbf{v}, \widetilde{\mathbf{v}} \in V$, $\mathbf{w}, \widetilde{\mathbf{w}} \in W$ defines an inner product on their Cartesian product $V \times W$.

Inner Products on Function Spaces

Inner products and norms on function spaces are essential in modern analysis and its applications, particularly Fourier analysis, boundary value problems, ordinary and partial differential equations, and numerical analysis. Let us introduce the most important examples.

EXAMPLE 3.4

Let $[a, b] \subset \mathbb{R}$ be a bounded closed interval. Consider the vector space $\mathrm{C}^0[a, b]$ consisting of all continuous scalar functions $f(x)$ defined for $a \le x \le b$. The integral of the product of two continuous functions

$$\langle f, g \rangle = \int_a^b f(x)\, g(x)\, dx \qquad (3.12)$$

defines an inner product on the vector space $\mathrm{C}^0[a, b]$, as we shall prove below. The associated norm is, according to the basic definition (3.7),

$$\| f \| = \sqrt{\int_a^b f(x)^2\, dx}\,, \qquad (3.13)$$

and is known as the L^2 *norm* of the function f over the interval $[a, b]$. The L^2 inner product and norm of functions can be viewed as the infinite-dimensional function space versions of the dot product and Euclidean norm of vectors in $\mathbb{R}^n$. The reason for the name L^2 will become clearer later on.

For example, if we take $[a, b] = \left[0, \tfrac{1}{2}\pi \right]$, then the L^2 inner product between $f(x) = \sin x$ and $g(x) = \cos x$ is equal to

$$\langle \sin x, \cos x \rangle = \int_0^{\pi/2} \sin x\, \cos x\, dx = \frac{1}{2} \sin^2 x \,\Bigg|_{x=0}^{\pi/2} = \frac{1}{2}.$$

Similarly, the norm of the function $\sin x$ is

$$\| \sin x \| = \sqrt{\int_0^{\pi/2} (\sin x)^2\, dx} = \sqrt{\frac{\pi}{4}}\,.$$

One must always be careful when evaluating function norms. For example, the constant function $c(x) \equiv 1$ has norm

$$\| 1 \| = \sqrt{\int_0^{\pi/2} 1^2 \, dx} = \sqrt{\frac{\pi}{2}} \,,$$

not 1 as you might have expected. We also note that the value of the norm depends upon which interval the integral is taken over. For instance, on the longer interval $[0, \pi]$,

$$\| 1 \| = \sqrt{\int_0^{\pi} 1^2 \, dx} = \sqrt{\pi} \,.$$

Thus, when dealing with the L^2 inner product or norm, one must always be careful to specify the function space, or, equivalently, the interval on which it is being evaluated.

Let us prove that formula (3.12) does, indeed, define an inner product. First, we need to check that $\langle f, g \rangle$ is well-defined. This follows because the product $f(x)\, g(x)$ of two continuous functions is also continuous, and hence its integral over a bounded interval is defined and finite. The symmetry requirement is immediate:

$$\langle f, g \rangle = \int_a^b f(x)\, g(x)\, dx = \langle g, f \rangle,$$

because multiplication of functions is commutative. The first bilinearity axiom

$$\langle c\, f + d\, g, h \rangle = c\, \langle f, h \rangle + d\, \langle g, h \rangle$$

amounts to the following elementary integral identity

$$\int_a^b \big[\, c\, f(x) + d\, g(x) \,\big]\, h(x)\, dx = c \int_a^b f(x)\, h(x)\, dx + d \int_a^b g(x)\, h(x)\, dx,$$

valid for arbitrary continuous functions f, g, h and scalars (constants) c, d. The second bilinearity axiom is proved similarly; alternatively, one can use symmetry to deduce it from the first as in Exercise 3.1.9. Finally, positivity requires that

$$\| f \|^2 = \langle f, f \rangle = \int_a^b f(x)^2\, dx \geq 0.$$

This is clear because $f(x)^2 \geq 0$, and the integral of a nonnegative function is nonnegative. Moreover, since the function $f(x)^2$ is continuous and nonnegative, its integral will vanish, $\int_a^b f(x)^2\, dx = 0$, if and only if $f(x) \equiv 0$ is the zero function, cf. Exercise 3.1.27. This completes the proof that (3.12) defines a *bona fide* inner product on the function space $C^0[a, b]$. ●

Remark: The L^2 inner product formula can also be applied to more general functions, but we have restricted our attention to continuous functions in order to avoid certain technical complications. The most general function space admitting this inner product is known as *Hilbert space*, which lies at the foundation of most of modern analysis, function theory and Fourier analysis, [36, 38, 52, 54], as well as providing the theoretical setting for all of quantum mechanics, [40]. Unfortunately, we cannot provide the mathematical details of the Hilbert space construction since it requires that you be familiar with measure theory and the Lebesgue integral, [54].

Warning: One does need to be extremely careful when trying to extend the L^2 inner product to other spaces of functions. Indeed, there are nonzero, discontinuous functions with zero "L^2 norm". For example, the non-zero function

$$f(x) = \begin{cases} 1, & x = 0, \\ 0, & \text{otherwise} \end{cases} \qquad \text{satisfies} \quad \| f \|^2 = \int_{-1}^{1} f(x)^2 \, dx = 0, \tag{3.14}$$

because any function that is zero except at finitely many (or even countably many) points has zero integral.

The L^2 inner product is but one of a vast number of possible inner products on function spaces. For example, one can also define weighted inner products on the space $C^0[a,b]$. The weights along the interval are specified by a (continuous) positive scalar function $w(x) > 0$. The corresponding *weighted inner product* and *norm* are

$$\langle f, g \rangle = \int_a^b f(x)\, g(x)\, w(x)\, dx, \qquad \| f \| = \sqrt{\int_a^b f(x)^2\, w(x)\, dx} \,. \tag{3.15}$$

The verification of the inner product axioms in this case is left as an exercise for the reader. As in the finite-dimensional version, weighted inner products are often used in statistics and data analysis, [17].

EXERCISES 3.1

3.1.19. For each of the given pairs of functions in $C^0[0,1]$, find their L^2 inner product $\langle f, g \rangle = \int_0^1 f(x)\, g(x)\, dx$ and their L^2 norms $\| f \|$, $\| g \|$:
(a) $f(x) = 1$, $g(x) = x$
(b) $f(x) = \cos 2\pi x$, $g(x) = \sin 2\pi x$
(c) $f(x) = x$, $g(x) = e^x$
(d) $f(x) = (x+1)^2$, $g(x) = \dfrac{1}{x+1}$

3.1.20. Let $f(x) = x$, $g(x) = 1 + x^2$. Compute $\langle f, g \rangle$, $\| f \|$, and $\| g \|$ for
(a) the L^2 inner product

$$\langle f, g \rangle = \int_0^1 f(x)\, g(x)\, dx,$$

(b) the L^2 inner product

$$\langle f, g \rangle = \int_{-1}^1 f(x)\, g(x)\, dx,$$

(c) the weighted inner product

$$\langle f, g \rangle = \int_0^1 f(x)\, g(x)\, x\, dx.$$

3.1.21. Which of the following formulas for $\langle f, g \rangle$ define inner products on the space $C^0[-1,1]$?
(a) $\displaystyle\int_{-1}^1 f(x)\, g(x)\, e^{-x}\, dx$
(b) $\displaystyle\int_{-1}^1 f(x)\, g(x)\, x\, dx$
(c) $\displaystyle\int_{-1}^1 f(x)\, g(x)\, (x+2)\, dx$
(d) $\displaystyle\int_{-1}^1 f(x)\, g(x)\, x^2\, dx$

3.1.22. Prove that

$$\langle f, g \rangle = \int_0^1 f(x)\, g(x)\, dx$$

does *not* define an inner product on the vector space $C^0[-1,1]$. Explain why this does not contradict the fact that it defines an inner product on the vector space $C^0[0,1]$. Does it define an inner product on the subspace $\mathcal{P}^{(n)} \subset C^0[-1,1]$ consisting of all polynomial functions?

3.1.23. Does either of the following define an inner product on $C^0[0,1]$?
(a) $\langle f, g \rangle = f(0)\, g(0) + f(1)\, g(1)$
(b) $\langle f, g \rangle = f(0)\, g(0) + f(1)\, g(1) + \int_0^1 f(x)\, g(x)\, dx$

3.1.24. Let $f(x)$ be a function, and $\| f \|$ its L^2 norm on $[a,b]$. Is $\| f^2 \| = \| f \|^2$? If yes, prove the statement. If no, give a counterexample.

3.1.25. Prove that

$$\langle f, g \rangle = \int_a^b \left[f(x)\, g(x) + f'(x)\, g'(x) \right] dx$$

defines an inner product on the space $C^1[a,b]$ of continuously differentiable functions on the interval $[a,b]$. Write out the corresponding norm, known as the *Sobolev H^1 norm*; it and its generalizations play an extremely important role in advanced mathematical analysis, [36].

3.1.26. Let $V = C^1[-1,1]$ denote the vector space of continuously differentiable functions for $-1 \le x \le 1$.

(a) Does the expression

$$\langle f, g \rangle = \int_{-1}^{1} f'(x)\, g'(x)\, dx$$

define an inner product on V?

(b) Answer the same question for the subspace $W = \{ f \in V \mid f(0) = 0 \}$ consisting of all continuously differentiable functions which vanish at 0.

◇ **3.1.27.** (a) Let $h(x) \ge 0$ be a continuous, non-negative function defined on an interval $[a,b]$. Prove that $\int_a^b h(x)\, dx = 0$ if and only if $h(x) \equiv 0$. *Hint*: Use the fact that $\int_c^d h(x)\, dx > 0$ if $h(x) > 0$ for $c \le x \le d$.

(b) Give an example that shows that this result is not valid if h is allowed to be discontinuous.

◇ **3.1.28.** (a) Prove the inner product axioms for the weighted inner product (3.15), assuming $w(x) > 0$ for all $a \le x \le b$.

(b) Explain why it does *not* define an inner product if w is continuous and $w(x_0) < 0$ for some $x_0 \in [a,b]$.

(c) If $w(x) \ge 0$ for $a \le x \le b$, does (3.15) define an inner product? *Hint*: Your answer may depend upon $w(x)$.

♡ **3.1.29.** Let $\Omega \subset \mathbb{R}^2$ be a bounded domain. Let $C^0(\Omega)$ denote the vector space consisting of all continuous, bounded real-valued functions $f(x,y)$ defined for $(x,y) \in \Omega$.

(a) Prove that if $f(x,y) \ge 0$ is continuous and $\iint_\Omega f(x,y)\, dx\, dy = 0$, then $f(x,y) \equiv 0$. *Hint*: Mimic Exercise 3.1.27.

(b) Use this result to prove that

$$\langle f, g \rangle = \iint_\Omega f(x,y)\, g(x,y)\, dx\, dy \qquad (3.16)$$

defines an inner product on $C^0(\Omega)$, called the L^2 inner product on the domain Ω. What is the corresponding norm?

3.1.30. Compute the L^2 inner product (3.16) and norms of the functions $f(x,y) \equiv 1$ and $g(x,y) = x^2 + y^2$, when

(a) $\Omega = \{ 0 \le x \le 1, 0 \le y \le 1 \}$ is the unit square;

(b) $\Omega = \{ x^2 + y^2 \le 1 \}$ is the unit disk. *Hint*: Use polar coordinates.

♡ **3.1.31.** Let V be the vector space consisting of all continuous, vector-valued functions

$$\mathbf{f}(x) = (f_1(x), f_2(x))^T$$

defined on the interval $0 \le x \le 1$.

(a) Prove that

$$\langle\!\langle \mathbf{f}, \mathbf{g} \rangle\!\rangle = \int_0^1 \left[f_1(x)\, g_1(x) + f_2(x)\, g_2(x) \right] dx$$

defines an inner product on V.

(b) Prove, more generally, that if $\langle \mathbf{v}, \mathbf{w} \rangle$ is any inner product on $\mathbb{R}^2$, then $\langle\!\langle \mathbf{f}, \mathbf{g} \rangle\!\rangle = \int_a^b \langle \mathbf{f}(x), \mathbf{g}(x) \rangle\, dx$ defines an inner product on V. (Part (a) corresponds to the dot product.)

(c) Use part (b) to prove that

$$\langle\!\langle \mathbf{f}, \mathbf{g} \rangle\!\rangle = \int_a^b \left[f_1(x)\, g_1(x) - f_1(x)\, g_2(x) - f_2(x)\, g_1(x) + 3 f_2(x)\, g_2(x) \right] dx$$

defines an inner product on V.

3.2 INEQUALITIES

There are two absolutely fundamental inequalities that are valid for *any* inner product on any vector space. The first is inspired by the geometric interpretation of the dot product on Euclidean space in terms of the angle between vectors. It is named after two of the founders of modern analysis, Augustin Cauchy and Herman Schwarz, who established it in the case of the L^2 inner product on function space.* The more familiar triangle inequality, that the length of any side of a triangle is bounded by the sum of the lengths of the other two sides is, in fact, an immediate

*Russians also give credit for its discovery to their compatriot Viktor Bunyakovskii, and, indeed, some authors append his name to the inequality.

consequence of the Cauchy–Schwarz inequality, and hence also valid for any norm based on an inner product.

We will present these two inequalities in their most general, abstract form, since this brings their essence into the spotlight. Specializing to different inner products and norms on both finite-dimensional and infinite-dimensional vector spaces leads to a wide variety of striking and useful inequalities.

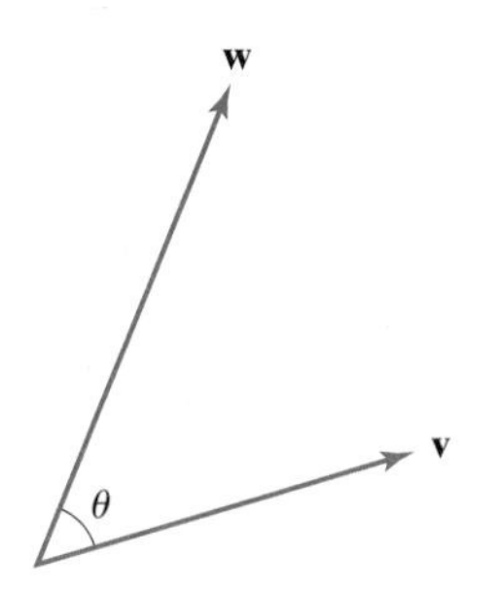

Figure 3.2 Angle between two vectors.

The Cauchy–Schwarz Inequality

In Euclidean geometry, the dot product between two vectors can be geometrically characterized by the equation

$$\mathbf{v} \cdot \mathbf{w} = \|\mathbf{v}\| \, \|\mathbf{w}\| \, \cos\theta, \tag{3.17}$$

where θ measures the angle between the vectors $\mathbf{v}$ and $\mathbf{w}$, as drawn in Figure 3.2. Since

$$|\cos\theta| \leq 1,$$

the absolute value of the dot product is bounded by the product of the lengths of the vectors:

$$|\mathbf{v} \cdot \mathbf{w}| \; \leq \; \|\mathbf{v}\| \, \|\mathbf{w}\|.$$

This is the simplest form of the general *Cauchy–Schwarz inequality*. We present a simple, algebraic proof that does not rely on the geometrical notions of length and angle and thus demonstrates its universal validity for *any* inner product.

THEOREM 3.5 Every inner product satisfies the Cauchy–Schwarz inequality

$$|\langle \mathbf{v}, \mathbf{w} \rangle| \; \leq \; \|\mathbf{v}\| \, \|\mathbf{w}\|, \qquad \text{for all} \quad \mathbf{v}, \mathbf{w} \in V. \tag{3.18}$$

Here, $\|\mathbf{v}\|$ is the associated norm, while $|\cdot|$ denotes absolute value of real numbers. Equality holds if and only if $\mathbf{v}$ and $\mathbf{w}$ are parallel vectors.

Proof The case when $\mathbf{w} = \mathbf{0}$ is trivial, since both sides of (3.18) are equal to 0. Thus, we may suppose $\mathbf{w} \neq \mathbf{0}$. Let $t \in \mathbb{R}$ be an arbitrary scalar. Using the three inner product axioms, we have

$$0 \leq \|\mathbf{v} + t\,\mathbf{w}\|^2 = \langle \mathbf{v} + t\,\mathbf{w}, \mathbf{v} + t\,\mathbf{w} \rangle = \|\mathbf{v}\|^2 + 2t\,\langle \mathbf{v}, \mathbf{w} \rangle + t^2\,\|\mathbf{w}\|^2, \tag{3.19}$$

with equality holding if and only if $\mathbf{v} = -\,t\,\mathbf{w}$—which requires $\mathbf{v}$ and $\mathbf{w}$ to be parallel vectors. We fix $\mathbf{v}$ and $\mathbf{w}$, and consider the right hand side of (3.19) as a quadratic function,

$$0 \leq p(t) = a\,t^2 + 2\,b\,t + c, \qquad \text{where} \quad a = \|\mathbf{w}\|^2, \quad b = \langle \mathbf{v}, \mathbf{w} \rangle, \quad c = \|\mathbf{v}\|^2,$$

of the scalar variable t. To get the maximum mileage out of the fact that $p(t) \geq 0$, let us look at where it assumes its minimum, which occurs when its derivative is zero:

$$p'(t) = 2\,a\,t + 2\,b = 0, \qquad \text{and so} \quad t \; = \; -\,\frac{b}{a} \; = \; -\,\frac{\langle \mathbf{v}, \mathbf{w} \rangle}{\|\mathbf{w}\|^2}\,.$$

Substituting this particular value of t into (3.19), we find

$$0 \; \leq \; \|\mathbf{v}\|^2 - 2\,\frac{\langle \mathbf{v}, \mathbf{w} \rangle^2}{\|\mathbf{w}\|^2} + \frac{\langle \mathbf{v}, \mathbf{w} \rangle^2}{\|\mathbf{w}\|^2} \; = \; \|\mathbf{v}\|^2 - \frac{\langle \mathbf{v}, \mathbf{w} \rangle^2}{\|\mathbf{w}\|^2}\,.$$

Rearranging this last inequality, we conclude that

$$\frac{\langle \mathbf{v}, \mathbf{w} \rangle^2}{\|\mathbf{w}\|^2} \; \leq \; \|\mathbf{v}\|^2, \qquad \text{or} \quad \langle \mathbf{v}, \mathbf{w} \rangle^2 \; \leq \; \|\mathbf{v}\|^2 \, \|\mathbf{w}\|^2.$$

Taking the (positive) square root of both sides of the final inequality completes the proof of the Cauchy–Schwarz inequality (3.18). ■

Given any inner product, we can use the quotient

$$\cos\theta = \frac{\langle \mathbf{v}, \mathbf{w} \rangle}{\|\mathbf{v}\| \, \|\mathbf{w}\|} \tag{3.20}$$

to define the "angle" between the vector space elements $\mathbf{v}, \mathbf{w} \in V$. The Cauchy–Schwarz inequality tells us that the ratio lies between -1 and $+1$, and hence the angle θ is well defined, and, in fact, unique if we restrict it to lie in the range $0 \le \theta \le \pi$.

For example, the vectors $\mathbf{v} = (1, 0, 1)^T$, $\mathbf{w} = (0, 1, 1)^T$ have dot product $\mathbf{v} \cdot \mathbf{w} = 1$ and norms $\|\mathbf{v}\| = \|\mathbf{w}\| = \sqrt{2}$. Hence the Euclidean angle between them is given by

$$\cos\theta = \frac{1}{\sqrt{2} \cdot \sqrt{2}} = \frac{1}{2}, \qquad \text{and so} \quad \theta = \tfrac{1}{3}\pi = 1.0472\ldots .$$

On the other hand, if we adopt the weighted inner product

$$\langle \mathbf{v}, \mathbf{w} \rangle = v_1 w_1 + 2 v_2 w_2 + 3 v_3 w_3,$$

then $\mathbf{v} \cdot \mathbf{w} = 3$, $\|\mathbf{v}\| = 2$, $\|\mathbf{w}\| = \sqrt{5}$, and hence their "weighted" angle becomes

$$\cos\theta = \frac{3}{2\sqrt{5}} = .67082\ldots, \qquad \text{with} \quad \theta = .835482\ldots .$$

Thus, the measurement of angle (and length) is dependent upon the choice of an underlying inner product.

Similarly, under the L^2 inner product on the interval $[0, 1]$, the "angle" θ between the polynomials $p(x) = x$ and $q(x) = x^2$ is given by

$$\cos\theta = \frac{\langle x, x^2 \rangle}{\|x\| \, \|x^2\|} = \frac{\displaystyle\int_0^1 x^3 \, dx}{\sqrt{\displaystyle\int_0^1 x^2 \, dx}\ \sqrt{\displaystyle\int_0^1 x^4 \, dx}} = \frac{\frac{1}{4}}{\sqrt{\frac{1}{3}}\ \sqrt{\frac{1}{5}}} = \sqrt{\frac{15}{16}},$$

so that $\theta = .25268$ radians.

Warning: You should not try to give this notion of angle between functions more significance than the formal definition warrants —it does not correspond to any "angular" properties of their graphs. Also, the value depends on the choice of inner product and the interval upon which it is being computed. For example, if we change to the L^2 inner product on the interval $[-1, 1]$, then

$$\langle x, x^2 \rangle = \int_{-1}^1 x^3 \, dx = 0.$$

Hence (3.20) becomes $\cos\theta = 0$, so the "angle" between x and x^2 is now $\theta = \frac{1}{2}\pi$.

EXERCISES 3.2

3.2.1. Verify the Cauchy–Schwarz inequality for each of the following pairs of vectors $\mathbf{v}, \mathbf{w}$, using the standard dot product, and then determine the angle between them:

(a) $(1, 2)^T, (-1, 2)^T$

(b) $(1, -1, 0)^T, (-1, 0, 1)^T$

(c) $(1, -1, 0)^T, (2, 2, 2)^T$

(d) $(1, -1, 1, 0)^T, (-2, 0, -1, 1)^T$

(e) $(2, 1, -2, -1)^T, (0, -1, 2, -1)^T$

3.2.2. (a) Find the Euclidean angle between the vectors $(1, 1, 1, 1)^T$ and $(1, 1, 1, -1)^T$ in $\mathbb{R}^4$.

(b) List the possible angles between $(1, 1, 1, 1)^T$ and $(a_1, a_2, a_3, a_4)^T$, where each a_i is either 1 or -1.

3.2.3. Prove that the points $(0, 0, 0)$, $(1, 1, 0)$, $(1, 0, 1)$, $(0, 1, 1)$ form the vertices of a regular tetrahedron, meaning that all sides have the same length. What is the common Euclidean angle between the edges? What is the angle between any two rays going from the center $\left(\frac{1}{2}, \frac{1}{2}, \frac{1}{2}\right)$ to the vertices? *Remark*: Methane molecules assume this geometric configuration, and the angle influences their chemistry.

3.2.4. Verify the Cauchy–Schwarz inequality for the vectors $\mathbf{v} = (1, 2)^T$, $\mathbf{w} = (1, -3)^T$, using

(a) the dot product

(b) the weighted inner product $\langle \mathbf{v}, \mathbf{w} \rangle = v_1 w_1 + 2 v_2 w_2$

(c) the inner product (3.9)

3.2.5. Verify the Cauchy–Schwarz inequality for the vectors $\mathbf{v} = (3, -1, 2)^T$, $\mathbf{w} = (1, -1, 1)^T$, using

(a) the dot product

(b) the weighted inner product $\langle \mathbf{v}, \mathbf{w} \rangle = v_1 w_1 + 2 v_2 w_2 + 3 v_3 w_3$

(c) the inner product

$$\langle \mathbf{v}, \mathbf{w} \rangle = \mathbf{v}^T \begin{pmatrix} 2 & -1 & 0 \\ -1 & 2 & -1 \\ 0 & -1 & 2 \end{pmatrix} \mathbf{w}$$

3.2.6. Use the Cauchy–Schwarz inequality to prove $(a \cos\theta + b \sin\theta)^2 \le a^2 + b^2$ for any θ, a, b.

3.2.7. Prove that

$$(a_1 + a_2 + \cdots + a_n)^2 \le n\,(a_1^2 + a_2^2 + \cdots + a_n^2)$$

for any real numbers $a_1, \ldots, a_n$. When does equality hold?

◇ **3.2.8.** *The Law of Cosines*: Prove that the formula

$$\|\mathbf{v} - \mathbf{w}\|^2 = \|\mathbf{v}\|^2 + \|\mathbf{w}\|^2 - 2\,\|\mathbf{v}\|\,\|\mathbf{w}\|\,\cos\theta, \tag{3.21}$$

where θ is the angle between $\mathbf{v}$ and $\mathbf{w}$, is valid in any inner product space.

◇ **3.2.9.** Explain why the inequality $\langle \mathbf{v}, \mathbf{w} \rangle \le \|\mathbf{v}\|\,\|\mathbf{w}\|$, obtained by omitting the absolute value sign on the left hand side of Cauchy–Schwarz, is valid.

◇ **3.2.10.** Show that one can determine the angle θ between $\mathbf{v}$ and $\mathbf{w}$ via the formula

$$\cos\theta = \frac{\|\mathbf{v} + \mathbf{w}\|^2 - \|\mathbf{v} - \mathbf{w}\|^2}{4\,\|\mathbf{v}\|\,\|\mathbf{w}\|}.$$

Draw a picture illustrating what is being measured.

♡ **3.2.11.** The *cross product* of two vectors in $\mathbb{R}^2$ is defined as the scalar

$$\mathbf{v} \times \mathbf{w} = v_1 w_2 - v_2 w_1 \qquad \text{for} \quad \mathbf{v} = (v_1, v_2)^T, \quad \mathbf{w} = (w_1, w_2)^T. \tag{3.22}$$

(a) Does the cross product define an inner product on $\mathbb{R}^2$? Carefully explain which axioms are valid and which are not.

(b) Prove that $\mathbf{v} \times \mathbf{w} = \|\mathbf{v}\|\,\|\mathbf{w}\|\,\sin\theta$, where θ denotes the angle from $\mathbf{v}$ to $\mathbf{w}$ as in Figure 3.2.

(c) Prove that $\mathbf{v} \times \mathbf{w} = 0$ if and only if $\mathbf{v}$ and $\mathbf{w}$ are parallel vectors.

(d) Show that $|\mathbf{v} \times \mathbf{w}|$ equals the area of the parallelogram defined by $\mathbf{v}$ and $\mathbf{w}$.

3.2.12. Verify the Cauchy–Schwarz inequality for the functions $f(x) = x$ and $g(x) = e^x$ with respect to

(a) the L^2 inner product on the interval $[0, 1]$,

(b) the L^2 inner product on $[-1, 1]$,

(c) the weighted inner product

$$\langle f, g \rangle = \int_0^1 f(x)\,g(x)\,e^{-x}\,dx.$$

3.2.13. Using the L^2 inner product on the interval $[0, \pi]$ find the angle between the functions

(a) 1 and $\cos x$

(b) 1 and $\sin x$

(c) $\cos x$ and $\sin x$

3.2.14. Verify the Cauchy–Schwarz inequality for the two particular functions appearing in Exercise 3.1.30 with respect to the L^2 inner product on

(a) the unit square;

(b) the unit disk.

Orthogonal Vectors

In Euclidean geometry, a particularly noteworthy configuration occurs when two vectors are *perpendicular*. Perpendicular vectors meet at a right angle, $\theta = \frac{1}{2}\pi$ or $\frac{3}{2}\pi$, with $\cos\theta = 0$. The angle formula (3.17) implies that the vectors $\mathbf{v}$, $\mathbf{w}$ are perpendicular if and only if their dot product vanishes: $\mathbf{v} \cdot \mathbf{w} = 0$. Perpendicularity is of interest in general inner product spaces, but, for historical reasons, has been given a more suggestive name.

Definition 3.6 Two elements $\mathbf{v}, \mathbf{w} \in V$ of an inner product space V are called *orthogonal* if their inner product vanishes: $\langle \mathbf{v}, \mathbf{w} \rangle = 0$.

In particular, the zero element is orthogonal to everything: $\langle \mathbf{0}, \mathbf{v} \rangle = 0$ for all $\mathbf{v} \in V$. Orthogonality is a remarkably powerful tool in all applications of linear algebra, and often serves to dramatically simplify many computations. We will devote all of Chapter 5 to a detailed exploration of its manifold implications.

EXAMPLE 3.7 The vectors $\mathbf{v} = (1, 2)^T$ and $\mathbf{w} = (6, -3)^T$ are orthogonal with respect to the Euclidean dot product in $\mathbb{R}^2$, since $\mathbf{v} \cdot \mathbf{w} = 1 \cdot 6 + 2 \cdot (-3) = 0$. We deduce that they meet at a right angle. However, these vectors are *not* orthogonal with respect to the weighted inner product (3.8):

$$\langle \mathbf{v}, \mathbf{w} \rangle = \left\langle \begin{pmatrix} 1 \\ 2 \end{pmatrix}, \begin{pmatrix} 6 \\ -3 \end{pmatrix} \right\rangle = 2 \cdot 1 \cdot 6 + 5 \cdot 2 \cdot (-3) = -18 \neq 0.$$

Thus, the property of orthogonality, like angles in general, depends upon which inner product is being used. ●

EXAMPLE 3.8 The polynomials $p(x) = x$ and $q(x) = x^2 - \frac{1}{2}$ are orthogonal with respect to the inner product $\langle p, q \rangle = \int_0^1 p(x)\, q(x)\, dx$ on the interval $[0, 1]$, since

$$\left\langle x, x^2 - \tfrac{1}{2} \right\rangle = \int_0^1 x\left(x^2 - \tfrac{1}{2}\right) dx = \int_0^1 \left(x^3 - \tfrac{1}{2}x\right) dx = 0.$$

They fail to be orthogonal on most other intervals. For example, on the interval $[0, 2]$,

$$\left\langle x, x^2 - \tfrac{1}{2} \right\rangle = \int_0^2 x\left(x^2 - \tfrac{1}{2}\right) dx = \int_0^2 \left(x^3 - \tfrac{1}{2}x\right) dx = 3. \qquad ●$$

EXERCISES 3.2

Note: Unless stated otherwise, the inner product is the standard dot product on $\mathbb{R}^n$.

3.2.15. (a) Find a so that $(2, a, -3)^T$ is orthogonal to $(-1, 3, -2)^T$.
(b) Is there any value of a for which $(2, a, -3)^T$ is parallel to $(-1, 3, -2)^T$?

3.2.16. Find all vectors in $\mathbb{R}^3$ that are orthogonal to both $(1, 2, 3)^T$ and $(-2, 0, 1)^T$.

3.2.17. Answer Exercises 3.2.15 and 3.2.16 using the weighted inner product $\langle \mathbf{v}, \mathbf{w} \rangle = 3\, v_1 w_1 + 2\, v_2 w_2 + v_3 w_3$.

3.2.18. Find all vectors in $\mathbb{R}^4$ that are orthogonal to both $(1, 2, 3, 4)^T$ and $(5, 6, 7, 8)^T$.

3.2.19. Determine a basis for the subspace $W \subset \mathbb{R}^4$ consisting of all vectors which are orthogonal to the vector $(1, 2, -1, 3)^T$.

3.2.20. Find three vectors $\mathbf{u}$, $\mathbf{v}$ and $\mathbf{w}$ in $\mathbb{R}^3$ such that $\mathbf{u}$ and $\mathbf{v}$ are orthogonal, $\mathbf{u}$ and $\mathbf{w}$ are orthogonal, but $\mathbf{v}$ and $\mathbf{w}$ are *not* orthogonal. Are your vectors linearly independent or linearly dependent? Can you find vectors of the opposite dependency satisfying the same conditions? Why or why not?

3.2.21. For what values of a, b are the vectors $(1, 1, a)^T$ and $(b, -1, 1)^T$ orthogonal
(a) with respect to the dot product?
(b) with respect to the weighted inner product of Exercise 3.2.17?

3.2.22. When is a vector orthogonal to itself?

◇ **3.2.23.** Prove that the only element $\mathbf{w}$ in an inner product space V that is orthogonal to every vector, so $\langle \mathbf{w}, \mathbf{v} \rangle = 0$ for all $\mathbf{v} \in V$, is the zero vector: $\mathbf{w} = \mathbf{0}$.

3.2.24. A vector with $\| \mathbf{v} \| = 1$ is known as a *unit vector*. Prove that if $\mathbf{v}, \mathbf{w}$ are both unit vectors, then $\mathbf{v} + \mathbf{w}$ and $\mathbf{v} - \mathbf{w}$ are orthogonal. Are they also unit vectors?

◇ **3.2.25.** Let V be an inner product space and $\mathbf{v} \in V$ a fixed element. Prove that the set of all vectors $\mathbf{w} \in V$ that are orthogonal to $\mathbf{v}$ forms a subspace of V.

3.2.26. (a) Show that the polynomials $p_1(x) = 1$, $p_2(x) = x - \frac{1}{2}$, $p_3(x) = x^2 - x + \frac{1}{6}$ are mutually orthogonal with respect to the L^2 inner product on the interval $[0, 1]$.

(b) Show that the functions $\sin n\pi x$, $n = 1, 2, 3, \ldots$, are mutually orthogonal with respect to the same inner product.

3.2.27. Find a non-zero quadratic polynomial that is orthogonal to both $p_1(x) = 1$ and $p_2(x) = x$ under the L^2 inner product on the interval $[-1, 1]$.

3.2.28. Find all quadratic polynomials that are orthogonal to the function e^x with respect to the L^2 inner product on the interval $[0, 1]$.

3.2.29. Determine all pairs among the functions 1, x, $\cos \pi x$, $\sin \pi x$, e^x, that are orthogonal with respect to the L^2 inner product on $[-1, 1]$.

3.2.30. Find two non-zero functions that are orthogonal with respect to the weighted inner product $\langle f, g \rangle = \int_0^1 f(x)\, g(x)\, x\, dx$.

The Triangle Inequality

The familiar triangle inequality states that the length of one side of a triangle is at most equal to the sum of the lengths of the other two sides. Referring to Figure 3.3, if the first two sides are represented by vectors **v** and **w**, then the third corresponds to their sum $\mathbf{v} + \mathbf{w}$. The triangle inequality turns out to be an elementary consequence of the Cauchy–Schwarz inequality, and hence is valid in *any* inner product space.

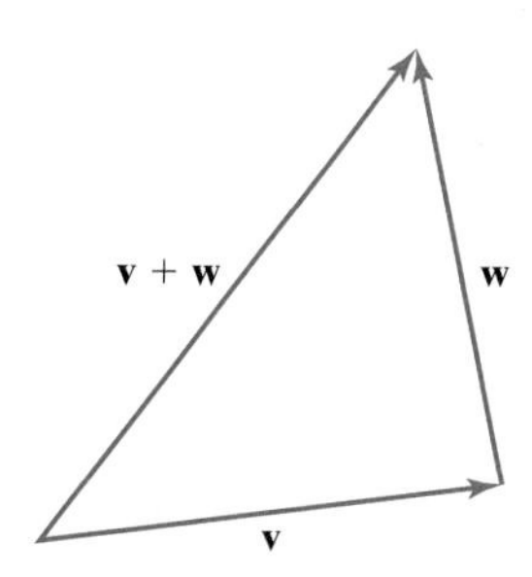

Figure 3.3 Triangle inequality.

THEOREM 3.9 The norm associated with an inner product satisfies the *triangle inequality*

$$\|\mathbf{v} + \mathbf{w}\| \le \|\mathbf{v}\| + \|\mathbf{w}\| \quad \text{for all} \quad \mathbf{v}, \mathbf{w} \in V. \tag{3.23}$$

Equality holds if and only if **v** and **w** are parallel vectors.

Proof We compute

$$\begin{aligned} \|\mathbf{v} + \mathbf{w}\|^2 &= \langle \mathbf{v} + \mathbf{w}, \mathbf{v} + \mathbf{w} \rangle = \|\mathbf{v}\|^2 + 2\, \langle \mathbf{v}, \mathbf{w} \rangle + \|\mathbf{w}\|^2 \\ &\le \|\mathbf{v}\|^2 + 2\, \|\mathbf{v}\|\, \|\mathbf{w}\| + \|\mathbf{w}\|^2 = \big(\|\mathbf{v}\| + \|\mathbf{w}\| \big)^2, \end{aligned}$$

where the middle inequality follows from Cauchy–Schwarz, cf. Exercise 3.2.9. Taking square roots of both sides and using positivity completes the proof. ■

EXAMPLE 3.10 The vectors

$$\mathbf{v} = \begin{pmatrix} 1 \\ 2 \\ -1 \end{pmatrix} \quad \text{and} \quad \mathbf{w} = \begin{pmatrix} 2 \\ 0 \\ 3 \end{pmatrix} \quad \text{sum to} \quad \mathbf{v} + \mathbf{w} = \begin{pmatrix} 3 \\ 2 \\ 2 \end{pmatrix}.$$

Their Euclidean norms are $\|\mathbf{v}\| = \sqrt{6}$ and $\|\mathbf{w}\| = \sqrt{13}$, while $\|\mathbf{v} + \mathbf{w}\| = \sqrt{17}$. The triangle inequality (3.23) in this case says $\sqrt{17} \le \sqrt{6} + \sqrt{13}$, which is valid. ●

EXAMPLE 3.11 Consider the functions $f(x) = x - 1$ and $g(x) = x^2 + 1$. Using the L^2 norm on the interval $[0, 1]$, we find

$$\|f\| = \sqrt{\int_0^1 (x-1)^2\,dx} = \sqrt{\frac{1}{3}},$$

$$\|g\| = \sqrt{\int_0^1 (x^2+1)^2\,dx} = \sqrt{\frac{28}{15}},$$

$$\|f+g\| = \sqrt{\int_0^1 (x^2+x)^2\,dx} = \sqrt{\frac{31}{30}}.$$

The triangle inequality requires $\sqrt{\frac{31}{30}} \le \sqrt{\frac{1}{3}} + \sqrt{\frac{28}{15}}$, which is true. ●

The Cauchy–Schwarz and triangle inequalities look much more impressive when written out in full detail. For the Euclidean dot product (3.1), they are

$$\left| \sum_{i=1}^{n} v_i\, w_i \right| \le \sqrt{\sum_{i=1}^{n} v_i^2}\ \sqrt{\sum_{i=1}^{n} w_i^2}\,,$$
$$\sqrt{\sum_{i=1}^{n} (v_i + w_i)^2} \le \sqrt{\sum_{i=1}^{n} v_i^2} + \sqrt{\sum_{i=1}^{n} w_i^2}\,. \tag{3.24}$$

Theorems 3.18 and 3.23 imply that these inequalities are valid for arbitrary real numbers $v_1, \ldots, v_n, w_1, \ldots, w_n$. For the L^2 inner product (3.13) on function space, they produce the following splendid integral inequalities:

$$\left| \int_a^b f(x)\,g(x)\,dx \right| \le \sqrt{\int_a^b f(x)^2\,dx}\ \sqrt{\int_a^b g(x)^2\,dx}\,,$$
$$\sqrt{\int_a^b \left[f(x)+g(x)\right]^2 dx} \le \sqrt{\int_a^b f(x)^2\,dx} + \sqrt{\int_a^b g(x)^2\,dx}\,, \tag{3.25}$$

which hold for arbitrary continuous (and, in fact, rather general) functions. The first of these is the original Cauchy–Schwarz inequality, whose proof appeared to be quite deep when it first appeared. Only after the abstract notion of an inner product space was properly formalized did its innate simplicity and generality become evident.

EXERCISES 3.2

3.2.31. Use the dot product on $\mathbb{R}^3$ to answer the following:

(a) Find the angle between the vectors $(1, 2, 3)$ and $(1, -1, 2)$.

(b) Verify the Cauchy–Schwarz and triangle inequalities for these two particular vectors.

(c) Find all vectors that are orthogonal to both of these vectors.

3.2.32. Verify the triangle inequality for each pair of vectors in Exercise 3.2.1.

3.2.33. Verify the triangle inequality for the vectors and inner products in Exercise 3.2.4.

3.2.34. Verify the triangle inequality for the functions in Exercise 3.2.12 for the indicated inner products.

3.2.35. Verify the triangle inequality for the two particular functions appearing in Exercise 3.1.30 with respect to the L^2 inner product on

(a) the unit square (b) the unit disk

3.2.36. Use the L^2 inner product

$$\langle f, g \rangle = \int_{-1}^{1} f(x)\, g(x)\, dx$$

to answer the following:

(a) Find the "angle" between the functions 1 and x. Are they orthogonal?

(b) Verify the Cauchy–Schwarz and triangle inequalities for these two functions.

(c) Find all quadratic polynomials $p(x) = a + b x + c x^2$ that are orthogonal to both of these functions.

3.2.37. (a) Write down the explicit formulae for the Cauchy–Schwarz and triangle inequalities based on the weighted inner product

$$\langle f, g \rangle = \int_{0}^{1} f(x)\, g(x)\, e^x\, dx.$$

(b) Verify that the inequalities hold when $f(x) = 1$, $g(x) = e^x$ by direct computation.

(c) What is the "angle" between these two functions in this inner product?

3.2.38. Answer Exercise 3.2.37 for the Sobolev H^1 inner product

$$\langle f, g \rangle = \int_{0}^{1} \big[\, f(x)\, g(x) + f'(x)\, g'(x) \,\big]\, dx,$$

cf. Exercise 3.1.25.

3.2.39. Prove that $\| \mathbf{v} - \mathbf{w} \| \ \geq \ | \ \| \mathbf{v} \| - \| \mathbf{w} \| \ |$. Interpret this result pictorially.

3.2.40. *True or false*: $\| \mathbf{w} \| \ \leq \ \| \mathbf{v} \| + \| \mathbf{v} + \mathbf{w} \|$ for any $\mathbf{v}, \mathbf{w} \in V$.

♡ **3.2.41.** (a) Prove that the space $\mathbb{R}^\infty$ consisting of all infinite sequences $\mathbf{x} = (x_1, x_2, x_3, \ldots)$ of real numbers $x_i \in \mathbb{R}$ forms a vector space.

(b) Prove that the set of all sequences $\mathbf{x}$ such that $\sum_{k=1}^{\infty} x_k^2 < \infty$ forms a subspace, commonly denoted $\ell^2 \subset \mathbb{R}^\infty$.

(c) Write down two examples of sequences $\mathbf{x}$ belonging to ℓ^2 and two that do not belong to ℓ^2.

(d) *True or false*: If $\mathbf{x} \in \ell^2$, then $x_k \to 0$ and $k \to \infty$.

(e) *True or false*: If $x_k \to 0$ as $k \to \infty$, then $\mathbf{x} \in \ell^2$.

(f) Let α be fixed, and let $\mathbf{x}$ be the sequence with $x_k = \alpha^k$. For which values of α is $\mathbf{x} \in \ell^2$?

(g) Answer part (f) when $x_k = k^\alpha$.

(h) Prove that

$$\langle \mathbf{x}, \mathbf{y} \rangle = \sum_{k=1}^{\infty} x_k\, y_k$$

defines an inner product on the vector space ℓ^2. What is the corresponding norm?

(i) Write out the Cauchy–Schwarz and triangle inequalities for the inner product space ℓ^2.

3.3 NORMS

Every inner product gives rise to a norm that can be used to measure the magnitude or length of the elements of the underlying vector space. However, not every norm that is used in analysis and applications arises from an inner product. To define a general norm on a vector space, we will extract those properties that do not directly rely on the inner product structure.

Definition 3.12 A *norm* on the vector space V assigns a real number $\| \mathbf{v} \|$ to each vector $\mathbf{v} \in V$, subject to the following axioms for every $\mathbf{v}, \mathbf{w} \in V$, and $c \in \mathbb{R}$.

(a) *Positivity*: $\| \mathbf{v} \| \geq 0$, with $\| \mathbf{v} \| = 0$ if and only if $\mathbf{v} = \mathbf{0}$.

(b) *Homogeneity*: $\| c \mathbf{v} \| = | c | \, \| \mathbf{v} \|$.

(c) *Triangle inequality*: $\| \mathbf{v} + \mathbf{w} \| \leq \| \mathbf{v} \| + \| \mathbf{w} \|$.

As we now know, every inner product gives rise to a norm. Indeed, positivity of the norm is one of the inner product axioms. The homogeneity property follows since

$$\| c\, \mathbf{v} \| \ = \ \sqrt{\langle c\, \mathbf{v}, c\, \mathbf{v} \rangle} \ = \ \sqrt{c^2\, \langle \mathbf{v}, \mathbf{v} \rangle} \ = \ | c | \, \sqrt{\langle \mathbf{v}, \mathbf{v} \rangle} \ = \ | c | \, \| \mathbf{v} \|.$$

Finally, the triangle inequality for an inner product norm was established in Theorem 3.9. Let us introduce some of the principal examples of norms that do not come from inner products.

First, let $V = \mathbb{R}^n$. The *1–norm* of a vector $\mathbf{v} = (v_1, v_2, \ldots, v_n)^T$ is defined as the sum of the absolute values of its entries:

$$\|\mathbf{v}\|_1 = |v_1| + |v_2| + \cdots + |v_n|. \tag{3.26}$$

The *max* or *∞–norm* is equal to its maximal entry (in absolute value):

$$\|\mathbf{v}\|_\infty = \max\{\, |v_1|, |v_2|, \ldots, |v_n| \,\}. \tag{3.27}$$

Verification of the positivity and homogeneity properties for these two norms is straightforward; the triangle inequality is a direct consequence of the elementary inequality

$$|a + b| \le |a| + |b|, \qquad a, b \in \mathbb{R},$$

for absolute values.

The Euclidean norm, 1–norm, and ∞–norm on $\mathbb{R}^n$ are just three representatives of the general *p–norm*

$$\|\mathbf{v}\|_p = \sqrt[p]{\sum_{i=1}^{n} |v_i|^p}\,. \tag{3.28}$$

This quantity defines a norm for any $1 \le p < \infty$. The ∞–norm is a limiting case of (3.28) as $p \to \infty$. Note that the Euclidean norm (3.3) is the 2–norm, and is often designated as such; it is the only p–norm which comes from an inner product. The positivity and homogeneity properties of the p–norm are not hard to establish. The triangle inequality, however, is not trivial; in detail, it reads

$$\sqrt[p]{\sum_{i=1}^{n} |v_i + w_i|^p} \le \sqrt[p]{\sum_{i=1}^{n} |v_i|^p} + \sqrt[p]{\sum_{i=1}^{n} |w_i|^p}\,, \tag{3.29}$$

and is known as *Minkowski's inequality*. A complete proof can be found in [38].

There are analogous norms on the space $C^0[a,b]$ of continuous functions on an interval $[a,b]$. Basically, one replaces the previous sums by integrals. Thus, the L^p–norm is defined as

$$\|f\|_p = \sqrt[p]{\int_a^b |f(x)|^p\, dx}\,. \tag{3.30}$$

In particular, the L^1 norm is given by integrating the absolute value of the function:

$$\|f\|_1 = \int_a^b |f(x)|\, dx. \tag{3.31}$$

The L^2 norm (3.13) appears as a special case, $p = 2$, and, again, is the only one arising from an inner product. The limiting L^∞ norm is defined by the maximum

$$\|f\|_\infty = \max\{\, |f(x)| \;:\; a \le x \le b \,\}. \tag{3.32}$$

The proof of the general triangle or Minkowski inequality for $p \ne 1, 2, \infty$ is again not trivial, [16, 38, 52].

EXAMPLE 3.13 Consider the polynomial $p(x) = 3x^2 - 2$ on the interval $-1 \le x \le 1$. Its L^2 norm is

$$\| p \|_2 = \sqrt{\int_{-1}^{1} (3x^2 - 2)^2 \, dx} = \sqrt{\frac{18}{5}} = 1.8974\ldots .$$

Its L^∞ norm is

$$\| p \|_\infty = \max \left\{ \, |3x^2 - 2| \; : \; -1 \le x \le 1 \, \right\} = 2,$$

with the maximum occurring at $x = 0$. Finally, its L^1 norm is

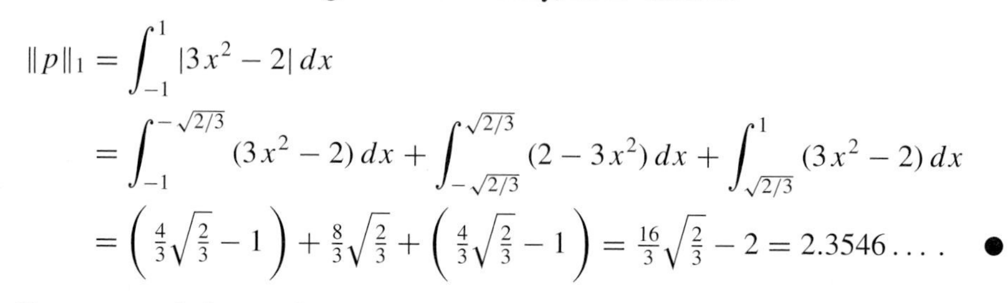

$$\begin{aligned} \| p \|_1 &= \int_{-1}^{1} |3x^2 - 2| \, dx \\ &= \int_{-1}^{-\sqrt{2/3}} (3x^2 - 2)\, dx + \int_{-\sqrt{2/3}}^{\sqrt{2/3}} (2 - 3x^2)\, dx + \int_{\sqrt{2/3}}^{1} (3x^2 - 2)\, dx \\ &= \left(\tfrac{4}{3}\sqrt{\tfrac{2}{3}} - 1 \right) + \tfrac{8}{3}\sqrt{\tfrac{2}{3}} + \left(\tfrac{4}{3}\sqrt{\tfrac{2}{3}} - 1 \right) = \tfrac{16}{3}\sqrt{\tfrac{2}{3}} - 2 = 2.3546\ldots . \end{aligned}$$ ●

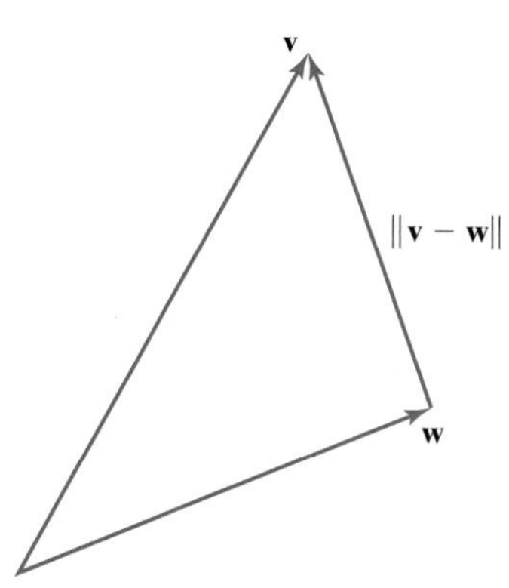

Figure 3.4 Distance between vectors.

Every norm defines a *distance* between vector space elements, namely

$$d(\mathbf{v}, \mathbf{w}) = \| \mathbf{v} - \mathbf{w} \|. \tag{3.33}$$

For the standard dot product norm, we recover the usual notion of distance between points in Euclidean space. Other types of norms produce alternative (and sometimes quite useful) notions of distance that are, nevertheless, subject to all the familiar properties:

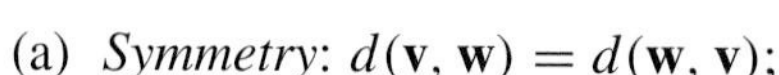

(a) *Symmetry*: $d(\mathbf{v}, \mathbf{w}) = d(\mathbf{w}, \mathbf{v})$;

(b) *Positivity*: $d(\mathbf{v}, \mathbf{w}) = 0$ if and only if $\mathbf{v} = \mathbf{w}$;

(c) *Triangle Inequality*: $d(\mathbf{v}, \mathbf{w}) \le d(\mathbf{v}, \mathbf{z}) + d(\mathbf{z}, \mathbf{w})$.

EXERCISES 3.3

3.3.1. Compute the 1, 2, 3 and ∞ norms of the vectors $\begin{pmatrix} 1 \\ 0 \end{pmatrix}, \begin{pmatrix} 0 \\ 1 \end{pmatrix}$. Verify the triangle inequality in each case.

3.3.2. Answer Exercise 3.3.1 for

(a) $\begin{pmatrix} 2 \\ -1 \end{pmatrix}, \begin{pmatrix} 1 \\ -2 \end{pmatrix}$ (b) $\begin{pmatrix} 1 \\ 0 \\ -1 \end{pmatrix}, \begin{pmatrix} -1 \\ 1 \\ 0 \end{pmatrix}$

(c) $\begin{pmatrix} 1 \\ -2 \\ -1 \end{pmatrix}, \begin{pmatrix} 2 \\ -1 \\ -3 \end{pmatrix}$

3.3.3. Which two of the vectors $\mathbf{u} = (-2, 2, 1)^T$, $\mathbf{v} = (1, 4, 1)^T$, $\mathbf{w} = (0, 0, -1)^T$ are closest to each other in distance for

(a) the Euclidean norm? (b) the ∞ norm?

(c) the 1 norm?

3.3.4. (a) Compute the L^∞ norm on $[0, 1]$ of the functions $f(x) = \frac{1}{3} - x$ and $g(x) = x - x^2$.

(b) Verify the triangle inequality for these two particular functions.

3.3.5. Answer Exercise 3.3.4 using the L^1 norm.

3.3.6. Which two of the functions $f(x) = 1$, $g(x) = x$, $h(x) = \sin \pi x$ are closest to each other on the interval $[0, 1]$ under

(a) the L^1 norm? (b) the L^2 norm?

(c) the L^∞ norm?

3.3.7. Consider the functions $f(x) = 1$ and $g(x) = x - \frac{3}{4}$ as elements of the vector space $C^0[0, 1]$. For each of the indicated norms, compute $\| f \|$, $\| g \|$, $\| f + g \|$, and verify the triangle inequality:

(a) the L^1 norm (b) the L^2 norm

(c) the L^3 norm (d) the L^∞ norm

3.3.8. Answer Exercise 3.3.7 when $f(x) = e^x$ and $g(x) = e^{-x}$.

3.3.9. Carefully prove that $\| (x, y)^T \| = |x| + 2\,|x - y|$ defines a norm on $\mathbb{R}^2$.

3.3.10. Prove that the following formulas define norms on $\mathbb{R}^2$:

(a) $\| \mathbf{v} \| = \sqrt{2\, v_1^2 + 3\, v_2^2}$

(b) $\| \mathbf{v} \| = \sqrt{2\, v_1^2 - v_1 v_2 + 2\, v_2^2}$

(c) $\| \mathbf{v} \| = 2\,| v_1 | + | v_2 |$

(d) $\| \mathbf{v} \| = \max\{\, 2\,| v_1 |, | v_2 | \,\}$

(e) $\| \mathbf{v} \| = \max\{\, | v_1 - v_2 |, | v_1 + v_2 | \,\}$

(f) $\| \mathbf{v} \| = | v_1 - v_2 | + | v_1 + v_2 |$

3.3.11. Which of the following formulas define norms on $\mathbb{R}^3$?

(a) $\| \mathbf{v} \| = \sqrt{2\, v_1^2 + v_2^2 + 3\, v_3^2}$

(b) $\| \mathbf{v} \| = \sqrt{v_1^2 + 2\, v_1 v_2 + v_2^2 + v_3^2}$

(c) $\| \mathbf{v} \| = \max\{\, | v_1 |, | v_2 |, | v_3 | \,\}$

(d) $\| \mathbf{v} \| = | v_1 - v_2 | + | v_2 - v_3 | + | v_3 - v_1 |$

(e) $\| \mathbf{v} \| = | v_1 | + \max\{\, | v_2 |, | v_3 | \,\}$

3.3.12. Prove that two parallel vectors $\mathbf{v}$ and $\mathbf{w}$ have the same norm if and only if $\mathbf{v} = \pm\mathbf{w}$.

3.3.13. *True or false*: If $\| \mathbf{v} + \mathbf{w} \| = \| \mathbf{v} \| + \| \mathbf{w} \|$, then $\mathbf{v}$, $\mathbf{w}$ are parallel vectors.

3.3.14. Prove that the ∞ norm on $\mathbb{R}^2$ does not come from an inner product. *Hint*: Look at Exercise 3.1.12.

3.3.15. Can formula (3.11) be used to define an inner product for

(a) the 1 norm $\| \mathbf{v} \|_1$ on $\mathbb{R}^2$?

(b) the ∞ norm $\| \mathbf{v} \|_\infty$ on $\mathbb{R}^2$?

◇ **3.3.16.** Prove that $\lim_{p \to \infty} \| \mathbf{v} \|_p = \| \mathbf{v} \|_\infty$ for any $\mathbf{v} \in \mathbb{R}^2$.

◇ **3.3.17.** Justify the triangle inequality for

(a) the L^1 norm (3.31);

(b) the L^∞ norm (3.32).

◇ **3.3.18.** Let $w(x) > 0$ for $a \le x \le b$ be a weight function.

(a) Prove that

$$\| f \|_{1,w} = \int_a^b | f(x) | \, w(x) \, dx$$

defines a norm on $C^0[\,a, b\,]$, called the *weighted* L^1 *norm*.

(b) Do the same for the *weighted* L^∞ *norm* $\| f \|_{\infty,w} = \max\{\, | f(x) | \, w(x) \; : \; a \le x \le b \,\}$.

3.3.19. Let $\| \cdot \|_1$ and $\| \cdot \|_2$ be two different norms on a vector space V.

(a) Prove that $\| \mathbf{v} \| = \max\{\, \| \mathbf{v} \|_1, \| \mathbf{v} \|_2 \,\}$ defines a norm on V.

(b) Does $\| \mathbf{v} \| = \min\{\, \| \mathbf{v} \|_1, \| \mathbf{v} \|_2 \,\}$ define a norm?

(c) Does the arithmetic mean

$$\| \mathbf{v} \| = \tfrac{1}{2} \big(\| \mathbf{v} \|_1 + \| \mathbf{v} \|_2 \big)$$

define a norm?

(d) Does the geometric mean

$$\| \mathbf{v} \| = \sqrt{\| \mathbf{v} \|_1 \, \| \mathbf{v} \|_2}$$

define a norm?

Unit Vectors

Let V be a fixed normed vector space. The elements $\mathbf{u} \in V$ that have unit norm, $\| \mathbf{u} \| = 1$, play a special role, and are known as *unit vectors* (or functions or elements). The following easy lemma shows how to construct a unit vector pointing in the same direction as any given nonzero vector.

Lemma 3.14 If $\mathbf{v} \ne \mathbf{0}$ is any nonzero vector, then the vector $\mathbf{u} = \mathbf{v} / \| \mathbf{v} \|$ obtained by dividing $\mathbf{v}$ by its norm is a unit vector parallel to $\mathbf{v}$.

Proof We compute, making use of the homogeneity property of the norm:

$$\| \mathbf{u} \| = \left\| \frac{\mathbf{v}}{\| \mathbf{v} \|} \right\| = \frac{\| \mathbf{v} \|}{\| \mathbf{v} \|} = 1. \qquad \blacksquare$$

EXAMPLE 3.15 The vector $\mathbf{v} = (1, -2)^T$ has length $\| \mathbf{v} \|_2 = \sqrt{5}$ with respect to the standard Euclidean norm. Therefore, the unit vector pointing in the same direction is

$$\mathbf{u} = \frac{\mathbf{v}}{\| \mathbf{v} \|_2} = \frac{1}{\sqrt{5}} \begin{pmatrix} 1 \\ -2 \end{pmatrix} = \begin{pmatrix} \frac{1}{\sqrt{5}} \\ -\frac{2}{\sqrt{5}} \end{pmatrix}.$$

On the other hand, for the 1 norm, $\|\mathbf{v}\|_1 = 3$, and so

$$\widetilde{\mathbf{u}} = \frac{\mathbf{v}}{\|\mathbf{v}\|_1} = \frac{1}{3}\begin{pmatrix} 1 \\ -2 \end{pmatrix} = \begin{pmatrix} \frac{1}{3} \\ -\frac{2}{3} \end{pmatrix}$$

is the unit vector parallel to $\mathbf{v}$ in the 1 norm. Finally, $\|\mathbf{v}\|_\infty = 2$, and hence the corresponding unit vector for the ∞ norm is

$$\widehat{\mathbf{u}} = \frac{\mathbf{v}}{\|\mathbf{v}\|_\infty} = \frac{1}{2}\begin{pmatrix} 1 \\ -2 \end{pmatrix} = \begin{pmatrix} \frac{1}{2} \\ -1 \end{pmatrix}.$$

Thus, the notion of unit vector will depend upon which norm is being used. ●

EXAMPLE 3.16 Similarly, on the interval $[0, 1]$, the quadratic polynomial $p(x) = x^2 - \frac{1}{2}$ has L^2 norm

$$\|p\|_2 = \sqrt{\int_0^1 \left(x^2 - \tfrac{1}{2}\right)^2 dx} = \sqrt{\int_0^1 \left(x^4 - x^2 + \tfrac{1}{4}\right) dx} = \sqrt{\frac{7}{60}}.$$

Therefore,

$$u(x) = \frac{p(x)}{\|p\|} = \sqrt{\tfrac{60}{7}}\, x^2 - \sqrt{\tfrac{15}{7}}$$

is a "unit polynomial", $\|u\|_2 = 1$, which is "parallel" to (or, more precisely, a scalar multiple of) the polynomial p. On the other hand, for the L^∞ norm,

$$\|p\|_\infty = \max\left\{ \left| x^2 - \tfrac{1}{2} \right| \;\middle|\; 0 \le x \le 1 \right\} = \tfrac{1}{2},$$

and hence, in this case $\tilde{u}(x) = 2\,p(x) = 2x^2 - 1$ is the corresponding unit polynomial. ●

The *unit sphere* for the given norm is defined as the set of all unit vectors

$$S_1 = \left\{ \|\mathbf{u}\| = 1 \right\}, \qquad \text{while} \qquad S_r = \left\{ \|\mathbf{u}\| = r \right\} \tag{3.34}$$

is the sphere of radius $r \ge 0$. Thus, the unit sphere for the Euclidean norm on $\mathbb{R}^n$ is the usual round sphere

$$S_1 = \left\{ \|\mathbf{x}\|^2 = x_1^2 + x_2^2 + \cdots + x_n^2 = 1 \right\}.$$

The unit sphere for the ∞ norm is the unit cube

$$S_1 = \left\{ \mathbf{x} \in \mathbb{R}^n \;\middle|\; \begin{array}{c} |x_i| \le 1, \quad i = 1, \ldots, n, \text{ and either} \\ x_1 = \pm 1 \text{ or } x_2 = \pm 1 \text{ or} \ldots \text{ or } x_n = \pm 1 \end{array} \right\}.$$

For the 1 norm,

$$S_1 = \left\{ \mathbf{x} \in \mathbb{R}^n \;\middle|\; |x_1| + |x_2| + \cdots + |x_n| = 1 \right\}$$

is the unit diamond in two dimensions, unit octahedron in three dimensions, and unit *cross polytope* in general. See Figure 3.5 for the two-dimensional pictures.

In all cases, the *unit ball* $B_1 = \left\{ \|\mathbf{u}\| \le 1 \right\}$ consists of all vectors of norm less than or equal to 1, and has the unit sphere as its boundary. If V is a finite-dimensional normed vector space, then the unit ball B_1 forms a *compact* subset, meaning that it is closed and bounded. This basic topological fact, which is *not* true in infinite-dimensional normed spaces, underscores the distinction between finite-dimensional vector analysis and the vastly more complicated infinite-dimensional realm.

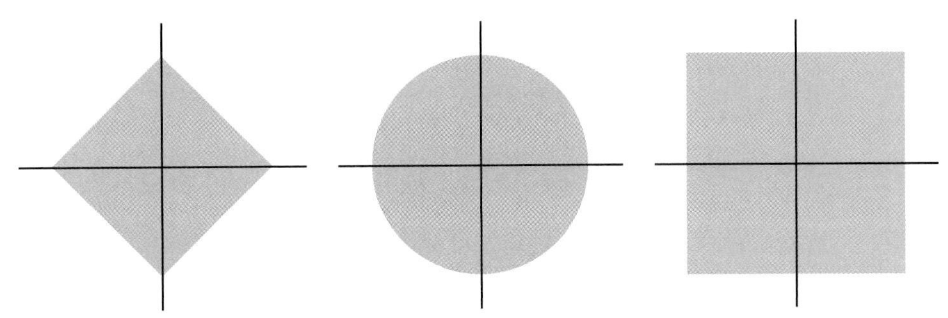

Figure 3.5 Unit balls and spheres for 1, 2 and ∞ norms in $\mathbb{R}^2$.

EXERCISES 3.3

3.3.20. Find a unit vector in the same direction as $\mathbf{v} = (1, 2, -3)^T$ for
(a) the Euclidean norm
(b) the weighted norm $\|\mathbf{v}\|^2 = 2\,v_1^2 + v_2^2 + \frac{1}{3}\,v_3^2$
(c) the 1 norm
(d) the ∞ norm
(e) the norm based on the inner product $2\,v_1\,w_1 - v_1\,w_2 - v_2\,w_1 + 2\,v_2\,w_2 - v_2\,w_3 - v_3\,w_2 + 2\,v_3\,w_3$

3.3.21. Show for any choice of given angles θ, ϕ and ψ, the following are unit vectors in the Euclidean norm:
(a) $(\cos\theta\cos\phi, \cos\theta\sin\phi, \sin\theta)^T$
(b) $\frac{1}{\sqrt{2}}\,(\cos\theta, \sin\theta, \cos\phi, \sin\phi)^T$
(c) $(\cos\theta\cos\phi\cos\psi, \cos\theta\cos\phi\sin\psi, \cos\theta\sin\phi, \sin\theta)^T$

3.3.22. How many unit vectors are parallel to a given vector $\mathbf{v} \neq \mathbf{0}$?
(a) 1 (b) 2
(c) 3 (d) ∞
(e) depends upon the norm.
Explain your choice of answer.

3.3.23. Plot the unit circle (sphere) for
(a) the weighted norm $\|\mathbf{v}\| = \sqrt{v_1^2 + 4\,v_2^2}$;
(b) the norm based on the inner product (3.9);
(c) the norm of Exercise 3.3.9.

3.3.24. Draw the unit circle for each norm in Exercise 3.3.10.

3.3.25. Sketch the unit sphere $S_1 \subset \mathbb{R}^3$ for
(a) the L^1 norm, (b) the L^∞ norm,
(c) the weighted norm $\|\mathbf{v}\|^2 = 2\,v_1^2 + v_2^2 + 3\,v_3^2$,
(d) the norm

$$\|\mathbf{v}\| = \max\{\,|v_1 + v_2|, |v_1 + v_3|, |v_2 + v_3|\,\}.$$

3.3.26. Let $\mathbf{v} \neq \mathbf{0}$ be any nonzero vector in a normed vector space V. Show how to construct a new norm on V that changes $\mathbf{v}$ into a unit vector.

3.3.27. *True or false*: Two norms on a vector space have the same unit sphere if and only if they are the same norm.

3.3.28. Find the unit function that is a constant multiple of the function $f(x) = x - \frac{1}{3}$ with respect to the
(a) L^1 norm on $[0, 1]$
(b) L^2 norm on $[0, 1]$
(c) L^∞ norm on $[0, 1]$
(d) L^1 norm on $[-1, 1]$
(e) L^2 norm on $[-1, 1]$
(f) L^∞ norm on $[-1, 1]$

3.3.29. For which norms is the constant function $f(x) \equiv 1$ a unit function?
(a) L^1 norm on $[0, 1]$
(b) L^2 norm on $[0, 1]$
(c) L^∞ norm on $[0, 1]$
(d) L^1 norm on $[-1, 1]$
(e) L^2 norm on $[-1, 1]$
(f) L^∞ norm on $[-1, 1]$
(g) L^1 norm on $\mathbb{R}$
(h) L^2 norm on $\mathbb{R}$
(i) L^∞ norm on $\mathbb{R}$

◇ **3.3.30.** A subset $S \subset \mathbb{R}^n$ is called *convex* if, for any $\mathbf{x}$, $\mathbf{y} \in S$, the line segment joining $\mathbf{x}$ to $\mathbf{y}$ is also in S, i.e., $t\,\mathbf{x} + (1-t)\,\mathbf{y} \in S$ for all $0 \le t \le 1$. Prove that the unit ball is a convex subset of a normed vector space. Is the unit sphere convex?

Equivalence of Norms

While there are many different types of norms, in a finite-dimensional vector space they are all more or less equivalent. "Equivalence" does not mean that they assume the same value, but rather that they are, in a certain sense, always close to one another, and so, for many analytical purposes, may be used interchangeably. As a consequence, we may be able to simplify the analysis of a problem by choosing a suitably adapted norm; examples can be found in Chapter 10.

THEOREM 3.17 Let $\|\cdot\|_1$ and $\|\cdot\|_2$ be any two norms on $\mathbb{R}^n$. Then there exist positive constants $c^\star, C^\star > 0$ such that

$$c^\star \|\mathbf{v}\|_1 \le \|\mathbf{v}\|_2 \le C^\star \|\mathbf{v}\|_1 \quad \text{for every} \quad \mathbf{v} \in \mathbb{R}^n. \tag{3.35}$$

Proof We just sketch the basic idea, leaving the details to a more rigorous real analysis course, [16, §7.6]. We begin by noting that a norm defines a continuous real-valued function $f(\mathbf{v}) = \|\mathbf{v}\|$ on $\mathbb{R}^n$. (Continuity is, in fact, a consequence of the triangle inequality.) Let $S_1 = \left\{ \|\mathbf{u}\|_1 = 1 \right\}$ denote the unit sphere of the first norm. Any continuous function defined on a compact set achieves both a maximum and a minimum value. Thus, restricting the second norm function to the unit sphere S_1 of the first norm, we can set

$$c^\star = \min\{ \|\mathbf{u}\|_2 \mid \mathbf{u} \in S_1 \}, \qquad C^\star = \max\{ \|\mathbf{u}\|_2 \mid \mathbf{u} \in S_1 \}. \tag{3.36}$$

Moreover, $0 < c^\star \le C^\star < \infty$, with equality holding if and only if the norms are the same. The minimum and maximum (3.36) will serve as the constants in the desired inequalities (3.35). Indeed, by definition,

$$c^\star \le \|\mathbf{u}\|_2 \le C^\star \quad \text{when} \quad \|\mathbf{u}\|_1 = 1, \tag{3.37}$$

which proves that (3.35) is valid for all unit vectors $\mathbf{v} = \mathbf{u} \in S_1$. To prove the inequalities in general, assume $\mathbf{v} \ne \mathbf{0}$. (The case $\mathbf{v} = \mathbf{0}$ is trivial.) Lemma 3.14 says that $\mathbf{u} = \mathbf{v}/\|\mathbf{v}\|_1 \in S_1$ is a unit vector in the first norm: $\|\mathbf{u}\|_1 = 1$. Moreover, by the homogeneity property of the norm, $\|\mathbf{u}\|_2 = \|\mathbf{v}\|_2/\|\mathbf{v}\|_1$. Substituting into (3.37) and clearing denominators completes the proof of (3.35). ■

EXAMPLE 3.18 For example, consider the Euclidean norm $\|\cdot\|_2$ and the max norm $\|\cdot\|_\infty$ on $\mathbb{R}^n$. According to (3.36), the bounding constants are found by minimizing and maximizing $\|\mathbf{u}\|_\infty = \max\{ |u_1|, \ldots, |u_n| \}$ over all unit vectors $\|\mathbf{u}\|_2 = 1$ on the (round) unit sphere. The maximal value is achieved at the poles $\pm\mathbf{e}_k$, with $\|\pm\mathbf{e}_k\|_\infty = C^\star = 1$ The minimal value is attained at the points $\left(\pm\frac{1}{\sqrt{n}}, \ldots, \pm\frac{1}{\sqrt{n}} \right)$, whereby $c^\star = \frac{1}{\sqrt{n}}$. Therefore,

$$\frac{1}{\sqrt{n}} \|\mathbf{v}\|_2 \le \|\mathbf{v}\|_\infty \le \|\mathbf{v}\|_2. \tag{3.38}$$

We can interpret these inequalities as follows. Suppose $\mathbf{v}$ is a vector lying on the unit sphere in the Euclidean norm, so $\|\mathbf{v}\|_2 = 1$. Then (3.38) tells us that its ∞ norm is bounded from above and below by $\frac{1}{\sqrt{n}} \le \|\mathbf{v}\|_\infty \le 1$. Therefore, the Euclidean unit sphere sits inside the ∞ norm unit sphere, and outside the ∞ norm sphere of radius $\frac{1}{\sqrt{n}}$. Figure 3.6 illustrates the two-dimensional situation: the unit circle is inside the unit square, and contains the square of size $\frac{1}{\sqrt{2}}$. ●

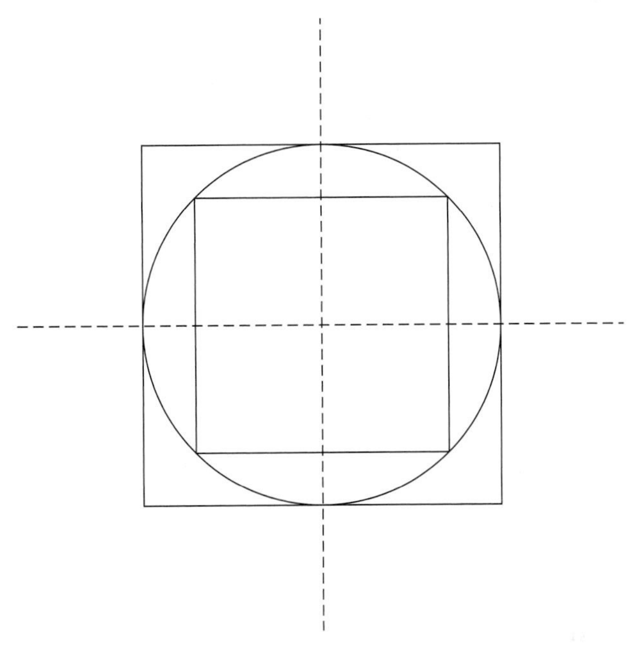

Figure 3.6 Equivalence of the ∞ and 2 norms.

One significant consequence of the equivalence of norms is that, in $\mathbb{R}^n$, convergence is independent of the norm. The following are all equivalent to the standard notion of convergence of a sequence $\mathbf{u}^{(1)}, \mathbf{u}^{(2)}, \mathbf{u}^{(3)}, \ldots$ of vectors in $\mathbb{R}^n$:

(a) the vectors converge: $\mathbf{u}^{(k)} \longrightarrow \mathbf{u}^\star$:

(b) the individual coordinates all converge: $u_i^{(k)} \longrightarrow u_i^\star$ for $i = 1, \ldots, n$.

(c) the difference in norms goes to zero: $\| \mathbf{u}^{(k)} - \mathbf{u}^\star \| \longrightarrow 0$.

The last version, known as *convergence in norm*, does not depend on which norm is chosen. Indeed, the inequality (3.35) implies that if one norm goes to zero, so does any other norm. A consequence is that all norms on $\mathbb{R}^n$ induce the same topology—convergence of sequences, notions of open and closed sets, and so on. None of this is true in infinite-dimensional function space! A rigorous development of the underlying topological and analytical properties of compactness, continuity, and convergence is beyond the scope of this course. The motivated student is encouraged to consult a text in real analysis, e.g., [16], to find the relevant definitions, theorems and proofs.

EXAMPLE 3.19

Consider the infinite-dimensional vector space $C^0[0,1]$ consisting of all continuous functions on the interval $[0,1]$. The functions

$$f_n(x) = \begin{cases} 1 - nx, & 0 \le x \le \frac{1}{n}, \\ 0, & \frac{1}{n} \le x \le 1, \end{cases}$$

have identical L^∞ norms

$$\| f_n \|_\infty = \sup \{ \, | f_n(x) | \mid 0 \le x \le 1 \, \} = 1.$$

On the other hand, their L^2 norm

$$\| f_n \|_2 = \sqrt{\int_0^1 f_n(x)^2 \, dx} = \sqrt{\int_0^{1/n} (1 - nx)^2 \, dx} = \frac{1}{\sqrt{3n}}$$

goes to zero as $n \to \infty$. This example shows that there *is no* constant $C^\star$ such that

$$\| f \|_\infty \le C^\star \| f \|_2$$

for all $f \in C^0[0,1]$. Thus, the L^∞ and L^2 norms on $C^0[0,1]$ are not equivalent— there exist functions that have unit L^∞ norm, but arbitrarily small L^2 norm. Similar comparative results can be established for the other function space norms. Analysis and topology on function space is intimately linked to the underlying choice of norm. ●

EXERCISES 3.3

3.3.31. Check the validity of the inequalities (3.38) for the particular vectors

(a) $(1,-1)^T$ (b) $(1,2,3)^T$
(c) $(1,1,1,1)^T$ (d) $(1,-1,-2,-1,1)^T$

3.3.32. Find all $\mathbf{v} \in \mathbb{R}^2$ such that

(a) $\|\mathbf{v}\|_1 = \|\mathbf{v}\|_\infty$ (b) $\|\mathbf{v}\|_1 = \|\mathbf{v}\|_2$
(c) $\|\mathbf{v}\|_2 = \|\mathbf{v}\|_\infty$ (d) $\|\mathbf{v}\|_\infty = \frac{1}{\sqrt{2}}\|\mathbf{v}\|_2$

3.3.33. How would you quantify the following statement: The norm of a vector is small if and only if all its entries are small.

3.3.34. Can you find an elementary proof of the inequalities $\|\mathbf{v}\|_\infty \le \|\mathbf{v}\|_2 \le \sqrt{n}\,\|\mathbf{v}\|_\infty$ for $\mathbf{v} \in \mathbb{R}^n$ directly from the formulas for the norms?

3.3.35. (i) Show the equivalence of the Euclidean norm and the 1 norm on $\mathbb{R}^n$ by proving $\|\mathbf{v}\|_2 \le \|\mathbf{v}\|_1 \le \sqrt{n}\,\|\mathbf{v}\|_2$.
(ii) Verify that the vectors in Exercise 3.3.31 satisfy both inequalities.
(iii) For which vectors $\mathbf{v} \in \mathbb{R}^n$ is
(a) $\|\mathbf{v}\|_2 = \|\mathbf{v}\|_1$?
(b) $\|\mathbf{v}\|_1 = \sqrt{n}\,\|\mathbf{v}\|_2$?

3.3.36. (a) Establish the equivalence inequalities (3.35) between the 1 and ∞ norms.
(b) Verify them for the vectors in Exercise 3.3.31.
(c) For which vectors $\mathbf{v} \in \mathbb{R}^n$ are your inequalities equality?

3.3.37. Let $\|\cdot\|_2$ denote the usual Euclidean norm on $\mathbb{R}^n$. Determine the constants in the norm equivalence inequalities $c^\star\,\|\mathbf{v}\| \le \|\mathbf{v}\|_2 \le C^\star\,\|\mathbf{v}\|$ for the following norms:

(a) the weighted norm $\|\mathbf{v}\| = \sqrt{2\,v_1^2 + 3\,v_2^2}$,
(b) the norm $\|\mathbf{v}\| = \max\big\{\,|v_1 + v_2|, |v_1 - v_2|\,\big\}$.

3.3.38. Let $\|\cdot\|$ be any norm on $\mathbb{R}^n$. Prove that there is a constant $C > 0$ such that the entries of any $\mathbf{v} = (v_1, v_2, \ldots, v_n)^T \in \mathbb{R}^n$ are all bounded, in absolute value, by $|v_i| \le C\,\|\mathbf{v}\|$.

◇ **3.3.39.** What does it mean if the constants defined in (3.36) are equal: $c^\star = C^\star$?

3.3.40. Prove that if $[a,b]$ is a bounded interval and $f \in C^0[a,b]$, then $\|f\|_2 \le \sqrt{b-a}\,\|f\|_\infty$.

♡ **3.3.41.** In this exercise, the indicated function norms are taken over all of $\mathbb{R}$.

(a) Let $f_n(x) = \begin{cases} 1, & -n \le x \le n, \\ 0 & \text{otherwise.} \end{cases}$ Prove that $\|f_n\|_\infty = 1$, but $\|f_n\|_2 \to \infty$ as $n \to \infty$.
(b) Explain why there is no constant C such that $\|f\|_2 \le C\,\|f\|_\infty$ for all functions f.
(c) Let $f_n(x) = \begin{cases} \sqrt{\dfrac{n}{2}}, & -\dfrac{1}{n} \le x \le \dfrac{1}{n}, \\ 0 & \text{otherwise.} \end{cases}$ Prove that $\|f_n\|_2 = 1$, but $\|f_n\|_\infty \to \infty$ as $n \to \infty$. Conclude that there is no constant C such that $\|f\|_\infty \le C\,\|f\|_2$.
(d) Construct similar examples that disprove the related inequalities
(i) $\|f\|_\infty \le C\,\|f\|_1$
(ii) $\|f\|_1 \le C\,\|f\|_2$
(iii) $\|f\|_2 \le C\,\|f\|_1$

♡ **3.3.42.** (a) Prove that the L^∞ and L^2 norms on the vector space $C^0[-1,1]$ are not equivalent. *Hint*: Look at Exercise 3.3.41 for ideas.
(b) Can you establish a bound in either direction, i.e., $\|f\|_\infty \le C\,\|f\|_2$ or $\|f\|_2 \le \widetilde{C}\,\|f\|_\infty$ for all $f \in C^0[-1,1]$ for some positive constant $C, \widetilde{C}$?
(c) Are the L^1 and L^∞ norms equivalent?

3.3.43. Suppose $\langle \mathbf{v}, \mathbf{w} \rangle_1$ and $\langle \mathbf{v}, \mathbf{w} \rangle_2$ are two inner products on the same vector space V. For which $\alpha, \beta \in \mathbb{R}$ is the linear combination $\langle \mathbf{v}, \mathbf{w} \rangle = \alpha\,\langle \mathbf{v}, \mathbf{w} \rangle_1 + \beta\,\langle \mathbf{v}, \mathbf{w} \rangle_2$ a legitimate inner product? *Hint*: The case when $\alpha, \beta \ge 0$ is easy. However, some negative values are also permitted, and your task is to decide which.

3.4 POSITIVE DEFINITE MATRICES

Let us now return to the study of inner products and fix our attention on the finite-dimensional situation. Our immediate goal is to determine the most general inner product that can be placed on the finite-dimensional vector space $\mathbb{R}^n$. The answer will lead us to the important class of positive definite matrices, which appear in a wide range of applications, including minimization problems, mechanics, electrical circuits, differential equations, statistics, and numerical methods. Moreover, their infinite-dimensional counterparts, positive definite linear operators, govern most boundary value problems in continuum physics and engineering.

Let $\langle \mathbf{x}, \mathbf{y} \rangle$ be an inner product between vectors $\mathbf{x} = (\, x_1 \; x_2 \; \ldots \; x_n \,)^T$ and $\mathbf{y} = (\, y_1 \; y_2 \; \ldots \; y_n \,)^T$ in $\mathbb{R}^n$. We begin by writing the vectors in terms of the standard basis vectors (2.17):

$$\begin{aligned} \mathbf{x} &= x_1 \mathbf{e}_1 + \cdots + x_n \mathbf{e}_n = \sum_{i=1}^{n} x_i \mathbf{e}_i, \\ \mathbf{y} &= y_1 \mathbf{e}_1 + \cdots + y_n \mathbf{e}_n = \sum_{j=1}^{n} y_j \mathbf{e}_j. \end{aligned} \tag{3.39}$$

To evaluate their inner product, we will appeal to the three basic axioms. We first employ bilinearity to expand

$$\langle \mathbf{x}, \mathbf{y} \rangle = \left\langle \sum_{i=1}^{n} x_i \mathbf{e}_i, \sum_{j=1}^{n} y_j \mathbf{e}_j \right\rangle = \sum_{i,j=1}^{n} x_i \, y_j \langle \mathbf{e}_i, \mathbf{e}_j \rangle.$$

Therefore,

$$\langle \mathbf{x}, \mathbf{y} \rangle = \sum_{i,j=1}^{n} k_{ij} \, x_i \, y_j = \mathbf{x}^T K \, \mathbf{y}, \tag{3.40}$$

where K denotes the $n \times n$ matrix of inner products of the basis vectors, with entries

$$k_{ij} = \langle \mathbf{e}_i, \mathbf{e}_j \rangle, \qquad i, j = 1, \ldots, n. \tag{3.41}$$

We conclude that any inner product must be expressed in the general *bilinear form* (3.40).

The two remaining inner product axioms will impose certain conditions on the inner product matrix K. Symmetry implies that

$$k_{ij} = \langle \mathbf{e}_i, \mathbf{e}_j \rangle = \langle \mathbf{e}_j, \mathbf{e}_i \rangle = k_{ji}, \qquad i, j = 1, \ldots, n.$$

Consequently, the inner product matrix K is symmetric:

$$K = K^T.$$

Conversely, symmetry of K ensures symmetry of the bilinear form:

$$\langle \mathbf{x}, \mathbf{y} \rangle = \mathbf{x}^T K \, \mathbf{y} = (\mathbf{x}^T K \, \mathbf{y})^T = \mathbf{y}^T K^T \mathbf{x} = \mathbf{y}^T K \, \mathbf{x} = \langle \mathbf{y}, \mathbf{x} \rangle,$$

where the second equality follows from the fact that the quantity is a scalar, and hence equals its transpose.

The final condition for an inner product is positivity. This requires that

$$\| \mathbf{x} \|^2 = \langle \mathbf{x}, \mathbf{x} \rangle = \mathbf{x}^T K \, \mathbf{x} = \sum_{i,j=1}^{n} k_{ij} \, x_i \, x_j \geq 0 \quad \text{for all} \quad \mathbf{x} \in \mathbb{R}^n, \tag{3.42}$$

with equality if and only if $\mathbf{x} = \mathbf{0}$. The precise meaning of this positivity condition on the matrix K is not as immediately evident, and so will be encapsulated in a definition.

Definition 3.20 An $n \times n$ matrix K is called *positive definite* if it is symmetric, $K^T = K$, and satisfies the positivity condition

$$\mathbf{x}^T K \mathbf{x} > 0 \quad \text{for all} \quad \mathbf{0} \neq \mathbf{x} \in \mathbb{R}^n. \tag{3.43}$$

We will sometimes write $K > 0$ to mean that K is a symmetric, positive definite matrix.

Warning: The condition $K > 0$ does *not* mean that all the entries of K are positive. There are many positive definite matrices that have some negative entries; see Example 3.22 below. Conversely, many symmetric matrices with all positive entries are not positive definite!

REMARK: Although some authors allow non-symmetric matrices to be designated as positive definite, we will *only* say that a matrix is positive definite when it is symmetric. But, to underscore our convention and remind the casual reader, we will often include the superfluous adjective "symmetric" when speaking of positive definite matrices.

Our preliminary analysis has resulted in the following characterization of inner products on a finite-dimensional vector space.

THEOREM 3.21 Every inner product on $\mathbb{R}^n$ is given by

$$\langle \mathbf{x}, \mathbf{y} \rangle = \mathbf{x}^T K \mathbf{y} \quad \text{for} \quad \mathbf{x}, \mathbf{y} \in \mathbb{R}^n, \tag{3.44}$$

where K is a symmetric, positive definite matrix.

Given any symmetric matrix K, the homogeneous quadratic polynomial

$$q(\mathbf{x}) = \mathbf{x}^T K \mathbf{x} = \sum_{i,j=1}^{n} k_{ij}\, x_i\, x_j, \tag{3.45}$$

is known as a *quadratic form** on $\mathbb{R}^n$. The quadratic form is called *positive definite* if

$$q(\mathbf{x}) > 0 \quad \text{for all} \quad 0 \neq \mathbf{x} \in \mathbb{R}^n. \tag{3.46}$$

Thus, a quadratic form is positive definite if and only if its coefficient matrix is.

EXAMPLE 3.22 Even though the symmetric matrix

$$K = \begin{pmatrix} 4 & -2 \\ -2 & 3 \end{pmatrix}$$

has two negative entries, it is, nevertheless, a positive definite matrix. Indeed, the corresponding quadratic form

$$q(\mathbf{x}) = \mathbf{x}^T K \mathbf{x} = 4x_1^2 - 4x_1 x_2 + 3x_2^2 = (2x_1 - x_2)^2 + 2x_2^2 \geq 0$$

is a sum of two non-negative quantities. Moreover, $q(\mathbf{x}) = 0$ if and only if both $2x_1 - x_2 = 0$ and $x_2 = 0$, which implies $x_1 = 0$ also. This proves $q(\mathbf{x}) > 0$ for all $\mathbf{x} \neq \mathbf{0}$, and hence K is indeed a positive definite matrix. The corresponding inner product on $\mathbb{R}^2$ is

$$\langle \mathbf{x}, \mathbf{y} \rangle = (\, x_1 \; x_2 \,) \begin{pmatrix} 4 & -2 \\ -2 & 3 \end{pmatrix} \begin{pmatrix} y_1 \\ y_2 \end{pmatrix} = 4x_1 y_1 - 2x_1 y_2 - 2x_2 y_1 + 3x_2 y_2.$$

*Exercise 3.4.20 shows that the coefficient matrix K in any quadratic form can be taken to be symmetric without any loss of generality.

On the other hand, despite the fact that

$$K = \begin{pmatrix} 1 & 2 \\ 2 & 1 \end{pmatrix}$$

has all positive entries, it is *not* a positive definite matrix. Indeed, writing out

$$q(\mathbf{x}) = \mathbf{x}^T K \mathbf{x} = x_1^2 + 4x_1x_2 + x_2^2,$$

we find, for instance, that $q(1, -1) = -2 < 0$, violating positivity. These two simple examples should be enough to convince the reader that the problem of determining whether a given symmetric matrix is or is not positive definite is not completely elementary. ●

EXAMPLE 3.23 By definition, a general symmetric 2×2 matrix

$$K = \begin{pmatrix} a & b \\ b & c \end{pmatrix}$$

is positive definite if and only if the associated quadratic form satisfies

$$q(\mathbf{x}) = a x_1^2 + 2b x_1 x_2 + c x_2^2 > 0 \quad \text{for all} \quad \mathbf{x} \neq \mathbf{0}. \tag{3.47}$$

Analytic geometry tells us that this is the case if and only if

$$a > 0, \qquad a c - b^2 > 0, \tag{3.48}$$

i.e., the quadratic form has positive leading coefficient and positive determinant (or negative discriminant). A direct proof of this well-known fact will appear shortly. ●

With a little practice, it is not difficult to read off the coefficient matrix K from the explicit formula for the quadratic form (3.45).

EXAMPLE 3.24 Consider the quadratic form

$$q(x, y, z) = x^2 + 4xy + 6y^2 - 2xz + 9z^2$$

depending upon three variables. The corresponding coefficient matrix is

$$K = \begin{pmatrix} 1 & 2 & -1 \\ 2 & 6 & 0 \\ -1 & 0 & 9 \end{pmatrix}$$

whereby

$$q(x, y, z) = (x \quad y \quad z) \begin{pmatrix} 1 & 2 & -1 \\ 2 & 6 & 0 \\ -1 & 0 & 9 \end{pmatrix} \begin{pmatrix} x \\ y \\ z \end{pmatrix}.$$

Note that the squared terms in q contribute directly to the diagonal entries of K, while the mixed terms are split in half to give the symmetric off-diagonal entries. As a challenge, the reader might wish to try proving that this particular matrix is positive definite by establishing positivity of the quadratic form: $q(x, y, z) > 0$ for all nonzero $(x, y, z)^T \in \mathbb{R}^3$. Later, we will devise a simple, systematic test for positive definiteness. ●

Slightly more generally, a quadratic form and its associated symmetric coefficient matrix are called *positive semi-definite* if

$$q(\mathbf{x}) = \mathbf{x}^T K \mathbf{x} \geq 0 \quad \text{for all} \quad \mathbf{x} \in \mathbb{R}^n. \tag{3.49}$$

A positive semi-definite matrix may have *null directions*, meaning non-zero vectors $\mathbf{z}$ such that $q(\mathbf{z}) = \mathbf{z}^T K \mathbf{z} = \mathbf{0}$. Clearly, any nonzero vector $\mathbf{z} \in \ker K$ that lies in the coefficient matrix's kernel defines a null direction, but there may be others. A positive definite matrix is not allowed to have null directions, and so $\ker K = \{\mathbf{0}\}$. Recalling Proposition 2.42, we deduce that all positive definite matrices are nonsingular. The converse, however, is *not* valid; many symmetric, nonsingular matrices fail to be positive definite.

Proposition 3.25 If K is positive definite, then K is nonsingular.

EXAMPLE 3.26 The matrix

$$K = \begin{pmatrix} 1 & -1 \\ -1 & 1 \end{pmatrix}$$

is positive semi-definite, but not positive definite. Indeed, the associated quadratic form

$$q(\mathbf{x}) = \mathbf{x}^T K \mathbf{x} = x_1^2 - 2x_1x_2 + x_2^2 = (x_1 - x_2)^2 \geq 0$$

is a perfect square, and so clearly non-negative. However, the elements of $\ker K$, namely the scalar multiples of the vector $(1, 1)^T$, define null directions: $q(c, c) = 0$. ●

In a similar fashion, a quadratic form $q(\mathbf{x}) = \mathbf{x}^T K \mathbf{x}$ and its associated symmetric matrix K are called *negative semi-definite* if $q(\mathbf{x}) \leq 0$ for all $\mathbf{x}$ and *negative definite* if $q(\mathbf{x}) < 0$ for all $\mathbf{x} \neq \mathbf{0}$. A quadratic form is called *indefinite* if it is neither positive nor negative semi-definite; equivalently, there exist points $\mathbf{x}_+$ where $q(\mathbf{x}_+) > 0$ and points $\mathbf{x}_-$ where $q(\mathbf{x}_-) < 0$. Details can be found in the exercises.

EXERCISES 3.4

3.4.1. Which of the following 2×2 matrices are positive definite?

(a) $\begin{pmatrix} 1 & 0 \\ 0 & 2 \end{pmatrix}$ (b) $\begin{pmatrix} 0 & 1 \\ 2 & 0 \end{pmatrix}$

(c) $\begin{pmatrix} 1 & 2 \\ 2 & 1 \end{pmatrix}$ (d) $\begin{pmatrix} 5 & 3 \\ 3 & -2 \end{pmatrix}$

(e) $\begin{pmatrix} 1 & -1 \\ -1 & 3 \end{pmatrix}$ (f) $\begin{pmatrix} 1 & 1 \\ -1 & 2 \end{pmatrix}$

In the positive definite cases, write down the formula for the associated inner product.

3.4.2. Let $K = \begin{pmatrix} 1 & 2 \\ 2 & 3 \end{pmatrix}$. Prove that the associated quadratic form $q(\mathbf{x}) = \mathbf{x}^T K \mathbf{x}$ is indefinite by finding a point $\mathbf{x}^+$ where $q(\mathbf{x}^+) > 0$ and a point $\mathbf{x}^-$ where $q(\mathbf{x}^-) < 0$.

◇ **3.4.3.** (a) Prove that a diagonal matrix

$$D = \operatorname{diag}(c_1, c_2, \ldots, c_n)$$

is positive definite if and only if all its diagonal entries are positive: $c_i > 0$.

(b) Write down and identify the associated inner product.

◇ **3.4.4.** (a) Show that every diagonal entry of a positive definite matrix must be positive.

(b) Write down a symmetric matrix with all positive diagonal entries that is not positive definite.

(c) Find a nonzero matrix with one or more zero diagonal entries that is positive semi-definite.

3.4.5. Write out the Cauchy–Schwarz and triangle inequalities for the inner product defined in Example 3.22.

3.4.6. Prove that if K is any positive definite matrix, then any positive scalar multiple cK, $c > 0$, is also positive definite.

◇ **3.4.7.** (a) Show that if K and L are positive definite matrices, so is $K + L$.

(b) Give an example of two matrices that are not positive definite whose sum is positive definite.

3.4.8. Find a pair of positive definite matrices K and L whose product KL is not positive definite.

3.4.9. Write down a nonsingular symmetric matrix that is not positive or negative definite.

◇ **3.4.10.** Let K be a nonsingular symmetric matrix.

(a) Show that $\mathbf{x}^T K^{-1} \mathbf{x} = \mathbf{y}^T K \mathbf{y}$, where $K\mathbf{y} = \mathbf{x}$.

(b) Prove that if K is positive definite, so is K^{-1}.

♢ **3.4.11.** Prove that an $n \times n$ symmetric matrix K is positive definite if and only if, for every $\mathbf{0} \neq \mathbf{v} \in \mathbb{R}^n$, the vectors $\mathbf{v}$ and $K\mathbf{v}$ meet at an acute Euclidean angle: $|\theta| < \frac{1}{2}\pi$.

♢ **3.4.12.** Prove that the inner product associated with a positive definite quadratic form $q(\mathbf{x})$ is given by the *polarization formula*

$$\langle \mathbf{x}, \mathbf{y} \rangle = \tfrac{1}{2}\big[\, q(\mathbf{x}+\mathbf{y}) - q(\mathbf{x}) - q(\mathbf{y}) \,\big].$$

3.4.13. (a) Is it possible for a quadratic form to be positive, $q(\mathbf{x}_+) > 0$, at only one point $\mathbf{x}_+ \in \mathbb{R}^n$?

(b) Under what conditions is $q(\mathbf{x}_0) = 0$ at only one point?

3.4.14. (a) Show that a symmetric matrix N is negative definite if and only if $K = -N$ is positive definite.

(b) Write down two explicit criteria that tell whether or not a 2×2 matrix

$$N = \begin{pmatrix} a & b \\ b & c \end{pmatrix}$$

is negative definite.

(c) Use your criteria to check whether

(i) $\begin{pmatrix} -1 & 1 \\ 1 & -2 \end{pmatrix}$ (ii) $\begin{pmatrix} -4 & -5 \\ -5 & -6 \end{pmatrix}$

(iii) $\begin{pmatrix} -3 & -1 \\ -1 & 2 \end{pmatrix}$

are negative definite.

3.4.15. Show that $\mathbf{x} = \begin{pmatrix} 1 \\ 1 \end{pmatrix}$ is a null direction for

$$K = \begin{pmatrix} 1 & -2 \\ -2 & 3 \end{pmatrix},$$

but $\mathbf{x} \notin \ker K$.

3.4.16. Explain why an indefinite quadratic form necessarily has a non-zero null direction.

3.4.17. Let $K = K^T$. *True or false*:

(a) If K admits a null direction, then $\ker K \neq \{\mathbf{0}\}$.

(b) If K has no null directions, then K is either positive or negative definite.

♢ **3.4.18.** In special relativity, light rays in Minkowski space-time $\mathbb{R}^n$ travel along the *light cone* which, by definition, consists of all null directions associated with an indefinite quadratic form $q(\mathbf{x}) = \mathbf{x}^T K \mathbf{x}$. Find and sketch a picture of the light cone when the coefficient matrix K is

(a) $\begin{pmatrix} 1 & 0 \\ 0 & -1 \end{pmatrix}$ (b) $\begin{pmatrix} 1 & 2 \\ 2 & 3 \end{pmatrix}$

(c) $\begin{pmatrix} 1 & 0 & 0 \\ 0 & -1 & 0 \\ 0 & 0 & -1 \end{pmatrix}$

Remark: In the physical universe [41] , space-time is $n = 4$-dimensional, and

$$K = \operatorname{diag}(1, -1, -1, -1).$$

♢ **3.4.19.** (a) Let K and L be symmetric $n \times n$ matrices. Prove that $\mathbf{x}^T K \mathbf{x} = \mathbf{x}^T L \mathbf{x}$ for all $\mathbf{x} \in \mathbb{R}^n$ if and only if $K = L$.

(b) Find an example of two non-symmetric matrices $K \neq L$ such that $\mathbf{x}^T K \mathbf{x} = \mathbf{x}^T L \mathbf{x}$ for all $\mathbf{x} \in \mathbb{R}^n$.

♢ **3.4.20.** Suppose

$$q(\mathbf{x}) = \mathbf{x}^T A \mathbf{x} = \sum_{i,j=1}^{n} a_{ij}\, x_i\, x_j$$

is a general quadratic form on $\mathbb{R}^n$, whose coefficient matrix A is not necessarily symmetric. Prove that $q(\mathbf{x}) = \mathbf{x}^T K \mathbf{x}$, where $K = \frac{1}{2}(A + A^T)$ is a symmetric matrix. Therefore, we do not lose any generality by restricting our discussion to quadratic forms that are constructed from symmetric matrices.

♢ **3.4.21.** A function $f(\mathbf{x})$ on $\mathbb{R}^n$ is called *homogeneous of degree k* if $f(c\,\mathbf{x}) = c^k f(\mathbf{x})$ for all scalars c.

(a) If $\mathbf{a} \in \mathbb{R}^n$ is a fixed vector, show that a *linear form* $\ell(\mathbf{x}) = \mathbf{a} \cdot \mathbf{x} = a_1 x_1 + \cdots + a_n x_n$ is homogeneous of degree 1.

(b) Show that a quadratic form

$$q(\mathbf{x}) = \mathbf{x}^T K \mathbf{x} = \sum_{i,j=1}^{n} k_{ij}\, x_i\, x_j$$

is homogeneous of degree 2.

(c) Find a homogeneous function of degree 2 on $\mathbb{R}^2$ that is not a quadratic form.

Gram Matrices

Symmetric matrices whose entries are given by inner products of elements of an inner product space will appear throughout this text. They are named after the nineteenth century Danish mathematician Jorgen Gram—not the metric mass unit!

Definition 3.27 Let V be an inner product space, and let $\mathbf{v}_1, \ldots, \mathbf{v}_n \in V$. The associated *Gram matrix*

$$K = \begin{pmatrix} \langle \mathbf{v}_1, \mathbf{v}_1 \rangle & \langle \mathbf{v}_1, \mathbf{v}_2 \rangle & \ldots & \langle \mathbf{v}_1, \mathbf{v}_n \rangle \\ \langle \mathbf{v}_2, \mathbf{v}_1 \rangle & \langle \mathbf{v}_2, \mathbf{v}_2 \rangle & \ldots & \langle \mathbf{v}_2, \mathbf{v}_n \rangle \\ \vdots & \vdots & \ddots & \vdots \\ \langle \mathbf{v}_n, \mathbf{v}_1 \rangle & \langle \mathbf{v}_n, \mathbf{v}_2 \rangle & \ldots & \langle \mathbf{v}_n, \mathbf{v}_n \rangle \end{pmatrix} \tag{3.50}$$

is the $n \times n$ matrix whose entries are the inner products between the selected vector space elements.

Symmetry of the inner product implies symmetry of the Gram matrix:

$$k_{ij} = \langle \mathbf{v}_i, \mathbf{v}_j \rangle = \langle \mathbf{v}_j, \mathbf{v}_i \rangle = k_{ji}, \quad \text{and hence} \quad K^T = K. \tag{3.51}$$

In fact, the most direct method for producing positive definite and semi-definite matrices is through the Gram matrix construction.

THEOREM 3.28 All Gram matrices are positive semi-definite. The Gram matrix (3.50) is positive definite if and only if $\mathbf{v}_1, \ldots, \mathbf{v}_n$ are linearly independent.

Proof To prove positive (semi-)definiteness of K, we need to examine the associated quadratic form

$$q(\mathbf{x}) = \mathbf{x}^T K \mathbf{x} = \sum_{i,j=1}^{n} k_{ij}\, x_i\, x_j.$$

Substituting the values (3.51) for the matrix entries, we find

$$q(\mathbf{x}) = \sum_{i,j=1}^{n} \langle \mathbf{v}_i, \mathbf{v}_j \rangle\, x_i\, x_j.$$

Bilinearity of the inner product on V implies that we can assemble this summation into a single inner product

$$q(\mathbf{x}) = \left\langle \sum_{i=1}^{n} x_i\, \mathbf{v}_i \,,\, \sum_{j=1}^{n} x_j\, \mathbf{v}_j \right\rangle = \langle \mathbf{v}, \mathbf{v} \rangle = \| \mathbf{v} \|^2 \geq 0, \quad \text{where} \quad \mathbf{v} = \sum_{i=1}^{n} x_i\, \mathbf{v}_i$$

lies in the subspace of V spanned by the given vectors. This immediately proves that K is positive semi-definite.

Moreover, $q(\mathbf{x}) = \| \mathbf{v} \|^2 > 0$ as long as $\mathbf{v} \neq \mathbf{0}$. If $\mathbf{v}_1, \ldots, \mathbf{v}_n$ are linearly independent,

$$\mathbf{v} = x_1\, \mathbf{v}_1 + \cdots + x_n\, \mathbf{v}_n = \mathbf{0} \quad \text{if and only if} \quad x_1 = \cdots = x_n = 0,$$

and hence $q(\mathbf{x}) = 0$ if and only if $\mathbf{x} = \mathbf{0}$. Thus, in this case, $q(\mathbf{x})$ and K are positive definite. ■

EXAMPLE 3.29 Consider the vectors

$$\mathbf{v}_1 = \begin{pmatrix} 1 \\ 2 \\ -1 \end{pmatrix}, \quad \mathbf{v}_2 = \begin{pmatrix} 3 \\ 0 \\ 6 \end{pmatrix}.$$

For the standard Euclidean dot product on $\mathbb{R}^3$, the Gram matrix is

$$K = \begin{pmatrix} \mathbf{v}_1 \cdot \mathbf{v}_1 & \mathbf{v}_1 \cdot \mathbf{v}_2 \\ \mathbf{v}_2 \cdot \mathbf{v}_1 & \mathbf{v}_2 \cdot \mathbf{v}_2 \end{pmatrix} = \begin{pmatrix} 6 & -3 \\ -3 & 45 \end{pmatrix}.$$

Since $\mathbf{v}_1, \mathbf{v}_2$ are linearly independent, $K > 0$. Positive definiteness implies that

$$q(x_1, x_2) = 6x_1^2 - 6x_1x_2 + 45x_2^2 > 0 \quad \text{for all} \quad (x_1, x_2) \neq \mathbf{0}.$$

Indeed, this can be checked directly, using the criteria in (3.48).

On the other hand, for the weighted inner product

$$\langle \mathbf{v}, \mathbf{w} \rangle = 3\, v_1 w_1 + 2\, v_2 w_2 + 5\, v_3 w_3, \tag{3.52}$$

the corresponding Gram matrix is

$$\widetilde{K} = \begin{pmatrix} \langle \mathbf{v}_1, \mathbf{v}_1 \rangle & \langle \mathbf{v}_1, \mathbf{v}_2 \rangle \\ \langle \mathbf{v}_2, \mathbf{v}_1 \rangle & \langle \mathbf{v}_2, \mathbf{v}_2 \rangle \end{pmatrix} = \begin{pmatrix} 16 & -21 \\ -21 & 207 \end{pmatrix}. \tag{3.53}$$

Since $\mathbf{v}_1, \mathbf{v}_2$ are still linearly independent (which, of course, does not depend upon which inner product is used), the matrix $\widetilde{K}$ is also positive definite. ●

In the case of the Euclidean dot product, the construction of the Gram matrix K can be directly implemented as follows. Given column vectors $\mathbf{v}_1, \ldots, \mathbf{v}_n \in \mathbb{R}^m$, let us form the $m \times n$ matrix $A = (\, \mathbf{v}_1 \; \mathbf{v}_2 \, \ldots \, \mathbf{v}_n \,)$. In view of the identification (3.2) between the dot product and multiplication of row and column vectors, the (i, j) entry of K is given as the product

$$k_{ij} = \mathbf{v}_i \cdot \mathbf{v}_j = \mathbf{v}_i^T \mathbf{v}_j$$

of the ith row of the transpose A^T with the jth column of A. In other words, the Gram matrix can be evaluated as a matrix product:

$$K = A^T A. \tag{3.54}$$

For the preceding Example 3.29,

$$A = \begin{pmatrix} 1 & 3 \\ 2 & 0 \\ -1 & 6 \end{pmatrix},$$

and so

$$K = A^T A = \begin{pmatrix} 1 & 2 & -1 \\ 3 & 0 & 6 \end{pmatrix} \begin{pmatrix} 1 & 3 \\ 2 & 0 \\ -1 & 6 \end{pmatrix} = \begin{pmatrix} 6 & -3 \\ -3 & 45 \end{pmatrix}.$$

Theorem 3.28 implies that the Gram matrix (3.54) is positive definite if and only if the columns of A are linearly independent vectors. This implies the following result.

Proposition 3.30 Given an $m \times n$ matrix A, the following are equivalent:

(a) The $n \times n$ Gram matrix $K = A^T A$ is positive definite.

(b) A has linearly independent columns.

(c) $\operatorname{rank} A = n \leq m$.

(d) $\ker A = \{0\}$.

Changing the underlying inner product will, of course, change the Gram matrix. As noted in Theorem 3.21, every inner product on $\mathbb{R}^m$ has the form

$$\langle \mathbf{v}, \mathbf{w} \rangle = \mathbf{v}^T C\, \mathbf{w} \quad \text{for} \quad \mathbf{v}, \mathbf{w} \in \mathbb{R}^m, \tag{3.55}$$

where $C > 0$ is a symmetric, positive definite $m \times m$ matrix. Therefore, given n vectors $\mathbf{v}_1, \ldots, \mathbf{v}_n \in \mathbb{R}^m$, the entries of the Gram matrix with respect to this inner product are

$$k_{ij} = \langle \mathbf{v}_i, \mathbf{v}_j \rangle = \mathbf{v}_i^T C\, \mathbf{v}_j.$$

If, as above, we assemble the column vectors into an $m \times n$ matrix $A = (\mathbf{v}_1\ \mathbf{v}_2\ \ldots\ \mathbf{v}_n)$, then the Gram matrix entry k_{ij} is obtained by multiplying the ith row of A^T by the jth column of the product matrix $C\,A$. Therefore, the Gram matrix based on the alternative inner product (3.55) is given by

$$K = A^T C\,A. \tag{3.56}$$

Theorem 3.28 immediately implies that K is positive definite—provided that A has rank n.

THEOREM 3.31 Suppose A is an $m \times n$ matrix with linearly independent columns. Suppose C is any positive definite $m \times m$ matrix. Then the matrix $K = A^T C\,A$ is a positive definite $n \times n$ matrix.

The Gram matrices constructed in (3.56) arise in a wide variety of applications, including least squares approximation theory, cf. Chapter 4, and mechanical structures and electrical circuits, cf. Chapters 6 and 9. In the majority of applications, $C = \operatorname{diag}(c_1, \ldots, c_m)$ is a diagonal positive definite matrix, which requires it to have strictly positive diagonal entries $c_i > 0$. This choice corresponds to a weighted inner product (3.10) on $\mathbb{R}^m$.

EXAMPLE 3.32 Returning to the situation of Example 3.29, the weighted inner product (3.52) corresponds to the diagonal positive definite matrix

$$C = \begin{pmatrix} 3 & 0 & 0 \\ 0 & 2 & 0 \\ 0 & 0 & 5 \end{pmatrix}.$$

Therefore, the weighted Gram matrix (3.56) based on the vectors

$$\begin{pmatrix} 1 \\ 2 \\ -1 \end{pmatrix}, \quad \begin{pmatrix} 3 \\ 0 \\ 6 \end{pmatrix}$$

is

$$\widetilde{K} = A^T C\,A = \begin{pmatrix} 1 & 2 & -1 \\ 3 & 0 & 6 \end{pmatrix} \begin{pmatrix} 3 & 0 & 0 \\ 0 & 2 & 0 \\ 0 & 0 & 5 \end{pmatrix} \begin{pmatrix} 1 & 3 \\ 2 & 0 \\ -1 & 6 \end{pmatrix} = \begin{pmatrix} 16 & -21 \\ -21 & 207 \end{pmatrix},$$

reproducing (3.53). ●

The Gram matrix construction is not restricted to finite-dimensional vector spaces, but also applies to inner products on function space. For instance, the covariance matrices arising in statistical analysis, [17], are of Gram matrix form. Here is a particularly important example.

EXAMPLE 3.33 Consider the vector space $\mathrm{C}^0[0,1]$ consisting of continuous functions on the interval $0 \le x \le 1$, equipped with the L^2 inner product $\langle f, g \rangle = \displaystyle\int_0^1 f(x)\,g(x)\,dx$. Let us construct the Gram matrix corresponding to the simple monomial functions

$1, x, x^2$. We compute the required inner products

$$\langle 1,1\rangle = \|1\|^2 = \int_0^1 dx = 1, \qquad \langle 1,x\rangle = \int_0^1 x\,dx = \frac{1}{2},$$
$$\langle x,x\rangle = \|x\|^2 = \int_0^1 x^2\,dx = \frac{1}{3}, \qquad \langle 1,x^2\rangle = \int_0^1 x^2\,dx = \frac{1}{3},$$
$$\langle x^2,x^2\rangle = \|x^2\|^2 = \int_0^1 x^4\,dx = \frac{1}{5}, \qquad \langle x,x^2\rangle = \int_0^1 x^3\,dx = \frac{1}{4}.$$

Therefore, the Gram matrix is

$$K = \begin{pmatrix} \langle 1,1\rangle & \langle 1,x\rangle & \langle 1,x^2\rangle \\ \langle x,1\rangle & \langle x,x\rangle & \langle x,x^2\rangle \\ \langle x^2,1\rangle & \langle x^2,x\rangle & \langle x^2,x^2\rangle \end{pmatrix} = \begin{pmatrix} 1 & \frac{1}{2} & \frac{1}{3} \\ \frac{1}{2} & \frac{1}{3} & \frac{1}{4} \\ \frac{1}{3} & \frac{1}{4} & \frac{1}{5} \end{pmatrix}.$$

As we know, the monomial functions $1, x, x^2$ are linearly independent, and so Theorem 3.28 immediately implies that the matrix K is positive definite. ●

The alert reader may recognize this particular Gram matrix as the 3×3 *Hilbert matrix* that we encountered in (1.70). More generally, the Gram matrix corresponding to the monomials $1, x, x^2, \ldots, x^n$ has entries

$$k_{ij} = \langle x^{i-1}, x^{j-1}\rangle = \int_0^1 x^{i+j-2}\,dt = \frac{1}{i+j-1}, \qquad i,j = 1,\ldots,n+1.$$

Therefore, the monomial Gram matrix is the $(n+1)\times(n+1)$ Hilbert matrix (1.70): $K = H_{n+1}$. As a consequence of Proposition 3.25 and Theorem 3.31, we have proved the following non-trivial result.

Proposition 3.34 The $n \times n$ Hilbert matrix H_n is positive definite. Consequently, H_n is a nonsingular matrix.

EXAMPLE 3.35 Let us construct the Gram matrix corresponding to the functions $1, \cos x, \sin x$ with respect to the inner product $\langle f,g\rangle = \int_{-\pi}^{\pi} f(x)\,g(x)\,dx$ on the interval $[-\pi,\pi]$. We compute the inner products

$$\langle 1,1\rangle = \|1\|^2 = \int_{-\pi}^{\pi} dx = 2\pi,$$
$$\langle 1,\cos x\rangle = \int_{-\pi}^{\pi} \cos x\,dx = 0,$$
$$\langle 1,\sin x\rangle = \int_{-\pi}^{\pi} \sin x\,dx = 0,$$
$$\langle \cos x,\cos x\rangle = \|\cos x\|^2 = \int_{-\pi}^{\pi} \cos^2 x\,dx = \pi,$$
$$\langle \sin x,\sin x\rangle = \|\sin x\|^2 = \int_{-\pi}^{\pi} \sin^2 x\,dx = \pi,$$
$$\langle \cos x,\sin x\rangle = \int_{-\pi}^{\pi} \cos x \sin x\,dx = 0.$$

Therefore, the Gram matrix is a simple diagonal matrix:

$$K = \begin{pmatrix} 2\pi & 0 & 0 \\ 0 & \pi & 0 \\ 0 & 0 & \pi \end{pmatrix}.$$

Positive definiteness of K is immediately evident. ●

If the columns of A are linearly dependent, then the associated Gram matrix is only positive semi-definite. In this case, the Gram matrix has nontrivial null directions $\mathbf{0} \neq \mathbf{v} \in \ker K = \ker A$.

Proposition 3.36 Let $K = A^T C A$ be the $n \times n$ Gram matrix constructed from an $m \times n$ matrix A and a positive definite $m \times m$ matrix $C > 0$. Then $\ker K = \ker A$, and hence $\operatorname{rank} K = \operatorname{rank} A$.

Proof Clearly, if $A\mathbf{x} = \mathbf{0}$ then $K\mathbf{x} = A^T C A \mathbf{x} = \mathbf{0}$, and so $\ker A \subset \ker K$. Conversely, if $K\mathbf{x} = \mathbf{0}$, then

$$0 = \mathbf{x}^T K \mathbf{x} = \mathbf{x}^T A^T C A \mathbf{x} = \mathbf{y}^T C \mathbf{y}, \qquad \text{where} \quad \mathbf{y} = A\mathbf{x}.$$

Since $C > 0$, this implies $\mathbf{y} = \mathbf{0}$, and hence $\mathbf{x} \in \ker A$. Finally, by Theorem 2.49, $\operatorname{rank} K = n - \dim\ker K = n - \dim\ker A = \operatorname{rank} A$. ■

EXERCISES 3.4

3.4.22. (a) Find the Gram matrix corresponding to each of the following sets of vectors using the Euclidean dot product on $\mathbb{R}^n$.

(i) $\begin{pmatrix}-1\\3\end{pmatrix}, \begin{pmatrix}0\\2\end{pmatrix}$

(ii) $\begin{pmatrix}1\\2\end{pmatrix}, \begin{pmatrix}-2\\3\end{pmatrix}, \begin{pmatrix}-1\\-1\end{pmatrix}$

(iii) $\begin{pmatrix}2\\1\\-1\end{pmatrix}, \begin{pmatrix}-3\\0\\2\end{pmatrix}$

(iv) $\begin{pmatrix}1\\1\\0\end{pmatrix}, \begin{pmatrix}1\\0\\1\end{pmatrix}, \begin{pmatrix}0\\1\\1\end{pmatrix}$

(v) $\begin{pmatrix}1\\-2\\2\end{pmatrix}, \begin{pmatrix}2\\-1\\1\end{pmatrix}, \begin{pmatrix}-1\\-1\\1\end{pmatrix}$

(vi) $\begin{pmatrix}1\\0\\-1\\0\end{pmatrix}, \begin{pmatrix}-1\\1\\0\\1\end{pmatrix}$

(vii) $\begin{pmatrix}1\\2\\3\\4\end{pmatrix}, \begin{pmatrix}-2\\1\\-4\\3\end{pmatrix}, \begin{pmatrix}-1\\3\\-1\\-2\end{pmatrix}$

(viii) $\begin{pmatrix}1\\0\\0\\1\end{pmatrix}, \begin{pmatrix}-2\\1\\0\\0\end{pmatrix}, \begin{pmatrix}-1\\0\\-1\\0\end{pmatrix}, \begin{pmatrix}0\\2\\-3\\0\end{pmatrix}$

(b) Which are positive definite?

(c) If the matrix is positive semi-definite, find all its null vectors.

3.4.23. Recompute the Gram matrices for cases (iii)–(v) in the previous exercise using the weighted inner product $\langle \mathbf{x}, \mathbf{y} \rangle = x_1 y_1 + 2 x_2 y_2 + 3 x_3 y_3$. Does this change its positive definiteness?

3.4.24. Recompute the Gram matrices for cases (vi)–(viii) in Exercise 3.4.22 for the weighted inner product $\langle \mathbf{x}, \mathbf{y} \rangle = x_1 y_1 + \frac{1}{2} x_2 y_2 + \frac{1}{3} x_3 y_3 + \frac{1}{4} x_4 y_4$.

3.4.25. Find the Gram matrix K for the functions $1, e^x, e^{2x}$ using the L^2 inner product on $[0, 1]$. Is K positive definite?

3.4.26. Answer Exercise 3.4.25 using the weighted inner product $\langle f, g \rangle = \int_0^1 f(x)\, g(x)\, e^{-x}\, dx$.

3.4.27. Find the Gram matrix K for the monomials $1, x, x^2, x^3$ using the L^2 inner product on $[-1, 1]$. Is K positive definite?

3.4.28. Answer Exercise 3.4.27 using the weighted inner product $\langle f, g \rangle = \int_{-1}^1 f(x)\, g(x)\,(1+x)\, dx$.

◇ **3.4.29.** Prove that every positive definite matrix K can be written as a Gram matrix.

◇ **3.4.30.** (a) Prove that if K is a positive definite matrix, then K^2 is also positive definite.

(b) More generally, prove that if $S = S^T$ is symmetric and nonsingular, then S^2 is positive definite.

◇ **3.4.31.** Let $K = A^T C A$, where $C > 0$. Prove that

(a) $\ker K = \operatorname{coker} K = \ker A$;

(b) $\operatorname{rng} K = \operatorname{corng} K = \operatorname{corng} A$.

3.4.32. Show that $\mathbf{z}$ is a null vector for the quadratic form $q(\mathbf{x}) = \mathbf{x}^T K \mathbf{x}$ based on the Gram matrix $K = A^T A$ if and only if $\mathbf{z} \in \ker K$.

3.4.33. Let A be an $m \times n$ matrix.

(a) Explain why the product $L = A A^T$ is a Gram matrix.

(b) Show that, even though they may be of different sizes, both Gram matrices $K = A^T A$ and $L = A A^T$ have the same rank.

(c) Under what conditions are both K and L positive definite?

3.4.34. Suppose K is the Gram matrix computed from $\mathbf{v}_1, \ldots, \mathbf{v}_n \in V$ relative to a given inner product. Let $\widetilde{K}$ be the Gram matrix for the same elements, but computed relative to a *different* inner product. Show that $K > 0$ if and only if $\widetilde{K} > 0$.

◇ **3.4.35.** Let $K_1 = A_1^T C_1 A_1$ and $K_2 = A_2^T C_2 A_2$ be any two $n \times n$ Gram matrices. Let $K = K_1 + K_2$.

(a) Show that if $K_1, K_2 > 0$ then $K > 0$.

(b) Give an example where K_1 and K_2 are not positive definite, but $K > 0$.

(c) Show that $K = A^T C A$ is also a Gram matrix. *Hint*: : A will have size $(m_1 + m_2) \times n$, where m_1 and m_2 are the number of rows in A_1, A_2, respectively.

3.5 COMPLETING THE SQUARE

Gram matrices furnish us with an almost inexhaustible supply of positive definite matrices. However, we still do not know how to test whether a given symmetric matrix is positive definite. As we shall soon see, the secret already appears in the particular computations in Examples 3.2 and 3.22.

You may recall the algebraic technique known as "completing the square", first in the derivation of the quadratic formula for the solution to

$$q(x) = a x^2 + 2 b x + c = 0, \tag{3.57}$$

and, later, in facilitating the integration of various types of rational and algebraic functions. The idea is to combine the first two terms in (3.57) as a perfect square, and so rewrite the quadratic function in the form

$$q(x) = a \left(x + \frac{b}{a} \right)^2 + \frac{a c - b^2}{a} = 0. \tag{3.58}$$

As a consequence,

$$\left(x + \frac{b}{a} \right)^2 = \frac{b^2 - a c}{a^2} .$$

The familiar quadratic formula

$$x = \frac{-b \pm \sqrt{b^2 - a c}}{a}$$

follows by taking the square root of both sides and then solving for x. The intermediate step (3.58), where we eliminate the linear term, is known as *completing the square*.

We can perform the same kind of manipulation on a homogeneous quadratic form

$$q(x_1, x_2) = a x_1^2 + 2 b x_1 x_2 + c x_2^2. \tag{3.59}$$

In this case, provided $a \neq 0$, completing the square amounts to writing

$$\begin{aligned} q(x_1, x_2) &= a x_1^2 + 2 b x_1 x_2 + c x_2^2 \\ &= a \left(x_1 + \frac{b}{a} x_2 \right)^2 + \frac{a c - b^2}{a} x_2^2 \\ &= a y_1^2 + \frac{a c - b^2}{a} y_2^2. \end{aligned} \tag{3.60}$$

The net result is to re-express $q(x_1, x_2)$ as a simpler sum of squares of the new variables

$$y_1 = x_1 + \frac{b}{a}\, x_2, \qquad y_2 = x_2. \tag{3.61}$$

It is not hard to see that the final expression in (3.60) is positive definite, as a function of y_1, y_2, if and only if both coefficients are positive:

$$a > 0, \qquad \frac{a\,c - b^2}{a} > 0. \tag{3.62}$$

Therefore, $q(x_1, x_2) \geq 0$, with equality if and only if $y_1 = y_2 = 0$, or, equivalently, $x_1 = x_2 = 0$. This conclusively proves that conditions (3.62) are necessary and sufficient for the quadratic form (3.59) to be positive definite.

Our goal is to adapt this simple idea to analyze the positivity of quadratic forms depending on more than two variables. To this end, let us rewrite the quadratic form identity (3.60) in matrix form. The original quadratic form (3.59) is

$$q(\mathbf{x}) = \mathbf{x}^T K\, \mathbf{x}, \qquad \text{where} \quad K = \begin{pmatrix} a & b \\ b & c \end{pmatrix}, \qquad \mathbf{x} = \begin{pmatrix} x_1 \\ x_2 \end{pmatrix}. \tag{3.63}$$

Similarly, the right hand side of (3.60) can be written as

$$\widehat{q}\,(\mathbf{y}) = \mathbf{y}^T D\, \mathbf{y}, \qquad \text{where} \quad D = \begin{pmatrix} a & 0 \\ 0 & \dfrac{a\,c - b^2}{a} \end{pmatrix}, \qquad \mathbf{y} = \begin{pmatrix} y_1 \\ y_2 \end{pmatrix}. \tag{3.64}$$

Anticipating the final result, the equations (3.61) connecting $\mathbf{x}$ and $\mathbf{y}$ can themselves be written in matrix form as

$$\mathbf{y} = L^T \mathbf{x} \quad \text{or} \quad \begin{pmatrix} y_1 \\ y_2 \end{pmatrix} = \begin{pmatrix} x_1 + \frac{b}{a} x_2 \\ x_2 \end{pmatrix}, \qquad \text{where} \quad L^T = \begin{pmatrix} 1 & \dfrac{b}{a} \\ 0 & 1 \end{pmatrix}.$$

Substituting into (3.64), we find

$$\mathbf{y}^T D\, \mathbf{y} = (L^T \mathbf{x})^T D\, (L^T \mathbf{x}) = \mathbf{x}^T L\, D\, L^T \mathbf{x} = \mathbf{x}^T K\, \mathbf{x}, \qquad \text{where} \quad K = L\, D\, L^T. \tag{3.65}$$

The result is the *same* factorization (1.59) of the coefficient matrix that we previously obtained via Gaussian Elimination. We are thus led to the realization that *completing the square is the same as the $L\,D\,L^T$ factorization of a symmetric matrix*!

Recall the definition of a regular matrix as one that can be reduced to upper triangular form without any row interchanges. Theorem 1.34 says that the regular symmetric matrices are precisely those that admit an $L\,D\,L^T$ factorization. The identity (3.65) is therefore valid for all regular $n \times n$ symmetric matrices, and shows how to write the associated quadratic form as a sum of squares:

$$q(\mathbf{x}) = \mathbf{x}^T K\, \mathbf{x} = \mathbf{y}^T D\, \mathbf{y} = d_1\, y_1^2 + \cdots + d_n\, y_n^2, \qquad \text{where} \quad \mathbf{y} = L^T \mathbf{x}. \tag{3.66}$$

The coefficients d_i are the diagonal entries of D, which are the pivots of K. Furthermore, the diagonal quadratic form is positive definite, $\mathbf{y}^T D\, \mathbf{y} > 0$ for all $\mathbf{y} \neq \mathbf{0}$ if and only if all the pivots are positive, $d_i > 0$. Invertibility of L^T tells us that $\mathbf{y} = \mathbf{0}$ if and only if $\mathbf{x} = \mathbf{0}$, and hence, positivity of the pivots is equivalent to positive definiteness of the original quadratic form: $q(\mathbf{x}) > 0$ for all $\mathbf{x} \neq 0$. We have thus almost proved the main result that completely characterizes positive definite matrices.

THEOREM 3.37 A symmetric matrix K is positive definite if and only if it is regular and has all positive pivots.

In other words, a square matrix K is positive definite if and only if it can be factored $K = L D L^T$, where L is special lower triangular and D is diagonal with all positive diagonal entries.

EXAMPLE 3.38 Consider the symmetric matrix

$$K = \begin{pmatrix} 1 & 2 & -1 \\ 2 & 6 & 0 \\ -1 & 0 & 9 \end{pmatrix}.$$

Gaussian Elimination produces the factors

$$L = \begin{pmatrix} 1 & 0 & 0 \\ 2 & 1 & 0 \\ -1 & 1 & 1 \end{pmatrix}, \qquad D = \begin{pmatrix} 1 & 0 & 0 \\ 0 & 2 & 0 \\ 0 & 0 & 6 \end{pmatrix}, \qquad L^T = \begin{pmatrix} 1 & 2 & -1 \\ 0 & 1 & 1 \\ 0 & 0 & 1 \end{pmatrix},$$

in its factorization $K = LDL^T$. Since the pivots—the diagonal entries 1, 2 and 6 in D — are all positive, Theorem 3.37 implies that K is positive definite, which means that the associated quadratic form satisfies

$$q(\mathbf{x}) = x_1^2 + 4x_1x_2 - 2x_1x_3 + 6x_2^2 + 9x_3^2 > 0, \qquad \text{for all} \quad \mathbf{x} = (x_1, x_2, x_3)^T \neq \mathbf{0}.$$

Indeed, the $L D L^T$ factorization implies that $q(\mathbf{x})$ can be explicitly written as a sum of squares:

$$q(\mathbf{x}) = x_1^2 + 4x_1x_2 - 2x_1x_3 + 6x_2^2 + 9x_3^2 = y_1^2 + 2y_2^2 + 6y_3^2, \tag{3.67}$$

where

$$y_1 = x_1 + 2x_2 - x_3, \qquad y_2 = x_2 + x_3, \qquad y_3 = x_3$$

are the entries of $\mathbf{y} = L^T\mathbf{x}$. Positivity of the coefficients of the y_i^2 (which are the pivots) implies that $q(\mathbf{x})$ is positive definite. ●

EXAMPLE 3.39 Let's test whether the matrix

$$K = \begin{pmatrix} 1 & 2 & 3 \\ 2 & 3 & 7 \\ 3 & 7 & 8 \end{pmatrix}$$

is positive definite. When we perform Gaussian Elimination, the second pivot turns out to be -1, which immediately implies that K is not positive definite—even though all its entries are positive. (The third pivot is 3, but this does not affect the conclusion; all it takes is one non-positive pivot to disqualify a matrix from being positive definite. Also, row interchanges aren't of any help, since we are not allowed to perform them when checking for positive definiteness.) This means that the associated quadratic form

$$q(\mathbf{x}) = x_1^2 + 4x_1x_2 + 6x_1x_3 + 3x_2^2 + 14x_2x_3 + 8x_3^2$$

assumes negative values at some points. For instance, $q(-2, 1, 0) = -1$. ●

A direct method for completing the square in a quadratic form goes as follows: The first step is to put all the terms involving x_1 in a suitable square, at the expense of introducing extra terms involving only the other variables. For instance, in the case of the quadratic form in (3.67), the terms involving x_1 can be written as

$$x_1^2 + 4x_1x_2 - 2x_1x_3 = (x_1 + 2x_2 - x_3)^2 - 4x_2^2 + 4x_2x_3 - x_3^2.$$

Therefore,

$$q(\mathbf{x}) = (x_1 + 2x_2 - x_3)^2 + 2x_2^2 + 4x_2x_3 + 8x_3^2 = (x_1 + 2x_2 - x_3)^2 + \widetilde{q}(x_2, x_3),$$

where

$$\widetilde{q}(x_2, x_3) = 2x_2^2 + 4x_2x_3 + 8x_3^2$$

is a quadratic form that only involves x_2, x_3. We then repeat the process, combining all the terms involving x_2 in the remaining quadratic form into a square, writing

$$\widetilde{q}(x_2, x_3) = 2(x_2 + x_3)^2 + 6x_3^2.$$

This gives the final form

$$q(\mathbf{x}) = (x_1 + 2x_2 - x_3)^2 + 2(x_2 + x_3)^2 + 6x_3^2,$$

which reproduces (3.67).

In general, as long as $k_{11} \neq 0$, we can write

$$\begin{aligned} q(\mathbf{x}) = \mathbf{x}^T K \mathbf{x} &= k_{11}x_1^2 + 2k_{12}x_1x_2 + \cdots + 2k_{1n}x_1x_n + k_{22}x_2^2 + \cdots + k_{nn}x_n^2 \\ &= k_{11}\left(x_1 + \frac{k_{12}}{k_{11}}x_2 + \cdots + \frac{k_{1n}}{k_{11}}x_n\right)^2 + \widetilde{q}(x_2, \ldots, x_n) \\ &= k_{11}(x_1 + l_{21}x_2 + \cdots + l_{n1}x_n)^2 + \widetilde{q}(x_2, \ldots, x_n), \end{aligned} \tag{3.68}$$

where

$$l_{21} = \frac{k_{21}}{k_{11}} = \frac{k_{12}}{k_{11}}, \quad \ldots \quad l_{n1} = \frac{k_{n1}}{k_{11}} = \frac{k_{1n}}{k_{11}}$$

are precisely the multiples appearing in the matrix L obtained from applying Gaussian Elimination to K, while

$$\widetilde{q}(x_2, \ldots, x_n) = \sum_{i,j=2}^{n} \widetilde{k}_{ij}\, x_i\, x_j$$

is a quadratic form involving one less variable. The entries of its symmetric coefficient matrix $\widetilde{K}$ are

$$\widetilde{k}_{ij} = \widetilde{k}_{ji} = k_{ij} - l_{j1}\,k_{1i} = k_{ij} - \frac{k_{1j}\,k_{1i}}{k_{11}}, \qquad i, j = 2, \ldots n,$$

which are exactly the same as the entries appearing below and to the right of the first pivot after applying the the first phase of the Gaussian Elimination process to K. In particular, the second pivot of K is the diagonal entry $\widetilde{k}_{22}$. We continue by applying the same procedure to the reduced quadratic form $\widetilde{q}(x_2, \ldots, x_n)$ and repeating until only the final variable remains. Completing the square at each stage reproduces the corresponding phase of the Gaussian Elimination process. The final result is our formula (3.66) rewriting the original quadratic form as a sum of squares whose coefficients are the pivots.

With this in hand, we can now complete the proof of Theorem 3.37. First, if the upper left entry k_{11}, namely the first pivot, is not strictly positive, then K cannot be positive definite because $q(\mathbf{e}_1) = \mathbf{e}_1^T K \mathbf{e}_1 = k_{11} \leq 0$. Otherwise, suppose $k_{11} > 0$ and so we can write $q(\mathbf{x})$ in the form (3.68). We claim that $q(\mathbf{x})$ is positive definite if and only if the reduced quadratic form $\widetilde{q}(x_2, \ldots, x_n)$ is positive definite. Indeed, if $\widetilde{q}$ is positive definite and $k_{11} > 0$, then $q(\mathbf{x})$ is the sum of two positive quantities, which simultaneously vanish if and only if $x_1 = x_2 = \cdots = x_n = 0$. On the other hand, suppose $\widetilde{q}(x_2^\star, \ldots, x_n^\star) \leq 0$ for some $x_2^\star, \ldots, x_n^\star$, not all zero. Setting $x_1^\star = -l_{21}x_2^\star - \cdots - l_{n1}x_n^\star$ makes the initial square term in (3.68) equal to 0, so

$$q(x_1^\star, x_2^\star, \ldots, x_n^\star) = \widetilde{q}(x_2^\star, \ldots, x_n^\star) \leq 0,$$

proving the claim. In particular, positive definiteness of $\widetilde{q}$ requires that the second pivot $\widetilde{k}_{22} > 0$. We then continue the reduction procedure outlined in the preceding paragraph; if a non-positive entry appears in the diagonal pivot position at any stage, the original quadratic form and matrix cannot be positive definite. On the other hand, finding all positive pivots (without using any row interchanges) will, in the absence of numerical errors, ensure positive definiteness. ■

EXERCISES 3.5

3.5.1. Decide whether the following matrices are positive definite:

(a) $\begin{pmatrix} 4 & -2 \\ -2 & 4 \end{pmatrix}$ (b) $\begin{pmatrix} 1 & 1 \\ 1 & 1 \end{pmatrix}$

(c) $\begin{pmatrix} 1 & 1 & 2 \\ 1 & 2 & 1 \\ 2 & 1 & 1 \end{pmatrix}$ (d) $\begin{pmatrix} 1 & 1 & 1 \\ 1 & 2 & -2 \\ 1 & -2 & 4 \end{pmatrix}$

(e) $\begin{pmatrix} 2 & 1 & 1 & 1 \\ 1 & 2 & 1 & 1 \\ 1 & 1 & 2 & 1 \\ 1 & 1 & 1 & 2 \end{pmatrix}$

(f) $\begin{pmatrix} -1 & 1 & 1 & 1 \\ 1 & -1 & 1 & 1 \\ 1 & 1 & -1 & 1 \\ 1 & 1 & 1 & -1 \end{pmatrix}$

3.5.2. Find an $L\,D\,L^T$ factorization of the following symmetric matrices. Which are positive definite?

(a) $\begin{pmatrix} 1 & 2 \\ 2 & 3 \end{pmatrix}$ (b) $\begin{pmatrix} 5 & -1 \\ -1 & 3 \end{pmatrix}$

(c) $\begin{pmatrix} 3 & -1 & 3 \\ -1 & 5 & 1 \\ 3 & 1 & 5 \end{pmatrix}$ (d) $\begin{pmatrix} -2 & 1 & -1 \\ 1 & -2 & 1 \\ -1 & 1 & -2 \end{pmatrix}$

(e) $\begin{pmatrix} 2 & 1 & -2 \\ 1 & 1 & -3 \\ -2 & -3 & 11 \end{pmatrix}$ (f) $\begin{pmatrix} 1 & 1 & 1 & 0 \\ 1 & 2 & 0 & 1 \\ 1 & 0 & 1 & 1 \\ 0 & 1 & 1 & 2 \end{pmatrix}$

(g) $\begin{pmatrix} 3 & 2 & 1 & 0 \\ 2 & 3 & 0 & 1 \\ 1 & 0 & 3 & 2 \\ 0 & 1 & 2 & 3 \end{pmatrix}$

(h) $\begin{pmatrix} 2 & 1 & -2 & 0 \\ 1 & 1 & -3 & 2 \\ -2 & -3 & 10 & -1 \\ 0 & 2 & -1 & 7 \end{pmatrix}$

3.5.3. (a) For which values of c is the matrix

$$A = \begin{pmatrix} 1 & 1 & 0 \\ 1 & c & 1 \\ 0 & 1 & 1 \end{pmatrix}$$

positive definite?

(b) For the particular value $c = 3$, carry out elimination to find the factorization $A = L\,D\,L^T$.

(c) Use your result from part (b) to rewrite the quadratic form $q(x, y, z) = x^2 + 2\,x\,y + 3\,y^2 + 2\,y\,z + z^2$ as a sum of squares.

(d) Explain how your result is related to the positive definiteness of A.

3.5.4. Write the quadratic form $q(\mathbf{x}) = x_1^2 + x_1 x_2 + 2x_2^2 - x_1 x_3 + 3x_3^2$ in the form $q(\mathbf{x}) = \mathbf{x}^T K\,\mathbf{x}$ for some symmetric matrix K. Is $q(\mathbf{x})$ positive definite?

3.5.5. Write the following quadratic forms on $\mathbb{R}^2$ as a sum of squares. Which are positive definite?

(a) $x^2 + 8\,x\,y + y^2$ (b) $x^2 - 4\,x\,y + 7\,y^2$

(c) $x^2 - 2\,x\,y - y^2$ (d) $x^2 + 6\,x\,y$

3.5.6. Prove that the following quadratic forms on $\mathbb{R}^3$ are positive definite by writing each as a sum of squares:

(a) $x^2 + 4\,x\,z + 3\,y^2 + 5\,z^2$

(b) $x^2 + 3\,x\,y + 3\,y^2 - 2\,x\,z + 8\,z^2$

(c) $2x_1^2 + x_1 x_2 - 2x_1 x_3 + 2x_2^2 - 2x_2 x_3 + 2x_3^2$

3.5.7. Write the following quadratic forms in matrix notation and determine if they are positive definite:

(a) $x^2 + 4\,x\,z + 2\,y^2 + 8\,y\,z + 12\,z^2$

(b) $3\,x^2 - 2\,y^2 - 8\,x\,y + x\,z + z^2$

(c) $x^2 + 2\,x\,y + 2\,y^2 - 4\,x\,z - 6\,y\,z + 6\,z^2$

(d) $3x_1^2 - x_2^2 + 5x_3^2 + 4x_1 x_2 - 7x_1 x_3 + 9x_2 x_3$

(e) $x_1^2 + 4x_1 x_2 - 2x_1 x_3 + 5x_2^2 - 2x_2 x_4 + 6x_3^2 - x_3 x_4 + 4x_4^2$

3.5.8. For what values of a, b and c is the quadratic form $x^2 + a\,x\,y + y^2 + b\,x\,z + c\,y\,z + z^2$ positive definite?

3.5.9. *True or false*: Every planar quadratic form $q(x, y) = a\,x^2 + 2\,b\,x\,y + c\,y^2$ can be written as a sum of squares.

3.5.10. (a) Prove that a positive definite matrix has positive determinant: $\det K > 0$.

(b) Show that a positive definite matrix has positive trace: $\operatorname{tr} K > 0$.

(c) Show that every 2×2 matrix with positive determinant and positive trace is positive definite.

(d) Find a symmetric 3×3 matrix with positive determinant and positive trace that is not positive definite.

3.5.11. (a) Prove that if K_1, K_2 are positive definite $n \times n$ matrices, then $K = \begin{pmatrix} K_1 & O \\ O & K_2 \end{pmatrix}$ is a positive definite $2n \times 2n$ matrix.
(b) Is the converse true?

3.5.12. Let $\| \cdot \|$ be any norm on $\mathbb{R}^n$.
(a) Show that $q(\mathbf{x})$ is a positive definite quadratic form if and only if $q(\mathbf{u}) > 0$ for all unit vectors, $\| \mathbf{u} \| = 1$.
(b) Prove that if $S = S^T$ is any symmetric matrix, then $K = S + c\,\mathrm{I} > 0$ is positive definite for $c \gg 0$ sufficiently large.

3.5.13. Prove that every symmetric matrix $S = K + N$ can be written as the sum of a positive definite matrix K and a negative definite matrix N. *Hint*: Use Exercise 3.5.12(b)

◇ **3.5.14.** (a) Prove that any regular symmetric matrix can be decomposed as a linear combination

$$K = d_1 \mathbf{l}_1 \mathbf{l}_1^T + d_2 \mathbf{l}_2 \mathbf{l}_2^T + \cdots + d_n \mathbf{l}_n \mathbf{l}_n^T \qquad (3.69)$$

of symmetric rank 1 matrices, as in Exercise 1.8.15, where $\mathbf{l}_1, \ldots, \mathbf{l}_n$ are the columns of the special lower triangular matrix L and $d_1, \ldots, d_n$ are the pivots, i.e., the diagonal entries of D. *Hint*: See Exercise 1.2.34.
(b) Decompose $\begin{pmatrix} 4 & -1 \\ -1 & 1 \end{pmatrix}$ and $\begin{pmatrix} 1 & 2 & 1 \\ 2 & 6 & 1 \\ 1 & 1 & 4 \end{pmatrix}$ in this manner.

♡ **3.5.15.** There is an alternative criterion for positive definiteness based on subdeterminants of the matrix. The 2×2 version already appears in (3.62).
(a) Prove that a 3×3 matrix

$$K = \begin{pmatrix} a & b & c \\ b & d & e \\ c & e & f \end{pmatrix}$$

is positive definite if and only if $a > 0$, $a d - b^2 > 0$, and $\det K > 0$.
(b) Prove the general version: an $n \times n$ matrix $K > 0$ is positive definite if and only if all the upper left square $k \times k$ subdeterminants are positive for $k = 1, \ldots, n$. *Hint*: See Exercise 1.9.19.

◇ **3.5.16.** Let K be a symmetric matrix. Prove that if a non-positive diagonal entry appears anywhere (not necessarily in the pivot position) in the matrix during Regular Gaussian Elimination, then K is not positive definite.

◇ **3.5.17.** Formulate a determinantal criterion similar to Exercise 3.5.15 for negative definite matrices. Write out the 2×2 and 3×3 cases explicitly.

3.5.18. *True or false*: A negative definite matrix must have negative trace and negative determinant.

The Cholesky Factorization

The identity (3.65) shows us how to write any regular quadratic form $q(\mathbf{x})$ as a linear combination of squares. We can push this result slightly further in the positive definite case. Since each pivot $d_i > 0$, we can write the diagonal quadratic form (3.66) as a sum of pure squares:

$$d_1 y_1^2 + \cdots + d_n y_n^2 = \left(\sqrt{d_1}\, y_1 \right)^2 + \cdots + \left(\sqrt{d_n}\, y_n \right)^2 = z_1^2 + \cdots + z_n^2,$$

where $z_i = \sqrt{d_i}\, y_i$. In matrix form, we are writing

$$\widehat{q}\,(\mathbf{y}) = \mathbf{y}^T D\, \mathbf{y} = \mathbf{z}^T \mathbf{z} = \| \mathbf{z} \|^2,$$

where

$$\mathbf{z} = S\, \mathbf{y}, \qquad \text{with} \qquad S = \operatorname{diag}\left(\sqrt{d_1}, \ldots, \sqrt{d_n} \right).$$

Since $D = S^2$, the matrix S can be thought of as a "square root" of the diagonal matrix D. Substituting back into (1.56), we deduce the *Cholesky factorization*

$$K = L\, D\, L^T = L\, S\, S^T L^T = M\, M^T, \qquad \text{where} \qquad M = L\, S \qquad (3.70)$$

of a positive definite matrix, first proposed by the early twentieth century French geographer André–Louis Cholesky for solving problems in geodesic surveying. Note that M is a lower triangular matrix with all positive entries, namely the square roots of the pivots $m_{ii} = \sqrt{d_i}$ on its diagonal. Applying the Cholesky factorization to the corresponding quadratic form produces

$$q(\mathbf{x}) = \mathbf{x}^T K\, \mathbf{x} = \mathbf{x}^T M\, M^T \mathbf{x} = \mathbf{z}^T \mathbf{z} = \| \mathbf{z} \|^2, \qquad \text{where} \qquad \mathbf{z} = M^T \mathbf{x}. \qquad (3.71)$$

We can interpret (3.71) as a change of variables from $\mathbf{x}$ to $\mathbf{z}$ that converts an arbitrary inner product norm, as defined by the square root of the positive definite quadratic form $q(\mathbf{x})$, into the standard Euclidean norm $\|\mathbf{z}\|$.

EXAMPLE 3.40 For the matrix

$$K = \begin{pmatrix} 1 & 2 & -1 \\ 2 & 6 & 0 \\ -1 & 0 & 9 \end{pmatrix}$$

considered in Example 3.38, the Cholesky formula (3.70) gives $K = M M^T$, where

$$M = L S = \begin{pmatrix} 1 & 0 & 0 \\ 2 & 1 & 0 \\ -1 & 1 & 1 \end{pmatrix} \begin{pmatrix} 1 & 0 & 0 \\ 0 & \sqrt{2} & 0 \\ 0 & 0 & \sqrt{6} \end{pmatrix} = \begin{pmatrix} 1 & 0 & 0 \\ 2 & \sqrt{2} & 0 \\ -1 & \sqrt{2} & \sqrt{6} \end{pmatrix}.$$

The associated quadratic function can then be written as a sum of pure squares:

$$q(\mathbf{x}) = x_1^2 + 4x_1x_2 - 2x_1x_3 + 6x_2^2 + 9x_3^2 = z_1^2 + z_2^2 + z_3^2,$$

where $\mathbf{z} = M^T\mathbf{x}$, or, explicitly,

$$z_1 = x_1 + 2x_2 - x_3, \qquad z_2 = \sqrt{2}\,x_2 + \sqrt{2}\,x_3, \qquad z_3 = \sqrt{6}\,x_3.$$ ●

EXERCISES 3.5

3.5.19. Find the Cholesky factorizations of the following matrices:

(a) $\begin{pmatrix} 3 & -2 \\ -2 & 2 \end{pmatrix}$ (b) $\begin{pmatrix} 4 & -12 \\ -12 & 45 \end{pmatrix}$

(c) $\begin{pmatrix} 1 & 1 & 1 \\ 1 & 2 & -2 \\ 1 & -2 & 14 \end{pmatrix}$ (d) $\begin{pmatrix} 2 & 1 & 1 \\ 1 & 2 & 1 \\ 1 & 1 & 2 \end{pmatrix}$

(e) $\begin{pmatrix} 2 & 1 & 0 & 0 \\ 1 & 2 & 1 & 0 \\ 0 & 1 & 2 & 1 \\ 0 & 0 & 1 & 2 \end{pmatrix}$

3.5.20. Which of the matrices in Exercise 3.5.1 have a Cholesky factorization? When valid, write out the factorization.

3.5.21. Write the following positive definite quadratic forms as a sum of pure squares, as in (3.71):

(a) $16x_1^2 + 25x_2^2$ (b) $x_1^2 - 2x_1x_2 + 4x_2^2$

(c) $5x_1^2 + 4x_1x_2 + 3x_2^2$

(d) $3x_1^2 - 2x_1x_2 - 2x_1x_3 + 2x_2^2 + 6x_3^2$

(e) $x_1^2 + x_1x_2 + x_2^2 + x_2x_3 + x_3^2$

(f) $4x_1^2 - 2x_1x_2 - 4x_1x_3 + \frac{1}{2}x_2^2 - x_2x_3 + 6x_3^2$

(g) $3x_1^2 + 2x_1x_2 + 3x_2^2 + 2x_2x_3 + 3x_3^2 + 2x_3x_4 + 3x_4^2$

3.6 COMPLEX VECTOR SPACES

Although physical applications ultimately require real answers, complex numbers and complex vector spaces play an extremely useful, if not essential, role in the intervening analysis. Particularly in the description of periodic phenomena, complex numbers and complex exponentials help to simplify complicated trigonometric formulae. Complex variable methods are ubiquitous in electrical engineering, Fourier analysis, potential theory, fluid mechanics, electromagnetism, and many other applied fields, [36, 59]. In quantum mechanics, the basic physical quantities are complex-valued wave functions, [40]. Moreover, the Schrödinger equation, which governs quantum dynamics, is an inherently complex partial differential equation.

In this section, we survey the main facts concerning complex numbers and complex vector spaces. Most of the constructions are entirely analogous to their real counterparts, and so will not be dwelled on at length. The one exception is the complex version of an inner product, which does introduce some novelties not found in its simpler real cousin.

Complex Numbers

Recall that a *complex number* is an expression of the form $z = x + \mathrm{i}\, y$, where $x, y \in \mathbb{R}$ are real and* $\mathrm{i} = \sqrt{-1}$. The set of all complex numbers (scalars) is denoted by $\mathbb{C}$. We call $x = \operatorname{Re} z$ the *real part* of z and $y = \operatorname{Im} z$ the *imaginary part* of $z = x + \mathrm{i}\, y$. (*Note*: The imaginary part is the real number y, *not* $\mathrm{i}\, y$.) A real number x is merely a complex number with zero imaginary part, $\operatorname{Im} z = 0$, and so we may regard $\mathbb{R} \subset \mathbb{C}$. Complex addition and multiplication are based on simple adaptations of the rules of real arithmetic to include the identity $\mathrm{i}^2 = -1$, and so

$$\begin{aligned} (x + \mathrm{i}\, y) + (u + \mathrm{i}\, v) &= (x + u) + \mathrm{i}\,(y + v), \\ (x + \mathrm{i}\, y)(u + \mathrm{i}\, v) &= (x\,u - y\,v) + \mathrm{i}\,(x\,v + y\,u). \end{aligned} \tag{3.72}$$

Complex numbers enjoy all the usual laws of real addition and multiplication, *including commutativity*: $z\,w = w\,z$.

We can identify a complex number $x + \mathrm{i}\, y$ with a vector $(x, y)^T \in \mathbb{R}^2$ in the real plane. For this reason, $\mathbb{C}$ is sometimes referred to as the *complex plane*. Complex addition (3.72) corresponds to vector addition, but complex multiplication does not have a readily identifiable vector counterpart.

Another useful operation on complex numbers is that of complex conjugation.

Definition 3.41 The *complex conjugate* of $z = x + \mathrm{i}\, y$ is $\overline{z} = x - \mathrm{i}\, y$, whereby $\operatorname{Re} \overline{z} = \operatorname{Re} z$, while $\operatorname{Im} \overline{z} = -\operatorname{Im} z$.

Geometrically, the complex conjugate of z is obtained by reflecting the corresponding vector through the real axis, as illustrated in Figure 3.7. In particular $\overline{z} = z$ if and only if z is real. Note that

$$\operatorname{Re} z = \frac{z + \overline{z}}{2}, \qquad \operatorname{Im} z = \frac{z - \overline{z}}{2\,\mathrm{i}}. \tag{3.73}$$

Complex conjugation is compatible with complex arithmetic:

$$\overline{z + w} = \overline{z} + \overline{w}, \qquad \overline{z\,w} = \overline{z}\;\overline{w}.$$

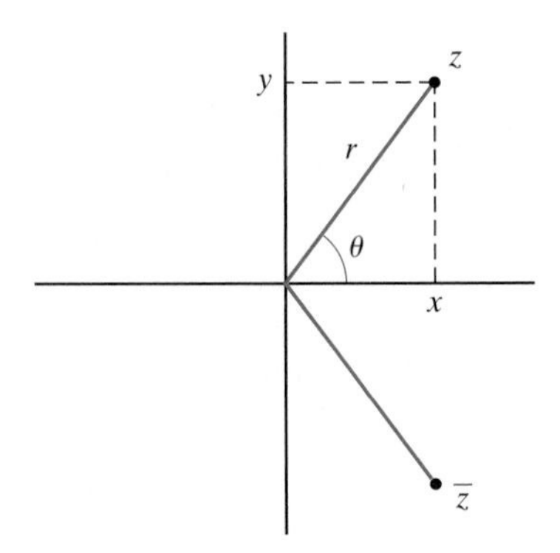

Figure 3.7 Complex numbers.

*To avoid confusion with the symbol for current, electrical engineers prefer to use j to indicate the imaginary unit.

In particular, the product of a complex number and its conjugate

$$z\,\overline{z} = (x + \mathrm{i}\,y)\,(x - \mathrm{i}\,y) = x^2 + y^2 \tag{3.74}$$

is real and non-negative. Its square root is known as the *modulus* or *norm* of the complex number $z = x + \mathrm{i}\,y$, and written

$$|z| = \sqrt{x^2 + y^2}. \tag{3.75}$$

Note that $|z| \geq 0$, with $|z| = 0$ if and only if $z = 0$. The modulus $|z|$ generalizes the absolute value of a real number, and coincides with the standard Euclidean norm in the $x\,y$–plane, which implies the validity of the triangle inequality

$$|z + w| \leq |z| + |w|. \tag{3.76}$$

Equation (3.74) can be rewritten in terms of the modulus as

$$z\,\overline{z} = |z|^2. \tag{3.77}$$

Rearranging the factors, we deduce the formula for the reciprocal of a nonzero complex number:

$$\frac{1}{z} = \frac{\overline{z}}{|z|^2}, \qquad z \neq 0, \qquad \text{or, equivalently,} \qquad \frac{1}{x + \mathrm{i}\,y} = \frac{x - \mathrm{i}\,y}{x^2 + y^2}. \tag{3.78}$$

The general formula for complex division,

$$\frac{w}{z} = \frac{w\,\overline{z}}{|z|^2} \qquad \text{or} \qquad \frac{u + \mathrm{i}\,v}{x + \mathrm{i}\,y} = \frac{(x\,u + y\,v) + \mathrm{i}\,(x\,v - y\,u)}{x^2 + y^2}, \tag{3.79}$$

is an immediate consequence.

The modulus of a complex number,

$$r = |z| = \sqrt{x^2 + y^2}\,,$$

is one component of its polar coordinate representation

$$x = r\cos\theta, \qquad y = r\sin\theta \qquad \text{or} \qquad z = r(\cos\theta + \mathrm{i}\,\sin\theta). \tag{3.80}$$

The polar angle, which measures the angle that the line connecting z to the origin makes with the horizontal axis, is known as the *phase*, and written

$$\theta = \operatorname{ph} z. \tag{3.81}$$

As such, the phase is only defined up to an integer multiple of 2π. The more common term for the angle is the *argument*, written $\arg z = \operatorname{ph} z$. However, we prefer to use "phase" throughout this text, in part to avoid confusion with the argument z of a function $f(z)$. We note that the modulus and phase of a product of complex numbers can be readily computed:

$$|z\,w| = |z|\,|w|, \qquad \operatorname{ph}(z\,w) = \operatorname{ph} z + \operatorname{ph} w. \tag{3.82}$$

Complex conjugation preserves the modulus, but negates the phase:

$$|\overline{z}| = |z|, \qquad \operatorname{ph}\overline{z} = -\operatorname{ph} z. \tag{3.83}$$

One of the most profound formulas in all of mathematics is *Euler's formula*

$$e^{\mathrm{i}\,\theta} = \cos\theta + \mathrm{i}\,\sin\theta, \tag{3.84}$$

relating the complex exponential with the real sine and cosine functions. It has a variety of mathematical justifications; see Exercise 3.6.23 for one that is based on comparing power series. Euler's formula can be used to compactly rewrite the polar form (3.80) of a complex number as

$$z = r\, e^{\mathrm{i}\,\theta} \quad \text{where} \quad r = |z|, \qquad \theta = \operatorname{ph} z. \tag{3.85}$$

The complex conjugation identity

$$e^{-\mathrm{i}\,\theta} = \cos(-\theta) + \mathrm{i}\,\sin(-\theta) = \cos\theta - \mathrm{i}\,\sin\theta = \overline{e^{\mathrm{i}\,\theta}}$$

permits us to express the basic trigonometric functions in terms of complex exponentials:

$$\cos\theta = \frac{e^{\mathrm{i}\,\theta} + e^{-\mathrm{i}\,\theta}}{2}, \qquad \sin\theta = \frac{e^{\mathrm{i}\,\theta} - e^{-\mathrm{i}\,\theta}}{2\,\mathrm{i}}. \tag{3.86}$$

These formulae are very useful when working with trigonometric identities and integrals.

The exponential of a general complex number is easily derived from the Euler formula and the standard properties of the exponential function—which carry over unaltered to the complex domain; thus,

$$e^{z} = e^{x+\mathrm{i}\,y} = e^{x}\, e^{\mathrm{i}\,y} = e^{x}\cos y + \mathrm{i}\, e^{x}\sin y. \tag{3.87}$$

Graphs of the real and imaginary parts of the complex exponential function appear in Figure 3.8. Note that $e^{2\pi\,\mathrm{i}} = 1$, and hence the exponential function is periodic

$$e^{z+2\pi\,\mathrm{i}} = e^{z} \tag{3.88}$$

with imaginary period $2\pi\,\mathrm{i}$—indicative of the periodicity of the trigonometric functions in Euler's formula.

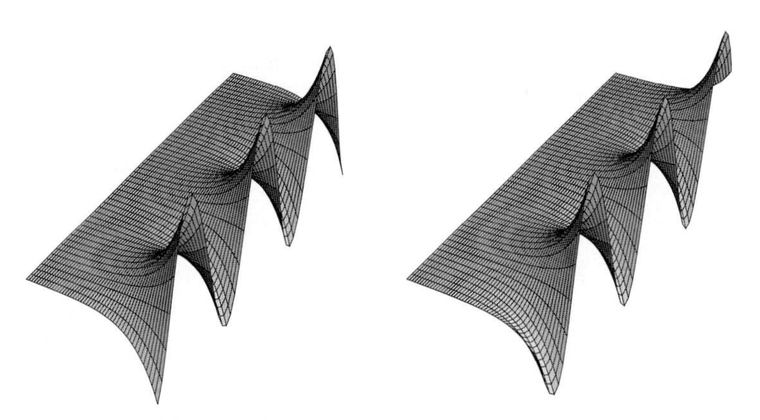

Figure 3.8 Real and imaginary parts of e^{z}.

EXERCISES 3.6

3.6.1. Write down a single equation that relates the five most important numbers in mathematics, which are $0, 1, e, \pi$ and i.

3.6.2. For any integer k, prove that $e^{k\pi\,\mathrm{i}} = (-1)^{k}$.

3.6.3. Is the formula $1^{z} = 1$ valid for all complex values of z?

3.6.4. What is wrong with the calculation $e^{2a\pi\,\mathrm{i}} = (e^{2\pi\,\mathrm{i}})^{a} = 1^{a} = 1$?

3.6.5. (a) Write i in phase–modulus form.
(b) Use this expression to find $\sqrt{\mathrm{i}}$, i.e., a complex number z such that $z^{2} = \mathrm{i}$. Can you find a second square root?
(c) Find explicit formulas for the three third roots and four fourth roots of i.

3.6.6. In Figure 3.7, where would you place the point $1/z$?

3.6.7. (a) If z moves counterclockwise around a circle of radius r in the complex plane, around which circle and in which direction does $w = 1/z$ move?

(b) What about $w = \overline{z}$?

(c) What if the circle is not centered at the origin?

◇ **3.6.8.** Show that $-|z| \leq \operatorname{Re} z \leq |z|$ and $-|z| \leq \operatorname{Im} z \leq |z|$.

◇ **3.6.9.** Prove that if φ is real, then $\operatorname{Re}(e^{\mathrm{i}\varphi} z) \leq |z|$, with equality if and only if $\varphi = -\operatorname{ph} z$.

3.6.10. Prove the identities in (3.82) and (3.83).

3.6.11. Prove $\operatorname{ph}(z/w) = \operatorname{ph} z - \operatorname{ph} w = \operatorname{ph}(z\,\overline{w})$ is equal to the angle between the vectors representing z and w.

3.6.12. The phase of a complex number $z = x + \mathrm{i}\, y$ is often written as $\operatorname{ph} z = \tan^{-1}(y/x)$. Explain why this formula is ambiguous, and does not uniquely define $\operatorname{ph} z$.

3.6.13. Show that if we identify the complex numbers z, w with vectors in the plane, then their Euclidean dot product is equal to $\operatorname{Re}(z\,\overline{w})$.

3.6.14. (a) Prove that the complex numbers z and w correspond to orthogonal vectors in $\mathbb{R}^2$ if and only if $\operatorname{Re} z\,\overline{w} = 0$.

(b) Prove that z and $\mathrm{i}\, z$ are always orthogonal.

3.6.15. Prove that $e^{z+w} = e^z e^w$. Conclude that $e^{mz} = (e^z)^m$ whenever m is an integer.

3.6.16. (a) Use the formula $e^{2\mathrm{i}\theta} = (e^{\mathrm{i}\theta})^2$ to deduce the well-known trigonometric identities for $\cos 2\theta$ and $\sin 2\theta$.

(b) Derive the corresponding identities for $\cos 3\theta$ and $\sin 3\theta$.

(c) Write down the explicit identities for $\cos m\theta$ and $\sin m\theta$ as polynomials in $\cos\theta$ and $\sin\theta$. *Hint*: Apply the Binomial Formula to $(e^{\mathrm{i}\theta})^m$.

◇ **3.6.17.** Use complex exponentials to prove the identity

$$\cos\theta - \cos\varphi = 2\cos\frac{\theta-\varphi}{2}\,\cos\frac{\theta+\varphi}{2}.$$

3.6.18. Prove that if $z = x + \mathrm{i}\, y$, then $|e^z| = e^x$, $\operatorname{ph} e^z = y$.

3.6.19. The formulas

$$\cos z = \frac{e^{\mathrm{i}z} + e^{-\mathrm{i}z}}{2} \quad \text{and} \quad \sin z = \frac{e^{\mathrm{i}z} - e^{-\mathrm{i}z}}{2\,\mathrm{i}}$$

serve to define the basic complex trigonometric functions. Write out the formulas for their real and imaginary parts in terms of $z = x + \mathrm{i}\, y$, and show that $\cos z$ and $\sin z$ reduce to their usual real forms when $z = x$ is real. What do they become when $z = \mathrm{i}\, y$ is purely imaginary?

3.6.20. The complex *hyperbolic functions* are defined as

$$\cosh z = \frac{e^z + e^{-z}}{2}, \qquad \sinh z = \frac{e^z - e^{-z}}{2}.$$

(a) Write out the formulas for their real and imaginary parts in terms of $z = x + \mathrm{i}\, y$.

(b) Prove that $\cos \mathrm{i}\, z = \cosh z$ and $\sin \mathrm{i}\, z = \mathrm{i}\sinh z$.

♡ **3.6.21.** Generalizing Example 2.17(c), by a *trigonometric polynomial* of *degree* $\leq n$, we mean a function

$$T(x) = \sum_{0 \leq j+k \leq n} c_{jk}\,(\cos\theta)^j\,(\sin\theta)^k$$

in the powers of the sine and cosine functions up to degree n.

(a) Use formula (3.86) to prove that any trigonometric polynomial of degree $\leq n$ can be written as a complex linear combination of the $2n+1$ complex exponentials $e^{-n\mathrm{i}\theta}, \ldots e^{-\mathrm{i}\theta}, e^{0\,\mathrm{i}\theta} = 1, e^{\mathrm{i}\theta}, e^{2\mathrm{i}\theta}, \ldots e^{n\,\mathrm{i}\theta}$.

(b) Prove that any trigonometric polynomial of degree $\leq n$ can be written as a real linear combination of the trigonometric functions $1, \cos\theta, \sin\theta, \cos 2\theta, \sin 2\theta, \ldots \cos n\theta, \sin n\theta$.

(c) Write out the following trigonometric polynomials in both of the preceding forms:

(i) $\cos^2\theta$ (ii) $\cos\theta\sin\theta$ (iii) $\cos^3\theta$ (iv) $\sin^4\theta$ (v) $\cos^2\theta\sin^2\theta$

◇ **3.6.22.** Write out the real and imaginary parts of the power function x^c with complex exponent $c = a + \mathrm{i}\, b \in \mathbb{C}$.

◇ **3.6.23.** Write the power series expansions for $e^{\mathrm{i}x}$. Prove that the real terms give the power series for $\cos x$, while the imaginary terms give that of $\sin x$. Use this identification to justify Euler's formula (3.84).

◇ **3.6.24.** The derivative of a complex-valued function $f(x) = u(x) + \mathrm{i}\, v(x)$, depending on a real variable x, is given by $f'(x) = u'(x) + \mathrm{i}\, v'(x)$.

(a) Prove that if $\lambda = \mu + \mathrm{i}\,\nu$ is any complex scalar, then

$$\frac{d}{dx}\, e^{\lambda x} = \lambda\, e^{\lambda x}.$$

(b) Prove, conversely,

$$\int_a^b e^{\lambda x}\, dx = \frac{1}{\lambda}\left(e^{\lambda b} - e^{\lambda a}\right)$$

provided $\lambda \neq 0$.

3.6.25. Use the complex trigonometric formulae (3.86) and Exercise 3.6.24 to evaluate the following trigonometric integrals:

(a) $\displaystyle\int \cos^2 x\, dx$ (b) $\displaystyle\int \sin^2 x\, dx$ (c) $\displaystyle\int \cos x \sin x\, dx$ (d) $\displaystyle\int \cos 3x \sin 5x\, dx$

How did you calculate them in first year calculus? If you're not convinced this method is easier, try the more complicated integrals

(e) $\displaystyle\int \cos^4 x\,dx$ (f) $\displaystyle\int \sin^4 x\,dx$

(g) $\displaystyle\int \cos^2 x \sin^2 x\,dx$

(h) $\displaystyle\int \cos 3x \sin 5x \cos 7x\,dx$

♠ **3.6.26.** Use a computer to graph the real and imaginary parts of the complex functions z^2 and $1/z$. Discuss what you see.

♠ **3.6.27.** Use a computer to graph $\operatorname{ph} z$ and $|z|$. Discuss what you observe.

Complex Vector Spaces and Inner Products

A *complex vector space* is defined in exactly the same manner as its real counterpart, cf. Definition 2.1, the only difference being that we replace real scalars by complex scalars. The most basic example is the n-dimensional complex vector space $\mathbb{C}^n$ consisting of all column vectors $\mathbf{z} = (z_1, z_2, \ldots, z_n)^T$ that have n complex entries $z_1, \ldots, z_n \in \mathbb{C}$. Verification of each of the vector space axioms is immediate.

We can write any complex vector $\mathbf{z} = \mathbf{x} + \mathrm{i}\,\mathbf{y} \in \mathbb{C}^n$ as a linear combination of two real vectors $\mathbf{x} = \operatorname{Re}\mathbf{z}$ and $\mathbf{y} = \operatorname{Im}\mathbf{z} \in \mathbb{R}^n$ called its *real* and *imaginary parts.* Its complex conjugate $\overline{\mathbf{z}} = \mathbf{x} - \mathrm{i}\,\mathbf{y}$ is obtained by taking the complex conjugates of its individual entries. Thus, for example, if

$$\mathbf{z} = \begin{pmatrix} 1+2\,\mathrm{i} \\ -3 \\ 5\,\mathrm{i} \end{pmatrix} = \begin{pmatrix} 1 \\ -3 \\ 0 \end{pmatrix} + \mathrm{i}\begin{pmatrix} 2 \\ 0 \\ 5 \end{pmatrix},$$

then

$$\overline{\mathbf{z}} = \begin{pmatrix} 1-2\,\mathrm{i} \\ -3 \\ -5\,\mathrm{i} \end{pmatrix} = \begin{pmatrix} 1 \\ -3 \\ 0 \end{pmatrix} - \mathrm{i}\begin{pmatrix} 2 \\ 0 \\ 5 \end{pmatrix}.$$

In particular, $\mathbf{z} \in \mathbb{R}^n \subset \mathbb{C}^n$ is a real vector if and only if $\mathbf{z} = \overline{\mathbf{z}}$.

Most of the vector space concepts we developed in the real domain, including span, linear independence, basis, and dimension, can be straightforwardly extended to the complex regime. The one exception is the concept of an inner product, which requires a little thought. In analysis, the primary applications of inner products and norms rely on the associated inequalities: Cauchy–Schwarz and triangle. But there is no natural ordering of the complex numbers, and so one *cannot* assign a meaning to a complex inequality like $z < w$. Inequalities only make sense in the real domain, and so the norm of a complex vector should still be a positive and real. With this in mind, the naïve idea of simply summing the squares of the entries of a complex vector will *not* define a norm on $\mathbb{C}^n$, since the result will typically be complex. Moreover, some nonzero complex vectors, e.g., $(1, \mathrm{i})^T$, would then have zero "norm".

The correct definition is modeled on the formula

$$|z| = \sqrt{z\,\overline{z}}$$

that defines the modulus of a complex scalar $z \in \mathbb{C}$. If, in analogy with the real definition (3.7), the quantity inside the square root should represent the inner product of z with itself, then we should define the "dot product" between two complex numbers to be

$$z \cdot w = z\,\overline{w}, \qquad \text{so that} \quad z \cdot z = z\,\overline{z} = |z|^2.$$

Writing out the formula when $z = x + \mathrm{i}\,y$ and $w = u + \mathrm{i}\,v$, we find

$$z \cdot w = z\,\overline{w} = (x + \mathrm{i}\,y)\,(u - \mathrm{i}\,v) = (x\,u + y\,v) + \mathrm{i}\,(y\,u - x\,v). \tag{3.89}$$

Thus, the dot product of two complex numbers is, in general, complex. The real part of $z \cdot w$ is, in fact, the Euclidean dot product between the corresponding vectors in $\mathbb{R}^2$, while its imaginary part is, interestingly, their scalar cross-product, cf. (3.22).

The vector version of this construction is named after the nineteenth century French mathematician Charles Hermite, and called the *Hermitian dot product* on $\mathbb{C}^n$. It has the explicit formula

$$\mathbf{z} \cdot \mathbf{w} = \mathbf{z}^T \overline{\mathbf{w}} = z_1 \overline{w}_1 + z_2 \overline{w}_2 + \cdots + z_n \overline{w}_n, \qquad \text{for} \quad \mathbf{z} = \begin{pmatrix} z_1 \\ \vdots \\ z_n \end{pmatrix}, \quad \mathbf{w} = \begin{pmatrix} w_1 \\ \vdots \\ w_n \end{pmatrix}. \tag{3.90}$$

Pay attention to the fact that we must apply complex conjugation to all the entries of the second vector. For example, if

$$\mathbf{z} = \begin{pmatrix} 1 + \mathrm{i} \\ 3 + 2\,\mathrm{i} \end{pmatrix}, \qquad \mathbf{w} = \begin{pmatrix} 1 + 2\,\mathrm{i} \\ \mathrm{i} \end{pmatrix},$$

then

$$\mathbf{z} \cdot \mathbf{w} = (1 + \mathrm{i})(1 - 2\,\mathrm{i}) + (3 + 2\,\mathrm{i})(-\,\mathrm{i}) = 5 - 4\,\mathrm{i}.$$

On the other hand,

$$\mathbf{w} \cdot \mathbf{z} = (1 + 2\,\mathrm{i})(1 - \mathrm{i}) + \mathrm{i}\,(3 - 2\,\mathrm{i}) = 5 + 4\,\mathrm{i},$$

and we conclude that the Hermitian dot product is *not* symmetric. Indeed, reversing the order of the vectors conjugates their dot product:

$$\mathbf{w} \cdot \mathbf{z} = \overline{\mathbf{z} \cdot \mathbf{w}}.$$

This is an unexpected complication, but it does have the desired effect that the induced norm, namely

$$0 \le \| \mathbf{z} \| = \sqrt{\mathbf{z} \cdot \mathbf{z}} = \sqrt{\mathbf{z}^T \overline{\mathbf{z}}} = \sqrt{|z_1|^2 + \cdots + |z_n|^2}, \tag{3.91}$$

is strictly positive for all $\mathbf{0} \ne \mathbf{z} \in \mathbb{C}^n$. For example, if

$$\mathbf{z} = \begin{pmatrix} 1 + 3\,\mathrm{i} \\ -2\,\mathrm{i} \\ -5 \end{pmatrix}, \qquad \text{then} \quad \| \mathbf{z} \| = \sqrt{|1 + 3\,\mathrm{i}|^2 + |-2\,\mathrm{i}|^2 + |-5|^2} = \sqrt{39}.$$

The Hermitian dot product is well behaved under complex vector addition:

$$(\mathbf{z} + \widetilde{\mathbf{z}}) \cdot \mathbf{w} = \mathbf{z} \cdot \mathbf{w} + \widetilde{\mathbf{z}} \cdot \mathbf{w}, \qquad \mathbf{z} \cdot (\mathbf{w} + \widetilde{\mathbf{w}}) = \mathbf{z} \cdot \mathbf{w} + \mathbf{z} \cdot \widetilde{\mathbf{w}}.$$

However, while complex scalar multiples can be extracted from the first vector without alteration, when they multiply the second vector, they emerge as complex conjugates:

$$(c\,\mathbf{z}) \cdot \mathbf{w} = c\,(\mathbf{z} \cdot \mathbf{w}), \qquad \mathbf{z} \cdot (c\,\mathbf{w}) = \overline{c}\,(\mathbf{z} \cdot \mathbf{w}), \qquad c \in \mathbb{C}.$$

Thus, the Hermitian dot product is not bilinear in the strict sense, but satisfies something that, for lack of a better name, is known as *sesquilinearity*.

The general definition of an inner product on a complex vector space is modeled on the preceding properties of the Hermitian dot product.

Definition 3.42 An *inner product* on the complex vector space V is a pairing that takes two vectors $\mathbf{v}, \mathbf{w} \in V$ and produces a complex number $\langle \mathbf{v}, \mathbf{w} \rangle \in \mathbb{C}$, subject to the following requirements, for $\mathbf{u}, \mathbf{v}, \mathbf{w} \in V$, and $c, d \in \mathbb{C}$:

(i) *Sesquilinearity*:

$$\begin{aligned} \langle c\,\mathbf{u} + d\,\mathbf{v}, \mathbf{w} \rangle &= c\,\langle \mathbf{u}, \mathbf{w} \rangle + d\,\langle \mathbf{v}, \mathbf{w} \rangle, \\ \langle \mathbf{u}, c\,\mathbf{v} + d\,\mathbf{w} \rangle &= \bar{c}\,\langle \mathbf{u}, \mathbf{v} \rangle + \bar{d}\,\langle \mathbf{u}, \mathbf{w} \rangle. \end{aligned} \tag{3.92}$$

(ii) *Conjugate Symmetry*:

$$\langle \mathbf{v}, \mathbf{w} \rangle = \overline{\langle \mathbf{w}, \mathbf{v} \rangle}. \tag{3.93}$$

(iii) *Positivity*:

$$\|\mathbf{v}\|^2 = \langle \mathbf{v}, \mathbf{v} \rangle \geq 0, \qquad \text{and} \qquad \langle \mathbf{v}, \mathbf{v} \rangle = 0 \qquad \text{if and only if} \qquad \mathbf{v} = \mathbf{0}. \tag{3.94}$$

Thus, when dealing with a complex inner product space, one must pay careful attention to the complex conjugate that appears when the second argument in the inner product is multiplied by a complex scalar, as well as the complex conjugate that appears when reversing the order of the two arguments. But, once this initial complication has been properly dealt with, the further properties of the inner product carry over directly from the real domain.

THEOREM 3.43 The Cauchy–Schwarz inequality,

$$|\langle \mathbf{v}, \mathbf{w} \rangle| \ \leq \ \|\mathbf{v}\|\,\|\mathbf{w}\|, \tag{3.95}$$

with $|\cdot|$ now denoting the complex modulus, and the triangle inequality

$$\|\mathbf{v} + \mathbf{w}\| \ \leq \ \|\mathbf{v}\| + \|\mathbf{w}\| \tag{3.96}$$

are both valid on any complex inner product space.

The proof of (3.95–96) is modeled on the real case, and the details are left to the reader.

EXAMPLE 3.44 The vectors $\mathbf{v} = (1 + \mathrm{i}\,, 2\mathrm{i}\,, -3)^T$, $\mathbf{w} = (2 - \mathrm{i}\,, 1, 2 + 2\,\mathrm{i}\,)^T$, satisfy

$$\begin{aligned} \|\mathbf{v}\| &= \sqrt{2+4+9} = \sqrt{15}, \\ \|\mathbf{w}\| &= \sqrt{5+1+8} = \sqrt{14}, \\ \mathbf{v} \cdot \mathbf{w} &= (1+\mathrm{i}\,)(2+\mathrm{i}\,) + 2\mathrm{i} + (-3)(2 - 2\,\mathrm{i}\,) = -5 + 11\,\mathrm{i}\,. \end{aligned}$$

Thus, the Cauchy–Schwarz inequality reads

$$|\langle \mathbf{v}, \mathbf{w} \rangle| = |-5 + 11\,\mathrm{i}\,| = \sqrt{146} \ \leq \ \sqrt{210} = \sqrt{15}\,\sqrt{14} = \|\mathbf{v}\|\,\|\mathbf{w}\|.$$

Similarly, the triangle inequality tells us that

$$\begin{aligned} \|\mathbf{v} + \mathbf{w}\| &= \|(3, 1 + 2\,\mathrm{i}\,, -1 + 2\,\mathrm{i}\,)^T\| \\ &= \sqrt{9+5+5} = \sqrt{19} \leq \sqrt{15} + \sqrt{14} = \|\mathbf{v}\| + \|\mathbf{w}\|. \end{aligned}$$

●

EXAMPLE 3.45 Let $\mathrm{C}^0[-\pi, \pi\,]$ denote the complex vector space consisting of all complex valued continuous functions $f(x) = u(x) + \mathrm{i}\,v(x)$ depending upon the *real* variable $-\pi \leq x \leq \pi$. The Hermitian L^2 inner product on $\mathrm{C}^0[-\pi, \pi\,]$ is defined as

$$\langle f, g \rangle = \int_{-\pi}^{\pi} f(x)\,\overline{g(x)}\,dx\,, \tag{3.97}$$

i.e., the integral of f times the complex conjugate of g, with corresponding norm

$$\| f \| = \sqrt{\int_{-\pi}^{\pi} |f(x)|^2 \, dx} = \sqrt{\int_{-\pi}^{\pi} \left[u(x)^2 + v(x)^2 \right] dx} \,. \tag{3.98}$$

The reader can verify that (3.97) satisfies the Hermitian inner product axioms.

In particular, if k, l are integers, then the inner product of the complex exponential functions $e^{\mathrm{i}kx}$ and $e^{\mathrm{i}lx}$ is

$$\begin{aligned} \langle e^{\mathrm{i}kx}, e^{\mathrm{i}lx} \rangle &= \int_{-\pi}^{\pi} e^{\mathrm{i}kx} e^{-\mathrm{i}lx} \, dx \\ &= \int_{-\pi}^{\pi} e^{\mathrm{i}(k-l)x} \, dx = \begin{cases} 2\pi, & k = l, \\ \left. \dfrac{e^{\mathrm{i}(k-l)x}}{\mathrm{i}(k-l)} \right|_{x=-\pi}^{\pi} = 0, & k \neq l. \end{cases} \end{aligned}$$

We conclude that when $k \neq l$, the complex exponentials $e^{\mathrm{i}kx}$ and $e^{\mathrm{i}lx}$ are orthogonal, since their inner product is zero. The complex formulation of Fourier analysis, [16, 47], is founded on this key example. ●

EXERCISES 3.6

3.6.28. Determine whether the indicated sets of complex vectors are linearly independent or dependent.

(a) $\begin{pmatrix} \mathrm{i} \\ 1 \end{pmatrix}, \begin{pmatrix} 1 \\ \mathrm{i} \end{pmatrix}$

(b) $\begin{pmatrix} 1+\mathrm{i} \\ 1 \end{pmatrix}, \begin{pmatrix} 2 \\ 1-\mathrm{i} \end{pmatrix}$

(c) $\begin{pmatrix} 1+3\,\mathrm{i} \\ 2-\mathrm{i} \end{pmatrix}, \begin{pmatrix} 2-3\,\mathrm{i} \\ 1-\mathrm{i} \end{pmatrix}$

(d) $\begin{pmatrix} -2+\mathrm{i} \\ \mathrm{i} \end{pmatrix}, \begin{pmatrix} 4-3\,\mathrm{i} \\ 1 \end{pmatrix}, \begin{pmatrix} 2\,\mathrm{i} \\ 1-5\,\mathrm{i} \end{pmatrix}$

(e) $\begin{pmatrix} 1+2\,\mathrm{i} \\ 2 \\ 0 \end{pmatrix}, \begin{pmatrix} 2 \\ 0 \\ 1-\mathrm{i} \end{pmatrix}$

(f) $\begin{pmatrix} 1 \\ 3\,\mathrm{i} \\ 2-\mathrm{i} \end{pmatrix}, \begin{pmatrix} 1+2\,\mathrm{i} \\ -3 \\ 0 \end{pmatrix}, \begin{pmatrix} 1-\mathrm{i} \\ -\mathrm{i} \\ 1 \end{pmatrix}$

(g) $\begin{pmatrix} 1+\mathrm{i} \\ 2-\mathrm{i} \\ 1 \end{pmatrix}, \begin{pmatrix} 1-\mathrm{i} \\ -3\,\mathrm{i} \\ 1-2\,\mathrm{i} \end{pmatrix}, \begin{pmatrix} -1+\mathrm{i} \\ 2+3\,\mathrm{i} \\ 1+2\,\mathrm{i} \end{pmatrix}$

3.6.29. (a) Determine whether the vectors

$$\mathbf{v}_1 = \begin{pmatrix} 1 \\ \mathrm{i} \\ 0 \end{pmatrix}, \quad \mathbf{v}_2 = \begin{pmatrix} 0 \\ 1+\mathrm{i} \\ 2 \end{pmatrix}, \quad \mathbf{v}_3 = \begin{pmatrix} -1+\mathrm{i} \\ 1+\mathrm{i} \\ -1 \end{pmatrix}$$

are linearly independent or linearly dependent.

(b) Do they form a basis of $\mathbb{C}^3$?

(c) Compute the Hermitian norm of each vector.

(d) Compute the Hermitian dot products between all different pairs. Which vectors are orthogonal?

(e) Do the vectors form an orthogonal or orthonormal basis of $\mathbb{C}^3$?

3.6.30. Find the dimension of and a basis for the following subspaces of $\mathbb{C}^3$:

(a) The set of all complex multiples of $(1, \mathrm{i}, 1-\mathrm{i})^T$.

(b) The plane $z_1 + \mathrm{i}\, z_2 + (1-\mathrm{i}) z_3 = 0$.

(c) The range of the matrix

$$A = \begin{pmatrix} 1 & \mathrm{i} & 2-\mathrm{i} \\ 2+\mathrm{i} & 1+3\,\mathrm{i} & -1-\mathrm{i} \end{pmatrix}.$$

(d) The kernel of the same matrix.

(e) The set of vectors that are orthogonal to $(1-\mathrm{i}, 2\,\mathrm{i}, 1+\mathrm{i})^T$.

3.6.31. *True or false*: The set of complex vectors of the form $\begin{pmatrix} z \\ \bar{z} \end{pmatrix}$ for $z \in \mathbb{C}$ forms a subspace of $\mathbb{C}^2$.

3.6.32. Find bases for the four fundamental subspaces associated with the complex matrices

(a) $\begin{pmatrix} \mathrm{i} & 2 \\ -1 & 2\,\mathrm{i} \end{pmatrix}$

(b) $\begin{pmatrix} 2 & -1+\mathrm{i} & 1-2\mathrm{i} \\ -4 & 3-\mathrm{i} & 1+\mathrm{i} \end{pmatrix}$

(c) $\begin{pmatrix} \mathrm{i} & -1 & 2-\mathrm{i} \\ -1+2\mathrm{i} & -2-\mathrm{i} & 3 \\ \mathrm{i} & -1 & 1+\mathrm{i} \end{pmatrix}$

3.6.33. Prove that $\mathbf{v} = \mathbf{x} + \mathrm{i}\,\mathbf{y}$ and $\bar{\mathbf{v}} = \mathbf{x} - \mathrm{i}\,\mathbf{y}$ are linearly independent complex vectors if and only if their real and imaginary parts $\mathbf{x}$ and $\mathbf{y}$ are linearly independent real vectors.

3.6.34. Prove that the space of complex $m \times n$ matrices forms a complex vector space. What is its dimension?

3.6.35. Determine which of the following are subspaces of the vector space consisting of all complex 2×2 matrices.

(a) All matrices with real diagonals.

(b) All matrices for which the sum of the diagonal entries is zero.

(c) All singular complex matrices.

(d) All matrices whose determinant is real.

(e) All matrices of the form $\begin{pmatrix} a & b \\ \bar{a} & \bar{b} \end{pmatrix}$, where $a, b \in \mathbb{C}$.

3.6.36. *True or false*: The set of all complex valued functions $u(x) = v(x) + \mathrm{i}\,w(x)$ with $u(0) = \mathrm{i}$ forms a subspace of the vector space of complex-valued functions.

3.6.37. Let V denote the complex vector space spanned by the functions 1, $e^{\mathrm{i}x}$ and $e^{-\mathrm{i}x}$, where x is a real variable. Which of the following functions belong to V?

(a) $\sin x$ (b) $\cos x - 2\,\mathrm{i}\,\sin x$

(c) $\cosh x$ (d) $\sin^2 \frac{1}{2}x$ (e) $\cos^2 x$

3.6.38. Prove that the following define Hermitian inner products on $\mathbb{C}^2$:

(a) $\langle \mathbf{v}, \mathbf{w} \rangle = v_1\,\overline{w}_1 + 2\,v_2\overline{w}_2$,

(b) $\langle \mathbf{v}, \mathbf{w} \rangle = v_1\,\overline{w}_1 + \mathrm{i}\,v_1\,\overline{w}_2 - \mathrm{i}\,v_2\,\overline{w}_1 + 2\,v_2\overline{w}_2$.

3.6.39. Which of the following define inner products on $\mathbb{C}^2$?

(a) $\langle \mathbf{v}, \mathbf{w} \rangle = v_1\,\overline{w}_1 + 2\,\mathrm{i}\,v_2\overline{w}_2$

(b) $\langle \mathbf{v}, \mathbf{w} \rangle = v_1\,w_1 + 2v_2\,w_2$

(c) $\langle \mathbf{v}, \mathbf{w} \rangle = v_1\,\overline{w}_2 + v_2\overline{w}_1$

(d) $\langle \mathbf{v}, \mathbf{w} \rangle = 2\,v_1\,\overline{w}_1 + v_1\,\overline{w}_2 + v_2\,\overline{w}_1 + 2\,v_2\overline{w}_2$

(e) $\langle \mathbf{v}, \mathbf{w} \rangle = 2\,v_1\,\overline{w}_1 + (1+\mathrm{i})\,v_1\,\overline{w}_2 + (1-\mathrm{i})\,v_2\,\overline{w}_1 + 3\,v_2\,\overline{w}_2$

◇ **3.6.40.** Let $A = A^T$ be a real symmetric $n \times n$ matrix. Show that $(A\,\mathbf{v}) \cdot \mathbf{w} = \mathbf{v} \cdot (A\,\mathbf{w})$ for all $\mathbf{v}, \mathbf{w} \in \mathbb{C}^n$.

3.6.41. Let $\mathbf{z} = \mathbf{x} + \mathrm{i}\,\mathbf{y} \in \mathbb{C}^n$.

(a) Prove that, for the Hermitian dot product, $\|\mathbf{z}\|^2 = \|\mathbf{x}\|^2 + \|\mathbf{y}\|^2$.

(b) Does this formula remain valid under a more general Hermitian inner product on $\mathbb{C}^n$?

◇ **3.6.42.** Let V be a complex inner product space. Prove that, for all $\mathbf{z}, \mathbf{w} \in V$,

(a) $\|\mathbf{z} + \mathbf{w}\|^2 = \|\mathbf{z}\|^2 + 2\,\mathrm{Re}\langle \mathbf{z}, \mathbf{w} \rangle + \|\mathbf{w}\|^2$;

(b) $\langle \mathbf{z}, \mathbf{w} \rangle = \frac{1}{4}\left(\|\mathbf{z} + \mathbf{w}\|^2 - \|\mathbf{z} - \mathbf{w}\|^2 + \mathrm{i}\,\|\mathbf{z} + \mathrm{i}\,\mathbf{w}\|^2 - \mathrm{i}\,\|\mathbf{z} - \mathrm{i}\,\mathbf{w}\|^2 \right)$.

3.6.43. (a) How would you define the angle between two elements of a complex inner product space?

(b) What is the angle between

$$(-1, 2-\mathrm{i}, -1+2\mathrm{i})^T$$

and

$$(-2-\mathrm{i}, -\mathrm{i}, 1-\mathrm{i})^T$$

relative to the Hermitian dot product?

3.6.44. Let $\mathbf{0} \neq \mathbf{v} \in \mathbb{C}^n$. Which scalar multiples $c\,\mathbf{v}$ have the same Hermitian norm as $\mathbf{v}$?

◇ **3.6.45.** Prove the Cauchy–Schwarz inequality (3.95) and the triangle inequality (3.96) for a general complex inner product. *Hint*: Use Exercises 3.6.8, 3.6.42.

3.6.46. (a) Formulate a general definition of a norm on a complex vector space.

(b) How would you define analogs of the 1 , 2 and ∞ norms on $\mathbb{C}^n$?

3.6.47. *Multiple choice*: Let V be a complex normed vector space. How many unit vectors are parallel to a given vector $\mathbf{0} \neq \mathbf{v} \in V$?

(a) none (b) 1 (c) 2

(d) 3 (e) ∞

(f) depends upon the vector;

(g) depends upon the norm.

Explain your answer.

◇ **3.6.48.** The *Hermitian adjoint* of a complex $m \times n$ matrix A is the complex conjugate of its transpose, written $A^\dagger = \overline{A^T} = \bar{A}^T$. For example, if

$$A = \begin{pmatrix} 1+\mathrm{i} & 2\mathrm{i} \\ -3 & 2-5\mathrm{i} \end{pmatrix}$$

then

$$A^\dagger = \begin{pmatrix} 1-\mathrm{i} & -3 \\ -2\mathrm{i} & 2+5\mathrm{i} \end{pmatrix}.$$

Prove

(a) $(A^\dagger)^\dagger = A$,

(b) $(z\,A + w\,B)^\dagger = \bar{z}\,A^\dagger + \overline{w}\,B^\dagger$ for $z, w \in \mathbb{C}$,

(c) $(AB)^\dagger = B^\dagger A^\dagger$.

◇ **3.6.49.** A complex matrix H is called *Hermitian* if it equals its Hermitian adjoint, $H^\dagger = H$, as defined in the preceding exercise.

(a) Prove that the diagonal entries of a Hermitian matrix are real.

(b) Prove that $(H\,\mathbf{z}) \cdot \mathbf{w} = \mathbf{z} \cdot (H\,\mathbf{w})$ for $\mathbf{z}, \mathbf{w} \in \mathbb{C}^n$.

(c) Prove that every Hermitian inner product on $\mathbb{C}^n$ has the form $\langle \mathbf{z}, \mathbf{w} \rangle = \mathbf{z}^T H\,\overline{\mathbf{w}}$ where H is an $n \times n$ positive definite Hermitian matrix.

(d) How would you verify positive definiteness of a complex matrix?

◇ **3.6.50.** Let $\mathbf{v}_1, \ldots, \mathbf{v}_n$ be elements of a complex inner product space. Let K denote the corresponding $n \times n$ *Gram matrix*, defined in the usual manner.

(a) Prove that K is a Hermitian matrix, as defined in Exercise 3.6.49.

(b) Prove that K is positive semi-definite, meaning $\mathbf{z}^T K\,\mathbf{z} \geq 0$ for all $\mathbf{z} \in \mathbb{C}^n$.

(c) Prove that K is positive definite if and only if $\mathbf{v}_1, \ldots, \mathbf{v}_n$ are linearly independent.

3.6.51. For each the following pairs of complex-valued functions,

(i) compute their L^2 norm and Hermitian inner product on the interval $[0, 1]$, and then

(ii) check the validity of the Cauchy–Schwarz and triangle inequalities.

(a) $1,\ e^{\,\mathrm{i}\,\pi x}$ (b) $x + \mathrm{i},\ x - \mathrm{i}$

(c) $\mathrm{i}\,x^2,\ (1 - 2\,\mathrm{i})x + 3\,\mathrm{i}$

3.6.52. Formulate conditions on a weight function $w(x)$ that guarantee that the weighted integral

$$\langle f, g \rangle = \int_a^b f(x)\,\overline{g(x)}\,w(x)\,dx$$

defines an inner product on the space of continuous complex-valued functions on $[a, b]$.

CHAPTER 4

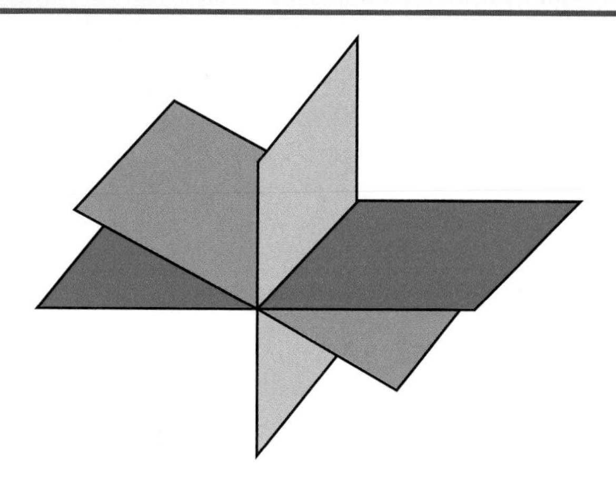

Minimization and Least Squares Approximation

Because Nature strives for efficiency, many systems arising in physical applications are founded on a minimization principle. For example, in a mechanical system, the stable equilibrium configurations minimize the potential energy. In an electrical circuit, the current adjusts itself to minimize the power. In optics and relativity, light rays follow the paths of minimal distance—the geodesics on the curved space-time. In data analysis, the most common way of fitting a function to prescribed data points is to minimize the least squares error, which serves to quantify the overall deviation between the data and the sampled function values. Solutions to most of the boundary value problems arising in applications to continuum mechanics and physics are also characterized by a minimization principle, which is then used to design the finite element numerical solution method. Optimization (minimization and maximization) is ubiquitous in control theory, engineering design and manufacturing, linear programming, econometrics, and most other fields of analysis.

This chapter introduces and solves the most basic mathematical minimization problem: a quadratic function depending on several variables. Assuming the quadratic coefficient matrix is positive definite, the minimizer can be found by solving an associated linear algebraic system. With the solution in hand, we are able to treat a wide range of applications, including, in this chapter, least squares fitting of data, along with interpolation and approximation of functions. Applications to equilibrium mechanics will form the focus of Chapter 6. Applications to the numerical solution of differential equations can be found in Chapter 11. In this text, we are only able to deal with the simplest quadratic minimization problems. Minimization of more complicated functions is of comparable significance, but relies on the nonlinear methods of multivariable calculus, and thus lies outside our scope.

4.1 MINIMIZATION PROBLEMS

Let us begin by introducing three important minimization problems—the first arising in physics, the second in analysis, and the third in geometry.

Equilibrium Mechanics

A fundamental principle of mechanics is that systems in equilibrium minimize potential energy. For example, a ball in a bowl will roll downhill unless it is sitting at the bottom, where its potential energy due to gravity is at a (local) minimum. In the simplest class of examples, the energy is a quadratic function, e.g.,

$$f(x, y) = 3x^2 - 2xy + 4y^2 + x - 2y + 1, \tag{4.1}$$

and one seeks the point $x = x^\star$, $y = y^\star$ (if one exists) at which $f(x^\star, y^\star)$ achieves its overall minimal value.

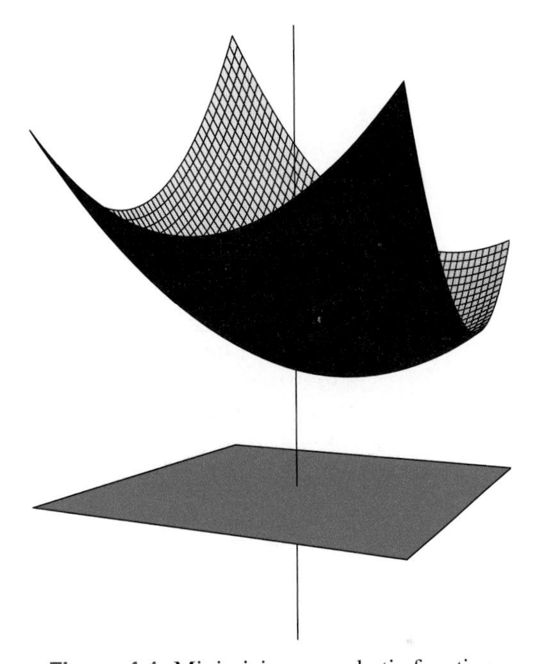

Figure 4.1 Minimizing a quadratic function.

Similarly, a pendulum will swing back and forth unless it is at the bottom of its arc, where potential energy is minimized. Actually, the pendulum has a second equilibrium position at the top of the arc, but this is rarely observed since it is an *unstable* equilibrium, meaning that any tiny movement will knock it off balance. Therefore, a better way of stating the principle is that *stable equilibria* are where the mechanical system (locally) minimizes potential energy. For the ball rolling on a curved surface, the local minima—the bottoms of valleys—are the stable equilibria, while the local maxima — the tops of hills—are unstable. Minimization principles serve to characterize the equilibrium configurations of a wide range of physical systems, including masses and springs, structures, electrical circuits, and even continuum models of solid mechanics and elasticity, fluid mechanics, electromagnetism, thermodynamics, and so on.

Solution of Equations

Suppose we wish to solve a system of equations

$$f_1(\mathbf{x}) = 0, \quad f_2(\mathbf{x}) = 0, \quad \ldots \quad f_m(\mathbf{x}) = 0, \tag{4.2}$$

where $\mathbf{x} = (x_1, \ldots, x_n) \in \mathbb{R}^n$. This system can be converted into a minimization problem in the following seemingly silly manner. Define

$$p(\mathbf{x}) = \left[f_1(\mathbf{x}) \right]^2 + \cdots + \left[f_m(\mathbf{x}) \right]^2. \tag{4.3}$$

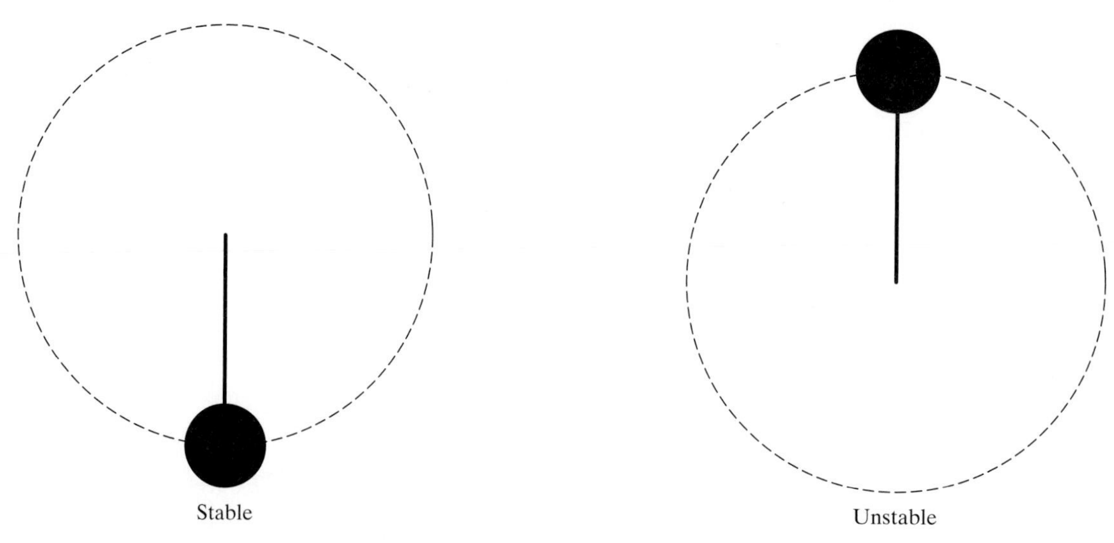

Figure 4.2 Equilibria of a pendulum.

Clearly, $p(\mathbf{x}) \geq 0$ for all $\mathbf{x}$. Moreover, $p(\mathbf{x}^\star) = 0$ if and only if each summand is zero, and hence $\mathbf{x} = \mathbf{x}^\star$ is a solution to (4.2). Therefore, the minimum value of $p(\mathbf{x})$ is zero, and the minimum is achieved if and only if we are at a solution to the original system of equations.

For us, the most important case is when we have a linear system

$$A\mathbf{x} = \mathbf{b} \tag{4.4}$$

consisting of m equations in n unknowns. In this case, the solutions may be obtained by minimizing the function

$$p(\mathbf{x}) = \|A\mathbf{x} - \mathbf{b}\|^2, \tag{4.5}$$

where $\|\cdot\|$ denotes the Euclidean norm on $\mathbb{R}^m$. Clearly $p(\mathbf{x})$ has a minimum value of 0, which is achieved if and only if $\mathbf{x}$ is a solution to the linear system (4.4). Of course, it is not clear that we have gained much, since we already know how to solve $A\mathbf{x} = \mathbf{b}$ by Gaussian Elimination. However, this artifice turns out to have profound consequences.

Suppose that the linear system (4.4) does *not* have a solution, i.e., $\mathbf{b}$ does not lie in the range of the matrix A. This situation is very typical when there are more equations than unknowns. Such problems arise in data fitting, when the measured data points are all supposed to lie on a straight line, say, but rarely do so exactly, due to experimental error. Although we know there is no exact solution to the system, we might still try to find an approximate solution—a vector $\mathbf{x}^\star$ that comes as close to solving the system as possible. One way to measure closeness is by looking at the magnitude of the *residual vector* $\mathbf{r} = A\mathbf{x} - \mathbf{b}$, i.e., the difference between the left and right hand sides of the system. The smaller its norm $\|\mathbf{r}\| = \|A\mathbf{x} - \mathbf{b}\|$, the better the attempted solution. The vector $\mathbf{x}^\star$ that minimizes the squared residual norm function (4.5) is known as the *least squares solution* to the linear system. As before, if the linear system (4.4) happens to have an actual solution, with $A\mathbf{x}^\star = \mathbf{b}$, then $\mathbf{x}^\star$ qualifies as the least squares solution too, since in this case $p(\mathbf{x}^\star) = 0$ achieves its absolute minimum. So least squares solutions include traditional solutions as special cases. While not the only possible approach to this issue, least squares is easiest to analyze and solve, and, hence, is often the method of choice for fitting functions to experimental data and performing statistical analysis.

The Closest Point

The following minimization problem arises in elementary geometry, although its practical implications cut a much wider swath. Given a point $\mathbf{b} \in \mathbb{R}^m$ and a subset $V \subset \mathbb{R}^m$, find the point $\mathbf{v}^\star \in V$ that is closest to $\mathbf{b}$. In other words, we seek to minimize the distance $d(\mathbf{v}, \mathbf{b}) = \|\mathbf{v} - \mathbf{b}\|$ over all possible $\mathbf{v} \in V$.

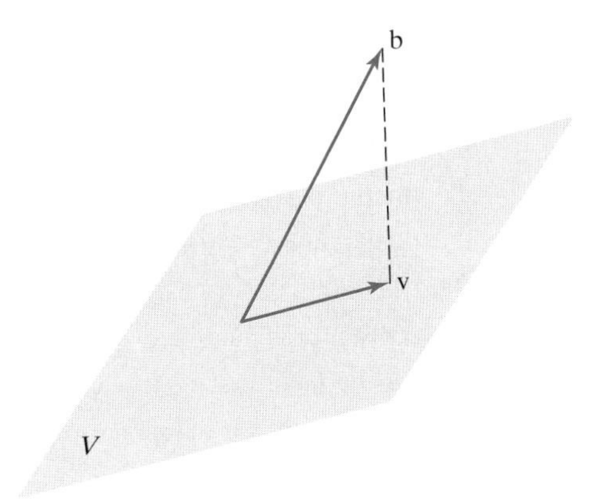

Figure 4.3 The closest point.

The simplest situation occurs when V is a subspace of $\mathbb{R}^m$. In this case, the closest point problem can, in fact, be reformulated as a least squares minimization problem. Let $\mathbf{v}_1, \ldots, \mathbf{v}_n$ be a basis for V. The general element $\mathbf{v} \in V$ is a linear combination of the basis vectors. Applying our handy matrix multiplication formula (2.13), we can write the subspace elements in the form

$$\mathbf{v} = x_1 \mathbf{v}_1 + \cdots + x_n \mathbf{v}_n = A\mathbf{x},$$

where $A = (\, \mathbf{v}_1 \; \mathbf{v}_2 \; \ldots \; \mathbf{v}_n \,)$ is the $m \times n$ matrix formed by the (column) basis vectors and $\mathbf{x} = (x_1, x_2, \ldots, x_n)^T$ are the coordinates of $\mathbf{v}$ relative to the chosen basis. In this manner, we can identify V with the range of A, i.e., the subspace spanned by its columns. Consequently, the closest point in V to $\mathbf{b}$ is found by minimizing

$$\|\mathbf{v} - \mathbf{b}\|^2 = \|A\mathbf{x} - \mathbf{b}\|^2$$

over all possible $\mathbf{x} \in \mathbb{R}^n$. But this is exactly the same as the least squares function (4.5)! Thus, if $\mathbf{x}^\star$ is the least squares solution to the system $A\mathbf{x} = \mathbf{b}$, then $\mathbf{v}^\star = A\mathbf{x}^\star$ is the closest point to $\mathbf{b}$ belonging to $V = \operatorname{rng} A$. In this way, we have established a profound and fertile connection between least squares solutions to linear systems and the geometrical problem of minimizing distances to subspaces.

All three of the preceding minimization problems are solved by the same underlying mathematical construction, which will be described in detail in Sections 4.2 and 4.3.

REMARK: In this book, we will concentrate on minimization problems. Maximizing a function $p(\mathbf{x})$ is the same as minimizing its negative $-\,p(\mathbf{x})$, and so can be easily handled by the same techniques.

EXERCISES 4.1

Note: Unless otherwise indicated, "distance" refers to the Euclidean norm.

4.1.1. Find the least squares solution to the pair of equations $3x = 1,\ 2x = -1$.

4.1.2. Find the minimizer of the function $f(x, y) = (3x - 2y + 1)^2 + (2x + y + 2)^2$.

4.1.3. Find the closest point or points to $\mathbf{b} = (-1, 2)^T$ that lie on

(a) the x-axis (b) the y-axis

(c) the line $y = x$ (d) the line $x + y = 0$

(e) the line $2x + y = 0$

4.1.4. Solve Exercise 4.1.3 when distance is measured in

(a) the ∞-norm (b) the 1-norm

4.1.5. Given $\mathbf{b} \in \mathbb{R}^2$, is the closest point on a line L unique when distance is measured in

(a) the Euclidean norm?

(b) the 1-norm? (c) the ∞-norm?

♡ **4.1.6.** Let $L \subset \mathbb{R}^2$ be a line through the origin, and let $\mathbf{b} \in \mathbb{R}^2$ be any point.

(a) Find a geometrical construction of the closest point $\mathbf{v} \in L$ to $\mathbf{b}$ when distance is measured in the standard Euclidean norm.

(b) Use your construction to prove that there is one and only one closest point.

(c) Show that if $\mathbf{0} \neq \mathbf{a} \in L$, then the distance equals

$$\frac{\sqrt{\|\mathbf{a}\|^2\,\|\mathbf{b}\|^2 - (\mathbf{a}\cdot\mathbf{b})^2}}{\|\mathbf{a}\|} = \frac{|\mathbf{a}\times\mathbf{b}|}{\|\mathbf{a}\|},$$

using the two-dimensional cross product (3.22).

4.1.7. Suppose $\mathbf{a}$ and $\mathbf{b}$ are unit vectors in $\mathbb{R}^2$. Show that the distance from $\mathbf{a}$ to the line through $\mathbf{b}$ is the same as the distance from $\mathbf{b}$ to the line through $\mathbf{a}$. Use a picture to explain why this holds. How is the distance related to the angle between the two vectors?

4.1.8. (a) Prove that the distance from the point $(x_0, y_0)^T$ to the line $ax + by = 0$ is

$$\frac{|a x_0 + b y_0|}{\sqrt{a^2 + b^2}}.$$

(b) What is the minimum distance to the line $ax + by + c = 0$?

♡ **4.1.9.** (a) Generalize Exercise 4.1.8 to find the distance between a point $(x_0, y_0, z_0)^T$ and the plane $ax + by + cz + d = 0$ in $\mathbb{R}^3$.

(b) Use your formula to compute the distance between $(1, 1, 1)^T$ and the plane $3x - 2y + z = 1$.

4.1.10. (a) Explain in detail why the minimizer of $\|\mathbf{v} - \mathbf{b}\|$ coincides with the minimizer of $\|\mathbf{v} - \mathbf{b}\|^2$.

(b) Find all scalar functions $F(x)$ for which the minimizer of $F(\|\mathbf{v} - \mathbf{b}\|)$ is the same as the minimizer of $\|\mathbf{v} - \mathbf{b}\|$.

4.1.11. (a) Explain why the problem of maximizing the distance from a point to a subspace does not have a solution.

(b) Can you formulate a situation where maximizing distance to a point leads to a problem with a solution?

4.2 MINIMIZATION OF QUADRATIC FUNCTIONS

The simplest algebraic equations are linear systems. As such, they must be thoroughly understood before venturing into the far more complicated nonlinear realm. For minimization problems, the starting point is the quadratic function. (Linear functions do not have minima—think of the function $f(x) = \alpha x + \beta$ whose graph is a straight line.) In this section, we shall see how the problem of minimizing a general quadratic function of n variables can be solved by linear algebra techniques.

Let us begin by reviewing the very simplest example—minimizing a scalar quadratic function

$$p(x) = a x^2 + 2 b x + c \tag{4.6}$$

over all possible values of $x \in \mathbb{R}$. If $a > 0$, then the graph of p is a parabola pointing upwards, and so there exists a unique minimum value. If $a < 0$, the parabola points downwards, and there is no minimum (although there is a maximum). If $a = 0$, the graph is a straight line, and there is neither minimum nor maximum—except in the trivial case when $b = 0$ also, and the function $p(x) = c$ is constant, with every

x qualifying as a minimum and a maximum. The three nontrivial possibilities are sketched in Figure 4.4.

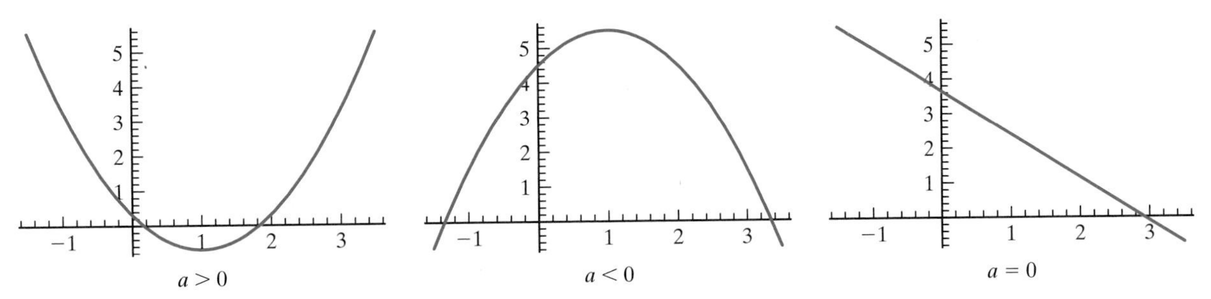

Figure 4.4 Parabolas.

In the case $a > 0$, the minimum can be found by calculus. The *critical points* of a function, which are candidates for minima (and maxima), are found by setting its derivative to zero. In this case, differentiating, and solving

$$p'(x) = 2ax + 2b = 0,$$

we conclude that the only possible minimum value occurs at

$$x^\star = -\frac{b}{a}, \qquad \text{where} \quad p(x^\star) = c - \frac{b^2}{a}. \tag{4.7}$$

Of course, one must check that this critical point is indeed a minimum, and not a maximum or inflection point. The second derivative test will show that $p''(x^\star) = 2a > 0$, and so $x^\star$ is at least a local minimum.

A more instructive approach to this problem—and one that only requires elementary algebra— is to "complete the square". As in (3.58), we rewrite

$$p(x) = a\left(x + \frac{b}{a}\right)^2 + \frac{ac - b^2}{a}. \tag{4.8}$$

If $a > 0$, then the first term is always ≥ 0, and, moreover, attains its minimum value 0 only at $x^\star = -b/a$. The second term is constant, and so is unaffected by the value of x. Thus, the global minimum of $p(x)$ is at $x^\star = -b/a$. Moreover, its minimal value equals the constant term, $p(x^\star) = (ac - b^2)/a$, thereby reconfirming and strengthening the calculus result in (4.7).

Now that we have the one-variable case firmly in hand, let us turn our attention to the more substantial problem of minimizing quadratic functions of several variables. Thus, we seek to minimize a (real) *quadratic function*

$$p(\mathbf{x}) = p(x_1, \ldots, x_n) = \sum_{i,j=1}^{n} k_{ij}\, x_i\, x_j - 2\sum_{i=1}^{n} f_i\, x_i + c, \tag{4.9}$$

depending on n variables $\mathbf{x} = (x_1, x_2, \ldots, x_n)^T \in \mathbb{R}^n$. The coefficients k_{ij}, f_i and c are all assumed to be real. Moreover, we can assume, without loss of generality, that the coefficients of the quadratic terms are symmetric: $k_{ij} = k_{ji}$. (See Exercise 3.4.20 for a justification.) Note that $p(\mathbf{x})$ is more general than a quadratic form (3.45) in that it also contains linear and constant terms. We seek a global minimum, and so the variables $\mathbf{x}$ are allowed to vary over all of $\mathbb{R}^n$. (Minimizing a quadratic function over a proper subset $\mathbf{x} \in S \subsetneq \mathbb{R}^n$ is a far more intricate problem, and will not be discussed here.)

Let us begin by rewriting the quadratic function (4.9) in a more compact matrix notation:

$$p(\mathbf{x}) = \mathbf{x}^T K \mathbf{x} - 2\mathbf{x}^T \mathbf{f} + c, \tag{4.10}$$

in which $K = (k_{ij})$ is a symmetric $n \times n$ matrix, $\mathbf{f}$ is a constant vector, and c is a constant scalar. We first note that in the simple scalar case (4.6), we needed to impose the condition that the quadratic coefficient a is *positive* in order to obtain a (unique) minimum. The corresponding condition for the multivariable case is that the quadratic coefficient matrix K be *positive definite*. This key assumption enables us to establish a general minimization criterion.

THEOREM 4.1 If K is a symmetric, positive definite matrix, then the quadratic function (4.10) has a unique minimizer, which is the solution to the linear system

$$K\mathbf{x} = \mathbf{f}, \qquad \text{namely} \quad \mathbf{x}^\star = K^{-1}\mathbf{f}. \tag{4.11}$$

The minimum value of $p(\mathbf{x})$ is equal to any of the following expressions:

$$p(\mathbf{x}^\star) = p(K^{-1}\mathbf{f}) = c - \mathbf{f}^T K^{-1}\mathbf{f} = c - \mathbf{f}^T \mathbf{x}^\star = c - (\mathbf{x}^\star)^T K \mathbf{x}^\star. \tag{4.12}$$

Proof First recall that, by Proposition 3.25, positive definiteness implies that K is a nonsingular matrix, and hence the linear system (4.11) does have a unique solution $\mathbf{x}^\star = K^{-1}\mathbf{f}$. Then, for any $\mathbf{x} \in \mathbb{R}^n$, we can write

$$\begin{aligned} p(\mathbf{x}) &= \mathbf{x}^T K \mathbf{x} - 2\mathbf{x}^T \mathbf{f} + c = \mathbf{x}^T K \mathbf{x} - 2\mathbf{x}^T K \mathbf{x}^\star + c \\ &= (\mathbf{x} - \mathbf{x}^\star)^T K (\mathbf{x} - \mathbf{x}^\star) + \left[c - (\mathbf{x}^\star)^T K \mathbf{x}^\star \right], \end{aligned} \tag{4.13}$$

where we used the symmetry of $K = K^T$ to identify $\mathbf{x}^T K \mathbf{x}^\star = (\mathbf{x}^\star)^T K \mathbf{x}$. The first term in the final expression has the form $\mathbf{y}^T K \mathbf{y}$, where $\mathbf{y} = \mathbf{x} - \mathbf{x}^\star$. Since we assumed that K is positive definite, we know that $\mathbf{y}^T K \mathbf{y} > 0$ for all $\mathbf{y} \neq \mathbf{0}$. Thus, the first term achieves its minimum value, namely 0, if and only if $\mathbf{0} = \mathbf{y} = \mathbf{x} - \mathbf{x}^\star$. Moreover, since $\mathbf{x}^\star$ is fixed, the second term does not depend on $\mathbf{x}$. Therefore, the minimum of $p(\mathbf{x})$ occurs at $\mathbf{x} = \mathbf{x}^\star$ and its minimum value $p(\mathbf{x}^\star)$ is equal to the constant term. The alternative expressions in (4.12) follow from simple substitutions. ■

EXAMPLE 4.2 Consider the problem of minimizing the quadratic function

$$p(x_1, x_2) = 4x_1^2 - 2x_1 x_2 + 3x_2^2 + 3x_1 - 2x_2 + 1$$

over all (real) x_1, x_2. We first write the function in our matrix form (4.10), so

$$p(x_1, x_2) = (\, x_1 \;\; x_2 \,) \begin{pmatrix} 4 & -1 \\ -1 & 3 \end{pmatrix} \begin{pmatrix} x_1 \\ x_2 \end{pmatrix} - 2\,(\, x_1 \;\; x_2 \,) \begin{pmatrix} -\frac{3}{2} \\ 1 \end{pmatrix} + 1,$$

whereby

$$K = \begin{pmatrix} 4 & -1 \\ -1 & 3 \end{pmatrix}, \qquad \mathbf{f} = \begin{pmatrix} -\frac{3}{2} \\ 1 \end{pmatrix}. \tag{4.14}$$

(Pay attention to the overall factor of -2 in front of the linear terms.) According to Theorem 4.1, to find the minimum we must solve the linear system

$$\begin{pmatrix} 4 & -1 \\ -1 & 3 \end{pmatrix} \begin{pmatrix} x_1 \\ x_2 \end{pmatrix} = \begin{pmatrix} -\frac{3}{2} \\ 1 \end{pmatrix}. \tag{4.15}$$

Applying the usual Gaussian Elimination algorithm, only one row operation is required to place the coefficient matrix in upper triangular form:

$$\left(\begin{array}{cc|c} 4 & -1 & -\frac{3}{2} \\ -1 & 3 & 1 \end{array}\right) \longmapsto \left(\begin{array}{cc|c} 4 & -1 & -\frac{3}{2} \\ 0 & \frac{11}{4} & \frac{5}{8} \end{array}\right).$$

The coefficient matrix is regular as no row interchanges were required, and its two pivots, namely $4, \frac{11}{4}$, are both positive. Thus, by Theorem 3.37, $K > 0$ and hence $p(x_1, x_2)$ really does have a minimum, obtained by applying Back Substitution to the reduced system:

$$\mathbf{x}^\star = \begin{pmatrix} x_1^\star \\ x_2^\star \end{pmatrix} = \begin{pmatrix} -\frac{7}{22} \\ \frac{5}{22} \end{pmatrix} \approx \begin{pmatrix} -.31818 \\ .22727 \end{pmatrix}. \tag{4.16}$$

The quickest way to compute the minimal value

$$p(\mathbf{x}^\star) = p\left(-\tfrac{7}{22}, \tfrac{5}{22}\right) = \tfrac{13}{44} \approx .29546$$

is to use the second formula in (4.12).

It is instructive to compare the algebraic solution method with the minimization procedure you learned in multi-variable calculus, [2, 58]. The *critical points* of $p(x_1, x_2)$ are found by setting both partial derivatives equal to zero:

$$\frac{\partial p}{\partial x_1} = 8x_1 - 2x_2 + 3 = 0, \qquad \frac{\partial p}{\partial x_2} = -2x_1 + 6x_2 - 2 = 0.$$

If we divide by an overall factor of 2, these are precisely the *same* linear equations we already constructed in (4.15). Thus, not surprisingly, the calculus approach leads to the same minimizer (4.16). To check whether $\mathbf{x}^\star$ is a (local) minimum, we need to apply the second derivative test. In the case of a function of several variables, this requires analyzing the *Hessian matrix*, which is the symmetric matrix of second order partial derivatives

$$H = \begin{pmatrix} \dfrac{\partial^2 p}{\partial x_1^2} & \dfrac{\partial^2 p}{\partial x_1 \partial x_2} \\ \dfrac{\partial^2 p}{\partial x_1 \partial x_2} & \dfrac{\partial^2 p}{\partial x_2^2} \end{pmatrix} = \begin{pmatrix} 8 & -2 \\ -2 & 6 \end{pmatrix} = 2K,$$

which is exactly twice the quadratic coefficient matrix (4.14). If the Hessian matrix is positive definite—which we already know in this case—then the critical point is indeed a (local) minimum.

Thus, the calculus and algebraic approaches to this minimization problem lead, as they must, to identical results. However, the algebraic method is *more* powerful, because it immediately produces the *unique*, *global* minimum, whereas, barring additional work, calculus can only guarantee that the critical point is a local minimum. Moreover, the proof of the calculus local minimization criterion—that the Hessian matrix be positive definite at the critical point—relies, in fact, on the algebraic solution to the quadratic minimization problem! In summary: minimization of quadratic functions is a problem in linear algebra, while minimizing more complicated functions requires the full force of multivariable calculus. ●

The most efficient method for producing a minimum of a quadratic function $p(\mathbf{x})$ on $\mathbb{R}^n$, then, is to first write out the symmetric coefficient matrix K and the vector $\mathbf{f}$. Solving the system $K\mathbf{x} = \mathbf{f}$ will produce the minimizer $\mathbf{x}^\star$ *provided* $K > 0$ — which should be checked during the course of the procedure using the criteria of

Theorem 3.37, that is, making sure that no row interchanges are used and all the pivots are positive.

EXAMPLE 4.3 Let us minimize the quadratic function

$$p(x, y, z) = x^2 + 2xy + xz + 2y^2 + yz + 2z^2 + 6y - 7z + 5.$$

This has the matrix form (4.10) with

$$K = \begin{pmatrix} 1 & 1 & \frac{1}{2} \\ 1 & 2 & \frac{1}{2} \\ \frac{1}{2} & \frac{1}{2} & 2 \end{pmatrix}, \qquad \mathbf{x} = \begin{pmatrix} x \\ y \\ z \end{pmatrix}, \qquad \mathbf{f} = \begin{pmatrix} 0 \\ -3 \\ \frac{7}{2} \end{pmatrix}, \qquad c = 5.$$

Gaussian Elimination produces the $L\,D\,L^T$ factorization

$$K = \begin{pmatrix} 1 & 1 & \frac{1}{2} \\ 1 & 2 & \frac{1}{2} \\ \frac{1}{2} & \frac{1}{2} & 2 \end{pmatrix} = \begin{pmatrix} 1 & 0 & 0 \\ 1 & 1 & 0 \\ \frac{1}{2} & 0 & 1 \end{pmatrix} \begin{pmatrix} 1 & 0 & 0 \\ 0 & 1 & 0 \\ 0 & 0 & \frac{7}{4} \end{pmatrix} \begin{pmatrix} 1 & 1 & \frac{1}{2} \\ 0 & 1 & 0 \\ 0 & 0 & 1 \end{pmatrix}.$$

The pivots, i.e., the diagonal entries of D, are all positive, and hence K is positive definite. Theorem 4.1 then guarantees that $p(x, y, z)$ has a unique minimizer, which is found by solving the linear system $K\mathbf{x} = \mathbf{f}$. The solution is then quickly obtained by forward and back substitution:

$$x^\star = 2, \ \mathbf{y}^\star = -3, \ \mathbf{z}^\star = 2, \qquad \text{with} \quad p(x^\star, y^\star, z^\star) = p(2, -3, 2) = -11. \quad \bullet$$

Theorem 4.1 solves the general quadratic minimization problem when the quadratic coefficient matrix is positive definite. If K is not positive definite, then the quadratic function (4.10) does not have a minimum, apart from one exceptional situation.

THEOREM 4.4 If K is positive definite, then the quadratic function $p(\mathbf{x}) = \mathbf{x}^T K\mathbf{x} - 2\mathbf{x}^T\mathbf{f} + c$ has a unique global minimizer $\mathbf{x}^\star$ satisfying $K\mathbf{x}^\star = \mathbf{f}$. If K is only positive semi-definite, and $\mathbf{f} \in \operatorname{rng} K$, then every solution to the linear system $K\mathbf{x}^\star = \mathbf{f}$ is a global minimum of $p(\mathbf{x})$, but the minimum is not unique since $p(\mathbf{x}^\star + \mathbf{z}) = p(\mathbf{x}^\star)$ for any null vector $\mathbf{z} \in \ker K$. In all other cases, $p(\mathbf{x})$ has no global minimum.

Proof The first part is merely a restatement of Theorem 4.1. The second part is proved by a similar computation, and is left to the reader. If K is not positive semi-definite, then one can find a vector $\mathbf{y}$ such that $a = \mathbf{y}^T K\mathbf{y} < 0$. If we set $\mathbf{x} = t\mathbf{y}$, then $p(\mathbf{x}) = p(t\mathbf{y}) = at^2 + 2bt + c$, with $b = \mathbf{y}^T\mathbf{f}$. Since $a < 0$, by choosing $|t| \gg 0$ sufficiently large, we can arrange that $p(t\mathbf{y}) \ll 0$ is an arbitrarily large negative quantity, and so p has no (finite) minimum value. The one remaining case—when K is positive semi-definite, but $\mathbf{f} \notin \operatorname{rng} K$—is deferred until Exercise 5.6.29. ■

EXERCISES 4.2

4.2.1. Find the minimum value of the function $f(x, y, z) = x^2 + 2xy + 3y^2 + 2yz + z^2 - 2x + 3z + 2$. How do you know that your answer is really the global minimum?

4.2.2. For the potential energy function in (4.1), where is the equilibrium position of the ball?

4.2.3. For each of the following quadratic functions, determine if there is a minimum. If so, find the minimizer and the minimum value for the function.

(a) $x^2 - 2xy + 4y^2 + x - 1$

(b) $3x^2 + 3xy + 3y^2 - 2x - 2y + 4$

(c) $x^2 + 5xy + 3y^2 + 2x - y$

(d) $x^2 + y^2 + yz + z^2 + x + y - z$

(e) $x^2 + xy - y^2 - yz + z^2 - 3$

(f) $x^2 + 5xz + y^2 - 2yz + z^2 + 2x - z - 3$

(g) $x^2 + xy + y^2 + yz + z^2 + zw + w^2 - 2x - w$

4.2.4. (a) For which numbers b (allowing both positive and negative numbers) is the matrix

$$A = \begin{pmatrix} 1 & b \\ b & 4 \end{pmatrix}$$

positive definite?

(b) Find the factorization $A = L D L^T$ when b is in the range for positive definiteness.

(c) Find the minimum value (depending on b; it might be finite or it might be $-\infty$) of the function $p(x, y) = x^2 + 2bxy + 4y^2 - 2y$.

4.2.5. For each matrix K, vector $\mathbf{f}$, and scalar c, write out the quadratic function $p(\mathbf{x})$ given by (4.10). Then either find the minimizer $\mathbf{x}^\star$ and minimum value $p(\mathbf{x}^\star)$, or explain why there is none.

(a) $K = \begin{pmatrix} 4 & -12 \\ -12 & 45 \end{pmatrix}$, $\mathbf{f} = \begin{pmatrix} -\frac{1}{2} \\ 2 \end{pmatrix}$, $c = 3$

(b) $K = \begin{pmatrix} 3 & 2 \\ 2 & 1 \end{pmatrix}$, $\mathbf{f} = \begin{pmatrix} 4 \\ 1 \end{pmatrix}$, $c = 0$

(c) $K = \begin{pmatrix} 3 & -1 & 1 \\ -1 & 2 & -1 \\ 1 & -1 & 3 \end{pmatrix}$, $\mathbf{f} = \begin{pmatrix} 1 \\ 0 \\ -2 \end{pmatrix}$, $c = -3$

(d) $K = \begin{pmatrix} 1 & 1 & 1 \\ 1 & 2 & -1 \\ 1 & -1 & 1 \end{pmatrix}$, $\mathbf{f} = \begin{pmatrix} -3 \\ -1 \\ 2 \end{pmatrix}$, $c = 1$

(e) $K = \begin{pmatrix} 1 & 1 & 0 & 0 \\ 1 & 2 & 1 & 0 \\ 0 & 1 & 3 & 1 \\ 0 & 0 & 1 & 4 \end{pmatrix}$, $\mathbf{f} = \begin{pmatrix} -1 \\ 2 \\ -3 \\ 4 \end{pmatrix}$, $c = 0$

4.2.6. Find the minimum value of the quadratic function

$$p(x_1, \ldots, x_n) = 4 \sum_{i=1}^{n} x_i^2 - 2 \sum_{i=1}^{n-1} x_i x_{i+1} + \sum_{i=1}^{n} x_i$$

for $n = 2$,3 and 4

4.2.7. Find the maximum value of the quadratic functions

(a) $-x^2 + 3xy - 5y^2 - x + 1$,

(b) $-2x^2 + 6xy - 3y^2 + 4x - 3y$.

4.2.8. Suppose K_1 and K_2 are positive definite $n \times n$ matrices. Suppose that, for $i = 1, 2$, the minimizer of $p_i(\mathbf{x}) = \mathbf{x}^T K_i \mathbf{x} - 2\mathbf{x}^T \mathbf{f}_i + c_i$, is $\mathbf{x}_i^\star$. Is the minimizer of $p(\mathbf{x}) = p_1(\mathbf{x}) + p_2(\mathbf{x})$ given by $\mathbf{x}^\star = \mathbf{x}_1^\star + \mathbf{x}_2^\star$? Prove or give a counterexample.

◇ **4.2.9.** Let $K > 0$. Prove that a quadratic function $p(\mathbf{x}) = \mathbf{x}^T K \mathbf{x} - 2\mathbf{x}^T \mathbf{f}$ without constant term has non-positive minimum value: $p(\mathbf{x}^\star) \le 0$. When is the minimum value equal to zero?

4.2.10. Let $q(\mathbf{x}) = \mathbf{x}^T A \mathbf{x}$ be a quadratic form. Prove that the minimum value of $q(\mathbf{x})$ is either 0 or $-\infty$.

4.2.11. Under what conditions does the affine function $p(\mathbf{x}) = \mathbf{x}^T \mathbf{f} + c$ have a minimum?

◇ **4.2.12.** Under what conditions does a quadratic function $p(\mathbf{x}) = \mathbf{x}^T K \mathbf{x} - 2\mathbf{x}^T \mathbf{f} + c$ have a finite global maximum? Explain how to find the maximizer and maximum value.

4.2.13. Why can't you minimize a complex-valued quadratic function?

4.3 LEAST SQUARES AND THE CLOSEST POINT

We are now in a position to solve our geometric problem of finding the element in a subspace that is closest to a given point.

PROBLEM Let V be a subspace of $\mathbb{R}^m$. Given $\mathbf{b} \in \mathbb{R}^m$, find $\mathbf{v}^\star \in V$ which minimizes $\|\mathbf{v} - \mathbf{b}\|$ over all possible $\mathbf{v} \in V$. The minimal distance $d^\star = \|\mathbf{v}^\star - \mathbf{b}\|$ to the closest point is called the *distance* from the point $\mathbf{b}$ to the subspace V.

Of course, if $\mathbf{b} \in V$ lies in the subspace, then the answer is easy: the closest point is $\mathbf{v}^\star = \mathbf{b}$ itself, and the distance from $\mathbf{b}$ to the subspace is zero. Thus, the problem only becomes interesting when $\mathbf{b} \notin V$.

Initially, you may assume that $\|\cdot\|$ denotes the usual Euclidean norm, and so we are dealing with the Euclidean distance between points. But it will be no more difficult to solve the closest point problem for *any* norm that arises from an inner product.

Warning: The method does *not* apply to more general norms not coming from inner products, such as the 1 norm or ∞ norm. In such cases, finding the closest point problem is a *nonlinear* minimization problem whose solution requires more sophisticated analytical techniques, [51, 59].

When solving the closest point problem, the goal is to minimize the squared distance

$$\|\mathbf{v}-\mathbf{b}\|^2 = \langle \mathbf{v}-\mathbf{b}, \mathbf{v}-\mathbf{b}\rangle = \|\mathbf{v}\|^2 - 2\,\langle \mathbf{v}, \mathbf{b}\rangle + \|\mathbf{b}\|^2 \tag{4.17}$$

over all possible $\mathbf{v}$ belonging to the subspace $V \subset \mathbb{R}^m$. Let us assume that we know a basis $\mathbf{v}_1, \ldots, \mathbf{v}_n$ of V, with $n = \dim V$. Then the most general vector in V is a linear combination

$$\mathbf{v} = x_1\,\mathbf{v}_1 + \cdots + x_n\,\mathbf{v}_n \tag{4.18}$$

of the basis vectors. We substitute the formula (4.18) for $\mathbf{v}$ into the squared distance function (4.17). As we shall see, the resulting expression is a quadratic function of the coefficients $\mathbf{x} = (x_1, x_2, \ldots, x_n)^T$, and so the minimum is provided by Theorem 4.1.

First, the quadratic terms come from expanding

$$\|\mathbf{v}\|^2 = \langle x_1\,\mathbf{v}_1 + \cdots + x_n\,\mathbf{v}_n, x_1\,\mathbf{v}_1 + \cdots + x_n\,\mathbf{v}_n\rangle = \sum_{i,j=1}^{n} x_i\,x_j\,\langle \mathbf{v}_i, \mathbf{v}_j\rangle. \tag{4.19}$$

Therefore,

$$\|\mathbf{v}\|^2 = \sum_{i,j=1}^{n} k_{ij}\,x_i\,x_j = \mathbf{x}^T K \mathbf{x},$$

where K is the symmetric $n \times n$ Gram matrix whose (i, j) entry is the inner product

$$k_{ij} = \langle \mathbf{v}_i, \mathbf{v}_j\rangle \tag{4.20}$$

between the basis vectors of our subspace. Similarly,

$$\langle \mathbf{v}, \mathbf{b}\rangle = \langle x_1\,\mathbf{v}_1 + \cdots + x_n\,\mathbf{v}_n, \mathbf{b}\rangle = \sum_{i=1}^{n} x_i\,\langle \mathbf{v}_i, \mathbf{b}\rangle,$$

and so

$$\langle \mathbf{v}, \mathbf{b}\rangle = \sum_{i=1}^{n} x_i\,f_i = \mathbf{x}^T \mathbf{f},$$

where $\mathbf{f} \in \mathbb{R}^n$ is the vector whose ith entry is the inner product

$$f_i = \langle \mathbf{v}_i, \mathbf{b}\rangle \tag{4.21}$$

between the point and the subspace's basis elements. Substituting back, we conclude that the squared distance function (4.17) reduces to the quadratic function

$$p(\mathbf{x}) = \mathbf{x}^T K \mathbf{x} - 2\mathbf{x}^T \mathbf{f} + c = \sum_{i,j=1}^{n} k_{ij}\,x_i\,x_j - 2\sum_{i=1}^{n} f_i\,x_i + c, \tag{4.22}$$

in which K and $\mathbf{f}$ are given in (4.20–21), while $c = \|\mathbf{b}\|^2$.

Since we assumed that the basis vectors $\mathbf{v}_1, \ldots, \mathbf{v}_n$ are linearly independent, Proposition 3.30 assures us that the associated Gram matrix $K = A^T A$ is positive definite. Therefore, we may directly apply our basic minimization Theorem 4.1 to solve the closest point problem.

THEOREM 4.5 Let $\mathbf{v}_1, \ldots, \mathbf{v}_n$ form a basis for the subspace $V \subset \mathbb{R}^m$. Given $\mathbf{b} \in \mathbb{R}^m$, the closest point $\mathbf{v}^\star = x_1^\star \mathbf{v}_1 + \cdots + x_n^\star \mathbf{v}_n \in V$ is prescribed by the solution $\mathbf{x}^\star = K^{-1}\mathbf{f}$ to the linear system

$$K\,\mathbf{x} = \mathbf{f}, \tag{4.23}$$

where K and $\mathbf{f}$ are given in (4.20–21). The distance between the point and the subspace is

$$d^\star = \|\mathbf{v}^\star - \mathbf{b}\| = \sqrt{\|\mathbf{b}\|^2 - \mathbf{f}^T\mathbf{x}^\star}\,. \tag{4.24}$$

When using the standard Euclidean inner product and norm on $\mathbb{R}^n$ to measure distance, the entries of the Gram matrix K and the vector $\mathbf{f}$ are given by dot products:

$$k_{ij} = \mathbf{v}_i \cdot \mathbf{v}_j = \mathbf{v}_i^T\mathbf{v}_j, \qquad f_i = \mathbf{v}_i \cdot \mathbf{b} = \mathbf{v}_i^T\mathbf{b}.$$

As in (3.54), both sets of equations can be combined into a single matrix equation. If $A = (\,\mathbf{v}_1\ \mathbf{v}_2\ \ldots\ \mathbf{v}_n\,)$ denotes the $m \times n$ matrix formed by the basis vectors, then

$$K = A^T A, \qquad \mathbf{f} = A^T\mathbf{b}, \qquad c = \|\mathbf{b}\|^2. \tag{4.25}$$

A direct derivation of these equations is instructive. Since, by formula (2.13),

$$\mathbf{v} = x_1\,\mathbf{v}_1 + \cdots + x_n\,\mathbf{v}_n = A\mathbf{x},$$

we find

$$\begin{aligned}\|\mathbf{v} - \mathbf{b}\|^2 &= \|A\mathbf{x} - \mathbf{b}\|^2 = (A\mathbf{x} - \mathbf{b})^T(A\mathbf{x} - \mathbf{b}) = (\mathbf{x}^T A^T - \mathbf{b}^T)(A\mathbf{x} - \mathbf{b})\\ &= \mathbf{x}^T A^T A\mathbf{x} - 2\,\mathbf{x}^T A^T\mathbf{b} + \mathbf{b}^T\mathbf{b} = \mathbf{x}^T K\mathbf{x} - 2\mathbf{x}^T\mathbf{f} + c,\end{aligned}$$

thereby justifying (4.25). (In the next to last equality, we equate the scalar quantities $\mathbf{b}^T A\mathbf{x} = (\mathbf{b}^T A\mathbf{x})^T = \mathbf{x}^T A^T\mathbf{b}$.)

If, instead of the Euclidean inner product, we adopt an alternative inner product $\langle \mathbf{v}, \mathbf{w} \rangle = \mathbf{v}^T C\,\mathbf{w}$ prescribed by a positive definite matrix $C > 0$, then the same computations produce

$$K = A^T C\,A, \qquad \mathbf{f} = A^T C\,\mathbf{b}, \qquad c = \|\mathbf{b}\|^2. \tag{4.26}$$

The weighted Gram matrix formula was previously derived in (3.56).

EXAMPLE 4.6 Let $V \subset \mathbb{R}^3$ be the plane spanned by

$$\mathbf{v}_1 = \begin{pmatrix} 1 \\ 2 \\ -1 \end{pmatrix}, \qquad \mathbf{v}_2 = \begin{pmatrix} 2 \\ -3 \\ -1 \end{pmatrix}.$$

Our goal is to find the point $\mathbf{v}^\star \in V$ closest to $\mathbf{b} = \begin{pmatrix} 1 \\ 0 \\ 0 \end{pmatrix}$, where distance is measured in the usual Euclidean norm. We combine the basis vectors to form the matrix

$$A = \begin{pmatrix} 1 & 2 \\ 2 & -3 \\ -1 & -1 \end{pmatrix}.$$

According to (4.25), the positive definite Gram matrix and associated vector are

$$K = A^T A = \begin{pmatrix} 6 & -3 \\ -3 & 14 \end{pmatrix}, \qquad \mathbf{f} = A^T\mathbf{b} = \begin{pmatrix} 1 \\ 2 \end{pmatrix}.$$

(Or, alternatively, these can be computed directly by taking inner products, as in (4.20–21).) We solve the linear system $K\mathbf{x} = \mathbf{f}$ for $\mathbf{x}^\star = K^{-1}\mathbf{f} = \left(\frac{4}{15}, \frac{1}{5}\right)^T$. Theorem 4.5 implies that the closest point is

$$\mathbf{v}^\star = A\mathbf{x}^\star = x_1^\star \mathbf{v}_1 + x_2^\star \mathbf{v}_2 = \begin{pmatrix} \frac{2}{3} \\ -\frac{1}{15} \\ -\frac{7}{15} \end{pmatrix} \approx \begin{pmatrix} .6667 \\ -.0667 \\ -.4667 \end{pmatrix}.$$

The distance from the point $\mathbf{b}$ to the plane is $d^\star = \|\mathbf{v}^\star - \mathbf{b}\| = \frac{1}{\sqrt{3}} \approx .5774$.

Suppose, on the other hand, that distance is measured in the weighted norm $\|\mathbf{v}\| = v_1^2 + \frac{1}{2}v_2^2 + \frac{1}{3}v_3^2$ corresponding to the positive definite diagonal matrix $C = \operatorname{diag}\left(1, \frac{1}{2}, \frac{1}{3}\right)$. In this case, we form the weighted Gram matrix and vector (4.26):

$$K = A^T C A = \begin{pmatrix} 1 & 2 & -1 \\ 2 & -3 & -1 \end{pmatrix} \begin{pmatrix} 1 & 0 & 0 \\ 0 & \frac{1}{2} & 0 \\ 0 & 0 & \frac{1}{3} \end{pmatrix} \begin{pmatrix} 1 & 2 \\ 2 & -3 \\ -1 & -1 \end{pmatrix} = \begin{pmatrix} \frac{10}{3} & -\frac{2}{3} \\ -\frac{2}{3} & \frac{53}{6} \end{pmatrix},$$

$$\mathbf{f} = A^T C \mathbf{b} = \begin{pmatrix} 1 & 2 & -1 \\ 2 & -3 & -1 \end{pmatrix} \begin{pmatrix} 1 & 0 & 0 \\ 0 & \frac{1}{2} & 0 \\ 0 & 0 & \frac{1}{3} \end{pmatrix} \begin{pmatrix} 1 \\ 0 \\ 0 \end{pmatrix} = \begin{pmatrix} 1 \\ 2 \end{pmatrix},$$

and so

$$\mathbf{x}^\star = K^{-1}\mathbf{f} \approx \begin{pmatrix} .3506 \\ .2529 \end{pmatrix}, \qquad \mathbf{v}^\star = A\mathbf{x}^\star \approx \begin{pmatrix} .8563 \\ -.0575 \\ -.6034 \end{pmatrix}.$$

Hence, the distance between the point and the subspace is measured in the weighted norm: $d^\star = \|\mathbf{v}^\star - \mathbf{b}\| \approx .3790$. ●

REMARK: The solution to the closest point problem given in Theorem 4.5 applies, as stated, to the more general case when $V \subset W$ is a finite-dimensional subspace of a general inner product space W. The underlying inner product space W can even be infinite-dimensional, which it is when dealing with least squares approximations in function space, to be described at the end of this chapter.

EXERCISES 4.3

Note: Unless otherwise indicated, "distance" refers to the Euclidean norm.

4.3.1. Find the closest point on the plane spanned by $(1, 2, -1)^T$, $(0, -1, 3)^T$ to the point $(1, 1, 1)^T$. What is the distance between the point and the plane?

4.3.2. Redo Exercise 4.3.1 using

(a) the weighted inner product $\langle \mathbf{v}, \mathbf{w} \rangle = 2v_1w_1 + 4v_2w_2 + 3v_3w_3$;

(b) the inner product $\langle \mathbf{v}, \mathbf{w} \rangle = \mathbf{v}^T C\mathbf{w}$ based on the positive definite matrix

$$C = \begin{pmatrix} 2 & -1 & 0 \\ -1 & 2 & -1 \\ 0 & -1 & 2 \end{pmatrix}.$$

4.3.3. Find the point on the plane $x + 2y - z = 0$ that is closest to $(0, 0, 1)^T$.

4.3.4. Let $\mathbf{b} = (3, 1, 2, 1)^T$. Find the closest point and the distance from $\mathbf{b}$ to the following subspaces:

(a) the line in the direction $(1, 1, 1, 1)^T$,

(b) the plane spanned by $(1,1,0,0)^T$ and $(0,0,1,1)^T$,

(c) the hyperplane spanned by $(1,0,0,0)^T$, $(0,1,0,0)^T$, $(0,0,1,0)^T$,

(d) the hyperplane defined by the equation $x+y+z+w=0$.

4.3.5. Find the closest point and the distance from $\mathbf{b}=(1,1,2,-2)^T$ to the subspace spanned by $(1,2,-1,0)^T$, $(0,1,-2,-1)^T$, $(1,0,3,2)^T$.

4.3.6. Redo Exercises 4.3.4, 4.3.5 using

(i) the weighted inner product $\langle \mathbf{v},\mathbf{w}\rangle = \frac{1}{2}v_1w_1 + v_2w_2 + \frac{1}{2}v_3w_3 + v_4w_4$

(ii) the inner product based on the positive definite matrix

$$C = \begin{pmatrix} 4 & -1 & 1 & 0 \\ -1 & 4 & -1 & 1 \\ 1 & -1 & 4 & -1 \\ 0 & 1 & -1 & 4 \end{pmatrix}$$

4.3.7. Find the vector $\mathbf{v}^\star \in \operatorname{span}\{(0,0,1,1),(2,1,1,1)\}$ that minimizes $\|\mathbf{v}-(0,3,1,2)\|$.

4.3.8. (a) Find the distance from the point $\mathbf{b}=(1,2,-1)^T$ to the plane $x-2y+z=0$.

(b) Find the distance to the plane $x-2y+z=3$. *Hint*: Move the point and the plane so that the plane goes through the origin.

♡ **4.3.9.** Let A be an $m\times n$ matrix with rank $A=n$.

(a) Prove that the matrix $P = A(A^TA)^{-1}A^T$ is a *projection matrix*, meaning that $P^2=P$, cf. Exercise 2.5.8.

(b) Construct the projection matrix corresponding to

(i) $A=\begin{pmatrix} 1 \\ -1 \end{pmatrix}$ (ii) $A=\begin{pmatrix} 1 & 2 \\ -1 & 1 \end{pmatrix}$

(iii) $A=\begin{pmatrix} 1 \\ 2 \\ -1 \end{pmatrix}$ (iv) $A=\begin{pmatrix} 1 & 0 \\ -1 & 2 \\ 0 & -1 \end{pmatrix}$

(c) Prove that P is symmetric.

(d) Prove that $\operatorname{rng} P = \operatorname{rng} A$.

(e) Show that $\mathbf{v}^\star = P\mathbf{b}$ is the closest point on the subspace $\operatorname{rng} A = \operatorname{rng} P$ to $\mathbf{b}$.

(f) Show that if A is nonsingular, then $P = \mathrm{I}$. How do you interpret this in light of part (e)?

◇ **4.3.10.** (a) Given a configuration of n points $\mathbf{a}_1,\ldots,\mathbf{a}_n$ in the plane, explain how to find the point $\mathbf{x}\in\mathbb{R}^2$ that minimizes the total squared distance $\sum_{i=1}^{n}\|\mathbf{x}-\mathbf{a}_i\|^2$.

(b) Apply your method when

(i) $\mathbf{a}_1=(1,3)$, $\mathbf{a}_2=(-2,5)$;

(ii) $\mathbf{a}_1=(0,0)$, $\mathbf{a}_2=(0,1)$, $\mathbf{a}_3=(1,0)$;

(iii) $\mathbf{a}_1=(0,0)$, $\mathbf{a}_2=(0,2)$, $\mathbf{a}_3=(1,2)$, $\mathbf{a}_4=(-2,-1)$.

4.3.11. Answer Exercise 4.3.10 when distance is measured in

(a) the weighted norm $\|\mathbf{x}\| = \sqrt{2x_1^2+3x_2^2}$;

(b) the norm based on the positive definite matrix $\begin{pmatrix} 3 & -1 \\ -1 & 2 \end{pmatrix}$.

4.3.12. Explain why the quantity inside the square root in (4.24) is always non-negative.

◇ **4.3.13.** Justify the formulae in (4.26).

Least Squares

As we first observed in Section 4.1, the solution to the closest point problem also solves the basic least squares minimization problem! Let us first officially define the notion of a (classical) least squares solution to a linear system.

Definition 4.7 A *least squares solution* to a linear system of equations

$$A\mathbf{x}=\mathbf{b} \tag{4.27}$$

is a vector $\mathbf{x}^\star\in\mathbb{R}^n$ that minimizes the Euclidean norm $\|A\mathbf{x}-\mathbf{b}\|$.

REMARK: Later, we will generalize the least squares method to more general weighted norms coming from inner products. However, for the time being we restrict our attention to the Euclidean version.

If the system (4.27) actually has a solution, then it is automatically the least squares solution. Thus, the concept of least squares solution is new only when the system does not have a solution, i.e., $\mathbf{b}$ does not lie in the range of A. We also want the least squares solution to be unique. As with an ordinary solution, this happens if and only if $\ker A = \{\mathbf{0}\}$, or, equivalently, the columns of A are linearly independent, or, equivalently, $\operatorname{rank} A = n$.

As before, to make the connection with the closest point problem, we identify the subspace $V = \operatorname{rng} A \subset \mathbb{R}^m$ as the range or column space of the matrix A. If the columns of A are linearly independent, then they form a basis for the range V. Since every element of the range can be written as $\mathbf{v} = A\mathbf{x}$, minimizing $\| A\mathbf{x} - \mathbf{b} \|$ is the same as minimizing the distance $\| \mathbf{v} - \mathbf{b} \|$ between the point and the subspace. The least squares solution $\mathbf{x}^\star$ to the minimization problem gives the closest point $\mathbf{v}^\star = A\mathbf{x}^\star$ in $V = \operatorname{rng} A$. Therefore, the least squares solution follows from Theorem 4.5. In the Euclidean case, we state the result more explicitly by using (4.25) to write out the linear system (4.23) and the minimal distance (4.24).

THEOREM 4.8 Assume that $\ker A = \{\mathbf{0}\}$. Set $K = A^T A$ and $\mathbf{f} = A^T \mathbf{b}$. Then the least squares solution to the linear system $A\mathbf{x} = \mathbf{b}$ is the unique solution $\mathbf{x}^\star$ to the so-called *normal equations*

$$K\mathbf{x} = \mathbf{f} \quad \text{or, explicitly,} \quad (A^T A)\mathbf{x} = A^T \mathbf{b}, \tag{4.28}$$

namely

$$\mathbf{x}^\star = (A^T A)^{-1} A^T \mathbf{b}. \tag{4.29}$$

The least squares error is

$$\| A\mathbf{x}^\star - \mathbf{b} \|^2 = \| \mathbf{b} \|^2 - \mathbf{f}^T \mathbf{x}^\star = \| \mathbf{b} \|^2 - \mathbf{b}^T A (A^T A)^{-1} A^T \mathbf{b}. \tag{4.30}$$

Note that the normal equations (4.28) can be simply obtained by multiplying the original system $A\mathbf{x} = \mathbf{b}$ on both sides by A^T. In particular, if A is square and invertible, then $(A^T A)^{-1} = A^{-1}(A^T)^{-1}$, and so (4.29) reduces to $\mathbf{x} = A^{-1}\mathbf{b}$, while the two terms in the error formula (4.30) cancel out, producing zero error. In the rectangular case—when inversion is *not* allowed—(4.29) gives a *new* formula for the solution to the linear system $A\mathbf{x} = \mathbf{b}$ whenever $\mathbf{b} \in \operatorname{rng} A$.

EXAMPLE 4.9 Consider the linear system

$$\begin{aligned} x_1 + 2x_2 \qquad &= 1 \\ 3x_1 - x_2 + x_3 &= 0 \\ -x_1 + 2x_2 + x_3 &= -1 \\ x_1 - x_2 - 2x_3 &= 2 \\ 2x_1 + x_2 - x_3 &= 2, \end{aligned}$$

consisting of 5 equations in 3 unknowns. The coefficient matrix and right hand side are

$$A = \begin{pmatrix} 1 & 2 & 0 \\ 3 & -1 & 1 \\ -1 & 2 & 1 \\ 1 & -1 & -2 \\ 2 & 1 & -1 \end{pmatrix}, \qquad \mathbf{b} = \begin{pmatrix} 1 \\ 0 \\ -1 \\ 2 \\ 2 \end{pmatrix}.$$

A direct application of Gaussian Elimination shows that $\mathbf{b} \notin \operatorname{rng} A$, and so the system is incompatible—it has no solution. Of course, to apply the least squares method, we are not required to check this in advance. If the system has a solution, it is the least squares solution too, and the least squares method will find it.

To form the normal equations (4.28), we compute

$$K = A^T A = \begin{pmatrix} 16 & -2 & -2 \\ -2 & 11 & 2 \\ -2 & 2 & 7 \end{pmatrix}, \qquad \mathbf{f} = A^T \mathbf{b} = \begin{pmatrix} 8 \\ 0 \\ -7 \end{pmatrix}.$$

Solving the 3×3 system $K\mathbf{x} = \mathbf{f}$ by Gaussian Elimination, we find

$$\mathbf{x} = K^{-1}\mathbf{f} \approx (.4119, .2482, -.9532)^T$$

to be the least squares solution to the system. The least squares error is

$$\| \mathbf{b} - A\mathbf{x}^\star \| \approx \| (-.0917, .0342, .1313, .0701, .0252)^T \| \approx .1799,$$

which is reasonably small—indicating that the system is, roughly speaking, not too incompatible. ●

EXERCISES 4.3

4.3.14. Find the least squares solutions to the following linear systems:

(a) $x + 2y = 1,\ 3x - y = 0,\ -x + 2y = 3,$

(b) $4x - 2y = 1,\ 2x + 3y = -4,\ x - 2y = -1,\ 2x + 2y = 2,$

(c) $2u + v - 2w = 1,\ 3u - 2w = 0,\ u - v + 3w = 2,$

(d) $x - z = -1,\ 2x - y + 3z = 1,\ y - 3z = 0,\ -5x + 2y + z = 3,$

(e) $x_1 + x_2 = 2,\ x_2 + x_4 = 1,\ x_1 + x_3 = 0,\ x_3 - x_4 = 1,\ x_1 - x_4 = 2.$

4.3.15. Find the least squares solution to the linear system $A\mathbf{x} = \mathbf{b}$ when

(a) $A = \begin{pmatrix} 1 \\ 2 \\ -1 \end{pmatrix}, \quad \mathbf{b} = \begin{pmatrix} 1 \\ 1 \\ 0 \end{pmatrix}$

(b) $A = \begin{pmatrix} 1 & 0 \\ 2 & -1 \\ 3 & 5 \end{pmatrix}, \quad \mathbf{b} = \begin{pmatrix} 1 \\ 3 \\ 7 \end{pmatrix}$

(c) $A = \begin{pmatrix} 2 & 1 & -1 \\ 1 & -2 & 0 \\ 3 & -1 & 1 \end{pmatrix}, \quad \mathbf{b} = \begin{pmatrix} 1 \\ 0 \\ 1 \end{pmatrix}$

(d) $A = \begin{pmatrix} 2 & 3 \\ 4 & -2 \\ 1 & 5 \\ 2 & 0 \end{pmatrix}, \quad \mathbf{b} = \begin{pmatrix} 2 \\ -1 \\ 1 \\ 3 \end{pmatrix}$

(e) $A = \begin{pmatrix} 2 & 1 & 4 \\ 1 & -2 & 1 \\ 1 & 0 & -3 \\ 5 & 2 & -2 \end{pmatrix}, \quad \mathbf{b} = \begin{pmatrix} 0 \\ 0 \\ 1 \\ 0 \end{pmatrix}$

4.3.16. Let

$$A = \begin{pmatrix} 3 & -3 & 1 \\ 2 & 4 & 1 \\ 1 & 2 & 1 \end{pmatrix}, \qquad \mathbf{b} = \begin{pmatrix} 6 \\ 5 \\ 4 \end{pmatrix}.$$

Prove, using Gaussian Elimination, that the linear system $A\mathbf{x} = \mathbf{b}$ has a unique solution. Show that the least squares solution (4.29) is the same. Explain why this is necessarily the case.

4.3.17. Given

$$A = \begin{pmatrix} 1 & 2 & -1 \\ 0 & -2 & 3 \\ 1 & 5 & -1 \\ -3 & 1 & 1 \end{pmatrix} \quad \text{and} \quad \mathbf{b} = \begin{pmatrix} 0 \\ 5 \\ 6 \\ 8 \end{pmatrix},$$

find the least squares solution to the system $A\mathbf{x} = \mathbf{b}$. What is the error? Interpret your result.

4.3.18. Suppose we are interested in solving a linear system $A\mathbf{x} = \mathbf{b}$ by the method of least squares when the coefficient matrix A has linearly dependent columns. Let $K\mathbf{x} = \mathbf{f}$, where $K = A^T C A$, $\mathbf{f} = A^T C \mathbf{b}$, be the corresponding normal equations.

(a) Prove that $\mathbf{f} \in \operatorname{rng} K$, and so the normal equations have a solution. *Hint*: Use Exercise 3.4.31.

(b) Prove that any solution to the normal equations minimizes the least squares error, and hence qualifies as a least squares solution to the original system.

(c) Explain why the least squares solution is not unique.

4.4 DATA FITTING AND INTERPOLATION

One of the most important applications of the least squares minimization process is to the fitting of data points. Suppose we are running an experiment in which we measure a certain time-dependent physical quantity. At time t_i we make the measurement y_i, and thereby obtain a set of, say, m data points

$$(t_1, y_1), \quad (t_2, y_2), \quad \ldots \quad (t_m, y_m). \tag{4.31}$$

Suppose our theory indicates that the data points are supposed to all lie on a single line

$$y = \alpha + \beta t, \tag{4.32}$$

whose precise form—meaning its coefficients α, β—is to be determined. For example, a police car is interested in clocking the speed of a vehicle by using measurements of its relative distance at several times. Assuming that the vehicle is traveling at constant speed, its position at time t will have the linear form (4.32), with β, the velocity, and α, the initial position, to be determined. Experimental error will almost inevitably make this impossible to achieve exactly, and so the problem is to find the straight line (4.32) that "best fits" the measured data and then use its slope to estimate the vehicle's velocity.

The *error* between the measured value y_i and the sample value predicted by the function (4.32) at $t = t_i$ is

$$e_i = y_i - (\alpha + \beta t_i), \qquad i = 1, \ldots, m.$$

We can write this system of equations in matrix form as

$$\mathbf{e} = \mathbf{y} - A\mathbf{x},$$

where

$$\mathbf{e} = \begin{pmatrix} e_1 \\ e_2 \\ \vdots \\ e_m \end{pmatrix}, \quad \mathbf{y} = \begin{pmatrix} y_1 \\ y_2 \\ \vdots \\ y_m \end{pmatrix}, \quad \text{while} \quad A = \begin{pmatrix} 1 & t_1 \\ 1 & t_2 \\ \vdots & \vdots \\ 1 & t_m \end{pmatrix}, \quad \mathbf{x} = \begin{pmatrix} \alpha \\ \beta \end{pmatrix}. \tag{4.33}$$

We call $\mathbf{e}$ the *error vector* and $\mathbf{y}$ the *data vector*. The coefficients α, β of our desired function (4.32) are the unknowns, forming the entries of the column vector $\mathbf{x}$.

If we could fit the data exactly, so $y_i = \alpha + \beta t_i$ for all i, then each $e_i = 0$, and we could solve $A\mathbf{x} = \mathbf{y}$ for the coefficients α, β. In the language of linear algebra, the data points all lie on a straight line if and only if $\mathbf{y} \in \operatorname{rng} A$. If the data points are not collinear, then we seek the straight line that minimizes the total squared error or Euclidean norm

$$\text{Error} = \|\mathbf{e}\| = \sqrt{e_1^2 + \cdots + e_m^2}\,.$$

Pictorially, referring to Figure 4.5, the errors are the vertical distances from the points to the line, and we are seeking to minimize the square root of the sum of the squares of the individual errors*, hence the term *least squares*. In other words,

*This choice of minimization may strike the reader as a little odd. Why not just minimize the sum of the absolute value of the errors, i.e., the 1 norm $\|\mathbf{e}\|_1 = |e_1| + \cdots + |e_n|$ of the error vector, or minimize the maximal error, i.e., the ∞ norm $\|\mathbf{e}\|_\infty = \max\{|e_1|, \ldots, |e_n|\}$? Or, even better, why minimize the vertical distance to the line? The perpendicular distance from each data point to the line, as computed in Exercise 4.1.8, might strike you as a better measure of error. The answer is that, although each of these alternative minimization criteria is interesting and potentially useful, they all lead to *nonlinear* minimization problems, and so are much harder to solve! The least squares minimization problem can

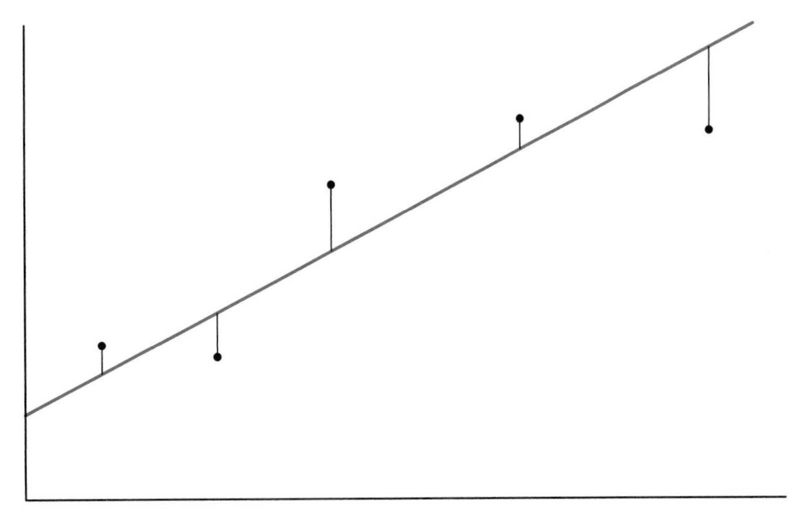

Figure 4.5 Least squares approximation of data by a straight line.

we are looking for the coefficient vector $\mathbf{x} = (\alpha, \beta)^T$ that minimizes the Euclidean norm of the error vector

$$\| \mathbf{e} \| = \| A\mathbf{x} - \mathbf{y} \|. \tag{4.34}$$

Thus, we recover the problem of characterizing the least squares solution to the linear system $A\mathbf{x} = \mathbf{y}$.

Theorem 4.8 prescribes the solution to this least squares minimization problem. We form the normal equations

$$(A^T A)\mathbf{x} = A^T \mathbf{y}, \qquad \text{with solution} \quad \mathbf{x}^\star = (A^T A)^{-1} A^T \mathbf{y}. \tag{4.35}$$

Invertibility of the Gram matrix $K = A^T A$ relies on the assumption that the matrix A has linearly independent columns. For the particular matrix in (4.33), linear independence of its two columns requires that not all the t_i's be equal, i.e., we must measure the data at at least two distinct times. Note that this restriction does not preclude measuring some of the data at the same time, e.g., by repeating the experiment. However, choosing *all* the t_i's to be the same is a silly data fitting problem. (Why?)

Under this assumption, we then compute

$$A^T A = \begin{pmatrix} 1 & 1 & \dots & 1 \\ t_1 & t_2 & \dots & t_m \end{pmatrix} \begin{pmatrix} 1 & t_1 \\ 1 & t_2 \\ \vdots & \vdots \\ 1 & t_m \end{pmatrix} = \begin{pmatrix} m & \sum t_i \\ \sum t_i & \sum (t_i)^2 \end{pmatrix} = m \begin{pmatrix} 1 & \overline{t} \\ \overline{t} & \overline{t^2} \end{pmatrix},$$

$$A^T \mathbf{y} = \begin{pmatrix} 1 & 1 & \dots & 1 \\ t_1 & t_2 & \dots & t_m \end{pmatrix} \begin{pmatrix} y_1 \\ y_2 \\ \vdots \\ y_m \end{pmatrix} = \begin{pmatrix} \sum y_i \\ \sum t_i\, y_i \end{pmatrix} = m \begin{pmatrix} \overline{y} \\ \overline{t\, y} \end{pmatrix}, \tag{4.36}$$

be solved by linear algebra, and so, purely on the grounds of simplicity, is the method of choice in most applications. Moreover, one needs to fully understand the linear problem before diving into more treacherous nonlinear waters.

where the overbars, namely

$$\overline{t} = \frac{1}{m}\sum_{i=1}^{m} t_i, \quad \overline{y} = \frac{1}{m}\sum_{i=1}^{m} y_i, \quad \overline{t^2} = \frac{1}{m}\sum_{i=1}^{m} t_i^2, \quad \overline{t\,y} = \frac{1}{m}\sum_{i=1}^{m} t_i\,y_i, \tag{4.37}$$

denote the *average* sample values of the indicated variables.

Warning: The average of a product is *not* equal to the product of the averages! In particular,

$$\overline{t^2} \neq (\overline{t})^2, \qquad \overline{t\,y} \neq \overline{t}\,\overline{y}.$$

Substituting (4.36) into the normal equations (4.35), and canceling the common factor of m, we find that we have only to solve a pair of linear equations

$$\alpha + \overline{t}\,\beta = \overline{y}, \qquad \overline{t}\,\alpha + \overline{t^2}\,\beta = \overline{t\,y},$$

for the coefficients:

$$\alpha = \overline{y} - \overline{t}\,\beta, \qquad \beta = \frac{\overline{t\,y} - \overline{t}\,\overline{y}}{\overline{t^2} - (\overline{t}\,)^2} = \frac{\sum (t_i - \overline{t}\,)\,y_i}{\sum (t_i - \overline{t}\,)^2}. \tag{4.38}$$

Therefore, the best (in the least squares sense) straight line that fits the given data is

$$y = \beta\,(t - \overline{t}\,) + \overline{y}, \tag{4.39}$$

where the line's slope β is given in (4.38).

EXAMPLE 4.10 Suppose the data points are given by the table

t_i	0	1	3	6
y_i	2	3	7	12

To find the least squares line, we construct

$$A = \begin{pmatrix} 1 & 0 \\ 1 & 1 \\ 1 & 3 \\ 1 & 6 \end{pmatrix}, \qquad A^T = \begin{pmatrix} 1 & 1 & 1 & 1 \\ 0 & 1 & 3 & 6 \end{pmatrix}, \qquad \mathbf{y} = \begin{pmatrix} 2 \\ 3 \\ 7 \\ 12 \end{pmatrix}.$$

Therefore

$$A^T A = \begin{pmatrix} 4 & 10 \\ 10 & 46 \end{pmatrix}, \qquad A^T\mathbf{y} = \begin{pmatrix} 24 \\ 96 \end{pmatrix}.$$

The normal equations (4.35) reduce to

$$4\alpha + 10\beta = 24, \qquad 10\alpha + 46\beta = 96, \qquad \text{so} \quad \alpha = \tfrac{12}{7}, \quad \beta = \tfrac{12}{7}.$$

Therefore, the best least squares fit to the data is the straight line

$$y = \tfrac{12}{7} + \tfrac{12}{7}\,t.$$

Alternatively, one can compute this formula directly from (4.38–39). As you can see in Figure 4.6, the least squares line does a fairly good job of approximating the data points. ●

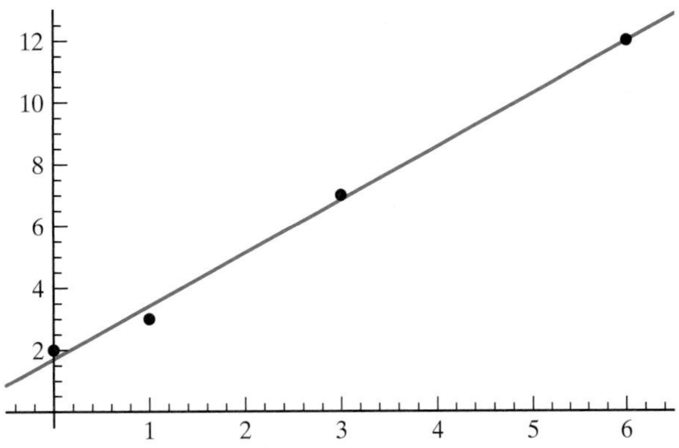

Figure 4.6 Least squares line.

EXAMPLE 4.11 Suppose we are given a sample of an unknown radioactive isotope. At time t_i we measure, using a Geiger counter, the amount m_i of radioactive material in the sample. The problem is to determine the initial amount of material along with the isotope's half-life. If the measurements were exact, we would have $m(t) = m_0 e^{\beta t}$, where $m_0 = m(0)$ is the initial mass, and $\beta < 0$ the decay rate. The half life is given by $t^\star = \beta^{-1} \log 2$; see Example 8.1 for additional details.

As it stands this is *not* a linear least squares problem. But it can be easily converted to the proper form by taking logarithms:

$$y(t) = \log m(t) = \log m_0 + \beta t = \alpha + \beta t.$$

We can thus do a linear least squares fit on the logarithms $y_i = \log m_i$ of the radioactive mass data at the measurement times t_i to determine the best values for β and $\alpha = \log m_0$. ●

EXERCISES 4.4

4.4.1. Find the straight line $y = \alpha + \beta t$ that best fits the following data in the least squares sense:

(a)

t_i	0	1	3	6
y_i	2	3	7	12

(b)

t_i	1	2	3	4	5
y_i	1	0	−2	−3	−3

(c)

t_i	−2	−1	0	1	2
y_i	−5	−3	−2	0	3

4.4.2. The proprietor of an internet travel company compiled the following data relating the annual profit of the firm to its annual advertising expenditure (both measured in thousands of dollars):

Annual advertising expenditure	12	14	17	21	26	30
Annual profit	60	70	90	100	100	120

(a) Determine the equation of the least squares line.

(b) Plot the data and the least squares line. Estimate the profit when the annual advertising budget is \$50,000.

(c) What about a \$100,000 budget?

4.4.3. The median price (in thousands of dollars) of existing homes in a certain metropolitan area from 1989 to 1999 was:

year	price
1989	86.4
1990	89.8
1991	92.8
1992	96.0
1993	99.6
1994	103.1
1995	106.3
1996	109.5
1997	113.3
1998	120.0
1999	129.5

(a) Find an equation of the least squares line for these data.

(b) Use the result to estimate the median price of a house in the year 2005, and the year 2010, assuming that the trend continues.

♡ **4.4.4.** The amount of waste (in millions of tons a day) generated in Lower Slobbovia from 1960 to 1995 was

year	amount
1960	86.0
1965	99.8
1970	115.8
1975	125.0
1980	132.6
1985	143.1
1990	156.3
1995	169.5

(a) Find the equation for the least squares line that best fits these data. Use the result to estimate the amount of waste in the year 2000, and in the year 2005.

(b) Redo your calculations using an exponential growth model $y = c e^{\alpha t}$.

(c) Which model do you think most accurately reflects the data? Why?

4.4.5. A 20-pound turkey that has been thawed at the room temperature of $72°$ is placed in the oven at 1:00 pm. The temperature of the turkey is observed in 20 minute intervals to be $79°$, $88°$, and $96°$. A turkey is cooked when its temperature reaches $165°$. How much longer do you need to wait until the turkey is done?

4.4.6. The amount of radium-224 in a sample was measured at the indicated times.

time in days	mg
0	100
1	82.7
2	68.3
3	56.5
4	46.7
5	38.6
6	31.9
7	26.4

(a) Estimate how much radium will be left after 10 days.

(b) If the sample is considered to be safe when the amount of radium is less than .01 mg, estimate how long the sample needs to be stored before it can be safely disposed.

4.4.7. A sample of lead-210 measured the following radioactivity data at the given times:

time in days	mg
0	10
4	8.8
8	7.8
10	7.3
14	6.4
18	6.4

(a) Estimate the half-life of the lead-210.

(b) How long until only 1% of the original amount remains?

4.4.8. The following table gives the population of the United States for the years 1900-1950.

year	population in millions
1900	76
1910	92
1920	106
1930	123
1940	131
1950	150

(a) Use an exponential growth model to predict the population in 2000, 2010, and 2050.

(b) The actual population for the year 2000 has recently been estimated to be 281 million. How does this affect your prediction for the year 2050?

4.4.9. For the data points

x	1	1	2	2	3	3
y	1	2	1	2	2	4
z	3	6	11	-2	0	3

(a) determine the best plane $z = a + b\,x + c\,y$ which fits the data in the least squares sense;

(b) how would you answer the question in part (a) if the plane is constrained to go through the point $x = 2$, $y = 2$, $z = 0$?

4.4.10. Show, by constructing explicit examples, that $\overline{t^2} \neq (\overline{t}\,)^2$ and $\overline{t\,y} \neq \overline{t}\,\overline{y}$. Can you find any data for which either equality is valid?

4.4.11. Given points $t_1, \ldots, t_m$, prove

$$\overline{t^2} - (\overline{t}\,)^2 = \frac{1}{m} \sum_{i=1}^{m} (t_i - \overline{t}\,)^2,$$

thereby justifying (4.38).

Polynomial Approximation and Interpolation

The basic least squares philosophy has a variety of different extensions, all interesting and all useful. First, we can replace the straight line (4.32) by a parabola defined by a quadratic function

$$y = \alpha + \beta\,t + \gamma\,t^2. \tag{4.40}$$

For example, Newton's theory of gravitation says that (in the absence of air resistance) a falling object obeys the parabolic law (4.40), where $\alpha = h_0$ is the initial height, $\beta = v_0$ is the initial velocity, and $\gamma = -\frac{1}{2}\,g$ is minus one half the gravitational constant. Suppose we observe a falling body on a new planet, and measure its height y_i at times t_i. Then we can approximate its initial height, initial velocity and gravitational acceleration by finding the parabola (4.40) that best fits the data. Again, we characterize the least squares fit by minimizing the sum of the squares of the individual errors $e_i = y_i - y(t_i)$.

The method can evidently be extended to a completely general polynomial function

$$y(t) = \alpha_0 + \alpha_1\,t + \cdots + \alpha_n\,t^n \tag{4.41}$$

of degree n. The total least squares error between the data and the sample values of the function is equal to

$$\|\mathbf{e}\|^2 = \sum_{i=1}^{m} \left[\,y_i - y(t_i)\,\right]^2 = \|\mathbf{y} - A\,\mathbf{x}\|^2, \tag{4.42}$$

where

$$A = \begin{pmatrix} 1 & t_1 & t_1^2 & \ldots & t_1^n \\ 1 & t_2 & t_2^2 & \ldots & t_2^n \\ \vdots & \vdots & \vdots & \ddots & \vdots \\ 1 & t_m & t_m^2 & \ldots & t_m^n \end{pmatrix}, \qquad \mathbf{x} = \begin{pmatrix} \alpha_0 \\ \alpha_1 \\ \alpha_2 \\ \vdots \\ \alpha_n \end{pmatrix}, \qquad \mathbf{y} = \begin{pmatrix} y_1 \\ y_2 \\ \vdots \\ y_m \end{pmatrix}. \tag{4.43}$$

The coefficient $m \times (n+1)$ coefficient matrix is known as a *Vandermonde matrix*, named after the eighteenth century French mathematician, scientist and musicologist Alexandre–Théophile Vandermonde—despite the fact that it appears nowhere in his four mathematical papers! In particular, if $m = n + 1$, then A is square, and so, assuming invertibility, we can solve $A\,\mathbf{x} = \mathbf{y}$ exactly. In other words, there is no error, and the solution is an *interpolating polynomial*, meaning that it fits the data exactly. A proof of the following result can be found at the end of this section.

Lemma 4.12 If $t_1, \ldots, t_{n+1}$ are distinct, $t_i \neq t_j$, then the $(n+1) \times (n+1)$ Vandermonde interpolation matrix (4.43) is nonsingular.

This result immediately implies the basic existence theorem for interpolating polynomials.

THEOREM 4.13 Let $t_1, \ldots, t_{n+1}$ be distinct sample points. Then, for any data $y_1, \ldots, y_{n+1}$, there exists a unique interpolating polynomial of degree $\leq n$ with the prescribed sample values $y(t_i) = y_i$ for all $i = 1, \ldots, n+1$.

Thus, two points will determine a unique interpolating line, three points a unique interpolating parabola, four points an interpolating cubic, and so on; see Figure 4.7.

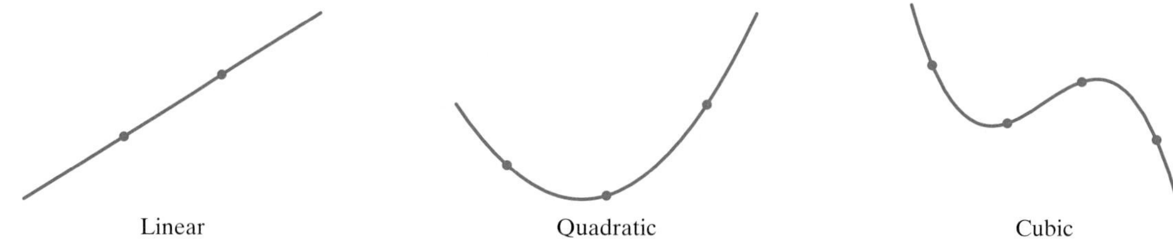

Figure 4.7 Interpolating polynomials.

The basic ideas of interpolation and least squares fitting of data can be applied to approximate complicated mathematical functions by much simpler polynomials. Such approximation schemes are used in all numerical computations. Your computer or calculator is only able to add, subtract, multiply and divide. Thus, when you ask it to compute $\sqrt{t}$ or e^t or $\cos t$ or any other non-rational function, the program must rely on an approximation scheme based on polynomials†. In the "dark ages" before computers, one would consult precomputed tables of values of the function at particular data points. If one needed a value at a nontabulated point, then some form of polynomial interpolation would be used to accurately approximate the intermediate value.

EXAMPLE 4.14 Suppose that we would like to compute reasonably accurate values for the exponential function e^t for values of t lying in the interval $0 \leq t \leq 1$ by approximating it by a quadratic polynomial

$$p(t) = \alpha + \beta t + \gamma t^2. \tag{4.44}$$

If we choose 3 points, say $t_1 = 0$, $t_2 = .5$, $t_3 = 1$, then there is a unique quadratic polynomial (4.44) that interpolates e^t at the data points, i.e.,

$$p(t_i) = e^{t_i} \quad \text{for} \quad i = 1, 2, 3.$$

In this case, the coefficient matrix (4.43), namely

$$A = \begin{pmatrix} 1 & 0 & 0 \\ 1 & .5 & .25 \\ 1 & 1 & 1 \end{pmatrix},$$

is nonsingular. Therefore, we can exactly solve the interpolation equations

$$A\mathbf{x} = \mathbf{y}, \quad \text{where} \quad \mathbf{y} = \begin{pmatrix} e^{t_1} \\ e^{t_2} \\ e^{t_3} \end{pmatrix} = \begin{pmatrix} 1. \\ 1.64872 \\ 2.71828 \end{pmatrix}$$

†Actually, one could also allow interpolation and approximation by rational functions, a subject known as *Padé approximation theory*, [3].

is the data vector, which we assume we already know. The solution

$$\mathbf{x} = \begin{pmatrix} \alpha \\ \beta \\ \gamma \end{pmatrix} = \begin{pmatrix} 1. \\ .876603 \\ .841679 \end{pmatrix}$$

yields the interpolating polynomial

$$p(t) = 1 + .876603\, t + .841679\, t^2. \tag{4.45}$$

It is the unique quadratic polynomial that agrees with e^t at the three specified data points. See Figure 4.8 for a comparison of the graphs; the first graph shows e^t, the second $p(t)$, and the third lays the two graphs on top of each other. Even with such a primitive interpolation scheme, the two functions are quite close. The maximum error or L^∞ norm of the difference is

$$\| e^t - p(t) \|_\infty = \max \left\{ \, | e^t - p(t) | \;\middle|\; 0 \le t \le 1 \, \right\} \approx .01442,$$

with the largest deviation occurring at $t \approx .796$. ●

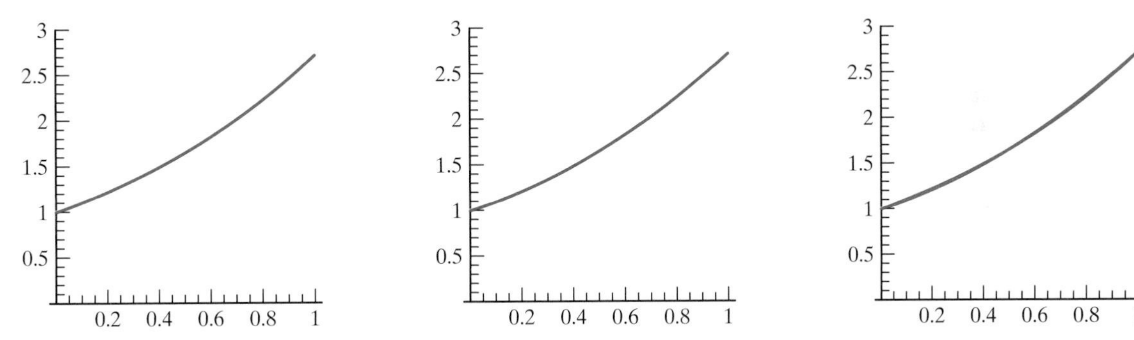

Figure 4.8 Quadratic interpolating polynomial for e^t.

There is, in fact, an explicit formula for the interpolating polynomial that is named after the influential eighteenth century Italo–French mathematician Joseph–Louis Lagrange. It relies on the basic superposition principle for solving inhomogeneous systems, cf. Theorem 2.44. Specifically, suppose we know the solutions $\mathbf{x}_1, \ldots, \mathbf{x}_{n+1}$ to the particular interpolation systems

$$A\mathbf{x}_k = \mathbf{e}_k, \qquad k = 1, \ldots, n+1, \tag{4.46}$$

where $\mathbf{e}_1, \ldots, \mathbf{e}_{n+1}$ are the standard basis vectors of $\mathbb{R}^{n+1}$. Then the solution to

$$A\mathbf{x} = \mathbf{y} = y_1 \mathbf{e}_1 + \cdots + y_{n+1} \mathbf{e}_{n+1}$$

is given by the superposition formula

$$\mathbf{x} = y_1 \mathbf{x}_1 + \cdots + y_{n+1} \mathbf{x}_{n+1}.$$

The particular interpolation equation (4.46) corresponds to the interpolation data $\mathbf{y} = \mathbf{e}_k$, meaning that $y_k = 1$, while $y_i = 0$ at all points t_i with $i \neq k$. If we can find the $n+1$ particular interpolating polynomials that realize this very special data, we can use superposition to construct the general interpolating polynomial.

THEOREM 4.15 Given distinct sample points $t_1, \ldots, t_{n+1}$, the kth *Lagrange interpolating polynomial* is given by

$$L_k(t) = \frac{(t - t_1) \cdots (t - t_{k-1})(t - t_{k+1}) \cdots (t - t_{n+1})}{(t_k - t_1) \cdots (t_k - t_{k-1})(t_k - t_{k+1}) \cdots (t_k - t_{n+1})}, \qquad k = 1, \ldots, n+1. \tag{4.47}$$

It is the unique polynomial of degree n that satisfies

$$L_k(t_i) = \begin{cases} 1, & i = k, \\ 0, & i \neq k, \end{cases} \qquad i, k = 1, \ldots, n+1. \tag{4.48}$$

Proof The uniqueness of the Lagrange interpolating polynomial is an immediate consequence of Theorem 4.13. To show that (4.47) is the correct formula, we note that when $t = t_i$ for any $i \neq k$, the factor $(t - t_i)$ in the numerator of $L_k(t)$ vanishes, while the denominator is not zero since the points are distinct. On the other hand, when $t = t_k$, the numerator and denominator are equal, and so $L_k(t_k) = 1$. ■

THEOREM 4.16 If $t_1, \ldots, t_{n+1}$ are distinct, then the polynomial of degree $\leq n$ that interpolates the associated data $y_1, \ldots, y_{n+1}$ is

$$p(t) = y_1 L_1(t) + \cdots + y_{n+1} L_{n+1}(t). \tag{4.49}$$

Proof We merely compute

$$p(t_k) = y_1 L_1(t_k) + \cdots + y_k L_k(t) + \cdots + y_{n+1} L_{n+1}(t_k) = y_k,$$

where, according to (4.48), every summand except the kth is zero. ■

EXAMPLE 4.17 For example, the three quadratic Lagrange interpolating polynomials for the values $t_1 = 0$, $t_2 = \frac{1}{2}$, $t_3 = 1$ used to interpolate e^t in Example 4.14 are

$$\begin{aligned} L_1(t) &= \frac{\left(t - \frac{1}{2}\right)(t-1)}{\left(0 - \frac{1}{2}\right)(0-1)} = 2t^2 - 3t + 1, \\ L_2(t) &= \frac{(t-0)(t-1)}{\left(\frac{1}{2} - 0\right)\left(\frac{1}{2} - 1\right)} = -4t^2 + 4t, \\ L_3(t) &= \frac{(t-0)\left(t - \frac{1}{2}\right)}{(1-0)\left(1 - \frac{1}{2}\right)} = 2t^2 - t. \end{aligned} \tag{4.50}$$

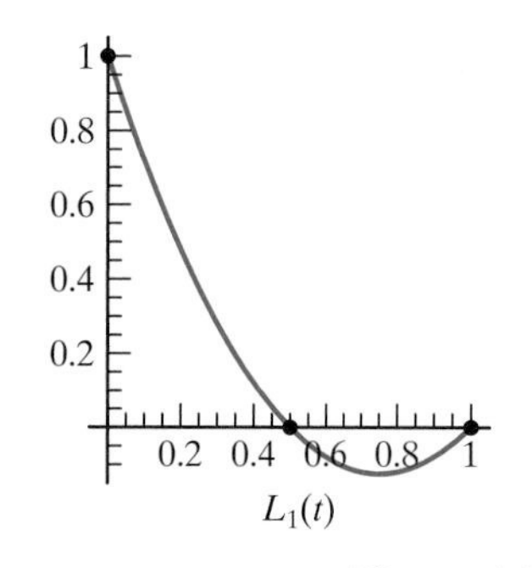

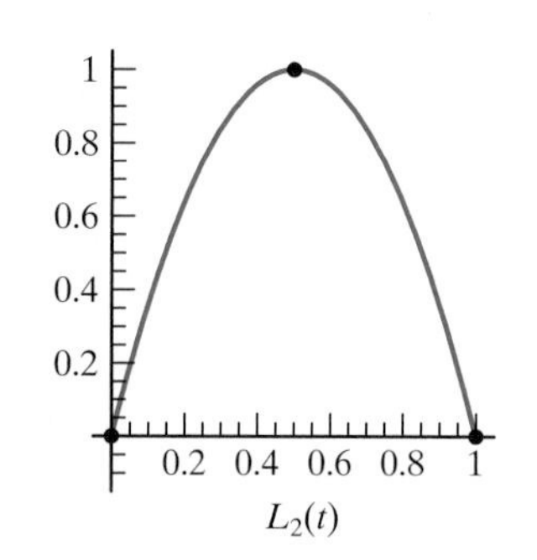

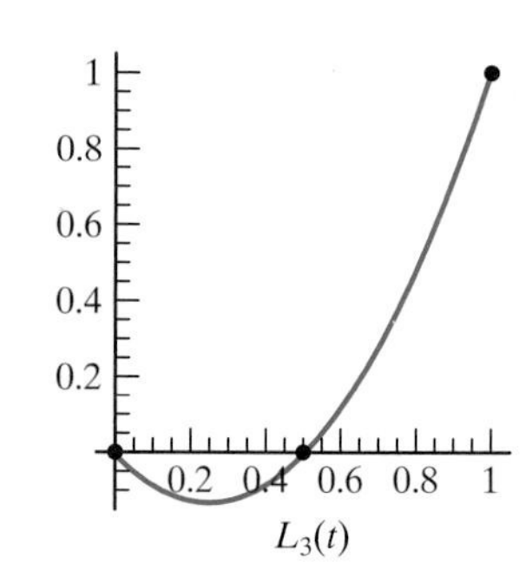

Figure 4.9 Lagrange interpolating polynomials for the points 0, .5, 1.

Thus, we can rewrite the quadratic interpolant (4.45) to e^t as

$$\begin{aligned} y(t) &= L_1(t) + e^{1/2} L_2(t) + e\, L_3(t) \\ &= (2t^2 - 3t + 1) + 1.64872(-4t^2 + 4t) + 2.71828(2t^2 - t). \end{aligned}$$

We stress that this is the *same* interpolating polynomial—we have merely rewritten it in the more transparent Lagrange form. ●

You might expect that the higher the degree, the more accurate the interpolating polynomial. This expectation turns out, unfortunately, not to be uniformly valid. While low degree interpolating polynomials are usually reasonable approximants to functions, high degree interpolants are not only more expensive to compute, but can be rather badly behaved, particularly near the ends of the interval. For example, Figure 4.10 displays the degree 2, 4 and 10 interpolating polynomials for the function $1/(1+t^2)$ on the interval $-3 \le t \le 3$ using equally spaced data points. Note the rather poor approximation of the function near the ends of the interval. Higher degree interpolants fare even worse, although the bad behavior becomes more and more concentrated near the endpoints. As a consequence, high degree polynomial interpolation tends not to be used in practical applications. Better alternatives rely on least squares approximants by low degree polynomials, to be described next, and interpolation by piecewise cubic splines, a topic that will be discussed in depth in Chapter 11.

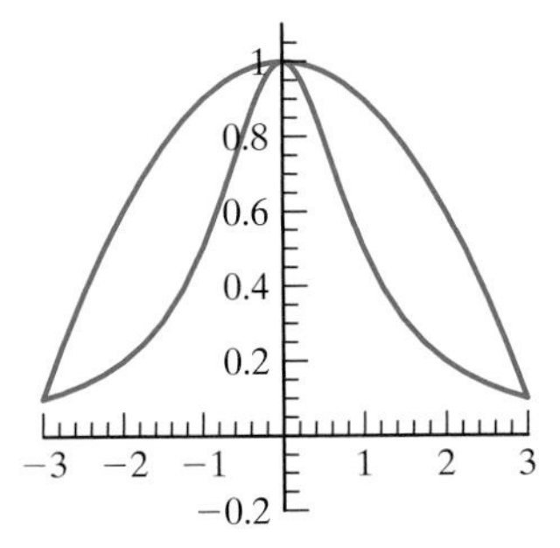

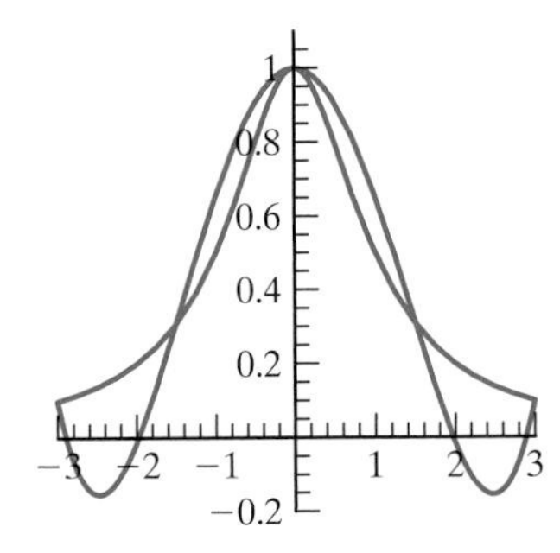

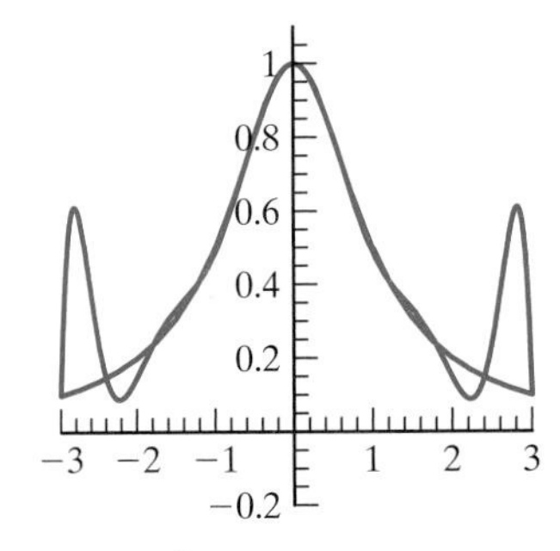

Figure 4.10 Degree 2, 4 and 10 interpolating polynomials for $1/(1+t^2)$.

If we have $m > n+1$ data points, then, usually, there is no degree n polynomial that fits all the data, and so we must switch over to a least squares approximation. The first requirement is that the associated $m \times (n+1)$ interpolation matrix (4.43) has rank $n+1$; this follows from Lemma 4.12 (coupled with Exercise 2.5.41), provided that at least $n+1$ of the values $t_1, \ldots, t_m$ are distinct. Thus, given data at $m \ge n+1$ different sample points $t_1, \ldots, t_m$, we can uniquely determine the best least squares polynomial of degree n that fits the data by solving the normal equations (4.35).

EXAMPLE 4.18

Let us return to the problem of approximating the exponential function e^t. If we use more than three data points, but still require a quadratic polynomial, then we can no longer interpolate exactly, and must devise a least squares approximant. For instance, using five equally spaced sample points

$$t_1 = 0, \quad t_2 = .25, \quad t_3 = .5, \quad t_4 = .75, \quad t_5 = 1,$$

the coefficient matrix and sampled data vector (4.43) are

$$A = \begin{pmatrix} 1 & 0 & 0 \\ 1 & .25 & .0625 \\ 1 & .5 & .25 \\ 1 & .75 & .5625 \\ 1 & 1 & 1 \end{pmatrix}, \qquad \mathbf{y} = \begin{pmatrix} 1. \\ 1.28403 \\ 1.64872 \\ 2.11700 \\ 2.71828 \end{pmatrix}.$$

The solution to the normal equations (4.28), with

$$K = A^T A = \begin{pmatrix} 5. & 2.5 & 1.875 \\ 2.5 & 1.875 & 1.5625 \\ 1.875 & 1.5625 & 1.38281 \end{pmatrix}, \qquad \mathbf{f} = A^T \mathbf{y} = \begin{pmatrix} 8.76803 \\ 5.45140 \\ 4.40153 \end{pmatrix},$$

is

$$\mathbf{x} = K^{-1}\mathbf{f} = (1.00514,\ .864277,\ .843538)^T.$$

This leads to the quadratic least squares approximant

$$p_2(t) = 1.00514 + .864277\, t + .843538\, t^2.$$

On the other hand, the quartic interpolating polynomial

$$p_4(t) = 1 + .998803\, t + .509787\, t^2 + .140276\, t^3 + .069416\, t^4$$

is found directly from the data values as above. The quadratic polynomial has a maximal error of $\approx .011$ over the interval $[0, 1]$—slightly better than the quadratic interpolant—while the quartic has a significantly smaller maximal error: $\approx .0000527$. (In this case, high degree interpolants are not ill behaved.) See Figure 4.11 for a comparison of the graphs, and Example 4.21 below for further discussion. ●

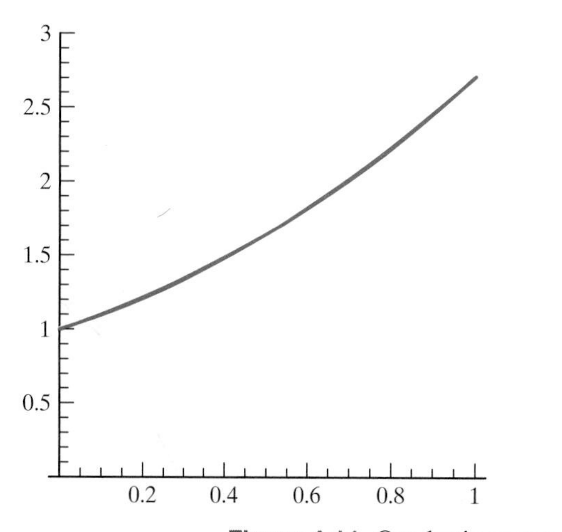

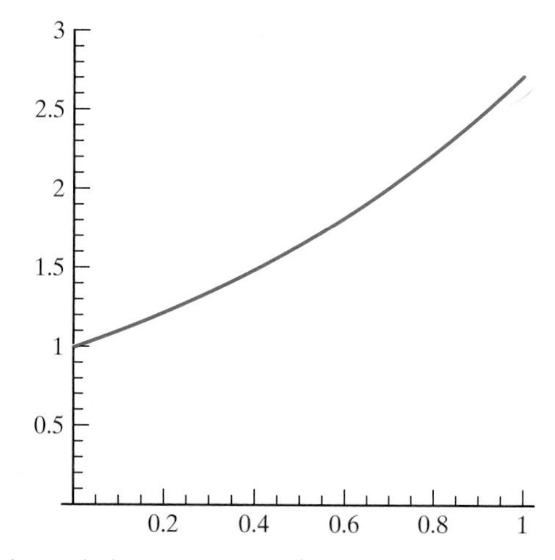

Figure 4.11 Quadratic approximant and quartic interpolant for e^t.

Proof of Lemma 4.12 We will establish the rather striking LU factorization of the transposed Vandermonde matrix $V = A^T$, which will immediately prove that, when $t_1, \ldots, t_{n+1}$ are distinct, both V and A are nonsingular matrices. The 4×4 case is instructive for understanding the general pattern. Applying regular Gaussian Elimination, we find the explicit LU factorization

$$\begin{pmatrix} 1 & 1 & 1 & 1 \\ t_1 & t_2 & t_3 & t_4 \\ t_1^2 & t_2^2 & t_3^2 & t_4^2 \\ t_1^3 & t_2^3 & t_3^3 & t_4^3 \end{pmatrix} = \begin{pmatrix} 1 & 0 & 0 & 0 \\ t_1 & 1 & 0 & 0 \\ t_1^2 & t_1 + t_2 & 1 & 0 \\ t_1^3 & t_1^2 + t_1 t_2 + t_2^2 & t_1 + t_2 + t_3 & 1 \end{pmatrix}$$
$$\begin{pmatrix} 1 & 1 & 1 & 1 \\ 0 & t_2 - t_1 & t_3 - t_1 & t_4 - t_1 \\ 0 & 0 & (t_3 - t_1)(t_3 - t_2) & (t_4 - t_1)(t_4 - t_2) \\ 0 & 0 & 0 & (t_4 - t_1)(t_4 - t_2)(t_4 - t_3) \end{pmatrix}.$$

In the general $(n+1) \times (n+1)$ case, the individual entries of the matrices appearing in factorization $V = LU$ are

$$
\begin{aligned}
v_{ij} &= t_j^{i-1}, && i, j = 1, \ldots, n+1, \\
\ell_{ij} &= \sum_{1 \le k_1 \le \cdots \le k_{i-j} \le j} t_{k_1} t_{k_2} \cdots t_{k_{i-j}}, && 1 \le j < i \le n+1, \\
\ell_{ii} &= 1, && i = 1, \ldots, n+1, \\
\ell_{ij} &= 0, && 1 \le i < j \le n+1, \\
u_{ij} &= \prod_{k=1}^{i} (t_j - t_k), && 1 < i \le j \le n+1, \\
u_{1j} &= 1, && j = 1, \ldots, n+1, \\
u_{ij} &= 0, && 1 \le j < i \le n+1.
\end{aligned}
\tag{4.51}
$$

Full details of the proof that $V = LU$ can be found in [25, 49]. (Surprisingly, as far as we know, these are the first places this factorization appears in the literature.) The entries of L lying below the diagonal are known as the *complete monomial polynomials* since ℓ_{ij} is obtained by summing, with unit coefficients, all monomials of degree $i - j$ in the j variables $t_1, \ldots, t_j$. The entries of U appearing on or above the diagonal are known as the *Newton difference polynomials*. In particular, if $t_1, \ldots, t_n$ are distinct, so $t_i \ne t_j$ for $i \ne j$, all entries of U lying on or above the diagonal are nonzero. In this case, V has all nonzero pivots, and is a regular, hence nonsingular matrix. ■

EXERCISES 4.4

4.4.12. For the following data values, construct the interpolating polynomial in Lagrange form:

(a)

t_i	−3	2
y_i	1	5

(b)

t_i	0	1	3
y_i	1	.5	.25

(c)

t_i	−1	0	1
y_i	1	2	−1

(d)

t_i	0	1	2	3
y_i	0	1	4	9

(e)

t_i	−2	−1	0	1	2
y_i	−1	−2	2	1	3

4.4.13. Find and graph the polynomial of minimal degree that passes through the following points:

(a) $(3, -1), (6, 5)$;

(b) $(0, 6), (-2, 4), (1, 10)$;

(c) $(-2, 3), (0, -1), (1, -3)$;

(d) $(0, -1), (1, 0), (-1, 2), (2, -1)$;

(e) $(0, -3), (-1, -3), (1, -1), (-2, 17), (2, 9)$.

4.4.14. Given

t_i	1	2	3
y_i	3	6	11

(a) find the best straight line $y = \alpha + \beta t$ that fits the data in the least squares sense;

(b) find the best parabola $y = \alpha + \beta t + \gamma t^2$ that fits the data. Interpret the error.

4.4.15. A student runs an experiment six times in an attempt to obtain an equation relating two physical quantities x and y. For $x = 1, 2, 4, 6, 8, 10$ units, the experiments result in corresponding y values of $3, 3, 4, 6, 7, 8$ units. Find and graph the following:

(a) the least squares line;

(b) the least squares quadratic polynomial;

(c) the interpolating polynomial.

(d) Which do you think is the most likely theoretical model for this data?

4.4.16. The table

time in sec	0	10	20	30
meters	4500	4300	3930	3000

measures the altitude of a falling parachutist. Predict how many seconds she can wait to open the parachute before reaching the minimum altitude of 1500 meters.

4.4.17. A missile is launched in your direction. Using a range finder, you measure its altitude at the times:

time in sec	altitude in meters
0	200
10	650
20	970
30	1200
40	1375
50	1130

How long until you have to run?

4.4.18. (a) Write down the Taylor polynomials of degrees 2 and 4 at $t = 0$ for the function $f(t) = e^t$.

(b) Compare their accuracy with the interpolating and least squares polynomials in Examples 4.14 and 4.18.

4.4.19. Given the known values of $\sin t$ at $t = 0°, 30°, 45°, 60°$, find the following approximations:

(a) the least squares linear polynomial,

(b) the least squares quadratic polynomial,

(c) the quadratic Taylor polynomial at $t = 0$,

(d) the interpolating polynomial.

(e) the cubic Taylor polynomial at $t = 0$.

Graph each approximation and discuss their accuracy.

4.4.20. (a) Find the least squares linear polynomial approximating $\sqrt{t}$ on $[0, 1]$, choosing six different exact data values.

(b) How much more accurate is the least squares quadratic polynomial based on the same data?

4.4.21. Find the quartic (degree 4) polynomial that exactly interpolates the function $\tan t$ at the five data points $t_0 = 0, t_1 = .25, t_2 = .5, t_3 = .75, t_4 = 1$. Compare the graphs of the two functions over $0 \le t \le \frac{1}{2}\pi$.

4.4.22. A table of logarithms contains the following entries:

t	1.0	2.0	3.0	4.0
$\log_{10} t$	0	.3010	.4771	.6021

Approximate $\log_{10} e$ by constructing an interpolating polynomial of

(a) degree two using the entries at $x = 1.0, 2.0$, and 3.0,

(b) degree three using all the entries.

4.4.23. Let $q(t)$ denote the quadratic interpolating polynomial that goes through the data points (t_0, y_0), (t_1, y_1), (t_2, y_2).

(a) Under what conditions does $q(t)$ have a minimum? A maximum?

(b) Show that the minimizing/maximizing value is at

$$t^\star = \frac{m_0\, s_1 - m_1\, s_0}{s_1 - s_0},$$

where

$$s_0 = \frac{y_1 - y_0}{t_1 - t_0}, \quad s_1 = \frac{y_2 - y_1}{t_2 - t_1},$$
$$m_0 = \frac{t_0 + t_1}{2}, \quad m_1 = \frac{t_1 + t_2}{2}.$$

What is $q(t^\star)$?

♠ **4.4.24.** Use $n + 1$ equally spaced data points to interpolate $f(t) = 1/(1 + t^2)$ on an interval $-a \le t \le a$ for $a = 1, 1.5, 2, 2.5, 3$, and $n = 2, 4, 10, 20$. Do all intervals exhibit the pathology illustrated in Figure 4.10? If not, how large can a be before the interpolants have poor approximation properties? What happens when the number of interpolation points is taken to be $n = 50$?

♠ **4.4.25.** Repeat Exercise 4.4.24 for the hyperbolic secant function $f(t) = \operatorname{sech} t = 1/\cosh t$.

4.4.26. Given A as in (4.43) with $m < n + 1$, how would you characterize those polynomials $p(t)$ whose coefficient vector $\mathbf{x} \in \ker A$?

4.4.27. (a) Give an example of an interpolating polynomial through $n + 1$ points that has degree $< n$.

(b) Can you explain, without referring to the explicit formulae, why all the Lagrange interpolating polynomials based on $n + 1$ points must have degree equal to n?

◇ **4.4.28.** (a) Prove that a polynomial $p(x) = a_0 + a_1 x + a_2 x^2 + \cdots + a_n x^n$ of degree $\le n$ vanishes at $n+1$ *distinct* points, so $p(x_1) = p(x_2) = \cdots = p(x_{n+1}) = 0$, if and only if $p(x) \equiv 0$ is the zero polynomial.

(b) Prove that the monomials $1, x, x^2, \ldots, x^n$ are linearly independent.

(c) Explain why a polynomial $p(x) \equiv 0$ if and only if all its coefficients $a_0 = a_1 = \cdots = a_n = 0$. *Hint*: Use Lemma 4.12 and Exercise 2.3.37.

◇ **4.4.29.** Prove the determinant formula

$$\det A = \prod_{1 \le i < j \le n+1} (t_i - t_j)$$

for the $(n + 1) \times (n + 1)$ Vandermonde matrix defined in (4.43).

4.4.30. Let $x_1, \ldots, x_n$ be distinct real numbers. Prove that the $n \times n$ matrix K with entries

$$k_{ij} = \frac{1 - (x_i x_j)^n}{1 - x_i x_j}$$

is positive definite.

♡ **4.4.31.** *Numerical Differentiation*: The most common numerical methods for approximating the derivatives of a function are based on interpolation. To approximate the kth derivative $f^{(k)}(x_0)$ at a point x_0, one replaces the function $f(x)$ by an interpolating polynomial $p_n(x)$ of degree $n \geq k$ based on the nearby points $x_0, \ldots, x_n$ (the point x_0 is almost always included as one of the interpolation points), leading to the approximation $f^{(k)}(x_0) \approx p_n^{(k)}(x_0)$. Use this method to construct numerical approximations to

(a) $f'(x)$ using a quadratic interpolating polynomial based on $x-h, x, x+h$.

(b) $f''(x)$ with the same quadratic polynomial.

(c) $f'(x)$ using a quadratic interpolating polynomial based on $x, x+h, x+2h$.

(d) $f'(x)$, $f''(x)$, $f'''(x)$ and $f^{(iv)}(x)$ using a quartic interpolating polynomial based on $x-2h, x-h, x, x+h, x+2h$.

(e) Test your methods by approximating the derivatives of e^x and $\tan x$ at $x=0$ with step sizes $h = \frac{1}{10}, \frac{1}{100}, \frac{1}{1000}, \frac{1}{10000}$. Discuss the accuracies you observe. Can the step size be arbitrarily small?

(f) Why do you need $n \geq k$?

♡ **4.4.32.** *Numerical Integration*: Most numerical methods for evaluating an integral $\int_a^b f(x)\,dx$ are based on interpolation. One chooses $n+1$ interpolation points $a \leq x_0 < x_1 < \cdots < x_n \leq b$ and replaces the integrand by its interpolating polynomial $p_n(x)$ of degree n, leading to the approximation

$$\int_a^b f(x)\,dx \approx \int_a^b p_n(x)\,dx,$$

where the polynomial integral can be done explicitly. Write down the following popular integration rules:

(a) *Trapezoid Rule*: $x_0 = a, x_1 = b$.

(b) *Simpson's Rule*: $x_0 = a, x_1 = \frac{1}{2}(a+b)$, $x_2 = b$.

(c) *Simpson's $\frac{3}{8}$ Rule*: $x_0 = a$, $x_1 = \frac{1}{3}(a+b)$, $x_2 = \frac{2}{3}(a+b)$, $x_3 = b$.

(d) *Midpoint Rule*: $x_0 = \frac{1}{2}(a+b)$.

(e) *Open Rule*: $x_0 = \frac{1}{3}(a+b)$, $x_1 = \frac{2}{3}(a+b)$.

Test your methods for accuracy on the following integrals:

(i) $\displaystyle\int_0^1 e^x\,dx$ (ii) $\displaystyle\int_0^\pi \sin x\,dx$

(iii) $\displaystyle\int_1^e \log x\,dx$ (iv) $\displaystyle\int_0^{\pi/2} \sqrt{x^3+1}\,dx$

Note: For more details on numerical differentiation and integration, you are encouraged to consult a basic numerical analysis text, e.g., [10].

Approximation and Interpolation by General Functions

There is nothing special about polynomial functions in the preceding approximation scheme. For example, suppose we were interested in finding the best trigonometric approximation

$$y = \alpha_1 \cos t + \alpha_2 \sin t$$

to a given set of data. Again, the least squares error takes the same form $\| \mathbf{y} - A\mathbf{x} \|^2$ as in (4.42), where

$$A = \begin{pmatrix} \cos t_1 & \sin t_1 \\ \cos t_2 & \sin t_2 \\ \vdots & \vdots \\ \cos t_m & \sin t_m \end{pmatrix}, \qquad \mathbf{x} = \begin{pmatrix} \alpha_1 \\ \alpha_2 \end{pmatrix}, \qquad \mathbf{y} = \begin{pmatrix} y_1 \\ \vdots \\ y_m \end{pmatrix}.$$

Thus, the columns of A are the sampled values of the functions $\cos t$, $\sin t$. The key is that the unspecified parameters—in this case α_1, α_2—occur *linearly* in the approximating function. Thus, the most general case is to approximate the data (4.31) by a linear combination

$$y(t) = \alpha_1 h_1(t) + \alpha_2 h_2(t) + \cdots + \alpha_n h_n(t)$$

of prescribed functions $h_1(x), \ldots, h_n(x)$. The least squares error is, as always, given by

$$\text{Error} = \sqrt{\sum_{i=1}^{m} \bigl(y_i - y(t_i) \bigr)^2} = \| \mathbf{y} - A\mathbf{x} \|,$$

where the sample matrix A, the vector of unknown coefficients $\mathbf{x}$, and the data vector $\mathbf{y}$ are

$$A = \begin{pmatrix} h_1(t_1) & h_2(t_1) & \ldots & h_n(t_1) \\ h_1(t_2) & h_2(t_2) & \ldots & h_n(t_2) \\ \vdots & \vdots & \ddots & \vdots \\ h_1(t_m) & h_2(t_m) & \ldots & h_n(t_m) \end{pmatrix}, \qquad \mathbf{x} = \begin{pmatrix} \alpha_1 \\ \alpha_2 \\ \vdots \\ \alpha_n \end{pmatrix}, \qquad \mathbf{y} = \begin{pmatrix} y_1 \\ y_2 \\ \vdots \\ y_m \end{pmatrix}. \tag{4.52}$$

If A is square and nonsingular, then we can find an interpolating function of the prescribed form by solving the linear system

$$A\mathbf{x} = \mathbf{y}. \tag{4.53}$$

A particularly important case is provided by the $2n + 1$ trigonometric functions

$$1, \quad \cos x, \quad \sin x, \quad \cos 2x, \quad \sin 2x, \quad \ldots \quad \cos nx, \quad \sin nx.$$

Interpolation on $2n + 1$ equally spaced data points on the interval $[0, 2\pi]$ leads to the Discrete Fourier Transform, used in signal processing, data transmission, and compression, and to be the focus of Section 5.7.

If there are more than n data points, then we cannot, in general, interpolate exactly, and must content ourselves with a least squares approximation that minimizes the error at the sample points as best it can. The least squares solution to the interpolation equations (4.53) is found by solving the associated normal equations $K\mathbf{x} = \mathbf{f}$, where the (i, j) entry of $K = A^T A$ is m times the average sample value of the product of $h_i(t)$ and $h_j(t)$, namely

$$k_{ij} = m\,\overline{h_i(t)\,h_j(t)} = \sum_{\kappa=1}^{m} h_i(t_\kappa)\,h_j(t_\kappa), \tag{4.54}$$

whereas the ith entry of $\mathbf{f} = A^T\mathbf{y}$ is

$$f_i = m\,\overline{h_i(t)\,y} = \sum_{\kappa=1}^{m} h_i(t_\kappa)\,y_\kappa. \tag{4.55}$$

The one issue is whether the columns of the sample matrix A are linearly independent. This is more subtle than the polynomial case covered by Lemma 4.12. Linear independence of the sampled function vectors is, in general, more restrictive than merely requiring the functions themselves to be linearly independent; see Exercise 2.3.37 for further details.

If the parameters do not occur linearly in the functional formula, then we cannot use linear algebra to effect a least squares approximation. For example, one cannot determine the frequency ω, the amplitude r, *and* the phase shift δ of the general trigonometric approximation

$$y = c_1 \cos \omega t + c_2 \sin \omega t = r \cos(\omega t + \delta)$$

that minimizes the least squares error at the sample points. Approximating data by such a function constitutes a *nonlinear* minimization problem.

EXERCISES 4.4

4.4.33. Given the data values

t_i	0	.5	1
y_i	1	.5	.25

construct the trigonometric function of the form

$$g(t) = a\cos\pi t + b\sin\pi t$$

that best approximates the data in the least squares sense.

4.4.34. Find the hyperbolic function $g(t) = a\cosh t + b\sinh t$ that best approximates the data in Exercise 4.4.33.

4.4.35. (a) Find the exponential function of the form $g(t) = a\,e^t + b\,e^{2t}$ that best approximates t^2 in the least squares sense based on the sample points 0, 1, 2, 3, 4.
(b) What is the least squares error?
(c) Compare the graphs on the interval $[0,4]$—where is the approximation the worst?
(d) How much better can you do by including a constant term in $g(t) = a\,e^t + b\,e^{2t} + c$?

4.4.36. (a) Find the best trigonometric approximation of the form $g(t) = r\cos(t+\delta)$ to t^2 using 5 and 9 equally spaced sample points on $[0,\pi]$.
(b) Can you answer the question for $g(t) = r_1\cos(t+\delta_1) + r_2\cos(2t+\delta_2)$?

4.4.37. A *trigonometric polynomial* of degree n is a function of the form

$$p(t) = a_0 + a_1\cos t + b_1\sin t + a_2\cos 2t + b_2\sin 2t + \cdots + a_n\cos nt + b_n\sin nt,$$

where $a_0, a_1, b_1, \ldots, a_n, b_n$ are the coefficients. Find the trigonometric polynomial of degree n that is the least squares approximation to the function $f(t) = 1/(1+t^2)$ on the interval $[-\pi,\pi]$ based on the k equally spaced data points

$$t_j = -\pi + \frac{2\pi j}{k}, \qquad j = 0,\ldots,k-1,$$

(omitting the right hand endpoint), when
(a) $n=1, k=4$ (b) $n=2, k=8$
(c) $n=2, k=16$ (d) $n=3, k=16$
(e) Compare the graphs of the trigonometric approximant and the function, and discuss.
(f) Why do we not include the right hand endpoint $t_k = \pi$?

♡ **4.4.38.** The *sinc functions* are defined as

$$S_0(x) = \frac{\sin(\pi x/h)}{\pi x/h},$$

while $S_j(x) = S_0(x - jh)$ whenever $h > 0$ and j is an integer. We will interpolate a function $f(x)$ at the mesh points $x_j = jh$, $j = 0,\ldots,n$, by a linear combination of sinc functions: $S(x) = c_0 S_0(x) + \cdots + c_n S_n(x)$. What are the coefficients c_j? Graph and discuss the accuracy of the sinc interpolant for the functions x^2 and $\frac{1}{2} - \left|x - \frac{1}{2}\right|$ on the interval $[0,1]$ using $h = .25, .1$ and $.025$.

Weighted Least Squares

Another extension to the basic least squares method is to introduce weights in the measurement of the error. Suppose some of the data is known to be more reliable or more significant than others. For example, measurements at an earlier time may be more accurate, or more critical to the data fitting problem, than later measurements. In that situation, we should penalize any errors in the earlier measurements and downplay errors in the later data.

In general, this requires the introduction of a positive weight $c_i > 0$ associated to each data point (t_i, y_i); the larger the weight, the more vital the error. For a straight line approximation $y = \alpha + \beta t$, the *weighted least squares error* is defined as

$$\text{Error} = \sqrt{\sum_{i=1}^m c_i e_i^2} = \sqrt{\sum_{i=1}^m c_i\left[y_i - (\alpha + \beta t_i)\right]^2}.$$

Let us rewrite this formula in matrix form. Let $C = \operatorname{diag}(c_1,\ldots,c_m)$ denote the diagonal *weight matrix*. Note that $C > 0$ is positive definite, since all the weights are positive. The least squares error,

$$\text{Error} = \sqrt{\mathbf{e}^T C\,\mathbf{e}} = \|\mathbf{e}\|,$$

is then the norm of the error vector $\mathbf{e}$ with respect to the weighted inner product

$$\langle \mathbf{v}, \mathbf{w} \rangle = \mathbf{v}^T C \, \mathbf{w}.$$

Since $\mathbf{e} = \mathbf{y} - A\,\mathbf{x}$,

$$\begin{aligned} \|\mathbf{e}\|^2 = \|A\,\mathbf{x} - \mathbf{y}\|^2 &= (A\,\mathbf{x} - \mathbf{y})^T C\,(A\,\mathbf{x} - \mathbf{y}) \\ &= \mathbf{x}^T A^T C\, A\,\mathbf{x} - 2\mathbf{x}^T A^T C\,\mathbf{y} + \mathbf{y}^T C\,\mathbf{y} = \mathbf{x}^T K\,\mathbf{x} - 2\mathbf{x}^T\mathbf{f} + c, \end{aligned} \tag{4.56}$$

where

$$K = A^T C\, A, \qquad \mathbf{f} = A^T C\,\mathbf{y}, \qquad c = \mathbf{y}^T C\,\mathbf{y} = \|\mathbf{y}\|^2.$$

Note that K is the weighted Gram matrix derived in (3.56), whose entries

$$k_{ij} = \langle \mathbf{v}_i, \mathbf{v}_j \rangle = \mathbf{v}_i^T C\,\mathbf{v}_j$$

are the weighted inner products between the column vectors $\mathbf{v}_1, \ldots, \mathbf{v}_n$ of A. Theorem 3.31 immediately implies that K is positive definite—provided A has linearly independent columns or, equivalently, has rank n.

THEOREM 4.19 Suppose A is an $m \times n$ matrix with linearly independent columns. Suppose $C > 0$ is any positive definite $m \times m$ matrix. Then, the quadratic function (4.56) giving the weighted least squares error has a unique minimizer, which is the solution to the *weighted normal equations*

$$A^T C\, A\,\mathbf{x} = A^T C\,\mathbf{y}, \qquad \text{so that} \quad \mathbf{x} = (A^T C\, A)^{-1} A^T C\,\mathbf{y}. \tag{4.57}$$

In brief, the weighted least squares solution is obtained by multiplying both sides of the original system $A\,\mathbf{x} = \mathbf{y}$ by the matrix $A^T C$. The derivation of this result allows $C > 0$ to be *any* positive definite matrix. In applications, the off-diagonal entries of C can be used to weight cross-correlation terms in the data, although this extra freedom is rarely used in practice.

EXAMPLE 4.20 In Example 4.10, we fit the following data

t_i	0	1	3	6
y_i	2	3	7	12
c_i	3	2	$\frac{1}{2}$	$\frac{1}{4}$

with an unweighted least squares line. Now we shall assign the weights listed in the last row of the table for the error at each sample point. Thus, errors in the first two data values carry more weight than the latter two. To find the weighted least squares line $y = \alpha + \beta\, t$ that best fits the data, we compute

$$A^T C\, A = \begin{pmatrix} 1 & 1 & 1 & 1 \\ 0 & 1 & 3 & 6 \end{pmatrix} \begin{pmatrix} 3 & 0 & 0 & 0 \\ 0 & 2 & 0 & 0 \\ 0 & 0 & \frac{1}{2} & 0 \\ 0 & 0 & 0 & \frac{1}{4} \end{pmatrix} \begin{pmatrix} 1 & 0 \\ 1 & 1 \\ 1 & 3 \\ 1 & 6 \end{pmatrix} = \begin{pmatrix} \frac{23}{4} & 5 \\ 5 & \frac{31}{2} \end{pmatrix},$$

$$A^T C\,\mathbf{y} = \begin{pmatrix} 1 & 1 & 1 & 1 \\ 0 & 1 & 3 & 6 \end{pmatrix} \begin{pmatrix} 3 & 0 & 0 & 0 \\ 0 & 2 & 0 & 0 \\ 0 & 0 & \frac{1}{2} & 0 \\ 0 & 0 & 0 & \frac{1}{4} \end{pmatrix} \begin{pmatrix} 2 \\ 3 \\ 7 \\ 12 \end{pmatrix} = \begin{pmatrix} \frac{37}{2} \\ \frac{69}{2} \end{pmatrix}.$$

Thus, the weighted normal equations (4.57) reduce to

$$\tfrac{23}{4}\alpha + 5\beta = \tfrac{37}{2}, \qquad 5\alpha + \tfrac{31}{2}\beta = \tfrac{69}{2}, \qquad \text{so} \quad \alpha = 1.7817, \qquad \beta = 1.6511.$$

Therefore, the least squares fit to the data under the given weights is

$$y = 1.7817 + 1.6511\,t,$$

as plotted in Figure 4.12. ●

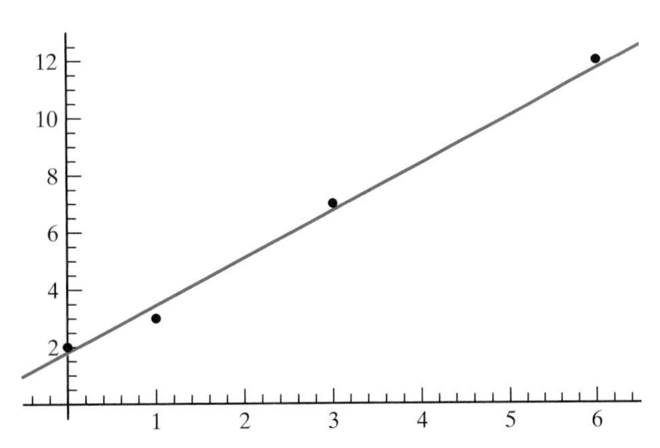

Figure 4.12 Weighted least squares line.

EXERCISES 4.4

4.4.39. Re-solve Exercise 4.4.14 using the respective weight 2, 1, .5 at the three data points.

4.4.40. Find the best linear least squares fit of the following data using the indicated weights:

(a)

t_i	1	2	3	4
y_i	.2	.4	.7	1.2
c_i	1	2	3	4

(b)

t_i	0	1	3	6
y_i	2	3	7	12
c_i	4	3	2	1

(c)

t_i	−2	−1	0	1	2
y_i	−5	−3	−2	0	3
c_i	2	1	.5	1	2

(d)

t_i	1	2	3	4	5
y_i	2	1.3	1.1	.8	.2
c_i	5	4	3	2	1

4.4.41. For the data points in Exercise 4.4.9, determine the plane $z = \alpha + \beta x + \gamma y$ that fits the data in the least squares sense when the errors are weighted according to the reciprocal of the distance of the point (x_i, y_i, z_i) from the origin.

◇ **4.4.42.** Write out an explicit formula for the weighted least squares error.

Least Squares Approximation in Function Spaces

So far, while we have used least squares minimization to interpolate and approximate known, complicated functions by simpler polynomials, we have only worried about the errors committed at a discrete, preassigned set of sample points. A more uniform approach would be to take into account the errors committed at *all* points in the interval of interest. This can be accomplished by replacing the discrete, finite-dimensional vector space norm on sample vectors by a continuous, infinite-dimensional function space norm in order to specify the least squares error that must be minimized over the entire interval.

More specifically, we let $V = C^0[a,b]$ denote the space of continuous functions on the bounded interval $[a,b]$ with L^2 inner product and norm

$$\langle f, g \rangle = \int_a^b f(t)\,g(t)\,dt, \qquad \| f \| = \sqrt{\int_a^b f(t)^2\,dt}\,. \tag{4.58}$$

Let $\mathcal{P}^{(n)}$ denote the subspace consisting of all polynomials of degree $\leq n$. For simplicity, we employ the standard monomial basis $1, t, t^2, \ldots, t^n$. We will be approximating a general function $f(t) \in \mathrm{C}^0[a, b]$ by a polynomial

$$p(t) = \alpha_0 + \alpha_1 t + \cdots + \alpha_n t^n \in \mathcal{P}^{(n)} \tag{4.59}$$

of degree at most n. The *error function* $e(t) = f(t) - p(t)$ measures the discrepancy between the function and its approximating polynomial at each t. Instead of summing the squares of the errors at a finite set of sample points, we go to a continuous limit that sums or, rather, integrates the squared errors of all points in the interval. Thus, the approximating polynomial will be characterized as the one that minimizes the L^2 *least squares error*:

$$\text{Error} = \| e \| = \| p - f \| = \sqrt{\int_a^b [\, p(t) - f(t)\,]^2 \, dt}\,. \tag{4.60}$$

To solve the minimization problem, we begin by substituting (4.59) and expanding, as in (4.19):

$$\begin{aligned} \| p - f \|^2 &= \left\| \sum_{i=0}^{n} \alpha_i t^i - f(t) \right\|^2 \\ &= \sum_{i,j=0}^{n} \alpha_i \alpha_j \langle t^i, t^j \rangle - 2 \sum_{i=0}^{n} \alpha_i \langle t^i, f(t) \rangle + \| f(t) \|^2. \end{aligned}$$

As a result, we are led to minimize the same sort of quadratic function

$$\mathbf{x}^T K \mathbf{x} - 2\mathbf{x}^T \mathbf{f} + c, \tag{4.61}$$

where $\mathbf{x} = (\alpha_0, \alpha_1, \ldots, \alpha_n)^T$ is the vector containing the unknown coefficients in the minimizing polynomial, while‡

$$k_{ij} = \langle t^i, t^j \rangle = \int_a^b t^{i+j} \, dt, \qquad f_i = \langle t^i, f \rangle = \int_a^b t^i f(t) \, dt, \tag{4.62}$$

are, as before, the Gram matrix K consisting of inner products between basis monomials along with the vector $\mathbf{f}$ of inner products between the monomials and the right hand side. The coefficients of the least squares minimizing polynomial are thus found by solving the associated normal equations $K\mathbf{x} = \mathbf{f}$.

EXAMPLE 4.21

Let us return to the problem of approximating the exponential function $f(t) = e^t$ by a quadratic polynomial on the interval $0 \leq t \leq 1$, but now with the least squares error being measured by the L^2 norm. Thus, we consider the subspace $\mathcal{P}^{(2)}$ consisting of all quadratic polynomials

$$p(t) = \alpha + \beta t + \gamma t^2.$$

Using the monomial basis $1, t, t^2$, the normal equations are

$$\begin{pmatrix} 1 & \frac{1}{2} & \frac{1}{3} \\ \frac{1}{2} & \frac{1}{3} & \frac{1}{4} \\ \frac{1}{3} & \frac{1}{4} & \frac{1}{5} \end{pmatrix} \begin{pmatrix} \alpha \\ \beta \\ \gamma \end{pmatrix} = \begin{pmatrix} e - 1 \\ 1 \\ e - 2 \end{pmatrix}.$$

‡*Warning*: Here, the indices i, j labeling the entries of the $(n+1) \times (n+1)$ matrix K and vectors $\mathbf{x}, \mathbf{f} \in \mathbb{R}^{n+1}$ range from 0 to n instead of 1 to $n+1$.

The coefficient matrix is the Gram matrix K consisting of the inner products

$$\langle t^i, t^j \rangle = \int_0^1 t^{i+j}\, dt = \frac{1}{i+j+1}$$

between basis monomials, while the right hand side is the vector of inner products

$$\langle t^i, e^t \rangle = \int_0^1 t^i\, e^t\, dt.$$

The solution to the normal system is computed to be

$$\begin{aligned} \alpha &= 39\,e - 105 \simeq 1.012991, \\ \beta &= -216\,e + 588 \simeq .851125, \\ \gamma &= 210\,e - 570 \simeq .839184, \end{aligned}$$

leading to the least squares quadratic approximant

$$p^\star(t) = 1.012991 + .851125\, t + .839184\, t^2, \tag{4.63}$$

plotted in Figure 4.13. The least squares error is

$$\| e^t - p^\star(t) \| \simeq .00527593.$$

The maximal error over the interval is measured by the L^∞ norm of the difference:

$$\| e^t - p^\star(t) \|_\infty = \max \left\{ \, | e^t - p^\star(t) | \;\middle|\; 0 \le t \le 1 \, \right\} \simeq .014981815,$$

with the maximum occurring at $t = 1$. Thus, the simple quadratic polynomial (4.63) will give a reasonable approximation to the first two decimal places in e^t on the entire interval $[\,0, 1\,]$. A more accurate approximation can be made by taking a higher degree polynomial, or by decreasing the length of the interval. ●

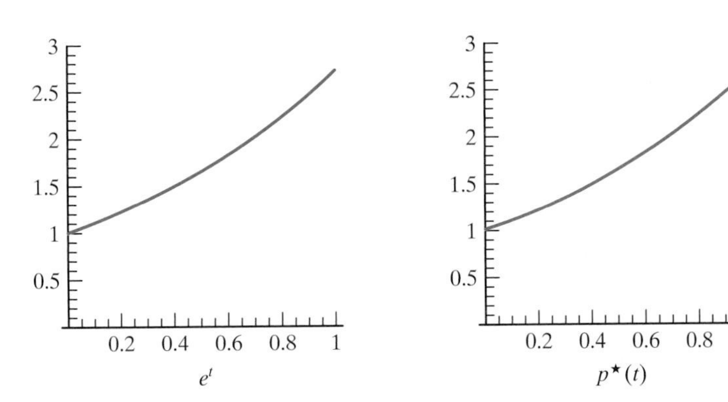

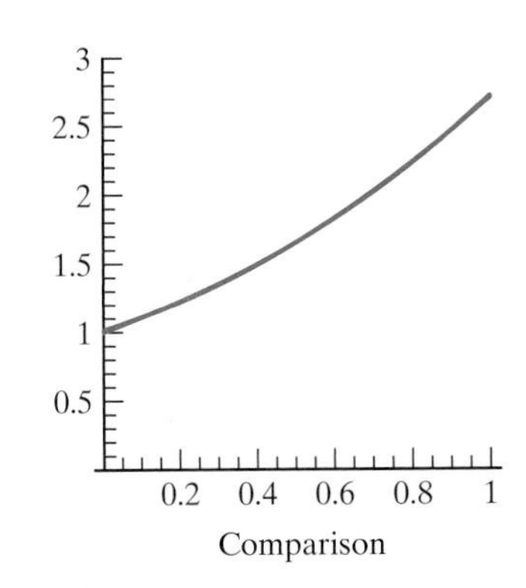

Figure 4.13 Quadratic least squares approximation of e^t.

REMARK: Although the least squares polynomial (4.63) minimizes the L^2 norm of the error, it does slightly worse with the L^∞ norm than the previous sample-based minimizer (4.45). The problem of finding the quadratic polynomial that minimizes the L^∞ norm is more difficult, and must be solved by nonlinear minimization techniques.

REMARK: As noted in Example 3.33, the Gram matrix for the simple monomial basis is the $n \times n$ Hilbert matrix (1.70). The ill conditioned nature of the Hilbert matrix and the consequential difficulty in accurately solving the normal equations complicate the practical numerical implementation of high degree least squares polynomial approximations. A better approach, based on an alternative orthogonal polynomial basis, will be discussed in in the ensuing chapter.

EXERCISES 4.4

4.4.43. Approximate the function $f(t) = \sqrt[3]{t}$ using the least squares method based on the L^2 norm on the interval $[0, 1]$ by

(a) a straight line (b) a parabola

(c) a cubic polynomial

4.4.44. Approximate the function $f(t) = \frac{1}{8}(2t-1)^3 + \frac{1}{4}$ by a quadratic polynomial on the interval $[-1, 1]$ using the least squares method based on the L^2 norm. Compare the graphs. Where is the error the largest?

4.4.45. For the function $f(t) = \sin t$ determine the best approximating linear and quadratic polynomials that minimize the least squares error based on the L^2 norm on $\left[0, \frac{1}{2}\pi\right]$.

4.4.46. Find the quartic (degree 4) polynomial that best approximates the function e^t on the interval $[0, 1]$ by minimizing the L^2 error (4.58).

♡ **4.4.47.** (a) Find the quadratic interpolant to $f(x) = x^5$ on the interval [0, 1] based on equally spaced data points.

(b) Find the quadratic least squares approximation based on the data points 0, .25, .5, .75, 1.

(c) Find the quadratic least squares approximation with respect to the L^2 norm.

(d) Discuss the strengths and weaknesses of each approximation.

4.4.48. Let $f(x) = x$. Find the trigonometric function of the form $g(x) = a + b\cos x + c\sin x$ that minimizes the L^2 error

$$\|g - f\| = \sqrt{\int_{-\pi}^{\pi} [g(x) - f(x)]^2 \, dx}\,.$$

◇ **4.4.49.** Let $g_1(t), \ldots, g_n(t)$ be prescribed, linearly independent functions. Explain how to best approximate a function $f(t)$ by a linear combination $c_1 g_1(t) + \cdots + c_n g_n(t)$ when the least squares error is measured in a weighted L^2 norm $\|f\|_w^2 = \int_a^b f(t)^2 \, w(t)\, dt$ with weight function $w(t) > 0$.

4.4.50. (a) Find the quadratic least squares approximation to $f(t) = t^5$ on the interval $[0, 1]$ with weights

(i) $w(t) = 1$ (ii) $w(t) = t$

(iii) $w(t) = e^{-t}$

(b) Compare the errors—which gives the best result over the entire interval?

4.4.51. Let $f_a(x) = \sqrt{\dfrac{a}{1 + a^4 x^2}}$. Prove that

(a) $\|f_a\|_2 = \sqrt{\dfrac{\pi}{a}}$, where $\|\cdot\|_2$ denotes the L^2 norm on $(-\infty, \infty)$.

(b) $\|f_a\|_\infty = \sqrt{a}$, where $\|\cdot\|_\infty$ denotes the L^∞ norm on $(-\infty, \infty)$.

(c) Use this example to explain why having a small least squares error does not necessarily mean that the functions are everywhere close.

4.4.52. Find the plane $z = \alpha + \beta x + \gamma y$ that best approximates the following functions on the square

$$S = \{-1 \le x \le 1, -1 \le y \le 1\}$$

using the L^2 norm

$$\|f\|^2 = \iint_S |f(x, y)|^2 \, dx \, dy$$

to measure the least squares error:

(a) $x^2 + y^2$ (b) $x^3 - y^3$

(c) $\sin \pi x \, \sin \pi y$

4.4.53. Find the radial polynomial $p(x, y) = a + br + cr^2$, where $r^2 = x^2 + y^2$, that best approximates the function $f(x, y) = x$ using the L^2 norm on the unit disk $D = \{r \le 1\}$ to measure the least squares error.

CHAPTER 5

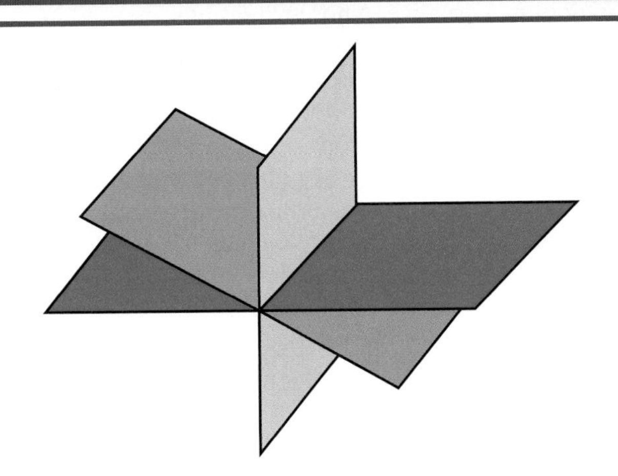

Orthogonality

Orthogonality is the mathematical formalization of the geometrical property of perpendicularity, as adapted to general inner product spaces. In linear algebra, bases consisting of mutually orthogonal elements play an essential role in theoretical developments, in a broad range of applications, and in the design of practical numerical algorithms. Computations become dramatically simpler and less prone to numerical instabilities when performed in orthogonal coordinate systems. Indeed, many large scale modern applications would be impractical, if not completely infeasible were it not for the dramatic simplifying power of orthogonality. The orthogonality of eigenvector and eigenfunction bases for symmetric matrices and self-adjoint boundary value problems is the key to understanding the dynamics of discrete and continuous mechanical, thermodynamical, and electrical systems, as well as all of Fourier analysis and signal processing.

The duly famous Gram–Schmidt process will convert an arbitrary basis of an inner product space into an orthogonal basis. In Euclidean space, the Gram–Schmidt process can be reinterpreted as a new kind of matrix factorization, in which a nonsingular matrix $A = Q\,R$ is written as the product of an orthogonal matrix Q and an upper triangular matrix R. The $Q\,R$ factorization and its generalizations are used in statistical data analysis as well the design of numerical algorithms for computing eigenvalues and eigenvectors. In function space, the Gram–Schmidt algorithm is employed to construct orthogonal polynomials and other useful systems of orthogonal functions.

Orthogonality is motivated by geometry, and orthogonal matrices are of fundamental importance in the mathematics of symmetry, in computer graphics and animation, and in image processing. The orthogonal projection of a point onto a subspace turns out to be the closest point or least squares minimizer. Moreover, when written in terms of an orthogonal basis for the subspace, the orthogonal projection has an elegant explicit formula that not only avoids setting up and solving the normal equations, but also offers numerical advantages over the direct approach to least squares minimization. Yet another important fact is that the four fundamental subspaces of a matrix come in mutually orthogonal pairs. This observation leads directly to a new characterization of the compatibility conditions for linear algebraic systems known as the Fredholm alternative, whose extensions are used in the analysis of boundary value problems and differential equations.

One of the most fertile applications of orthogonal bases is in signal processing. Fourier analysis decomposes a signal into its simple periodic components—sines and cosines—which form an orthogonal system of functions. Modern digital media, such as CD's, DVD's and MP3's, are based on discrete data obtained by sampling the physical signal. The Discrete Fourier Transform uses orthogonality to decompose the sampled signal vector into a linear combination of sampled trigonometric functions (or, more correctly, complex exponentials). Basic data compression and noise removal algorithms are applied to the discrete Fourier coefficients, acting on the observation that noise tends to accumulate in the high frequency Fourier modes. Section 5.7 is devoted to the basics of discrete Fourier analysis, culminating in the remarkable Fast Fourier Transform, a key algorithm in modern signal processing and numerical analysis.

5.1 ORTHOGONAL BASES

Let V be a fixed real* inner product space. Recall that two elements $\mathbf{v}, \mathbf{w} \in V$ are called *orthogonal* if their inner product vanishes: $\langle \mathbf{v}, \mathbf{w} \rangle = 0$. In the case of vectors in Euclidean space, orthogonality under the dot product means that they meet at a right angle.

A particularly important configuration is when V admits a basis consisting of mutually orthogonal elements.

Definition 5.1 A basis $\mathbf{u}_1, \ldots, \mathbf{u}_n$ of V is called *orthogonal* if $\langle \mathbf{u}_i, \mathbf{u}_j \rangle = 0$ for all $i \neq j$. The basis is called *orthonormal* if, in addition, each vector has unit length: $\| \mathbf{u}_i \| = 1$, for all $i = 1, \ldots, n$.

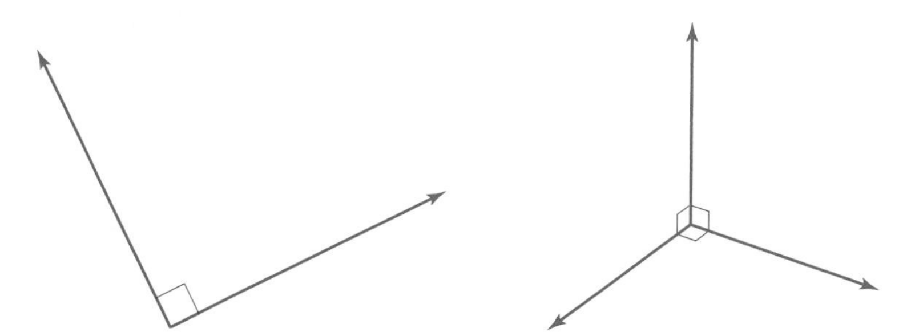

Figure 5.1 Orthonormal bases in $\mathbb{R}^2$ and $\mathbb{R}^3$.

*The methods can be adapted more or less straightforwardly to complex inner product spaces. The main complication, as noted in Section 3.6, is that we need to be careful with the order of vectors appearing in the non-symmetric complex inner products. In this chapter, we will write the inner product formulas in the proper order so that they retain their validity in complex vector spaces.

For the Euclidean space $\mathbb{R}^n$ equipped with the standard dot product, the simplest example of an orthonormal basis is the standard basis

$$\mathbf{e}_1 = \begin{pmatrix} 1 \\ 0 \\ 0 \\ \vdots \\ 0 \\ 0 \end{pmatrix}, \quad \mathbf{e}_2 = \begin{pmatrix} 0 \\ 1 \\ 0 \\ \vdots \\ 0 \\ 0 \end{pmatrix}, \quad \ldots \quad \mathbf{e}_n = \begin{pmatrix} 0 \\ 0 \\ 0 \\ \vdots \\ 0 \\ 1 \end{pmatrix}.$$

Orthogonality follows because $\mathbf{e}_i \cdot \mathbf{e}_j = 0$, for $i \neq j$, while $\|\mathbf{e}_i\| = 1$ implies normality.

Since a basis cannot contain the zero vector, there is an easy way to convert an orthogonal basis to an orthonormal basis. Namely, we replace each basis vector with a unit vector pointing in the same direction, as in Lemma 3.14.

Lemma 5.2 If $\mathbf{v}_1, \ldots, \mathbf{v}_n$ is an orthogonal basis of a vector space V, then the normalized vectors $\mathbf{u}_i = \mathbf{v}_i / \|\mathbf{v}_i\|$, $i = 1, \ldots, n$, form an orthonormal basis.

EXAMPLE 5.3 The vectors

$$\mathbf{v}_1 = \begin{pmatrix} 1 \\ 2 \\ -1 \end{pmatrix}, \quad \mathbf{v}_2 = \begin{pmatrix} 0 \\ 1 \\ 2 \end{pmatrix}, \quad \mathbf{v}_3 = \begin{pmatrix} 5 \\ -2 \\ 1 \end{pmatrix},$$

are easily seen to form a basis of $\mathbb{R}^3$. Moreover, they are mutually perpendicular, $\mathbf{v}_1 \cdot \mathbf{v}_2 = \mathbf{v}_1 \cdot \mathbf{v}_3 = \mathbf{v}_2 \cdot \mathbf{v}_3 = 0$, and so form an orthogonal basis with respect to the standard dot product on $\mathbb{R}^3$. When we divide each orthogonal basis vector by its length, the result is the orthonormal basis

$$\mathbf{u}_1 = \frac{1}{\sqrt{6}} \begin{pmatrix} 1 \\ 2 \\ -1 \end{pmatrix} = \begin{pmatrix} \frac{1}{\sqrt{6}} \\ \frac{2}{\sqrt{6}} \\ -\frac{1}{\sqrt{6}} \end{pmatrix},$$

$$\mathbf{u}_2 = \frac{1}{\sqrt{5}} \begin{pmatrix} 0 \\ 1 \\ 2 \end{pmatrix} = \begin{pmatrix} 0 \\ \frac{1}{\sqrt{5}} \\ \frac{2}{\sqrt{5}} \end{pmatrix},$$

$$\mathbf{u}_3 = \frac{1}{\sqrt{30}} \begin{pmatrix} 5 \\ -2 \\ 1 \end{pmatrix} = \begin{pmatrix} \frac{5}{\sqrt{30}} \\ -\frac{2}{\sqrt{30}} \\ \frac{1}{\sqrt{30}} \end{pmatrix},$$

satisfying $\mathbf{u}_1 \cdot \mathbf{u}_2 = \mathbf{u}_1 \cdot \mathbf{u}_3 = \mathbf{u}_2 \cdot \mathbf{u}_3 = 0$ and $\|\mathbf{u}_1\| = \|\mathbf{u}_2\| = \|\mathbf{u}_3\| = 1$. The appearance of square roots in the elements of an orthonormal basis is fairly typical. ●

A useful observation is that any orthogonal collection of nonzero vectors is automatically linearly independent.

Proposition 5.4 If $\mathbf{v}_1, \ldots, \mathbf{v}_k \in V$ are nonzero, mutually orthogonal elements, so $\mathbf{v}_i \neq \mathbf{0}$ and $\langle \mathbf{v}_i, \mathbf{v}_j \rangle = 0$ for all $i \neq j$, then they are linearly independent.

Proof Suppose

$$c_1 \mathbf{v}_1 + \cdots + c_k \mathbf{v}_k = \mathbf{0}.$$

Let us take the inner product of this equation with any $\mathbf{v}_i$. Using linearity of the inner product and orthogonality, we compute

$$\begin{aligned} 0 &= \langle c_1 \mathbf{v}_1 + \cdots + c_k \mathbf{v}_k , \mathbf{v}_i \rangle \\ &= c_1 \langle \mathbf{v}_1 , \mathbf{v}_i \rangle + \cdots + c_k \langle \mathbf{v}_k , \mathbf{v}_i \rangle \\ &= c_i \langle \mathbf{v}_i , \mathbf{v}_i \rangle = c_i \, \| \mathbf{v}_i \|^2 . \end{aligned}$$

Therefore, provided $\mathbf{v}_i \neq \mathbf{0}$, we conclude that the coefficient $c_i = 0$. Since this holds for all $i = 1, \ldots , k$, the linear independence of $\mathbf{v}_1, \ldots , \mathbf{v}_k$ follows. ■

As a direct corollary, we infer that any collection of nonzero orthogonal vectors forms a basis for its span.

THEOREM 5.5 Suppose $\mathbf{v}_1, \ldots , \mathbf{v}_n \in V$ are nonzero, mutually orthogonal elements of an inner product space V. Then $\mathbf{v}_1, \ldots , \mathbf{v}_n$ form an orthogonal basis for their span $W = \operatorname{span}\{\mathbf{v}_1, \ldots , \mathbf{v}_n\} \subset V$, which is therefore a subspace of dimension $n = \dim W$. In particular, if $\dim V = n$, then $\mathbf{v}_1, \ldots , \mathbf{v}_n$ form a orthogonal basis for V.

Orthogonality is also of profound significance for function spaces. Here is a relatively simple example.

EXAMPLE 5.6 Consider the vector space $\mathcal{P}^{(2)}$ consisting of all quadratic polynomials $p(x) = \alpha + \beta x + \gamma x^2$, equipped with the L^2 inner product and norm

$$\langle p , q \rangle = \int_0^1 p(x)\, q(x)\, dx, \qquad \| p \| = \sqrt{\langle p , p \rangle} = \sqrt{\int_0^1 p(x)^2 \, dx} \,.$$

The standard monomials $1, x, x^2$ do *not* form an orthogonal basis. Indeed,

$$\langle 1 , x \rangle = \tfrac{1}{2}, \qquad \langle 1 , x^2 \rangle = \tfrac{1}{3}, \qquad \langle x , x^2 \rangle = \tfrac{1}{4}.$$

One orthogonal basis of $\mathcal{P}^{(2)}$ is provided by following polynomials:

$$p_1(x) = 1, \qquad p_2(x) = x - \tfrac{1}{2}, \qquad p_3(x) = x^2 - x + \tfrac{1}{6}. \tag{5.1}$$

Indeed, one easily verifies that $\langle p_1 , p_2 \rangle = \langle p_1 , p_3 \rangle = \langle p_2 , p_3 \rangle = 0$, while

$$\| p_1 \| = 1, \qquad \| p_2 \| = \frac{1}{\sqrt{12}} = \frac{1}{2\sqrt{3}}, \qquad \| p_3 \| = \frac{1}{\sqrt{180}} = \frac{1}{6\sqrt{5}}.$$

The corresponding orthonormal basis is found by dividing each orthogonal basis element by its norm:

$$u_1(x) = 1, \qquad u_2(x) = \sqrt{3}\,(2x - 1), \qquad u_3(x) = \sqrt{5}\left(6x^2 - 6x + 1\right).$$

In Section 5.4 below, we will learn how to systematically construct such orthogonal systems of polynomials. ●

EXERCISES 5.1

5.1.1. Let $\mathbb{R}^2$ have the standard dot product. Classify the following pairs of vectors as

(i) basis,

(ii) orthogonal basis, and/or

(iii) orthonormal basis:

(a) $\mathbf{v}_1 = \begin{pmatrix} -1 \\ 2 \end{pmatrix}, \mathbf{v}_2 = \begin{pmatrix} 2 \\ 1 \end{pmatrix}$

(b) $\mathbf{v}_1 = \begin{pmatrix} \frac{1}{\sqrt{2}} \\ \frac{1}{\sqrt{2}} \end{pmatrix}, \mathbf{v}_2 = \begin{pmatrix} -\frac{1}{\sqrt{2}} \\ \frac{1}{\sqrt{2}} \end{pmatrix}$

(c) $\mathbf{v}_1 = \begin{pmatrix} -1 \\ -1 \end{pmatrix}, \mathbf{v}_2 = \begin{pmatrix} 2 \\ 2 \end{pmatrix}$

(d) $\mathbf{v}_1 = \begin{pmatrix} 2 \\ 3 \end{pmatrix}, \mathbf{v}_2 = \begin{pmatrix} 1 \\ -6 \end{pmatrix}$

(e) $\mathbf{v}_1 = \begin{pmatrix} -1 \\ 0 \end{pmatrix}, \mathbf{v}_2 = \begin{pmatrix} 0 \\ 3 \end{pmatrix}$

(f) $\mathbf{v}_1 = \begin{pmatrix} \frac{3}{5} \\ \frac{4}{5} \end{pmatrix}, \mathbf{v}_2 = \begin{pmatrix} -\frac{4}{5} \\ \frac{3}{5} \end{pmatrix}$

5.1.2. Let $\mathbb{R}^3$ have the standard dot product. Classify the following sets of vectors as

(i) basis,

(ii) orthogonal basis, and/or

(iii) orthonormal basis:

(a) $\begin{pmatrix} 1 \\ 1 \\ 0 \end{pmatrix}, \begin{pmatrix} 1 \\ -1 \\ 1 \end{pmatrix}, \begin{pmatrix} 0 \\ 1 \\ 1 \end{pmatrix}$

(b) $\begin{pmatrix} -\frac{4}{13} \\ \frac{3}{5} \\ -\frac{48}{65} \end{pmatrix}, \begin{pmatrix} \frac{12}{13} \\ 0 \\ -\frac{5}{13} \end{pmatrix}, \begin{pmatrix} \frac{3}{13} \\ \frac{4}{5} \\ \frac{36}{65} \end{pmatrix}$

(c) $\begin{pmatrix} 0 \\ \frac{1}{\sqrt{2}} \\ -\frac{1}{\sqrt{2}} \end{pmatrix}, \begin{pmatrix} -\frac{1}{\sqrt{2}} \\ 0 \\ \frac{1}{\sqrt{2}} \end{pmatrix}, \begin{pmatrix} \frac{1}{\sqrt{2}} \\ -\frac{1}{\sqrt{2}} \\ 0 \end{pmatrix}$

5.1.3. Repeat Exercise 5.1.1, but use the weighted inner product $\langle \mathbf{v}, \mathbf{w} \rangle = v_1 w_1 + \frac{1}{9} v_2 w_2$ instead of the dot product.

5.1.4. Show that the standard basis vectors $\mathbf{e}_1, \mathbf{e}_2, \mathbf{e}_3$ form an orthogonal basis with respect to the weighted inner product $\langle \mathbf{v}, \mathbf{w} \rangle = v_1 w_1 + 2 v_2 w_2 + 3 v_3 w_3$ on $\mathbb{R}^3$. Find an orthonormal basis for this inner product space.

5.1.5. Find all values of a so that the vectors

$$\begin{pmatrix} a \\ 1 \end{pmatrix}, \quad \begin{pmatrix} -a \\ 1 \end{pmatrix}$$

form an orthogonal basis of $\mathbb{R}^2$ under

(a) the dot product;

(b) the weighted inner product $\langle \mathbf{v}, \mathbf{w} \rangle = 3 v_1 w_1 + 2 v_2 w_2$;

(c) the inner product prescribed by the positive definite matrix $K = \begin{pmatrix} 2 & -1 \\ -1 & 3 \end{pmatrix}$.

5.1.6. Find all possible values of a and b in the inner product $\langle \mathbf{v}, \mathbf{w} \rangle = a\, v_1 w_1 + b\, v_2 w_2$ that make the vectors $(1, 2)^T, (-1, 1)^T$, an orthogonal basis in $\mathbb{R}^2$.

5.1.7. Answer Exercise 5.1.6 for the vectors

(a) $(2, 3)^T, (-2, 2)^T$ (b) $(1, 4)^T, (2, 1)^T$

5.1.8. Find an inner product such that the vectors $(-1, 2)^T$ and $(1, 2)^T$ form an orthonormal basis of $\mathbb{R}^2$.

5.1.9. *True or false*: If $\mathbf{v}_1, \mathbf{v}_2, \mathbf{v}_3$ are a basis for $\mathbb{R}^3$, then they form an orthogonal basis under some appropriately weighted inner product $\langle \mathbf{v}, \mathbf{w} \rangle = a\, v_1 w_1 + b\, v_2 w_2 + c\, v_3 w_3$.

♡ **5.1.10.** The *cross product* between vectors in $\mathbb{R}^3$ is defined by the formula

$$\mathbf{v} \times \mathbf{w} = \begin{pmatrix} v_2 w_3 - v_3 w_2 \\ v_3 w_1 - v_1 w_3 \\ v_1 w_2 - v_2 w_1 \end{pmatrix}, \qquad (5.2)$$

where

$$\mathbf{v} = \begin{pmatrix} v_1 \\ v_2 \\ v_3 \end{pmatrix}, \quad \mathbf{w} = \begin{pmatrix} w_1 \\ w_2 \\ w_3 \end{pmatrix}.$$

(a) Show that $\mathbf{u} = \mathbf{v} \times \mathbf{w}$ is orthogonal, under the dot product, to both $\mathbf{v}$ and $\mathbf{w}$.

(b) Show that $\mathbf{v} \times \mathbf{w} = \mathbf{0}$ if and only if $\mathbf{v}$ and $\mathbf{w}$ are parallel.

(c) Prove that if $\mathbf{v}, \mathbf{w} \in \mathbb{R}^3$ are orthogonal nonzero vectors, then $\mathbf{u} = \mathbf{v} \times \mathbf{w}, \mathbf{v}, \mathbf{w}$ form an orthogonal basis of $\mathbb{R}^3$.

(d) *True or false*: If $\mathbf{v}, \mathbf{w} \in \mathbb{R}^3$ are orthogonal unit vectors, then $\mathbf{v}, \mathbf{w}$ and $\mathbf{u} = \mathbf{v} \times \mathbf{w}$ form an orthonormal basis of $\mathbb{R}^3$.

◇ **5.1.11.** Prove that every orthonormal basis of $\mathbb{R}^2$ under the standard dot product has the form

$$\mathbf{u}_1 = \begin{pmatrix} \cos\theta \\ \sin\theta \end{pmatrix} \quad \text{and} \quad \mathbf{u}_2 = \begin{pmatrix} \sin\theta \\ \pm\cos\theta \end{pmatrix}$$

for some $0 \leq \theta < 2\pi$ and some choice of $\pm$ sign.

◇ **5.1.12.** Given angles θ, φ, ψ, prove that the vectors

$$\mathbf{u}_1 = \begin{pmatrix} \cos\psi\cos\varphi - \cos\theta\sin\varphi\sin\psi \\ -\sin\psi\cos\varphi - \cos\theta\sin\varphi\cos\psi \\ \sin\theta\sin\varphi \end{pmatrix},$$

$$\mathbf{u}_2 = \begin{pmatrix} \cos\psi\sin\varphi + \cos\theta\cos\varphi\sin\psi \\ -\sin\psi\sin\varphi + \cos\theta\cos\varphi\cos\psi \\ -\sin\theta\cos\varphi \end{pmatrix},$$

$$\mathbf{u}_3 = \begin{pmatrix} \sin\theta\sin\varphi \\ \sin\theta\cos\varphi \\ \cos\theta \end{pmatrix},$$

form an orthonormal basis of $\mathbb{R}^3$ under the standard dot product. *Remark*: It can be proved, [26, p. 147], that *every* orthonormal basis of $\mathbb{R}^3$ has the form $\mathbf{u}_1, \mathbf{u}_2, \pm\mathbf{u}_3$ for some choice of angles θ, φ, ψ.

♡ **5.1.13.** (a) Show that $\mathbf{v}_1, \ldots, \mathbf{v}_n$ form an orthonormal basis of $\mathbb{R}^n$ for the inner product $\langle \mathbf{v}, \mathbf{w} \rangle = \mathbf{v}^T K \mathbf{w}$ for $K > 0$ if and only if $A^T K A = \mathrm{I}$ where $A = (\, \mathbf{v}_1\ \mathbf{v}_2 \ldots \mathbf{v}_n\,)$.

(b) Prove that any basis of $\mathbb{R}^n$ is an orthonormal basis with respect to some inner product. Is the inner product uniquely determined?

(c) Find the inner product on $\mathbb{R}^2$ that makes $\mathbf{v}_1 = (1, 1)^T$, $\mathbf{v}_2 = (2, 3)^T$ into an orthonormal basis.

(d) Find the inner product on $\mathbb{R}^3$ that makes $\mathbf{v}_1 = (1, 1, 1)^T$, $\mathbf{v}_2 = (1, 1, 2)^T$, $\mathbf{v}_3 = (1, 2, 3)^T$ into an orthonormal basis.

5.1.14. Describe all orthonormal bases of $\mathbb{R}^2$ for the inner products

(a) $\langle \mathbf{v}, \mathbf{w} \rangle = \mathbf{v}^T \begin{pmatrix} 1 & 0 \\ 0 & 2 \end{pmatrix} \mathbf{w}$;

(b) $\langle \mathbf{v}, \mathbf{w} \rangle = \mathbf{v}^T \begin{pmatrix} 1 & -1 \\ -1 & 2 \end{pmatrix} \mathbf{w}$.

5.1.15. Let $\mathbf{v}$ and $\mathbf{w}$ be elements of an inner product space. Prove that $\|\mathbf{v} + \mathbf{w}\|^2 = \|\mathbf{v}\|^2 + \|\mathbf{w}\|^2$ if and only if $\mathbf{v}, \mathbf{w}$ are orthogonal. Explain why this formula can be viewed as the generalization of the Pythagorean Theorem.

5.1.16. Prove that if $\mathbf{v}_1, \mathbf{v}_2$ form a basis of an inner product space V and $\|\mathbf{v}_1\| = \|\mathbf{v}_2\|$ then $\mathbf{v}_1 + \mathbf{v}_2$ and $\mathbf{v}_1 - \mathbf{v}_2$ form an orthogonal basis of V.

5.1.17. Suppose $\mathbf{v}_1, \ldots, \mathbf{v}_k$ are nonzero mutually orthogonal elements of an inner product space V. Write down their Gram matrix. Why is it nonsingular?

5.1.18. Let $V = \mathcal{P}^{(1)}$ be the vector space consisting of linear polynomials $p(t) = a\,t + b$.

(a) Carefully explain why

$$\langle p, q \rangle = \int_0^1 t\, p(t)\, q(t)\, dt$$

defines an inner product on V.

(b) Find all polynomials $p(t) = a\,t + b \in V$ that are orthogonal to $p_1(t) = 1$ based on this inner product.

(c) Use part (b) to construct an orthonormal basis of V for this inner product.

(d) Find an orthonormal basis of the space $\mathcal{P}^{(2)}$ of quadratic polynomials for the same inner product. *Hint*: First find a quadratic polynomial that is orthogonal to the basis you constructed in part (c).

5.1.19. Explain why the functions $\cos x$, $\sin x$ form an orthogonal basis for the space of solutions to the differential equation $y'' + y = 0$ under the L^2 inner product on $[\,-\pi, \pi\,]$.

5.1.20. Do the functions $e^{x/2}, e^{-x/2}$ form an orthogonal basis for the space of solutions to the differential equation $4\,y'' - y = 0$ under the L^2 inner product on $[\,0, 1\,]$? If not, can you find an orthogonal basis of the solution space?

Computations in Orthogonal Bases

What are the advantages of orthogonal and orthonormal bases? Once one has a basis of a vector space, a key issue is how to express other elements as linear combinations of the basis elements—that is, to find their *coordinates* in the prescribed basis. In general, this is not so easy, since it requires solving a system of linear equations, cf. (2.22). In high dimensional situations arising in applications, computing the solution may require a considerable, if not infeasible amount of time and effort.

However, if the basis is orthogonal, or, even better, orthonormal, then the change of basis computation requires almost no work. This is the crucial insight underlying the efficacy of both discrete and continuous Fourier analysis, least squares approximations and statistical analysis of large data sets, signal, image and video processing, and a multitude of other applications, both classical and modern.

THEOREM 5.7 Let $\mathbf{u}_1, \ldots, \mathbf{u}_n$ be an orthonormal basis for an inner product space V. Then one can write any element $\mathbf{v} \in V$ as a linear combination

$$\mathbf{v} = c_1 \mathbf{u}_1 + \cdots + c_n \mathbf{u}_n, \tag{5.3}$$

in which its *coordinates*

$$c_i = \langle \mathbf{v}, \mathbf{u}_i \rangle, \qquad i = 1, \ldots, n, \tag{5.4}$$

are explicitly given as inner products. Moreover, its norm

$$\|\mathbf{v}\| = \sqrt{c_1^2 + \cdots + c_n^2} = \sqrt{\sum_{i=1}^{n} \langle \mathbf{v}, \mathbf{u}_i \rangle^2} \tag{5.5}$$

is the square root of the sum of the squares of its orthonormal basis coordinates.

Proof Let us compute the inner product of (5.3) with one of the basis vectors. Using the orthonormality conditions

$$\langle \mathbf{u}_i, \mathbf{u}_j \rangle = \begin{cases} 0 & i \neq j, \\ 1 & i = j, \end{cases} \tag{5.6}$$

and bilinearity of the inner product, we find

$$\langle \mathbf{v}, \mathbf{u}_i \rangle = \left\langle \sum_{j=1}^{n} c_j \mathbf{u}_j, \mathbf{u}_i \right\rangle = \sum_{j=1}^{n} c_j \langle \mathbf{u}_j, \mathbf{u}_i \rangle = c_i \|\mathbf{u}_i\|^2 = c_i.$$

To prove formula (5.5), we similarly expand

$$\|\mathbf{v}\|^2 = \langle \mathbf{v}, \mathbf{v} \rangle = \sum_{i,j=1}^{n} c_i c_j \langle \mathbf{u}_i, \mathbf{u}_j \rangle = \sum_{i=1}^{n} c_i^2,$$

again making use of the orthonormality of the basis elements. ∎

EXAMPLE 5.8 Let us rewrite the vector $\mathbf{v} = (1, 1, 1)^T$ in terms of the orthonormal basis

$$\mathbf{u}_1 = \begin{pmatrix} \frac{1}{\sqrt{6}} \\ \frac{2}{\sqrt{6}} \\ -\frac{1}{\sqrt{6}} \end{pmatrix}, \qquad \mathbf{u}_2 = \begin{pmatrix} 0 \\ \frac{1}{\sqrt{5}} \\ \frac{2}{\sqrt{5}} \end{pmatrix}, \qquad \mathbf{u}_3 = \begin{pmatrix} \frac{5}{\sqrt{30}} \\ -\frac{2}{\sqrt{30}} \\ \frac{1}{\sqrt{30}} \end{pmatrix},$$

constructed in Example 5.3. Computing the dot products

$$\mathbf{v} \cdot \mathbf{u}_1 = \frac{2}{\sqrt{6}}, \qquad \mathbf{v} \cdot \mathbf{u}_2 = \frac{3}{\sqrt{5}}, \qquad \mathbf{v} \cdot \mathbf{u}_3 = \frac{4}{\sqrt{30}},$$

we immediately conclude that

$$\mathbf{v} = \frac{2}{\sqrt{6}} \mathbf{u}_1 + \frac{3}{\sqrt{5}} \mathbf{u}_2 + \frac{4}{\sqrt{30}} \mathbf{u}_3.$$

Needless to say, a direct computation based on solving the associated linear system, as in Chapter 2, is more tedious. ●

While passage from an orthogonal basis to its orthonormal version is elementary—one simply divides each basis element by its norm—we shall often find it more convenient to work directly with the unnormalized version. The next result provides the corresponding formula expressing a vector in terms of an orthogonal, but not necessarily orthonormal basis. The proof proceeds exactly as in the orthonormal case, and details are left to the reader.

THEOREM 5.9 If $\mathbf{v}_1, \ldots, \mathbf{v}_n$ form an orthogonal basis, then the corresponding coordinates of a vector

$$\mathbf{v} = a_1 \mathbf{v}_1 + \cdots + a_n \mathbf{v}_n \quad \text{are given by} \quad a_i = \frac{\langle \mathbf{v}, \mathbf{v}_i \rangle}{\| \mathbf{v}_i \|^2}. \tag{5.7}$$

In this case, its norm can be computed using the formula

$$\| \mathbf{v} \|^2 = \sum_{i=1}^{n} a_i^2 \, \| \mathbf{v}_i \|^2 = \sum_{i=1}^{n} \left(\frac{\langle \mathbf{v}, \mathbf{v}_i \rangle}{\| \mathbf{v}_i \|} \right)^2. \tag{5.8}$$

Equation (5.7), along with its orthonormal simplification (5.4), is one of the most useful formulas we shall establish, and applications will appear repeatedly throughout the sequel.

EXAMPLE 5.10 The wavelet basis

$$\mathbf{v}_1 = \begin{pmatrix} 1 \\ 1 \\ 1 \\ 1 \end{pmatrix}, \quad \mathbf{v}_2 = \begin{pmatrix} 1 \\ 1 \\ -1 \\ -1 \end{pmatrix}, \quad \mathbf{v}_3 = \begin{pmatrix} 1 \\ -1 \\ 0 \\ 0 \end{pmatrix}, \quad \mathbf{v}_4 = \begin{pmatrix} 0 \\ 0 \\ 1 \\ -1 \end{pmatrix}, \tag{5.9}$$

introduced in Example 2.35 is, in fact, an orthogonal basis of $\mathbb{R}^4$. The norms are

$$\| \mathbf{v}_1 \| = 2, \qquad \| \mathbf{v}_2 \| = 2, \qquad \| \mathbf{v}_3 \| = \sqrt{2}, \qquad \| \mathbf{v}_4 \| = \sqrt{2}.$$

Therefore, using (5.7), we can readily express any vector as a linear combination of the wavelet basis vectors. For example,

$$\mathbf{v} = \begin{pmatrix} 4 \\ -2 \\ 1 \\ 5 \end{pmatrix} = 2\,\mathbf{v}_1 - \mathbf{v}_2 + 3\,\mathbf{v}_3 - 2\,\mathbf{v}_4,$$

where the wavelet coordinates are computed directly by

$$\frac{\langle \mathbf{v}, \mathbf{v}_1 \rangle}{\| \mathbf{v}_1 \|^2} = \frac{8}{4} = 2, \qquad \frac{\langle \mathbf{v}, \mathbf{v}_2 \rangle}{\| \mathbf{v}_2 \|^2} = \frac{-4}{4} = -1,$$

$$\frac{\langle \mathbf{v}, \mathbf{v}_3 \rangle}{\| \mathbf{v}_3 \|^2} = \frac{6}{2} = 3, \qquad \frac{\langle \mathbf{v}, \mathbf{v}_4 \rangle}{\| \mathbf{v}_4 \|^2} = \frac{-4}{2} = -2.$$

This is clearly quicker than solving the linear system, as we did in Example 2.35. Finally, we note that

$$\begin{aligned} 46 = \| \mathbf{v} \|^2 &= 2^2 \, \| \mathbf{v}_1 \|^2 + (-1)^2 \, \| \mathbf{v}_2 \|^2 + 3^2 \, \| \mathbf{v}_3 \|^2 + (-2)^2 \, \| \mathbf{v}_4 \|^2 \\ &= 4 \cdot 4 + 1 \cdot 4 + 9 \cdot 2 + 4 \cdot 2, \end{aligned}$$

in conformity with (5.8). ●

EXAMPLE 5.11 The same formulae are equally valid for orthogonal bases in function spaces. For example, to express a quadratic polynomial

$$p(x) = c_1 \, p_1(x) + c_2 \, p_2(x) + c_3 \, p_3(x) = c_1 + c_2 \left(x - \tfrac{1}{2} \right) + c_3 \left(x^2 - x + \tfrac{1}{6} \right)$$

in terms of the orthogonal basis (5.1), we merely compute the L^2 inner product integrals

$$c_1 = \frac{\langle p, p_1 \rangle}{\| p_1 \|^2} = \int_0^1 p(x)\, dx,$$

$$c_2 = \frac{\langle p, p_2 \rangle}{\| p_2 \|^2} = 12 \int_0^1 p(x)\left(x - \tfrac{1}{2}\right) dx,$$

$$c_3 = \frac{\langle p, p_3 \rangle}{\| p_3 \|^2} = 180 \int_0^1 p(x)\left(x^2 - x + \tfrac{1}{6}\right) dx.$$

Thus, for example, the coefficients for $p(x) = x^2 + x + 1$ are

$$c_1 = \int_0^1 (x^2 + x + 1)\, dx = \tfrac{11}{6},$$

$$c_2 = 12 \int_0^1 (x^2 + x + 1)\left(x - \tfrac{1}{2}\right) dx = 2,$$

$$c_3 = 180 \int_0^1 (x^2 + x + 1)\left(x^2 - x + \tfrac{1}{6}\right) dx = 1,$$

and so

$$p(x) = x^2 + x + 1 = \tfrac{11}{6} + 2\left(x - \tfrac{1}{2}\right) + \left(x^2 - x + \tfrac{1}{6}\right). \qquad \bullet$$

EXAMPLE 5.12 Perhaps the most important example of an orthogonal basis is provided by the basic trigonometric functions. Let $\mathcal{T}^{(n)}$ denote the vector space consisting of all *trigonometric polynomials*

$$T(x) = \sum_{0 \le j+k \le n} a_{jk} (\sin x)^j (\cos x)^k \tag{5.10}$$

of *degree* $\le n$. The individual monomials $(\sin x)^j (\cos x)^k$ span $\mathcal{T}^n$, but, as we saw in Example 2.20, they do not form a basis owing to identities stemming from the basic trigonometric formula $\cos^2 x + \sin^2 x = 1$. Exercise 3.6.21 introduced a more convenient spanning set consisting of the $2n + 1$ functions

$$1, \quad \cos x, \quad \sin x, \quad \cos 2x, \quad \sin 2x, \quad \ldots \quad \cos nx, \quad \sin nx. \tag{5.11}$$

Let us prove that these functions form an orthogonal basis of $\mathcal{T}^n$ with respect to the L^2 inner product and norm:

$$\langle f, g \rangle = \int_{-\pi}^{\pi} f(x)\, g(x)\, dx, \qquad \| f \|^2 = \int_{-\pi}^{\pi} f(x)^2\, dx. \tag{5.12}$$

The elementary integration formulae

$$\begin{aligned}
&\int_{-\pi}^{\pi} \cos kx \, \cos lx \, dx = \begin{cases} 0, & k \ne l, \\ 2\pi, & k = l = 0, \\ \pi, & k = l \ne 0, \end{cases} \\
&\int_{-\pi}^{\pi} \sin kx \, \sin lx \, dx = \begin{cases} 0, & k \ne l, \\ \pi, & k = l \ne 0, \end{cases} \\
&\int_{-\pi}^{\pi} \cos kx \, \sin lx \, dx = 0,
\end{aligned} \tag{5.13}$$

which are valid for all nonnegative integers $k, l \geq 0$, imply the orthogonality relations

$$\langle \cos kx, \cos lx \rangle = \langle \sin kx, \sin lx \rangle = 0, \quad k \neq l, \quad \langle \cos kx, \sin lx \rangle = 0,$$
$$\| \cos kx \| = \| \sin kx \| = \sqrt{\pi}, \quad k \neq 0, \quad \| 1 \| = \sqrt{2\pi}. \tag{5.14}$$

Theorem 5.5 now assures us that the functions (5.11) form a basis for $\mathcal{T}^n$. One consequence is that $\dim \mathcal{T}^n = 2n+1$—a fact that is not so easy to establish directly.

Orthogonality of the trigonometric functions (5.11) means that we can compute the coefficients $a_0, \ldots, a_n, b_1, \ldots, b_n$ of any trigonometric polynomial

$$p(x) = a_0 + \sum_{k=1}^{n} \left(a_k \cos kx + b_k \sin kx \right) \tag{5.15}$$

by an explicit integration formula. Namely,

$$a_0 = \frac{\langle f, 1 \rangle}{\| 1 \|^2} = \frac{1}{2\pi} \int_{-\pi}^{\pi} f(x)\, dx,$$
$$a_k = \frac{\langle f, \cos kx \rangle}{\| \cos kx \|^2} = \frac{1}{\pi} \int_{-\pi}^{\pi} f(x) \cos kx\, dx, \tag{5.16}$$
$$b_k = \frac{\langle f, \sin kx \rangle}{\| \sin kx \|^2} = \frac{1}{\pi} \int_{-\pi}^{\pi} f(x) \sin kx\, dx, \quad k \geq 1.$$

These formulae play an essential role in the theory and applications of Fourier series, [16, 59]. ●

EXERCISES 5.1

5.1.21. (a) Prove that

$$\mathbf{v}_1 = \begin{pmatrix} \frac{3}{5} \\ 0 \\ \frac{4}{5} \end{pmatrix}, \quad \mathbf{v}_2 = \begin{pmatrix} -\frac{4}{13} \\ \frac{12}{13} \\ \frac{3}{13} \end{pmatrix}, \quad \mathbf{v}_3 = \begin{pmatrix} -\frac{48}{65} \\ -\frac{5}{13} \\ \frac{36}{65} \end{pmatrix}$$

form an orthonormal basis for $\mathbb{R}^3$ for the usual dot product.

(b) Find the coordinates of $\mathbf{v} = (1, 1, 1)^T$ relative to this basis.

(c) Verify formula (5.5) in this particular case.

♡ **5.1.22.** (a) Prove that the vectors

$$\mathbf{v}_1 = \begin{pmatrix} 1 \\ 1 \\ 1 \end{pmatrix}, \quad \mathbf{v}_2 = \begin{pmatrix} 1 \\ 1 \\ -2 \end{pmatrix}, \quad \mathbf{v}_3 = \begin{pmatrix} -1 \\ 1 \\ 0 \end{pmatrix}$$

form an orthogonal basis of $\mathbb{R}^3$ with the dot product.

(b) Use orthogonality to write the vector $\mathbf{v} = (1, 2, 3)^T$ as a linear combination of $\mathbf{v}_1, \mathbf{v}_2, \mathbf{v}_3$.

(c) Verify the formula (5.8) for $\| \mathbf{v} \|$.

(d) Construct an orthonormal basis, using the given vectors.

(e) Write $\mathbf{v}$ as a linear combination of the orthonormal basis, and verify (5.5).

5.1.23. Let $\mathbb{R}^2$ have the inner product defined by the positive definite matrix

$$K = \begin{pmatrix} 2 & -1 \\ -1 & 3 \end{pmatrix}.$$

(a) Show that $\mathbf{v}_1 = (1, 1)^T$, $\mathbf{v}_2 = (-2, 1)^T$ form an orthogonal basis.

(b) Write the vector $\mathbf{v} = (3, 2)^T$ as a linear combination of $\mathbf{v}_1, \mathbf{v}_2$ using the orthogonality formula (5.7).

(c) Verify the formula (5.8) for $\| \mathbf{v} \|$.

(d) Find an orthonormal basis $\mathbf{u}_1, \mathbf{u}_2$ for this inner product.

(e) Write $\mathbf{v}$ as a linear combination of the orthonormal basis, and verify (5.5).

5.1.24. Find an example that demonstrates why equation (5.5) is not valid for a non-orthonormal basis.

5.1.25. Use orthogonality to write the polynomials $1, x$ and x^2 as linear combinations of the orthogonal basis (5.1).

5.1.26. (a) Prove that the polynomials

$$P_0(t) = 1, \quad P_1(t) = t,$$
$$P_2(t) = t^2 - \tfrac{1}{3}, \quad P_3(t) = t^3 - \tfrac{3}{5}t$$

form an orthogonal basis for the vector space $\mathcal{P}^3$ of cubic polynomials for the L^2 inner product

$$\langle f, g \rangle = \int_{-1}^{1} f(t)\, g(t)\, dt.$$

(b) Find an orthonormal basis of $\mathcal{P}^3$.

(c) Write t^3 as a linear combination of P_0, P_1, P_2, P_3 using the orthogonal basis formula (5.7).

5.1.27. (a) Prove that the polynomials

$$P_0(t) = 1, \quad P_1(t) = t - \tfrac{2}{3},$$
$$P_2(t) = t^2 - \tfrac{6}{5}t + \tfrac{3}{10}$$

form an orthogonal basis for $\mathcal{P}^2$ with respect to the weighted inner product

$$\langle f, g \rangle = \int_{0}^{1} f(t)\, g(t)\, t\, dt.$$

(b) Find the corresponding orthonormal basis.

(c) Write t^2 as a linear combination of P_0, P_1, P_2 using the orthogonal basis formula (5.7).

5.1.28. Write the following trigonometric polynomials in terms of the basis functions (5.11):

(a) $\cos^2 x$ (b) $\cos x \sin x$

(c) $\sin^3 x$ (d) $\cos^2 x \sin^3 x$

(e) $\cos^4 x$

Hint: You can use complex exponentials to simplify the inner product integrals.

5.1.29. Write down an orthonormal basis of the space of trigonometric polynomials $\mathcal{T}^{(n)}$ with respect to the L^2 inner product

$$\langle f, g \rangle = \int_{-\pi}^{\pi} f(x)\, g(x)\, dx.$$

◇ **5.1.30.** Show that the $2n+1$ complex exponentials e^{ikx} for $k = -n, -n+1, \ldots, -1, 0, 1, \ldots, n$, form an orthonormal basis for the space of complex-valued trigonometric polynomials under the Hermitian inner product

$$\langle f, g \rangle = \frac{1}{2\pi}\int_{-\pi}^{\pi} f(x)\, \overline{g(x)}\, dx.$$

◇ **5.1.31.** Prove the trigonometric integral identities (5.13). *Hint*: You can either use a trigonometric summation identity, or, if you can't remember the right one, use Euler's formula (3.86) to rewrite sine and cosine as combinations of complex exponentials.

◇ **5.1.32.** Fill in the complete details of the proof of Theorem 5.9.

5.2 THE GRAM–SCHMIDT PROCESS

Once we become convinced of the utility of orthogonal and orthonormal bases, a natural question arises: How can we construct them? A practical algorithm was first discovered by the French mathematician Pierre–Simon Laplace in the eighteenth century. Today the algorithm is known as the *Gram–Schmidt process*, after its rediscovery by Gram, whom we already met in Chapter 3, and Erhard Schmidt, a nineteenth century German mathematician. The Gram–Schmidt process is one of the premier algorithms of applied and computational linear algebra.

Let V denote a finite-dimensional inner product space. (To begin with, you might wish to think of V as a subspace of $\mathbb{R}^m$, equipped with the standard Euclidean dot product, although the algorithm will be formulated in complete generality.) We assume that we already know some basis $\mathbf{w}_1, \ldots, \mathbf{w}_n$ of V, where $n = \dim V$. Our goal is to use this information to construct an orthogonal basis $\mathbf{v}_1, \ldots, \mathbf{v}_n$.

We will construct the orthogonal basis elements one by one. Since initially we are not worrying about normality, there are no conditions on the first orthogonal basis element $\mathbf{v}_1$, and so there is no harm in choosing

$$\mathbf{v}_1 = \mathbf{w}_1.$$

Note that $\mathbf{v}_1 \neq \mathbf{0}$, since $\mathbf{w}_1$ appears in the original basis. The second basis vector must be orthogonal to the first: $\langle \mathbf{v}_2, \mathbf{v}_1 \rangle = 0$. Let us try to arrange this by subtracting a suitable multiple of $\mathbf{v}_1$, and set

$$\mathbf{v}_2 = \mathbf{w}_2 - c\,\mathbf{v}_1,$$

where c is a scalar to be determined. The orthogonality condition

$$0 = \langle \mathbf{v}_2, \mathbf{v}_1 \rangle = \langle \mathbf{w}_2, \mathbf{v}_1 \rangle - c \, \langle \mathbf{v}_1, \mathbf{v}_1 \rangle = \langle \mathbf{w}_2, \mathbf{v}_1 \rangle - c \, \| \mathbf{v}_1 \|^2$$

requires that $c = \langle \mathbf{w}_2, \mathbf{v}_1 \rangle / \| \mathbf{v}_1 \|^2$, and therefore

$$\mathbf{v}_2 = \mathbf{w}_2 - \frac{\langle \mathbf{w}_2, \mathbf{v}_1 \rangle}{\| \mathbf{v}_1 \|^2} \, \mathbf{v}_1. \tag{5.17}$$

Linear independence of $\mathbf{v}_1 = \mathbf{w}_1$ and $\mathbf{w}_2$ ensures that $\mathbf{v}_2 \neq \mathbf{0}$.

Next, we construct

$$\mathbf{v}_3 = \mathbf{w}_3 - c_1 \, \mathbf{v}_1 - c_2 \, \mathbf{v}_2$$

by subtracting suitable multiples of the first two orthogonal basis elements from $\mathbf{w}_3$. We want $\mathbf{v}_3$ to be orthogonal to both $\mathbf{v}_1$ and $\mathbf{v}_2$. Since we already arranged that $\langle \mathbf{v}_1, \mathbf{v}_2 \rangle = 0$, this requires

$$0 = \langle \mathbf{v}_3, \mathbf{v}_1 \rangle = \langle \mathbf{w}_3, \mathbf{v}_1 \rangle - c_1 \, \langle \mathbf{v}_1, \mathbf{v}_1 \rangle, \qquad 0 = \langle \mathbf{v}_3, \mathbf{v}_2 \rangle = \langle \mathbf{w}_3, \mathbf{v}_2 \rangle - c_2 \, \langle \mathbf{v}_2, \mathbf{v}_2 \rangle,$$

and hence

$$c_1 = \frac{\langle \mathbf{w}_3, \mathbf{v}_1 \rangle}{\| \mathbf{v}_1 \|^2}, \qquad c_2 = \frac{\langle \mathbf{w}_3, \mathbf{v}_2 \rangle}{\| \mathbf{v}_2 \|^2}.$$

Therefore, the next orthogonal basis vector is given by the formula

$$\mathbf{v}_3 = \mathbf{w}_3 - \frac{\langle \mathbf{w}_3, \mathbf{v}_1 \rangle}{\| \mathbf{v}_1 \|^2} \, \mathbf{v}_1 - \frac{\langle \mathbf{w}_3, \mathbf{v}_2 \rangle}{\| \mathbf{v}_2 \|^2} \, \mathbf{v}_2.$$

Since $\mathbf{v}_1$ and $\mathbf{v}_2$ are linear combinations of $\mathbf{w}_1$ and $\mathbf{w}_2$, we must have $\mathbf{v}_3 \neq \mathbf{0}$, as otherwise this would imply that $\mathbf{w}_1, \mathbf{w}_2, \mathbf{w}_3$ are linearly dependent, and hence could not come from a basis.

Continuing in the same manner, suppose we have already constructed the mutually orthogonal vectors $\mathbf{v}_1, \ldots, \mathbf{v}_{k-1}$ as linear combinations of $\mathbf{w}_1, \ldots, \mathbf{w}_{k-1}$. The next orthogonal basis element $\mathbf{v}_k$ will be obtained from $\mathbf{w}_k$ by subtracting off a suitable linear combination of the previous orthogonal basis elements:

$$\mathbf{v}_k = \mathbf{w}_k - c_1 \, \mathbf{v}_1 - \cdots - c_{k-1} \, \mathbf{v}_{k-1}.$$

Since $\mathbf{v}_1, \ldots, \mathbf{v}_{k-1}$ are already orthogonal, the orthogonality constraint

$$0 = \langle \mathbf{v}_k, \mathbf{v}_j \rangle = \langle \mathbf{w}_k, \mathbf{v}_j \rangle - c_j \, \langle \mathbf{v}_j, \mathbf{v}_j \rangle$$

requires

$$c_j = \frac{\langle \mathbf{w}_k, \mathbf{v}_j \rangle}{\| \mathbf{v}_j \|^2} \quad \text{for} \quad j = 1, \ldots, k-1. \tag{5.18}$$

In this fashion, we establish the general *Gram–Schmidt formula*

$$\mathbf{v}_k = \mathbf{w}_k - \sum_{j=1}^{k-1} \frac{\langle \mathbf{w}_k, \mathbf{v}_j \rangle}{\| \mathbf{v}_j \|^2} \, \mathbf{v}_j, \qquad k = 1, \ldots, n. \tag{5.19}$$

The Gram–Schmidt process (5.19) defines an explicit, recursive procedure for constructing the orthogonal basis vectors $\mathbf{v}_1, \ldots, \mathbf{v}_n$. If we are actually after an orthonormal basis $\mathbf{u}_1, \ldots, \mathbf{u}_n$, we merely normalize the resulting orthogonal basis vectors, setting $\mathbf{u}_k = \mathbf{v}_k / \| \mathbf{v}_k \|$ for each $k = 1, \ldots, n$.

EXAMPLE 5.13 The vectors

$$\mathbf{w}_1 = \begin{pmatrix} 1 \\ 1 \\ -1 \end{pmatrix}, \qquad \mathbf{w}_2 = \begin{pmatrix} 1 \\ 0 \\ 2 \end{pmatrix}, \qquad \mathbf{w}_3 = \begin{pmatrix} 2 \\ -2 \\ 3 \end{pmatrix}, \tag{5.20}$$

are readily seen to form a basis* of $\mathbb{R}^3$. To construct an orthogonal basis (with respect to the standard dot product) using the Gram–Schmidt procedure, we begin by setting

$$\mathbf{v}_1 = \mathbf{w}_1 = \begin{pmatrix} 1 \\ 1 \\ -1 \end{pmatrix}.$$

The next basis vector is

$$\mathbf{v}_2 = \mathbf{w}_2 - \frac{\mathbf{w}_2 \cdot \mathbf{v}_1}{\|\mathbf{v}_1\|^2}\,\mathbf{v}_1 = \begin{pmatrix} 1 \\ 0 \\ 2 \end{pmatrix} - \frac{-1}{3}\begin{pmatrix} 1 \\ 1 \\ -1 \end{pmatrix} = \begin{pmatrix} \frac{4}{3} \\ \frac{1}{3} \\ \frac{5}{3} \end{pmatrix}.$$

The last orthogonal basis vector is

$$\begin{aligned} \mathbf{v}_3 &= \mathbf{w}_3 - \frac{\mathbf{w}_3 \cdot \mathbf{v}_1}{\|\mathbf{v}_1\|^2}\,\mathbf{v}_1 - \frac{\mathbf{w}_3 \cdot \mathbf{v}_2}{\|\mathbf{v}_2\|^2}\,\mathbf{v}_2 \\ &= \begin{pmatrix} 2 \\ -2 \\ 3 \end{pmatrix} - \frac{-3}{3}\begin{pmatrix} 1 \\ 1 \\ -1 \end{pmatrix} - \frac{7}{\frac{14}{3}}\begin{pmatrix} \frac{4}{3} \\ \frac{1}{3} \\ \frac{5}{3} \end{pmatrix} = \begin{pmatrix} 1 \\ -\frac{3}{2} \\ -\frac{1}{2} \end{pmatrix}. \end{aligned}$$

The reader can easily validate the orthogonality of $\mathbf{v}_1, \mathbf{v}_2, \mathbf{v}_3$.

An orthonormal basis is obtained by dividing each vector by its length. Since

$$\|\mathbf{v}_1\| = \sqrt{3}, \qquad \|\mathbf{v}_2\| = \sqrt{\frac{14}{3}}, \qquad \|\mathbf{v}_3\| = \sqrt{\frac{7}{2}}.$$

we produce the corresponding orthonormal basis vectors

$$\mathbf{u}_1 = \begin{pmatrix} \frac{1}{\sqrt{3}} \\ \frac{1}{\sqrt{3}} \\ -\frac{1}{\sqrt{3}} \end{pmatrix}, \qquad \mathbf{u}_2 = \begin{pmatrix} \frac{4}{\sqrt{42}} \\ \frac{1}{\sqrt{42}} \\ \frac{5}{\sqrt{42}} \end{pmatrix}, \qquad \mathbf{u}_3 = \begin{pmatrix} \frac{2}{\sqrt{14}} \\ -\frac{3}{\sqrt{14}} \\ -\frac{1}{\sqrt{14}} \end{pmatrix}. \tag{5.21}$$

●

EXAMPLE 5.14 Here is a typical problem: find an orthonormal basis, with respect to the dot product, for the subspace $V \subset \mathbb{R}^4$ consisting of all vectors which are orthogonal to the given vector $\mathbf{a} = (1, 2, -1, -3)^T$. The first task is to find a basis for the subspace. Now, a vector $\mathbf{x} = (x_1, x_2, x_3, x_4)^T$ is orthogonal to $\mathbf{a}$ if and only if

$$\mathbf{x} \cdot \mathbf{a} = x_1 + 2x_2 - x_3 - 3x_4 = 0.$$

*This will, in fact, be a consequence of the successful completion of the Gram–Schmidt process and does not need to be checked in advance. If the given vectors were not linearly independent, then eventually one of the Gram–Schmidt vectors would vanish, $\mathbf{v}_k = \mathbf{0}$, and the process will break down.

Solving this homogeneous linear system by the usual method, we find that the free variables are x_2, x_3, x_4, and so a (non-orthogonal) basis for the subspace is

$$\mathbf{w}_1 = \begin{pmatrix} -2 \\ 1 \\ 0 \\ 0 \end{pmatrix}, \quad \mathbf{w}_2 = \begin{pmatrix} 1 \\ 0 \\ 1 \\ 0 \end{pmatrix}, \quad \mathbf{w}_3 = \begin{pmatrix} 3 \\ 0 \\ 0 \\ 1 \end{pmatrix}.$$

To obtain an orthogonal basis, we apply the Gram–Schmidt process. First,

$$\mathbf{v}_1 = \mathbf{w}_1 = \begin{pmatrix} -2 \\ 1 \\ 0 \\ 0 \end{pmatrix}.$$

The next element is

$$\mathbf{v}_2 = \mathbf{w}_2 - \frac{\mathbf{w}_2 \cdot \mathbf{v}_1}{\|\mathbf{v}_1\|^2}\, \mathbf{v}_1 = \begin{pmatrix} 1 \\ 0 \\ 1 \\ 0 \end{pmatrix} - \frac{-2}{5} \begin{pmatrix} -2 \\ 1 \\ 0 \\ 0 \end{pmatrix} = \begin{pmatrix} \frac{1}{5} \\ \frac{2}{5} \\ 1 \\ 0 \end{pmatrix}.$$

The last element of our orthogonal basis is

$$\begin{aligned} \mathbf{v}_3 &= \mathbf{w}_3 - \frac{\mathbf{w}_3 \cdot \mathbf{v}_1}{\|\mathbf{v}_1\|^2}\, \mathbf{v}_1 - \frac{\mathbf{w}_3 \cdot \mathbf{v}_2}{\|\mathbf{v}_2\|^2}\, \mathbf{v}_2 \\ &= \begin{pmatrix} 3 \\ 0 \\ 0 \\ 1 \end{pmatrix} - \frac{-6}{5} \begin{pmatrix} -2 \\ 1 \\ 0 \\ 0 \end{pmatrix} - \frac{\frac{3}{5}}{\frac{6}{5}} \begin{pmatrix} \frac{1}{5} \\ \frac{2}{5} \\ 1 \\ 0 \end{pmatrix} = \begin{pmatrix} \frac{1}{2} \\ 1 \\ -\frac{1}{2} \\ 1 \end{pmatrix}. \end{aligned}$$

An orthonormal basis can then be obtained by dividing each $\mathbf{v}_i$ by its length:

$$\mathbf{u}_1 = \begin{pmatrix} -\frac{2}{\sqrt{5}} \\ \frac{1}{\sqrt{5}} \\ 0 \\ 0 \end{pmatrix}, \quad \mathbf{u}_2 = \begin{pmatrix} \frac{1}{\sqrt{30}} \\ \frac{2}{\sqrt{30}} \\ \frac{5}{\sqrt{30}} \\ 0 \end{pmatrix}, \quad \mathbf{u}_3 = \begin{pmatrix} \frac{1}{\sqrt{10}} \\ \frac{2}{\sqrt{10}} \\ -\frac{1}{\sqrt{10}} \\ \frac{2}{\sqrt{10}} \end{pmatrix}. \tag{5.22}$$

●

The Gram–Schmidt procedure has one final important consequence. According to Theorem 2.29, every finite-dimensional vector space (except $\{\mathbf{0}\}$) admits a basis. Given an inner product, the Gram–Schmidt process enables one to construct an orthogonal and even orthonormal basis of the space. Therefore, we have, in fact, implemented a constructive proof of the existence of orthogonal and orthonormal bases of finite-dimensional inner product spaces.

THEOREM 5.15 Every non-zero finite-dimensional inner product space has an orthonormal basis.

In fact, if its dimension is > 1, then the inner product space has infinitely many orthonormal bases.

EXERCISES 5.2

Note: For exercise #1–8 use the Euclidean dot product on $\mathbb{R}^n$.

5.2.1. Use the Gram–Schmidt process to determine an orthonormal basis for $\mathbb{R}^3$ starting with the following sets of vectors:

(a) $\begin{pmatrix} 1\\0\\1 \end{pmatrix}, \begin{pmatrix} 1\\1\\1 \end{pmatrix}, \begin{pmatrix} -1\\2\\1 \end{pmatrix}$

(b) $\begin{pmatrix} 1\\1\\0 \end{pmatrix}, \begin{pmatrix} 0\\1\\-1 \end{pmatrix}, \begin{pmatrix} 1\\0\\-1 \end{pmatrix}$

(c) $\begin{pmatrix} 1\\2\\3 \end{pmatrix}, \begin{pmatrix} 4\\5\\0 \end{pmatrix}, \begin{pmatrix} 2\\3\\-1 \end{pmatrix}$

5.2.2. Use the Gram–Schmidt process to construct an orthonormal basis for $\mathbb{R}^4$ starting with the following sets of vectors:

(a) $(1,0,1,0)^T, (0,1,0,-1)^T, (1,0,0,1)^T, (1,1,1,1)^T$

(b) $(1,0,0,1)^T, (4,1,0,0)^T, (1,0,2,1)^T, (0,2,0,1)^T$.

5.2.3. Try the Gram–Schmidt procedure on the vectors

$$\begin{pmatrix} 1\\-1\\0\\1 \end{pmatrix}, \begin{pmatrix} 0\\-1\\1\\2 \end{pmatrix}, \begin{pmatrix} 2\\-1\\-1\\0 \end{pmatrix}, \begin{pmatrix} 2\\2\\-2\\1 \end{pmatrix}.$$

What happens? Can you explain why you are unable to complete the algorithm?

5.2.4. Use the Gram–Schmidt process to construct an orthonormal basis for the following subspaces of $\mathbb{R}^3$:

(a) the plane spanned by $(0,2,1)^T, (1,-2,-1)^T$;

(b) the plane defined by the equation $2x - y + 3z = 0$;

(c) the set of all vectors orthogonal to $(1,-1,-2)^T$.

5.2.5. Find an orthogonal basis of the subspace spanned by the vectors $\mathbf{w}_1 = (1,-1,-1,1,1)^T$, $\mathbf{w}_2 = (2,1,4,-4,2)^T$, and $\mathbf{w}_3 = (5,-4,-3,7,1)^T$.

5.2.6. Find an orthonormal basis for the following subspaces of $\mathbb{R}^4$:

(a) the span of the vectors

$$\begin{pmatrix} 1\\1\\-1\\0 \end{pmatrix}, \begin{pmatrix} -1\\0\\1\\1 \end{pmatrix}, \begin{pmatrix} 2\\-1\\2\\1 \end{pmatrix}$$

(b) the kernel of the matrix $\begin{pmatrix} 2 & 1 & 0 & -1\\ 3 & 2 & -1 & -1 \end{pmatrix}$

(c) the corange of the preceding matrix

(d) the range of the matrix $\begin{pmatrix} 1 & -2 & 2\\ 2 & -4 & 1\\ 0 & 0 & -1\\ -2 & 4 & 5 \end{pmatrix}$

(e) the cokernel of the preceding matrix

(f) the set of all vectors orthogonal to $(1,1,-1,-1)^T$

5.2.7. Find orthonormal bases for the four fundamental subspaces associated with the following matrices:

(a) $\begin{pmatrix} 1 & -1\\ -3 & 3 \end{pmatrix}$ (b) $\begin{pmatrix} -1 & 0 & 2\\ 1 & 1 & -1\\ 0 & 1 & 1 \end{pmatrix}$

(c) $\begin{pmatrix} 1 & 0 & 1 & 0\\ 1 & 1 & 1 & 1\\ -1 & 2 & 0 & 1 \end{pmatrix}$

(d) $\begin{pmatrix} 1 & 2 & 1\\ 0 & -2 & 1\\ -1 & 0 & -2\\ 1 & -2 & 3 \end{pmatrix}$

5.2.8. Redo Exercise 5.2.1 using

(i) the weighted inner product $\langle \mathbf{v}, \mathbf{w} \rangle = 3 v_1 w_1 + 2 v_2 w_2 + v_3 w_3$;

(ii) the inner product induced by the positive definite matrix

$$K = \begin{pmatrix} 2 & -1 & 0\\ -1 & 2 & -1\\ 0 & -1 & 2 \end{pmatrix}.$$

5.2.9. Construct an orthonormal basis of $\mathbb{R}^2$ for the nonstandard inner products

(a) $\langle \mathbf{x}, \mathbf{y} \rangle = \mathbf{x}^T \begin{pmatrix} 3 & 0\\ 0 & 5 \end{pmatrix} \mathbf{y}$,

(b) $\langle \mathbf{x}, \mathbf{y} \rangle = \mathbf{x}^T \begin{pmatrix} 4 & -1\\ -1 & 1 \end{pmatrix} \mathbf{y}$,

(c) $\langle \mathbf{x}, \mathbf{y} \rangle = \mathbf{x}^T \begin{pmatrix} 2 & 1\\ 1 & 3 \end{pmatrix} \mathbf{y}$.

5.2.10. Construct an orthonormal basis for $\mathbb{R}^3$ with respect to the inner products defined by the following positive definite matrices:

(a) $\begin{pmatrix} 4 & -2 & 0\\ -2 & 3 & -1\\ 0 & -1 & 2 \end{pmatrix}$ (b) $\begin{pmatrix} 3 & -1 & 1\\ -1 & 4 & -2\\ 1 & -2 & 4 \end{pmatrix}$

5.2.11. (a) How many orthonormal bases does $\mathbb{R}$ have?

(b) What about $\mathbb{R}^2$?

(c) Does your answer change if you use a different inner product?

◇ **5.2.12.** Verify that the Gram–Schmidt formulae (5.19) also produce an orthogonal basis of a complex vector space under a Hermitian inner product.

5.2.13. (a) Apply the complex Gram–Schmidt algorithm from Exercise 5.2.12 to produce an orthonormal basis starting with the vectors $(1+\mathrm{i}, 1-\mathrm{i})^T, (1-2\mathrm{i}, 5\mathrm{i})^T \in \mathbb{C}^2$.

(b) Do the same for $(1+\mathrm{i}, 1-\mathrm{i}, 2-\mathrm{i})^T$, $(1+2\mathrm{i}, -2\mathrm{i}, 2-\mathrm{i})^T, (1, 1-2\mathrm{i}, \mathrm{i})^T \in \mathbb{C}^3$.

5.2.14. Use the complex Gram–Schmidt algorithm from Exercise 5.2.12 to construct orthonormal bases for

(a) the subspace spanned by $(1-\mathrm{i}, 1, 0)^T, (0, 3-\mathrm{i}, 2\mathrm{i})^T$;

(b) the set of solutions to $(2-\mathrm{i})x - 2\mathrm{i}\,y + (1-2\mathrm{i})z = 0$;

(c) the subspace spanned by $(-\mathrm{i}, 1, -1, \mathrm{i})^T$, $(0, 2\mathrm{i}, 1-\mathrm{i}, -1+\mathrm{i})^T$, $(1, \mathrm{i}, -\mathrm{i}, 1-2\mathrm{i})^T$.

5.2.15. *True or false*: Reordering the original basis before starting the Gram–Schmidt process leads to the same orthogonal basis.

◇ **5.2.16.** Suppose that $V \subsetneq \mathbb{R}^n$ is a proper subspace, and $\mathbf{u}_1, \ldots, \mathbf{u}_m$ forms an orthonormal basis of V. Prove that there exist vectors $\mathbf{u}_{m+1}, \ldots, \mathbf{u}_n \in \mathbb{R}^n \setminus V$ such that the complete collection $\mathbf{u}_1, \ldots, \mathbf{u}_n$ forms an orthonormal basis for $\mathbb{R}^n$. *Hint*: Begin with Exercise 2.4.24.

Modifications of the Gram–Schmidt Process

With the basic Gram–Schmidt algorithm now in hand, it is worth looking at a couple of reformulations that have both practical and theoretical advantages. The first can be used to directly construct the orthonormal basis vectors $\mathbf{u}_1, \ldots, \mathbf{u}_n$ from the basis $\mathbf{w}_1, \ldots, \mathbf{w}_n$.

We begin by replacing each orthogonal basis vector in the basic Gram–Schmidt formula (5.19) by its normalized version $\mathbf{u}_j = \mathbf{v}_j / \| \mathbf{v}_j \|$. The original basis vectors can be expressed in terms of the orthonormal basis via a "triangular" system

$$\begin{aligned}
\mathbf{w}_1 &= r_{11}\,\mathbf{u}_1, \\
\mathbf{w}_2 &= r_{12}\,\mathbf{u}_1 + r_{22}\,\mathbf{u}_2, \\
\mathbf{w}_3 &= r_{13}\,\mathbf{u}_1 + r_{23}\,\mathbf{u}_2 + r_{33}\,\mathbf{u}_3, \\
&\;\;\vdots \qquad\quad \vdots \qquad\quad \vdots \qquad\quad \ddots \\
\mathbf{w}_n &= r_{1n}\,\mathbf{u}_1 + r_{2n}\,\mathbf{u}_2 + \cdots + r_{nn}\,\mathbf{u}_n.
\end{aligned} \tag{5.23}$$

The coefficients r_{ij} can, in fact, be computed directly from these formulae. Indeed, taking the inner product of the equation for $\mathbf{w}_j$ with the orthonormal basis vector $\mathbf{u}_i$ for $i \le j$, we find, in view of the orthonormality constraints (5.6),

$$\langle \mathbf{w}_j, \mathbf{u}_i \rangle = \langle r_{1j}\,\mathbf{u}_1 + \cdots + r_{jj}\,\mathbf{u}_j, \mathbf{u}_i \rangle = r_{1j}\,\langle \mathbf{u}_1, \mathbf{u}_i \rangle + \cdots + r_{jj}\,\langle \mathbf{u}_n, \mathbf{u}_i \rangle = r_{ij},$$

and hence

$$r_{ij} = \langle \mathbf{w}_j, \mathbf{u}_i \rangle. \tag{5.24}$$

On the other hand, according to (5.5),

$$\| \mathbf{w}_j \|^2 = \| r_{1j}\,\mathbf{u}_1 + \cdots + r_{jj}\,\mathbf{u}_j \|^2 = r_{1j}^2 + \cdots + r_{j-1,j}^2 + r_{jj}^2. \tag{5.25}$$

The pair of equations (5.24–25) can be rearranged to devise a recursive procedure to compute the orthonormal basis. At stage j, we assume that we have already constructed $\mathbf{u}_1, \ldots, \mathbf{u}_{j-1}$. We then compute†

$$r_{ij} = \langle \mathbf{w}_j, \mathbf{u}_i \rangle, \qquad \text{for each} \quad i = 1, \ldots, j-1. \tag{5.26}$$

†When $j = 1$, there is nothing to do.

We obtain the next orthonormal basis vector $\mathbf{u}_j$ by computing

$$r_{jj} = \sqrt{\|\mathbf{w}_j\|^2 - r_{1j}^2 - \cdots - r_{j-1,j}^2}\,, \qquad \mathbf{u}_j = \frac{\mathbf{w}_j - r_{1j}\,\mathbf{u}_1 - \cdots - r_{j-1,j}\,\mathbf{u}_{j-1}}{r_{jj}}\,. \tag{5.27}$$

Running through the formulae (5.26–27) for $j = 1, \ldots, n$ leads to the *same* orthonormal basis $\mathbf{u}_1, \ldots, \mathbf{u}_n$ as the previous version of the Gram–Schmidt procedure.

EXAMPLE 5.16 Let us apply the revised algorithm to the vectors

$$\mathbf{w}_1 = \begin{pmatrix} 1 \\ 1 \\ -1 \end{pmatrix}, \qquad \mathbf{w}_2 = \begin{pmatrix} 1 \\ 0 \\ 2 \end{pmatrix}, \qquad \mathbf{w}_3 = \begin{pmatrix} 2 \\ -2 \\ 3 \end{pmatrix},$$

of Example 5.13. To begin, we set

$$r_{11} = \|\mathbf{w}_1\| = \sqrt{3}\,, \qquad \mathbf{u}_1 = \frac{\mathbf{w}_1}{r_{11}} = \begin{pmatrix} \frac{1}{\sqrt{3}} \\ \frac{1}{\sqrt{3}} \\ -\frac{1}{\sqrt{3}} \end{pmatrix}.$$

The next step is to compute

$$r_{12} = \langle \mathbf{w}_2, \mathbf{u}_1 \rangle = -\frac{1}{\sqrt{3}}\,, \qquad r_{22} = \sqrt{\|\mathbf{w}_2\|^2 - r_{12}^2} = \sqrt{\frac{14}{3}}\,,$$

$$\mathbf{u}_2 = \frac{\mathbf{w}_2 - r_{12}\mathbf{u}_1}{r_{22}} = \begin{pmatrix} \frac{4}{\sqrt{42}} \\ \frac{1}{\sqrt{42}} \\ \frac{5}{\sqrt{42}} \end{pmatrix}.$$

The final step yields

$$r_{13} = \langle \mathbf{w}_3, \mathbf{u}_1 \rangle = -\sqrt{3}\,, \qquad r_{23} = \langle \mathbf{w}_3, \mathbf{u}_2 \rangle = \sqrt{\frac{21}{2}}\,,$$

$$r_{33} = \sqrt{\|\mathbf{w}_3\|^2 - r_{13}^2 - r_{23}^2} = \sqrt{\frac{7}{2}}\,, \qquad \mathbf{u}_3 = \frac{\mathbf{w}_3 - r_{13}\,\mathbf{u}_1 - r_{23}\,\mathbf{u}_2}{r_{33}} = \begin{pmatrix} \frac{2}{\sqrt{14}} \\ -\frac{3}{\sqrt{14}} \\ -\frac{1}{\sqrt{14}} \end{pmatrix}.$$

As advertised, the result is the same orthonormal basis vectors that we found in Example 5.13. ●

For hand computations, the original version (5.19) of the Gram–Schmidt process is slightly easier—even if one does ultimately want an orthonormal basis—since it avoids the square roots that are ubiquitous in the orthonormal version (5.26–27). On the other hand, for numerical implementation on a computer, the orthonormal version is a bit faster, as it involves fewer arithmetic operations.

However, in practical, large scale computations, both versions of the Gram–Schmidt process suffer from a serious flaw. They are subject to numerical instabilities, and so accumulating round-off errors may seriously corrupt the computations, leading to inaccurate, non-orthogonal vectors. Fortunately, there is a simple rearrangement of the calculation that obviates this difficulty and leads to the numerically robust algorithm that is most often used in practice, [18, 32, 51]. The idea is to treat the vectors simultaneously rather than sequentially, making full use of the orthonormal basis vectors as they arise. More specifically, the algorithm begins as before—we take $\mathbf{u}_1 = \mathbf{w}_1 / \|\mathbf{w}_1\|$. We then subtract off the appropriate multiples of $\mathbf{u}_1$ from *all* of the remaining basis vectors so as to arrange their orthogonality to $\mathbf{u}_1$. This is accomplished by setting

$$\mathbf{w}_k^{(2)} = \mathbf{w}_k - \langle \mathbf{w}_k, \mathbf{u}_1 \rangle \, \mathbf{u}_1 \quad \text{for} \quad k = 2, \ldots, n.$$

The second orthonormal basis vector $\mathbf{u}_2 = \mathbf{w}_2^{(2)} / \|\mathbf{w}_2^{(2)}\|$ is then obtained by normalizing. We next modify the remaining $\mathbf{w}_3^{(2)}, \ldots, \mathbf{w}_n^{(2)}$ to produce vectors

$$\mathbf{w}_k^{(3)} = \mathbf{w}_k^{(2)} - \langle \mathbf{w}_k^{(2)}, \mathbf{u}_2 \rangle \, \mathbf{u}_2, \qquad k = 3, \ldots, n,$$

that are orthogonal to both $\mathbf{u}_1$ and $\mathbf{u}_2$. Then $\mathbf{u}_3 = \mathbf{w}_3^{(3)} / \|\mathbf{w}_3^{(3)}\|$ is the next orthonormal basis element, and the process continues. The full algorithm starts with the initial basis vectors $\mathbf{w}_j = \mathbf{w}_j^{(1)}$, $j = 1, \ldots, n$, and then recursively computes

$$\mathbf{u}_j = \frac{\mathbf{w}_j^{(j)}}{\|\mathbf{w}_j^{(j)}\|}, \qquad \mathbf{w}_k^{(j+1)} = \mathbf{w}_k^{(j)} - \langle \mathbf{w}_k^{(j)}, \mathbf{u}_j \rangle \, \mathbf{u}_j, \qquad \begin{array}{l} j = 1, \ldots, n, \\ k = j+1, \ldots, n. \end{array} \tag{5.28}$$

(In the final phase, when $j = n$, the second formula is no longer needed.) The result is a numerically stable computation of the *same* orthonormal basis vectors $\mathbf{u}_1, \ldots, \mathbf{u}_n$.

EXAMPLE 5.17 Let us apply the stable Gram–Schmidt process (5.28) to the basis vectors

$$\mathbf{w}_1^{(1)} = \mathbf{w}_1 = \begin{pmatrix} 2 \\ 2 \\ -1 \end{pmatrix}, \quad \mathbf{w}_2^{(1)} = \mathbf{w}_2 = \begin{pmatrix} 0 \\ 4 \\ -1 \end{pmatrix}, \quad \mathbf{w}_3^{(1)} = \mathbf{w}_3 = \begin{pmatrix} 1 \\ 2 \\ -3 \end{pmatrix}.$$

The first orthonormal basis vector is

$$\mathbf{u}_1 = \frac{\mathbf{w}_1^{(1)}}{\|\mathbf{w}_1^{(1)}\|} = \begin{pmatrix} \frac{2}{3} \\ \frac{2}{3} \\ -\frac{1}{3} \end{pmatrix}.$$

Next, we compute

$$\mathbf{w}_2^{(2)} = \mathbf{w}_2^{(1)} - \langle \mathbf{w}_2^{(1)}, \mathbf{u}_1 \rangle \, \mathbf{u}_1 = \begin{pmatrix} -2 \\ 2 \\ 0 \end{pmatrix},$$

$$\mathbf{w}_3^{(2)} = \mathbf{w}_3^{(1)} - \langle \mathbf{w}_3^{(1)}, \mathbf{u}_1 \rangle \, \mathbf{u}_1 = \begin{pmatrix} -1 \\ 0 \\ -2 \end{pmatrix}.$$

The second orthonormal basis vector is

$$\mathbf{u}_2 = \frac{\mathbf{w}_2^{(2)}}{\|\mathbf{w}_2^{(2)}\|} = \begin{pmatrix} -\frac{1}{\sqrt{2}} \\ \frac{1}{\sqrt{2}} \\ 0 \end{pmatrix}.$$

Finally,

$$\mathbf{w}_3^{(3)} = \mathbf{w}_3^{(2)} - \langle \mathbf{w}_3^{(2)}, \mathbf{u}_2 \rangle \, \mathbf{u}_2 = \begin{pmatrix} -\frac{1}{2} \\ -\frac{1}{2} \\ -2 \end{pmatrix}, \qquad \mathbf{u}_3 = \frac{\mathbf{w}_3^{(3)}}{\| \mathbf{w}_3^{(3)} \|} = \begin{pmatrix} -\frac{\sqrt{2}}{6} \\ -\frac{\sqrt{2}}{6} \\ -\frac{2\sqrt{2}}{3} \end{pmatrix}.$$

The resulting vectors $\mathbf{u}_1, \mathbf{u}_2, \mathbf{u}_3$ form the desired orthonormal basis. ●

EXERCISES 5.2

5.2.17. Use the modified Gram-Schmidt process (5.26–27) to produce orthonormal bases for the spaces spanned by the following vectors:

(a) $\begin{pmatrix} 0 \\ 1 \\ 1 \end{pmatrix}, \begin{pmatrix} 1 \\ 0 \\ 1 \end{pmatrix}, \begin{pmatrix} 2 \\ 1 \\ 0 \end{pmatrix}$

(b) $\begin{pmatrix} 1 \\ 1 \\ -1 \\ 0 \end{pmatrix}, \begin{pmatrix} -1 \\ 0 \\ 1 \\ 1 \end{pmatrix}, \begin{pmatrix} 2 \\ -1 \\ 2 \\ 1 \end{pmatrix}$

(c) $\begin{pmatrix} 2 \\ 1 \\ 3 \\ 1 \\ 0 \end{pmatrix}, \begin{pmatrix} 0 \\ -1 \\ 2 \\ -1 \\ 1 \end{pmatrix}, \begin{pmatrix} 1 \\ 2 \\ -1 \\ 0 \\ 1 \end{pmatrix}$

(d) $\begin{pmatrix} 0 \\ 1 \\ 0 \\ 1 \\ 0 \end{pmatrix}, \begin{pmatrix} 1 \\ 0 \\ 1 \\ 1 \\ 0 \end{pmatrix}, \begin{pmatrix} 1 \\ -1 \\ 0 \\ 1 \\ -1 \end{pmatrix}, \begin{pmatrix} 1 \\ 0 \\ -1 \\ 0 \\ 1 \end{pmatrix}$

5.2.18. Repeat Exercise 5.2.17 using the numerically stable algorithm (5.28) and check that you get the same result. Which of the two algorithms was easier for you to implement?

5.2.19. Redo each of the exercises in the preceding subsection by implementing the numerically stable Gram-Schmidt process (5.28) instead, and verify that you end up with the same orthonormal basis.

◇ **5.2.20.** Prove that (5.28) does indeed produce an orthonormal basis. Explain why the result is the same orthonormal basis as the ordinary Gram–Schmidt method.

5.2.21. Let $\mathbf{w}_j^{(j)}$ be the vectors in the stable Gram–Schmidt algorithm (5.28). Prove that the coefficients in (5.23) are given by $r_{ii} = \| \mathbf{w}_i^{(i)} \|$, and $r_{ij} = \langle \mathbf{w}_j^{(i)}, \mathbf{u}_i \rangle$ for $i < j$.

5.3 ORTHOGONAL MATRICES

Matrices whose columns form an orthonormal basis of $\mathbb{R}^n$ relative to the standard Euclidean dot product play a distinguished role. Such "orthogonal matrices" appear in a wide range of applications in geometry, physics, quantum mechanics, crystallography, partial differential equations, symmetry theory, and special functions. Rotational motions of bodies in three-dimensional space are described by orthogonal matrices, and hence they lie at the foundations of rigid body mechanics, including satellite and underwater vehicle motions, as well as three-dimensional computer graphics and animation. Furthermore, orthogonal matrices are an essential ingredient in one of the most important methods of numerical linear algebra: the QR algorithm for computing eigenvalues of matrices, to be presented in Section 10.6.

Definition 5.18 A square matrix Q is called an *orthogonal matrix* if it satisfies

$$Q^T Q = \mathrm{I}. \tag{5.29}$$

The orthogonality condition implies that one can easily invert an orthogonal matrix:

$$Q^{-1} = Q^T. \tag{5.30}$$

In fact, the two conditions are equivalent, and hence a matrix is orthogonal if and only if its inverse is equal to its transpose. The second important characterization of orthogonal matrices relates them directly to orthonormal bases.

Proposition 5.19 A matrix Q is orthogonal if and only if its columns form an orthonormal basis with respect to the Euclidean dot product on $\mathbb{R}^n$.

Proof Let $\mathbf{u}_1, \ldots, \mathbf{u}_n$ be the columns of Q. Then $\mathbf{u}_1^T, \ldots, \mathbf{u}_n^T$ are the rows of the transposed matrix Q^T. The (i, j) entry of the product $Q^T Q$ is given as the product of the ith row of Q^T times the jth column of Q. Thus, the orthogonality requirement (5.29) implies

$$\mathbf{u}_i \cdot \mathbf{u}_j = \mathbf{u}_i^T \mathbf{u}_j = \begin{cases} 1, & i = j, \\ 0, & i \neq j, \end{cases}$$

which are precisely the conditions (5.6) for $\mathbf{u}_1, \ldots, \mathbf{u}_n$ to form an orthonormal basis. ■

Warning: Technically, we should be referring to an "orthonormal" matrix, not an "orthogonal" matrix. But the terminology is so standard throughout mathematics that we have no choice but to adopt it here. There is no commonly accepted name for a matrix whose columns form an orthogonal but not orthonormal basis.

EXAMPLE 5.20 A 2×2 matrix $Q = \begin{pmatrix} a & b \\ c & d \end{pmatrix}$ is orthogonal if and only if its columns

$$\mathbf{u}_1 = \begin{pmatrix} a \\ c \end{pmatrix}, \qquad \mathbf{u}_2 = \begin{pmatrix} b \\ d \end{pmatrix},$$

form an orthonormal basis of $\mathbb{R}^2$. Equivalently, the requirement

$$Q^T Q = \begin{pmatrix} a & c \\ b & d \end{pmatrix} \begin{pmatrix} a & b \\ c & d \end{pmatrix} = \begin{pmatrix} a^2 + c^2 & a\,c + b\,d \\ a\,c + b\,d & b^2 + d^2 \end{pmatrix} = \begin{pmatrix} 1 & 0 \\ 0 & 1 \end{pmatrix},$$

implies that its entries must satisfy the algebraic equations

$$a^2 + c^2 = 1, \qquad a\,c + b\,d = 0, \qquad b^2 + d^2 = 1.$$

The first and last equations say that the points $(a, c)^T$ and $(b, d)^T$ lie on the unit circle in $\mathbb{R}^2$, and so

$$a = \cos\theta, \qquad c = \sin\theta, \qquad b = \cos\psi, \qquad d = \sin\psi,$$

for some choice of angles θ, ψ. The remaining orthogonality condition is

$$0 = a\,c + b\,d = \cos\theta\ \cos\psi + \sin\theta\ \sin\psi = \cos(\theta - \psi),$$

which implies that θ and ψ differ by a right angle: $\psi = \theta \pm \frac{1}{2}\pi$. The $\pm$ sign leads to two cases:

$$b = -\sin\theta, \qquad d = \cos\theta, \qquad \text{or} \qquad b = \sin\theta, \qquad d = -\cos\theta.$$

As a result, every 2×2 orthogonal matrix has one of two possible forms

$$\begin{pmatrix} \cos\theta & -\sin\theta \\ \sin\theta & \cos\theta \end{pmatrix} \quad \text{or} \quad \begin{pmatrix} \cos\theta & \sin\theta \\ \sin\theta & -\cos\theta \end{pmatrix}, \qquad \text{where} \quad 0 \le \theta < 2\pi. \tag{5.31}$$

The corresponding orthonormal bases are illustrated in Figure 5.2. The former is a right handed basis, as defined in Exercise 2.4.7, and can be obtained from the standard basis $\mathbf{e}_1, \mathbf{e}_2$ by a rotation through angle θ, while the latter has the opposite, reflected orientation. ●

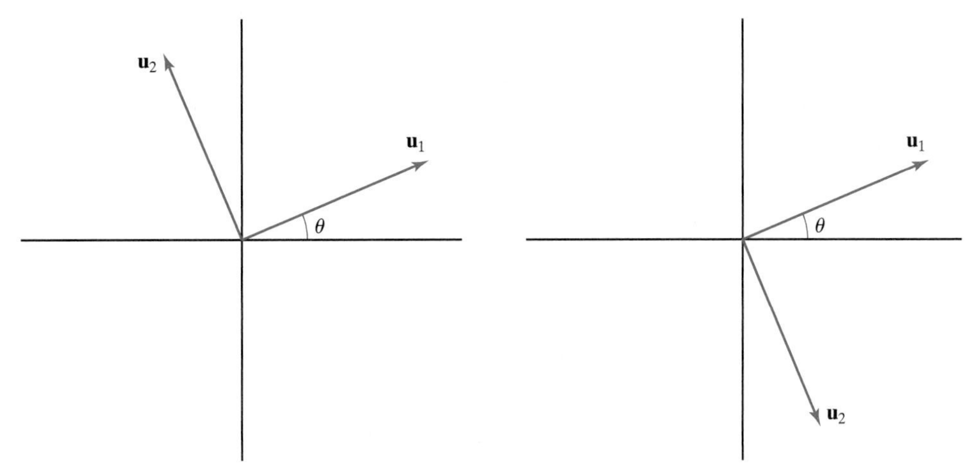

Figure 5.2 Orthonormal bases in $\mathbb{R}^2$.

EXAMPLE 5.21 A 3×3 orthogonal matrix $Q = (\mathbf{u}_1\ \mathbf{u}_2\ \mathbf{u}_3)$ is prescribed by 3 mutually perpendicular vectors of unit length in $\mathbb{R}^3$. For instance, the orthonormal basis constructed in (5.21) corresponds to the orthogonal matrix

$$Q = \begin{pmatrix} \frac{1}{\sqrt{3}} & \frac{4}{\sqrt{42}} & \frac{2}{\sqrt{14}} \\ \frac{1}{\sqrt{3}} & \frac{1}{\sqrt{42}} & -\frac{3}{\sqrt{14}} \\ -\frac{1}{\sqrt{3}} & \frac{5}{\sqrt{42}} & -\frac{1}{\sqrt{14}} \end{pmatrix}.$$

A complete list of 3×3 orthogonal matrices can be found in Exercises 5.3.4 and 5.3.5. ●

Lemma 5.22 An orthogonal matrix has determinant $\det Q = \pm 1$.

Proof Taking the determinant of (5.29), and using the determinantal formulae (1.82), (1.84), shows that

$$1 = \det \mathrm{I} = \det(Q^T Q) = \det Q^T \det Q = (\det Q)^2,$$

which immediately proves the lemma. ■

An orthogonal matrix is called *proper* or *special* if it has determinant $+1$. Geometrically, the columns of a proper orthogonal matrix form a right-handed basis of $\mathbb{R}^n$, as defined in Exercise 2.4.7. An *improper* orthogonal matrix, with determinant -1, corresponds to a left handed basis that lives in a mirror image world.

Proposition 5.23 The product of two orthogonal matrices is also orthogonal.

Proof If $Q_1^T Q_1 = \mathrm{I} = Q_2^T Q_2$, then

$$(Q_1 Q_2)^T (Q_1 Q_2) = Q_2^T Q_1^T Q_1 Q_2 = Q_2^T Q_2 = \mathrm{I},$$

and so the product $Q_1 Q_2$ is also orthogonal. ■

This property says that the set of all orthogonal matrices forms a *group**. The *orthogonal group* lies at the foundation of everyday Euclidean geometry, as well as rigid body mechanics, atomic structure and chemistry, computer graphics and animation, and many other areas.

EXERCISES 5.3

5.3.1. Determine which of the following matrices are
(i) orthogonal;
(ii) proper orthogonal.

(a) $\begin{pmatrix} 1 & 1 \\ -1 & 1 \end{pmatrix}$ (b) $\begin{pmatrix} \frac{12}{13} & \frac{5}{13} \\ -\frac{5}{13} & \frac{12}{13} \end{pmatrix}$

(c) $\begin{pmatrix} 0 & 1 & 0 \\ -1 & 0 & 0 \\ 0 & 0 & -1 \end{pmatrix}$

(d) $\begin{pmatrix} -\frac{1}{3} & \frac{2}{3} & \frac{2}{3} \\ \frac{2}{3} & -\frac{1}{3} & \frac{2}{3} \\ \frac{2}{3} & \frac{2}{3} & -\frac{1}{3} \end{pmatrix}$

(e) $\begin{pmatrix} \frac{1}{2} & \frac{1}{3} & \frac{1}{4} \\ \frac{1}{3} & \frac{1}{4} & \frac{1}{5} \\ \frac{1}{4} & \frac{1}{5} & \frac{1}{6} \end{pmatrix}$

(f) $\begin{pmatrix} \frac{3}{5} & 0 & \frac{4}{5} \\ -\frac{4}{13} & \frac{12}{13} & \frac{3}{13} \\ -\frac{48}{65} & -\frac{5}{13} & \frac{36}{65} \end{pmatrix}$

(g) $\begin{pmatrix} \frac{2}{3} & -\frac{\sqrt{2}}{6} & \frac{\sqrt{2}}{2} \\ -\frac{2}{3} & \frac{\sqrt{2}}{6} & \frac{\sqrt{2}}{2} \\ \frac{1}{3} & \frac{2\sqrt{2}}{3} & 0 \end{pmatrix}$

5.3.2. (a) Show that

$$R = \begin{pmatrix} 1 & 0 & 0 \\ 0 & 0 & 1 \\ 0 & 1 & 0 \end{pmatrix},$$

a reflection matrix, and

$$Q = \begin{pmatrix} \cos\theta & \sin\theta & 0 \\ -\sin\theta & \cos\theta & 0 \\ 0 & 0 & 1 \end{pmatrix},$$

representing a rotation by angle θ around the z axis, are both orthogonal.

(b) Verify that the products $R\,Q$ and $Q\,R$ are also orthogonal.

(c) Which of the preceding matrices, R, Q, $R\,Q$, $Q\,R$ are proper orthogonal?

5.3.3. *True or false*:

(a) If Q is an improper 2×2 orthogonal matrix, then $Q^2 = \mathrm{I}$.

(b) If Q is an improper 3×3 orthogonal matrix, then $Q^2 = \mathrm{I}$.

5.3.4. (a) Prove that, for any θ, φ, ψ,

$$Q = \begin{pmatrix} \cos\varphi\,\cos\psi - \cos\theta\,\sin\varphi\,\sin\psi & \sin\varphi\,\cos\psi + \cos\theta\,\cos\varphi\,\sin\psi & \sin\theta\,\sin\psi \\ -\cos\varphi\,\sin\psi - \cos\theta\,\sin\varphi\,\cos\psi & -\sin\varphi\,\sin\psi + \cos\theta\,\cos\varphi\,\cos\psi & \sin\theta\,\cos\psi \\ \sin\theta\,\sin\varphi & -\sin\theta\,\cos\varphi & \cos\theta \end{pmatrix}$$

is a proper orthogonal matrix.

(b) Write down a formula for Q^{-1}.

Remark: It can be shown that every proper orthogonal matrix can be parameterized in this manner; θ, φ, ψ are known as the *Euler angles*, and play an important role in applications in mechanics and geometry, [26, p. 147].

*The precise mathematical definition of a group can be found in Exercise 5.3.24. Although they will not play a significant role in this text, groups underlie the mathematical formalization of symmetry and, as such, form one of the most fundamental concepts in advanced mathematics and its applications, particularly quantum mechanics and modern theoretical physics, [40]. Indeed, according to the mathematician Felix Klein, [69], all geometry is based on group theory.

5.3.5. (a) Show that if $y_1^2 + y_2^2 + y_3^2 + y_4^2 = 1$, then the matrix

$$Q = \begin{pmatrix} y_1^2 + y_2^2 - y_3^2 - y_4^2 & 2\,(y_2\,y_3 + y_1\,y_4) & 2\,(y_2\,y_4 - y_1\,y_3) \\ 2\,(y_2\,y_3 - y_1\,y_4) & y_1^2 - y_2^2 + y_3^2 - y_4^2 & 2\,(y_3\,y_4 + y_1\,y_2) \\ 2\,(y_2\,y_4 + y_1\,y_3) & 2\,(y_3\,y_4 - y_1\,y_2) & y_1^2 - y_2^2 - y_3^2 + y_4^2 \end{pmatrix}$$

is a proper orthogonal matrix. The numbers y_1, y_2, y_3, y_4 are known as *Cayley–Klein parameters*.

(b) Write down a formula for Q^{-1}.

(c) Prove the formulas

$$y_1 = \cos\frac{\varphi+\psi}{2}\,\cos\frac{\theta}{2}, \quad y_2 = \cos\frac{\varphi-\psi}{2}\,\sin\frac{\theta}{2}, \quad y_3 = \sin\frac{\varphi-\psi}{2}\,\sin\frac{\theta}{2}, \quad y_4 = \sin\frac{\varphi+\psi}{2}\,\cos\frac{\theta}{2},$$

relating the Cayley–Klein parameters and the Euler angles of Exercise 5.3.4, [26, §4–5].

5.3.6. Show that if Q is a proper orthogonal matrix, and R is obtained from Q by interchanging two rows, then R is an improper orthogonal matrix.

5.3.7. Show that the product of two proper orthogonal matrices is also proper orthogonal. What can you say about the product of two improper orthogonal matrices? What about an improper times a proper orthogonal matrix?

◇ **5.3.8.** (a) Prove that the transpose of an orthogonal matrix is also orthogonal.

(b) Explain why the rows of an $n \times n$ orthogonal matrix also form an orthonormal basis of $\mathbb{R}^n$.

5.3.9. Prove that the inverse of an orthogonal matrix is orthogonal.

5.3.10. *True or false*:

(a) A matrix whose columns form an orthogonal basis of $\mathbb{R}^n$ is an orthogonal matrix.

(b) A matrix whose rows form an orthonormal basis of $\mathbb{R}^n$ is an orthogonal matrix.

(c) An orthogonal matrix is symmetric if and only if it is a diagonal matrix.

5.3.11. Write down all diagonal $n \times n$ orthogonal matrices.

5.3.12. Prove that an upper triangular matrix U is orthogonal if and only if U is a diagonal matrix. What are its diagonal entries?

5.3.13. (a) Show that the elementary row operation matrix corresponding to the interchange of two rows is an improper orthogonal matrix.

(b) Are there any other orthogonal elementary matrices?

5.3.14. *True or false*: Applying an elementary row operation to an orthogonal matrix produces an orthogonal matrix.

5.3.15. (a) Prove that every permutation matrix is orthogonal.

(b) How many permutation matrices of a given size are proper orthogonal?

◇ **5.3.16.** (a) Prove that if Q is an orthogonal matrix, then $\| Q\,\mathbf{x} \| = \| \mathbf{x} \|$ for any vector $\mathbf{x} \in \mathbb{R}^n$, where $\|\cdot\|$ denotes the standard Euclidean norm.

(b) Prove the converse: if $\| Q\,\mathbf{x} \| = \| \mathbf{x} \|$ for all $\mathbf{x} \in \mathbb{R}^n$, then Q is an orthogonal matrix.

♡ **5.3.17.** (a) Show that if $\mathbf{u} \in \mathbb{R}^n$ is a unit vector, then the $n \times n$ matrix $Q = \mathrm{I} - 2\,\mathbf{u}\,\mathbf{u}^T$ is an orthogonal matrix, known as an *elementary reflection* or *Householder matrix*.

(b) Write down the elementary reflection matrices corresponding to the following unit vectors:

(i) $(1, 0)^T$ (ii) $\left(\frac{3}{5}, \frac{4}{5}\right)^T$

(iii) $(0, 1, 0)^T$ (iv) $\left(\frac{1}{\sqrt{2}}, 0, -\frac{1}{\sqrt{2}}\right)^T$

(c) Prove that

(i) Q is symmetric

(ii) $Q^{-1} = Q$ (iii) $Q\,\mathbf{u} = -\,\mathbf{u}$

(d) Find all vectors fixed by an elementary reflection matrix, i.e., $Q\,\mathbf{v} = \mathbf{v}$—first for the matrices in part (b), and then in general.

◇ **5.3.18.** Show that if $A^T = -\,A$ is any skew-symmetric matrix, then its *Cayley Transform*

$$Q = (\mathrm{I} - A)^{-1}(\mathrm{I} + A)$$

is an orthogonal matrix. Can you prove that $\mathrm{I} - A$ is always invertible?

5.3.19. (a) Suppose S is an $n \times n$ matrix whose columns form an orthogonal, but not orthonormal, basis of $\mathbb{R}^n$.

(b) Find a formula for S^{-1} mimicking the formula $Q^{-1} = Q^T$ for an orthogonal matrix.

(c) Use your formula to determine the inverse of the wavelet matrix W whose columns form the orthogonal wavelet basis (5.9) of $\mathbb{R}^4$.

◇ **5.3.20.** Let $\mathbf{v}_1, \ldots, \mathbf{v}_n$ and $\mathbf{w}_1, \ldots, \mathbf{w}_n$ be two sets of linearly independent vectors in $\mathbb{R}^n$. Show that all their dot products are the same, so $\mathbf{v}_i \cdot \mathbf{v}_j = \mathbf{w}_i \cdot \mathbf{w}_j$ for all $i, j = 1, \ldots, n$, if and only if there is an orthogonal matrix Q such that $\mathbf{w}_i = Q\,\mathbf{v}_i$ for all $i = 1, \ldots, n$.

5.3.21. Suppose $\mathbf{u}_1, \ldots, \mathbf{u}_k$ form an orthonormal set of vectors in $\mathbb{R}^n$ with $k < n$. Let $Q = (\mathbf{u}_1\ \mathbf{u}_2 \ldots \mathbf{u}_k)$ denote the $n \times k$ matrix whose columns are the orthonormal vectors.

(a) Prove that $Q^T Q = \mathrm{I}_k$. (b) Is $Q Q^T = \mathrm{I}_n$?

5.3.22. (a) Let A be an $m \times n$ matrix whose columns are nonzero, mutually orthogonal vectors in $\mathbb{R}^m$.

(b) Explain why $m \geq n$.

(c) Prove that $A^T A$ is a diagonal matrix. What are the diagonal entries?

(d) Is $A A^T$ diagonal?

◇ **5.3.23.** Let $K > 0$ be a positive definite $n \times n$ matrix. Prove that an $n \times n$ matrix S satisfies $S^T K S = \mathrm{I}$ if and only if the columns of S form an orthonormal basis of $\mathbb{R}^n$ with respect to the inner product $\langle \mathbf{v}, \mathbf{w} \rangle = \mathbf{v}^T K \mathbf{w}$.

♡ **5.3.24.** A set of $n \times n$ matrices $G \subset \mathcal{M}_{n \times n}$ is said to form a *group* if

(1) whenever $A, B \in G$, so is the product $A B \in G$, and

(2) whenever $A \in G$, then A is nonsingular, and $A^{-1} \in G$.

(a) Show that $\mathrm{I} \in G$.

(b) Prove that the following sets of $n \times n$ matrices form a group:

(i) all nonsingular matrices;
(ii) all nonsingular upper triangular matrices;
(iii) all matrices of determinant 1;
(iv) all orthogonal matrices;
(v) all proper orthogonal matrices;
(vi) all permutation matrices;
(vii) all 2×2 matrices with integer entries and determinant 1.

(c) Explain why the set of all nonsingular 2×2 matrices with integer entries does not form a group.

(d) Does the set of positive definite matrices form a group?

♡ **5.3.25.** *Unitary Matrices*: A complex, square matrix U is called *unitary* if it satisfies $U^\dagger U = \mathrm{I}$, where $U^\dagger = \overline{U^T}$ denotes the *Hermitian transpose* in which one first transposes and then takes complex conjugates of all entries.

(a) Show that U is a unitary matrix if and only if $U^{-1} = U^\dagger$.

(b) Show that the following matrices are unitary and compute their inverses:

(i) $\begin{pmatrix} \frac{1}{\sqrt{2}} & \frac{\mathrm{i}}{\sqrt{2}} \\ \frac{\mathrm{i}}{\sqrt{2}} & \frac{1}{\sqrt{2}} \end{pmatrix}$

(ii) $\begin{pmatrix} \frac{1}{\sqrt{3}} & \frac{1}{\sqrt{3}} & \frac{1}{\sqrt{3}} \\ \frac{1}{\sqrt{3}} & -\frac{1}{2\sqrt{3}} + \frac{\mathrm{i}}{2} & -\frac{1}{2\sqrt{3}} - \frac{\mathrm{i}}{2} \\ \frac{1}{\sqrt{3}} & -\frac{1}{2\sqrt{3}} - \frac{\mathrm{i}}{2} & -\frac{1}{2\sqrt{3}} + \frac{\mathrm{i}}{2} \end{pmatrix}$

(iii) $\begin{pmatrix} \frac{1}{2} & \frac{1}{2} & \frac{1}{2} & \frac{1}{2} \\ \frac{1}{2} & \frac{\mathrm{i}}{2} & -\frac{1}{2} & -\frac{\mathrm{i}}{2} \\ \frac{1}{2} & -\frac{1}{2} & \frac{1}{2} & -\frac{1}{2} \\ \frac{1}{2} & -\frac{\mathrm{i}}{2} & -\frac{1}{2} & \frac{\mathrm{i}}{2} \end{pmatrix}$

(c) Are the following matrices unitary?

(i) $\begin{pmatrix} 2 & 1+2\,\mathrm{i} \\ 1-2\,\mathrm{i} & 3 \end{pmatrix}$

(ii) $\frac{1}{5}\begin{pmatrix} -1+2\,\mathrm{i} & -4-2\,\mathrm{i} \\ 2-4\,\mathrm{i} & -2-\mathrm{i} \end{pmatrix}$

(iii) $\begin{pmatrix} \frac{12}{13} & \frac{5}{13} \\ \frac{5}{13} & -\frac{12}{13} \end{pmatrix}$

(d) Show that U is a unitary matrix if and only if its columns form an orthonormal basis of $\mathbb{C}^n$ with respect to the Hermitian dot product.

(e) Prove that the set of unitary matrices forms a group, as defined in Exercise 5.3.24.

The *QR* Factorization

The Gram–Schmidt procedure for orthonormalizing bases of $\mathbb{R}^n$ can be reinterpreted as a matrix factorization. This is more subtle than the $L\,U$ factorization that resulted from Gaussian Elimination, but is of comparable significance, and is used in a broad range of applications in mathematics, statistics, physics, engineering, and numerical analysis.

Let $\mathbf{w}_1, \ldots, \mathbf{w}_n$ be a basis of $\mathbb{R}^n$, and let $\mathbf{u}_1, \ldots, \mathbf{u}_n$ be the corresponding orthonormal basis that results from any one of the three implementations of the Gram–Schmidt process. We assemble both sets of column vectors to form nonsingular $n \times n$

matrices

$$A = (\mathbf{w}_1 \quad \mathbf{w}_2 \quad \ldots \quad \mathbf{w}_n), \qquad Q = (\mathbf{u}_1 \quad \mathbf{u}_2 \quad \ldots \quad \mathbf{u}_n).$$

Since the $\mathbf{u}_i$ form an orthonormal basis, Q is an orthogonal matrix. In view of the matrix multiplication formula (2.13), the Gram–Schmidt equations (5.23) can be recast into an equivalent matrix form:

$$A = Q\,R, \quad \text{where} \quad R = \begin{pmatrix} r_{11} & r_{12} & \ldots & r_{1n} \\ 0 & r_{22} & \ldots & r_{2n} \\ \vdots & \vdots & \ddots & \vdots \\ 0 & 0 & \ldots & r_{nn} \end{pmatrix} \tag{5.32}$$

is an upper triangular matrix whose entries are the coefficients in (5.26–27). Since the Gram–Schmidt process works on any basis, the only requirement on the matrix A is that its columns form a basis of $\mathbb{R}^n$, and hence A can be any nonsingular matrix. We have therefore established the celebrated *$Q\,R$ factorization* of nonsingular matrices.

THEOREM 5.24 Any nonsingular matrix A can be factored, $A = Q\,R$, into the product of an orthogonal matrix Q and an upper triangular matrix R. The factorization is unique if all the diagonal entries of R are assumed to be positive.

The proof of uniqueness is relegated to Exercise 5.3.32.

EXAMPLE 5.25 The columns of the matrix

$$A = \begin{pmatrix} 1 & 1 & 2 \\ 1 & 0 & -2 \\ -1 & 2 & 3 \end{pmatrix}$$

are the same as the basis vectors considered in Example 5.16. The orthonormal basis (5.21) constructed using the Gram–Schmidt algorithm leads to the orthogonal and upper triangular matrices

$$Q = \begin{pmatrix} \frac{1}{\sqrt{3}} & \frac{4}{\sqrt{42}} & \frac{2}{\sqrt{14}} \\ \frac{1}{\sqrt{3}} & \frac{1}{\sqrt{42}} & -\frac{3}{\sqrt{14}} \\ -\frac{1}{\sqrt{3}} & \frac{5}{\sqrt{42}} & -\frac{1}{\sqrt{14}} \end{pmatrix}, \qquad R = \begin{pmatrix} \sqrt{3} & -\frac{1}{\sqrt{3}} & -\sqrt{3} \\ 0 & \frac{\sqrt{14}}{\sqrt{3}} & \frac{\sqrt{21}}{\sqrt{2}} \\ 0 & 0 & \frac{\sqrt{7}}{\sqrt{2}} \end{pmatrix}. \tag{5.33}$$

The reader may wish to verify that, indeed, $A = Q\,R$. ●

While any of the three implementations of the Gram–Schmidt algorithm will produce the $Q\,R$ factorization of a given matrix $A = (\mathbf{w}_1\ \mathbf{w}_2\ \ldots\ \mathbf{w}_n)$, the stable version, as encoded in equations (5.28), is the one to use in practical computations, as it is the least likely to fail due to numerical artifacts produced by round-off errors. The accompanying pseudocode program reformulates the algorithm purely in terms of the matrix entries a_{ij} of A. During the course of the algorithm, the entries of the matrix A are successively overwritten; the final result is the orthogonal matrix Q appearing in place of A. The entries r_{ij} of R must be stored separately.

QR Factorization of a Matrix A

```
start
    for j = 1 to n
        set r_jj = sqrt(a_1j^2 + ··· + a_nj^2)
        if r_jj = 0, stop; print "A has linearly dependent columns"
            else for i = 1 to n
                set a_ij = a_ij / r_jj
            next i
        for k = j + 1 to n
            set r_jk = a_1j a_1k + ··· + a_nj a_nk
            for i = 1 to n
                set a_ik = a_ik − a_ij r_jk
            next i
        next k
    next j
end
```

EXAMPLE 5.26 Let us factor the matrix

$$A = \begin{pmatrix} 2 & 1 & 0 & 0 \\ 1 & 2 & 1 & 0 \\ 0 & 1 & 2 & 1 \\ 0 & 0 & 1 & 2 \end{pmatrix}$$

using the numerically stable QR algorithm. As in the program, we work directly on the matrix A, gradually changing it into orthogonal form. In the first loop, we set $r_{11} = \sqrt{5}$ to be the norm of the first column vector of A. We then normalize the first column by dividing by r_{11}; the resulting matrix is

$$\begin{pmatrix} \frac{2}{\sqrt{5}} & 1 & 0 & 0 \\ \frac{1}{\sqrt{5}} & 2 & 1 & 0 \\ 0 & 1 & 2 & 1 \\ 0 & 0 & 1 & 2 \end{pmatrix}.$$

The next entries $r_{12} = \frac{4}{\sqrt{5}}$, $r_{13} = \frac{1}{\sqrt{5}}$, $r_{14} = 0$, are obtained by taking the dot products of the first column with the other three columns. For $j = 1, 2, 3$, we subtract r_{1j} times the first column from the jth column; the result

$$\begin{pmatrix} \frac{2}{\sqrt{5}} & -\frac{3}{5} & -\frac{2}{5} & 0 \\ \frac{1}{\sqrt{5}} & \frac{6}{5} & \frac{4}{5} & 0 \\ 0 & 1 & 2 & 1 \\ 0 & 0 & 1 & 2 \end{pmatrix}$$

is a matrix whose first column is normalized to have unit length, and whose second, third and fourth columns are orthogonal to it. In the next loop, we normalize the second column by dividing by its norm $r_{22} = \sqrt{\frac{14}{5}}$, and so obtain the matrix

$$\begin{pmatrix} \frac{2}{\sqrt{5}} & -\frac{3}{\sqrt{70}} & -\frac{2}{5} & 0 \\ \frac{1}{\sqrt{5}} & \frac{6}{\sqrt{70}} & \frac{4}{5} & 0 \\ 0 & \frac{5}{\sqrt{70}} & 2 & 1 \\ 0 & 0 & 1 & 2 \end{pmatrix}.$$

We then take dot products of the second column with the remaining two columns to produce $r_{23} = \frac{16}{\sqrt{70}}$, $r_{24} = \frac{5}{\sqrt{70}}$. Subtracting these multiples of the second column from the third and fourth columns, we obtain

$$\begin{pmatrix} \frac{2}{\sqrt{5}} & -\frac{3}{\sqrt{70}} & \frac{2}{7} & \frac{3}{14} \\ \frac{1}{\sqrt{5}} & \frac{6}{\sqrt{70}} & -\frac{4}{7} & -\frac{3}{7} \\ 0 & \frac{5}{\sqrt{70}} & \frac{6}{7} & \frac{9}{14} \\ 0 & 0 & 1 & 2 \end{pmatrix},$$

which now has its first two columns orthonormalized, and orthogonal to the last two columns. We then normalize the third column by dividing by $r_{33} = \sqrt{\frac{15}{7}}$, and so

$$\begin{pmatrix} \frac{2}{\sqrt{5}} & -\frac{3}{\sqrt{70}} & \frac{2}{\sqrt{105}} & \frac{3}{14} \\ \frac{1}{\sqrt{5}} & \frac{6}{\sqrt{70}} & -\frac{4}{\sqrt{105}} & -\frac{3}{7} \\ 0 & \frac{5}{\sqrt{70}} & \frac{6}{\sqrt{105}} & \frac{9}{14} \\ 0 & 0 & \frac{7}{\sqrt{105}} & 2 \end{pmatrix}.$$

Finally, we subtract $r_{34} = \frac{20}{\sqrt{105}}$ times the third column from the fourth column. Dividing the resulting fourth column by its norm $r_{44} = \sqrt{\frac{5}{6}}$ results in the final formulas,

$$Q = \begin{pmatrix} \frac{2}{\sqrt{5}} & -\frac{3}{\sqrt{70}} & \frac{2}{\sqrt{105}} & -\frac{1}{\sqrt{30}} \\ \frac{1}{\sqrt{5}} & \frac{6}{\sqrt{70}} & -\frac{4}{\sqrt{105}} & \frac{2}{\sqrt{30}} \\ 0 & \frac{5}{\sqrt{70}} & \frac{6}{\sqrt{105}} & -\frac{3}{\sqrt{30}} \\ 0 & 0 & \frac{7}{\sqrt{105}} & \frac{4}{\sqrt{30}} \end{pmatrix}, \qquad R = \begin{pmatrix} \sqrt{5} & \frac{4}{\sqrt{5}} & \frac{1}{\sqrt{5}} & 0 \\ 0 & \frac{\sqrt{14}}{\sqrt{5}} & \frac{16}{\sqrt{70}} & \frac{5}{\sqrt{70}} \\ 0 & 0 & \frac{\sqrt{15}}{\sqrt{7}} & \frac{20}{\sqrt{105}} \\ 0 & 0 & 0 & \frac{\sqrt{5}}{\sqrt{6}} \end{pmatrix},$$

for the $A = Q R$ factorization. ●

Ill-Conditioned Systems and Householder's Method

The $Q R$ factorization can be employed as an alternative to Gaussian Elimination to solve linear systems. Indeed, the system

$$A\mathbf{x} = \mathbf{b} \quad \text{becomes} \quad Q R \mathbf{x} = \mathbf{b}, \quad \text{and hence} \quad R\mathbf{x} = Q^T\mathbf{b}, \tag{5.34}$$

because $Q^{-1} = Q^T$ is an orthogonal matrix. Since R is upper triangular, the latter system can be solved for $\mathbf{x}$ by Back Substitution. The resulting algorithm, while more expensive to compute, does offer some numerical advantages over traditional Gaussian Elimination, as it is less prone to inaccuracies resulting from ill-conditioning.

EXAMPLE 5.27 Let us apply the $A = QR$ factorization

$$\begin{pmatrix} 1 & 1 & 2 \\ 1 & 0 & -2 \\ -1 & 2 & 3 \end{pmatrix} = \begin{pmatrix} \frac{1}{\sqrt{3}} & \frac{4}{\sqrt{42}} & \frac{2}{\sqrt{14}} \\ \frac{1}{\sqrt{3}} & \frac{1}{\sqrt{42}} & -\frac{3}{\sqrt{14}} \\ -\frac{1}{\sqrt{3}} & \frac{5}{\sqrt{42}} & -\frac{1}{\sqrt{14}} \end{pmatrix} \begin{pmatrix} \sqrt{3} & -\frac{1}{\sqrt{3}} & -\sqrt{3} \\ 0 & \frac{\sqrt{14}}{\sqrt{3}} & \frac{\sqrt{21}}{\sqrt{2}} \\ 0 & 0 & \frac{\sqrt{7}}{\sqrt{2}} \end{pmatrix}$$

that we found in Example 5.25 to solve the linear system $A\mathbf{x} = (0, -4, 5)^T$. We first compute

$$Q^T\mathbf{b} = \begin{pmatrix} \frac{1}{\sqrt{3}} & \frac{1}{\sqrt{3}} & -\frac{1}{\sqrt{3}} \\ \frac{4}{\sqrt{42}} & \frac{1}{\sqrt{42}} & \frac{5}{\sqrt{42}} \\ \frac{2}{\sqrt{14}} & -\frac{3}{\sqrt{14}} & -\frac{1}{\sqrt{14}} \end{pmatrix} \begin{pmatrix} 0 \\ -4 \\ 5 \end{pmatrix} = \begin{pmatrix} -3\sqrt{3} \\ \frac{\sqrt{21}}{\sqrt{2}} \\ \frac{\sqrt{7}}{\sqrt{2}} \end{pmatrix}.$$

We then solve the upper triangular system

$$R\,\mathbf{x} = \begin{pmatrix} \sqrt{3} & -\frac{1}{\sqrt{3}} & -\sqrt{3} \\ 0 & \frac{\sqrt{14}}{\sqrt{3}} & \frac{\sqrt{21}}{\sqrt{2}} \\ 0 & 0 & \frac{\sqrt{7}}{\sqrt{2}} \end{pmatrix} \begin{pmatrix} x \\ y \\ z \end{pmatrix} = \begin{pmatrix} -3\sqrt{3} \\ \frac{\sqrt{21}}{\sqrt{2}} \\ \frac{\sqrt{7}}{\sqrt{2}} \end{pmatrix}$$

by Back Substitution, leading to the solution $\mathbf{x} = (-2, 0, 1)^T$. ●

In computing the QR factorization of a mildly ill-conditioned matrix, one should employ the stable version (5.28) of the Gram–Schmidt process. However, more recalcitrant matrices require a completely different approach to the factorization, as formulated by the mid-twentieth century American mathematician Alston Householder. His idea was to use a sequence of simple orthogonal matrices to gradually convert the matrix into upper triangular form.

Consider the *Householder* or *elementary reflection matrix*

$$H = \mathrm{I} - 2\,\mathbf{u}\,\mathbf{u}^T \tag{5.35}$$

in which $\mathbf{u}$ is a unit vector (in the Euclidean norm). The matrix H represents a reflection of vectors through the subspace $\mathbf{u}^\perp = \{\, \mathbf{v} \mid \mathbf{v}\cdot\mathbf{u} = 0 \,\}$ of vectors orthogonal to $\mathbf{u}$, as illustrated in Figure 5.3. According to Exercise 5.3.17, H is a symmetric orthogonal matrix, and so

$$H^T = H, \qquad H^2 = \mathrm{I}, \qquad H^{-1} = H. \tag{5.36}$$

The proof is straightforward: symmetry is immediate, while

$$H\,H^T = H^2 = (\mathrm{I} - 2\,\mathbf{u}\,\mathbf{u}^T)\,(\mathrm{I} - 2\,\mathbf{u}\,\mathbf{u}^T) = \mathrm{I} - 4\,\mathbf{u}\,\mathbf{u}^T + 4\,\mathbf{u}\,(\mathbf{u}^T\mathbf{u})\,\mathbf{u}^T = \mathrm{I}$$

since, by assumption, $\mathbf{u}^T\mathbf{u} = \|\mathbf{u}\|^2 = 1$. By suitably forming the unit vector $\mathbf{u}$, we can construct an elementary reflection matrix that interchanges any two vectors of the same length.

Lemma 5.28 Let $\mathbf{v}, \mathbf{w} \in \mathbb{R}^n$ with $\|\mathbf{v}\| = \|\mathbf{w}\|$. Set $\mathbf{u} = (\mathbf{v} - \mathbf{w})/\|\mathbf{v} - \mathbf{w}\|$ and let $H = \mathrm{I} - 2\,\mathbf{u}\,\mathbf{u}^T$ be the corresponding elementary reflection matrix. Then $H\,\mathbf{v} = \mathbf{w}$ and $H\,\mathbf{w} = \mathbf{v}$.

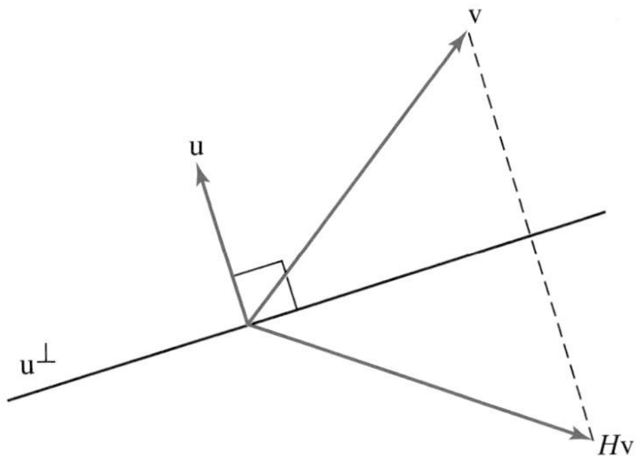

Figure 5.3 Elementary reflection matrix.

Proof Keeping in mind that $\mathbf{x}$ and $\mathbf{y}$ have the same Euclidean norm, we compute

$$
\begin{aligned}
H\,\mathbf{v} = (\mathrm{I} - 2\,\mathbf{u}\,\mathbf{u}^T)\,\mathbf{v} &= \mathbf{v} - 2\,\frac{(\mathbf{v}-\mathbf{w})(\mathbf{v}-\mathbf{w})^T\mathbf{v}}{\|\mathbf{v}-\mathbf{w}\|^2} \\
&= \mathbf{v} - 2\,\frac{(\mathbf{v}-\mathbf{w})\left(\|\mathbf{v}\|^2 - \mathbf{w}\cdot\mathbf{v}\right)}{2\,\|\mathbf{v}\|^2 - 2\,\mathbf{v}\cdot\mathbf{w}} \\
&= \mathbf{v} - (\mathbf{v}-\mathbf{w}) = \mathbf{w}.
\end{aligned}
$$

The proof of the second equation is similar. ■

In the first phase of Householder's method, we introduce the elementary reflection matrix that maps the first column $\mathbf{v}_1$ of the matrix A to a multiple of the first basis vector, namely $\mathbf{w}_1 = \|\mathbf{v}_1\|\,\mathbf{e}_1$, noting that $\|\mathbf{v}_1\| = \|\mathbf{w}_1\|$. We define the first Householder vector and corresponding elementary reflection matrix as

$$
\mathbf{u}_1 = \frac{\mathbf{v}_1 - \|\mathbf{v}_1\|\,\mathbf{e}_1}{\|\mathbf{v}_1 - \|\mathbf{v}_1\|\,\mathbf{e}_1\|}, \qquad H_1 = \mathrm{I} - 2\,\mathbf{u}_1\,\mathbf{u}_1^T.
$$

(If $\mathbf{v}_1 = c\,\mathbf{e}_1$ is already in the desired form, then we set $\mathbf{u}_1 = \mathbf{0}$ and $H_1 = \mathrm{I}$.) Since, by the lemma, $H_1\mathbf{v}_1 = \mathbf{w}_1$, when we multiply A on the left by H_1, we obtain a matrix

$$
A_2 = H_1\,A = \begin{pmatrix}
r_{11} & \widetilde{a}_{12} & \widetilde{a}_{13} & \ldots & \widetilde{a}_{1n} \\
0 & \widetilde{a}_{22} & \widetilde{a}_{23} & \ldots & \widetilde{a}_{2n} \\
0 & \widetilde{a}_{32} & \widetilde{a}_{33} & \ldots & \widetilde{a}_{3n} \\
\vdots & \vdots & \vdots & \ddots & \vdots \\
0 & \widetilde{a}_{n2} & \widetilde{a}_{n3} & \ldots & \widetilde{a}_{nn}
\end{pmatrix}
$$

whose first column is in the desired upper triangular form.

In the next phase, we construct a second elementary reflection matrix to make all the entries below the diagonal in the second column of A_2 zero, keeping in mind that, at the same time, we should not mess up the first column. The latter requirement tells us that the vector used for the reflection should have a zero in its first entry. The correct choice is to set

$$
\widetilde{\mathbf{v}}_2 = (0, \widetilde{a}_{22}, \widetilde{a}_{32}, \ldots, \widetilde{a}_{n2})^T, \qquad \mathbf{u}_2 = \frac{\widetilde{\mathbf{v}}_2 - \|\widetilde{\mathbf{v}}_2\|\,\mathbf{e}_2}{\|\widetilde{\mathbf{v}}_2 - \|\widetilde{\mathbf{v}}_2\|\,\mathbf{e}_2\|}, \qquad H_2 = \mathrm{I} - 2\,\mathbf{u}_2\,\mathbf{u}_2^T.
$$

(As before, if $\tilde{\mathbf{v}}_2 = c\,\mathbf{e}_2$, then $\mathbf{u}_2 = \mathbf{0}$ and $H_2 = \mathrm{I}$.) The net effect is

$$A_3 = H_2\,A_2 = \begin{pmatrix} r_{11} & r_{12} & \widehat{a}_{13} & \dots & \widehat{a}_{1n} \\ 0 & r_{22} & \widehat{a}_{23} & \dots & \widehat{a}_{2n} \\ 0 & 0 & \widehat{a}_{33} & \dots & \widehat{a}_{3n} \\ \vdots & \vdots & \vdots & \ddots & \vdots \\ 0 & 0 & \widehat{a}_{n3} & \dots & \widehat{a}_{nn} \end{pmatrix}.$$

and now the first two columns are in upper triangular form.

The process continues; at the kth stage, we are dealing with a matrix A_k whose first $k-1$ columns coincide with the first k columns of the eventual upper triangular matrix R. Let $\widehat{\mathbf{v}}_k$ denote the vector obtained from the kth column of A_k by setting its initial $k-1$ entries equal to 0. We define the kth Householder vector and corresponding elementary reflection matrix by

$$\mathbf{w}_k = \widehat{\mathbf{v}}_k - \|\widehat{\mathbf{v}}_k\|\,\mathbf{e}_k, \qquad \mathbf{u}_k = \begin{cases} \mathbf{w}_k/\|\mathbf{w}_k\|, & \mathbf{w}_k \neq \mathbf{0}, \\ \mathbf{0}, & \mathbf{w}_k = \mathbf{0}, \end{cases} \qquad \begin{aligned} H_k &= \mathrm{I} - 2\,\mathbf{u}_k\,\mathbf{u}_k^T, \\ A_{k+1} &= H_k\,A_k. \end{aligned} \tag{5.37}$$

The process is completed after $n-1$ steps, and the final result is

$$R = H_{n-1}A_{n-1} = H_{n-1}H_{n-2}\cdots H_1 A = Q^T A, \qquad \text{where} \qquad Q = H_1 H_2 \cdots H_{n-1}$$

is an orthogonal matrix, since it is the product of orthogonal matrices, cf. Proposition 5.23. In this manner, we have reproduced a[†] $Q\,R$ factorization of

$$A = Q\,R = H_1 H_2 \cdots H_{n-1} R. \tag{5.38}$$

EXAMPLE 5.29 Let us implement Householder's Method on the particular matrix

$$A = \begin{pmatrix} 1 & 1 & 2 \\ 1 & 0 & -2 \\ -1 & 2 & 3 \end{pmatrix}$$

considered earlier in Example 5.25. The first Householder vector

$$\widehat{\mathbf{v}}_1 = \begin{pmatrix} 1 \\ 1 \\ -1 \end{pmatrix} - \sqrt{3}\begin{pmatrix} 1 \\ 0 \\ 0 \end{pmatrix} = \begin{pmatrix} -.7321 \\ 1 \\ -1 \end{pmatrix}$$

leads to the elementary reflection matrix

$$H_1 = \begin{pmatrix} .5774 & .5774 & -.5774 \\ .5774 & .2113 & .7887 \\ -.5774 & .7887 & .2113 \end{pmatrix},$$

whereby

$$A_2 = H_1 A = \begin{pmatrix} 1.7321 & -.5774 & -1.7321 \\ 0 & 2.1547 & 3.0981 \\ 0 & -.1547 & -2.0981 \end{pmatrix}.$$

[†] The upper triangular matrix R may not have positive diagonal entries; if desired, this can be easily fixed by changing the signs of the appropriate columns of Q.

To construct the second and final Householder matrix, we start with the second column of A_2 and then set the first entry to 0; the resulting Householder vector is

$$\widehat{\mathbf{v}}_2 = \begin{pmatrix} 0 \\ 2.1547 \\ -.1547 \end{pmatrix} - 2.1603 \begin{pmatrix} 0 \\ 1 \\ 0 \end{pmatrix} = \begin{pmatrix} 0 \\ -.0055 \\ -.1547 \end{pmatrix}.$$

Therefore,

$$H_2 = \begin{pmatrix} 1 & 0 & 0 \\ 0 & .9974 & -.0716 \\ 0 & -.0716 & -.9974 \end{pmatrix},$$

and so

$$R = H_2 A_2 = \begin{pmatrix} 1.7321 & -.5774 & -1.7321 \\ 0 & 2.1603 & 3.2404 \\ 0 & 0 & 1.8708 \end{pmatrix}$$

is the upper triangular matrix in the QR decomposition of A. The orthogonal matrix Q is obtained by multiplying the reflection matrices:

$$Q = H_1 H_2 = \begin{pmatrix} .5774 & .6172 & .5345 \\ .5774 & .1543 & -.8018 \\ -.5774 & .7715 & -.2673 \end{pmatrix},$$

which reconfirms the previous factorization (5.33). ●

REMARK: If the purpose of the QR factorization is to solve a linear system via (5.34), it is not necessary to explicitly multiply out the Householder matrices to form Q; we merely need to store the corresponding unit Householder vectors $\mathbf{u}_1, \ldots, \mathbf{u}_{n-1}$. The solution to

$$A\mathbf{x} = QR\mathbf{x} = \mathbf{b} \quad \text{can be found by solving} \quad R\mathbf{x} = H_{n-1}H_{n-2}\cdots H_1\mathbf{b} \qquad (5.39)$$

by Back Substitution. This is the method of choice for moderately ill-conditioned systems. Severe ill-conditioning will defeat even this ingenious approach, and the resolution of such systems can be an extreme challenge.

EXERCISES 5.3

5.3.26. Write down the QR matrix factorization corresponding to the vectors in Example 5.17.

5.3.27. Find the QR factorization of the following matrices:

(a) $\begin{pmatrix} 1 & -3 \\ 2 & 1 \end{pmatrix}$ (b) $\begin{pmatrix} 4 & 3 \\ 3 & 2 \end{pmatrix}$

(c) $\begin{pmatrix} 2 & 1 & -1 \\ 0 & 1 & 3 \\ -1 & -1 & 1 \end{pmatrix}$ (d) $\begin{pmatrix} 0 & 1 & 2 \\ -1 & 1 & 1 \\ -1 & 1 & 3 \end{pmatrix}$

(e) $\begin{pmatrix} 0 & 0 & 2 \\ 0 & 4 & 1 \\ -1 & 0 & 1 \end{pmatrix}$ (f) $\begin{pmatrix} 1 & 1 & 1 & 1 \\ 1 & 2 & 1 & 0 \\ 1 & 1 & 2 & 1 \\ 1 & 0 & 1 & 1 \end{pmatrix}$

5.3.28. For each of the following linear systems:
(a) find the QR factorization of the coefficient matrix, and then
(b) use your factorization to solve the system.

(i) $\begin{pmatrix} 1 & 2 \\ -1 & 3 \end{pmatrix}\begin{pmatrix} x \\ y \end{pmatrix} = \begin{pmatrix} -1 \\ 2 \end{pmatrix}$

(ii) $\begin{pmatrix} 2 & 1 & -1 \\ 1 & 0 & 2 \\ 2 & -1 & 3 \end{pmatrix}\begin{pmatrix} x \\ y \\ z \end{pmatrix} = \begin{pmatrix} 2 \\ -1 \\ 0 \end{pmatrix}$

(iii) $\begin{pmatrix} 1 & 1 & 0 \\ -1 & 0 & 1 \\ 0 & -1 & 1 \end{pmatrix}\begin{pmatrix} x \\ y \\ z \end{pmatrix} = \begin{pmatrix} 0 \\ 1 \\ 0 \end{pmatrix}$

♠ **5.3.29.** Use the numerically stable version of the Gram–Schmidt process to find the QR factorizations of the 3×3, 4×4 and 5×5 versions of the tridiagonal matrix that has 4's along the diagonal and 1's on the sub- and super-diagonals, as in Example 1.37.

5.3.30. Use Householder's Method to solve Exercises 5.3.27 and 5.3.29.

♡ **5.3.31.** (a) How many arithmetic operations are required to compute the QR factorization of an $n \times n$ matrix?

(b) How many additional operations are needed to utilize the factorization to solve a linear system $A\mathbf{x} = \mathbf{b}$ via (5.34)?

(c) Compare the amount of computational effort with standard Gaussian Elimination.

◇ **5.3.32.** Prove that the QR factorization of a matrix is unique if all the diagonal entries of R are assumed to be positive. *Hint*: Use Exercise 5.3.12.

♡ **5.3.33.** Suppose A is an $m \times n$ matrix with rank $A = n$.

(a) Show that applying the Gram–Schmidt algorithm to the columns of A produces an orthonormal basis for rng A.

(b) Prove that this is equivalent to the matrix factorization $A = QR$, where Q is an $m \times n$ matrix with orthonormal columns, while R is a nonsingular $n \times n$ upper triangular matrix.

(c) Show that the QR program in the text also works for rectangular, $m \times n$, matrices as stated, the only modification being that the row indices i run from 1 to m.

(d) Apply this method to factor

(i) $\begin{pmatrix} 1 & -1 \\ 2 & 3 \\ 0 & 2 \end{pmatrix}$ (ii) $\begin{pmatrix} -3 & 2 \\ 1 & -1 \\ 4 & 1 \end{pmatrix}$, (iii) $\begin{pmatrix} -1 & 1 \\ 1 & -2 \\ -1 & -3 \\ 0 & 5 \end{pmatrix}$ (iv) $\begin{pmatrix} 0 & 1 & 2 \\ -3 & 1 & -1 \\ -1 & 0 & -2 \\ 1 & 1 & -2 \end{pmatrix}$

(e) Explain what happens if rank $A < n$.

♡ **5.3.34.** (a) According to Exercise 5.2.12, the Gram–Schmidt process can also be applied to produce orthonormal bases of complex vector spaces. In the case of $\mathbb{C}^n$, explain how this is equivalent to the factorization of a nonsingular complex matrix $A = UR$ into the product of a unitary matrix U (see Exercise 5.3.25) and a nonsingular upper triangular matrix R.

(b) Factor the following complex matrices into unitary times upper triangular:

(i) $\begin{pmatrix} \mathrm{i} & 1 \\ -1 & 2\,\mathrm{i} \end{pmatrix}$ (ii) $\begin{pmatrix} 1+\mathrm{i} & 2-\mathrm{i} \\ 1-\mathrm{i} & -\mathrm{i} \end{pmatrix}$

(iii) $\begin{pmatrix} \mathrm{i} & 1 & 0 \\ 1 & \mathrm{i} & 1 \\ 0 & 1 & \mathrm{i} \end{pmatrix}$

(iv) $\begin{pmatrix} \mathrm{i} & 1 & -\mathrm{i} \\ 1-\mathrm{i} & 0 & 1+\mathrm{i} \\ -1 & 2+3\,\mathrm{i} & 1 \end{pmatrix}$

(c) What can you say about uniqueness of the factorization?

5.3.35. Write out a pseudocode program to implement Householder's Method. The input should be an $n \times n$ matrix A and the output should be the Householder unit vectors $\mathbf{u}_1, \ldots, \mathbf{u}_{n-1}$ and the upper triangular matrix R. Test your code on one of the examples in Exercises 5.3.26–28.

5.4 ORTHOGONAL POLYNOMIALS

Orthogonal and orthonormal bases play, if anything, an even more essential role in function spaces. Unlike the Euclidean space $\mathbb{R}^n$, most of the obvious bases of a typical (finite dimensional) function space are not orthogonal with respect to any natural inner product. Thus, the computation of an orthonormal basis of functions is a critical step towards simplification of the analysis. The Gram–Schmidt algorithm, in any of the above formulations, can be successfully applied to construct suitably orthogonal functions. The most important examples are the classical orthogonal polynomials that arise in approximation and interpolation theory. Other orthogonal systems of functions play starring roles in Fourier analysis and its generalizations, in quantum mechanics, in the solution of partial differential equations by separation of variables, and a host of further applications in mathematics, physics, engineering, and elsewhere.

The Legendre Polynomials

We shall construct an orthonormal basis for the vector space $\mathcal{P}^{(n)}$ of polynomials of degree $\leq n$. For definiteness, the construction will be based on the L^2 inner product

$$\langle p, q \rangle = \int_{-1}^{1} p(t)\, q(t)\, dt \tag{5.40}$$

on the interval $[-1, 1]$. The underlying method will work on any other bounded interval, as well as for weighted inner products, but (5.40) is of particular importance. We shall apply the Gram–Schmidt orthogonalization process to the elementary, but non-orthogonal monomial basis $1, t, t^2, \ldots t^n$. Because

$$\langle t^k, t^l \rangle = \int_{-1}^{1} t^{k+l}\, dt = \begin{cases} \dfrac{2}{k+l+1}, & k+l \text{ even}, \\ 0, & k+l \text{ odd}, \end{cases} \tag{5.41}$$

odd degree monomials are orthogonal to those of even degree, but that is all. We will use $q_0(t), q_1(t), \ldots, q_n(t)$ to denote the resulting orthogonal polynomials. We begin by setting

$$q_0(t) = 1, \qquad \text{with} \quad \| q_0 \|^2 = \int_{-1}^{1} q_0(t)^2\, dt = 2.$$

According to formula (5.17), the next orthogonal basis polynomial is

$$q_1(t) = t \; - \; \frac{\langle t, q_0 \rangle}{\| q_0 \|^2}\, q_0(t) = t, \qquad \text{with} \quad \| q_1 \|^2 = \tfrac{2}{3}\,.$$

In general, the Gram–Schmidt formula (5.19) says we should define

$$q_k(t) \; = \; t^k \; - \; \sum_{j=0}^{k-1} \frac{\langle t^k, q_j \rangle}{\| q_j \|^2}\, q_j(t) \qquad \text{for} \quad k = 1, 2, \ldots\,.$$

We can thus recursively compute the next few orthogonal polynomials:

$$\begin{aligned} q_2(t) &= t^2 - \tfrac{1}{3}, & \| q_2 \|^2 &= \tfrac{8}{45}\,, \\ q_3(t) &= t^3 - \tfrac{3}{5}\, t, & \| q_3 \|^2 &= \tfrac{8}{175}\,, \\ q_4(t) &= t^4 - \tfrac{6}{7}\, t^2 + \tfrac{3}{35}\,, & \| q_4 \|^2 &= \tfrac{128}{11025}\,, \end{aligned} \tag{5.42}$$

and so on. The reader can verify that they satisfy the orthogonality conditions

$$\langle q_i, q_j \rangle = \int_{-1}^{1} q_i(t)\, q_j(t)\, dt = 0, \qquad i \neq j.$$

The resulting polynomials $q_0, q_1, q_2, \ldots$ are known as the *monic* Legendre polynomials*, in honor of the 18th century French mathematician Adrien–Marie Legendre who first used them for studying Newtonian gravitation. Since the first n, namely $q_0, \ldots, q_{n-1}$ span the subspace $\mathcal{P}^{(n-1)}$ of polynomials of degree $\leq n-1$, the next one, q_n, can be characterized as the unique monic polynomial that is orthogonal to every polynomial of degree $\leq n-1$:

$$\langle t^k, q_n \rangle = 0, \qquad k = 0, \ldots, n-1. \tag{5.43}$$

Since the monic Legendre polynomials form a basis for the space of polynomials, we can uniquely rewrite any polynomial of degree n as a linear combination:

$$p(t) = c_0\, q_0(t) + c_1\, q_1(t) + \cdots + c_n\, q_n(t). \tag{5.44}$$

---

*A polynomial is called *monic* if its leading coefficient is equal to 1.

In view of the general orthogonality formula (5.7), the coefficients are simply given by inner products

$$c_k = \frac{\langle p, q_k \rangle}{\| q_k \|^2} = \frac{1}{\| q_k \|^2} \int_{-1}^{1} p(t)\, q_k(t)\, dt, \qquad k = 0, \ldots, n. \tag{5.45}$$

For example,

$$t^4 = q_4(t) + \tfrac{6}{7} q_2(t) + \tfrac{1}{5} q_0(t) = \left(t^4 - \tfrac{6}{7} t^2 + \tfrac{3}{35}\right) + \tfrac{6}{7}\left(t^2 - \tfrac{1}{3}\right) + \tfrac{1}{5},$$

where the coefficients can be obtained either directly or via (5.45):

$$c_4 = \frac{11025}{128} \int_{-1}^{1} t^4\, q_4(t)\, dt = 1, \qquad c_3 = \frac{175}{8} \int_{-1}^{1} t^4\, q_3(t)\, dt = 0, \qquad \text{and so on.}$$

The classical *Legendre polynomials* are certain scalar multiples, namely

$$P_k(t) = \frac{(2k)!}{2^k\, (k!)^2}\, q_k(t), \qquad k = 0, 1, 2, \ldots, \tag{5.46}$$

and so also define a system of orthogonal polynomials. The multiple is fixed by the requirement that

$$P_k(1) = 1, \tag{5.47}$$

which is not so important here, but does play a role in other applications. The first few classical Legendre polynomials are

$$\begin{aligned}
P_0(t) &= 1, & \| P_0 \|^2 &= 2,\\
P_1(t) &= t, & \| P_1 \|^2 &= \tfrac{2}{3},\\
P_2(t) &= \tfrac{3}{2} t^2 - \tfrac{1}{2}, & \| P_2 \|^2 &= \tfrac{2}{5},\\
P_3(t) &= \tfrac{5}{2} t^3 - \tfrac{3}{2} t, & \| P_3 \|^2 &= \tfrac{2}{7},\\
P_4(t) &= \tfrac{35}{8} t^4 - \tfrac{15}{4} t^2 + \tfrac{3}{8}, & \| P_4 \|^2 &= \tfrac{2}{9},\\
P_5(t) &= \tfrac{63}{8} t^5 - \tfrac{35}{4} t^3 + \tfrac{15}{8} t. & \| P_5 \|^2 &= \tfrac{2}{11},\\
P_6(t) &= \tfrac{231}{16} t^6 - \tfrac{315}{16} t^4 + \tfrac{105}{16} t^2 - \tfrac{5}{16}, & \| P_6 \|^2 &= \tfrac{2}{13},
\end{aligned}$$

and are graphed in Figure 5.4. There is, in fact, an explicit formula for the Legendre polynomials, due to the early nineteenth century Portuguese mathematician Olinde Rodrigues.

THEOREM 5.30 The *Rodrigues formula* for the classical Legendre polynomials is

$$P_k(t) = \frac{1}{2^k\, k!} \frac{d^k}{dt^k} (t^2 - 1)^k, \qquad \| P_k \| = \sqrt{\frac{2}{2k+1}}, \qquad k = 0, 1, 2, \ldots. \tag{5.48}$$

Thus, for example,

$$P_4(t) = \frac{1}{16 \cdot 4!} \frac{d^4}{dt^4} (t^2 - 1)^4 = \frac{1}{384} \frac{d^4}{dt^4} (t^2 - 1)^4 = \tfrac{35}{8} t^4 - \tfrac{15}{4} t^2 + \tfrac{3}{8}.$$

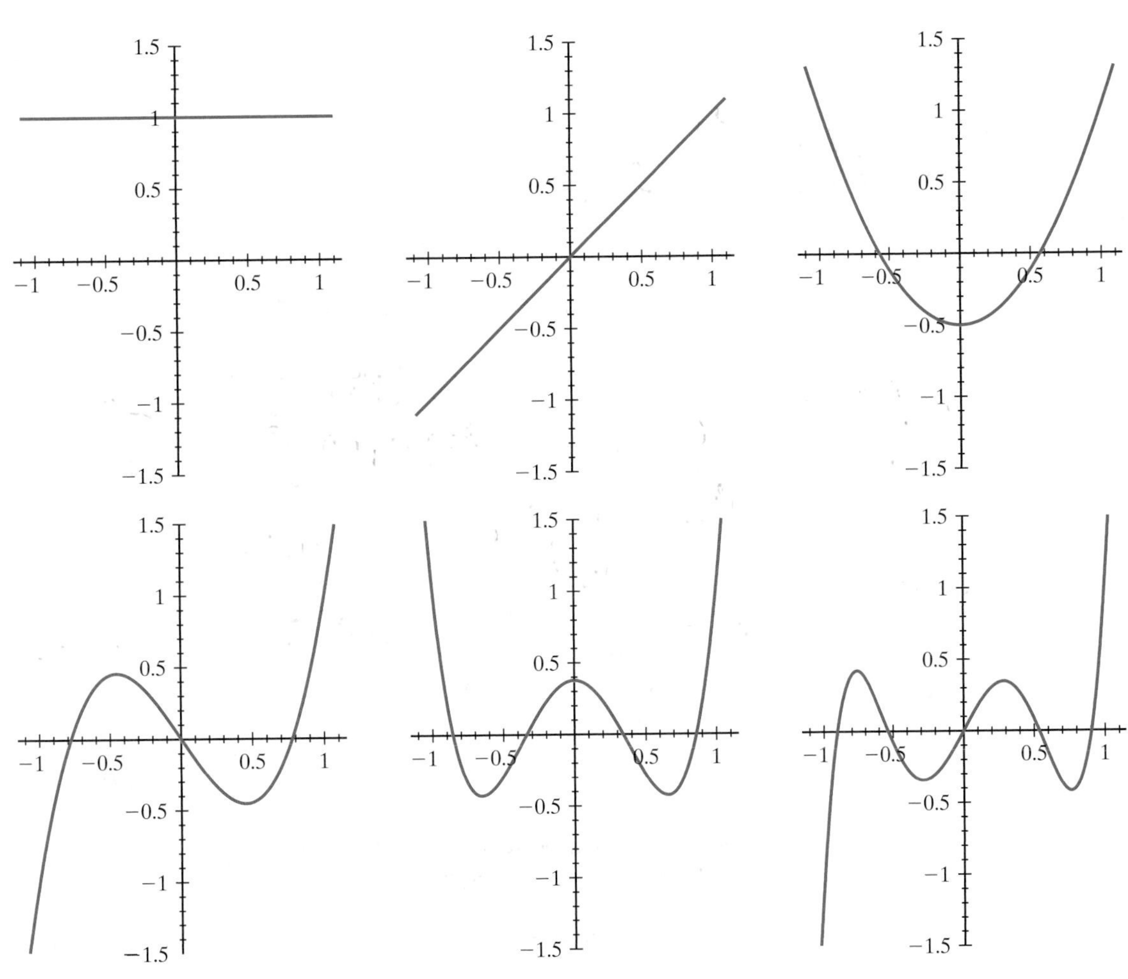

Figure 5.4 The Legendre polynomials $P_0(t), \ldots, P_5(t)$.

Proof of Theorem 5.30 Let

$$R_{j,k}(t) = \frac{d^j}{dt^j}\,(t^2-1)^k, \tag{5.49}$$

which is evidently a polynomial of degree $2k-j$. In particular, the Rodrigues formula (5.48) claims that $P_k(t)$ is a multiple of $R_{k,k}(t)$. Note that

$$\frac{d}{dt} R_{j,k}(t) = R_{j+1,k}(t). \tag{5.50}$$

Moreover,

$$R_{j,k}(1) = 0 = R_{j,k}(-1) \qquad \text{whenever} \qquad j < k, \tag{5.51}$$

since, by the product rule, differentiating $(t^2-1)^k$ a total of $j < k$ times still leaves at least one factor of t^2-1 in each summand, which therefore vanishes at $t = \pm 1$. In order to complete the proof, let us establish:

Lemma 5.31 If $j \le k$, then the polynomial $R_{j,k}(t)$ is orthogonal to all polynomials of degree $\le j-1$.

Proof In other words,

$$\langle t^i, R_{j,k} \rangle = \int_{-1}^{1} t^i R_{j,k}(t)\, dt = 0, \qquad \text{for all} \quad 0 \le i < j \le k. \tag{5.52}$$

Since $j > 0$, we use (5.50) to write $R_{j,k}(t) = R'_{j-1,k}(t)$. Integrating by parts,

$$\begin{aligned} \langle t^i, R_{j,k} \rangle &= \int_{-1}^{1} t^i R'_{j-1,k}(t)\, dt \\ &= i\, t^i R_{j-1,k}(t) - i \int_{-1}^{1} t^{i-1} R_{j-1,k}(t)\, dt \\ &= -i\, \langle t^{i-1}, R_{j-1,k} \rangle, \end{aligned}$$

where the boundary terms vanish owing to (5.51). In particular, setting $i = 0$ proves $\langle 1, R_{j,k} \rangle = 0$ for all $j > 0$. We then repeat the process, and, eventually, for any $j > i$,

$$\begin{aligned} \langle t^i, R_{j,k} \rangle &= -i\, \langle t^{i-1}, R_{j-1,k} \rangle = i(i-1)\, \langle t^{i-2}, R_{j-2,k} \rangle \\ &= \cdots = (-1)^i\, i\,(i-1) \cdots 3 \cdot 2\, \langle 1, R_{j-i,k} \rangle = 0, \end{aligned}$$

completing the proof. ■

In particular, $R_{k,k}(t)$ is a polynomial of degree k which is orthogonal to every polynomial of degree $\le k - 1$. By our earlier remarks, this implies that it must be a constant multiple,

$$R_{k,k}(t) = c_k\, P_k(t)$$

of the kth Legendre polynomial. To determine c_k, we need only compare the leading terms:

$$R_{k,k}(t) = \frac{d^k}{dt^k}\,(t^2 - 1)^k = \frac{d^k}{dt^k}\,(t^{2k} + \cdots) = \frac{(2k)!}{k!}\, t^k + \cdots,$$

while

$$P_k(t) = \frac{(2k)!}{2^k\,(k!)^2}\, t^{2k} + \cdots.$$

We conclude that $c_k = 2^k\, k!$, which proves (5.48). The proof of the formula for $\| P_k \|$ can be found in Exercise 5.4.9. ■

The Legendre polynomials play an important role in many aspects of applied mathematics, including numerical analysis, least squares approximation of functions, and solution of partial differential equations.

EXERCISES 5.4

5.4.1. Write the following polynomials as linear combinations of monic Legendre polynomials. Use orthogonality to compute the coefficients:
(a) t^3 (b) $t^4 + t^2$ (c) $7t^4 + 2t^3 - t$

5.4.2. (a) Find the monic Legendre polynomial of degree 5 using the Gram–Schmidt process. Check your answer by using the Rodrigues formula.
(b) Use orthogonality to write t^5 as linear combinations of Legendre polynomials.
(c) Repeat the exercise for degree 6.

◇ **5.4.3.** (a) Explain why q_n is the unique monic polynomial that satisfies (5.43).
(b) Use this characterization to directly construct $q_5(t)$.

5.4.4. Prove that the even (odd) degree Legendre polynomials are even (odd) functions of t.

5.4.5. Prove that if $p(t) = p(-t)$ is an even polynomial, then all the odd order coefficients $c_{2j+1} = 0$ in its Legendre expansion (5.44) vanish.

5.4.6. Write out an explicit Rodrigues-type formula for the monic Legendre polynomial $q_k(t)$ and its norm.

5.4.7. Write out an explicit Rodrigues-type formula for an *orthonormal basis* $Q_0(t), \ldots, Q_n(t)$ for the space of polynomials of degree $\leq n$ under the inner product (5.40).

◇ **5.4.8.** Use the Rodrigues formula to prove (5.47). *Hint*: Write $(t^2-1)^k = (t-1)^k\,(t+1)^k$.

♡ **5.4.9.** A proof of the formula in (5.48) for the norm of the Legendre polynomial is based on the following steps.

(a) First, prove that

$$\| R_{k,k} \|^2 = (-1)^k\,(2k)! \int_{-1}^{1} (t^2-1)^k\,dt$$

by a repeated integration by parts.

(b) Second, prove that

$$\int_{-1}^{1} (t^2-1)^k\,dt = (-1)^k\,\frac{2^{2k+1}\,(k!)^2}{(2k+1)!}$$

by using the change of variables $t = \cos\theta$ in the integral. The resulting trigonometric integral can be done by another repeated integration by parts.

(c) Finally, use the Rodrigues formula to complete the proof.

♡ **5.4.10.** (a) Find the roots, $P_n(t) = 0$, of the Legendre polynomials P_2, P_3 and P_4.

(b) Prove that for $0 \leq j \leq k$, the polynomial $R_{j,k}(t)$ defined in (5.49) has roots of order $k-j$ at $t = \pm 1$, and j additional simple roots lying between -1 and 1. *Hint*: Use induction on j and Rolle's Theorem from calculus, [2, 58].

(c) Conclude that all k roots of the Legendre polynomial $P_k(t)$ are real, simple and lie in the interval $-1 < t < 1$.

Other Systems of Orthogonal Polynomials

The standard Legendre polynomials form an orthogonal system with respect to the L^2 inner product on the interval $[-1, 1]$. Dealing with any other interval, or, more generally, a weighted inner product, leads to a different, suitably adapted collection of orthogonal polynomials. In all cases, applying the Gram–Schmidt process to the standard monomials $1, t, t^2, t^3, \ldots$ will produce the desired orthogonal system.

EXAMPLE 5.32 In this example, we construct orthogonal polynomials for the weighted inner product

$$\langle f, g \rangle = \int_0^\infty f(t)\,g(t)\,e^{-t}\,dt \tag{5.53}$$

on the interval $[0, \infty)$. A straightforward integration by parts proves that

$$\int_0^\infty t^k\,e^{-t}\,dt = k!, \qquad \text{and hence} \quad \langle t^i, t^j \rangle = (i+j)!, \quad \| t^i \|^2 = (2i)! \tag{5.54}$$

We apply the Gram–Schmidt process to construct a system of orthogonal polynomials for this inner product. The first few are

$$\begin{aligned}
q_0(t) &= 1, && \| q_0 \|^2 = 1, \\
q_1(t) &= t - \frac{\langle t, q_0 \rangle}{\| q_0 \|^2}\,q_0(t) = t - 1, && \| q_1 \|^2 = 1, \\
q_2(t) &= t^2 - \frac{\langle t^2, q_0 \rangle}{\| q_0 \|^2}\,q_0(t) - \frac{\langle t^2, q_1 \rangle}{\| q_1 \|^2}\,q_1(t) && \\
&= t^2 - 4t + 2, && \| q_2 \|^2 = 4 \\
q_3(t) &= t^3 - 9t^2 + 18t - 6, && \| q_3 \|^2 = 36.
\end{aligned} \tag{5.55}$$

The resulting orthogonal polynomials are known as the (monic) *Laguerre polynomials*, named after the nineteenth century French mathematician Edmond Laguerre. ●

In some cases, a change of variables may be used to relate systems of orthogonal polynomials and thereby circumvent the Gram–Schmidt computation. Suppose, for instance, that our goal is to construct an orthogonal system of polynomials for the L^2 inner product

$$\langle\langle f, g \rangle\rangle = \int_a^b f(t)\, g(t)\, dt$$

on the interval $[a, b]$. The key remark is that we can map the interval $[-1, 1]$ to $[a, b]$ by a simple change of variables of the form $s = \alpha + \beta t$. Specifically,

$$s = \frac{2t - b - a}{b - a} \quad \text{will change} \quad a \le t \le b \quad \text{to} \quad -1 \le s \le 1. \tag{5.56}$$

It therefore changes functions $F(s)$, $G(s)$, defined for $-1 \le s \le 1$, into functions

$$f(t) = F\left(\frac{2t - b - a}{b - a}\right), \qquad g(t) = G\left(\frac{2t - b - a}{b - a}\right), \tag{5.57}$$

defined for $a \le t \le b$. Moreover, when integrating, $ds = \dfrac{2}{b - a}\, dt$, and so the inner products are related by

$$\begin{aligned} \langle f, g \rangle = \int_a^b f(t)\, g(t)\, dt &= \int_a^b F\left(\frac{2t - b - a}{b - a}\right) G\left(\frac{2t - b - a}{b - a}\right) dt \\ &= \int_{-1}^1 F(s)\, G(s)\, \frac{b - a}{2}\, ds = \frac{b - a}{2} \langle F, G \rangle, \end{aligned} \tag{5.58}$$

where the final L^2 inner product is over the interval $[-1, 1]$. In particular, the change of variables maintains orthogonality, while rescaling the norms:

$$\langle f, g \rangle = 0 \quad \text{if and only if} \quad \langle F, G \rangle = 0, \qquad \text{while} \quad \| f \| = \sqrt{\frac{b - a}{2}}\, \| F \|. \tag{5.59}$$

Moreover, if $F(s)$ is a polynomial of degree n in s, then $f(t)$ is a polynomial of degree n in t and vice versa. Let us apply these observations to the Legendre polynomials:

Proposition 5.33 The transformed Legendre polynomials

$$\widetilde{P}_k(t) = P_k\left(\frac{2t - b - a}{b - a}\right), \qquad \| \widetilde{P}_k \| = \sqrt{\frac{b - a}{2k + 1}}, \qquad k = 0, 1, 2, \ldots, \tag{5.60}$$

form an orthogonal system of polynomials with respect to the L^2 inner product on the interval $[a, b]$.

EXAMPLE 5.34 Consider the L^2 inner product

$$\langle\langle f, g \rangle\rangle = \int_0^1 f(t)\, g(t)\, dt.$$

The map $s = 2t - 1$ will change $0 \le t \le 1$ to $-1 \le s \le 1$. According to Proposition 5.33, this change of variables will convert the Legendre polynomials $P_k(s)$ into an orthogonal system of polynomials on $[0, 1]$, namely

$$\widetilde{P}_k(t) = P_k(2t - 1), \qquad \text{with corresponding } L^2 \text{ norms} \quad \| \widetilde{P}_k \| = \sqrt{\frac{1}{2k + 1}}.$$

The first few are

$$\begin{aligned}
\widetilde{P}_0(t) &= 1, \\
\widetilde{P}_1(t) &= 2t - 1, \\
\widetilde{P}_2(t) &= 6t^2 - 6t + 1, \\
\widetilde{P}_3(t) &= 20t^3 - 30t^2 + 12t - 1, \\
\widetilde{P}_4(t) &= 70t^4 - 140t^3 + 90t^2 - 20t + 1, \\
\widetilde{P}_5(t) &= 252t^5 - 630t^4 + 560t^3 - 210t^2 + 30t - 1.
\end{aligned} \tag{5.61}$$

One can, as an alternative, derive these formulae through a direct application of the Gram–Schmidt process. ●

EXERCISES 5.4

5.4.11. Construct polynomials P_0, P_1, P_2, and P_3 of degree 0, 1, 2, and 3, respectively, which are orthogonal with respect to the inner products

(a) $\langle f, g\rangle = \displaystyle\int_1^2 f(t)\, g(t)\, dt$

(b) $\langle f, g\rangle = \displaystyle\int_0^1 f(t)\, g(t)\, t\, dt$

(c) $\langle f, g\rangle = \displaystyle\int_{-1}^1 f(t)\, g(t)\, t^2\, dt$

(d) $\langle f, g\rangle = \displaystyle\int_{-\infty}^\infty f(t)\, g(t)\, e^{-|t|}\, dt$

5.4.12. Find the first four orthogonal polynomials on the interval $[0, 1]$ for the weighted L^2 inner product with weight $w(t) = t^2$.

5.4.13. Write down an orthogonal basis for vector space $\mathcal{P}^{(5)}$ of quintic polynomials under the inner product

$$\langle f, g\rangle = \int_{-2}^2 f(t)\, g(t)\, dt.$$

5.4.14. Use the Gram–Schmidt process based on the L^2 inner product on $[0, 1]$ to construct a system of orthogonal polynomials of degree ≤ 4. Verify that your polynomials are multiples of the modified Legendre polynomials found in Example 5.34.

5.4.15. Find the first four orthogonal polynomials under the Sobolev H^1 inner product

$$\langle f, g\rangle = \int_{-1}^1 \left[\, f(x)\, g(x) + f'(x)\, g'(x)\,\right] dx,$$

cf. Exercise 3.1.25.

◇ **5.4.16.** Prove the formula for $\| \widetilde{P}_k \|$ in (5.60).

5.4.17. Find the monic Laguerre polynomials of degrees 4 and 5 and their norms.

◇ **5.4.18.** Prove the integration formula (5.54).

◇ **5.4.19.** The *Hermite polynomials* are orthogonal with respect to the inner product

$$\langle f, g\rangle = \int_{-\infty}^\infty f(t)\, g(t)\, e^{-t^2}\, dt.$$

Find the first five monic Hermite polynomials. *Hint*: $\displaystyle\int_{-\infty}^\infty e^{-t^2}\, dt = \sqrt{\pi}$.

♡ **5.4.20.** The *Chebyshev polynomials*:

(a) Prove that

$$T_n(t) = \cos(n \arccos t), \qquad n = 0, 1, 2, \ldots$$

form a system of orthogonal polynomials under the weighted inner product

$$\langle f, g\rangle = \int_{-1}^1 \frac{f(t)\, g(t)\, dt}{\sqrt{1 - t^2}}. \tag{5.62}$$

(b) What is $\| T_n \|$?

(c) Write out the formulae for $T_0(t), \ldots, T_6(t)$ and plot their graphs.

5.4.21. Does the Gram–Schmidt process for the inner product (5.62) lead to the Chebyshev polynomials $T_n(t)$ defined in the preceding exercise? Explain why or why not.

5.4.22. Find two functions that form an orthogonal basis for the space of the solutions to the differential equation $y'' - 3y' + 2y = 0$ under the L^2 inner product on $[0, 1]$.

5.4.23. Find an orthogonal basis for the space of the solutions to the differential equation $y''' - y'' + y' - y = 0$ for the L^2 inner product on $[-\pi, \pi]$.

5.4.24. Explain how to adapt the numerically stable Gram–Schmidt method in (5.28) to construct a system of orthogonal polynomials. Test your algorithm on one of the preceding exercises.

♡ **5.4.25.** In this exercise, we investigate the effect of more general changes of variables on orthogonal polynomials.

(a) Prove that $t = 2s^2 - 1$ defines a one-to-one map from the interval $0 \le s \le 1$ to the interval $-1 \le t \le 1$.

(b) Let $p_k(t)$ denote the monic Legendre polynomials, which are orthogonal on $-1 \le t \le 1$. Show that $q_k(s) = p_k(2s^2 - 1)$ defines a polynomial. Write out the cases $k = 0, 1, 2, 3$ explicitly.

(c) Are the polynomials $q_k(s)$ orthogonal under the L^2 inner product on $[0, 1]$? If not, do they retain any sort of orthogonality property? *Hint*: What happens to the L^2 inner product on $[-1, 1]$ under the change of variables?

5.4.26. (a) Show that the change of variables $s = e^{-t}$ maps the Laguerre inner product (5.53) to the standard L^2 inner product on $[0, 1]$. However, explain why this does *not* allow you to change Legendre polynomials into Laguerre polynomials.

(b) Describe the functions resulting from applying the change of variables to the modified Legendre polynomials (5.61) and their orthogonality properties.

(c) Describe the functions results from applying the inverse change of variables to the Laguerre polynomials (5.55) and their orthogonality properties.

5.5 ORTHOGONAL PROJECTIONS AND LEAST SQUARES

In Chapter 4, we solved, and then discovered the significance of, the problem of finding the point on a subspace that lies closest to a prescribed point. In this section, we shall explore an important geometrical interpretation of our solution: the closest point is the *orthogonal projection* of the prescribed point onto the subspace. Furthermore, if we adopt an orthogonal, or, even better, orthonormal basis for the subspace, then the closest point can be easily determined through an elegant, explicit formula. In this manner, orthogonality provides an efficient shortcut that effectively bypasses the tedious solving of the normal equations. The resulting orthogonal projection formula streamlines the solution of a wide range of applied problems, including least squares minimization and data fitting.

Orthogonal Projection

We begin by characterizing the orthogonal projection of a vector onto a subspace. Throughout this section, $W \subset V$ will be a finite-dimensional subspace of a real inner product space. The inner product space V is allowed to be infinite-dimensional. But, to facilitate your geometric intuition, you may initially want to view W as a subspace of Euclidean space $V = \mathbb{R}^m$ equipped with the ordinary dot product.

Definition 5.35 A vector $\mathbf{z} \in V$ is said to be *orthogonal* to the subspace $W \subset V$ if it is orthogonal to every vector in W, so $\langle \mathbf{z}, \mathbf{w} \rangle = 0$ for all $\mathbf{w} \in W$.

Given a basis (or, more generally, a spanning set) $\mathbf{w}_1, \ldots, \mathbf{w}_n$ for W, we note that $\mathbf{z}$ is orthogonal to W if and only if it is orthogonal to every basis vector: $\langle \mathbf{z}, \mathbf{w}_i \rangle = 0$ for $i = 1, \ldots, n$. Indeed, any other vector in W has the form $\mathbf{w} = c_1 \mathbf{w}_1 + \cdots + c_n \mathbf{w}_n$, and hence, by linearity, $\langle \mathbf{z}, \mathbf{w} \rangle = c_1 \langle \mathbf{z}, \mathbf{w}_1 \rangle + \cdots + c_n \langle \mathbf{z}, \mathbf{w}_n \rangle = 0$, as required.

Definition 5.36 The *orthogonal projection* of $\mathbf{v}$ onto the subspace W is the element $\mathbf{w} \in W$ that makes the difference $\mathbf{z} = \mathbf{v} - \mathbf{w}$ orthogonal to W.

The geometric configuration underlying orthogonal projection can be seen in Figure 5.5. As we shall see, the orthogonal projection is unique. The explicit construction is greatly simplified by taking a orthonormal basis of the subspace, which, if necessary, can be arranged by applying the Gram–Schmidt process to a known basis.

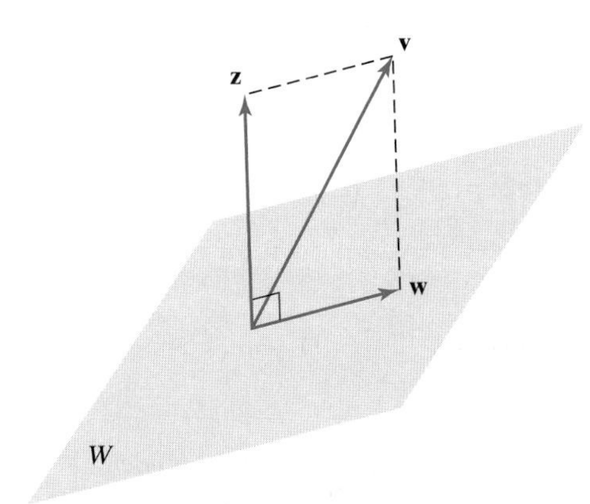

Figure 5.5 The orthogonal projection of a vector onto a subspace.

(The direct construction of the orthogonal projection in terms of a non-orthogonal basis appears in Exercise 5.5.26.)

THEOREM 5.37 Let $\mathbf{u}_1, \ldots, \mathbf{u}_n$ be an orthonormal basis for the subspace $W \subset V$. Then the orthogonal projection of a vector $\mathbf{v} \in V$ onto W is

$$\mathbf{w} = c_1 \mathbf{u}_1 + \cdots + c_n \mathbf{u}_n \quad \text{where} \quad c_i = \langle \mathbf{v}, \mathbf{u}_i \rangle, \quad i = 1, \ldots, n. \tag{5.63}$$

Proof First, since $\mathbf{u}_1, \ldots, \mathbf{u}_n$ form a basis of the subspace, the orthogonal projection element must be some linear combination: $\mathbf{w} = c_1 \mathbf{u}_1 + \cdots + c_n \mathbf{u}_n$. Definition 5.36 requires that the difference $\mathbf{z} = \mathbf{v} - \mathbf{w}$ be orthogonal to W, and, as noted above, it suffices to check orthogonality to the basis vectors. By our orthonormality assumption,

$$\begin{aligned} 0 = \langle \mathbf{z}, \mathbf{u}_i \rangle &= \langle \mathbf{v}, \mathbf{u}_i \rangle - \langle \mathbf{w}, \mathbf{u}_i \rangle = \langle \mathbf{v}, \mathbf{u}_i \rangle - \langle c_1 \mathbf{u}_1 + \cdots + c_n \mathbf{u}_n, \mathbf{u}_i \rangle \\ &= \langle \mathbf{v}, \mathbf{u}_i \rangle - c_1 \langle \mathbf{u}_1, \mathbf{u}_i \rangle - \cdots - c_n \langle \mathbf{u}_n, \mathbf{u}_i \rangle = \langle \mathbf{v}, \mathbf{u}_i \rangle - c_i. \end{aligned}$$

We deduce that the coefficients $c_i = \langle \mathbf{v}, \mathbf{u}_i \rangle$ of the orthogonal projection $\mathbf{w}$ are uniquely prescribed by the orthogonality requirement, which also proves its uniqueness. ■

More generally, if we employ an orthogonal basis $\mathbf{v}_1, \ldots, \mathbf{v}_n$ for the subspace W, then the same argument demonstrates that the orthogonal projection of $\mathbf{v}$ onto W is given by

$$\mathbf{w} = a_1 \mathbf{v}_1 + \cdots + a_n \mathbf{v}_n, \quad \text{where} \quad a_i = \frac{\langle \mathbf{v}, \mathbf{v}_i \rangle}{\| \mathbf{v}_i \|^2}, \qquad i = 1, \ldots, n. \tag{5.64}$$

Of course, we could equally well replace the orthogonal basis by the orthonormal basis obtained by dividing each vector by its length: $\mathbf{u}_i = \mathbf{v}_i / \| \mathbf{v}_i \|$. The reader should be able to prove that the two formulae (5.63–64) for the orthogonal projection yield the same vector $\mathbf{w}$.

EXAMPLE 5.38 Consider the plane $W \subset \mathbb{R}^3$ spanned by the orthogonal vectors

$$\mathbf{v}_1 = \begin{pmatrix} 1 \\ -2 \\ 1 \end{pmatrix}, \qquad \mathbf{v}_2 = \begin{pmatrix} 1 \\ 1 \\ 1 \end{pmatrix}.$$

According to formula (5.64), the orthogonal projection of $\mathbf{v} = (1, 0, 0)^T$ onto W is

$$\mathbf{w} = \frac{\langle \mathbf{v}, \mathbf{v}_1 \rangle}{\|\mathbf{v}_1\|^2}\, \mathbf{v}_1 + \frac{\langle \mathbf{v}, \mathbf{v}_2 \rangle}{\|\mathbf{v}_2\|^2}\, \mathbf{v}_2 = \frac{1}{6}\begin{pmatrix} 1 \\ -2 \\ 1 \end{pmatrix} + \frac{1}{3}\begin{pmatrix} 1 \\ 1 \\ 1 \end{pmatrix} = \begin{pmatrix} \frac{1}{2} \\ 0 \\ \frac{1}{2} \end{pmatrix}.$$

Alternatively, we can replace $\mathbf{v}_1, \mathbf{v}_2$ by the orthonormal basis

$$\mathbf{u}_1 = \frac{\mathbf{v}_1}{\|\mathbf{v}_1\|} = \begin{pmatrix} \frac{1}{\sqrt{6}} \\ -\frac{2}{\sqrt{6}} \\ \frac{1}{\sqrt{6}} \end{pmatrix}, \qquad \mathbf{u}_2 = \frac{\mathbf{v}_2}{\|\mathbf{v}_2\|} = \begin{pmatrix} \frac{1}{\sqrt{3}} \\ \frac{1}{\sqrt{3}} \\ \frac{1}{\sqrt{3}} \end{pmatrix}.$$

Then, using the orthonormal version (5.63),

$$\mathbf{w} = \langle \mathbf{v}, \mathbf{u}_1 \rangle\, \mathbf{u}_1 + \langle \mathbf{v}, \mathbf{u}_2 \rangle\, \mathbf{u}_2 = \frac{1}{\sqrt{6}}\begin{pmatrix} \frac{1}{\sqrt{6}} \\ -\frac{2}{\sqrt{6}} \\ \frac{1}{\sqrt{6}} \end{pmatrix} + \frac{1}{\sqrt{3}}\begin{pmatrix} \frac{1}{\sqrt{3}} \\ \frac{1}{\sqrt{3}} \\ \frac{1}{\sqrt{3}} \end{pmatrix} = \begin{pmatrix} \frac{1}{2} \\ 0 \\ \frac{1}{2} \end{pmatrix}.$$

The answer is, of course, the same. As the reader may notice, while the theoretical formula is simpler when written in an orthonormal basis, for hand computations the orthogonal basis version avoids dealing with square roots. (Of course, when performing the computation on a computer, this is not a significant issue.) ●

An intriguing observation is that the coefficients in the orthogonal projection formulae (5.63), (5.64) coincide with the formulae (5.4), (5.7) for writing a vector in terms of an orthonormal or orthogonal basis. Indeed, if $\mathbf{v}$ were an element of W, then it would coincide with its orthogonal projection, $\mathbf{w} = \mathbf{v}$. (Why?) As a result, the orthogonal projection formulae include the orthogonal basis formulae as a special case.

It is also worth noting that the *same* formulae occur in the Gram–Schmidt algorithm, cf. (5.19). This observation leads to a useful geometric interpretation for the Gram–Schmidt construction. For each $k = 1, \ldots, n$, let

$$V_k = \operatorname{span}\{\mathbf{w}_1, \ldots, \mathbf{w}_k\} = \operatorname{span}\{\mathbf{v}_1, \ldots, \mathbf{v}_k\} = \operatorname{span}\{\mathbf{u}_1, \ldots, \mathbf{u}_k\} \tag{5.65}$$

denote the k-dimensional subspace spanned by the first k basis elements. The basic Gram–Schmidt formula (5.19) can be rewritten in the form $\mathbf{v}_k = \mathbf{w}_k - \mathbf{y}_k$, where $\mathbf{y}_k$ is the orthogonal projection of $\mathbf{w}_k$ onto the subspace V_{k-1}. The resulting vector $\mathbf{v}_k$ is, by construction, orthogonal to the subspace, and hence orthogonal to all of the previous basis elements, which serves to rejustify the Gram–Schmidt procedure.

EXERCISES 5.5

Note: Use the Euclidean dot product and norm unless otherwise specified.

5.5.1. Determine which of the vectors

$$\mathbf{v}_1 = \begin{pmatrix} 1 \\ 1 \\ 0 \end{pmatrix}, \quad \mathbf{v}_2 = \begin{pmatrix} -2 \\ 2 \\ 2 \end{pmatrix},$$

$$\mathbf{v}_3 = \begin{pmatrix} 2 \\ -1 \\ -3 \end{pmatrix}, \quad \mathbf{v}_4 = \begin{pmatrix} -1 \\ 3 \\ 4 \end{pmatrix},$$

is orthogonal to

(a) the line spanned by $\begin{pmatrix} 1 \\ 3 \\ -2 \end{pmatrix}$

(b) the plane spanned by $\begin{pmatrix} 1 \\ -1 \\ 1 \end{pmatrix}, \begin{pmatrix} 2 \\ 1 \\ 1 \end{pmatrix}$

(c) the plane defined by $x - y - z = 0$

(d) the kernel of the matrix $\begin{pmatrix} 1 & -1 & -1 \\ 3 & -2 & -4 \end{pmatrix}$

(e) the range of the matrix $\begin{pmatrix} -3 & 1 \\ 3 & -1 \\ -1 & 0 \end{pmatrix}$

(f) the cokernel of the matrix $\begin{pmatrix} -1 & 0 & 3 \\ 2 & 1 & -2 \\ 3 & 1 & -5 \end{pmatrix}$

5.5.2. Find the orthogonal projection of the vector $\mathbf{v} = (1, 1, 1)^T$ onto the following subspaces, using the indicated orthonormal/orthogonal bases:

(a) the line in the direction $\left(-\frac{1}{\sqrt{3}}, \frac{1}{\sqrt{3}}, \frac{1}{\sqrt{3}}\right)^T$

(b) the line spanned by $(2, -1, 3)^T$

(c) the plane spanned by $(1, 1, 0)^T, (-2, 2, 1)^T$

(d) the plane spanned by $\left(-\frac{3}{5}, \frac{4}{5}, 0\right)^T, \left(\frac{4}{13}, \frac{3}{13}, -\frac{12}{13}\right)^T$

5.5.3. Find the orthogonal projection of the vector $\begin{pmatrix} 1 \\ 2 \\ 3 \end{pmatrix}$ onto the range of $\begin{pmatrix} 3 & 2 \\ 2 & -2 \\ 1 & -2 \end{pmatrix}$.

5.5.4. Find the orthogonal projection of the vector $\mathbf{v} = (1, 3, -1)^T$ onto the plane spanned by $(-1, 2, 1)^T, (2, 1, -3)^T$ by first using the Gram–Schmidt process to construct an orthogonal basis.

5.5.5. Find the orthogonal projection of

$$\mathbf{v} = (1, 2, -1, 2)^T$$

onto the following subspaces:

(a) the span of $\begin{pmatrix} 1 \\ -1 \\ 2 \\ 1 \end{pmatrix}, \begin{pmatrix} 2 \\ 1 \\ 0 \\ -1 \end{pmatrix}$

(b) the range of the matrix $\begin{pmatrix} 1 & 2 \\ -1 & 1 \\ 0 & 3 \\ -1 & 1 \end{pmatrix}$

(c) the kernel of the matrix $\begin{pmatrix} 1 & -1 & 0 & 1 \\ -2 & 1 & 1 & 0 \end{pmatrix}$

(d) the subspace orthogonal to $\mathbf{a} = (1, -1, 0, 1)^T$

Warning: Make sure you have an orthogonal basis before applying formula (5.64)!

5.5.6. Redo Exercise 5.5.2 using

(i) the weighted inner product $\langle \mathbf{v}, \mathbf{w} \rangle = 2 v_1 w_1 + 2 v_2 w_2 + v_3 w_3$,

(ii) the inner product induced by the positive definite matrix

$$K = \begin{pmatrix} 2 & -1 & 0 \\ -1 & 2 & -1 \\ 0 & -1 & 2 \end{pmatrix}.$$

5.5.7. Find the orthogonal projection of $\mathbf{v} = (1, 2, -1, 2)^T$ onto the span of $(1, -1, 2, 5)^T$ and $(2, 1, 0, -1)^T$ using the weighted inner product $\langle \mathbf{v}, \mathbf{w} \rangle = 4 v_1 w_1 + 3 v_2 w_2 + 2 v_3 w_3 + v_4 w_4$.

♡ **5.5.8.** Let $\mathbf{u}_1, \ldots, \mathbf{u}_k$ be an orthonormal basis for the subspace $W \subset \mathbb{R}^m$. Let $A = (\mathbf{u}_1\ \mathbf{u}_2 \ldots \mathbf{u}_k)$ be the $m \times k$ matrix whose columns are the orthonormal basis vectors, and define $P = A A^T$ to be the corresponding *projection matrix*.

(a) Given $\mathbf{v} \in \mathbb{R}^n$, prove that its orthogonal projection $\mathbf{w} \in W$ is given by matrix multiplication: $\mathbf{w} = P\mathbf{v}$.

(b) Write out the projection matrix corresponding to the subspaces spanned by

(i) $\begin{pmatrix} \frac{1}{\sqrt{2}} \\ \frac{1}{\sqrt{2}} \end{pmatrix}$ (ii) $\begin{pmatrix} \frac{2}{3} \\ -\frac{2}{3} \\ \frac{1}{3} \end{pmatrix}$

(iii) $\begin{pmatrix} \frac{1}{\sqrt{6}} \\ -\frac{2}{\sqrt{6}} \\ \frac{1}{\sqrt{6}} \end{pmatrix}, \begin{pmatrix} \frac{1}{\sqrt{3}} \\ \frac{1}{\sqrt{3}} \\ \frac{1}{\sqrt{3}} \end{pmatrix}$

(iv) $\begin{pmatrix} \frac{1}{3} \\ -\frac{2}{3} \\ \frac{1}{3} \\ 0 \end{pmatrix}, \begin{pmatrix} 0 \\ \frac{1}{3} \\ \frac{2}{3} \\ -\frac{1}{3} \end{pmatrix}$

(v) $\begin{pmatrix} \frac{1}{2} \\ \frac{1}{2} \\ \frac{1}{2} \\ -\frac{1}{2} \end{pmatrix}, \begin{pmatrix} \frac{1}{2} \\ -\frac{1}{2} \\ \frac{1}{2} \\ \frac{1}{2} \end{pmatrix}, \begin{pmatrix} \frac{1}{2} \\ \frac{1}{2} \\ -\frac{1}{2} \\ \frac{1}{2} \end{pmatrix}$

(c) Prove that $P = P^T$ is symmetric.

(d) Prove that $P^2 = P$. Give a geometrical explanation of this fact.

(e) Prove that rank $P = k$.

5.5.9. (a) Prove that the set of all vectors orthogonal to a given subspace $V \subset \mathbb{R}^m$ forms a subspace.

(b) Find a basis for the set of all vectors in $\mathbb{R}^4$ that are orthogonal to the subspace spanned by $(1, 2, 0, -1)^T$, $(2, 0, 3, 1)^T$.

Orthogonal Least Squares

Now we make an important connection: The orthogonal projection of a vector onto a subspace is also the least squares vector—the closest point in the subspace!

THEOREM 5.39 Let $W \subset V$ be a finite-dimensional subspace of an inner product space. Given a vector $\mathbf{v} \in V$, the closest point or least squares minimizer $\mathbf{w} \in W$ is the same as the orthogonal projection of $\mathbf{v}$ onto W.

Proof Let $\mathbf{w} \in W$ be the orthogonal projection of $\mathbf{v}$ onto the subspace, which requires that the difference $\mathbf{z} = \mathbf{v} - \mathbf{w}$ be orthogonal to W. Suppose $\widetilde{\mathbf{w}} \in W$ is any other vector in the subspace. Then, the squared distance from $\mathbf{v}$ to $\widetilde{\mathbf{w}}$ is

$$\|\mathbf{v} - \widetilde{\mathbf{w}}\|^2 = \|\mathbf{w} + \mathbf{z} - \widetilde{\mathbf{w}}\|^2 = \|\mathbf{w} - \widetilde{\mathbf{w}}\|^2 + 2\langle \mathbf{w} - \widetilde{\mathbf{w}}, \mathbf{z}\rangle + \|\mathbf{z}\|^2 = \|\mathbf{w} - \widetilde{\mathbf{w}}\|^2 + \|\mathbf{z}\|^2.$$

The inner product term $\langle \mathbf{w} - \widetilde{\mathbf{w}}, \mathbf{z}\rangle = 0$ vanishes because $\mathbf{z}$ is orthogonal to every vector in W, including $\mathbf{w} - \widetilde{\mathbf{w}}$. Since $\mathbf{w}$ is uniquely prescribed by the vector $\mathbf{v}$, so is $\mathbf{z} = \mathbf{v} - \mathbf{w}$, and hence the second term $\|\mathbf{z}\|^2$ does not change with the choice of the point $\widetilde{\mathbf{w}} \in W$. As a result, $\|\mathbf{v} - \widetilde{\mathbf{w}}\|^2$ will be minimized if and only if $\|\mathbf{w} - \widetilde{\mathbf{w}}\|^2$ is minimized. Since $\widetilde{\mathbf{w}} \in W$ is allowed to be any element of the subspace W, the minimal value $\|\mathbf{w} - \widetilde{\mathbf{w}}\|^2 = 0$ occurs when $\widetilde{\mathbf{w}} = \mathbf{w}$. This implies that the closest point $\widetilde{\mathbf{w}}$ coincides with the orthogonal projection $\mathbf{w}$. ■

In particular, if we are supplied with an orthonormal or orthogonal basis of our subspace, then we can easily compute the closest least squares point $\mathbf{w} \in W$ to $\mathbf{v}$ using our orthogonal projection formulae (5.63) or (5.64). In this way, orthogonal bases have a dramatic simplifying effect on the least squares approximation formulae. Not only do they completely avoid the construction and solution of the normal equations, the resulting computational algorithm for determining the least squares solution is significantly more efficient; see Exercises 5.5.22, 24 for details.

EXAMPLE 5.40 Consider the three-dimensional subspace $W \subset \mathbb{R}^4$ spanned by the orthogonal* vectors $\mathbf{v}_1 = (1, -1, 2, 0)^T$, $\mathbf{v}_2 = (0, 2, 1, -2)^T$, $\mathbf{v}_3 = (1, 1, 0, 1)^T$. Our task is to solve the the least squares problem of finding the closest point $\mathbf{w} \in W$ to the vector $\mathbf{v} = (1, 2, 2, 1)^T$. Since the spanning vectors are orthogonal (but not orthonormal), we can use the orthogonal projection formula (5.64) to find $\mathbf{w} = a_1 \mathbf{v}_1 + a_2 \mathbf{v}_2 + a_3 \mathbf{v}_3$,

*We use the ordinary Euclidean norm on $\mathbb{R}^4$ throughout this example.

with

$$a_1 = \frac{\langle \mathbf{v}, \mathbf{v}_1 \rangle}{\| \mathbf{v}_1 \|^2} = \frac{3}{6} = \frac{1}{2},$$

$$a_2 = \frac{\langle \mathbf{v}, \mathbf{v}_2 \rangle}{\| \mathbf{v}_2 \|^2} = \frac{4}{9},$$

$$a_3 = \frac{\langle \mathbf{v}, \mathbf{v}_3 \rangle}{\| \mathbf{v}_3 \|^2} = \frac{4}{3}.$$

Thus, the closest point to $\mathbf{v}$ in the given subspace is

$$\mathbf{w} = \tfrac{1}{2}\mathbf{v}_1 + \tfrac{4}{9}\mathbf{v}_2 + \tfrac{4}{3}\mathbf{v}_3 = \left(\tfrac{11}{6}, \tfrac{31}{18}, \tfrac{13}{9}, \tfrac{4}{9}\right)^T.$$

We further note that, in accordance with the orthogonal projection property, the vector

$$\mathbf{z} = \mathbf{v} - \mathbf{w} = \left(-\tfrac{5}{6}, \tfrac{5}{18}, \tfrac{5}{9}, \tfrac{5}{9}\right)^T$$

is orthogonal to $\mathbf{v}_1, \mathbf{v}_2, \mathbf{v}_3$ and hence to the entire subspace. ●

Even when we only know a non-orthogonal basis for the subspace, it may still be a good strategy to first apply the Gram–Schmidt process in order to replace it by an orthogonal or even orthonormal basis, and then apply the orthogonal projection formulae (5.63, 64) to calculate the least squares point. Not only does this simplify the final computation, it will often avoid the numerical inaccuracies due to ill-conditioning that can afflict the direct solution to the normal equations (4.28). The following example illustrates this alternative procedure.

EXAMPLE 5.41 Let us return to the problem, solved in Example 4.6, of finding the closest point on the plane V spanned by $\mathbf{w}_1 = (1, 2, -1)^T$, $\mathbf{w}_2 = (2, -3, -1)^T$ to the point $\mathbf{b} = (1, 0, 0)^T$. We proceed by first using the Gram–Schmidt process to compute an orthogonal basis

$$\mathbf{v}_1 = \mathbf{w}_1 = \begin{pmatrix} 1 \\ 2 \\ -1 \end{pmatrix}, \qquad \mathbf{v}_2 = \mathbf{w}_2 - \frac{\mathbf{w}_2 \cdot \mathbf{v}_1}{\| \mathbf{v}_1 \|^2}\, \mathbf{w}_1 = \begin{pmatrix} \frac{5}{2} \\ -2 \\ -\frac{3}{2} \end{pmatrix},$$

for our subspace. As a result, we can use the orthogonal projection (5.64) to produce the closest point

$$\mathbf{v}^\star = \frac{\mathbf{b} \cdot \mathbf{v}_1}{\| \mathbf{v}_1 \|^2}\, \mathbf{v}_1 + \frac{\mathbf{b} \cdot \mathbf{v}_2}{\| \mathbf{v}_2 \|^2}\, \mathbf{v}_2 = \begin{pmatrix} \frac{2}{3} \\ -\frac{1}{15} \\ -\frac{7}{15} \end{pmatrix},$$

reconfirming our earlier result. ●

Let us now revisit the problem, described in Section 4.4, of approximating experimental data by a least squares minimization procedure. The required calculations are significantly simplified by the introduction of an orthogonal basis of the least squares subspace. Given sample points $t_1, \ldots, t_m$, let

$$\mathbf{t}_k = \left(t_1^k, t_2^k, \ldots, t_m^k\right)^T, \qquad k = 0, 1, 2, \ldots,$$

be the vector obtained by sampling the monomial t^k. More generally, sampling a polynomial, i.e., a linear combination of monomials

$$y = p(t) = \alpha_0 + \alpha_1 t + \cdots + \alpha_n t^n \tag{5.66}$$

results in the self-same linear combination

$$\mathbf{p} = (p(t_1), \ldots, p(t_n))^T = \alpha_0 \mathbf{t}_0 + \alpha_1 \mathbf{t}_1 + \cdots + \alpha_n \mathbf{t}_n \tag{5.67}$$

of monomial sample vectors. Thus, all sampled polynomial vectors belong to the subspace $W = \operatorname{span}\{\mathbf{t}_0, \ldots, \mathbf{t}_n\} \subset \mathbb{R}^m$ spanned by the monomial sample vectors.

Let $\mathbf{y} = (y_1, y_2, \ldots, y_m)^T$ be data that has been measured at the sample points. The polynomial least squares approximation is, by definition, the polynomial $y = p(t)$ whose sample vector $\mathbf{p}$ is the closest point in the subspace W, which is the same as the orthogonal projection of the data vector $\mathbf{y}$ onto W. But the monomial sample vectors $\mathbf{t}_0, \ldots, \mathbf{t}_n$ are not orthogonal, and so a direct approach requires solving the normal equations (4.35) for the least squares coefficients $\alpha_0, \ldots, \alpha_n$.

An alternative approach is to first apply the Gram–Schmidt process to construct an orthogonal basis for the subspace W, from which the least squares coefficients are then found by simply taking inner products. Let us adopt the rescaled version

$$\langle \mathbf{v}, \mathbf{w} \rangle = \frac{1}{m} \sum_{i=1}^{m} v_i w_i = \overline{v\, w} \tag{5.68}$$

of the standard dot product† on $\mathbb{R}^m$. If $\mathbf{v}, \mathbf{w}$ represent the sample vectors corresponding to the functions $v(t), w(t)$, then their inner product $\langle \mathbf{v}, \mathbf{w} \rangle$ is equal to the average value of the product function $v(t)\, w(t)$ on the m sample points. In particular, the inner product between our "monomial" basis vectors corresponding to sampling t^k and t^l is

$$\langle \mathbf{t}_k, \mathbf{t}_l \rangle = \frac{1}{m} \sum_{i=1}^{m} t_i^k\, t_i^l = \frac{1}{m} \sum_{i=1}^{m} t_i^{k+l} = \overline{t^{k+l}}, \tag{5.69}$$

which is the averaged sample value of the monomial t^{k+l}.

To keep the formulae reasonably simple, let us further assume‡ that the sample points are evenly spaced and symmetric about 0. The first requirement is that $t_i - t_{i-1} = h$ is fixed, while the second means that if t_i is a sample point, so is $-t_i$. An example would be the seven sample points $-3, -2, -1, 0, 1, 2, 3$. As a consequence of these two assumptions, the averaged sample values of the odd powers of t vanish: $\overline{t^{2i+1}} = 0$. Hence, by (5.69), the sample vectors $\mathbf{t}_k$ and $\mathbf{t}_l$ are orthogonal whenever $k + l$ is odd.

Applying the Gram–Schmidt algorithm to $\mathbf{t}_0, \mathbf{t}_1, \mathbf{t}_2, \ldots$ produces the orthogonal basis vectors $\mathbf{q}_0, \mathbf{q}_1, \mathbf{q}_2, \ldots$. Each

$$\mathbf{q}_k = (q_k(t_1), \ldots, q_k(t_m))^T = c_{k0}\, \mathbf{t}_0 + c_{k1}\, \mathbf{t}_1 + \cdots + c_{kk}\, \mathbf{t}_k \tag{5.70}$$

can be interpreted as the sample vector for a certain degree k interpolating polynomial

$$q_k(t) = c_{k0} + c_{k1}\, t + \cdots + c_{kk}\, t^k.$$

The first few of these polynomials, along with their corresponding orthogonal sam-

†For weighted least squares, we would use an appropriately weighted inner product.

‡The method works without this restriction, but the formulas become more unwieldy. See Exercise 5.5.20 for details.

ple vectors, follow:

$$\begin{aligned} q_0(t) &= 1, & \mathbf{q}_0 &= \mathbf{t}_0, & \|\mathbf{q}_0\|^2 &= 1, \\ q_1(t) &= t, & \mathbf{q}_1 &= \mathbf{t}_1, & \|\mathbf{q}_1\|^2 &= \overline{t^2}, \\ q_2(t) &= t^2 - \overline{t^2}, & \mathbf{q}_2 &= \mathbf{t}_2 - \overline{t^2}\,\mathbf{t}_0, & \|\mathbf{q}_2\|^2 &= \overline{t^4} - \left(\overline{t^2}\right)^2, \\ q_3(t) &= t^3 - \frac{\overline{t^4}}{\overline{t^2}}\,t, & \mathbf{q}_3 &= \mathbf{t}_3 - \frac{\overline{t^4}}{\overline{t^2}}\,\mathbf{t}_1, & \|\mathbf{q}_3\|^2 &= \overline{t^6} - \frac{\left(\overline{t^4}\right)^2}{\overline{t^2}}. \end{aligned} \tag{5.71}$$

With these in hand, the least squares approximating polynomial of degree n to the given data vector $\mathbf{y}$ is given by a linear combination

$$p(t) = a_0\, q_0(t) + a_1\, q_1(t) + a_2\, q_2(t) + \cdots + a_n\, q_n(t). \tag{5.72}$$

The coefficients can now be obtained directly through the orthogonality formulae (5.64), and so

$$a_k = \frac{\langle \mathbf{q}_k, \mathbf{y} \rangle}{\|\mathbf{q}_k\|^2} = \frac{\overline{q_k\, y}}{\overline{q_k^2}}. \tag{5.73}$$

Thus, once we have set up the orthogonal basis, we no longer need to solve any linear system to construct the least squares approximation.

An additional advantage of the orthogonal basis is that, unlike the direct method, the formulae (5.73) for the least squares coefficients *do not depend on the degree of the approximating polynomial.* As a result, one can readily increase the degree, and, presumably, the accuracy, of the approximant without having to recompute any of the lower degree terms. For instance, if a quadratic polynomial $p_2(t) = a_0 + a_1\, q_1(t) + a_2\, q_2(t)$ is insufficiently accurate, the cubic least squares approximant $p_3(t) = p_2(t) + a_3\, q_3(t)$ can be constructed *without* having to recompute the quadratic coefficients a_0, a_1, a_2. This doesn't work when using the non-orthogonal monomials, all of whose coefficients will be affected by increasing the degree of the approximating polynomial.

EXAMPLE 5.42 Consider the following tabulated sample values:

t_i	-3	-2	-1	0	1	2	3
y_i	-1.4	-1.3	$-.6$	$.1$	$.9$	1.8	2.9

To compute polynomial least squares fits of degrees $1, 2$ and 3, we begin by computing the polynomials (5.71), which for the given sample points t_i are

$$\begin{aligned} &q_0(t) = 1, && q_1(t) = t, && q_2(t) = t^2 - 4, && q_3(t) = t^3 - 7t, \\ &\|\mathbf{q}_0\|^2 = 1, && \|\mathbf{q}_1\|^2 = 4, && \|\mathbf{q}_2\|^2 = 12, && \|\mathbf{q}_3\|^2 = \tfrac{216}{7}. \end{aligned}$$

Thus, to four decimal places, the coefficients for the least squares approximation (5.72) are

$$\begin{aligned} a_0 &= \langle \mathbf{q}_0, \mathbf{y} \rangle = .3429, & a_1 &= \tfrac{1}{4}\langle \mathbf{q}_1, \mathbf{y} \rangle = .7357, \\ a_2 &= \tfrac{1}{12}\langle \mathbf{q}_2, \mathbf{y} \rangle = .0738, & a_3 &= \tfrac{7}{216}\langle \mathbf{q}_3, \mathbf{y} \rangle = -.0083. \end{aligned}$$

To obtain the best linear approximation, we use

$$p_1(t) = a_0\, q_0(t) + a_1\, q_1(t) = .3429 + .7357\, t,$$

with a least squares error of .7081. Similarly, the quadratic and cubic least squares approximations are

$$\begin{aligned} p_2(t) &= .3429 + .7357\, t + .0738\,(t^2 - 4), \\ p_3(t) &= .3429 + .7357\, t + .0738\,(t^2 - 4) - .0083\,(t^3 - 7t), \end{aligned}$$

with respective least squares errors .2093 and .1697 at the sample points. Note particularly that the lower order coefficients do not change as we increase the degree. A plot of the three approximations appears in Figure 5.6. The cubic term does not significantly increase the accuracy of the approximation, and so this data probably comes from sampling a quadratic function. ●

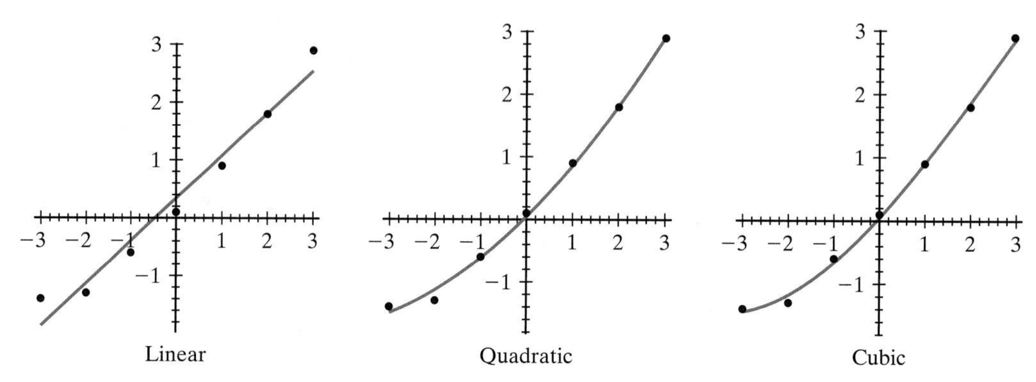

Figure 5.6 Least squares data approximations.

EXERCISES 5.5

Note: Use the Euclidean distance unless otherwise specified.

5.5.10. Find the closest point to the vector $\mathbf{b} = (1, 0, 2)^T$ belonging the two-dimensional subspace spanned by the orthogonal vectors $\mathbf{v}_1 = (1, -1, 1)^T$, $\mathbf{v}_2 = (-1, 1, 2)^T$.

5.5.11. Find the least squares solutions to the following linear systems. *Hint*: Check orthogonality of the columns of the coefficient matrix.

(a) $\begin{pmatrix} 1 & -1 \\ 2 & 2 \\ 3 & -1 \end{pmatrix} \begin{pmatrix} x \\ y \end{pmatrix} = \begin{pmatrix} 1 \\ 0 \\ -1 \end{pmatrix}$

(b) $\begin{pmatrix} 3 & -1 \\ 0 & 2 \\ -2 & 1 \\ 1 & 5 \end{pmatrix} \begin{pmatrix} x \\ y \end{pmatrix} = \begin{pmatrix} 2 \\ 1 \\ -1 \\ 1 \end{pmatrix}$

(c) $\begin{pmatrix} 1 & -1 & -1 \\ 1 & 3 & 2 \\ -2 & 1 & 0 \\ 1 & 0 & -1 \\ 0 & 7 & -1 \end{pmatrix} \begin{pmatrix} x \\ y \\ z \end{pmatrix} = \begin{pmatrix} -1 \\ 0 \\ 1 \\ -1 \\ 0 \end{pmatrix}$

5.5.12. Let $\mathbf{b} = (0, 3, 1, 2)^T$. Find the vector $\mathbf{v} \in \operatorname{span}\{(0, 0, 1, 1)^T, (2, 1, 1, -1)^T\}$ such that $\|\mathbf{v} - \mathbf{b}\|$ is minimized.

5.5.13. Find the closest point to $\mathbf{b} = (1, 2, -1, 3)^T$ in the subspace $W = \operatorname{span}\left\{(1, 0, 2, 1)^T, (1, 1, 0, -1)^T, (2, 0, 1, -1)^T\right\}$ by first constructing an orthogonal basis of W and then applying the orthogonal projection formula (5.64).

5.5.14. Repeat Exercise 5.5.13 using the weighted norm $\|\mathbf{v}\| = v_1^2 + 2\, v_2^2 + v_3^2 + 3\, v_4^2$.

5.5.15. Use the orthogonal sample vectors (5.71) to find the best polynomial least squares fits of degree 1, 2 and 3 for the following sets of data:

(a)

t_i	−2	−1	0	1	2
y_i	7	11	13	18	21

(b)

t_i	−3	−2	−1	0	1	2	3
y_i	−2.7	−2.1	−.5	.5	1.2	2.4	3.2

(c)

t_i	−3	−2	−1	0	1	2	3
y_i	60	80	90	100	120	120	130

◇ **5.5.16.** (a) Verify the orthogonality of the sample polynomial vectors in (5.71).

(b) Construct the next orthogonal sample polynomial $q_4(t)$ and the norm of its sample vector.

(c) Use your result to compute the quartic least squares approximation for the data in Example 5.42.

5.5.17. Use the result of Exercise 5.5.16 to find the best approximating polynomial of degree 4 to the data in Exercise 5.5.15.

5.5.18. Justify the fact that the orthogonal sample vector $\mathbf{q}_k$ in (5.70) is a linear combination of only the first k monomial sample vectors.

♡ **5.5.19.** The formulas (5.71) only apply when the sample times are symmetric around 0. When the sample points $t_1, \ldots, t_n$ are equally spaced, so $t_{i+1} - t_i = h$ for all $i = 1, \ldots, n-1$, then there is a simple trick to convert the least squares problem into a symmetric form.

(a) Show that the translated sample points $s_i = t_i - \bar{t}$, where

$$\bar{t} = \frac{1}{n} \sum_{i=1}^{n} t_i$$

is the average, are symmetric around 0.

(b) Suppose $q(s)$ is the least squares polynomial for the data points (s_i, y_i). Prove that $p(t) = q(t - \bar{t})$ is the least squares polynomial for the original data (t_i, y_i).

(c) Apply this method to find the least squares polynomials of degrees 1 and 2 for the following data:

t_i	1	2	3	4	5	6
y_i	−8	−6	−4	−1	1	3

◇ **5.5.20.** Construct the first three orthogonal basis elements for sample points $t_1, \ldots, t_m$ that are in general position.

5.5.21. John knows that the least squares solution to $A\mathbf{x} = \mathbf{b}$ can be identified with the closest point on the subspace rng A spanned by the columns of the coefficient matrix. Therefore, he tries to find the solution by first orthonormalizing the columns using Gram–Schmidt, and then finding the least squares coefficients by the orthonormal basis formula (5.63). To his surprise, he does not get the same solution! Can you explain the source of his difficulty. How can you use his solution to obtain the proper least squares solution $\mathbf{x}$? Check your algorithm with the system that we treated in Example 4.9.

◇ **5.5.22.** Let A be an $m \times n$ matrix with $\ker A = \{\mathbf{0}\}$. Suppose that we use the Gram–Schmidt algorithm to factor $A = QR$ as in Exercise 5.3.33. Prove that the least squares solution to the linear system $A\mathbf{x} = \mathbf{b}$ is found by solving the triangular system $R\mathbf{x} = Q^T\mathbf{b}$ by Back Substitution.

5.5.23. Apply the method in Exercise 5.5.22 to find the least squares solutions to the systems in Exercise 4.3.14.

◇ **5.5.24.** Which is the more efficient algorithm: direct least squares based on solving the normal equations by Gaussian Elimination, or using Gram–Schmidt orthonormalization and then solving the resulting triangular system by Back Substitution as in Exercise 5.5.22? Justify your answer.

◇ **5.5.25.** (a) Find a formula for the least squares error (4.30) in terms of an orthonormal basis of the subspace.

(b) Generalize your formula to the case of an orthogonal basis.

♡ **5.5.26.** Let $\mathbf{w}_1, \ldots, \mathbf{w}_n$ be any basis of the subspace $W \subset \mathbb{R}^m$. Let $A = (\mathbf{w}_1, \ldots, \mathbf{w}_n)$ be the $m \times n$ matrix whose columns are the basis vectors, so that $W = \operatorname{rng} A$ and $\operatorname{rank} A = n$. Let $P = A(A^TA)^{-1}A^T$ be the corresponding *projection matrix*, as defined in Exercise 2.5.8.

(a) Prove that the orthogonal projection of $\mathbf{v} \in \mathbb{R}^n$ onto $\mathbf{w} \in W$ is obtained by multiplying by the projection matrix: $\mathbf{w} = P\mathbf{v}$.

(b) Explain why Exercise 5.5.8 is a special case of this result.

(c) Show that if $A = QR$ is the factorization of Exercise 5.3.33, then $P = QQ^T$. Why is $P \neq \mathrm{I}$?

5.5.27. Use the projection matrix method of Exercise 5.5.26 to find the orthogonal projection of $\mathbf{v} = (1, 0, 0, 0)^T$ onto the range of the following matrices:

(a) $\begin{pmatrix} 5 \\ -5 \\ -7 \\ 1 \end{pmatrix}$ (b) $\begin{pmatrix} 1 & 0 \\ -1 & 2 \\ 0 & -1 \\ 1 & 2 \end{pmatrix}$

(c) $\begin{pmatrix} 2 & -1 \\ -3 & 1 \\ 1 & -2 \\ 1 & 2 \end{pmatrix}$

(d) $\begin{pmatrix} 0 & 1 & -1 \\ 0 & -1 & 2 \\ 1 & 1 & 1 \\ -2 & -1 & 0 \end{pmatrix}$

Orthogonal Polynomials and Least Squares

In a similar fashion, the orthogonality of Legendre polynomials and their relatives serves to simplify the construction of least squares approximants in function space. Suppose, for instance, that our goal is to approximate the exponential function e^t by a polynomial on the interval $-1 \le t \le 1$, where the least squares error is measured using the standard L^2 norm. We will write the best least squares approximant as a linear combination of the Legendre polynomials,

$$\begin{aligned} p(t) &= a_0 P_0(t) + a_1 P_1(t) + \cdots + a_n P_n(t) \\ &= a_0 + a_1 t + a_2 \left(\tfrac{3}{2} t^2 - \tfrac{1}{2}\right) + \cdots . \end{aligned} \tag{5.74}$$

By orthogonality, the least squares coefficients can be immediately computed by the inner product formula (5.64), so, by the Rodrigues formula (5.48),

$$a_k = \frac{\langle e^t, P_k \rangle}{\| P_k \|^2} = \frac{2k+1}{2} \int_{-1}^{1} e^t P_k(t)\, dt. \tag{5.75}$$

For example, the quadratic approximation is given by the first three terms in (5.74), whose coefficients are

$$\begin{aligned} a_0 &= \frac{1}{2} \int_{-1}^{1} e^t\, dt = \frac{1}{2}\left(e - \frac{1}{e}\right) \simeq 1.175201, \\ a_1 &= \frac{3}{2} \int_{-1}^{1} t\, e^t\, dt = \frac{3}{e} \simeq 1.103638, \\ a_2 &= \frac{5}{2} \int_{-1}^{1} \left(\tfrac{3}{2} t^2 - \tfrac{1}{2}\right) e^t\, dt = \frac{5}{2}\left(e - \frac{7}{e}\right) \simeq .357814. \end{aligned}$$

Graphs appear in Figure 5.7; the first shows e^t, the second its quadratic approximant

$$e^t \approx 1.175201 + 1.103638\, t + .357814 \left(\tfrac{3}{2} t^2 - \tfrac{1}{2}\right), \tag{5.76}$$

and the third compares the two by laying them on top of each other.

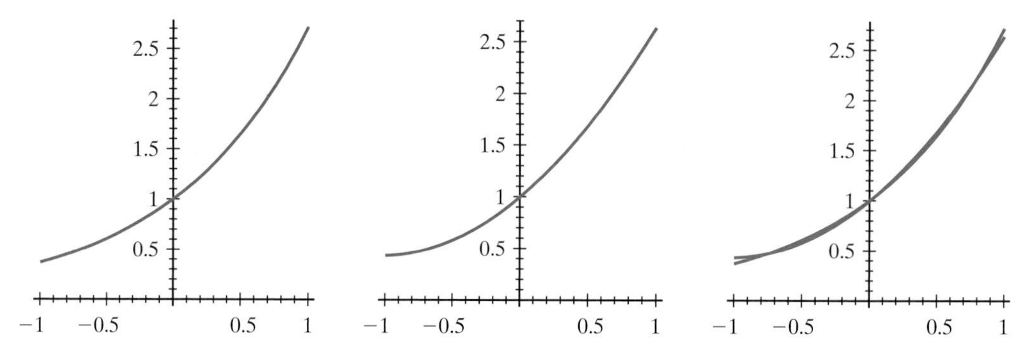

Figure 5.7 Quadratic least squares approximation to e^t.

As in the discrete case, there are two major advantages of the orthogonal Legendre polynomials over the direct approach presented in Example 4.21. First, we do not need to solve any linear systems of equations since the required coefficients (5.75) are found by direct integration. Indeed, the coefficient matrix for polynomial least squares approximation based on the monomial basis is some variant of the notoriously ill-conditioned Hilbert matrix, (1.70), and the computation of an accurate solution can be tricky. Our precomputation of an orthogonal system of polynomials has successfully circumvented the ill-conditioned normal system.

The second advantage is that the coefficients a_k do *not* depend on the degree of the approximating polynomial. Thus, if the quadratic approximation (5.76) is insufficiently accurate, we merely append the cubic correction

$$a_3 P_3(t) = a_3 \left(\tfrac{5}{2} t^3 - \tfrac{3}{2} t\right),$$

where

$$a_3 = \frac{7}{2} \int_{-1}^{1} \left(\tfrac{5}{2} t^3 - \tfrac{3}{2} t\right) e^t\, dt = \frac{7}{2}\left(37\, e - \frac{5}{e}\right) \simeq .070456.$$

Unlike the earlier method, there is no need to recompute the coefficients a_0, a_1, a_2. And, if we desire yet further accuracy, we need only compute the next one or two coefficients.

To exploit orthogonality, each interval and norm requires the construction of a corresponding system of orthogonal polynomials. Let us reconsider Example 4.21, in which we used the method of least squares to approximate e^t based on the L^2 norm on $[0, 1]$. Here, the ordinary Legendre polynomials are no longer orthogonal, and so we must use the rescaled Legendre polynomials (5.61) instead. Thus, the quadratic least squares approximant can be written as

$$\begin{aligned} p(t) &= a_0 + a_1 \widetilde{P}_1(t) + a_2 \widetilde{P}_2(t) \\ &= 1.718282 + .845155\,(2t-1) + .139864\,(6t^2-6t+1) \\ &= 1.012991 + .851125\,t + .839184\,t^2, \end{aligned}$$

where the coefficients

$$a_k = \frac{\langle e^t, \widetilde{P}_k \rangle}{\| \widetilde{P}_k \|^2} = (2k+1) \int_0^1 \widetilde{P}_k(t)\, e^t\, dt$$

are found by direct integration:

$$\begin{aligned} a_0 &= \int_0^1 e^t\, dt = e - 1 \simeq 1.718282, \\ a_1 &= 3 \int_0^1 (2t-1)\, e^t\, dt = 3(3-e) \simeq .845155, \\ a_2 &= 5 \int_0^1 (6t^2-6t+1) e^t\, dt = 5(7e-19) \simeq .139864. \end{aligned}$$

It is worth emphasizing that this is the *same* approximating polynomial as we computed in (4.63). The use of an orthogonal system of polynomials merely streamlines the computation.

EXERCISES 5.5

5.5.28. Use the Legendre polynomials to find the best (a) quadratic, and (b) cubic approximation to t^4, based on the L^2 norm on $[-1, 1]$.

5.5.29. Repeat Exercise 5.5.28 using the L^2 norm on $[0, 1]$.

5.5.30. Find the best cubic approximation to $f(t) = e^t$ based on the L^2 norm on $[0, 1]$.

5.5.31. Find the (a) linear (b) quadratic, (c) cubic polynomials $q(t)$ that minimize the following integral:

$$\int_0^1 [q(t) - t^3]^2\, dt.$$

What is the minimum value in each case?

5.5.32. Find the best quadratic and cubic approximations for $\sin t$ for the L^2 norm on $[0, \pi]$ by using an orthogonal basis. Graph your results and estimate the maximal error.

♠ **5.5.33.** Answer Exercise 5.5.30 when $f(t) = \sin t$. Use a computer to numerically evaluate the integrals.

♠ **5.5.34.** Find the degree 6 least squares polynomial approximation to e^t on the interval $[-1, 1]$ under the L^2 norm.

5.5.35. (a) Use the polynomials and weighted norm from Exercise 5.4.12 to find the quadratic least squares approximation to $f(t) = 1/t$. In what sense is your quadratic approximation "best"? (b) Now find the best approximating cubic polynomial. (c) Compare the graphs of the quadratic and cubic approximants with the original function and discuss what you observe.

♠ **5.5.36.** Use the Laguerre polynomials (5.55) to find the quadratic and cubic polynomial least squares approximation to $f(t) = \tan^{-1} t$ relative to the weighted inner product (5.53). Use a computer to evaluate the coefficients. Graph your result and discuss what you observe.

5.6 ORTHOGONAL SUBSPACES

We now extend the notion of orthogonality from individual elements to entire subspaces of an inner product space V.

Definition 5.43 Two subspaces $W, Z \subset V$ are called *orthogonal* if every vector in W is orthogonal to every vector in Z.

In other words, W and Z are orthogonal subspaces if and only if $\langle \mathbf{w}, \mathbf{z} \rangle = 0$ for every $\mathbf{w} \in W$ and $\mathbf{z} \in Z$. In practice, one only needs to check orthogonality of basis elements, or, more generally, spanning sets.

Lemma 5.44 If $\mathbf{w}_1, \ldots, \mathbf{w}_k$ span W and $\mathbf{z}_1, \ldots, \mathbf{z}_l$ span Z, then W and Z are orthogonal subspaces if and only if $\langle \mathbf{w}_i, \mathbf{z}_j \rangle = 0$ for all $i = 1, \ldots, k$ and $j = 1, \ldots, l$.

EXAMPLE 5.45 Let $V = \mathbb{R}^3$ have the ordinary dot product. Then the plane $W \subset \mathbb{R}^3$ defined by the equation $2x - y + 3z = 0$ is orthogonal to the line Z spanned by its normal vector $\mathbf{n} = (2, -1, 3)^T$. Indeed, every $\mathbf{w} = (x, y, z)^T \in W$ satisfies the orthogonality condition $\mathbf{w} \cdot \mathbf{n} = 2x - y + 3z = 0$, which is just the equation for the plane. ●

EXAMPLE 5.46 Let W be the span of $\mathbf{w}_1 = (1, -2, 0, 1)^T$, $\mathbf{w}_2 = (3, -5, 2, 1)^T$, and Z the span of $\mathbf{z}_1 = (3, 2, 0, 1)^T$, $\mathbf{z}_2 = (1, 0, -1, -1)^T$. We find $\mathbf{w}_1 \cdot \mathbf{z}_1 = \mathbf{w}_1 \cdot \mathbf{z}_2 = \mathbf{w}_2 \cdot \mathbf{z}_1 = \mathbf{w}_2 \cdot \mathbf{z}_2 = 0$, and so W and Z are orthogonal subspaces of $\mathbb{R}^4$ under the Euclidean dot product. ●

Definition 5.47 The *orthogonal complement* to a subspace $W \subset V$, denoted* $W^\perp$, is defined as the set of all vectors which are orthogonal to W:

$$W^\perp = \{ \mathbf{v} \in V \mid \langle \mathbf{v}, \mathbf{w} \rangle = 0 \text{ for all } \mathbf{w} \in W \}. \tag{5.77}$$

One easily checks that the orthogonal complement $W^\perp$ is also a subspace. Moreover, $W \cap W^\perp = \{\mathbf{0}\}$. (Why?) Keep in mind that the orthogonal complement will depend upon which inner product is being used.

EXAMPLE 5.48 Let $W = \{ (t, 2t, 3t)^T \mid t \in \mathbb{R} \}$ be the line (one-dimensional subspace) in the direction of the vector $\mathbf{w}_1 = (1, 2, 3)^T \in \mathbb{R}^3$. Under the dot product, its orthogonal complement $W^\perp$ is the plane passing through the origin having normal vector $\mathbf{w}_1$, as sketched in Figure 5.8. In other words, $\mathbf{z} = (x, y, z)^T \in W^\perp$ if and only if

$$\mathbf{z} \cdot \mathbf{w}_1 = x + 2y + 3z = 0. \tag{5.78}$$

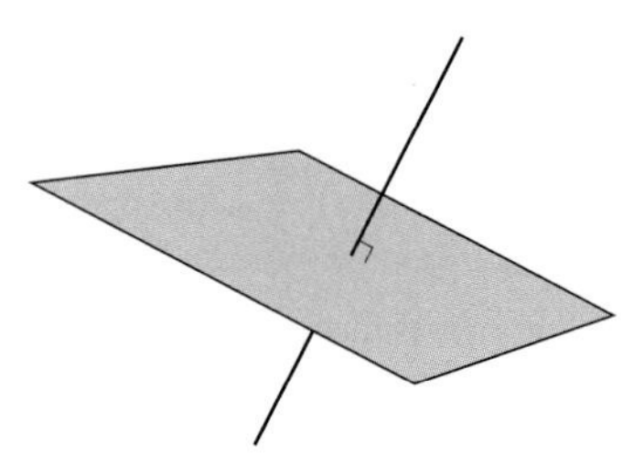

Figure 5.8 Orthogonal complement to a line.

Thus, $W^\perp$ is characterized as the solution space to the homogeneous linear equation (5.78), or, equivalently, the kernel of the 1×3 matrix $A = \mathbf{w}_1^T = (\,1 \quad 2 \quad 3\,)$. We

*And usually pronounced "W perp"

can write the general solution in the form

$$\mathbf{z} = \begin{pmatrix} -2y - 3z \\ y \\ z \end{pmatrix} = y \begin{pmatrix} -2 \\ 1 \\ 0 \end{pmatrix} + z \begin{pmatrix} -3 \\ 0 \\ 1 \end{pmatrix} = y\,\mathbf{z}_1 + z\,\mathbf{z}_2,$$

where y, z are the free variables. The indicated vectors $\mathbf{z}_1 = (-2, 1, 0)^T$, $\mathbf{z}_2 = (-3, 0, 1)^T$, form a (non-orthogonal) basis for the orthogonal complement $W^\perp$. ●

Proposition 5.49 Suppose that $W \subset V$ is a finite-dimensional subspace of an inner product space. Then every vector $\mathbf{v} \in V$ can be uniquely decomposed into $\mathbf{v} = \mathbf{w} + \mathbf{z}$, where $\mathbf{w} \in W$ and $\mathbf{z} \in W^\perp$.

Proof We let $\mathbf{w} \in W$ be the orthogonal projection of $\mathbf{v}$ onto W. Then $\mathbf{z} = \mathbf{v} - \mathbf{w}$ is, by definition, orthogonal to W and hence belongs to $W^\perp$. Note that $\mathbf{z}$ can be viewed as the orthogonal projection of $\mathbf{v}$ onto the complementary subspace $W^\perp$ (provided it is finite-dimensional). If we are given two such decompositions, $\mathbf{v} = \mathbf{w} + \mathbf{z} = \widetilde{\mathbf{w}} + \widetilde{\mathbf{z}}$, then $\mathbf{w} - \widetilde{\mathbf{w}} = \widetilde{\mathbf{z}} - \mathbf{z}$. The left hand side of this equation lies in W while the right hand side belongs to $W^\perp$. But, as we already noted, the only vector that belongs to both W and $W^\perp$ is the zero vector. Thus, $\mathbf{w} - \widetilde{\mathbf{w}} = \mathbf{0} = \widetilde{\mathbf{z}} - \mathbf{z}$, so $\mathbf{w} = \widetilde{\mathbf{w}}$ and $\mathbf{z} = \widetilde{\mathbf{z}}$, which proves uniqueness. ■

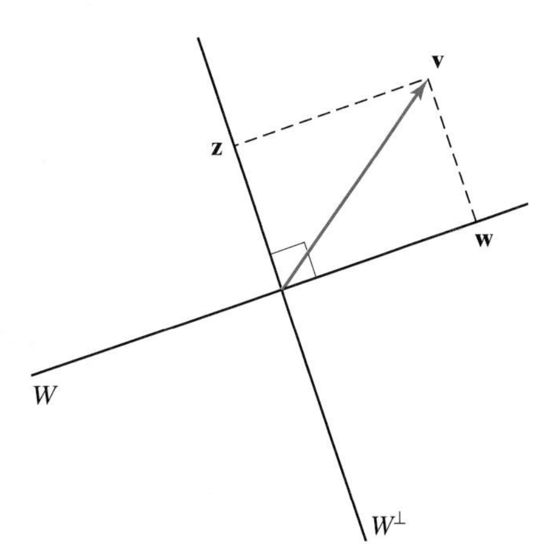

Figure 5.9 Orthogonal decomposition of a vector.

As a direct consequence of Exercise 2.4.26, we conclude that, in a finite-dimensional inner product space, a subspace and its orthogonal complement have complementary dimensions:

Proposition 5.50 If $\dim W = m$ and $\dim V = n$, then $\dim W^\perp = n - m$.

EXAMPLE 5.51 Return to the situation described in Example 5.48. Let us decompose the vector $\mathbf{v} = (1, 0, 0)^T \in \mathbb{R}^3$ into a sum $\mathbf{v} = \mathbf{w} + \mathbf{z}$ of a vector $\mathbf{w}$ lying in the line W and a vector $\mathbf{z}$ belonging to its orthogonal plane $W^\perp$, defined by (5.78). Each is obtained by an orthogonal projection onto the subspace in question, but we only need to compute one of the two directly, since the second can be obtained by subtracting the first from $\mathbf{v}$.

Orthogonal projection onto a one-dimensional subspace is easy since any basis is, trivially, an orthogonal basis. Thus, the projection of $\mathbf{v}$ onto the line spanned by

$$\mathbf{w}_1 = (1,2,3)^T \quad \text{is} \quad \mathbf{w} = \frac{\langle \mathbf{v}, \mathbf{w}_1 \rangle}{\| \mathbf{w}_1 \|^2}\, \mathbf{w}_1 = \left(\tfrac{1}{14}, \tfrac{2}{14}, \tfrac{3}{14} \right)^T.$$

The component in $W^\perp$ is then obtained by subtraction:

$$\mathbf{z} = \mathbf{v} - \mathbf{w} = \left(\tfrac{13}{14}, -\tfrac{2}{14}, -\tfrac{3}{14} \right)^T.$$

Alternatively, one can obtain $\mathbf{z}$ directly by orthogonal projection onto the plane $W^\perp$. But you need to be careful: the basis found in Example 5.48 is not orthogonal, and so you will need to set up and solve the normal equations to find the closest point $\mathbf{z}$. Or, you can first convert to an orthogonal basis by a single Gram–Schmidt step, and then use the orthogonal projection formula (5.64). All three methods lead to the same vector $\mathbf{z} \in W^\perp$. ●

EXAMPLE 5.52 Let $W \subset \mathbb{R}^4$ be the two-dimensional subspace spanned by the orthogonal vectors $\mathbf{w}_1 = (1,1,0,1)^T$ and $\mathbf{w}_2 = (1,1,1,-2)^T$. Its orthogonal complement $W^\perp$ (with respect to the Euclidean dot product) is the set of all vectors $\mathbf{v} = (x,y,z,w)^T$ that satisfy the linear system

$$\mathbf{v} \cdot \mathbf{w}_1 = x + y + w = 0, \qquad \mathbf{v} \cdot \mathbf{w}_2 = x + y + z - 2w = 0.$$

Applying the usual algorithm—the free variables are y and w—we find that the solution space is spanned by

$$\mathbf{z}_1 = (-1,1,0,0)^T, \qquad \mathbf{z}_2 = (-1,0,3,1)^T,$$

which form a non-orthogonal basis for $W^\perp$. An orthogonal basis

$$\mathbf{y}_1 = \mathbf{z}_1 = (-1,1,0,0)^T, \qquad \mathbf{y}_2 = \mathbf{z}_2 - \tfrac{1}{2}\mathbf{z}_1 = \left(-\tfrac{1}{2}, -\tfrac{1}{2}, 3, 1 \right)^T,$$

for $W^\perp$ is obtained by a single Gram–Schmidt step. To decompose the vector $\mathbf{v} = (1,0,0,0)^T = \mathbf{w} + \mathbf{z}$, say, we compute the two orthogonal projections:

$$\begin{aligned} \mathbf{w} &= \tfrac{1}{3}\mathbf{w}_1 + \tfrac{1}{7}\mathbf{w}_2 = \left(\tfrac{10}{21}, \tfrac{10}{21}, \tfrac{1}{7}, \tfrac{1}{21} \right)^T \in W, \\ \mathbf{z} &= \mathbf{v} - \mathbf{w} = -\tfrac{1}{2}\mathbf{y}_1 - \tfrac{1}{21}\mathbf{y}_2 = \left(\tfrac{11}{21}, -\tfrac{10}{21}, -\tfrac{1}{7}, -\tfrac{1}{21} \right)^T \in W^\perp. \end{aligned}$$ ●

Proposition 5.53 If W is a finite-dimensional subspace of an inner product space, then $(W^\perp)^\perp = W$.

This result is a direct corollary of the orthogonal decomposition derived in Proposition 5.49.

Warning: Propositions 5.49 and 5.53 are *not* necessarily true for infinite-dimensional subspaces. If $\dim W = \infty$, one can only assert that $W \subseteq (W^\perp)^\perp$. For example, it can be shown, [16, Exercise 10.2.D], that on any bounded interval $[a,b]$ the orthogonal complement to the subspace of all polynomials $\mathcal{P}^{(\infty)} \subset \mathrm{C}^0[a,b]$ with respect to the L^2 inner product is trivial: $(\mathcal{P}^{(\infty)})^\perp = \{0\}$. This means that the only continuous function which satisfies

$$\langle x^n, f(x) \rangle = \int_a^b x^n f(x)\, dx = 0 \quad \text{for all} \quad n = 0,1,2,\ldots$$

is the zero function $f(x) \equiv 0$. But the orthogonal complement of $\{0\}$ is the entire space, and so $((\mathcal{P}^{(\infty)})^\perp)^\perp = \mathrm{C}^0[a,b] \neq \mathcal{P}^{(\infty)}$.

The difference is that, in infinite-dimensional function space, a proper subspace $W \subsetneq V$ can be *dense*,* whereas in finite dimensions, every proper subspace is a "thin" subset that only occupies an infinitesimal fraction of the entire vector space. This seeming paradox turns out to be the reason for the success of numerical approximation schemes in function space, such as the finite element method.

EXERCISES 5.6

Note: In Exercises 5.6.1–4, use the Euclidean dot product.

5.6.1. Find the orthogonal complement $W^\perp$ to the subspaces $W \subset \mathbb{R}^3$ spanned by the indicated vectors. What is the dimension of $W^\perp$ in each case?

(a) $\begin{pmatrix} 3 \\ -1 \\ 1 \end{pmatrix}$ (b) $\begin{pmatrix} 1 \\ 2 \\ 3 \end{pmatrix}, \begin{pmatrix} 2 \\ 0 \\ 1 \end{pmatrix}$

(c) $\begin{pmatrix} 1 \\ 2 \\ 3 \end{pmatrix}, \begin{pmatrix} 2 \\ 4 \\ 6 \end{pmatrix}$

(d) $\begin{pmatrix} 0 \\ 1 \\ -1 \end{pmatrix}, \begin{pmatrix} -2 \\ 3 \\ 1 \end{pmatrix}, \begin{pmatrix} -1 \\ 2 \\ 0 \end{pmatrix}$

(e) $\begin{pmatrix} 1 \\ 1 \\ 0 \end{pmatrix}, \begin{pmatrix} 1 \\ 0 \\ 1 \end{pmatrix}, \begin{pmatrix} 0 \\ 1 \\ 1 \end{pmatrix}$

5.6.2. Find a basis for the orthogonal complement to the following subspaces of $\mathbb{R}^3$:

(a) the plane $3x + 4y - 5z = 0$

(b) the line in the direction $(-2, 1, 3)^T$

(c) the range of the matrix

$$\begin{pmatrix} 1 & 2 & -1 & 3 \\ -2 & 0 & 2 & 1 \\ -1 & 2 & 1 & 4 \end{pmatrix}$$

(d) the cokernel of the same matrix

5.6.3. Find a basis for the orthogonal complement to the following subspaces of $\mathbb{R}^4$:

(a) the set of solutions to $-x + 3y - 2z + w = 0$

(b) the subspace spanned by $(1, 2, -1, 3)^T$, $(-2, 0, 1, -2)^T$, $(-1, 2, 0, 1)^T$

(c) the kernel of the matrix in Exercise 5.6.2c

(d) the corange of the same matrix.

5.6.4. Decompose each of the following vectors with respect to the indicated subspace as $\mathbf{v} = \mathbf{w} + \mathbf{z}$, where $\mathbf{w} \in W$, $\mathbf{z} \in W^\perp$.

(a) $\mathbf{v} = \begin{pmatrix} 1 \\ 2 \end{pmatrix}$, $W = \operatorname{span}\left\{\begin{pmatrix} -3 \\ 1 \end{pmatrix}\right\}$

(b) $\mathbf{v} = \begin{pmatrix} 1 \\ 2 \\ -1 \end{pmatrix}$, $W = \operatorname{span}\left\{\begin{pmatrix} -3 \\ 2 \\ 1 \end{pmatrix}, \begin{pmatrix} -1 \\ 0 \\ 5 \end{pmatrix}\right\}$

(c) $\mathbf{v} = \begin{pmatrix} 1 \\ 0 \\ 0 \end{pmatrix}$, $W = \ker \begin{pmatrix} 1 & 2 & -1 \\ 2 & 0 & 2 \end{pmatrix}$

(d) $\mathbf{v} = \begin{pmatrix} 1 \\ 0 \\ 0 \end{pmatrix}$, $W = \operatorname{rng} \begin{pmatrix} 1 & 0 & 1 \\ -2 & -1 & 0 \\ 1 & 3 & -5 \end{pmatrix}$

(e) $\mathbf{v} = \begin{pmatrix} 1 \\ 0 \\ 0 \\ 1 \end{pmatrix}$, $W = \ker \begin{pmatrix} 1 & 0 & 0 & 2 \\ -2 & -1 & 1 & -3 \end{pmatrix}$

5.6.5. Redo Exercise 5.6.1 using the weighted inner product $\langle \mathbf{v}, \mathbf{w} \rangle = v_1 w_1 + 2 v_2 w_2 + 3 v_3 w_3$ instead of the dot product.

5.6.6. Redo Example 5.52 using the weighted inner product $\langle \mathbf{v}, \mathbf{w} \rangle = v_1 w_1 + 2 v_2 w_2 + 3 v_3 w_3 + 4 v_4 w_4$ instead of the dot product.

5.6.7. Let $V = \mathcal{P}^{(4)}$ denote the space of quartic polynomials, with the L^2 inner product

$$\langle p, q \rangle = \int_{-1}^{1} p(x)\, q(x)\, dx.$$

Let $W = \mathcal{P}^2$ be the subspace of quadratic polynomials.

(a) Write down the conditions that a polynomial $p \in \mathcal{P}^{(4)}$ must satisfy in order to belong to the orthogonal complement $W^\perp$.

(b) Find a basis for and the dimension of $W^\perp$.

(c) Find an orthogonal basis for $W^\perp$.

◇ **5.6.8.** Prove that the orthogonal complement $W^\perp$ to a subspace $W \subset V$ is itself a subspace.

*In general, a subset $W \subset V$ of a normed vector space is *dense* if, for every $\mathbf{v} \in V$, and every $\varepsilon > 0$, one can find $\mathbf{w} \in W$ with $\|\mathbf{v} - \mathbf{w}\| < \varepsilon$. The Weierstrass approximation theorem, [16], tells us that the polynomials form a dense subspace of the space of continuous functions, and underlies the proof of the result mentioned in the preceding paragraph.

5.6.9. Let $W \subset V$. Prove that

(a) $W \cap W^{\perp} = \{\mathbf{0}\}$ (b) $W \subseteq (W^{\perp})^{\perp}$

5.6.10. Let V be an inner product space. Prove that

(a) $V^{\perp} = \{\mathbf{0}\}$ (b) $\{\mathbf{0}\}^{\perp} = V$

5.6.11. Show that if $W_1 \subset W_2$ are finite dimensional subspaces of an inner product space, then $W_1^{\perp} \supset W_2^{\perp}$.

5.6.12. (a) Show that if $W, Z \subset \mathbb{R}^n$ are complementary subspaces, then $W^{\perp}$ and $Z^{\perp}$ are also complementary subspaces.

(b) Sketch a picture illustrating this result when W and Z are lines in $\mathbb{R}^2$.

◇ **5.6.13.** Fill in the details of the proof of Proposition 5.53.

5.6.14. Prove Lemma 5.44.

◇ **5.6.15.** Let $W \subset V$ with $\dim V = n$. Suppose $\mathbf{w}_1, \ldots, \mathbf{w}_m$ is an orthogonal basis for W and $\mathbf{w}_{m+1}, \ldots, \mathbf{w}_n$ is an orthogonal basis for $W^{\perp}$.

(a) Prove that the combination $\mathbf{w}_1, \ldots, \mathbf{w}_n$ forms an orthogonal basis of V.

(b) Show that if $\mathbf{v} = c_1 \mathbf{w}_1 + \cdots + c_n \mathbf{w}_n$ is any vector in V, then its orthogonal decomposition $\mathbf{v} = \mathbf{w} + \mathbf{z}$ is given by $\mathbf{w} = c_1 \mathbf{w}_1 + \cdots + c_m \mathbf{w}_m \in W$ and $\mathbf{z} = c_{m+1} \mathbf{w}_{m+1} + \cdots + c_n \mathbf{w}_n \in W^{\perp}$.

♡ **5.6.16.** Consider the subspace $W = \{ u(a) = 0 = u(b) \}$ of the vector space $C^0[a, b]$ with the usual L^2 inner product.

(a) Show that W has a complementary subspace of dimension 2.

(b) Prove that there does not exist an orthogonal complement to W. Thus, an infinite-dimensional subspace may not admit an orthogonal complement!

Orthogonality of the Fundamental Matrix Subspaces and the Fredholm Alternative

In Chapter 2, we introduced the four fundamental subspaces associated with an $m \times n$ matrix A. According to the Fundamental Theorem 2.49, the first two, the kernel (or null space) and the corange (or row space), are subspaces of $\mathbb{R}^n$ having complementary dimensions. The second two, the cokernel (or left null space) and the range (or column space), are subspaces of $\mathbb{R}^m$, also of complementary dimensions. In fact, more than this is true—the paired subspaces are orthogonal complements with respect to the standard Euclidean dot product!

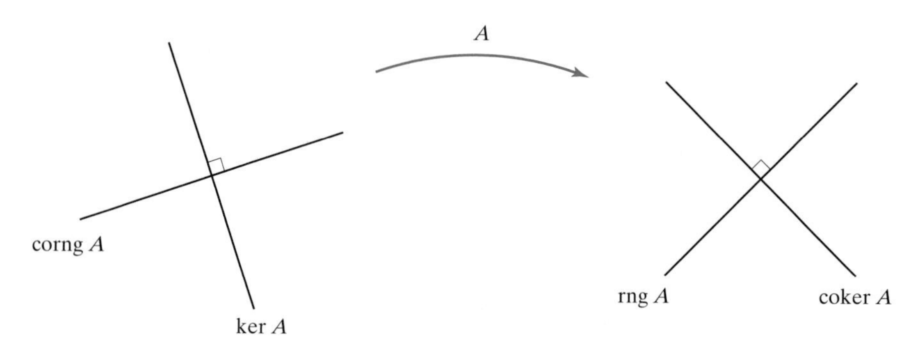

Figure 5.10 The fundamental matrix subspaces.

THEOREM 5.54 Let A be a real $m \times n$ matrix. Then its kernel and corange are orthogonal complements as subspaces of $\mathbb{R}^n$ under the dot product, while its cokernel and range are orthogonal complements in $\mathbb{R}^m$, also under the dot product:

$$\ker A = (\operatorname{corng} A)^{\perp} \subset \mathbb{R}^n, \qquad \operatorname{coker} A = (\operatorname{rng} A)^{\perp} \subset \mathbb{R}^m. \tag{5.79}$$

Proof A vector $\mathbf{x} \in \mathbb{R}^n$ lies in $\ker A$ if and only if $A\mathbf{x} = \mathbf{0}$. According to the rules of matrix multiplication, the ith entry of $A\mathbf{x}$ equals the vector product of the ith row $\mathbf{r}_i^T$ of A and $\mathbf{x}$. But this product vanishes, $\mathbf{r}_i^T \mathbf{x} = \mathbf{r}_i \cdot \mathbf{x} = 0$, if and only if $\mathbf{x}$ is orthogonal to $\mathbf{r}_i$. Therefore, $\mathbf{x} \in \ker A$ if and only if $\mathbf{x}$ is orthogonal to all the rows of A. Since the rows span $\operatorname{corng} A$, this is equivalent to $\mathbf{x}$ lying in its orthogonal complement $(\operatorname{corng} A)^{\perp}$, which proves the first statement. Orthogonality of the range and cokernel follows by the same argument applied to the transposed matrix A^T. ■

Combining Theorems 2.49 and 5.54, we deduce the following important characterization of compatible linear systems.

THEOREM 5.55 The linear system $A\mathbf{x} = \mathbf{b}$ has a solution if and only if $\mathbf{b}$ is orthogonal to the cokernel of A.

Indeed, the system has a solution if and only if the right hand side belongs to the range of the coefficient matrix, $\mathbf{b} \in \operatorname{rng} A$, which, by (5.79), requires that $\mathbf{b}$ be orthogonal to its cokernel. Thus, the compatibility conditions for the linear system $A\mathbf{x} = \mathbf{b}$ can be written in the form

$$\mathbf{y} \cdot \mathbf{b} = 0 \quad \text{for every } \mathbf{y} \text{ satisfying} \quad A^T\mathbf{y} = \mathbf{0}. \tag{5.80}$$

In practice, one only needs to check orthogonality of $\mathbf{b}$ with respect to a basis $\mathbf{y}_1, \ldots, \mathbf{y}_{m-r}$ of the cokernel, leading to a system of $m - r$ compatibility constraints

$$\mathbf{y}_i \cdot \mathbf{b} = 0, \qquad i = 1, \ldots, m - r. \tag{5.81}$$

Here $r = \operatorname{rank} A$ denotes the rank of the coefficient matrix, and so $m - r$ is also the number of all zero rows in the row echelon form of A. Hence, (5.81) contains precisely the same number of constraints as would be derived using Gaussian Elimination.

Theorem 5.55 is known as the *Fredholm alternative*, named after the Swedish mathematician Ivar Fredholm. Fredholm's primary interest was in solving linear integral equations, but his compatibility criterion was recognized to be a general property of linear systems, including linear algebraic systems, linear differential equations, linear boundary value problems, and so on.

EXAMPLE 5.56 In Example 2.40, we analyzed the linear system $A\mathbf{x} = \mathbf{b}$ with coefficient matrix

$$A = \begin{pmatrix} 1 & 0 & -1 \\ 0 & 1 & -2 \\ 1 & -2 & 3 \end{pmatrix}.$$

Using direct Gaussian Elimination, we were led to a single compatibility condition, namely $-b_1 + 2b_2 + b_3 = 0$, required for the system to have a solution. We now understand the meaning behind this equation: it is telling us that the right hand side $\mathbf{b}$ must be orthogonal to the cokernel of A. The cokernel is determined by solving the homogeneous adjoint system $A^T\mathbf{y} = \mathbf{0}$, and is the line spanned by the vector $\mathbf{y}_1 = (-1, 2, 1)^T$. Thus, the compatibility condition requires that $\mathbf{b}$ be orthogonal to $\mathbf{y}_1$, in accordance with the Fredholm alternative (5.81). ●

EXAMPLE 5.57 Let us determine the compatibility conditions for the linear system

$$\begin{aligned} x_1 - x_2 + 3x_3 &= b_1, & 2x_1 + 3x_2 + x_3 &= b_3, \\ -x_1 + 2x_2 - 4x_3 &= b_2, & x_1 + 2x_3 &= b_4, \end{aligned}$$

by computing the cokernel of its coefficient matrix

$$A = \begin{pmatrix} 1 & -1 & 3 \\ -1 & 2 & -4 \\ 2 & 3 & 1 \\ 1 & 0 & 2 \end{pmatrix}.$$

We need to solve the homogeneous adjoint system $A^T \mathbf{y} = \mathbf{0}$, namely

$$y_1 - y_2 + 2\,y_3 + y_4 = 0, \qquad -\,y_1 + 2\,y_2 + 3\,y_3 = 0, \qquad 3\,y_1 - 4\,y_2 + y_3 + 2\,y_4 = 0.$$

Applying Gaussian Elimination, we deduce the general solution

$$\mathbf{y} = y_3\,(-7, -5, 1, 0)^T + y_4\,(-2, -1, 0, 1)^T$$

is a linear combination (whose coefficients are the free variables) of the two basis vectors for $\operatorname{coker} A$. Thus, the Fredholm compatibility conditions (5.81) are obtained by taking their dot products with the right hand side of the original system:

$$-7\,b_1 - 5\,b_2 + b_3 = 0, \qquad -2\,b_1 - b_2 + b_4 = 0.$$

The reader can check that these are indeed the same compatibility conditions that result from a direct Gaussian Elimination on the augmented matrix $\left(A \mid \mathbf{b} \right)$. ●

REMARK: Vice versa, rather than solving to homogeneous adjoint system, we can use Gaussian Elimination on the augmented matrix $\left(A \mid \mathbf{b} \right)$ to determine $m - r$ basis vectors $\mathbf{y}_1, \ldots, \mathbf{y}_{m-r}$ for $\operatorname{coker} A$. They are formed from the coefficients of $b_1, \ldots, b_m$ in the $m - r$ consistency conditions $\mathbf{y}_i \cdot \mathbf{b} = 0$ for $i = 1, \ldots m - r$, arising from the all zero rows in the reduced row echelon form.

We are now very close to a full understanding of the fascinating geometry that lurks behind the simple algebraic operation of multiplying a vector $\mathbf{x} \in \mathbb{R}^n$ by an $m \times n$ matrix, resulting in a vector $\mathbf{b} = A\mathbf{x} \in \mathbb{R}^m$. Since the kernel and corange of A are orthogonal complements in the domain space $\mathbb{R}^n$, Proposition 5.50 tells us that we can uniquely decompose $\mathbf{x} = \mathbf{w} + \mathbf{z}$ where $\mathbf{w} \in \operatorname{corng} A$, while $\mathbf{z} \in \ker A$. Since $A\mathbf{z} = \mathbf{0}$, we have

$$\mathbf{b} = A\mathbf{x} = A(\mathbf{w} + \mathbf{z}) = A\mathbf{w}.$$

Therefore, we can regard multiplication by A as a combination of two operations:

(i) The first is an orthogonal projection onto the corange taking $\mathbf{x}$ to $\mathbf{w}$.

(ii) The second maps a vector in $\operatorname{corng} A \subset \mathbb{R}^n$ to a vector in $\operatorname{rng} A \subset \mathbb{R}^m$, taking the orthogonal projection $\mathbf{w}$ to the image vector $\mathbf{b} = A\mathbf{w} = A\mathbf{x}$.

Moreover, if A has rank r, then both $\operatorname{rng} A$ and $\operatorname{corng} A$ are r-dimensional subspaces, albeit of different vector spaces. Each vector $\mathbf{b} \in \operatorname{rng} A$ corresponds to a *unique* vector $\mathbf{w} \in \operatorname{corng} A$. Indeed, if $\mathbf{w}, \widetilde{\mathbf{w}} \in \operatorname{corng} A$ satisfy $\mathbf{b} = A\mathbf{w} = A\widetilde{\mathbf{w}}$, then $A(\mathbf{w} - \widetilde{\mathbf{w}}) = \mathbf{0}$, and hence $\mathbf{w} - \widetilde{\mathbf{w}} \in \ker A$. But, since the kernel and the corange are complementary subspaces, the only vector that belongs to both is the zero vector, and hence $\mathbf{w} = \widetilde{\mathbf{w}}$. In this manner, we have proved the first part of the following result; the second is left as Exercise 5.6.26.

Proposition 5.58 Multiplication by an $m \times n$ matrix A of rank r defines a one-to-one correspondence between the r-dimensional subspaces $\operatorname{corng} A \subset \mathbb{R}^n$ and $\operatorname{rng} A \subset \mathbb{R}^m$. Moreover, if $\mathbf{v}_1, \ldots, \mathbf{v}_r$ forms a basis of $\operatorname{corng} A$ then their images $A\mathbf{v}_1, \ldots, A\mathbf{v}_r$ form a basis for $\operatorname{rng} A$.

In summary, the linear system $A\mathbf{x} = \mathbf{b}$ has a solution if and only if $\mathbf{b} \in \operatorname{rng} A$, or, equivalently, is orthogonal to every vector $\mathbf{y} \in \operatorname{coker} A$. If the compatibility conditions hold, then the system has a *unique* solution $\mathbf{w} \in \operatorname{corng} A$ that, by the definition of the corange, is a linear combination of the *rows* of A. The general solution to the system is $\mathbf{x} = \mathbf{w} + \mathbf{z}$ where $\mathbf{w}$ is the particular solution belonging to the corange, while $\mathbf{z} \in \ker A$ is an arbitrary element of the kernel.

THEOREM 5.59 A compatible linear system $A\mathbf{x} = \mathbf{b}$ with $\mathbf{b} \in \operatorname{rng} A = (\operatorname{coker} A)^\perp$ has a unique solution $\mathbf{w} \in \operatorname{corng} A$ satisfying $A\mathbf{w} = \mathbf{b}$. The general solution is $\mathbf{x} = \mathbf{w} + \mathbf{z}$ where $\mathbf{z} \in \ker A$. The particular solution $\mathbf{w}$ is distinguished by the fact that it has the smallest Euclidean norm of all possible solutions: $\|\mathbf{w}\| \le \|\mathbf{x}\|$ whenever $A\mathbf{x} = \mathbf{b}$.

Proof We have already established all but the last statement. Since the corange and kernel are orthogonal subspaces, the norm of a general solution $\mathbf{x} = \mathbf{w} + \mathbf{z}$ is

$$\|\mathbf{x}\|^2 = \|\mathbf{w}+\mathbf{z}\|^2 = \|\mathbf{w}\|^2 + 2\,\mathbf{w}\cdot\mathbf{z} + \|\mathbf{z}\|^2 = \|\mathbf{w}\|^2 + \|\mathbf{z}\|^2 \ge \|\mathbf{w}\|^2,$$

with equality if and only if $\mathbf{z} = \mathbf{0}$. ■

In practice, to determine the unique minimum norm solution to a compatible linear system, we invoke the orthogonality of the corange and kernel of the coefficient matrix. Thus, if $\mathbf{z}_1, \ldots, \mathbf{z}_{n-r}$ form a basis for $\ker A$, then the minimum norm solution $\mathbf{x} = \mathbf{w} \in \operatorname{corng} A$ is obtained by solving the enlarged system

$$A\mathbf{x} = \mathbf{b}, \qquad \mathbf{z}_1^T\mathbf{x} = 0, \qquad \ldots \qquad \mathbf{z}_{n-r}^T\mathbf{x} = 0. \tag{5.82}$$

The associated $(m+n-r) \times n$ coefficient matrix is simply obtained by appending the (transposed) kernel vectors to the original matrix A. The resulting matrix is guaranteed to have maximum rank n, and so, assuming $\mathbf{b} \in \operatorname{rng} A$, the enlarged system has a unique solution, which is the minimum norm solution to the original system $A\mathbf{x} = \mathbf{b}$.

EXAMPLE 5.60 Consider the linear system

$$\begin{pmatrix} 1 & -1 & 2 & -2 \\ 0 & 1 & -2 & 1 \\ 1 & 3 & -5 & 2 \\ 5 & -1 & 9 & -6 \end{pmatrix}\begin{pmatrix} x \\ y \\ z \\ w \end{pmatrix} = \begin{pmatrix} -1 \\ 1 \\ 4 \\ 6 \end{pmatrix}. \tag{5.83}$$

Applying the usual Gaussian Elimination algorithm, we discover that the coefficient matrix has rank 3, and its kernel is spanned by the single vector $\mathbf{z}_1 = (1, -1, 0, 1)^T$. The system itself is compatible; indeed, the right hand side is orthogonal to the basis cokernel vector $(2, 24, -7, 1)^T$, and so satisfies the Fredholm condition (5.81). The general solution to the linear system is $\mathbf{x} = (t, 3-t, 1, t)^T$, where $t = w$ is the free variable.

To find the solution of minimum Euclidean norm, we can apply the algorithm described in the previous paragraph.† Thus, we supplement the original system by the constraint

$$\begin{pmatrix} 1 & -1 & 0 & 1 \end{pmatrix}\begin{pmatrix} x \\ y \\ z \\ w \end{pmatrix} = x - y + w = 0 \tag{5.84}$$

that the solution be orthogonal to the kernel basis vector. Solving the combined linear system (5.83, 84) leads to the unique solution $\mathbf{x} = \mathbf{w} = (1, 2, 1, 1)^T$, obtained by setting the free variable $t = 1$. Let us check that its norm is indeed the smallest among all solutions to the original system:

$$\|\mathbf{w}\| = \sqrt{7} \le \|\mathbf{x}\| = \|(t, 3-t, 1, t)^T\| = \sqrt{3t^2 - 6t + 10}\,,$$

† An alternative is to orthogonally project the general solution onto the corange. The result is the same.

where the quadratic function inside the square root achieves its minimum value at $t = 1$. It is further distinguished as the only solution that can be expressed as a linear combination of the rows of the coefficient matrix:

$$\begin{aligned}\mathbf{w}^T &= (\,1,\ 2,\ 1,\ 1\,)\\ &= -4\,(\,1,\ -1,\ 2,\ -2\,) - 17\,(\,0,\ 1,\ -2,\ 1\,) + 5\,(\,1,\ 3,\ -5,\ 2\,).\end{aligned}$$ ●

EXERCISES 5.6

5.6.17. For each of the following matrices A,

(i) find a basis for each of the four fundamental subspaces;

(ii) verify that the range and cokernel are orthogonal complements;

(iii) verify that the corange and kernel are orthogonal complements:

(a) $\begin{pmatrix} 1 & -2 \\ 2 & -4 \end{pmatrix}$ (b) $\begin{pmatrix} 5 & 0 \\ 1 & 2 \\ 0 & 2 \end{pmatrix}$

(c) $\begin{pmatrix} 0 & 1 & 2 \\ -1 & 0 & -3 \\ -2 & 3 & 0 \end{pmatrix}$

(d) $\begin{pmatrix} 1 & 2 & 0 & 1 \\ -1 & 1 & 3 & 1 \\ 0 & 3 & 3 & 2 \end{pmatrix}$

(e) $\begin{pmatrix} 3 & 1 & 4 & 2 & 7 \\ 1 & 1 & 2 & 0 & 3 \\ 5 & 2 & 7 & 3 & 12 \end{pmatrix}$

(f) $\begin{pmatrix} 1 & 3 & 0 & -2 \\ -2 & 1 & 2 & 3 \\ -3 & 5 & 4 & 4 \\ 1 & -4 & -2 & -1 \end{pmatrix}$

(g) $\begin{pmatrix} -1 & 2 & 2 & -1 \\ 2 & -4 & -5 & 2 \\ -3 & 6 & 2 & -3 \\ 1 & -2 & -3 & 1 \\ -2 & 4 & -5 & -2 \end{pmatrix}$

5.6.18. For each of the following matrices, use Gaussian elimination on the augmented matrix $\left(A \mid \mathbf{b} \right)$ to determine a basis for its cokernel:

(a) $\begin{pmatrix} 9 & -6 \\ 6 & -4 \end{pmatrix}$ (b) $\begin{pmatrix} 1 & 3 \\ 2 & 6 \\ -3 & -9 \end{pmatrix}$

(c) $\begin{pmatrix} 1 & 1 & 3 \\ -1 & 1 & -2 \\ -1 & 3 & 6 \end{pmatrix}$ (d) $\begin{pmatrix} 1 & -2 & -2 \\ 0 & -1 & 3 \\ 2 & -5 & -1 \\ -2 & 2 & 10 \end{pmatrix}$

5.6.19. Let $A = \begin{pmatrix} 1 & -2 & 2 & -1 \\ -2 & 4 & -3 & 5 \\ -1 & 2 & 0 & 7 \end{pmatrix}$.

(a) Find a basis for corng A.

(b) Use Proposition 5.58 to find a basis of rng A.

(c) Write each column of A as a linear combination of the basis vectors you found in part (b).

5.6.20. Write down the compatibility conditions on the following systems of linear equations by first computing a basis for the cokernel of the coefficient matrix.

(a) $2x + y = a,\ x + 4y = b,\ -3x + 2y = c$;

(b) $x + 2y + 3z = a,\ -x + 5y - 2z = b,\ 2x - 3y + 5z = c$;

(c) $x_1 + 2x_2 + 3x_3 = b_1,\ x_2 + 2x_3 = b_2,\ 3x_1 + 5x_2 + 7x_3 = b_3,\ -2x_1 + x_2 - 4x_3 = b_4$;

(d) $x - 3y + 2z + w = a,\ 4x - 2y + 2z + 3w = b,\ 5x - 5y + 4z + 4w = c,\ 2x + 4y - 2z + w = d$.

5.6.21. For each of the following $m \times n$ matrices, decompose the first standard basis vector $\mathbf{e}_1 = \mathbf{w} + \mathbf{z} \in \mathbb{R}^n$, where $\mathbf{w} \in$ corng A and $\mathbf{z} \in$ ker A. Verify your answer by expressing $\mathbf{w}$ as a linear combination of the rows of A.

(a) $\begin{pmatrix} 1 & -2 & 1 \\ 2 & -3 & 2 \end{pmatrix}$ (b) $\begin{pmatrix} 1 & 1 & 2 \\ -1 & 0 & -1 \\ -2 & -1 & -3 \end{pmatrix}$

(c) $\begin{pmatrix} 1 & -1 & 0 & 3 \\ 2 & 1 & 3 & 3 \\ 1 & 2 & 3 & 0 \end{pmatrix}$

(d) $\begin{pmatrix} -1 & 1 & 1 & -1 & 2 \\ -3 & 2 & -1 & -2 & 0 \end{pmatrix}$

5.6.22. For each of the following linear systems

(i) verify compatibility using the Fredholm alternative,

(ii) find the general solution, and

(iii) find the solution of minimum Euclidean norm.

(a) $2x - 4y = -6,\ -x + 2y = 3$

(b) $2x + 3y = -1,\ 3x + 7y = 1,\ -3x + 2y = 8$

(c) $6x - 3y + 9z = 12,\ 2x - y + 3z = 4$

(d) $x+3y+5z=3,\ -x+4y+9z=11,\ 2x+3y+4z=0$

(e) $x_1-3x_2+7x_3=-8,\ 2x_1+x_2=5,\ 4x_1-3x_2+10x_3=-5,\ -2x_1+2x_2-6x_3=4$

(f) $x-y+2z+3w=5,\ 3x-3y+5z+7w=13,\ -2x+2y+z+4w=0$

5.6.23. Let $A=\begin{pmatrix} 1 & -1 & 0 & 2 \\ 2 & -2 & 0 & 4 \\ -1 & 1 & 1 & -1 \\ 0 & 0 & 2 & 2 \end{pmatrix}$.

(a) Find an orthogonal basis for corng A.

(b) Find an orthogonal basis for ker A.

(c) If you combine your bases from part (a) and (b), do you get an orthogonal basis of $\mathbb{R}^4$? Why or why not?

5.6.24. Show that if $A=A^T$ is a symmetric matrix, then $A\mathbf{x}=\mathbf{b}$ has a solution if and only if $\mathbf{b}$ is orthogonal to ker A.

◇ **5.6.25.** Suppose $\mathbf{v}_1,\ldots,\mathbf{v}_n$ span a subspace $V\subset\mathbb{R}^m$. Prove that $\mathbf{w}$ is orthogonal to V if and only if $\mathbf{w}\in\operatorname{coker}A$ where $A=(\,\mathbf{v}_1\ \mathbf{v}_2\ \ldots\ \mathbf{v}_n\,)$ is the matrix with the indicated columns.

◇ **5.6.26.** Prove that if $\mathbf{v}_1,\ldots,\mathbf{v}_r$ are a basis of corng A then their images $A\mathbf{v}_1,\ldots,A\mathbf{v}_r$ are a basis for rng A.

5.6.27. *True or false*: The standard algorithm for finding a basis for ker A will always produce an orthogonal basis.

5.6.28. *True or false*: The minimal norm solution to $A\mathbf{x}=\mathbf{b}$ is obtained by setting all the free variables to zero.

◇ **5.6.29.** Prove that if K is a positive semi-definite matrix, and $\mathbf{f}\notin\operatorname{rng}K$, then the quadratic function $p(\mathbf{x})=\mathbf{x}^TK\mathbf{x}-2\mathbf{x}^T\mathbf{f}+c$ has no minimum value. *Hint*: Try looking at vectors $\mathbf{x}\in\ker K$.

5.6.30. Is Theorem 5.54 true as stated for complex matrices? If not, can you formulate a similar theorem that is true? What is the Fredholm alternative for complex matrices?

5.7 DISCRETE FOURIER ANALYSIS AND THE FAST FOURIER TRANSFORM

In modern digital media— audio, still images or video—continuous signals are sampled at discrete time intervals before being processed. Fourier analysis decomposes the sampled signal into its fundamental periodic constituents—sines and cosines, or, more conveniently, complex exponentials. The crucial fact, upon which all of modern signal processing is based, is that the sampled complex exponentials form an orthogonal basis. The section introduces the Discrete Fourier Transform, and concludes with an introduction to the Fast Fourier Transform, an efficient algorithm for computing the discrete Fourier representation and reconstructing the signal from its Fourier coefficients.

We will concentrate on the one-dimensional version here. Let $f(x)$ be a function representing the signal, defined on an interval $a\le x\le b$. Our computer can only store its measured values at a finite number of *sample points* $a\le x_0<x_1<\cdots<x_n\le b$. In the simplest and, by far, the most common case, the sample points are equally spaced, and so

$$x_j=a+j\,h,\qquad j=0,\ldots,n,\qquad \text{where}\quad h=\frac{b-a}{n}$$

indicates the sample rate. In signal processing applications, x represents time instead of space, and the x_j are the times at which we sample the signal $f(x)$. Sample rates can be very high, e.g., every 10–20 milliseconds in current speech recognition systems.

For simplicity, we adopt the "standard" interval of $0\le x\le 2\pi$, and the n equally spaced sample points*

$$x_0=0,\quad x_1=\frac{2\pi}{n},\quad x_2=\frac{4\pi}{n},\quad\ldots\quad x_j=\frac{2j\pi}{n},\quad\ldots\quad x_{n-1}=\frac{2(n-1)\pi}{n}. \tag{5.85}$$

(Signals defined on other intervals can be handled by simply rescaling the interval to have length 2π.) Sampling a (complex-valued) signal or function $f(x)$ produces

*We will find it convenient to omit the final sample point $x_n=2\pi$ from consideration.

the *sample vector*

$$\mathbf{f} = (f_0, f_1, \ldots, f_{n-1})^T = (f(x_0), f(x_1), \ldots, f(x_{n-1}))^T,$$

where

$$f_j = f(x_j) = f\left(\frac{2j\pi}{n}\right). \tag{5.86}$$

Sampling cannot distinguish between functions that have the same values at all of the sample points—from the sampler's point of view they are identical. For example, the periodic complex exponential function

$$f(x) = e^{\mathrm{i}nx} = \cos nx + \mathrm{i}\sin nx$$

has sampled values

$$f_j = f\left(\frac{2j\pi}{n}\right) = \exp\left(\mathrm{i}n\,\frac{2j\pi}{n}\right) = e^{2j\pi\mathrm{i}} = 1 \quad \text{for all} \quad j = 0, \ldots, n-1,$$

and hence is indistinguishable from the constant function $c(x) \equiv 1$—both lead to the *same* sample vector $(1, 1, \ldots, 1)^T$. This has the important implication that sampling at n equally spaced sample points *cannot* detect periodic signals of frequency n. More generally, the two complex exponential signals

$$e^{\mathrm{i}(k+n)x} \quad \text{and} \quad e^{\mathrm{i}kx}$$

are also indistinguishable when sampled. This has the important consequence that we need only use the first n periodic complex exponential functions

$$f_0(x) = 1, \quad f_1(x) = e^{\mathrm{i}x}, \quad f_2(x) = e^{2\mathrm{i}x}, \quad \ldots \quad f_{n-1}(x) = e^{(n-1)\mathrm{i}x}, \tag{5.87}$$

in order to represent any 2π periodic sampled signal. In particular, exponentials $e^{-\mathrm{i}kx}$ of "negative" frequency can all be converted into positive versions, namely $e^{\mathrm{i}(n-k)x}$, by the same sampling argument. For example,

$$e^{-\mathrm{i}x} = \cos x - \mathrm{i}\sin x \quad \text{and} \quad e^{(n-1)\mathrm{i}x} = \cos(n-1)x + \mathrm{i}\sin(n-1)x$$

have identical values on the sample points (5.85). However, off of the sample points, they are quite different; the former is slowly varying, while the latter represents a high frequency oscillation. In Figure 5.11, we compare $e^{-\mathrm{i}x}$ and $e^{7\mathrm{i}x}$ when there

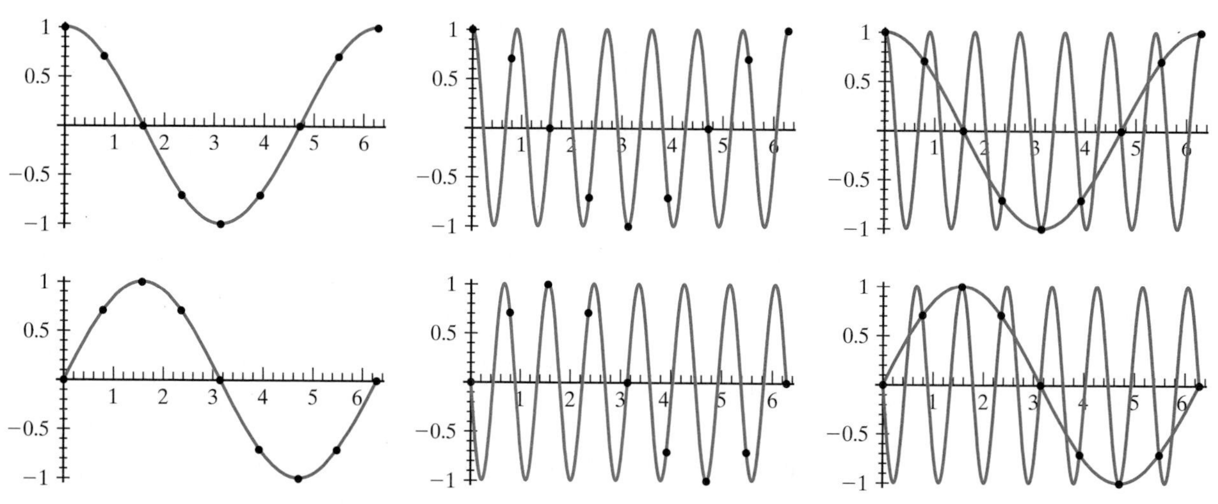

Figure 5.11 Sampling $e^{-\mathrm{i}x}$ and $e^{7\mathrm{i}x}$ on $n = 8$ sample points.

are $n = 8$ sample values, indicated by the dots on the graphs. The top row compares the real parts, $\cos x$ and $\cos 7x$, while the bottom row compares the imaginary parts, $\sin x$ and $-\sin 7x$. Note that both functions have the same pattern of sample values, even though their overall behavior is strikingly different.

This effect is commonly referred to as *aliasing*†. If you view a moving particle under a stroboscopic light that flashes only eight times, you would be unable to determine which of the two graphs the particle was following. Aliasing is the cause of a well-known artifact in movies: spoked wheels can appear to be rotating backwards when our brain interprets the discretization of the high frequency forward motion imposed by the frames of the film as an equivalently discretized low frequency motion in reverse. Aliasing also has important implications for the design of music CD's. We must sample an audio signal at a sufficiently high rate that all audible frequencies can be adequately represented. In fact, human appreciation of music also relies on inaudible high frequency tones, and so a much higher sample rate is actually used in commercial CD design. But the sample rate that was selected remains controversial; hi fi aficionados complain that it was not set high enough to fully reproduce the musical quality of an analog LP record!

The *discrete Fourier representation* decomposes a sampled function $f(x)$ into a linear combination of complex exponentials. Since we cannot distinguish sampled exponentials of frequency higher than n, we only need consider a finite linear combination

$$f(x) \sim p(x) = c_0 + c_1 e^{\,\mathrm{i}\,x} + c_2 e^{2\,\mathrm{i}\,x} + \cdots + c_{n-1} e^{(n-1)\,\mathrm{i}\,x} = \sum_{k=0}^{n-1} c_k\, e^{\,\mathrm{i}\,k x} \tag{5.88}$$

of the first n exponentials (5.87). The symbol $\sim$ in (5.88) means that the function $f(x)$ and the sum $p(x)$ agree on the sample points:

$$f(x_j) = p(x_j), \qquad j = 0, \ldots, n-1. \tag{5.89}$$

Therefore, $p(x)$ can be viewed as a (complex-valued) *interpolating trigonometric polynomial* of degree $\leq n-1$ for the sample data $f_j = f(x_j)$.

REMARK: If $f(x)$ is real, then $p(x)$ is also real on the sample points, but may very well be complex-valued in between. To avoid this unsatisfying state of affairs, we will usually discard its imaginary component, and regard the real part of $p(x)$ as "the" interpolating trigonometric polynomial. On the other hand, sticking with a purely real construction unnecessarily complicates the analysis, and so we will retain the complex exponential form (5.88) of the discrete Fourier sum.

Since we are working in the finite-dimensional vector space $\mathbb{C}^n$ throughout, we may reformulate the discrete Fourier series in vectorial form. Sampling the basic exponentials (5.87) produces the complex vectors

$$\begin{aligned} \boldsymbol{\omega}_k &= \left(e^{\,\mathrm{i}\,k x_0}, e^{\,\mathrm{i}\,k x_1}, e^{\,\mathrm{i}\,k x_2}, \ldots, e^{\,\mathrm{i}\,k x_n} \right)^T \\ &= \left(1, e^{2k\pi\,\mathrm{i}/n}, e^{4k\pi\,\mathrm{i}/n}, \ldots, e^{2(n-1)k\pi\,\mathrm{i}/n} \right)^T, \end{aligned} \qquad k = 0, \ldots, n-1. \tag{5.90}$$

The interpolation conditions (5.89) can be recast in the equivalent vector form

$$\mathbf{f} = c_0\, \boldsymbol{\omega}_0 + c_1\, \boldsymbol{\omega}_1 + \cdots + c_{n-1}\, \boldsymbol{\omega}_{n-1}. \tag{5.91}$$

†In computer graphics, the term "aliasing" is used in a much broader sense that covers a variety of artifacts introduced by discretization—particularly, the jagged appearance of lines and smooth curves on a digital monitor.

In other words, to compute the discrete Fourier coefficients $c_0, \ldots, c_{n-1}$ of f, all we need to do is rewrite its sample vector $\mathbf{f}$ as a linear combination of the sampled exponential vectors $\boldsymbol{\omega}_0, \ldots, \boldsymbol{\omega}_{n-1}$.

Now, the absolutely crucial property is the orthonormality of the basis elements $\boldsymbol{\omega}_0, \ldots, \boldsymbol{\omega}_{n-1}$. Were it not for the power of orthogonality, Fourier analysis might have remained a mere mathematical curiosity, rather than today's indispensable tool.

Proposition 5.61 The sampled exponential vectors $\boldsymbol{\omega}_0, \ldots, \boldsymbol{\omega}_{n-1}$ form an orthonormal basis of $\mathbb{C}^n$ with respect to the inner product

$$\langle \mathbf{f}, \mathbf{g} \rangle = \frac{1}{n} \sum_{j=0}^{n-1} f_j \, \overline{g_j} = \frac{1}{n} \sum_{j=0}^{n-1} f(x_j) \, \overline{g(x_j)}, \qquad \mathbf{f}, \mathbf{g} \in \mathbb{C}^n. \tag{5.92}$$

REMARK: The inner product (5.92) is a rescaled version of the standard Hermitian dot product (3.90) between complex vectors. We can interpret the inner product between the sample vectors $\mathbf{f}$, $\mathbf{g}$ as the *average* of the sampled values of the product signal $f(x)\,\overline{g(x)}$.

Proof The crux of the matter relies on properties of the remarkable complex numbers

$$\zeta_n = e^{2\pi\,\mathrm{i}/n} = \cos\frac{2\pi}{n} + \mathrm{i}\,\sin\frac{2\pi}{n}, \qquad \text{where} \quad n = 1, 2, 3, \ldots. \tag{5.93}$$

Particular cases include

$$\zeta_2 = -1, \quad \zeta_3 = -\tfrac{\sqrt{3}}{2} + \tfrac{1}{2}\,\mathrm{i}, \quad \zeta_4 = \mathrm{i}, \quad \text{and} \quad \zeta_8 = \tfrac{\sqrt{2}}{2} + \tfrac{\sqrt{2}}{2}\,\mathrm{i}. \tag{5.94}$$

The nth power of ζ_n is

$$\zeta_n^n = \left(e^{2\pi\,\mathrm{i}/n}\right)^n = e^{2\pi\,\mathrm{i}} = 1,$$

and hence ζ_n is one of the complex *nth roots of unity*: $\zeta_n = \sqrt[n]{1}$. There are, in fact, n different complex nth roots of 1, including 1 itself, namely the powers of ζ_n:

$$\zeta_n^k = e^{2k\pi\,\mathrm{i}/n} = \cos\frac{2k\pi}{n} + \mathrm{i}\,\sin\frac{2k\pi}{n}, \qquad k = 0, \ldots, n-1. \tag{5.95}$$

Since it generates all the others, ζ_n is known as the *primitive nth root of unity*. Geometrically, the nth roots (5.95) lie on the vertices of a regular unit n–gon in the complex plane; see Figure 5.12. The primitive root ζ_n is the first vertex we encounter as we go around the n–gon in a counterclockwise direction, starting at 1. Continuing around, the other roots appear in their natural order $\zeta_n^2, \zeta_n^3, \ldots, \zeta_n^{n-1}$, and finishing back at $\zeta_n^n = 1$. The complex conjugate of ζ_n is the "last" nth root

$$e^{-2\pi\,\mathrm{i}/n} = \overline{\zeta}_n = \frac{1}{\zeta_n} = \zeta_n^{n-1} = e^{2(n-1)\pi\,\mathrm{i}/n}. \tag{5.96}$$

The complex numbers (5.95) are a complete set of roots of the polynomial $z^n - 1$, which can therefore be factored:

$$z^n - 1 = (z-1)(z-\zeta_n)(z-\zeta_n^2)\cdots(z-\zeta_n^{n-1}).$$

On the other hand, elementary algebra provides us with the real factorization

$$z^n - 1 = (z-1)(1 + z + z^2 + \cdots + z^{n-1}).$$

Comparing the two, we conclude that

$$1 + z + z^2 + \cdots + z^{n-1} = (z-\zeta_n)(z-\zeta_n^2)\cdots(z-\zeta_n^{n-1}).$$

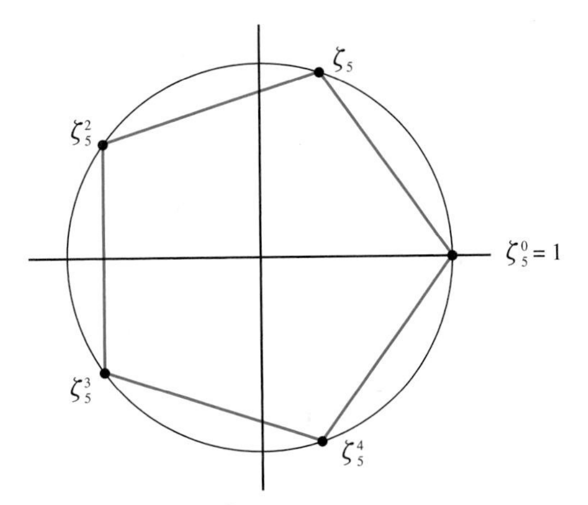

Figure 5.12 The fifth roots of unity.

Substituting $z = \zeta_n^k$ into both sides of this identity, we deduce the useful formula

$$1 + \zeta_n^k + \zeta_n^{2k} + \cdots + \zeta_n^{(n-1)k} = \begin{cases} n, & k = 0, \\ 0, & 0 < k < n. \end{cases} \tag{5.97}$$

Since $\zeta_n^{n+k} = \zeta_n^k$, this formula can easily be extended to general integers k; the sum is equal to n if n evenly divides k and is 0 otherwise.

Now, let us apply what we've learned to prove Proposition 5.61. First, in view of (5.95), the sampled exponential vectors (5.90) can all be written in terms of the nth roots of unity:

$$\omega_k = \left(1, \zeta_n^k, \zeta_n^{2k}, \zeta_n^{3k}, \ldots, \zeta_n^{(n-1)k}\right)^T, \qquad k = 0, \ldots, n-1. \tag{5.98}$$

Therefore, applying (5.96, 97), we conclude that

$$\langle \omega_k, \omega_l \rangle = \frac{1}{n} \sum_{j=0}^{n-1} \zeta_n^{jk} \, \overline{\zeta_n^{jl}} = \frac{1}{n} \sum_{j=0}^{n-1} \zeta_n^{j(k-l)} = \begin{cases} 1, & k = l, \\ 0, & k \neq l, \end{cases} \qquad 0 \leq k, l < n,$$

which establishes orthonormality of the sampled exponential vectors. ■

Orthonormality of the basis vectors implies that we can immediately compute the Fourier coefficients in the discrete Fourier sum (5.88) by taking inner products:

$$c_k = \langle \mathbf{f}, \omega_k \rangle = \frac{1}{n} \sum_{j=0}^{n-1} f_j \, \overline{e^{\,\mathrm{i}\,k x_j}} = \frac{1}{n} \sum_{j=0}^{n-1} f_j \, e^{-\mathrm{i}\,k x_j} = \frac{1}{n} \sum_{j=0}^{n-1} \zeta_n^{-jk} f_j. \tag{5.99}$$

In other words, the discrete Fourier coefficient c_k is obtained by averaging the sampled values of the product function $f(x)\, e^{-\mathrm{i}\,k x}$. The passage from a signal to its Fourier coefficients is known as the *Discrete Fourier Transform* or DFT for short. The reverse procedure of reconstructing a signal from its discrete Fourier coefficients via the sum (5.88) (or (5.91)) is known as the *Inverse Discrete Fourier Transform* or IDFT.

EXAMPLE 5.62 If $n = 4$, then $\zeta_4 = \mathrm{i}$. The corresponding sampled exponential vectors

$$\omega_0 = \begin{pmatrix} 1 \\ 1 \\ 1 \\ 1 \end{pmatrix}, \qquad \omega_1 = \begin{pmatrix} 1 \\ \mathrm{i} \\ -1 \\ -\mathrm{i} \end{pmatrix}, \qquad \omega_2 = \begin{pmatrix} 1 \\ -1 \\ 1 \\ -1 \end{pmatrix}, \qquad \omega_3 = \begin{pmatrix} 1 \\ -\mathrm{i} \\ -1 \\ \mathrm{i} \end{pmatrix},$$

form an orthonormal basis of $\mathbb{C}^4$ with respect to the averaged Hermitian dot product

$$\langle \mathbf{v}, \mathbf{w} \rangle = \tfrac{1}{4}\left(v_0\,\overline{w_0} + v_1\,\overline{w_1} + v_2\,\overline{w_2} + v_3\,\overline{w_3} \right),$$

where $\mathbf{v} = (v_0, v_1, v_2, v_3)^T$, $\mathbf{w} = (w_0, w_1, w_2, w_3)^T$.

Given the sampled function values

$$f_0 = f(0), \quad f_1 = f\left(\tfrac{1}{2}\pi\right), \quad f_2 = f(\pi), \quad f_3 = f\left(\tfrac{3}{2}\pi\right),$$

we construct the discrete Fourier representation

$$\mathbf{f} = c_0\,\boldsymbol{\omega}_0 + c_1\,\boldsymbol{\omega}_1 + c_2\,\boldsymbol{\omega}_2 + c_3\,\boldsymbol{\omega}_3, \tag{5.100}$$

where

$$c_0 = \langle \mathbf{f}, \boldsymbol{\omega}_0 \rangle = \tfrac{1}{4}(f_0 + f_1 + f_2 + f_3), \quad c_1 = \langle \mathbf{f}, \boldsymbol{\omega}_1 \rangle = \tfrac{1}{4}(f_0 - \mathrm{i}\, f_1 - f_2 + \mathrm{i}\, f_3),$$
$$c_2 = \langle \mathbf{f}, \boldsymbol{\omega}_2 \rangle = \tfrac{1}{4}(f_0 - f_1 + f_2 - f_3), \quad c_3 = \langle \mathbf{f}, \boldsymbol{\omega}_3 \rangle = \tfrac{1}{4}(f_0 + \mathrm{i}\, f_1 - f_2 - \mathrm{i}\, f_3).$$

We interpret this decomposition as the complex exponential interpolant

$$f(x) \sim p(x) = c_0 + c_1\, e^{\mathrm{i}\,x} + c_2\, e^{2\mathrm{i}\,x} + c_3\, e^{3\mathrm{i}\,x}$$

that agrees with $f(x)$ on the sample points.

For instance, if

$$f(x) = 2\pi x - x^2,$$

then

$$f_0 = 0., \quad f_1 = 7.4022, \quad f_2 = 9.8696, \quad f_3 = 7.4022,$$

and hence

$$c_0 = 6.1685, \quad c_1 = -2.4674, \quad c_2 = -1.2337, \quad c_3 = -2.4674.$$

Therefore, the interpolating trigonometric polynomial is given by the real part of

$$p(x) = 6.1685 - 2.4674\, e^{\mathrm{i}\,x} - 1.2337\, e^{2\mathrm{i}\,x} - 2.4674\, e^{3\mathrm{i}\,x}, \tag{5.101}$$

namely,

$$\operatorname{Re} p(x) = 6.1685 - 2.4674\,\cos x - 1.2337\,\cos 2x - 2.4674\,\cos 3x. \tag{5.102}$$

In Figure 5.13 we compare the function, with the interpolation points indicated, and discrete Fourier representations (5.102) for both $n = 4$ and $n = 16$ points. The resulting graphs point out a significant difficulty with the Discrete Fourier Transform as developed so far. While the trigonometric polynomials do indeed correctly match the sampled function values, their pronounced oscillatory behavior makes them completely unsuitable for interpolation away from the sample points. ●

However, this difficulty can be rectified by being a little more clever. The problem is that we have not been paying sufficient attention to the frequencies that are represented in the Fourier sum. Indeed, the graphs in Figure 5.13 might remind you of our earlier observation that, due to aliasing, low and high frequency exponentials can have the same sample data, but differ wildly in between the sample points. While the first half of the summands in (5.88) represent relatively low frequencies, the second half do not, and can be replaced by equivalent lower frequency, and hence less oscillatory exponentials. Namely, if $0 < k \le \frac{1}{2} n$, then $e^{-\mathrm{i}\,k x}$ and $e^{\mathrm{i}\,(n-k) x}$ have the same sample values, but the former is of lower frequency than the latter. Thus, for interpolatory purposes, we should replace the second half of the summands in

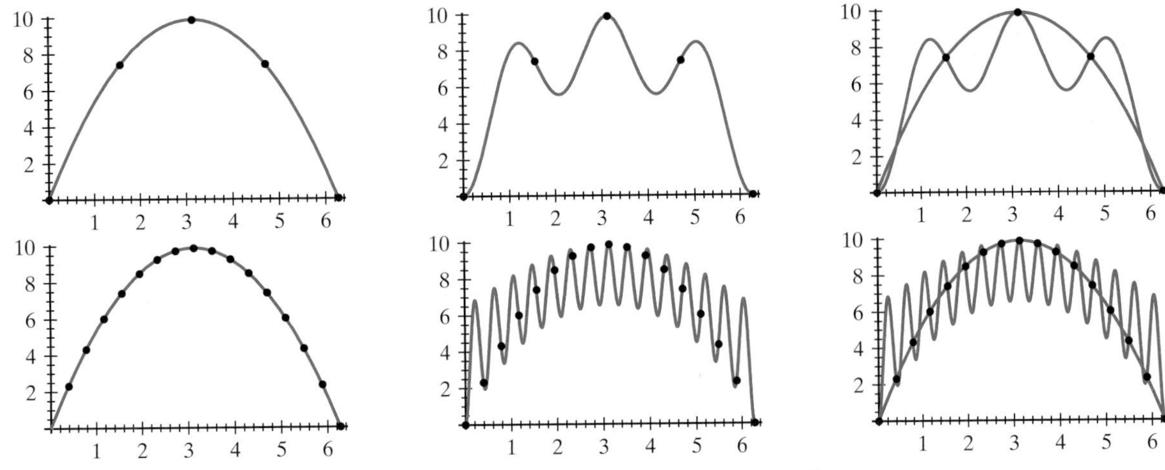

Figure 5.13 The discrete fourier representation of $x^2 - 2\pi x$.

the Fourier sum (5.88) by their low frequency alternatives. If $n = 2m+1$ is odd, then we take

$$\begin{aligned}\widehat{p}(x) &= c_{-m}\, e^{-\mathrm{i}\, m x} + \cdots + c_{-1}\, e^{-\mathrm{i}\, x} + c_0 + c_1\, e^{\mathrm{i}\, x} + \cdots + c_m\, e^{\mathrm{i}\, m x} \\ &= \sum_{k=-m}^{m} c_k\, e^{\mathrm{i}\, k x}\end{aligned} \tag{5.103}$$

as the equivalent low frequency interpolant. If $n = 2m$ is even—which is the most common case occurring in applications—then

$$\begin{aligned}\widehat{p}(x) &= c_{-m}\, e^{-\mathrm{i}\, m x} + \cdots + c_{-1}\, e^{-\mathrm{i}\, x} + c_0 + c_1\, e^{\mathrm{i}\, x} + \cdots + c_{m-1}\, e^{\mathrm{i}\,(m-1) x} \\ &= \sum_{k=-m}^{m-1} c_k\, e^{\mathrm{i}\, k x}\end{aligned} \tag{5.104}$$

will be our choice. (It is a matter of personal taste whether to use $e^{-\mathrm{i}\, m x}$ or $e^{\mathrm{i}\, m x}$ to represent the highest frequency term.) In both cases, the Fourier coefficients with negative indices are the same as their high frequency alternatives:

$$c_{-k} = c_{n-k} = \langle \mathbf{f}, \boldsymbol{\omega}_{n-k} \rangle = \langle \mathbf{f}, \boldsymbol{\omega}_{-k} \rangle, \tag{5.105}$$

where $\boldsymbol{\omega}_{-k} = \boldsymbol{\omega}_{n-k}$ is the sample vector for $e^{-\mathrm{i}\, k x} \sim e^{\mathrm{i}\,(n-k) x}$.

Returning to the previous example, for interpolating purposes, we should replace (5.101) by the equivalent low frequency interpolant

$$\widehat{p}(x) = -1.2337\, e^{-2\mathrm{i}\, x} - 2.4674\, e^{-\mathrm{i}\, x} + 6.1685 - 2.4674\, e^{\mathrm{i}\, x}, \tag{5.106}$$

with real part

$$\operatorname{Re} \widehat{p}(x) = 6.1685 - 4.9348 \cos x - 1.2337 \cos 2x.$$

Graphs of the $n = 4$ and 16 low frequency trigonometric interpolants can be seen in Figure 5.14. Thus, by utilizing only the lowest frequency exponentials, we successfully suppress the aliasing artifacts, resulting in a quite reasonable trigonometric interpolant to the given function.

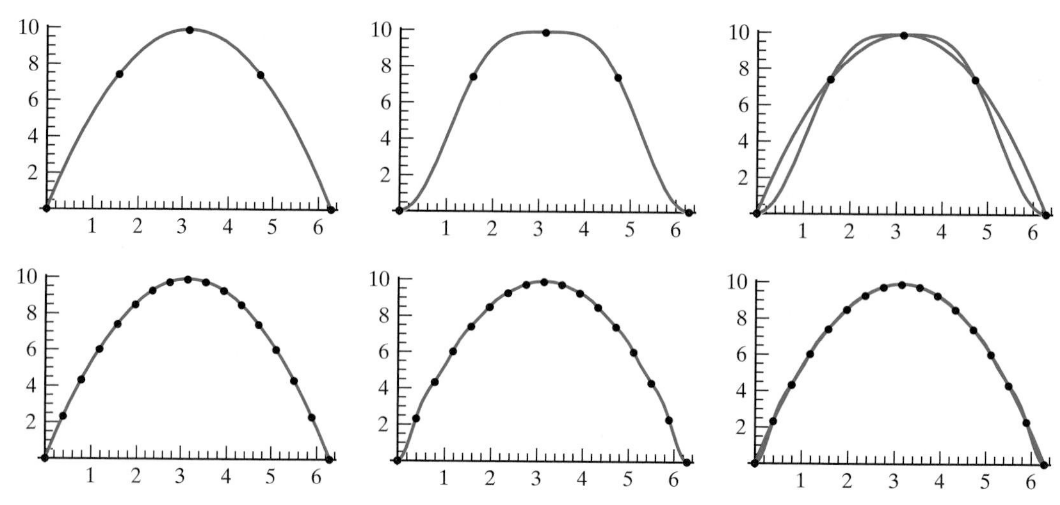

Figure 5.14 The low frequency discrete fourier representation of $x^2 - 2\pi x$.

REMARK: The low frequency version also serves to unravel the reality of the Fourier representation of a real function $f(x)$. Since $\omega_{-k} = \overline{\omega_k}$, formula (5.105) implies that $c_{-k} = \overline{c_k}$, and so the common frequency terms

$$c_{-k}\, e^{-i k x} + c_k\, e^{i k x} = a_k \cos k x + b_k \sin k x$$

add up to a real trigonometric function. Therefore, the odd n interpolant (5.103) is a real trigonometric polynomial, whereas in the even version (5.104) only the highest frequency term $c_{-m}\, e^{-i m x}$ produces a complex term—which is, in fact, 0 on the sample points.

EXERCISES 5.7

5.7.1. Find

(i) the discrete Fourier coefficients,

(ii) the low frequency trigonometric interpolant,

for the following functions using the indicated number of sample points:

(a) $\sin x,\ n = 4$ (b) $|x - \pi|,\ n = 6$

(c) $f(x) = \begin{cases} 1, & x \le 2, \\ 0, & x > 2, \end{cases}\quad n = 6$

(d) $\operatorname{sign}(x - \pi),\ n = 8$

5.7.2. Find

(i) the sample values, and

(ii) the trigonometric interpolant

corresponding to the following discrete Fourier coefficients:

(a) $c_{-1} = c_1 = 1,\ c_0 = 0$

(b) $c_{-2} = c_0 = c_2 = 1,\ c_{-1} = c_1 = -1$

(c) $c_{-2} = c_0 = c_1 = 2,\ c_{-1} = c_2 = 0$

(d) $c_0 = c_2 = c_4 = 1,\ c_1 = c_3 = c_5 = -1$

♠ **5.7.3.** Let $f(x) = x$. Compute its discrete Fourier coefficients based on $n = 4, 8$ and 16 sample points. Then, plot $f(x)$ along with the resulting (real) trigonometric interpolants and discuss their accuracy.

♠ **5.7.4.** Answer Exercise 5.7.3 for the functions

(a) x^2 (b) $(x - \pi)^2$

(c) $\sin x$ (d) $\cos \frac{1}{2} x$

(e) $\begin{cases} 1, & \frac{1}{2}\pi \le x \le \frac{3}{2}\pi, \\ 0 & \text{otherwise} \end{cases}$

(f) $\begin{cases} x, & 0 \le x \le \pi, \\ 2\pi - x, & \pi \le x \le 2\pi. \end{cases}$

5.7.5. (a) Draw a picture of the complex plane with the complex solutions to $z^6 = 1$ marked.

(b) What is the exact formula (no trigonometric functions allowed) for the primitive sixth root of unity ζ_6?

(c) Verify explicitly that $1 + \zeta_6 + \zeta_6^2 + \zeta_6^3 + \zeta_6^4 + \zeta_6^5 = 0$.

(d) Give a geometrical explanation of this identity.

◇ **5.7.6.** (a) Explain in detail why the nth roots of 1 lie on the vertices of a regular n–gon. What is the angle between two consecutive sides?

(b) Explain why this is also true for the nth roots of any non-zero complex number $z \neq 0$.

(c) Sketch a picture of the hexagon corresponding to $\sqrt[6]{z}$ for a given $z \neq 0$.

◇ **5.7.7.** In general, an nth root of unit ζ is called *primitive* if all the nth roots of unity are obtained by raising it to successive powers: $1, \zeta, \zeta^2, \zeta^3, \ldots$.

(a) Find all primitive

(i) fourth (ii) fifth

(iii) ninth roots of unity

(b) Can you characterize all the primitive nth roots of unity?

5.7.8. (a) In Example 5.62, the $n = 4$ discrete Fourier coefficients of the function $f(x) = 2\pi x - x^2$ were found to be real. Is this true when $n = 16$? For general n?

(b) What property of a function $f(x)$ will guarantee that its Fourier coefficients are real?

♡ **5.7.9.** Let $\mathbf{c} = (c_0, c_1, \ldots, c_{n-1})^T \in \mathbb{C}^n$ be the vector of discrete Fourier coefficients corresponding to the sample vector $\mathbf{f} = (f_0, f_1, \ldots, f_{n-1})^T$.

(a) Explain why the sampled signal $\mathbf{f} = F_n\,\mathbf{c}$ can be reconstructed by multiplying its Fourier coefficient vector by an $n \times n$ matrix F_n. Write down F_2, F_3, F_4, and F_8. What is the general formula for the entries of F_n?

(b) Prove that, in general,

$$F_n^{-1} = \frac{1}{n}\,F_n^{\dagger} = \frac{1}{n}\,\overline{F}_n^{\,T},$$

where † denotes the Hermitian transpose defined in Exercise 5.3.25.

(c) Prove that

$$U_n = \frac{1}{\sqrt{n}}\,F_n$$

is a unitary matrix, i.e., $U_n^{-1} = U_n^{\dagger}$.

Compression and Noise Removal

In a typical experimental signal, noise primarily affects the high frequency modes, while the authentic features tend to appear in the low frequencies. Think of the hiss and static you hear on an AM radio station or a low quality audio recording. Thus, a very simple, but effective, method for denoising a corrupted signal is to decompose it into its Fourier modes, as in (5.88), and then discard the high frequency constituents. A similar idea underlies the DolbyⓉ recording system used on most movie soundtracks: during the recording process, the high frequency modes are artificially boosted, so that scaling them back when showing the movie in the theater has the effect of eliminating much of the extraneous noise. The one design issue is the specification of a cut-off between low and high frequency, that is, between signal and noise. This choice will depend upon the properties of the measured signal, and is left to the discretion of the signal processor.

A correct implementation of the denoising procedure is facilitated by using the unaliased forms (5.103, 104) of the trigonometric interpolant, in which the low frequency summands only appear when $|k|$ is small. In this version, to eliminate high frequency components, we replace the full summation by

$$q_l(x) = \sum_{k=-l}^{l} c_k\, e^{\,i\,k\,x}, \tag{5.107}$$

where $l < \frac{1}{2}(n+1)$ specifies the selected cut-off frequency between signal and noise. The $2l+1 \ll n$ low frequency Fourier modes retained in (5.107) will, in favorable situations, capture the essential features of the original signal while simultaneously eliminating the high frequency noise.

In Figure 5.15 we display a sample signal followed by the same signal corrupted by adding in random noise. We use $n = 2^8 = 256$ sample points in the discrete Fourier representation, and to remove the noise, we retain only the $2l+1 = 11$ lowest frequency modes. In other words, instead of all $n = 512$ Fourier coefficients $c_{-256}, \ldots, c_{-1}, c_0, c_1, \ldots, c_{255}$, we only compute the 11 lowest order ones $c_{-5}, \ldots, c_5$. Summing up just those 11 exponentials produces the denoised signal

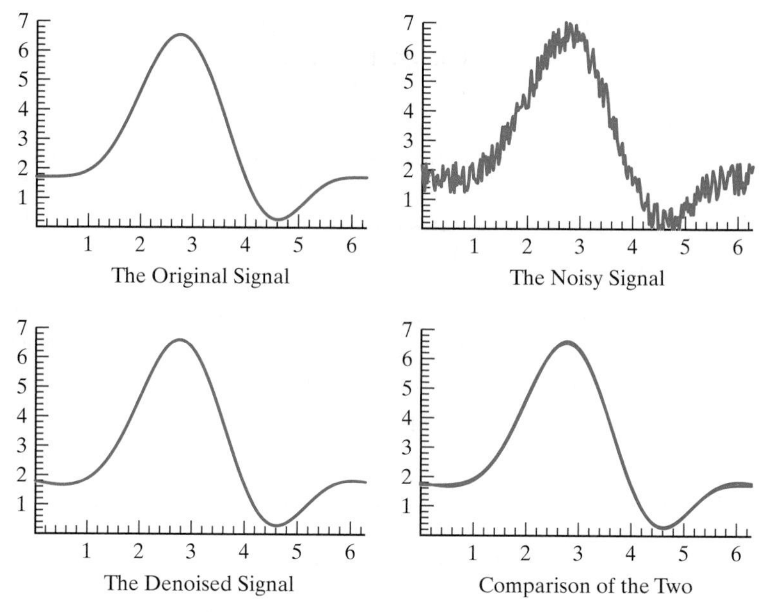

Figure 5.15 Denoising a signal.

$q(x) = c_{-5}\,e^{-5\,i\,x} + \cdots + c_5\,e^{5\,i\,x}$. To compare, we plot both the original signal and the denoised version on the same graph. In this case, the maximal deviation is less than .15 over the entire interval $[0, 2\pi]$.

The same idea underlies many data compression algorithms for audio recordings, digital images and, particularly, video. The goal is efficient storage and/or transmission of the signal. As before, we expect all the important features to be contained in the low frequency constituents, and so discarding the high frequency terms will, in favorable situations, not lead to any noticeable degradation of the signal or image. Thus, to compress a signal (and, simultaneously, remove high frequency noise), we retain only its low frequency discrete Fourier coefficients. The signal is reconstructed by summing the associated truncated discrete Fourier series (5.107). A mathematical justification of Fourier-based compression algorithms relies on the fact that the Fourier coefficients of smooth functions tend rapidly to zero—the smoother the function, the faster the decay rate. Thus, the small high frequency Fourier coefficients will be of negligible importance.

In Figure 5.16, the previous signal is compressed by retaining, respectively, $2l + 1 = 21$ and $2l + 1 = 7$ Fourier coefficients only instead of all $n = 512$ that would be required for complete accuracy. For the case of moderate compression, the maximal deviation between the signal and the compressed version is less

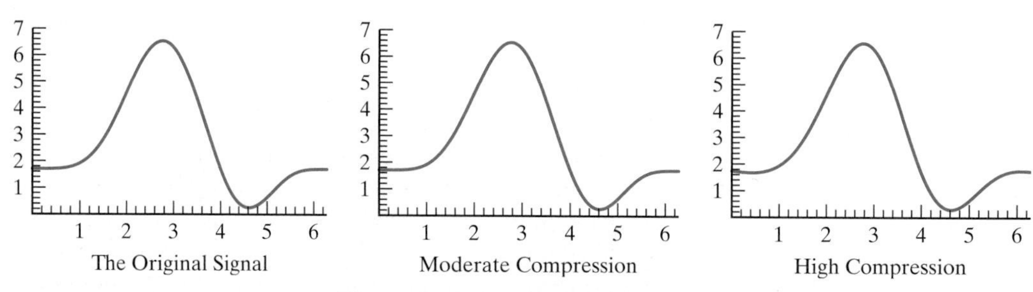

Figure 5.16 Compressing a signal.

than 1.5×10^{-4} over the entire interval, while even the highly compressed version deviates at most .05 from the original signal. Of course, the lack of any fine scale features in this particular signal means that a very high compression can be achieved—the more complicated or detailed the original signal, the more Fourier modes need to be retained for accurate reproduction.

EXERCISES 5.7

♠ **5.7.10.** Construct the discrete Fourier coefficients for

$$f(x) = \begin{cases} -x, & 0 \le x \le \frac{1}{3}\pi, \\ x - \frac{2}{3}\pi, & \frac{1}{3}\pi \le x \le \frac{4}{3}\pi, \\ -x + 2\pi, & \frac{4}{3}\pi \le x \le 2\pi \end{cases}$$

based on $n = 128$ sample points. Then graph the reconstructed function when using the data compression algorithm that retains only the 11 and 21 lowest frequency modes. Discuss what you observe.

♠ **5.7.11.** Answer Exercise 5.7.10 when $f(x) =$

(a) x (b) $x^2(2\pi - x)^2$

(c) $\begin{cases} \sin x, & 0 \le x \le \pi, \\ 0, & \pi \le x \le 2\pi. \end{cases}$

♣ **5.7.12.** Let $q_l(x)$ denote the trigonometric polynomial (5.107) obtained by summing the first $2l + 1$ discrete Fourier modes. Suppose the criterion for compression of a signal $f(x)$ is that $\| f - q_l \|_\infty = \max\{\, | f(x) - q_l(x) | \mid 0 \le x \le 2\pi \,\} < \varepsilon$. For the particular function in Exercise 5.7.10, how large do you need to choose k when $\varepsilon = .1$? $\varepsilon = .01$? $\varepsilon = .001$?

♣ **5.7.13.** Let $f(x) = x(2\pi - x)$ be sampled on $n = 128$ equally spaced points between 0 and 2π. Use a random number generator with $-1 \le r_j \le 1$ to add noise by replacing each sample value $f_j = f(x_j)$ by $g_j = f_j + \varepsilon\, r_j$. Investigate, for different values of ε, how many discrete Fourier modes are required to reconstruct a reasonable denoised approximation to the original signal.

♠ **5.7.14.** The signal in Figure 5.15 was obtained from the explicit formula

$$-\frac{1}{5}\left(\frac{x(2\pi - x)}{10}\right)^5 (x+1.5)(x+2.5)(x-4)+1.7.$$

Noise was added by using a random number generator. Experiment with different intensities of noise and different numbers of sample points and discuss what you observe.

♣ **5.7.15.** If we use the original form (5.88) of the discrete Fourier representation, we might be tempted to denoise/compress the signal by only retaining the first $0 \le k \le l$ terms in the sum. Test this method on the signal in Exercise 5.7.10 and discuss what you observe.

5.7.16. *True or false*: If $f(x)$ is real, the compressed/denoised signal (5.107) is a real trigonometric polynomial.

The Fast Fourier Transform

While one may admire an algorithm for its intrinsic beauty, in the real world, the bottom line is always efficiency of implementation: the less total computation, the faster the processing, and hence the more extensive the range of applications. Orthogonality is the first and most important feature of many practical linear algebra algorithms, and is *the* critical feature of Fourier analysis. Still, even the power of orthogonality reaches its limits when it comes to dealing with truly large scale problems such as three-dimensional medical imaging or video processing. In the early 1960's, James Cooley and John Tukey, [14], discovered‡ a much more efficient approach to the Discrete Fourier Transform, exploiting the rather special structure of the sampled exponential vectors. The resulting algorithm is known as the *Fast Fourier Transform*, often abbreviated FFT, and its discovery launched the modern revolution in digital signal and data processing, [7, 8].

In general, computing all the discrete Fourier coefficients (5.99) of an n times sampled signal requires a total of n^2 complex multiplications and $n^2 - n$ complex

‡In fact, the key ideas can be found in Gauss' hand computations in the early 1800's, but his insight was not fully appreciated until modern computers arrived on the scene.

additions. Note also that each complex addition

$$z + w = (x + \mathrm{i}\, y) + (u + \mathrm{i}\, v) = (x + u) + \mathrm{i}\,(y + v) \tag{5.108}$$

generally requires two real additions, while each complex multiplication

$$z\, w = (x + \mathrm{i}\, y)\,(u + \mathrm{i}\, v) = (x\, u - y\, v) + \mathrm{i}\,(x\, v + y\, u) \tag{5.109}$$

requires 4 real multiplications and 2 real additions, or, by employing the alternative formula

$$x\, v + y\, u = (x + y)\,(u + v) - x\, u - y\, v \tag{5.110}$$

for the imaginary part, 3 real multiplications and 5 real additions. (The choice of formula (5.109) or (5.110) will depend upon the processor's relative speeds of multiplication and addition.) Similarly, given the Fourier coefficients $c_0, \ldots, c_{n-1}$, reconstruction of the sampled signal via (5.88) requires $n^2 - n$ complex multiplications and $n^2 - n$ complex additions. As a result, both computations become quite labor intensive for large n. Extending these ideas to multi-dimensional data only exacerbates the problem.

In order to explain the method without undue complication, we return to the original, aliased form of the discrete Fourier representation (5.88). (Once one understands how the FFT works, one can easily adapt the algorithm to the low frequency version (5.104).) The seminal observation is that if the number of sample points

$$n = 2m$$

is even, then the primitive mth root of unity $\zeta_m = \sqrt[m]{1}$ equals the square of the primitive nth root:

$$\zeta_m = \zeta_n^2.$$

We use this fact to split the summation (5.99) for the order n discrete Fourier coefficients into two parts, collecting together the even and the odd powers of ζ_n^k:

$$\begin{aligned}
c_k &= \frac{1}{n}\left(f_0 + f_1\, \zeta_n^{-k} + f_2\, \zeta_n^{-2k} + \cdots + f_{n-1}\, \zeta_n^{-(n-1)k} \right) \\
&= \frac{1}{n}\left(f_0 + f_2\, \zeta_n^{-2k} + f_4\, \zeta_n^{-4k} + \cdots + f_{2m-2}\, \zeta_n^{-(2m-2)k} \right) \\
&\quad + \zeta_n^{-k}\, \frac{1}{n}\left(f_1 + f_3\, \zeta_n^{-2k} + f_5\, \zeta_n^{-4k} + \cdots + f_{2m-1}\, \zeta_n^{-(2m-2)k} \right) \\
&= \frac{1}{2}\left\{ \frac{1}{m}\left(f_0 + f_2\, \zeta_m^{-k} + f_4\, \zeta_m^{-2k} + \cdots + f_{2m-2}\, \zeta_m^{-(m-1)k} \right) \right\} \\
&\quad + \frac{\zeta_n^{-k}}{2}\left\{ \frac{1}{m}\left(f_1 + f_3\, \zeta_m^{-k} + f_5\, \zeta_m^{-2k} + \cdots + f_{2m-1}\, \zeta_m^{-(m-1)k} \right) \right\}.
\end{aligned} \tag{5.111}$$

Now, observe that the expressions in braces are the order m Fourier coefficients for the sample data

$$\begin{aligned}
\mathbf{f}^e &= (f_0, f_2, f_4, \ldots, f_{2m-2})^T = (f(x_0), f(x_2), f(x_4), \ldots, f(x_{2m-2}))^T, \\
\mathbf{f}^o &= (f_1, f_3, f_5, \ldots, f_{2m-1})^T = (f(x_1), f(x_3), f(x_5), \ldots, f(x_{2m-1}))^T.
\end{aligned} \tag{5.112}$$

Note that $\mathbf{f}^e$ is obtained by sampling $f(x)$ on the *even* sample points x_{2j}, while $\mathbf{f}^o$ is obtained by sampling the same function $f(x)$, but now at the *odd* sample points

x_{2j+1}. In other words, we are splitting the original sampled signal into two "half-sampled" signals obtained by sampling on every other point. The even and odd Fourier coefficients are

$$c_k^e = \frac{1}{m}\left(f_0 + f_2\,\zeta_m^{-k} + f_4\,\zeta_m^{-2k} + \cdots + f_{2m-2}\,\zeta_m^{-(m-1)k}\right), \qquad k = 0, \ldots, m-1.$$
$$c_k^o = \frac{1}{m}\left(f_1 + f_3\,\zeta_m^{-k} + f_5\,\zeta_m^{-2k} + \cdots + f_{2m-1}\,\zeta_m^{-(m-1)k}\right), \tag{5.113}$$

Since they contain just m data values, both the even and odd samples require only m distinct Fourier coefficients, and we adopt the identification

$$c_{k+m}^e = c_k^e, \qquad c_{k+m}^o = c_k^o, \qquad k = 0, \ldots, m-1. \tag{5.114}$$

Therefore, the order $n = 2m$ discrete Fourier coefficients (5.111) can be constructed from a pair of order m discrete Fourier coefficients via

$$c_k = \tfrac{1}{2}\left(c_k^e + \zeta_n^{-k}\,c_k^o\right), \qquad k = 0, \ldots, n-1. \tag{5.115}$$

Now if $m = 2l$ is also even, then we can play the same game on the order m Fourier coefficients (5.113), reconstructing each of them from a pair of order l discrete Fourier coefficients—obtained by sampling the signal at every fourth point. If $n = 2^r$ is a power of 2, then this game can be played all the way back to the start, beginning with the trivial order 1 discrete Fourier representation, which just samples the function at a single point. The result is the desired algorithm. After some rearrangement of the basic steps, we arrive at the Fast Fourier Transform, which we now present in its final form.

We begin with a sampled signal on $n = 2^r$ sample points. To efficiently program the Fast Fourier Transform, it helps to write out each index $0 \le j < 2^r$ in its binary (as opposed to decimal) representation

$$j = j_{r-1}\,j_{r-2}\,\ldots\,j_2\,j_1\,j_0, \qquad \text{where} \quad j_\nu = 0 \text{ or } 1; \tag{5.116}$$

the notation is shorthand for its r digit binary expansion

$$j = j_0 + 2\,j_1 + 4\,j_2 + 8\,j_3 + \cdots + 2^{r-1}\,j_{r-1}.$$

We then define the *bit reversal* map

$$\rho(j_{r-1}\,j_{r-2}\,\ldots\,j_2\,j_1\,j_0) = j_0\,j_1\,j_2\,\ldots\,j_{r-2}\,j_{r-1}. \tag{5.117}$$

For instance, if $r = 5$, and $j = 13$, with 5 digit binary representation 01101, then $\rho(j) = 22$ has the reversed binary representation 10110. Note especially that the bit reversal map $\rho = \rho_r$ depends upon the original choice of $r = \log_2 n$.

Secondly, for each $0 \le k < r$, define the maps

$$\begin{aligned} \alpha_k(j) &= j_{r-1}\,\ldots\,j_{k+1}\,0\,j_{k-1}\,\ldots\,j_0, \\ \beta_k(j) &= j_{r-1}\,\ldots\,j_{k+1}\,1\,j_{k-1}\,\ldots\,j_0 = \alpha_k(j) + 2^k, \end{aligned} \qquad \text{for} \quad j = j_{r-1}\,j_{r-2}\,\ldots\,j_1\,j_0. \tag{5.118}$$

In other words, $\alpha_k(j)$ sets the kth binary digit of j to 0, while $\beta_k(j)$ sets it to 1. In the preceding example, $\alpha_2(13) = 9$, with binary form 01001, while $\beta_2(13) = 13$ with binary form 01101. The bit operations (5.117, 118) are especially easy to implement on modern binary computers.

Given a sampled signal $f_0, \ldots, f_{n-1}$, its discrete Fourier coefficients $c_0, \ldots, c_{n-1}$ are computed by the following iterative algorithm:

$$c_j^{(0)} = f_{\rho(j)}, \qquad j = 0, \ldots, n-1,$$
$$c_j^{(k+1)} = \tfrac{1}{2}\left(c_{\alpha_k(j)}^{(k)} + \zeta_{2^{k+1}}^{-j}\, c_{\beta_k(j)}^{(k)} \right), \qquad k = 0, \ldots, r-1, \tag{5.119}$$

in which $\zeta_{2^{k+1}}$ is the primitive 2^{k+1} root of unity. The final output of the iterative procedure, namely

$$c_j = c_j^{(r)}, \qquad j = 0, \ldots, n-1, \tag{5.120}$$

are the discrete Fourier coefficients of our signal. The preprocessing step of the algorithm, where we define $c_j^{(0)}$, produces a more convenient rearrangement of the sample values. The subsequent steps successively combine the Fourier coefficients of the appropriate even and odd sampled subsignals together, reproducing (5.111) in a different notation. The following example should help make the overall process clearer.

EXAMPLE 5.63 Consider the case $r = 3$, and so our signal has $n = 2^3 = 8$ sampled values $f_0, f_1, \ldots, f_7$. We begin the process by rearranging the sample values

$$c_0^{(0)} = f_0, \quad c_1^{(0)} = f_4, \quad c_2^{(0)} = f_2, \quad c_3^{(0)} = f_6,$$
$$c_4^{(0)} = f_1, \quad c_5^{(0)} = f_5, \quad c_6^{(0)} = f_3, \quad c_7^{(0)} = f_7,$$

in the order specified by the bit reversal map ρ. For instance $\rho(3) = 6$, or, in binary notation, $\rho(011) = 110$.

The first stage of the iteration is based on $\zeta_2 = -1$. Equation (5.119) gives

$$c_0^{(1)} = \tfrac{1}{2}(c_0^{(0)} + c_1^{(0)}), \qquad c_1^{(1)} = \tfrac{1}{2}(c_0^{(0)} - c_1^{(0)}),$$
$$c_2^{(1)} = \tfrac{1}{2}(c_2^{(0)} + c_3^{(0)}), \qquad c_3^{(1)} = \tfrac{1}{2}(c_2^{(0)} - c_3^{(0)}),$$
$$c_4^{(1)} = \tfrac{1}{2}(c_4^{(0)} + c_5^{(0)}), \qquad c_5^{(1)} = \tfrac{1}{2}(c_4^{(0)} - c_5^{(0)}),$$
$$c_6^{(1)} = \tfrac{1}{2}(c_6^{(0)} + c_7^{(0)}), \qquad c_7^{(1)} = \tfrac{1}{2}(c_6^{(0)} - c_7^{(0)}),$$

where we combine successive pairs of the rearranged sample values. The second stage of the iteration has $k = 1$ with $\zeta_4 = \mathrm{i}$. We find

$$c_0^{(2)} = \tfrac{1}{2}(c_0^{(1)} + c_2^{(1)}), \qquad c_1^{(2)} = \tfrac{1}{2}(c_1^{(1)} - \mathrm{i}\, c_3^{(1)}),$$
$$c_2^{(2)} = \tfrac{1}{2}(c_0^{(1)} - c_2^{(1)}), \qquad c_3^{(2)} = \tfrac{1}{2}(c_1^{(1)} + \mathrm{i}\, c_3^{(1)}),$$
$$c_4^{(2)} = \tfrac{1}{2}(c_4^{(1)} + c_6^{(1)}), \qquad c_5^{(2)} = \tfrac{1}{2}(c_5^{(1)} - \mathrm{i}\, c_7^{(1)}),$$
$$c_6^{(2)} = \tfrac{1}{2}(c_4^{(1)} - c_6^{(1)}), \qquad c_7^{(2)} = \tfrac{1}{2}(c_5^{(1)} + \mathrm{i}\, c_7^{(1)}).$$

Note that the indices of the combined pairs of coefficients differ by 2. In the last step, where $k = 2$ and $\zeta_8 = \frac{\sqrt{2}}{2}(1 + \mathrm{i})$, we combine coefficients whose indices differ by $4 = 2^2$; the final output

$$c_0 = c_0^{(3)} = \tfrac{1}{2}(c_0^{(2)} + c_4^{(2)}), \qquad c_1 = c_1^{(3)} = \tfrac{1}{2}\left(c_1^{(2)} + \tfrac{\sqrt{2}}{2}(1 - \mathrm{i})\, c_5^{(2)}\right),$$
$$c_2 = c_2^{(3)} = \tfrac{1}{2}\left(c_2^{(2)} - \mathrm{i}\, c_6^{(2)}\right), \qquad c_3 = c_3^{(3)} = \tfrac{1}{2}\left(c_3^{(2)} - \tfrac{\sqrt{2}}{2}(1 + \mathrm{i})\, c_7^{(2)}\right),$$
$$c_4 = c_4^{(3)} = \tfrac{1}{2}(c_0^{(2)} - c_4^{(2)}), \qquad c_5 = c_5^{(3)} = \tfrac{1}{2}\left(c_1^{(2)} - \tfrac{\sqrt{2}}{2}(1 - \mathrm{i})\, c_5^{(2)}\right),$$
$$c_6 = c_6^{(3)} = \tfrac{1}{2}\left(c_2^{(2)} + \mathrm{i}\, c_6^{(2)}\right), \qquad c_7 = c_7^{(3)} = \tfrac{1}{2}\left(c_3^{(2)} + \tfrac{\sqrt{2}}{2}(1 + \mathrm{i})\, c_7^{(2)}\right),$$

is the complete set of discrete Fourier coefficients. ●

Let us count the number of arithmetic operations required in the Fast Fourier Transform algorithm. At each stage in the computation, we must perform $n = 2^r$ complex additions/subtractions and the same number of complex multiplications. (Actually, the number of multiplications is slightly smaller since multiplications by ± 1 and $\pm\,\mathrm{i}$ are extremely simple. However, this does not significantly alter the final operations count.) There are $r = \log_2 n$ stages, and so we require a total of $r\,n = n \log_2 n$ complex additions/subtractions and the same number of multiplications. Now, when n is large, $n \log_2 n$ is *significantly* smaller than n^2, which is the number of operations required for the direct algorithm. For instance, if $n = 2^{10} = 1,024$, then $n^2 = 1,048,576$, while $n \log_2 n = 10,240$—a net savings of 99%. As a result, many large scale computations that would be intractable using the direct approach are immediately brought into the realm of feasibility. This is the reason why all modern implementations of the Discrete Fourier Transform are based on the FFT algorithm and its variants.

The reconstruction of the signal from the discrete Fourier coefficients $c_0, \ldots, c_{n-1}$ is speeded up in exactly the same manner. The only differences are that we replace $\zeta_n^{-1} = \overline{\zeta_n}$ by ζ_n, and drop the factors of $\frac{1}{2}$ since there is no need to divide by n in the final result (5.88). Therefore, we apply the slightly modified iterative procedure

$$\begin{aligned} f_j^{(0)} &= c_{\rho(j)}, & j &= 0, \ldots, n-1, \\ f_j^{(k+1)} &= f_{\alpha_k(j)}^{(k)} + \zeta_{2^{k+1}}^{\,j}\, f_{\beta_k(j)}^{(k)}, & k &= 0, \ldots, r-1, \end{aligned} \tag{5.121}$$

and finish with

$$f(x_j) = f_j = f_j^{(r)}, \qquad j = 0, \ldots, n-1. \tag{5.122}$$

EXAMPLE 5.64 The reconstruction formulae in the case of $n = 8 = 2^3$ Fourier coefficients $c_0, \ldots, c_7$, which were computed in Example 5.63, can be implemented as follows. First, we rearrange the Fourier coefficients in bit reversed order:

$$\begin{aligned} f_0^{(0)} &= c_0, & f_1^{(0)} &= c_4, & f_2^{(0)} &= c_2, & f_3^{(0)} &= c_6, \\ f_4^{(0)} &= c_1, & f_5^{(0)} &= c_5, & f_6^{(0)} &= c_3, & f_7^{(0)} &= c_7, \end{aligned}$$

Then we begin combining them in successive pairs:

$$\begin{aligned} f_0^{(1)} &= f_0^{(0)} + f_1^{(0)}, & f_1^{(1)} &= f_0^{(0)} - f_1^{(0)}, \\ f_2^{(1)} &= f_2^{(0)} + f_3^{(0)}, & f_3^{(1)} &= f_2^{(0)} - f_3^{(0)}, \\ f_4^{(1)} &= f_4^{(0)} + f_5^{(0)}, & f_5^{(1)} &= f_4^{(0)} - f_5^{(0)}, \\ f_6^{(1)} &= f_6^{(0)} + f_7^{(0)}, & f_7^{(1)} &= f_6^{(0)} - f_7^{(0)}. \end{aligned}$$

Next,

$$\begin{aligned} f_0^{(2)} &= f_0^{(1)} + f_2^{(1)}, & f_1^{(2)} &= f_1^{(1)} + \mathrm{i}\, f_3^{(1)}, \\ f_2^{(2)} &= f_0^{(1)} - f_2^{(1)}, & f_3^{(2)} &= f_1^{(1)} - \mathrm{i}\, f_3^{(1)}, \\ f_4^{(2)} &= f_4^{(1)} + f_6^{(1)}, & f_5^{(2)} &= f_5^{(1)} + \mathrm{i}\, f_7^{(1)}, \\ f_6^{(2)} &= f_4^{(1)} - f_6^{(1)}, & f_7^{(2)} &= f_5^{(1)} - \mathrm{i}\, f_7^{(1)}. \end{aligned}$$

Finally, the sampled signal values are

$$
\begin{aligned}
f(x_0) &= f_0^{(3)} = f_0^{(2)} + f_4^{(2)},\\
f(x_1) &= f_1^{(3)} = f_1^{(2)} + \tfrac{\sqrt{2}}{2}\,(1+\mathrm{i})\, f_5^{(2)},\\
f(x_2) &= f_2^{(3)} = f_2^{(2)} + \mathrm{i}\, f_6^{(2)},\\
f(x_3) &= f_3^{(3)} = f_3^{(2)} - \tfrac{\sqrt{2}}{2}\,(1-\mathrm{i})\, f_7^{(2)},\\
f(x_4) &= f_4^{(3)} = f_0^{(2)} - f_4^{(2)},\\
f(x_5) &= f_5^{(3)} = f_1^{(2)} - \tfrac{\sqrt{2}}{2}\,(1+\mathrm{i})\, f_5^{(2)},\\
f(x_6) &= f_6^{(3)} = f_2^{(2)} - \mathrm{i}\, f_6^{(2)},\\
f(x_7) &= f_7^{(3)} = f_3^{(2)} + \tfrac{\sqrt{2}}{2}\,(1-\mathrm{i})\, f_7^{(2)}
\end{aligned}
$$

●

EXERCISES 5.7

♠ **5.7.17.** Use the Fast Fourier Transform to find the discrete Fourier coefficients for the the following functions using the indicated number of sample points. Carefully indicate each step in your analysis.

(a) $\dfrac{x}{\pi}$, $n = 4$ (b) $\sin x$, $n = 8$

(c) $|x - \pi|$, $n = 8$ (d) $\operatorname{sign}(x - \pi)$, $n = 16$

♠ **5.7.18.** Use the Inverse Fast Fourier Transform to reassemble the sampled function data corresponding to the following discrete Fourier coefficients. Carefully indicate each step in your analysis.

(a) $c_0 = c_2 = 1$, $c_1 = c_3 = -1$

(b) $c_0 = c_1 = c_4 = 2$, $c_2 = c_6 = 0$, $c_3 = c_5 = c_7 = -1$

♡ **5.7.19.** In this exercise, we show how the Fast Fourier Transform is equivalent to a certain matrix factorization. Let

$$\mathbf{c} = (c_0, c_1 \ldots, c_7)^T$$

be vector of Fourier coefficients, and let

$$\mathbf{f}^{(k)} = (f_0^{(k)}, f_1^{(k)}, \ldots, f_7^{(k)})^T, \quad k = 0, 1, 2, 3,$$

be vectors containing the coefficients defined in the reconstruction algorithm Example 5.64.

(a) Show that

$$\mathbf{f}^{(0)} = M_0\mathbf{c}, \qquad \mathbf{f}^{(1)} = M_1\mathbf{f}^{(0)},$$
$$\mathbf{f}^{(2)} = M_2\mathbf{f}^{(1)}, \quad \mathbf{f} = \mathbf{f}^{(3)} = M_3\mathbf{f}^{(2)},$$

where M_0, M_1, M_2, M_3 are 8×8 matrices. Write down their explicit forms.

(b) Explain why the matrix product $F_8 = M_3M_2M_1M_0$ reproduces the Fourier matrix derived in Exercise 5.7.9. Check the factorization directly.

(c) Write down the corresponding matrix factorization for the direct algorithm of Example 5.63.

CHAPTER

6

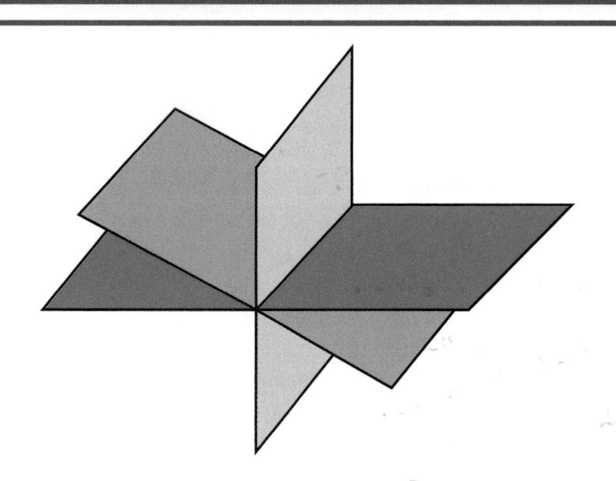

Equilibrium

In this chapter, we will apply what we have learned so far to the analysis of equilibrium configurations and stability of mechanical structures and electrical networks. Both physical problems fit into a common, and surprisingly general, mathematical framework. The physical laws of equilibrium mechanics and circuits lead to linear algebraic systems whose coefficient matrix is of positive (semi-)definite Gram form. The positive definite cases correspond to stable structures and networks, which can support any applied forcing or external current. The result is a unique, stable equilibrium solution that can be characterized by an energy minimization principle. On the other hand, systems with semi-definite coefficient matrices model unstable structures and networks that are unable to remain in equilibrium except under very special configurations of external forces. In the case of mechanical structures, the instabilities are of two types: rigid motions, in which the structure moves while maintaining its overall geometrical shape, and mechanisms, in which it spontaneously deforms in the absence of any applied force. In Chapter 11, we will discover that the same linear algebra framework, but now reformulated for infinite-dimensional function space, also characterizes the boundary value problems modeling the equilibria of continuous media, including bars, beams, solid bodies, and many other systems arising throughout physics and engineering.

The starting point is a linear chain of masses interconnected by springs and constrained to move only in the longitudinal direction. Our general mathematical framework is already manifest in this rather simple mechanical system. In the second section, we discuss simple electrical networks consisting of resistors, current sources and/or batteries, interconnected by a network of wires. Finally, we treat small (so as to remain in a linear modeling regime) displacements of two- and three-dimensional structures constructed out of elastic bars. In all cases, we only consider the equilibrium solutions. Dynamical (time-varying) processes for each of these physical systems are governed by systems of linear ordinary differential equations, to be formulated and analyzed in Chapter 9.

6.1 SPRINGS AND MASSES

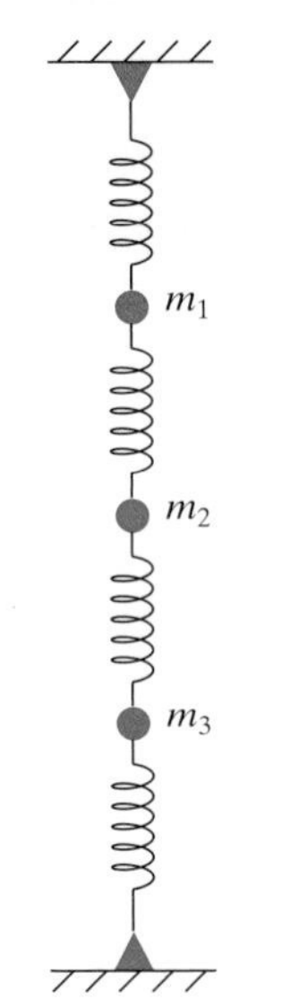

Figure 6.1 A mass–spring chain with fixed ends.

A *mass–spring chain* consists of n masses $m_1, m_2, \cdots m_n$ arranged in a straight line. Each mass is connected to its immediate neighbor(s) by springs. Moreover, the chain may be connected at one or both ends to a fixed support by a spring. At first, for specificity, let us look at the case when both ends of the chain are attached to unmoving supports, as illustrated in Figure 6.1

We assume that the masses are arranged in a vertical line, and order them from top to bottom. For simplicity, we will only allow the masses to move in the vertical direction, that is, we restrict to a one-dimensional motion. (Section 6.3 deals with the more complicated two- and three-dimensional situations.)

If we subject some or all of the masses to an external force, e.g., gravity, then the system will move* to a new equilibrium position. The resulting position of the ith mass is measured by its *displacement* u_i from its original position, which, since we are only allowing vertical motion, is a scalar quantity. Referring to Figure 6.1, we use the convention that $u_i > 0$ if the mass has moved downwards, and $u_i < 0$ if it has moved upwards. Our goal is to determine the new equilibrium configuration of the chain under the prescribed forcing, that is, to set up and solve a system of equations for the displacements $u_1, \ldots, u_n$.

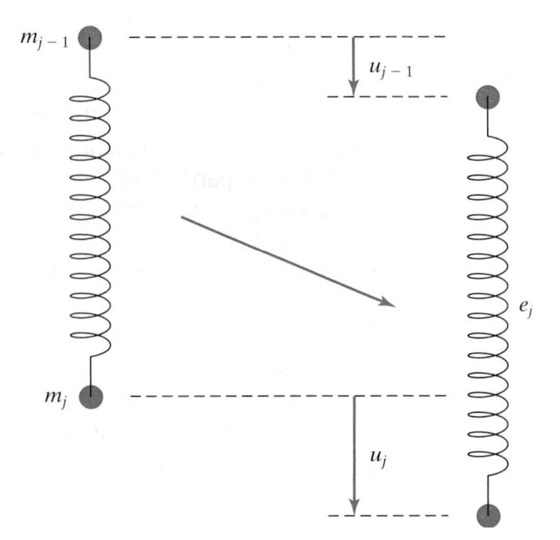

Figure 6.2 Elongation of a spring.

Let e_j denote the *elongation* of the jth spring, which connects mass m_{j-1} to mass m_j. By "elongation", we mean how far the spring has been stretched, so that $e_j > 0$ if the spring is longer than its reference length, while $e_j < 0$ if the spring has been compressed. The elongations of the internal springs can be determined directly from the displacements of the masses at each end according to the geometric formula

$$e_j = u_j - u_{j-1}, \qquad j = 2, \ldots, n, \tag{6.1}$$

while, for the top and bottom springs,

$$e_1 = u_1, \qquad e_{n+1} = -u_n, \tag{6.2}$$

*The differential equations governing its dynamical behavior during the motion will be the subject of Chapter 9. Damping or frictional effects will cause the system to eventually settle down into a stable equilibrium configuration, if such exists.

since the supports are not allowed to move. We write the elongation equations (6.1)-(6.2) in matrix form

$$\mathbf{e} = A\mathbf{u}, \tag{6.3}$$

where $\mathbf{e} = \begin{pmatrix} e_1 \\ \vdots \\ e_{n+1} \end{pmatrix}$ is the *elongation vector*, $\mathbf{u} = \begin{pmatrix} u_1 \\ \vdots \\ u_n \end{pmatrix}$ is the *displacement vector*, and the coefficient matrix

$$A = \begin{pmatrix} 1 & & & & & \\ -1 & 1 & & & & \\ & -1 & 1 & & & \\ & & -1 & 1 & & \\ & & & \ddots & \ddots & \\ & & & & -1 & 1 \\ & & & & & -1 \end{pmatrix} \tag{6.4}$$

has size $(n+1) \times n$, with only its non-zero entries being indicated. We refer to A as the *reduced incidence matrix*[†] for the mass–spring chain. The incidence matrix effectively encodes the underlying geometry of the mass–spring chain, including the "boundary conditions" at the top and the bottom.

The next step is to relate the elongation e_j experienced by the jth spring to its internal force y_j. This is the basic *constitutive assumption*, that relates geometry to kinematics. In the present case, we suppose that the springs are not stretched (or compressed) particularly far. Under this assumption, *Hooke's Law*, named in honor of the seventeenth century English scientist and inventor Robert Hooke, states that the internal force is directly proportional to the elongation —the more you stretch a spring, the more it tries to pull you back. Thus,

$$y_j = c_j\, e_j, \tag{6.5}$$

where the constant of proportionality $c_j > 0$ measures the spring's *stiffness*. Hard springs have large stiffness and so takes a large force to stretch, whereas soft springs have a small, but still positive, stiffness. We will also write the constitutive equations (6.5) in matrix form

$$\mathbf{y} = C\,\mathbf{e}, \tag{6.6}$$

where

$$\mathbf{y} = \begin{pmatrix} y_1 \\ y_2 \\ \vdots \\ y_{n+1} \end{pmatrix}, \qquad C = \begin{pmatrix} c_1 & & & \\ & c_2 & & \\ & & \ddots & \\ & & & c_{n+1} \end{pmatrix}$$

are the internal force vector and the matrix of spring stiffnesses. Note particularly that C is a diagonal matrix, and, more importantly, positive definite, since all its diagonal entries are strictly positive.

Finally, the forces must balance if the system is to remain in equilibrium. In this simplified model, the external forces act only on the masses and not on the springs. Let f_i denote the external force on the ith mass m_i. We also measure force in the downwards direction, so $f_i > 0$ means the force is pulling the ith mass downwards. (In particular, gravity would induce a positive force on each mass.) If

[†] The connection with the incidence matrix of a graph will become evident in Section 6.2.

the ith spring is stretched, it will exert an upwards force on m_i, while if the $(i+1)$st spring is stretched, it will pull m_i downwards. Therefore, the balance of forces on m_i requires that

$$f_i = y_i - y_{i+1}. \tag{6.7}$$

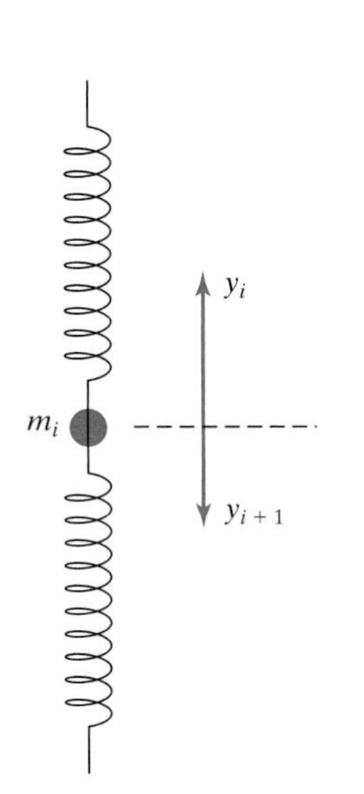

Figure 6.3 Force balance.

The vectorial form of the force balance law is

$$\mathbf{f} = A^T \mathbf{y}, \tag{6.8}$$

where $\mathbf{f} = (f_1, \ldots, f_n)^T$. The remarkable fact is that the force balance coefficient matrix

$$A^T = \begin{pmatrix} 1 & -1 & & & & \\ & 1 & -1 & & & \\ & & 1 & -1 & & \\ & & & 1 & -1 & \\ & & & & \ddots & \ddots \\ & & & & & 1 \; -1 \end{pmatrix} \tag{6.9}$$

is the *transpose* of the reduced incidence matrix (6.4) for the chain. This connection between geometry and force balance turns out to be of almost universal applicability, and is the reason underlying the positivity of the final coefficient matrix in the resulting system of equilibrium equations.

Summarizing, the basic geometrical and physical properties of our mechanical system lead us to the equilibrium equations relating its displacements $\mathbf{u}$, elongations $\mathbf{e}$, internal forces $\mathbf{y}$, and external forces $\mathbf{f}$:

$$\mathbf{e} = A\,\mathbf{u}, \qquad \mathbf{y} = C\,\mathbf{e}, \qquad \mathbf{f} = A^T \mathbf{y}. \tag{6.10}$$

These equations can be combined into a single linear system

$$K\mathbf{u} = \mathbf{f}, \qquad \text{where} \qquad K = A^T C\, A \tag{6.11}$$

is called the *stiffness matrix* associated with the entire mass–spring chain. In the particular case under consideration,

$$K = \begin{pmatrix} c_1 + c_2 & -c_2 & & & & & \\ -c_2 & c_2 + c_3 & -c_3 & & & & \\ & -c_3 & c_3 + c_4 & -c_4 & & & \\ & & -c_4 & c_4 + c_5 & -c_5 & & \\ & & & \ddots & \ddots & \ddots & \\ & & & & -c_{n-1} & c_{n-1} + c_n & -c_n \\ & & & & & -c_n & c_n + c_{n+1} \end{pmatrix} \tag{6.12}$$

has a very simple symmetric, tridiagonal form. As such, we can use the tridiagonal solution algorithm of Section 1.7 to rapidly solve the linear system (6.11) for the displacements of the masses.

EXAMPLE 6.1

Let us consider the particular case of $n = 3$ masses connected by identical springs with unit spring constant. Thus, $c_1 = c_2 = c_3 = c_4 = 1$ and $C = \operatorname{diag}(1, 1, 1, 1) = \mathrm{I}$ is the 4×4 identity matrix. The 3×3 stiffness matrix is then

$$K = A^T A = \begin{pmatrix} 1 & -1 & 0 & 0 \\ 0 & 1 & -1 & 0 \\ 0 & 0 & 1 & -1 \end{pmatrix} \begin{pmatrix} 1 & 0 & 0 \\ -1 & 1 & 0 \\ 0 & -1 & 1 \\ 0 & 0 & -1 \end{pmatrix} = \begin{pmatrix} 2 & -1 & 0 \\ -1 & 2 & -1 \\ 0 & -1 & 2 \end{pmatrix}.$$

A straightforward Gaussian Elimination produces the $K = L\,D\,L^T$ factorization

$$\begin{pmatrix} 2 & -1 & 0 \\ -1 & 2 & -1 \\ 0 & -1 & 2 \end{pmatrix} = \begin{pmatrix} 1 & 0 & 0 \\ -\frac{1}{2} & 1 & 0 \\ 0 & -\frac{2}{3} & 1 \end{pmatrix} \begin{pmatrix} 2 & 0 & 0 \\ 0 & \frac{3}{2} & 0 \\ 0 & 0 & \frac{4}{3} \end{pmatrix} \begin{pmatrix} 1 & -\frac{1}{2} & 0 \\ 0 & 1 & -\frac{2}{3} \\ 0 & 0 & 1 \end{pmatrix}. \tag{6.13}$$

With this in hand, we can solve the basic equilibrium equations $K\mathbf{u} = \mathbf{f}$ by the usual Forward and Back Substitution algorithm. ●

REMARK: Even though we construct $K = A^T C\,A$ and then factor it as $K = L\,D\,L^T$, there is no direct algorithm to get from A and C to L and D, which, typically, are matrices of different sizes.

Suppose, for example, we pull the middle mass downwards with a unit force, so $f_2 = 1$ while $f_1 = f_3 = 0$. Then $\mathbf{f} = (0, 1, 0)^T$, and the solution to the equilibrium equations (6.11) is $\mathbf{u} = \left(\frac{1}{2}, 1, \frac{1}{2}\right)^T$, whose entries prescribe the mass displacements. Observe that all three masses have moved down, with the middle mass moving twice as far as the other two. The corresponding spring elongations and internal forces are obtained by matrix multiplication

$$\mathbf{y} = \mathbf{e} = A\mathbf{u} = \left(\tfrac{1}{2}, \tfrac{1}{2}, -\tfrac{1}{2}, -\tfrac{1}{2}\right)^T.$$

Thus, the top two springs are elongated, while the bottom two are compressed, all by an equal amount.

Similarly, if all the masses are equal, $m_1 = m_2 = m_3 = m$, then the solution under a constant downwards gravitational force $\mathbf{f} = (m\,g, m\,g, m\,g)^T$ of magnitude g is

$$\mathbf{u} = K^{-1}\begin{pmatrix} m\,g \\ m\,g \\ m\,g \end{pmatrix} = \begin{pmatrix} \frac{3}{2}m\,g \\ 2m\,g \\ \frac{3}{2}m\,g \end{pmatrix},$$

and

$$\mathbf{y} = \mathbf{e} = A\mathbf{u} = \left(\tfrac{3}{2}m\,g, \tfrac{1}{2}m\,g, -\tfrac{1}{2}m\,g, -\tfrac{3}{2}m\,g\right)^T.$$

Now, the middle mass has only moved 33% farther than the others, whereas the top and bottom spring are experiencing three times as much elongation/compression as the middle two springs.

An important observation is that we *cannot* determine the internal forces $\mathbf{y}$ or elongations $\mathbf{e}$ directly from the force balance law (6.8) because the transposed matrix A^T is not square, and so the system $\mathbf{f} = A^T\mathbf{y}$ does not have a unique solution. We must first compute the displacements $\mathbf{u}$ by solving the full equilibrium equations (6.11), and then use the resulting displacements to reconstruct the elongations and internal forces. Such systems are referred to as *statically indeterminate*.

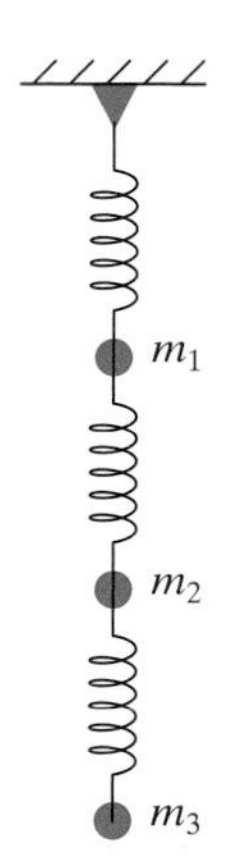

Figure 6.4 A mass–spring chain with one free end.

The behavior of the system will depend upon both the forcing and the boundary conditions. Suppose, by way of contrast, that we only fix the top of the chain to a support, and leave the bottom mass hanging freely, as in Figure 6.4. The geometric relation between the displacements and the elongations has the same form (6.3) as before, but the reduced incidence matrix is slightly altered:

$$A = \begin{pmatrix} 1 & & & & & \\ -1 & 1 & & & & \\ & -1 & 1 & & & \\ & & -1 & 1 & & \\ & & & \ddots & \ddots & \\ & & & & -1 & 1 \end{pmatrix}. \tag{6.14}$$

This matrix has size $n \times n$ and is obtained from the preceding example (6.4) by eliminating the last row corresponding to the missing bottom spring. The constitutive equations are still governed by Hooke's law $\mathbf{y} = C\mathbf{e}$, as in (6.6), with $C = \operatorname{diag}(c_1, \ldots, c_n)$ the $n \times n$ diagonal matrix of spring stiffnesses. Finally, the force balance equations are also found to have the same general form $\mathbf{f} = A^T\mathbf{y}$ as in (6.8), but with the transpose of the revised incidence matrix (6.14). In conclusion, the equilibrium equations $K\mathbf{x} = \mathbf{f}$ have an identical form (6.11), based on the revised stiffness matrix

$$K = A^T C A = \begin{pmatrix} c_1+c_2 & -c_2 & & & & & \\ -c_2 & c_2+c_3 & -c_3 & & & & \\ & -c_3 & c_3+c_4 & -c_4 & & & \\ & & -c_4 & c_4+c_5 & -c_5 & & \\ & & & \ddots & \ddots & \ddots & \\ & & & & -c_{n-1} & c_{n-1}+c_n & -c_n \\ & & & & & -c_n & c_n \end{pmatrix} \tag{6.15}$$

Only the bottom right entry differs from the fixed end matrix (6.12).

This system is called *statically determinate* because the incidence matrix A is square and nonsingular, and so it is possible to solve the force balance law (6.8) directly for the internal forces $\mathbf{y} = A^{-1}\mathbf{f}$ without having to solve the full equilibrium equations for the displacements $\mathbf{u}$ before computing the internal forces $\mathbf{y} = C\,A\,\mathbf{u}$.

EXAMPLE 6.2 For a three mass chain with one free end and equal unit spring constants $c_1 = c_2 = c_3 = 1$, the stiffness matrix is

$$K = A^T A = \begin{pmatrix} 1 & -1 & 0 \\ 0 & 1 & -1 \\ 0 & 0 & 1 \end{pmatrix} \begin{pmatrix} 1 & 0 & 0 \\ -1 & 1 & 0 \\ 0 & -1 & 1 \end{pmatrix} = \begin{pmatrix} 2 & -1 & 0 \\ -1 & 2 & -1 \\ 0 & -1 & 1 \end{pmatrix}.$$

Pulling the middle mass downwards with a unit force, whereby $\mathbf{f} = (0, 1, 0)^T$, results in the displacements

$$\mathbf{u} = K^{-1}\mathbf{f} = \begin{pmatrix} 1 \\ 2 \\ 2 \end{pmatrix}, \qquad \text{so that} \quad \mathbf{y} = \mathbf{e} = A\mathbf{u} = \begin{pmatrix} 1 \\ 1 \\ 0 \end{pmatrix}.$$

In this configuration, the bottom two masses have moved by the same amount, which is twice as far as the top mass. Because we are only pulling on the middle mass, the bottom spring hangs free and experiences no elongation, whereas the top two springs are stretched by the same amount.

Similarly, for a chain of equal masses subject to a constant downwards gravitational force $\mathbf{f} = (m\,g, m\,g, m\,g)^T$, the equilibrium position is

$$\mathbf{u} = K^{-1}\begin{pmatrix} m\,g \\ m\,g \\ m\,g \end{pmatrix} = \begin{pmatrix} 3m\,g \\ 5m\,g \\ 6m\,g \end{pmatrix}, \qquad \text{and} \quad \mathbf{y} = \mathbf{e} = A\mathbf{u} = \begin{pmatrix} 3m\,g \\ 2m\,g \\ m\,g \end{pmatrix}.$$

Note how much farther the masses have moved now that the restraining influence of the bottom support has been removed. The top spring is experiencing the most elongation, and is thus the most likely to break, because it must support all three masses. ●

EXERCISES 6.1

6.1.1. A mass–spring chain consists of two masses connected to two fixed supports. The spring constants are $c_1 = c_3 = 1$ and $c_2 = 2$.

(a) Find the stiffness matrix K.

(b) Solve the equilibrium equations $K\mathbf{u} = \mathbf{f}$ when $\mathbf{f} = (4, 3)^T$.

(c) Which mass moved the farthest?

(d) Which spring has been stretched the most? Compressed the most?

6.1.2. Solve Exercise 6.1.1 when the first and second spring are interchanged, $c_1 = 2$, $c_2 = c_3 = 1$. Which of your conclusions changed?

6.1.3. Redo Exercises 6.1.1–2 when the bottom support and spring are removed.

6.1.4. A mass–spring chain consists of four masses suspended between two fixed supports. The spring stiffnesses are $c_1 = 1, c_2 = \frac{1}{2}, c_3 = \frac{2}{3}, c_4 = \frac{1}{2}, c_5 = 1$.

(a) Determine the equilibrium positions of the masses and the elongations of the springs when the external force is $\mathbf{f} = (0, 1, 1, 0)^T$. Is your solution unique?

(b) Suppose we only fix the top support. Solve the problem with the same data and compare your results.

6.1.5. (a) Show that, in a mass-spring chain with two fixed ends, under any external force, the average elongation of the springs is zero:

$$\frac{1}{n+1}(e_1 + \cdots + e_{n+1}) = 0.$$

(b) What can you say about the average elongation of the springs in a chain with one fixed end?

◇ **6.1.6.** Suppose we subject the ith mass (and no others) in a chain to a unit force, and then measure the resulting displacement of the jth mass. Prove that this is the *same* as the displacement of the ith mass when the chain is subject to a unit force on the jth mass. *Hint*: See Exercise 1.6.20.

♣ **6.1.7.** Find the displacements $u_1, u_2, \ldots, u_{100}$ of 100 masses connected in a row by identical springs, with spring constant $c = 1$. Consider the following three types of force functions:

(a) Constant force: $f_1 = \cdots = f_{100} = .01$;

(b) Linear force: $f_i = .0002\, i$;

(c) Quadratic force: $f_i = 6 \cdot 10^{-6}\, i\,(100 - i)$.

Also consider two different boundary conditions at the bottom:

(i) Spring 101 connects the last mass to a support.

(ii) Mass 100 hangs free at the end of the line of springs.

Graph the displacements and elongations in all six cases. Discuss your results; in particular, comment on whether they agree with your physical intuition.

6.1.8. (a) Suppose you are given three springs with respective stiffnesses $c = 1, c' = 2, c'' = 3$. In what order should you connect them to three masses and a top support so that the bottom mass goes down the farthest under a uniform gravitational force?

(b) Answer Exercise 6.1.8 when the springs connect two masses to both top and bottom supports.

♣ **6.1.9.** Generalizing Exercise 6.1.8, suppose you are given n different springs.

(a) In which order should you connect them to n masses and a top support so that the bottom mass goes down the farthest under a uniform gravitational force? Does your answer depend upon the relative sizes of the spring constants?

(b) Answer the same question when the springs connect $n - 1$ masses to both top and bottom supports.

6.1.10. Find the $L D L^T$ factorization of an $n \times n$ tridiagonal matrix whose diagonal entries are all equal to 2 and whose sub- and super-diagonal entries are all equal to -1. *Hint*: Start with the 3×3 case (6.13), and then analyze a slightly larger one to spot the pattern.

♡ **6.1.11.** In a statically indeterminate situation, the equations $A^T\mathbf{y} = \mathbf{f}$ do not have a unique solution for the internal forces $\mathbf{y}$ in terms of the external forces $\mathbf{f}$.

(a) Prove that, nevertheless, if $C = \mathrm{I}$, the internal forces are the *unique* solution of minimal Euclidean norm, as given by Theorem 5.59.

(b) Use this method to directly find the internal force for the system in Example 6.1. Make sure that your values agree with those in the example.

Positive Definiteness and the Minimization Principle

You may have already observed that the stiffness matrix $K = A^T C A$ of a mass–spring chain has the form of a Gram matrix, cf. (3.56), for the weighted inner product $\langle \mathbf{v}, \mathbf{w} \rangle = \mathbf{v}^T C\, \mathbf{w}$ induced by the diagonal matrix of spring stiffnesses. Moreover, since A has linearly independent columns (which should be checked), and C is positive definite, Theorem 3.31 tells us that the stiffness matrix is positive definite: $K > 0$. In particular, Theorem 3.37 guarantees that K is nonsingular, and hence the linear system (6.11) has a unique solution $\mathbf{u} = K^{-1}\mathbf{f}$. We can therefore conclude that the mass–spring chain assumes a unique equilibrium position under an arbitrary external force. However, one must keep in mind that this is a mathematical result and may not hold in all physical situations. Indeed, we should anticipate that a very large force will take us outside the regime covered by the linear Hooke's law relation (6.5), and render our simple mathematical model physically irrelevant.

According to Theorem 4.1, when the coefficient matrix of a linear system is positive definite, the equilibrium solution can be characterized by a minimization principle. For mass-spring chains, the quadratic function to be minimized has a physical interpretation: it is the potential energy of the system. Nature is parsimonious with her energy, so a physical system seeks out an energy-minimizing equilibrium configuration. Energy minimization principles are of almost universal validity, and can often be advantageously used for the construction of mathematical models, as well as their solutions, both analytical and numerical.

The energy function to be minimized can be determined directly from physical principles. For a mass–spring chain, the potential energy of the ith mass equals the product of the applied force times the displacement: $-f_i\,u_i$. The minus sign is the result of our convention that a positive displacement $u_i > 0$ means that the mass has moved down, and hence decreased its potential energy. Thus, the total potential energy due to external forcing on all the masses in the chain is

$$-\sum_{i=1}^{n} f_i\,u_i = -\mathbf{u}^T\mathbf{f}.$$

Next, we calculate the internal energy of the system. In a single spring elongated by an amount e, the work done by the internal forces $y = c\,e$ is stored as potential energy, and so is calculated by integrating the force over the elongated distance:

$$\int_0^e y\,de = \int_0^e c\,e\,de = \tfrac{1}{2}\,c\,e^2.$$

Totalling the contributions from each spring, we find the internal spring energy to be

$$\frac{1}{2}\sum_{i=1}^{n} c_i\,e_i^2 = \tfrac{1}{2}\,\mathbf{e}^T C\,\mathbf{e} = \tfrac{1}{2}\,\mathbf{u}^T A^T C A\,\mathbf{u} = \tfrac{1}{2}\,\mathbf{u}^T K\,\mathbf{u},$$

where we used the incidence equation $\mathbf{e} = A\,\mathbf{u}$ relating elongation and displacement. Therefore, the total potential energy is

$$p(\mathbf{u}) = \tfrac{1}{2}\,\mathbf{u}^T K\,\mathbf{u} - \mathbf{u}^T\mathbf{f}. \tag{6.16}$$

Since $K > 0$, Theorem 4.1 implies that this quadratic function has a unique minimizer that satisfies the equilibrium equation $K\,\mathbf{u} = \mathbf{f}$.

EXAMPLE 6.3 For the three mass chain with two fixed ends described in Example 6.1, the potential energy function (6.16) has the explicit form

$$p(\mathbf{u}) = \frac{1}{2}\,(\,u_1 \quad u_2 \quad u_3\,)\begin{pmatrix} 2 & -1 & 0 \\ -1 & 2 & -1 \\ 0 & -1 & 2 \end{pmatrix}\begin{pmatrix} u_1 \\ u_2 \\ u_3 \end{pmatrix} - (\,u_1 \quad u_2 \quad u_3\,)\begin{pmatrix} f_1 \\ f_2 \\ f_3 \end{pmatrix}$$
$$= u_1^2 - u_1\,u_2 + u_2^2 - u_2\,u_3 + u_3^2 - f_1\,u_1 - f_2\,u_2 - f_3\,u_3,$$

where $\mathbf{f} = (f_1, f_2, f_3)^T$ is the external forcing. The minimizer of this particular quadratic function gives the equilibrium displacements $\mathbf{u} = (u_1, u_2, u_3)^T$ of the three masses. ●

EXERCISES 6.1

6.1.12. Prove directly that the stiffness matrices in Examples 6.1 and 6.2 are positive definite.

6.1.13. Write down the potential energy for the following mass-spring chains with identical unit springs when subject to a uniform gravitational force:

(a) three identical masses connected to only a top support.

(b) four identical masses connected to top and bottom supports.

(c) four identical masses connected only to a top support.

6.1.14. (a) Find the total potential energy of the equilibrium configuration of the mass–spring chain in Exercise 6.1.1.

(b) Test the minimum principle by substituting three other possible displacements of the masses and checking that they all have larger potential energy.

6.1.15. Answer Exercise 6.1.14 for the mass–spring chain in Exercise 6.1.4.

6.1.16. Describe the mass-spring chains that gives rise to the following potential energy functions, and find their equilibrium configuration:

(a) $3u_1^2 - 4u_1u_2 + 3u_2^2 + u_1 - 3u_2$

(b) $5u_1^2 - 6u_1u_2 + 3u_2^2 + 2u_2$

(c) $2u_1^2 - 3u_1u_2 + 4u_2^2 - 5u_2u_3 + \frac{5}{2}u_3^2 - u_1 - u_2 + u_3$

(d) $2u_1^2 - u_1u_2 + u_2^2 - u_2u_3 + u_3^2 - u_3u_4 + 2u_4^2 + u_1 - 2u_3$

6.1.17. Explain why the columns of the reduced incidence matrices (6.4) and (6.14) are linearly independent.

6.1.18. Suppose that, when subject to a nonzero external force $\mathbf{f} \neq \mathbf{0}$, a mass-spring chain has equilibrium position $\mathbf{u}^\star$. Prove that the potential energy is strictly negative at equilibrium: $p(\mathbf{u}^\star) < 0$.

♡ **6.1.19.** Return to the situation investigated in Exercise 6.1.8. How should you arrange the springs in order to minimize the potential energy in the resulting mass–spring chain?

6.1.20. *True or false*: The potential energy function uniquely determines the mass/spring chain.

6.2 ELECTRICAL NETWORKS

By an electrical *network*, we mean a collection of (insulated) wires that are joined together at their ends. The junctions connecting the ends of one or more wires are called *nodes*. Mathematically, we can view any such electrical network as a graph, the wires being the edges and the nodes the vertices. To begin with, we assume that there are no electrical devices (batteries, inductors, capacitors, etc.) in the network and so the only impediments to the current flowing through the network are the resistances in the wires. As we shall see, resistance (or, rather, its reciprocal) plays a very similar role to spring stiffness. We shall feed a current into the network at one or more of the nodes, and would like to determine how the induced current flows through the wires. The basic equations governing the equilibrium voltages and currents in such a network follow from the three fundamental laws of electricity, named after the pioneering nineteenth century German physicists Gustav Kirchhoff and Georg Ohm., two of the founders of electric circuit theory.

Voltage is defined as the electromotive force that moves electrons through a wire. An individual wire's voltage is determined by the difference in the voltage potentials at its two ends—just as the gravitational force on a mass is induced by a difference in gravitational potential. To quantify voltage, we need to fix an orientation for the wire. A positive voltage will mean that the electrons move in the chosen direction, while under a negative voltage they move in reverse. The original choice of orientation is arbitrary, but once assigned will pin down the sign conventions to be used by voltages, currents, etc. To this end, we draw a digraph to represent the network, whose edges represent wires and whose vertices represent nodes. Each edge is assigned an orientation that indicates the wire's starting and ending nodes. A simple example consisting of five wires joined at four different nodes can be seen in Figure 6.5. The arrows indicate the selected directions for the wires, while the wavy lines are the standard electrical symbols for resistance.

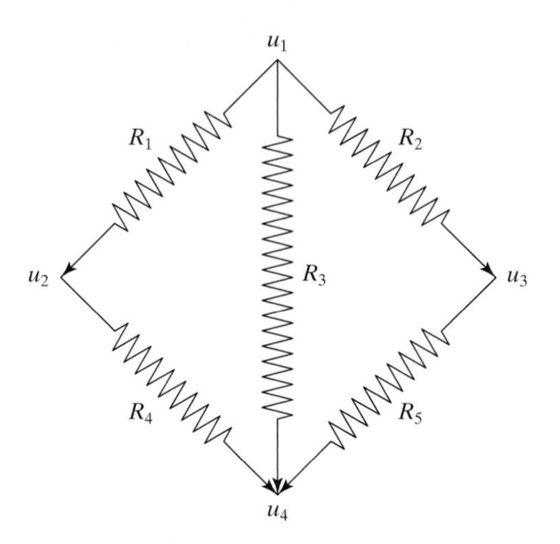

Figure 6.5 A simple electrical network.

In an electrical network, each node will have a voltage potential, denoted u_i. If wire k starts at node i and ends at node j under its assigned orientation, then its voltage v_k equals the difference between the voltage potentials at its ends:

$$v_k = u_i - u_j. \tag{6.17}$$

Note that $v_k > 0$ if $u_i > u_j$, indicating that the electrons flow from the starting node i to the ending node j. In our particular illustrative example, the five wires have respective voltages

$$v_1 = u_1 - u_2, \quad v_2 = u_1 - u_3, \quad v_3 = u_1 - u_4, \quad v_4 = u_2 - u_4, \quad v_5 = u_3 - u_4.$$

Let us rewrite this linear system of equations in vector form

$$\mathbf{v} = A\mathbf{u}, \tag{6.18}$$

where

$$A = \begin{pmatrix} 1 & -1 & 0 & 0 \\ 1 & 0 & -1 & 0 \\ 1 & 0 & 0 & -1 \\ 0 & 1 & 0 & -1 \\ 0 & 0 & 1 & -1 \end{pmatrix}. \tag{6.19}$$

The alert reader will recognize the *incidence matrix* (2.44) for the digraph defined by the network. This is true in general—*the voltages along the wires of an electrical network are related to the potentials at the nodes by a linear system of the form* (6.18), *in which A is the incidence matrix of the network digraph*. The rows of the incidence matrix are indexed by the wires, and the columns by the nodes. Each row of the matrix A has a single $+1$ in the column indexed by the starting node, and a single -1 in the column of the ending node.

Kirchhoff's Voltage Law states that the sum of the voltages around each closed circuit in the network is zero. For example, in the network under consideration, summing the voltages around the left-hand triangular circuit gives

$$v_1 + v_4 - v_3 = (u_1 - u_2) + (u_2 - u_4) - (u_1 - u_4) = 0.$$

Note that v_3 appears with a minus sign since we must traverse wire 3 in the opposite direction to its assigned orientation when going around the circuit in the counterclockwise direction. The voltage law is a direct consequence of (6.18). Indeed, as discussed in Section 2.6, the circuits can be identified with vectors $\ell \in \operatorname{coker} A = \ker A^T$ in the cokernel of the incidence matrix, and so

$$\ell \cdot \mathbf{v} = \ell^T \mathbf{v} = \ell^T A \mathbf{u} = 0. \tag{6.20}$$

Therefore, orthogonality of the voltage vector $\mathbf{v}$ to the circuit vector ℓ is the mathematical formalization of Kirchhoff's Voltage Law.

Given a prescribed set of voltages $\mathbf{v}$ along the wires, can one find corresponding voltage potentials $\mathbf{u}$ at the nodes? To answer this question, we need to solve $\mathbf{v} = A\mathbf{u}$, which requires $\mathbf{v} \in \operatorname{rng} A$. According to the Fredholm Alternative Theorem 5.55, the necessary and sufficient condition for this to hold is that $\mathbf{v}$ be orthogonal to $\operatorname{coker} A$. Theorem 2.53 says that the cokernel of an incidence matrix is spanned by the circuit vectors, and so $\mathbf{v}$ is a possible set of voltages if and only if $\mathbf{v}$ is orthogonal to all the circuit vectors $\ell \in \operatorname{coker} A$, i.e., the Voltage Law is necessary and sufficient for the given voltages to be physically realizable in the network.

Kirchhoff's Law is related to the topology of the network—how the different wires are connected together. *Ohm's Law* is a constitutive relation, indicating what the wires are made of. The resistance along a wire (including any added resistors) prescribes the relation between voltage and current or the rate of flow of electric charge. The law reads

$$v_k = R_k \, y_k, \tag{6.21}$$

where v_k is the voltage, R_k is the resistance, and y_k (often denoted I_k in the engineering literature) denotes the current along wire k. Thus, for a fixed voltage, the larger the resistance of the wire, the smaller the current that flows through it. The direction of the current is also prescribed by our choice of orientation of the wire, so that $y_k > 0$ if the current is flowing from the starting to the ending node. We combine the individual equations (6.21) into a single vector equation

$$\mathbf{v} = R\, \mathbf{y}, \tag{6.22}$$

where the *resistance matrix* $R = \operatorname{diag}(R_1, \dots, R_n) > 0$ is diagonal and positive definite. We shall, in analogy with (6.6), replace (6.22) by the inverse relationship

$$\mathbf{y} = C\, \mathbf{v}, \tag{6.23}$$

where $C = R^{-1}$ is the *conductance matrix*, again diagonal, positive definite, whose entries are the *conductances* $c_k = 1/R_k$ of the wires. For the particular network in Figure 6.5,

$$C = \begin{pmatrix} c_1 & 0 & 0 & 0 & 0 \\ 0 & c_2 & 0 & 0 & 0 \\ 0 & 0 & c_3 & 0 & 0 \\ 0 & 0 & 0 & c_4 & 0 \\ 0 & 0 & 0 & 0 & c_5 \end{pmatrix} = \begin{pmatrix} 1/R_1 & 0 & 0 & 0 & 0 \\ 0 & 1/R_2 & 0 & 0 & 0 \\ 0 & 0 & 1/R_3 & 0 & 0 \\ 0 & 0 & 0 & 1/R_4 & 0 \\ 0 & 0 & 0 & 0 & 1/R_5 \end{pmatrix}. \tag{6.24}$$

Finally, we stipulate that electric current is not allowed to accumulate at any node, i.e., every electron which arrives at a node must leave along one of the wires.

Let $y_k, y_l, \ldots, y_m$ denote the currents along all the wires $k, l, \ldots, m$ that meet at node i in the network, and f_i an external current source, if any, applied at node i. *Kirchhoff's Current Law* requires that the net current leaving the node along the wires equals the external current coming into the node, and so

$$\pm\, y_k \,\pm\, y_l \,\pm\, \cdots \,\pm\, y_m = f_i. \tag{6.25}$$

Each $\pm$ sign is determined by the orientation of the wire, with $+$ if node i is its starting node or $-$ if it is its ending node.

In our particular example, suppose that we send a 1 amp current source into the first node. Then Kirchhoff's Current Law requires

$$y_1 + y_2 + y_3 = 1, \qquad -y_1 + y_4 = 0, \qquad -y_2 + y_5 = 0, \qquad -y_3 - y_4 - y_5 = 0,$$

the four equations corresponding to the four nodes in our network. The vector form of this linear system is

$$A^T \mathbf{y} = \mathbf{f}, \tag{6.26}$$

where $\mathbf{y} = (y_1, y_2, y_3, y_4, y_5)^T$ are the currents along the five wires, and $\mathbf{f} = (1, 0, 0, 0)^T$ represents the current sources at the four nodes. The coefficient matrix

$$A^T = \begin{pmatrix} 1 & 1 & 1 & 0 & 0 \\ -1 & 0 & 0 & 1 & 0 \\ 0 & -1 & 0 & 0 & 1 \\ 0 & 0 & -1 & -1 & -1 \end{pmatrix} \tag{6.27}$$

is the *transpose* of the incidence matrix (6.19). As in the mass–spring chain, this is a remarkable general fact, and follows directly from Kirchhoff's two laws. *The coefficient matrix for the Current Law is the transpose of the incidence matrix for the Voltage Law.*

Let us assemble the full system of equilibrium equations:

$$\mathbf{v} = A\,\mathbf{u}, \qquad \mathbf{y} = C\,\mathbf{v}, \qquad \mathbf{f} = A^T \mathbf{y}. \tag{6.28}$$

Remarkably, we arrive at a system of linear relations that has an identical form to the mass–spring chain system (6.10). As before, they combine into a single linear system

$$K\,\mathbf{u} = \mathbf{f}, \qquad \text{where} \qquad K = A^T C\, A \tag{6.29}$$

is known as the *resistivity matrix* associated with the network. In our particular example, combining (6.19), (6.24), (6.27) produces the resistivity matrix

$$K = A^T C\, A = \begin{pmatrix} c_1 + c_2 + c_3 & -c_1 & -c_2 & -c_3 \\ -c_1 & c_1 + c_4 & 0 & -c_4 \\ -c_2 & 0 & c_2 + c_5 & -c_5 \\ -c_3 & -c_4 & -c_5 & c_3 + c_4 + c_5 \end{pmatrix} \tag{6.30}$$

whose entries depend on the conductances of the five wires in the network.

REMARK: There is a simple pattern to the resistivity matrix, evident in (6.30). The diagonal entries k_{ii} equal the sum of the conductances of all the wires having node i at one end. The non-zero off-diagonal entries k_{ij}, $i \neq j$, equal $-c_k$, the conductance of the wire* joining node i to node j, while $k_{ij} = 0$ if there is no wire joining the two nodes.

*This assumes that there is only one wire joining the two nodes.

Consider the case when all the wires in our network have equal unit resistance, and so $c_k = 1/R_k = 1$ for $k = 1, \ldots, 5$. Then the resistivity matrix is

$$K = \begin{pmatrix} 3 & -1 & -1 & -1 \\ -1 & 2 & 0 & -1 \\ -1 & 0 & 2 & -1 \\ -1 & -1 & -1 & 3 \end{pmatrix}. \qquad (6.31)$$

However, trying to solve the linear system (6.29), we run into an immediate difficulty: *there is no solution*! The matrix (6.31) is *not* positive definite—it is a singular matrix. Moreover, the particular current source vector $\mathbf{f} = (1, 0, 0, 0)^T$ does not lie in the range of K. Something is clearly amiss.

Before getting discouraged, let us sit back and use a little physical intuition. We are trying to put a 1 amp current into the network at node 1. Where can the electrons go? The answer is nowhere—they are trapped in the network and, as they accumulate, something drastic will happen—sparks will fly! This is clearly an unstable situation, and so the fact that the equilibrium equations do not have a solution is trying to tell us that the physical system cannot remain in a steady state. The physics rescues the mathematics, or, vice versa, the mathematics elucidates the underlying physical processes.

In order to achieve equilibrium in an electrical network, we must remove as much current as we put in. Thus, if we feed a 1 amp current into node 1, then we must extract a total of 1 amp's worth of current from the other nodes. In other words, the sum of all the external current sources must vanish:

$$f_1 + f_2 + \cdots + f_n = 0,$$

and so there is no net current being fed into the network. Suppose we also extract a 1 amp current from node 4; then the modified current source vector $\mathbf{f} = (1, 0, 0, -1)^T$ does indeed lie in the range of K, as you can check, and the equilibrium system (6.29) has a solution. Fine ...

But we are not out of the woods yet. As we know, if a linear system has a singular coefficient matrix, then either it has no solutions—the case we already rejected — or it has infinitely many solutions—the case we are considering now. In the particular network under consideration, the general solution to the linear system

$$\begin{pmatrix} 3 & -1 & -1 & -1 \\ -1 & 2 & 0 & -1 \\ -1 & 0 & 2 & -1 \\ -1 & -1 & -1 & 3 \end{pmatrix} \begin{pmatrix} u_1 \\ u_2 \\ u_3 \\ u_4 \end{pmatrix} = \begin{pmatrix} 1 \\ 0 \\ 0 \\ -1 \end{pmatrix}$$

is found by Gaussian Elimination:

$$\mathbf{u} = \begin{pmatrix} \frac{1}{2} + t \\ \frac{1}{4} + t \\ \frac{1}{4} + t \\ t \end{pmatrix} = \begin{pmatrix} \frac{1}{2} \\ \frac{1}{4} \\ \frac{1}{4} \\ 0 \end{pmatrix} + t \begin{pmatrix} 1 \\ 1 \\ 1 \\ 1 \end{pmatrix}, \qquad (6.32)$$

where $t = u_4$ is the free variable. The resulting nodal voltage potentials

$$u_1 = \tfrac{1}{2} + t, \quad u_2 = \tfrac{1}{4} + t, \quad u_3 = \tfrac{1}{4} + t, \quad u_4 = t,$$

depend on a free parameter t.

The ambiguity arises because voltage potential is a mathematical abstraction that cannot be measured directly; only relative potential differences have physical import. To resolve the inherent ambiguity, we need to assign a baseline value for the voltage potentials. In terrestrial electricity, the Earth is assumed to have zero potential. Specifying a particular node to have zero potential is physically equivalent to grounding that node. For our example, suppose we ground node 4 by setting $u_4 = 0$. This fixes the free variable $t = 0$ in our solution (6.32), and so uniquely specifies all the other voltage potentials: $u_1 = \frac{1}{2}$, $u_2 = \frac{1}{4}$, $u_3 = \frac{1}{4}$, $u_4 = 0$.

On the other hand, even without specification of a baseline potential level, the corresponding physical voltages and currents along the wires are uniquely specified. In our example, computing $\mathbf{y} = \mathbf{v} = A\,\mathbf{u}$ gives

$$y_1 = v_1 = \tfrac{1}{4}, \quad y_2 = v_2 = \tfrac{1}{4}, \quad y_3 = v_3 = \tfrac{1}{2}, \quad y_4 = v_4 = \tfrac{1}{4}, \quad y_5 = v_5 = \tfrac{1}{4},$$

independent of the value of t in (6.32). Thus, the nonuniqueness of the voltage potential solution $\mathbf{u}$ is not an essential difficulty. All physical quantities that we can measure—currents and voltages—*are* uniquely specified by the solution to the equilibrium system.

REMARK: Although they have no real physical meaning, we *cannot* dispense with the nonmeasurable (and nonunique) voltage potentials $\mathbf{u}$. Most networks are *statically indeterminate*, since their incidence matrix is rectangular and not invertible, so the linear system $A^T\mathbf{y} = \mathbf{f}$ cannot be solved directly for the currents in terms of the voltage sources since the system does not have a unique solution. Only by first solving the full equilibrium system (6.29) for the potentials, and then using the relation $\mathbf{y} = C\,A\,\mathbf{u}$ between the potentials and the currents, can we determine their actual values.

Let us analyze what is going on in the context of our general mathematical framework. Proposition 3.30 says that the resistivity matrix $K = A^T C A$ is a positive semi-definite Gram matrix, which is positive definite (and hence nonsingular) if and only if A has linearly independent columns, or, equivalently, $\ker A = \{\mathbf{0}\}$. But Proposition 2.51 says that the incidence matrix A of a directed graph *never* has a trivial kernel. Therefore, the resistivity matrix K is only positive semi-definite, and hence singular. If the network is connected, then $\ker A = \ker K = \operatorname{coker} K$ is one-dimensional, spanned by the vector $\mathbf{z} = (1, 1, 1, \ldots, 1)^T$. According to the Fredholm Alternative Theorem 5.55, the fundamental network equation $K\mathbf{u} = \mathbf{f}$ has a solution if and only if $\mathbf{f}$ is orthogonal to $\operatorname{coker} K$, and so the current source vector must satisfy

$$\mathbf{z} \cdot \mathbf{f} = f_1 + f_2 + \cdots + f_n = 0, \tag{6.33}$$

as we already observed. Therefore, the linear algebra reconfirms our physical intuition: a connected network admits an equilibrium configuration, obtained by solving (6.29), if and only if the nodal current sources add up to zero, i.e., there is no net influx of current into the network.

Grounding one of the nodes is equivalent to nullifying the value of its voltage potential: $u_i = 0$. This variable is now fixed, and can be safely eliminated from our system. To accomplish this, we let $A^\star$ denote the $m \times (n-1)$ matrix obtained by deleting the ith column from A. For example, grounding node 4 in our sample network, so $u_4 = 0$, allows us to erase the fourth column of the incidence matrix (6.19), leading to the *reduced incidence matrix*

$$A^\star = \begin{pmatrix} 1 & -1 & 0 \\ 1 & 0 & -1 \\ 1 & 0 & 0 \\ 0 & 1 & 0 \\ 0 & 0 & 1 \end{pmatrix}. \tag{6.34}$$

The key observation is that $A^\star$ has trivial kernel, $\ker A^\star = \{\mathbf{0}\}$, and therefore the reduced network resistivity matrix

$$K^\star = (A^\star)^T C\, A^\star = \begin{pmatrix} c_1 + c_2 + c_3 & -c_1 & -c_2 \\ -c_1 & c_1 + c_4 & 0 \\ -c_2 & 0 & c_2 + c_5 \end{pmatrix} \tag{6.35}$$

is positive definite. Note that we can obtain $K^\star$ directly from K in (6.30) by deleting both its fourth row and fourth column. Let $\mathbf{f}^\star = (1, 0, 0)^T$ denote the reduced current source vector obtained by deleting the fourth entry from $\mathbf{f}$. Then the reduced linear system is

$$K^\star \mathbf{u}^\star = \mathbf{f}^\star, \tag{6.36}$$

where $\mathbf{u}^\star = (u_1, u_2, u_3)^T$ is the reduced voltage potential vector. Positive definiteness of $K^\star$ implies that (6.36) has a unique solution $\mathbf{u}^\star$, from which we can reconstruct the voltages $\mathbf{v} = A^\star \mathbf{u}^\star$ and currents $\mathbf{y} = C\,\mathbf{v} = C A^\star \mathbf{u}^\star$ along the wires. In our example, if all the wires have unit resistance, then the reduced system (6.36) is

$$\begin{pmatrix} 3 & -1 & -1 \\ -1 & 2 & 0 \\ -1 & 0 & 2 \end{pmatrix} \begin{pmatrix} u_1 \\ u_2 \\ u_3 \end{pmatrix} = \begin{pmatrix} 1 \\ 0 \\ 0 \end{pmatrix},$$

and has unique solution $\mathbf{u}^\star = \left(\frac{1}{2}, \frac{1}{4}, \frac{1}{4}\right)^T$. The voltage potentials are

$$u_1 = \tfrac{1}{2}, \qquad u_2 = \tfrac{1}{4}, \qquad u_3 = \tfrac{1}{4}, \qquad u_4 = 0,$$

and correspond to the earlier solution (6.32) when $t = 0$. The corresponding voltages and currents along the wires are the same as before.

Batteries, Power, and the Electrical–Mechanical Correspondence

So far, we have only considered the effect of current sources at the nodes. Suppose now that the network contains one or more batteries. Each *battery* serves as a voltage source along a wire, and we let b_k denote the voltage of a battery connected to wire k. The sign of b_k indicates the relative orientation of the battery's terminals with respect to the wire, with $b_k > 0$ if the current from the battery runs in the same direction as our chosen orientation of the wire. The battery's voltage is included in the voltage balance equation (6.17):

$$v_k = u_i - u_j + b_k.$$

The corresponding vector equation (6.18) becomes

$$\mathbf{v} = A\,\mathbf{u} + \mathbf{b}, \tag{6.37}$$

where $\mathbf{b} = (b_1, b_2, \ldots, b_m)^T$ is the *battery vector* whose entries are indexed by the wires. (If there is no battery on wire k, the corresponding entry is $b_k = 0$.) The remaining two equations are as before, so $\mathbf{y} = C\,\mathbf{v}$ are the currents in the wires, and, in the absence of external current sources, Kirchhoff's Current Law implies $A^T \mathbf{y} = \mathbf{0}$. Using the modified formula (6.37) for the voltages, these combine into the following equilibrium system:

$$K\,\mathbf{u} = A^T C\, A\,\mathbf{u} = -\,A^T C\,\mathbf{b}. \tag{6.38}$$

Thus, interestingly, the voltage potentials satisfy the normal equations (4.57) that characterize the least squares solution to system $A\mathbf{u} = -\,\mathbf{b}$ for the weighted norm

$$\| \mathbf{v} \| = \sqrt{\mathbf{v}^T C\, \mathbf{v}} \tag{6.39}$$

determined by the network's conductance matrix C. It is a striking fact that Nature solves a least squares problem in order to make the weighted norm of the voltages $\mathbf{v}$ as small as possible.

Batteries have exactly the same effect on the voltage potentials as if we imposed the current source vector

$$\mathbf{f} = -A^T C\,\mathbf{b}. \tag{6.40}$$

Namely, placing battery of voltage b_k on wire k is exactly the same as introducing additional current sources of $-c_k\, b_k$ at the starting node and $c_k\, b_k$ at the ending node. Note that the induced current vector $\mathbf{f} \in \operatorname{corng} A = \operatorname{rng} K$ (see Exercise 3.4.31) continues to satisfy the network constraint (6.33). Vice versa, a system of allowed current sources $\mathbf{f} \in \operatorname{rng} K$ has the same effect as any collection of batteries $\mathbf{b}$ that satisfies (6.40).

In the absence of external current sources, a network with batteries always admits a solution for the voltage potentials and currents. Although the currents are uniquely determined, the voltage potentials are not. As before, to eliminate the ambiguity, we can ground one of the nodes and use the reduced incidence matrix $A^\star$ and reduced current source vector $\mathbf{f}^\star$ obtained by eliminating the column/entry corresponding to the grounded node. The complete details are left to the interested reader.

EXAMPLE 6.4

Consider an electrical network running along the sides of a cube, where each wire contains a 2 ohm resistor and there is a 9 volt battery source on one wire. The problem is to determine how much current flows through the wire directly opposite the battery. Orienting the wires and numbering them as indicated in Figure 6.6, the incidence matrix is

$$A = \begin{pmatrix} 1 & -1 & 0 & 0 & 0 & 0 & 0 & 0 \\ 1 & 0 & -1 & 0 & 0 & 0 & 0 & 0 \\ 1 & 0 & 0 & -1 & 0 & 0 & 0 & 0 \\ 0 & 1 & 0 & 0 & -1 & 0 & 0 & 0 \\ 0 & 1 & 0 & 0 & 0 & -1 & 0 & 0 \\ 0 & 0 & 1 & 0 & -1 & 0 & 0 & 0 \\ 0 & 0 & 1 & 0 & 0 & 0 & -1 & 0 \\ 0 & 0 & 0 & 1 & 0 & -1 & 0 & 0 \\ 0 & 0 & 0 & 1 & 0 & 0 & -1 & 0 \\ 0 & 0 & 0 & 0 & 1 & 0 & 0 & -1 \\ 0 & 0 & 0 & 0 & 0 & 1 & 0 & -1 \\ 0 & 0 & 0 & 0 & 0 & 0 & 1 & -1 \end{pmatrix}.$$

We connect the battery along wire 1 and measure the resulting current along wire 12. To avoid the ambiguity in the voltage potentials, we ground the last node and erase the final column from A to obtain the reduced incidence matrix $A^\star$. Since the resistance matrix R has all 2's along the diagonal, the conductance matrix is $C = \frac{1}{2}\,\mathrm{I}$. Therefore, the network resistivity matrix is

$$K^\star = (A^\star)^T C A^\star = \tfrac{1}{2}\,(A^\star)^T A^\star = \frac{1}{2}\begin{pmatrix} 3 & -1 & -1 & -1 & 0 & 0 & 0 \\ -1 & 3 & 0 & 0 & -1 & -1 & 0 \\ -1 & 0 & 3 & 0 & -1 & 0 & -1 \\ -1 & 0 & 0 & 3 & 0 & -1 & -1 \\ 0 & -1 & -1 & 0 & 3 & 0 & 0 \\ 0 & -1 & 0 & -1 & 0 & 3 & 0 \\ 0 & 0 & -1 & -1 & 0 & 0 & 3 \end{pmatrix}.$$

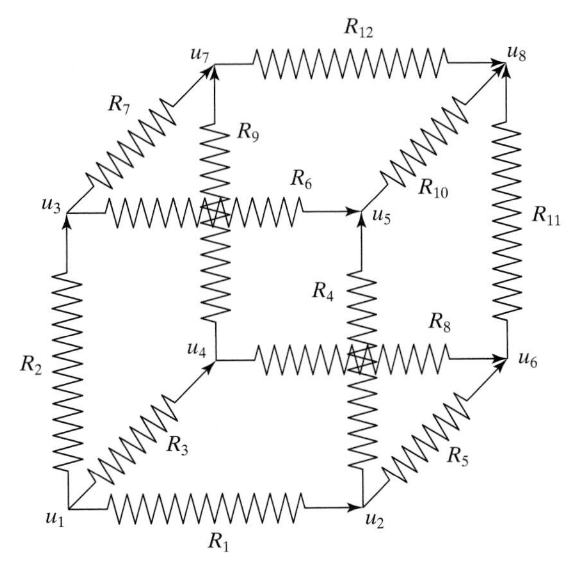

Figure 6.6 Cubical electrical network.

The reduced current source vector corresponding to the battery

$$\mathbf{b} = (9, 0, 0, 0, 0, 0, 0, 0, 0, 0, 0, 0)^T$$

along the first wire is

$$\mathbf{f}^\star = -(A^\star)^T C\,\mathbf{b} = \left(-\tfrac{9}{2}, \tfrac{9}{2}, 0, 0, 0, 0, 0\right)^T.$$

Solving the resulting linear system $K^\star \mathbf{u}^\star = \mathbf{f}^\star$ by Gaussian Elimination yields the voltage potentials

$$\mathbf{u}^\star = \left(-3, \tfrac{9}{4}, -\tfrac{9}{8}, -\tfrac{9}{8}, \tfrac{3}{8}, \tfrac{3}{8}, -\tfrac{3}{4}\right)^T.$$

Thus, the induced currents along the sides of the cube are

$$\begin{aligned}\mathbf{y} &= C\,\mathbf{v} = C\,(A^\star \mathbf{u}^\star + \mathbf{b}) \\ &= \left(\tfrac{15}{8}, -\tfrac{15}{16}, -\tfrac{15}{16}, \tfrac{15}{16}, \tfrac{15}{16}, -\tfrac{3}{4}, -\tfrac{3}{16}, -\tfrac{3}{4}, -\tfrac{3}{16}, \tfrac{3}{16}, \tfrac{3}{16}, -\tfrac{3}{8}\right)^T.\end{aligned}$$

In particular, the current on the wire that is opposite the battery is $y_{12} = -\tfrac{3}{8}$, flowing in the opposite direction to its orientation. The largest current flows through the battery wire, while wires 7, 9, 10 and 11 transmit the least. ●

As with a mass–spring chain, the voltage potentials in such a resistive electrical network can be characterized by a minimization principle. The *power* in a single conducting wire is defined as the product of its current y_j and voltage v_j,

$$P_j = y_j\, v_j = R_j\, y_j^2 = c_j\, v_j^2, \tag{6.41}$$

where R_j is the resistance, $c_j = 1/R_j$ the conductance, and we are using Ohm's Law (6.21) to relate voltage and current, [44]. Physically, the power quantifies the rate at which electrical energy is converted into heat by the wire's resistance. Summing over all wires in the system, the internal power† of the network

$$P_{int} = \sum_j P_j = \sum_j c_j\, v_j^2 = \|\mathbf{v}\|^2$$

is identified as the square of the weighted norm (6.39).

† So far, we have not considered the effect of batteries or current sources on the network.

Consider a network that contains batteries, but no external current sources. Summing over all the wires in the network, the total power due to internal and external sources can be identified as the product of the current and voltage vectors

$$\begin{aligned} P &= y_1 v_1 + \cdots + y_m v_m = \mathbf{y}^T \mathbf{v} = \mathbf{v}^T C \mathbf{v} \\ &= (A\mathbf{u} + \mathbf{b})^T C (A\mathbf{u} + \mathbf{b}) = \mathbf{u}^T A^T C A \mathbf{u} + 2\mathbf{u}^T A^T C \mathbf{b} + \mathbf{b}^T C \mathbf{b}, \end{aligned}$$

and is thus a quadratic function of the voltage potentials, which we rewrite in our usual form[‡]

$$\tfrac{1}{2} P = p(\mathbf{u}) = \tfrac{1}{2} \mathbf{u}^T K \mathbf{u} - \mathbf{u}^T \mathbf{f} + c, \tag{6.42}$$

where $K = A^T C A$ is the network resistivity matrix, while $\mathbf{f} = -A^T C \mathbf{b}$ are the equivalent current sources at the nodes (6.40) that correspond to the batteries. The last term $c = \frac{1}{2} \mathbf{b}^T C \mathbf{b}$ is one half the internal power of the batteries, and is not affected by the currents/voltages in the wires. In deriving (6.42), we have ignored external current sources at the nodes. By the preceding discussion, external current sources can be viewed as an equivalent collection of batteries, and so will contribute to the linear terms $\mathbf{u}^T \mathbf{f}$ in the power, which will then represent the combined effect of all batteries and external current sources.

In general, the resistivity matrix K is only positive semi-definite, and so the quadratic power function (6.42) does not, in general, possess a minimizer. As argued above, to ensure equilibrium, we need to ground one or more of the nodes. The resulting reduced power function

$$p(\mathbf{u}^\star) = \tfrac{1}{2} (\mathbf{u}^\star)^T K^\star \mathbf{u}^\star - (\mathbf{u}^\star)^T \mathbf{f}^\star, \tag{6.43}$$

has a positive definite coefficient matrix: $K^\star > 0$. Its unique minimizer is the voltage potential solution $\mathbf{u}^\star$ to the reduced linear system (6.36). We conclude that the electrical network adjusts itself so as to *minimize the power or total energy loss* throughout the network. As in mechanics, Nature solves a minimization problem in an effort to conserve energy.

Structures	Variables	Networks
Displacements	$\mathbf{u}$	Voltage Potentials
Prestressed bars/springs	$\mathbf{b}$	Batteries
Elongations[†]	$\mathbf{v} = A\mathbf{u} + \mathbf{b}$	Voltages
Spring stiffnesses	C	Conductivities
Internal Forces	$\mathbf{y} = C\mathbf{v}$	Currents
External forcing	$\mathbf{f} = A^T \mathbf{y}$	Current sources
Stiffness matrix	$K = A^T C A$	Resistivity matrix
Potential energy	$p(\mathbf{u}) = \frac{1}{2}\mathbf{u}^T K \mathbf{u} - \mathbf{u}^T \mathbf{f}$	$\frac{1}{2} \times$ Power

We have now discovered the remarkable correspondence between the equilibrium equations for electrical networks (6.10) and those of mass–spring chains (6.28). This *Electrical–Mechanical Correspondence* is summarized in the above table. In the following section, we will see that the analogy extends to more general mechanical structures. In Chapter 11, we will discover that, suitably interpreted, the

[‡] For alternating currents, the power is reduced by a factor of $\frac{1}{2}$, so $p(\mathbf{u})$ equals the power.

[†] Here, we use $\mathbf{v}$ instead of $\mathbf{e}$ to represent elongation.

theory continues to apply in the continuous regime, and subsumes solid mechanics, fluid mechanics, electrostatics, and many other physical systems in a universal mathematical framework!

EXERCISES 6.2

6.2.1. Draw the electrical networks corresponding to the following incidence matrices.

(a) $\begin{pmatrix} 1 & 0 & -1 & 0 \\ 0 & 1 & 0 & -1 \\ -1 & 1 & 0 & 0 \\ 0 & 0 & 1 & -1 \end{pmatrix}$

(b) $\begin{pmatrix} 0 & 0 & 1 & -1 \\ 1 & 0 & 0 & -1 \\ 0 & -1 & 1 & 0 \\ 1 & -1 & 0 & 0 \end{pmatrix}$

(c) $\begin{pmatrix} 0 & 1 & 0 & 0 & -1 \\ -1 & 0 & 1 & 0 & 0 \\ 0 & 0 & 0 & -1 & 1 \\ 0 & -1 & 1 & 0 & 0 \end{pmatrix}$

(d) $\begin{pmatrix} -1 & 0 & 1 & 0 & 0 \\ 0 & -1 & 0 & 1 & 0 \\ 1 & -1 & 0 & 0 & 0 \\ 0 & 0 & 0 & -1 & 1 \\ 0 & 0 & -1 & 0 & 1 \end{pmatrix}$

(e) $\begin{pmatrix} 0 & -1 & 0 & 1 & 0 & 0 & 0 \\ 0 & 0 & -1 & 0 & 1 & 0 & 0 \\ 1 & 0 & -1 & 0 & 0 & 0 & 0 \\ 1 & -1 & 0 & 0 & 0 & 0 & 0 \\ 0 & 0 & 0 & -1 & 0 & 1 & 0 \\ 0 & 0 & 0 & 0 & -1 & 0 & 1 \\ 0 & 0 & -1 & 0 & 0 & 1 & 0 \\ 0 & 0 & 0 & 0 & 0 & 1 & -1 \end{pmatrix}$

6.2.2. Suppose that all wires in the illustrated network have unit resistivity.

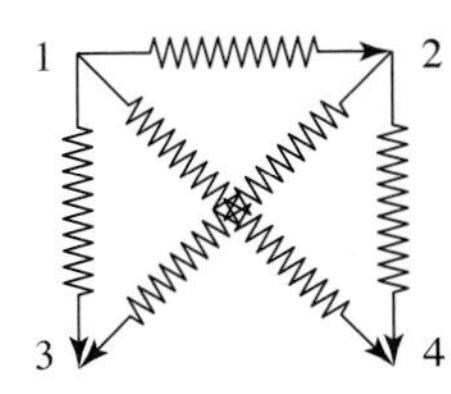

(a) Write down the incidence matrix A.

(b) Write down the equilibrium system for the network when node 4 is grounded and there is a current source of magnitude 3 at node 1.

(c) Solve the system for the voltage potentials at the ungrounded nodes.

(d) If you connect a light bulb to the network, which wire should you connect it to so that it shines the brightest?

6.2.3. What happens in the network in Figure 6.5 if we ground both nodes 3 and 4? Set up and solve the system and compare the currents for the two cases.

6.2.4. (a) Write down the incidence matrix A for the illustrated electrical network.

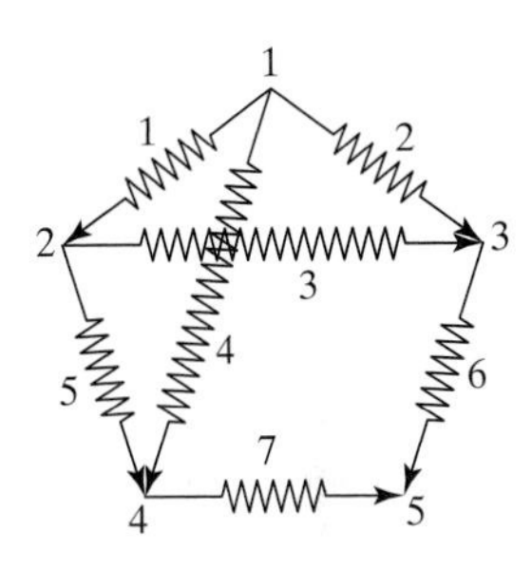

(b) Suppose all the wires contain unit resistors, except for $R_4 = 2$. Let there be a unit current source at node 1, and assume node 5 is grounded. Find the voltage potentials at the nodes and the currents through the wires.

(c) Which wire would shock you the most?

6.2.5. Answer Exercise 6.2.4 if, instead of the current source, you put a 1.5 volt battery on wire 1.

♠ **6.2.6.** Consider an electrical network running along the sides of a tetrahedron.

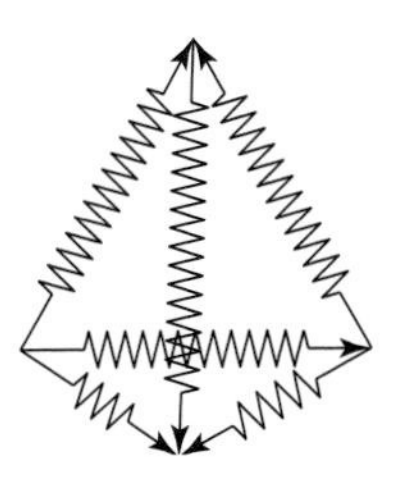

Suppose that each wire contains a 3 ohm resistor and there is a 10 volt battery source on one wire. Determine how much current flows through the wire directly opposite the battery.

♠ **6.2.7.** Now suppose that each wire in the tetrahedral network in Exercise 6.2.6 contains a 1 ohm resistor and there are two 5 volt battery sources located on two non-adjacent wires. Determine how much current flows through the wires in the network.

♠ **6.2.8.** (a) How do the currents change if the resistances in the wires in the cubical network in Example 6.4 are all equal to 1 ohm?

(b) What if wire k has resistance $R_k = k$ ohms?

♣ **6.2.9.** Suppose you are given six resistors with respective resistances 1, 2, 3, 4, 5 and 6. How should you connect them in a tetrahedral network (one resistor per wire) so that a light bulb on the wire opposite the battery burns the brightest?

♣ **6.2.10.** The nodes in an electrical network lie on the vertices $\left(\frac{i}{n}, \frac{j}{n}\right)$ for $-n \leq i, j \leq n$ in a square grid centered at the origin; the wires run along the grid lines. The boundary nodes, when x or $y = \pm 1$, are all grounded. A unit current source is introduced at the origin.

(a) Compute the potentials at the nodes and currents along the wires for $n = 2, 3, 4$.

(b) Investigate and compare the solutions for large n, i.e., as the grid size becomes small. Do you detect any form of limiting behavior?

6.2.11. Show that, in a network with all unit resistors, the currents $\mathbf{y}$ can be characterized as the unique solution to the Kirchhoff equations $A^T \mathbf{y} = \mathbf{f}$ of minimum Euclidean norm.

6.2.12. (a) Find the voltage potentials at all the nodes and the currents along the wires of the following trees if the bottom node is grounded and a unit current source is introduced at the top node.

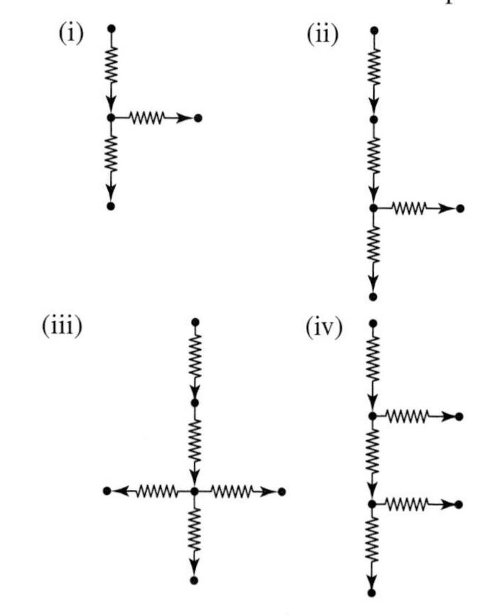

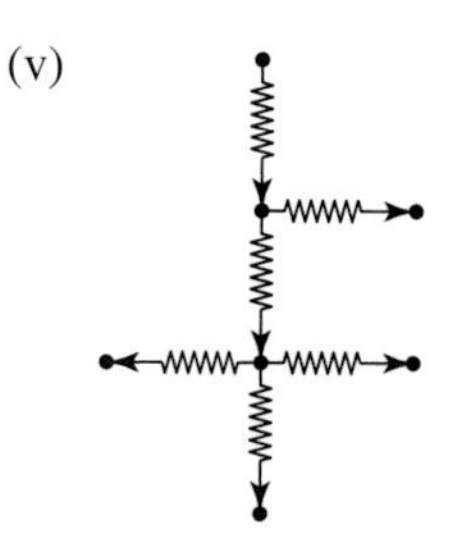

(b) Can you make any general predictions about electrical currents in trees?

6.2.13. A node in a tree is called *terminating* if it has only one edge. Repeat the preceding exercise when all terminating nodes except for the top one are grounded.

6.2.14. Suppose the graph of an electrical network is a tree, as in Exercise 2.6.8. Show that if one of the nodes in the tree is grounded, the system is statically determinate.

6.2.15. *True or false*:

(a) The nodal voltage potentials in a network with a battery $\mathbf{b}$ are the same as in the same network with the current sources $\mathbf{f} = -A^T C\,\mathbf{b}$.

(b) Are the currents the same?

6.2.16. Suppose two wires in a network join the *same* pair of nodes. Explain why their effect on the rest of the network is the same as a single wire whose conductance $c = c_1 + c_2$ is the sum of the individual conductances. How are the resistances related?

6.2.17. (a) Write down the equilibrium equations for a network that contains both batteries and current sources.

(b) Formulate a general superposition principle for such situations.

◇ **6.2.18.** Prove that the voltage potential at node i due to a unit current source at node j is the *same* as the voltage potential at node j due to a unit current source at node i. Can you give a physical explanation of this *reciprocity relation*?

6.2.19. What is the analog of condition (6.33) for a disconnected graph?

6.3 STRUCTURES

A *structure* (sometimes called a *truss*) is a mathematical idealization of a framework for a building. Think of a skyscraper when just the I-beams are connected together—before the walls, floors, ceilings, roof and ornamentation are added. An ideal structure is constructed of elastic bars connected at *joints*. By a *bar*, we mean a straight, rigid rod that can be (slightly) elongated, but not bent. (Beams, which are allowed to bend, are more complicated, and we defer their treatment until Section 11.4.) When a bar is stretched, it obeys Hooke's law (at least in the linear regime we are modeling) and so, for all practical purposes, behaves like a spring with a very large stiffness. As a result, a structure can be regarded as a two- or three-dimensional generalization of a mass–spring chain.

The joints will allow the bar to rotate in any direction. Of course, this is an idealization; in a building, the rivets and bolts will prevent rotation to a significant degree. However, under moderate stress—for example, if the wind is blowing on our skyscraper, the bolts can only be expected to keep the structure connected, and the rotational motion will induce stresses on the joints which must be taken into account when designing the structure. Of course, under extreme stress, the structure will collapse—a disaster that its designers must avoid. The purpose of this section is to derive conditions that will guarantee that a structure is rigidly stable under moderate forcing, or, alternatively, understand the processes that might lead to its collapse.

The first order of business is to understand how an individual bar reacts to motion. We have already encountered the basic idea in our treatment of springs. The key complication here is that the ends of the bar are not restricted to a single direction of motion, but can move in either two or three-dimensional space. We use d to denote the dimension of the underlying space. In the $d = 1$ dimensional case, the structure reduces to a mass–spring chain that we analyzed in the Section 6.1. Here we concentrate on structures in $d = 2$ and 3 dimensions.

Consider an unstressed bar with one end at position $\mathbf{a}_1 \in \mathbb{R}^d$ and the other end at position $\mathbf{a}_2 \in \mathbb{R}^d$. In $d = 2$ dimensions, we write $\mathbf{a}_i = (a_i, b_i)^T$, while in $d = 3$-dimensional space $\mathbf{a}_i = (a_i, b_i, c_i)^T$. The length of the bar is $L = \|\mathbf{a}_1 - \mathbf{a}_2\|$, where we use the standard Euclidean norm to measure distance on $\mathbb{R}^d$ throughout this section.

Suppose we move the ends of the bar a little, sending $\mathbf{a}_i$ to $\mathbf{b}_i = \mathbf{a}_i + \varepsilon\,\mathbf{u}_i$ and, simultaneously, $\mathbf{a}_j$ to $\mathbf{b}_j = \mathbf{a}_j + \varepsilon\,\mathbf{u}_j$, as in Figure 6.7. The vectors $\mathbf{u}_i, \mathbf{u}_j \in \mathbb{R}^d$ indicate the respective direction of displacement of the two ends, and we use ε to represent the relative magnitude of the displacement. How much has this motion stretched the bar? Since we are assuming that the bar can't bend, the length of the displaced bar is

$$
\begin{aligned}
L + e &= \|\mathbf{b}_i - \mathbf{b}_j\| = \|(\mathbf{a}_i + \varepsilon\,\mathbf{u}_i) - (\mathbf{a}_j + \varepsilon\,\mathbf{u}_j)\| \\
&= \|(\mathbf{a}_i - \mathbf{a}_j) + \varepsilon\,(\mathbf{u}_i - \mathbf{u}_j)\| \qquad (6.44) \\
&= \sqrt{\|\mathbf{a}_i - \mathbf{a}_j\|^2 + 2\,\varepsilon\,(\mathbf{a}_i - \mathbf{a}_j)\cdot(\mathbf{u}_i - \mathbf{u}_j) + \varepsilon^2\,\|\mathbf{u}_i - \mathbf{u}_j\|^2}
\end{aligned}
$$

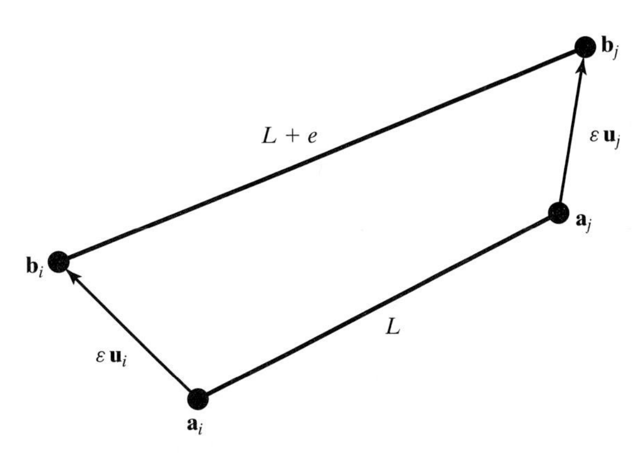

Figure 6.7 Displacement of a bar.

The difference between the new length and the original length, namely

$$e = \sqrt{\|\mathbf{a}_i - \mathbf{a}_j\|^2 + 2\,\varepsilon\,(\mathbf{a}_i - \mathbf{a}_j)\cdot(\mathbf{u}_i - \mathbf{u}_j) + \varepsilon^2\,\|\mathbf{u}_i - \mathbf{u}_j\|^2} \; - \; \|\mathbf{a}_i - \mathbf{a}_j\|, \tag{6.45}$$

is, by definition, the bar's *elongation*.

If the underlying dimension d is 2 or more, the elongation (6.45) is a *nonlinear* function of the displacement vectors $\mathbf{u}_i$, $\mathbf{u}_j$. Thus, an exact, geometrical treatment of structures in equilibrium requires dealing with complicated nonlinear systems of equations. In some situations, e.g., the design of robotic mechanisms, [43], analysis of the nonlinear system is crucial, but this lies beyond the scope of this text. However, in many practical situations, the displacements are fairly small, so $|\varepsilon| \ll 1$. For example, when a building moves, the lengths of bars are in meters, but the displacements are, barring catastrophes, typically in centimeters if not millimeters. In such situations, we can replace the geometrically exact elongation by a much simpler linear approximation.

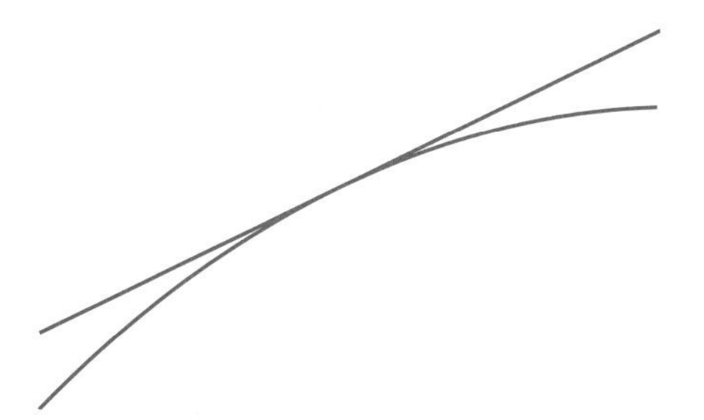

Figure 6.8 Tangent line approximation.

As you learned in calculus, the most basic linear approximation to a nonlinear function $g(\varepsilon)$ near $\varepsilon = 0$ is given by its tangent line or linear Taylor polynomial

$$g(\varepsilon) \approx g(0) + g'(0)\,\varepsilon, \qquad |\varepsilon| \ll 1, \tag{6.46}$$

as sketched in Figure 6.8. In the case of small displacements of a bar, the elongation (6.45) is a square root function of the particular form

$$g(\varepsilon) = \sqrt{a^2 + 2\,\varepsilon\,b + \varepsilon^2\,c^2} \; - a,$$

where

$$a = \|\mathbf{a}_i - \mathbf{a}_j\|, \qquad b = (\mathbf{a}_i - \mathbf{a}_j)\cdot(\mathbf{u}_i - \mathbf{u}_j), \qquad c = \|\mathbf{u}_i - \mathbf{u}_j\|,$$

are independent of ε. Since $g(0) = 0$ and $g'(0) = \dfrac{b}{a}$, the linear approximation (6.46) is

$$\sqrt{a^2 + 2\,\varepsilon\,b + \varepsilon^2\,c^2} \; - a \approx \varepsilon\,\frac{b}{a} \qquad \text{for} \quad |\varepsilon| \ll 1.$$

In this manner, we arrive at the linear approximation to the bar's elongation

$$e \approx \varepsilon\,\frac{(\mathbf{a}_i - \mathbf{a}_j)\cdot(\mathbf{u}_i - \mathbf{u}_j)}{\|\mathbf{a}_i - \mathbf{a}_j\|} = \mathbf{n}\cdot(\varepsilon\,\mathbf{u}_i - \varepsilon\,\mathbf{u}_j), \qquad \text{where} \quad \mathbf{n} = \frac{\mathbf{a}_i - \mathbf{a}_j}{\|\mathbf{a}_i - \mathbf{a}_j\|}$$

is the unit vector, $\|\mathbf{n}\| = 1$, that points in the direction of the bar from node j to node i.

The factor ε was merely a device used to derive the linear approximation. It can now be safely discarded, so that the displacement of the ith node is now $\mathbf{u}_i$ instead of $\varepsilon\,\mathbf{u}_i$, and we assume $\|\mathbf{u}_i\|$ is small. If bar k connects node i to node j, then its (approximate) elongation is equal to

$$e_k = \mathbf{n}_k \cdot (\mathbf{u}_i - \mathbf{u}_j) = \mathbf{n}_k \cdot \mathbf{u}_i - \mathbf{n}_k \cdot \mathbf{u}_j, \qquad \text{where} \qquad \mathbf{n}_k = \frac{\mathbf{a}_i - \mathbf{a}_j}{\|\mathbf{a}_i - \mathbf{a}_j\|}. \tag{6.47}$$

The elongation e_k is the sum of two terms: the first, $\mathbf{n}_k \cdot \mathbf{u}_i$, is the component of the displacement vector for node i in the direction of the unit vector $\mathbf{n}_k$ that points along the bar *towards* node i, whereas the second, $-\,\mathbf{n}_k \cdot \mathbf{u}_j$, is the component of the displacement vector for node j in the direction of the unit vector $-\,\mathbf{n}_k$ that points in the opposite direction along the bar *towards* node j; see Figure 6.9. Their sum equals the total elongation of the bar.

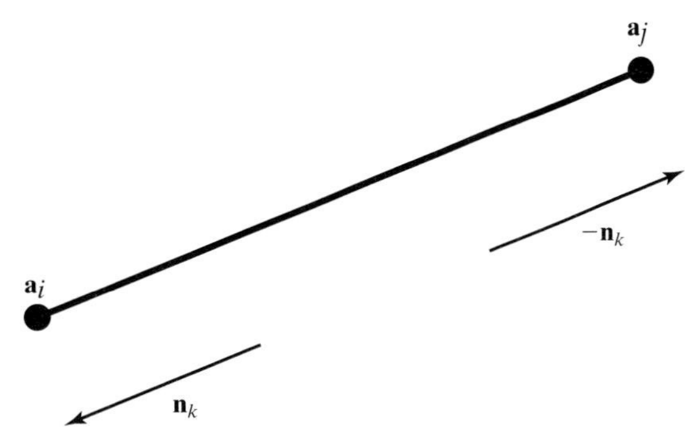

Figure 6.9 Unit vectors for a bar.

We assemble all the linear equations (6.47) relating nodal displacements to bar elongations in matrix form

$$\mathbf{e} = A\mathbf{u}. \tag{6.48}$$

Here

$$\mathbf{e} = \begin{pmatrix} e_1 \\ \vdots \\ e_m \end{pmatrix} \in \mathbb{R}^m$$

is the vector of elongations, while

$$\mathbf{u} = \begin{pmatrix} \mathbf{u}_1 \\ \vdots \\ \mathbf{u}_n \end{pmatrix} \in \mathbb{R}^{dn}$$

is the vector of displacements. Each $\mathbf{u}_i \in \mathbb{R}^d$ is itself a column vector with d entries, and so $\mathbf{u}$ has a total of $d\,n$ entries. For example, in the planar case $d = 2$, we have $\mathbf{u}_i = \begin{pmatrix} x_i \\ y_i \end{pmatrix}$, since each node's displacement has both an x and y component, and so

$$\mathbf{u} = \begin{pmatrix} \mathbf{u}_1 \\ \vdots \\ \mathbf{u}_n \end{pmatrix} = \begin{pmatrix} x_1 \\ y_1 \\ x_2 \\ y_2 \\ \vdots \\ x_n \\ y_n \end{pmatrix} \in \mathbb{R}^{2n}.$$

In three dimensions, $d = 3$, we have $\mathbf{u}_i = (x_i, y_i, z_i)^T$, and so each node will contribute three components to the displacement vector

$$\mathbf{u} = (x_1, y_1, z_1, x_2, y_2, z_2, \ldots, x_n, y_n, z_n)^T \in \mathbb{R}^{3n}.$$

The *incidence matrix* A connecting the displacements and elongations will be of size $m \times (dn)$. The kth row of A will have (at most) $2d$ nonzero entries. The entries in the d slots corresponding to node i will be the components of the (transposed) unit bar vector $\mathbf{n}_k^T$ pointing towards node i, as given in (6.47), while the entries in the d slots corresponding to node j will be the components of its negative $-\mathbf{n}_k^T$, which is the unit bar vector pointing towards node j. All other entries are 0. The general formulation is best appreciated by working through an explicit example.

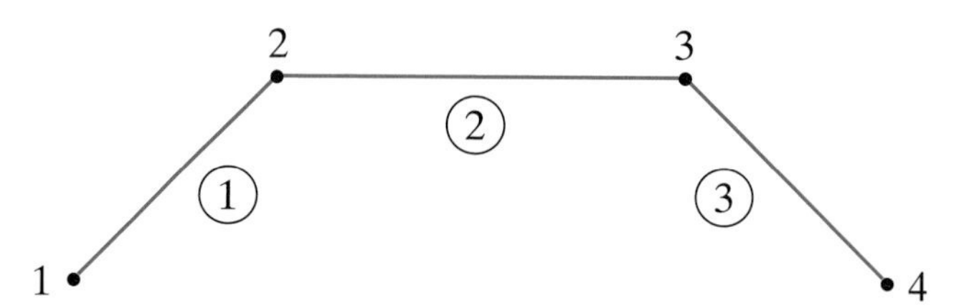

Figure 6.10 Three bar planar structure.

EXAMPLE 6.5 Consider the planar structure pictured in Figure 6.10. The four nodes are at positions

$$\mathbf{a}_1 = (0, 0)^T, \quad \mathbf{a}_2 = (1, 1)^T, \quad \mathbf{a}_3 = (3, 1)^T, \quad \mathbf{a}_4 = (4, 0)^T,$$

so the two side bars are at $45°$ angles and the center bar is horizontal. Implementing our construction, the associated incidence matrix is

$$A = \left(\begin{array}{cc|cc|cc|cc} -\frac{1}{\sqrt{2}} & -\frac{1}{\sqrt{2}} & \frac{1}{\sqrt{2}} & \frac{1}{\sqrt{2}} & 0 & 0 & 0 & 0 \\ 0 & 0 & -1 & 0 & 1 & 0 & 0 & 0 \\ 0 & 0 & 0 & 0 & -\frac{1}{\sqrt{2}} & \frac{1}{\sqrt{2}} & \frac{1}{\sqrt{2}} & -\frac{1}{\sqrt{2}} \end{array}\right). \tag{6.49}$$

The three rows of A refer to the three bars in our structure. The columns come in pairs, as indicated by the vertical lines in the matrix: the first two columns refer to the x and y displacements of the first node; the third and fourth columns refer to the second node, and so on. The first two entries of the first row of A indicate the unit vector

$$\mathbf{n}_1 = \frac{\mathbf{a}_1 - \mathbf{a}_2}{\|\mathbf{a}_1 - \mathbf{a}_2\|} = \left(-\frac{1}{\sqrt{2}}, -\frac{1}{\sqrt{2}}\right)^T$$

that points along the first bar towards the first node, while the third and fourth entries have the opposite signs, and form the unit vector

$$-\mathbf{n}_1 = \frac{\mathbf{a}_2 - \mathbf{a}_1}{\|\mathbf{a}_2 - \mathbf{a}_1\|} = \left(\frac{1}{\sqrt{2}}, \frac{1}{\sqrt{2}}\right)^T$$

along the same bar that points in the opposite direction — towards the second node. The remaining entries are zero because the first bar only connects the first two nodes. Similarly, the unit vector along the second bar pointing towards node 2 is

$$\mathbf{n}_2 = \frac{\mathbf{a}_2 - \mathbf{a}_3}{\|\mathbf{a}_2 - \mathbf{a}_3\|} = (-1, 0)^T,$$

and this gives the third and fourth entries of the second row of A; the fifth and sixth entries are their negatives, corresponding to the unit vector $-\mathbf{n}_2$ pointing towards node 3. The last row is constructed from the unit vectors along bar #3 in the same fashion. ●

REMARK: Interestingly, the incidence matrix for a structure only depends on the directions of the bars and not their lengths. This is analogous to the fact that the incidence matrix for an electrical network only depends on the connectivity properties of the wires and not on their overall lengths. One can regard the incidence matrix for a structure as a kind of d–dimensional generalization of the incidence matrix for a directed graph.

The next phase of our procedure is to introduce the constitutive relations for the bars that determine their internal forces or stresses. As we remarked at the beginning of the section, each bar is viewed as a hard spring, subject to a linear Hooke's law equation

$$y_k = c_k\, e_k \tag{6.50}$$

that relates its elongation e_k to its internal force y_k. The bar stiffness $c_k > 0$ is a positive scalar, and so $y_k > 0$ if the bar is in tension, while $y_k < 0$ if the bar is compressed. We write (6.50) in matrix form

$$\mathbf{y} = C\,\mathbf{e},$$

where $C = \operatorname{diag}(c_1, \ldots, c_m) > 0$ is a diagonal, positive definite matrix.

Finally, we need to balance the forces at each node in order to achieve equilibrium. If bar k terminates at node i, then it exerts a force $-\,y_k\,\mathbf{n}_k$ on the node, where $\mathbf{n}_k$ is the unit vector pointing towards the node in the direction of the bar, as in (6.47). The minus sign comes from physics: if the bar is under tension, so $y_k > 0$, then it is trying to contract back to its unstressed state, and so will pull the node towards it—in the opposite direction to $\mathbf{n}_k$—while a bar in compression will push the node away. In addition, we may have an externally applied force vector, denoted by $\mathbf{f}_i$, on node i, which might be some combination of gravity, weights, mechanical forces, and so on. (In this admittedly simplified model, external forces only act on the nodes.) Force balance at equilibrium requires that all the nodal forces, external and internal, cancel; thus,

$$\mathbf{f}_i + \sum_k (-\,y_k\,\mathbf{n}_k) = 0, \qquad \text{or} \qquad \sum_k y_k\,\mathbf{n}_k = \mathbf{f}_i,$$

where the sum is over all the bars that are attached to node i. The matrix form of the force balance equations is (and this should no longer come as a surprise)

$$\mathbf{f} = A^T\mathbf{y}, \tag{6.51}$$

where A^T is the transpose of the incidence matrix, and

$$\mathbf{f} = \begin{pmatrix} \mathbf{f}_1 \\ \vdots \\ \mathbf{f}_n \end{pmatrix} \in \mathbb{R}^{dn}$$

is the vector containing all external force on the nodes. Putting everything together, (6.48–51), i.e.,

$$\mathbf{e} = A\mathbf{u}, \quad \mathbf{y} = C\,\mathbf{e}, \quad \mathbf{f} = A^T\mathbf{y},$$

once again we are led to a by now familiar linear system of equations:

$$K\mathbf{u} = \mathbf{f}, \qquad \text{where} \qquad K = A^T C\, A \tag{6.52}$$

is the stiffness matrix for our structure.

The stiffness matrix K is a positive (semi-)definite Gram matrix (3.56) associated with the weighted inner product on the space of elongations prescribed by the diagonal matrix C. As we know, K will be positive definite if and only if the kernel of the incidence matrix is trivial: $\ker A = \{\mathbf{0}\}$. However, the preceding example does not enjoy this property because we have not tied down (or "grounded") our structure. In essence, we are considering a structure floating in outer space, which is free to move around in any direction. Each rigid motion of the structure will correspond to an element of the kernel of its incidence matrix, and thereby preclude positive definiteness of its stiffness matrix.

EXAMPLE 6.6

Consider a planar space station in the shape of a unit equilateral triangle, as in Figure 6.11. Placing the nodes at positions

$$\mathbf{a}_1 = \left(\tfrac{1}{2}, \tfrac{\sqrt{3}}{2}\right)^T, \qquad \mathbf{a}_2 = (1,0)^T, \qquad \mathbf{a}_3 = (0,0)^T,$$

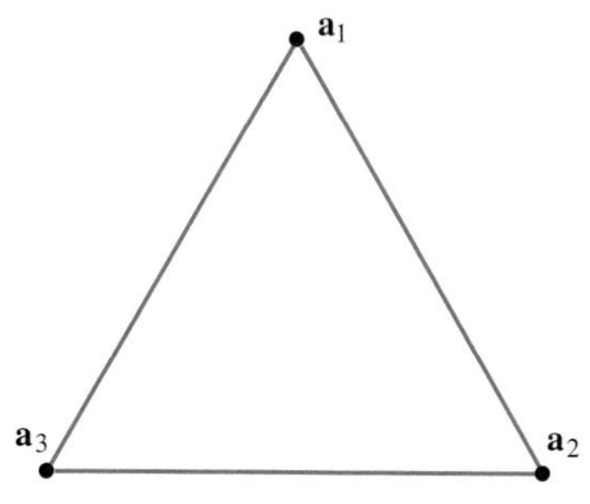

Figure 6.11 A triangular structure.

we use the preceding algorithm to construct the incidence matrix

$$A = \left(\begin{array}{cc|cc|cc} -\frac{1}{2} & \frac{\sqrt{3}}{2} & \frac{1}{2} & -\frac{\sqrt{3}}{2} & 0 & 0 \\ \frac{1}{2} & \frac{\sqrt{3}}{2} & 0 & 0 & -\frac{1}{2} & -\frac{\sqrt{3}}{2} \\ 0 & 0 & 1 & 0 & -1 & 0 \end{array}\right)$$

whose rows are indexed by the bars, and whose columns are indexed in pairs by the three nodes. The kernel of A is three-dimensional, with basis

$$\mathbf{z}_1 = \begin{pmatrix} 1 \\ 0 \\ 1 \\ 0 \\ 1 \\ 0 \end{pmatrix}, \qquad \mathbf{z}_2 = \begin{pmatrix} 0 \\ 1 \\ 0 \\ 1 \\ 0 \\ 1 \end{pmatrix}, \qquad \mathbf{z}_3 = \begin{pmatrix} -\frac{\sqrt{3}}{2} \\ \frac{1}{2} \\ 0 \\ 1 \\ 0 \\ 0 \end{pmatrix}. \tag{6.53}$$

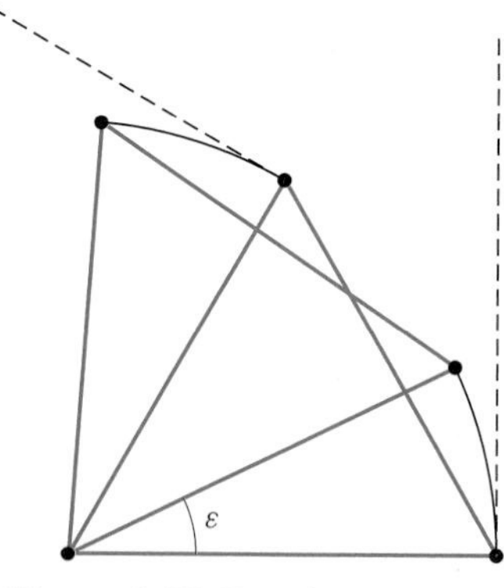

Figure 6.12 Rotating a space station.

We claim that these three displacement vectors represent three different planar rigid motions: the first two correspond to translations, and the third to a rotation.

The translations are easy to discern. Translating the space station in a horizontal direction means that we move all three nodes the same amount, and so the displacements are $\mathbf{u}_1 = \mathbf{u}_2 = \mathbf{u}_3 = \mathbf{a}$ for some fixed vector $\mathbf{a}$. In particular, a rigid unit horizontal translation has $\mathbf{a} = \mathbf{e}_1 = (1,0)^T$, and corresponds to the first kernel basis vector. Similarly, a unit vertical translation of all three nodes corresponds to $\mathbf{a} = \mathbf{e}_2 = (0,1)^T$, and corresponds to the second kernel basis vector. Any other translation is a linear combination of these two. Translations do not alter the lengths of any of the bars, and so do not induce any stress in the structure.

The rotations are a little more subtle, owing to the linear approximation that we used to compute the elongations. Referring to Figure 6.12, rotating the space station through a small angle ε around the node $\mathbf{a}_3 = (0,0)^T$ will move the other two nodes to positions

$$\mathbf{b}_1 = \begin{pmatrix} \frac{1}{2}\cos\varepsilon - \frac{\sqrt{3}}{2}\sin\varepsilon \\ \frac{1}{2}\sin\varepsilon + \frac{\sqrt{3}}{2}\cos\varepsilon \end{pmatrix}, \qquad \mathbf{b}_2 = \begin{pmatrix} \cos\varepsilon \\ \sin\varepsilon \end{pmatrix}, \qquad \mathbf{b}_3 = \begin{pmatrix} 0 \\ 0 \end{pmatrix}. \tag{6.54}$$

However, the corresponding displacements

$$\mathbf{u}_1 = \mathbf{b}_1 - \mathbf{a}_1 = \begin{pmatrix} \frac{1}{2}(\cos\varepsilon - 1) - \frac{\sqrt{3}}{2}\sin\varepsilon \\ \frac{1}{2}\sin\varepsilon + \frac{\sqrt{3}}{2}(\cos\varepsilon - 1) \end{pmatrix},$$
$$\mathbf{u}_2 = \mathbf{b}_2 - \mathbf{a}_2 = \begin{pmatrix} \cos\varepsilon - 1 \\ \sin\varepsilon \end{pmatrix}, \qquad \mathbf{u}_3 = \mathbf{b}_3 - \mathbf{a}_3 = \begin{pmatrix} 0 \\ 0 \end{pmatrix}, \tag{6.55}$$

do *not* combine into a vector that belongs to $\ker A$. The problem is that, under a rotation, the nodes move along circles, while the kernel displacements $\mathbf{u} = \varepsilon\,\mathbf{z} \in \ker A$ correspond to straight line motion! In order to maintain consistency, we must adopt a similar linear approximation of the nonlinear circular motion of the nodes. Thus, we replace the nonlinear displacements $\mathbf{u}_j(\varepsilon)$ in (6.55) by their linear tangent approximations* $\varepsilon\,\mathbf{u}_j'(0)$, so

$$\mathbf{u}_1 \approx \varepsilon \begin{pmatrix} -\frac{\sqrt{3}}{2} \\ \frac{1}{2} \end{pmatrix}, \qquad \mathbf{u}_2 \approx \varepsilon \begin{pmatrix} 0 \\ 1 \end{pmatrix}, \qquad \mathbf{u}_3 = \begin{pmatrix} 0 \\ 0 \end{pmatrix}.$$

The resulting displacements *do* combine to produce the displacement vector

$$\mathbf{u} = \varepsilon\left(-\tfrac{\sqrt{3}}{2},\ \tfrac{1}{2},\ 0,\ 1,\ 0,\ 0\right)^T = \varepsilon\,\mathbf{z}_3$$

that moves the space station in the direction of the third kernel basis vector. Thus, as claimed, $\mathbf{z}_3$ represents the linear approximation to a rigid rotation around the first node.

Remarkably, the rotations around the other two nodes, although distinct nonlinear motions, can be linearly approximated by particular combinations of the three kernel basis elements $\mathbf{z}_1, \mathbf{z}_2, \mathbf{z}_3$, and so already appear in our description of $\ker A$. For example, the displacement vector

$$\mathbf{u} = \varepsilon\left(\tfrac{\sqrt{3}}{2}\,\mathbf{z}_1 + \tfrac{1}{2}\,\mathbf{z}_2 - \mathbf{z}_3\right) = \varepsilon\left(0,\ 0,\ \tfrac{\sqrt{3}}{2},\ -\tfrac{1}{2},\ \tfrac{\sqrt{3}}{2},\ \tfrac{1}{2}\right)^T \tag{6.56}$$

represents the linear approximation to a rigid rotation around the first node. We conclude that the three-dimensional kernel of the incidence matrix represents the sum total of all possible rigid motions of the space station, or, more correctly, their linear approximations.

Which types of forces will maintain the space station in equilibrium? This will happen if and only if we can solve the force balance equations

$$A^T\mathbf{y} = \mathbf{f} \tag{6.57}$$

for the internal forces $\mathbf{y}$. The Fredholm Alternative Theorem 5.55 implies that the system (6.57) has a solution if and only if $\mathbf{f}$ is orthogonal to $\operatorname{coker} A^T = \ker A$. Therefore, $\mathbf{f} = (f_1, g_1, f_2, g_2, f_3, g_3)^T$ must be orthogonal to the kernel basis vectors (6.53), and so must satisfy the three linear constraints

$$\begin{aligned} \mathbf{z}_1 \cdot \mathbf{f} &= f_1 + f_2 + f_3 = 0, \\ \mathbf{z}_2 \cdot \mathbf{f} &= g_1 + g_2 + g_3 = 0, \\ \mathbf{z}_3 \cdot \mathbf{f} &= \tfrac{\sqrt{3}}{2}\,f_1 + \tfrac{1}{2}\,g_1 + g_3 = 0. \end{aligned} \tag{6.58}$$

The first requires that there is no net horizontal force on the space station. The second requires no net vertical force. The last constraint requires that the *moment* of the forces around the first node vanishes. The vanishing of the force moments

*Note that $\mathbf{u}_j(0) = \mathbf{0}$.

around each of the other two nodes is a consequence of these three conditions, since the associated kernel vectors can be expressed as linear combinations of the three basis elements. The corresponding physical requirements are clear. If there is a net horizontal or vertical force, the space station will rigidly translate in that direction; if there is a non-zero force moment, the station will rigidly rotate. In any event, unless the force balance constraints (6.58) are satisfied, the space station cannot remain in equilibrium. A freely floating space station is an unstable structure that can easily be set into motion with a tiny external force.

Since there are three independent rigid motions, we must impose three constraints on the structure in order to fully stabilize it under general external forcing. "Grounding" one of the nodes, i.e., preventing it from moving by attaching it to a fixed support, will serve to eliminate the two translational instabilities. For example, setting $\mathbf{u}_3 = \mathbf{0}$ has the effect of fixing the third node of the space station to a support. With this specification, we can eliminate the variables associated with that node, and thereby delete the corresponding columns of the incidence matrix—leaving the *reduced incidence matrix*

$$A^\star = \begin{pmatrix} \frac{1}{2} & \frac{\sqrt{3}}{2} & 0 & 0 \\ -\frac{1}{2} & \frac{\sqrt{3}}{2} & \frac{1}{2} & -\frac{\sqrt{3}}{2} \\ 0 & 0 & 1 & 0 \end{pmatrix}.$$

The kernel of $A^\star$ is one-dimensional, spanned by the single vector

$$\mathbf{z}_3^\star = \left(\tfrac{\sqrt{3}}{2},\ \tfrac{1}{2},\ 0,\ 1 \right)^T,$$

which corresponds to (the linear approximation of) the rotations around the fixed node. To prevent the structure from rotating, we can also fix the second node, by further requiring $\mathbf{u}_2 = \mathbf{0}$. This serve to also eliminate the third and fourth columns of the original incidence matrix. The resulting "doubly reduced" incidence matrix

$$A^{\star\star} = \begin{pmatrix} \frac{1}{2} & \frac{\sqrt{3}}{2} \\ -\frac{1}{2} & \frac{\sqrt{3}}{2} \\ 0 & 0 \end{pmatrix}$$

has trivial kernel: $\ker A^{\star\star} = \{\mathbf{0}\}$. Therefore, the corresponding reduced stiffness matrix

$$K^{\star\star} = (A^{\star\star})^T A^{\star\star} = \begin{pmatrix} \frac{1}{2} & -\frac{1}{2} & 0 \\ \frac{\sqrt{3}}{2} & \frac{\sqrt{3}}{2} & 0 \end{pmatrix} \begin{pmatrix} \frac{1}{2} & \frac{\sqrt{3}}{2} \\ -\frac{1}{2} & \frac{\sqrt{3}}{2} \\ 0 & 0 \end{pmatrix} = \begin{pmatrix} \frac{1}{2} & 0 \\ 0 & \frac{3}{2} \end{pmatrix}$$

is positive definite. A planar triangle with two fixed nodes is a stable structure, which can now support an arbitrary external forcing on the remaining free node. (Forces on the fixed nodes have no effect since they are no longer allowed to move.) ●

In general, a planar structure without any fixed nodes will have at least a three-dimensional kernel, corresponding to the rigid motions of translations and (linear approximations to) rotations. To stabilize the structure, one must fix two (non-coincident) nodes. A three-dimensional structure that is not tied to any fixed supports will admit 6 independent rigid motions in its kernel. Three of these correspond to rigid translations in the three coordinate directions, while the other three correspond to linear approximations to the rigid rotations around the three coordinate axes. To eliminate the rigid motion instabilities of the structure, we need to fix three non-collinear nodes. Indeed, fixing one node will eliminate translations; fixing two nodes will still leave the rotations around the axis through the fixed nodes. Details can be found in the exercises.

Even after attaching a sufficient number of nodes to fixed supports so as to eliminate all possible rigid motions, there may still remain nonzero vectors in the kernel of the reduced incidence matrix of the structure. These indicate additional instabilities in which the shape of the structure can deform without any applied force. Such non-rigid motions are known as *mechanisms* of the structure. Since a mechanism moves the nodes without elongating any of the bars, it does not induce any internal forces. A structure that admits a mechanism is unstable—even tiny external forces may provoke a large motion.

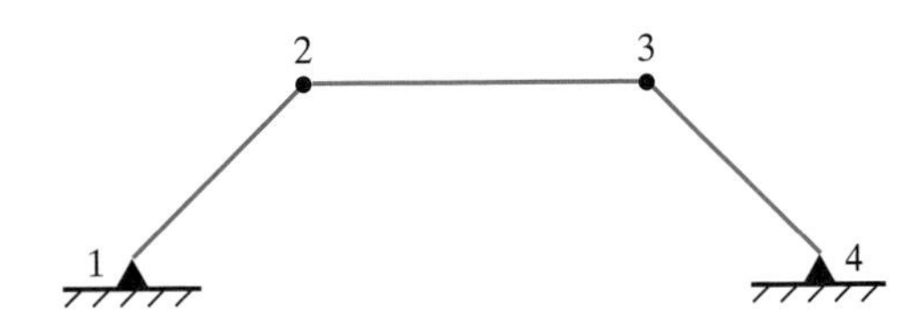

Figure 6.13 Three bar structure with fixed supports.

EXAMPLE 6.7

Consider the three bar structure of Example 6.5, but now with its two ends attached to supports, as pictured in Figure 6.13. Since we are fixing nodes 1 and 4, we set $\mathbf{u}_1 = \mathbf{u}_4 = \mathbf{0}$. Hence, we should remove the first and last column pairs from the incidence matrix (6.49), leading to the reduced incidence matrix

$$A^\star = \begin{pmatrix} \frac{1}{\sqrt{2}} & \frac{1}{\sqrt{2}} & 0 & 0 \\ -1 & 0 & 1 & 0 \\ 0 & 0 & -\frac{1}{\sqrt{2}} & \frac{1}{\sqrt{2}} \end{pmatrix}.$$

The structure no longer admits any rigid motions. However, the kernel of $A^\star$ is one-dimensional, spanned by reduced displacement vector $\mathbf{z}^\star = (1, -1, 1, 1)^T$, which corresponds to the unstable mechanism that displaces the second node in the direction $\mathbf{u}_2 = (1, -1)^T$ and the third node in the direction $\mathbf{u}_3 = (1, 1)^T$. Geometrically, then, $\mathbf{z}^\star$ represents the displacement where node 2 moves down and to the right at a $45°$ angle, while node 3 moves simultaneously up and to the right at a $45°$ angle; the result of the mechanism is sketched in Figure 6.14. This mechanism does not alter the lengths of the three bars (at least in our linear approximation regime) and so requires no net force to be set into motion.

As with the rigid motions of the space station, an external forcing vector $\mathbf{f}^\star$ will maintain equilibrium only when it lies in the corange of $A^\star$, and hence, by the Fredholm Alternative, must be orthogonal to all the mechanisms in $\ker A^\star$. Thus, the nodal forces $\mathbf{f}_2 = (f_2, g_2)^T$ and $\mathbf{f}_3 = (f_3, g_3)^T$ must satisfy the balance law

$$\mathbf{z}^\star \cdot \mathbf{f}^\star = f_2 - g_2 + f_3 + g_3 = 0.$$

If this fails, the equilibrium equation has no solution, and the structure will be set into motion. For example, a uniform horizontal force $f_2 = f_3 = 1$, $g_2 = g_3 = 0$ will induce the mechanism, whereas a uniform vertical force, $f_2 = f_3 = 0$, $g_2 =$

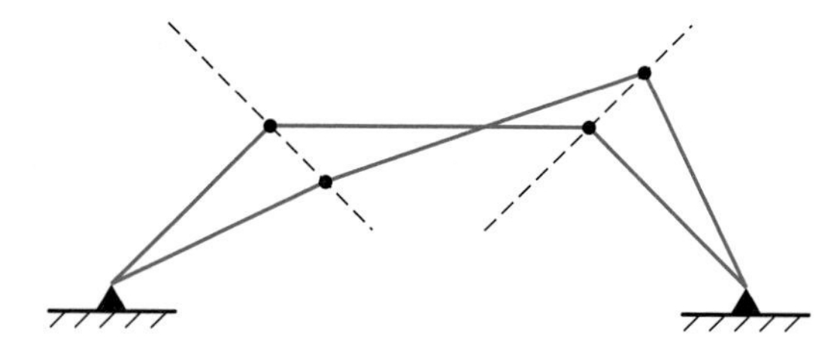

Figure 6.14 Unstable mechanism of the three bar structure.

$g_3 = 1$ will maintain equilibrium. In the latter case, the equilibrium equations

$$K^\star \mathbf{u}^\star = \mathbf{f}^\star, \quad \text{where} \quad K^\star = (A^\star)^T A^\star = \begin{pmatrix} \frac{3}{2} & \frac{1}{2} & -1 & 0 \\ \frac{1}{2} & \frac{1}{2} & 0 & 0 \\ -1 & 0 & \frac{3}{2} & -\frac{1}{2} \\ 0 & 0 & -\frac{1}{2} & \frac{1}{2} \end{pmatrix},$$

have an indeterminate solution

$$\mathbf{u}^\star = (-3, 5, -2, 0)^T + t\,(1, -1, 1, 1)^T,$$

since we can add in any element of $\ker K^\star = \ker A^\star$. In other words, the equilibrium position is not unique, since the structure can still be displaced in the direction of the unstable mechanism while maintaining the overall force balance. On the other hand, the elongations and internal forces

$$\mathbf{y} = \mathbf{e} = A^\star \mathbf{u}^\star = \left(-\sqrt{2},\ -1,\ -\sqrt{2}\ \right)^T,$$

are well-defined, indicating that, under our stabilizing uniform vertical force, all three bars are compressed, with the two diagonals experiencing 41.4% more compression than the horizontal bar. ●

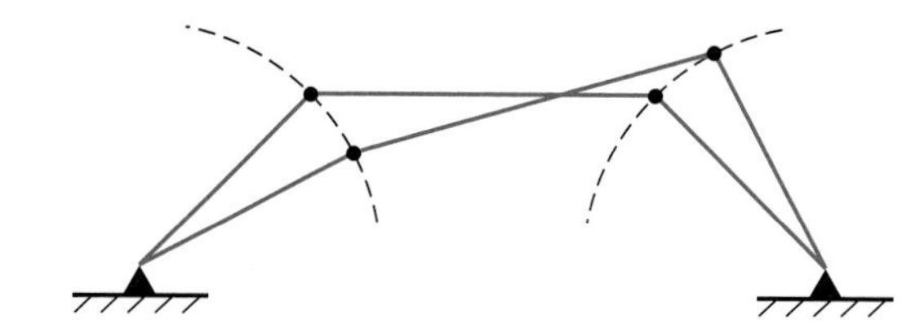

Figure 6.15 Nonlinear mechanism of the three bar structure.

REMARK: Just like the rigid rotations, the mechanisms described here are linear approximations to the actual nonlinear motions. In a physical structure, the vertices will move along curves whose tangents at the initial configuration are the directions indicated by the mechanism vector. In the linear approximation illustrated in Figure 6.14, the lengths of the bars will change slightly. In the true nonlinear mechanism, illustrated in Figure 6.15, the nodes must move along circles so as to rigidly preserve the lengths of all three bars. In certain cases, a structure can admit a linear mechanism, but one that cannot be physically realized due to the nonlinear constraints imposed by the geometrical configurations of the bars. Nevertheless, such a structure is at best borderline stable, and should not be used in any real-world constructions.

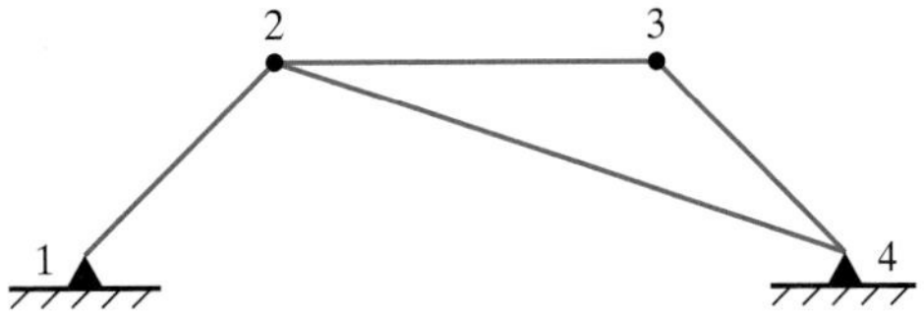

Figure 6.16 Reinforced planar structure.

We can always stabilize a structure by first fixing nodes to eliminate rigid motions, and then adding in a sufficient number of extra bars to prevent mechanisms. In the preceding example, suppose we attach an additional bar connecting nodes 2 and 4, leading to the reinforced structure in Figure 6.16. The revised incidence matrix is

$$A = \left(\begin{array}{cc|cc|cc|cc} -\frac{1}{\sqrt{2}} & -\frac{1}{\sqrt{2}} & \frac{1}{\sqrt{2}} & \frac{1}{\sqrt{2}} & 0 & 0 & 0 & 0 \\ 0 & 0 & -1 & 0 & 1 & 0 & 0 & 0 \\ 0 & 0 & 0 & 0 & -\frac{1}{\sqrt{2}} & \frac{1}{\sqrt{2}} & \frac{1}{\sqrt{2}} & -\frac{1}{\sqrt{2}} \\ 0 & 0 & -\frac{3}{\sqrt{10}} & \frac{1}{\sqrt{10}} & 0 & 0 & \frac{3}{\sqrt{10}} & -\frac{1}{\sqrt{10}} \end{array}\right)$$

and is obtained from (6.49) by appending another row representing the added bar. When nodes 1 and 4 are fixed, the reduced incidence matrix

$$A^\star = \begin{pmatrix} \frac{1}{\sqrt{2}} & \frac{1}{\sqrt{2}} & 0 & 0 \\ -1 & 0 & 1 & 0 \\ 0 & 0 & -\frac{1}{\sqrt{2}} & \frac{1}{\sqrt{2}} \\ -\frac{3}{\sqrt{10}} & \frac{1}{\sqrt{10}} & 0 & 0 \end{pmatrix}$$

has trivial kernel, $\ker A^\star = \{\mathbf{0}\}$, and hence the reinforced structure is stable. It admits no mechanisms, and can support any configuration of forces (within reason — mathematically the structure will support an arbitrarily large external force, but very large forces will take us outside the linear regime described by the model, and the structure may be crushed).

This particular case is *statically determinate* owing to the fact that the incidence matrix is square and nonsingular, which implies that one can solve the force balance equations (6.57) directly for the internal forces. For instance, a uniform downwards vertical force $f_2 = f_3 = 0$, $g_2 = g_3 = -1$, e.g., gravity, will produce the internal forces

$$y_1 = -\sqrt{2}, \quad y_2 = -1, \quad y_3 = -\sqrt{2}, \quad y_4 = 0$$

indicating that bars 1, 2 and 3 are compressed, while, interestingly, the reinforcing bar 4 remains unchanged in length and hence experiences no internal force. Assuming the bars are all of the same material, and taking the elastic constant to be 1, so $C = \mathrm{I}$, then the reduced stiffness matrix is

$$K^\star = (A^\star)^T A^\star = \begin{pmatrix} \frac{12}{5} & \frac{1}{5} & -1 & 0 \\ \frac{1}{5} & \frac{3}{5} & 0 & 0 \\ -1 & 0 & \frac{3}{2} & -\frac{1}{2} \\ 0 & 0 & -\frac{1}{2} & \frac{1}{2} \end{pmatrix}.$$

The solution to the reduced equilibrium equations is

$$\mathbf{u}^\star = \left(-\tfrac{1}{2}, -\tfrac{3}{2}, -\tfrac{3}{2}, -\tfrac{7}{2}\right)^T, \quad \text{so} \quad \mathbf{u}_2 = \left(-\tfrac{1}{2}, -\tfrac{3}{2}\right)^T, \quad \mathbf{u}_3 = \left(-\tfrac{3}{2}, -\tfrac{7}{2}\right)^T.$$

give the displacements of the two nodes under the applied force. Both are moving down and to the left, with node 3 moving relatively farther owing to its lack of reinforcement.

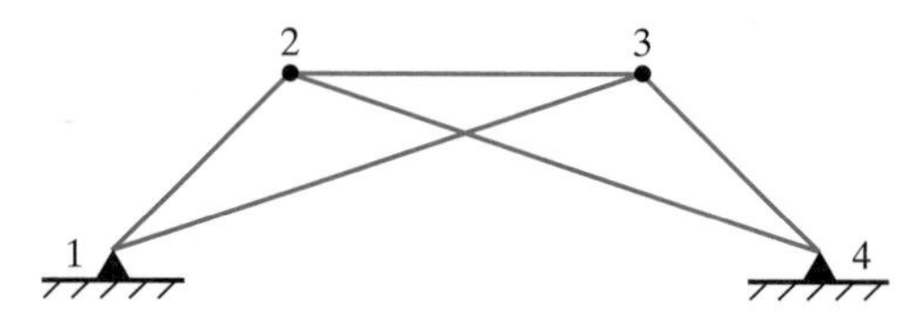

Figure 6.17 Doubly reinforced planar structure.

Suppose we reinforce the structure yet further by adding in a bar connecting nodes 1 and 3, as in Figure 6.17. The resulting reduced incidence matrix

$$A^\star = \begin{pmatrix} \frac{1}{\sqrt{2}} & \frac{1}{\sqrt{2}} & 0 & 0 \\ -1 & 0 & 1 & 0 \\ 0 & 0 & -\frac{1}{\sqrt{2}} & \frac{1}{\sqrt{2}} \\ -\frac{3}{\sqrt{10}} & \frac{1}{\sqrt{10}} & 0 & 0 \\ 0 & 0 & \frac{3}{\sqrt{10}} & \frac{1}{\sqrt{10}} \end{pmatrix}$$

again has trivial kernel, $\ker A^\star = \{\mathbf{0}\}$, and hence the structure is stable. Indeed, adding in extra bars to a stable structure cannot cause it to lose stability. (In matrix language, appending additional rows to a matrix cannot increase the size of its kernel, cf. Exercise 2.5.10.) Since the incidence matrix is rectangular, the structure is now *statically indeterminate*, and we cannot determine the internal forces without first solving the full equilibrium equations (6.52) for the displacements. The stiffness matrix is

$$K^\star = (A^\star)^T A^\star = \begin{pmatrix} \frac{12}{5} & \frac{1}{5} & -1 & 0 \\ \frac{1}{5} & \frac{3}{5} & 0 & 0 \\ -1 & 0 & \frac{12}{5} & -\frac{1}{5} \\ 0 & 0 & -\frac{1}{5} & \frac{3}{5} \end{pmatrix}.$$

Under the same uniform vertical force, the displacement

$$\mathbf{u}^\star = \left(\tfrac{1}{10}, -\tfrac{17}{10}, -\tfrac{1}{10}, -\tfrac{17}{10}\right)^T$$

indicates that the free nodes now move symmetrically down and towards the center of the structure. The internal forces on the bars are

$$y_1 = -\tfrac{4}{5}\sqrt{2}, \quad y_2 = -\tfrac{1}{5}, \quad y_3 = -\tfrac{4}{5}\sqrt{2}, \quad y_4 = -\sqrt{\tfrac{2}{5}}, \quad y_5 = -\sqrt{\tfrac{2}{5}}.$$

All five bars are now experiencing compression, with the two outside bars being the most stressed. This relatively simple computation should already indicate to the practicing construction engineer which of the bars in the structure are more likely to collapse under an applied external force.

Summarizing our discussion, we have established the following fundamental result characterizing the stability and equilibrium of structures.

THEOREM 6.8 A structure is stable, and will maintain an equilibrium under arbitrary external forcing, if and only if its reduced incidence matrix $A^\star$ has linearly independent columns, or, equivalently, $\ker A^\star = \{\mathbf{0}\}$. More generally, an external force $\mathbf{f}^\star$ on a structure will maintain equilibrium if and only if $\mathbf{f}^\star \in \operatorname{corng} A^\star = (\ker A^\star)^\perp$, which requires that the external force be orthogonal to all rigid motions and all mechanisms admitted by the structure.

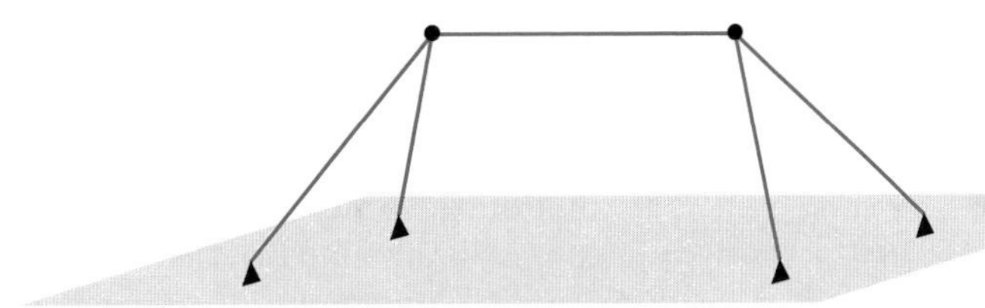

Figure 6.18 A swing set.

EXAMPLE 6.9 A swing set is to be constructed, consisting of two diagonal supports at each end joined by a horizontal cross bar. Is this configuration stable, i.e., can a child swing on it without it collapsing? The movable joints are at positions

$$\mathbf{a}_1 = (1, 1, 3)^T, \quad \mathbf{a}_2 = (4, 1, 3)^T,$$

while the four fixed supports are at

$$\mathbf{a}_3 = (0, 0, 0)^T, \quad \mathbf{a}_4 = (0, 2, 0)^T, \quad \mathbf{a}_5 = (5, 0, 0)^T, \quad \mathbf{a}_6 = (5, 2, 0)^T.$$

The reduced incidence matrix for the structure is calculated in the usual manner:

$$A^\star = \left(\begin{array}{ccc|ccc} \frac{1}{\sqrt{11}} & \frac{1}{\sqrt{11}} & \frac{3}{\sqrt{11}} & 0 & 0 & 0 \\ \frac{1}{\sqrt{11}} & -\frac{1}{\sqrt{11}} & \frac{3}{\sqrt{11}} & 0 & 0 & 0 \\ -1 & 0 & 0 & 1 & 0 & 0 \\ 0 & 0 & 0 & -\frac{1}{\sqrt{11}} & \frac{1}{\sqrt{11}} & \frac{3}{\sqrt{11}} \\ 0 & 0 & 0 & -\frac{1}{\sqrt{11}} & -\frac{1}{\sqrt{11}} & \frac{3}{\sqrt{11}} \end{array}\right)$$

For instance, the first three entries contained in the first row refer to the unit vector

$$\mathbf{n}_1 = \frac{\mathbf{a}_1 - \mathbf{a}_3}{\|\mathbf{a}_1 - \mathbf{a}_3\|}$$

in the direction of the bar going from $\mathbf{a}_3$ to $\mathbf{a}_1$. Suppose the three bars have the same stiffness $c_1 = \cdots = c_5 = 1$, so the reduced stiffness matrix for the structure is

$$K^\star = (A^\star)^T A^\star = \begin{pmatrix} \frac{13}{11} & 0 & \frac{6}{11} & -1 & 0 & 0 \\ 0 & \frac{2}{11} & 0 & 0 & 0 & 0 \\ \frac{6}{11} & 0 & \frac{18}{11} & 0 & 0 & 0 \\ -1 & 0 & 0 & \frac{13}{11} & 0 & -\frac{6}{11} \\ 0 & 0 & 0 & 0 & \frac{2}{11} & 0 \\ 0 & 0 & 0 & -\frac{6}{11} & 0 & \frac{18}{11} \end{pmatrix}.$$

Solving $A^\star \mathbf{z}^\star = \mathbf{0}$, we find $\ker A^\star = \ker K^\star$ is one-dimensional, spanned by

$$\mathbf{z}^\star = (3, 0, -1, 3, 0, 1)^T.$$

This indicates a mechanism that causes the swing set to collapse: the first node moves down and to the right, while the second node moves up and to the right, the horizontal motion being three times as large as the vertical. The swing set can only support forces $\mathbf{f}_1 = (f_1, g_1, h_1)^T$, $\mathbf{f}_2 = (f_2, g_2, h_2)^T$ on the free nodes whose combined force vector $\mathbf{f}^\star$ is orthogonal to the mechanism vector, and so

$$3\,(f_1 + f_2) - h_1 + h_2 = 0.$$

Otherwise, a reinforcing bar, say from node 1 to node 6 (although this will interfere with the swinging!) or another bar connecting one of the nodes to a new ground support, will be required to completely stabilize the swing.

For a uniform downwards unit vertical force, $\mathbf{f}^\star = (0, 0, -1, 0, 0, -1)^T$, a particular solution to (6.11) is $\mathbf{u}^\star = \left(\frac{13}{6}, 0, -\frac{4}{3}, \frac{11}{6}, 0, 0\right)^T$ and the general solution $\mathbf{u} = \mathbf{u}^\star + t\,\mathbf{z}^\star$ is obtained by adding in an arbitrary element of the kernel. The resulting forces/elongations are uniquely determined,

$$\mathbf{y} = \mathbf{e} = A^\star \mathbf{u} = A^\star \mathbf{u}^\star = \left(-\frac{\sqrt{11}}{6},\ -\frac{\sqrt{11}}{6},\ -\frac{1}{3},\ -\frac{\sqrt{11}}{6},\ -\frac{\sqrt{11}}{6}\right)^T,$$

so that every bar is compressed, the middle one experiencing slightly more than half the stress of the outer supports.

Figure 6.19 Reinforced swing set.

If we add in two vertical supports at the nodes, as in Figure 6.19, then the corresponding reduced incidence matrix

$$A^\star = \begin{pmatrix} \frac{1}{\sqrt{11}} & \frac{1}{\sqrt{11}} & \frac{3}{\sqrt{11}} & 0 & 0 & 0 \\ \frac{1}{\sqrt{11}} & -\frac{1}{\sqrt{11}} & \frac{3}{\sqrt{11}} & 0 & 0 & 0 \\ -1 & 0 & 0 & 1 & 0 & 0 \\ 0 & 0 & 0 & -\frac{1}{\sqrt{11}} & \frac{1}{\sqrt{11}} & \frac{3}{\sqrt{11}} \\ 0 & 0 & 0 & -\frac{1}{\sqrt{11}} & -\frac{1}{\sqrt{11}} & \frac{3}{\sqrt{11}} \\ 0 & 0 & 1 & 0 & 0 & 0 \\ 0 & 0 & 0 & 0 & 0 & 1 \end{pmatrix}$$

has trivial kernel, indicating stabilization of the structure. The reduced stiffness matrix

$$K^\star = \begin{pmatrix} \frac{13}{11} & 0 & \frac{6}{11} & -1 & 0 & 0 \\ 0 & \frac{2}{11} & 0 & 0 & 0 & 0 \\ \frac{6}{11} & 0 & \frac{29}{11} & 0 & 0 & 0 \\ -1 & 0 & 0 & \frac{13}{11} & 0 & -\frac{6}{11} \\ 0 & 0 & 0 & 0 & \frac{2}{11} & 0 \\ 0 & 0 & 0 & -\frac{6}{11} & 0 & \frac{29}{11} \end{pmatrix}$$

is only slightly different than before, but this is enough to make it positive definite, $K^\star > 0$, and so allow arbitrary external forcing without collapse. Under the same uniform vertical force, the internal forces are

$$\mathbf{y} = \mathbf{e} = A^\star \mathbf{u} = \left(-\tfrac{\sqrt{11}}{10},\ -\tfrac{\sqrt{11}}{10},\ -\tfrac{1}{5},\ -\tfrac{\sqrt{11}}{10},\ -\tfrac{\sqrt{11}}{10},\ -\tfrac{2}{5},\ -\tfrac{2}{5} \right)^T.$$

Note the overall reductions in stress in the original bars; the two reinforcing vertical bars are now experiencing the largest compression. ●

Further developments in the mathematical analysis of structures can be found in [24, 59].

EXERCISES 6.3

6.3.1. If a bar in a structure compresses 2 cm under a force of 5 Newtons applied to a node, how far will it compress under a force of 20 Newtons applied at the same node?

6.3.2. An individual bar in a structure experiences a stress of 3 under a unit horizontal force applied to all the nodes and a stress of -2 under a unit vertical force applied to all nodes. What combinations of horizontal and vertical forces will make the bar stress-free?

6.3.3. (a) For the reinforced structure illustrated in Figure 6.16, determine the displacements of the nodes and the stresses in the bars under a uniform horizontal force, and interpret physically.

(b) Answer the same question for the doubly reinforced structure in Figure 6.17.

6.3.4. Discuss the effect of a uniform horizontal force in the direction of the horizontal bar on the swing set and its reinforced version in Example 6.9.

♡ **6.3.5.** All the bars in the illustrated square planar structure have unit stiffness.

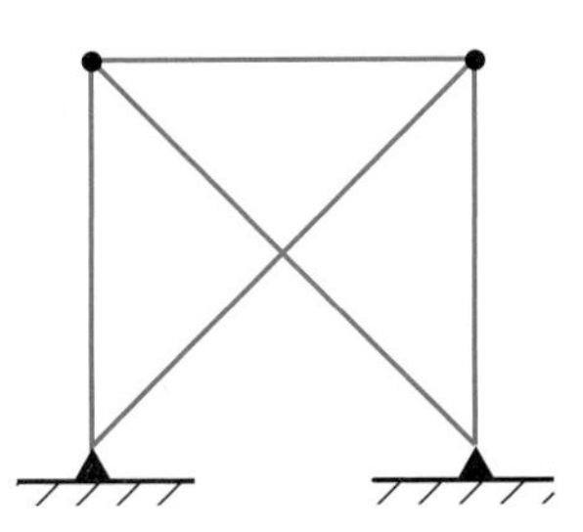

(a) Write down the reduced incidence matrix A.

(b) Write down the equilibrium equations for the structure when subjected to external forces at the free nodes.

(c) Is the structure stable? statically determinate? Explain in detail.

(d) Find a set of external forces with the property that the upper left node moves horizontally, while the upper right node stays in place. Which bar is under the most stress?

♡ **6.3.6.** In the square structure of Exercise 6.3.5, the diagonal struts simply cross each other. We could also try joining them at an additional central node. Compare the stresses in the two structures under a uniform horizontal and a uniform vertical force at the two upper nodes, and discuss what you observe.

6.3.7. Write down the reduced incidence matrix A for the pictured structure with 4 bars and 2 fixed supports. The width and the height of the vertical sides is 1 unit, while the top node is 1.5 units above the base.

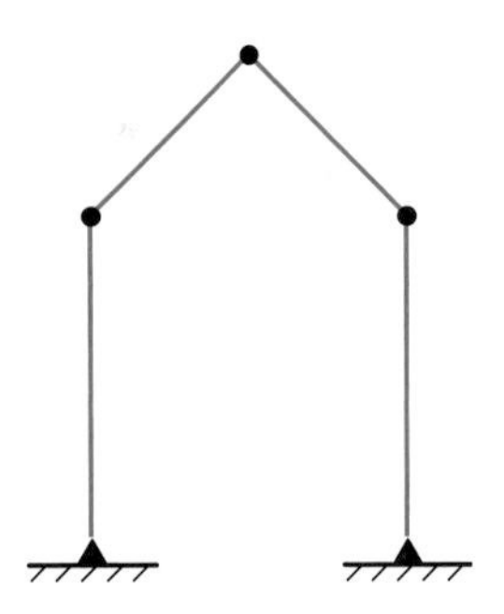

(a) Predict the number of independent solutions to $A\mathbf{u} = \mathbf{0}$, and then solve to describe them both numerically and geometrically.

(b) What condition(s) must be imposed on the external forces to maintain equilibrium in the structure?

(c) Add in just enough additional bars so that the resulting reinforced structure has only the trivial solution to $A\mathbf{u} = \mathbf{0}$. Is your reinforced structure stable?

♡ **6.3.8.** Consider the two-dimensional "house" constructed out of bars, as in the accompanying picture. The bottom nodes are fixed. The width of the house is 3 units, the height of the vertical sides 1 unit, and the peak is 1.5 units above the base.

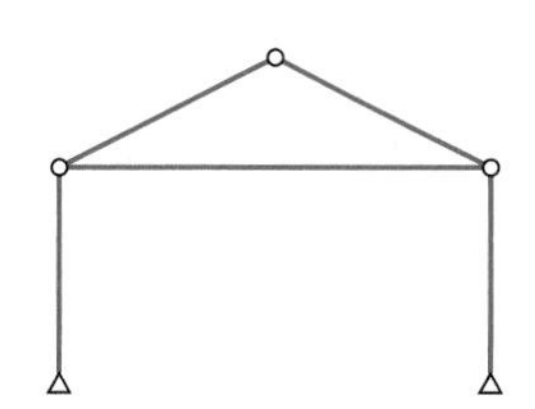

(a) Determine the reduced incidence matrix A for this structure.

(b) How many distinct modes of instability are there? Describe them geometrically, and indicate whether they are mechanisms or rigid motions.

(c) Suppose we apply a combination of forces to each non-fixed node in the structure. Determine conditions such that the structure can support the forces. Write down an explicit nonzero set of external forces that satisfy these conditions, and compute the corresponding elongations of the individual bars. Which bar is under the most stress?

(d) Add in a *minimal* number of bars so that the resulting structure can support any force. Before starting, decide, from general principles, how many bars you need to add.

(e) With your new stable configuration, use the same force as before, and recompute the forces on the individual bars. Which bar now has the most stress? How much have you reduced the maximal stress in your reinforced building?

♣ **6.3.9.** Answer Exercise 6.3.8 for the illustrated two- and three-dimensional houses. In the two-dimensional case, the width and total height of the vertical bars is 2 units, and the peak is an additional .5 unit higher. In the three-dimensional house, the width and vertical heights are equal to 1 unit, the length is 3 units, while the peaks are 1.5 units above the base.

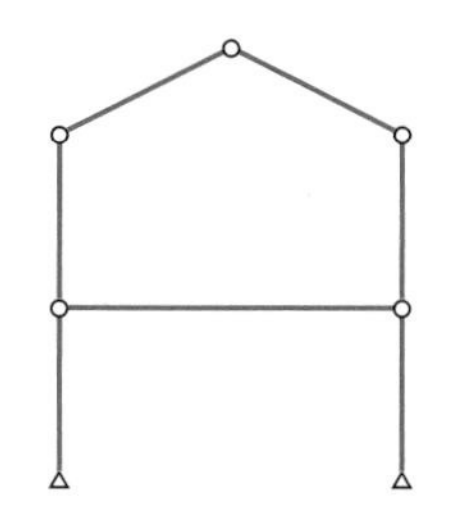

♡ **6.3.10.** Consider a structure consisting of three bars joined in a vertical line hanging from a top support.

(a) Write down the equilibrium equations for this system when only forces and displacements in the vertical direction are allowed, i.e., a one-dimensional structure. Is the problem statically determinate, statically indeterminate or unstable? If the latter, describe all possible mechanisms and the constraints on the forces required to maintain equilibrium.

(b) Answer part (a) when the structure is two-dimensional, i.e., is allowed to move in a plane.

(c) Answer the same question for the fully three-dimensional version.

♡ **6.3.11.** A *mass–spring ring* consists of n masses connected in a circle by n identical springs, and the masses are only allowed to move in the angular direction.

(a) Derive the equations of equilibrium.

(b) Discuss stability, and characterize the external forces that will maintain equilibrium.

(c) Find such a set of nonzero external forces in the case of a four mass ring and solve the equilibrium equations. What does the nonuniqueness of the solution represent?

♣ **6.3.12.** A space station is built in the shape of a three-dimensional *simplex* whose nodes are at the positions $\mathbf{0}, \mathbf{e}_1, \mathbf{e}_2, \mathbf{e}_3 \in \mathbb{R}^3$, and each pair of nodes is connected by a bar.

(a) Sketch the space station and find its incidence matrix A.

(b) Show that $\ker A$ is six-dimensional, and find a basis.

(c) Explain which three basis vectors correspond to rigid translations.

(d) Find three basis vectors that correspond to linear approximations to rotations around the three coordinate axes.

(e) Suppose the bars all have unit stiffness. Compute the full stiffness matrix for the space station.

(f) What constraints on external forces at the four nodes are required to maintain equilibrium? Can you interpret them physically?

(g) How many nodes do you need to fix to stabilize the structure?

(h) Suppose you fix the three nodes in the xy plane. How much internal force does each bar experience under a unit vertical force on the upper vertex?

♣ **6.3.13.** Suppose a space station is built in the shape of a regular tetrahedron with all sides of unit length. Answer all questions in Exercise 6.3.12.

6.3.14. *True or false*: If a structure constructed out of bars with identical stiffnesses is stable, then the same structure constructed out of bars with differing stiffnesses is also stable.

6.3.15. *True or false*: If a structure is statically indeterminate, then every non-zero applied force will result in

(a) one or more nodes having a non-zero displacement;

(b) one or more bars having a non-zero elongation.

6.3.16. A structure in $\mathbb{R}^3$ has n movable nodes, admits no rigid motions, and is statically determinate.

(a) How many bars must it have?

(b) Find an example with $n = 3$.

◇ **6.3.17.** Prove that if we apply a unit force to node i in a structure and measure the displacement of node j in the direction of the force, then we obtain the same value if we apply the force to node j and measure the displacement at node i in the same direction. *Hint*: First, solve Exercise 6.1.6.

6.3.18. *True or false*: A structure in $\mathbb{R}^3$ will admit no rigid motions if and only if at least 3 nodes are fixed.

6.3.19. Suppose all bars have unit stiffness. Explain why the internal forces in a structure form the solution of minimal Euclidean norm among all solutions to $A^T \mathbf{y} = \mathbf{f}$.

◇ **6.3.20.** Let A be the reduced incidence matrix for a structure and C the diagonal bar stiffness matrix. Suppose $\mathbf{f}$ is a set of external forces that maintains equilibrium of the structure.

(a) Prove that $\mathbf{f} = A^T C \mathbf{g}$ for some $\mathbf{g}$.

(b) Prove that an allowable displacement $\mathbf{u}$ is a least squares solution to the system $A\mathbf{u} = \mathbf{g}$ with respect to the weighted norm $\|\mathbf{v}\|^2 = \mathbf{v}^T C \mathbf{v}$.

♡ **6.3.21.** Suppose an *unstable* structure admits no rigid motions—only mechanisms. Let $\mathbf{f}$ be an external force on the structure that maintains equilibrium. Suppose that you stabilize the structure by adding in the *minimal* number of reinforcing bars. Prove that the given force $\mathbf{f}$ induces the *same* stresses in the original bars, while the reinforcing bars experience no stress. Are the displacements necessarily the same? Does the result continue to hold when more reinforcing bars are added to the structure? *Hint*: Use Exercise 6.3.20.

♡ **6.3.22.** When a node is fixed to a *roller*, it is only permitted to move along a straight line—the direction of the roller. Consider the three bar structure in Example 6.5. Suppose node 1 is fixed, but node 4 is attached to a roller that only permits it to move in the horizontal direction.

(a) Construct the reduced incidence matrix and the equilibrium equations in this situation. You should have a system of 5 equations in 5 unknowns—the horizontal and vertical displacements of nodes 2 and 3 and the horizontal displacement of node 4.

(b) Is your structure stable? If not, how many rigid motions and how many mechanisms does it permit?

♡ **6.3.23.** Answer Exercise 6.3.22 when the roller at node 4 only allows it to move in the vertical direction.

6.3.24. Redo Exercises 6.3.22–23 for the reinforced structure in Figure 6.16.

6.3.25. (a) Suppose that we fix one node in a planar structure and put a second node on a roller. Does the structure admit any rigid motions?

(b) How many rollers are needed to prevent all rigid motions in a three-dimensional structure? Are there any restrictions on the directions of the rollers?

CHAPTER 7

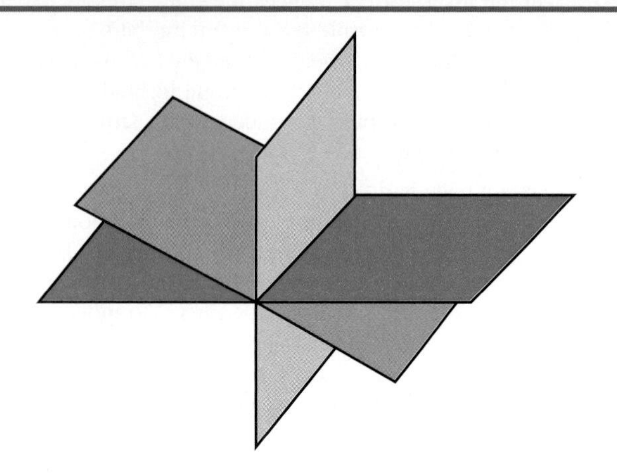

Linearity

We began this book by learning how to systematically solve systems of linear algebraic equations. This "elementary" problem formed our launching pad for developing the fundamentals of linear algebra. In its initial form, matrices and vectors were the primary focus of our study, but the theory was developed in a sufficiently general and abstract form that it can be immediately used in many other useful situations—particularly infinite-dimensional function spaces. Indeed, applied mathematics deals, not just with algebraic equations, but also with differential equations, difference equations, integral equations, differential delay equations, control systems, and many other types of system—not all of which, unfortunately, can be adequately developed in this introductory text. It is now time to assemble what we have learned about linear matrix systems and place the results in a suitably general framework that will lead to insight into the key principles that govern all linear systems arising in mathematics and its applications.

The most basic underlying object of linear systems theory is the vector space, and we have already seen that the elements of vector spaces can be vectors, or functions, or even vector-valued functions. The seminal ideas of span, linear independence, basis, and dimension are equally applicable and equally vital in more general contexts, particularly function spaces. Just as vectors in Euclidean space are prototypes for elements of general vector spaces, matrices are also prototypes for more general objects, known as *linear functions*. Linear functions are also known as linear maps or, when dealing with function spaces, linear operators, and include linear differential operators, linear integral operators, function evaluation, and many other basic operations. Generalized functions, such as the delta function, are, in fact, most properly formulated as linear operators on a suitable function space; see Chapter 11. Linear operators on infinite-dimensional function spaces are the basic objects of quantum mechanics. Each quantum mechanical observable (mass, energy, momentum) is formulated as a linear operator on an infinite-dimensional Hilbert space—the space of wave functions or states of the system, [40]. It is remarkable that quantum mechanics is an entirely linear theory, whereas classical and relativistic mechanics are inherently nonlinear. The holy grail of modern physics—the unification of general relativity and quantum mechanics—is to resolve the apparent incompatibility of the microscopic linear and macroscopic nonlinear physical regimes.

In geometry, linear functions are interpreted as linear transformations of space (or space-time), and, as such, lie at the foundations of motion of bodies, such as satellites and planets, computer graphics and games, and the mathematical formulation of symmetry. Many familiar geometrical transformations, including rotations, scalings, reflections, projections, and shears, are linear. But including translational motions requires a slight extension of linearity, known as an affine transformation.

Linear functions form the simplest class of functions on vector spaces, and must be thoroughly understood before any serious progress can be made in the vastly more complicated nonlinear world. Indeed, nonlinear functions are often approximated by linear functions, generalizing the calculus approximation of a scalar function by its tangent line. This linearization process is applied to nonlinear functions of several variables studied in multivariable calculus, as well as the nonlinear systems arising in physics and mechanics, which can often be well approximated by linear differential equations.

A linear system is just an equation formed by a linear function. The most basic linear system is a system of linear algebraic equations. Linear systems theory includes linear differential equations, linear boundary value problems, linear integral equations, and so on, all in a common conceptual framework. The fundamental ideas of linear superposition and the relation between the solutions to inhomogeneous and homogeneous systems are universally applicable to all linear systems. You have no doubt encountered many of these concepts in your study of elementary ordinary differential equations. In this text, they have already appeared in our discussion of the solutions to linear algebraic systems, and will be seen yet again in the context of boundary value problems.

7.1 LINEAR FUNCTIONS

We begin our study of linear functions with the basic definition. For simplicity, we shall concentrate on real linear functions between real vector spaces. Extending the concepts and constructions to complex linear functions on complex vector spaces is not difficult, and will be dealt with in due course.

Definition 7.1 Let V and W be real vector spaces. A function $L\colon V \to W$ is called *linear* if it obeys two basic rules:

$$L[\mathbf{v}+\mathbf{w}] = L[\mathbf{v}] + L[\mathbf{w}], \qquad L[c\,\mathbf{v}] = c\,L[\mathbf{v}], \tag{7.1}$$

for all $\mathbf{v}, \mathbf{w} \in V$ and all scalars c. We will call V the *domain space* and W the *target space** for L.

In particular, setting $c = 0$ in the second condition implies that a linear function always maps the zero element in V to the zero element in W, so

$$L[\mathbf{0}] = \mathbf{0}. \tag{7.2}$$

(But keep in mind that the two $\mathbf{0}$'s in this equation represent different zero elements—the left hand one is in V, but the right hand side is in W.) We can readily combine the two defining conditions (7.1) into a single rule

$$L[c\,\mathbf{v} + d\,\mathbf{w}] = c\,L[\mathbf{v}] + d\,L[\mathbf{w}], \qquad \text{for all} \quad \mathbf{v}, \mathbf{w} \in V, \quad c, d \in \mathbb{R}, \tag{7.3}$$

*The term "target" is used here to avoid later confusion with the range of L, which, in general, is a subspace of the target vector space W.

that characterizes linearity of a function L. An easy induction proves that a linear function respects linear combinations, so

$$L[c_1\mathbf{v}_1 + \cdots + c_k\mathbf{v}_k] = c_1 L[\mathbf{v}_1] + \cdots + c_k L[\mathbf{v}_k] \tag{7.4}$$

for any $c_1, \ldots, c_k \in \mathbb{R}$ and $\mathbf{v}_1, \ldots, \mathbf{v}_k \in V$.

The interchangeable terms *linear map*, *linear operator*, and, when $V = W$, *linear transformation* are all commonly used as alternatives to "linear function", depending on the circumstances and taste of the author. The term "linear operator" is particularly useful when the underlying vector space is a function space, so as to avoid confusing the two different uses of the word "function". As usual, we will often refer to the elements of a vector space as "vectors", even though they might be functions or matrices or something else, depending upon the context.

EXAMPLE 7.2

The simplest linear function is the zero function $\mathrm{O}[\mathbf{v}] \equiv \mathbf{0}$ which maps every element $\mathbf{v} \in V$ to the zero vector in W. Note that, in view of (7.2), this is the *only* constant linear function; a nonzero constant function is *not*, despite its evident simplicity, linear. Another simple but important linear function is the identity function $\mathrm{I} = \mathrm{I}_V : V \to V$ which leaves every vector unchanged: $\mathrm{I}[\mathbf{v}] = \mathbf{v}$. Slightly more generally, the operation of scalar multiplication $M_a[\mathbf{v}] = a\,\mathbf{v}$ by a fixed scalar $a \in \mathbb{R}$ defines a linear function from V to itself, with $M_0 = \mathrm{O}$, the zero function from V to itself, and $M_1 = \mathrm{I}$, the identity function on V, appearing as special cases. ●

EXAMPLE 7.3

Suppose $V = \mathbb{R}$. We claim that every linear function $L: \mathbb{R} \to \mathbb{R}$ has the form

$$y = L[x] = a x,$$

for some constant a. Therefore, the only scalar linear functions are those whose graph is a straight line passing through the origin. To prove this, we write $x \in \mathbb{R}$ as a scalar product $x = x \cdot 1$. Then, by the second property in (7.1),

$$L[x] = L[x \cdot 1] = x \cdot L[1] = a x, \qquad \text{where} \quad a = L[1],$$

as claimed. ●

Warning: Even though the graph of the function

$$y = a x + b, \tag{7.5}$$

is a straight line, it is *not* a linear function—unless $b = 0$, so the line goes through the origin. The proper mathematical name for a function of the form (7.5) is an *affine function*; see Definition 7.21 below.

EXAMPLE 7.4

Let $V = \mathbb{R}^n$ and $W = \mathbb{R}^m$. Let A be an $m \times n$ matrix. Then the function $L[\mathbf{v}] = A\mathbf{v}$ given by matrix multiplication is easily seen to be a linear function. Indeed, the requirements (7.1) reduce to the basic distributivity and scalar multiplication properties of matrix multiplication:

$$A(\mathbf{v} + \mathbf{w}) = A\mathbf{v} + A\mathbf{w}, \qquad A(c\,\mathbf{v}) = c\,A\mathbf{v}, \qquad \text{for all} \quad \mathbf{v}, \mathbf{w} \in \mathbb{R}^n, \quad c \in \mathbb{R}.$$

In fact, *every* linear function between two Euclidean spaces has this form. ●

THEOREM 7.5

Every linear function $L: \mathbb{R}^n \to \mathbb{R}^m$ is given by matrix multiplication, $L[\mathbf{v}] = A\mathbf{v}$, where A is an $m \times n$ matrix.

Warning: Pay attention to the order of m and n. While A has size $m \times n$, the linear function L goes *from* $\mathbb{R}^n$ *to* $\mathbb{R}^m$.

Proof The key idea is to look at what the linear function does to the basis vectors. Let $\mathbf{e}_1, \ldots, \mathbf{e}_n$ be the standard basis of $\mathbb{R}^n$, and let $\widehat{\mathbf{e}}_1, \ldots, \widehat{\mathbf{e}}_m$ be the standard basis of $\mathbb{R}^m$. (We temporarily place hats on the latter to avoid confusing the two.) Since $L[\mathbf{e}_j] \in \mathbb{R}^m$, we can write it as a linear combination of the latter basis vectors:

$$L[\mathbf{e}_j] = \mathbf{a}_j = \begin{pmatrix} a_{1j} \\ a_{2j} \\ \vdots \\ a_{mj} \end{pmatrix} = a_{1j}\widehat{\mathbf{e}}_1 + a_{2j}\widehat{\mathbf{e}}_2 + \cdots + a_{mj}\widehat{\mathbf{e}}_m, \qquad j = 1, \ldots, n. \tag{7.6}$$

Let us construct the $m \times n$ matrix

$$A = (\, \mathbf{a}_1\ \mathbf{a}_2 \ldots \mathbf{a}_n \,) = \begin{pmatrix} a_{11} & a_{12} & \cdots & a_{1n} \\ a_{21} & a_{22} & \cdots & a_{2n} \\ \vdots & \vdots & \ddots & \vdots \\ a_{m1} & a_{m2} & \cdots & a_{mn} \end{pmatrix} \tag{7.7}$$

whose columns are the image vectors (7.6). Using (7.4), we then compute the effect of L on a general vector $\mathbf{v} = (v_1, v_2, \ldots, v_n)^T \in \mathbb{R}^n$:

$$\begin{aligned} L[\mathbf{v}] &= L[\, v_1\mathbf{e}_1 + \cdots + v_n\mathbf{e}_n \,] \\ &= v_1 L[\mathbf{e}_1] + \cdots + v_n L[\mathbf{e}_n] \\ &= v_1\mathbf{a}_1 + \cdots + v_n\mathbf{a}_n = A\mathbf{v}. \end{aligned}$$

The final equality follows from our basic formula (2.13) connecting matrix multiplication and linear combinations. We conclude that the vector $L[\mathbf{v}]$ coincides with the vector $A\mathbf{v}$ obtained by multiplying $\mathbf{v}$ by the coefficient matrix A. ■

The proof of Theorem 7.5 shows us how to construct the matrix representative of a given linear function $L\colon \mathbb{R}^n \to \mathbb{R}^m$. We merely assemble the image column vectors $\mathbf{a}_1 = L[\mathbf{e}_1], \ldots, \mathbf{a}_n = L[\mathbf{e}_n]$ into an $m \times n$ matrix A.

The two basic linearity conditions (7.1) have a simple geometrical interpretation. Since vector addition is the same as completing the parallelogram sketched in Figure 7.1, the first linearity condition requires that L map parallelograms to parallelograms. The second linearity condition says that if we stretch a vector by a factor c, then its image under L must also be stretched by the same amount. Thus, one can often detect linearity by simply looking at the geometry of the function.

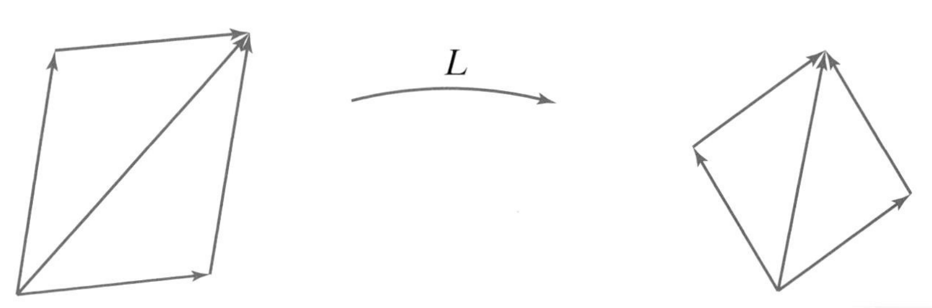

Figure 7.1 Linear function on Euclidean space.

EXAMPLE 7.6 As a specific example, consider the function $R_\theta\colon \mathbb{R}^2 \to \mathbb{R}^2$ that rotates the vectors in the plane around the origin by a specified angle θ. This geometric transformation clearly preserves parallelograms—see Figure 7.2. It also respects stretching of vectors, and hence defines a linear function. In order to find its matrix representative, we need to find out where the basis vectors $\mathbf{e}_1, \mathbf{e}_2$ are mapped. Referring to

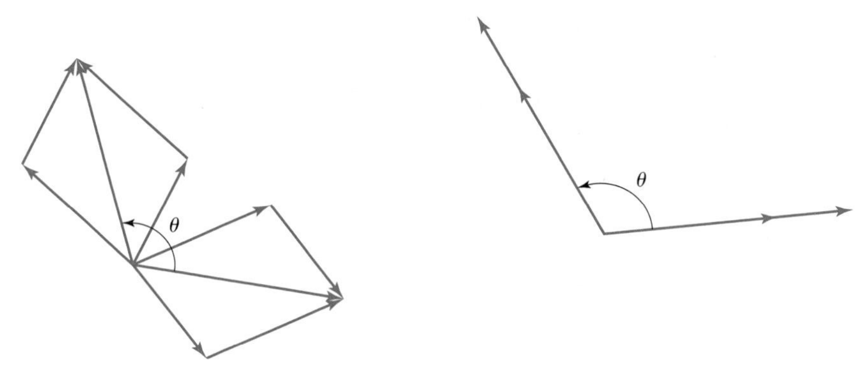

Figure 7.2 Linearity of rotations.

Figure 7.3, we have

$$R_\theta[\,\mathbf{e}_1\,] = (\cos\theta)\,\mathbf{e}_1 + (\sin\theta)\,\mathbf{e}_2 = \begin{pmatrix}\cos\theta\\ \sin\theta\end{pmatrix},$$

$$R_\theta[\,\mathbf{e}_2\,] = -\,(\sin\theta)\,\mathbf{e}_1 + (\cos\theta)\,\mathbf{e}_2 = \begin{pmatrix}-\sin\theta\\ \cos\theta\end{pmatrix}.$$

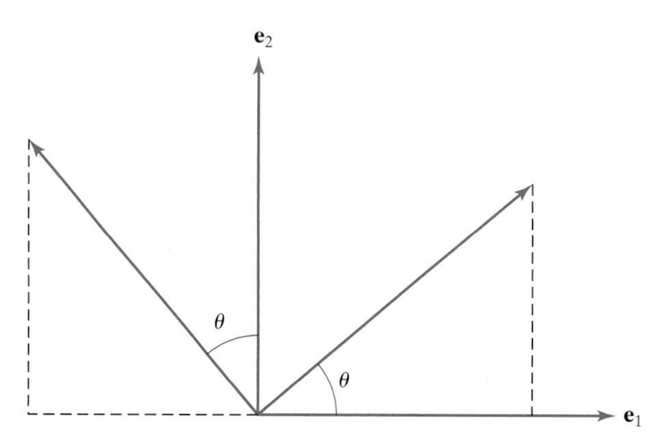

Figure 7.3 Rotation in $\mathbb{R}^2$.

According to the general recipe (7.7), we assemble these two column vectors to obtain the matrix form of the rotation transformation, and so

$$R_\theta[\,\mathbf{v}\,] = A_\theta\,\mathbf{v}, \qquad \text{where} \qquad A_\theta = \begin{pmatrix}\cos\theta & -\sin\theta\\ \sin\theta & \cos\theta\end{pmatrix}. \tag{7.8}$$

Therefore, rotating a vector $\mathbf{v} = \begin{pmatrix}x\\ y\end{pmatrix}$ through angle θ produces the vector

$$\widehat{\mathbf{v}} = R_\theta[\,\mathbf{v}\,] = A_\theta\,\mathbf{v} = \begin{pmatrix}\cos\theta & -\sin\theta\\ \sin\theta & \cos\theta\end{pmatrix}\begin{pmatrix}x\\ y\end{pmatrix} = \begin{pmatrix}x\cos\theta - y\sin\theta\\ x\sin\theta + y\cos\theta\end{pmatrix}$$

with coordinates

$$\hat{x} = x\cos\theta - y\sin\theta, \qquad \hat{y} = x\sin\theta + y\cos\theta.$$

These formulae can be proved directly, but, in fact, are a consequence of the underlying linearity of rotations. ●

EXERCISES 7.1

7.1.1. Which of the following functions $F: \mathbb{R}^3 \to \mathbb{R}$ are linear?
(a) $F(x,y,z) = x$ (b) $F(x,y,z) = y - 2$
(c) $F(x,y,z) = x + y + 3$
(d) $F(x,y,z) = x - y - z$
(e) $F(x,y,z) = x\,y\,z$
(f) $F(x,y,z) = x^2 - y^2 + z^2$
(g) $F(x,y,z) = e^{x-y+z}$

7.1.2. Which of the following functions $F: \mathbb{R}^2 \to \mathbb{R}^2$ are linear?
(a) $F\begin{pmatrix} x \\ y \end{pmatrix} = \begin{pmatrix} x - y \\ x + y \end{pmatrix}$
(b) $F\begin{pmatrix} x \\ y \end{pmatrix} = \begin{pmatrix} x + y + 1 \\ x - y - 1 \end{pmatrix}$
(c) $F\begin{pmatrix} x \\ y \end{pmatrix} = \begin{pmatrix} x\,y \\ x - y \end{pmatrix}$
(d) $F\begin{pmatrix} x \\ y \end{pmatrix} = \begin{pmatrix} 3\,y \\ 2\,x \end{pmatrix}$
(e) $F\begin{pmatrix} x \\ y \end{pmatrix} = \begin{pmatrix} x^2 + y^2 \\ x^2 - y^2 \end{pmatrix}$
(f) $F\begin{pmatrix} x \\ y \end{pmatrix} = \begin{pmatrix} y - 3\,x \\ x \end{pmatrix}$

7.1.3. Explain why the following functions $F: \mathbb{R}^2 \to \mathbb{R}^2$ are not linear.
(a) $\begin{pmatrix} x + 2 \\ x + y \end{pmatrix}$ (b) $\begin{pmatrix} x^2 \\ y^2 \end{pmatrix}$
(c) $\begin{pmatrix} |y| \\ |x| \end{pmatrix}$ (d) $\begin{pmatrix} \sin(x+y) \\ x - y \end{pmatrix}$
(e) $\begin{pmatrix} x + e^y \\ 2x + y \end{pmatrix}$

7.1.4. Explain why the translation function $T: \mathbb{R}^2 \to \mathbb{R}^2$, defined by

$$T\begin{pmatrix} x \\ y \end{pmatrix} = \begin{pmatrix} x + a \\ y + b \end{pmatrix}$$

for fixed a, b, is almost never linear. Precisely when is it linear?

7.1.5. Find a matrix representation for the following linear transformations on $\mathbb{R}^3$:
(a) A counterclockwise rotation by $90°$ around the z–axis.
(b) A clockwise rotation by $60°$ around the x–axis.
(c) Reflection through the (x,y)-plane.
(d) Counterclockwise rotation by $120°$ around the line $x = y = z$.
(e) Rotation by $180°$ around the line $x = y = z$.
(f) Orthogonal projection onto the (x,y) plane.
(g) Orthogonal projection onto the plane $x - y + 2z = 0$.

7.1.6. Find a linear function $L: \mathbb{R}^2 \to \mathbb{R}$ such that

$$L\begin{pmatrix} 1 \\ 1 \end{pmatrix} = 2 \quad \text{and} \quad L\begin{pmatrix} 1 \\ -1 \end{pmatrix} = 3.$$

Is it unique?

7.1.7. Find a linear function $L: \mathbb{R}^2 \to \mathbb{R}^2$ such that

$$L\begin{pmatrix} 1 \\ 2 \end{pmatrix} = \begin{pmatrix} 2 \\ -1 \end{pmatrix} \quad \text{and} \quad L\begin{pmatrix} 2 \\ 1 \end{pmatrix} = \begin{pmatrix} 0 \\ -1 \end{pmatrix}.$$

7.1.8. Under what conditions does there exist a linear function $L: \mathbb{R}^2 \to \mathbb{R}^2$ such that

$$L\begin{pmatrix} x_1 \\ y_1 \end{pmatrix} = \begin{pmatrix} a_1 \\ b_1 \end{pmatrix} \quad \text{and} \quad L\begin{pmatrix} x_2 \\ y_2 \end{pmatrix} = \begin{pmatrix} a_2 \\ b_2 \end{pmatrix}?$$

Under what conditions is L uniquely defined? In the latter case, write down the matrix representation of L.

7.1.9. Can you construct a linear function $L: \mathbb{R}^3 \to \mathbb{R}$ such that

$$L\begin{pmatrix} 1 \\ -1 \\ 0 \end{pmatrix} = 1, \quad L\begin{pmatrix} 1 \\ 0 \\ -1 \end{pmatrix} = 4$$

and

$$L\begin{pmatrix} 0 \\ 1 \\ -1 \end{pmatrix} = -2?$$

If yes, find one. If not, explain why not.

◇ **7.1.10.** Let $\mathbf{a} = (a,b,c)^T \in \mathbb{R}^3$ be a fixed vector. Prove that the cross product map $L_{\mathbf{a}}[\mathbf{v}] = \mathbf{a} \times \mathbf{v}$, as defined in (5.2), is linear, and find its matrix representative.

7.1.11. Is the Euclidean norm function $N(\mathbf{v}) = \|\mathbf{v}\|$, for $\mathbf{v} \in \mathbb{R}^n$, linear?

7.1.12. *True or false*: The quadratic form $Q(\mathbf{v}) = \mathbf{v}^T K \mathbf{v}$ defined by a symmetric $n \times n$ matrix K defines a linear function $Q: \mathbb{R}^n \to \mathbb{R}$.

7.1.13. Let V be a vector space. Prove that every linear function $L: \mathbb{R} \to V$ has the form $L[x] = x\,\mathbf{b}$ for $x \in \mathbb{R}$, where $\mathbf{b} \in V$ is a fixed vector.

7.1.14. Let

$$A = \begin{pmatrix} a & b \\ c & d \end{pmatrix}, \qquad B = \begin{pmatrix} p & q \\ r & s \end{pmatrix}$$

be fixed 2×2 matrices. For each of the following functions, prove that $L\colon \mathcal{M}_{2\times 2} \to \mathcal{M}_{2\times 2}$ defines a linear map, and then find its matrix representative with respect to the standard basis

$$\begin{pmatrix} 1 & 0 \\ 0 & 0 \end{pmatrix}, \begin{pmatrix} 0 & 1 \\ 0 & 0 \end{pmatrix}, \begin{pmatrix} 0 & 0 \\ 1 & 0 \end{pmatrix}, \begin{pmatrix} 0 & 0 \\ 0 & 1 \end{pmatrix}$$

of $\mathcal{M}_{2\times 2}$.

(a) $L[X] = AX$ (b) $R[X] = XB$

(c) $K[X] = AXB$

7.1.15. The domain space of the following functions is the space of $n \times n$ real matrices A. Which are linear? What is the target space in each case?

(a) $L[A] = 3A$ (b) $L[A] = \mathrm{I} - A$

(c) $L[A] = A^T$ (d) $L[A] = A^{-1}$

(e) $L[A] = \det A$ (f) $L[A] = \operatorname{tr} A$

(g) $L[A] = \operatorname{diag} A$, i.e., the diagonal entries of A

(h) $L[A] = A\mathbf{v}$ where $\mathbf{v} \in \mathbb{R}^n$ is a fixed vector

(i) $L[A] = \mathbf{v}^T A\mathbf{v}$ where $\mathbf{v} \in \mathbb{R}^n$ is a fixed vector

◇ **7.1.16.** (a) Prove that L is linear if and only if it satisfies (7.3).

(b) Use induction to prove that L satisfies (7.4).

◇ **7.1.17.** Let $\mathbf{v}_1, \ldots, \mathbf{v}_n$ be a basis of V and $\mathbf{w}_1, \ldots, \mathbf{w}_n$ be any vectors in W. Show that there is a unique linear function $L\colon V \to W$ such that $L[\mathbf{v}_i] = \mathbf{w}_i$, $i = 1, \ldots, n$.

♡ **7.1.18.** *Bilinear Functions*: Let V, W, Z be vector spaces. A function that takes any pair of vectors $\mathbf{v} \in V$ and $\mathbf{w} \in W$ to a vector $\mathbf{z} = B(\mathbf{v}, \mathbf{w}) \in Z$ is called *bilinear* if, for each fixed $\mathbf{w}$, it is a linear function of $\mathbf{v}$, so $B(c\,\mathbf{v} + d\,\widetilde{\mathbf{v}}, \mathbf{w}) = c\,B(\mathbf{v}, \mathbf{w}) + d\,B(\widetilde{\mathbf{v}}, \mathbf{w})$, and, for each fixed $\mathbf{v}$, it is a linear function of $\mathbf{w}$, so $B(\mathbf{v}, c\,\mathbf{w} + d\,\widetilde{\mathbf{w}}) = c\,B(\mathbf{v}, \mathbf{w}) + d\,B(\mathbf{v}, \widetilde{\mathbf{w}})$. Thus, $B\colon V \times W \to Z$ defines a function on the Cartesian product space $V \times W$, as defined in Exercise 2.1.13.

(a) Show that $B(\mathbf{v}, \mathbf{w}) = v_1 w_1 - 2 v_2 w_2$ is a bilinear function from $\mathbb{R}^2 \times \mathbb{R}^2$ to $\mathbb{R}$.

(b) Show that $B(\mathbf{v}, \mathbf{w}) = 2 v_1 w_2 - 3 v_2 w_3$ is a bilinear function from $\mathbb{R}^2 \times \mathbb{R}^3$ to $\mathbb{R}$.

(c) Show that if V is an inner product space, then $B(\mathbf{v}, \mathbf{w}) = \langle \mathbf{v}, \mathbf{w} \rangle$ defines a bilinear function $B\colon V \times V \to \mathbb{R}$.

(d) Show that if A is any $m \times n$ matrix, then $B(\mathbf{v}, \mathbf{w}) = \mathbf{v}^T A\mathbf{w}$ defines a bilinear function $B\colon \mathbb{R}^m \times \mathbb{R}^n \to \mathbb{R}$.

(e) Show that every bilinear function $B\colon \mathbb{R}^m \times \mathbb{R}^n \to \mathbb{R}$ arises in this way.

(f) Show that the vector-valued function $B\colon \mathbb{R}^m \times \mathbb{R}^n \to \mathbb{R}^k$ defines a bilinear function if and only if each of its entries is a bilinear function $B\colon \mathbb{R}^m \times \mathbb{R}^n \to \mathbb{R}$.

(g) *True or false*: A bilinear function $B\colon V \times W \to Z$ defines a linear function on the Cartesian product space.

Linear Operators

So far, we have concentrated on linear functions on Euclidean space, and discovered that they are all represented by matrices. For function spaces, there is a much wider variety of linear operators available, and a complete classification is out of the question. Let us look at some of the main representative examples that arise in applications.

EXAMPLE 7.7

(a) Recall that $\mathrm{C}^0[a, b]$ denotes the vector space consisting of all continuous functions on the interval $[a, b]$. Evaluation of the function at a point, namely $L[f] = f(x_0)$, defines a linear operator $L\colon \mathrm{C}^0[a, b] \to \mathbb{R}$, because

$$L[c\,f + d\,g] = (c\,f + d\,g)(x_0) = c\,f(x_0) + d\,g(x_0) = c\,L[f] + d\,L[g]$$

for any functions $f, g \in \mathrm{C}^0[a, b]$ and scalars (constants) c, d.

(b) Another real-valued linear function is the integration operator

$$I[f] = \int_a^b f(x)\,dx. \tag{7.9}$$

Linearity of I is an immediate consequence of the basic integration identity

$$\int_a^b \big[\, c\,f(x) + d\,g(x) \,\big]\,dx = c\int_a^b f(x)\,dx \; + \; d\int_a^b g(x)\,dx,$$

which is valid for arbitrary integrable—which includes continuous—functions f, g and constants c, d.

(c) We have already seen that multiplication of functions by a fixed scalar

$$M_c[f(x)] = c\,f(x)$$

defines a linear map $M_c\colon C^0[a,b] \to C^0[a,b]$; the particular case $c = 1$ reduces to the identity transformation $\mathrm{I} = M_1$. More generally, if $a(x) \in C^0[a,b]$ is a fixed continuous function, then the operation

$$M_a[f(x)] = a(x)\,f(x)$$

of multiplication by a also defines a linear transformation $M_a\colon C^0[a,b] \to C^0[a,b]$.

(d) Another important linear transformation is the indefinite integral

$$J[f] = \int_a^x f(y)\,dy. \tag{7.10}$$

According to the Fundamental Theorem of Calculus, [2, 58], the integral of a continuous function is continuously differentiable. Therefore, $J\colon C^0[a,b] \to C^1[a,b]$ defines a linear operator from the space of continuous functions to the space of continuously differentiable functions.

(e) Vice versa, differentiation of functions is also a linear operation. To be precise, since not every continuous function can be differentiated, we take the domain space to be the vector space $C^1[a,b]$ of continuously differentiable functions on the interval $[a,b]$. The derivative operator

$$D[f] = f' \tag{7.11}$$

defines a linear operator $D\colon C^1[a,b] \to C^0[a,b]$. This follows from the elementary differentiation formula

$$D[c\,f + d\,g] = (c\,f + d\,g)' = c\,f' + d\,g' = c\,D[f] + d\,D[g],$$

valid whenever c, d are constant. ●

EXERCISES 7.1

7.1.19. Which of the following define linear operators on the vector space $C^1(\mathbb{R})$ of continuously differentiable scalar functions? What is the target space?

(a) $L[f] = f(0) + f(1)$

(b) $L[f] = f(0)\,f(1)$ (c) $L[f] = f'(1)$

(d) $L[f] = f'(3) - f(2)$

(e) $L[f] = x^2 f(x)$ (f) $L[f] = f(x+2)$

(g) $L[f] = f(x) + 2$ (h) $L[f] = f'(2x)$

(i) $L[f] = f'(x^2)$

(j) $L[f] = f(x)\sin x - f'(x)\cos x$

(k) $L[f] = 2\log f(0)$

(l) $L[f] = \displaystyle\int_0^1 e^y f(y)\,dy$

(m) $L[f] = \displaystyle\int_0^1 |f(y)|\,dy$

(n) $L[f] = \displaystyle\int_{x-1}^{x+1} f(y)\,dy$

(o) $L[f] = \displaystyle\int_x^{x^2} \frac{f(y)}{y}\,dy$

(p) $L[f] = \displaystyle\int_0^{f(x)} y\,dy$

(q) $L[f] = \displaystyle\int_0^x y^2 f'(y)\,dy$

(r) $L[f] = \displaystyle\int_{-1}^1 [f(y) - f(0)]\,dy$

(s) $L[f] = \displaystyle\int_{-1}^x [f(y) - y]\,dy$.

7.1.20. *True or false*: The average or mean

$$A[f] = \frac{1}{b-a}\int_a^b f(x)\,dx$$

of a function on the interval $[a,b]$ defines a linear operator $A\colon C^0[a,b] \to \mathbb{R}$.

7.1.21. Prove that multiplication $M_h[f(x)] = h(x)\,f(x)$ by a fixed function $h \in C^n[a,b]$ defines a linear operator $M_h\colon C^n[a,b] \to C^n[a,b]$. Which result from calculus do you need to complete the proof?

7.1.22. Show that if $w(x)$ is any fixed continuous function, then the weighted integral

$$I_w[f] = \int_a^b f(x)\,w(x)\,dx$$

defines a linear operator $I_w\colon C^0[a,b] \to \mathbb{R}$.

7.1.23. (a) Show that the partial derivatives

$$\partial_x[f] = \frac{\partial f}{\partial x} \quad \text{and} \quad \partial_y[f] = \frac{\partial f}{\partial y}$$

both define linear operators on the space of continuously differentiable functions $f(x,y)$.

(b) For which values of a, b, c, d is the map

$$L[f] = a\,\frac{\partial f}{\partial x} + b\,\frac{\partial f}{\partial y} + c\,f + d$$

linear?

7.1.24. Prove that the Laplacian operator

$$\Delta[f] = \frac{\partial^2 f}{\partial x^2} + \frac{\partial^2 f}{\partial y^2}$$

defines a linear function on the vector space of twice continuously differentiable functions $f(x,y)$.

7.1.25. Show that the gradient $G[f] = \nabla f$ defines a linear operator from the space of continuously differentiable scalar-valued functions $f\colon \mathbb{R}^2 \to \mathbb{R}$ to the space of continuous vector fields $\mathbf{v}\colon \mathbb{R}^2 \to \mathbb{R}^2$.

7.1.26. Prove that, on $\mathbb{R}^3$, the gradient, curl and divergence all define linear operators. Be precise in your description of the domain space and the target space in each case.

The Space of Linear Functions

Given vector spaces V, W, we use $\mathcal{L}(V,W)$ to denote the set of all* linear functions $L\colon V \to W$. We claim that $\mathcal{L}(V,W)$ is itself a vector space. We add two linear functions $L, M \in \mathcal{L}(V,W)$ in the same way we add general functions: $(L+M)[\mathbf{v}] = L[\mathbf{v}] + M[\mathbf{v}]$. You should check that $L+M$ satisfies the linear function axioms (7.1), provided that L and M do. Similarly, multiplication of a linear function by a scalar $c \in \mathbb{R}$ is defined so that $(c\,L)[\mathbf{v}] = c\,L[\mathbf{v}]$, again producing a linear function. The verification that $\mathcal{L}(V,W)$ satisfies the basic vector space axioms of Definition 2.1 is left to the reader.

In particular, if $V = \mathbb{R}^n$ and $W = \mathbb{R}^m$, then Theorem 7.5 implies that we can identify $\mathcal{L}(\mathbb{R}^n, \mathbb{R}^m)$ with the space $\mathcal{M}_{m\times n}$ of all $m \times n$ matrices. Addition of linear functions corresponds to matrix addition, while scalar multiplication coincides with the usual scalar multiplication of matrices. Therefore, the space of all $m\times n$ matrices forms a vector space—a fact we already knew. A basis for $\mathcal{M}_{m\times n}$ is given by the $m\,n$ matrices E_{ij}, $1 \le i \le m$, $1 \le j \le n$, which have a single 1 in the (i,j) position and zeros everywhere else. Therefore, the dimension of $\mathcal{M}_{m\times n}$ is $m\,n$. Note that E_{ij} corresponds to the specific linear transformation mapping $\mathbf{e}_j \in \mathbb{R}^n$ to $\widehat{\mathbf{e}}_i \in \mathbb{R}^m$ and every other $\mathbf{e}_k \in \mathbb{R}^n$, $k \ne j$, to $\mathbf{0}$.

EXAMPLE 7.8 The space of linear transformations of the plane, $\mathcal{L}(\mathbb{R}^2, \mathbb{R}^2)$, is identified with the space $\mathcal{M}_{2\times 2}$ of 2×2 matrices $A = \begin{pmatrix} a & b \\ c & d \end{pmatrix}$. The standard basis of $\mathcal{M}_{2\times 2}$ consists of the $4 = 2 \cdot 2$ matrices

$$E_{11} = \begin{pmatrix} 1 & 0 \\ 0 & 0 \end{pmatrix}, \quad E_{12} = \begin{pmatrix} 0 & 1 \\ 0 & 0 \end{pmatrix}, \quad E_{21} = \begin{pmatrix} 0 & 0 \\ 1 & 0 \end{pmatrix}, \quad E_{22} = \begin{pmatrix} 0 & 0 \\ 0 & 1 \end{pmatrix}.$$

*In infinite-dimensional situations, one usually imposes additional restrictions, e.g., continuity or boundedness of the linear operators. We can safely relegate these more subtle distinctions to a more advanced treatment of the subject. See [38, 52] for a full discussion of the rather sophisticated analytical details, which play an important role in serious quantum mechanical applications.

Indeed, we can uniquely write any other matrix

$$A = \begin{pmatrix} a & b \\ c & d \end{pmatrix} = a\,E_{11} + b\,E_{12} + c\,E_{21} + d\,E_{22},$$

as a linear combination of these four basis matrices. ●

A particularly important case is when the target space of the linear functions is $\mathbb{R}$.

Definition 7.9 The *dual space* to a vector space V is defined as the vector space $V^* = \mathcal{L}(V, \mathbb{R})$ consisting of all real-valued linear functions $L\colon V \to \mathbb{R}$.

If $V = \mathbb{R}^n$, then, by Theorem 7.5, every linear function $L\colon \mathbb{R}^n \to \mathbb{R}$ is given by multiplication by a $1 \times n$ matrix, i.e., a row vector. Explicitly,

$$L[\mathbf{v}] = \mathbf{a}\,\mathbf{v} = a_1 v_1 + \cdots + a_n v_n, \qquad \text{where} \qquad \mathbf{a} = (\, a_1\; a_2 \,\ldots\, a_n \,), \qquad \mathbf{v} = \begin{pmatrix} v_1 \\ \vdots \\ v_n \end{pmatrix}.$$

Therefore, we can identify the dual space $(\mathbb{R}^n)^*$ with the space of *row* vectors with n entries. In light of this observation, the distinction between row vectors and column vectors is now seen to be much more sophisticated than mere semantics or notation. Row vectors should more properly be viewed as real-valued linear functions— the *dual* objects to column vectors.

The *standard dual basis* $\boldsymbol{\varepsilon}_1, \ldots, \boldsymbol{\varepsilon}_n$ of $(\mathbb{R}^n)^*$ consists of the standard row basis vectors; namely, $\boldsymbol{\varepsilon}_j$ is the row vector with 1 in the jth slot and zeros elsewhere. The jth dual basis element defines the linear function

$$E_j[\mathbf{v}] = \boldsymbol{\varepsilon}_j\,\mathbf{v} = v_j$$

that picks off the jth coordinate of $\mathbf{v}$—with respect to the original basis $\mathbf{e}_1, \ldots, \mathbf{e}_n$. Thus, the dimension of $V = \mathbb{R}^n$ and its dual $(\mathbb{R}^n)^*$ are both equal to n.

An inner product structure provides a mechanism for identifying a vector space and its dual. However, it should be borne in mind that this identification will depend upon the choice of inner product.

THEOREM 7.10 Let V be a finite-dimensional real inner product space. Then every linear function $L\colon V \to \mathbb{R}$ is given by an inner product with a fixed vector $\mathbf{a} \in V$:

$$L[\mathbf{v}] = \langle \mathbf{a}, \mathbf{v} \rangle \tag{7.12}$$

Proof Let $\mathbf{u}_1, \ldots, \mathbf{u}_n$ be an orthonormal basis of V. (If necessary, we can use the Gram–Schmidt process to generate such a basis.) If we write $\mathbf{v} = x_1\mathbf{u}_1 + \cdots + x_n\mathbf{u}_n$, then, by linearity,

$$L[\mathbf{v}] = x_1\,L[\mathbf{u}_1] + \cdots + x_n\,L[\mathbf{u}_n] = a_1 x_1 + \cdots + a_n x_n, \qquad \text{where} \quad a_i = L[\mathbf{u}_i\,].$$

On the other hand, if we write $\mathbf{a} = a_1\,\mathbf{u}_1 + \cdots + a_n\,\mathbf{u}_n$, then, by orthonormality of the basis,

$$\langle \mathbf{a}, \mathbf{v} \rangle = \sum_{i,j=1}^{n} a_i\,x_j\,\langle \mathbf{u}_i, \mathbf{u}_j \rangle = a_1 x_1 + \cdots + a_n x_n.$$

Thus, equation (7.12) holds, which completes the proof. ■

REMARK: In the particular case when $V = \mathbb{R}^n$ is endowed with the standard dot product, then Theorem 7.10 identifies a row vector representing a linear function with the corresponding column vector obtained by transposition $\mathbf{a} \mapsto \mathbf{a}^T$. Thus, the naïve identification of a row and a column vector is, in fact, an indication of a much more subtle phenomenon that relies on the identification of $\mathbb{R}^n$ with its dual vector space based on the Euclidean inner product. Alternative inner products will lead to alternative, more complicated, identifications of row and column vectors; see Exercise 7.1.31 for details.

Important: Theorem 7.10 is *not* true if V is infinite-dimensional. This fact will have important repercussions for the analysis of the differential equations of continuum mechanics, which will lead us immediately into the much deeper waters of generalized function theory. Details will be deferred until Section 11.2.

EXERCISES 7.1

7.1.27. Write down a basis for and dimension of the linear function spaces
(a) $\mathcal{L}(\mathbb{R}^3, \mathbb{R})$ (b) $\mathcal{L}(\mathbb{R}^2, \mathbb{R}^2)$
(c) $\mathcal{L}(\mathbb{R}^m, \mathbb{R}^n)$ (d) $\mathcal{L}(\mathcal{P}^{(3)}, \mathbb{R})$
(e) $\mathcal{L}(\mathcal{P}^{(2)}, \mathbb{R}^2)$ (f) $\mathcal{L}(\mathcal{P}^{(2)}, \mathcal{P}^{(2)})$
Here $\mathcal{P}^{(n)}$ is the space of polynomials of degree $\leq n$.

7.1.28. *True or false*: The set of linear transformations $L: \mathbb{R}^2 \to \mathbb{R}^2$ such that

$$L\begin{pmatrix}1\\0\end{pmatrix} = \begin{pmatrix}0\\0\end{pmatrix}$$

forms a subspace of $\mathcal{L}(\mathbb{R}^2, \mathbb{R}^2)$. If true, what is its dimension?

7.1.29. *True or false*: The set of linear transformations $L: \mathbb{R}^3 \to \mathbb{R}^3$ such that

$$L\begin{pmatrix}0\\1\\0\end{pmatrix} = \begin{pmatrix}0\\1\\0\end{pmatrix}$$

forms a subspace of $\mathcal{L}(\mathbb{R}^3, \mathbb{R}^3)$. If true, what is its dimension?

7.1.30. Consider the linear function $L: \mathbb{R}^3 \to \mathbb{R}$ defined by $L(x, y, z) = 3x - y + 2z$. Write down the vector $\mathbf{a} \in \mathbb{R}^3$ such that $L[\mathbf{v}] = \langle \mathbf{a}, \mathbf{v} \rangle$ when the inner product is
(a) the Euclidean dot product,
(b) the weighted inner product $\langle \mathbf{v}, \mathbf{w} \rangle = v_1 w_1 + 2 v_2 w_2 + 3 v_3 w_3$,
(c) the inner product defined by the positive definite matrix

$$K = \begin{pmatrix} 2 & -1 & 0 \\ -1 & 2 & 1 \\ 0 & 1 & 2 \end{pmatrix}.$$

◇ **7.1.31.** Let $\mathbb{R}^n$ be equipped with the inner product $\langle \mathbf{v}, \mathbf{w} \rangle = \mathbf{v}^T K \mathbf{w}$. Let $L[\mathbf{v}] = \mathbf{r}\,\mathbf{v}$ where $\mathbf{r}$ is a row vector of size $1 \times n$.
(a) Find a formula for the column vector $\mathbf{a}$ such that (7.12) holds for the linear function $L: \mathbb{R}^n \to \mathbb{R}$.
(b) Illustrate your result when $\mathbf{r} = (2, -1)$, using
(i) the dot product
(ii) the weighted inner product $\langle \mathbf{v}, \mathbf{w} \rangle = 3 v_1 w_1 + 2 v_2 w_2$,
(iii) the inner product induced by

$$K = \begin{pmatrix} 2 & -1 \\ -1 & 3 \end{pmatrix}.$$

♡ **7.1.32.** *Dual Bases*: Given a basis $\mathbf{v}_1, \ldots, \mathbf{v}_n$ of V, the *dual basis* $L_1, \ldots, L_n$ of V^* consists of the linear functions uniquely defined by the requirements

$$L_i(\mathbf{v}_j) = \begin{cases} 1 & i = j, \\ 0, & i \neq j. \end{cases}$$

(a) Show that $L_i[\mathbf{v}] = x_i$ gives the ith coordinate of a vector $\mathbf{v} = x_1 \mathbf{v}_1 + \cdots + x_n \mathbf{v}_n$ with respect to the given basis.
(b) Prove that the dual basis is indeed a basis for the dual vector space.
(c) Prove that if $V = \mathbb{R}^n$ and $A = (\mathbf{v}_1\ \mathbf{v}_2 \ldots \mathbf{v}_n)$ is the $n \times n$ matrix whose columns are the basis vectors, then the rows of the inverse matrix A^{-1} can be identified as the corresponding dual basis of $(\mathbb{R}^n)^*$.

7.1.33. Use Exercise 7.1.32(c) to find the dual basis for:
(a) $\mathbf{v}_1 = \begin{pmatrix}1\\1\end{pmatrix}, \mathbf{v}_2 = \begin{pmatrix}1\\-1\end{pmatrix}$

(b) $\mathbf{v}_1 = \begin{pmatrix} 1 \\ 2 \end{pmatrix}$, $\mathbf{v}_2 = \begin{pmatrix} 3 \\ -1 \end{pmatrix}$

(c) $\mathbf{v}_1 = \begin{pmatrix} 1 \\ 1 \\ 0 \end{pmatrix}$, $\mathbf{v}_2 = \begin{pmatrix} 1 \\ 0 \\ 1 \end{pmatrix}$, $\mathbf{v}_3 = \begin{pmatrix} 0 \\ 1 \\ 1 \end{pmatrix}$

(d) $\mathbf{v}_1 = \begin{pmatrix} 1 \\ 2 \\ -3 \end{pmatrix}$, $\mathbf{v}_2 = \begin{pmatrix} 0 \\ -3 \\ 1 \end{pmatrix}$, $\mathbf{v}_3 = \begin{pmatrix} -1 \\ 2 \\ 2 \end{pmatrix}$

(e) $\mathbf{v}_1 = \begin{pmatrix} 1 \\ 1 \\ 0 \\ 0 \end{pmatrix}$, $\mathbf{v}_2 = \begin{pmatrix} 0 \\ 1 \\ 1 \\ 0 \end{pmatrix}$, $\mathbf{v}_3 = \begin{pmatrix} 0 \\ 0 \\ 1 \\ 1 \end{pmatrix}$, $\mathbf{v}_4 = \begin{pmatrix} 1 \\ -1 \\ 1 \\ 2 \end{pmatrix}$

7.1.34. Let $\mathcal{P}^{(2)}$ denote the space of quadratic polynomials equipped with the L^2 inner product

$$\langle p, q \rangle = \int_0^1 p(x)\, q(x)\, dx.$$

Find the polynomial q that represents the following linear functions, i.e., such that $L[\,p\,] = \langle q, p \rangle$:

(a) $L[\,p\,] = p(0)$ (b) $L[\,p\,] = \frac{1}{2}\, p'(1)$

(c) $L[\,p\,] = \displaystyle\int_0^1 p(x)\, dx$

(d) $L[\,p\,] = \displaystyle\int_{-1}^1 p(x)\, dx$

7.1.35. Find the dual basis, as defined in Exercise 7.1.32, for the monomial basis of $\mathcal{P}^{(2)}$ with respect to the L^2 inner product

$$\langle p, q \rangle = \int_0^1 p(x)\, q(x)\, dx.$$

7.1.36. Write out a proof of Theorem 7.10 that does not rely on finding an orthonormal basis.

Composition

Besides adding and multiplying by scalars, one can also compose linear functions.

Lemma 7.11 Let V, W, Z be vector spaces. If $L\colon V \to W$ and $M\colon W \to Z$ are linear functions, then the composite function $M \circ L\colon V \to Z$, defined by $(M \circ L)[\mathbf{v}] = M[L[\mathbf{v}]]$ is linear.

Proof This is straightforward:

$$\begin{aligned}(M \circ L)[c\,\mathbf{v} + d\,\mathbf{w}] &= M[L[c\,\mathbf{v} + d\,\mathbf{w}]] \\ &= M[c\,L[\mathbf{v}] + d\,L[\mathbf{w}]] \\ &= c\,M[L[\mathbf{v}]] + d\,M[L[\mathbf{w}]] \\ &= c\,(M \circ L)[\mathbf{v}] + d\,(M \circ L)[\mathbf{w}],\end{aligned}$$

where we used, successively, the linearity of L and then of M. ■

For example, if $L[\mathbf{v}] = A\mathbf{v}$ maps $\mathbb{R}^n$ to $\mathbb{R}^m$, and $M[\mathbf{w}] = B\,\mathbf{w}$ maps $\mathbb{R}^m$ to $\mathbb{R}^l$, so that A is an $m \times n$ matrix and B is a $l \times m$ matrix, then

$$(M \circ L)[\mathbf{v}] = M[L[\mathbf{v}]] = B(A\mathbf{v}) = (B\,A)\,\mathbf{v},$$

and hence the composition $M \circ L\colon \mathbb{R}^n \to \mathbb{R}^l$ corresponds to the $l \times n$ product matrix BA. In other words, on Euclidean space, *composition of linear functions is the same as matrix multiplication*. And, like matrix multiplication, composition of (linear) functions is not, in general, commutative.

EXAMPLE 7.12 Composing two rotations results in another rotation: $R_\varphi \circ R_\theta = R_{\varphi+\theta}$. In other words, if we first rotate by angle θ and then by angle φ, the net effect is rotation by angle $\varphi + \theta$. On the matrix level of (7.8), this implies that

$$\begin{pmatrix} \cos\varphi & -\sin\varphi \\ \sin\varphi & \cos\varphi \end{pmatrix} \begin{pmatrix} \cos\theta & -\sin\theta \\ \sin\theta & \cos\theta \end{pmatrix} = A_\varphi A_\theta$$

$$= A_{\varphi+\theta} = \begin{pmatrix} \cos(\varphi+\theta) & -\sin(\varphi+\theta) \\ \sin(\varphi+\theta) & \cos(\varphi+\theta) \end{pmatrix}.$$

Multiplying out the left hand side, we deduce the well-known trigonometric addition formulae

$$\cos(\varphi+\theta) = \cos\varphi\,\cos\theta - \sin\varphi\,\sin\theta,$$
$$\sin(\varphi+\theta) = \cos\varphi\,\sin\theta + \sin\varphi\,\cos\theta.$$

In fact, this constitutes a *bona fide* proof of these two trigonometric identities! ●

EXAMPLE 7.13 One can build up more sophisticated linear operators on function space by adding and composing simpler ones. In particular, higher order derivative operators are obtained by composing the derivative operator D, defined in (7.11), with itself. For example,

$$D^2[f] = D \circ D[f] = D[f'] = f''$$

defines the second derivative operator. One needs to exercise due care about the domain of definition, since not every function is differentiable. In general, the kth order derivative

$$D^k[f] = f^{(k)}(x)$$

defines a linear operator

$$D^k : \mathrm{C}^n[a,b] \longrightarrow \mathrm{C}^{n-k}[a,b] \quad \text{for any } n \ge k,$$

obtained by composing D with itself k times.

If we further compose D^k with the linear operation of multiplication by a fixed function $a(x)$ we obtain the linear operator $(a\,D^k)[f] = a(x)\,f^{(k)}(x)$. Finally, a general *linear ordinary differential operator* of order n

$$L = a_n(x)\,D^n + a_{n-1}(x)\,D^{n-1} + \cdots + a_1(x)\,D + a_0(x) \tag{7.13}$$

is obtained by summing such operators. If the coefficient functions $a_0(x), \ldots, a_n(x)$ are continuous, then

$$L[u] = a_n(x)\frac{d^n u}{dx^n} + a_{n-1}(x)\frac{d^{n-1}u}{dx^{n-1}} + \cdots + a_1(x)\frac{du}{dx} + a_0(x)u \tag{7.14}$$

defines a linear operator from $\mathrm{C}^n[a,b]$ to $\mathrm{C}^0[a,b]$. The most important case—but certainly not the only one arising in applications—is when the coefficients $a_i(x) = c_i$ are all constant. ●

EXERCISES 7.1

7.1.37. For each of the following pairs of linear functions $S, T\colon \mathbb{R}^2 \to \mathbb{R}^2$, describe the compositions $S \circ T$ and $T \circ S$. Do the functions commute?

(a) S = counterclockwise rotation by $60°$; T = clockwise rotation by $120°$

(b) S = reflection in the line $y = x$; T = rotation by $180°$

(c) S = reflection in the x axis; T = reflection in the y axis

(d) S = reflection in the line $y = x$; T = reflection in the line $y = 2x$

(e) S = orthogonal projection on the x–axis; T = orthogonal projection on the y–axis

(f) S = orthogonal projection on the x–axis; T = orthogonal projection on the line $y = x$

(g) S = orthogonal projection on the x–axis; T = rotation by 180°

(h) S = orthogonal projection on the x–axis; T = counterclockwise rotation by 90°

(i) S = orthogonal projection on the line $y = -2x$; T = reflection in the line $y = x$

7.1.38. Find a matrix representative for the linear functions

(a) $L\colon \mathbb{R}^2 \to \mathbb{R}^2$ that maps $\mathbf{e}_1$ to $\begin{pmatrix} 1 \\ -3 \end{pmatrix}$ and $\mathbf{e}_2$ to $\begin{pmatrix} -1 \\ 2 \end{pmatrix}$;

(b) $M\colon \mathbb{R}^2 \to \mathbb{R}^2$ that takes $\mathbf{e}_1$ to $\begin{pmatrix} -1 \\ -3 \end{pmatrix}$ and $\mathbf{e}_2$ to $\begin{pmatrix} 0 \\ 2 \end{pmatrix}$; and

(c) $N\colon \mathbb{R}^2 \to \mathbb{R}^2$ that takes $\begin{pmatrix} 1 \\ -3 \end{pmatrix}$ to $\begin{pmatrix} -1 \\ -3 \end{pmatrix}$ and $\begin{pmatrix} -1 \\ 2 \end{pmatrix}$ to $\begin{pmatrix} 0 \\ 2 \end{pmatrix}$.

(d) Explain why $M = N \circ L$.

(e) Verify part (d) by multiplying the matrix representatives in the proper order.

7.1.39. On the vector space $\mathbb{R}^3$, let R denote counterclockwise rotation around the x axis by 90° and S counterclockwise rotation around the z axis by 90°.

(a) Find matrix representatives for R and S.

(b) Show that $R \circ S \neq S \circ R$. Explain what happens to the standard basis vectors under the two compositions.

(c) Give an experimental demonstration of the noncommutativity of R and S by physically rotating a solid object, e.g., this book, in the prescribed manners.

7.1.40. Let P denote orthogonal projection of $\mathbb{R}^3$ onto the plane $V = \{z = x + y\}$ and Q denote orthogonal projection onto the plane $W = \{z = x - y\}$. Is the composition $R = Q \circ P$ the same as orthogonal projection onto the line $L = V \cap W$? Verify your conclusion by computing the matrix representatives of P, Q, and R.

7.1.41. (a) Write the linear operator $L[f(x)] = f'(b)$ as a composition of two linear functions. Do your linear functions commute?

(b) For which values of a, b, c, d, e is $L[f(x)] = a\, f'(b) + c\, f(d) + e$ a linear function?

7.1.42. Let $L = x\,D + 1$, and $M = D - x$ be differential operators. Find $L \circ M$ and $M \circ L$. Do the differential operators commute?

7.1.43. (a) Explain why the differential operator $L = D \circ M_a \circ D$ obtained by composing the linear operators of differentiation $D[f(x)] = f'(x)$ and multiplication $M_a[f(x)] = a(x)\, f(x)$ by a fixed function $a(x)$ defines a linear operator.

(b) Re-express L as a linear differential operator of the form (7.14).

◇ **7.1.44.** (a) Show that composition of linear functions is associative: $(L \circ M) \circ N = L \circ (M \circ N)$. Be precise about the domain and target spaces involved.

(b) How do you know the result is a linear function?

(c) Explain why this proves associativity of matrix multiplication.

7.1.45. Show that the space of constant coefficient linear differential operators of order $\leq n$ forms a vector space. Determine its dimension by exhibiting a basis.

7.1.46. Show that if $p(x, y)$ is any polynomial, then $L = p(\partial_x, \partial_y)$ defines a linear, constant coefficient partial differential operator. For example, if $p(x, y) = x^2 + y^2$, then $L = \partial_x^2 + \partial_y^2$ is the Laplacian operator

$$\Delta[f] = \frac{\partial^2 f}{\partial x^2} + \frac{\partial^2 f}{\partial y^2}.$$

♡ **7.1.47.** The *commutator* of two linear transformations $L, M\colon V \to V$ on a vector space V is defined as

$$K = [L, M] = L \circ M - M \circ L. \qquad (7.15)$$

(a) Prove that the commutator K is a linear transformation.

(b) Prove that L and M commute if and only if $[L, M] = \mathrm{O}$.

(c) Compute the commutators of the linear transformations defined by the following pairs of matrices:

(i) $\begin{pmatrix} 1 & 1 \\ 0 & 1 \end{pmatrix}$, $\begin{pmatrix} -1 & 0 \\ 1 & 2 \end{pmatrix}$

(ii) $\begin{pmatrix} 0 & 1 \\ -1 & 0 \end{pmatrix}$, $\begin{pmatrix} 1 & 0 \\ 0 & -1 \end{pmatrix}$

(iii) $\begin{pmatrix} 1 & 1 & 0 \\ 1 & 0 & 1 \\ 0 & 1 & 1 \end{pmatrix}$, $\begin{pmatrix} -1 & 0 & 0 \\ 0 & 1 & 0 \\ 0 & 0 & -1 \end{pmatrix}$

(d) Prove the *Jacobi identity*

$$\big[[L, M], N\big] + \big[[N, L], M\big] + \big[[M, N], L\big] = \mathrm{O} \qquad (7.16)$$

is valid for any three linear transformations.

(e) Verify the Jacobi identity for the first three matrices in part (b).

(f) Prove that the commutator $B(L, M) = [L, M]$ defines a bilinear map on $\mathcal{L}(V, V)$, cf. Exercise 7.1.18.

◇ **7.1.48.** (a) In (one-dimensional) quantum mechanics, the differentiation operator $P[f(x)] = f'(x)$ represents the *momentum* of a particle, while the operator $Q[f(x)] = x f(x)$ of multiplication by the function x represents its *position*. Prove that the position and momentum operators satisfy the *Heisenberg Commutation Relations* $[P, Q] = P \circ Q - Q \circ P = \mathrm{I}$.

(b) Prove that there are no matrices P, Q that satisfy the Heisenberg Commutation Relations. *Hint*: Use Exercise 1.2.32.
Remark: The noncommutativity of quantum mechanical observables lies at the heart of the Uncertainty Principle. The result in part (b) is one of the main reasons why quantum mechanics must be an intrinsically infinite-dimensional theory.

♡ **7.1.49.** Let $\mathcal{D}^{(1)}$ denote the set of all first order linear differential operators $L = p(x)\,D + q(x)$ where p, q are polynomials.

(a) Prove that $\mathcal{D}^{(1)}$ is a vector space. Is it finite-dimensional or infinite-dimensional?

(b) Prove that the commutator (7.15) of $L, M \in \mathcal{D}^{(1)}$ is a first order differential operator $[L, M] \in \mathcal{D}^{(1)}$ by writing out an explicit formula.

(c) Verify the Jacobi identity (7.16) for the first order differential operators $L = D$, $M = x\,D + 1$, and $N = x^2 D + 2x$.

7.1.50. Do the conclusions of Exercise 7.1.49(a–b) hold for the space $\mathcal{D}^{(2)}$ of second order differential operators $L = p(x)\,D^2 + q(x)\,D + r(x)$ where p, q, r are polynomials?

Inverses

The inverse of a linear function is defined in direct analogy with the Definition 1.13 of the inverse of a (square) matrix.

Definition 7.14 Let $L\colon V \to W$ be a linear function. If $M\colon W \to V$ is a function such that both compositions

$$L \circ M = \mathrm{I}_W, \qquad M \circ L = \mathrm{I}_V \tag{7.17}$$

are equal to the identity function, then we call M the *inverse* of L and write $M = L^{-1}$.

The two conditions (7.17) require

$$L[M[\mathbf{w}]] = \mathbf{w} \quad \text{for all} \quad \mathbf{w} \in W, \qquad \text{and} \qquad M[L[\mathbf{v}]] = \mathbf{v} \quad \text{for all} \quad \mathbf{v} \in V.$$

In Exercise 7.1.55, you are asked to prove that, when it exists, the inverse is unique. Of course, if $M = L^{-1}$ is the inverse of L, then $L = M^{-1}$ is the inverse of M since the conditions are symmetric, and, in such cases, $(L^{-1})^{-1} = L$.

Lemma 7.15 If it exists, the inverse of a linear function is also a linear function.

Proof Let L, M satisfy the conditions of Definition 7.14. Given $\mathbf{w}, \widetilde{\mathbf{w}} \in W$, we note

$$\mathbf{w} = (L \circ M)[\mathbf{w}] = L[\mathbf{v}], \qquad \widetilde{\mathbf{w}} = (L \circ M)[\widetilde{\mathbf{w}}] = L[\widetilde{\mathbf{v}}],$$

where

$$\mathbf{v} = M[\mathbf{w}], \qquad \widetilde{\mathbf{v}} = M[\widetilde{\mathbf{w}}].$$

Therefore, given scalars c, d, and using only the linearity of L,

$$\begin{aligned} M[c\,\mathbf{w} + d\,\widetilde{\mathbf{w}}] &= M[c\,L[\mathbf{v}] + d\,L[\widetilde{\mathbf{v}}]] \\ &= (M \circ L)[c\,\mathbf{v} + d\,\widetilde{\mathbf{v}}] \\ &= c\,\mathbf{v} + d\,\widetilde{\mathbf{v}} \\ &= c\,M[\mathbf{w}] + d\,M[\widetilde{\mathbf{w}}], \end{aligned}$$

proving linearity of M. ■

If $V = \mathbb{R}^n$, $W = \mathbb{R}^m$, so that L and M are given by matrix multiplication, by A and B respectively, then the conditions (7.17) reduce to the usual conditions

$$A\,B = \mathrm{I}, \qquad B\,A = \mathrm{I}$$

for matrix inversion, cf. (1.36). Therefore, $B = A^{-1}$ is the inverse matrix. In particular, for $L\colon \mathbb{R}^m \to \mathbb{R}^n$ to have an inverse, we must have $m = n$ and its coefficient matrix A must be nonsingular.

The invertibility of linear transformations on infinite-dimensional function spaces is more subtle. Here is a familiar example from calculus.

EXAMPLE 7.16 The Fundamental Theorem of Calculus says, roughly, that differentiation

$$D[\,f\,] = f'$$

and (indefinite) integration

$$J[\,f\,] = \int_a^x f(y)\,dy$$

are "inverse" operations. More precisely, the derivative of the indefinite integral of f is equal to f, and hence

$$D[\,J[\,f(x)\,]\,] = \frac{d}{dx}\int_a^x f(y)\,dy = f(x).$$

In other words, the composition $D \circ J = \mathrm{I}_{\mathrm{C}^0[a,b]}$ defines the identity operator on the function space $\mathrm{C}^0[a,b]$. On the other hand, if we integrate the derivative of a continuously differentiable function $f \in \mathrm{C}^1[a,b]$, we obtain

$$J[\,D[\,f(x)\,]\,] = J[\,f'(x)\,] = \int_a^x f'(y)\,dy = f(x) - f(a).$$

Therefore

$$J[\,D[\,f(x)\,]\,] = f(x) - f(a), \qquad \text{and so} \qquad J \circ D \neq \mathrm{I}_{\mathrm{C}^1[a,b]}$$

is *not* the identity operator. In other words, differentiation, D, is a left inverse for integration, J, but not a right inverse!

If we restrict D to the subspace $V = \{\, f \mid f(a) = 0 \,\} \subset \mathrm{C}^1[a,b]$ consisting of all continuously differentiable functions that vanish at the left hand endpoint, then $J\colon \mathrm{C}^0[a,b] \to V$, and $D\colon V \to \mathrm{C}^0[a,b]$ are, by the preceding argument, inverse linear operators: $D \circ J = \mathrm{I}_{\mathrm{C}^0[a,b]}$, and $J \circ D = \mathrm{I}_V$. Note that $V \subsetneq \mathrm{C}^1[a,b] \subsetneq \mathrm{C}^0[a,b]$. Thus, we discover the curious and disconcerting infinite-dimensional phenomenon that J defines a one-to-one, invertible, linear map from a vector space $\mathrm{C}^0[a,b]$ to a proper subspace $V \subsetneq \mathrm{C}^0[a,b]$. This paradoxical situation *cannot* occur in finite dimensions. A linear map $L\colon \mathbb{R}^n \to \mathbb{R}^n$ can only be invertible when its range is the entire space—because it represents multiplication by a nonsingular square matrix. ●

EXERCISES 7.1

7.1.51. Determine which of the following linear functions $L\colon \mathbb{R}^2 \to \mathbb{R}^2$ has an inverse, and, if so, describe it:

(a) the scaling transformation that doubles the length of each vector;

(b) clockwise rotation by 45°;

(c) reflection through the y axis;

(d) orthogonal projection onto the line $y = x$;

(e) the shearing transformation defined by the matrix $\begin{pmatrix} 1 & 2 \\ 0 & 1 \end{pmatrix}$.

7.1.52. For each of the linear functions in Exercise 7.1.51, write down its matrix representative, the matrix representative of its inverse, and verify that the matrices are mutual inverses.

7.1.53. Let $L: \mathbb{R}^2 \to \mathbb{R}^2$ be the linear function such that $L[\mathbf{e}_1] = (1,-1)^T$, $L[\mathbf{e}_2] = (3,-2)^T$. Find $L^{-1}[\mathbf{e}_1]$ and $L^{-1}[\mathbf{e}_2]$.

7.1.54. Let $L: \mathbb{R}^3 \to \mathbb{R}^3$ be the linear function such that $L[\mathbf{e}_1] = (2,1,-1)^T$, $L[\mathbf{e}_2] = (1,2,1)^T$, $L[\mathbf{e}_3] = (-1,2,2)^T$. Find $L^{-1}[\mathbf{e}_1]$, $L^{-1}[\mathbf{e}_2]$ and $L^{-1}[\mathbf{e}_3]$.

◇ **7.1.55.** Prove that the inverse of a linear transformation is unique; i.e., given L, there is at most one linear transformation M that can satisfy (7.17).

♡ **7.1.56.** Suppose $\mathbf{v}_1, \ldots, \mathbf{v}_n$ is a basis for V and $\mathbf{w}_1, \ldots, \mathbf{w}_n$ a basis for W.

(a) Prove that there is a unique linear function $L: V \to W$ such that $L[\mathbf{v}_i] = \mathbf{w}_i$ for $i = 1, \ldots, n$.

(b) Prove that L is invertible.

(c) If $V = W = \mathbb{R}^n$, find a formula for the matrix representative of the linear functions L and L^{-1}.

(d) Apply your construction to produce a linear function that takes:

(i) $\mathbf{v}_1 = \begin{pmatrix} 1 \\ 0 \end{pmatrix}, \mathbf{v}_2 = \begin{pmatrix} 0 \\ 1 \end{pmatrix}$ to $\mathbf{w}_1 = \begin{pmatrix} 3 \\ 1 \end{pmatrix}, \mathbf{w}_2 = \begin{pmatrix} 5 \\ 2 \end{pmatrix}$,

(ii) $\mathbf{v}_1 = \begin{pmatrix} 1 \\ 2 \end{pmatrix}, \mathbf{v}_2 = \begin{pmatrix} 2 \\ 1 \end{pmatrix}$ to $\mathbf{w}_1 = \begin{pmatrix} 1 \\ -1 \end{pmatrix}, \mathbf{w}_2 = \begin{pmatrix} 1 \\ 1 \end{pmatrix}$,

(iii) $\mathbf{v}_1 = \begin{pmatrix} 0 \\ 1 \\ 1 \end{pmatrix}, \mathbf{v}_2 = \begin{pmatrix} 1 \\ 0 \\ 1 \end{pmatrix}, \mathbf{v}_3 = \begin{pmatrix} 1 \\ 1 \\ 0 \end{pmatrix}$ to $\mathbf{w}_1 = \begin{pmatrix} 1 \\ 0 \\ 0 \end{pmatrix}, \mathbf{w}_2 = \begin{pmatrix} 0 \\ 1 \\ 0 \end{pmatrix}, \mathbf{w}_3 = \begin{pmatrix} 0 \\ 0 \\ 1 \end{pmatrix}$.

7.1.57. Suppose $V, W \subset \mathbb{R}^n$ are subspaces of the same dimension. Prove that there is an invertible linear function $L: \mathbb{R}^n \to \mathbb{R}^n$ that takes V to W. *Hint*: Use Exercise 7.1.56.

7.1.58. Give an example of a matrix with a left inverse, but not a right inverse. Is your left inverse unique?

◇ **7.1.59.** Let $L: V \to W$ be a linear function. Suppose $M, N: W \to V$ satisfy $L \circ M = \mathrm{I}_V = N \circ L$. Prove that $M = N = L^{-1}$. Thus, a linear function may have only a left or a right inverse, but if it has both, then they must be the same.

7.1.60. (a) Prove that $L[p] = p' + p$ defines an invertible linear map on the space $\mathcal{P}^{(2)}$ of quadratic polynomials. Find a formula for its inverse.

(b) Does the derivative $D[p] = p'$ have either a left or a right inverse on $\mathcal{P}^{(2)}$?

♡ **7.1.61.** (a) Show that the set of all functions of the form $f(x) = (a x^2 + b x + c)\, e^x$ for $a, b, c, \in \mathbb{R}$ forms a vector space. What is its dimension?

(b) Show that the derivative $D[f(x)] = f'(x)$ defines an invertible linear transformation on this vector space, and determine its inverse.

(c) Generalize your result in part (b) to the infinite-dimensional vector space consisting of all functions of the form $p(x)\, e^x$, where $p(x)$ is an arbitrary polynomial.

7.2 LINEAR TRANSFORMATIONS

Consider a linear function $L: \mathbb{R}^n \to \mathbb{R}^n$ that maps n-dimensional Euclidean space to itself. The function L maps a point $\mathbf{x} \in \mathbb{R}^n$ to its image point $L[\mathbf{x}] = A\mathbf{x}$, where A is its $n \times n$ matrix representative. As such, it can be assigned a geometrical interpretation that leads to further insight into the nature and scope of linear functions on Euclidean space. The geometrically inspired term *linear transformation* is often used to refer to such linear functions. The two-, three- and four-dimensional (viewing time as the fourth dimension) cases have particular relevance to our physical universe. Many of the notable maps that appear in geometry, computer graphics, elasticity, symmetry, crystallography, and Einstein's special relativity, to name a few, are defined by linear transformations.

Most of the important classes of linear transformations already appear in the two-dimensional case. Every linear function $L: \mathbb{R}^2 \to \mathbb{R}^2$ has the form

$$L\begin{pmatrix} x \\ y \end{pmatrix} = \begin{pmatrix} a x + b y \\ c x + d y \end{pmatrix}, \quad \text{where} \quad A = \begin{pmatrix} a & b \\ c & d \end{pmatrix} \tag{7.18}$$

is an arbitrary 2×2 matrix. We have already encountered the *rotation matrices*

$$R_\theta = \begin{pmatrix} \cos\theta & -\sin\theta \\ \sin\theta & \cos\theta \end{pmatrix}, \tag{7.19}$$

whose effect is to rotate every vector in $\mathbb{R}^2$ through an angle θ; in Figure 7.4 we illustrate the effect on a couple of square regions in the plane. Planar rotations coincide with 2×2 proper orthogonal matrices, meaning matrices Q that satisfy

$$Q^T Q = \mathrm{I}, \qquad \det Q = +1. \tag{7.20}$$

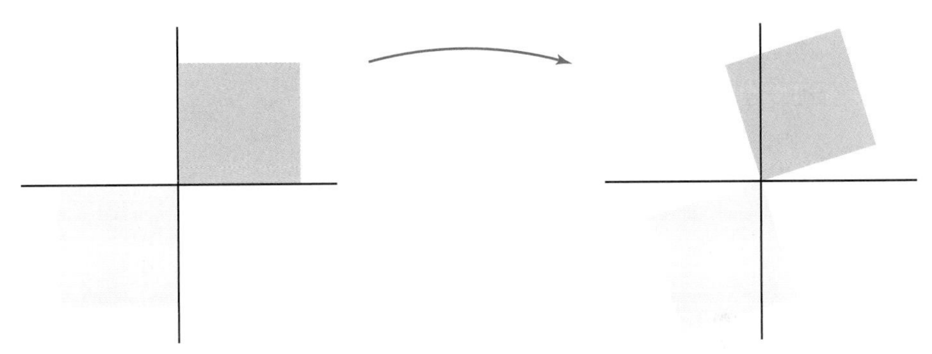

Figure 7.4 Rotation.

The improper orthogonal matrices, i.e., those with determinant -1, define *reflections*. For example, the matrix

$$A = \begin{pmatrix} -1 & 0 \\ 0 & 1 \end{pmatrix} \quad \text{corresponds to the linear transformation} \quad L\begin{pmatrix} x \\ y \end{pmatrix} = \begin{pmatrix} -x \\ y \end{pmatrix}, \tag{7.21}$$

which reflects the plane through the y axis. It can be visualized by thinking of the y axis as a mirror, as illustrated in Figure 7.5.

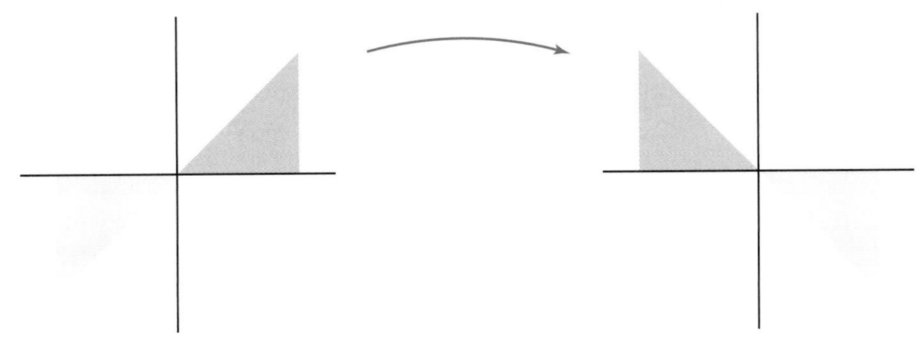

Figure 7.5 Reflection through the y axis.

Another simple example is the improper orthogonal matrix

$$R = \begin{pmatrix} 0 & 1 \\ 1 & 0 \end{pmatrix}. \quad \text{The corresponding linear transformation} \quad L\begin{pmatrix} x \\ y \end{pmatrix} = \begin{pmatrix} y \\ x \end{pmatrix} \tag{7.22}$$

is a reflection through the diagonal line $y = x$, as illustrated in Figure 7.6.

A similar classification of orthogonal matrices carries over to three-dimensional (and even higher dimensional) space. The proper orthogonal matrices correspond to

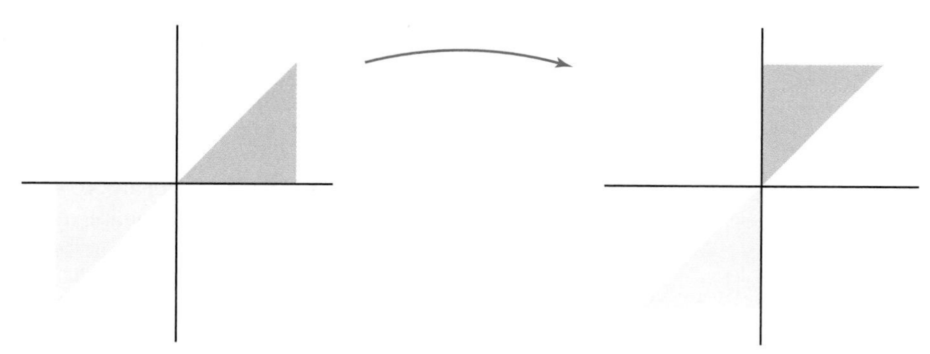

Figure 7.6 Reflection through the diagonal.

rotations and the improper to reflections, or, more generally, reflections combined with rotations. For example, the proper orthogonal matrix

$$Z_\theta = \begin{pmatrix} \cos\theta & -\sin\theta & 0 \\ \sin\theta & \cos\theta & 0 \\ 0 & 0 & 1 \end{pmatrix} \tag{7.23}$$

corresponds to a counterclockwise rotation through an angle θ around the z–axis, while

$$Y_\varphi = \begin{pmatrix} \cos\varphi & 0 & -\sin\varphi \\ 0 & 1 & 0 \\ \sin\varphi & 0 & \cos\varphi \end{pmatrix} \tag{7.24}$$

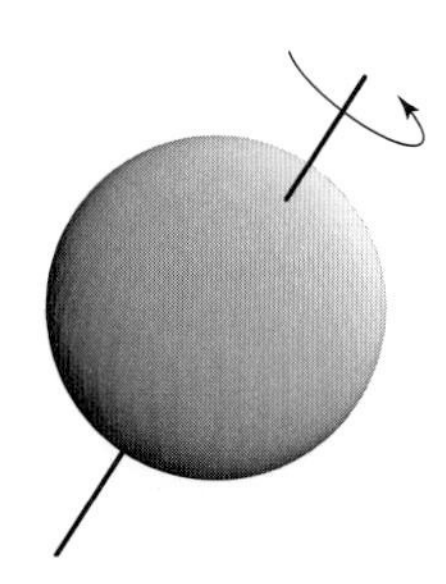

Figure 7.7 A three-dimensional rotation.

corresponds to a clockwise rotation through an angle φ around the y–axis. In general, a proper orthogonal matrix $Q = (\mathbf{u}_1\ \mathbf{u}_2\ \mathbf{u}_3)$ with columns $\mathbf{u}_i = Q\,\mathbf{e}_i$ corresponds to the rotation in which the standard basis vectors $\mathbf{e}_1, \mathbf{e}_2, \mathbf{e}_3$ are rotated to new positions given by the orthonormal basis $\mathbf{u}_1, \mathbf{u}_2, \mathbf{u}_3$. It can be shown—see Exercise 8.2.43—that *every* 3×3 orthogonal matrix corresponds to a rotation around a line through the origin in $\mathbb{R}^3$—the axis of the rotation, as sketched in Figure 7.7.

Since the product of two (proper) orthogonal matrices is also (proper) orthogonal, the composition of two rotations is also a rotation. Unlike the planar case, the order in which the rotations are performed is important! Multiplication of $n \times n$ orthogonal matrices is *not* commutative when $n \geq 3$. For example, rotating first around the z–axis and then rotating around the y–axis does *not* have the same effect as first rotating around the y–axis and then around the z–axis. If you don't believe this, try it out with a solid object such as this book. Rotate through 90°, say, around each axis; the final configuration of the book will depend upon the order in which you do the rotations. Then prove this mathematically by showing that the two rotation matrices (7.23, 24) do not commute.

Other important linear transformations arise from elementary matrices. First, the elementary matrices corresponding to the third type of row operations—multiplying a row by a scalar— correspond to simple stretching transformations. For example, if

$$A = \begin{pmatrix} 2 & 0 \\ 0 & 1 \end{pmatrix}, \qquad \text{then the linear transformation} \qquad L\begin{pmatrix} x \\ y \end{pmatrix} = \begin{pmatrix} 2x \\ y \end{pmatrix}$$

has the effect of stretching along the x axis by a factor of 2; see Figure 7.8. A negative diagonal entry corresponds to a reflection followed by a stretch. For example, the elementary matrix

$$\begin{pmatrix} -2 & 0 \\ 0 & 1 \end{pmatrix} = \begin{pmatrix} 2 & 0 \\ 0 & 1 \end{pmatrix} \begin{pmatrix} -1 & 0 \\ 0 & 1 \end{pmatrix}$$

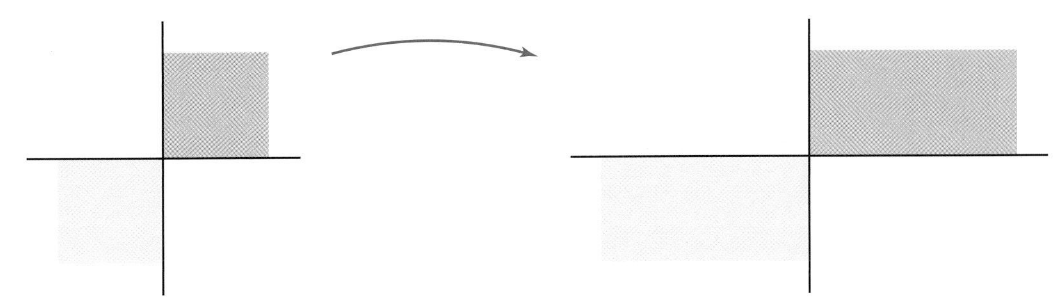

Figure 7.8 Stretch along the x–axis.

corresponds to a reflection through the y axis followed by a stretch along the x axis. In this case, the order of these operations is immaterial since the matrices commute.

In the 2×2 case, there is only one type of elementary row interchange, namely the matrix (7.22) that corresponds to a reflection through the diagonal $y = x$.

The elementary matrices of Type #1 correspond to *shearing transformations* of the plane. For example, the matrix

$$\begin{pmatrix} 1 & 2 \\ 0 & 1 \end{pmatrix} \quad \text{represents the linear transformation} \quad L\begin{pmatrix} x \\ y \end{pmatrix} = \begin{pmatrix} x + 2y \\ y \end{pmatrix},$$

which has the effect of shearing the plane along the x–axis. The constant 2 will be called the *shear factor*, and can be either positive or negative. Under the shearing transformation, each point moves parallel to the x axis by an amount proportional to its (signed) distance from the axis. Similarly, the elementary matrix

$$\begin{pmatrix} 1 & 0 \\ -3 & 1 \end{pmatrix} \quad \text{represents the linear transformation} \quad L\begin{pmatrix} x \\ y \end{pmatrix} = \begin{pmatrix} x \\ y - 3x \end{pmatrix},$$

which is a shear along the y axis of magnitude -3. As illustrated in Figure 7.9, shears map rectangles to parallelograms; distances are altered, but areas are unchanged.

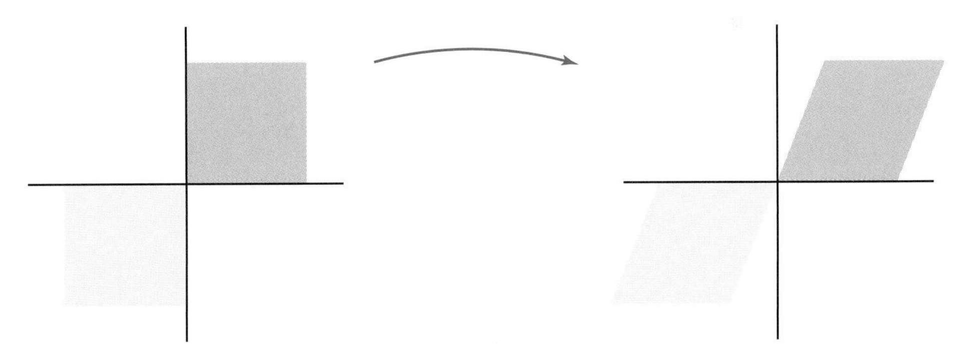

Figure 7.9 Shear in the x direction.

All of the preceding linear maps are invertible, and so are represented by nonsingular matrices. Besides the zero map/matrix, which sends every point $\mathbf{x} \in \mathbb{R}^2$ to the origin, the simplest singular map is

$$\begin{pmatrix} 1 & 0 \\ 0 & 0 \end{pmatrix} \quad \text{corresponding to the linear transformation} \quad L\begin{pmatrix} x \\ y \end{pmatrix} = \begin{pmatrix} x \\ 0 \end{pmatrix},$$

which defines the orthogonal projection of the vector $(x, y)^T$ onto the x–axis. Other rank one matrices represent various kinds of projections from the plane to a line through the origin; see Exercise 7.2.16 for details.

A similar classification of linear maps can be established in higher dimensions. The linear transformations constructed from elementary matrices can be built up from the following four basic types:

(i) a stretch in a single coordinate direction,
(ii) a reflection through a coordinate plane*,
(iii) a reflection through a diagonal plane,
(iv) a shear along a coordinate axis.

Moreover, we already proved, (1.45), that every nonsingular matrix can be written as a product of elementary matrices. This has the remarkable consequence that *every* invertible linear transformation can be constructed from a sequence of elementary stretches, reflections, and shears. In addition, there is one further, non-invertible type of basic linear transformation:

(v) an orthogonal projection onto a lower dimensional subspace.

All linear transformations of $\mathbb{R}^n$ can be built up, albeit non-uniquely, as a composition of these five basic types.

EXAMPLE 7.17 Consider the matrix

$$A = \begin{pmatrix} \frac{\sqrt{3}}{2} & -\frac{1}{2} \\ \frac{1}{2} & \frac{\sqrt{3}}{2} \end{pmatrix}$$

corresponding to a plane rotation through $\theta = 30^\circ$, cf. (7.19). Rotations are not elementary linear transformations. To express this particular rotation as a product of elementary matrices, we need to perform the Gauss-Jordan Elimination procedure to reduce it to the identity matrix. Let us indicate the basic steps:

$$E_1 = \begin{pmatrix} 1 & 0 \\ -\frac{1}{\sqrt{3}} & 1 \end{pmatrix}, \qquad E_1 A = \begin{pmatrix} \frac{\sqrt{3}}{2} & -\frac{1}{2} \\ 0 & \frac{2}{\sqrt{3}} \end{pmatrix},$$

$$E_2 = \begin{pmatrix} 1 & 0 \\ 0 & \frac{\sqrt{3}}{2} \end{pmatrix}, \qquad E_2 E_1 A = \begin{pmatrix} \frac{\sqrt{3}}{2} & -\frac{1}{2} \\ 0 & 1 \end{pmatrix},$$

$$E_3 = \begin{pmatrix} \frac{2}{\sqrt{3}} & 0 \\ 0 & 1 \end{pmatrix}, \qquad E_3 E_2 E_1 A = \begin{pmatrix} 1 & -\frac{1}{\sqrt{3}} \\ 0 & 1 \end{pmatrix},$$

$$E_4 = \begin{pmatrix} 1 & \frac{1}{\sqrt{3}} \\ 0 & 1 \end{pmatrix}, \qquad E_4 E_3 E_2 E_1 A = \mathrm{I} = \begin{pmatrix} 1 & 0 \\ 0 & 1 \end{pmatrix}.$$

We conclude that

$$\begin{pmatrix} \frac{\sqrt{3}}{2} & -\frac{1}{2} \\ \frac{1}{2} & \frac{\sqrt{3}}{2} \end{pmatrix} = A = E_1^{-1} E_2^{-1} E_3^{-1} E_4^{-1}$$

$$= \begin{pmatrix} 1 & 0 \\ \frac{1}{\sqrt{3}} & 1 \end{pmatrix} \begin{pmatrix} 1 & 0 \\ 0 & \frac{2}{\sqrt{3}} \end{pmatrix} \begin{pmatrix} \frac{\sqrt{3}}{2} & 0 \\ 0 & 1 \end{pmatrix} \begin{pmatrix} 1 & -\frac{1}{\sqrt{3}} \\ 0 & 1 \end{pmatrix}.$$

*In n-dimensional space, this should read "hyperplane", i.e., a subspace of dimension $n - 1$.

As a result, a $30°$ rotation can be effected by composing the following elementary transformations in the prescribed order, bearing in mind that the last matrix in the product will act first on the vector $\mathbf{x}$:

(1) First, a shear in the x–direction with shear factor $-\frac{1}{\sqrt{3}}$,

(2) Then a stretch (or, rather, a contraction) in the direction of the x–axis by a factor of $\frac{\sqrt{3}}{2}$,

(3) Then a stretch in the y-direction by the reciprocal factor $\frac{2}{\sqrt{3}}$,

(4) Finally, a shear in the direction of the y–axis with shear factor $\frac{1}{\sqrt{3}}$.

The fact that this combination of elementary transformations results in a pure rotation is surprising and non-obvious. ●

EXERCISES 7.2

7.2.1. For each of the following linear transformations $L: \mathbb{R}^2 \to \mathbb{R}^2$, find a matrix representative, and then describe its effect on

(i) the x–axis;

(ii) the unit square $S = \{0 \le x, y \le 1\}$;

(iii) the unit disk $D = \{x^2 + y^2 \le 1\}$.

(a) counterclockwise rotation by $45°$.

(b) rotation by $180°$;

(c) reflection in the line $y = 2x$;

(d) shear along the y axis of magnitude 2;

(e) shear along the line $x = y$ of magnitude 3;

(f) orthogonal projection on the line $y = 2x$.

7.2.2. Draw the parallelogram spanned by the vectors $\begin{pmatrix} 1 \\ 2 \end{pmatrix}$ and $\begin{pmatrix} 3 \\ 1 \end{pmatrix}$. Then draw its image under the linear transformations defined by the following matrices:

(a) $\begin{pmatrix} 1 & 0 \\ -1 & 1 \end{pmatrix}$ (b) $\begin{pmatrix} 0 & 1 \\ 1 & 0 \end{pmatrix}$

(c) $\begin{pmatrix} 1 & 2 \\ -1 & 4 \end{pmatrix}$ (d) $\begin{pmatrix} \frac{1}{\sqrt{2}} & -\frac{1}{\sqrt{2}} \\ \frac{1}{\sqrt{2}} & \frac{1}{\sqrt{2}} \end{pmatrix}$

(e) $\begin{pmatrix} -1 & -2 \\ 2 & 1 \end{pmatrix}$ (f) $\begin{pmatrix} \frac{1}{2} & -\frac{1}{2} \\ -\frac{1}{2} & \frac{1}{2} \end{pmatrix}$

(g) $\begin{pmatrix} 2 & -1 \\ -4 & 2 \end{pmatrix}$

7.2.3. Let L be the linear transformation represented by the matrix $\begin{pmatrix} 0 & 1 \\ -1 & 0 \end{pmatrix}$. Show that $L^2 = L \circ L$ is rotation by $180°$. Is L itself a rotation or a reflection?

7.2.4. Let L be the linear transformation determined by $\begin{pmatrix} 0 & 1 \\ 1 & 0 \end{pmatrix}$. Show $L^2 = \mathrm{I}$, and interpret geometrically.

7.2.5. What is the geometric interpretation of the linear transformation with matrix $A = \begin{pmatrix} 1 & 0 \\ 2 & -1 \end{pmatrix}$? Use this to explain why $A^2 = \mathrm{I}$.

7.2.6. Describe the image of the line ℓ that goes through the points $\begin{pmatrix} -2 \\ 1 \end{pmatrix}$, $\begin{pmatrix} 1 \\ -2 \end{pmatrix}$ under the linear transformation $\begin{pmatrix} 2 & 3 \\ -1 & 0 \end{pmatrix}$.

7.2.7. Find a linear transformation that maps the unit circle $x^2 + y^2 = 1$ to the ellipse $\frac{1}{4}x^2 + \frac{1}{9}y^2 = 1$. Is your answer unique?

7.2.8. Find a linear transformation that maps the unit sphere $x^2 + y^2 + z^2 = 1$ to the ellipsoid $x^2 + \frac{1}{4}y^2 + \frac{1}{16}z^2 = 1$.

7.2.9. *True or false*: A linear transformation $L: \mathbb{R}^2 \to \mathbb{R}^2$ maps

(a) straight lines to straight lines

(b) triangles to triangles

(c) squares to squares

(d) circles to circles

(e) ellipses to ellipses

◇ **7.2.10.** (a) Prove that the linear transformation associated with the improper orthogonal matrix

$$\begin{pmatrix} \cos\theta & \sin\theta \\ \sin\theta & -\cos\theta \end{pmatrix}$$

is a reflection through the line that makes an angle $\frac{1}{2}\theta$ with the x–axis.

(b) Show that the composition of two such reflections, with angles θ, φ, is a rotation. What is the angle of the rotation? Does the composition depend upon the order of the two reflections?

◇ **7.2.11.** Let $L \subset \mathbb{R}^2$ be a line through the origin in the direction of the unit vector $\mathbf{u}$.

(a) Prove that the matrix representative of reflection through L is $R = 2\mathbf{u}\mathbf{u}^T - \mathrm{I}$.

(b) Find the corresponding formula for reflection through a line in the direction of a general nonzero vector $\mathbf{v} \neq \mathbf{0}$.

(c) Determine the matrix representative for reflection through the line in the direction

(i) $(1,0)^T$ (ii) $\left(\frac{3}{5}, -\frac{4}{5}\right)^T$
(iii) $(1,1)^T$ (iv) $(2,-3)^T$

7.2.12. Decompose the following matrices into a product of elementary matrices. Then interpret each of the factors as a linear transformation.

(a) $\begin{pmatrix} 0 & 2 \\ -3 & 1 \end{pmatrix}$ (b) $\begin{pmatrix} 1 & 1 \\ -1 & 1 \end{pmatrix}$

(c) $\begin{pmatrix} 3 & 1 \\ 1 & 2 \end{pmatrix}$ (d) $\begin{pmatrix} 1 & 1 & 0 \\ 1 & 0 & 1 \\ 0 & 1 & 1 \end{pmatrix}$

(e) $\begin{pmatrix} 1 & 2 & 0 \\ 2 & 4 & 1 \\ 2 & 1 & 1 \end{pmatrix}$

7.2.13. (a) Prove that

$$\begin{pmatrix} \cos\theta & -\sin\theta \\ \sin\theta & \cos\theta \end{pmatrix} = \begin{pmatrix} 1 & a \\ 0 & 1 \end{pmatrix}\begin{pmatrix} 1 & 0 \\ b & 1 \end{pmatrix}\begin{pmatrix} 1 & a \\ 0 & 1 \end{pmatrix},$$

where $a = -\tan\frac{1}{2}\theta$ and $b = \sin\theta$.

(b) Is the factorization valid for all values of θ?

(c) Interpret the factorization geometrically.

Remark: The factored version is less prone to numerical errors due to round-off, and so can be used when extremely accurate numerical computations involving rotations are required.

7.2.14. (a) Find the matrix in $\mathbb{R}^3$ that corresponds to a counterclockwise rotation around the x–axis through an angle 60°.

(b) Write it as a product of elementary matrices, and interpret each of the factors.

7.2.15. Determine the matrix representative for orthogonal projection $P : \mathbb{R}^2 \to \mathbb{R}^2$ on the line through the origin in the direction

(a) $(1,0)^T$ (b) $(1,1)^T$ (c) $(2,-3)^T$

◇ **7.2.16.** (a) Prove that any 2×2 matrix of rank 1 can be written in the form $A = \mathbf{u}\mathbf{v}^T$ where $\mathbf{u}, \mathbf{v} \in \mathbb{R}^2$ are non-zero column vectors.

(b) Which rank one matrices correspond to orthogonal projection onto a one-dimensional subspace of $\mathbb{R}^2$?

7.2.17. Give a geometrical interpretation of the linear transformations on $\mathbb{R}^3$ defined by each of the six 3×3 permutation matrices (1.30).

7.2.18. Write down the 3×3 matrix X_ψ representing a clockwise rotation in $\mathbb{R}^3$ around the x-axis by angle ψ.

7.2.19. Explain why the linear map defined by $-\mathrm{I}$ defines a rotation in two-dimensional space, but a reflection in three-dimensional space.

◇ **7.2.20.** Let $\mathbf{u} = (u_1, u_2, u_3)^T \in \mathbb{R}^3$ be a unit vector. Show that $Q_\pi = 2\mathbf{u}\mathbf{u}^T - \mathrm{I}$ represents rotation around the axis $\mathbf{u}$ through an angle π.

◇ **7.2.21.** Let $\mathbf{u} \in \mathbb{R}^3$ be a unit vector.

(a) Explain why the *elementary reflection matrix* $R = \mathrm{I} - 2\mathbf{u}\mathbf{u}^T$ represents a reflection through the plane orthogonal to $\mathbf{u}$.

(b) Prove that R is an orthogonal matrix. Is it proper or improper?

(c) Write out R when $\mathbf{u} =$

(i) $\left(\frac{3}{5}, 0, -\frac{4}{5}\right)^T$ (ii) $\left(\frac{3}{13}, \frac{4}{13}, -\frac{12}{13}\right)^T$
(iii) $\left(\frac{1}{\sqrt{6}}, -\frac{2}{\sqrt{6}}, \frac{1}{\sqrt{6}}\right)^T$

(d) Give a geometrical explanation why $Q_\pi = -R$ represents the rotation of Exercise 7.2.20.

◇ **7.2.22.** Let $\mathbf{a} \in \mathbb{R}^3$ be a fixed vector and let Q be any 3×3 rotation matrix such that $Q\,\mathbf{a} = \mathbf{e}_3$.

(a) Show that, using the notation of (7.23), $R_\theta = Q^T Z_\theta\, Q$ represents rotation around $\mathbf{a}$ by angle θ.

(b) Verify this formula in the case $\mathbf{a} = \mathbf{e}_2$ by comparing with (7.24).

♡ **7.2.23.** *Quaternions*: The *skew field* $\mathbb{H}$ of quaternions can be identified with the vector space $\mathbb{R}^4$ equipped with a *noncommutative* multiplication operation. The standard basis vectors $\mathbf{e}_1, \mathbf{e}_2, \mathbf{e}_3, \mathbf{e}_4$ are traditionally denoted by the letters $1, \mathrm{i}, \mathrm{j}, \mathrm{k}$; the vector $(a,b,c,d)^T \in \mathbb{R}^4$ corresponds to the quaternion $q = a + b\,\mathrm{i} + c\,\mathrm{j} + d\,\mathrm{k}$. Quaternion addition coincides with vector addition. Quaternion multiplication is defined so that

$$\begin{aligned} 1\,q &= q = q\,1, \\ \mathrm{i}^2 &= \mathrm{j}^2 = \mathrm{k}^2 = -1, \\ \mathrm{i}\,\mathrm{j} &= \mathrm{k} = -\mathrm{j}\,\mathrm{i}, \\ \mathrm{i}\,\mathrm{k} &= -\mathrm{j} = -\mathrm{k}\,\mathrm{i}, \\ \mathrm{j}\,\mathrm{k} &= \mathrm{i} = -\mathrm{k}\,\mathrm{j}, \end{aligned}$$

along with the distributive laws

$$(q+r)s = qs + rs, \qquad q(r+s) = qr + qs,$$

for all $q, r, s \in \mathbb{H}$.

(a) Compute the following quaternion products:

(i) $\mathrm{j}\,(2 - 3\,\mathrm{j} + \mathrm{k})$

(ii) $(1 + \mathrm{i})(1 - 2\,\mathrm{i} + \mathrm{j})$

(iii) $(1 + \mathrm{i} - \mathrm{j} - 3\,\mathrm{k})^2$

(iv) $(2 + 2\,\mathrm{i} + 3\,\mathrm{j} - \mathrm{k})(2 - 2\,\mathrm{i} - 3\,\mathrm{j} + \mathrm{k})$

(b) The *conjugate* of the quaternion $q = a + b\,\mathrm{i} + c\,\mathrm{j} + d\,\mathrm{k}$ is defined to be $\overline{q} = a - b\,\mathrm{i} - c\,\mathrm{j} - d\,\mathrm{k}$. Prove that $q\,\overline{q} = \|q\|^2 = \overline{q}\,q$, where $\|\cdot\|$ is the usual Euclidean norm on $\mathbb{R}^4$.

(c) Prove that quaternion multiplication is associative.

(d) Let $q = a + b\,\mathrm{i} + c\,\mathrm{j} + d\,\mathrm{k} \in \mathbb{H}$ be a fixed quaternion. Show that $L_q[r] = qr$ and $R_q[r] = rq$ define linear transformations on the vector space $\mathbb{H} \simeq \mathbb{R}^4$. Write down their 4×4 matrix representatives, and observe that they are not the same since quaternion multiplication is not commutative.

(e) Show that L_q and R_q are orthogonal matrices if $\|q\|^2 = a^2 + b^2 + c^2 + d^2 = 1$.

(f) We can identify a quaternion $q = b\,\mathrm{i} + c\,\mathrm{j} + d\,\mathrm{k}$ with zero real part, $a = 0$, with a vector $\mathbf{q} = (b, c, d)^T \in \mathbb{R}^3$. Show that, in this case, the quaternion product $qr = \mathbf{q} \times \mathbf{r} - \mathbf{q} \cdot \mathbf{r}$ can be identified with the difference between the cross and dot product of the two vectors. Which vector identities result from the associativity of quaternion multiplication?

Remark: The *quaternions* were discovered by the Irish mathematician William Rowan Hamilton in 1843. Much of our modern vector calculus notation is of quaternionic origin, [15].

Change of Basis

Sometimes a linear transformation represents an elementary geometrical transformation, but this is not evident because the matrix happens to be written in the "wrong" coordinates. The characterization of linear functions from $\mathbb{R}^n$ to $\mathbb{R}^m$ as multiplication by $m \times n$ matrices in Theorem 7.5 relies on using the standard bases for both the domain and target spaces. In many cases, these bases are not particularly well-adapted to the linear transformation in question, and one can often gain additional insight by adopting more suitable bases. To this end, we first need to understand how to rewrite a linear transformation in terms of a new basis.

The following result says that, in *any* basis, a linear function on finite-dimensional vector spaces can always be realized by matrix multiplication of the coordinates. But bear in mind that the particular matrix representative will depend upon the choice of bases.

THEOREM 7.18 Let $L\colon V \to W$ be a linear function. Suppose V has basis $\mathbf{v}_1, \ldots, \mathbf{v}_n$ and W has basis $\mathbf{w}_1, \ldots, \mathbf{w}_m$. We can write

$$\mathbf{v} = x_1\mathbf{v}_1 + \cdots + x_n\mathbf{v}_n \in V, \qquad \mathbf{w} = y_1\mathbf{w}_1 + \cdots + y_m\mathbf{w}_m \in W,$$

where $\mathbf{x} = (x_1, x_2, \ldots, x_n)^T$ are the coordinates of $\mathbf{v}$ relative to the basis of V and $\mathbf{y} = (y_1, y_2, \ldots, y_m)^T$ are those of $\mathbf{w}$ relative to the basis of W. Then, in these coordinates, the linear function $\mathbf{w} = L[\mathbf{v}]$ is given by multiplication by an $m \times n$ matrix B, so $\mathbf{y} = B\mathbf{x}$.

Proof We mimic the proof of Theorem 7.5, replacing the standard basis vectors by more general basis vectors. In other words, we will apply L to the basis vectors of V and express the result as a linear combination of the basis vectors in W. Specifically, we write

$$L[\mathbf{v}_j] = \sum_{i=1}^{m} b_{ij}\,\mathbf{w}_i.$$

The coefficients b_{ij} form the entries of the desired coefficient matrix. Indeed, by linearity,

$$\begin{aligned} L[\mathbf{v}] &= L[x_1 \mathbf{v}_1 + \cdots + x_n \mathbf{v}_n] \\ &= x_1 L[\mathbf{v}_1] + \cdots + x_n L[\mathbf{v}_n] = \sum_{i=1}^{m} \left(\sum_{j=1}^{n} b_{ij} x_j \right) \mathbf{w}_i, \end{aligned}$$

and so $y_i = \sum_{j=1}^{n} b_{ij} x_j$ as claimed. ■

Suppose that the linear transformation $L: \mathbb{R}^n \to \mathbb{R}^m$ is represented by a certain $m \times n$ matrix A relative to the standard bases $\mathbf{e}_1, \ldots, \mathbf{e}_n$ and $\widehat{\mathbf{e}}_1, \ldots, \widehat{\mathbf{e}}_m$ of the domain and target spaces. If we introduce alternative bases for $\mathbb{R}^n$ and $\mathbb{R}^m$ then the *same* linear transformation may have a completely different matrix representation. Therefore, different matrices may represent the same underlying linear transformation, with respect to different bases of its domain and target spaces.

EXAMPLE 7.19 Consider the linear transformation

$$L(x_1, x_2) = \begin{pmatrix} x_1 - x_2 \\ 2x_1 + 4x_2 \end{pmatrix}$$

which we write in the standard, Cartesian coordinates x, y on $\mathbb{R}^2$. The corresponding coefficient matrix

$$A = \begin{pmatrix} 1 & -1 \\ 2 & 4 \end{pmatrix} \tag{7.25}$$

is the matrix representation of L—relative to the standard basis $\mathbf{e}_1, \mathbf{e}_2$ of $\mathbb{R}^2$. This means that

$$L[\mathbf{e}_1] = \begin{pmatrix} 1 \\ 2 \end{pmatrix} = \mathbf{e}_1 + 2\,\mathbf{e}_2, \qquad L[\mathbf{e}_2] = \begin{pmatrix} -1 \\ 4 \end{pmatrix} = -\,\mathbf{e}_1 + 4\,\mathbf{e}_2.$$

Let us see what happens if we replace the standard basis by the alternative basis

$$\mathbf{v}_1 = \begin{pmatrix} 1 \\ -1 \end{pmatrix}, \qquad \mathbf{v}_2 = \begin{pmatrix} 1 \\ -2 \end{pmatrix}.$$

What is the corresponding matrix formulation of the same linear transformation? According to the recipe of Theorem 7.18, we must compute

$$L[\mathbf{v}_1] = \begin{pmatrix} 2 \\ -2 \end{pmatrix} = 2\,\mathbf{v}_1, \qquad L[\mathbf{v}_2] = \begin{pmatrix} 3 \\ -6 \end{pmatrix} = 3\,\mathbf{v}_2.$$

The linear transformation acts by stretching in the direction $\mathbf{v}_1$ by a factor of 2 and simultaneously stretching in the direction $\mathbf{v}_2$ by a factor of 3. Therefore, the matrix form of L with respect to this new basis is the diagonal matrix

$$D = \begin{pmatrix} 2 & 0 \\ 0 & 3 \end{pmatrix}. \tag{7.26}$$

In general,

$$L[a\,\mathbf{v}_1 + b\,\mathbf{v}_2] = 2a\,\mathbf{v}_1 + 3b\,\mathbf{v}_2,$$

and the effect is to multiply the new basis coordinates $\mathbf{a} = (a, b)^T$ by the diagonal matrix D. Both (7.25) and (7.26) represent the *same* linear transformation—the former in the standard basis and the latter in the new basis. The hidden geometry of this linear transformation is thereby exposed through an inspired choice of basis. The secret behind such well-adapted bases will be revealed in Chapter 8. ●

How does one effect a change of basis in general? According to (2.22), if $\mathbf{v}_1, \ldots, \mathbf{v}_n$ form a basis of $\mathbb{R}^n$, then the coordinates $\mathbf{y} = (y_1, y_2, \ldots, y_n)^T$ of a vector

$$(x_1, x_2, \ldots, x_n)^T = \mathbf{x} = y_1 \mathbf{v}_1 + y_2 \mathbf{v}_2 + \cdots + y_n \mathbf{v}_n$$

are found by solving the linear system

$$S\,\mathbf{y} = \mathbf{x}, \qquad \text{where} \qquad S = (\,\mathbf{v}_1\ \mathbf{v}_2\ \ldots\ \mathbf{v}_n\,) \tag{7.27}$$

is the nonsingular $n \times n$ matrix whose columns are the basis vectors.

Consider first a linear transformation $L\colon \mathbb{R}^n \to \mathbb{R}^n$ from $\mathbb{R}^n$ to itself. When written in terms of the standard basis, $L[\mathbf{x}] = A\,\mathbf{x}$ has a certain $n \times n$ coefficient matrix A. To change to the new basis $\mathbf{v}_1, \ldots, \mathbf{v}_n$, we use (7.27) to rewrite the standard $\mathbf{x}$ coordinates in terms of the new $\mathbf{y}$ coordinates. We also need to find the coordinates $\mathbf{g}$ of a range vector $\mathbf{f} = A\,\mathbf{x}$ with respect to the new basis. By the same reasoning that led to (7.27), its new coordinates are found by solving the linear system $\mathbf{f} = S\,\mathbf{g}$. Therefore, the new target coordinates are expressed in terms of the new domain coordinates via

$$\mathbf{g} = S^{-1}\mathbf{f} = S^{-1}A\,\mathbf{x} = S^{-1}A\,S\,\mathbf{y} = B\,\mathbf{y}.$$

We conclude that, in the new basis $\mathbf{v}_1, \ldots, \mathbf{v}_n$, the matrix form of our linear transformation is

$$B = S^{-1}A\,S, \qquad \text{where} \qquad S = (\,\mathbf{v}_1\ \mathbf{v}_2\ \ldots\ \mathbf{v}_n\,). \tag{7.28}$$

Two matrices A and B that are related by such an equation for some nonsingular matrix S are called *similar*. Similar matrices represent the *same* linear transformation, but relative to *different* bases of the underlying vector space $\mathbb{R}^n$.

EXAMPLE 7.19 ***Continued***

Returning to the preceding example, we assemble the new basis vectors to form the change of basis matrix

$$S = \begin{pmatrix} 1 & 1 \\ -1 & -2 \end{pmatrix},$$

and verify that

$$S^{-1}A\,S = \begin{pmatrix} 2 & 1 \\ -1 & -1 \end{pmatrix}\begin{pmatrix} 1 & -1 \\ 2 & 4 \end{pmatrix}\begin{pmatrix} 1 & 1 \\ -1 & -2 \end{pmatrix} = \begin{pmatrix} 2 & 0 \\ 0 & 3 \end{pmatrix} = D,$$

reconfirming our earlier computation. ●

More generally, a linear transformation $L\colon \mathbb{R}^n \to \mathbb{R}^m$ is represented by an $m \times n$ matrix A with respect to the standard bases on both the domain and target spaces. What happens if we introduce a new basis $\mathbf{v}_1, \ldots, \mathbf{v}_n$ on the domain space $\mathbb{R}^n$ *and* a new basis $\mathbf{w}_1, \ldots, \mathbf{w}_m$ on the target space $\mathbb{R}^m$? Arguing as above, we conclude that the matrix representative of L with respect to these new bases is given by

$$B = T^{-1}A\,S, \tag{7.29}$$

where $S = (\,\mathbf{v}_1\ \mathbf{v}_2\ \ldots\ \mathbf{v}_n\,)$ is the domain basis matrix, while $T = (\,\mathbf{w}_1\ \mathbf{w}_2\ \ldots\ \mathbf{w}_m\,)$ is the range basis matrix.

In particular, suppose that the linear transformation has rank $r = \dim \operatorname{rng} A = \dim \operatorname{corng} A$. Let us choose a basis $\mathbf{v}_1, \ldots, \mathbf{v}_n$ of $\mathbb{R}^n$ such that $\mathbf{v}_1, \ldots, \mathbf{v}_r$ form a basis of $\operatorname{corng} A$ while $\mathbf{v}_{r+1}, \ldots, \mathbf{v}_n$ form a basis for $\ker A = (\operatorname{corng} A)^\perp$. According to Proposition 5.58, the image vectors $\mathbf{w}_1 = L[\mathbf{v}_1], \ldots, \mathbf{w}_r = L[\mathbf{v}_r]$, form a basis for $\operatorname{rng} A$, while $L[\mathbf{v}_{r+1}] = \cdots = L[\mathbf{v}_n] = \mathbf{0}$. We further choose a basis $\mathbf{w}_{r+1}, \ldots, \mathbf{w}_m$ for $\operatorname{coker} A = (\operatorname{rng} A)^\perp$, and note that the combination $\mathbf{w}_1, \ldots, \mathbf{w}_m$

forms a basis for $\mathbb{R}^m$. The matrix form of L relative to these two adapted bases is simply

$$B = T^{-1} A\, S = \begin{pmatrix} \mathrm{I}_r & \mathrm{O} \\ \mathrm{O} & \mathrm{O} \end{pmatrix} = \begin{pmatrix} 1 & 0 & 0 & \ldots & 0 & 0 & \ldots & 0 \\ 0 & 1 & 0 & \ldots & 0 & 0 & \ldots & 0 \\ 0 & 0 & 1 & \ldots & 0 & 0 & \ldots & 0 \\ \vdots & \vdots & \vdots & \ddots & \vdots & \vdots & \ddots & \vdots \\ 0 & 0 & 0 & \ldots & 1 & 0 & \ldots & 0 \\ 0 & 0 & 0 & \ldots & 0 & 0 & \ldots & 0 \\ \vdots & \vdots & \vdots & \ddots & \vdots & \vdots & \ddots & \vdots \\ 0 & 0 & 0 & \ldots & 0 & 0 & \ldots & 0 \end{pmatrix}. \tag{7.30}$$

In this matrix, the first r columns have a single 1 in the diagonal slot, indicating that the first r basis vectors of the domain space are mapped to the first r basis vectors of the target space while the last $n-r$ columns are all zero, indicating that the last $n-r$ basis vectors in the domain are all mapped to $\mathbf{0}$. Thus, by a suitable choice of bases on both the domain and target spaces, any linear transformation has an extremely simple *canonical form* that only depends upon its rank.

EXAMPLE 7.20 According to the illustrative example following Theorem 2.49, the matrix

$$A = \begin{pmatrix} 2 & -1 & 1 & 2 \\ -8 & 4 & -6 & -4 \\ 4 & -2 & 3 & 2 \end{pmatrix}$$

has rank 2. Based on those calculations, we choose the domain space basis

$$\mathbf{v}_1 = \begin{pmatrix} 2 \\ -1 \\ 1 \\ 2 \end{pmatrix}, \quad \mathbf{v}_2 = \begin{pmatrix} 0 \\ 0 \\ -2 \\ 4 \end{pmatrix}, \quad \mathbf{v}_3 = \begin{pmatrix} \frac{1}{2} \\ 1 \\ 0 \\ 0 \end{pmatrix}, \quad \mathbf{v}_4 = \begin{pmatrix} -2 \\ 0 \\ 2 \\ 1 \end{pmatrix},$$

noting that $\mathbf{v}_1, \mathbf{v}_2$ are a basis for corng A, while $\mathbf{v}_3, \mathbf{v}_4$ are a basis for ker A. For our basis of the target space, we first compute $\mathbf{w}_1 = A\mathbf{v}_1$ and $\mathbf{w}_2 = A\mathbf{v}_2$, which form a basis for rng A. We supplement these by the single basis vector $\mathbf{w}_3$ for coker A, and so

$$\mathbf{w}_1 = \begin{pmatrix} 10 \\ -34 \\ 17 \end{pmatrix}, \quad \mathbf{w}_2 = \begin{pmatrix} 6 \\ -4 \\ 2 \end{pmatrix}, \quad \mathbf{w}_3 = \begin{pmatrix} 0 \\ \frac{1}{2} \\ 1 \end{pmatrix}.$$

In terms of these two bases, the canonical matrix form of this particular linear function is

$$B = T^{-1} A\, S = \begin{pmatrix} 1 & 0 & 0 & 0 \\ 0 & 1 & 0 & 0 \\ 0 & 0 & 0 & 0 \end{pmatrix},$$

where the bases are assembled to form the matrices

$$S = \begin{pmatrix} 2 & 0 & \frac{1}{2} & -2 \\ -1 & 0 & 1 & 0 \\ 1 & -2 & 0 & 2 \\ 2 & 4 & 0 & 1 \end{pmatrix}, \qquad T = \begin{pmatrix} 10 & 6 & 0 \\ -34 & -4 & \frac{1}{2} \\ 17 & 2 & 1 \end{pmatrix}.$$

●

EXERCISES 7.2

7.2.24. Find the matrix form of the linear transformation
$$L(x, y) = \begin{pmatrix} x - 4y \\ -2x + 3y \end{pmatrix}$$
with respect to the following bases of $\mathbb{R}^2$:

(a) $\begin{pmatrix} 1 \\ 0 \end{pmatrix}, \begin{pmatrix} 0 \\ 1 \end{pmatrix}$ (b) $\begin{pmatrix} 2 \\ 0 \end{pmatrix}, \begin{pmatrix} 0 \\ 3 \end{pmatrix}$

(c) $\begin{pmatrix} 1 \\ 1 \end{pmatrix}, \begin{pmatrix} -1 \\ 1 \end{pmatrix}$ (d) $\begin{pmatrix} 2 \\ 1 \end{pmatrix}, \begin{pmatrix} -1 \\ 1 \end{pmatrix}$

(e) $\begin{pmatrix} 3 \\ 2 \end{pmatrix}, \begin{pmatrix} 2 \\ 3 \end{pmatrix}$

7.2.25. Find the matrix form of
$$L[\mathbf{x}] = \begin{pmatrix} -3 & 2 & 2 \\ -3 & 1 & 3 \\ -1 & 2 & 0 \end{pmatrix} \mathbf{x}$$
with respect to the following bases of $\mathbb{R}^3$:

(a) $\begin{pmatrix} 2 \\ 0 \\ 0 \end{pmatrix}, \begin{pmatrix} 0 \\ -1 \\ 0 \end{pmatrix}, \begin{pmatrix} 0 \\ 0 \\ -2 \end{pmatrix}$

(b) $\begin{pmatrix} 1 \\ 0 \\ 1 \end{pmatrix}, \begin{pmatrix} 0 \\ -1 \\ 1 \end{pmatrix}, \begin{pmatrix} 1 \\ 1 \\ 1 \end{pmatrix}$

(c) $\begin{pmatrix} 2 \\ 1 \\ 2 \end{pmatrix}, \begin{pmatrix} 0 \\ 1 \\ -1 \end{pmatrix}, \begin{pmatrix} 1 \\ -2 \\ 1 \end{pmatrix}$

7.2.26. Find bases of the domain and target spaces that place the following matrices in the canonical form (7.30). Use (7.29) to check your answer.

(a) $\begin{pmatrix} 1 & 2 \\ 2 & 1 \end{pmatrix}$ (b) $\begin{pmatrix} 1 & -3 & 4 \\ -2 & 6 & -8 \end{pmatrix}$

(c) $\begin{pmatrix} 2 & 3 \\ 0 & 4 \\ -1 & 1 \end{pmatrix}$ (d) $\begin{pmatrix} 1 & 2 & 1 \\ 1 & -1 & -1 \\ 2 & 1 & 0 \end{pmatrix}$

(e) $\begin{pmatrix} 1 & 3 & 0 & 1 \\ 2 & 6 & 1 & -2 \\ -1 & -3 & -1 & 3 \\ 0 & 0 & -1 & 4 \end{pmatrix}$

7.2.27. (a) Show that every invertible linear function $L: \mathbb{R}^n \to \mathbb{R}^n$ can be represented by the identity matrix by choosing appropriate (and not necessarily the same) bases on the domain and target spaces.

(b) Which linear transformations are represented by the identity matrix when the domain and target space are required to have the same basis?

(c) Find bases for which the following linear transformations on $\mathbb{R}^2$ are represented by the identity matrix:

(i) the scaling map $S[\mathbf{x}] = 2\mathbf{x}$;

(ii) counterclockwise rotation by $45°$;

(iii) the shear $\begin{pmatrix} 1 & 0 \\ -2 & 1 \end{pmatrix}$.

◇ **7.2.28.** Suppose A is an $m \times n$ matrix.

(a) Let $\mathbf{v}_1, \ldots, \mathbf{v}_n$ be a basis of $\mathbb{R}^n$, and $A\mathbf{v}_i = \mathbf{w}_i \in \mathbb{R}^m$, for $i = 1, \ldots, n$. Prove that the vectors $\mathbf{v}_1, \ldots, \mathbf{v}_n, \mathbf{w}_1, \ldots, \mathbf{w}_n$, serve to uniquely specify A.

(b) Write down a formula for A.

◇ **7.2.29.** Suppose a linear transformation $L: \mathbb{R}^n \to \mathbb{R}^n$ is represented by a symmetric matrix with respect to the standard basis $\mathbf{e}_1, \ldots, \mathbf{e}_n$.

(a) Prove that its matrix representative with respect to any orthonormal basis $\mathbf{u}_1, \ldots, \mathbf{u}_n$ is symmetric.

(b) Is it symmetric when expressed in terms of a non-orthonormal basis?

7.2.30. In this exercise, we show that any inner product $\langle \cdot, \cdot \rangle$ on $\mathbb{R}^n$ can be reduced to the dot product when expressed in a suitably adapted basis.

(a) Specifically, prove that there exists a basis $\mathbf{v}_1, \ldots, \mathbf{v}_n$ of $\mathbb{R}^n$ such that
$$\langle \mathbf{x}, \mathbf{y} \rangle = \sum_{i=1}^{n} c_i d_i = \mathbf{c} \cdot \mathbf{d}$$
where $\mathbf{c} = (c_1, c_2, \ldots, c_n)^T$ are the coordinates of $\mathbf{x}$ and $\mathbf{d} = (d_1, d_2, \ldots, d_n)^T$ those of $\mathbf{y}$ with respect to the basis. Is the basis uniquely determined?

(b) Find bases that reduce the following inner products to the dot product on $\mathbb{R}^2$:

(i) $\langle \mathbf{v}, \mathbf{w} \rangle = 2v_1w_1 + 3v_2w_2$,

(ii) $\langle \mathbf{v}, \mathbf{w} \rangle = v_1w_1 - v_1w_2 - v_2w_1 + 3v_2w_2$.

7.3 AFFINE TRANSFORMATIONS AND ISOMETRIES

Not every transformation of importance in geometrical applications arises as a linear function. A simple example is a *translation*, where all the points in $\mathbb{R}^n$ are moved in the same direction by a common distance. The function that accomplishes this is

$$T[\mathbf{x}] = \mathbf{x} + \mathbf{b}, \quad \mathbf{x} \in \mathbb{R}^n, \tag{7.31}$$

where $\mathbf{b} \in \mathbb{R}^n$ is a fixed vector that determines the direction and the distance that the points are translated. Except in the trivial case $\mathbf{b} = \mathbf{0}$, the translation T is *not* a linear function because

$$T[\mathbf{x}+\mathbf{y}] = \mathbf{x}+\mathbf{y}+\mathbf{b} \neq T[\mathbf{x}] + T[\mathbf{y}] = \mathbf{x}+\mathbf{y}+2\mathbf{b}.$$

Or, even more simply, we note that $T[\mathbf{0}] = \mathbf{b}$, which must be $\mathbf{0}$ if T is to be linear.

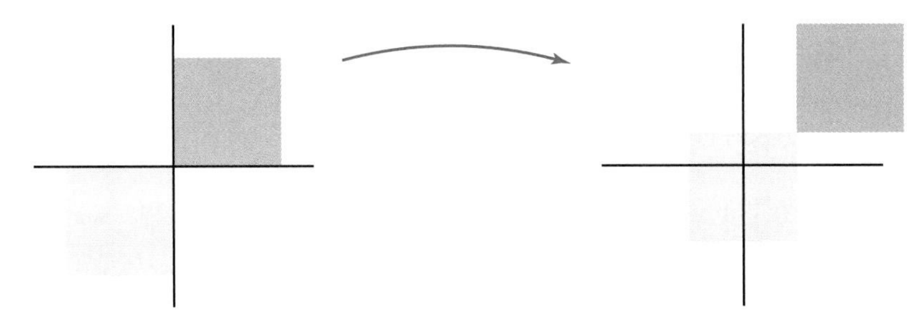

Figure 7.10 Translation.

Combining translations and linear functions leads us to an important class of geometrical transformations.

Definition 7.21 A function $F : \mathbb{R}^n \to \mathbb{R}^n$ of the form

$$F[\mathbf{x}] = A\mathbf{x} + \mathbf{b}, \tag{7.32}$$

where A is an $m \times n$ matrix and $\mathbf{b} \in \mathbb{R}^n$ a fixed vector, is called an *affine function*.

In general, $F[\mathbf{x}]$ is an affine function if and only if $L[\mathbf{x}] = F[\mathbf{x}] - F[\mathbf{0}]$ is a linear function. In the particular case (7.32), $F[\mathbf{0}] = \mathbf{b}$, and so $L[\mathbf{x}] = A\mathbf{x}$.

For example, every affine function from $\mathbb{R}$ to $\mathbb{R}$ has the form

$$f(x) = \alpha\, x + \beta. \tag{7.33}$$

As mentioned earlier, even though the graph of $f(x)$ is a straight line, f is *not* a linear function—unless $\beta = 0$, and the line goes through the origin. Thus, to be technically correct, we should refer to (7.33) as a *scalar affine function*.

EXAMPLE 7.22 The affine function

$$F(x,y) = \begin{pmatrix} 0 & -1 \\ 1 & 0 \end{pmatrix} \begin{pmatrix} x \\ y \end{pmatrix} + \begin{pmatrix} 1 \\ -2 \end{pmatrix} = \begin{pmatrix} -y+1 \\ x-2 \end{pmatrix}$$

has the effect of first rotating the plane $\mathbb{R}^2$ by 90° about the origin, and then translating by the vector $(1, -2)^T$. The reader may enjoy proving that this combination has the same effect as just rotating the plane through an angle of 90° centered at the point $\left(\frac{3}{4}, -\frac{1}{2}\right)$. For details, see Exercise 7.3.13. ●

Note that the affine function (7.32) can be obtained by composing a linear function $L[\mathbf{x}] = A\mathbf{x}$ with a translation $T[\mathbf{x}] = \mathbf{x} + \mathbf{b}$, so $F = T \circ L$. The order of composition is important, since $G = L \circ T$ defines the slightly different affine function $G[\mathbf{x}] = A(\mathbf{x} + \mathbf{b}) = A\mathbf{x} + \mathbf{c}$ where $\mathbf{c} = A\mathbf{b}$. More generally, the composition of any two affine functions is again an affine function. Specifically, given $F[\mathbf{x}] = A\mathbf{x} + \mathbf{a}$, $G[\mathbf{y}] = B\,\mathbf{y} + \mathbf{b}$, then

$$(G \circ F)[\mathbf{x}] = G[\,F[\mathbf{x}]\,] = G[\,A\mathbf{x} + \mathbf{a}\,] = B\,(A\mathbf{x} + \mathbf{a}) + \mathbf{b} = C\,\mathbf{x} + \mathbf{c}, \qquad \text{where} \quad C = B\,A, \quad \mathbf{c} = B\,\mathbf{a} + \mathbf{b}. \tag{7.34}$$

Note that the coefficient matrix of the composition is the product of the coefficient matrices, but the resulting vector of translation is *not* the sum of the two translation vectors.

EXERCISES 7.3

7.3.1. Describe the image of

(i) the x-axis

(ii) the unit disk $x^2 + y^2 \leq 1$

(iii) the unit square $0 \leq x, y \leq 1$

under the following affine transformations:

(a) $T_1\begin{pmatrix} x \\ y \end{pmatrix} = \begin{pmatrix} x \\ y \end{pmatrix} + \begin{pmatrix} 2 \\ -1 \end{pmatrix}$

(b) $T_2\begin{pmatrix} x \\ y \end{pmatrix} = \begin{pmatrix} 3 & 0 \\ 0 & 2 \end{pmatrix}\begin{pmatrix} x \\ y \end{pmatrix} + \begin{pmatrix} -1 \\ 0 \end{pmatrix}$

(c) $T_3\begin{pmatrix} x \\ y \end{pmatrix} = \begin{pmatrix} 1 & 2 \\ 0 & 1 \end{pmatrix}\begin{pmatrix} x \\ y \end{pmatrix} + \begin{pmatrix} 1 \\ 2 \end{pmatrix}$

(d) $T_4\begin{pmatrix} x \\ y \end{pmatrix} = \begin{pmatrix} 0 & 1 \\ -1 & 0 \end{pmatrix}\begin{pmatrix} x \\ y \end{pmatrix} + \begin{pmatrix} 1 \\ 0 \end{pmatrix}$

(e) $T_5\begin{pmatrix} x \\ y \end{pmatrix} = \begin{pmatrix} .6 & .8 \\ -.8 & .6 \end{pmatrix}\begin{pmatrix} x \\ y \end{pmatrix} + \begin{pmatrix} -3 \\ 2 \end{pmatrix}$

(f) $T_6\begin{pmatrix} x \\ y \end{pmatrix} = \begin{pmatrix} \frac{1}{2} & \frac{1}{2} \\ \frac{1}{2} & \frac{1}{2} \end{pmatrix}\begin{pmatrix} x \\ y \end{pmatrix} + \begin{pmatrix} 1 \\ 0 \end{pmatrix}$

(g) $T_7\begin{pmatrix} x \\ y \end{pmatrix} = \begin{pmatrix} 1 & 1 \\ -1 & 1 \end{pmatrix}\begin{pmatrix} x \\ y \end{pmatrix} + \begin{pmatrix} 2 \\ -3 \end{pmatrix}$

(h) $T_8\begin{pmatrix} x \\ y \end{pmatrix} = \begin{pmatrix} 2 & 1 \\ -2 & -1 \end{pmatrix}\begin{pmatrix} x \\ y \end{pmatrix} + \begin{pmatrix} 1 \\ 1 \end{pmatrix}$

7.3.2. Using the affine functions in Exercise 7.3.1, write out the following compositions and verify that they satisfy (7.34):

(a) $T_3 \circ T_4$ (b) $T_4 \circ T_3$ (c) $T_3 \circ T_6$

(d) $T_6 \circ T_3$ (e) $T_7 \circ T_8$ (f) $T_8 \circ T_7$

7.3.3. *True or false*: An affine transformation takes

(a) straight lines to straight lines

(b) triangles to triangles

(c) squares to squares

(d) circles to circles

(e) ellipses to ellipses

7.3.4. Describe the image of the triangle with vertices $(-1, 0)$, $(1, 0)$, $(0, 2)$ under the affine transformation

$$T\begin{pmatrix} x \\ y \end{pmatrix} = \begin{pmatrix} 4 & -1 \\ 2 & 5 \end{pmatrix}\begin{pmatrix} x \\ y \end{pmatrix} + \begin{pmatrix} 3 \\ -4 \end{pmatrix}.$$

7.3.5. Under what conditions is the composition of two affine functions

(a) a translation? (b) a linear function?

7.3.6. (a) Under what conditions does an affine function have an inverse?

(b) Is the inverse an affine function? If so, find a formula for its matrix and vector constituents.

(c) Find the inverse, when it exists, of the affine functions in Exercise 7.3.1.

◇ **7.3.7.** Let $\mathbf{v}_1, \ldots, \mathbf{v}_n$ be a basis for $\mathbb{R}^n$.

(a) Show that any affine function $F[\mathbf{x}] = A\mathbf{x} + \mathbf{b}$ on $\mathbb{R}^n$ is uniquely determined by the $n + 1$ vectors $\mathbf{w}_0 = F[\mathbf{0}], \mathbf{w}_1 = F[\mathbf{v}_1], \ldots, \mathbf{w}_n = F[\mathbf{v}_n]$.

(b) Find the formula for A and $\mathbf{b}$ when $\mathbf{v}_1 = \mathbf{e}_1, \ldots, \mathbf{v}_n = \mathbf{e}_n$ are the standard basis vectors.

(c) Find the formula for $A, \mathbf{b}$ for a general basis $\mathbf{v}_1, \ldots, \mathbf{v}_n$.

7.3.8. Show that the space of all affine functions on $\mathbb{R}^n$ forms a vector space. What is its dimension?

◇ **7.3.9.** In this exercise, we establish a useful matrix representation for affine functions. We identify $\mathbb{R}^n$ with the n-dimensional affine subspace (as in Exercise 2.2.30)

$$V_n = \{\, (\mathbf{x}, 1)^T = (x_1, \ldots, x_n, 1)^T \,\} \subset \mathbb{R}^{n+1}$$

consisting of vectors whose last coordinate is fixed at $x_{n+1} = 1$.

(a) Show that multiplication of vectors

$$\begin{pmatrix} \mathbf{x} \\ 1 \end{pmatrix} \in V_n$$

by the $(n+1) \times (n+1)$ *affine matrix*

$$\begin{pmatrix} A & \mathbf{b} \\ \mathbf{0} & 1 \end{pmatrix}$$

coincides with the action (7.32) of an affine function on $\mathbf{x} \in \mathbb{R}^n$.

(b) Prove that the composition law (7.34) for affine functions corresponds to multiplication of their affine matrices.

(c) Define the inverse of an affine function in the evident manner, and show that it corresponds to the inverse affine matrix.

Isometry

A transformation that preserves distance is known as an *isometry*. (The mathematical term *metric* refers to the underlying norm or distance on the space; thus, "isometric" translates as "distance preserving".) In Euclidean geometry, the isometries consist of the *rigid motions*—translations and rotations—along with reflections.

Definition 7.23 A function $F: V \to V$ is called an *isometry* on a normed vector space if it preserves distance, meaning

$$d(F[\mathbf{v}], F[\mathbf{w}]) = d(\mathbf{v}, \mathbf{w}) \quad \text{for all} \quad \mathbf{v}, \mathbf{w} \in V. \tag{7.35}$$

Since the distance between points is just the norm of the vector connecting them, $d(\mathbf{v}, \mathbf{w}) = \| \mathbf{v} - \mathbf{w} \|$, cf. (3.33), the isometry condition (7.35) can be restated as

$$\| \, F[\mathbf{v}] - F[\mathbf{w}] \, \| = \| \mathbf{v} - \mathbf{w} \| \quad \text{for all} \quad \mathbf{v}, \mathbf{w} \in V. \tag{7.36}$$

Clearly, any translation

$$T[\mathbf{v}] = \mathbf{v} + \mathbf{a}, \qquad \text{where} \quad \mathbf{a} \in V \quad \text{is a fixed vector}$$

defines an isometry, since $T[\mathbf{v}] - T[\mathbf{w}] = \mathbf{v} - \mathbf{w}$. A linear transformation $L: V \to V$ defines an isometry if and only if

$$\| \, L[\mathbf{v}] \, \| = \| \mathbf{v} \| \quad \text{for all} \quad \mathbf{v} \in V, \tag{7.37}$$

because, by linearity,

$$\| L[\mathbf{v}] - L[\mathbf{w}] \| = \| L[\mathbf{v} - \mathbf{w}] \| = \| \mathbf{v} - \mathbf{w} \|.$$

A similar computation proves that an affine transformation $F[\mathbf{v}] = L[\mathbf{v}] + \mathbf{a}$ is an isometry if and only if its linear part $L[\mathbf{v}]$ is.

The simplest class of isometries are the translations

$$T[\mathbf{x}] = \mathbf{x} + \mathbf{b} \tag{7.38}$$

in a fixed direction $\mathbf{b}$. For the standard Euclidean norm on $V = \mathbb{R}^n$, the linear isometries consist of rotations and reflections. As we shall prove, both are characterized by orthogonal matrices:

$$L[\mathbf{x}] = Q\,\mathbf{x}, \qquad \text{where} \quad Q^T Q = \mathrm{I}. \tag{7.39}$$

The *proper isometries* correspond to the rotations, with $\det Q = +1$, and can be realized as physical motions; *improper isometries*, with $\det Q = -1$, are then obtained by reflection in a mirror.

Proposition 7.24 A linear transformation $L[\mathbf{x}] = Q\,\mathbf{x}$ defines a Euclidean isometry of $\mathbb{R}^n$ if and only if Q is an orthogonal matrix.

Proof The linear isometry condition (7.37) requires that

$$\| Q\mathbf{x} \|^2 = (Q\mathbf{x})^T Q\mathbf{x} = \mathbf{x}^T Q^T Q\mathbf{x} = \mathbf{x}^T\mathbf{x} = \|\mathbf{x}\|^2 \quad \text{for all} \quad \mathbf{x} \in \mathbb{R}^n.$$

According to Exercise 5.3.16, this holds if and only if $Q^T Q = \mathrm{I}$, which is precisely the condition (5.29) that Q be an orthogonal matrix. ■

It can be proved, [70], that the most general Euclidean isometry of $\mathbb{R}^n$ is an affine transformation, and hence of the form $F[\mathbf{x}] = Q\mathbf{x} + \mathbf{b}$, where Q is an orthogonal matrix and $\mathbf{b}$ is a constant vector. Therefore, every Euclidean isometry is a combination of translations, rotations and reflections.

In the two-dimensional case, the proper linear isometries $R[\mathbf{x}] = Q\mathbf{x}$ with $\det Q = 1$ represent rotations around the origin. More generally, a rotation of the plane around a center at $\mathbf{c}$ is represented by the affine isometry

$$R[\mathbf{x}] = Q(\mathbf{x} - \mathbf{c}) + \mathbf{c} = Q\mathbf{x} + \mathbf{b}, \quad \text{where} \quad \mathbf{b} = (\mathrm{I} - Q)\mathbf{c}, \tag{7.40}$$

and where Q is a rotation matrix. In Exercise 7.3.13, we ask you to prove that every plane isometry is either a translation or a rotation around a center.

In three-dimensional space, both translations (7.38) and rotations around a center (7.40) continue to define proper isometries. There is one additional type, representing the motion of a point on the head of a screw. A *screw motion* is an affine map of the form

$$S[\mathbf{x}] = Q\mathbf{x} + \mathbf{a}, \tag{7.41}$$

where the 3×3 orthogonal matrix Q represents a rotation through an angle θ around a fixed axis $\mathbf{a}$, which is also the direction of the translation term. The result is indicated in Figure 7.11. For example,

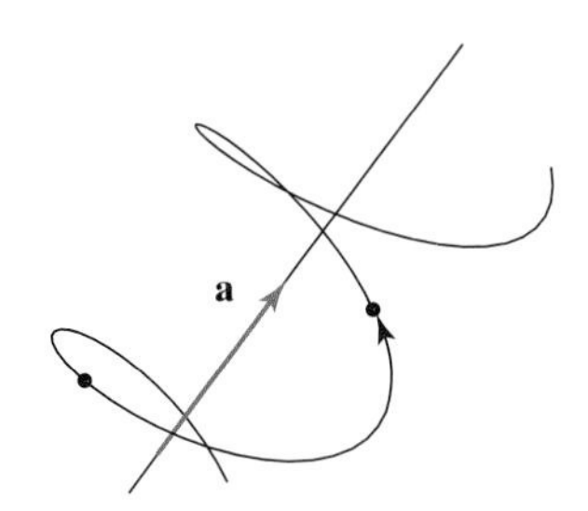

Figure 7.11 A screw.

$$S_\theta \begin{pmatrix} x \\ y \\ z \end{pmatrix} = \begin{pmatrix} \cos\theta & -\sin\theta & 0 \\ \sin\theta & \cos\theta & 0 \\ 0 & 0 & 1 \end{pmatrix} \begin{pmatrix} x \\ y \\ z \end{pmatrix} + \begin{pmatrix} 0 \\ 0 \\ a \end{pmatrix}$$

represents a vertical screw along the z–axis through an angle θ by a distance a. In Exercise 8.2.46 you are asked to prove that every proper isometry of $\mathbb{R}^3$ is either a translation, a rotation, or a screw motion.

The isometries of $\mathbb{R}^2$ and $\mathbb{R}^3$ are indispensable for understanding of how physical objects move in three-dimensional space. Basic computer graphics and animation require efficient implementation of rigid isometries in three-dimensional space and their compositions—coupled with appropriate (nonlinear) perspective maps prescribing the projection of three-dimensional objects onto a two-dimensional viewing screen, [12, 55].

EXERCISES 7.3

Note: All exercises are based on the Euclidean norm unless otherwise noted.

7.3.10. Which of the indicated maps $\mathbf{F}(x, y)$ define isometries of the Euclidean plane?

(a) $\begin{pmatrix} y \\ -x \end{pmatrix}$ (b) $\begin{pmatrix} x-2 \\ y-1 \end{pmatrix}$

(c) $\begin{pmatrix} x-y+1 \\ x+2 \end{pmatrix}$ (d) $\dfrac{1}{\sqrt{2}}\begin{pmatrix} x+y-3 \\ x+y-2 \end{pmatrix}$

(e) $\dfrac{1}{5}\begin{pmatrix} 3x+4y \\ -4x+3y+1 \end{pmatrix}$

7.3.11. *True or false*: The map $L[\mathbf{x}] = -\mathbf{x}$ for $\mathbf{x} \in \mathbb{R}^n$ defines

(a) an isometry (b) a rotation

7.3.12. Prove that the planar affine isometry

$$F\begin{pmatrix} x \\ y \end{pmatrix} = \begin{pmatrix} -y+1 \\ x-2 \end{pmatrix}$$

represents a rotation through an angle of 90° around the center $\left(\frac{3}{2}, -\frac{1}{2}\right)^T$.

◇ **7.3.13.** Prove that every *proper* affine plane isometry $F[\mathbf{x}] = Q\mathbf{x} + \mathbf{b}$ of $\mathbb{R}^2$, where $\det Q = 1$, is either

(a) a translation, or

(b) a rotation (7.40) centered at some point $\mathbf{c} \in \mathbb{R}^2$.

Hint: Use Exercise 1.5.7.

7.3.14. Compute both compositions $F \circ G$ and $G \circ F$ of the following affine functions on $\mathbb{R}^2$. Which pairs commute?

(a) F = counterclockwise rotation around the origin by 45°; G = translation in the y direction by 3 units.

(b) F = counterclockwise rotation around the point $(1,1)^T$ by 30°; G = counterclockwise rotation around the point $(-2,1)^T$ by 90°.

(c) F = reflection through the line $y = x+1$; G = rotation around $(1,1)^T$ by 180°.

♡ **7.3.15.** In $\mathbb{R}^2$, show the following:

(a) The composition of two affine isometries is another affine isometry.

(b) The composition of two translations is another translation.

(c) The composition of a translation and a rotation (not necessarily centered at the origin) in either order is a rotation.

(d) The composition of two plane rotations is either another rotation or a translation. What is the condition for the latter possibility?

(e) Any plane translation can be written as the composition of two rotations.

7.3.16. Let ℓ be a line in $\mathbb{R}^2$. A *glide reflection* is an affine map on $\mathbb{R}^2$ composed of a translation in the direction of ℓ by a distance d followed by a reflection through ℓ.

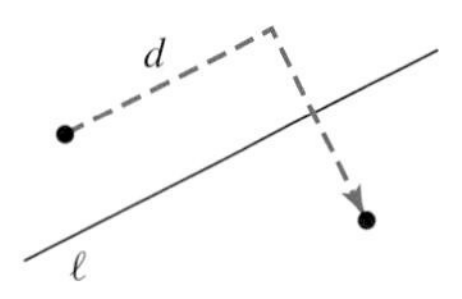

Find the formula for a glide reflection along

(a) the x axis by a distance 2,

(b) the line $y = x$ by a distance 3 in the direction of increasing x,

(c) the line $x + y = 1$ by a distance 2 in the direction of increasing x.

◇ **7.3.17.** Let ℓ be the line in the direction of the unit vector $\mathbf{u}$ through the point $\mathbf{a}$.

(a) Write down the formula for the affine map defining the reflection through the line ℓ. *Hint*: Use Exercise 7.2.11.

(b) Write down the formula for the glide reflection, as defined in Exercise 7.3.16, along ℓ by a distance d in the direction of $\mathbf{u}$.

(c) Prove that every improper affine plane isometry is either a reflection or a glide reflection. *Hint*: Use Exercise 7.2.10.

♡ **7.3.18.** A set of $n+1$ points $\mathbf{a}_0, \ldots, \mathbf{a}_n \in \mathbb{R}^n$ is said to be *in general position* if the differences $\mathbf{a}_i - \mathbf{a}_j$ span $\mathbb{R}^n$.

(a) Show that the points are in general position if and only if they do not all lie in a proper affine subspace $A \subsetneq \mathbb{R}^n$, cf. Exercise 2.2.30.

(b) Let $\mathbf{a}_0, \ldots, \mathbf{a}_n$ and $\mathbf{b}_0, \ldots, \mathbf{b}_n$ be two sets in general position. Show that there is an isometry $F\colon \mathbb{R}^n \to \mathbb{R}^n$ such that $F[\mathbf{a}_i] = \mathbf{b}_i$ for all $i = 0, \ldots, n$, if and only if their interpoint distances agree: $\|\mathbf{a}_i - \mathbf{a}_j\| = \|\mathbf{b}_i - \mathbf{b}_j\|$ for all $0 \le i < j \le n$. *Hint*: Use Exercise 5.3.20.

◇ **7.3.19.** Suppose that V is an inner product space and $L\colon V \to V$ is an isometry, so $\|L[\mathbf{v}]\| = \|\mathbf{v}\|$ for all $\mathbf{v} \in V$. Prove that L also preserves the inner product: $\langle L[\mathbf{v}], L[\mathbf{w}]\rangle = \langle \mathbf{v}, \mathbf{w}\rangle$. *Hint*: Look at $\|L[\mathbf{v}+\mathbf{w}]\|^2$.

◇ **7.3.20.** Let V be a normed vector space. Prove that a linear map $L\colon V \to V$ defines an isometry of V for the given norm if and only if it maps the unit sphere $S_1 = \{\,\|\mathbf{u}\| = 1\,\}$ to itself:

$$L[S_1] = \{\, L[\mathbf{u}] \mid \mathbf{u} \in S_1 \,\} = S_1.$$

7.3.21. (a) List all linear and affine isometries of $\mathbb{R}^2$ with respect to the ∞ norm. *Hint*: Use Exercise 7.3.20.

(b) Can you generalize your results to $\mathbb{R}^3$?

7.3.22. Answer Exercise 7.3.21 for the 1 norm.

♡ **7.3.23.** A matrix of the form

$$H = \begin{pmatrix} \cosh\alpha & \sinh\alpha \\ \sinh\alpha & \cosh\alpha \end{pmatrix}$$

for $\alpha \in \mathbb{R}$ defines a *hyperbolic rotation* of $\mathbb{R}^2$.

(a) Prove that all hyperbolic rotations preserve the indefinite quadratic form $q(\mathbf{x}) = x^2 - y^2$ in the sense that $q(H\mathbf{x}) = q(\mathbf{x})$ for all $\mathbf{x} = (x, y)^T \in \mathbb{R}^2$. Observe that ordinary rotations preserve circles $x^2 + y^2 = a$, while hyperbolic rotations preserve hyperbolas $x^2 - y^2 = a$.

(b) Are there any other affine transformations of $\mathbb{R}^2$ that preserve the quadratic form $q(\mathbf{x})$?

Remark: The four-dimensional version of this construction, i.e., affine maps preserving the indefinite

Minkowski form $t^2 - x^2 - y^2 - z^2$, forms the geometrical foundation for Einstein's theory of special relativity, [41].

♡ **7.3.24.** Let $\ell \subset \mathbb{R}^2$ be a line, and $\mathbf{p} \notin \ell$ a point. A *perspective map* takes a point $\mathbf{x} \in \mathbb{R}^2$ to the point $\mathbf{q} \in \ell$ that is the intersection of ℓ with the line going through $\mathbf{p}$ and $\mathbf{x}$. If the line is parallel to ℓ then the map is not defined. Find the formula for the perspective map when

(a) ℓ is the x axis and $\mathbf{p} = (0, 1)^T$,

(b) ℓ is the line $y = x$ and $\mathbf{p} = (1, 0)^T$.

Is either map affine? An isometry?

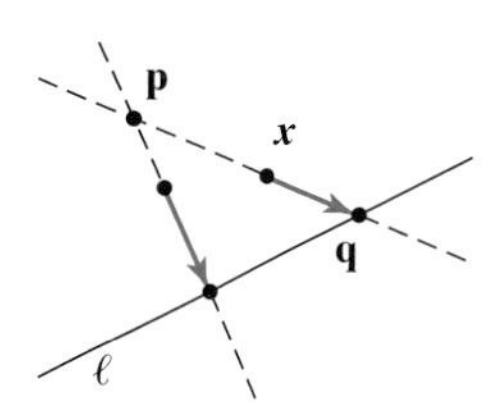

Remark: Mapping three-dimensional objects onto a two-dimensional screen (or your retina) is based on perspective maps, which are thus of fundamental importance in art, optics, computer vision, computer graphics and animation, and computer games.

7.4 LINEAR SYSTEMS

The abstract notion of a linear system serves to unify, in a common conceptual framework, linear systems of algebraic equations, linear differential equations, both ordinary and partial, linear boundary value problems, linear integral equations, linear control systems, and a huge variety of other linear systems that appear in all aspects of mathematics and its applications. The idea is simply to replace matrix multiplication by a general linear function. Many of the structural results we learned in the matrix context have, when suitably formulated, direct counterparts in these more general frameworks. The result is a unified understanding of the basic properties and nature of solutions to all such linear systems.

Definition 7.25 A *linear system* is an equation of the form

$$L[\mathbf{u}] = \mathbf{f}, \tag{7.42}$$

in which $L\colon U \to V$ is a linear function between vector spaces, the right hand side $\mathbf{f} \in V$ is an element of the target space, while the desired solution $\mathbf{u} \in U$ belongs to the domain space. The system is *homogeneous* if $\mathbf{f} = \mathbf{0}$; otherwise, it is called *inhomogeneous*.

EXAMPLE 7.26 If $U = \mathbb{R}^n$ and $V = \mathbb{R}^m$, then, according to Theorem 7.5, every linear function $L\colon \mathbb{R}^n \to \mathbb{R}^m$ is given by matrix multiplication: $L[\mathbf{u}] = A\mathbf{u}$. Therefore, in this particular case, every linear system is a matrix system, namely $A\mathbf{u} = \mathbf{f}$. ●

EXAMPLE 7.27 A *linear ordinary differential equation* takes the form $L[u] = f$, where L is an nth order linear differential operator of the form (7.13), and the right hand side is, say, a continuous function. Written out, the differential equation takes the familiar form

$$L[u] = a_n(x)\frac{d^n u}{dx^n} + a_{n-1}(x)\frac{d^{n-1} u}{dx^{n-1}} + \cdots + a_1(x)\frac{du}{dx} + a_0(x)u = f(x). \tag{7.43}$$

You should already have some familiarity with solving the constant coefficient case as covered, for instance, in [6, 19]. ●

EXAMPLE 7.28 Let $K(x, y)$ be a function of two variables that is continuous for all $a \le x, y \le b$. Then the integral

$$I_K[u] = \int_a^b K(x, y)\, u(y)\, dy$$

defines a linear operator $I_K\colon \mathrm{C}^0[a, b] \to \mathrm{C}^0[a, b]$, known as an *integral transform*. Important examples include the Fourier and Laplace transforms, [36, 59]. Finding the inverse transform requires solving a *linear integral equation* $I_K[u] = f$, which has the explicit form

$$\int_a^b K(x, y)\, u(y)\, dy = f(x).$$

EXAMPLE 7.29 We can combine linear maps to form more complicated, "mixed" types of linear systems. For example, consider a typical initial value problem

$$u'' + u' - 2u = x, \qquad u(0) = 1, \qquad u'(0) = -1, \tag{7.44}$$

for a scalar unknown function $u(x)$. The differential equation can be written as a linear system

$$L[u] = x, \quad \text{where} \quad L[u] = (D^2 + D - 2)[u] = u'' + u' - 2u$$

is a linear, constant coefficient differential operator. Further,

$$M[u] = \begin{pmatrix} L[u] \\ u(0) \\ u'(0) \end{pmatrix} = \begin{pmatrix} u''(x) + u'(x) - 2u(x) \\ u(0) \\ u'(0) \end{pmatrix}$$

defines a linear map whose domain is the space $U = \mathrm{C}^2$ of twice continuously differentiable functions $u(x)$, and whose range is the vector space V consisting of all triples*

$$\mathbf{v} = \begin{pmatrix} f(x) \\ a \\ b \end{pmatrix},$$

where $f \in \mathrm{C}^0$ is a continuous function and $a, b \in \mathbb{R}$ are real constants. You should convince yourself that V is indeed a vector space under the evident addition and scalar multiplication operations. In this way, we can write the initial value problem (7.44) in linear systems form as $M[u] = \mathbf{f}$, where $\mathbf{f} = (x, 1, -1)^T$.

A similar construction applies to linear boundary value problems. For example, the boundary value problem

$$u'' + u = e^x, \qquad u(0) = 1, \qquad u(1) = 2,$$

is in the form of a linear system

$$B[u] = \mathbf{f}, \quad \text{where} \quad B[u] = \begin{pmatrix} u''(x) + u(x) \\ u(0) \\ u(1) \end{pmatrix}, \quad \mathbf{f} = \begin{pmatrix} e^x \\ 1 \\ 2 \end{pmatrix}.$$

Note that $B\colon \mathrm{C}^2 \to V$ defines a linear map having the same domain and target spaces as the initial value problem map M.

EXERCISES 7.4

7.4.1. Place each of the following linear systems in the form (7.42). Carefully describe the linear function, its domain space, its target space, and the right hand side of the system. Which systems are homogeneous?
(a) $3x + 5 = 0$ (b) $x = y + z$
(c) $a = 2b - 3,\ b = c - 1$
(d) $3(p - 2) = 2(q - 3),\ p + q = 0$
(e) $u' + 3xu = 0$ (f) $u' + 3x = 0$

*This is a particular case of the general Cartesian product construction between vector spaces; here $V = \mathrm{C}^0 \times \mathbb{R}^2$. See Exercise 2.1.13 for details.

(g) $u' = u, \ u(0) = 1$

(h) $u'' - u = e^x, \ u(0) = 3\,u(1)$

(i) $u'' + x^2 u = 3x, \ u(0) = 1, \ u'(0) = 0$

(j) $u' = v, \ v' = 2u$

(k) $u'' - v'' = 2u - v, \ u(0) = v(0), \ u(1) = v(1)$

(l) $u(x) = 1 - 3\int_0^x u(y)\,dy$

(m) $\int_0^\infty u(t)\,e^{-st}\,dt = 1 + s^2$

(n) $\int_0^1 u(x)\,dx = u\left(\frac{1}{2}\right)$

(o) $\int_0^1 u(y)\,dy = \int_0^1 y\,v(y)\,dy$

(p) $\dfrac{\partial u}{\partial t} + 2\,\dfrac{\partial u}{\partial x} = 1$

(q) $\dfrac{\partial u}{\partial x} = \dfrac{\partial v}{\partial y}, \ \dfrac{\partial u}{\partial y} = -\,\dfrac{\partial v}{\partial x}$

(r) $-\,\dfrac{\partial^2 u}{\partial x^2} - \dfrac{\partial^2 u}{\partial y^2} = x^2 + y^2 - 1$

7.4.2. The Fredholm Alternative of Theorem 5.55 first appeared in the study of what are now known as *Fredholm integral equations*:

$$u(x) + \int_a^b K(x,y)\,u(y)\,dy = f(x),$$

in which $K(x,y)$ and $f(x)$ are prescribed continuous functions. Explain how the integral equation is a linear system; i.e., describe the linear map L, its domain and target spaces, and prove linearity.

7.4.3. Answer Exercise 7.4.2 for the *Volterra integral equation*

$$u(t) + \int_a^t K(t,s)\,u(s)\,ds = f(t),$$

where $a \le t \le b$.

7.4.4. (a) Prove that the solution to the linear integral equation

$$u(t) = a + \int_0^t k(s)\,u(s)\,ds$$

solves the linear initial value problem $du/dt = k(t)\,u(t), \quad u(0) = a$.

(b) Use part (a) to solve the following integral equations

(i) $u(t) = 2 - \int_0^t u(s)\,ds$,

(ii) $u(t) = 1 + 2\int_1^t s\,u(s)\,ds$,

(iii) $u(t) = 3 + \int_0^t e^s\,u(s)\,ds$.

7.4.5. *True or false*: If $F[\mathbf{x}]$ is an affine function on $\mathbb{R}^n$, then the equation $F[\mathbf{x}] = \mathbf{c}$ defines a linear system.

The Superposition Principle

Before attempting to tackle general inhomogeneous linear systems, we should look first at the homogeneous version. The most important fact is that homogeneous linear systems admit a superposition principle that allows one to construct new solutions from known solutions. Recall that the word "superposition" refers to taking linear combinations of solutions.

Consider a general homogeneous linear system

$$L[\mathbf{z}] = \mathbf{0}, \qquad (7.45)$$

where $L\colon U \to V$ is a linear function. If we are given two solutions, say $\mathbf{z}_1$ and $\mathbf{z}_2$, meaning that

$$L[\mathbf{z}_1] = \mathbf{0}, \qquad L[\mathbf{z}_2] = \mathbf{0},$$

then their sum $\mathbf{z}_1 + \mathbf{z}_2$ is automatically a solution, since, in view of the linearity of L,

$$L[\mathbf{z}_1 + \mathbf{z}_2] = L[\mathbf{z}_1] + L[\mathbf{z}_2] = \mathbf{0} + \mathbf{0} = \mathbf{0}.$$

Similarly, given a solution $\mathbf{z}$ and any scalar c, the scalar multiple $c\,\mathbf{z}$ is automatically a solution, since

$$L[c\,\mathbf{z}] = c\,L[\mathbf{z}] = c\,\mathbf{0} = \mathbf{0}.$$

Combining these two elementary observations, we can now state the general *superposition principle*. The proof is an immediate consequence of formula (7.4).

THEOREM 7.30 If $\mathbf{z}_1, \ldots, \mathbf{z}_k$ are all solutions to the same homogeneous linear system $L[\mathbf{z}] = \mathbf{0}$, then any linear combination $c_1 \mathbf{z}_1 + \cdots + c_k \mathbf{z}_k$ is also a solution.

As with matrices, we call the solution space to the homogeneous linear system (7.45) the *kernel* of the linear function L. The superposition principle implies that the kernel always forms a subspace.

Proposition 7.31 If $L \colon U \to V$ is a linear function, then its *kernel*

$$\ker L = \{\, \mathbf{z} \in U \mid L[\mathbf{z}] = 0 \,\} \subset U \tag{7.46}$$

forms a subspace of the domain space U.

As we know, in the case of linear matrix systems, the kernel can be explicitly determined by applying the usual Gaussian Elimination algorithm. To solve more general homogeneous linear systems, e.g., linear differential equations, one must develop appropriate analytical solution techniques.

EXAMPLE 7.32 Consider the second order linear differential operator

$$L = D^2 - 2D - 3, \tag{7.47}$$

which maps the function $u(x)$ to the function

$$L[u] = (D^2 - 2D - 3)[u] = u'' - 2u' - 3u.$$

The associated homogeneous system takes the form of a homogeneous, linear, constant coefficient second order ordinary differential equation

$$L[u] = u'' - 2u' - 3u = 0. \tag{7.48}$$

In accordance with the standard solution method, we plug the exponential *ansatz*†

$$u = e^{\lambda x}$$

into the equation. The result is

$$\begin{aligned} L[e^{\lambda x}] &= D^2[e^{\lambda x}] - 2D[e^{\lambda x}] - 3e^{\lambda x} \\ &= \lambda^2 e^{\lambda x} - 2\lambda e^{\lambda x} - 3e^{\lambda x} = (\lambda^2 - 2\lambda - 3)e^{\lambda x}. \end{aligned}$$

Therefore, $u = e^{\lambda x}$ is a solution if and only if λ satisfies the *characteristic equation*

$$0 = \lambda^2 - 2\lambda - 3 = (\lambda - 3)(\lambda + 1).$$

The two roots are $\lambda_1 = 3$, $\lambda_2 = -1$, and hence

$$u_1(x) = e^{3x}, \qquad u_2(x) = e^{-x}, \tag{7.49}$$

are two linearly independent solutions of (7.48). According to the general superposition principle, every linear combination

$$u(x) = c_1 u_1(x) + c_2 u_2(x) = c_1 e^{3x} + c_2 e^{-x}$$

of these two basic solutions is also a solution, for any choice of constants c_1, c_2. In fact, this two-parameter family constitutes the most general solution to the ordinary differential equation (7.48); this is a consequence of Theorem 7.34 below. Thus, the kernel of the second order differential operator (7.47) is two-dimensional, with basis given by the independent exponential solutions (7.49). ●

---

†The German word *ansatz* (plural *ansätze*) refers to the method of finding a solution to a complicated equation by guessing the solution's form in advance. Typically, one is not clever enough to guess the precise solution, and so the ansatz will have one or more free parameters—in this case the constant exponent λ—that, with some luck, can be rigged up to fulfill the requirements imposed by the equation. Thus, a reasonable English translation of "ansatz" is "inspired guess".

In general, the solution space to an nth order homogeneous linear ordinary differential equation

$$L[u] = a_n(x) \frac{d^n u}{dx^n} + a_{n-1}(x) \frac{d^{n-1} u}{dx^{n-1}} + \cdots + a_1(x) \frac{du}{dx} + a_0(x) u = 0 \tag{7.50}$$

forms a subspace of the vector space $C^n(a,b)$ of n times continuously differentiable functions defined on an open interval[‡] $a < x < b$, since it is just the kernel of a linear differential operator $L\colon C^n(a,b) \to C^0(a,b)$. This implies that linear combinations of solutions are also solutions. To determine the number of solutions, or, more precisely, the dimension of the solution space, we need to impose some mild restrictions on the differential operator.

Definition 7.33 The differential operator L is called *nonsingular* on an open interval (a,b) if all its coefficients $a_n(x), \ldots, a_0(x) \in C^0(a,b)$ are continuous functions and its *leading coefficient* does not vanish: $a_n(x) \neq 0$ for all $a < x < b$.

The basic existence and uniqueness theorems governing nonsingular homogeneous linear ordinary differential equations can be reformulated as a characterization of the dimension of the solution space.

THEOREM 7.34 The kernel of a nonsingular nth order ordinary differential operator forms an n-dimensional subspace $\ker L \subset C^n(a,b)$.

A proof of this theorem can be found in [6, 28]. The fact that the kernel has dimension n means that it has a basis consisting of n linearly independent solutions $u_1(x), \ldots, u_n(x) \in C^n(a,b)$ with the property that every solution to the homogeneous differential equation (7.50) is given by a linear combination

$$u(x) = c_1 u_1(x) + \cdots + c_n u_n(x),$$

where $c_1, \ldots, c_n$ are arbitrary constants. Therefore, once we find n linearly independent solutions of an nth order homogeneous linear ordinary differential equation, we can immediately write down its most general solution.

The condition that the leading coefficient $a_n(x) \neq 0$ is essential. Points where $a_n(x) = 0$ are known as *singular points*. Singular points show up in many applications, but must be treated separately and with care, [6, 19]. Of course, if the coefficients are constant then there is nothing to worry about — either the leading coefficient is nonzero, $a_n \neq 0$, or the differential equation is, in fact, of lower order than advertised. Here is the prototypical example of an ordinary differential equation with a singular point.

EXAMPLE 7.35 A second order *Euler differential equation* takes the form

$$E[u] = a x^2 u'' + b x u' + c u = 0, \tag{7.51}$$

where $a \neq 0$ and b, c are constants, and $E = a x^2 D^2 + b x D + c$ is a second order differential operator. Instead of the exponential solution ansatz used in the constant coefficient case, Euler equations are solved by using a power ansatz

$$u(x) = x^r \tag{7.52}$$

‡We allow a and/or b to be infinite.

with unknown exponent r. Substituting into the differential equation, we find

$$\begin{aligned} E[x^r] &= a x^2 D^2[x^r] + b x D[x^r] + c x^r \\ &= a r (r-1) x^r + b r x^r + c x^r \\ &= [a r (r-1) + b r + c] x^r. \end{aligned}$$

Thus, x^r is a solution if and only if r satisfies the *characteristic equation*

$$a r (r-1) + b r + c = a r^2 + (b-a) r + c = 0. \tag{7.53}$$

If the characteristic equation has two distinct real roots, $r_1 \neq r_2$, then we obtain two linearly independent solutions $u_1(x) = x^{r_1}$ and $u_2(x) = x^{r_2}$, and so the general (real) solution to (7.51) has the form

$$u(x) = c_1 |x|^{r_1} + c_2 |x|^{r_2}. \tag{7.54}$$

(The absolute values are usually needed to ensure that the solutions remain real when $x < 0$.) The other cases — repeated roots and complex roots—will be discussed below.

The Euler equation has a singular point at $x = 0$, where its leading coefficient vanishes. Theorem 7.34 assures us that the differential equation has a two-dimensional solution space on any interval not containing the singular point. However, predicting the number of solutions that remain continuously differentiable at $x = 0$ is not so easy, since it depends on the values of the exponents r_1 and r_2. For instance, the case

$$x^2 u'' - 3 x u' + 3 u = 0 \quad \text{has solution} \quad u = c_1 x + c_2 x^3,$$

which forms a two-dimensional subspace of $\mathrm{C}^0(\mathbb{R})$. However,

$$x^2 u'' + x u' - u = 0 \quad \text{has solution} \quad u = c_1 x + \frac{c_2}{x},$$

and only the multiples of the first solution x are continuous at $x = 0$. Therefore, the solutions that are continuous everywhere form only a one-dimensional subspace of $\mathrm{C}^0(\mathbb{R})$. Finally,

$$x^2 u'' + 5 x u' + 3 u = 0 \quad \text{has solution} \quad u = \frac{c_1}{x} + \frac{c_2}{x^3}.$$

In this case, there are no nontrivial solutions $u(x) \not\equiv 0$ that are continuous at $x = 0$, and so the space of solutions defined on all of $\mathbb{R}$ is zero-dimensional. ●

The superposition principle is equally valid in the study of homogeneous linear partial differential equations. Here is a particularly noteworthy example.

EXAMPLE 7.36 Consider the *Laplace equation*

$$\Delta[u] = \frac{\partial^2 u}{\partial x^2} + \frac{\partial^2 u}{\partial y^2} = 0 \tag{7.55}$$

for a function $u(x, y)$ defined on a domain $\Omega \subset \mathbb{R}^2$. The Laplace equation is named after the influential French mathematician Pierre–Simon Laplace, and is *the* most important partial differential equation. Its applications range over almost all fields of mathematics, physics, and engineering, including complex analysis, geometry, fluid mechanics, electromagnetism, elasticity, thermodynamics, and quantum mechanics.

The Laplace equation is a homogeneous linear partial differential equation corresponding to the partial differential operator $\Delta = \partial_x^2 + \partial_y^2$ known as the *Laplacian*. Linearity can either be proved directly, or by noting that Δ is built up from the basic linear partial derivative operators ∂_x, ∂_y by the processes of composition and addition, as detailed in Exercise 7.1.46. Solutions to the Laplace equation are known as *harmonic functions*.

Unlike homogeneous linear ordinary differential equations, there are an infinite number of linearly independent solutions to the Laplace equation. Examples include the trigonometric/exponential solutions

$$e^{\omega x} \cos \omega y, \quad e^{\omega x} \sin \omega y, \quad e^{\omega y} \cos \omega x, \quad e^{\omega y} \sin \omega y,$$

where ω is *any* real constant. There are also infinitely many independent *harmonic polynomial* solutions, the first few of which are

$$1, \quad x, \quad y, \quad x^2 - y^2, \quad x\,y, \quad x^3 - 3\,x\,y^2, \quad \ldots$$

The reader might enjoy finding some more polynomial solutions and trying to spot the pattern. (The answer will appear shortly.) As usual, we can build up more complicated solutions by taking general linear combinations of these particular ones; for instance, $u(x, y) = 1 - 4\,x\,y + 2\,e^{3x} \cos 3\,y$ is automatically a solution. ●

EXERCISES 7.4

7.4.6. Solve the following homogeneous linear ordinary differential equations. What is the dimension of the solution space?

(a) $u'' - 4\,u = 0$ (b) $u'' - 6\,u' + 8\,u = 0$

(c) $u''' - 9\,u' = 0$

(d) $u'''' + 4\,u''' - u'' - 16\,u' - 12\,u = 0$

7.4.7. Define $L[\,y\,] = y'' + y$.

(a) Prove directly from the definition that $L\colon C^2[a, b] \to C^0[a, b]$ is a linear transformation.

(b) Determine $\ker L$.

7.4.8. Answer Exercise 7.4.7 when $L = 3\,D^2 - 2\,D - 5$.

7.4.9. Consider the linear differential equation $y''' + 5\,y'' + 3\,y' - 9\,y = 0$.

(a) Write the equation in the form $L[\,y\,] = 0$ for a differential operator $L = p(D)$.

(b) Find a basis for $\ker L$, and then write out the general solution to the differential equation.

7.4.10. The following functions are solutions to a constant coefficient homogeneous scalar ordinary differential equation.

(i) Determine the least possible order of the differential equation, and

(ii) write down an appropriate differential equation.

(a) $e^{2x} + e^{-3x}$ (b) $1 + e^{-x}$

(c) $x\,e^x$ (d) $e^x + 2\,e^{2x} + 3\,e^{3x}$

7.4.11. Solve the following Euler differential equations:

(a) $x^2 u'' + 5\,x\,u' - 5\,u = 0$

(b) $2\,x^2 u'' - x\,u' - 2\,u = 0$

(c) $x^2 u'' - u = 0$

(d) $x^2 u'' + x\,u' - 3\,u = 0$

(e) $3\,x^2 u'' - 5\,x\,u' - 3\,u = 0$

(f) $\dfrac{d^2u}{dx^2} + \dfrac{2}{x}\dfrac{du}{dx} = 0$

7.4.12. Solve the third order Euler differential equation $x^3 u''' + 2\,x^2 u'' - 3\,x\,u' + 3\,u = 0$ by using the power ansatz (7.52). What is the dimension of the solution space for $x > 0$? For all x?

7.4.13. (i) Show that if $u(x)$ solves the Euler equation

$$a\,x^2 \frac{d^2u}{dx^2} + b\,x\,\frac{du}{dx} + c\,u = 0,$$

then $v(t) = u(e^t)$ solves a linear, constant coefficient differential equation.

(ii) Use this alternative technique to solve the Euler differential equations in Exercise 7.4.11.

◇ **7.4.14.** (a) Use the method in Exercise 7.4.13 to solve an Euler equation whose characteristic equation has a double root $r_1 = r_2 = r$. Solve the specific equations

(i) $x^2 u'' - x\,u' + u = 0$

(ii) $\dfrac{d^2u}{dx^2} + \dfrac{1}{x}\dfrac{du}{dx} = 0.$

7.4.15. Show that if $u(x)$ solves $x\,u'' + 2\,u' - 4\,x\,u = 0$, then $v(x) = x\,u(x)$ solves a linear, constant coefficient equation. Use this to find the general solution to the given differential equation. Which of your solutions are continuous at the singular point $x = 0$? Differentiable?

7.4.16. Let $S \subset \mathbb{R}$ be an open subset (i.e., a union of open intervals). *True or false*: $\ker D$ is a one-dimensional subspace of $\mathrm{C}^1(S)$.

7.4.17. Show that

$$\log(x^2 + y^2) \quad \text{and} \quad \frac{x}{x^2 + y^2}$$

are harmonic functions, that is, solutions of the two-dimensional Laplace equation.

7.4.18. Find all solutions $u = f(r)$ of the two-dimensional Laplace equation that depend only on the radial coordinate $r = \sqrt{x^2 + y^2}$. Do these solutions form a vector space? If so, what is its dimension?

7.4.19. Find all (real) solutions to the two-dimensional Laplace equation of the form $u = \log p(x, y)$ where $p(x, y)$ is a quadratic polynomial. Do these solutions form a vector space? If so, what is its dimension?

♡ **7.4.20.** (a) Show that the function $e^x \cos y$ is a solution to the two-dimensional Laplace equation.

(b) Show that its quadratic Taylor polynomial at $x = y = 0$ is harmonic.

(c) What about its degree 3 Taylor polynomial?

(d) Can you state a general theorem?

(e) Test your result by looking at the Taylor polynomials of the harmonic function $\log\left[(x-1)^2 + y^2 \right]$.

7.4.21. (a) Find a basis for, and the dimension of the vector space consisting of all quadratic polynomial solutions of the three-dimensional Laplace equation

$$\frac{\partial^2 u}{\partial x^2} + \frac{\partial^2 u}{\partial y^2} + \frac{\partial^2 u}{\partial z^2} = 0.$$

(b) Do the same for the homogeneous cubic polynomial solutions.

7.4.22. Find all solutions $u = f(r)$ of the three-dimensional Laplace equation

$$\frac{\partial^2 u}{\partial x^2} + \frac{\partial^2 u}{\partial y^2} + \frac{\partial^2 u}{\partial z^2} = 0$$

that depend only on the radial coordinate

$$r = \sqrt{x^2 + y^2 + z^2}.$$

Do these solutions form a vector space? If so, what is its dimension?

7.4.23. Let L, M be linear functions.

(a) Prove that $\ker(L \circ M) \supseteq \ker M$.

(b) Find an example where $\ker(L \circ M) \neq \ker M$.

Inhomogeneous Systems

Now we turn our attention to inhomogeneous linear systems

$$L[\mathbf{u}] = \mathbf{f}, \tag{7.56}$$

where $L\colon U \to V$ is a linear function, $\mathbf{f} \in V$, and the desired solution $\mathbf{u} \in U$. Unless $\mathbf{f} = \mathbf{0}$, the solution space to (7.56) is *not* a subspace of U. (Why?) Here, the crucial question is existence—is there a solution to the system? In contrast, for the homogeneous system $L[\mathbf{z}] = \mathbf{0}$, existence is not an issue, since $\mathbf{0}$ is always a solution. The key question for homogeneous systems is uniqueness: either $\ker L = \{\mathbf{0}\}$, in which case $\mathbf{0}$ is the only solution, or $\ker L \neq \{\mathbf{0}\}$, in which case there are infinitely many nontrivial solutions $\mathbf{0} \neq \mathbf{z} \in \ker L$.

In the matrix case, the compatibility of an inhomogeneous system $A\mathbf{x} = \mathbf{b}$—which was required for the existence of a solution—led to the general definition of the range of a matrix, which we copy verbatim for linear functions.

Definition 7.37 The *range* of a linear function $L\colon U \to V$ is the subspace

$$\operatorname{rng} L = \{\, L[\mathbf{u}] \mid \mathbf{u} \in U \,\} \subset V.$$

The proof that $\operatorname{rng} L$ is a subspace of the target space is straightforward: If $\mathbf{f} = L[\mathbf{u}]$ and $\mathbf{g} = L[\mathbf{v}]$ are any two elements of the range, so is any linear combination, since, by linearity

$$c\,\mathbf{f} + d\,\mathbf{g} = c\,L[\mathbf{u}] + d\,L[\mathbf{v}] = L[c\,\mathbf{u} + d\,\mathbf{v}] \in \operatorname{rng} L.$$

For example, if $L[\mathbf{u}] = A\mathbf{u}$ is given by multiplication by an $m \times n$ matrix, then its range is the subspace $\operatorname{rng} L = \operatorname{rng} A \subset \mathbb{R}^m$ spanned by the columns of A—the *column space* of the coefficient matrix. When L is a linear differential operator, or more general linear operator, characterizing its range can be a much more challenging problem.

The fundamental theorem regarding solutions to inhomogeneous linear equations exactly mimics our earlier result, Theorem 2.39, for matrix systems.

THEOREM 7.38 Let $L: U \to V$ be a linear function. Let $\mathbf{f} \in V$. Then the inhomogeneous linear system

$$L[\mathbf{u}] = \mathbf{f} \tag{7.57}$$

has a solution if and only if $\mathbf{f} \in \operatorname{rng} L$. In this case, the general solution to the system has the form

$$\mathbf{u} = \mathbf{u}^\star + \mathbf{z} \tag{7.58}$$

where $\mathbf{u}^\star$ is a particular solution, so $L[\mathbf{u}^\star] = \mathbf{f}$, and $\mathbf{z}$ is any element of $\ker L$, i.e., a solution to the corresponding homogeneous system

$$L[\mathbf{z}] = \mathbf{0}. \tag{7.59}$$

Proof We merely repeat the proof of Theorem 2.39. The existence condition $\mathbf{f} \in \operatorname{rng} L$ is an immediate consequence of the definition of the range. Suppose $\mathbf{u}^\star$ is a particular solution to (7.57). If $\mathbf{z}$ is a solution to (7.59), then, by linearity,

$$L[\mathbf{u}^\star + \mathbf{z}] = L[\mathbf{u}^\star] + L[\mathbf{z}] = \mathbf{f} + \mathbf{0} = \mathbf{f},$$

and hence $\mathbf{u}^\star + \mathbf{z}$ is also a solution to (7.57). To show that every solution has this form, let $\mathbf{u}$ be a second solution, so that $L[\mathbf{u}] = \mathbf{f}$. Setting $\mathbf{z} = \mathbf{u} - \mathbf{u}^\star$, we find

$$L[\mathbf{z}] = L[\mathbf{u} - \mathbf{u}^\star] = L[\mathbf{u}] - L[\mathbf{u}^\star] = \mathbf{f} - \mathbf{f} = \mathbf{0}.$$

Therefore $\mathbf{z} \in \ker L$, and so $\mathbf{u}$ has the proper form (7.58). ■

REMARK: In physical systems, the inhomogeneity $\mathbf{f}$ typically corresponds to an external force. The decomposition formula (7.58) states that its effect on the linear system can be viewed as a direct combination of a specific response $\mathbf{u}^\star$ to the forcing and the system's internal, unencumbered motion, as represented by the homogeneous solution $\mathbf{z}$. Examples of this important principle appear throughout the book.

Corollary 7.39 The inhomogeneous linear system (7.57) has a *unique* solution if and only if $\mathbf{f} \in \operatorname{rng} L$ and $\ker L = \{\mathbf{0}\}$.

Therefore, to prove that a linear system has a unique solution, we first need to prove an *existence result* that there is at least one solution, which requires the right hand side $\mathbf{f}$ to lie in the range of the operator L, and then a *uniqueness result*, that the only solution to the homogeneous system $L[\mathbf{z}] = \mathbf{0}$ is the trivial zero solution $\mathbf{z} = \mathbf{0}$. Observe that whenever an inhomogeneous system $L[\mathbf{u}] = \mathbf{f}$ has a unique solution, then any other inhomogeneous system $L[\mathbf{u}] = \mathbf{g}$ that is defined by the *same* linear function also has a unique solution, provided $\mathbf{g} \in \operatorname{rng} L$. In other words, uniqueness does not depend upon the external forcing — although existence might.

EXAMPLE 7.40 Consider the inhomogeneous linear second order differential equation

$$u'' + u' - 2u = x.$$

Note that this can be written in the linear system form

$$L[u] = x, \quad \text{where} \quad L = D^2 + D - 2$$

is a linear second order differential operator. The kernel of the differential operator L is found by solving the associated homogeneous linear equation

$$L[z] = z'' + z' - 2z = 0.$$

Applying the usual solution method, we find that the homogeneous differential equation has a two-dimensional solution space, with basis functions

$$z_1(x) = e^{-2x}, \quad z_2(x) = e^x.$$

Therefore, the general element of $\ker L$ is a linear combination

$$z(x) = c_1 z_1(x) + c_2 z_2(x) = c_1 e^{-2x} + c_2 e^x.$$

To find a particular solution to the inhomogeneous differential equation, we rely on the method of *undetermined coefficients*.[§] We introduce the solution ansatz $u = ax + b$, and compute

$$L[u] = L[ax + b] = a - 2(ax + b) = -2ax + (a - 2b) = x,$$

Equating the coefficients of x and 1, and then solving for $a = -\frac{1}{2}$, $b = -\frac{1}{4}$, we deduce that

$$u^\star(x) = -\tfrac{1}{2}x - \tfrac{1}{4}$$

is a particular solution to the inhomogeneous differential equation. Theorem 7.38 then says that the general solution is

$$u(x) = u^\star(x) + z(x) = -\tfrac{1}{2}x - \tfrac{1}{4} + c_1 e^{-2x} + c_2 e^x.$$ ●

EXAMPLE 7.41 By inspection, we see that

$$u(x, y) = -\tfrac{1}{2}\sin(x + y)$$

is a solution to the particular *Poisson equation*

$$\frac{\partial^2 u}{\partial x^2} + \frac{\partial^2 u}{\partial y^2} = \sin(x + y). \tag{7.60}$$

Theorem 7.38 implies that *every* solution to this inhomogeneous version of the Laplace equation (7.55) takes the form

$$u(x, y) = -\tfrac{1}{2}\sin(x + y) + z(x, y),$$

where $z(x, y)$ is an arbitrary harmonic function, i.e., solution to the homogeneous Laplace equation. ●

EXAMPLE 7.42 Let us solve the second order linear boundary value problem

$$u'' + u = x, \quad u(0) = 0, \quad u(\pi) = 0. \tag{7.61}$$

As with initial value problems, the first step is to solve the differential equation. To this end, we first solve the corresponding homogeneous differential equation $z'' + z = 0$. The usual method (see [6] or Example 7.50 below) shows that $\cos x$ and $\sin x$ form a basis for its solution space. The method of undetermined coefficients

---

[§]One could also employ the method of *variation of parameters*, although usually the undetermined coefficient method, when applicable, is the more straightforward of the two. Details can be found in most ordinary differential equations texts, including [6, 19].

then produces the particular solution $u^\star(x) = x$ to the inhomogeneous differential equation, and so its general solution is

$$u(x) = x + c_1 \cos x + c_2 \sin x. \tag{7.62}$$

The next step is to see whether any solutions also satisfy the boundary conditions. Plugging formula (7.62) into the boundary conditions yields

$$u(0) = c_1 = 0, \qquad u(\pi) = \pi - c_1 = 0.$$

However, these two conditions are incompatible, and so there is *no* solution to the linear system (7.61). The function $f(x) = x$ does not lie in the range of the differential operator $L[u] = u'' + u$ when u is subjected to the boundary conditions. Or, to state it another way, $(x, 0, 0)^T$ does not belong to the range of the linear operator $M[u] = \big(u'' + u, u(0), u(\pi)\big)^T$ defining the boundary value problem.

On the other hand, if we slightly modify the inhomogeneity, the boundary value problem

$$u'' + u = x - \tfrac{1}{2}\pi, \qquad u(0) = 0, \qquad u(\pi) = 0. \tag{7.63}$$

does admit a solution, but it fails to be unique. Applying the preceding solution techniques, we find that

$$u(x) = x - \tfrac{1}{2}\pi + \tfrac{1}{2}\pi \cos x + c \sin x$$

solves the system for *any* choice of constant c, and so the boundary value problem admits infinitely many solutions. Observe that

$$z(x) = \sin x$$

forms a basis for the kernel or solution space of the corresponding homogeneous boundary value problem

$$z'' + z = 0, \qquad z(0) = 0, \qquad z(\pi) = 0,$$

while

$$u^\star(x) = x - \tfrac{1}{2}\pi + \tfrac{1}{2}\pi \cos x$$

represents a particular solution to the inhomogeneous system. Thus, $u(x) = u^\star(x) + z(x)$, in conformity with the general formula (7.58).

Incidentally, if we modify the interval of definition, considering

$$u'' + u = f(x), \qquad u(0) = 0, \qquad u\left(\tfrac{1}{2}\pi\right) = 0, \tag{7.64}$$

then the homogeneous boundary value problem, with $f(x) \equiv 0$, has only the trivial solution, and the inhomogeneous system admits a unique solution for *any* inhomogeneity $f(x)$. For example, if $f(x) = x$, then

$$u(x) = x - \tfrac{1}{2}\pi \sin x \tag{7.65}$$

is the unique solution to the resulting boundary value problem (7.64).

This example highlights some crucial differences between boundary value problems and initial value problems for ordinary differential equations. Nonsingular initial value problems have a unique solution for any set of initial conditions. Boundary value problems have more of the flavor of linear algebraic systems, either possessing a unique solution for all possible inhomogeneities, or admitting either no solution or infinitely many solutions, depending on the right hand side. An interesting question is how to characterize the inhomogeneities $f(x)$ that admit a solution, i.e., lie in the range of the associated linear operator. We will explore these issues in depth in Chapter 11. ●

EXERCISES 7.4

7.4.24. For each of the following inhomogeneous systems, determine whether the right hand side lies in the range of the coefficient matrix, and, if so, write out the general solution, clearly identifying the particular solution and the kernel element.

(a) $\begin{pmatrix} 1 & -1 \\ 3 & -3 \end{pmatrix} \mathbf{x} = \begin{pmatrix} 1 \\ 2 \end{pmatrix}$

(b) $\begin{pmatrix} 2 & 1 & 4 \\ -1 & 2 & 1 \end{pmatrix} \mathbf{x} = \begin{pmatrix} 1 \\ 2 \end{pmatrix}$

(c) $\begin{pmatrix} 1 & 2 & -1 \\ 2 & 0 & 1 \\ 1 & -2 & 2 \end{pmatrix} \mathbf{x} = \begin{pmatrix} 0 \\ 3 \\ 3 \end{pmatrix}$

(d) $\begin{pmatrix} -1 & 3 & 0 & 2 \\ 2 & -6 & 1 & -1 \\ -3 & 9 & -2 & 0 \end{pmatrix} \mathbf{x} = \begin{pmatrix} 2 \\ -2 \\ 2 \end{pmatrix}$

(e) $\begin{pmatrix} -2 & 1 \\ -2 & 3 \\ 3 & -5 \end{pmatrix} \mathbf{x} = \begin{pmatrix} 1 \\ 0 \\ -1 \end{pmatrix}$

(f) $\begin{pmatrix} 1 & 0 & -1 & 0 \\ 0 & 1 & 1 & -1 \\ 1 & 1 & 0 & -1 \\ 1 & 2 & 1 & -2 \end{pmatrix} \mathbf{x} = \begin{pmatrix} -2 \\ 1 \\ -1 \\ 0 \end{pmatrix}$

7.4.25. Which of the following systems has a unique solution?

(a) $\begin{pmatrix} 3 & 1 \\ -1 & -1 \\ 2 & 0 \end{pmatrix} \begin{pmatrix} x \\ y \end{pmatrix} = \begin{pmatrix} 0 \\ 2 \\ 2 \end{pmatrix}$

(b) $\begin{pmatrix} 1 & 2 & -1 \\ -2 & 3 & 0 \end{pmatrix} \begin{pmatrix} x \\ y \\ z \end{pmatrix} = \begin{pmatrix} 1 \\ 2 \end{pmatrix}$

(c) $\begin{pmatrix} 2 & 1 & -1 \\ 0 & -3 & -3 \\ 2 & 0 & -2 \end{pmatrix} \begin{pmatrix} u \\ v \\ w \end{pmatrix} = \begin{pmatrix} 3 \\ -1 \\ 5 \end{pmatrix}$

(d) $\begin{pmatrix} 1 & 4 & -1 \\ 1 & 3 & -3 \\ 2 & 3 & -2 \end{pmatrix} \begin{pmatrix} u \\ v \\ w \end{pmatrix} = \begin{pmatrix} -2 \\ -1 \\ 1 \end{pmatrix}$

(e) $\begin{pmatrix} 1 & -4 & 2 & 0 \\ 3 & -9 & 5 & 0 \\ 1 & 1 & 0 & -2 \end{pmatrix} \begin{pmatrix} p \\ q \\ r \\ s \end{pmatrix} = \begin{pmatrix} 0 \\ 1 \\ 2 \end{pmatrix}$

7.4.26. Solve the following inhomogeneous linear ordinary differential equations:

(a) $u' - 4u = x - 3$

(b) $5u'' - 4u' + 4u = e^x \cos x$,

(c) $u'' - 3u' = e^{3x}$

7.4.27. Solve the following initial value problems:

(a) $u' + 3u = e^x$, $u(1) = 0$

(b) $u'' + 4u = 1$, $u(\pi) = u'(\pi) = 0$

(c) $u'' - u' - 2u = e^x + e^{-x}$, $u(0) = u'(0) = 0$

(d) $u'' + 2u' + 5u = \sin x$, $u(0) = 1$, $u'(0) = 0$

(e) $u''' - u'' + u' - u = x$, $u(0) = 0$, $u'(0) = 1$, $u''(0) = 0$

7.4.28. Write down all solutions to the following boundary value problems. Label your answer as

(i) unique solution,

(ii) no solution,

(iii) infinitely many solutions.

(a) $u'' + 2u = 2x$, $u(0) = 0$, $u(\pi) = 0$

(b) $u'' + 4u = \cos x$, $u(-\pi) = 0$, $u(\pi) = 1$

(c) $u'' - 2u' + u = x - 2$, $u(0) = -1$, $u(1) = 1$

(d) $u'' + 2u' + 2u = 1$, $u(0) = \frac{1}{2}$, $u(\pi) = \frac{1}{2}$

(e) $u'' - 3u' + 2u = 4x$, $u(0) = 0$, $u(1) = 0$

(f) $x^2 u'' + x u' - u = 0$, $u(0) = 1$, $u(1) = 0$

(g) $x^2 u'' - 6u = 0$, $u(1) = 1$, $u(2) = -1$

(h) $x^2 u'' - 2x u' + 2u = 0$, $u(0) = 0$, $u(1) = 1$

7.4.29. Solve the following inhomogeneous Euler equations using either variation of parameters or the change of variables method discussed in Exercise 7.4.13:

(a) $x^2 u'' + x u' - u = x$

(b) $x^2 u'' - 2x u' + 2u = \log x$

(c) $x^2 u'' - 3x u' - 5u = 3x - 5$

7.4.30. Let L, M be linear functions.

(a) Prove that $\operatorname{rng}(L \circ M) \subseteq \operatorname{rng} L$.

(b) Give an example where $\operatorname{rng}(L \circ M) \neq \operatorname{rng} L$.

◇ **7.4.31.** Let $L \colon U \to V$ be a linear function, and let $W \subset U$ be a subspace of the domain space.

(a) Prove that $Y = \{ L[\mathbf{w}] \mid \mathbf{w} \in W \} \subset \operatorname{rng} L \subset V$ is a subspace of the range.

(b) Prove that $\dim Y \leq \dim W$. Conclude that a linear transformation can never increase the dimension of a subspace.

◇ **7.4.32.** The subspace W of a vector space V is said to be an *invariant subspace* under the linear transformation $L \colon V \to V$ if $L[\mathbf{w}] \in W$ whenever $\mathbf{w} \in W$. Prove that $\ker L$ and $\operatorname{rng} L$ are both invariant subspaces.

7.4.33. Find all its invariant subspaces $W \subset \mathbb{R}^2$, as defined in Exercise 7.4.32, of the following linear transformations $L \colon \mathbb{R}^2 \to \mathbb{R}^2$:

(a) the scaling transformation $(2x, 3y)^T$

(b) the shear $(x + 3y, y)^T$

(c) counterclockwise rotation by an angle $0 \leq \theta < 2\pi$

◇ **7.4.34.** (a) Show that if $L\colon V \to V$ is linear and $\ker L \neq \{\mathbf{0}\}$, then L is not invertible.
(b) Show that if $\operatorname{rng} L \neq V$, then L is not invertible.
(c) Give an example of a linear map with $\ker L = \{\mathbf{0}\}$ which is not invertible. *Hint*: First explain why your example must be on an infinite-dimensional vector space.

Superposition Principles for Inhomogeneous Systems

The *superposition principle* for inhomogeneous linear systems allows us to combine different inhomogeneities—provided that we do not change the underlying linear operator. The result is a straightforward generalization of the matrix version described in Theorem 2.44.

THEOREM 7.43 Let $L\colon U \to V$ be a prescribed linear function. Suppose that, for each $i = 1, \ldots, k$, we know a particular solution $\mathbf{u}_i^\star$ to the inhomogeneous linear system $L[\mathbf{u}] = \mathbf{f}_i$ for some $\mathbf{f}_i \in \operatorname{rng} L$. Then, given scalars $c_1, \ldots, c_k$, a particular solution to the combined inhomogeneous system

$$L[\mathbf{u}] = c_1 \mathbf{f}_1 + \cdots + c_k \mathbf{f}_k \tag{7.66}$$

is the corresponding linear combination

$$\mathbf{u}^\star = c_1 \mathbf{u}_1^\star + \cdots + c_k \mathbf{u}_k^\star \tag{7.67}$$

of particular solutions. The general solution to the inhomogeneous system (7.66) is

$$\mathbf{u} = \mathbf{u}^\star + \mathbf{z} = c_1 \mathbf{u}_1^\star + \cdots + c_k \mathbf{u}_k^\star + \mathbf{z}, \tag{7.68}$$

where $\mathbf{z} \in \ker L$ is an arbitrary solution to the associated homogeneous system $L[\mathbf{z}] = \mathbf{0}$.

The proof is an easy consequence of linearity, and left to the reader. In physical terms, the superposition principle can be interpreted as follows. If we know the response of a linear physical system to several different external forces, represented by $\mathbf{f}_1, \ldots, \mathbf{f}_k$, then the response of the system to a linear combination of these forces is just the self-same linear combination of the individual responses. The homogeneous solution $\mathbf{z}$ represents an internal motion that the system acquires independent of any external forcing. Superposition relies on the linearity of the system, and so is always applicable in quantum mechanics, which is an inherently linear theory. But, in classical and relativistic mechanics, superposition is valid only in the linear approximation regime governing small motions/displacements/etc. Large scale motions of the fully nonlinear physical system are much more subtle, and combinations of external forces may lead to unexpected results.

EXAMPLE 7.44 In Example 7.42, we found that a particular solution to the linear differential equation

$$u'' + u = x \quad \text{is} \quad u_1^\star = x.$$

The method of undetermined coefficients is used to solve the inhomogeneous equation

$$u'' + u = \cos x.$$

Since $\cos x$ and $\sin x$ are already solutions to the homogeneous equation, we must use the solution ansatz

$$u = a\,x \cos x + b\,x \sin x,$$

which, when substituted into the differential equation, produces the particular solution

$$u_2^\star = -\tfrac{1}{2}\,x \sin x.$$

Therefore, by the superposition principle, the combined inhomogeneous system

$$u'' + u = 3x - 2\cos x$$

has a particular solution

$$u^\star = 3u_1^\star - 2u_2^\star = 3x + x\sin x.$$

The general solution is obtained by appending an arbitrary solution to the homogeneous equation:

$$u = 3x + x\sin x + c_1\cos x + c_2\sin x. \qquad \bullet$$

EXAMPLE 7.45 Consider the boundary value problem

$$u'' + u = x, \quad u(0) = 2, \quad u\left(\tfrac{1}{2}\pi\right) = -1, \tag{7.69}$$

which is a modification of (7.64) with inhomogeneous boundary conditions. The superposition principle applies here, and allows us to decouple the inhomogeneity due to the forcing from the inhomogeneity due to the boundary conditions. We decompose the right hand side, written in vectorial form, into simpler constituents¶

$$\begin{pmatrix} x \\ 2 \\ -1 \end{pmatrix} = \begin{pmatrix} x \\ 0 \\ 0 \end{pmatrix} + 2\begin{pmatrix} 0 \\ 1 \\ 0 \end{pmatrix} - \begin{pmatrix} 0 \\ 0 \\ 1 \end{pmatrix}.$$

The first vector on the right hand side corresponds to the preceding boundary value problem (7.64), whose solution was found in (7.65). The second and third vectors correspond to the unforced boundary value problems

$$u'' + u = 0, \quad u(0) = 1, \quad u\left(\tfrac{1}{2}\pi\right) = 0,$$

and

$$u'' + u = 0, \quad u(0) = 0, \quad u\left(\tfrac{1}{2}\pi\right) = 1,$$

with respective solutions $u(x) = \cos x$ and $u(x) = \sin x$. Therefore, the solution to the combined boundary value problem (7.69) is the same linear combination of these individual solutions:

$$u(x) = \left(x - \tfrac{1}{2}\pi\sin x\right) + 2\cos x - \sin x = x + 2\cos x - \left(1 + \tfrac{1}{2}\pi\right)\sin x.$$

The solution is unique because the corresponding homogeneous boundary value problem

$$z'' + z = 0, \quad z(0) = 0, \quad z\left(\tfrac{1}{2}\pi\right) = 0,$$

has only the trivial solution $z(x) \equiv 0$, as you can verify. $\bullet$

EXERCISES 7.4

7.4.35. Use superposition to solve the following inhomogeneous ordinary differential equations:

(a) $u' + 2u = 1 + \cos x$

(b) $u'' - 9u = x + \sin x$

(c) $9u'' - 18u' + 10u = 1 + e^x\cos x$

(d) $u'' + u' - 2u = \sinh x$, where

$$\sinh x = \tfrac{1}{2}(e^x - e^{-x})$$

(e) $u''' + 9u' = 1 + e^{3x}$

7.4.36. Consider the differential equation

$$x u'' - (x+1)u' + u = 0.$$

¶ *Warning*: When writing out a linear combination, make sure the scalars are *constants*. Writing the first summand as $x\,(1,0,0)^T$ will lead to an *incorrect* application of the superposition principle.

Suppose we know the solution to the initial value problem $u(1) = 2$, $u'(1) = 1$ is $u(x) = x+1$, while the solution to the initial value problem $u(1) = 1$, $u'(1) = 1$ is $u(x) = e^{x-1}$.

(a) What is the solution to the initial value problem $u(1) = 3$, $u'(1) = -2$?

(b) What is the general solution to the differential equation?

7.4.37. Consider the differential equation $4x\,u''+2u'+u = 0$. Given that $\cos\sqrt{x}$ solves the boundary value problem $u\left(\frac{1}{4}\pi^2\right) = 0$, $u(\pi^2) = -1$, and $\sin\sqrt{x}$ solves the boundary value problem $u\left(\frac{1}{4}\pi^2\right) = 1$, $u(\pi^2) = 0$, write down the solution to the boundary value problem $u\left(\frac{1}{4}\pi^2\right) = -3$, $u(\pi^2) = 7$.

7.4.38. Consider the differential equation $u'' + x\,u = 2$. Suppose you know solutions to the two boundary value problems $u(0) = 1$, $u(1) = 0$ and $u(0) = 0$, $u(1) = 1$. List all possible boundary value problems you can solve using superposition.

7.4.39. Solve the following boundary value problems by using superposition:

(a) $u'' + 9u = x$, $u(0) = 1$, $u'(\pi) = 0$

(b) $u'' - 8u' + 16u = e^{4x}$, $u(0) = 1$, $u(1) = 0$

(c) $u'' + 4u = \sin 3x$, $u'(0) = 0$, $u(2\pi) = 3$

(d) $u'' - 2u' + u = 1 + e^x$, $u'(0) = -1$, $u'(1) = 1$

7.4.40. Given that $x^2 + y^2$ solves the Poisson equation

$$\frac{\partial^2 u}{\partial x^2} + \frac{\partial^2 u}{\partial y^2} = 4,$$

while $x^4 + y^4$ solves

$$\frac{\partial^2 u}{\partial x^2} + \frac{\partial^2 u}{\partial y^2} = 12(x^2 + y^2),$$

write down a solution to

$$\frac{\partial^2 u}{\partial x^2} + \frac{\partial^2 u}{\partial y^2} = 1 + x^2 + y^2.$$

♡ **7.4.41.** *Reduction of Order*: Suppose you know one solution $u_1(x)$ to the second order homogeneous differential equation $u'' + a(x)u' + b(x)u = 0$.

(a) Show that if $u(x) = v(x)\,u_1(x)$ is any other solution, then $w(x) = v'(x)$ satisfies a first order differential equation.

(b) Use reduction of order to find the general solution to the following equations, based on the indicated solution:

(i) $u'' - 2u' + u = 0$, $u_1(x) = e^x$

(ii) $x\,u'' + (x-1)u' - u = 0$, $u_1(x) = x - 1$

(iii) $u''+4x\,u'+(4x^2+2)u = 0$, $u_1(x) = e^{-x^2}$

(iv) $u'' - (x^2+1)u = 0$, $u_1(x) = e^{x^2/2}$

◇ **7.4.42.** Write out the details of the proof of Theorem 7.43.

Complex Solutions to Real Systems

As we know, solutions to a linear, homogeneous, constant coefficient ordinary differential equation are found by substituting an exponential ansatz, which effectively reduces the differential equation to the polynomial characteristic equation. Complex roots of the characteristic equation yield complex exponential solutions. But, if the equation is real, then the real and imaginary parts of the complex solutions are automatically real solutions. This solution technique is a particular case of a general principle for producing real solutions to real linear systems from, typically, simpler complex solutions. To work, the method requires us to impose some additional structure on the complex vector spaces involved.

Definition 7.46 A complex vector space V is called *conjugated* if it admits an operation of *complex conjugation* taking $\mathbf{u} \in V$ to $\overline{\mathbf{u}} \in V$ that is compatible with scalar multiplication, meaning that if $\mathbf{u} \in V$ and $\lambda \in \mathbb{C}$, then $\overline{\lambda\,\mathbf{u}} = \overline{\lambda}\,\overline{\mathbf{u}}$.

The simplest example of a conjugated vector space is $\mathbb{C}^n$. The complex conjugate of a vector $\mathbf{u} = (u_1, u_2, \ldots, u_n)^T$ is obtained by conjugating all its entries: $\overline{\mathbf{u}} = (\overline{u}_1, \overline{u}_2, \ldots, \overline{u}_n)^T$. Thus,

$$\mathbf{u} = \mathbf{v} + \mathrm{i}\,\mathbf{w}, \qquad \overline{\mathbf{u}} = \mathbf{v} - \mathrm{i}\,\mathbf{w},$$
$$\text{where} \quad \mathbf{v} = \operatorname{Re}\mathbf{u} = \frac{\mathbf{u}+\overline{\mathbf{u}}}{2}, \qquad \mathbf{w} = \operatorname{Im}\mathbf{u} = \frac{\mathbf{u}-\overline{\mathbf{u}}}{2\,\mathrm{i}}, \tag{7.70}$$

are the real and imaginary parts of $\mathbf{u} \in \mathbb{C}^n$. For example, if

$$\mathbf{u} = \begin{pmatrix} 1 - 2\,\mathrm{i} \\ 3\,\mathrm{i} \\ 5 \end{pmatrix} = \begin{pmatrix} 1 \\ 0 \\ 5 \end{pmatrix} + \mathrm{i} \begin{pmatrix} -2 \\ 3 \\ 0 \end{pmatrix},$$

then

$$\overline{\mathbf{u}} = \begin{pmatrix} 1 + 2\,\mathrm{i} \\ -3\,\mathrm{i} \\ 5 \end{pmatrix} = \begin{pmatrix} 1 \\ 0 \\ 5 \end{pmatrix} - \mathrm{i} \begin{pmatrix} -2 \\ 3 \\ 0 \end{pmatrix}.$$

The other prototypical example of a conjugated vector space is the space of complex-valued functions $f(x) = r(x) + \mathrm{i}\,s(x)$ defined on the interval $a \le x \le b$. The complex conjugate function is $\overline{f}(x) = \overline{f(x)} = r(x) - \mathrm{i}\,s(x)$. Thus, the complex conjugate of $e^{(1+3\,\mathrm{i})x} = e^x \cos 3x + \mathrm{i}\, e^x \sin 3x$ is

$$\overline{e^{(1+3\,\mathrm{i})x}} = e^{(1-3\,\mathrm{i})x} = e^x \cos 3x - \mathrm{i}\, e^x \sin 3x.$$

An element $\mathbf{v} \in V$ of a conjugated vector space is called *real* if $\overline{\mathbf{v}} = \mathbf{v}$. One easily checks that the real and imaginary parts of a general element, as defined by (7.70), are both real elements.

Warning: Not all subspaces of a conjugated vector space are conjugated. For example, the one-dimensional subspace of $\mathbb{C}^2$ spanned by $\mathbf{v}_1 = (1, 2)^T$ is conjugated. Indeed, the complex conjugate of a general element $c\,\mathbf{v}_1 = (c, 2c)^T$ is $(\overline{c}, 2\,\overline{c})^T = \overline{c}\,\mathbf{v}_1$ which also belongs to the subspace. On the other hand, the subspace spanned by $(1, \mathrm{i})^T$ is *not* conjugated because the complex conjugate of the element $(c, \mathrm{i}\,c)^T$ is $(\overline{c}, -\mathrm{i}\,\overline{c})^T$, which does not belong to the subspace unless $c = 0$. In Exercise 7.4.52 you are asked to prove that a subspace $V \subset \mathbb{C}^n$ is conjugated if and only if it has a basis $\mathbf{v}_1, \ldots, \mathbf{v}_k$ consisting entirely of real vectors. While conjugated subspaces play a role in certain applications, in practice we will only deal with $\mathbb{C}^n$ and the entire space of complex-valued functions, and so can suppress most of these somewhat technical details.

Definition 7.47 A linear function $L\colon U \to V$ between conjugated vector spaces is called *real* if it commutes with complex conjugation:

$$L[\,\overline{\mathbf{u}}\,] = \overline{L[\,\mathbf{u}\,]}. \tag{7.71}$$

For example, any linear function $L\colon \mathbb{C}^n \to \mathbb{C}^m$ is given by multiplication by an $m \times n$ matrix: $L[\mathbf{u}] = A\mathbf{u}$. The function is real if and only if A is a real matrix. Similarly, a differential operator (7.13) is real if and only if its coefficients are real-valued functions.

THEOREM 7.48 If $L[\mathbf{u}] = \mathbf{0}$ is a real homogeneous linear system and $\mathbf{u} = \mathbf{v} + \mathrm{i}\,\mathbf{w}$ is a complex solution, then its complex conjugate $\overline{\mathbf{u}} = \mathbf{v} - \mathrm{i}\,\mathbf{w}$ is also a solution. Moreover, both the real and imaginary parts, $\mathbf{v}$ and $\mathbf{w}$, of a complex solution are real solutions.

Proof First note that, by reality, $L[\,\overline{\mathbf{u}}\,] = \overline{L[\,\mathbf{u}\,]} = \mathbf{0}$ whenever $L[\mathbf{u}] = \mathbf{0}$, and hence the complex conjugate $\overline{\mathbf{u}}$ of any solution is also a solution. Therefore, by linear superposition, $\mathbf{v} = \operatorname{Re} \mathbf{u} = \frac{1}{2}(\mathbf{u} + \overline{\mathbf{u}})$ and $\mathbf{w} = \operatorname{Im} \mathbf{u} = \frac{1}{2\,\mathrm{i}}(\mathbf{u} - \overline{\mathbf{u}})$ are also solutions. ■

EXAMPLE 7.49 The real linear matrix system

$$\begin{pmatrix} 2 & -1 & 3 & 0 \\ -2 & 1 & 1 & 2 \end{pmatrix} \begin{pmatrix} x \\ y \\ z \\ w \end{pmatrix} = \begin{pmatrix} 0 \\ 0 \end{pmatrix}$$

has a complex solution

$$\mathbf{u} = \begin{pmatrix} -1-3\,\mathrm{i} \\ 1 \\ 1+2\,\mathrm{i} \\ -2-4\,\mathrm{i} \end{pmatrix} = \begin{pmatrix} -1 \\ 1 \\ 1 \\ -2 \end{pmatrix} + \mathrm{i} \begin{pmatrix} -3 \\ 0 \\ 2 \\ -4 \end{pmatrix}.$$

Since the coefficient matrix is real, the real and imaginary parts,

$$\mathbf{v} = (-1, 1, 1, -2)^T, \qquad \mathbf{w} = (-3, 0, 2, -4)^T,$$

are both solutions of the system.

On the other hand, the complex linear system

$$\begin{pmatrix} 2 & -2\,\mathrm{i} & \mathrm{i} & 0 \\ 1+\mathrm{i} & 0 & -2-\mathrm{i} & 1 \end{pmatrix} \begin{pmatrix} x \\ y \\ z \\ w \end{pmatrix} = \begin{pmatrix} 0 \\ 0 \end{pmatrix}$$

has the complex solution

$$\mathbf{u} = \begin{pmatrix} 1-\mathrm{i} \\ -\mathrm{i} \\ 2 \\ 2+2\,\mathrm{i} \end{pmatrix} = \begin{pmatrix} 1 \\ 0 \\ 2 \\ 2 \end{pmatrix} + \mathrm{i} \begin{pmatrix} -1 \\ -1 \\ 0 \\ 2 \end{pmatrix}.$$

However, neither its real nor its imaginary part is a solution to the system. ●

EXAMPLE 7.50 Consider the real ordinary differential equation

$$u'' + 2u' + 5u = 0.$$

To solve it, we use the usual exponential ansatz $u = e^{\lambda x}$, leading to the characteristic equation

$$\lambda^2 + 2\lambda + 5 = 0.$$

There are two roots,

$$\lambda_1 = -1 + 2\,\mathrm{i}, \qquad \lambda_2 = -1 - 2\,\mathrm{i},$$

leading, via Euler's formula (3.84), to the complex solutions

$$u_1(x) = e^{(-1+2\,\mathrm{i})x} = e^{-x}\cos 2x + \mathrm{i}\, e^{-x}\sin 2x,$$
$$u_2(x) = e^{(-1-2\,\mathrm{i})x} = e^{-x}\cos 2x - \mathrm{i}\, e^{-x}\sin 2x.$$

The complex conjugate of the first solution is the second, in accordance with Theorem 7.48. Moreover, the real and imaginary parts of the two solutions

$$v(x) = e^{-x}\cos 2x, \; w(x) = e^{-x}\sin 2x,$$

are individual real solutions. Since the solution space is two-dimensional, the general solution is a linear combination

$$u(x) = c_1\, e^{-x}\cos 2x + c_2\, e^{-x}\sin 2x,$$

of the two linearly independent real solutions. ●

EXAMPLE 7.51 Consider the real second order Euler differential equation

$$L[u] = x^2 u'' + 7x u' + 13 u = 0.$$

The roots of the associated characteristic equation

$$r(r-1) + 7r + 13 = r^2 + 6r + 13 = 0$$

are complex: $r = -3 \pm 2\,\mathrm{i}$, and the resulting solutions $x^{-3+2\mathrm{i}}$, $x^{-3-2\mathrm{i}}$ are complex conjugate powers. Using Euler's formula (3.84), we write each of them in real and imaginary form, e.g.,

$$x^{-3+2\mathrm{i}} = x^{-3} e^{2\mathrm{i}\log x} = x^{-3}\cos(2\log x) + \mathrm{i}\, x^{-3}\sin(2\log x).$$

Again, by Theorem 7.48, the real and imaginary parts of the complex solution are by themselves real solutions to the equation. Therefore, the general real solution to this differential equation is

$$u(x) = c_1 x^{-3}\cos(2\log x) + c_2 x^{-3}\sin(2\log x).$$

●

EXAMPLE 7.52 The complex monomial

$$u(x, y) = (x + \mathrm{i}\, y)^n$$

is a solution to the Laplace equation

$$\frac{\partial^2 u}{\partial x^2} + \frac{\partial^2 u}{\partial y^2} = 0$$

because, by the chain rule,

$$\frac{\partial^2 u}{\partial x^2} = n(n-1)(x + \mathrm{i}\, y)^{n-2},$$

$$\frac{\partial^2 u}{\partial y^2} = n(n-1)\,\mathrm{i}^2 (x + \mathrm{i}\, y)^{n-2} = -n(n-1)(x + \mathrm{i}\, y)^{n-2}.$$

Since the Laplace operator is real, Theorem 7.48 implies that the real and imaginary parts of this complex solution are real solutions. The resulting real solutions are the harmonic polynomials introduced in Example 7.36.

Knowing this, it is relatively easy to find the explicit formulae for the harmonic polynomials. We appeal to the Binomial Formula

$$(a+b)^n = \sum_{i=0}^{n} \binom{n}{k} x^k y^{n-k}, \quad \text{where} \quad \binom{n}{k} = \frac{n!}{k!\,(n-k)!} \tag{7.72}$$

is the standard notation for the *binomial coefficients*. Since $\mathrm{i}^2 = -1$, $\mathrm{i}^3 = -\mathrm{i}$, $\mathrm{i}^4 = 1$, etc.,

$$\begin{aligned}(x + \mathrm{i}\, y)^n &= x^n + n x^{n-1}(\mathrm{i}\, y) + \binom{n}{2} x^{n-2}(\mathrm{i}\, y)^2 + \binom{n}{3} x^{n-3}(\mathrm{i}\, y)^3 + \cdots + (\mathrm{i}\, y)^n \\ &= x^n + \mathrm{i}\, n x^{n-1} y - \binom{n}{2} x^{n-2} y^2 - \mathrm{i}\binom{n}{3} x^{n-3} y^3 + \cdots.\end{aligned}$$

Separating the real and imaginary terms, we find the explicit formulae

$$\begin{aligned}\operatorname{Re}(x + \mathrm{i}\, y)^n &= x^n - \binom{n}{2} x^{n-2} y^2 + \binom{n}{4} x^{n-4} y^4 + \cdots, \\ \operatorname{Im}(x + \mathrm{i}\, y)^n &= n x^{n-1} y - \binom{n}{3} x^{n-3} y^3 + \binom{n}{5} x^{n-5} y^5 + \cdots,\end{aligned} \tag{7.73}$$

for the two independent harmonic polynomials of degree n. The first few of these polynomials were described in Example 7.36. In fact, it can be proved that the most general solution to the Laplace equation can be written as a convergent infinite series in the basic harmonic polynomials, cf. [47, 68]. ●

EXERCISES 7.4

7.4.43. Can you find a complex matrix A such that $\ker A \neq \{\mathbf{0}\}$ and the real and imaginary parts of every complex solution to $A\mathbf{u} = \mathbf{0}$ are also solutions?

7.4.44. Find the general real solution to the following homogeneous differential equations:

(a) $u'' + 4u = 0$ (b) $u'' + 6u' + 10u = 0$

(c) $2u''' + 3u' - 5u = 0$ (d) $u'''' + u = 0$

(e) $u'''' + 13u'' + 36u = 0$

(f) $x^2 u'' - x u' + 3u = 0$

(g) $x^3 u''' + x^2 u'' + 3 x u' - 8u = 0$

7.4.45. The following functions are solutions to a real constant coefficient homogeneous scalar ordinary differential equation.

(i) Determine the least possible order of the differential equation, and

(ii) write down an appropriate differential equation.

(a) $e^{-x}\sin$ (b) $3x$ (c) $x\sin x$

(d) $1 + x e^{-x}\cos 2x$

(e) $\sin x + \cos 2x,\ \sin x + x^2\cos x$

7.4.46. Find the general solution to the following complex ordinary differential equations. Verify that, in these cases, the real and imaginary parts of a complex solution are *not* real solutions.

(a) $u' + \mathrm{i}\,u = 0$

(b) $u'' - \mathrm{i}\,u' + (\mathrm{i} - 1)u = 0$

(c) $u'' - \mathrm{i}\,u = 0$

7.4.47. (a) Write down the explicit formulas for the harmonic polynomials of degree 4 and check that they are indeed solutions to the Laplace equation.

(b) Prove that every homogeneous polynomial solution of degree 4 is a linear combination of the two basic harmonic polynomials.

♡ **7.4.48.** (a) Show that $u(t,x) = e^{-k^2 t + \mathrm{i} k x}$ is a complex solution to the *heat equation*

$$\frac{\partial u}{\partial t} = \frac{\partial^2 u}{\partial x^2}$$

for any real constant k.

(b) Write down another complex solution by using complex conjugation.

(c) Find two independent real solutions to the heat equation.

(d) Can k be complex? If so, what real solutions are produced?

(e) Which of your solutions decay to zero as $t \to \infty$?

7.4.49. Find all complex exponential solutions $u(t,x) = e^{\omega t + k x}$ of the *beam equation*

$$\frac{\partial^2 u}{\partial t^2} = \frac{\partial^4 u}{\partial x^4}.$$

How many different real solutions can you produce?

7.4.50. Which of the following sets of vectors span conjugated subspaces of $\mathbb{C}^3$?

(a) $\begin{pmatrix} 1 \\ -1 \\ 2 \end{pmatrix}$ (b) $\begin{pmatrix} 1 \\ -\mathrm{i} \\ 2\mathrm{i} \end{pmatrix}$

(c) $\begin{pmatrix} 1 \\ 0 \\ 3 \end{pmatrix}, \begin{pmatrix} 1 \\ 1 \\ -1 \end{pmatrix}$ (d) $\begin{pmatrix} 1 \\ 0 \\ \mathrm{i} \end{pmatrix}, \begin{pmatrix} \mathrm{i} \\ 1 \\ 0 \end{pmatrix}$

(e) $\begin{pmatrix} \mathrm{i} \\ 1 \\ 0 \end{pmatrix}, \begin{pmatrix} 1 \\ 0 \\ -\mathrm{i} \end{pmatrix}, \begin{pmatrix} 0 \\ 1 \\ \mathrm{i} \end{pmatrix}$

◇ **7.4.51.** Prove that the real and imaginary parts of a general element of a conjugated vector space, as defined by (7.70), are both real elements.

◇ **7.4.52.** Prove that a subspace $V \subset \mathbb{C}^n$ is conjugated if and only if it admits a basis all of whose elements are real.

◇ **7.4.53.** Prove that a linear function $L\colon \mathbb{C}^n \to \mathbb{C}^m$ is real if and only if $L[\mathbf{u}] = A\mathbf{u}$ where A is a real $m \times n$ matrix.

◇ **7.4.54.** Prove that if $L[\mathbf{u}] = \mathbf{f}$ is a real inhomogeneous linear system with real right hand side $\mathbf{f}$, and $\mathbf{u} = \mathbf{v} + \mathrm{i}\,\mathbf{w}$ is a complex solution, then its real part $\mathbf{v}$ is a solution to the system, $L[\mathbf{v}] = \mathbf{f}$, while its imaginary part $\mathbf{w}$ solves the homogeneous system $L[\mathbf{w}] = \mathbf{0}$.

7.4.55. (a) Show that $\mathbf{u}$ solves the complex homogeneous linear system $L[\mathbf{u}] = \mathbf{0}$ if and only if its complex conjugate $\mathbf{v} = \overline{\mathbf{u}}$ solves the complex conjugate system $\overline{L}[\mathbf{v}] = \mathbf{0}$.

(b) Solve

$$\begin{pmatrix} 1-\mathrm{i} & 1 & 1-2\,\mathrm{i} \\ 2 & 1+\mathrm{i} & 3-\mathrm{i} \end{pmatrix}\mathbf{u} = \mathbf{0},$$

and then use your result to write down the solution to the system

$$\begin{pmatrix} 1+\mathrm{i} & 1 & 1+2\,\mathrm{i} \\ 2 & 1-\mathrm{i} & 3+\mathrm{i} \end{pmatrix}\mathbf{u} = \mathbf{0}.$$

◇ **7.4.56.** Let $\mathbf{u} = \mathbf{x} + \mathrm{i}\,\mathbf{y}$ be a complex solution to a real linear system. Under what conditions are its real and imaginary parts $\mathbf{x}, \mathbf{y}$ linearly independent real solutions?

7.5 ADJOINTS

Sections 2.5 and 5.6 revealed the importance of the adjoint system $A^T\mathbf{y} = \mathbf{f}$ in the analysis of systems of linear algebraic equations $A\mathbf{x} = \mathbf{b}$. Two of the four fundamental matrix subspaces are based on the transposed matrix. While the $m \times n$ matrix A defines a linear function from $\mathbb{R}^n$ to $\mathbb{R}^m$, its transpose, A^T, has size $n \times m$ and hence characterizes a linear function in the *reverse* direction, from $\mathbb{R}^m$ to $\mathbb{R}^n$.

As with most basic concepts for linear algebraic systems, the adjoint system and transpose operation on the coefficient matrix are the prototypes of a more general construction that is valid for general linear functions. However, it is not immediately obvious how to "transpose" a more general linear operator $L[\,u\,]$, e.g., a differential operator acting on function space. In this section, we shall introduce the abstract concept of the *adjoint* of a linear function that generalizes the transpose operation on matrices. Unfortunately, most of the interesting examples must be deferred until we develop additional analytical tools; see Chapter 11. You may wish to wait until then before reading through this final section.

The adjoint (and transpose) relies on an inner product structure on both the domain and target spaces. For simplicity, we restrict our attention to real inner product spaces, leaving the complex version to the interested reader. Thus, we begin with a linear function $L\colon U \to V$ that maps an inner product space U to a second inner product space V. We distinguish the inner products on U and V (which may be different even when U and V are the same vector space) by using a single angle bracket

$$\langle \mathbf{u}, \widetilde{\mathbf{u}} \rangle \quad \text{to denote the inner product between} \quad \mathbf{u}, \widetilde{\mathbf{u}} \in U,$$

and a double angle bracket

$$\langle\!\langle \mathbf{v}, \widetilde{\mathbf{v}} \rangle\!\rangle \quad \text{to denote the inner product between} \quad \mathbf{v}, \widetilde{\mathbf{v}} \in V.$$

Once inner products on both the domain and target spaces are prescribed, the abstract definition of the adjoint of a linear function can be formulated.

Definition 7.53 Let U, V be inner product spaces, and let $L\colon U \to V$ be a linear function. The *adjoint* of L is the function $L^*\colon V \to U$ that satisfies

$$\langle\!\langle L[\mathbf{u}], \mathbf{v} \rangle\!\rangle = \langle \mathbf{u}, L^*[\mathbf{v}] \rangle \quad \text{for all} \quad \mathbf{u} \in U, \quad \mathbf{v} \in V. \tag{7.74}$$

Note that the adjoint function goes in the *opposite* direction to L, just like the transposed matrix. Also, the left hand side of equation (7.74) indicates the inner product on V, while the right hand side is the inner product on U—which is where the respective vectors live. In infinite-dimensional situations, the adjoint may not exist. But if it does, then it is uniquely determined by (7.74); see Exercise 7.5.7.

REMARK: Technically, (7.74) only serves to define the "formal adjoint" of the linear operator L. For the infinite-dimensional function spaces arising in analysis, a true adjoint must satisfy certain additional analytical requirements, [38, 52]. However, for pedagogical reasons, it is better to suppress such advanced analytical complications in this introductory treatment.

Lemma 7.54 The adjoint of a linear function is a linear function.

Proof Given $\mathbf{u} \in U$, $\mathbf{v}, \mathbf{w} \in V$, and scalars $c, d \in \mathbb{R}$, we find

$$\begin{aligned}\langle \mathbf{u}, L^*[c\,\mathbf{v} + d\,\mathbf{w}] \rangle &= \langle\!\langle L[\mathbf{u}], c\,\mathbf{v} + d\,\mathbf{w} \rangle\!\rangle \\ &= c\,\langle\!\langle L[\mathbf{u}], \mathbf{v} \rangle\!\rangle + d\,\langle\!\langle L[\mathbf{u}], \mathbf{w} \rangle\!\rangle \\ &= c\,\langle \mathbf{u}, L^*[\mathbf{v}] \rangle + d\,\langle \mathbf{u}, L^*[\mathbf{w}] \rangle \\ &= \langle \mathbf{u}, c\,L^*[\mathbf{v}] + d\,L^*[\mathbf{w}] \rangle.\end{aligned}$$

Since this holds for all $\mathbf{u} \in U$, we must have

$$L^*[c\,\mathbf{v} + d\,\mathbf{w}] = c\,L^*[\mathbf{v}] + d\,L^*[\mathbf{w}],$$

proving linearity. ■

EXAMPLE 7.55 Let us first show how the defining equation (7.74) for the adjoint leads directly to the transpose of a matrix. Let $L\colon \mathbb{R}^n \to \mathbb{R}^m$ be the linear function $L[\mathbf{v}] = A\mathbf{v}$ defined by multiplication by the $m \times n$ matrix A. Then $L^*\colon \mathbb{R}^m \to \mathbb{R}^n$ is linear, and so is represented by matrix multiplication, $L^*[\mathbf{v}] = A^*\mathbf{v}$, by an $n \times m$ matrix A^*. We impose the ordinary Euclidean dot products

$$\begin{aligned}\langle \mathbf{u}, \widetilde{\mathbf{u}} \rangle &= \mathbf{u} \cdot \widetilde{\mathbf{u}} = \mathbf{u}^T \widetilde{\mathbf{u}}, \qquad \mathbf{u}, \widetilde{\mathbf{u}} \in \mathbb{R}^n, \\ \langle\!\langle \mathbf{v}, \widetilde{\mathbf{v}} \rangle\!\rangle &= \mathbf{v} \cdot \widetilde{\mathbf{v}} = \mathbf{v}^T \widetilde{\mathbf{v}}, \qquad \mathbf{v}, \widetilde{\mathbf{v}} \in \mathbb{R}^m,\end{aligned}$$

as our inner products on both $\mathbb{R}^n$ and $\mathbb{R}^m$. Evaluation of both sides of the adjoint identity (7.74) yields

$$\begin{aligned}\langle\!\langle L[\mathbf{u}], \mathbf{v} \rangle\!\rangle &= \langle\!\langle A\mathbf{u}, \mathbf{v} \rangle\!\rangle = (A\mathbf{u})^T \mathbf{v} = \mathbf{u}^T A^T \mathbf{v}, \\ \langle \mathbf{u}, L^*[\mathbf{v}] \rangle &= \langle \mathbf{u}, A^* \mathbf{v} \rangle = \mathbf{u}^T A^* \mathbf{v}.\end{aligned} \tag{7.75}$$

Since these expressions must agree for all $\mathbf{u}, \mathbf{v}$, the matrix A^* representing L^* is equal to the transposed matrix A^T; see Exercise 1.6.13. Therefore, *the adjoint of a matrix with respect to the Euclidean inner product is its transpose*: $A^* = A^T$. ●

EXAMPLE 7.56 Let us now adopt different, weighted inner products on the domain and target spaces for the linear map $L\colon \mathbb{R}^n \to \mathbb{R}^m$ given by $L[\mathbf{u}] = A\mathbf{v}$. Suppose that

(i) the inner product on the domain space $\mathbb{R}^n$ is given by $\langle \mathbf{u}, \widetilde{\mathbf{u}} \rangle = \mathbf{u}^T M \widetilde{\mathbf{u}}$, while

(ii) the inner product on the target space $\mathbb{R}^m$ is given by $\langle\!\langle \mathbf{v}, \widetilde{\mathbf{v}} \rangle\!\rangle = \mathbf{v}^T C \widetilde{\mathbf{v}}$.

Here M and C are positive definite matrices of respective sizes $n \times n$ and $m \times m$. Then, in place of (7.75), we have

$$\langle\!\langle A\mathbf{u}, \mathbf{v} \rangle\!\rangle = (A\mathbf{u})^T C\,\mathbf{v} = \mathbf{u}^T A^T C\,\mathbf{v}, \qquad \langle \mathbf{u}, A^*\mathbf{v} \rangle = \mathbf{u}^T M A^*\mathbf{v}.$$

Equating these expressions, we deduce that $A^T C = M A^*$. Therefore, the *weighted adjoint* of the matrix A is given by the more complicated formula

$$A^* = M^{-1} A^T C. \tag{7.76}$$

In mechanical applications, M plays the role of the mass matrix, and explicitly appears in the dynamical systems to be studied in Chapter 9. In particular, suppose A is square, defining a linear transformation $L\colon \mathbb{R}^n \to \mathbb{R}^n$. If we adopt the same inner product $\langle \mathbf{v}, \widetilde{\mathbf{v}} \rangle = \mathbf{v}^T C \widetilde{\mathbf{v}}$ on both the domain and target spaces $\mathbb{R}^n$, then its adjoint matrix $A^* = C^{-1} A^T C$ is similar to its transpose. ●

Everything that we learned about transposes can be reinterpreted in the more general language of adjoints. First, applying the adjoint operation twice returns you to where you began; this is an immediate consequence of the defining equation (7.74).

Proposition 7.57 The adjoint of the adjoint of L is just $L = (L^*)^*$.

The next result generalizes the fact, (1.53), that the transpose of the product of two matrices is the product of the transposes, in the reverse order.

Proposition 7.58 If $L\colon U \to V$ and $M\colon V \to W$ have respective adjoints $L^*\colon V \to U$ and $M^*\colon W \to V$, then the composite linear function $M \circ L\colon U \to W$ has adjoint $(M \circ L)^* = L^* \circ M^*$, which maps W to U.

Proof Let $\langle \mathbf{u}, \widetilde{\mathbf{u}} \rangle$, $\langle\!\langle \mathbf{v}, \widetilde{\mathbf{v}} \rangle\!\rangle$, $\langle\!\langle\!\langle \mathbf{w}, \widetilde{\mathbf{w}} \rangle\!\rangle\!\rangle$, denote, respectively, the inner products on U, V, W. For $\mathbf{u} \in U$, $\mathbf{w} \in W$, we compute using the definition (7.74) repeatedly:

$$\begin{aligned} \langle \mathbf{u}, (M \circ L)^*[\mathbf{w}] \rangle &= \langle\!\langle\!\langle M \circ L[\mathbf{u}], \mathbf{w} \rangle\!\rangle\!\rangle \\ &= \langle\!\langle\!\langle M[L[\mathbf{u}]], \mathbf{w} \rangle\!\rangle\!\rangle \\ &= \langle\!\langle L[\mathbf{u}], M^*[\mathbf{w}] \rangle\!\rangle \\ &= \langle \mathbf{u}, L^*[M^*[\mathbf{w}]] \rangle \\ &= \langle \mathbf{u}, L^* \circ M^*[\mathbf{w}] \rangle. \end{aligned}$$

Since this holds for all $\mathbf{u}$ and $\mathbf{w}$, the identification follows. ■

In this chapter, we have only been able to actually compute adjoints in the finite-dimensional situation, when the linear functions are given by matrix multiplication. The more challenging case of adjoints of linear operators on function spaces, e.g., differential operators appearing in boundary value problems, will be the focus of Section 11.3.

EXERCISES 7.5

7.5.1. Choose one from the following list of inner products on $\mathbb{R}^2$. Then find the adjoint of

$$A = \begin{pmatrix} 1 & 2 \\ -1 & 3 \end{pmatrix}$$

when your inner product is used on both its domain and target space.

(a) the Euclidean dot product

(b) the weighted inner product $\langle \mathbf{v}, \mathbf{w} \rangle = 2\, v_1 w_1 + 3\, v_1 w_1$

(c) the inner product $\langle \mathbf{v}, \mathbf{w} \rangle = \mathbf{v}^T K \mathbf{w}$ defined by the positive definite matrix

$$K = \begin{pmatrix} 2 & -1 \\ -1 & 4 \end{pmatrix}$$

7.5.2. From the list in Exercise 7.5.1, choose different inner products on the domain and target space, and then determine the adjoint of the matrix A.

7.5.3. Choose one from the following list of inner products on $\mathbb{R}^3$ for both the domain and target space, and find the adjoint of

$$A = \begin{pmatrix} 1 & 1 & 0 \\ -1 & 0 & 1 \\ 0 & -1 & 2 \end{pmatrix}:$$

(a) the Euclidean dot product on $\mathbb{R}^3$

(b) the weighted inner product $\langle \mathbf{v}, \mathbf{w} \rangle = v_1 w_1 + 2\, v_2 w_2 + 3\, v_3 w_3$

(c) the inner product $\langle \mathbf{v}, \mathbf{w} \rangle = \mathbf{v}^T K \mathbf{w}$ defined by the positive definite matrix

$$K = \begin{pmatrix} 2 & 1 & 0 \\ 1 & 2 & 1 \\ 0 & 1 & 2 \end{pmatrix}$$

7.5.4. From the list in Exercise 7.5.3, choose different inner products on the domain and target space, and then compute the adjoint of the matrix A.

7.5.5. Choose an inner product on $\mathbb{R}^2$ from the list in Exercise 7.5.1, and an inner product on $\mathbb{R}^3$ from the list in Exercise 7.5.3, and then compute the adjoint of

$$A = \begin{pmatrix} 1 & 3 \\ 0 & 2 \\ -1 & 1 \end{pmatrix}.$$

7.5.6. Let $\mathcal{P}^{(2)}$ be the space of quadratic polynomials equipped with the inner product

$$\langle p, q \rangle = \int_0^1 p(x)\, q(x)\, dx.$$

Find the adjoint of the derivative operator $D[\,p\,] = p'$ acting on $\mathcal{P}^{(2)}$.

◇ **7.5.7.** Prove that, if it exists, the adjoint of a linear function is uniquely determined by (7.74).

◇ **7.5.8.** Prove that

(a) $(L + M)^* = L^* + M^*$

(b) $(c\,L)^* = c\,L^*$ for $c \in \mathbb{R}$

(c) $(L^*)^* = L$

(d) $(L^{-1})^* = (L^*)^{-1}$

Self–Adjoint and Positive Definite Linear Functions

Throughout this section, U will be a fixed inner product space. We will show how to generalize the notions of symmetric and positive definite matrices to linear operators on U in a natural fashion. First, we define the analog of a symmetric matrix.

Definition 7.59 A linear function $K: U \to U$ is called *self-adjoint* if $K^* = K$. A self-adjoint linear function is *positive definite*, written $K > 0$, if

$$\langle \mathbf{u}, K[\mathbf{u}] \rangle > 0 \quad \text{for all} \quad \mathbf{0} \neq \mathbf{u} \in U. \tag{7.77}$$

In particular, if $K > 0$ then $\ker K = \{\mathbf{0}\}$. (Why?) Thus, a positive definite linear system $K[\mathbf{u}] = \mathbf{f}$ with $\mathbf{f} \in \operatorname{rng} K$ must have a unique solution. The next result generalizes our basic observation that the Gram matrices $K = A^T A$ and $A^T C A$, cf. (3.54), (3.56), are symmetric and positive (semi-)definite.

THEOREM 7.60 Let $L: U \to V$ be a linear map between inner product spaces with adjoint $L^*: V \to U$. Then the composite map $K = L^* \circ L: U \to U$ is self-adjoint. Moreover, K is positive definite if and only if $\ker L = \{\mathbf{0}\}$.

Proof First, by Propositions 7.58 and 7.57,

$$K^* = (L^* \circ L)^* = L^* \circ (L^*)^* = L^* \circ L = K,$$

proving self-adjointness. Furthermore, for $\mathbf{v} \in U$, the inner product

$$\langle \mathbf{v}, K[\mathbf{v}] \rangle = \langle \mathbf{v}, L^*[L[\mathbf{v}]] \rangle = \langle L[\mathbf{v}], L[\mathbf{v}] \rangle = \| L[\mathbf{v}] \|^2 > 0$$

is strictly positive, provided that $L[\mathbf{v}] \neq \mathbf{0}$. Thus, if $\ker L = \{\mathbf{0}\}$, then the positivity condition (7.77) holds, and conversely. ■

Let us specialize to the case of a linear function $L: \mathbb{R}^n \to \mathbb{R}^m$ that is represented by the $m \times n$ matrix A. When the Euclidean dot product is used on the two spaces, the adjoint L^* is represented by the transpose A^T, and hence the map $K = L^* \circ L$ has matrix representation $A^T A$. Therefore, in this case Theorem 7.60 reduces to our earlier Proposition 3.30 governing the positive definiteness of the Gram matrix product $A^T A$. If we change the inner product on the target space to $\langle\langle \mathbf{w}, \widetilde{\mathbf{w}} \rangle\rangle = \mathbf{w}^T C\, \widetilde{\mathbf{w}}$, then L^* is represented by $A^T C$, and hence $K = L^* \circ L$ has matrix form

$A^T C A$, which is the general symmetric, positive definite Gram matrix constructed in (3.56) that underlay our development of the equations of equilibrium in Chapter 6. Finally, if we replace the dot product on the domain space $\mathbb{R}^n$ by the alternative inner product $\langle \mathbf{v}, \widetilde{\mathbf{v}} \rangle = \mathbf{v}^T M \widetilde{\mathbf{v}}$, then, according to (7.76), the adjoint of L has matrix form

$$A^* = M^{-1} A^T C, \quad \text{and therefore} \quad K = A^* A = M^{-1} A^T C A \tag{7.78}$$

is a self-adjoint, positive (semi-)definite matrix with respect to the weighted inner product on $\mathbb{R}^n$ prescribed by the positive definite matrix M. In this case, the positive definite, self-adjoint operator K is *no longer represented by a symmetric matrix*. So, we did not quite tell the truth when we said we would only allow symmetric matrices to be positive definite—we really meant only self-adjoint matrices.

General self-adjoint matrices will be important in our discussion of the vibrations of mass-spring chains that have unequal masses. Extensions of these constructions to differential operators underlies the analysis of the boundary value problems of continuum mechanics, to be studied in Chapter 11.

EXERCISES 7.5

7.5.9. Show that the following linear transformations of $\mathbb{R}^2$ are self-adjoint with respect to the Euclidean dot product:
(a) rotation through the angle $\theta = \pi$;
(b) reflection about the line $y = x$.
(c) The scaling map $S[\mathbf{x}] = 3\mathbf{x}$;
(d) orthogonal projection onto the line $y = x$.

◇ **7.5.10.** Let M be a positive definite matrix. Show that $A\colon \mathbb{R}^n \to \mathbb{R}^n$ is self-adjoint with respect to the inner product $\langle \mathbf{v}, \mathbf{w} \rangle = \mathbf{v}^T M \mathbf{w}$ if and only if $M A$ is a symmetric matrix.

7.5.11. Prove that

$$A = \begin{pmatrix} 6 & 3 \\ 2 & 4 \end{pmatrix}$$

is self-adjoint with respect to the weighted inner product $\langle \mathbf{v}, \mathbf{w} \rangle = 2 v_1 w_1 + 3 v_2 w_2$. *Hint*: Use the criterion in Exercise 7.5.10.

7.5.12. Consider the weighted inner product $\langle \mathbf{v}, \mathbf{w} \rangle = v_1 w_1 + \frac{1}{2} v_2 w_2 + \frac{1}{3} v_3 w_3$ on $\mathbb{R}^3$.
(a) What are the conditions on the entries of a 3×3 matrix A in order that it be self-adjoint? *Hint*: Use the criterion in Exercise 7.5.10.
(b) Write down an example of a non-diagonal self-adjoint matrix.

7.5.13. Answer Exercise 7.5.12 for the inner product based on

$$\begin{pmatrix} 2 & -1 & 0 \\ -1 & 2 & -1 \\ 0 & -1 & 2 \end{pmatrix}.$$

7.5.14. *True or false*: The identity transformation is self-adjoint for any inner product on the underlying vector space.

7.5.15. *True or false*: A diagonal matrix is self-adjoint for any inner product on $\mathbb{R}^n$.

7.5.16. Suppose $L\colon U \to U$ has an adjoint $L^*\colon U \to U$.
(a) Show that $L + L^*$ is self-adjoint.
(b) Show that $L \circ L^*$ is self-adjoint.

◇ **7.5.17.** (a) Suppose $K, M\colon U \to U$ are self-adjoint linear functions on an inner product space U. Prove that $\langle K[\mathbf{u}], \mathbf{u} \rangle = \langle M[\mathbf{u}], \mathbf{u} \rangle$ for all $\mathbf{u} \in U$ if and only if $K = M$.
(b) Explain why this result is false if the self-adjointness hypothesis is dropped.

7.5.18. Prove that if $L\colon U \to U$ is an invertible linear transformation on an inner product space U, then the following three statements are equivalent:
(a) $\langle L[\mathbf{u}], L[\mathbf{v}] \rangle = \langle \mathbf{u}, \mathbf{v} \rangle$ for all $\mathbf{u}, \mathbf{v} \in U$
(b) $\| L[\mathbf{u}] \| = \| \mathbf{u} \|$ for all $\mathbf{u} \in U$
(c) $L^* = L^{-1}$. *Hint*: Use Exercise 7.5.17

7.5.19. (a) Prove that the operation $M_a[u(x)] = a(x)\, u(x)$ of multiplication by a fixed continuous function $a(x)$ defines a self-adjoint linear operator on the function space $C^0[a, b]$ with respect to the L^2 inner product.
(b) Is M_a also self-adjoint with respect to the weighted inner product

$$\langle\langle f, g \rangle\rangle = \int_a^b f(x)\, g(x)\, w(x)\, dx?$$

♡ **7.5.20.** A linear transformation $S\colon U \to U$ is called *skew-adjoint* if $S^* = -S$.
(a) Prove that a skew-symmetric matrix is skew-adjoint with respect to the standard dot product on $\mathbb{R}^n$.

(b) Under what conditions is $S[\mathbf{x}] = A\mathbf{x}$ skew-adjoint with respect to the inner product $\langle \mathbf{x}, \mathbf{y} \rangle = \mathbf{x}^T M \mathbf{y}$ on $\mathbb{R}^n$?

(c) Let $L : U \to U$ have an adjoint L^*. Prove that $L - L^*$ is skew-adjoint.

(d) Explain why every linear operator $L : U \to U$ that has an adjoint L^* can be written as the sum of a self-adjoint and a skew-adjoint operator.

◇ **7.5.21.** Let $L_1 : U \to V_1$ and $L_2 : U \to V_2$ be linear maps between inner product spaces, with V_1, V_2 not necessarily the same. Let $K_1 = L_1^* \circ L_1$, $K_2 = L_2^* \circ L_2$. Show that the sum $K = K_1 + K_2$ can be written as a self-adjoint combination $K = L^* \circ L$ for some linear operator L. *Hint*: See Exercise 3.4.35 for the matrix case.

Minimization

In Chapter 4, we learned how the solution to a linear algebraic system $K\mathbf{u} = \mathbf{f}$ with positive definite coefficient matrix K can be characterized as the unique minimizer for the quadratic function $p(\mathbf{u}) = \frac{1}{2}\mathbf{u}^T K \mathbf{u} - \mathbf{u}^T \mathbf{f}$. There is an analogous minimization principle that characterizes the solutions to linear systems defined by positive definite linear operators. This general result is of tremendous importance in analysis of boundary value problems for differential equations, for both physical and mathematical reasons, and also inspires the finite element numerical solution algorithm. Details will appear in Chapter 11.

We restrict our attention to real linear operators on real vector spaces in this section.

THEOREM 7.61 Let $K : U \to U$ be a positive definite operator on a real inner product space U. If $\mathbf{f} \in \operatorname{rng} K$, then the quadratic function

$$p(\mathbf{u}) = \tfrac{1}{2}\langle \mathbf{u}, K[\mathbf{u}] \rangle - \langle \mathbf{u}, \mathbf{f} \rangle \tag{7.79}$$

has a unique minimizer $\mathbf{u} = \mathbf{u}^\star$, which is the solution to the linear system $K[\mathbf{u}] = \mathbf{f}$.

Proof The proof mimics that of its matrix counterpart in Theorem 4.1. Our assumption that $\mathbf{f} \in \operatorname{rng} K$ implies that there is a $\mathbf{u}^\star \in U$ such that $K[\mathbf{u}^\star] = \mathbf{f}$. Thus, we can write

$$\begin{aligned} p(\mathbf{u}) &= \tfrac{1}{2}\langle \mathbf{u}, K[\mathbf{u}] \rangle - \langle \mathbf{u}, K[\mathbf{u}^\star] \rangle \\ &= \tfrac{1}{2}\langle \mathbf{u} - \mathbf{u}^\star, K[\mathbf{u} - \mathbf{u}^\star] \rangle - \tfrac{1}{2}\langle \mathbf{u}^\star, K[\mathbf{u}^\star] \rangle. \end{aligned} \tag{7.80}$$

where we used linearity, along with the fact that K is self-adjoint to identify the terms $\langle \mathbf{u}, K[\mathbf{u}^\star] \rangle = \langle \mathbf{u}^\star, K[\mathbf{u}] \rangle$. Since $K > 0$, the first term on the right hand side of (7.80) is always ≥ 0; moreover it equals its minimal value 0 if and only if $\mathbf{u} = \mathbf{u}^\star$. On the other hand, the second term does not depend upon $\mathbf{u}$ at all, and hence is unaffected by variations in $\mathbf{u}$. Therefore, to minimize $p(\mathbf{u})$, we must make the first term as small as possible, which is accomplished by setting $\mathbf{u} = \mathbf{u}^\star$. ∎

REMARK: For linear functions given by matrix multiplication, positive definiteness automatically implies invertibility, and so the linear system $K\mathbf{u} = \mathbf{f}$ has a solution for every right hand side. This is not so immediate when K is a positive definite operator on an infinite-dimensional function space. Therefore, the existence of a solution or minimizer is a significant issue. And, in fact, many modern analytical existence results rely on the determination of suitable minimization principles. On the other hand, once existence is assured, uniqueness follows immediately from the positive definiteness of the operator K.

THEOREM 7.62 Suppose $L\colon U \to V$ is a linear map between inner product spaces with $\ker L = \{\mathbf{0}\}$ and adjoint map $L^*\colon V \to U$. Let $K = L^* \circ L\colon U \to U$ be the associated positive definite operator. If $\mathbf{f} \in \operatorname{rng} K$, then the quadratic function

$$p(\mathbf{u}) = \tfrac{1}{2}\,\| L[\mathbf{u}] \|^2 - \langle \mathbf{u}, \mathbf{f} \rangle \tag{7.81}$$

has a unique minimizer $\mathbf{u}^\star$, which is the solution to the linear system $K[\mathbf{u}^\star] = \mathbf{f}$.

Proof It suffices to note that the quadratic term in (7.79) can be written in the alternative form

$$\langle \mathbf{u}, K[\mathbf{u}] \rangle = \langle \mathbf{u}, L^*[L[\mathbf{u}]] \rangle = \langle\!\langle L[\mathbf{u}], L[\mathbf{u}] \rangle\!\rangle = \| L[\mathbf{u}] \|^2.$$

Thus, (7.81) reduces to the quadratic function of the form (7.79) with $K = L^* \circ L$, and so Theorem 7.62 follows directly from Theorem 7.61. ■

Warning: In (7.81), the first term $\| L[\mathbf{u}] \|^2$ is computed using the norm based on the inner product on V, while the second term $\langle \mathbf{u}, \mathbf{f} \rangle$ employs the inner product on U.

EXAMPLE 7.63 For a general positive definite matrix (7.78), the quadratic function (7.81) is computed with respect to the alternative inner product $\langle \mathbf{u}, \widetilde{\mathbf{u}} \rangle = \mathbf{u}^T M \widetilde{\mathbf{u}}$, so

$$\begin{aligned} p(\mathbf{u}) &= \tfrac{1}{2}\,\| A\mathbf{u} \|^2 - \langle \mathbf{u}, \mathbf{f} \rangle = \tfrac{1}{2}\,(A\mathbf{u})^T C A\mathbf{u} - \mathbf{u}^T M \mathbf{f} \\ &= \tfrac{1}{2}\,\mathbf{u}^T (A^T C A)\mathbf{u} - \mathbf{u}^T (M\mathbf{f}). \end{aligned}$$

Theorem 7.62 tells us that the minimizer of the quadratic function is the solution to

$$A^T C A\mathbf{u} = M\mathbf{f}, \qquad \text{which we rewrite as} \qquad K\mathbf{u} = M^{-1} A^T C A\mathbf{u} = \mathbf{f}.$$

This conclusion also follows from our earlier finite-dimensional minimization Theorem 4.1. ●

This section is really a preview of things to come, since the full implications will require us to acquire more analytical expertise. In Chapter 11, we will find that the most important minimization principles that characterize solutions to the linear boundary value problems of physics and engineering all arise through this remarkably general, mathematical construction.

EXERCISES 7.5

7.5.22. Find the minimum value of the quadratic function

$$p(\mathbf{u}) = \frac{1}{2}\,\mathbf{u}^T \begin{pmatrix} 3 & -2 \\ -2 & 3 \end{pmatrix} \mathbf{u} - \mathbf{u}^T \begin{pmatrix} 1 \\ -1 \end{pmatrix}$$

for $\mathbf{u} \in \mathbb{R}^2$.

7.5.23. Minimize the quadratic function

$$p(\mathbf{u}) = \frac{1}{2}\,\mathbf{u}^T \begin{pmatrix} 2 & -1 & 0 \\ -1 & 4 & -2 \\ 0 & -2 & 3 \end{pmatrix} \mathbf{u} - \mathbf{u}^T \begin{pmatrix} 2 \\ 0 \\ -1 \end{pmatrix}$$

for $\mathbf{u} \in \mathbb{R}^3$.

7.5.24. Minimize $\| (2x - y, x + y)^T \|^2 - 6x$ over all x, y, where $\| \cdot \|$ denotes the Euclidean norm on $\mathbb{R}^2$.

7.5.25. Answer Exercise 7.5.24 for

(a) the weighted norm $\| (x, y)^T \| = \sqrt{2x^2 + 3y^2}$;

(b) the norm based on $\begin{pmatrix} 2 & -1 \\ -1 & 1 \end{pmatrix}$;

(c) the norm based on $\begin{pmatrix} 3 & 1 \\ 1 & 3 \end{pmatrix}$.

7.5.26. Let

$$L(x, y) = \begin{pmatrix} x - 2y \\ x + y \\ -x + 3y \end{pmatrix} \quad \text{and} \quad \mathbf{f} = \begin{pmatrix} 1 \\ 0 \end{pmatrix}.$$

Minimize $p(\mathbf{x}) = \frac{1}{2}\,\| L[\mathbf{x}] \|^2 - \langle \mathbf{x}, \mathbf{f} \rangle$ using

(a) the Euclidean inner products and norms on both $\mathbb{R}^2$ and $\mathbb{R}^3$

(b) the Euclidean inner product on $\mathbb{R}^2$ and the weighted norm $\|\mathbf{w}\| = \sqrt{w_1^2 + 2\,w_2^2 + 3\,w_3^2}$ on $\mathbb{R}^3$

(c) the inner product given by

$$\begin{pmatrix} 2 & -1 \\ -1 & 2 \end{pmatrix}$$

on $\mathbb{R}^2$ and the Euclidean norm on $\mathbb{R}^3$

(d) the inner product given by

$$\begin{pmatrix} 2 & -1 \\ -1 & 2 \end{pmatrix}$$

on $\mathbb{R}^2$ and the weighted norm

$$\|\mathbf{w}\| = \sqrt{w_1^2 + 2\,w_2^2 + 3\,w_3^2}$$

on $\mathbb{R}^3$

7.5.27. Find the minimum distance between the point $(1, 0, 0)^T$ and the plane $x+y-z = 0$ when distance is measured in

(a) the Euclidean norm

(b) the weighted norm $\|\mathbf{w}\| = \sqrt{w_1^2 + 2\,w_2^2 + 3\,w_3^2}$

(c) the norm based on the positive definite matrix

$$\begin{pmatrix} 3 & -1 & 1 \\ -1 & 2 & -1 \\ 1 & -1 & 3 \end{pmatrix}$$

◇ **7.5.28.** How would you modify the statement of Theorem 7.62 if $\ker L \neq \{\mathbf{0}\}$?

CHAPTER 8

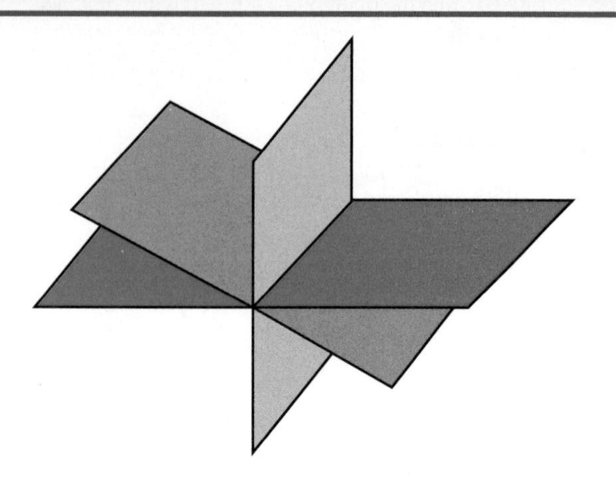

Eigenvalues

So far, our physical applications of linear algebra have concentrated on statics: unchanging equilibrium configurations of mass–spring chains, circuits, and structures, all modeled by linear systems of algebraic equations. It is now time to set the universe in motion. In general, a *dynamical system* refers to the (differential) equations governing the time-varying motion of a physical system, be it mechanical, electrical, chemical, fluid, thermodynamical, biological, financial, ... Our immediate goal is to solve the simplest class of dynamical models, which are first order autonomous linear systems of ordinary differential equations.

We begin with a very quick review of the scalar case, whose solutions are simple exponential functions. This inspires us to try to solve a vector-valued linear system by substituting a similar exponential solution formula. We are immediately led to the system of algebraic equations that define the eigenvalues and eigenvectors of the coefficient matrix. Thus, before we can make any progress in our study of differential equations, we need to learn about eigenvalues and eigenvectors, and that is the purpose of the present chapter. Dynamical systems are used to motivate the subject, but serious applications will be deferred until Chapter 9. Additional applications of eigenvalues and eigenvectors to iterative systems, stochastic processes, and numerical solution algorithms for linear algebraic systems form the focus of Chapter 10.

Each square matrix has a collection of one or more complex scalars, called eigenvalues, and associated vectors, called eigenvectors. From a geometrical viewpoint, the matrix defines a linear transformation on Euclidean space; the eigenvectors indicate the directions of pure stretch and the eigenvalues the extent of stretching. Most matrices are complete, meaning that their (complex) eigenvectors form a basis of the underlying vector space. When expressed in terms of its eigenvector basis, the matrix representing the linear transformation assumes a very simple diagonal form, facilitating the detailed analysis of its properties. A particularly important class are the symmetric matrices, whose eigenvectors form an orthogonal basis of $\mathbb{R}^n$; in fact, this is the most common way for orthogonal bases to appear. Incomplete matrices are trickier, and we relegate their non-diagonal Schur decomposition and Jordan canonical form to the final section.

A non-square matrix A does not have eigenvalues. In their place, one uses the square roots of the eigenvalues of the associated square Gram matrix $K = A^T A$, which are called singular values of the original matrix. Singular values underlie the method of principal component analysis, which appears in an increasingly broad range of modern applications, including statistics, data mining, image processing, semantics, language and speech recognition, and learning theory, to name a few. The corresponding singular value decomposition (SVD) supplies the final details for our understanding of the remarkable geometric structure governing matrix multiplication. The singular value decomposition is used to define the pseudoinverse of a matrix, which provides a mechanism for "inverting" non-square and singular matrices, and an alternative construction of least squares solutions to general linear systems.

REMARK: The numerical computation of eigenvalues and eigenvectors is a challenging issue, and must be be deferred until Section 10.6. Unless you are prepared to consult that section in advance, solving the computer-based problems in this chapter will require access to computer software that can accurately compute eigenvalues and eigenvectors.

8.1 SIMPLE DYNAMICAL SYSTEMS

Our new goal is to solve and analyze the simplest class of dynamical systems, namely those modeled by first order linear systems of ordinary differential equations. We begin with a thorough review of the scalar case, including a complete investigation into the stability of their equilibria—in preparation for the general situation to be treated in depth in Chapter 9. Readers who are not interested in such motivational material may skip ahead to Section 8.2 without incurring any penalty.

Scalar Ordinary Differential Equations

Consider the elementary scalar ordinary differential equation

$$\frac{du}{dt} = a\,u. \tag{8.1}$$

Here $a \in \mathbb{R}$ is a real constant, while the unknown $u(t)$ is a scalar function. As you learned in first year calculus, [6, 19, 58], the general solution to (8.1) is an exponential function

$$u(t) = c\,e^{a\,t}. \tag{8.2}$$

The integration constant c is uniquely determined by a single *initial condition*

$$u(t_0) = b \tag{8.3}$$

imposed at an initial time t_0. Substituting $t = t_0$ into the solution formula (8.2),

$$u(t_0) = c\,e^{a\,t_0} = b, \qquad \text{and so} \qquad c = b\,e^{-a\,t_0}.$$

We conclude that

$$u(t) = b\,e^{a(t-t_0)} \tag{8.4}$$

is the unique solution to the scalar initial value problem (8.1, 3).

EXAMPLE 8.1 The radioactive decay of an isotope, say uranium–238, is governed by the differential equation

$$\frac{du}{dt} = -\gamma\, u. \tag{8.5}$$

Here $u(t)$ denotes the amount of the isotope remaining at time t; the coefficient $\gamma > 0$ governs the decay rate. The solution is an exponentially decaying function $u(t) = c\, e^{-\gamma t}$, where $c = u(0)$ is the initial amount of radioactive material.

The isotope's *half-life* $t_\star$ is the time it takes for half of a sample to decay, that is when $u(t_\star) = \frac{1}{2} u(0)$. To determine $t_\star$, we solve the algebraic equation

$$e^{-\gamma t_\star} = \frac{1}{2}, \qquad \text{so that} \qquad t_\star = \frac{\log 2}{\gamma}. \tag{8.6}$$

●

Returning to the general situation, let us make some elementary, but pertinent, observations about this simplest linear dynamical system. First of all, since the equation is homogeneous, the zero function $u(t) \equiv 0$ (corresponding to $c = 0$ in the solution (8.2)) is a constant solution, known as an *equilibrium solution*, or *fixed point*, since it does not depend on t. If the coefficient $a > 0$ is positive, then the solutions (8.2) are exponentially growing (in absolute value) as $t \to +\infty$. This implies that the zero equilibrium solution is *unstable*. The initial condition $u(t_0) = 0$ produces the zero solution, but if we make a tiny error (either physical, numerical, or mathematical) in the initial data, say $u(t_0) = \varepsilon$, then the solution $u(t) = \varepsilon\, e^{a(t-t_0)}$ will eventually be far away from equilibrium. More generally, any two solutions with very close, but not equal, initial data will eventually become arbitrarily far apart: $|u_1(t) - u_2(t)| \to \infty$ as $t \to \infty$. One consequence is an inherent difficulty in accurately computing the long time behavior of the solution, since small numerical errors may eventually have very large effects.

On the other hand, if $a < 0$, the solutions are exponentially decaying in time. In this case, the zero solution is *stable*, since a small change in the initial data will have a negligible effect on the solution. In fact, the zero solution is *globally asymptotically stable*. The phrase "asymptotically stable" implies that solutions that start out near equilibrium eventually return; more specifically, if $u(t_0) = \varepsilon$ is small, then $u(t) \to 0$ as $t \to \infty$. "Globally" implies that all solutions, no matter how large the initial data, return to equilibrium. In fact, for a linear system, the stability of an equilibrium solution is inevitably a global phenomenon.

The borderline case is when $a = 0$. Then all the solutions to (8.1) are constant. In this case, the zero solution is *stable*—indeed, *globally stable*—but not asymptotically stable. The solution to the initial value problem $u(t_0) = \varepsilon$ is $u(t) \equiv \varepsilon$. Therefore, a solution that starts out near equilibrium will remain near, but will not asymptotically return. The three qualitatively different possibilities are illustrated in Figure 8.1.

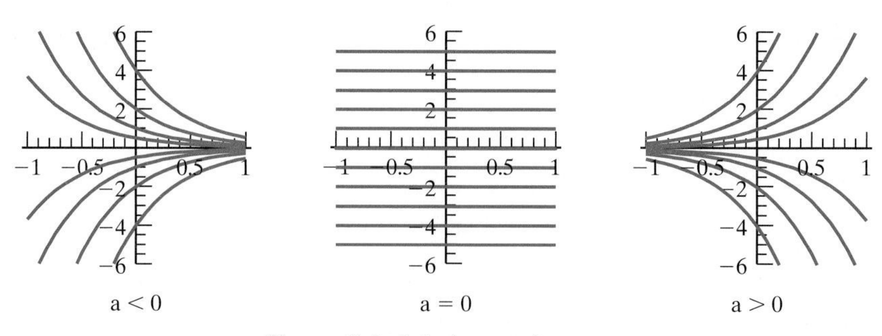

Figure 8.1 Solutions to $\dot{u} = a\, u$.

EXERCISES 8.1

8.1.1. Solve the following initial value problems:

(a) $\dfrac{du}{dt} = 5u, \ u(0) = -3,$

(b) $\dfrac{du}{dt} = 2u, \ u(1) = 3,$

(c) $\dfrac{du}{dt} = -3u, \ u(-1) = 1.$

8.1.2. Suppose a radioactive material has a half-life of 100 years. What is the decay rate γ? Starting with an initial sample of 100 grams, how much will be left after 10 years? 100 years? 1,000 years?

8.1.3. Carbon-14 has a half-life of 5730 years. Human skeletal fragments discovered in a cave are analyzed and found to have only 6.24% of the carbon-14 that living tissue would have. How old are the remains?

8.1.4. Prove that if $t_\star$ is the half-life of a radioactive material, then $u(n t_\star) = 2^{-n} u(0)$. Explain the meaning of this equation in your own words.

8.1.5. A bacteria colony grows according to the equation $du/dt = 1.3\, u$. How long until the colony doubles? quadruples? If the initial population is 2, how long until the population reaches 2 million?

8.1.6. Deer in Northern Minnesota reproduce according to the linear differential equation

$$\frac{du}{dt} = .27\, u$$

where t is measured in years. If the initial population is $u(0) = 5{,}000$ and the environment can sustain at most 1,000,000 deer, how long until the deer run out of resources?

◇ **8.1.7.** Consider the inhomogeneous differential equation

$$\frac{du}{dt} = a\, u + b,$$

where a, b are constants.

(a) Show that $u_\star = -b/a$ is a constant equilibrium solution.

(b) Solve the differential equation. *Hint*: Look at the differential equation satisfied by $v = u - u_\star$.

(c) Discuss the stability of the equilibrium solution $u_\star$.

8.1.8. Use the method of Exercise 8.1.7 to solve the following initial value problems:

(a) $\dfrac{du}{dt} = 2u - 1, \ u(0) = 1,$

(b) $\dfrac{du}{dt} = 5u + 15, \ u(1) = -3,$

(c) $\dfrac{du}{dt} = -3u + 6, \ u(2) = -1.$

8.1.9. The radioactive waste from a nuclear reactor has a half-life of 1000 years. Waste is continually produced at the rate of 5 tons per year and stored in a dump site.

(a) Set up an inhomogeneous differential equation, of the form in Exercise 8.1.7, to model the amount of radioactive waste.

(b) Determine whether the amount of radioactive material at the dump increases indefinitely, decreases to zero, or eventually stabilizes at some fixed amount.

(c) Starting with a brand new site, how long will it be until the dump contains 100 tons of radioactive material?

♡ **8.1.10.** Suppose that hunters are allowed to shoot a fixed number of the Northern Minnesota deer in Exercise 8.1.6 each year.

(a) Explain why the population model takes the form

$$\frac{du}{dt} = .27\, u - b,$$

where b is the number killed yearly. (Ignore the seasonal aspects of hunting.)

(b) If $b = 1{,}000$, how long until the deer run out of resources? *Hint*: See Exercise 8.1.7.

(c) What is the maximal rate at which deer can be hunted without causing their extinction?

◇ **8.1.11.** (a) Prove that if $u_1(t)$ and $u_2(t)$ are any two distinct solutions to

$$\frac{du}{dt} = a\, u$$

with $a > 0$, then $|u_1(t) - u_2(t)| \to \infty$ as $t \to \infty$.

(b) If $a = .02$ and $u_1(0) = .1$, $u_2(0) = .05$, how long do you have to wait until $|u_1(t) - u_2(t)| > 1{,}000$?

8.1.12. (a) Write down the exact solution to the initial value problem $du/dt = \frac{2}{7}\, u,\ u(0) = \frac{1}{3}$.

(b) Suppose you make the approximation $u(0) = .3333$. At what point does your solution differ from the true solution by 1 unit? by 1000?

(c) Answer the same question if you also approximate the coefficient in the differential equation by

$$\frac{du}{dt} = .2857\, u.$$

◇ **8.1.13.** Let a be complex. Prove that $u(t) = c\, e^{a t}$ is the (complex) solution to our scalar ordinary differential equation (8.1). Describe the asymptotic behavior of the solution as $t \to \infty$, and the stability properties of the zero equilibrium solution.

First Order Dynamical Systems

The simplest class of *dynamical systems* consists of n first order ordinary differential equations

$$\frac{du_1}{dt} = f_1(t, u_1, \ldots, u_n), \qquad \ldots \qquad \frac{du_n}{dt} = f_n(t, u_1, \ldots, u_n).$$

The unknowns are n scalar functions $u_1(t), \ldots, u_n(t)$ depending on the scalar variable $t \in \mathbb{R}$, which we usually view as time. We will often write the system in the equivalent vector form

$$\frac{d\mathbf{u}}{dt} = \mathbf{f}(t, \mathbf{u}). \tag{8.7}$$

A vector-valued solution $\mathbf{u}(t) = (u_1(t), \ldots, u_n(t))^T$ serves to parametrize a curve in $\mathbb{R}^n$, called a *solution trajectory*. A dynamical system is called *autonomous* if the time variable t does not appear explicitly on the right hand side, and so has the form

$$\frac{d\mathbf{u}}{dt} = \mathbf{f}(\mathbf{u}). \tag{8.8}$$

Dynamical systems of ordinary differential equations appear in an astonishing variety of applications, including physics, chemistry, astronomy, biology, economics, etc., and have been the focus of intense research activity since the very first days of the calculus.

In this text, we shall concentrate most of our attention on the very simplest case: a homogeneous, linear, autonomous dynamical system

$$\frac{d\mathbf{u}}{dt} = A\mathbf{u}, \tag{8.9}$$

in which A is a constant $n \times n$ matrix. In full detail, the system has the form

$$\begin{aligned} \frac{du_1}{dt} &= a_{11} u_1 + a_{12} u_2 + \cdots + a_{1n} u_n, \\ \frac{du_2}{dt} &= a_{21} u_1 + a_{22} u_2 + \cdots + a_{2n} u_n, \\ &\vdots \qquad\qquad\qquad \vdots \\ \frac{du_n}{dt} &= a_{n1} u_1 + a_{n2} u_2 + \cdots + a_{nn} u_n. \end{aligned} \tag{8.10}$$

In the autonomous case, which is the only type to be treated in depth here, the coefficients a_{ij} are assumed to be (real) constants. We seek not only to develop basic solution techniques, but to also to understand their qualitative and quantitative behavior.

Drawing our inspiration from the exponential solution (8.2) in the scalar case, let us investigate whether the vector system admits any solutions of a similar exponential form

$$\mathbf{u}(t) = e^{\lambda t}\, \mathbf{v}. \tag{8.11}$$

We assume that λ is a constant scalar, so $e^{\lambda t}$ is the usual scalar exponential function, while $\mathbf{v} \in \mathbb{R}^n$ is a constant vector. In other words, the components $u_i(t) = v_i\, e^{\lambda t}$ of our desired solution are assumed to be constant multiples of the *same* exponential function. Since $\mathbf{v}$ is constant, the derivative of $\mathbf{u}(t)$ is easily found:

$$\frac{d\mathbf{u}}{dt} = \frac{d}{dt}\left(e^{\lambda t}\, \mathbf{v}\right) = \lambda\, e^{\lambda t}\, \mathbf{v}.$$

REMARK: In general, if $\mathbf{v}$ is an eigenvector of A for the eigenvalue λ, then so is any nonzero scalar multiple of $\mathbf{v}$. In practice, we only distinguish linearly independent eigenvectors. Thus, in this example, we shall say "$\mathbf{v}_1 = (1, 1)^T$ is *the* eigenvector corresponding to the eigenvalue $\lambda_1 = 4$", when we really mean that the eigenvectors for $\lambda_1 = 4$ consist of all nonzero scalar multiples of $\mathbf{v}_1$.

Similarly, for the second eigenvalue $\lambda_2 = 2$, the eigenvector equation is

$$(A - 2\,\mathrm{I})\,\mathbf{v} = \begin{pmatrix} 1 & 1 \\ 1 & 1 \end{pmatrix} \begin{pmatrix} x \\ y \end{pmatrix} = \begin{pmatrix} 0 \\ 0 \end{pmatrix}.$$

The solution $(-a, a)^T = a\,(-1, 1)^T$ is the set of scalar multiples of the eigenvector $\mathbf{v}_2 = (-1, 1)^T$. Therefore, the complete list of eigenvalues and eigenvectors (up to scalar multiple) for this particular matrix is

$$\lambda_1 = 4, \quad \mathbf{v}_1 = \begin{pmatrix} 1 \\ 1 \end{pmatrix}, \qquad \lambda_2 = 2, \quad \mathbf{v}_2 = \begin{pmatrix} -1 \\ 1 \end{pmatrix}.$$

EXAMPLE 8.6 Consider the 3×3 matrix

$$A = \begin{pmatrix} 0 & -1 & -1 \\ 1 & 2 & 1 \\ 1 & 1 & 2 \end{pmatrix}.$$

Using the formula (1.86) for a 3×3 determinant, we compute the characteristic equation

$$\begin{aligned} 0 = \det(A - \lambda\,\mathrm{I}) &= \det \begin{pmatrix} -\lambda & -1 & -1 \\ 1 & 2-\lambda & 1 \\ 1 & 1 & 2-\lambda \end{pmatrix} \\ &= (-\lambda)(2-\lambda)^2 + (-1)\cdot 1 \cdot 1 + (-1)\cdot 1 \cdot 1 - \\ &\quad - 1 \cdot (2-\lambda)(-1) - 1 \cdot 1 \cdot (-\lambda) - (2-\lambda)\cdot 1 \cdot (-1) \\ &= -\lambda^3 + 4\lambda^2 - 5\lambda + 2. \end{aligned}$$

The resulting cubic polynomial can be factored:

$$-\lambda^3 + 4\lambda^2 - 5\lambda + 2 = -(\lambda - 1)^2(\lambda - 2) = 0.$$

Most 3×3 matrices have three different eigenvalues, but this particular one has only two: $\lambda_1 = 1$, which is called a *double eigenvalue* since it is a double root of the characteristic equation, along with a *simple eigenvalue* $\lambda_2 = 2$.

The eigenvector equation (8.13) for the double eigenvalue $\lambda_1 = 1$ is

$$(A - \mathrm{I})\mathbf{v} = \begin{pmatrix} -1 & -1 & -1 \\ 1 & 1 & 1 \\ 1 & 1 & 1 \end{pmatrix} \begin{pmatrix} x \\ y \\ z \end{pmatrix} = \begin{pmatrix} 0 \\ 0 \\ 0 \end{pmatrix}.$$

The general solution to this homogeneous linear system

$$\mathbf{v} = \begin{pmatrix} -a-b \\ a \\ b \end{pmatrix} = a \begin{pmatrix} -1 \\ 1 \\ 0 \end{pmatrix} + b \begin{pmatrix} -1 \\ 0 \\ 1 \end{pmatrix}$$

depends upon two free variables: $y = a$ and $z = b$. Any nonzero solution forms a valid eigenvector for the eigenvalue $\lambda_1 = 1$, and so the general eigenvector is any non-zero linear combination of the two "basis eigenvectors" $\mathbf{v}_1 = (-1, 1, 0)^T$, $\widehat{\mathbf{v}}_1 = (-1, 0, 1)^T$.

On the other hand, the eigenvector equation for the simple eigenvalue $\lambda_2 = 2$ is

$$(A - 2\,\mathrm{I})\mathbf{v} = \begin{pmatrix} -2 & -1 & -1 \\ 1 & 0 & 1 \\ 1 & 1 & 0 \end{pmatrix} \begin{pmatrix} x \\ y \\ z \end{pmatrix} = \begin{pmatrix} 0 \\ 0 \\ 0 \end{pmatrix}.$$

The general solution

$$\mathbf{v} = \begin{pmatrix} -a \\ a \\ a \end{pmatrix} = a \begin{pmatrix} -1 \\ 1 \\ 1 \end{pmatrix}$$

consists of all scalar multiples of the eigenvector $\mathbf{v}_2 = (-1, 1, 1)^T$.

In summary, the eigenvalues and (basis) eigenvectors for this matrix are

$$\begin{aligned} \lambda_1 &= 1, \quad \mathbf{v}_1 = \begin{pmatrix} -1 \\ 1 \\ 0 \end{pmatrix}, \quad \widehat{\mathbf{v}}_1 = \begin{pmatrix} -1 \\ 0 \\ 1 \end{pmatrix}, \\ \lambda_2 &= 2, \quad \mathbf{v}_2 = \begin{pmatrix} -1 \\ 1 \\ 1 \end{pmatrix}. \end{aligned} \tag{8.15}$$

●

In general, given a real eigenvalue λ, the corresponding *eigenspace* $V_\lambda \subset \mathbb{R}^n$ is the subspace spanned by all its eigenvectors. Equivalently, the eigenspace is the kernel

$$V_\lambda = \ker(A - \lambda\,\mathrm{I}). \tag{8.16}$$

In particular, $\lambda \in \mathbb{R}$ is an eigenvalue if and only if $V_\lambda \neq \{\mathbf{0}\}$ is a nontrivial subspace, and then every nonzero element of V_λ is a corresponding eigenvector. The most economical way to indicate each eigenspace is by writing out a basis, as in (8.15) with $\mathbf{v}_1, \widehat{\mathbf{v}}_1$ giving a basis for V_1, while $\mathbf{v}_2$ is a basis for V_2.

EXAMPLE 8.7 The characteristic equation of the matrix

$$A = \begin{pmatrix} 1 & 2 & 1 \\ 1 & -1 & 1 \\ 2 & 0 & 1 \end{pmatrix}$$

is

$$0 = \det(A - \lambda\,\mathrm{I}) = -\lambda^3 + \lambda^2 + 5\lambda + 3 = -(\lambda + 1)^2(\lambda - 3).$$

Again, there is a double eigenvalue $\lambda_1 = -1$ and a simple eigenvalue $\lambda_2 = 3$. However, in this case the matrix

$$A - \lambda_1\,\mathrm{I} = A + \mathrm{I} = \begin{pmatrix} 2 & 2 & 1 \\ 1 & 0 & 1 \\ 2 & 0 & 2 \end{pmatrix}$$

has only a one-dimensional kernel, spanned by $\mathbf{v}_1 = (2, -1, -2)^T$. Thus, even though λ_1 is a double eigenvalue, it only admits a one-dimensional eigenspace. The list of eigenvalues and eigenvectors is, in a sense, incomplete:

$$\lambda_1 = -1, \quad \mathbf{v}_1 = \begin{pmatrix} 2 \\ -1 \\ -2 \end{pmatrix}, \qquad \lambda_2 = 3, \quad \mathbf{v}_2 = \begin{pmatrix} 2 \\ 1 \\ 2 \end{pmatrix}.$$

●

EXAMPLE 8.8 Finally, consider the matrix

$$A = \begin{pmatrix} 1 & 2 & 0 \\ 0 & 1 & -2 \\ 2 & 2 & -1 \end{pmatrix}.$$

The characteristic equation is

$$0 = \det(A - \lambda\, \mathrm{I}) = -\lambda^3 + \lambda^2 - 3\lambda - 5 = -\,(\lambda + 1)\,(\lambda^2 - 2\lambda + 5).$$

The linear factor yields the eigenvalue -1. The quadratic factor leads to two complex roots, $1 + 2\,\mathrm{i}$ and $1 - 2\,\mathrm{i}$, which can be obtained via the quadratic formula. Hence A has one real and two complex eigenvalues:

$$\lambda_1 = -1, \qquad \lambda_2 = 1 + 2\,\mathrm{i}, \qquad \lambda_3 = 1 - 2\,\mathrm{i}.$$

Solving the associated linear system, the real eigenvalue is found to have corresponding eigenvector $\mathbf{v}_1 = (-1, 1, 1)^T$.

Complex eigenvalues are as important as real eigenvalues, and we need to be able to handle them too. To find the corresponding eigenvectors, which will also be complex, we need to solve the usual eigenvalue equation (8.13), which is now a complex homogeneous linear system. For example, the eigenvector(s) for $\lambda_2 = 1 + 2\,\mathrm{i}$ are found by solving

$$\bigl[\,A - (1 + 2\,\mathrm{i})\,\mathrm{I}\,\bigr]\mathbf{v} = \begin{pmatrix} -2\,\mathrm{i} & 2 & 0 \\ 0 & -2\,\mathrm{i} & -2 \\ 2 & 2 & -2 - 2\,\mathrm{i} \end{pmatrix} \begin{pmatrix} x \\ y \\ z \end{pmatrix} = \begin{pmatrix} 0 \\ 0 \\ 0 \end{pmatrix}.$$

This linear system can be solved by Gaussian Elimination (with complex pivots). A simpler strategy is to work directly: the first equation $-2\,\mathrm{i}\,x + 2\,y = 0$ tells us that $y = \mathrm{i}\,x$, while the second equation $-2\,\mathrm{i}\,y - 2z = 0$ says $z = -\mathrm{i}\,y = x$. If we trust our calculations so far, we do not need to solve the final equation $2x + 2\,y + (-2 - 2\,\mathrm{i})z = 0$, since we know that the coefficient matrix is singular and hence this equation must be a consequence of the first two. (However, it does serve as a useful check on our work.) So, the general solution $\mathbf{v} = (x,\, \mathrm{i}\,x,\, x)^T$ is an arbitrary constant multiple of the complex eigenvector $\mathbf{v}_2 = (1,\, \mathrm{i}\,,\, 1)^T$. The eigenvector equation for $\lambda_3 = 1 - 2\,\mathrm{i}$ is similarly solved for the third eigenvector $\mathbf{v}_3 = (1, -\mathrm{i}\,, 1)^T$.

Summarizing, the matrix under consideration has three complex eigenvalues and three corresponding eigenvectors, each unique up to (complex) scalar multiple:

$$\lambda_1 = -1, \qquad \lambda_2 = 1 + 2\,\mathrm{i}, \qquad \lambda_3 = 1 - 2\,\mathrm{i},$$

$$\mathbf{v}_1 = \begin{pmatrix} -1 \\ 1 \\ 1 \end{pmatrix}, \qquad \mathbf{v}_2 = \begin{pmatrix} 1 \\ \mathrm{i} \\ 1 \end{pmatrix}, \qquad \mathbf{v}_3 = \begin{pmatrix} 1 \\ -\,\mathrm{i} \\ 1 \end{pmatrix}.$$

Note that the third complex eigenvalue is the complex conjugate of the second, and the eigenvectors are similarly related. This is indicative of a general fact for real matrices: ●

Proposition 8.9 If A is a real matrix with a complex eigenvalue $\lambda = \mu + \mathrm{i}\,\nu$ and corresponding complex eigenvector $\mathbf{v} = \mathbf{x} + \mathrm{i}\,\mathbf{y}$, then the complex conjugate $\overline{\lambda} = \mu - \mathrm{i}\,\nu$ is also an eigenvalue with complex conjugate eigenvector $\overline{\mathbf{v}} = \mathbf{x} - \mathrm{i}\,\mathbf{y}$.

Proof First take complex conjugates of the eigenvalue equation (8.12):

$$\overline{A}\,\overline{\mathbf{v}} = \overline{A\mathbf{v}} = \overline{\lambda\mathbf{v}} = \overline{\lambda}\,\overline{\mathbf{v}}.$$

Using the fact that a real matrix is unaffected by conjugation, so $\overline{A} = A$, we conclude $A\,\overline{\mathbf{v}} = \overline{\lambda}\,\overline{\mathbf{v}}$, which is the equation for the eigenvalue $\overline{\lambda}$ and eigenvector $\overline{\mathbf{v}}$. ■

As a consequence, when dealing with real matrices, we only need to compute the eigenvectors for *one* of each complex conjugate pair of eigenvalues. This observation effectively halves the amount of work in the unfortunate event that we are confronted with complex eigenvalues.

The *eigenspace* associated with a complex eigenvalue λ is the subspace $V_\lambda \subset \mathbb{C}^n$ spanned by the associated eigenvectors. One might also consider complex eigenvectors associated with a real eigenvalue, but this doesn't add anything to the picture— they are merely complex linear combinations of the real eigenvalues. Thus, we only introduce complex eigenvectors when dealing with genuinely complex eigenvalues.

REMARK: The reader may recall that we said one should never use determinants in practical computations. So why have we reverted to using determinants to find eigenvalues? The truthful answer is that the practical computation of eigenvalues and eigenvectors *never* resorts to the characteristic equation! The method is fraught with numerical traps and inefficiencies when

(a) computing the determinant leading to the characteristic equation, then

(b) solving the resulting polynomial equation, which is itself a nontrivial numerical problem[‡], [10, 51], and, finally,

(c) solving each of the resulting linear eigenvector systems. Worse, if we only know an approximation $\widetilde{\lambda}$ to the true eigenvalue λ, the approximate eigenvector system $(A - \widetilde{\lambda}\,)\mathbf{v} = \mathbf{0}$ will almost certainly have a nonsingular coefficient matrix, and hence only admits the trivial solution $\mathbf{v} = \mathbf{0}$—which does not even qualify as an eigenvector!

Nevertheless, the characteristic equation does give us important theoretical insight into the structure of the eigenvalues of a matrix, and can be used when dealing with small matrices, e.g., 2×2 and 3×3, presuming exact arithmetic is employed. Numerical algorithms for computing eigenvalues and eigenvectors are based on completely different ideas, and will be deferred until Section 10.6.

EXERCISES 8.2

8.2.1. Find the eigenvalues and eigenvectors of the following matrices:

(a) $\begin{pmatrix} 1 & -2 \\ -2 & 1 \end{pmatrix}$ (b) $\begin{pmatrix} 1 & -\frac{2}{3} \\ \frac{1}{2} & -\frac{1}{6} \end{pmatrix}$

(c) $\begin{pmatrix} 3 & 1 \\ -1 & 1 \end{pmatrix}$ (d) $\begin{pmatrix} 1 & 2 \\ -1 & 1 \end{pmatrix}$

(e) $\begin{pmatrix} 3 & -1 & 0 \\ -1 & 2 & -1 \\ 0 & -1 & 3 \end{pmatrix}$ (f) $\begin{pmatrix} -1 & -1 & 4 \\ 1 & 3 & -2 \\ 1 & 1 & -1 \end{pmatrix}$

(g) $\begin{pmatrix} 1 & -3 & 11 \\ 2 & -6 & 16 \\ 1 & -3 & 7 \end{pmatrix}$ (h) $\begin{pmatrix} 2 & -1 & -1 \\ -2 & 1 & 1 \\ 1 & 0 & 1 \end{pmatrix}$

(i) $\begin{pmatrix} -4 & -4 & 2 \\ 3 & 4 & -1 \\ -3 & -2 & 3 \end{pmatrix}$ (j) $\begin{pmatrix} 3 & 4 & 0 & 0 \\ 4 & 3 & 0 & 0 \\ 0 & 0 & 1 & 3 \\ 0 & 0 & 4 & 5 \end{pmatrix}$

(k) $\begin{pmatrix} 4 & 0 & 0 & 0 \\ 1 & 3 & 0 & 0 \\ -1 & 1 & 2 & 0 \\ 1 & -1 & 1 & 1 \end{pmatrix}$

[‡]In fact, one effective numerical strategy for finding the roots of a polynomial is to turn the procedure on its head, and calculate the eigenvalues of a matrix whose characteristic equation is the polynomial in question! See [51] for details.

8.2.2. (a) Find the eigenvalues of the rotation matrix

$$R_\theta = \begin{pmatrix} \cos\theta & -\sin\theta \\ \sin\theta & \cos\theta \end{pmatrix}.$$

(b) For what values of θ are the eigenvalues real?

(c) Explain why your answer gives an immediate solution to Exercise 1.5.7c.

8.2.3. Answer Exercise 8.2.2a for the reflection matrix

$$F_\theta = \begin{pmatrix} \cos\theta & \sin\theta \\ \sin\theta & -\cos\theta \end{pmatrix}.$$

8.2.4. Write down

(a) a 2×2 matrix that has 0 as one of its eigenvalues and $(1,2)^T$ as a corresponding eigenvector;

(b) a 3×3 matrix that has $(1,2,3)^T$ as an eigenvector for the eigenvalue -1.

8.2.5. (a) Write out the characteristic equation for the matrix

$$\begin{pmatrix} 0 & 1 & 0 \\ 0 & 0 & 1 \\ \alpha & \beta & \gamma \end{pmatrix}.$$

(b) Show that, given any 3 numbers a, b, and c, there is a 3×3 matrix with characteristic equation $-\lambda^3 + a\lambda^2 + b\lambda + c = 0$.

8.2.6. Find the eigenvalues and eigenvectors of the cross product matrix

$$A = \begin{pmatrix} 0 & c & -b \\ -c & 0 & a \\ b & -a & 0 \end{pmatrix}.$$

8.2.7. Find all eigenvalues and eigenvectors of the following complex matrices:

(a) $\begin{pmatrix} i & 1 \\ 0 & -1+i \end{pmatrix}$ (b) $\begin{pmatrix} 2 & i \\ -i & -2 \end{pmatrix}$

(c) $\begin{pmatrix} i-2 & i+1 \\ i+2 & i-1 \end{pmatrix}$

(d) $\begin{pmatrix} 1+i & -1-i & 1-i \\ 2 & -2-i & 2-2i \\ -1 & 1+i & -1+2i \end{pmatrix}$

8.2.8. Find all eigenvalues and eigenvectors of

(a) the $n \times n$ zero matrix O;

(b) the $n \times n$ identity matrix I.

8.2.9. Find the eigenvalues and eigenvectors of an $n \times n$ matrix with every entry equal to 1. *Hint*: Try with $n = 2, 3$, and then generalize.

◇ **8.2.10.** Let A be a given square matrix.

(a) Explain in detail why any nonzero scalar multiple of an eigenvector of A is also an eigenvector.

(b) Show that any nonzero linear combination of two eigenvectors $\mathbf{v}, \mathbf{w}$ corresponding to the *same* eigenvalue is also an eigenvector.

(c) Prove that a linear combination $c\,\mathbf{v} + d\,\mathbf{w}$, with $c, d \neq 0$, of two eigenvectors corresponding to *different* eigenvalues is never an eigenvector.

8.2.11. *True or false*: If $\mathbf{v}$ is a real eigenvector of a real matrix A, then a nonzero complex multiple $\mathbf{w} = c\,\mathbf{v}$ for $c \in \mathbb{C}$ is a complex eigenvector of A.

◇ **8.2.12.** Let λ be a real eigenvalue of the real $n \times n$ matrix A, and $\mathbf{v}_1, \ldots, \mathbf{v}_k$ a basis for the associated eigenspace V_λ. Suppose $\mathbf{w} \in \mathbb{C}^n$ is a complex eigenvector, so $A\mathbf{w} = \lambda\mathbf{w}$. Prove that $\mathbf{w} = c_1\mathbf{v}_1 + \cdots + c_k\mathbf{v}_k$ is a complex linear combination of the real eigenspace basis. *Hint*: Look at the real and imaginary parts of the eigenvector equation.

◇ **8.2.13.** Define the *shift map* $S\colon \mathbb{C}^n \to \mathbb{C}^n$ by

$$S\,(v_1, v_2, \ldots, v_{n-1}, v_n)^T = (v_2, v_3, \ldots, v_n, v_1)^T.$$

(a) Prove that S is a linear map, and write down its matrix representation A.

(b) Prove that A is an orthogonal matrix.

(c) Prove that the sampled exponential vectors $\boldsymbol{\omega}_0, \ldots, \boldsymbol{\omega}_{n-1}$ in (5.90) form an eigenvector basis of A. What are the eigenvalues?

Basic Properties of Eigenvalues

If A is an $n \times n$ matrix, then its *characteristic polynomial* is

$$p_A(\lambda) = \det(A - \lambda\,\mathrm{I}) = c_n\lambda^n + c_{n-1}\lambda^{n-1} + \cdots + c_1\lambda + c_0. \tag{8.17}$$

The fact that $p_A(\lambda)$ is a polynomial of degree n is a consequence of the general determinantal formula (1.85). Indeed, every term is prescribed by a permutation π of the rows of the matrix, and equals plus or minus a product of n distinct matrix entries including one from each row and one from each column. The term corresponding to the identity permutation is obtained by multiplying the diagonal entries together, which, in this case, is

$$\begin{aligned}(a_{11}-\lambda)(a_{22}-\lambda)\cdots(a_{nn}-\lambda) \\ = (-1)^n\lambda^n + (-1)^{n-1}\big(a_{11}+a_{22}+\cdots+a_{nn}\big)\lambda^{n-1} + \cdots.\end{aligned} \tag{8.18}$$

All of the other terms have at most $n-2$ diagonal factors $a_{ii}-\lambda$, and so are polynomials of degree $\le n-2$ in λ. Thus, (8.18) is the only summand containing the monomials λ^n and λ^{n-1}, and so their respective coefficients are

$$c_n = (-1)^n, \quad c_{n-1} = (-1)^{n-1}(a_{11} + a_{22} + \cdots + a_{nn}) = (-1)^{n-1}\operatorname{tr} A, \tag{8.19}$$

where $\operatorname{tr} A$, the sum of its diagonal entries, is called the *trace* of the matrix A. The other coefficients $c_{n-2}, \ldots, c_1, c_0$ in (8.17) are more complicated combinations of the entries of A. However, setting $\lambda = 0$ implies

$$p_A(0) = \det A = c_0,$$

and hence the constant term in the characteristic polynomial equals the determinant of the matrix. In particular, if $A = \begin{pmatrix} a & b \\ c & d \end{pmatrix}$ is a 2×2 matrix, its characteristic polynomial has the explicit form

$$\begin{aligned} p_A(\lambda) = \det(A - \lambda\, \mathrm{I}) &= \det \begin{pmatrix} a-\lambda & b \\ c & d-\lambda \end{pmatrix} \\ &= \lambda^2 - (a+d)\lambda + (ad-bc) \\ &= \lambda^2 - (\operatorname{tr} A)\lambda + (\det A). \end{aligned} \tag{8.20}$$

As a result of these considerations, the characteristic equation of an $n \times n$ matrix A is a polynomial equation of degree n. According to the Fundamental Theorem of Algebra, [22], every (complex) polynomial of degree $n \ge 1$ can be completely factored, and so we can write the characteristic polynomial in factored form:

$$p_A(\lambda) = (-1)^n(\lambda - \lambda_1)(\lambda - \lambda_2)\cdots(\lambda - \lambda_n). \tag{8.21}$$

The complex numbers $\lambda_1, \ldots, \lambda_n$, some of which may be repeated, are the *roots* of the characteristic equation $p_A(\lambda) = 0$, and hence the eigenvalues of the matrix A. Therefore, we immediately conclude:

THEOREM 8.10 An $n \times n$ matrix A has at least one and at most n distinct complex eigenvalues.

Most $n \times n$ matrices—meaning those for which the characteristic polynomial factors into n *distinct* factors—have *exactly* n complex eigenvalues. More generally, an eigenvalue λ_j is said to have *multiplicity* m if the factor $(\lambda - \lambda_j)$ appears exactly m times in the factorization (8.21) of the characteristic polynomial. An eigenvalue is *simple* if it has multiplicity 1. In particular, A has n distinct eigenvalues if and only if all its eigenvalues are simple. In all cases, when the repeated eigenvalues are counted in accordance with their multiplicity, every $n \times n$ matrix has a total of n, possibly repeated, eigenvalues.

An example of a matrix with just one eigenvalue, of multiplicity n, is the $n \times n$ identity matrix I, whose only eigenvalue is $\lambda = 1$. In this case, *every* nonzero vector in $\mathbb{R}^n$ is an eigenvector of the identity matrix, and so the eigenspace is all of $\mathbb{R}^n$. At the other extreme, the "bidiagonal" *Jordan block matrix*[§]

$$J_a = \begin{pmatrix} a & 1 & & & & \\ & a & 1 & & & \\ & & a & 1 & & \\ & & & \ddots & \ddots & \\ & & & & a & 1 \\ & & & & & a \end{pmatrix} \tag{8.22}$$

[§] All non-displayed entries are zero.

also has only one eigenvalue, $\lambda = a$, again of multiplicity n. But in this case, J_a has only one eigenvector (up to scalar multiple), which is the first standard basis vector $\mathbf{e}_1$, and so its eigenspace is one-dimensional.

REMARK: If λ is a complex eigenvalue of multiplicity k for the real matrix A, then its complex conjugate $\overline{\lambda}$ also has multiplicity k. This is because complex conjugate roots of a real polynomial necessarily appear with identical multiplicities.

REMARK: If $n \leq 4$, then one can, in fact, write down an explicit formula for the solution to a polynomial equation of degree n, and hence explicit (but not particularly helpful) formulae for the eigenvalues of general 2×2, 3×3 and 4×4 matrices. As soon as $n \geq 5$, there is no explicit formula (at least in terms of radicals), and so one must usually resort to numerical approximations. This remarkable and deep algebraic result, which is a consequence of the symmetries of the polynomial's roots, was proved by the young Norwegian mathematician Niels Henrik Abel in the early part of the nineteenth century, [22].

If we explicitly multiply out the factored product (8.21) and equate the result to the characteristic polynomial (8.17), we find that its coefficients $c_0, c_1, \ldots c_{n-1}$ can be written as certain polynomials of the roots, known as the *elementary symmetric polynomials*. The first and last are of particular importance:

$$c_0 = \lambda_1\, \lambda_2 \cdots \lambda_n, \qquad c_{n-1} = (-1)^{n-1}\,(\lambda_1 + \lambda_2 + \cdots + \lambda_n). \tag{8.23}$$

Comparison with our previous formulae for the coefficients c_0 and c_{n-1} leads to the following useful result.

Proposition 8.11 The sum of the eigenvalues of a matrix equals its trace:

$$\lambda_1 + \lambda_2 + \cdots + \lambda_n = \operatorname{tr} A = a_{11} + a_{22} + \cdots + a_{nn}. \tag{8.24}$$

The product of the eigenvalues equals its determinant:

$$\lambda_1\, \lambda_2 \cdots \lambda_n = \det A. \tag{8.25}$$

REMARK: For repeated eigenvalues, one must add or multiply them in the formulae (8.24–25) according to their multiplicity.

EXAMPLE 8.12 The matrix

$$A = \begin{pmatrix} 1 & 2 & 1 \\ 1 & -1 & 1 \\ 2 & 0 & 1 \end{pmatrix}$$

considered in Example 8.7 has trace and determinant

$$\operatorname{tr} A = 1, \qquad \det A = 3,$$

which fix, respectively, the coefficient of λ^2 and the constant term in its characteristic equation. This matrix has two distinct eigenvalues: -1, which is a double eigenvalue, and 3, which is simple. For this particular matrix, formulae (8.24–25) become

$$1 = \operatorname{tr} A = (-1) + (-1) + 3, \qquad 3 = \det A = (-1)(-1)\,3.$$

Note that the double eigenvalue contributes twice to the sum and to the product. ●

EXERCISES 8.2

8.2.14. (a) Compute the eigenvalues and corresponding eigenvectors of

$$A = \begin{pmatrix} 1 & 4 & 4 \\ 3 & -1 & 0 \\ 0 & 2 & 3 \end{pmatrix}.$$

(b) Compute the trace of A and check that it equals the sum of the eigenvalues.

(c) Find the determinant of A and check that it is equal to to the product of the eigenvalues.

8.2.15. Verify the trace and determinant formulae (8.24–25) for the matrices in Exercise 8.2.1.

8.2.16. (a) Find the explicit formula for the characteristic polynomial $\det(A - \lambda \mathrm{I}) = -\lambda^3 + a\lambda^2 - b\lambda + c$ of a general 3×3 matrix. Verify that $a = \operatorname{tr} A$, $c = \det A$. What is the formula for b?

(b) Prove that if A has eigenvalues $\lambda_1, \lambda_2, \lambda_3$, then $a = \operatorname{tr} A = \lambda_1 + \lambda_2 + \lambda_3$, $b = \lambda_1 \lambda_2 + \lambda_1 \lambda_3 + \lambda_2 \lambda_3$, $c = \det A = \lambda_1 \lambda_2 \lambda_3$.

8.2.17. Prove that the eigenvalues of an upper triangular (or lower triangular) matrix are its diagonal entries.

◇ **8.2.18.** Let J_a be the $n \times n$ Jordan block matrix (8.22). Prove that its only eigenvalue is $\lambda = a$ and the only eigenvectors are the nonzero scalar multiples of the standard basis vector $\mathbf{e}_1$.

◇ **8.2.19.** Suppose that λ is an eigenvalue of A.

(a) Prove that $c\lambda$ is an eigenvalue of the scalar multiple cA.

(b) Prove that $\lambda + d$ is an eigenvalue of $A + d\,\mathrm{I}$.

(c) More generally, $c\lambda + d$ is an eigenvalue of $B = cA + d\,\mathrm{I}$ for scalars c, d.

8.2.20. Show that if λ is an eigenvalue of A, then λ^2 is an eigenvalue of A^2.

8.2.21. *True or false*:

(a) If λ is an eigenvalue of both A and B, then it is an eigenvalue of the sum $A + B$.

(b) If $\mathbf{v}$ is an eigenvector of both A and B, then it is an eigenvector of $A + B$.

8.2.22. *True or false*: If λ is an eigenvalue of A and μ is an eigenvalue of B, then $\lambda\mu$ is an eigenvalue of the matrix product $C = AB$.

◇ **8.2.23.** Let A and B be $n \times n$ matrices. Prove that the matrix products AB and BA have the same eigenvalues. *Hint*: How should the eigenvectors be related?

◇ **8.2.24.** (a) Prove that if $\lambda \neq 0$ is a nonzero eigenvalue of A, then $1/\lambda$ is an eigenvalue of A^{-1}.

(b) What happens if A has 0 as an eigenvalue?

◇ **8.2.25.** (a) Prove that if $\det A > 1$, then A has at least one eigenvalue with $|\lambda| > 1$.

(b) If $|\det A| < 1$, are all eigenvalues $|\lambda| < 1$? Prove or find a counter example.

8.2.26. Prove that A is a singular matrix if and only if 0 is an eigenvalue.

8.2.27. Prove that *every* nonzero vector $\mathbf{0} \neq \mathbf{v} \in \mathbb{R}^n$ is an eigenvector of A if and only if A is a scalar multiple of the identity matrix.

8.2.28. How many unit eigenvectors correspond to a given eigenvalue of a matrix?

8.2.29. *True or false*:

(a) Performing an elementary row operation of type #1 does not change the eigenvalues of a matrix.

(b) Interchanging two rows of a matrix changes the sign of its eigenvalues.

(c) Multiplying one row of a matrix by a scalar multiplies one of its eigenvalues by the same scalar.

8.2.30. (a) *True or false*: If $\lambda_1, \mathbf{v}_1$ and $\lambda_2, \mathbf{v}_2$ solve the eigenvalue equation (8.12) for a given matrix A, so does $\lambda_1 + \lambda_2$, $\mathbf{v}_1 + \mathbf{v}_2$.

(b) Explain what this has to do with linearity.

8.2.31. An *elementary reflection matrix* has the form $Q = \mathrm{I} - 2\mathbf{u}\mathbf{u}^T$, where $\mathbf{u} \in \mathbb{R}^n$ is a unit vector.

(a) Find the eigenvalues and eigenvectors for the elementary reflection matrices corresponding to the following unit vectors:

(i) $\begin{pmatrix} 1 \\ 0 \end{pmatrix}$ (ii) $\begin{pmatrix} \frac{3}{5} \\ \frac{4}{5} \end{pmatrix}$

(iii) $\begin{pmatrix} 0 \\ 1 \\ 0 \end{pmatrix}$ (iv) $\begin{pmatrix} \frac{1}{\sqrt{2}} \\ 0 \\ -\frac{1}{\sqrt{2}} \end{pmatrix}$

(b) What are the eigenvalues and eigenvectors of a general elementary reflection matrix?

◇ **8.2.32.** Let A and B be similar matrices, so $B = S^{-1} A S$ for some nonsingular matrix S.

(a) Prove that A and B have the same characteristic polynomial: $p_B(\lambda) = p_A(\lambda)$.

(b) Explain why similar matrices have the same eigenvalues. Do they have the same eigenvectors? If not, how are their eigenvectors related?

(c) Prove that the converse is not true by showing that

$$\begin{pmatrix} 2 & 0 \\ 0 & 2 \end{pmatrix} \quad \text{and} \quad \begin{pmatrix} 1 & 1 \\ -1 & 3 \end{pmatrix}$$

have the same eigenvalues, but are *not* similar.

8.2.33. Let A be a nonsingular $n \times n$ matrix with characteristic polynomial $p_A(\lambda)$.

(a) Explain how to construct the characteristic polynomial $p_{A^{-1}}(\lambda)$ of its inverse directly from $p_A(\lambda)$.

(b) Check your result when $A =$

(i) $\begin{pmatrix} 1 & 2 \\ 3 & 4 \end{pmatrix}$ (ii) $\begin{pmatrix} 1 & 4 & 4 \\ -2 & -1 & 0 \\ 0 & 2 & 3 \end{pmatrix}$

♡ **8.2.34.** A square matrix A is called *nilpotent* if $A^k = \mathrm{O}$ for some $k \geq 1$.

(a) Prove that the only eigenvalue of a nilpotent matrix is 0. (The converse is also true; see Exercise 8.6.18.)

(b) Find examples where $A^{k-1} \neq \mathrm{O}$ but $A^k = \mathrm{O}$ when $k = 2, 3$, and in general.

♡ **8.2.35.** (a) Prove that every eigenvalue of a matrix A is also an eigenvalue of its transpose A^T.

(b) Do they have the same eigenvectors? Prove that if $\mathbf{v}$ is an eigenvector of A with eigenvalue λ and $\mathbf{w}$ is an eigenvector of A^T with a different eigenvalue $\mu \neq \lambda$, then $\mathbf{v}$ and $\mathbf{w}$ are orthogonal vectors with respect to the dot product.

(c) Illustrate this result when

(i) $A = \begin{pmatrix} 0 & -1 \\ 2 & 3 \end{pmatrix}$

(ii) $A = \begin{pmatrix} 5 & -4 & 2 \\ 5 & -4 & 1 \\ -2 & 2 & -3 \end{pmatrix}$

8.2.36. (a) Prove that every real 3×3 matrix has at least one real eigenvalue.

(b) Find a real 4×4 matrix with no real eigenvalues.

(c) Can you find a real 5×5 matrix with no real eigenvalues?

8.2.37. (a) Show that if A is a matrix such that $A^4 = \mathrm{I}$, then the only possible eigenvalues of A are $1, -1, \mathrm{i}$ and $-\mathrm{i}$.

(b) Give an example of a real matrix that has all four numbers as eigenvalues.

8.2.38. A *projection matrix* satisfies $P^2 = P$. Find all eigenvalues and eigenvectors of P.

8.2.39. *True or false*: All the eigenvalues of an $n \times n$ permutation matrix are real.

8.2.40. (a) Show that if all the row sums of A are equal to 1, then A has 1 as an eigenvalue.

(b) Suppose all the column sums of A are equal to 1. Does the same result hold? *Hint*: Use Exercise 8.2.35.

8.2.41. Let Q be an orthogonal matrix.

(a) Prove that if λ is an eigenvalue, then so is $1/\lambda$.

(b) Prove that all its eigenvalues are complex numbers of modulus $|\lambda| = 1$. In particular, the only possible real eigenvalues of an orthogonal matrix are ± 1.

(c) Suppose $\mathbf{v} = \mathbf{x} + \mathrm{i}\,\mathbf{y}$ is a complex eigenvector corresponding to a non-real eigenvalue. Prove that its real and imaginary parts are orthogonal vectors having the same Euclidean norm.

◇ **8.2.42.** (a) Prove that every 3×3 proper orthogonal matrix has $+1$ as an eigenvalue.

(b) *True or false*: An improper 3×3 orthogonal matrix has -1 as an eigenvalue.

◇ **8.2.43.** (a) Show that the linear transformation defined by a 3×3 proper orthogonal matrix corresponds to rotating through an angle around a line through the origin in $\mathbb{R}^3$—the axis of the rotation. *Hint*: Use Exercise 8.2.42(a).

(b) Find the axis and angle of rotation of the orthogonal matrix

$$\begin{pmatrix} \frac{3}{5} & 0 & \frac{4}{5} \\ -\frac{4}{13} & \frac{12}{13} & \frac{3}{13} \\ -\frac{48}{65} & -\frac{5}{13} & \frac{36}{65} \end{pmatrix}.$$

8.2.44. Find all invariant subspaces, cf. Exercise 7.4.32, of a rotation in $\mathbb{R}^3$.

8.2.45. Suppose Q is an orthogonal matrix.

(a) Prove that $K = 2\,\mathrm{I} - Q - Q^T$ is a positive semi-definite matrix.

(b) Under what conditions is $K > 0$?

◇ **8.2.46.** Prove that every *proper* affine isometry $F(\mathbf{x}) = Q\,\mathbf{x} + \mathbf{b}$ of $\mathbb{R}^3$, where $\det Q = +1$, is one of the following:

(a) a *translation* $\mathbf{x} + \mathbf{b}$,

(b) a *rotation* centered at some point of $\mathbb{R}^3$, or

(c) a *screw* consisting of a rotation around an axis followed by a translation in the direction of the axis. *Hint*: Use Exercise 8.2.43.

◇ **8.2.47.** Let M_n be the $n \times n$ tridiagonal matrix whose diagonal entries are all equal to 0 and whose sub- and super-diagonal entries all equal 1.

(a) Find the eigenvalues and eigenvectors of M_2 and M_3 directly.

(b) Prove that the eigenvalues and eigenvectors of M_n are explicitly given by

$$\lambda_k = 2\cos\frac{k\pi}{n+1},$$

$$\mathbf{v}_k = \left(\sin\frac{k\pi}{n+1}, \ \sin\frac{2k\pi}{n+1}, \ \ldots, \ \sin\frac{nk\pi}{n+1} \right),$$

for $k = 1, \ldots, n$. How do you know that there are no other eigenvalues?

◇ **8.2.48.** Let a, b be fixed scalars. Determine the eigenvalues and eigenvectors of the $n \times n$ tridiagonal matrix with all diagonal entries equal to a and all sub- and super-diagonal entries equal to b. *Hint*: See Exercises 8.2.19, 8.2.47.

♡ **8.2.49.** Find a formula for the eigenvalues of the tricirculant $n \times n$ matrix Z_n that has 1's on the sub- and super-diagonals as well as its $(1, n)$ and $(n, 1)$ entries, while all other entries are 0. *Hint*: Use Exercise 8.2.47 as a guide.

8.2.50. Let A be an $n \times n$ matrix with eigenvalues $\lambda_1, \ldots, \lambda_k$, and B an $m \times m$ matrix with eigenvalues $\mu_1, \ldots, \mu_l$. Show that the $(m+n) \times (m+n)$ block diagonal matrix

$$D = \begin{pmatrix} A & \mathrm{O} \\ \mathrm{O} & B \end{pmatrix}$$

has eigenvalues $\lambda_1, \ldots, \lambda_k, \mu_1, \ldots, \mu_l$ and no others. How are the eigenvectors related?

♡ **8.2.51.** Let

$$A = \begin{pmatrix} a & b \\ c & d \end{pmatrix}$$

be a 2×2 matrix.

(a) Prove that A satisfies its own characteristic equation, meaning $p_A(A) = A^2 - (\operatorname{tr} A)\, A + (\det A)\, \mathrm{I} = \mathrm{O}$. *Remark*: This result is a special case of the *Cayley–Hamilton Theorem*, to be developed in Exercise 8.6.20.

(b) Prove the inverse formula

$$A^{-1} = \frac{(\operatorname{tr} A)\, \mathrm{I} - A}{\det A}$$

when $\det A \neq 0$.

(c) Check the Cayley–Hamilton and inverse formulas when

$$A = \begin{pmatrix} 2 & 1 \\ -3 & 2 \end{pmatrix}.$$

♡ **8.2.52.** *Deflation*: Suppose A has eigenvalue λ and corresponding eigenvector $\mathbf{v}$.

(a) Let $\mathbf{b}$ be any vector. Prove that the matrix $B = A - \mathbf{v}\,\mathbf{b}^T$ also has $\mathbf{v}$ as an eigenvector, now with eigenvalue $\lambda - \beta$ where $\beta = \mathbf{v} \cdot \mathbf{b}$.

(b) Prove that if $\mu \neq \lambda - \beta$ is any other eigenvalue of A, then it is also an eigenvalue of B. *Hint*: Look for an eigenvector of the form $\mathbf{w} + c\,\mathbf{v}$, where $\mathbf{w}$ is the eigenvector of A.

(c) Given a nonsingular matrix A with eigenvalues $\lambda_1, \lambda_2, \ldots, \lambda_n$ and $\lambda_1 \neq \lambda_j$, $j \geq 2$, explain how to construct a *deflated* matrix B whose eigenvalues are $0, \lambda_2, \ldots, \lambda_n$.

(d) Try out your method on the matrices

$$\begin{pmatrix} 3 & 3 \\ 1 & 5 \end{pmatrix} \quad \text{and} \quad \begin{pmatrix} 3 & -1 & 0 \\ -1 & 2 & -1 \\ 0 & -1 & 3 \end{pmatrix}.$$

8.3 EIGENVECTOR BASES AND DIAGONALIZATION

Most of the vector space bases that play a distinguished role in applications are assembled from the eigenvectors of a particular matrix. In this section, we show that the eigenvectors of any "complete" matrix automatically form a basis for $\mathbb{R}^n$ or, in the complex case, $\mathbb{C}^n$. In the following subsection, we use the eigenvector basis to rewrite the linear transformation determined by the matrix in a simple diagonal form. The most important cases—symmetric and positive definite matrices—will be treated in the following section.

The first task is to show that eigenvectors corresponding to distinct eigenvalues are automatically linearly independent.

Lemma 8.13 If $\lambda_1, \ldots, \lambda_k$ are *distinct* eigenvalues of the same matrix A, then the corresponding eigenvectors $\mathbf{v}_1, \ldots, \mathbf{v}_k$ are linearly independent.

Proof The result is proved by induction on the number of eigenvalues. The case $k = 1$ is immediate since an eigenvector cannot be zero. Assume that we know the result is valid for $k - 1$ eigenvalues. Suppose we have a vanishing linear combination:

$$c_1\, \mathbf{v}_1 + \cdots + c_{k-1}\, \mathbf{v}_{k-1} + c_k\, \mathbf{v}_k = \mathbf{0}. \tag{8.26}$$

Let us multiply this equation by the matrix A:

$$A\big(c_1\,\mathbf{v}_1 + \cdots + c_{k-1}\,\mathbf{v}_{k-1} + c_k\,\mathbf{v}_k\big) = c_1\,A\,\mathbf{v}_1 + \cdots + c_{k-1}\,A\,\mathbf{v}_{k-1} + c_k\,A\,\mathbf{v}_k$$
$$= c_1\,\lambda_1\,\mathbf{v}_1 + \cdots + c_{k-1}\,\lambda_{k-1}\,\mathbf{v}_{k-1} + c_k\,\lambda_k\,\mathbf{v}_k = \mathbf{0}.$$

On the other hand, if we multiply the original equation (8.26) by λ_k, we also have

$$c_1\,\lambda_k\,\mathbf{v}_1 + \cdots + c_{k-1}\,\lambda_k\,\mathbf{v}_{k-1} + c_k\,\lambda_k\,\mathbf{v}_k = \mathbf{0}.$$

Subtracting this from the previous equation, the final terms cancel and we are left with the equation

$$c_1(\lambda_1 - \lambda_k)\mathbf{v}_1 + \cdots + c_{k-1}(\lambda_{k-1} - \lambda_k)\mathbf{v}_{k-1} = \mathbf{0}.$$

This is a vanishing linear combination of the first $k-1$ eigenvectors, and so, by our induction hypothesis, can only happen if all the coefficients are zero:

$$c_1(\lambda_1 - \lambda_k) = 0, \qquad \ldots \qquad c_{k-1}(\lambda_{k-1} - \lambda_k) = 0.$$

The eigenvalues were assumed to be distinct, so $\lambda_j \neq \lambda_k$ when $j \neq k$. Consequently, $c_1 = \cdots = c_{k-1} = 0$. Substituting these values back into (8.26), we find $c_k\,\mathbf{v}_k = \mathbf{0}$, and so $c_k = 0$ also, since the eigenvector $\mathbf{v}_k \neq \mathbf{0}$. Thus we have proved that (8.26) holds if and only if $c_1 = \cdots = c_k = 0$, which implies the linear independence of the eigenvectors $\mathbf{v}_1, \ldots, \mathbf{v}_k$. This completes the induction step. ■

The most important consequence of this result is when a matrix has the maximum allotment of eigenvalues.

THEOREM 8.14 If the $n \times n$ real matrix A has n distinct real eigenvalues $\lambda_1, \ldots, \lambda_n$, then the corresponding real eigenvectors $\mathbf{v}_1, \ldots, \mathbf{v}_n$ form a basis of $\mathbb{R}^n$. If A (which may now be either a real or a complex matrix) has n distinct complex eigenvalues, then the corresponding eigenvectors $\mathbf{v}_1, \ldots, \mathbf{v}_n$ form a basis of $\mathbb{C}^n$.

For instance, the 2×2 matrix in Example 8.5 has two distinct real eigenvalues, and its two independent eigenvectors form a basis of $\mathbb{R}^2$. The 3×3 matrix in Example 8.8 has three distinct complex eigenvalues, and its eigenvectors form a basis for $\mathbb{C}^3$. If a matrix has multiple eigenvalues, then there may or may not be an eigenvector basis of $\mathbb{R}^n$ (or $\mathbb{C}^n$). The matrix in Example 8.6 admits an eigenvector basis, whereas the matrix in Example 8.7 does not. In general, it can be proved* that the dimension of the eigenspace is less than or equal to the eigenvalue's multiplicity. In particular, every simple eigenvalue has a one-dimensional eigenspace, and hence, up to scalar multiple, only one associated eigenvector.

Definition 8.15 An eigenvalue λ of a matrix A is called *complete* if the corresponding eigenspace $V_\lambda = \ker(A - \lambda\,\mathrm{I})$ has the same dimension as its multiplicity. The matrix A is *complete* if all its eigenvalues are.

Note that a simple eigenvalue is automatically complete, since its eigenspace is the one-dimensional subspace spanned by the corresponding eigenvector. Thus, only multiple eigenvalues can cause a matrix to be incomplete.

---

*This follows from Theorem 8.50 below.

REMARK: The multiplicity of an eigenvalue λ_i is sometimes referred to as its *algebraic multiplicity*. The dimension of the eigenspace V_λ is its *geometric multiplicity*, and so completeness requires that the two multiplicities are equal. The word "complete" is not completely standard; other common terms for such matrices are *perfect*, *semi-simple* and, as discussed shortly, *diagonalizable*.

THEOREM 8.16 An $n \times n$ real or complex matrix A is complete if and only if its eigenvectors span $\mathbb{C}^n$. In particular, any $n \times n$ matrix that has n distinct eigenvalues is complete.

Or, stated another way, a matrix is complete if and only if its eigenvectors can be used to form a basis of $\mathbb{C}^n$. Most matrices are complete. Incomplete $n \times n$ matrices, which have fewer than n linearly independent complex eigenvectors, are less pleasant to deal with, and we relegate most of the messy details to Section 8.6.

REMARK: We already noted that complex eigenvectors of a real matrix always appear in conjugate pairs: $\mathbf{v} = \mathbf{x} \pm \mathrm{i}\,\mathbf{y}$. If the matrix is complete, then it can be shown that its real eigenvectors combined with the real and imaginary parts of its complex conjugate eigenvectors form a real basis for $\mathbb{R}^n$. For instance, the complex eigenvectors of the 3×3 matrix appearing in Example 8.8 are

$$\begin{pmatrix} 1 \\ 0 \\ 1 \end{pmatrix} \pm \mathrm{i} \begin{pmatrix} 0 \\ 1 \\ 0 \end{pmatrix}.$$

The vectors

$$\begin{pmatrix} -1 \\ 1 \\ 1 \end{pmatrix}, \quad \begin{pmatrix} 1 \\ 0 \\ 1 \end{pmatrix}, \quad \begin{pmatrix} 0 \\ 1 \\ 0 \end{pmatrix}$$

consisting of its real eigenvector and the real and imaginary parts of its complex eigenvectors, form a basis for $\mathbb{R}^3$.

EXERCISES 8.3

8.3.1. Which of the following are complete eigenvalues for the indicated matrix? What is the dimension of the associated eigenspace?

(a) $3, \begin{pmatrix} 2 & 1 \\ 1 & 2 \end{pmatrix}$ (b) $2, \begin{pmatrix} 2 & 1 \\ 0 & 2 \end{pmatrix}$

(c) $1, \begin{pmatrix} 0 & 0 & -1 \\ 1 & 1 & 0 \\ 1 & 0 & 0 \end{pmatrix}$ (d) $1, \begin{pmatrix} 1 & 1 & -1 \\ 1 & 1 & 0 \\ 1 & 1 & 2 \end{pmatrix}$

(e) $-1, \begin{pmatrix} -1 & -4 & -4 \\ 0 & -1 & 0 \\ 0 & 4 & 3 \end{pmatrix}$

(f) $-\mathrm{i}, \begin{pmatrix} -\mathrm{i} & 1 & 0 \\ -\mathrm{i} & 1 & -1 \\ 0 & 0 & -\mathrm{i} \end{pmatrix}$

(g) $-2, \begin{pmatrix} 1 & 0 & -1 & 1 \\ 0 & 1 & 0 & 1 \\ -1 & 1 & 1 & 0 \\ 1 & 0 & 0 & 1 \end{pmatrix}$

(h) $1, \begin{pmatrix} 1 & -1 & -1 & -1 & -1 \\ 0 & 1 & -1 & -1 & -1 \\ 0 & 0 & 1 & -1 & -1 \\ 0 & 0 & 0 & 1 & -1 \\ 0 & 0 & 0 & 0 & 1 \end{pmatrix}$

8.3.2. Find the eigenvalues and a basis for the each of the eigenspaces of the following matrices. Which are complete?

(a) $\begin{pmatrix} 4 & -4 \\ 1 & 0 \end{pmatrix}$ (b) $\begin{pmatrix} 6 & -8 \\ 4 & -6 \end{pmatrix}$

(c) $\begin{pmatrix} 3 & -2 \\ 4 & -1 \end{pmatrix}$ (d) $\begin{pmatrix} \mathrm{i} & -1 \\ 1 & \mathrm{i} \end{pmatrix}$

(e) $\begin{pmatrix} 4 & -1 & -1 \\ 0 & 3 & 0 \\ 1 & -1 & 2 \end{pmatrix}$

(f) $\begin{pmatrix} -6 & 0 & -8 \\ -4 & 2 & -4 \\ 4 & 0 & 6 \end{pmatrix}$

(g) $\begin{pmatrix} -2 & 1 & -1 \\ 5 & -3 & 6 \\ 5 & -1 & 4 \end{pmatrix}$

(h) $\begin{pmatrix} 1 & 0 & 0 & 0 \\ 0 & 1 & 0 & 0 \\ -1 & 1 & -1 & 0 \\ 1 & 0 & -1 & 0 \end{pmatrix}$

(i) $\begin{pmatrix} -1 & 0 & 1 & 2 \\ 0 & 1 & 0 & 1 \\ -1 & -4 & 1 & -2 \\ 0 & 1 & 0 & 1 \end{pmatrix}$

8.3.3. Which of the following matrices admit eigenvector bases of $\mathbb{R}^n$? For those that do, exhibit such a basis. If not, what is the dimension of the subspace of $\mathbb{R}^n$ spanned by the eigenvectors?

(a) $\begin{pmatrix} 1 & 3 \\ 3 & 1 \end{pmatrix}$ (b) $\begin{pmatrix} 1 & 3 \\ -3 & 1 \end{pmatrix}$

(c) $\begin{pmatrix} 1 & 3 \\ 0 & 1 \end{pmatrix}$ (d) $\begin{pmatrix} 1 & -2 & 0 \\ 0 & -1 & 0 \\ 4 & -4 & -1 \end{pmatrix}$

(e) $\begin{pmatrix} 1 & -2 & 0 \\ 0 & -1 & 0 \\ 0 & -4 & -1 \end{pmatrix}$ (f) $\begin{pmatrix} 2 & 0 & 0 \\ 1 & -1 & 1 \\ 2 & 1 & -1 \end{pmatrix}$

(g) $\begin{pmatrix} 0 & 0 & -1 \\ 0 & 1 & 0 \\ 1 & 0 & 0 \end{pmatrix}$

(h) $\begin{pmatrix} 0 & 0 & -1 & 1 \\ 0 & -1 & 0 & 1 \\ 1 & 0 & -1 & 0 \\ 1 & 0 & 1 & 0 \end{pmatrix}$

8.3.4. Answer Exercise 8.3.3 with $\mathbb{R}^n$ replaced by $\mathbb{C}^n$.

8.3.5. (a) Give an example of a 3×3 matrix that only has 1 as an eigenvalue, and has only one linearly independent eigenvector.

(b) Give an example that has two linearly independent eigenvectors.

8.3.6. *True or false*:

(a) Every diagonal matrix is complete.

(b) Every upper triangular matrix is complete.

8.3.7. Prove that if A is a complete matrix, so is $c\,A + d\,\mathrm{I}$, where c, d are any scalars. *Hint*: Use Exercise 8.2.19.

8.3.8. (a) Prove that if A is complete, so is A^2.

(b) Give an example of an incomplete matrix A such that A^2 is complete.

◇ **8.3.9.** Suppose $\mathbf{v}_1, \ldots, \mathbf{v}_n$ forms an eigenvector basis for the complete matrix A, with $\lambda_1, \ldots, \lambda_n$ the corresponding eigenvalues. Prove that *every* eigenvalue of A is one of the $\lambda_1, \ldots, \lambda_n$.

8.3.10. (a) Prove that if λ is an eigenvalue of A, then λ^n is an eigenvalue of A^n.

(b) State and prove a converse if A is complete. *Hint*: Use Exercise 8.3.9. (The completeness hypothesis is not essential, but this is harder, relying on the Jordan canonical form.)

◇ **8.3.11.** Show that if A is complete, then any similar matrix $B = S^{-1} A\, S$ is also complete.

8.3.12. Let U be an upper triangular matrix with all its diagonal entries equal. Prove that U is complete if and only if U is a diagonal matrix.

8.3.13. Show that each eigenspace of an $n \times n$ matrix A is an invariant subspace, as defined in Exercise 7.4.32.

◇ **8.3.14.** (a) Prove that if $\mathbf{x} \pm \mathrm{i}\,\mathbf{y}$ is a complex conjugate pair of eigenvectors of a real matrix A corresponding to complex conjugate eigenvalues $\mu \pm \mathrm{i}\,\nu$ with $\nu \neq 0$, then $\mathbf{x}$ and $\mathbf{y}$ are linearly independent real vectors.

(b) More generally, if $\mathbf{v}_j = \mathbf{x}_j \pm \mathrm{i}\,\mathbf{y}_j$, $j = 1, \ldots, k$ are complex conjugate pairs of eigenvectors corresponding to *distinct* pairs of complex conjugate eigenvalues $\mu_j \pm \mathrm{i}\,\nu_j$, $\nu_j \neq 0$, then the real vectors $\mathbf{x}_1, \ldots, \mathbf{x}_k, \mathbf{y}_1, \ldots, \mathbf{y}_k$ are linearly independent.

Diagonalization

Let $L\colon \mathbb{R}^n \to \mathbb{R}^n$ be a linear transformation on n-dimensional Euclidean space. As we know, cf. Theorem 7.5, $L[\mathbf{x}] = A\mathbf{x}$ is prescribed by multiplication by an $n \times n$ matrix A. However, the matrix representing a given linear transformation will depend upon the choice of basis for the underlying vector space $\mathbb{R}^n$. Linear transformations having a complicated matrix representation in terms of the standard basis $\mathbf{e}_1, \ldots, \mathbf{e}_n$ may be considerably simplified by choosing a suitably adapted basis $\mathbf{v}_1, \ldots, \mathbf{v}_n$. We are now in a position to understand how to effect such a simplification.

For example, the linear transformation

$$L\begin{pmatrix} x \\ y \end{pmatrix} = \begin{pmatrix} x - y \\ 2x + 4y \end{pmatrix}$$

studied in Example 7.19 is represented by the matrix $A = \begin{pmatrix} 1 & -1 \\ 2 & 4 \end{pmatrix}$—when expressed in terms of the standard basis of $\mathbb{R}^2$. In terms of the alternative basis

$$\mathbf{v}_1 = \begin{pmatrix} 1 \\ -1 \end{pmatrix}, \qquad \mathbf{v}_2 = \begin{pmatrix} 1 \\ -2 \end{pmatrix},$$

it is represented by the diagonal matrix $\begin{pmatrix} 2 & 0 \\ 0 & 3 \end{pmatrix}$, indicating it has a simple stretching action on the new basis vectors:

$$A\mathbf{v}_1 = 2\,\mathbf{v}_1, \qquad A\mathbf{v}_2 = 3\,\mathbf{v}_2.$$

Now we can understand the reason for this simplification. *The new basis consists of the two eigenvectors of the matrix A.* This observation is indicative of a general fact: representing a linear transformation in terms of an eigenvector basis has the effect of changing its matrix representative into a simple diagonal form—thereby *diagonalizing* the original coefficient matrix.

According to (7.28), if $\mathbf{v}_1, \ldots, \mathbf{v}_n$ form a basis of $\mathbb{R}^n$, then the corresponding matrix representative of the linear transformation $L[\mathbf{v}] = A\mathbf{v}$ is given by the similar matrix $B = S^{-1} A\, S$, where $S = (\,\mathbf{v}_1\ \mathbf{v}_2\ \ldots\ \mathbf{v}_n\,)^T$ is the matrix whose columns are the basis vectors. In the preceding example,

$$S = \begin{pmatrix} 1 & 1 \\ -1 & -2 \end{pmatrix},$$

and hence

$$S^{-1} A\, S = \begin{pmatrix} 2 & 1 \\ -1 & -1 \end{pmatrix}\begin{pmatrix} 1 & -1 \\ 2 & 4 \end{pmatrix}\begin{pmatrix} 1 & 1 \\ -1 & -2 \end{pmatrix} = \begin{pmatrix} 2 & 0 \\ 0 & 3 \end{pmatrix}.$$

Definition 8.17 A square matrix A is called *diagonalizable* if there exists a nonsingular matrix S and a diagonal matrix $\Lambda = \operatorname{diag}(\lambda_1, \ldots, \lambda_n)$ such that

$$S^{-1} A\, S = \Lambda, \quad \text{or, equivalently} \quad A = S\,\Lambda\, S^{-1}. \tag{8.27}$$

A diagonal matrix represents a linear transformation that simultaneously stretches[†] in the direction of the basis vectors. Thus, every diagonalizable matrix represents an elementary combination of (complex) stretching transformations.

To understand the diagonalization equation (8.27), we rewrite it in the equivalent form

$$A\, S = S\,\Lambda. \tag{8.28}$$

Using the columnwise action (1.11) of matrix multiplication, one easily sees that the kth column of this $n \times n$ matrix equation is given by

$$A\,\mathbf{v}_k = \lambda_k \mathbf{v}_k,$$

[†] A negative diagonal entry represents the combination of a reflection and stretch. Complex entries indicate complex "stretching" transformations. See Section 7.2 for details.

where $\mathbf{v}_k$ denotes the kth column of S. Therefore, the columns of S are necessarily eigenvectors, and the entries of the diagonal matrix Λ are the corresponding eigenvalues! And, as a result, a diagonalizable matrix A must have n linearly independent eigenvectors, i.e., an eigenvector basis, to form the columns of the nonsingular diagonalizing matrix S. Since the diagonal form Λ contains the eigenvalues along its diagonal, it is uniquely determined up to a permutation of its entries.

Now, as we know, not every matrix has an eigenvector basis. Moreover, even when it exists, the eigenvector basis may be complex, in which case S is a complex matrix, and the entries of the diagonal matrix Λ are the complex eigenvalues. Thus, we should distinguish between complete matrices that are diagonalizable over the complex numbers and the more restrictive class of matrices that can be diagonalized by a real matrix S.

THEOREM 8.18 A matrix is complex diagonalizable if and only if it is complete. A matrix is real diagonalizable if and only if it is complete and has all real eigenvalues.

EXAMPLE 8.19 The 3×3 matrix

$$A = \begin{pmatrix} 0 & -1 & -1 \\ 1 & 2 & 1 \\ 1 & 1 & 2 \end{pmatrix}$$

considered in Example 8.5 has eigenvector basis

$$\mathbf{v}_1 = \begin{pmatrix} -1 \\ 1 \\ 0 \end{pmatrix}, \qquad \mathbf{v}_2 = \begin{pmatrix} -1 \\ 0 \\ 1 \end{pmatrix}, \qquad \mathbf{v}_3 = \begin{pmatrix} -1 \\ 1 \\ 1 \end{pmatrix}.$$

We assemble these to form the eigenvector matrix

$$S = \begin{pmatrix} -1 & -1 & -1 \\ 1 & 0 & 1 \\ 0 & 1 & 1 \end{pmatrix}, \qquad \text{whereby} \quad S^{-1} = \begin{pmatrix} -1 & 0 & -1 \\ -1 & -1 & 0 \\ 1 & 1 & 1 \end{pmatrix}.$$

The diagonalization equation (8.27) becomes

$$\begin{aligned} S^{-1} A\, S &= \begin{pmatrix} -1 & 0 & -1 \\ -1 & -1 & 0 \\ 1 & 1 & 1 \end{pmatrix} \begin{pmatrix} 0 & -1 & -1 \\ 1 & 2 & 1 \\ 1 & 1 & 2 \end{pmatrix} \begin{pmatrix} -1 & -1 & -1 \\ 1 & 0 & 1 \\ 0 & 1 & 1 \end{pmatrix} \\ &= \begin{pmatrix} 1 & 0 & 0 \\ 0 & 1 & 0 \\ 0 & 0 & 2 \end{pmatrix} = \Lambda, \end{aligned}$$

with the eigenvalues of A appearing along the diagonal of Λ, in the same order as the eigenvectors. ●

REMARK: If a matrix is not complete, then it cannot be diagonalized. A simple example is a matrix of the form

$$A = \begin{pmatrix} 1 & c \\ 0 & 1 \end{pmatrix}$$

with $c \neq 0$, which represents a shear in the direction of the x axis. Incomplete matrices will be the subject of the Section 8.6.

EXERCISES 8.3

8.3.15. Diagonalize the following matrices:

(a) $\begin{pmatrix} 3 & -9 \\ 2 & -6 \end{pmatrix}$ (b) $\begin{pmatrix} 5 & -4 \\ 2 & -1 \end{pmatrix}$

(c) $\begin{pmatrix} -4 & -2 \\ 5 & 2 \end{pmatrix}$ (d) $\begin{pmatrix} -2 & 3 & 1 \\ 0 & 1 & -1 \\ 0 & 0 & 3 \end{pmatrix}$

(e) $\begin{pmatrix} 8 & 0 & -3 \\ -3 & 0 & -1 \\ 3 & 0 & -2 \end{pmatrix}$

(f) $\begin{pmatrix} 3 & 3 & 5 \\ 5 & 6 & 5 \\ -5 & -8 & -7 \end{pmatrix}$

(g) $\begin{pmatrix} 2 & 5 & 5 \\ 0 & 2 & 0 \\ 0 & -5 & -3 \end{pmatrix}$

(h) $\begin{pmatrix} 1 & 0 & -1 & 1 \\ 0 & 2 & -1 & 1 \\ 0 & 0 & -1 & 0 \\ 0 & 0 & 0 & -2 \end{pmatrix}$

(i) $\begin{pmatrix} 0 & 0 & 1 & 0 \\ 0 & 0 & 0 & 1 \\ 1 & 0 & 0 & 0 \\ 0 & 1 & 0 & 0 \end{pmatrix}$

(j) $\begin{pmatrix} 2 & 1 & -1 & 0 \\ -3 & -2 & 0 & 1 \\ 0 & 0 & 1 & -2 \\ 0 & 0 & 1 & -1 \end{pmatrix}$

8.3.16. Diagonalize the *Fibonacci matrix* $F = \begin{pmatrix} 1 & 1 \\ 1 & 0 \end{pmatrix}$.

8.3.17. Diagonalize the matrix $\begin{pmatrix} 0 & -1 \\ 1 & 0 \end{pmatrix}$ of rotation through $90°$. How would you interpret the result?

8.3.18. Diagonalize the rotation matrices

(a) $\begin{pmatrix} 0 & -1 & 0 \\ 1 & 0 & 0 \\ 0 & 0 & 1 \end{pmatrix}$ (b) $\begin{pmatrix} \frac{5}{13} & 0 & \frac{12}{13} \\ 0 & 1 & 0 \\ -\frac{12}{13} & 0 & \frac{5}{13} \end{pmatrix}$

8.3.19. Which of the following matrices have real diagonal forms?

(a) $\begin{pmatrix} -2 & 1 \\ 4 & 1 \end{pmatrix}$ (b) $\begin{pmatrix} 1 & 2 \\ -3 & 1 \end{pmatrix}$

(c) $\begin{pmatrix} 0 & 1 & 0 \\ -1 & 0 & 1 \\ 1 & 1 & 0 \end{pmatrix}$ (d) $\begin{pmatrix} 0 & 3 & 2 \\ -1 & 1 & -1 \\ 1 & -3 & -1 \end{pmatrix}$

(e) $\begin{pmatrix} 3 & -8 & 2 \\ -1 & 2 & 2 \\ 1 & -4 & 2 \end{pmatrix}$

(f) $\begin{pmatrix} 1 & 0 & 0 & 0 \\ 0 & 1 & 0 & 0 \\ 0 & 0 & 1 & 0 \\ -1 & -1 & -1 & -1 \end{pmatrix}$

8.3.20. Diagonalize the following complex matrices:

(a) $\begin{pmatrix} i & 1 \\ 1 & i \end{pmatrix}$ (b) $\begin{pmatrix} 1-i & 0 \\ i & 2+i \end{pmatrix}$

(c) $\begin{pmatrix} 2-i & 2+i \\ 3-i & 1+i \end{pmatrix}$ (d) $\begin{pmatrix} -i & 0 & 1 \\ -i & 1 & -1 \\ 1 & 0 & -i \end{pmatrix}$

8.3.21. Write down a real matrix that has

(a) eigenvalues $-1, 3$ and corresponding eigenvectors $\begin{pmatrix} -1 \\ 2 \end{pmatrix}, \begin{pmatrix} 1 \\ 1 \end{pmatrix}$

(b) eigenvalues $0, 2, -2$ and associated eigenvectors $\begin{pmatrix} -1 \\ 1 \\ 0 \end{pmatrix}, \begin{pmatrix} 2 \\ -1 \\ 1 \end{pmatrix}, \begin{pmatrix} 0 \\ 1 \\ 3 \end{pmatrix}$

(c) an eigenvalue of 3 and corresponding eigenvectors $\begin{pmatrix} 2 \\ -3 \end{pmatrix}, \begin{pmatrix} 1 \\ 2 \end{pmatrix}$

(d) an eigenvalue $-1 + 2i$ and corresponding eigenvector $\begin{pmatrix} 1+i \\ 3i \end{pmatrix}$

(e) an eigenvalue -2 and corresponding eigenvector $\begin{pmatrix} 2 \\ 0 \\ -1 \end{pmatrix}$

(f) an eigenvalue $3 + i$ and corresponding eigenvector $\begin{pmatrix} 1 \\ 2i \\ -1-i \end{pmatrix}$

8.3.22. A matrix A has eigenvalues -1 and 2 and associated eigenvectors $\begin{pmatrix} 1 \\ 2 \end{pmatrix}$ and $\begin{pmatrix} 2 \\ 3 \end{pmatrix}$. Write down the matrix form of the linear transformation $L[\mathbf{u}] = A\mathbf{u}$ in terms of

(a) the standard basis $\mathbf{e}_1, \mathbf{e}_2$;

(b) the basis consisting of its eigenvectors;

(c) the basis $\begin{pmatrix} 1 \\ 1 \end{pmatrix}, \begin{pmatrix} 3 \\ 4 \end{pmatrix}$.

◇ **8.3.23.** Prove that two complete matrices A, B have the same eigenvalues (with multiplicities) if and only if they are similar, i.e., $B = S^{-1} A S$ for some nonsingular matrix S.

8.3.24. Let B be obtained from A by permuting both its rows and columns using the same permutation π, so $b_{ij} = a_{\pi(i),\pi(j)}$. Prove that A and B have the same eigenvalues. How are their eigenvectors related?

8.3.25. *True or false*: If A is a complete upper triangular matrix, then it has an upper triangular eigenvector matrix S.

8.3.26. Suppose the $n \times n$ matrix A is diagonalizable. How many different diagonal forms does it have?

8.3.27. Characterize all complete matrices that are their own inverses: $A^{-1} = A$. Write down a non-diagonal example.

♡ **8.3.28.** Two $n \times n$ matrices A, B are said to be *simultaneously diagonalizable* if there is a nonsingular matrix S such that both $S^{-1} A S$ and $S^{-1} B S$ are diagonal matrices.

(a) Show that simultaneously diagonalizable matrices commute: $A B = B A$.

(b) Prove that the converse is valid, provided that one of the matrices has no multiple eigenvalues.

(c) Is every pair of commuting matrices simultaneously diagonalizable?

8.4 EIGENVALUES OF SYMMETRIC MATRICES

Fortunately, the matrices that arise in most applications are complete and, in fact, possess some additional structure that ameliorates the calculation of their eigenvalues and eigenvectors. The most important class are the symmetric, including positive definite, matrices. In fact, not only are the eigenvalues of a symmetric matrix necessarily real, the eigenvectors always form an *orthogonal basis* of the underlying Euclidean space, enjoying all the wonderful properties we studied in Chapter 5. In fact, this is by far the most common way for orthogonal bases to appear—as the eigenvector bases of symmetric matrices. Let us state this important result, but defer its proof until the end of the section.

THEOREM 8.20 Let $A = A^T$ be a real symmetric $n \times n$ matrix. Then

(a) All the eigenvalues of A are real.

(b) Eigenvectors corresponding to distinct eigenvalues are orthogonal.

(c) There is an orthonormal basis of $\mathbb{R}^n$ consisting of n eigenvectors of A.

In particular, all symmetric matrices are complete.

REMARK: Orthogonality is with respect to the standard dot product on $\mathbb{R}^n$. As we noted in Section 7.5, the transpose or adjoint operation is intimately connected with the Euclidean dot product. An analogous result holds for more general self-adjoint linear transformations on $\mathbb{R}^n$; see Exercise 8.4.10 for details.

EXAMPLE 8.21 The 2×2 matrix

$$A = \begin{pmatrix} 3 & 1 \\ 1 & 3 \end{pmatrix}$$

considered in Example 8.5 is symmetric, and so has real eigenvalues $\lambda_1 = 4$ and $\lambda_2 = 2$. You can easily check that the corresponding eigenvectors $\mathbf{v}_1 = (1, 1)^T$ and $\mathbf{v}_2 = (-1, 1)^T$ are orthogonal: $\mathbf{v}_1 \cdot \mathbf{v}_2 = 0$, and hence form an orthogonal basis of $\mathbb{R}^2$. The orthonormal eigenvector basis promised by Theorem 8.20 is obtained by dividing each eigenvector by its Euclidean norm:

$$\mathbf{u}_1 = \begin{pmatrix} \frac{1}{\sqrt{2}} \\ \frac{1}{\sqrt{2}} \end{pmatrix}, \qquad \mathbf{u}_2 = \begin{pmatrix} -\frac{1}{\sqrt{2}} \\ \frac{1}{\sqrt{2}} \end{pmatrix}.$$

●

EXAMPLE 8.22 Consider the symmetric matrix

$$A = \begin{pmatrix} 5 & -4 & 2 \\ -4 & 5 & 2 \\ 2 & 2 & -1 \end{pmatrix}.$$

A straightforward computation produces its eigenvalues and eigenvectors:

$$\lambda_1 = 9, \qquad \lambda_2 = 3, \qquad \lambda_3 = -3,$$

$$\mathbf{v}_1 = \begin{pmatrix} 1 \\ -1 \\ 0 \end{pmatrix}, \quad \mathbf{v}_2 = \begin{pmatrix} 1 \\ 1 \\ 1 \end{pmatrix}, \quad \mathbf{v}_3 = \begin{pmatrix} 1 \\ 1 \\ -2 \end{pmatrix}.$$

As the reader can check, the eigenvectors form an orthogonal basis of $\mathbb{R}^3$. An orthonormal basis is provided by the unit eigenvectors

$$\mathbf{u}_1 = \begin{pmatrix} \frac{1}{\sqrt{2}} \\ -\frac{1}{\sqrt{2}} \\ 0 \end{pmatrix}, \quad \mathbf{u}_2 = \begin{pmatrix} \frac{1}{\sqrt{3}} \\ \frac{1}{\sqrt{3}} \\ \frac{1}{\sqrt{3}} \end{pmatrix}, \quad \mathbf{u}_3 = \begin{pmatrix} \frac{1}{\sqrt{6}} \\ \frac{1}{\sqrt{6}} \\ -\frac{2}{\sqrt{6}} \end{pmatrix}.$$ ●

In particular, the eigenvalues of a symmetric matrix can be used to test its positive definiteness.

THEOREM 8.23 A symmetric matrix $K = K^T$ is positive definite if and only if all of its eigenvalues are strictly positive.

Proof First, if $K > 0$, then, by definition, $\mathbf{x}^T K \mathbf{x} > 0$ for all nonzero vectors $\mathbf{x} \in \mathbb{R}^n$. In particular, if $\mathbf{x} = \mathbf{v} \neq \mathbf{0}$ is an eigenvector with (necessarily real) eigenvalue λ, then

$$0 < \mathbf{v}^T K \mathbf{v} = \mathbf{v}^T (\lambda \mathbf{v}) = \lambda \|\mathbf{v}\|^2, \tag{8.29}$$

which immediately proves that $\lambda > 0$. Conversely, suppose K has all positive eigenvalues. Let $\mathbf{u}_1, \ldots, \mathbf{u}_n$ be the orthonormal eigenvector basis of $\mathbb{R}^n$ guaranteed by Theorem 8.20, with $K \mathbf{u}_j = \lambda_j \mathbf{u}_j$. Then, writing

$$\mathbf{x} = c_1 \mathbf{u}_1 + \cdots + c_n \mathbf{u}_n, \qquad \text{we find} \qquad K\mathbf{x} = c_1 \lambda_1 \mathbf{u}_1 + \cdots + c_n \lambda_n \mathbf{u}_n.$$

Therefore, using the orthonormality of the eigenvectors, for any $\mathbf{x} \neq \mathbf{0}$,

$$\begin{aligned} \mathbf{x}^T K \mathbf{x} &= (c_1 \mathbf{u}_1^T + \cdots + c_n \mathbf{u}_n^T)(c_1 \lambda_1 \mathbf{u}_1 + \cdots + c_n \lambda_n \mathbf{u}_n) \\ &= \lambda_1 c_1^2 + \cdots + \lambda_n c_n^2 > 0 \end{aligned}$$

since only $\mathbf{x} = \mathbf{0}$ has coordinates $c_1 = \cdots = c_n = 0$. This proves that K is positive definite. ■

REMARK: The same proof shows that K is positive semi-definite if and only if all its eigenvalues satisfy $\lambda \geq 0$. A positive semi-definite matrix that is not positive definite admits a zero eigenvalue and one or more *null eigenvectors*, i.e., solutions to $K\mathbf{v} = \mathbf{0}$. Every nonzero element $\mathbf{0} \neq \mathbf{v} \in \ker K$ of its kernel is a null eigenvector.

EXAMPLE 8.24 Consider the symmetric matrix

$$K = \begin{pmatrix} 8 & 0 & 1 \\ 0 & 8 & 1 \\ 1 & 1 & 7 \end{pmatrix}.$$

Its characteristic equation is

$$\det(K - \lambda I) = -\lambda^3 + 23\lambda^2 - 174\lambda + 432 = -(\lambda - 9)(\lambda - 8)(\lambda - 6),$$

and so its eigenvalues are 9, 8, and 6. Since they are all positive, K is a positive definite matrix. The associated eigenvectors are

$$\lambda_1 = 9, \quad \mathbf{v}_1 = \begin{pmatrix} 1 \\ 1 \\ 1 \end{pmatrix},$$

$$\lambda_2 = 8, \quad \mathbf{v}_2 = \begin{pmatrix} -1 \\ 1 \\ 0 \end{pmatrix},$$

$$\lambda_3 = 6, \quad \mathbf{v}_3 = \begin{pmatrix} -1 \\ -1 \\ 2 \end{pmatrix}.$$

Note that the eigenvectors form an orthogonal basis of $\mathbb{R}^3$, as guaranteed by Theorem 8.20. As usual, we can construct an corresponding orthonormal eigenvector basis

$$\mathbf{u}_1 = \begin{pmatrix} \frac{1}{\sqrt{3}} \\ \frac{1}{\sqrt{3}} \\ \frac{1}{\sqrt{3}} \end{pmatrix}, \quad \mathbf{u}_2 = \begin{pmatrix} -\frac{1}{\sqrt{2}} \\ \frac{1}{\sqrt{2}} \\ 0 \end{pmatrix}, \quad \mathbf{u}_3 = \begin{pmatrix} -\frac{1}{\sqrt{6}} \\ -\frac{1}{\sqrt{6}} \\ \frac{2}{\sqrt{6}} \end{pmatrix},$$

by dividing each eigenvector by its norm. ●

Proof of Theorem 8.20 First recall that (see Exercise 3.6.40) if $A = A^T$ is real, symmetric, then

$$(A\mathbf{v}) \cdot \mathbf{w} = \mathbf{v} \cdot (A\mathbf{w}) \quad \text{for all} \quad \mathbf{v}, \mathbf{w} \in \mathbb{C}^n, \tag{8.30}$$

where $\cdot$ indicates the Euclidean dot product when the vectors are real and, more generally, the Hermitian dot product $\mathbf{v} \cdot \mathbf{w} = \mathbf{v}^T \overline{\mathbf{w}}$ when they are complex.

To prove property (a), suppose λ is a complex eigenvalue with complex eigenvector $\mathbf{v} \in \mathbb{C}^n$. Consider the Hermitian dot product of the complex vectors $A\mathbf{v}$ and $\mathbf{v}$:

$$(A\mathbf{v}) \cdot \mathbf{v} = (\lambda \mathbf{v}) \cdot \mathbf{v} = \lambda \, \|\mathbf{v}\|^2.$$

On the other hand, by (8.30),

$$(A\mathbf{v}) \cdot \mathbf{v} = \mathbf{v} \cdot (A\mathbf{v}) = \mathbf{v} \cdot (\lambda \mathbf{v}) = \mathbf{v}^T \, \overline{\lambda \mathbf{v}} = \overline{\lambda} \, \|\mathbf{v}\|^2.$$

Equating these two expressions, we deduce

$$\overline{\lambda} \, \|\mathbf{v}\|^2 = \lambda \, \|\mathbf{v}\|^2.$$

Since $\mathbf{v} \neq \mathbf{0}$, as it is an eigenvector, we conclude that $\overline{\lambda} = \lambda$, proving that the eigenvalue λ is real.

To prove (b), suppose

$$A\mathbf{v} = \lambda \mathbf{v}, \qquad A\mathbf{w} = \mu \mathbf{w},$$

where $\lambda \neq \mu$ are distinct real eigenvalues. Then, again by (8.30),

$$\lambda \mathbf{v} \cdot \mathbf{w} = (A\mathbf{v}) \cdot \mathbf{w} = \mathbf{v} \cdot (A\mathbf{w}) = \mathbf{v} \cdot (\mu \mathbf{w}) = \mu \mathbf{v} \cdot \mathbf{w},$$

and hence

$$(\lambda - \mu) \mathbf{v} \cdot \mathbf{w} = 0.$$

Since $\lambda \neq \mu$, this implies that $\mathbf{v} \cdot \mathbf{w} = 0$ and hence the eigenvectors $\mathbf{v}, \mathbf{w}$ are orthogonal.

Finally, the proof of (c) is easy if all the eigenvalues of A are distinct. Theorem 8.14 implies that the eigenvectors form a basis of $\mathbb{R}^n$, and part (b) proves they are orthogonal. (An alternative proof starts with orthogonality, and then applies Proposition 5.4 to prove that the eigenvectors form a basis.) To obtain an orthonormal basis, we merely divide the eigenvectors by their lengths: $\mathbf{u}_k = \mathbf{v}_k / \|\mathbf{v}_k\|$, as in Lemma 5.2.

To prove (c) in general, we proceed by induction on the size n of the matrix A. To start, the case of a 1×1 matrix is trivial. (Why?) Next, suppose A has size $n \times n$. We know that A has at least one eigenvalue, λ_1, which is necessarily real. Let $\mathbf{v}_1$ be the associated eigenvector. Let

$$V^{\perp} = \left\{ \mathbf{w} \in \mathbb{R}^n \;\middle|\; \mathbf{v}_1 \cdot \mathbf{w} = 0 \right\}$$

denote the orthogonal complement to the eigenspace V_{λ_1}—the set of all vectors orthogonal to the first eigenvector. Proposition 5.50 implies that $\dim V^{\perp} = n - 1$, and so we can choose an orthonormal basis $\mathbf{y}_1, \ldots, \mathbf{y}_{n-1}$. Now, if $\mathbf{w}$ is any vector in $V^{\perp}$, so is $A\mathbf{w}$, since, by (8.30),

$$\mathbf{v}_1 \cdot (A\mathbf{w}) = (A\,\mathbf{v}_1) \cdot \mathbf{w} = \lambda_1\, \mathbf{v}_1 \cdot \mathbf{w} = 0.$$

Thus, A defines a linear transformation on $V^{\perp}$ that is represented by an $(n-1) \times (n-1)$ matrix with respect to the chosen orthonormal basis $\mathbf{y}_1, \ldots, \mathbf{y}_{n-1}$. In Exercise 8.4.11 you are asked to prove that the representing matrix is symmetric. Our induction hypothesis then implies that there is an orthonormal basis of $V^{\perp}$ consisting of eigenvectors $\mathbf{u}_2, \ldots, \mathbf{u}_n$ of A. Appending the unit eigenvector $\mathbf{u}_1 = \mathbf{v}_1 / \|\mathbf{v}_1\|$ to this collection will complete the orthonormal basis of $\mathbb{R}^n$. ∎

Proposition 8.25 Let $A = A^T$ be an $n \times n$ symmetric matrix. Let $\mathbf{v}_1, \ldots, \mathbf{v}_n$ be an orthogonal eigenvector basis such that $\mathbf{v}_1, \ldots, \mathbf{v}_r$ correspond to nonzero eigenvalues, while $\mathbf{v}_{r+1}, \ldots, \mathbf{v}_n$ are null eigenvectors corresponding to the zero eigenvalue (if any). Then $r = \operatorname{rank} A$; the non-null eigenvectors $\mathbf{v}_1, \ldots, \mathbf{v}_r$ form an orthogonal basis for $\operatorname{rng} A = \operatorname{corng} A$, while the null eigenvectors $\mathbf{v}_{r+1}, \ldots, \mathbf{v}_n$ form an orthogonal basis for $\ker A = \operatorname{coker} A$.

Proof The zero eigenspace coincides with the kernel, $V_0 = \ker A$. Thus, the linearly independent null eigenvectors form a basis for $\ker A$, which has dimension $n - r$ where $r = \operatorname{rank} A$. Moreover, the remaining r non-null eigenvectors are orthogonal to the null eigenvectors. Therefore, they must form a basis for the kernel's orthogonal complement, namely $\operatorname{corng} A$. ∎

EXERCISES 8.4

8.4.1. Find the eigenvalues and an orthonormal eigenvector basis for the following symmetric matrices:

(a) $\begin{pmatrix} 2 & 6 \\ 6 & -7 \end{pmatrix}$ (b) $\begin{pmatrix} 5 & -2 \\ -2 & 5 \end{pmatrix}$

(c) $\begin{pmatrix} 2 & -1 \\ -1 & 5 \end{pmatrix}$ (d) $\begin{pmatrix} 1 & 0 & 4 \\ 0 & 1 & 3 \\ 4 & 3 & 1 \end{pmatrix}$

(e) $\begin{pmatrix} 6 & -4 & 1 \\ -4 & 6 & -1 \\ 1 & -1 & 11 \end{pmatrix}$

8.4.2. Determine whether the following symmetric matrices are positive definite by computing their eigenvalues. Validate your conclusions by using the methods from Chapter 4.

(a) $\begin{pmatrix} 2 & -2 \\ -2 & 3 \end{pmatrix}$ (b) $\begin{pmatrix} -2 & 3 \\ 3 & 6 \end{pmatrix}$

(c) $\begin{pmatrix} 1 & -1 & 0 \\ -1 & 2 & -1 \\ 0 & -1 & 1 \end{pmatrix}$

(d) $\begin{pmatrix} 4 & -1 & -2 \\ -1 & 4 & -1 \\ -2 & -1 & 4 \end{pmatrix}$

8.4.3. Prove that a symmetric matrix is negative definite if and only if all its eigenvalues are negative.

8.4.4. How many orthonormal eigenvector bases does a symmetric $n \times n$ matrix have?

8.4.5. Let $A = \begin{pmatrix} a & b \\ c & d \end{pmatrix}$.

(a) Write down necessary and sufficient conditions on the entries a, b, c, d that ensures that A has only real eigenvalues.

(b) Verify that all symmetric 2×2 matrices satisfy your conditions.

(c) Write down a non-symmetric matrix that satisfies your conditions.

♡ **8.4.6.** Let $A^T = -A$ be a real, skew-symmetric $n \times n$ matrix.

(a) Prove that the only possible real eigenvalue of A is $\lambda = 0$.

(b) More generally, prove that all eigenvalues λ of A are purely imaginary, i.e., $\operatorname{Re}\lambda = 0$.

(c) Explain why 0 is an eigenvalue of A whenever n is odd.

(d) Explain why, if $n = 3$, the eigenvalues of $A \neq \mathrm{O}$ are 0, $\mathrm{i}\,\omega$, $-\mathrm{i}\,\omega$, for some real $\omega \neq 0$.

(e) Verify these facts for the particular matrices

(i) $\begin{pmatrix} 0 & -2 \\ 2 & 0 \end{pmatrix}$ (ii) $\begin{pmatrix} 0 & 3 & 0 \\ -3 & 0 & -4 \\ 0 & 4 & 0 \end{pmatrix}$

(iii) $\begin{pmatrix} 0 & 1 & -1 \\ -1 & 0 & -1 \\ 1 & 1 & 0 \end{pmatrix}$

(iv) $\begin{pmatrix} 0 & 0 & 2 & 0 \\ 0 & 0 & 0 & -3 \\ -2 & 0 & 0 & 0 \\ 0 & 3 & 0 & 0 \end{pmatrix}$

♡ **8.4.7.** (a) Prove that every eigenvalue of a Hermitian matrix A, satisfying $A^T = \overline{A}$ as in Exercise 3.6.49, is real.

(b) Show that the eigenvectors corresponding to distinct eigenvalues are orthogonal under the Hermitian dot product on $\mathbb{C}^n$.

(c) Find the eigenvalues and eigenvectors of the following Hermitian matrices, and verify orthogonality:

(i) $\begin{pmatrix} 2 & \mathrm{i} \\ -\mathrm{i} & -2 \end{pmatrix}$ (ii) $\begin{pmatrix} 3 & 2-\mathrm{i} \\ 2+\mathrm{i} & -1 \end{pmatrix}$

(iii) $\begin{pmatrix} 0 & \mathrm{i} & 0 \\ -\mathrm{i} & 0 & \mathrm{i} \\ 0 & -\mathrm{i} & 0 \end{pmatrix}$

♡ **8.4.8.** Let $M > 0$ be a fixed positive definite $n \times n$ matrix. A nonzero vector $\mathbf{v} \neq \mathbf{0}$ is called a *generalized eigenvector* of the $n \times n$ matrix K if

$$K\,\mathbf{v} = \lambda\, M\,\mathbf{v}, \qquad \mathbf{v} \neq \mathbf{0}, \tag{8.31}$$

where the scalar λ is the corresponding *generalized eigenvalue*.

(a) Prove that λ is a generalized eigenvalue of the matrix K if and only if it is an ordinary eigenvalue of the matrix $M^{-1}K$. How are the eigenvectors related?

(b) Now suppose K is a symmetric matrix. Prove that its generalized eigenvalues are all real. *Hint*: First explain why this does *not* follow from part (a). Instead mimic the proof of part (a) of Theorem 8.20, using the weighted Hermitian inner product $\langle \mathbf{v}, \mathbf{w} \rangle = \mathbf{v}^T M\,\overline{\mathbf{w}}$ in place of the dot product.

(c) Show that if $K > 0$, then its generalized eigenvalues are all positive: $\lambda > 0$.

(d) Prove that the eigenvectors corresponding to different generalized eigenvalues are orthogonal under the weighted inner product $\langle \mathbf{v}, \mathbf{w} \rangle = \mathbf{v}^T M\,\mathbf{w}$.

(e) Show that, if the matrix pair K, M has n distinct generalized eigenvalues, then the eigenvectors form an orthogonal basis for $\mathbb{R}^n$. *Remark*: One can, by mimicking the proof of part (c) of Theorem 8.20, show that this holds even when there are repeated generalized eigenvalues.

8.4.9. Compute the generalized eigenvalues and eigenvectors, as in (8.31), for the following matrix pairs. Verify orthogonality of the eigenvectors under the appropriate inner product.

(a) $K = \begin{pmatrix} 3 & -1 \\ -1 & 2 \end{pmatrix}$, $M = \begin{pmatrix} 2 & 0 \\ 0 & 3 \end{pmatrix}$

(b) $K = \begin{pmatrix} 3 & 1 \\ 1 & 1 \end{pmatrix}$, $M = \begin{pmatrix} 2 & 0 \\ 0 & 1 \end{pmatrix}$

(c) $K = \begin{pmatrix} 2 & -1 \\ -1 & 4 \end{pmatrix}$, $M = \begin{pmatrix} 2 & -1 \\ -1 & 1 \end{pmatrix}$

(d) $K = \begin{pmatrix} 6 & -8 & 3 \\ -8 & 24 & -6 \\ 3 & -6 & 99 \end{pmatrix}$, $M = \begin{pmatrix} 1 & 0 & 0 \\ 0 & 4 & 0 \\ 0 & 0 & 9 \end{pmatrix}$

(e) $K = \begin{pmatrix} 1 & 2 & 0 \\ 2 & 8 & 2 \\ 0 & 2 & 1 \end{pmatrix}$, $M = \begin{pmatrix} 1 & 1 & 0 \\ 1 & 3 & 1 \\ 0 & 1 & 1 \end{pmatrix}$

(f) $K = \begin{pmatrix} 5 & 3 & -5 \\ 3 & 3 & -1 \\ -5 & -1 & 9 \end{pmatrix}$

$M = \begin{pmatrix} 3 & 2 & -3 \\ 2 & 2 & -1 \\ -3 & -1 & 5 \end{pmatrix}$

◇ **8.4.10.** Let $L = L^*: \mathbb{R}^n \to \mathbb{R}^n$ be a self-adjoint linear transformation with respect to the inner product $\langle \cdot , \cdot \rangle$. Prove that all its eigenvalues are real and the eigenvectors are orthogonal. *Hint*: Mimic the proof of Theorem 8.20, replacing the dot product by the given inner product.

◇ **8.4.11.** Using the set-up in the proof of Theorem 8.20, explain why

$$A\mathbf{y}_i = \sum_{j=1}^{n-1} b_{ij}\,\mathbf{y}_j.$$

Then prove that $b_{ij} = b_{ji}$, justifying the statement that the restriction of A to $V^\perp$ is represented by a symmetric matrix.

♡ **8.4.12.** The *difference map* $\Delta: \mathbb{C}^n \to \mathbb{C}^n$ is defined as $\Delta = S - \mathrm{I}$, where S is the shift map of Exercise 8.2.13.

(a) Write down the matrix corresponding to Δ.

(b) Prove that the sampled exponential vectors $\boldsymbol{\omega}_0, \ldots, \boldsymbol{\omega}_{n-1}$ from (5.90) form an eigenvector basis of Δ. What are the eigenvalues?

(c) Prove that $K = \Delta^T \Delta$ has the same eigenvectors as Δ. What are its eigenvalues? Is K positive definite?

(d) According to Theorem 8.20 the eigenvectors of a symmetric matrix are real and orthogonal. Use this to explain the orthogonality of the sampled exponential vectors. But, why aren't they real?

♡ **8.4.13.** An $n \times n$ *circulant matrix* has the form

$$C = \begin{pmatrix} c_0 & c_1 & c_2 & c_3 & \ldots & c_{n-1} \\ c_{n-1} & c_0 & c_1 & c_2 & \ldots & c_{n-2} \\ c_{n-2} & c_{n-1} & c_0 & c_1 & \ldots & c_{n-3} \\ \vdots & \vdots & \vdots & \vdots & \ddots & \vdots \\ c_1 & c_2 & c_3 & c_4 & \ldots & c_0 \end{pmatrix},$$

in which the entries of each succeeding row are obtained by moving all the previous row's entries one slot to the right, the last entry moving to the front.

(a) Check that the shift matrix S of Exercise 8.2.13, the difference matrix Δ, and its symmetric product K of Exercise 8.4.12 are all circulant matrices.

(b) Prove that the sampled exponential vectors $\boldsymbol{\omega}_0, \ldots, \boldsymbol{\omega}_{n-1}$, cf. (5.90) are eigenvectors of C. Thus, all circulant matrices have *the same eigenvectors*! What are the eigenvalues?

(c) Prove that $F_n^{-1} C F_n = \Lambda$ where F_n is the Fourier matrix in Exercise 5.7.9 and Λ is the diagonal matrix with the eigenvalues of C along the diagonal.

(d) Find the eigenvalues and eigenvectors of the following circulant matrices:

(i) $\begin{pmatrix} 1 & 2 \\ 2 & 1 \end{pmatrix}$ (ii) $\begin{pmatrix} 1 & 2 & 3 \\ 3 & 1 & 2 \\ 2 & 3 & 1 \end{pmatrix}$

(iii) $\begin{pmatrix} 1 & -1 & -1 & 1 \\ 1 & 1 & -1 & -1 \\ -1 & 1 & 1 & -1 \\ -1 & -1 & 1 & 1 \end{pmatrix}$

(iv) $\begin{pmatrix} 2 & -1 & 0 & -1 \\ -1 & 2 & -1 & 0 \\ 0 & -1 & 2 & -1 \\ -1 & 0 & -1 & 2 \end{pmatrix}$

(e) Find the eigenvalues of the tricirculant matrices in Exercise 1.7.13. Can you find a general formula for the $n \times n$ version? Explain why the eigenvalues must be real and positive. Does your formula reflect this fact?

(f) Which of the preceding matrices are invertible? Write down a general criterion for checking the invertibility of circulant matrices.

The Spectral Theorem

Every real, symmetric matrix admits an eigenvector basis, and hence is diagonalizable. Moreover, since we can choose eigenvectors that form an orthonormal basis, the diagonalizing matrix takes a particularly simple form. Recall that an $n \times n$ matrix Q is called *orthogonal* if and only if its columns form an orthonormal basis of $\mathbb{R}^n$. Alternatively, one characterizes orthogonal matrices by the condition $Q^{-1} = Q^T$, as in Definition 5.18.

Using the orthonormal eigenvector basis in the diagonalization (8.27) results in the so-called *spectral factorization* of the symmetric matrix.

THEOREM 8.26 Let A be a real, symmetric matrix. Then there exists an orthogonal matrix Q such that

$$A = Q\,\Lambda\,Q^{-1} = Q\,\Lambda\,Q^T, \tag{8.32}$$

where Λ is a real diagonal matrix. The eigenvalues of A appear on the diagonal of Λ, while the columns of Q are the corresponding orthonormal eigenvectors.

REMARK: The term "spectrum" refers to the eigenvalues of a matrix or, more generally, a linear operator. The terminology is motivated by physics. The spectral energy lines of atoms, molecules and nuclei are characterized as the eigenvalues of the governing quantum mechanical Schrödinger operator, [40]. The *Spectral Theorem* 8.26 is the finite-dimensional version for the decomposition of quantum mechanical linear operators into their spectral eigenstates.

Warning: The spectral factorization $A = Q\,\Lambda\,Q^T$ and the Gaussian factorization $A = L\,D\,L^T$ of a regular symmetric matrix, cf. (1.56), are completely different. In particular, the eigenvalues are *not* the pivots, so $\Lambda \neq D$.

The spectral factorization (8.32) provides us with an alternative means of diagonalizing the associated quadratic form $q(\mathbf{x}) = \mathbf{x}^T A\,\mathbf{x}$, i.e., of completing the square. We write

$$q(\mathbf{x}) = \mathbf{x}^T A\,\mathbf{x} = \mathbf{x}^T\,Q\,\Lambda\,Q^T\,\mathbf{x} = \mathbf{y}^T\,\Lambda\,\mathbf{y} = \sum_{i=1}^{n} \lambda_i\,y_i^2, \tag{8.33}$$

where $\mathbf{y} = Q^T\mathbf{x} = Q^{-1}\mathbf{x}$ form the coordinates of $\mathbf{x}$ with respect to the orthonormal eigenvector basis of A. In particular, $q(\mathbf{x}) > 0$ for all $\mathbf{x} \neq \mathbf{0}$ and so A is positive definite if and only if each eigenvalue $\lambda_i > 0$ is strictly positive, reconfirming Theorem 8.23.

EXAMPLE 8.27 For the 2×2 matrix $A = \begin{pmatrix} 3 & 1 \\ 1 & 3 \end{pmatrix}$ considered in Example 8.21, the orthonormal eigenvectors produce the diagonalizing orthogonal matrix

$$Q = \begin{pmatrix} \frac{1}{\sqrt{2}} & -\frac{1}{\sqrt{2}} \\ \frac{1}{\sqrt{2}} & \frac{1}{\sqrt{2}} \end{pmatrix}.$$

The reader can validate the resulting spectral factorization

$$\begin{pmatrix} 3 & 1 \\ 1 & 3 \end{pmatrix} = A = Q\,\Lambda\,Q^T = \begin{pmatrix} \frac{1}{\sqrt{2}} & -\frac{1}{\sqrt{2}} \\ \frac{1}{\sqrt{2}} & \frac{1}{\sqrt{2}} \end{pmatrix} \begin{pmatrix} 4 & 0 \\ 0 & 2 \end{pmatrix} \begin{pmatrix} \frac{1}{\sqrt{2}} & \frac{1}{\sqrt{2}} \\ -\frac{1}{\sqrt{2}} & \frac{1}{\sqrt{2}} \end{pmatrix}.$$

According to (8.33), the associated quadratic form is diagonalized as

$$q(\mathbf{x}) = 3x_1^2 + 2x_1x_2 + 3x_2^2 = 4y_1^2 + 2y_2^2,$$

where $\mathbf{y} = Q^T\mathbf{x}$, i.e.,

$$y_1 = \frac{x_1 + x_2}{\sqrt{2}}, \qquad y_2 = \frac{-x_1 + x_2}{\sqrt{2}}.$$

You can always choose Q to be a proper orthogonal matrix, so $\det Q = 1$, since an improper orthogonal matrix can be made proper by multiplying one of its columns by -1, which does not affect its status as an eigenvector matrix. Since a proper orthogonal matrix Q represents a rigid rotation of $\mathbb{R}^n$, the diagonalization of a symmetric matrix can be interpreted as a rotation of the coordinate system that makes the orthogonal eigenvectors line up along the coordinate axes. Therefore, a linear transformation $L[\mathbf{x}] = A\mathbf{x}$ represented by a positive definite matrix A can be regarded as a combination of stretches in n mutually orthogonal directions. For instance, in elasticity, the stress tensor of a deformed body is represented by a positive definite matrix. Its eigenvalues are known as the *principal stretches* and its eigenvectors the *principal directions* of the elastic deformation. A good way to visualize this is to consider the effect of the linear transformation on the unit (Euclidean) sphere $S_1 = \{ \|\mathbf{x}\| = 1 \}$. Stretching the sphere in mutually orthogonal directions will map it to an ellipsoid $E = L[S_1] = \{ A\mathbf{x} \mid \|\mathbf{x}\| = 1 \}$ whose principal axes are aligned with the directions of stretch; see Figure 8.2 for the two-dimensional case.

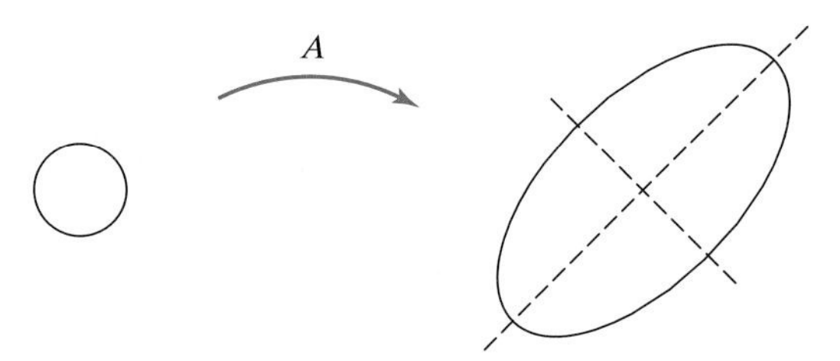

Figure 8.2 Stretching a circle into an ellipse.

EXERCISES 8.4

8.4.14. Write out the spectral factorization of the following matrices:

(a) $\begin{pmatrix} -3 & 4 \\ 4 & 3 \end{pmatrix}$ (b) $\begin{pmatrix} 2 & -1 \\ -1 & 4 \end{pmatrix}$

(c) $\begin{pmatrix} 1 & 1 & 0 \\ 1 & 2 & 1 \\ 0 & 1 & 1 \end{pmatrix}$ (d) $\begin{pmatrix} 3 & -1 & -1 \\ -1 & 2 & 0 \\ -1 & 0 & 2 \end{pmatrix}$

8.4.15. Write out the spectral factorization of the matrices listed in Exercise 8.4.1.

8.4.16. Construct a symmetric matrix with the following eigenvectors and eigenvalues, or explain why none exists:

(a) $\lambda_1 = 1,\quad \mathbf{v}_1 = \left(\frac{3}{5}, \frac{4}{5}\right)^T,$
$\lambda_2 = 3,\quad \mathbf{v}_2 = \left(-\frac{4}{5}, \frac{3}{5}\right)^T$

(b) $\lambda_1 = -2,\quad \mathbf{v}_1 = (1, -1)^T,$
$\lambda_2 = 1,\quad \mathbf{v}_2 = (1, 1)^T$

(c) $\lambda_1 = 3,\quad \mathbf{v}_1 = (2, -1)^T,$
$\lambda_2 = -1,\quad \mathbf{v}_2 = (-1, 2)^T$

(d) $\lambda_1 = 2,\quad \mathbf{v}_1 = (2, 1)^T,$
$\lambda_2 = 2,\quad \mathbf{v}_2 = (1, 2)^T$

8.4.17. Use the spectral factorization to diagonalize the following quadratic forms:

(a) $x^2 - 3xy + 5y^2$
(b) $3x^2 + 4xy + 6y^2$
(c) $x^2 + 8xz + y^2 + 6yz + z^2$
(d) $\frac{3}{2}x^2 - xy - xz + y^2 + z^2$
(e) $6x^2 - 8xy + 2xz + 6y^2 - 2yz + 11z^2$

♡ **8.4.18.** Find the eigenvalues and eigenvectors of the matrix

$$A = \begin{pmatrix} 2 & 1 & -1 \\ 1 & 2 & 1 \\ -1 & 1 & 2 \end{pmatrix}.$$

(a) Use the eigenvalues to compute the determinant of A.
(b) Is A positive definite? Why or why not?
(c) Find an orthonormal eigenvector basis of $\mathbb{R}^3$ determined by A or explain why none exists.
(d) Write out the spectral factorization of A if possible.
(e) Use orthogonality to write the vector $(1, 0, 0)^T$ as a linear combination of eigenvectors of A.

◇ **8.4.19.** Let $\mathbf{u}_1, \ldots, \mathbf{u}_n$ be an orthonormal basis of $\mathbb{R}^n$. Prove that it forms an eigenvector basis for some symmetric $n \times n$ matrix A. Can you characterize all such matrices?

8.4.20. *True or false*: A matrix with a real orthonormal eigenvector basis is symmetric.

8.4.21. Prove that any quadratic form can be written as

$$\mathbf{x}^T A\,\mathbf{x} = \|\mathbf{x}\|^2 \left(\sum_{i=1}^{n} \lambda_i \cos^2 \theta_i \right),$$

where λ_i are the eigenvalues of A and $\theta_i = \measuredangle\,(\mathbf{x}, \mathbf{v}_i)$ denotes the angle between $\mathbf{x}$ and the ith eigenvector.

8.4.22. An elastic body has stress tensor

$$T = \begin{pmatrix} 3 & 1 & 2 \\ 1 & 3 & 1 \\ 2 & 1 & 3 \end{pmatrix}.$$

Find the principal stretches and principal directions of stretch.

◇ **8.4.23.** Given a solid body spinning around its center of mass, the eigenvectors of its positive definite *inertia tensor* prescribe three mutually orthogonal *principal directions* of rotation, while the corresponding eigenvalues are the *moments of inertia*. Given the inertia tensor

$$T = \begin{pmatrix} 2 & 1 & 0 \\ 1 & 3 & 1 \\ 0 & 1 & 2 \end{pmatrix},$$

find the principal directions and moments of inertia.

◇ **8.4.24.** Let K be a positive definite 2×2 matrix.

(a) Explain why the quadratic equation $\mathbf{x}^T K\,\mathbf{x} = 1$ defines an ellipse. Prove that its principal axes are the eigenvectors of K, and the semi-axes are the reciprocals of the square roots of the eigenvalues.

(b) Graph and describe the following curves:

(i) $x^2 + 4\,y^2 = 1$
(ii) $x^2 + x\,y + y^2 = 1$
(iii) $3\,x^2 + 2\,x\,y + y^2 = 1$

(c) What sort of curve(s) does $\mathbf{x}^T K\,\mathbf{x} = 1$ describe if K is not positive definite?

◇ **8.4.25.** Let K be a positive definite 3×3 matrix.

(a) Prove that the quadratic equation $\mathbf{x}^T K\,\mathbf{x} = 1$ defines an ellipsoid in $\mathbb{R}^3$. What are its principal axes and semi-axes?

(b) Describe the surface defined by the quadratic equation $11\,x^2 - 8\,x\,y + 20\,y^2 - 10\,x\,z + 8\,y\,z + 11\,z^2 = 1$.

8.4.26. Prove that $A = A^T$ has a repeated eigenvalue if and only if it commutes, $A\,J = J\,A$, with a nonzero skew-symmetric matrix: $J^T = -J \neq \mathrm{O}$. *Hint*: First prove this when A is a diagonal matrix.

◇ **8.4.27.** (a) Prove that every positive definite matrix K has a unique positive definite *square root*, i.e., a matrix $B > 0$ satisfying $B^2 = K$.

(b) Find the positive definite square roots of the following matrices:

(i) $\begin{pmatrix} 2 & 1 \\ 1 & 2 \end{pmatrix}$ (ii) $\begin{pmatrix} 3 & -1 \\ -1 & 1 \end{pmatrix}$

(iii) $\begin{pmatrix} 2 & 0 & 0 \\ 0 & 5 & 0 \\ 0 & 0 & 9 \end{pmatrix}$

(iv) $\begin{pmatrix} 6 & -4 & 1 \\ -4 & 6 & -1 \\ 1 & -1 & 11 \end{pmatrix}$

8.4.28. Find all positive definite orthogonal matrices.

◇ **8.4.29.** *The Polar Decomposition*: Prove that every invertible matrix A has a *polar decomposition*, written $A = Q\,B$, into the product of an orthogonal matrix Q and a positive definite matrix $B > 0$. Show that if $\det A > 0$, then Q is a proper orthogonal matrix. *Hint*: Look at the Gram matrix $K = A^T A$ and use Exercise 8.4.27.
Remark: In mechanics, if A represents the deformation of a body, then Q represents a rotation, while B represents a stretching along the orthogonal eigendirections of K. Thus, any linear deformation of an elastic body can be decomposed into a pure stretching transformation followed by a rotation.

8.4.30. Find the polar decompositions $A = Q\,B$, as defined in Exercise 8.4.29, of the following matrices:

(a) $\begin{pmatrix} 0 & 1 \\ 2 & 0 \end{pmatrix}$ (b) $\begin{pmatrix} 2 & -3 \\ 1 & 6 \end{pmatrix}$

(c) $\begin{pmatrix} 1 & 2 \\ 0 & 1 \end{pmatrix}$ (d) $\begin{pmatrix} 0 & -3 & 8 \\ 1 & 0 & 0 \\ 0 & 4 & 6 \end{pmatrix}$

(e) $\begin{pmatrix} 1 & 0 & 1 \\ 1 & -2 & 0 \\ 1 & 1 & 0 \end{pmatrix}$

♡ **8.4.31.** *The Spectral Decomposition*:

(i) Let A be a symmetric matrix with eigenvalues $\lambda_1, \ldots, \lambda_n$ and corresponding orthonormal eigenvectors $\mathbf{u}_1, \ldots, \mathbf{u}_n$. Let $P_k = \mathbf{u}_k\,\mathbf{u}_k^T$ be the orthogonal projection matrix onto the eigenline spanned by $\mathbf{u}_k$, as defined in Exercise 5.5.8. Prove that the spectral factorization (8.32) can be rewritten as

$$\begin{aligned} A &= \lambda_1\,P_1 + \lambda_2\,P_2 + \cdots + \lambda_n\,P_n \\ &= \lambda_1\,\mathbf{u}_1\,\mathbf{u}_1^T + \lambda_2\,\mathbf{u}_2\,\mathbf{u}_2^T + \cdots + \lambda_n\,\mathbf{u}_n\,\mathbf{u}_n^T, \end{aligned} \tag{8.34}$$

expressing A as a linear combination of projection matrices.

(ii) Write out the *spectral decomposition* (8.34) for the matrices in Exercise 8.4.14.

◇ **8.4.32.** *The Spectral Theorem for Hermitian Matrices.* Prove that any complex Hermitian matrix can be factored as $H = U \Lambda U^\dagger$ where U is a unitary matrix and Λ is a real diagonal matrix. *Hint*: See Exercises 5.3.25, 8.4.7.

8.4.33. Find the spectral factorization, as in Exercise 8.4.32, of the following Hermitian matrices:

(a) $\begin{pmatrix} 3 & 2\,\mathrm{i} \\ -2\,\mathrm{i} & 6 \end{pmatrix}$ (b) $\begin{pmatrix} 6 & 1-2\,\mathrm{i} \\ 1+2\,\mathrm{i} & 2 \end{pmatrix}$

(c) $\begin{pmatrix} -1 & 5\,\mathrm{i} & -4 \\ -5\,\mathrm{i} & -1 & 4\,\mathrm{i} \\ -4 & -4\,\mathrm{i} & 8 \end{pmatrix}$

Optimization Principles for Eigenvalues

As we learned in Chapter 4, the solution to a linear system with positive definite coefficient matrix can be characterized by a minimization principle. Thus, it should come as no surprise that eigenvalues of positive definite, and even more general symmetric matrices, can also be characterized by some sort of optimization procedure. A number of basic numerical algorithms for computing eigenvalues of matrices and, subsequently, of differential operators are based on such optimization principles.

First, consider the relatively simple case of a diagonal matrix

$$\Lambda = \operatorname{diag}(\lambda_1, \ldots, \lambda_n).$$

We assume that the diagonal entries, which are the *same* as the eigenvalues, appear in decreasing order,

$$\lambda_1 \geq \lambda_2 \geq \cdots \geq \lambda_n, \tag{8.35}$$

so λ_1 is the largest eigenvalue, while λ_n is the smallest. The effect of Λ on a vector $\mathbf{y} = (y_1, y_2, \ldots, y_n)^T \in \mathbb{R}^n$ is to multiply its entries by the diagonal eigenvalues: $\Lambda\,\mathbf{y} = (\lambda_1 y_1, \lambda_2 y_2, \ldots, \lambda_n y_n)^T$. In other words, the linear transformation represented by the coefficient matrix Λ has the effect of stretching* the ith coordinate direction by the factor λ_i. In particular, the maximal stretch occurs in the $\mathbf{e}_1$ direction, with factor λ_1, while the minimal (or largest negative) stretch occurs in the $\mathbf{e}_n$ direction, with factor λ_n. The germ of the optimization principles for characterizing the extreme eigenvalues is contained in this geometrical observation.

Let us turn our attention to the associated quadratic form

$$q(\mathbf{y}) = \mathbf{y}^T \Lambda\,\mathbf{y} = \lambda_1 y_1^2 + \lambda_2 y_2^2 + \cdots + \lambda_n y_n^2. \tag{8.36}$$

Note that $q(t\,\mathbf{e}_1) = \lambda_1 t^2$, and hence if $\lambda_1 > 0$, then $q(\mathbf{y})$ has no maximum; on the other hand, if $\lambda_1 \leq 0$, so all eigenvalues are non-positive, then $q(\mathbf{y}) \leq 0$ for all $\mathbf{y}$, and its maximal value is $q(\mathbf{0}) = 0$. Thus, in either case, a strict maximization of $q(\mathbf{y})$ is of scant help.

Suppose, however, that we continue in our quest to maximize $q(\mathbf{y})$, but now restrict $\mathbf{y}$ to be a unit vector (in the Euclidean norm), so

$$\|\mathbf{y}\|^2 = y_1^2 + \cdots + y_n^2 = 1.$$

In view of (8.35),

$$\begin{aligned} q(\mathbf{y}) &= \lambda_1 y_1^2 + \lambda_2 y_2^2 + \cdots + \lambda_n y_n^2 \\ &\leq \lambda_1 y_1^2 + \lambda_1 y_2^2 + \cdots + \lambda_1 y_n^2 \\ &= \lambda_1 \big(y_1^2 + \cdots + y_n^2 \big) \\ &= \lambda_1. \end{aligned}$$

*If $\lambda_i < 0$, then the effect is to stretch and reflect.

Moreover, $q(\mathbf{e}_1) = \lambda_1$. We conclude that the maximal value of $q(\mathbf{y})$ over all unit vectors *is* the largest eigenvalue of Λ:

$$\lambda_1 = \max\{\, q(\mathbf{y}) \mid \|\mathbf{y}\| = 1 \,\}.$$

By the same reasoning, its minimal value equals the smallest eigenvalue:

$$\lambda_n = \min\{\, q(\mathbf{y}) \mid \|\mathbf{y}\| = 1 \,\}.$$

Thus, we can characterize the two extreme eigenvalues by optimization principles, albeit of a slightly different character than we treated in Chapter 4.

Now suppose A is any symmetric matrix. We use its spectral factorization (8.32) to diagonalize the associated quadratic form

$$q(\mathbf{x}) = \mathbf{x}^T A\,\mathbf{x} = \mathbf{x}^T Q\,\Lambda\,Q^T\mathbf{x} = \mathbf{y}^T \Lambda\,\mathbf{y}, \qquad \text{where} \quad \mathbf{y} = Q^T\mathbf{x} = Q^{-1}\mathbf{x},$$

as in (8.33). According to the preceding discussion, the maximum of $\mathbf{y}^T\Lambda\,\mathbf{y}$ over all unit vectors $\|\mathbf{y}\| = 1$ is the largest eigenvalue λ_1 of Λ, which is the *same* as the largest eigenvalue of A. Moreover, since Q is an orthogonal matrix, Proposition 7.24 tell us that it maps unit vectors to unit vectors:

$$1 = \|\mathbf{y}\| = \|Q^T\mathbf{x}\| = \|\mathbf{x}\|,$$

and so the maximum of $q(\mathbf{x})$ over all unit vectors $\|\mathbf{x}\| = 1$ is the same maximum eigenvalue λ_1. Similar reasoning applies to the smallest eigenvalue λ_n. In this fashion, we have established the basic optimization principles for the extreme eigenvalues of a symmetric matrix.

THEOREM 8.28 If A is a symmetric matrix, then

$$\lambda_1 = \max\left\{\, \mathbf{x}^T A\,\mathbf{x} \;\middle|\; \|\mathbf{x}\| = 1 \,\right\}, \qquad \lambda_n = \min\left\{\, \mathbf{x}^T A\,\mathbf{x} \;\middle|\; \|\mathbf{x}\| = 1 \,\right\}, \tag{8.37}$$

are, respectively its largest and smallest eigenvalues. The maximal value is achieved when $\mathbf{x} = \pm\mathbf{u}_1$ is one of the unit eigenvectors corresponding to the largest eigenvalue; similarly, the minimal value is at $\mathbf{x} = \pm\mathbf{u}_n$.

REMARK: In multivariable calculus, the eigenvalue λ plays the role of a *Lagrange multiplier* for the constrained optimization problem, [2, 58, 59].

EXAMPLE 8.29 The problem is to maximize the value of the quadratic form

$$q(x, y) = 3x^2 + 2xy + 3y^2$$

for all x, y lying on the unit circle $x^2 + y^2 = 1$. This maximization problem is precisely of form (8.37). The symmetric coefficient matrix for the quadratic form is

$$A = \begin{pmatrix} 3 & 1 \\ 1 & 3 \end{pmatrix},$$

whose eigenvalues are, according to Example 8.5, $\lambda_1 = 4$ and $\lambda_2 = 2$. Theorem 8.28 implies that the maximum is the largest eigenvalue, and hence equal to 4, while the minimum is the smallest eigenvalue, and hence equal to 2. Thus, evaluating $q(x, y)$ on the unit eigenvectors, we conclude that

$$q\left(-\tfrac{1}{\sqrt{2}}, \tfrac{1}{\sqrt{2}}\right) = 2 \le q(x, y) \le 4 = q\left(-\tfrac{1}{\sqrt{2}}, \tfrac{1}{\sqrt{2}}\right) \qquad \text{for all} \quad x^2 + y^2 = 1. \qquad \bullet$$

In practical applications, the restriction of the quadratic form to unit vectors may not be particularly convenient. We can, however, rephrase the eigenvalue optimization principles in a form that utilizes general vectors. If $\mathbf{v} \neq \mathbf{0}$ is any nonzero vector, then $\mathbf{x} = \mathbf{v}/\|\mathbf{v}\|$ is a unit vector. Substituting this expression for $\mathbf{x}$ in the quadratic form $q(\mathbf{x}) = \mathbf{x}^T A\mathbf{x}$ leads to the following optimization principles for the extreme eigenvalues of a symmetric matrix:

$$\lambda_1 = \max \left\{ \frac{\mathbf{v}^T A \mathbf{v}}{\|\mathbf{v}\|^2} \;\middle|\; \mathbf{v} \neq \mathbf{0} \right\}, \qquad \lambda_n = \min \left\{ \frac{\mathbf{v}^T A \mathbf{v}}{\|\mathbf{v}\|^2} \;\middle|\; \mathbf{v} \neq \mathbf{0} \right\}. \tag{8.38}$$

Thus, we replace optimization of a quadratic polynomial over the unit sphere by optimization of a rational function over all of $\mathbb{R}^n \setminus \{\mathbf{0}\}$. For instance, referring back to Example 8.29, the maximum value of

$$r(x,y) = \frac{3x^2 + 2xy + 3y^2}{x^2 + y^2} \qquad \text{for all} \quad \begin{pmatrix} x \\ y \end{pmatrix} \neq \begin{pmatrix} 0 \\ 0 \end{pmatrix}$$

is equal to 4, the same maximal eigenvalue of the corresponding coefficient matrix.

What about characterizing one of the intermediate eigenvalues? Then we need to be a little more sophisticated in designing the optimization principle. To motivate the construction, look first at the diagonal case. If we restrict the quadratic form (8.36) to vectors $\widetilde{\mathbf{y}} = (0, y_2, \ldots, y_n)^T$ whose first component is zero, we obtain

$$q(\widetilde{\mathbf{y}}) = q(0, y_2, \ldots, y_n) = \lambda_2 y_2^2 + \cdots + \lambda_n y_n^2.$$

The maximum value of $q(\widetilde{\mathbf{y}})$ over all such $\widetilde{\mathbf{y}}$ of norm 1 is, by the same reasoning, the second largest eigenvalue λ_2. Moreover, $\widetilde{\mathbf{y}} \cdot \mathbf{e}_1 = 0$, and so $\widetilde{\mathbf{y}}$ can be characterized as being orthogonal to the first standard basis vector, which also happens to be the eigenvector of Λ corresponding to the eigenvalue λ_1. Thus, to find the second eigenvalue, we maximize the quadratic form over all unit vectors that are orthogonal to the first eigenvector. Similarly, if we want to find the jth largest eigenvalue λ_j, we maximize $q(\widehat{\mathbf{y}})$ over all unit vectors $\widehat{\mathbf{y}}$ whose first $j-1$ components vanish, $y_1 = \cdots = y_{j-1} = 0$, or, stated geometrically, over all vectors $\widehat{\mathbf{y}}$ such that $\|\widehat{\mathbf{y}}\| = 1$ and $\widehat{\mathbf{y}} \cdot \mathbf{e}_1 = \cdots = \widehat{\mathbf{y}} \cdot \mathbf{e}_{j-1} = 0$, that is, over all unit vectors orthogonal to the first $j-1$ eigenvectors of Λ.

A similar reasoning based on the Spectral Theorem 8.26 and the orthogonality of eigenvectors of symmetric matrices leads to the general result.

THEOREM 8.30 Let A be a symmetric matrix with eigenvalues $\lambda_1 \geq \lambda_2 \geq \cdots \geq \lambda_n$ and corresponding orthogonal eigenvectors $\mathbf{v}_1, \ldots, \mathbf{v}_n$. Then the maximal value of the quadratic form $q(\mathbf{x}) = \mathbf{x}^T A\mathbf{x}$ over all unit vectors that are orthogonal to the first $j-1$ eigenvectors is its jth eigenvalue:

$$\lambda_j = \max \left\{ \mathbf{x}^T A \mathbf{x} \;\middle|\; \|\mathbf{x}\| = 1, \;\; \mathbf{x} \cdot \mathbf{v}_1 = \cdots = \mathbf{x} \cdot \mathbf{v}_{j-1} = 0 \right\}. \tag{8.39}$$

Thus, at least in principle, one can compute the eigenvalues and eigenvectors of a symmetric matrix by the following recursive procedure. First, find the largest eigenvalue λ_1 by the basic maximization principle (8.37) and its associated eigenvector $\mathbf{v}_1$ by solving the eigenvector system (8.13). The next largest eigenvalue λ_2 is then characterized by the constrained maximization principle (8.39), and so on. Although of theoretical interest, this algorithm is of somewhat limited value in practical numerical computations.

EXERCISES 8.4

8.4.34. Find the minimum and maximum values of the quadratic form $5x^2 + 4xy + 5y^2$ where x, y are subject to the constraint $x^2 + y^2 = 1$.

8.4.35. Find the minimum and maximum values of the quadratic form $2x^2 + xy + 2xz + 2y^2 + 2z^2$ where x, y, z are required to satisfy $x^2 + y^2 + z^2 = 1$.

8.4.36. Write down and solve an optimization principle characterizing the largest and smallest eigenvalue of the following positive definite matrices:

(a) $\begin{pmatrix} 2 & -1 \\ -1 & 3 \end{pmatrix}$ (b) $\begin{pmatrix} 4 & 1 \\ 1 & 4 \end{pmatrix}$

(c) $\begin{pmatrix} 6 & -4 & 1 \\ -4 & 6 & -1 \\ 1 & -1 & 11 \end{pmatrix}$

(d) $\begin{pmatrix} 4 & -1 & -2 \\ -1 & 4 & -1 \\ -2 & -1 & 4 \end{pmatrix}$

8.4.37. Write down a maximization principle that characterizes the middle eigenvalue of the matrices in parts (c–d) of Exercise 8.4.36.

8.4.38. What are the minimum and maximum values of the following rational functions:

(a) $\dfrac{3x^2 - 2y^2}{x^2 + y^2}$ (b) $\dfrac{x^2 - 3xy + y^2}{x^2 + y^2}$

(c) $\dfrac{3x^2 + xy + 5y^2}{x^2 + y^2}$

(d) $\dfrac{2x^2 + xy + 3xz + 2y^2 + 2z^2}{x^2 + y^2 + z^2}$

8.4.39. Find the minimum and maximum values of

$$q(\mathbf{x}) = \sum_{i=1}^{n-1} x_i x_{i+1}$$

for $\|\mathbf{x}\|^2 = 1$. *Hint*: See Exercise 8.2.47.

8.4.40. Suppose $K > 0$. What is the maximum value of $q(\mathbf{x}) = \mathbf{x}^T K \mathbf{x}$ when $\mathbf{x}$ is constrained to a sphere of radius $\|\mathbf{x}\| = r$?

8.4.41. Let $K > 0$. Prove the product formula

$$\max\left\{ \mathbf{x}^T K \mathbf{x} \;\middle|\; \|\mathbf{x}\| = 1 \right\} \cdot \min\left\{ \mathbf{x}^T K^{-1} \mathbf{x} \;\middle|\; \|\mathbf{x}\| = 1 \right\} = 1.$$

◇ **8.4.42.** Write out the details in the proof of Theorem 8.30.

◇ **8.4.43.** Reformulate Theorem 8.30 as a minimum principle for intermediate eigenvalues.

8.4.44. Under the set-up of Theorem 8.30, explain why

$$\lambda_j = \max\left\{ \frac{\mathbf{v}^T K \mathbf{v}}{\|\mathbf{v}\|^2} \;\middle|\; \begin{array}{l} \mathbf{v} \neq \mathbf{0}, \\ \mathbf{v} \cdot \mathbf{v}_1 = \cdots \\ \quad = \mathbf{v} \cdot \mathbf{v}_{j-1} = 0 \end{array} \right\}.$$

♡ **8.4.45.** (a) Let K, M be positive definite $n \times n$ matrices and $\lambda_1 \geq \cdots \geq \lambda_n$ be their generalized eigenvalues, as in Exercise 8.4.9. Prove that that the largest generalized eigenvalue can be characterized by the maximum principle

$$\lambda_1 = \max\left\{ \mathbf{x}^T K \mathbf{x} \;\middle|\; \mathbf{x}^T M \mathbf{x} = 1 \right\}.$$

Hint: Use Exercise 8.4.27.

(b) Prove the alternative maximum principle

$$\lambda_1 = \max\left\{ \frac{\mathbf{x}^T K \mathbf{x}}{\mathbf{x}^T M \mathbf{x}} \;\middle|\; \mathbf{x} \neq \mathbf{0} \right\}.$$

(c) How would you characterize the smallest generalized eigenvalue?

(d) An intermediate generalized eigenvalue?

8.4.46. Use Exercise 8.4.45 to find the minimum and maximum of the following rational functions:

(a) $\dfrac{3x^2 + 2y^2}{4x^2 + 5y^2}$ (b) $\dfrac{x^2 - xy + 2y^2}{2x^2 - xy + y^2}$

(c) $\dfrac{2x^2 + 3y^2 + z^2}{x^2 + 3y^2 + 2z^2}$

(d) $\dfrac{2x^2 + 6xy + 11y^2 + 6yz + 2z^2}{x^2 + 2xy + 3y^2 + 2yz + z^2}$

8.4.47. Let A be a complete square matrix, not necessarily symmetric, with all positive eigenvalues. Is the quadratic form $q(\mathbf{x}) = \mathbf{x}^T A \mathbf{x} > 0$ for all $\mathbf{x} \neq \mathbf{0}$?

8.5 SINGULAR VALUES

We have already indicated the central role played by the eigenvalues and eigenvectors of a square matrix in both theory and applications. Much more evidence to this effect will appear in the ensuing chapters. Alas, rectangular matrices do not have eigenvalues (why?), and so, at first glance, do not appear to possess any quantities of comparable significance. But you no doubt recall that our earlier treatment of least squares minimization problems, as well as the equilibrium equations for structures and circuits, made essential use of the symmetric, positive semi-definite *square* Gram matrix $K = A^T A$—which can be naturally formed even when A is not square. Perhaps the eigenvalues of K might play a comparably important role for general matrices. Since they are not easily related to the eigenvalues of A—which, in the non-square case, don't even exist — we shall endow them with a new name.

Definition 8.31 The *singular values* $\sigma_1, \ldots, \sigma_r$ of an $m \times n$ matrix A are the positive square roots, $\sigma_i = \sqrt{\lambda_i} > 0$, of the nonzero eigenvalues of the associated Gram matrix $K = A^T A$. The corresponding eigenvectors of K are known as the *singular vectors* of A.

Since K is necessarily positive semi-definite, its eigenvalues are always non-negative, $\lambda_i \geq 0$, which justifies the positivity of the singular values of A—independently of whether A itself has positive, negative, or even complex eigenvalues—or is rectangular and has no eigenvalues at all. The standard convention is to label the singular values in decreasing order, so that $\sigma_1 \geq \sigma_2 \geq \cdots \geq \sigma_r > 0$. Thus, σ_1 will always denote the largest or *dominant* singular value. If $K = A^T A$ has repeated eigenvalues, the singular values of A are repeated with the same multiplicities. As we will see, the number r of singular values is always equal to the rank of the matrix.

Warning: Many texts include the zero eigenvalues of K as singular values of A. We find this to be somewhat less convenient, but you should be aware of the differences in the two conventions.

EXAMPLE 8.32 Let $A = \begin{pmatrix} 3 & 5 \\ 4 & 0 \end{pmatrix}$. The associated Gram matrix

$$K = A^T A = \begin{pmatrix} 25 & 15 \\ 15 & 25 \end{pmatrix}$$

has eigenvalues

$$\lambda_1 = 40, \qquad \lambda_2 = 10,$$

and corresponding eigenvectors

$$\mathbf{v}_1 = \begin{pmatrix} 1 \\ 1 \end{pmatrix}, \qquad \mathbf{v}_2 = \begin{pmatrix} 1 \\ -1 \end{pmatrix}.$$

Thus, the singular values of A are $\sigma_1 = \sqrt{40} \approx 6.3246$ and $\sigma_2 = \sqrt{10} \approx 3.1623$, with $\mathbf{v}_1, \mathbf{v}_2$ being the singular vectors. Note that the singular values are *not* the same as its eigenvalues, which are $\lambda_1 = \frac{1}{2}(3 + \sqrt{89}) \approx 6.2170$ and $\lambda_2 = \frac{1}{2}(3 - \sqrt{89}) \approx -3.2170$—nor are the singular vectors eigenvectors of A. ●

Only in the special case of symmetric matrices is there a direct connection between the singular values and the eigenvalues.

Proposition 8.33 If $A = A^T$ is a symmetric matrix, its singular values are the absolute values of its nonzero eigenvalues: $\sigma_i = |\lambda_i| > 0$; its singular vectors coincide with the associated non-null eigenvectors.

Proof When A is symmetric, $K = A^T A = A^2$. So, if $A\mathbf{v} = \lambda\,\mathbf{v}$, then $K\,\mathbf{v} = A^2\,\mathbf{v} = \lambda^2\,\mathbf{v}$. Thus, every eigenvector $\mathbf{v}$ of A is also an eigenvector of K with eigenvalue λ^2. Therefore, the eigenvector basis of A is also an eigenvector basis for K, and hence also forms a complete system of singular vectors for A. ■

The generalization of the spectral factorization (8.32) to non-symmetric matrices is known as the *singular value decomposition*, commonly abbreviated SVD. Unlike the spectral factorization, the singular value decomposition applies to arbitrary (nonzero) real rectangular matrices.

THEOREM 8.34 Any nonzero real $m \times n$ matrix A of rank $r > 0$ can be factored,

$$A = P\,\Sigma\,Q^T, \tag{8.40}$$

into the product of an $m \times r$ matrix P with orthonormal* columns, so $P^T P = \mathrm{I}$, the $r \times r$ diagonal matrix $\Sigma = \operatorname{diag}(\sigma_1, \ldots, \sigma_r)$ that has the singular values of A as its diagonal entries, and an $r \times n$ matrix Q^T with orthonormal rows, so $Q^T Q = \mathrm{I}$.

Proof Let's begin by rewriting the desired factorization (8.40) as

$$A\,Q = P\,\Sigma.$$

The individual columns of this matrix equation are the vector equations

$$A\,\mathbf{q}_i = \sigma_i\,\mathbf{p}_i, \qquad i = 1, \ldots, r, \tag{8.41}$$

relating the orthonormal columns of $Q = (\,\mathbf{q}_1\ \mathbf{q}_2\ \ldots\ \mathbf{q}_r\,)$ to the orthonormal columns of $P = (\,\mathbf{p}_1\ \mathbf{p}_2\ \ldots\ \mathbf{p}_r\,)$. Thus, our goal is to find vectors $\mathbf{p}_1, \ldots, \mathbf{p}_r, \mathbf{q}_1, \ldots, \mathbf{q}_r$, that satisfy (8.41). To this end, we let $\mathbf{q}_1, \ldots, \mathbf{q}_r$ be orthonormal eigenvectors of the Gram matrix $K = A^T A$ corresponding to the non-zero eigenvalues, which, according to Proposition 8.25 form a basis for $\operatorname{rng} K = \operatorname{corng} A$, of dimension $r = \operatorname{rank} K = \operatorname{rank} A$. Thus, by the definition of the singular values,

$$A^T A\,\mathbf{q}_i = K\,\mathbf{q}_i = \lambda_i\,\mathbf{q}_i = \sigma_i^2\,\mathbf{q}_i, \qquad i = 1, \ldots, r.$$

We claim that the image vectors $\mathbf{w}_i = A\,\mathbf{q}_i$ are automatically orthogonal. Indeed, in view of the orthonormality of the $\mathbf{q}_i$,

$$\begin{aligned} \mathbf{w}_i \cdot \mathbf{w}_j = \mathbf{w}_i^T \mathbf{w}_j = (A\,\mathbf{q}_i)^T A\,\mathbf{q}_j &= \mathbf{q}_i^T A^T A\,\mathbf{q}_j = \sigma_j^2\,\mathbf{q}_i^T\,\mathbf{q}_j \\ &= \sigma_j^2\,\mathbf{q}_i \cdot \mathbf{q}_j = \begin{cases} 0, & i \neq j, \\ \sigma_i^2, & i = j. \end{cases} \end{aligned}$$

Consequently, $\mathbf{w}_1, \ldots, \mathbf{w}_r$ form an orthogonal system of vectors having respective norms

$$\|\mathbf{w}_i\| = \sqrt{\mathbf{w}_i \cdot \mathbf{w}_i} = \sigma_i.$$

We conclude that the corresponding unit vectors

$$\mathbf{p}_i = \frac{\mathbf{w}_i}{\sigma_i} = \frac{A\,\mathbf{q}_i}{\sigma_i}, \qquad i = 1, \ldots, r, \tag{8.42}$$

form an orthonormal set of vectors satisfying the required equations (8.41). ■

EXAMPLE 8.35 For the matrix $A = \begin{pmatrix} 3 & 5 \\ 4 & 0 \end{pmatrix}$ considered in Example 8.32, the orthonormal eigenvector basis of

$$K = A^T A = \begin{pmatrix} 25 & 15 \\ 15 & 25 \end{pmatrix}$$

is given by the unit singular vectors

$$\mathbf{q}_1 = \begin{pmatrix} \frac{1}{\sqrt{2}} \\ \frac{1}{\sqrt{2}} \end{pmatrix} \qquad \text{and} \qquad \mathbf{q}_2 = \begin{pmatrix} -\frac{1}{\sqrt{2}} \\ \frac{1}{\sqrt{2}} \end{pmatrix}.$$

---

*Throughout this section, we exclusively use the Euclidean dot product and norm.

Thus,

$$Q = \begin{pmatrix} \frac{1}{\sqrt{2}} & -\frac{1}{\sqrt{2}} \\ \frac{1}{\sqrt{2}} & \frac{1}{\sqrt{2}} \end{pmatrix}.$$

On the other hand, according to (8.42),

$$\mathbf{p}_1 = \frac{A\,\mathbf{q}_1}{\sigma_1} = \frac{1}{\sqrt{40}} \begin{pmatrix} 4\sqrt{2} \\ 2\sqrt{2} \end{pmatrix} = \begin{pmatrix} \frac{2}{\sqrt{5}} \\ \frac{1}{\sqrt{5}} \end{pmatrix},$$

$$\mathbf{p}_2 = \frac{A\,\mathbf{q}_2}{\sigma_2} = \frac{1}{\sqrt{10}} \begin{pmatrix} \sqrt{2} \\ -2\sqrt{2} \end{pmatrix} = \begin{pmatrix} \frac{1}{\sqrt{5}} \\ -\frac{2}{\sqrt{5}} \end{pmatrix},$$

and thus

$$P = \begin{pmatrix} \frac{2}{\sqrt{5}} & \frac{1}{\sqrt{5}} \\ \frac{1}{\sqrt{5}} & -\frac{2}{\sqrt{5}} \end{pmatrix}.$$

You may wish to validate the resulting singular value factorization

$$A = \begin{pmatrix} 3 & 5 \\ 4 & 0 \end{pmatrix} = \begin{pmatrix} \frac{2}{\sqrt{5}} & \frac{1}{\sqrt{5}} \\ \frac{1}{\sqrt{5}} & -\frac{2}{\sqrt{5}} \end{pmatrix} \begin{pmatrix} \sqrt{40} & 0 \\ 0 & \sqrt{10} \end{pmatrix} \begin{pmatrix} \frac{1}{\sqrt{2}} & \frac{1}{\sqrt{2}} \\ -\frac{1}{\sqrt{2}} & \frac{1}{\sqrt{2}} \end{pmatrix} = P\,\Sigma\,Q^T.$$

●

The singular value decomposition is telling us some interesting new geometrical information concerning matrix multiplication, further supplementing the discussion begun in Section 2.5 and continued in Section 5.6.

Proposition 8.36 Given the singular value decomposition $A = P\,\Sigma\,Q^T$, the columns $\mathbf{q}_1, \ldots, \mathbf{q}_r$ of Q form an orthonormal basis for corng A, while the columns $\mathbf{p}_1, \ldots, \mathbf{p}_r$ of P form an orthonormal basis for rng A.

Proof The first part of the proposition was proved during the course of the proof of Theorem 8.34. Moreover, since $\mathbf{p}_i = A(\mathbf{q}_i/\sigma_i) \in \operatorname{rng} A$ are $r = \operatorname{rank} A = \dim \operatorname{rng} A$ nonzero orthogonal vectors, they must form an orthogonal basis for the range. ■

If A is a nonsingular $n \times n$ matrix, then the matrices P, Σ, Q in its singular value decomposition (8.40) are all of size $n \times n$. Interpreting them as linear transformations of $\mathbb{R}^n$, the two orthogonal matrices represent rigid rotations/reflections, while the diagonal matrix Σ represents a combination of simple stretches, by an amount given by the singular values, in the orthogonal coordinate directions. Thus, every invertible linear transformation on $\mathbb{R}^n$ can be decomposed into a rotation/reflection Q^T, followed by the orthogonal stretching transformations represented by Σ, followed by another rotation/reflection P.

In the more general rectangular case, the matrix Q^T represents a projection from $\mathbb{R}^n$ to corng A, the matrix Σ continues to represent stretching transformations within this r-dimensional subspace, while P maps the result to rng $A \subset \mathbb{R}^m$. We already noted in Section 5.6 that the linear transformation $L\colon \mathbb{R}^n \to \mathbb{R}^m$ defined by matrix multiplication, $L[\mathbf{x}] = A\,\mathbf{x}$, can be interpreted as a projection from $\mathbb{R}^n$ to corng A followed by an invertible map from corng A to rng A. The singular value decomposition tells us that not only is the latter map invertible, it is simply a combination of stretches in the r mutually orthogonal singular directions $\mathbf{q}_1, \ldots, \mathbf{q}_r$, whose magnitudes equal the nonzero singular values. In this way, we have at last reached a complete understanding of the subtle geometry underlying the simple operation of matrix multiplication.

Condition Number, Rank, and Principal Component Analysis

The singular values not only provide a pretty geometric interpretation of the action of the matrix, they also play a key role in modern computational algorithms. The relative magnitudes of the singular values can be used to distinguish well-behaved linear systems from ill-conditioned systems which are much trickier to solve accurately. Since the number of singular values equals the matrix's rank, an $n \times n$ matrix with fewer than n singular values is singular. For the same reason, a square matrix with one or more very small singular values should be considered to be close to singular. The potential difficulty of accurately solving a linear algebraic system with coefficient matrix A is traditionally quantified as follows.

Definition 8.37 The *condition number* of a nonsingular $n \times n$ matrix is the ratio between its largest and smallest singular value: $\kappa(A) = \sigma_1/\sigma_n$.

If A is singular, it is said to have condition number ∞. A matrix with a very large condition number is said to be *ill-conditioned*; in practice, this occurs when the condition number is larger than the reciprocal of the machine's precision, e.g., 10^7 for typical single precision arithmetic. As the name implies, it is much harder to solve a linear system $A\mathbf{x} = \mathbf{b}$ when its coefficient matrix is ill-conditioned.

Determining the rank of a large matrix can be a numerical challenge. Small numerical errors in the entries can have an unpredictable effect. For example, the matrix

$$A = \begin{pmatrix} 1 & 1 & -1 \\ 2 & 2 & -2 \\ 3 & 3 & -3 \end{pmatrix}$$

has rank $r = 1$, but a tiny change, e.g.

$$\widetilde{A} = \begin{pmatrix} 1.00001 & 1. & -1. \\ 2. & 2.00001 & -2. \\ 3. & 3. & -3.00001 \end{pmatrix},$$

will produce a nonsingular matrix with rank $r = 3$. The latter matrix, however, is very close to singular, and this is highlighted by its singular values, which are $\sigma_1 \approx 6.48075$ while $\sigma_2 \approx \sigma_3 \approx .000001$. The fact that the second and third singular values are very small indicates that $\widetilde{A}$ is very close to a matrix of rank 1 and should be viewed as a numerical (or experimental) perturbation of such a matrix. Thus, an effective practical method for computing the rank of a matrix is to first assign a threshold, e.g., 10^{-5}, for singular values, and then treat any small singular value lying below the threshold as if it were zero.

This idea underlies the method of *Principal Component Analysis* that is assuming an increasingly visible role in modern statistics, data mining, imaging, speech recognition, semantics, and a variety of other fields, [35]. The singular vectors associated with the larger singular values indicate the *principal components* of the matrix, while small singular values indicate relatively unimportant directions. In applications, the columns of the matrix A represent the data vectors, which are normalized to have mean $\mathbf{0}$. The corresponding Gram matrix $K = A^T A$ can be identified as the associated *covariance matrix*, [17]. Its eigenvectors are the principal components that serve to indicate directions of correlation and clustering in the data.

The Pseudoinverse

The singular value decomposition enables us to substantially generalize the concept of a matrix inverse. The pseudoinverse was first defined by the American mathematician Eliakim Moore in the 1920's and rediscovered by the influential British mathematician and physicist Roger Penrose in the 1950's, and often has their names attached.

Definition 8.38 The *pseudoinverse* of a nonzero $m \times n$ matrix with singular value decomposition $A = P\,\Sigma\,Q^T$ is the $n \times m$ matrix $A^+ = Q\,\Sigma^{-1}P^T$.

Note that the latter equation is the singular value decomposition of the pseudoinverse A^+, and hence its nonzero singular values are the reciprocals of the nonzero singular values of A. The only matrix without a pseudoinverse is the zero matrix O. If A is a non-singular square matrix, then its pseudoinverse agrees with its ordinary inverse. Indeed, since both P and Q are square, orthogonal matrices in this case,

$$A^{-1} = (P\,\Sigma\,Q^T)^{-1} = (Q^{-1})^T\,\Sigma^{-1}P^{-1} = Q\,\Sigma^{-1}P^T = A^+,$$

where we used the fact that the inverse of an orthogonal matrix is equal to its transpose. More generally, if A has linearly independent columns, or, equivalently, $\ker A = \{\mathbf{0}\}$, then we can bypass the singular value decomposition to compute its pseudoinverse.

Lemma 8.39 Let A be an $m \times n$ matrix of rank n. Then

$$A^+ = (A^TA)^{-1}A^T. \tag{8.43}$$

Proof Replacing A by its singular value decomposition (8.40), we find

$$A^TA = (P\,\Sigma\,Q^T)^T(P\,\Sigma\,Q^T) = Q\,\Sigma\,P^TP\,\Sigma\,Q^T = Q\,\Sigma^2\,Q^T, \tag{8.44}$$

since $\Sigma = \Sigma^T$ is a diagonal matrix, while $P^TP = \mathrm{I}$ since the columns of P are orthonormal. This is merely the spectral factorization of the Gram matrix A^TA—that we in fact already knew from the original definition of the singular values and vectors. Now if A has rank n, then Q is an $n \times n$ orthogonal matrix. Therefore,

$$(A^TA)^{-1}A^T = (Q\,\Sigma^{-2}\,Q^T)(Q\,\Sigma P^T) = Q\,\Sigma^{-1}P^T = A^+. \qquad \blacksquare$$

If A is square and nonsingular, then, as we know, the solution to the linear system $A\mathbf{x} = \mathbf{b}$ is given by $\mathbf{x}^\star = A^{-1}\mathbf{b}$. For a general coefficient matrix, the vector $\mathbf{x}^\star = A^+\mathbf{b}$ obtained by applying the pseudoinverse to the right hand side plays a distinguished role—it is the *least squares solution* to the system under the Euclidean norm.

THEOREM 8.40 Consider the linear system $A\mathbf{x} = \mathbf{b}$. Let $\mathbf{x}^\star = A^+\mathbf{b}$, where A^+ is the pseudoinverse of A. If $\ker A = \{\mathbf{0}\}$, then $\mathbf{x}^\star$ is the (Euclidean) least squares solution to the linear system. If, more generally, $\ker A \neq \{\mathbf{0}\}$, then $\mathbf{x}^\star \in \operatorname{corng} A$ is the least squares solution of minimal Euclidean norm among all vectors that minimize the least squares error $\|A\mathbf{x} - \mathbf{b}\|$.

Proof To show that $\mathbf{x}^\star = A^+\mathbf{b}$ is the least squares solution to the system, we must check that it satisfies the normal equations $A^T A\mathbf{x}^\star = A^T\mathbf{b}$. If $\operatorname{rank} A = n$, so $A^T A$ is nonsingular, this follows immediately from (8.43). More generally, using (8.44), the definition of the pseudoinverse, and the fact that Q has orthonormal columns, so $Q^T Q = \mathrm{I}$,

$$A^T A\mathbf{x}^\star = A^T A\, A^+\mathbf{b} = (Q\,\Sigma^2\,Q^T)(Q\,\Sigma^{-1}P^T)\mathbf{b} = Q\,\Sigma\,P^T\mathbf{b} = A^T\mathbf{b}.$$

This proves that $\mathbf{x}^\star$ solves the normal equations, and hence, by Theorem 4.8 and Exercise 4.3.18, minimizes the least squares error. Moreover,

$$\mathbf{x}^\star = A^+\mathbf{b} = Q\,\Sigma^{-1}\,P^T\mathbf{b} = Q\,\mathbf{c} = c_1\,\mathbf{q}_1 + \cdots + c_r\,\mathbf{q}_r,$$

where

$$\mathbf{c} = (c_1, \ldots c_r)^T = \Sigma^{-1}P^T\mathbf{b}.$$

Thus, $\mathbf{x}^\star$ is a linear combination of the singular vectors, and hence, by Proposition 8.36 and Theorem 5.59, $\mathbf{x}^\star \in \operatorname{corng} A$ is the solution with minimum norm; the most general least squares solution has the form $\mathbf{x} = \mathbf{x}^\star + \mathbf{z}$. ■

EXAMPLE 8.41 Let us use the pseudoinverse to solve the linear system $A\,\mathbf{x} = \mathbf{b}$, with

$$A = \begin{pmatrix} 1 & 2 & -1 \\ 3 & -4 & 1 \\ -1 & 3 & -1 \\ 2 & -1 & 0 \end{pmatrix}, \qquad \mathbf{b} = \begin{pmatrix} 1 \\ 0 \\ -1 \\ 2 \end{pmatrix}.$$

In this case, $\ker A \neq \{\mathbf{0}\}$, and so we are not able use the simpler formula (8.43); thus, we begin by establishing the singular value decomposition of A. The corresponding Gram matrix

$$K = \begin{pmatrix} 15 & -15 & 3 \\ -15 & 30 & -9 \\ 3 & -9 & 3 \end{pmatrix}$$

has eigenvalues and eigenvectors

$$\lambda_1 = 24 + 3\sqrt{34} = 41.4929, \quad \lambda_2 = 24 - 3\sqrt{34} = 6.5071, \quad \lambda_3 = 0,$$

$$\mathbf{v}_1 = \begin{pmatrix} 2.1324 \\ -3.5662 \\ 1. \end{pmatrix}, \qquad \mathbf{v}_2 = \begin{pmatrix} -2.5324 \\ -1.2338 \\ 1. \end{pmatrix}, \qquad \mathbf{v}_3 = \begin{pmatrix} 1 \\ 2 \\ 5 \end{pmatrix}.$$

The singular values are the square roots of the positive eigenvalues, and so

$$\sigma_1 = \sqrt{\lambda_1} = 6.4415, \qquad \sigma_2 = \sqrt{\lambda_2} = 2.5509,$$

which are used to construct the diagonal singular value matrix

$$\Sigma = \begin{pmatrix} 6.4415 & 0 \\ 0 & 2.5509 \end{pmatrix}.$$

Note that A has rank 2 because it has just two singular values. The first two eigenvectors of K are the singular vectors of A, and we use the normalized (unit) singular vectors to form the columns of

$$Q = (\mathbf{q}_1\ \mathbf{q}_2) = \begin{pmatrix} .4990 & -.8472 \\ -.8344 & -.4128 \\ .2340 & .3345 \end{pmatrix}.$$

Next, we apply A to the singular vectors and divide by the corresponding singular value, as in (8.42); the resulting vectors

$$\mathbf{p}_1 = \frac{A\mathbf{q}_1}{\sigma_1} = \begin{pmatrix} -.2180 \\ .7869 \\ -.5024 \\ .2845 \end{pmatrix}, \qquad \mathbf{p}_2 = \frac{A\mathbf{q}_2}{\sigma_2} = \begin{pmatrix} -.7869 \\ -.2180 \\ -.2845 \\ -.5024 \end{pmatrix},$$

will form the orthonormal columns of

$$P = (\mathbf{p}_1, \mathbf{p}_2) = \begin{pmatrix} -.2180 & -.7869 \\ .7869 & -.2180 \\ -.5024 & -.2845 \\ .2845 & -.5024 \end{pmatrix},$$

and, as you can verify

$$\begin{pmatrix} -.2180 & .7869 & -.5024 & .2845 \\ -.7869 & -.2180 & -.2845 & -.5024 \end{pmatrix} \begin{pmatrix} 6.4415 & 0 \\ 0 & 2.5509 \end{pmatrix} \begin{pmatrix} .4990 & -.8472 \\ -.8344 & -.4128 \\ .2340 & .3345 \end{pmatrix}$$

is the singular value decomposition of our coefficient matrix: $A = P^T \Sigma\, Q$. Its pseudoinverse is immediately computed:

$$A^+ = Q^T \Sigma^{-1} P = \begin{pmatrix} .2444 & .1333 & .0556 & .1889 \\ .1556 & -.0667 & .1111 & .0444 \\ -.1111 & 0 & -.0556 & -.0556 \end{pmatrix}.$$

Finally, the least squares solution to the original linear system of minimal Euclidean norm is

$$\mathbf{x}^\star = A^+\mathbf{b} = \begin{pmatrix} .5667 \\ .1333 \\ -.1667 \end{pmatrix}. \qquad \bullet$$

When forming the pseudoinverse of an ill conditioned matrix, its very small singular values lead to very large entries in Σ^{-1}, which can induce numerical inaccuracies when computing the ordinary or, more generally, least squares solution $\mathbf{x}^\star = A^+\mathbf{b}$ to the linear system. A common and effective strategy is to eliminate all singular values below a specified cut-off, replacing $A = P\,\Sigma\,Q^T$ by $\widetilde{A} = \widetilde{P}\,\widetilde{\Sigma}\,\widetilde{Q}^T$, where $\widetilde{\Sigma}$ is the $k \times k$ diagonal matrix containing only the non-small singular values, while $\widetilde{P}$, $\widetilde{Q}$ are obtained by extracting the first k columns of P, Q, respectively. Applying the regularized pseudoinverse $\widetilde{A}^+ = \widetilde{Q}\,\widetilde{\Sigma}^{-1}\widetilde{P}^T$ to solve for $\mathbf{x}^\star = \widetilde{A}^+\mathbf{b}$ will, in favorable situations, effectively circumvent the effects of ill-conditioning.

Finally, we note that practical numerical algorithms for computing singular values and the singular value decomposition can be found in [27, 51].

EXERCISES 8.5

8.5.1. Find the singular values of the following matrices:
(a) $\begin{pmatrix} 1 & 1 \\ 0 & 2 \end{pmatrix}$ (b) $\begin{pmatrix} 0 & 1 \\ -1 & 0 \end{pmatrix}$ (c) $\begin{pmatrix} 1 & -2 \\ -3 & 6 \end{pmatrix}$ (d) $\begin{pmatrix} 2 & 0 & 0 \\ 0 & 3 & 0 \end{pmatrix}$ (e) $\begin{pmatrix} 2 & 1 & 0 & -1 \\ 0 & -1 & 1 & 1 \end{pmatrix}$ (f) $\begin{pmatrix} 1 & -1 & 0 \\ -1 & 2 & -1 \\ 0 & -1 & 1 \end{pmatrix}$

8.5.2. Write out the singular value decomposition (8.40) of the matrices in Exercise 8.5.1.

8.5.3. (a) Construct the singular value decomposition of the shear matrix $A = \begin{pmatrix} 1 & 1 \\ 0 & 1 \end{pmatrix}$.

(b) Explain how a shear can be realized as a combination of a rotation, a stretch, followed by a second rotation.

8.5.4. Find the condition number of the following matrices. Which would you characterize as ill-conditioned?

(a) $\begin{pmatrix} 2 & -1 \\ -3 & 1 \end{pmatrix}$ (b) $\begin{pmatrix} -.999 & .341 \\ -1.001 & .388 \end{pmatrix}$

(c) $\begin{pmatrix} 1 & 2 \\ 1.001 & 1.9997 \end{pmatrix}$ (d) $\begin{pmatrix} -1 & 3 & 4 \\ 2 & 10 & 6 \\ 1 & 2 & -3 \end{pmatrix}$

(e) $\begin{pmatrix} 72 & 96 & 103 \\ 42 & 55 & 59 \\ 67 & 95 & 102 \end{pmatrix}$

(f) $\begin{pmatrix} 5 & 7 & 6 & 5 \\ 7 & 10 & 8 & 7 \\ 6 & 8 & 10 & 9 \\ 5 & 7 & 9 & 10 \end{pmatrix}$

♠ **8.5.5.** Solve the following systems of equations using Gaussian Elimination with three-digit rounding arithmetic. Is your answer a reasonable approximation to the exact solution? Compare the accuracy of your answers with the condition number of the coefficient matrix, and discuss the implications of ill-conditioning.

(a) $\begin{aligned} 1000x + 999y &= 1 \\ 554x + 555y &= -1 \end{aligned}$

(b) $\begin{aligned} 97x + 175y + 83z &= 1 \\ 44x + 78y + 37z &= 1 \\ 52x + 97y + 46z &= 1 \end{aligned}$

(c) $\begin{aligned} 3.001x + 2.999y + 5z &= 1 \\ -x + 1.002y - 2.999z &= 2 \\ 2.002x + 4y + 2z &= 1.002 \end{aligned}$

♠ **8.5.6.** Compute the singular values and condition numbers of the 2×2, 3×3, and 4×4 Hilbert matrices. What is the smallest Hilbert matrix with condition number larger than 10^6?

8.5.7. (a) What are the singular values of a $1 \times n$ matrix?

(b) Write down its singular value decomposition.

(c) Write down its pseudoinverse.

8.5.8. Answer Exercise 8.5.7 for an $m \times 1$ matrix.

8.5.9. *True or false*: Every matrix has at least one singular value.

8.5.10. Explain why the singular values of A are the same as the nonzero eigenvalues of the positive definite square root matrix $S = \sqrt{A^T A}$, defined in Exercise 8.4.27.

8.5.11. *True or false*: The singular values of A^T are the same as the singular values of A.

◇ **8.5.12.** Prove that if A is square, nonsingular, then the singular values of A^{-1} are the reciprocals of the singular values of A. How are their condition numbers related?

◇ **8.5.13.** Let A be a nonsingular square matrix.

(a) Prove that the product of the singular values of A equals the absolute value of its determinant: $\sigma_1 \sigma_2 \cdots \sigma_n = |\det A|$.

(b) Does their sum equal the absolute value of the trace: $\sigma_1 + \cdots + \sigma_n = |\operatorname{tr} A|$?

(c) Show that if $|\det A| < 10^{-k}$, then its minimal singular value satisfies $\sigma_n < 10^{-k/n}$.

(d) *True or false*: A matrix whose determinant is very small is ill-conditioned.

(e) Construct an ill-conditioned matrix with $\det A = 1$.

8.5.14. *True or false*: If $\det A > 1$, then A is not ill-conditioned.

8.5.15. *True or false*: If A is a symmetric matrix, then its singular values are the same as its eigenvalues.

8.5.16. *True or false*: If U is an upper triangular matrix whose diagonal entries are all positive, then its singular values are the same as its diagonal entries.

8.5.17. *True or false*: The singular values of A^2 are the squares σ_i^2 of the singular values of A.

8.5.18. *True or false*: If $B = S^{-1} A S$ are similar matrices, then A and B have the same singular values.

♡ **8.5.19.** Let A be a nonsingular 2×2 matrix with singular value decomposition $A = P \Sigma Q^T$ and singular values $\sigma_1 \geq \sigma_2 > 0$.

(a) Prove that the image of the unit (Euclidean) circle under the linear transformation defined by A is an ellipse, $E = \{ A\mathbf{x} \mid \|\mathbf{x}\| = 1 \}$, whose principal axes are the columns $\mathbf{p}_1, \mathbf{p}_2$ of P, and whose corresponding semi-axes are the singular values σ_1, σ_2.

(b) Show that if A is symmetric, then the ellipse's principal axes are the eigenvectors of A and the semi-axes are the absolute values of its eigenvalues.

(c) Prove that the area of E equals $\pi\, |\det A|$.

(d) Find the principal axes, semi-axes, and area of the ellipses defined by

(i) $\begin{pmatrix} 0 & 1 \\ 1 & 1 \end{pmatrix}$ (ii) $\begin{pmatrix} 2 & 1 \\ -1 & 2 \end{pmatrix}$

(iii) $\begin{pmatrix} 5 & -4 \\ 0 & -3 \end{pmatrix}$

(e) What happens if A is singular?

8.5.20. Let

$$A = \begin{pmatrix} 2 & -3 \\ -3 & 10 \end{pmatrix}.$$

Write down the equation for the ellipse $E = \{ A\mathbf{x} \mid \|\mathbf{x}\| = 1 \}$ and draw a picture. What are its principal axes? Its semi-axes? Its area?

8.5.21. Let

$$A = \begin{pmatrix} 6 & -4 & 1 \\ -4 & 6 & -1 \\ 1 & -1 & 11 \end{pmatrix},$$

and let $E = \{ \mathbf{y} = A\mathbf{x} \mid \|\mathbf{x}\| = 1 \}$ be the image of the unit Euclidean sphere under the linear map induced by A.

(a) Explain why E is an ellipsoid and write down its equation.

(b) What are its principal axes and their lengths—the semi-axes of the ellipsoid?

(c) What is the volume of the solid ellipsoidal domain enclosed by E?

◇ **8.5.22.** *Optimization Principles for Singular Values*: Let A be any nonzero $m \times n$ matrix. Prove that

(a) $\sigma_1 = \max\{ \|A\mathbf{u}\| \mid \|\mathbf{u}\| = 1 \}$.

(b) Is the minimum the smallest singular value?

(c) Can you design an optimization principle for the intermediate singular values?

◇ **8.5.23.** Let A be a square matrix. Prove that its maximal eigenvalue is smaller than its maximal singular value: $\max |\lambda_i| \leq \max \sigma_i$. *Hint*: Use Exercise 8.5.22.

8.5.24. Let A be a nonsingular square matrix. Prove that

$$\kappa(A) = \frac{\max\{ \|A\mathbf{u}\| \mid \|\mathbf{u}\| = 1 \}}{\min\{ \|A\mathbf{u}\| \mid \|\mathbf{u}\| = 1 \}}.$$

8.5.25. Find the pseudoinverse of the following matrices:

(a) $\begin{pmatrix} 1 & -1 \\ -3 & 3 \end{pmatrix}$ (b) $\begin{pmatrix} 1 & -2 \\ 2 & 1 \end{pmatrix}$

(c) $\begin{pmatrix} 2 & 0 \\ 0 & -1 \\ 0 & 0 \end{pmatrix}$ (d) $\begin{pmatrix} 0 & 0 & 1 \\ 0 & -1 & 0 \\ 0 & 0 & 0 \end{pmatrix}$

(e) $\begin{pmatrix} 1 & -1 & 1 \\ -2 & 2 & -2 \end{pmatrix}$

(f) $\begin{pmatrix} 1 & 3 \\ 2 & 6 \\ 3 & 9 \end{pmatrix}$ (g) $\begin{pmatrix} 1 & 2 & 0 \\ 0 & 1 & 1 \\ 1 & 1 & -1 \end{pmatrix}$

8.5.26. Use the pseudoinverse to find the least squares solution of minimal norm to the following linear systems:

(a) $\begin{aligned} x + y &= 1 \\ 3x + 3y &= -2 \end{aligned}$ (b) $\begin{aligned} x + y + z &= 5 \\ 2x - y + z &= 2 \end{aligned}$

(c) $\begin{aligned} x - 3y &= 2 \\ 2x + y &= -1 \\ x + y &= 0 \end{aligned}$

♡ **8.5.27.** Prove that the pseudoinverse satisfies the following identities:

(a) $(A^+)^+ = A$ (b) $A A^+ A = A$

(c) $A^+ A A^+ = A^+$ (d) $(A A^+)^T = A A^+$

(e) $(A^+ A)^T = A^+ A$

8.5.28. Suppose $\mathbf{b} \in \operatorname{rng} A$ and $\ker A = \{\mathbf{0}\}$. Prove that $\mathbf{x}^\star = A^+\mathbf{b}$ is the unique solution to the linear system $A\mathbf{x} = \mathbf{b}$. What if $\ker A \neq \{\mathbf{0}\}$?

8.6 INCOMPLETE MATRICES

Unfortunately, not all square matrices are complete. Matrices that do not have enough (complex) eigenvectors to form a basis are considerably less pleasant to work with. However, as they occasionally appear in applications, it is worth learning how to handle them. There are two approaches: the first, named after the twentieth century Russian/German mathematician Issai Schur, is a generalization of the spectral theorem, and converts an arbitrary square matrix into a similar, upper triangular matrix with the eigenvalues along the diagonal. Thus, although not every matrix can be diagonalized, they can all be "triangularized". Applications of the Schur decomposition, including the numerical computation of eigenvalues, can be found in [18].

The second approach, due to the nineteenth century French mathematician Camille Jordan,* shows how to supplement the eigenvectors of an incomplete matrix in order to obtain a basis in which the matrix assumes a simple, but now

*No relation to Wilhelm Jordan of Gauss–Jordan fame.

non-diagonal canonical (meaning distinguished) form. Applications of the Jordan canonical form will appear in our study of systems of ordinary differential equations with incomplete coefficient matrices. We remark that the two subsections are completely independent of one another.

The Schur Decomposition

As noted above, the Schur decomposition is used to convert a square matrix into similar upper triangular matrix. The similarity transformation can be chosen to be represented by a unitary matrix—a complex generalization of an orthogonal matrix.

Definition 8.42 A complex, square matrix U is called *unitary* if it satisfies

$$U^\dagger U = \mathrm{I}, \qquad \text{where} \quad U^\dagger = \overline{U^T} \tag{8.45}$$

denotes the *Hermitian transpose* in which one first transposes and then takes complex conjugates of all entries.

Thus, U is unitary if and only if its inverse equals its Hermitian transpose: $U^{-1} = U^\dagger$. For example,

$$U = \begin{pmatrix} \frac{\mathrm{i}}{\sqrt{2}} & -\frac{1}{\sqrt{2}} \\ \frac{1}{\sqrt{2}} & -\frac{\mathrm{i}}{\sqrt{2}} \end{pmatrix}$$

is a 2×2 unitary matrix, since

$$U^{-1} = U^\dagger = \begin{pmatrix} -\frac{\mathrm{i}}{\sqrt{2}} & \frac{1}{\sqrt{2}} \\ -\frac{1}{\sqrt{2}} & \frac{\mathrm{i}}{\sqrt{2}} \end{pmatrix},$$

as you can easily check.

The (i, j) entry of the defining equation (8.45) is the Hermitian dot product between the ith and jth columns of U, and hence U is an $n \times n$ unitary matrix if and only if its columns form an orthonormal basis of $\mathbb{C}^n$. In particular, a real matrix is unitary if and only if it is an orthogonal matrix. The next result is proved in the same fashion as Proposition 5.23.

Proposition 8.43 If U_1 and U_2 are $n \times n$ unitary matrices, so is their product $U_1 U_2$.

The *Schur decomposition* states that every square matrix is unitarily similar to an upper triangular matrix. The method of proof provides a recursive algorithm for constructing the decomposition.

THEOREM 8.44 Let A be an $n \times n$ matrix, either real or complex. Then there exists a unitary matrix U and an upper triangular matrix Δ such that

$$A = U\,\Delta\,U^\dagger = U\,\Delta\,U^{-1}. \tag{8.46}$$

The diagonal entries of Δ are the eigenvalues of A.

Warning: The Schur decomposition (8.46) is not unique. As the method of proof makes clear, there are many inequivalent choices for the matrices U and Δ.

REMARK: If all eigenvalues of A are real, then $U = Q$ can be chosen to be a (real) orthogonal matrix, and Δ is also a real matrix. This follows from the construction outlined in the proof.

Proof The proof proceeds by induction on the size of A. According to Theorem 8.10, A has at least one, possibly complex, eigenvalue λ_1. Let $\mathbf{u}_1 \in \mathbb{C}^n$ be a corresponding unit eigenvector, so its Hermitian norm is $\|\mathbf{u}_1\| = 1$. Let U_1 be an $n \times n$ unitary matrix whose first column is the unit eigenvector $\mathbf{u}_1$. In practice, U_1 can be constructed by applying the Gram–Schmidt process to any basis of $\mathbb{C}^n$ whose first element is the eigenvector $\mathbf{u}_1$. The eigenvector equation $A\mathbf{u}_1 = \lambda_1 \mathbf{u}_1$ forms the first column of the matrix product equation

$$A U_1 = U_1 B, \quad \text{where} \quad B = U_1^\dagger A U_1 = \begin{pmatrix} \lambda_1 & * \\ \mathbf{0} & C \end{pmatrix},$$

with C an $(n-1) \times (n-1)$ matrix and $*$ indicating a row vector. By our induction hypothesis, there is an $(n-1) \times (n-1)$ unitary matrix V such that $V^\dagger C V = \Gamma$ is an upper triangular $(n-1) \times (n-1)$ matrix. Set

$$U_2 = \begin{pmatrix} 1 & \mathbf{0} \\ \mathbf{0} & V \end{pmatrix}.$$

It is easily checked that U_2 is also unitary, and, moreover

$$U_2^\dagger B U_2 = \begin{pmatrix} \lambda_1 & * \\ \mathbf{0} & \Gamma \end{pmatrix} = \Delta$$

is upper triangular. Therefore, the unitary product matrix $U = U_1 U_2$ yields the desired result:

$$U^\dagger A U = (U_1 U_2)^\dagger A U_1 U_2 = U_2^\dagger U_1^\dagger A U_1 U_2 = U_2^\dagger B U_2 = \Delta,$$

which, since $U^{-1} = U^\dagger$, establishes the Schur decomposition (8.46). Finally, since A and Δ are similar matrices, they have the same eigenvalues, which justifies the final statement of the theorem. ■

EXAMPLE 8.45 The matrix

$$A = \begin{pmatrix} 6 & 4 & -3 \\ -4 & -2 & 2 \\ 4 & 4 & -2 \end{pmatrix}$$

has a simple eigenvalue of 2, with eigenvector $\mathbf{v}_1 = (-1, 1, 0)^T$, and a double eigenvalue of 0, with only one independent eigenvector $\mathbf{v}_2 = (1, 0, 2)^T$. Thus A is incomplete, and so not diagonalizable. To construct a Schur decomposition, we begin with the first eigenvector $\mathbf{v}_1$ and apply the Gram–Schmidt process to the basis $\mathbf{v}_1, \mathbf{e}_2, \mathbf{e}_3$ to obtain the orthogonal matrix

$$U_1 = \begin{pmatrix} -\frac{1}{\sqrt{2}} & \frac{1}{\sqrt{2}} & 0 \\ \frac{1}{\sqrt{2}} & \frac{1}{\sqrt{2}} & 0 \\ 0 & 0 & 1 \end{pmatrix}.$$

The resulting similar matrix*

$$B = U_1^T A U_1 = \begin{pmatrix} 2 & -8 & \frac{\sqrt{5}}{2} \\ 0 & 2 & -\frac{1}{\sqrt{2}} \\ 0 & 4\sqrt{2} & -2 \end{pmatrix}$$

*Since all matrices are real in this example, the Hermitian transpose † reduces to the ordinary transpose ^{T}.

has its first column in upper triangular form. To continue the procedure, we extract the lower 2×2 submatrix

$$C = \begin{pmatrix} 2 & -\frac{1}{\sqrt{2}} \\ 4\sqrt{2} & -2 \end{pmatrix},$$

and find that it has a single (incomplete) eigenvalue 0, with unit eigenvector

$$\begin{pmatrix} \frac{1}{3} \\ \frac{2\sqrt{2}}{3} \end{pmatrix}.$$

The corresponding orthogonal matrix

$$V = \begin{pmatrix} \frac{1}{3} & \frac{2\sqrt{2}}{3} \\ \frac{2\sqrt{2}}{3} & -\frac{1}{3} \end{pmatrix}$$

will convert C to upper triangular form

$$V^T C\, V = \begin{pmatrix} 0 & \frac{9}{\sqrt{2}} \\ 0 & 0 \end{pmatrix}.$$

Therefore,

$$U_2 = \begin{pmatrix} 1 & 0 & 0 \\ 0 & \frac{1}{3} & \frac{2\sqrt{2}}{3} \\ 0 & \frac{2\sqrt{2}}{3} & -\frac{1}{3} \end{pmatrix}$$

will complete the conversion of the original matrix into upper triangular form

$$\Delta = U_2^T B\, U_2 = U^T A\, U = \begin{pmatrix} 2 & \frac{2}{3} & -\frac{37}{3\sqrt{2}} \\ 0 & 0 & \frac{9}{\sqrt{2}} \\ 0 & 0 & 0 \end{pmatrix},$$

where

$$U = U_1 U_2 = \begin{pmatrix} -\frac{1}{\sqrt{2}} & \frac{1}{3\sqrt{2}} & \frac{2}{3} \\ \frac{1}{\sqrt{2}} & \frac{1}{3\sqrt{2}} & \frac{2}{3} \\ 0 & \frac{2\sqrt{2}}{3} & -\frac{1}{3} \end{pmatrix}$$

is the desired orthogonal (unitary) matrix. Use of a computer to carry out the detailed calculations is essential in most examples. ●

EXERCISES 8.6

8.6.1. Establish a Schur Decomposition for the following matrices:

(a) $\begin{pmatrix} 1 & -1 \\ 1 & 3 \end{pmatrix}$ (b) $\begin{pmatrix} 1 & -2 \\ -2 & 1 \end{pmatrix}$ (c) $\begin{pmatrix} 8 & 9 \\ -6 & -7 \end{pmatrix}$ (d) $\begin{pmatrix} 1 & 5 \\ -2 & -1 \end{pmatrix}$

(e) $\begin{pmatrix} 2 & -1 & 2 \\ -2 & 3 & -1 \\ -6 & 6 & -5 \end{pmatrix}$ (f) $\begin{pmatrix} 0 & 2 & -1 \\ -1 & -1 & 1 \\ -1 & 0 & 0 \end{pmatrix}$

8.6.2. Show that a real unitary matrix is an orthogonal matrix.

◇ **8.6.3.** Prove Proposition 8.43.

◇ **8.6.4.** Write out a new proof of the Spectral Theorem 8.26 based on the Schur Decomposition.

♡ **8.6.5.** A complex matrix A is called *normal* if it commutes with its Hermitian transpose: $A^\dagger A = A A^\dagger$.

(a) Show that every real symmetric matrix is normal.

(b) Show that every unitary matrix is normal.

(c) Show that every real orthogonal matrix is normal.

(d) Show that an upper triangular matrix is normal if and only if it is diagonal.

(e) Show that the eigenvectors of a normal matrix form an orthogonal basis of $\mathbb{C}^n$ under the Hermitian dot product.

(f) Show that the converse is true: a matrix has an orthogonal eigenvector basis of $\mathbb{C}^n$ if and only if it is normal. *Hint*: Use the Schur Decomposition.

(g) How can you tell when a real matrix has a real orthonormal eigenvector basis?

The Jordan Canonical Form

We now turn to the more sophisticated Jordan canonical form. Throughout this section, A will be an $n \times n$ matrix, with either real or complex entries. We let $\lambda_1, \ldots, \lambda_k$ denote the *distinct* eigenvalues of A. We recall that Theorem 8.10 guarantees that every matrix has at least one (complex) eigenvalue, so $k \geq 1$. Moreover, we are assuming that $k < n$, as otherwise A would be complete.

Definition 8.46 A *Jordan chain* of length j is a sequence of non-zero vectors $\mathbf{w}_1, \ldots, \mathbf{w}_j \in \mathbb{C}^m$ that satisfies

$$A\mathbf{w}_1 = \lambda\,\mathbf{w}_1, \qquad A\mathbf{w}_i = \lambda\,\mathbf{w}_i + \mathbf{w}_{i-1}, \qquad i = 2, \ldots, j, \tag{8.47}$$

where λ is an eigenvalue of A.

Note that the initial vector $\mathbf{w}_1$ in a Jordan chain is a genuine eigenvector, and so Jordan chains only exist when λ is an eigenvalue. The rest, $\mathbf{w}_2, \ldots, \mathbf{w}_j$, are *generalized eigenvectors*, in accordance with the following definition.

Definition 8.47 A nonzero vector $\mathbf{w} \neq \mathbf{0}$ that satisfies

$$(A - \lambda\,\mathrm{I})^k\,\mathbf{w} = \mathbf{0} \tag{8.48}$$

for some $k > 0$ and $\lambda \in \mathbb{C}$ is called a *generalized eigenvector* of the matrix A.

Note that every ordinary eigenvector is automatically a generalized eigenvector, since we can just take $k = 1$ in (8.48); but the converse is not necessarily valid. We shall call the minimal value of k for which (8.48) holds the *index* of the generalized eigenvector. Thus, an ordinary eigenvector is a generalized eigenvector of index 1. Since $A - \lambda\,\mathrm{I}$ is nonsingular whenever λ is not an eigenvalue of A, its kth power $(A - \lambda\,\mathrm{I})^k$ is also nonsingular. Therefore, generalized eigenvectors can only exist when λ is an ordinary eigenvalue of A—there are no additional "generalized eigenvalues".

Lemma 8.48 The ith vector $\mathbf{w}_i$ in a Jordan chain (8.47) is a generalized eigenvector of index i.

Proof By definition, $(A - \lambda\,\mathrm{I})\,\mathbf{w}_1 = \mathbf{0}$, and so $\mathbf{w}_1$ is an eigenvector. Next, we have $(A - \lambda\,\mathrm{I})\,\mathbf{w}_2 = \mathbf{w}_1$, and so $(A - \lambda\,\mathrm{I})^2\,\mathbf{w}_2 = (A - \lambda\,\mathrm{I})\,\mathbf{w}_1 = \mathbf{0}$. Thus, $\mathbf{w}_2$ a generalized eigenvector of index 2. A simple induction proves that $(A - \lambda\,\mathrm{I})^i\,\mathbf{w}_i = \mathbf{0}$. ■

EXAMPLE 8.49 Consider the 3×3 Jordan block

$$A = \begin{pmatrix} 2 & 1 & 0 \\ 0 & 2 & 1 \\ 0 & 0 & 2 \end{pmatrix}.$$

The only eigenvalue is

$$\lambda = 2, \quad \text{and} \quad A - 2\,\mathrm{I} = \begin{pmatrix} 0 & 1 & 0 \\ 0 & 0 & 1 \\ 0 & 0 & 0 \end{pmatrix}.$$

We claim that the standard basis vectors $\mathbf{e}_1$, $\mathbf{e}_2$ and $\mathbf{e}_3$ form a Jordan chain. Indeed, $A\,\mathbf{e}_1 = 2\,\mathbf{e}_1$, and hence $\mathbf{e}_1 \in \ker(A - 2\,\mathrm{I})$ is a genuine eigenvector. Furthermore, $A\,\mathbf{e}_2 = 2\,\mathbf{e}_2 + \mathbf{e}_1$, and $A\,\mathbf{e}_3 = 2\,\mathbf{e}_3 + \mathbf{e}_2$, as you can easily check. Thus, $\mathbf{e}_1$, $\mathbf{e}_2$ and $\mathbf{e}_3$ satisfy the Jordan chain equations (8.47) for the eigenvalue $\lambda = 2$. Note that $\mathbf{e}_2$ lies in the kernel of

$$(A - 2\,\mathrm{I})^2 = \begin{pmatrix} 0 & 0 & 1 \\ 0 & 0 & 0 \\ 0 & 0 & 0 \end{pmatrix},$$

and so is a generalized eigenvector of index 2. Indeed, every vector of the form $\mathbf{w} = a\,\mathbf{e}_1 + b\,\mathbf{e}_2$ with $b \neq 0$ is a generalized eigenvector of index 2. (When $b = 0$, $a \neq 0$, the vector $\mathbf{w} = a\,\mathbf{e}_1$ is an ordinary eigenvector of index 1.) Finally, $(A - 2\,\mathrm{I})^3 = \mathrm{O}$, and so every nonzero vector $\mathbf{0} \neq \mathbf{v} \in \mathbb{R}^3$, including $\mathbf{e}_3$, is a generalized eigenvector of index 3 (or less). ●

A basis of $\mathbb{R}^n$ or $\mathbb{C}^n$ is called a *Jordan basis* for the matrix A if it consists of one or more Jordan chains that have no elements in common. Thus, for the Jordan matrix in Example 8.49, the standard basis $\mathbf{e}_1, \mathbf{e}_2, \mathbf{e}_3$ is, in fact, a Jordan basis. An eigenvector basis qualifies as a Jordan basis, since each eigenvector belongs to a Jordan chain of length 1. Jordan bases are the desired extension of eigenvector bases, and every square matrix has one. The proof of the *Jordan Basis Theorem* will appear at the end of this section.

THEOREM 8.50 Every $n \times n$ matrix admits a Jordan basis of $\mathbb{C}^n$. The first elements of the Jordan chains form a maximal system of linearly independent eigenvectors. Moreover, the number of generalized eigenvectors in the Jordan basis that belong to the Jordan chains associated with the eigenvalue λ is the same as the eigenvalue's multiplicity.

EXAMPLE 8.51 Consider the matrix

$$A = \begin{pmatrix} -1 & 0 & 1 & 0 & 0 \\ -2 & 2 & -4 & 1 & 1 \\ -1 & 0 & -3 & 0 & 0 \\ -4 & -1 & 3 & 1 & 0 \\ 4 & 0 & 2 & -1 & 0 \end{pmatrix}.$$

With some work, its characteristic equation is found to be

$$\begin{aligned} p_A(\lambda) = \det(A - \lambda\,\mathrm{I}) &= \lambda^5 + \lambda^4 - 5\,\lambda^3 - \lambda^2 + 8\,\lambda - 4 \\ &= (\lambda - 1)^3(\lambda + 2)^2 = 0, \end{aligned}$$

and hence A has two eigenvalues: $\lambda_1 = 1$, which is a triple eigenvalue, and $\lambda_2 = -2$, which is double. Solving the associated homogeneous systems $(A - \lambda_j\,\mathrm{I})\mathbf{v} = \mathbf{0}$, we discover that, up to constant multiple, there are only two eigenvectors:

$$\mathbf{v}_1 = (0, 0, 0, -1, 1)^T \quad \text{for} \quad \lambda_1 = 1$$

and, anticipating our final numbering,

$$\mathbf{v}_4 = (-1, 1, 1, -2, 0)^T \quad \text{for} \quad \lambda_2 = -2.$$

Thus, A is far from complete.

To construct a Jordan basis, we first note that since A has 2 linearly independent eigenvectors, the Jordan basis will contain two Jordan chains: the one associated with the triple eigenvalue $\lambda_1 = 1$ has length 3, while $\lambda_2 = -2$ leads to a Jordan chain of length 2. To construct the former, we need to first solve the system $(A - \mathrm{I})\mathbf{w} = \mathbf{v}_1$. Note that the coefficient matrix is singular—it must be since 1 is an eigenvalue—and the general solution is $\mathbf{w} = \mathbf{v}_2 + t\,\mathbf{v}_1$ where $\mathbf{v}_2 = (0, 1, 0, 0, -1)^T$, and t is the free variable. The appearance of an arbitrary multiple of the eigenvector $\mathbf{v}_1$ in the solution is not unexpected; indeed, the kernel of $A - \mathrm{I}$ is the eigenspace for $\lambda_1 = 1$. We can choose any solution, e.g., $\mathbf{v}_2$ as the second element in the Jordan chain. To find the last element of the chain, we solve $(A - \mathrm{I})\mathbf{w} = \mathbf{v}_2$ to find $\mathbf{w} = \mathbf{v}_3 + t\,\mathbf{v}_1$ where $\mathbf{v}_3 = (0, 0, 0, 1, 0)^T$ can be used as the last element of this Jordan chain. Similarly, to construct the Jordan chain for the second eigenvalue, we solve $(A + 2\,\mathrm{I})\mathbf{w} = \mathbf{v}_4$ and find $\mathbf{w} = \mathbf{v}_5 + t\,\mathbf{v}_4$ where $\mathbf{v}_5 = (-1, 0, 0, -2, 1)^T$. Thus, the desired Jordan basis is

$$\mathbf{v}_1 = \begin{pmatrix} 0 \\ 0 \\ 0 \\ -1 \\ 1 \end{pmatrix}, \quad \mathbf{v}_2 = \begin{pmatrix} 0 \\ 1 \\ 0 \\ 0 \\ -1 \end{pmatrix}, \quad \mathbf{v}_3 = \begin{pmatrix} 0 \\ 0 \\ 0 \\ 1 \\ 0 \end{pmatrix}, \quad \mathbf{v}_4 = \begin{pmatrix} -1 \\ 1 \\ 1 \\ -2 \\ 0 \end{pmatrix}, \quad \mathbf{v}_5 = \begin{pmatrix} -1 \\ 0 \\ 0 \\ -2 \\ 1 \end{pmatrix},$$

with

$$A\mathbf{v}_1 = \mathbf{v}_1, \quad A\mathbf{v}_2 = \mathbf{v}_2 + \mathbf{v}_1, \quad A\mathbf{v}_3 = \mathbf{v}_3 + \mathbf{v}_2,$$
$$A\mathbf{v}_4 = -2\,\mathbf{v}_4, \quad A\mathbf{v}_5 = -2\,\mathbf{v}_5 + \mathbf{v}_4.$$

Just as an eigenvector basis diagonalizes a complete matrix, a Jordan basis provides a particularly simple form for an incomplete matrix, known as the *Jordan canonical form*.

Definition 8.52 An $n \times n$ matrix of the form

$$J_{\lambda,n} = \begin{pmatrix} \lambda & 1 & & & & \\ & \lambda & 1 & & & \\ & & \lambda & 1 & & \\ & & & \ddots & \ddots & \\ & & & & \lambda & 1 \\ & & & & & \lambda \end{pmatrix}, \qquad (8.49)$$

in which λ is a real or complex number, is known as a *Jordan block*.

In particular, a 1×1 Jordan block is merely a scalar $J_{\lambda,1} = \lambda$. Since every matrix has at least one (complex) eigenvector—see Theorem 8.10— the Jordan block matrices have the least possible number of eigenvectors.

Lemma 8.53 The $n \times n$ Jordan block matrix $J_{\lambda,n}$ has a single eigenvalue, λ, and a single independent eigenvector, $\mathbf{e}_1$. The standard basis vectors $\mathbf{e}_1, \ldots, \mathbf{e}_n$ form a Jordan chain for $J_{\lambda,n}$.

Definition 8.54 A *Jordan matrix* is a square matrix of block diagonal form

$$J = \operatorname{diag}(J_{\lambda_1,n_1}, J_{\lambda_2,n_2}, \ldots, J_{\lambda_k,n_k}) = \begin{pmatrix} J_{\lambda_1,n_1} & & & \\ & J_{\lambda_2,n_2} & & \\ & & \ddots & \\ & & & J_{\lambda_k,n_k} \end{pmatrix}, \tag{8.50}$$

in which one or more Jordan blocks, not necessarily of the same size, lie along the diagonal, while all off-diagonal blocks are zero.

Note that the only non-zero entries in a Jordan matrix are those on the diagonal, which can have any complex value, and those on the superdiagonal, which are either 1 or 0. The positions of the superdiagonal 1's uniquely prescribes the Jordan blocks. For example, the 6×6 matrices

$$\begin{pmatrix} 1 & 0 & 0 & 0 & 0 & 0 \\ 0 & 2 & 0 & 0 & 0 & 0 \\ 0 & 0 & 3 & 0 & 0 & 0 \\ 0 & 0 & 0 & 3 & 0 & 0 \\ 0 & 0 & 0 & 0 & 2 & 0 \\ 0 & 0 & 0 & 0 & 0 & 1 \end{pmatrix}, \quad \begin{pmatrix} -1 & 1 & 0 & 0 & 0 & 0 \\ 0 & -1 & 1 & 0 & 0 & 0 \\ 0 & 0 & -1 & 1 & 0 & 0 \\ 0 & 0 & 0 & -1 & 0 & 0 \\ 0 & 0 & 0 & 0 & -1 & 1 \\ 0 & 0 & 0 & 0 & 0 & -1 \end{pmatrix},$$

$$\begin{pmatrix} 0 & 1 & 0 & 0 & 0 & 0 \\ 0 & 0 & 0 & 0 & 0 & 0 \\ 0 & 0 & 1 & 1 & 0 & 0 \\ 0 & 0 & 0 & 1 & 0 & 0 \\ 0 & 0 & 0 & 0 & 2 & 1 \\ 0 & 0 & 0 & 0 & 0 & 2 \end{pmatrix},$$

are all Jordan matrices: the first is a diagonal matrix, consisting of 6 distinct 1×1 Jordan blocks; the second has a 4×4 Jordan block followed by a 2×2 block that happen to have the same diagonal entries; the last has three 2×2 Jordan blocks.

As a direct corollary of Lemma 8.53 combined with the matrix's block structure, cf. Exercise 8.2.50, we obtain a complete classification of the eigenvectors and eigenvalues of a Jordan matrix.

Lemma 8.55 The Jordan matrix (8.50) has eigenvalues $\lambda_1, \ldots, \lambda_k$. The standard basis vectors $\mathbf{e}_1, \ldots, \mathbf{e}_n$ form a Jordan basis, which is partitioned into nonoverlapping Jordan chains labeled by the Jordan blocks.

Thus, in the preceding examples of Jordan matrices, the first has three double eigenvalues, 1 and 2 and 3, and corresponding linearly independent eigenvectors $\mathbf{e}_1, \mathbf{e}_6$, and $\mathbf{e}_2, \mathbf{e}_5$, and $\mathbf{e}_3, \mathbf{e}_4$, each of which belongs to a Jordan chain of length 1. The second matrix has only one eigenvalue, -1, but two independent eigenvectors $\mathbf{e}_1, \mathbf{e}_5$, and hence two Jordan chains, namely $\mathbf{e}_1, \mathbf{e}_2, \mathbf{e}_3, \mathbf{e}_4$, and $\mathbf{e}_5, \mathbf{e}_6$. The last has three eigenvalues $0, 1, 2$, three eigenvectors $\mathbf{e}_1, \mathbf{e}_3, \mathbf{e}_5$, and three Jordan chains of length 2: $\mathbf{e}_1, \mathbf{e}_2$, and $\mathbf{e}_3, \mathbf{e}_4$, and $\mathbf{e}_5, \mathbf{e}_6$. In particular, the only complete Jordan matrices are the diagonal matrices, all of whose Jordan blocks are of size 1×1.

The Jordan canonical form follows straightforwardly from the Jordan Basis Theorem 8.50.

THEOREM 8.56 Let A be an $n \times n$ real or complex matrix. Let $S = (\mathbf{w}_1\ \mathbf{w}_2 \ldots \mathbf{w}_n)$ be the matrix whose columns are a Jordan basis of A. Then S places A in *Jordan canonical form*

$$S^{-1} A\, S = J = \operatorname{diag}(J_{\lambda_1,n_1}, J_{\lambda_2,n_2}, \ldots, J_{\lambda_k,n_k}), \qquad \text{or, equivalently,} \qquad A = S\, J\, S^{-1}. \tag{8.51}$$

The diagonal entries of the similar Jordan matrix J are the eigenvalues of A. In particular, A is complete (diagonalizable) if and only if every Jordan block is of size 1×1 or, equivalently, all Jordan chains are of length 1. The Jordan canonical form of A is uniquely determined up to a permutation of the diagonal Jordan blocks.

For instance, the matrix

$$A = \begin{pmatrix} -1 & 0 & 1 & 0 & 0 \\ -2 & 2 & -4 & 1 & 1 \\ -1 & 0 & -3 & 0 & 0 \\ -4 & -1 & 3 & 1 & 0 \\ 4 & 0 & 2 & -1 & 0 \end{pmatrix}$$

considered in Example 8.51 has the following Jordan basis matrix and Jordan canonical form

$$S = \begin{pmatrix} 0 & 0 & 0 & -1 & -1 \\ 0 & 1 & 0 & 1 & 0 \\ 0 & 0 & 0 & 1 & 0 \\ -1 & 0 & 1 & -2 & -2 \\ 1 & -1 & 0 & 0 & 1 \end{pmatrix}, \qquad J = S^{-1} A\, S = \begin{pmatrix} 1 & 1 & 0 & 0 & 0 \\ 0 & 1 & 1 & 0 & 0 \\ 0 & 0 & 1 & 0 & 0 \\ 0 & 0 & 0 & -2 & 1 \\ 0 & 0 & 0 & 0 & -2 \end{pmatrix}.$$

Finally, to prove the Jordan Basis Theorem 8.50, we begin with a simple lemma.

Lemma 8.57 If $\mathbf{v}_1, \ldots, \mathbf{v}_n$ forms a Jordan basis for the matrix A, it also forms a Jordan basis for $B = A - c\,\mathrm{I}$, for any scalar c.

Proof We note that the eigenvalues of B are of the form $\lambda - c$, where λ is an eigenvalue of A. Moreover, given a Jordan chain $\mathbf{w}_1, \ldots, \mathbf{w}_j$ of A, we have

$$B\,\mathbf{w}_1 = (\lambda - c)\,\mathbf{w}_1, \qquad B\,\mathbf{w}_i = (\lambda - c)\,\mathbf{w}_i + \mathbf{w}_{i-1}, \qquad i = 2, \ldots, j,$$

so $\mathbf{w}_1, \ldots, \mathbf{w}_j$ is also a Jordan chain for B corresponding to the eigenvalue $\lambda - c$. ■

The proof of Theorem 8.50 will be done by induction on the size n of the matrix. The case $n = 1$ is trivial, since any nonzero element of $\mathbb{C}$ is a Jordan basis for a 1×1 matrix. To perform the induction step, we assume that the result is valid for all matrices of size $\leq n - 1$. Let A be an $n \times n$ matrix. According to Theorem 8.10, A has at least one complex eigenvalue λ_1. Let $B = A - \lambda_1 \mathrm{I}$. Since λ_1 is an eigenvalue of A, we know that 0 is an eigenvalue of B. This means that $\ker B \neq \{\mathbf{0}\}$, and so $r = \operatorname{rank} B < n$. Moreover, by Lemma 8.57, any Jordan basis of B is also a Jordan basis for A, and so we can concentrate all our attention on the singular matrix B from now on.

We note that $W = \operatorname{rng} B \subset \mathbb{C}^n$ is an invariant subspace, i.e., $B\,\mathbf{w} \in W$ whenever $\mathbf{w} \in W$, cf. Exercise 7.4.32. Moreover, since B is singular, $\dim W = r = \operatorname{rank} B < n$. Thus, by fixing a basis of W, we can realize the restriction $B\colon W \to W$ as multiplication by an $r \times r$ matrix. The fact that $r < n$ allows us to invoke the induction hypothesis and deduce the existence of a Jordan basis $\mathbf{w}_1, \ldots, \mathbf{w}_r \in W \subset \mathbb{C}^n$ for the action of B on the subspace W. Our goal is to complete this collection to a full Jordan basis on $\mathbb{C}^n$.

To this end, we append two additional kinds of vectors. Suppose that the Jordan basis of W contains k null Jordan chains associated with its zero eigenvalue. Each null Jordan chain consists of vectors $\mathbf{w}_1, \ldots, \mathbf{w}_j \in W$ satisfying

$$B\mathbf{w}_1 = \mathbf{0}, \qquad B\mathbf{w}_2 = \mathbf{w}_1, \qquad \ldots \qquad B\mathbf{w}_j = \mathbf{w}_{j-1}.$$

The number of null Jordan chains is equal to the number of linearly independent null eigenvectors of B that belong to $W = \operatorname{rng} B$, that is $k = \dim(\ker B \cap \operatorname{rng} B)$. To each such null Jordan chain, we append a vector $\mathbf{w}_{j+1} \in \mathbb{C}^n$ such that $B\mathbf{w}_{j+1} = \mathbf{w}_j$, noting that $\mathbf{w}_{j+1}$ exists because $\mathbf{w}_j \in \operatorname{rng} B$. We deduce that $\mathbf{w}_1, \ldots, \mathbf{w}_{j+1} \in \mathbb{C}^n$ forms a null Jordan chain, of length $j+1$. Having extended all the null Jordan chains in W, the resulting collection consists of $r+k$ vectors in $\mathbb{C}^n$ arranged in nonoverlapping Jordan chains. To complete to a basis, we append $n-r-k$ additional linearly independent null vectors $\mathbf{z}_1, \ldots, \mathbf{z}_{n-r-k} \in \ker B \setminus \operatorname{rng} B$ that lie outside its range. Since $B\mathbf{z}_j = \mathbf{0}$, each $\mathbf{z}_j$ forms a null Jordan chain of length 1. We claim that the complete collection consisting of the non-null Jordan chains in W, the k extended null chains, and the additional null vectors $\mathbf{z}_1, \ldots, \mathbf{z}_{n-r-k}$, forms the desired Jordan basis. By construction, it consists of nonoverlapping Jordan chains. The only remaining issue is proving that the vectors are linearly independent; this is left as a challenge for the reader in the final Exercise 8.6.21. ■

EXERCISES 8.6

8.6.6. For each of the following Jordan matrices, identify the Jordan blocks. Write down the eigenvalues, the eigenvectors, and the Jordan basis. Clearly identify the Jordan chains.

(a) $\begin{pmatrix} 2 & 1 \\ 0 & 2 \end{pmatrix}$ (b) $\begin{pmatrix} -3 & 0 \\ 0 & 6 \end{pmatrix}$

(c) $\begin{pmatrix} 1 & 0 & 0 \\ 0 & 1 & 1 \\ 0 & 0 & 1 \end{pmatrix}$ (d) $\begin{pmatrix} 0 & 1 & 0 \\ 0 & 0 & 1 \\ 0 & 0 & 0 \end{pmatrix}$

(e) $\begin{pmatrix} 4 & 0 & 0 & 0 \\ 0 & 3 & 1 & 0 \\ 0 & 0 & 3 & 0 \\ 0 & 0 & 0 & 2 \end{pmatrix}$

8.6.7. Write down all possible 4×4 Jordan matrices that only have an eigenvalue 2.

8.6.8. Write down all possible 3×3 Jordan matrices that have eigenvalues 2 and 5 (and no others).

8.6.9. Find Jordan bases and the Jordan canonical form for the following matrices:

(a) $\begin{pmatrix} 2 & 3 \\ 0 & 2 \end{pmatrix}$ (b) $\begin{pmatrix} -1 & -1 \\ 4 & -5 \end{pmatrix}$

(c) $\begin{pmatrix} 1 & 1 & 1 \\ 0 & 1 & 1 \\ 0 & 0 & 1 \end{pmatrix}$ (d) $\begin{pmatrix} -3 & 1 & 0 \\ 1 & -3 & -1 \\ 0 & 1 & -3 \end{pmatrix}$

(e) $\begin{pmatrix} -1 & 1 & 1 \\ -2 & -2 & -2 \\ 1 & -1 & -1 \end{pmatrix}$

(f) $\begin{pmatrix} 2 & -1 & 1 & 2 \\ 0 & 2 & 0 & 1 \\ 0 & 0 & 2 & -1 \\ 0 & 0 & 0 & 2 \end{pmatrix}$

8.6.10. Write down a formula for the inverse of a Jordan block matrix. *Hint*: Try some small examples first to help figure out the pattern.

8.6.11. *True or false*: If A is complete, every generalized eigenvector is an ordinary eigenvector.

◇ **8.6.12.** Suppose you know all eigenvalues of a matrix as well as their algebraic and geometric multiplicities. Can you determine the matrix's Jordan canonical form?

8.6.13. *True or false*: If $\mathbf{w}_1, \ldots, \mathbf{w}_j$ is a Jordan chain for a matrix A, so are the scalar multiples $c\,\mathbf{w}_1, \ldots, c\,\mathbf{w}_j$ for any $c \neq 0$.

8.6.14. *True or false*: If A has Jordan canonical form J, then A^2 has Jordan canonical J^2.

◇ **8.6.15.** (a) Give an example of a matrix A such that A^2 has an eigenvector that is *not* an eigenvector of A.

(b) Show that, in general, every eigenvalue of A^2 is the square of an eigenvalue of A.

8.6.16. Let A and B be $n \times n$ matrices. According to Exercise 8.2.23, the matrix products $A\,B$ and $B\,A$ have the same eigenvalues. Do they have the same Jordan form?

◇ **8.6.17.** Prove Lemma 8.53.

◇ **8.6.18.** (a) Prove that a Jordan block matrix $J_{0,n}$ with zero diagonal entries is nilpotent, as in Exercise 1.3.13.

(b) Prove that a Jordan matrix is nilpotent if and only if all its diagonal entries are zero.

(c) Prove that a matrix is nilpotent if and only if its Jordan canonical form is nilpotent.

(d) Explain why a matrix is nilpotent if and only if its only eigenvalue is 0.

8.6.19. Let J be a Jordan matrix.

(a) Prove that J^k is a complete matrix for some $k \geq 1$ if and only if either J is diagonal, or J is nilpotent with $J^k = \mathrm{O}$.

(b) Suppose that A is an incomplete matrix such that A^k is complete for some $k \geq 2$. Prove that $A^k = \mathrm{O}$. (A simpler version of this problem appears in Exercise 8.3.8.)

♡ **8.6.20.** The *Cayley-Hamilton Theorem*: Let $p_A(\lambda) = \det(A - \lambda\, \mathrm{I})$ be the characteristic polynomial of A.

(a) Prove that if D is a diagonal matrix, then[†] $p_D(D) = \mathrm{O}$. *Hint*: Leave $p_D(\lambda)$ in factored form.

(b) Prove that if A is complete, then $p_A(A) = \mathrm{O}$.

(c) Prove that if J is a Jordan block, then $p_J(J) = \mathrm{O}$.

(d) Prove that this also holds if J is a Jordan matrix.

(e) Prove that any square matrix satisfies its own characteristic equation: $p_A(A) = \mathrm{O}$.

◇ **8.6.21.** Prove that the n vectors constructed in the proof of Theorem 8.50 are linearly independent and hence form a Jordan basis. *Hint*: Suppose that some linear combination vanishes. Apply B to the equation, and then use the fact that we started with a Jordan basis for $W = \operatorname{rng} B$.

[†] See Exercise 1.2.35 for the basics of matrix polynomials.

CHAPTER

9

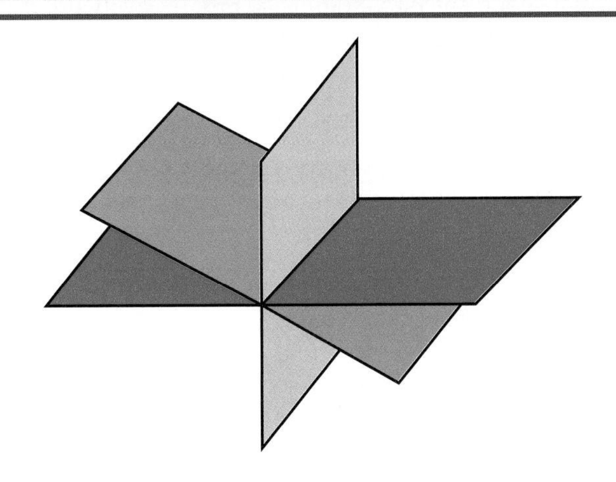

Linear Dynamical Systems

In this chapter, we will analyze the simplest class of finite dimensional dynamical systems, governed by first and second order linear systems of ordinary differential equations. Such systems, whose unvarying equilibria were the subject of Chapter 6, include the dynamical motions of mass–spring chains and structures, and the time-varying voltages and currents in simple electrical circuits. Dynamics of continuous media, including fluids, solids and gases, are modeled by infinite-dimensional dynamical systems described by partial differential equations, and will not be treated in this text, nor will we venture into the vastly more complicated realm of nonlinear dynamics.

Chapter 8 developed the basic mathematical tools—eigenvalues and eigenvectors—used in the analysis of linear systems of ordinary differential equations. For a first order system, the resulting *eigensolutions* describe the basic modes of exponential growth, decay, or periodic behavior. In particular, the stability of the equilibrium solution is determined by the eigenvalues. Most of the phenomenology inherent to linear dynamics can already be observed in the two-dimensional situation, and we devote Section 9.3 to a complete description of first order planar linear systems. In Section 9.4, we re-interpret the solution to a first order system in terms of the matrix exponential, which is defined by analogy with the usual scalar exponential function. Matrix exponentials are particularly effective for solving inhomogeneous or forced linear systems, and also appear in applications to geometry, computer graphics and animation, and theoretical physics.

As a consequence of Newton's laws of motion, mechanical vibrations are modeled by second order dynamical systems. For stable configurations with no frictional damping, the eigensolutions constitute the system's normal modes, each periodically vibrating with its associated fundamental frequency. The full dynamics is obtained by linear superposition of the periodic normal modes, but the resulting solution is, typically, no longer periodic. Such quasi-periodic motion may seem quite erratic—even though mathematically it is merely a combination of finitely many simple periodic motions. When subjected to an external periodic forcing, the system usually remains in a quasi-periodic motion that superimposes a periodic response onto its own internal vibrations. However, attempting to force the system at one of its natural frequencies may induce a resonant vibration, of progressively unbounded amplitude, resulting in a catastrophic breakdown of the physical system. In contrast, frictional effects serve to damp out the quasi-periodic vibrations and similarly help mitigate the dangers of resonance.

9.1 BASIC SOLUTION TECHNIQUES

Our initial focus will be on systems

$$\frac{d\mathbf{u}}{dt} = A\,\mathbf{u} \tag{9.1}$$

consisting of n first order linear ordinary differential equations in the n unknowns $\mathbf{u}(t) = (u_1(t), \ldots, u_n(t))^T \in \mathbb{R}^n$. In an *autonomous* system, the time variable does not appear explicitly, and so the coefficient matrix A, of size $n \times n$, is a constant real* matrix. Non-autonomous systems, in which $A(t)$ is time-dependent, are considerably more difficult to analyze, and we refer the reader to a more advanced text such as [28].

As we saw in Section 8.1, a vector-valued exponential function

$$\mathbf{u}(t) = e^{\lambda t}\,\mathbf{v},$$

in which λ is a constant scalar and $\mathbf{v}$ a constant vector, describes a solution to (9.1) if and only if

$$A\mathbf{v} = \lambda\,\mathbf{v}.$$

Hence, assuming $\mathbf{v} \neq \mathbf{0}$, the scalar λ must be an eigenvalue of A and $\mathbf{v}$ the corresponding eigenvector. The resulting exponential function will be called an *eigensolution* of the linear system. Since the system is linear and homogeneous, linear superposition allows us to combine the basic eigensolutions to form more general solutions.

If the coefficient matrix A is complete, then, by definition, its eigenvectors $\mathbf{v}_1$, $\mathbf{v}_2, \ldots, \mathbf{v}_n$ form a basis. The corresponding eigensolutions

$$\mathbf{u}_1(t) = e^{\lambda_1 t}\,\mathbf{v}_1, \qquad \ldots \qquad \mathbf{u}_n(t) = e^{\lambda_n t}\,\mathbf{v}_n,$$

will form a basis for the solution space to the system. Hence, the general solution to a first order linear system with complete coefficient matrix has the form

$$\mathbf{u}(t) = c_1\mathbf{u}_1(t) + \cdots + c_n\mathbf{u}_n(t) = c_1\,e^{\lambda_1 t}\,\mathbf{v}_1 + \cdots + c_n\,e^{\lambda_n t}\,\mathbf{v}_n, \tag{9.2}$$

where $c_1, \ldots, c_n$ are arbitrary constants, which are uniquely prescribed by the initial conditions

$$\mathbf{u}(t_0) = \mathbf{u}_0. \tag{9.3}$$

This all follows from the basic existence and uniqueness theorem for ordinary differential equations, which will be discussed shortly.

EXAMPLE 9.1 Let us solve the coupled pair of ordinary differential equations

$$\frac{du}{dt} = 3u + v, \qquad \frac{dv}{dt} = u + 3v.$$

*Extending the solution techniques to complex systems with complex coefficient matrices is straightforward, but will not be treated here.

We first write the system in matrix form (9.1) with unknown $\mathbf{u}(t) = \begin{pmatrix} u(t) \\ v(t) \end{pmatrix}$ and coefficient matrix $A = \begin{pmatrix} 3 & 1 \\ 1 & 3 \end{pmatrix}$. According to Example 8.5, the eigenvalues and eigenvectors of A are

$$\lambda_1 = 4, \quad \mathbf{v}_1 = \begin{pmatrix} 1 \\ 1 \end{pmatrix}, \qquad \lambda_2 = 2, \quad \mathbf{v}_2 = \begin{pmatrix} -1 \\ 1 \end{pmatrix}.$$

Both eigenvalues are simple, and so A is a complete matrix. The resulting eigensolutions

$$\mathbf{u}_1(t) = e^{4t}\begin{pmatrix} 1 \\ 1 \end{pmatrix} = \begin{pmatrix} e^{4t} \\ e^{4t} \end{pmatrix}, \quad \mathbf{u}_2(t) = e^{2t}\begin{pmatrix} -1 \\ 1 \end{pmatrix} = \begin{pmatrix} -e^{2t} \\ e^{2t} \end{pmatrix},$$

form a basis of the solution space, so the general solution is a linear combination

$$\mathbf{u}(t) = c_1 e^{4t}\begin{pmatrix} 1 \\ 1 \end{pmatrix} + c_2 e^{2t}\begin{pmatrix} -1 \\ 1 \end{pmatrix} = \begin{pmatrix} c_1 e^{4t} - c_2 e^{2t} \\ c_1 e^{4t} + c_2 e^{2t} \end{pmatrix}.$$

Hence,

$$u(t) = c_1 e^{4t} - c_2 e^{2t}, \qquad v(t) = c_1 e^{4t} + c_2 e^{2t},$$

in which c_1, c_2 are arbitrary constants. ●

The Phase Plane

A wide variety of physical systems are modeled by second order ordinary differential equations. Your first course on ordinary differential equations, e.g., [6, 19], covered the basic solution technique for constant coefficient scalar equations, which we quickly review in the context of an example.

EXAMPLE 9.2 To solve the differential equation

$$\frac{d^2u}{dt^2} + \frac{du}{dt} - 6u = 0, \tag{9.4}$$

we begin with the exponential ansatz†

$$u(t) = e^{\lambda t},$$

where the constant factor λ is to be determined. Substituting into the differential equation leads immediately to the *characteristic equation*

$$\lambda^2 + \lambda - 6 = 0, \quad \text{with roots} \quad \lambda_1 = 2, \quad \lambda_2 = -3.$$

Therefore, e^{2t} and e^{-3t} are individual solutions. Since the equation is second order, Theorem 7.34 implies that they form a basis for the two-dimensional solution space, and hence the general solution can be written as a linear combination

$$u(t) = c_1 e^{2t} + c_2 e^{-3t}, \tag{9.5}$$

where c_1, c_2 are arbitrary constants. ●

There is a standard trick to convert a second order equation

$$\frac{d^2u}{dt^2} + \alpha\,\frac{du}{dt} + \beta\,u = 0 \tag{9.6}$$

† See the footnote on p. 366 for an explanation of the term "ansatz", a.k.a. "inspired guess".

into a first order system. One introduces the so-called *phase plane variables*[‡]

$$u_1 = u, \qquad u_2 = \dot{u} = \frac{du}{dt}. \tag{9.7}$$

Assuming α, β are constants, the phase plane variables satisfy

$$\frac{du_1}{dt} = \frac{du}{dt} = u_2, \qquad \frac{du_2}{dt} = \frac{d^2u}{dt^2} = -\beta\, u - \alpha\, \frac{du}{dt} = -\beta\, u_1 - \alpha\, u_2.$$

In this manner, the second order equation (9.6) is converted into a first order system

$$\dot{u} = A\,\mathbf{u}, \qquad \text{where} \quad \mathbf{u}(t) = \begin{pmatrix} u_1(t) \\ u_2(t) \end{pmatrix}, \qquad A = \begin{pmatrix} 0 & 1 \\ -\beta & -\alpha \end{pmatrix}. \tag{9.8}$$

Every solution $u(t)$ to the second order equation yields a solution

$$\mathbf{u}(t) = \begin{pmatrix} u(t) \\ \dot{u}(t) \end{pmatrix}$$

to the system (9.8). Vice versa, if $\mathbf{u}(t) = (u_1(t), u_2(t))^T$ is any solution to (9.8), then its first component $u(t) = u_1(t)$ defines a solution to the original scalar equation (9.6). We conclude that the two are completely equivalent, in the sense that solving one will immediately resolve the other.

The variables $(u_1, u_2)^T = (u, \dot{u})^T$ serve as coordinates in the *phase plane* $\mathbb{R}^2$. The solutions $\mathbf{u}(t)$ parametrize curves in the phase plane, known as the solution *trajectories* or *orbits*. In particular, the equilibrium solution $\mathbf{u}(t) \equiv \mathbf{0}$ remains fixed at the origin, and so its trajectory is a single point. All other solutions describe genuine curves, whose tangent direction at a point $\mathbf{u}$ is prescribed by the right hand side of the differential equation, namely $\dot{u} = A\,\mathbf{u}$. The collection of all possible solution trajectories is called the *phase portrait* of the system. An important fact is that, in an autonomous first order system, *the phase plane trajectories never cross*. This striking property, which is also valid for nonlinear systems, is a consequence of the uniqueness properties of solutions, [6, 28]. Thus, the phase portrait consists of a family of non-intersecting curves which, when combined with the equilibrium points, fill out the entire phase plane. The direction of motion along a trajectory is indicated by a small arrow; nearby trajectories are traversed in the same direction. The one feature that is not so easily pictured in the phase portrait is the continuously varying speed at which the solution moves along its phase curve. Plotting this requires a more complicated three-dimensional plot using time as the third coordinate.

EXAMPLE 9.2 ***Continued***

For the second order equation (9.4), the equivalent phase plane system is

$$\frac{d\mathbf{u}}{dt} = \begin{pmatrix} 0 & 1 \\ 6 & -1 \end{pmatrix}\mathbf{u}, \qquad \text{or, in full detail,} \qquad \begin{aligned} \dot{u}_1 &= u_2 \\ \dot{u}_2 &= 6\,u_1 - u_2. \end{aligned} \tag{9.9}$$

Our previous solution formula (9.5) implies that the solution to the phase plane system (9.9) is given by

$$u_1(t) = u(t) = c_1\, e^{2t} + c_2\, e^{-3t}, \qquad u_2(t) = \frac{du}{dt} = 2\,c_1\, e^{2t} - 3\,c_2\, e^{-3t},$$

and hence

$$\mathbf{u}(t) = \begin{pmatrix} c_1\, \mathbf{e}^{2t} + c_2\, e^{-3t} \\ 2\,c_1\, e^{2t} - 3\,c_2\, e^{-3t} \end{pmatrix} = c_1 \begin{pmatrix} e^{2t} \\ 2\,e^{2t} \end{pmatrix} + c_2 \begin{pmatrix} e^{-3t} \\ -3\,e^{-3t} \end{pmatrix} \tag{9.10}$$

[‡]We will often use dots as a shorthand notation for time derivatives.

A sketch of the phase portrait, indicating several representative trajectories, appears in Figure 9.1. The solutions with $c_2 = 0$ go out to ∞ along the rays in the direction $(1, 2)^T$, whereas those with $c_1 = 0$ come in to the origin along rays in the direction $(1, -3)^T$. All other non-equilibrium solutions move along hyperbolic trajectories that asymptote to these rays in, respectively, forward and backward time. ●

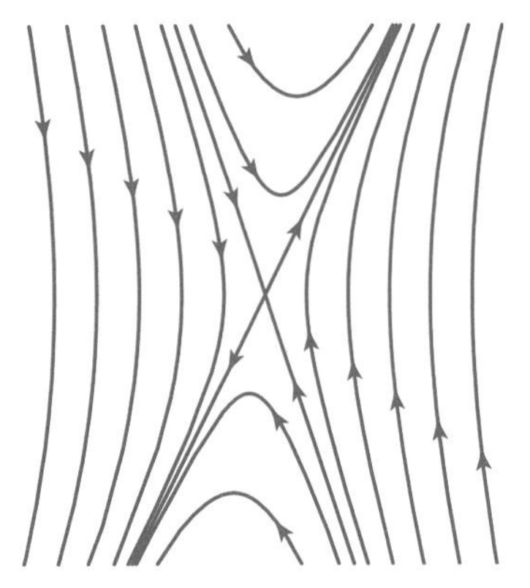

Figure 9.1 Phase portrait of $\dot{u}_1 = u_2,\ \dot{u}_2 = 6u_1 - u_2$.

With some practice, one learns to understand the temporal behavior of the solution by studying its phase plane trajectory. We will investigate the qualitative and quantitative behavior of phase plane systems in depth in Section 9.3.

EXERCISES 9.1

9.1.1. Choose one or more of the following differential equations, and then:

(a) Solve the equation directly.

(b) Write down its phase plane equivalent, and the general solution to the phase plane system.

(c) Plot at least four representative trajectories to illustrate the phase portrait.

(d) Choose two trajectories in your phase portrait and graph the corresponding solution curves $u(t)$. Explain in your own words how the orbit and the solution graph are related.

(i) $\ddot{u} + 4u = 0$ (ii) $\ddot{u} - 4u = 0$

(iii) $\ddot{u} + 2\dot{u} + u = 0$ (iv) $\ddot{u} + 4\dot{u} + 3u = 0$

(v) $\ddot{u} - 2\dot{u} + 10u = 0$

9.1.2. (a) Convert the third order equation

$$\frac{d^3u}{dt^3} + 3\frac{d^2u}{dt^2} + 4\frac{du}{dt} + 12u = 0$$

into a first order system in three variables by setting $u_1 = u,\ u_2 = \dot{u},\ u_3 = \ddot{u}$.

(b) Solve the equation directly, and then use this to write down the general solution to your first order system.

(c) What is the dimension of the solution space?

9.1.3. Convert the second order coupled system of ordinary differential equations $\ddot{u} = a\dot{u} + b\dot{v} + cu + dv$, $\dot{v} = p\dot{u} + q\dot{v} + ru + sv$, into a first order system involving four variables.

9.1.4. *True or false*: The phase plane trajectories (9.10) for $(c_1, c_2)^T \neq \mathbf{0}$ are hyperbolas.

◇ **9.1.5.** (a) Show that if $\mathbf{u}(t)$ solves $\dot{\mathbf{u}} = A\mathbf{u}$, then its *time reversal*, defined as $\mathbf{v}(t) = \mathbf{u}(-t)$, solves $\dot{\mathbf{v}} = B\mathbf{v}$, where $B = -A$.

(b) Explain why the two systems have the same phase portraits, but the direction of motion along the trajectories is reversed.

(c) Apply time reversal to the system(s) you derived in Exercise 9.1.1.

(d) What is the effect of time reversal on the original second order equation?

9.1.6. (a) Show that if $\mathbf{u}(t)$ solves $\dot{\mathbf{u}} = A\mathbf{u}$, then $\mathbf{v}(t) = \mathbf{u}(2t)$ solves $\dot{\mathbf{v}} = B\mathbf{v}$, where $B = 2A$.

(b) How are the solution trajectories of the two systems related?

◇ **9.1.7.** Let A be a constant $n \times n$ matrix, and suppose $\mathbf{u}(t)$ is a solution to the system

$$\frac{d\mathbf{u}}{dt} = A\mathbf{u}.$$

(a) Show that all derivatives

$$\frac{d^k\mathbf{u}}{dt^k}, \qquad k = 1, 2, \ldots,$$

are also solutions.

(b) Show that

$$\frac{d^k\mathbf{u}}{dt^k} = A^k\mathbf{u}.$$

9.1.8. *True or false*: Each solution to a phase plane system moves at a constant speed along its trajectory.

♠ **9.1.9.** Use a three-dimensional graphics package to plot solution curves $(t, u_1(t), u_2(t))^T$ of the phase plane systems in Exercise 9.1.1. Discuss their shape and explain how they are related to the phase plane trajectories.

♡ **9.1.10.** A first order linear system $\dot{u} = a\,u + b\,v$, $\dot{v} = c\,u + d\,v$, can be converted into a single second order differential equation by the following device. Assuming that $b \neq 0$, solve the system for v and $\dot{v}$ in terms of u and $\dot{u}$. Then differentiate your equation for v with respect to t, and eliminate $\dot{v}$ from the resulting pair of equations. The result is a second order ordinary differential equation for $u(t)$.

(a) Write out the second order equation in terms of the coefficients a, b, c, d of the first order system.

(b) Show that there is a one-to-one correspondence between solutions of the system and solutions of the scalar differential equation.

(c) Use this method to solve the following linear systems, and sketch the resulting phase portraits.

(i) $\dot{u} = v,\ \dot{v} = -u$

(ii) $\dot{u} = 2u + 5v,\ \dot{v} = -u$

(iii) $\dot{u} = 4u - v,\ \dot{v} = 6u - 3v$

(iv) $\dot{u} = u + v,\ \dot{v} = u - v$

(v) $\dot{u} = v,\ \dot{v} = 0$

(d) Show how to obtain a second order equation satisfied by $v(t)$ by an analogous device. Are the second order equations for u and for v the same?

(e) Discuss how you might proceed if $b = 0$.

Existence and Uniqueness

Before delving further into our subject, it will help to briefly summarize the basic existence and uniqueness theorems as they apply to linear systems of ordinary differential equations. Even though we will only study the constant coefficient case in detail, these results are equally applicable to non-autonomous systems, and so—but only in this subsection—we allow the coefficient matrix to depend continuously on t. A key fact is that a system of n first order ordinary differential equations requires n initial conditions—one for each variable—in order to uniquely specify its solution. More specifically:

THEOREM 9.3 Let $A(t)$ be an $n \times n$ matrix and $\mathbf{f}(t)$ an n component column vector of continuous functions on the interval $a < t < b$. Set an initial time $a < t_0 < b$ and an initial vector $\mathbf{b} \in \mathbb{R}^n$. Then the *initial value problem*

$$\frac{d\mathbf{u}}{dt} = A(t)\,\mathbf{u} + \mathbf{f}(t), \qquad \mathbf{u}(t_0) = \mathbf{b}, \tag{9.11}$$

admits a unique solution $\mathbf{u}(t)$ that is defined for all $a < t < b$.

For completeness, we have included an inhomogeneous forcing term $\mathbf{f}(t)$ in the system. We will not prove Theorem 9.3, which is a direct consequences of the more general existence and uniqueness theorem for nonlinear systems of ordinary differential equations. Full details can be found in most texts on ordinary differential equations, including [6, 28]. In the homogeneous case, when $\mathbf{f}(t) \equiv \mathbf{0}$, uniqueness of solutions implies that the solution with zero initial conditions, $\mathbf{u}(t_0) = \mathbf{0}$, is the trivial zero solution: $\mathbf{u}(t) \equiv \mathbf{0}$ for *all* t. In other words, if one starts at an equilibrium, one remains there for all time. Moreover, one can never arrive at equilibrium in a finite amount of time, since if $\mathbf{u}(t_1) = \mathbf{0}$, then, again by uniqueness, $\mathbf{u}(t) \equiv \mathbf{0}$ for all $t < t_1$ (and $\geq t_1$, too).

Uniqueness has another important consequence: linear independence of solutions needs only be checked at a single point.

Lemma 9.4 The solutions $\mathbf{u}_1(t), \ldots, \mathbf{u}_k(t)$ to a first order homogeneous linear system $\dot{\mathbf{u}} = A(t)\,\mathbf{u}$ are linearly independent if and only if their initial values $\mathbf{u}_1(t_0), \ldots, \mathbf{u}_k(t_0)$ are linearly independent vectors in $\mathbb{R}^n$.

Proof If the solutions were linearly dependent, one could find (constant) scalars $c_1, \ldots, c_k$, not all zero, such that

$$\mathbf{u}(t) = c_1\,\mathbf{u}_1(t) + \cdots + c_k\,\mathbf{u}_k(t) \equiv \mathbf{0}. \tag{9.12}$$

The equation holds, in particular, at $t = t_0$,

$$\mathbf{u}(t_0) = c_1\,\mathbf{u}_1(t_0) + \cdots + c_k\,\mathbf{u}_k(t_0) = \mathbf{0}. \tag{9.13}$$

This immediately proves linear dependence of the initial vectors. Conversely, if the initial values are linearly dependent, then (9.13) holds for some $c_1, \ldots, c_k$, not all zero. Linear superposition implies that the self-same linear combination $\mathbf{u}(t) = c_1\mathbf{u}_1(t) + \cdots + c_k\mathbf{u}_k(t)$ is a solution to the system, with zero initial condition. By uniqueness, $\mathbf{u}(t) \equiv \mathbf{0}$ for all t, and so (9.12) holds, proving linear dependence of the solutions. ■

Warning: This result is *not* true if the functions are not solutions to a *first order* linear system. For example,

$$\mathbf{u}_1(t) = \begin{pmatrix} 1 \\ t \end{pmatrix}, \qquad \mathbf{u}_2(t) = \begin{pmatrix} \cos t \\ \sin t \end{pmatrix},$$

are linearly independent vector-valued functions, but, at time $t = 0$, the vectors

$$\mathbf{u}_1(0) = \begin{pmatrix} 1 \\ 0 \end{pmatrix} = \mathbf{u}_2(0)$$

are linearly dependent. Even worse,

$$\mathbf{u}_1(t) = \begin{pmatrix} 1 \\ t \end{pmatrix}, \qquad \mathbf{u}_2(t) = \begin{pmatrix} t \\ t^2 \end{pmatrix},$$

define linearly dependent vectors at every fixed value of t. Nevertheless, as vector-valued functions, they are linearly independent. (Why?) In view of Lemma 9.4, neither pair of vector-valued functions can be solutions to a common first order homogeneous linear system.

The next result tells us how many different solutions are required in order to construct the general solution by linear superposition.

THEOREM 9.5 Let $\mathbf{u}_1(t), \ldots, \mathbf{u}_n(t)$ be n linearly independent solutions to the homogeneous system of n first order linear ordinary differential equations $\dot{\mathbf{u}} = A(t)\,\mathbf{u}$. Then the general solution is a linear combination $\mathbf{u}(t) = c_1\mathbf{u}_1(t) + \cdots + c_n\mathbf{u}_n(t)$ depending on n arbitrary constants $c_1, \ldots, c_n$.

Proof If we have n linearly independent solutions $\mathbf{u}_1(t), \ldots, \mathbf{u}_n(t)$, then Lemma 9.4 implies that, at the initial time t_0, the vectors $\mathbf{u}_1(t_0), \ldots, \mathbf{u}_n(t_0)$ are linearly independent, and hence form a basis for $\mathbb{R}^n$. This means that we can express any initial condition

$$\mathbf{u}(t_0) = \mathbf{b} = c_1\mathbf{u}_1(t_0) + \cdots + c_n\mathbf{u}_n(t_0)$$

as a linear combination of the initial vectors. Superposition and uniqueness of solutions implies that the corresponding solution to the initial value problem (9.11) is given by the same linear combination

$$\mathbf{u}(t) = c_1 \mathbf{u}_1(t) + \cdots + c_n \mathbf{u}_n(t).$$

We conclude that every solution to the ordinary differential equation can be written in the prescribed form. ■

Complete Systems

Thus, given a system of n homogeneous linear equations, the immediate goal is to find n linearly independent solutions. Each eigenvalue λ and eigenvector $\mathbf{v}$ leads to an exponential eigensolution $\mathbf{u}(t) = e^{\lambda t}\,\mathbf{v}$. The eigensolutions will be linearly independent if and only if the eigenvectors are—this follows directly from Lemma 9.4. Thus, if the $n \times n$ matrix admits an eigenvector basis, i.e., it is complete, then we have the requisite number of solutions, and hence have solved the differential equation.

THEOREM 9.6 If the $n \times n$ matrix A is complete, then the general (complex) solution to the autonomous linear system $\dot{\mathbf{u}} = A\,\mathbf{u}$ is given by

$$\mathbf{u}(t) = c_1 e^{\lambda_1 t}\,\mathbf{v}_1 + \cdots + c_n e^{\lambda_n t}\,\mathbf{v}_n, \tag{9.14}$$

where $\mathbf{v}_1, \ldots, \mathbf{v}_n$ are the eigenvector basis, $\lambda_1, \ldots, \lambda_n$ the corresponding eigenvalues. The constants $c_1, \ldots, c_n$ are uniquely specified by the initial conditions $\mathbf{u}(t_0) = \mathbf{b}$.

Proof Since the eigenvectors are linearly independent, the eigensolutions define linearly independent vectors $\mathbf{u}_k(0) = \mathbf{v}_k$ at time $t = 0$. Lemma 9.4 implies that the eigensolutions $\mathbf{u}_k(t)$ are, indeed, linearly independent. Hence, the result is an immediate consequence of Theorem 9.5. ■

EXAMPLE 9.7 Let us solve the initial value problem

$$\begin{aligned} \dot{u}_1 &= -2u_1 + u_2, \qquad & u_1(0) &= 3 \\ \dot{u}_2 &= 2u_1 - 3u_2, \qquad & u_2(0) &= 0. \end{aligned}$$

The coefficient matrix is

$$A = \begin{pmatrix} -2 & 1 \\ 2 & -3 \end{pmatrix}.$$

A straightforward computation produces its eigenvalues and eigenvectors:

$$\lambda_1 = -4, \quad \mathbf{v}_1 = \begin{pmatrix} 1 \\ -2 \end{pmatrix}, \qquad \lambda_2 = -1, \quad \mathbf{v}_2 = \begin{pmatrix} 1 \\ 1 \end{pmatrix}.$$

Theorem 9.6 assures us that the corresponding eigensolutions

$$\mathbf{u}_1(t) = e^{-4t}\begin{pmatrix} 1 \\ -2 \end{pmatrix}, \quad \mathbf{u}_2(t) = e^{-t}\begin{pmatrix} 1 \\ 1 \end{pmatrix},$$

form a basis for the two-dimensional solution space. The general solution is an arbitrary linear combination

$$\mathbf{u}(t) = \begin{pmatrix} u_1(t) \\ u_2(t) \end{pmatrix} = c_1 e^{-4t}\begin{pmatrix} 1 \\ -2 \end{pmatrix} + c_2 e^{-t}\begin{pmatrix} 1 \\ 1 \end{pmatrix} = \begin{pmatrix} c_1 e^{-4t} + c_2 \mathbf{e}^{-t} \\ -2c_1 \mathbf{e}^{-4t} + c_2 e^{-t} \end{pmatrix},$$

where c_1, c_2 are constant scalars. Once we have the general solution in hand, the final step is to determine the values of c_1, c_2 so as to satisfy the initial conditions. Evaluating the solution at $t = 0$, we find that we need to solve the linear system

$$c_1 + c_2 = 3, \qquad -2c_1 + c_2 = 0,$$

for $c_1 = 1$, $c_2 = 2$. Thus, the (unique) solution to the initial value problem is

$$u_1(t) = e^{-4t} + 2e^{-t}, \qquad u_2(t) = -2e^{-4t} + 2e^{-t}. \tag{9.15}$$

Note that both components of the solution decay exponentially fast to 0 as $t \to \infty$. ●

EXAMPLE 9.8 Consider the linear initial value problem

$$\begin{aligned} \dot u_1 &= u_1 + 2u_2, & u_1(0) &= 2\\ \dot u_2 &= u_2 - 2u_3, & u_2(0) &= -1\\ \dot u_3 &= 2u_1 + 2u_2 - u_3, & u_3(0) &= -2. \end{aligned}$$

The coefficient matrix is

$$A = \begin{pmatrix} 1 & 2 & 0\\ 0 & 1 & -2\\ 2 & 2 & -1 \end{pmatrix}.$$

In Example 8.8, we computed its eigenvalues and eigenvectors:

$$\lambda_1 = -1, \qquad \lambda_2 = 1 + 2\,\mathrm{i}, \quad \lambda_3 = 1 - 2\,\mathrm{i},$$

$$\mathbf{v}_1 = \begin{pmatrix} -1\\ 1\\ 1 \end{pmatrix}, \quad \mathbf{v}_2 = \begin{pmatrix} 1\\ \mathrm{i}\\ 1 \end{pmatrix}, \quad \mathbf{v}_3 = \begin{pmatrix} 1\\ -\mathrm{i}\\ 1 \end{pmatrix}.$$

The corresponding eigensolutions to the system are

$$\mathbf{u}_1(t) = e^{-t}\begin{pmatrix} -1\\ 1\\ 1 \end{pmatrix}, \quad \widehat{\mathbf{u}}_2(t) = e^{(1+2\mathrm{i})t}\begin{pmatrix} 1\\ \mathrm{i}\\ 1 \end{pmatrix}, \quad \widehat{\mathbf{u}}_3(t) = e^{(1-2\mathrm{i})t}\begin{pmatrix} 1\\ -\mathrm{i}\\ 1 \end{pmatrix}.$$

The first solution is real, but the second and third, while perfectly valid solutions, are complex-valued, and hence not as convenient to work with if, as in most applications, we are ultimately after real functions. But, since the underlying linear system is real, the general reality principle of Theorem 7.48 tells us that any complex solution can be broken up into its real and imaginary parts, each of which is a *real* solution. Here, applying Euler's formula (3.84) to the complex exponential, we find

$$\begin{aligned} \widehat{\mathbf{u}}_2(t) = e^{(1+2\mathrm{i})t}\begin{pmatrix} 1\\ \mathrm{i}\\ 1 \end{pmatrix} &= \left(e^t \cos 2t + \mathrm{i}\, e^t \sin 2t\right)\begin{pmatrix} 1\\ \mathrm{i}\\ 1 \end{pmatrix}\\ &= \begin{pmatrix} e^t \cos 2t\\ -e^t \sin 2t\\ e^t \cos 2t \end{pmatrix} + \mathrm{i}\begin{pmatrix} e^t \sin 2t\\ e^t \cos 2t\\ e^t \sin 2t \end{pmatrix}. \end{aligned}$$

The final two vector-valued functions are independent real solutions, as you can readily check. In this manner, we have produced three linearly independent real solutions

$$\mathbf{u}_1(t) = \begin{pmatrix} -e^{-t}\\ e^{-t}\\ e^{-t} \end{pmatrix}, \quad \mathbf{u}_2(t) = \begin{pmatrix} e^t \cos 2t\\ -e^t \sin 2t\\ e^t \cos 2t \end{pmatrix}, \quad \mathbf{u}_3(t) = \begin{pmatrix} e^t \sin 2t\\ e^t \cos 2t\\ e^t \sin 2t \end{pmatrix},$$

which, by Theorem 9.5, form a basis for the three-dimensional solution space to our system. The general solution can be written as a linear combination:

$$\mathbf{u}(t) = c_1\,\mathbf{u}_1(t) + c_2\,\mathbf{u}_2(t) + c_3\,\mathbf{u}_3(t) = \begin{pmatrix} -c_1\,e^{-t} + c_2\,e^{t}\cos 2t + c_3\,e^{t}\sin 2t \\ c_1\,e^{-t} - c_2\,e^{t}\sin 2t + c_3\,e^{t}\cos 2t \\ c_1\,e^{-t} + c_2\,e^{t}\cos 2t + c_3\,e^{t}\sin 2t \end{pmatrix}.$$

The constants c_1, c_2, c_3 are uniquely prescribed by imposing initial conditions. In our case, the solution satisfying

$$\mathbf{u}(0) = \begin{pmatrix} -c_1 + c_2 \\ c_1 + c_3 \\ c_1 + c_2 \end{pmatrix} = \begin{pmatrix} 2 \\ -1 \\ -2 \end{pmatrix} \qquad \text{results in} \qquad \begin{aligned} c_1 &= -2, \\ c_2 &= 0, \\ c_3 &= 1. \end{aligned}$$

Thus, the solution to the original initial value problem is

$$\begin{aligned} u_1(t) &= 2\,e^{-t} + e^{t}\sin 2t, \\ u_2(t) &= -2\,e^{-t} + e^{t}\cos 2t, \\ u_3(t) &= -2\,e^{-t} + e^{t}\sin 2t. \end{aligned}$$

Incidentally, the third complex eigensolution also produces two real solutions, but these reproduce the ones we have already listed since it is the complex conjugate of the second eigensolution, and so $\widehat{\mathbf{u}}_3(t) = \mathbf{u}_2(t) - \mathrm{i}\,\mathbf{u}_3(t)$. In general, when solving real systems, you only need to deal with one eigenvalue from each complex conjugate pair to construct a complete system of real solutions. ●

EXERCISES 9.1

9.1.11. Find the solution to the system of differential equations

$$\frac{du}{dt} = 3\,u + 4\,v, \qquad \frac{dv}{dt} = 4\,u - 3\,v,$$

with initial conditions $u(0) = 3$ and $v(0) = -2$.

9.1.12. Find the general real solution to the following systems of differential equations:

(a) $\dot u_1 = u_1 + 9\,u_2$, $\dot u_2 = u_1 + 3\,u_2$ (b) $\dot x_1 = 4\,x_1 + 3\,x_2$, $\dot x_2 = 3\,x_1 - 4\,x_2$

(c) $\dot y_1 = y_1 - y_2$, $\dot y_2 = 2\,y_1 + 3\,y_2$ (d) $\dot y_1 = y_2$, $\dot y_2 = 3\,y_1 + 2\,y_3$, $\dot y_3 = -\,y_2$

(e) $\dot x_1 = 3\,x_1 - 8\,x_2 + 2\,x_3$, $\dot x_2 = -\,x_1 + 2\,x_2 + 2\,x_3$, $\dot x_3 = x_1 - 4\,x_2 + 2\,x_3$

9.1.13. Solve the following initial value problems:

(a) $\dfrac{d\mathbf{u}}{dt} = \begin{pmatrix} 0 & 2 \\ 2 & 0 \end{pmatrix}\mathbf{u}, \quad \mathbf{u}(1) = \begin{pmatrix} 1 \\ 0 \end{pmatrix}$

(b) $\dfrac{d\mathbf{u}}{dt} = \begin{pmatrix} 1 & -2 \\ -2 & 1 \end{pmatrix}\mathbf{u}, \quad \mathbf{u}(0) = \begin{pmatrix} -2 \\ 4 \end{pmatrix}$

(c) $\dfrac{d\mathbf{u}}{dt} = \begin{pmatrix} 1 & 2 \\ -1 & 1 \end{pmatrix}\mathbf{u}, \quad \mathbf{u}(0) = \begin{pmatrix} 1 \\ 0 \end{pmatrix}$

(d) $\dfrac{d\mathbf{u}}{dt} = \begin{pmatrix} -1 & 3 & -3 \\ 2 & 2 & -7 \\ 0 & 3 & -4 \end{pmatrix}\mathbf{u}, \quad \mathbf{u}(0) = \begin{pmatrix} 1 \\ 0 \\ 0 \end{pmatrix}$

(e) $\dfrac{d\mathbf{u}}{dt} = \begin{pmatrix} 2 & 1 & -6 \\ -1 & 0 & 4 \\ 0 & -1 & -2 \end{pmatrix}\mathbf{u}, \quad \mathbf{u}(\pi) = \begin{pmatrix} 2 \\ -1 \\ -1 \end{pmatrix}$

(f) $\dfrac{d\mathbf{u}}{dt} = \begin{pmatrix} 0 & 0 & 1 & 0 \\ 0 & 0 & 0 & 2 \\ 1 & 0 & 0 & 0 \\ 0 & 2 & 0 & 0 \end{pmatrix}\mathbf{u}, \quad \mathbf{u}(2) = \begin{pmatrix} 1 \\ 0 \\ 0 \\ 1 \end{pmatrix}$

(g) $\dfrac{d\mathbf{u}}{dt} = \begin{pmatrix} 2 & 1 & -1 & 0 \\ -3 & -2 & 0 & 1 \\ 0 & 0 & 1 & -2 \\ 0 & 0 & 1 & -1 \end{pmatrix}\mathbf{u}, \quad \mathbf{u}(0) = \begin{pmatrix} 1 \\ -1 \\ 2 \\ 1 \end{pmatrix}$

9.1.14. (a) Find the solution to the system

$$\frac{dx}{dt} = -x + y, \qquad \frac{dy}{dt} = -x - y,$$

that has initial conditions $x(0) = 1,\ y(0) = 0$.

(b) Sketch a phase portrait of the system that shows several typical solution trajectories, including the solution you found in part (a). Clearly indicate the direction of increasing t on your curves.

9.1.15. A planar steady state fluid flow has velocity vector $\mathbf{v} = (2x - 3y, x - y)^T$. The motion of the fluid is described by the differential equation

$$\frac{d\mathbf{x}}{dt} = \mathbf{v}.$$

A floating object starts out at the point $(1, 1)^T$. Find its position after 1 time unit.

9.1.16. A steady state fluid flow has velocity vector $\mathbf{v} = (-2y, 2x, z)^T$. Describe the motion of the fluid particles as governed by the differential equation

$$\frac{d\mathbf{x}}{dt} = \mathbf{v}.$$

9.1.17. Solve the initial value problem

$$\frac{d\mathbf{u}}{dt} = \begin{pmatrix} -6 & 1 \\ 1 & -6 \end{pmatrix} \mathbf{u}, \qquad \mathbf{u}(0) = \begin{pmatrix} 1 \\ 2 \end{pmatrix}.$$

Explain how orthogonality can help.

9.1.18. (a) Find the eigenvalues and eigenvectors of

$$K = \begin{pmatrix} 1 & -1 & 0 \\ -1 & 2 & -1 \\ 0 & -1 & 1 \end{pmatrix}.$$

(b) Verify that the eigenvectors are mutually orthogonal. Based on part (a), is K positive definite, positive semi-definite or indefinite?

(c) Solve the initial value problem

$$\frac{d\mathbf{u}}{dt} = K\,\mathbf{u}, \qquad \mathbf{u}(0) = \begin{pmatrix} 1 \\ 2 \\ -1 \end{pmatrix},$$

using orthogonality to simplify the computations.

9.1.19. Demonstrate that one can also solve the initial value problem in Example 9.8 by writing the solution as a complex linear combination of the complex eigensolutions, and then using the initial conditions to specify the coefficients.

9.1.20. Determine whether the following vector-valued functions are linearly dependent or linearly independent:

(a) $\begin{pmatrix} 1 \\ t \end{pmatrix}, \begin{pmatrix} -t \\ 1 \end{pmatrix}$ (b) $\begin{pmatrix} 1+t \\ t \end{pmatrix}, \begin{pmatrix} 1-t^2 \\ t-t^2 \end{pmatrix}$

(c) $\begin{pmatrix} 1 \\ t \end{pmatrix}, \begin{pmatrix} t \\ 2 \end{pmatrix}, \begin{pmatrix} -t \\ t \end{pmatrix}$

(d) $\begin{pmatrix} e^{-t} \\ -e^{t} \end{pmatrix}, \begin{pmatrix} -e^{-t} \\ e^{t} \end{pmatrix}$

(e) $\begin{pmatrix} e^{2t}\cos 3t \\ -e^{2t}\sin 3t \end{pmatrix}, \begin{pmatrix} e^{2t}\sin 3t \\ e^{2t}\cos 3t \end{pmatrix}$

(f) $\begin{pmatrix} \cos 3t \\ \sin 3t \end{pmatrix}, \begin{pmatrix} \sin 3t \\ \cos 3t \end{pmatrix}$

(g) $\begin{pmatrix} 1 \\ t \\ 1-t \end{pmatrix}, \begin{pmatrix} 0 \\ -2 \\ 2 \end{pmatrix}, \begin{pmatrix} 3 \\ 1+3t \\ 2-3t \end{pmatrix}$

(h) $\begin{pmatrix} e^{t} \\ -e^{t} \\ e^{t} \end{pmatrix}, \begin{pmatrix} e^{t} \\ e^{t} \\ -e^{t} \end{pmatrix}, \begin{pmatrix} -e^{t} \\ e^{t} \\ e^{t} \end{pmatrix}$

(i) $\begin{pmatrix} e^{t} \\ t\,e^{t} \\ t^2 e^{t} \end{pmatrix}, \begin{pmatrix} t^2 e^{t} \\ e^{t} \\ t\,e^{t} \end{pmatrix}, \begin{pmatrix} t\,e^{t} \\ t^2 e^{t} \\ e^{t} \end{pmatrix}, \begin{pmatrix} e^{t} \\ e^{t} \\ e^{t} \end{pmatrix}$

◇ **9.1.21.** Let A be a constant matrix. Suppose $\mathbf{u}(t)$ solves the initial value problem $\dot{\mathbf{u}} = A\,\mathbf{u}$, $\mathbf{u}(0) = \mathbf{b}$. Prove that the solution to the initial value problem $\dot{\mathbf{u}} = A\,\mathbf{u}$, $\mathbf{u}(t_0) = \mathbf{b}$, is equal to $\widetilde{\mathbf{u}}(t) = \mathbf{u}(t - t_0)$. How are the solution trajectories related?

9.1.22. Suppose $\mathbf{u}(t)$ and $\widetilde{\mathbf{u}}(t)$ both solve the linear system $\dot{u} = A\,\mathbf{u}$.

(a) Suppose they have the same value $\mathbf{u}(t_1) = \widetilde{\mathbf{u}}(t_1)$ at any one time t_1. Show that they are, in fact, the same solution: $\mathbf{u}(t) = \widetilde{\mathbf{u}}(t)$ for all t.

(b) What happens if $\mathbf{u}(t_1) = \widetilde{\mathbf{u}}(t_2)$ for some $t_1 \neq t_2$. *Hint*: See Exercise 9.1.21.

9.1.23. Prove that the general solution to a linear system $\dot{\mathbf{u}} = \Lambda\,\mathbf{u}$ with diagonal coefficient matrix $\Lambda = \operatorname{diag}(\lambda_1, \ldots, \lambda_n)$ is given by $\mathbf{u}(t) = \left(c_1 e^{\lambda_1 t}, \ldots, c_n e^{\lambda_n t}\right)^T$.

9.1.24. Show that if $\mathbf{u}(t)$ is a solution to $\dot{\mathbf{u}} = A\mathbf{u}$, and S is a constant, nonsingular matrix of the same size as A, then $\mathbf{v}(t) = S\,\mathbf{u}(t)$ solves the linear system $\dot{\mathbf{v}} = B\mathbf{v}$, where $B = S\,A\,S^{-1}$ is similar to A.

◇ **9.1.25.** (i) Combine Exercises 9.1.23–24 to show that if $A = S\,\Lambda\,S^{-1}$ is diagonalizable, then the solution to $\dot{\mathbf{u}} = A\,\mathbf{u}$ can be written as $\mathbf{u}(t) = S\left(c_1 e^{\lambda_1 t}, \ldots, c_n e^{\lambda_n t}\right)^T$, where $\lambda_1, \ldots, \lambda_n$ are its eigenvalues and $S = (\,\mathbf{v}_1\ \mathbf{v}_2 \ldots \mathbf{v}_n\,)$ is the corresponding matrix of eigenvectors.

(ii) Write the general solution to the systems in Exercise 9.1.13 in this form.

The General Case

Summarizing the preceding subsection, if the coefficient matrix of a homogeneous, autonomous first order linear system is complete, then the eigensolutions form a (complex) basis for the solution space. In the incomplete cases, the formulae for the basis solutions are a little more intricate, and involve polynomials as well as (complex) exponentials. Readers who did not cover Section 8.6 are advised to skip ahead to Section 9.2; only Theorem 9.13, which summarizes the key features, will be used in the sequel.

EXAMPLE 9.9 The simplest incomplete case arises as the phase plane equivalent of a scalar ordinary differential equation whose characteristic equation has a repeated root. For example, to directly solve the second order equation

$$\frac{d^2u}{dt^2} - 2\,\frac{du}{dt} + u = 0,$$

we substitute the usual exponential ansatz $u = e^{\lambda t}$, leading to the characteristic equation

$$\lambda^2 - 2\lambda + 1 = 0.$$

There is only one double root, $\lambda = 1$, and hence, up to scalar multiple, only one exponential solution $u_1(t) = e^t$. For a scalar ordinary differential equation, the second "missing" solution is obtained by just multiplying by t, so that $u_2(t) = t\,e^t$. As a result, the general solution is

$$u(t) = c_1 u_1(t) + c_2 u_2(t) = c_1 e^t + c_2 t\,e^t.$$

The equivalent phase plane system is

$$\frac{d\mathbf{u}}{dt} = \begin{pmatrix} 0 & 1 \\ -1 & 2 \end{pmatrix}\mathbf{u}, \qquad \text{where} \qquad \mathbf{u}(t) = \begin{pmatrix} u(t) \\ \dot u(t) \end{pmatrix}.$$

Note that the coefficient matrix is incomplete—it has $\lambda = 1$ as a double eigenvalue, but only one independent eigenvector, namely $\mathbf{v} = (1,1)^T$. The two linearly independent solutions to the phase plane system can be constructed from the two solutions to the scalar equation. Thus,

$$\mathbf{u}_1(t) = \begin{pmatrix} e^t \\ e^t \end{pmatrix}, \qquad \mathbf{u}_2(t) = \begin{pmatrix} t\,e^t \\ t\,e^t + e^t \end{pmatrix}$$

form a basis for the two-dimensional solution space. The first is an eigensolution, while the second includes an additional polynomial factor. Observe that, unlike the scalar case, the second solution $\mathbf{u}_2$ is *not* obtained from the first by merely multiplying by t. Unfortunately, incomplete systems are not that easily handled. ●

In general, the eigenvectors of an incomplete matrix fail to form a basis, and, as noted in Section 8.6, must be extended to a Jordan basis. Thus, the key step is to describe the solutions associated with a Jordan chain.

Lemma 9.10 Suppose $\mathbf{w}_1, \ldots, \mathbf{w}_k$ form a Jordan chain of length k for the eigenvalue λ of the matrix A. Then there are k linearly independent solutions to the corresponding first order system $\dot{\mathbf{u}} = A\mathbf{u}$ having the form

$$\mathbf{u}_1(t) = e^{\lambda t}\mathbf{w}_1, \quad \mathbf{u}_2(t) = e^{\lambda t}(t\,\mathbf{w}_1 + \mathbf{w}_2), \quad \mathbf{u}_3(t) = e^{\lambda t}\big(\tfrac{1}{2}t^2\,\mathbf{w}_1 + t\,\mathbf{w}_2 + \mathbf{w}_3\big),$$

and, in general,

$$\mathbf{u}_j(t) = e^{\lambda t}\sum_{i=1}^{j}\frac{t^{j-i}}{(j-i)!}\,\mathbf{w}_i, \qquad 1 \le j \le k. \tag{9.16}$$

The proof is by direct substitution of the formulae (9.16) into the differential equation, invoking the defining relations (8.47) of the Jordan chain as needed. The details are left to the reader. If λ is a complex eigenvalue, then the Jordan chain solutions (9.16) will involve complex exponentials. As usual, they can be split into their real and imaginary parts, which, provided that A is a real matrix, are independent real solutions.

EXAMPLE 9.11 The coefficient matrix of the system

$$\frac{d\mathbf{u}}{dt} = \begin{pmatrix} -1 & 0 & 1 & 0 & 0 \\ -2 & 2 & -4 & 1 & 1 \\ -1 & 0 & -3 & 0 & 0 \\ -4 & -1 & 3 & 1 & 0 \\ 4 & 0 & 2 & -1 & 0 \end{pmatrix} \mathbf{u}$$

is incomplete; it has only 2 linearly independent eigenvectors associated with the eigenvalues 1 and -2. Using the Jordan basis computed in Example 8.51, we produce the following 5 linearly independent solutions:

$$\mathbf{u}_1(t) = e^t\, \mathbf{v}_1, \quad \mathbf{u}_2(t) = e^t\,(t\,\mathbf{v}_1 + \mathbf{v}_2), \quad \mathbf{u}_3(t) = e^t\,\left(\tfrac{1}{2}t^2\mathbf{v}_1 + t\,\mathbf{v}_2 + \mathbf{v}_3\right),$$
$$\mathbf{u}_4(t) = e^{-2t}\,\mathbf{v}_4, \qquad \mathbf{u}_5(t) = e^{-2t}\,(t\,\mathbf{v}_4 + \mathbf{v}_5),$$

or, explicitly,

$$\begin{pmatrix} 0 \\ 0 \\ 0 \\ -e^t \\ e^t \end{pmatrix}, \quad \begin{pmatrix} 0 \\ e^t \\ 0 \\ -t\,e^t \\ (-1+t)\,e^t \end{pmatrix}, \quad \begin{pmatrix} 0 \\ t\,e^t \\ 0 \\ \left(1-\frac{1}{2}t^2\right)e^t \\ \left(-t+\frac{1}{2}t^2\right)e^t \end{pmatrix},$$

$$\begin{pmatrix} -e^{-2t} \\ e^{-2t} \\ e^{-2t} \\ -2e^{-2t} \\ 0 \end{pmatrix}, \quad \begin{pmatrix} -(1+t)\,e^{-2t} \\ t\,e^{-2t} \\ t\,e^{-2t} \\ -2(1+t)\,e^{-2t} \\ e^{-2t} \end{pmatrix}.$$

The first three are associated with the $\lambda_1 = 1$ Jordan chain, the last two with the $\lambda_2 = -2$ chain. The eigensolutions are the pure exponentials $\mathbf{u}_1(t)$, $\mathbf{u}_4(t)$. The general solution to the system is an arbitrary linear combination of these five basis solutions. ●

Proposition 9.12 Let A be an $n \times n$ matrix. Then the Jordan chain solutions (9.16) constructed from a Jordan basis of A form a basis for the n-dimensional solution space for the corresponding linear system $\dot{\mathbf{u}} = A\,\mathbf{u}$.

The proof of linear independence of the Jordan chain solutions is reasonably straightforward, [28, 33].

While the precise analytical details prove to be quite messy, many important qualitative features can be readily gleaned from the general structure of the solution formulae (9.16). The following result describes the principal classes of solutions of homogeneous autonomous linear systems of ordinary differential equations.

THEOREM 9.13 Let A be a real $n \times n$ matrix. Every real solution to the linear system $\dot{u} = A\mathbf{u}$ is a linear combination of n linearly independent solutions appearing in the following four classes:

(1) If λ is a complete real eigenvalue of multiplicity m, then there exist m linearly independent solutions of the form

$$\mathbf{u}_k(t) = e^{\lambda t}\, \mathbf{v}_k, \qquad k = 1, \ldots, m,$$

where $\mathbf{v}_1, \ldots, \mathbf{v}_m$ are linearly independent eigenvectors.

(2) If $\mu \pm \mathrm{i}\,\nu$ form a pair of complete complex conjugate eigenvalues of multiplicity m, then there exist $2m$ linearly independent real solutions of the forms

$$\begin{aligned} \mathbf{u}_k(t) &= e^{\mu t}\big[\cos(\nu t)\,\mathbf{w}_k - \sin(\nu t)\,\mathbf{z}_k\,\big], \\ \widehat{\mathbf{u}}_k(t) &= e^{\mu t}\big[\sin(\nu t)\,\mathbf{w}_k + \cos(\nu t)\,\mathbf{z}_k\,\big], \end{aligned} \qquad k = 1, \ldots, m,$$

where $\mathbf{v}_k = \mathbf{w}_k \pm \mathrm{i}\,\mathbf{z}_k$ are the associated complex conjugate eigenvectors.

(3) If λ is an incomplete real eigenvalue of multiplicity m and r is the dimension of the eigenspace V_λ, then there exist m linearly independent solutions of the form

$$\mathbf{u}_k(t) = e^{\lambda t}\, \mathbf{p}_k(t), \qquad k = 1, \ldots, m,$$

where $\mathbf{p}_k(t)$ is a vector of polynomials of degree $\leq m - r$.

(4) If $\lambda = \mu \pm \mathrm{i}\,\nu$ form a pair of incomplete complex conjugate eigenvalues of multiplicity m and r is the common dimension of the two eigenspaces, then there exist $2m$ linearly independent real solutions

$$\begin{aligned} \mathbf{u}_k(t) &= e^{\mu t}\,\big[\cos(\nu t)\,\mathbf{p}_k(t) - \sin(\nu t)\,\mathbf{q}_k(t)\,\big], \\ \widehat{\mathbf{u}}_k(t) &= e^{\mu t}\,\big[\sin(\nu t)\,\mathbf{p}_k(t) + \cos(\nu t)\,\mathbf{q}_k(t)\,\big], \end{aligned} \qquad k = 1, \ldots, m,$$

where $\mathbf{p}_k(t), \mathbf{q}_k(t)$ are vectors of polynomials of degree $\leq m - r$.

As a result, every real solution to a homogeneous linear system of ordinary differential equations is a vector-valued function whose entries are linear combinations of functions of the particular form $t^k\, e^{\mu t} \cos \nu t$ and $t^k\, e^{\mu t} \sin \nu t$, i.e., sums of products of exponentials, trigonometric functions and polynomials. The exponents μ are the real parts of the eigenvalues of the coefficient matrix; the trigonometric frequencies ν are the imaginary parts of the eigenvalues; nonconstant polynomials appear only if the matrix is incomplete.

EXERCISES 9.1

9.1.26. Find the general solution to the linear system

$$\frac{d\mathbf{u}}{dt} = A\,\mathbf{u}$$

for the following incomplete coefficient matrices:

(a) $\begin{pmatrix} 2 & 1 \\ 0 & 2 \end{pmatrix}$ (b) $\begin{pmatrix} 2 & -1 \\ 9 & -4 \end{pmatrix}$

(c) $\begin{pmatrix} -1 & -1 \\ 4 & -5 \end{pmatrix}$ (d) $\begin{pmatrix} 4 & -1 & -3 \\ -2 & 1 & 2 \\ 5 & -1 & -4 \end{pmatrix}$

(e) $\begin{pmatrix} -3 & 1 & 0 \\ 1 & -3 & -1 \\ 0 & 1 & -3 \end{pmatrix}$ (f) $\begin{pmatrix} -1 & 1 & 1 \\ 0 & -1 & 1 \\ 0 & 0 & -1 \end{pmatrix}$

(g) $\begin{pmatrix} 3 & 1 & 1 & 1 \\ 0 & -1 & 0 & 1 \\ 0 & 0 & 3 & 1 \\ 0 & 0 & 0 & -1 \end{pmatrix}$

(h) $\begin{pmatrix} 0 & 1 & 1 & 0 \\ -1 & 0 & 0 & 1 \\ 0 & 0 & 0 & 1 \\ 0 & 0 & -1 & 0 \end{pmatrix}$

9.1.27. Find a first order system of ordinary differential equations that has the indicated vector-valued function as a solution:

(a) $\begin{pmatrix} e^t + e^{2t} \\ 2e^t \end{pmatrix}$ (b) $\begin{pmatrix} e^{-t}\cos 3t \\ -3e^{-t}\sin 3t \end{pmatrix}$

(c) $\begin{pmatrix} 1 \\ t-1 \end{pmatrix}$ (d) $\begin{pmatrix} \sin 2t - \cos 2t \\ \sin 2t + 3\cos 2t \end{pmatrix}$

(e) $\begin{pmatrix} e^{2t} \\ e^{-3t} \\ e^{2t} - e^{-3t} \end{pmatrix}$ (f) $\begin{pmatrix} \sin t \\ \cos t \\ 1 \end{pmatrix}$

(g) $\begin{pmatrix} t \\ 1-t^2 \\ 1+t \end{pmatrix}$ (h) $\begin{pmatrix} e^t \sin t \\ 2e^t \cos t \\ e^t \sin t \end{pmatrix}$

9.1.28. Which sets of functions in Exercise 9.1.20 can be solutions to a common first order, homogeneous, constant coefficient linear system of ordinary differential equations? If so, find a system they satisfy; if not, explain why not.

9.1.29. Solve the third order equation

$$\frac{d^3u}{dt^3} + 3\,\frac{d^2u}{dt^2} + 4\,\frac{du}{dt} + 12u = 0$$

by converting it into a first order system. Compare your answer with what you found in Exercise 9.1.2.

9.1.30. Solve the second order coupled system of ordinary differential equations $\ddot{u} = \dot{u} + u - v$, $\ddot{v} = \dot{v} - u + v$, by converting it into a first order system involving four variables.

9.1.31. Suppose that $\mathbf{u}(t) \in \mathbb{R}^n$ is a polynomial solution to the constant coefficient linear system $\dot{\mathbf{u}} = A\mathbf{u}$. What is the maximal possible degree of $\mathbf{u}(t)$? What can you say about A when $\mathbf{u}(t)$ has maximal degree?

◇ **9.1.32.** (a) Under the assumption that $\mathbf{u}_1, \ldots, \mathbf{u}_k$ form a Jordan chain for the coefficient matrix A, prove that the functions (9.16) are solutions to the differential equation $\dot{\mathbf{u}} = A\,\mathbf{u}$.

(b) Prove that they are linearly independent.

9.1.33. (a) Explain how to solve the inhomogeneous ordinary differential equation

$$\frac{d\mathbf{u}}{dt} = A\,\mathbf{u} + \mathbf{b}$$

when $\mathbf{b}$ is a constant vector belonging to rng A. *Hint*: Look at $\mathbf{v}(t) = \mathbf{u}(t) - \mathbf{u}^\star$ where $\mathbf{u}^\star$ is an equilibrium solution.

(b) Use your method to solve

(i) $\dfrac{du}{dt} = u - 3v + 1, \quad \dfrac{dv}{dt} = -u - v$

(ii) $\dfrac{du}{dt} = 4v + 2, \quad \dfrac{dv}{dt} = -u - 3.$

9.2 STABILITY OF LINEAR SYSTEMS

With the general solution formulae in hand, we are now ready to study the qualitative features of first order linear dynamical systems. Our primary focus will be on stability properties of the equilibrium solution(s). The starting point is a simple calculus lemma, whose proof is left to the reader.

Lemma 9.14 Let μ, ν be real and $k \geq 0$. A function of the form

$$f(t) = t^k e^{\mu t}\cos \nu t \quad \text{or} \quad t^k e^{\mu t}\sin \nu t \tag{9.17}$$

will decay to zero for large t, so $\lim_{t\to\infty} f(t) = 0$, if and only if $\mu < 0$. The function remains bounded, so $|f(t)| \leq C$ for some constant C, for all $t \geq 0$ if and only if either $\mu < 0$, or $\mu = 0$ and $k = 0$.

Loosely put, exponential decay will always overwhelm polynomial growth, while trigonometric functions remain neutrally bounded. Now, in the solution to our linear system, the functions (9.17) come from the eigenvalues $\lambda = \mu + \mathrm{i}\,\nu$ of the coefficient matrix. The lemma implies that the asymptotic behavior of the solutions, and hence the stability of the system, depends on the sign of $\mu = \operatorname{Re}\lambda$. If $\mu < 0$, then the solutions decay to zero at an exponential rate as $t \to \infty$. If $\mu > 0$, then the solutions become unbounded as $t \to \infty$. In the borderline case $\mu = 0$, the solutions remain bounded, provided that they don't involve any powers of t.

Asymptotic stability of the equilibrium zero solution requires that all other solutions tend to $\mathbf{0}$ as $t \to \infty$, and hence all the eigenvalues must satisfy $\mu = \operatorname{Re}\lambda < 0$. Or, stated another way, all eigenvalues must lie in the *left half plane*—the subset of $\mathbb{C}$ to the left of the imaginary axis, as sketched in Figure 9.2. In this manner, we have demonstrated the fundamental asymptotic stability criterion for linear systems.

Figure 9.2 The left half plane.

THEOREM 9.15 A first order autonomous homogeneous linear system of ordinary differential equations $\dot{\mathbf{u}} = A\,\mathbf{u}$ has an asymptotically stable zero solution if and only if all the eigenvalues of its coefficient matrix A lie in the left half plane: $\operatorname{Re}\lambda < 0$. If A has one or more eigenvalues with positive real part, $\operatorname{Re}\lambda > 0$, then the zero solution is unstable.

EXAMPLE 9.16 Consider the system

$$\frac{du}{dt} = 2u - 6v + w, \qquad \frac{dv}{dt} = 3u - 3v - w, \qquad \frac{dw}{dt} = 3u - v - 3w.$$

The coefficient matrix

$$A = \begin{pmatrix} 2 & -6 & 1 \\ 3 & -3 & -1 \\ 3 & -1 & -3 \end{pmatrix}$$

is found to have eigenvalues

$$\lambda_1 = -2, \quad \lambda_2 = -1 + \mathrm{i}\sqrt{6}, \quad \lambda_3 = -1 - \mathrm{i}\sqrt{6},$$

with respective real parts $-2, -1, -1$. The Stability Theorem 9.15 implies that the equilibrium solution $u_\star \equiv v_\star \equiv w_\star \equiv 0$ is asymptotically stable. Indeed, every solution involves the functions e^{-2t}, $e^{-t}\cos\sqrt{6}\,t$, and $e^{-t}\sin\sqrt{6}\,t$, all of which decay to 0 at an exponential rate. The latter two have the slowest decay rate, and so most solutions to the linear system go to $\mathbf{0}$ in proportion to e^{-t}, i.e., at an exponential rate determined by the eigenvalue with the least negative real part.

The final statement is a special case of the following general result, whose proof is left to the reader. ●

Proposition 9.17 If $\mathbf{u}(t)$ is any solution to $\dot{\mathbf{u}} = A\mathbf{u}$, then $\|\mathbf{u}(t)\| \leq C\,e^{at}$ for all $t \geq t_0$, for any $a > a^\star = \max\{\operatorname{Re}\lambda \mid \lambda \text{ is an eigenvalue of } A\}$, where the constant $C > 0$ depends on the solution and choice of norm. If the eigenvalue(s) achieving the maximum are complete, then one can set $a = a^\star$.

A particularly important class of systems are the linear *gradient flows*

$$\frac{d\mathbf{u}}{dt} = -K\mathbf{u}, \tag{9.18}$$

in which K is a symmetric, positive definite matrix. According to Theorem 8.23, all the eigenvalues of K are real and positive, and so the eigenvalues of the negative definite coefficient matrix $-K$ for the gradient flow system (9.18) are real and negative. Applying Theorem 9.15, we conclude that the zero solution to any gradient flow system (9.18) with negative definite coefficient matrix $-K$ is asymptotically stable.

EXAMPLE 9.18 Applying the test we learned in Chapter 3, the matrix

$$K = \begin{pmatrix} 1 & 1 \\ 1 & 5 \end{pmatrix}$$

is seen to be positive definite. The associated gradient flow is

$$\frac{du}{dt} = -u - v, \qquad \frac{dv}{dt} = -u - 5v. \tag{9.19}$$

The eigenvalues and eigenvectors of

$$-K = \begin{pmatrix} -1 & -1 \\ -1 & -5 \end{pmatrix}$$

are

$$\lambda_1 = -3 + \sqrt{5}, \quad \mathbf{v}_1 = \begin{pmatrix} 1 \\ 2 - \sqrt{5} \end{pmatrix},$$

$$\lambda_2 = -3 - \sqrt{5}, \quad \mathbf{v}_2 = \begin{pmatrix} 1 \\ 2 + \sqrt{5} \end{pmatrix}.$$

Therefore, the general solution to the system is

$$\mathbf{u}(t) = c_1\, e^{(-3+\sqrt{5})t} \begin{pmatrix} 1 \\ 2 - \sqrt{5} \end{pmatrix} + c_2\, e^{(-3-\sqrt{5})t} \begin{pmatrix} 1 \\ 2 + \sqrt{5} \end{pmatrix},$$

or, in components,

$$u(t) = c_1\, e^{(-3+\sqrt{5})t} + c_2\, e^{(-3-\sqrt{5})t},$$
$$v(t) = c_1\,(2 - \sqrt{5})\, e^{(-3+\sqrt{5})t} + c_2\,(2 + \sqrt{5})\, e^{(-3-\sqrt{5})t}.$$

All solutions tend to zero as $t \to \infty$ at the exponential rate prescribed by the least negative eigenvalue, which is $-3 + \sqrt{5} \approx -.7639$. This confirms the asymptotic stability of the gradient flow. ●

The reason for the term "gradient flow" is that the vector field $-K\mathbf{u}$ appearing on the right hand side of (9.18) is, in fact, the negative of the gradient of the quadratic function

$$q(\mathbf{u}) = \tfrac{1}{2}\mathbf{u}^T K\mathbf{u} = \frac{1}{2}\sum_{i,j=1}^{n} k_{ij}\,u_i\,u_j, \qquad \text{so that} \qquad \nabla q(\mathbf{u}) = K\mathbf{u}. \tag{9.20}$$

Thus, we can write (9.18) as

$$\frac{d\mathbf{u}}{dt} = -\nabla q(\mathbf{u}). \tag{9.21}$$

For the particular system (9.19),

$$q(u,v) = \tfrac{1}{2}\begin{pmatrix} u & v \end{pmatrix}\begin{pmatrix} 1 & 1 \\ 1 & 5 \end{pmatrix}\begin{pmatrix} u \\ v \end{pmatrix} = \tfrac{1}{2}u^2 + u\,v + \tfrac{5}{2}v^2,$$

and so the gradient flow is given by

$$\frac{du}{dt} = -\frac{\partial q}{\partial u} = -u - v, \qquad \frac{dv}{dt} = -\frac{\partial q}{\partial v} = -u - 5v.$$

The gradient ∇q of a function q points in the direction of its steepest increase, while its negative $-\nabla q$ points in the direction of steepest decrease, [2, 58]. Thus, the solutions to the gradient flow system (9.21) will decrease $q(\mathbf{u})$ as rapidly as possible, tending to its minimum at $\mathbf{u}^\star = \mathbf{0}$. For instance, if $q(u,v)$ represents the height of a hill at position (u,v), then the solutions to (9.21) are the paths of steepest descent followed by, say, water flowing down the hill (provided we ignore inertial effects). In physical applications, the quadratic function (9.20) often represents the potential energy in the system, and the gradient flow models the natural behavior of systems that seek to minimize their energy.

EXAMPLE 9.19 Let us solve the first order system

$$\frac{du}{dt} = -8u - w, \qquad \frac{dv}{dt} = -8v - w, \qquad \frac{dw}{dt} = -u - v - 7w,$$

subject to initial conditions

$$u(0) = 1, \qquad v(0) = -3, \qquad w(0) = 2.$$

The coefficient matrix is

$$\begin{pmatrix} -8 & 0 & -1 \\ 0 & -8 & -1 \\ -1 & -1 & -7 \end{pmatrix} = -\begin{pmatrix} 8 & 0 & 1 \\ 0 & 8 & 1 \\ 1 & 1 & 7 \end{pmatrix} = -K,$$

which is minus the positive definite matrix analyzed in Example 8.24. Using the computed eigenvalues and eigenvectors, we conclude that the general solution has the form

$$\mathbf{u}(t) = \begin{pmatrix} u(t) \\ v(t) \\ w(t) \end{pmatrix} = c_1 e^{-9t}\begin{pmatrix} 1 \\ 1 \\ 1 \end{pmatrix} + c_2 e^{-8t}\begin{pmatrix} -1 \\ 1 \\ 0 \end{pmatrix} + c_3 e^{-6t}\begin{pmatrix} -1 \\ -1 \\ 2 \end{pmatrix}.$$

The coefficients c_1, c_2, c_3 are prescribed by the initial data:

$$\mathbf{u}(0) = \begin{pmatrix} 1 \\ -3 \\ 2 \end{pmatrix} = c_1\begin{pmatrix} 1 \\ 1 \\ 1 \end{pmatrix} + c_2\begin{pmatrix} -1 \\ 1 \\ 0 \end{pmatrix} + c_3\begin{pmatrix} -1 \\ -1 \\ 2 \end{pmatrix} = c_1\mathbf{v}_1 + c_2\mathbf{v}_2 + c_3\mathbf{v}_3.$$

Rather than solve this linear system directly, we make use of the fact that the matrix is symmetric, and hence its eigenvectors $\mathbf{v}_1, \mathbf{v}_2, \mathbf{v}_3$ form an orthogonal basis. Thus, we can apply the orthogonal basis formula (5.7) to compute

$$c_1 = \frac{\langle \mathbf{u}(0), \mathbf{v}_1 \rangle}{\| \mathbf{v}_1 \|^2} = 0,$$

$$c_2 = \frac{\langle \mathbf{u}(0), \mathbf{v}_2 \rangle}{\| \mathbf{v}_2 \|^2} = \frac{-4}{2} = -2,$$

$$c_3 = \frac{\langle \mathbf{u}(0), \mathbf{v}_3 \rangle}{\| \mathbf{v}_3 \|^2} = \frac{6}{6} = 1.$$

We conclude that the solution to the initial value problem is

$$\mathbf{u}(t) = \begin{pmatrix} -e^{-6t} + 2e^{-8t} \\ -e^{-6t} - 2e^{-8t} \\ 2e^{-6t} \end{pmatrix}.$$

The solution decays exponentially fast to $\mathbf{0}$ at a rate of -6, which is the largest (least negative) eigenvalue of the coefficient matrix. ●

Asymptotic stability implies that the solutions return to equilibrium; *stability* only requires them to stay nearby. The appropriate eigenvalue criterion is readily established.

THEOREM 9.20 A first order linear, homogeneous, constant-coefficient system of ordinary differential equations (9.1) has a stable zero solution if and only if all the eigenvalues satisfy $\operatorname{Re}\lambda \leq 0$, and, moreover, any eigenvalue lying on the imaginary axis, $\operatorname{Re}\lambda = 0$, is complete, meaning that it has as many independent eigenvectors as its multiplicity.

Proof The proof is the same as before, based on Theorem 9.13 and the decay properties in Lemma 9.14. All the eigenvalues with negative real part lead to exponentially decaying solutions—even if they are incomplete. If the coefficient matrix has a complete zero eigenvalue, then the corresponding eigensolutions are constant, comprising a set of basis vectors for its kernel, which suffices to maintain stability. On the other hand, if 0 is an incomplete eigenvalue, then the associated Jordan chain solutions involve nonconstant polynomials, and become unbounded as $t \to \pm\infty$. Similarly, if a purely imaginary eigenvalue is complete, then the associated solutions only involve trigonometric functions, and hence remain bounded. On the other hand, solutions associated with incomplete purely imaginary eigenvalues contain polynomials in t multiplying sines and cosines, and hence cannot remain bounded. ■

EXAMPLE 9.21 A planar *Hamiltonian system* takes the form

$$\frac{du}{dt} = \frac{\partial H}{\partial v}, \qquad \frac{dv}{dt} = -\frac{\partial H}{\partial u}, \tag{9.22}$$

where $H(u, v)$ is known as the *Hamiltonian function*. If

$$H(u, v) = \tfrac{1}{2}\, a\, u^2 + b\, u\, v + \tfrac{1}{2}\, c\, v^2 \tag{9.23}$$

is a quadratic form, then the Hamiltonian system

$$\dot{u} = b\, u + c\, v, \qquad \dot{v} = -a\, u - b\, v, \tag{9.24}$$

is homogeneous, linear, with coefficient matrix

$$A = \begin{pmatrix} b & c \\ -a & -b \end{pmatrix}.$$

The associated characteristic equation is

$$\det(A - \lambda\,\mathrm{I}) = \lambda^2 + (a\,c - b^2) = 0.$$

If H is positive or negative definite, then $a\,c - b^2 > 0$, and so the eigenvalues are purely imaginary: $\lambda = \pm\,\mathrm{i}\,\sqrt{a\,c - b^2}$ and complete since they are simple. Thus, the stability criterion of Theorem 9.20 holds and we conclude that planar Hamiltonian systems with a definite Hamiltonian function are stable. On the other hand, if H is indefinite, then the coefficient matrix has one positive and one negative eigenvalue, and hence the Hamiltonian system is unstable. ●

REMARK: The equations of classical mechanics, such as motion of masses under gravitational attraction, can all be formulated as Hamiltonian systems. The Hamiltonian function represents the energy of the system. The Hamiltonian formulation is a crucial first step in the physical process of quantizing the classical equations in order to determine the quantum mechanical equations of motion, [40].

EXERCISES 9.2

9.2.1. Classify the following systems according to whether the origin is

(i) asymptotically stable,

(ii) stable, or

(iii) unstable

(a) $\dfrac{du}{dt} = -2u - v, \quad \dfrac{dv}{dt} = u - 2v$

(b) $\dfrac{du}{dt} = 2u - 5v, \quad \dfrac{dv}{dt} = u - v$

(c) $\dfrac{du}{dt} = -u - 2v, \quad \dfrac{dv}{dt} = 2u - 5v$

(d) $\dfrac{du}{dt} = -2v, \quad \dfrac{dv}{dt} = 8u$

(e) $\dfrac{du}{dt} = -2u - v + w, \quad \dfrac{dv}{dt} = -u - 2v + w, \quad \dfrac{dw}{dt} = -3u - 3v + 2w$

(f) $\dfrac{du}{dt} = -u - 2v, \quad \dfrac{dv}{dt} = 6u + 3v - 4w, \quad \dfrac{dw}{dt} = 4u - 3w$

(g) $\dfrac{du}{dt} = 2u - v + 3w, \quad \dfrac{dv}{dt} = u - v + w, \quad \dfrac{dw}{dt} = -4u + v - 5w$

(h) $\dfrac{du}{dt} = u + v - w, \quad \dfrac{dv}{dt} = -2u - 3v + 3w, \quad \dfrac{dw}{dt} = -v + w$

9.2.2. Write out the formula for the general real solution to the system in Example 9.16 and verify its stability.

9.2.3. Write out and solve the gradient flow system corresponding to the following quadratic forms:

(a) $u^2 + v^2$ (b) $u\,v$

(c) $4u^2 - 2u\,v + v^2$

(d) $2u^2 - u\,v - 2u\,w + 2v^2 - v\,w + 2w^2$

9.2.4. Write out and solve the Hamiltonian systems corresponding to the first three quadratic forms in Exercise 9.2.3. Which of them are stable?

9.2.5. Which of the following 2×2 systems are gradient flows? Which are Hamiltonian systems? In each case, discuss the stability of the zero solution.

(a) $\dot u = -2u + v, \quad \dot v = u - 2v$ (b) $\dot u = u - 2v, \quad \dot v = -2u + v$

(c) $\dot u = v, \quad \dot v = u$ (d) $\dot u = -v, \quad \dot v = u$

(e) $\dot u = -u - 2v, \quad \dot v = -2u - v$

9.2.6. *True or false*: A nonzero linear 2×2 gradient flow cannot be a Hamiltonian flow.

9.2.7. (a) Show that the matrix

$$A = \begin{pmatrix} 0 & 1 & 1 & 0 \\ -1 & 0 & 0 & 1 \\ 0 & 0 & 0 & 1 \\ 0 & 0 & -1 & 0 \end{pmatrix}$$

has $\lambda = \pm\,\mathrm{i}$ as incomplete complex conjugate eigenvalues.

(b) Find the general real solution to $\dot{\mathbf{u}} = A\,\mathbf{u}$.

(c) Explain the behavior of a typical solution. Why is the zero solution not stable?

9.2.8. Let A be a real 3×3 matrix, and assume that the linear system $\dot{\mathbf{u}} = A\,\mathbf{u}$ has a periodic solution of period P. Prove that every periodic solution of the system has period P. What other types of solutions can there be? Is the zero solution necessarily stable?

9.2.9. Are the conclusions of Exercise 9.2.8 valid when A is a 4×4 matrix?

9.2.10. Let A be a real 5×5 matrix, and assume that A has eigenvalues $\mathrm{i}, -\mathrm{i}, -2, -1$ (and no others). Is the zero solution to the linear system $\dot{\mathbf{u}} = A\,\mathbf{u}$ necessarily stable? Explain. Does your answer change if A is 6×6?

9.2.11. *True or false*: The system $\dot{\mathbf{u}} = -H_n\,\mathbf{u}$, where H_n is the $n \times n$ Hilbert matrix (1.70), is asymptotically stable.

9.2.12. *True or false*: If K is positive semi-definite, then the zero solution to $\dot{\mathbf{u}} = -K\,\mathbf{u}$ is stable.

9.2.13. Let $\mathbf{u}(t)$ solve $\dot{\mathbf{u}} = A\mathbf{u}$. Let $\mathbf{v}(t) = \mathbf{u}(-t)$ be its time reversal.

(a) Write down the linear system $\dot{\mathbf{v}} = B\mathbf{v}$ satisfied by $\mathbf{v}(t)$. Then classify the following statements as *true* or *false*. Explain your answers.

(b) If $\dot{\mathbf{u}} = A\,\mathbf{u}$ is asymptotically stable, then $\dot{\mathbf{v}} = B\mathbf{v}$ is unstable.

(c) If $\dot{\mathbf{u}} = A\,\mathbf{u}$ is unstable, then $\dot{\mathbf{v}} = B\mathbf{v}$ is asymptotically stable.

(d) If $\dot{\mathbf{u}} = A\,\mathbf{u}$ is stable, then $\dot{\mathbf{v}} = B\mathbf{v}$ is stable.

9.2.14. *True or false*: If A is a symmetric matrix, then the system $\dot{\mathbf{u}} = -A^2\mathbf{u}$ has an asymptotically stable equilibrium solution.

9.2.15. *True or false*:

(a) If $\operatorname{tr} A > 0$, then the system $\dot{\mathbf{u}} = A\,\mathbf{u}$ is unstable.

(b) If $\det A > 0$, then the system $\dot{\mathbf{u}} = A\,\mathbf{u}$ is unstable.

9.2.16. Consider the differential equation $\dot{\mathbf{u}} = -K\mathbf{u}$, where K is positive semi-definite.

(a) Find all equilibrium solutions.

(b) Prove that all non-constant solutions decay exponentially fast to some equilibrium. What is the decay rate?

(c) Is the origin

(i) stable,

(ii) asymptotically stable, or

(iii) unstable?

(d) Prove that, as $t \to \infty$, the solution $\mathbf{u}(t)$ converges to the orthogonal projection of its initial vector $\mathbf{a} = \mathbf{u}(0)$ onto $\ker K$.

9.2.17. (a) Let $H(u, v) = a\,u^2 + b\,u\,v + c\,v^2$ be a quadratic function. Prove that the non-equilibrium trajectories of the associated Hamiltonian system and those of the gradient flow are mutually orthogonal, i.e., they always intersect at right angles.

(b) Verify this result for

(i) $u^2 + 3\,v^2$ (ii) $u\,v$

by drawing representative trajectories of both systems on the same graph.

9.2.18. *True or false*: If the Hamiltonian system for $H(u, v)$ is stable, then the corresponding gradient flow $\dot{u} = -\nabla H$ is stable.

9.2.19. Suppose that $\mathbf{u}(t)$ satisfies the gradient flow system (9.21).

(a) Prove that $\dfrac{d}{dt}\,q(\mathbf{u}) = -\,\|K\,\mathbf{u}\|^2$.

(b) Explain why if $\mathbf{u}(t)$ is any nonconstant solution to the gradient flow, then $q(\mathbf{u}(t))$ is a strictly decreasing function of t, thus quantifying how fast a gradient flow decreases energy.

♡ **9.2.20.** The law of *conservation of energy* states that the energy in a Hamiltonian system is constant on solutions.

(a) Prove that if $\mathbf{u}(t)$ satisfies the Hamiltonian system (9.22), then $H(\mathbf{u}(t)) = c$ is a constant, and hence solutions $\mathbf{u}(t)$ move along the level sets of the Hamiltonian or energy function. Explain how the value of c is determined by the initial conditions.

(b) Plot the level curves of the particular Hamiltonian function $H(u, v) = u^2 - 2\,u\,v + 2\,v^2$ and verify that they coincide with the solution trajectories.

◇ **9.2.21.** Prove Lemma 9.14.

◇ **9.2.22.** Prove Proposition 9.17.

9.3 TWO-DIMENSIONAL SYSTEMS

The two-dimensional case is particularly instructive, since it is easy to analyze, but already manifests most of the key phenomena to be found in higher dimensions. Moreover, the solutions can be easily pictured and their behavior understood through their phase portraits. In this section, we will present a complete classification of the possible qualitative behaviors of real, planar linear dynamical systems.

Setting $\mathbf{u}(t) = (u(t), v(t))^T$, a first order planar homogeneous linear system has the explicit form

$$\frac{du}{dt} = a\,u + b\,v, \qquad \frac{dv}{dt} = c\,u + d\,v, \tag{9.25}$$

where $A = \begin{pmatrix} a & b \\ c & d \end{pmatrix}$ is the (constant) coefficient matrix. As in Section 9.1, we will refer to the (u, v)–plane as the *phase plane*. In particular, the phase plane equivalents (9.8) of second order scalar equations form a subclass.

According to (8.20), the characteristic equation for the given 2×2 matrix is

$$\det(A - \lambda\,\mathrm{I}) = \lambda^2 - \tau\,\lambda + \delta = 0, \tag{9.26}$$

where

$$\tau = \operatorname{tr} A = a + d, \qquad \delta = \det A = a\,d - b\,c, \tag{9.27}$$

are, respectively, the trace and the determinant of A. The eigenvalues, and hence the nature of the solutions, is almost entirely determined by these two quantities. The sign of the *discriminant*

$$\Delta = \tau^2 - 4\,\delta = (\operatorname{tr} A)^2 - 4\,\det A = (a - d)^2 + 4\,b\,c \tag{9.28}$$

determines whether the eigenvalues

$$\lambda = \frac{\tau \pm \sqrt{\Delta}}{2} \tag{9.29}$$

are real or complex, and thereby plays a key role in the classification.

Let us summarize the different possibilities as distinguished by their qualitative behavior. Each category will be illustrated by a representative phase portrait, which displays several typical solution trajectories in the phase plane. The complete portrait gallery of planar systems can be found in Figure 9.3.

Distinct Real Eigenvalues

The coefficient matrix A has two distinct real eigenvalues $\lambda_1 < \lambda_2$ if and only if the discriminant is positive: $\Delta > 0$. In this case, the solutions take the exponential form

$$\mathbf{u}(t) = c_1\,e^{\lambda_1 t}\,\mathbf{v}_1 + c_2\,e^{\lambda_2 t}\,\mathbf{v}_2, \tag{9.30}$$

where $\mathbf{v}_1, \mathbf{v}_2$ are the eigenvectors and c_1, c_2 are arbitrary constants, to be determined by the initial conditions. The asymptotic behavior of the solutions is governed by the size of the eigenvalues. Let $V_k = \{c\,\mathbf{v}_k\}$, $k = 1, 2$, denote the "eigenlines", i.e., the one-dimensional eigenspaces.

There are five qualitatively different cases, depending upon the signs of the two eigenvalues. These are listed by their descriptive name, followed by the required conditions on the discriminant, trace, and determinant of the coefficient matrix.

Ia. *Stable Node*: $\Delta > 0, \qquad \operatorname{tr} A < 0, \qquad \det A > 0.$

If $\lambda_1 < \lambda_2 < 0$ are both negative, then $\mathbf{0}$ is an asymptotically *stable node*. The solutions all tend to $\mathbf{0}$ as $t \to \infty$. Since the first exponential $e^{\lambda_1 t}$ decreases much faster than the second $e^{\lambda_2 t}$, the first term in the solution (9.30) will soon become negligible, and hence $\mathbf{u}(t) \approx c_2\,e^{\lambda_2 t}\,\mathbf{v}_2$ when t is large. Thus, solutions with $c_2 \neq 0$ will arrive at the origin along curves tangent to the eigenline V_2, including those with $c_1 = 0$ which move directly along the eigenline. The solutions with $c_2 = 0$ come in to the origin along the eigenline V_1, at a faster rate. Conversely, as $t \to -\infty$, all solutions become unbounded: $\|\mathbf{u}(t)\| \to \infty$. In this case, the first exponential

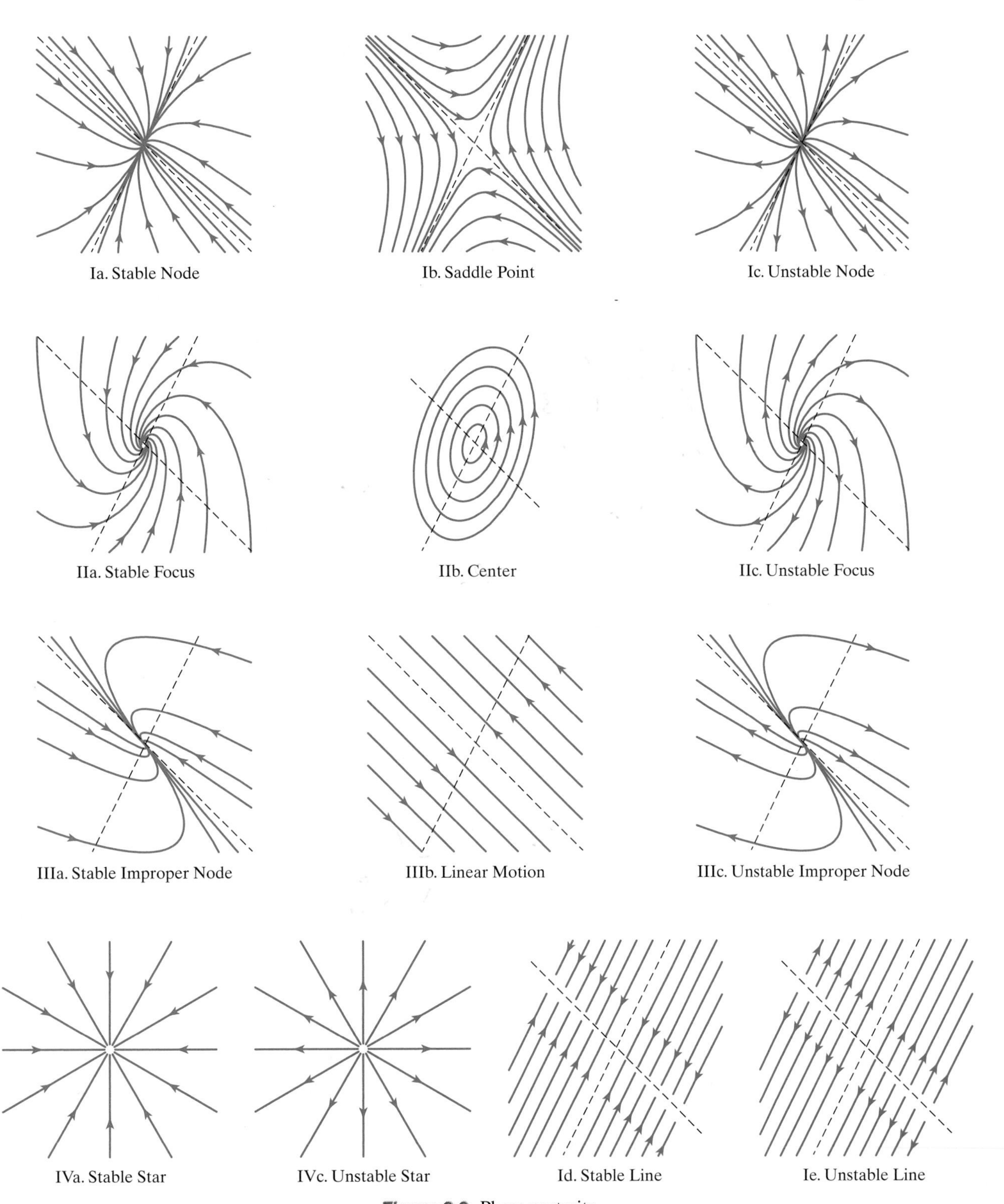

Figure 9.3 Phase portraits.

grows faster than the second, and so $\mathbf{u}(t) \approx c_1 e^{\lambda_1 t} \mathbf{v}_1$ for $t \ll 0$. In other words, as they escape to ∞, the solution trajectories become more and more parallel to the eigenline V_1—except for those with $c_1 = 0$ that remain in the eigenline V_2.

Ib. *Saddle Point*: $\Delta > 0, \qquad \det A < 0.$

If $\lambda_1 < 0 < \lambda_2$, then $\mathbf{0}$ is an unstable *saddle point*. Solutions (9.30) with $c_2 = 0$ start out on the eigenline V_1 and go in to $\mathbf{0}$ as $t \to \infty$, while solutions with $c_1 = 0$ start on V_2 and go to $\mathbf{0}$ as $t \to -\infty$. All other solutions become unbounded at both large positive and large negative times. As $t \to +\infty$, the solutions approach the *unstable eigenline* V_2, while as $t \to -\infty$, they asymptote to the *stable eigenline* V_1.

Ic. *Unstable Node*: $\Delta > 0, \qquad \operatorname{tr} A > 0, \qquad \det A > 0.$

If the eigenvalues $0 < \lambda_1 < \lambda_2$ are both positive, then $\mathbf{0}$ is an *unstable node*. The phase portrait is the same as that of a stable node, but the solution trajectories are traversed in the opposite direction. Time reversal $t \to -t$ will convert an unstable node into a stable node and vice versa. Thus, in the unstable case, the solutions all tend to the origin as $t \to -\infty$ and become unbounded as $t \to \infty$. Except for the eigensolutions, they asymptote to V_1 as $t \to -\infty$, and become parallel to V_2 as $t \to \infty$.

Id. *Stable Line*: $\Delta > 0, \qquad \operatorname{tr} A < 0, \qquad \det A = 0.$

If $\lambda_1 < \lambda_2 = 0$, then every point on the eigenline V_2 associated with the zero eigenvalue is a stable equilibrium point. The other solutions move along straight lines parallel to V_1, coming in to one of the equilibrium points on V_2 as $t \to \infty$.

Ie. *Unstable Line*: $\Delta > 0, \qquad \operatorname{tr} A > 0, \qquad \det A = 0.$

This is merely the time reversal of a stable line. If $0 = \lambda_1 < \lambda_2$, then every point on the eigenline V_1 is an equilibrium. The other solutions moves off to ∞ along straight lines parallel to V_2 as $t \to \infty$, and tend to an equilibrium on V_1 as $t \to -\infty$.

Complex Conjugate Eigenvalues

The coefficient matrix A has two complex conjugate eigenvalues

$$\lambda = \mu \pm \mathrm{i}\,\nu, \qquad \text{where} \qquad \mu = \tfrac{1}{2}\,\tau = \tfrac{1}{2}\operatorname{tr} A, \qquad \nu = \sqrt{-\Delta},$$

if and only if its discriminant is negative: $\Delta < 0$. In this case, the real solutions can be written in the phase–amplitude form

$$\mathbf{u}(t) = r\, e^{\mu t}\, [\, \cos(\nu t - \sigma)\, \mathbf{w} + \sin(\nu t - \sigma)\, \mathbf{z}\,], \tag{9.31}$$

where $\mathbf{w} \pm \mathrm{i}\,\mathbf{z}$ are the complex eigenvectors. As noted in Exercise 8.3.14, the real vectors $\mathbf{w}, \mathbf{z}$ are always linearly independent. The amplitude r and phase shift σ are uniquely prescribed by the initial conditions. There are three subcases, depending upon the sign of the real part μ, or, equivalently, the sign of the trace of A.

IIa. *Stable Focus*: $\Delta < 0, \qquad \operatorname{tr} A < 0.$

If $\mu < 0$, then $\mathbf{0}$ is an asymptotically *stable focus*. As $t \to \infty$, the solutions all spiral in to $\mathbf{0}$ with "frequency" ν —meaning it takes time $2\pi/\nu$ for the solution to go once around* the origin. On the other hand, as $t \to -\infty$, the solutions spiral off to ∞ at the same overall frequency.

IIb. *Center*: $\Delta < 0, \qquad \operatorname{tr} A = 0.$

If $\mu = 0$, then $\mathbf{0}$ is a *center*. The solutions all move periodically around elliptical orbits, with common frequency ν and hence period $2\pi/\nu$. In particular, solutions

*But keep in mind that these solutions are not periodic.

that start out near $\mathbf{0}$ stay nearby, and hence a center is a stable, but not asymptotically stable, equilibrium.

IIc. *Unstable Focus*: $\Delta < 0, \qquad \operatorname{tr} A > 0.$

If $\mu > 0$, then $\mathbf{0}$ is an *unstable focus*. The phase portrait is the time reversal of a stable focus, with solutions spiraling off to ∞ as $t \to \infty$ and in to the origin as $t \to -\infty$, again with a common "frequency" ν.

Incomplete Double Real Eigenvalue

The coefficient matrix has a double real eigenvalue $\lambda = \frac{1}{2}\tau = \frac{1}{2} \operatorname{tr} A$ if and only if the discriminant vanishes: $\Delta = 0$. The formula for the solutions depends on whether the eigenvalue λ is complete or not. If λ is an incomplete eigenvalue, admitting only one independent eigenvector $\mathbf{v}$, then the solutions are no longer given by simple exponentials. The general solution formula is

$$\mathbf{u}(t) = (c_1 + c_2 t)e^{\lambda t}\,\mathbf{v} + c_2\, e^{\lambda t}\,\mathbf{w}, \tag{9.32}$$

where $(A - \lambda\, \mathrm{I})\mathbf{w} = \mathbf{v}$, and so $\mathbf{v}, \mathbf{w}$ form a Jordan chain for the coefficient matrix. We let $V = \{c\,\mathbf{v}\}$ denote the eigenline associated with the genuine eigenvector $\mathbf{v}$.

IIIa. *Stable Improper Node*: $\Delta = 0, \qquad \operatorname{tr} A < 0, \qquad A \neq \lambda\,\mathrm{I}.$

If $\lambda < 0$ then $\mathbf{0}$ is an asymptotically *stable improper node*. Since $t\, e^{\lambda t}$ is larger than $e^{\lambda t}$ for $t > 1$, the solutions $\mathbf{u}(t) \approx c_2 t\, e^{\lambda t}$ tend to $\mathbf{0}$ as $t \to \infty$ along a curve that is tangent to the eigenline V. Similarly, as $t \to -\infty$, the solutions go off to ∞ in the opposite direction from their approach, becoming more and more parallel to the eigenline.

IIIb. *Linear Motion*: $\Delta = 0, \qquad \operatorname{tr} A = 0, \qquad A \neq \lambda\,\mathrm{I}.$

If $\lambda = 0$, then every point on the eigenline V is an unstable equilibrium point. Every other solution is a linear polynomial in t, and so moves along a straight line parallel to V, going off to ∞ in either direction.

IIIc. *Unstable Improper Node*: $\Delta = 0, \qquad \operatorname{tr} A > 0, \qquad A \neq \lambda\,\mathrm{I}.$

If $\lambda > 0$, then $\mathbf{0}$ is an *unstable improper node*. The phase portrait is the time reversal of the stable improper node. Solutions go off to ∞ as t increases, becoming progressively more parallel to the eigenline, and tend to the origin tangent to the eigenline as $t \to -\infty$.

Complete Double Real Eigenvalue

In this case, *every* vector in $\mathbb{R}^2$ is an eigenvector, and so the real solutions take the form $\mathbf{u}(t) = e^{\lambda t}\,\mathbf{v}$, where $\mathbf{v}$ is an *arbitrary* constant vector. In fact, this case occurs if and only if $A = \lambda\,\mathrm{I}$ is a scalar multiple of the identity matrix.

IVa. *Stable Star*: $A = \lambda\,\mathrm{I}, \qquad \lambda < 0.$

If $\lambda < 0$ then $\mathbf{0}$ is an asymptotically stable star. The solution trajectories are the rays coming in to the origin, and the solutions go to $\mathbf{0}$ at an exponential rate as $t \to \infty$.

IVb. *Trivial*: $A = \mathrm{O}.$

If $\lambda = 0$, then the only possibility is $A = \mathrm{O}$. Now every solution is constant and every point is a (stable) equilibrium point. Nothing happens! This is the only case not pictured in Figure 9.3.

IVc. *Unstable Star*: $A = \lambda \mathrm{I}, \qquad \lambda > 0.$

If $\lambda > 0$, then $\mathbf{0}$ is an unstable star. The phase portrait is the time reversal of the stable star, and so the solutions move out along rays as $t \to \infty$, while tending to $\mathbf{0}$ as $t \to -\infty$.

Figure 9.4 summarizes the different possibilities, as prescribed by the trace and determinant of the coefficient matrix. The horizontal axis indicates the value of $\tau = \operatorname{tr} A$, while the vertical axis refers to $\delta = \det A$. Points on the parabola $\tau^2 = 4\,\delta$ represent the cases with vanishing discriminant $\Delta = 0$, and correspond to either stars or improper nodes—except for the origin which is either linear motion or trivial. All the asymptotically stable cases lie in the shaded upper left quadrant where $\operatorname{tr} A < 0$ and $\det A > 0$. The borderline points are either stable centers, when $\operatorname{tr} A = 0$, $\det A > 0$, or stable lines, when $\operatorname{tr} A < 0$, $\det A = 0$, or the origin, which may or may not be stable depending upon whether A is the zero matrix or not. All other values for the trace and determinant result in unstable equilibria.

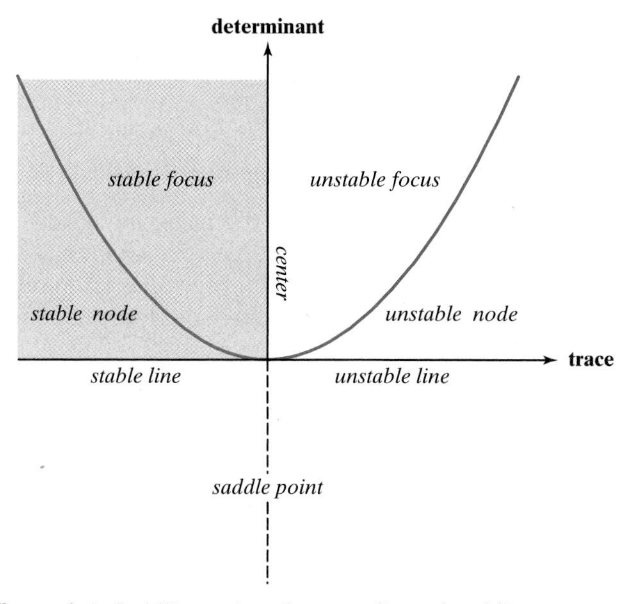

Figure 9.4 Stability regions for two–dimensional linear systems.

Proposition 9.22 Let τ, δ denote, respectively, the trace and determinant of the coefficient matrix A of a homogeneous, linear, autonomous planar system of first order ordinary differential equations. Then the system is

(i) asymptotically stable if and only if $\delta > 0$ and $\tau < 0$;

(ii) stable if and only if $\delta \geq 0$, $\tau \leq 0$, and, if $\delta = \tau = 0$, $A = \mathrm{O}$.

REMARK: Time reversal $t \to -t$ changes the sign of the coefficient matrix $A \to -A$, and hence the sign of its trace, $\tau \to -\tau$, while the determinant $\delta = \det A = \det(-A)$ is unchanged. Thus, the effect is to reflect Figure 9.4 through the vertical axis, interchanging the stable nodes and spirals with their unstable counterparts, while taking saddle points to saddle points.

In physical applications, the parameters occurring in the dynamical system are usually not known exactly, and so the real dynamics may, in fact, be governed by a slight perturbation of the mathematical model. Thus, it is important to know which systems are *structurally stable*, meaning that their basic qualitative features are preserved under sufficiently small changes in the coefficients.

Now, a small perturbation will alter the coefficient matrix slightly, and hence shift its trace and determinant by a comparably small amount. The net effect is to slightly perturb its eigenvalues. (A more in depth analysis can be used to quantify the effects.) Therefore, the question of structural stability reduces to whether the eigenvalues have moved sufficiently far to send the system into a different stability regime. Asymptotically stable systems remain asymptotically stable since a sufficiently small perturbation will not alter the signs of the real parts of its eigenvalues. For a similar reason, unstable systems remain unstable under small perturbations. On the other hand, a borderline stable system—either a center or the trivial system—might become either asymptotically stable or unstable, even under a minuscule perturbation. Such results continues to hold, at least locally, even under suitably small nonlinear perturbations.

Structural stability requires a bit more, since the overall phase portrait should not significantly change. A system in any of the open regions in the Stability Figure 9.4, i.e., a stable or unstable focus, a stable or unstable node, or a saddle point, is structurally stable, whereas a system that lies on the parabola $\tau^2 = 4\,\delta$, or the horizontal axis, or the positive vertical axis, e.g., an improper node, a stable line, etc., is not, since a small perturbation can easily kick it into either of the adjoining regions. Thus, structural stability requires that the eigenvalues be distinct, $\lambda_i \neq \lambda_j$, and have non-zero real part: $\operatorname{Re}\lambda \neq 0$. This final result remains valid for linear systems in higher dimensions, [28, 33].

EXERCISES 9.3

9.3.1. For each the following:

(a) Write the system as $\dot{\mathbf{u}} = A\,\mathbf{u}$.

(b) Find the eigenvalues and eigenvectors of A.

(c) Find the general real solution of the system.

(d) Draw the phase portrait, indicating its type and stability properties:

(i) $\dot u_1 = -u_2,\quad \dot u_2 = 9\,u_1$

(ii) $\dot u_1 = 2\,u_1 - 3\,u_2,\quad \dot u_2 = u_1 - u_2$

(iii) $\dot u_1 = 3\,u_1 - 2\,u_2,\quad \dot u_2 = 2\,u_1 - 2\,u_2$

9.3.2. For each of the following systems

(i) $\dot{\mathbf{u}} = \begin{pmatrix} 2 & -1 \\ 3 & -2 \end{pmatrix}\mathbf{u}$

(ii) $\dot{\mathbf{u}} = \begin{pmatrix} 1 & -1 \\ 5 & -3 \end{pmatrix}\mathbf{u}$

(iii) $\dot{\mathbf{u}} = \begin{pmatrix} -3 & 5/2 \\ -5/2 & 2 \end{pmatrix}\mathbf{u}$

(a) Find the general real solution.

(b) Using the solution formulas obtained in part (a), plot several trajectories of each system. On your graphs, identify the eigenlines (if relevant), and the direction of increasing t on the trajectories.

(c) Write down the type and stability properties of the system.

9.3.3. Classify the following systems, and sketch their phase portraits.

(a) $\dfrac{du}{dt} = -u + 4\,v,\quad \dfrac{dv}{dt} = u - 2\,v$

(b) $\dfrac{du}{dt} = -2\,u + v,\quad \dfrac{dv}{dt} = u - 4\,v$

(c) $\dfrac{du}{dt} = 5\,u + 4\,v,\quad \dfrac{dv}{dt} = u + 2\,v$

(d) $\dfrac{du}{dt} = -3\,u - 2\,v,\quad \dfrac{dv}{dt} = 3\,u + 2\,v$

9.3.4. Sketch the phase portrait for the following systems:

(a) $\dot u_1 = u_1 - 3\,u_2$, $\dot u_2 = -3\,u_1 + u_2$

(b) $\dot u_1 = 3\,u_1 - 4\,u_2$, $\dot u_2 = u_1 - u_2$

(c) $\dot u_1 = u_1 + u_2$, $\dot u_2 = 4\,u_1 - 2\,u_2$

(d) $\dot u_1 = u_1 + u_2$, $\dot u_2 = u_2$

(e) $\dot u_1 = \frac{3}{2}\,u_1 + \frac{5}{2}\,u_2$, $\dot u_2 = \frac{5}{2}\,u_1 - \frac{3}{2}\,u_2$

9.3.5. (a) Solve the initial value problem

$$\frac{d\mathbf{u}}{dt} = \begin{pmatrix} -1 & 2 \\ -1 & -3 \end{pmatrix}\mathbf{u}, \qquad \mathbf{u}(0) = \begin{pmatrix} 1 \\ 3 \end{pmatrix}.$$

(b) Sketch a picture of your solution curve $\mathbf{u}(t)$, indicating the direction of motion.

(c) Is the origin

(i) stable?

(ii) asymptotically stable?

(iii) unstable? (iv) none of these?

Justify your answer.

◇ **9.3.6.** Justify the solution formulas (9.31) and (9.32).

9.3.7. Which of the 14 possible two-dimensional phase portraits can occur for the phase plane equivalent (9.8) of a second order scalar ordinary differential equation?

9.4 MATRIX EXPONENTIALS

So far, our focus has been on vector-valued solutions $\mathbf{u}(t)$ to linear systems of ordinary differential equations

$$\frac{d\mathbf{u}}{dt} = A\mathbf{u}. \tag{9.33}$$

An evident, and, in fact, useful generalization is to look for *matrix solutions*. Specifically, we seek a matrix-valued function $U(t)$ that satisfies the corresponding *matrix differential equation*

$$\frac{dU}{dt} = A\,U(t). \tag{9.34}$$

As with vectors, we compute the derivative of $U(t)$ by differentiating its individual entries. If A is an $n \times n$ matrix, compatibility of matrix multiplication requires that $U(t)$ be of size $n \times k$ for some k. Since matrix multiplication acts column-wise, the individual columns of the matrix solution $U(t) = (\,\mathbf{u}_1(t) \ldots \mathbf{u}_k(t)\,)$ must solve the original vector system (9.33). Thus, a matrix solution is merely a convenient way of collecting together several different vector solutions. The most important case is when $U(t)$ is a square matrix, of size $n \times n$, and so consists of n vector solutions to the system.

EXAMPLE 9.23 According to Example 9.7, the vector-valued functions

$$\mathbf{u}_1(t) = \begin{pmatrix} e^{-4t} \\ -2e^{-4t} \end{pmatrix}, \qquad \mathbf{u}_2(t) = \begin{pmatrix} e^{-t} \\ e^{-t} \end{pmatrix},$$

are both solutions to the linear system

$$\frac{d\mathbf{u}}{dt} = \begin{pmatrix} -2 & 1 \\ 2 & -3 \end{pmatrix}\mathbf{u}.$$

They can be combined to form the matrix solution

$$U(t) = \begin{pmatrix} e^{-4t} & e^{-t} \\ -2e^{-4t} & e^{-t} \end{pmatrix} \quad \text{satisfying} \quad \frac{dU}{dt} = \begin{pmatrix} -2 & 1 \\ 2 & -3 \end{pmatrix} U.$$

Indeed, by direct calculation

$$\begin{aligned} \frac{dU}{dt} &= \begin{pmatrix} -4e^{-4t} & -e^{-t} \\ 8e^{-4t} & -e^{-t} \end{pmatrix} \\ &= \begin{pmatrix} -2 & 1 \\ 2 & -3 \end{pmatrix}\begin{pmatrix} e^{-4t} & e^{-t} \\ -2e^{-4t} & e^{-t} \end{pmatrix} \\ &= \begin{pmatrix} -2 & 1 \\ 2 & -3 \end{pmatrix} U. \end{aligned}$$

●

The existence and uniqueness theorems are readily adapted to matrix differential equations, and imply that there is a unique matrix solution to the system (9.34) that has initial conditions

$$U(t_0) = B, \tag{9.35}$$

where B is an $n \times k$ matrix. Note that the jth column $\mathbf{u}_j(t)$ of the matrix solution $U(t)$ satisfies the initial value problem

$$\frac{d\mathbf{u}_j}{dt} = A\,\mathbf{u}_j, \qquad \mathbf{u}_j(t_0) = \mathbf{b}_j,$$

where $\mathbf{b}_j$ denotes the jth column of B.

In the scalar case, the solution to the particular initial value problem

$$\frac{du}{dt} = a\,u, \qquad u(0) = 1,$$

is the ordinary exponential function $u(t) = e^{t\,a}$. Knowing this, we can write down the solution for a more general initial condition

$$u(t_0) = b \quad \text{as} \quad u(t) = b\,e^{(t-t_0)\,a}.$$

Let us formulate an analogous initial value problem for a linear system. Recall that, for matrices, the role of the multiplicative unit 1 is played by the identity matrix I. This inspires the following definition.

Definition 9.24 Let A be a square $n \times n$ matrix. The *matrix exponential*

$$U(t) = e^{t\,A} = \exp(t\,A) \tag{9.36}$$

is the unique $n \times n$ matrix solution to the initial value problem

$$\frac{dU}{dt} = A\,U, \qquad U(0) = \mathrm{I}. \tag{9.37}$$

The matrix exponential $e^{t\,A}$ turns out to enjoy almost all the properties you might expect from its scalar counterpart. First, it is defined for all $t \in \mathbb{R}$, and all $n \times n$ matrices. We can rewrite the defining properties (9.37) in the more suggestive form

$$\frac{d}{dt}\,e^{t\,A} = A\,e^{t\,A}, \qquad e^{0\,A} = \mathrm{I}. \tag{9.38}$$

As in the scalar case, once we know the matrix exponential, we are in a position to solve the general initial value problem.

Lemma 9.25 Let A be an $n \times n$ matrix. For any $n \times k$ matrix B, the solution to the initial value problem

$$\frac{dU}{dt} = A\,U, \qquad U(t_0) = B, \qquad \text{is} \qquad U(t) = e^{(t-t_0)\,A}\,B. \tag{9.39}$$

Proof Since B is a constant matrix,

$$\frac{dU}{dt} = \frac{d}{dt}\left[\,e^{(t-t_0)\,A}\,B\,\right] = A\,e^{(t-t_0)\,A}\,B = A\,U,$$

where we applied the chain rule and the first property (9.38). Thus, $U(t)$ is a matrix solution to the system. Moreover, by the second property in (9.38),

$$U(0) = e^{0\,A}\,B = \mathrm{I}\,B = B$$

has the correct initial conditions. ■

REMARK: The computation used in the proof is a special instance of the general *Leibniz rule*

$$\frac{d}{dt}\big[\,M(t)\,N(t)\,\big] = \frac{dM(t)}{dt}\,N(t) + M(t)\,\frac{dN(t)}{dt} \tag{9.40}$$

for the derivative of the product of (compatible) matrix-valued functions $M(t)$ and $N(t)$. The reader is asked to prove this formula in Exercise 9.4.18.

In particular, the solution to the original vector initial value problem

$$\frac{d\mathbf{u}}{dt} = A\,\mathbf{u}, \qquad \mathbf{u}(t_0) = \mathbf{b},$$

can be written in terms of the matrix exponential:

$$\mathbf{u}(t) = e^{(t-t_0)\,A}\,\mathbf{b}. \tag{9.41}$$

Thus, the matrix exponential provides us with an alternative formula for the solution of autonomous homogeneous first order linear systems, providing us with valuable new insight.

The next step is to find an algorithm for computing the matrix exponential. The solution formula (9.39) gives the hint. Suppose $U(t)$ is any $n \times n$ matrix solution. Then, by uniqueness, $U(t) = e^{t\,A}\,U(0)$, and hence, provided that $U(0)$ is a nonsingular matrix,

$$e^{t\,A} = U(t)\,U(0)^{-1}. \tag{9.42}$$

Thus, to construct the exponential of an $n \times n$ matrix A, you first need to find a basis of n linearly independent solutions $\mathbf{u}_1(t), \ldots, \mathbf{u}_n(t)$ to the linear system $\dot{\mathbf{u}} = A\mathbf{u}$ using the eigenvalues and eigenvectors, or, in the incomplete case, the Jordan chains. The resulting $n \times n$ matrix solution $U(t) = \big(\,\mathbf{u}_1(t) \ldots \mathbf{u}_n(t)\,\big)$ is then used to produce $e^{t\,A}$ via (9.42).

EXAMPLE 9.26 For the matrix

$$A = \begin{pmatrix} -2 & 1 \\ 2 & -3 \end{pmatrix}$$

considered in Example 9.23, we already constructed the nonsingular matrix solution

$$U(t) = \begin{pmatrix} e^{-4t} & e^{-t} \\ -2e^{-4t} & e^{-t} \end{pmatrix}.$$

Therefore, by (9.42), its matrix exponential is

$$\begin{aligned} e^{t\,A} &= U(t)\,U(0)^{-1} \\ &= \begin{pmatrix} e^{-4t} & e^{-t} \\ -2e^{-4t} & e^{-t} \end{pmatrix} \begin{pmatrix} 1 & 1 \\ -2 & 1 \end{pmatrix}^{-1} \\ &= \begin{pmatrix} \frac{1}{3}e^{-4t} + \frac{2}{3}e^{-t} & -\frac{1}{3}e^{-4t} + \frac{1}{3}e^{-t} \\ -\frac{2}{3}e^{-4t} + \frac{2}{3}e^{-t} & \frac{2}{3}e^{-4t} + \frac{1}{3}e^{-t} \end{pmatrix}. \end{aligned}$$

In particular, we obtain $e^{A} = \exp A$ by setting $t = 1$ in this formula:

$$\exp\begin{pmatrix} -2 & 1 \\ 2 & -3 \end{pmatrix} = \begin{pmatrix} \frac{1}{3}e^{-4} + \frac{2}{3}e^{-1} & -\frac{1}{3}e^{-4} + \frac{1}{3}e^{-1} \\ -\frac{2}{3}e^{-4} + \frac{2}{3}e^{-1} & \frac{2}{3}e^{-4} + \frac{1}{3}e^{-1} \end{pmatrix}.$$

Observe that the matrix exponential is *not* obtained by exponentiating the individual matrix entries.

To solve the initial value problem

$$\frac{d\mathbf{u}}{dt} = \begin{pmatrix} -2 & 1 \\ 2 & -3 \end{pmatrix} \mathbf{u}, \qquad \mathbf{u}(0) = \mathbf{b} = \begin{pmatrix} 3 \\ 0 \end{pmatrix},$$

we appeal to formula (9.39), whence

$$\begin{aligned} \mathbf{u}(t) &= e^{tA}\,\mathbf{b} \\ &= \begin{pmatrix} \frac{1}{3}e^{-4t} + \frac{2}{3}e^{-t} & -\frac{1}{3}e^{-4t} + \frac{1}{3}e^{-t} \\ -\frac{2}{3}e^{-4t} + \frac{2}{3}e^{-t} & \frac{2}{3}e^{-4t} + \frac{1}{3}e^{-t} \end{pmatrix} \begin{pmatrix} 3 \\ 0 \end{pmatrix} \\ &= \begin{pmatrix} e^{-4t} + 2e^{-t} \\ -2e^{-4t} + 2e^{-t} \end{pmatrix}. \end{aligned}$$

This reproduces our earlier solution (9.15). ●

EXAMPLE 9.27 Suppose $A = \begin{pmatrix} -1 & -2 \\ 2 & -1 \end{pmatrix}$. Its characteristic equation

$$\det(A - \lambda\,\mathrm{I}) = \lambda^2 + 2\lambda + 5 = 0 \qquad \text{has roots} \qquad \lambda = -1 \pm 2\,\mathrm{i},$$

which are thus the eigenvalues. The corresponding eigenvectors are $\mathbf{v} = \begin{pmatrix} \pm\mathrm{i} \\ 1 \end{pmatrix}$, leading to the complex conjugate solutions

$$\mathbf{u}_1(t) = \begin{pmatrix} \mathrm{i}\,e^{(-1+2\mathrm{i})t} \\ e^{(-1+2\mathrm{i})t} \end{pmatrix}, \qquad \mathbf{u}_2(t) = \begin{pmatrix} -\mathrm{i}\,e^{(-1-2\mathrm{i})t} \\ e^{(-1-2\mathrm{i})t} \end{pmatrix}.$$

We assemble them to form the (complex) matrix solution

$$U(t) = \begin{pmatrix} \mathrm{i}\,e^{(-1+2\mathrm{i})t} & -\mathrm{i}\,e^{(-1-2\mathrm{i})t} \\ e^{(-1+2\mathrm{i})t} & e^{(-1-2\mathrm{i})t} \end{pmatrix}.$$

The corresponding matrix exponential is, therefore,

$$\begin{aligned} e^{tA} = U(t)\,U(0)^{-1} &= \begin{pmatrix} \mathrm{i}\,e^{(-1+2\mathrm{i})t} & -\mathrm{i}\,e^{(-1-2\mathrm{i})t} \\ e^{(-1+2\mathrm{i})t} & e^{(-1-2\mathrm{i})t} \end{pmatrix} \begin{pmatrix} \mathrm{i} & -\mathrm{i} \\ 1 & 1 \end{pmatrix}^{-1} \\ &= \begin{pmatrix} \dfrac{e^{(-1+2\mathrm{i})t} + e^{(-1-2\mathrm{i})t}}{2} & \dfrac{-e^{(-1+2\mathrm{i})t} + e^{(-1-2\mathrm{i})t}}{2\,\mathrm{i}} \\ \dfrac{e^{(-1+2\mathrm{i})t} - e^{(-1-2\mathrm{i})t}}{2\,\mathrm{i}} & \dfrac{e^{(-1+2\mathrm{i})t} + e^{(-1-2\mathrm{i})t}}{2} \end{pmatrix} \\ &= \begin{pmatrix} e^{-t}\cos 2t & -e^{-t}\sin 2t \\ e^{-t}\sin 2t & e^{-t}\cos 2t \end{pmatrix}. \end{aligned}$$

Note that the final expression for the matrix exponential is real, as it must be since A is a real matrix. (See Exercise 9.4.19.) Also note that it wasn't necessary to find the real solutions to construct the matrix exponential—although this would have also worked and yielded the same result. Indeed, the two columns of e^{tA} form a basis for the space of (real) solutions to the linear system $\dot{\mathbf{u}} = A\,\mathbf{u}$. ●

Let us finish by listing some further important properties of the matrix exponential, all of which are direct analogs of the usual scalar exponential function. First, the *multiplicative property* says that

$$e^{(t+s)A} = e^{tA}\, e^{sA}, \qquad \text{for any} \quad s, t \in \mathbb{R}. \tag{9.43}$$

In particular, if we set $s = -t$, the left hand side of (9.43) reduces to the identity matrix, and hence

$$e^{-tA} = \left(e^{tA}\right)^{-1}. \tag{9.44}$$

As a consequence, for any A and any $t \in \mathbb{R}$, the exponential e^{tA} is a nonsingular matrix.

Warning: In general,

$$e^{t(A+B)} \neq e^{tA}\, e^{tB}. \tag{9.45}$$

Indeed, according to the discussion after (9.57) below, the left and right hand sides of (9.45) are equal for all t if and only if $A B = B A$—that is, A and B are commuting matrices.

While the matrix exponential can be painful to compute, there is a simple formula for its determinant in terms of the trace of the generating matrix.

Lemma 9.28 For any square matrix, $\det e^{tA} = e^{t \operatorname{tr} A}$.

Proof According to Exercise 9.4.21, if A has eigenvalues $\lambda_1, \ldots, \lambda_n$, then e^{tA} has eigenvalues* $e^{t\lambda_1}, \ldots, e^{t\lambda_n}$. Moreover, using (8.25), the determinant of $\det e^{tA}$ is the product of its eigenvalues, and so

$$\det e^{tA} = e^{t\lambda_1}\, e^{t\lambda_2} \cdots e^{t\lambda_n} = e^{t(\lambda_1+\lambda_2+\cdots+\lambda_n)} = e^{t \operatorname{tr} A},$$

where, by (8.24), we identify the sum of the eigenvalues as the trace of A. ■

For instance, the matrix

$$A = \begin{pmatrix} -2 & 1 \\ 2 & -3 \end{pmatrix}$$

considered in Example 9.26 has $\operatorname{tr} A = (-2) + (-3) = -5$, and hence $\det e^{tA} = e^{-5t}$, as you can easily check.

Finally, we note that the standard exponential series is also valid for matrices:

$$e^{tA} = \sum_{n=0}^{\infty} \frac{t^n}{n!} A^n = \mathrm{I} + tA + \frac{t^2}{2} A^2 + \frac{t^3}{6} A^3 + \cdots. \tag{9.46}$$

The proof that the series converges will be deferred until we have had a chance to discuss matrix norms in Section 10.3. Once convergence is established, proving that the exponential series satisfies the defining initial value problem (9.38) is straightforward:

$$\frac{d}{dt} \sum_{n=0}^{\infty} \frac{t^n}{n!} A^n = \sum_{n=1}^{\infty} \frac{t^{n-1}}{(n-1)!} A^n = \sum_{n=0}^{\infty} \frac{t^n}{n!} A^{n+1} = A \sum_{n=0}^{\infty} \frac{t^n}{n!} A^n.$$

Moreover, at $t = 0$ the sum collapses to the identity matrix: $\mathrm{I} = e^{0A}$. Thus, formula (9.46) follows from the uniqueness of solutions to the matrix initial value problem.

*In Exercise 9.4.29, you are asked to prove that repeated eigenvalues have the same multiplicities.

EXERCISES 9.4

9.4.1. Find the exponentials e^{tA} of the following 2×2 matrices:

(a) $\begin{pmatrix} 2 & -1 \\ 4 & -3 \end{pmatrix}$ (b) $\begin{pmatrix} 0 & 1 \\ 1 & 0 \end{pmatrix}$

(c) $\begin{pmatrix} 0 & -1 \\ 1 & 0 \end{pmatrix}$ (d) $\begin{pmatrix} 0 & 1 \\ 0 & 0 \end{pmatrix}$

(e) $\begin{pmatrix} -1 & 2 \\ -5 & 5 \end{pmatrix}$ (f) $\begin{pmatrix} 1 & 2 \\ -2 & -1 \end{pmatrix}$

9.4.2. Determine the matrix exponential e^{tA} for the following matrices:

(a) $\begin{pmatrix} 0 & 0 & 0 \\ 2 & 0 & 1 \\ 0 & -1 & 0 \end{pmatrix}$ (b) $\begin{pmatrix} 3 & -1 & 0 \\ -1 & 2 & -1 \\ 0 & -1 & 3 \end{pmatrix}$

(c) $\begin{pmatrix} -1 & 1 & 1 \\ -2 & -2 & -2 \\ 1 & -1 & -1 \end{pmatrix}$ (d) $\begin{pmatrix} 0 & 0 & 1 \\ 1 & 0 & 0 \\ 0 & 1 & 0 \end{pmatrix}$

9.4.3. Verify the determinant formula of Lemma 9.28 for the matrices in Exercises 9.4.1 and 9.4.2.

9.4.4. Find e^A when $A =$

(a) $\begin{pmatrix} 5 & -2 \\ -2 & 5 \end{pmatrix}$ (b) $\begin{pmatrix} 1 & -2 \\ 1 & 1 \end{pmatrix}$

(c) $\begin{pmatrix} 2 & -1 \\ 4 & -2 \end{pmatrix}$ (d) $\begin{pmatrix} 1 & 0 & 0 \\ 0 & -2 & 0 \\ 0 & 0 & -5 \end{pmatrix}$

(e) $\begin{pmatrix} 0 & 1 & -2 \\ -1 & 0 & 2 \\ 2 & -2 & 0 \end{pmatrix}$

9.4.5. Solve the indicated initial value problems by first exponentiating the coefficient matrix and then applying formula (9.41):

(a) $\dfrac{d\mathbf{u}}{dt} = \begin{pmatrix} 0 & -1 \\ 1 & 0 \end{pmatrix}\mathbf{u}, \ \mathbf{u}(0) = \begin{pmatrix} 1 \\ -2 \end{pmatrix}$

(b) $\dfrac{d\mathbf{u}}{dt} = \begin{pmatrix} 3 & -6 \\ 4 & -7 \end{pmatrix}\mathbf{u}, \ \mathbf{u}(0) = \begin{pmatrix} -1 \\ 1 \end{pmatrix}$

(c) $\dfrac{d\mathbf{u}}{dt} = \begin{pmatrix} -9 & -6 & 6 \\ 8 & 5 & -6 \\ -2 & 1 & 3 \end{pmatrix}\mathbf{u}, \ \mathbf{u}(0) = \begin{pmatrix} 0 \\ 1 \\ 0 \end{pmatrix}$

9.4.6. What is $e^{t\mathrm{O}}$ when O is the $n \times n$ zero matrix?

9.4.7. Find all matrices A such that $e^{tA} = \mathrm{O}$.

9.4.8. Let

$$A = \begin{pmatrix} 0 & -2\pi \\ 2\pi & 0 \end{pmatrix}.$$

Show that $e^A = \mathrm{I}$.

9.4.9. (a) Let A be a 2×2 matrix such that $\operatorname{tr} A = 0$ and $\delta = \sqrt{\det A} > 0$. Prove that

$$e^A = (\cos\delta)\,\mathrm{I} + \frac{\sin\delta}{\delta}\,A.$$

Hint: Use Exercise 8.2.51.

(b) Establish a similar formula when $\det A < 0$.

(c) What if $\det A = 0$?

◇ **9.4.10.** Explain in detail why the columns of e^{tA} form a basis for the solution space to the system $\dot{\mathbf{u}} = A\,\mathbf{u}$.

9.4.11. *True or false*:

(a) $e^{A^{-1}} = \left(e^A\right)^{-1}$ (b) $e^{A+A^{-1}} = e^A\, e^{A^{-1}}$

◇ **9.4.12.** Prove formula (9.43). *Hint*: Fix s and prove that, as functions of t, both sides of the equation define matrix solutions with the same initial conditions. Then use uniqueness.

9.4.13. Prove that A commutes with its exponential: $A\, e^{tA} = e^{tA} A$. *Hint*: Prove that both are matrix solutions to $\dot{U} = A\,U$ with the same initial conditions.

9.4.14. Prove that $e^{t(A-\lambda\,\mathrm{I})} = e^{-t\lambda}\, e^{tA}$ by showing that both sides are matrix solutions to the same initial value problem.

◇ **9.4.15.** (a) Prove that the exponential of the transpose of a matrix is the transpose of its exponential: $e^{tA^T} = (e^{tA})^T$.

(b) What does this imply about the solutions to the linear systems $\dot{\mathbf{u}} = A\,\mathbf{u}$ and $\mathbf{v} = A^T\mathbf{v}$?

◇ **9.4.16.** Prove that if $A = S\,B\,S^{-1}$ are similar matrices, then so are their exponentials: $e^{tA} = S\,e^{tB}\,S^{-1}$.

◇ **9.4.17.** Diagonalization provides an alternative method for computing the exponential of a complete matrix.

(a) First show that if

$$D = \operatorname{diag}(d_1, \ldots, d_n)$$

is a diagonal matrix, so is

$$e^{tD} = \operatorname{diag}(e^{t\,d_1}, \ldots, e^{t\,d_n}).$$

(b) Second, prove that if $A = S\,D\,S^{-1}$ is diagonalizable, so is $e^{tA} = S\,e^{tD}\,S^{-1}$.

(c) When possible, use diagonalization to compute the exponentials of the matrices in Exercises 9.4.1–2.

◇ **9.4.18.** Justify the matrix Leibniz rule (9.40) using the formula for matrix multiplication.

◇ **9.4.19.** Let A be a real matrix.

(a) Explain why e^A is a real matrix.

(b) Prove that $\det e^A > 0$.

9.4.20. Show that $\operatorname{tr} A = 0$ if and only if $\det e^{tA} = 1$ for all t.

◇ **9.4.21.** Prove that if λ is an eigenvalue of A, then $e^{t\lambda}$ is an eigenvalue of e^{tA}. What is the eigenvector?

9.4.22. Show that the origin is an asymptotically stable equilibrium solution to $\dot{\mathbf{u}} = A\,\mathbf{u}$ if and only if $\lim_{t \to \infty} e^{t A} = \mathrm{O}$.

9.4.23. Let A be a real square matrix and e^A its exponential. Under what conditions does the linear system $\dot{\mathbf{u}} = e^A\,\mathbf{u}$ have an asymptotically stable equilibrium solution?

9.4.24. Prove that if $U(t)$ is any matrix solution to

$$\frac{dU}{dt} = A\,U,$$

so is $\widetilde{U}(t) = U(t)\,C$ where C is any constant matrix (of compatible size).

◇ **9.4.25.** (a) Show that $U(t)$ satisfies the matrix differential equation $\dot{U} = U\,B$ if and only if $U(t) = C\,e^{t\,B}$, where $C = U(0)$.

(b) Show that if $U(0)$ is nonsingular, then $U(t)$ also satisfies a matrix differential equation of the form $\dot{U} = A\,U$. Is $A = B$? *Hint*: Use Exercise 9.4.16.

9.4.26. (a) Suppose $\mathbf{u}_1(t), \ldots, \mathbf{u}_n(t)$ are vector-valued functions whose values at each point t are linearly independent vectors in $\mathbb{R}^n$. Show that they form a basis for the solution space of a homogeneous constant coefficient linear system $\dot{\mathbf{u}} = A\,\mathbf{u}$ if and only if each $d\mathbf{u}_j/dt$ is a linear combination of $\mathbf{u}_1(t), \ldots, \mathbf{u}_n(t)$. *Hint*: Use Exercise 9.4.25.

(b) Show that a function $\mathbf{u}(t)$ belongs to the solution space of a homogeneous constant coefficient linear system $\dot{\mathbf{u}} = A\,\mathbf{u}$ if and only if

$$\frac{d^n\mathbf{u}}{dt^n}$$

is a linear combination of

$$\mathbf{u}, \quad \frac{d\mathbf{u}}{dt}, \quad \ldots, \quad \frac{d^{n-1}\mathbf{u}}{dt^{n-1}}.$$

Hint: Use Exercise 9.1.7.

9.4.27. Prove that if $A = \begin{pmatrix} B & \mathrm{O} \\ \mathrm{O} & C \end{pmatrix}$ is a block diagonal matrix, then so is $e^{t A} = \begin{pmatrix} e^{t B} & \mathrm{O} \\ \mathrm{O} & e^{t C} \end{pmatrix}$.

◇ **9.4.28.** (a) Prove that if $J_{0,n}$ is an $n \times n$ Jordan block matrix with 0 diagonal entries, cf. (8.49), then

$$e^{t\,J_{0,n}} = \begin{pmatrix} 1 & t & \dfrac{t^2}{2} & \dfrac{t^3}{6} & \ldots & \dfrac{t^n}{n!} \\ 0 & 1 & t & \dfrac{t^2}{2} & \ldots & \dfrac{t^{n-1}}{(n-1)!} \\ 0 & 0 & 1 & t & \ldots & \dfrac{t^{n-2}}{(n-2)!} \\ \vdots & \vdots & \vdots & \ddots & \ddots & \vdots \\ 0 & 0 & 0 & \ldots & 1 & t \\ 0 & 0 & 0 & \ldots & 0 & 1 \end{pmatrix}.$$

(b) Determine the exponential of a general Jordan block matrix $J_{\lambda,n}$. *Hint*: Use Exercise 9.4.14.

(c) Explain how you can use the Jordan canonical form to compute the exponential of a matrix. *Hint*: Use Exercise 9.4.27.

◇ **9.4.29.** Prove that if λ is an eigenvalue of A with multiplicity k, then $e^{t\lambda}$ is an eigenvalue of $e^{t A}$ with the same multiplicity. *Hint*: Combine the Jordan canonical form (8.51) with Exercises 9.4.17, 28.

♡ **9.4.30.** By a (natural) *logarithm* of a matrix B we mean a matrix A such that $e^A = B$.

(a) Explain why only nonsingular matrices can have a logarithm.

(b) Comparing Exercises 9.4.6–8, explain why the matrix logarithm is not unique.

(c) Find all real logarithms of the 2×2 identity matrix

$$\mathrm{I} = \begin{pmatrix} 1 & 0 \\ 0 & 1 \end{pmatrix}.$$

Hint: Use Exercise 9.4.21.

9.4.31. *True or false*: The solution to the non-autonomous initial value problem $\dot{\mathbf{u}} = A(t)\,\mathbf{u}$, $\mathbf{u}(0) = \mathbf{b}$, is $\mathbf{u}(t) = e^{\int_0^t A(s)\,ds}\,\mathbf{b}$.

Inhomogeneous Linear Systems

We now direct our attention to inhomogeneous linear systems of ordinary differential equations. For simplicity, we consider only first order† systems of the form

$$\frac{d\mathbf{u}}{dt} = A\,\mathbf{u} + \mathbf{f}(t), \tag{9.47}$$

in which A is a constant $n \times n$ matrix and $\mathbf{f}(t)$ is a vector of functions that can be interpreted as a collection of external forces acting on the system. According

†Higher order systems can, as remarked earlier, always be converted into first order systems involving additional variables.

Theorem 7.38, the solution to the inhomogeneous system will have the general form

$$\mathbf{u}(t) = \mathbf{u}^\star(t) + \mathbf{z}(t)$$

where $\mathbf{u}^\star(t)$ is a particular solution, representing a response to the forcing, while $\mathbf{z}(t)$ is a solution to the corresponding homogeneous system $\dot{\mathbf{z}} = A\,\mathbf{z}$, representing the system's internal motion. Since we now know how to find the solution $\mathbf{z}(t)$ to the homogeneous system, the only task is to find one particular solution to the inhomogeneous system.

In your first course on ordinary differential equations, you probably encountered a method known as *variation of parameters* for constructing particular solutions of inhomogeneous scalar ordinary differential equations, [6]. The method can be readily adapted to first order systems. Recall that, in the scalar case, to solve the inhomogeneous equation

$$\frac{du}{dt} = a\,u + f(t), \qquad \text{we set} \quad u(t) = e^{t\,a}\, v(t),$$

where the function $v(t)$ is to be determined. Differentiating,

$$\frac{du}{dt} = a\,e^{t\,a}\, v(t) + e^{t\,a}\,\frac{dv}{dt} = a\,u + e^{t\,a}\,\frac{dv}{dt}\,.$$

Therefore, $u(t)$ satisfies the differential equation if and only if

$$\frac{dv}{dt} = e^{-t\,a}\, f(t).$$

Since the right hand side is known, $v(t)$ can be immediately found by a direct integration.

The method is easily adapted to the vector-valued situation. We replace the scalar exponential by the exponential of the coefficient matrix, setting

$$\mathbf{u}(t) = e^{t\,A}\,\mathbf{v}(t). \tag{9.48}$$

Combining the product rule for matrix multiplication (9.40) with (9.38), we find

$$\frac{d\mathbf{u}}{dt} = A\,e^{t\,A}\,\mathbf{v} + e^{t\,A}\,\frac{d\mathbf{v}}{dt} = A\,\mathbf{u} + e^{t\,A}\,\frac{d\mathbf{v}}{dt}\,.$$

Comparing with the differential equation (9.47), we conclude that

$$\frac{d\mathbf{v}}{dt} = e^{-t\,A}\,\mathbf{f}(t).$$

The function $\mathbf{v}(t)$ can then be found by integration:[‡]

$$\mathbf{v}(t) = \mathbf{v}(t_0) + \int_{t_0}^{t} e^{-s\,A}\,\mathbf{f}(s)\,ds, \qquad \text{where} \quad \mathbf{v}(t_0) = e^{-t_0\,A}\,\mathbf{u}(t_0). \tag{9.49}$$

Substituting back into (9.48) leads to a general formula for the solution to the inhomogeneous linear system.

THEOREM 9.29 The solution to the initial value problem $d\mathbf{u}/dt = A\mathbf{u} + \mathbf{f}(t)$, $\mathbf{u}(t_0) = \mathbf{b}$, is

$$\mathbf{u}(t) = e^{(t-t_0)\,A}\,\mathbf{b} + \int_{t_0}^{t} e^{(t-s)\,A}\,\mathbf{f}(s)\,ds. \tag{9.50}$$

In the solution formula, the integral term can be viewed as the particular solution $\mathbf{u}^\star(t)$, while the first summand, $\mathbf{z}(t) = e^{(t-t_0)\,A}\,\mathbf{b}$ for $\mathbf{b} \in \mathbb{R}^n$, constitutes the general solution to the homogeneous system.

[‡]As with differentiation, vector- and matrix-valued functions are integrated entry-wise.

EXAMPLE 9.30 Our goal is to solve the initial value problem

$$\dot u_1 = 2u_1 - u_2, \qquad u_1(0) = 1,$$
$$\dot u_2 = 4u_1 - 3u_2 + e^t, \qquad u_2(0) = 0. \tag{9.51}$$

The first step is to determine eigenvalues and eigenvectors of the coefficient matrix:

$$A = \begin{pmatrix} 2 & -1 \\ 4 & -3 \end{pmatrix} \quad \text{so} \quad \lambda_1 = 1, \quad \mathbf{v}_1 = \begin{pmatrix} 1 \\ 1 \end{pmatrix}, \quad \lambda_2 = -2, \quad \mathbf{v}_2 = \begin{pmatrix} 1 \\ 4 \end{pmatrix}.$$

The resulting eigensolutions form the columns of the nonsingular matrix solution

$$U(t) = \begin{pmatrix} e^t & e^{-2t} \\ e^t & 4e^{-2t} \end{pmatrix},$$

so

$$e^{tA} = U(t)\,U(0)^{-1} = \begin{pmatrix} \frac43 e^t - \frac13 e^{-2t} & -\frac13 e^t + \frac13 e^{-2t} \\ \frac43 e^t - \frac43 e^{-2t} & -\frac13 e^t + \frac43 e^{-2t} \end{pmatrix}.$$

The two constituents of the solution formula (9.50) are

$$e^{tA}\mathbf{b} = \begin{pmatrix} \frac43 e^t - \frac13 e^{-2t} & -\frac13 e^t + \frac13 e^{-2t} \\ \frac43 e^t - \frac43 e^{-2t} & -\frac13 e^t + \frac43 e^{-2t} \end{pmatrix}\begin{pmatrix} 1 \\ 0 \end{pmatrix} = \begin{pmatrix} \frac43 e^t - \frac13 e^{-2t} \\ \frac43 e^t - \frac43 e^{-2t} \end{pmatrix},$$

which is the solution to the homogeneous system for the given nonzero initial conditions, and

$$\int_0^t e^{(t-s)A}\mathbf{f}(s)\,ds = \int_0^t \begin{pmatrix} \frac43 e^{t-s} - \frac13 e^{-2(t-s)} & -\frac13 e^{t-s} + \frac13 e^{-2(t-s)} \\ \frac43 e^{t-s} - \frac43 e^{-2(t-s)} & -\frac13 e^{t-s} + \frac43 e^{-2(t-s)} \end{pmatrix}\begin{pmatrix} 0 \\ e^s \end{pmatrix} ds$$
$$= \int_0^t \begin{pmatrix} -\frac13 e^t + \frac13 e^{-2t+3s} \\ -\frac13 e^t + \frac43 e^{-2t+3s} \end{pmatrix} ds$$
$$= \begin{pmatrix} -\frac13 t e^t + \frac19 (e^t - e^{-2t}) \\ -\frac13 t e^t + \frac49 (e^t - e^{-2t}) \end{pmatrix},$$

which is the particular solution to the inhomogeneous system that satisfies the homogeneous initial conditions $\mathbf{u}(0) = \mathbf{0}$. The solution to our initial value problem is their sum:

$$\mathbf{u}(t) = \begin{pmatrix} -\frac13 t e^t + \frac{13}{9} e^t - \frac49 e^{-2t} \\ -\frac13 t e^t + \frac{16}{9} e^t - \frac{16}{9} e^{-2t} \end{pmatrix}. \qquad \blacksquare$$

EXERCISES 9.4

9.4.32. Solve the following initial value problems:

(a) $\dot u_1 = 2u_1 - u_2, \quad u_1(0) = 0$
$\dot u_2 = 4u_1 - 3u_2 + e^{2t}, \quad u_2(0) = 0$

(b) $\dot u_1 = -u_1 + 2u_2 + e^t, \quad u_1(1) = 1$
$\dot u_2 = 2u_1 - u_2 + e^t, \quad u_2(1) = 1$

(c) $\dot u_1 = -u_2, \quad u_1(0) = 0$
$\dot u_2 = 4u_1 + \cos t, \quad u_2(0) = 1$

(d) $\dot u = 3u + v + 1, \quad u(1) = 1$
$\dot v = 4u + t, \quad v(1) = -1$

(e) $\dot p = p + q + t, \quad p(0) = 0$
$\dot q = -p - q + t, \quad q(0) = 0$

9.4.33. Solve the following initial value problems:
(a) $\dot u_1 = -2u_2 + 2u_3, \quad u_1(0) = 1$
$\dot u_2 = -u_1 + u_2 - 2u_3 + t, \quad u_2(0) = 0$
$\dot u_3 = -3u_1 + u_2 - 2u_3 + 1, \quad u_3(0) = 0$
(b) $\dot u_1 = u_1 - 2u_2, \quad u_1(0) = -1$
$\dot u_2 = -u_2 + e^{-t}, \quad u_2(0) = 0$
$\dot u_3 = 4u_1 - 4u_2 - u_3, \quad u_3(0) = -1$

9.4.34. Suppose that λ is *not* an eigenvalue of A. Show that the inhomogeneous system $\dot u = A\mathbf{u} + e^{\lambda t}\,\mathbf{v}$ has a solution of the form $\mathbf{u}^\star(t) = e^{\lambda t}\,\mathbf{w}$, where $\mathbf{w}$ is a constant vector. What is the general solution?

9.4.35. (a) Write down an integral formula for the solution to the initial value problem

$$\frac{d\mathbf{u}}{dt} = A\mathbf{u} + \mathbf{b}, \qquad \mathbf{u}(0) = \mathbf{0},$$

where $\mathbf{b}$ is a *constant* vector.

(b) Suppose $\mathbf{b} \in \operatorname{rng} A$. Do you recover the solution you found in Exercise 9.1.33?

Applications in Geometry

Matrix exponentials are an effective tool for understanding the linear transformations that appear in geometry and group theory, [70], quantum mechanics, [40], computer graphics and animation, [12, 55], computer vision, [56], and the symmetry analysis of differential equations, [13, 46]. We will only be able to scratch the surface of this useful and active area of mathematics.

Let A be an $n \times n$ matrix. For each $t \in \mathbb{R}$, the corresponding exponential e^{tA} is itself an $n \times n$ matrix and thus defines a linear transformation on the vector space $\mathbb{R}^n$:

$$L_t[\mathbf{x}] = e^{tA}\mathbf{x} \quad \text{for} \quad \mathbf{x} \in \mathbb{R}^n.$$

In this manner, each square matrix A generates a family of invertible linear transformations, parametrized by $t \in \mathbb{R}$. The resulting linear transformations are not arbitrary, but are subject the following three rules:

$$L_t \circ L_s = L_{t+s} = L_s \circ L_t, \qquad L_0 = \mathrm{I}, \qquad L_{-t} = L_t^{-1}. \tag{9.52}$$

These are merely restatements of three of the basic matrix exponential properties listed in (9.38, 43, 44). In particular, every transformation in the family commutes with every other one.

In geometry, the family of transformations $L_t = e^{tA}$ is said to form a *one-parameter group*,[§] [46], and A is referred to as its *infinitesimal generator*. Indeed, by the series formula (9.38) for the matrix exponential,

$$\begin{aligned} L_t[\mathbf{x}] = e^{tA}\mathbf{x} &= \left(\mathrm{I} + tA + \tfrac{1}{2}t^2A^2 + \cdots\right)\mathbf{x} \\ &= \mathbf{x} + tA\mathbf{x} + \tfrac{1}{2}t^2A^2\mathbf{x} + \cdots . \end{aligned} \tag{9.53}$$

When t is small, we can truncate the exponential series and approximate the transformation by the linear map

$$F_t[\mathbf{x}] = (\mathrm{I} + tA)\mathbf{x} = \mathbf{x} + tA\mathbf{x} \tag{9.54}$$

defined by the infinitesimal generator. We already made use of such approximations when we discussed the rigid motions and mechanisms of structures in Chapter 6. As t varies, the group transformations (9.53) typically move a point $\mathbf{x}$ along a curved trajectory. Under the first order approximation (9.54), the point $\mathbf{x}$ moves along a straight line in the direction $\mathbf{b} = A\mathbf{x}$, namely the tangent line to the curved trajectory. Thus, *the infinitesimal generator of a one-parameter group determines the tangent line approximation to the nonlinear group motion.*

[§] See also Exercise 5.3.24 for the general definition of a group.

Most of the linear transformations of interest in applications arise in this fashion. Let's look briefly at a few basic examples.

(a) When
$$A = \begin{pmatrix} 0 & 1 \\ 0 & 0 \end{pmatrix},$$
then
$$e^{tA} = \begin{pmatrix} 1 & t \\ 0 & 1 \end{pmatrix}$$
represents a shearing transformation. The group laws (9.52) imply that the composition of a shear of magnitude s and a shear of magnitude t is another shear of magnitude $s+t$.

(b) When
$$A = \begin{pmatrix} 1 & 0 \\ 0 & 1 \end{pmatrix},$$
then
$$e^{tA} = \begin{pmatrix} e^t & 0 \\ 0 & e^t \end{pmatrix}$$
represents a uniform scaling transformation. Composition and inverses of such scaling transformations are also scalings.

(c) When
$$A = \begin{pmatrix} 1 & 0 \\ 0 & -1 \end{pmatrix},$$
then
$$e^{tA} = \begin{pmatrix} e^t & 0 \\ 0 & e^{-t} \end{pmatrix}$$
which, for $t > 0$, represents a stretch in the x direction and a contraction in the y direction.

(d) When
$$A = \begin{pmatrix} 0 & -1 \\ 1 & 0 \end{pmatrix},$$
then
$$e^{tA} = \begin{pmatrix} \cos t & -\sin t \\ \sin t & \cos t \end{pmatrix}$$
is the matrix for a plane rotation, around the origin, by angle t. The group laws (9.52) say that the composition of a rotation through angle s followed by a rotation through angle t is a rotation through angle $s+t$, as previously noted in Example 7.12. Also, the inverse of a rotation through angle t is a rotation through angle $-t$.

Observe that the infinitesimal generator of the one-parameter group of plane rotations is a 2×2 skew-symmetric matrix. This turns out to be a general fact: rotations in higher dimensions are also generated by skew-symmetric matrices.

Lemma 9.31 If $A^T = -A$ is a skew-symmetric matrix, then $Q(t) = e^{tA}$ is a proper orthogonal matrix.

Proof According to equation (9.44) and Exercise 9.4.15,
$$Q(t)^{-1} = e^{-tA} = e^{tA^T} = \left(e^{tA}\right)^T = Q(t)^T,$$
which proves orthogonality. Properness, $\det Q = +1$, follows from Lemma 9.28. ■

With some more work, it can be shown that every proper orthogonal matrix is the exponential of some skew-symmetric matrix. Thus, the $\frac{1}{2}n(n-1)$ dimensional vector space of $n \times n$ skew symmetric matrices generates the group of rotations in n-dimensional Euclidean space. In the three-dimensional case, the three matrices A_x, A_y, A_z listed below form a basis and serve to generate, respectively, the one-parameter groups of counterclockwise rotations around the x, y and z axes:

$$
\begin{aligned}
A_x &= \begin{pmatrix} 0 & 0 & 0 \\ 0 & 0 & -1 \\ 0 & 1 & 0 \end{pmatrix}, & e^{t A_x} &= \begin{pmatrix} 1 & 0 & 0 \\ 0 & \cos t & -\sin t \\ 0 & \sin t & \cos t \end{pmatrix}, \\
A_y &= \begin{pmatrix} 0 & 0 & 1 \\ 0 & 0 & 0 \\ -1 & 0 & 0 \end{pmatrix}, & e^{t A_y} &= \begin{pmatrix} \cos t & 0 & \sin t \\ 0 & 1 & 0 \\ -\sin t & 0 & \cos t \end{pmatrix}, \\
A_z &= \begin{pmatrix} 0 & -1 & 0 \\ 1 & 0 & 0 \\ 0 & 0 & 0 \end{pmatrix}, & e^{t A_z} &= \begin{pmatrix} \cos t & -\sin t & 0 \\ \cos t & \sin t & 0 \\ 0 & 0 & 1 \end{pmatrix}.
\end{aligned}
\tag{9.55}
$$

Since every other skew-symmetric matrix can be expressed as a linear combination of A_x, A_y and A_z, every rotation can, in a sense, be generated by these three basic types. This reconfirms our earlier observations concerning the number of rigid motions (rotations and translations) experienced by an unattached structure; see Section 6.3 for details.

In the three-dimensional case, it can be shown that every non-zero skew-symmetric 3×3 matrix A is singular, with one-dimensional kernel. Let $\mathbf{0} \neq \mathbf{v} \in \ker A$ be the null eigenvector. Then the matrix exponentials $e^{t A}$ form the one-parameter group of rotations around the axis defined by $\mathbf{v}$. For instance, referring to (9.55), $\ker A_x$ is spanned by $\mathbf{e}_1 = (1, 0, 0)^T$, reconfirming that it generates the rotations around the x axis. Details can be found in Exercise 9.4.39.

Noncommutativity of linear transformations is reflected in the noncommutativity of their infinitesimal generators. Recall, (1.12), that the *commutator* of two $n \times n$ matrices A, B is

$$
[\, A, B \,] = A B - B A. \tag{9.56}
$$

Thus, A and B commute if and only if $[\, A, B \,] = \mathrm{O}$. We use the exponential series (9.46) to evaluate the commutator of the corresponding matrix exponentials:

$$
\begin{aligned}
\left[e^{t A}, e^{t B} \right] &= e^{t A} e^{t B} - e^{t B} e^{t A} \\
&= \left(\mathrm{I} + t A + \tfrac{1}{2} t^2 A^2 + \cdots \right)\left(\mathrm{I} + t B + \tfrac{1}{2} t^2 B^2 + \cdots \right) \\
&\quad - \left(\mathrm{I} + t B + \tfrac{1}{2} t^2 B^2 + \cdots \right)\left(\mathrm{I} + t A + \tfrac{1}{2} t^2 A^2 + \cdots \right) \\
&= t^2 (A B - B A) + \cdots = t^2 [\, A, B \,] + \cdots .
\end{aligned}
\tag{9.57}
$$

In particular, if the groups commute, then $[\, A, B \,] = \mathrm{O}$. The converse is also true, since if $A B = B A$ then all terms in the two series commute, and hence the matrix exponentials also commute.

Proposition 9.32 The matrix exponentials $e^{t A}$ and $e^{t B}$ commute for all t if and only if $A B = B A$.

In particular, the non-commutativity of three-dimensional rotations follows from the non-commutativity of their infinitesimal skew-symmetric generators. For instance, the commutator of the generators of rotations around the x and y axes is the generator of rotations around the z axis: $\left[A_x, A_y \right] = A_z$, since

$$\begin{pmatrix} 0 & 0 & 0 \\ 0 & 0 & -1 \\ 0 & 1 & 0 \end{pmatrix} \begin{pmatrix} 0 & 0 & 1 \\ 0 & 0 & 0 \\ -1 & 0 & 0 \end{pmatrix} - \begin{pmatrix} 0 & 0 & 1 \\ 0 & 0 & 0 \\ -1 & 0 & 0 \end{pmatrix} \begin{pmatrix} 0 & 0 & 0 \\ 0 & 0 & -1 \\ 0 & 1 & 0 \end{pmatrix}$$
$$= \begin{pmatrix} 0 & -1 & 0 \\ 1 & 0 & 0 \\ 0 & 0 & 0 \end{pmatrix}.$$

Hence, to a first (or, more correctly, second) order approximation, the difference between x and y rotations is, interestingly, a z rotation.

EXERCISES 9.4

9.4.36. Find the one-parameter groups generated by the following matrices and interpret geometrically: What are the trajectories? What are the fixed points?

(a) $\begin{pmatrix} 2 & 0 \\ 0 & 0 \end{pmatrix}$ (b) $\begin{pmatrix} 0 & 0 \\ 1 & 0 \end{pmatrix}$

(c) $\begin{pmatrix} 0 & 3 \\ -3 & 0 \end{pmatrix}$ (d) $\begin{pmatrix} 0 & -1 \\ 4 & 0 \end{pmatrix}$

(e) $\begin{pmatrix} 0 & 1 \\ 1 & 0 \end{pmatrix}$

9.4.37. Write down the one-parameter groups generated by the following matrices and interpret. What are the trajectories? What are the fixed points?

(a) $\begin{pmatrix} 2 & 0 & 0 \\ 0 & 1 & 0 \\ 0 & 0 & 0 \end{pmatrix}$ (b) $\begin{pmatrix} 0 & 0 & 1 \\ 0 & 0 & 0 \\ 0 & 0 & 0 \end{pmatrix}$

(c) $\begin{pmatrix} 0 & 0 & -2 \\ 0 & 0 & 0 \\ 2 & 0 & 0 \end{pmatrix}$ (d) $\begin{pmatrix} 0 & 1 & 0 \\ -1 & 0 & 0 \\ 0 & 0 & 1 \end{pmatrix}$

(e) $\begin{pmatrix} 0 & 0 & 1 \\ 0 & 0 & 0 \\ 1 & 0 & 0 \end{pmatrix}$

9.4.38. (a) Find the one-parameter group of rotations generated by the skew-symmetric matrix

$$A = \begin{pmatrix} 0 & 1 & 1 \\ -1 & 0 & -1 \\ -1 & 1 & 0 \end{pmatrix}.$$

(b) As noted above, e^{tA} represents a family of rotations around a fixed axis in $\mathbb{R}^3$. What is the axis?

♡ **9.4.39.** Let $\mathbf{0} \neq \mathbf{v} \in \mathbb{R}^3$.

(a) Show that the cross product $L_{\mathbf{v}}[\mathbf{x}] = \mathbf{v} \times \mathbf{x}$ defines a linear transformation on $\mathbb{R}^3$.

(b) Find the 3×3 matrix representative $A_{\mathbf{v}}$ of $L_{\mathbf{v}}$ and show that it is skew-symmetric.

(c) Show that every non-zero skew-symmetric 3×3 matrix defines such a cross product map.

(d) Show that $\ker A_{\mathbf{v}}$ is spanned by $\mathbf{v}$.

(e) Justify the fact that the matrix exponentials $e^{t A_{\mathbf{v}}}$ are rotations around the axis $\mathbf{v}$. Thus, the cross product with a vector serves as the infinitesimal generator of the one-parameter group of rotations around $\mathbf{v}$.

♡ **9.4.40.** Let

$$A = \begin{pmatrix} 0 & -1 & 0 \\ 1 & 0 & 0 \\ 0 & 0 & 0 \end{pmatrix}, \qquad \mathbf{b} = \begin{pmatrix} 0 \\ 0 \\ 1 \end{pmatrix}.$$

(a) Show that the solution to the linear system $\dot{\mathbf{x}} = A\mathbf{x}$ represents a rotation of $\mathbb{R}^3$ around the z axis. What is the trajectory of a point $\mathbf{x}_0$?

(b) Show that the solution to the inhomogeneous system $\dot{\mathbf{x}} = A\mathbf{x} + \mathbf{b}$ represents a screw motion of $\mathbb{R}^3$ around the z axis. What is the trajectory of a point $\mathbf{x}_0$?

(c) More generally, given $\mathbf{0} \neq \mathbf{a} \in \mathbb{R}^3$, show that the solution to $\dot{\mathbf{x}} = \mathbf{a} \times \mathbf{x} + \mathbf{a}$ represents a family of screw motions along the axis $\mathbf{a}$.

♡ **9.4.41.** Given a unit vector $\|\mathbf{u}\| = 1$ in $\mathbb{R}^3$, let $A = A_{\mathbf{u}}$ be the corresponding skew-symmetric 3×3 matrix that satisfies $A\mathbf{x} = \mathbf{u} \times \mathbf{x}$, as in Exercise 9.4.39.

(a) Prove the *Euler–Rodrigues formula* $e^{tA} = \mathrm{I} + (\sin t) A + (1 - \cos t) A^2$. *Hint*: Use the matrix exponential series (9.46).

(b) Show that $e^{tA} = \mathrm{I}$ if and only if t is an integer multiple of 2π.

(c) Generalize parts (a) and (b) to a non-unit vector $\mathbf{v} \neq \mathbf{0}$.

9.4.42. Choose two of the groups in Exercise 9.4.36 or 9.4.37, and determine whether or not they commute by looking at their infinitesimal generators. Then verify your conclusion by directly computing the commutator of the corresponding matrix exponentials.

9.4.43. (a) Prove that the commutator of two upper triangular matrices is upper triangular.

(b) Prove that the commutator of two skew symmetric matrices is skew symmetric.

(c) Is the commutator of two symmetric matrices symmetric?

◇ **9.4.44.** Prove that the *Jacobi identity* $[[A, B], C] + [[C, A], B] + [[B, C], A] = \mathrm{O}$ is valid for any three $n \times n$ matrices.

9.4.45. Let A be an $n \times n$ matrix whose last row has all zero entries. Prove that the last row of e^{tA} is $\mathbf{e}_n^T = (0, \ldots, 0, 1)$.

9.4.46. Let

$$A = \begin{pmatrix} B & \mathbf{c} \\ \mathbf{0} & 0 \end{pmatrix}$$

be in block form, where B is an $n \times n$ matrix, $\mathbf{c} \in \mathbb{R}^n$, while $\mathbf{0}$ denotes the zero row vector with n entries. Show that its matrix exponential is also in block form

$$e^{tA} = \begin{pmatrix} e^{tB} & \mathbf{f}(t) \\ \mathbf{0} & 1 \end{pmatrix}.$$

Can you find a formula for $\mathbf{f}(t)$?

◇ **9.4.47.** According to Exercise 7.3.9, any $(n+1) \times (n+1)$ matrix of the block form

$$\begin{pmatrix} A & \mathbf{b} \\ \mathbf{0} & 1 \end{pmatrix}$$

in which A is an $n \times n$ matrix and $\mathbf{b} \in \mathbb{R}^n$ can be identified with the affine transformation $F[\mathbf{x}] = A\mathbf{x} + \mathbf{a}$ on $\mathbb{R}^n$. Exercise 9.4.46 shows that every matrix in the one-parameter group e^{tB} generated by

$$B = \begin{pmatrix} A & \mathbf{b} \\ \mathbf{0} & 0 \end{pmatrix}$$

has such a form, and hence we can identify e^{tB} as a family of affine maps on $\mathbb{R}^n$. Describe the affine transformations of $\mathbb{R}^2$ generated by the following matrices:

(a) $\begin{pmatrix} 0 & 0 & 1 \\ 0 & 0 & 0 \\ 0 & 0 & 0 \end{pmatrix}$ (b) $\begin{pmatrix} 1 & 0 & 0 \\ 0 & -2 & 0 \\ 0 & 0 & 0 \end{pmatrix}$

(c) $\begin{pmatrix} 0 & -1 & 0 \\ 1 & 0 & 1 \\ 0 & 0 & 0 \end{pmatrix}$ (d) $\begin{pmatrix} 1 & 0 & 1 \\ 0 & -1 & -2 \\ 0 & 0 & 0 \end{pmatrix}$

9.5 DYNAMICS OF STRUCTURES

Chapter 6 was concerned with the equilibrium configurations of mass-spring chains and, more generally, structures constructed out of elastic bars. We are now able to undertake an analysis of their dynamical motions, which are governed by second order systems of ordinary differential equations. As in the first order case, the eigenvalues of the coefficient matrix play an essential role in both the explicit solution formulae and the system's qualitative behavior.

Let us begin with a mass-spring chain consisting of n masses $m_1, \ldots, m_n$ connected together in a row and, possibly, to top and bottom supports by springs. As in Section 6.1, we restrict our attention to purely one-dimensional motion of the masses in the direction of the chain. Thus the collective motion of the chain is prescribed by the displacement vector $\mathbf{u}(t) = (u_1(t), \ldots, u_n(t))^T$ whose ith entry represents the displacement from equilibrium of the ith mass. Since we are now interested in dynamics, the displacements are allowed to depend on time, t.

The motion of each mass is subject to Newton's Second Law:

$$\text{Force} = \text{Mass} \times \text{Acceleration}. \tag{9.58}$$

The acceleration of the ith mass is the second derivative $\ddot{u}_i = d^2 u_i / dt^2$ of its displacement, so the right hand sides of Newton's Law is $m_i \ddot{u}_i$. These form the entries of the vector $M\ddot{\mathbf{u}}$ obtained by multiplying the acceleration vector by the diagonal, positive definite mass matrix $M = \operatorname{diag}(m_1, \ldots, m_n)$. Keep in mind that the masses of the springs are assumed to be negligible in this model.

If, to begin with, we assume that there are no frictional effects, then the force exerted on each mass is the difference between the external force, if any, and the internal force due to the elongations of its two connecting springs. According to (6.11), the internal forces are the entries of the product $K\mathbf{u}$, where $K = A^T C A$ is the stiffness matrix, constructed from the chain's (reduced) incidence matrix A and the diagonal matrix of spring constants C. Thus, Newton's law immediately leads to the linear system of second order differential equations

$$M \frac{d^2\mathbf{u}}{dt^2} = \mathbf{f}(t) - K\mathbf{u} \tag{9.59}$$

governing the dynamical motions of the masses under a possibly time-dependent external force $\mathbf{f}(t)$. Such systems are also used to model the undamped dynamical motion of structures as well as resistanceless (superconducting) electrical circuits. As always, the first order of business is to analyze the homogeneous system

$$M \frac{d^2\mathbf{u}}{dt^2} + K\mathbf{u} = \mathbf{0} \tag{9.60}$$

modeling the unforced motions of the physical apparatus.

EXAMPLE 9.33 The simplest case is that of a single mass connected to a fixed support by a spring. Assuming no external force, the dynamical system (9.60) reduces to a homogeneous second order scalar equation

$$m \frac{d^2u}{dt^2} + k\,u = 0, \tag{9.61}$$

in which $m > 0$ is the mass, while $k > 0$ is the spring's stiffness. The general solution to (9.61) is

$$u(t) = c_1 \cos\omega t + c_2 \sin\omega t = r\cos(\omega t - \delta), \quad \text{where} \quad \omega = \sqrt{\frac{k}{m}} \tag{9.62}$$

is the natural frequency of vibration. In the second expression, we have used the phase-amplitude equation (2.7) to rewrite the solution as a single cosine with an amplitude $r = \sqrt{c_1^2 + c_2^2}$ and phase shift $\delta = \tan^{-1} c_2/c_1$. Thus, the mass' motion is periodic, with period $P = 2\pi/\omega$. The stiffer the spring or the lighter the mass, the faster the vibrations. Take note of the square root in the frequency formula; quadrupling the mass only slows down the vibrations by a factor of two.

The constants c_1, c_2—or their phase-amplitude counterparts r, δ—are determined by the initial conditions. Physically, we need to specify both an initial position and an initial velocity

$$u(t_0) = a, \qquad \dot{u}(t_0) = b, \tag{9.63}$$

in order to uniquely prescribe the subsequent motion of the system. The resulting solution is most conveniently written in the form

$$u(t) = a\cos\omega\,(t - t_0) + \frac{b}{\omega}\sin\omega\,(t - t_0) = r\cos\big[\,\omega\,(t - t_0) - \delta\,\big] \tag{9.64}$$

with amplitude $r = \sqrt{a^2 + \dfrac{b^2}{\omega^2}}$ and phase shift $\delta = \tan^{-1}\dfrac{b}{a\,\omega}$. A typical solution is plotted in Figure 9.5. ●

Let us turn to a more general second order system. To begin with, let us assume that the masses are all the same and equal to 1 (in some appropriate units), so that (9.60) reduces to

$$\frac{d^2\mathbf{u}}{dt^2} + K\mathbf{u} = \mathbf{0}. \tag{9.65}$$

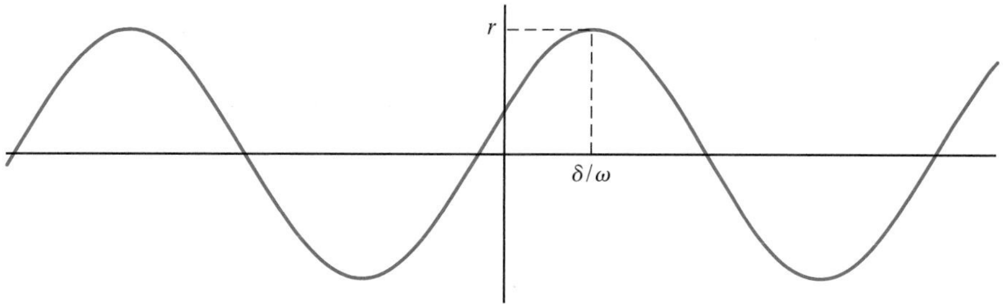

Figure 9.5 Vibration of a mass.

Inspired by the form of the solution of the scalar equation, let us try a trigonometric ansatz for the solution

$$\mathbf{u}(t) = \cos(\omega\, t)\, \mathbf{v}, \tag{9.66}$$

in which ω is a constant scalar and $\mathbf{v} \neq \mathbf{0}$ a constant vector. Differentiating, we find

$$\frac{d\mathbf{u}}{dt} = -\,\omega\, \sin(\omega\, t)\, \mathbf{v}, \qquad \frac{d^2\mathbf{u}}{dt^2} = -\,\omega^2\, \cos(\omega\, t)\, \mathbf{v},$$

whereas

$$K\,\mathbf{u} = \cos(\omega\, t)\; K\,\mathbf{v},$$

since the cosine factor is a scalar. Therefore, (9.66) will solve the second order system (9.65) if and only if

$$K\,\mathbf{v} = \omega^2\, \mathbf{v}. \tag{9.67}$$

The result is an eigenvector equation $K\,\mathbf{v} = \lambda\,\mathbf{v}$ for the stiffness matrix, with eigenvector $\mathbf{v} \neq \mathbf{0}$ and eigenvalue

$$\lambda = \omega^2. \tag{9.68}$$

Now, the scalar equation has both cosine and sine solutions. By the same token, the ansatz $\mathbf{u}(t) = \sin(\omega\, t)\, \mathbf{v}$ leads to the *same* eigenvector equation (9.67). We conclude that each (positive) eigenvalue leads to two different periodic trigonometric solutions. Summarizing:

Lemma 9.34 If $\mathbf{v}$ is an eigenvector of the matrix K with eigenvalue $\lambda = \omega^2$, then $\mathbf{u}(t) = \cos(\omega\, t)\mathbf{v}$ and $\mathbf{u}(t) = \sin(\omega\, t)\, \mathbf{v}$ are both solutions to the homogeneous second order system $\ddot{\mathbf{u}} + K\,\mathbf{u} = \mathbf{0}$.

Stable Structures

Let us begin with the motion of a stable mass/spring chain or structure, of the type introduced in Section 6.3. According to Theorem 6.8, stability requires that the reduced stiffness matrix be positive definite: $K > 0$. Theorem 8.23 says that all the eigenvalues of K are strictly positive, $\lambda_i > 0$, which is good, since it implies that the vibrational frequencies $\omega_i = \sqrt{\lambda_i}$ are all real. Moreover, positive definite matrices are always complete, and so K possesses an orthogonal eigenvector basis $\mathbf{v}_1, \ldots, \mathbf{v}_n$ of $\mathbb{R}^n$ corresponding to its eigenvalues $\lambda_1, \ldots, \lambda_n$, listed in accordance with their multiplicities. This yields a total $2n$ linearly independent trigonometric eigensolutions, namely

$$\begin{aligned} \mathbf{u}_i(t) &= \cos(\omega_i\, t\,)\, \mathbf{v}_i = \cos\!\left(\sqrt{\lambda_i}\, t\right) \mathbf{v}_i, \\ \widetilde{\mathbf{u}}_i(t) &= \sin(\omega_i\, t\,)\, \mathbf{v}_i = \sin\!\left(\sqrt{\lambda_i}\, t\right) \mathbf{v}_i, \end{aligned} \qquad i = 1, \ldots, n, \tag{9.69}$$

which is precisely the number required by the general existence and uniqueness theorems for linear ordinary differential equations. The general solution to (9.65) is an arbitrary linear combination of the eigensolutions:

$$\mathbf{u}(t) = \sum_{i=1}^{n} \left[c_i \cos(\omega_i t) + d_i \sin(\omega_i t) \right] \mathbf{v}_i = \sum_{i=1}^{n} r_i \cos(\omega_i t - \delta_i)\, \mathbf{v}_i. \tag{9.70}$$

The $2n$ coefficients c_i, d_i—or their phase–amplitude counterparts $r_i > 0$, and $0 \le \delta_i < 2\pi$—are uniquely determined by the initial conditions. As in (9.63), we need to specify both the initial positions and initial velocities of all the masses; this requires a total of $2n$ initial conditions

$$\mathbf{u}(t_0) = \mathbf{a}, \qquad \dot{\mathbf{u}}(t_0) = \mathbf{b}. \tag{9.71}$$

Suppose $t_0 = 0$; then substituting the solution formula (9.70) into the initial conditions, we find

$$\mathbf{u}(0) = \sum_{i=1}^{n} c_i \mathbf{v}_i = \mathbf{a}, \qquad \dot{\mathbf{u}}(0) = \sum_{i=1}^{n} \omega_i d_i \mathbf{v}_i = \mathbf{b}.$$

Since the eigenvectors are orthogonal, the coefficients are immediately found by our orthogonal basis formula (5.7), whence

$$c_i = \frac{\langle \mathbf{a}, \mathbf{v}_i \rangle}{\|\mathbf{v}_i\|^2}, \qquad d_i = \frac{\langle \mathbf{b}, \mathbf{v}_i \rangle}{\omega_i \|\mathbf{v}_i\|^2}. \tag{9.72}$$

The eigensolutions (9.69) are also known as the *normal modes of vibration* of the system, and the $\omega_i = \sqrt{\lambda_i}$ its *natural frequencies*, which are the *square roots of the eigenvalues of the stiffness matrix*. Each eigensolution is a periodic, vector-valued function of period $P_i = 2\pi/\omega_i$. Linear combinations of such periodic functions are called *quasi-periodic*, because they are *not*, typically, periodic!

A simple example is provided by the family of functions

$$f(t) = \cos t + \cos \omega t.$$

If $\omega = p/q$ is a rational number, then $f(t)$ is a periodic function, since $f(t + 2\pi q) = f(t)$, where $2\pi q$ is the minimal period, provided that p and q have no common factors. However, if ω is an irrational number, then $f(t)$ is not periodic, and will never precisely repeat itself. You are encouraged to carefully inspect the graphs in Figure 9.6. The first is periodic—can you spot where it begins to repeat?—whereas the second is only quasi-periodic and never quite succeeds in reproducing

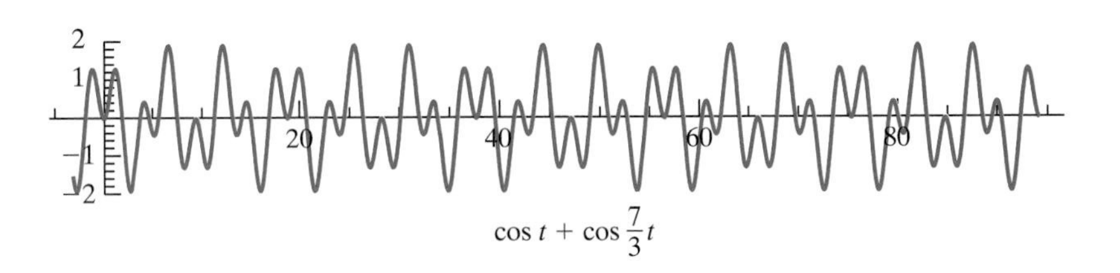

$\cos t + \cos \frac{7}{3} t$

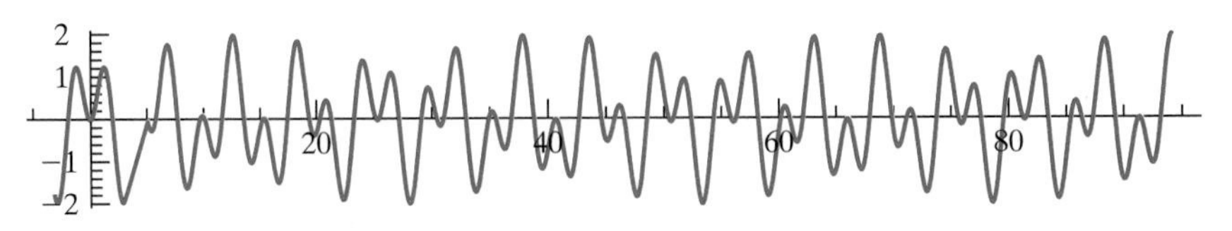

$\cos t + \cos \sqrt{5}\, t$

Figure 9.6 Periodic and quasi–periodic functions.

its behavior. The general solution (9.70) to the vibrational system is quasi-periodic, but only periodic when all the frequency ratios ω_i/ω_j are rational numbers. To the uninitiated, such quasi-periodic motions may appear to be rather chaotic,* even though they are built from a few simple periodic constituents. Most structures and circuits exhibit quasi-periodic vibrational motions. Let us analyze a couple of simple examples.

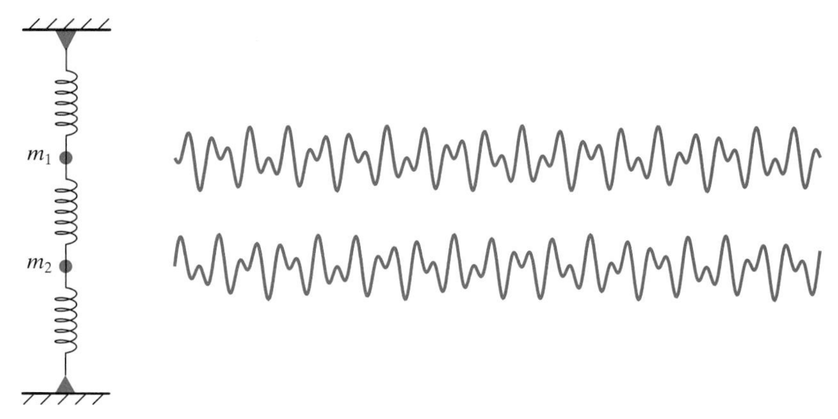

Figure 9.7 Motion of a double mass-spring chain with fixed supports.

EXAMPLE 9.35 Consider a chain consisting of two equal unit masses connected to top and bottom supports by three springs, as in Figure 9.7. If the spring constants are c_1, c_2, c_3 (labeled from top to bottom), then the stiffness matrix is

$$K = \begin{pmatrix} 1 & -1 & 0 \\ 0 & 1 & -1 \end{pmatrix} \begin{pmatrix} c_1 & 0 & 0 \\ 0 & c_2 & 0 \\ 0 & 0 & c_3 \end{pmatrix} \begin{pmatrix} 1 & 0 \\ -1 & 1 \\ 0 & -1 \end{pmatrix} = \begin{pmatrix} c_1 + c_2 & -c_2 \\ -c_2 & c_2 + c_3 \end{pmatrix}.$$

The eigenvalues and eigenvectors of K will prescribe the normal modes and vibrational frequencies of our two–mass chain.

Let us look in detail when the springs are identical, and choose our units so that $c_1 = c_2 = c_3 = 1$. The resulting stiffness matrix

$$K = \begin{pmatrix} 2 & -1 \\ -1 & 2 \end{pmatrix}$$

has eigenvalues and eigenvectors

$$\lambda_1 = 1, \quad \mathbf{v}_1 = \begin{pmatrix} 1 \\ 1 \end{pmatrix}, \qquad \lambda_2 = 3, \quad \mathbf{v}_2 = \begin{pmatrix} -1 \\ 1 \end{pmatrix}.$$

The general solution to the system is then

$$\mathbf{u}(t) = r_1 \cos(t - \delta_1) \begin{pmatrix} 1 \\ 1 \end{pmatrix} + r_2 \cos(\sqrt{3}\, t - \delta_2) \begin{pmatrix} -1 \\ 1 \end{pmatrix}.$$

The first summand is the normal mode vibrating at the relatively slow frequency $\omega_1 = 1$, with the two masses moving in tandem. The second normal mode vibrates faster, with frequency $\omega_2 = \sqrt{3} \approx 1.73205$, in which the two masses move in opposing directions. The general motion is a linear combination of these two normal modes. Since the frequency ratio $\omega_2/\omega_1 = \sqrt{3}$ is irrational, the motion is quasi-periodic. The system never quite returns to its initial configuration—unless

*This is *not* true chaos, which is is an inherently nonlinear phenomenon, [42].

it happens to be vibrating in one of the normal modes. A graph of some typical displacements of the masses is plotted in Figure 9.7.

If we eliminate the bottom spring, so the masses are just hanging from the top support as in Figure 9.8, then the reduced incidence matrix

$$A = \begin{pmatrix} 1 & -1 \\ 0 & 1 \end{pmatrix}$$

loses its last row. Assuming that the springs have unit stiffnesses $c_1 = c_2 = 1$, the corresponding stiffness matrix is

$$K = A^T A = \begin{pmatrix} 1 & -1 \\ 0 & 1 \end{pmatrix} \begin{pmatrix} 1 & 0 \\ -1 & 1 \end{pmatrix} = \begin{pmatrix} 2 & -1 \\ -1 & 1 \end{pmatrix}.$$

The eigenvalues and eigenvectors are

$$\lambda_1 = \frac{3-\sqrt{5}}{2}, \quad \mathbf{v}_1 = \begin{pmatrix} 1 \\ \frac{\sqrt{5}+1}{2} \end{pmatrix}, \quad \lambda_2 = \frac{3+\sqrt{5}}{2}, \quad \mathbf{v}_2 = \begin{pmatrix} 1 \\ -\frac{\sqrt{5}-1}{2} \end{pmatrix}.$$

The general solution to the system is the quasi-periodic linear combination

$$\mathbf{u}(t) = r_1 \cos\left(\tfrac{\sqrt{5}-1}{2}\, t - \delta_1\right) \begin{pmatrix} 1 \\ \frac{\sqrt{5}+1}{2} \end{pmatrix} + r_2 \cos\left(\tfrac{\sqrt{5}+1}{2}\, t - \delta_2\right) \begin{pmatrix} 1 \\ -\frac{\sqrt{5}-1}{2} \end{pmatrix}.$$

The slower normal mode, with frequency

$$\omega_1 = \sqrt{\tfrac{3-\sqrt{5}}{2}} = \tfrac{\sqrt{5}-1}{2} \approx .61803,$$

has the masses moving in tandem, with the bottom mass moving proportionally

$$\tfrac{\sqrt{5}+1}{2} \approx 1.61803$$

farther. The faster normal mode, with frequency

$$\omega_2 = \sqrt{\tfrac{3+\sqrt{5}}{2}} = \tfrac{\sqrt{5}+1}{2} \approx 1.61803,$$

has the masses moving in opposing directions, with the top mass experiencing the larger displacement. Thus, removing the bottom support has caused both modes to vibrate slower. A typical solution is plotted in Figure 9.8. ●

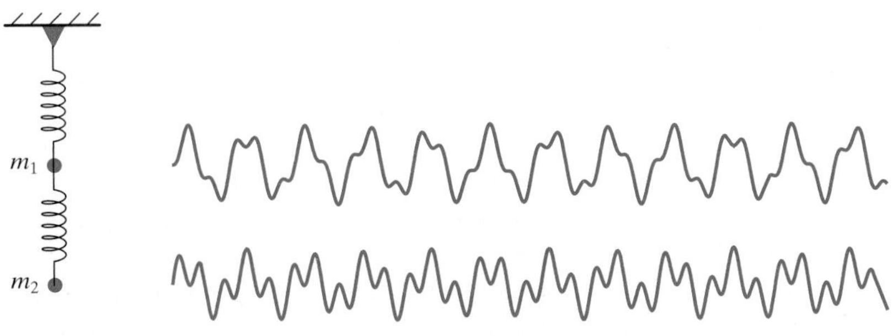

Figure 9.8 Motion of a double mass/spring chain with one free end.

EXAMPLE 9.36 Consider a three mass-spring chain, with unit springs and masses, and both ends attached to fixed supports. The stiffness matrix

$$K = \begin{pmatrix} 2 & -1 & 0 \\ -1 & 2 & -1 \\ 0 & -1 & 2 \end{pmatrix}$$

has eigenvalues and eigenvectors

$$\lambda_1 = 2 - \sqrt{2}, \quad \lambda_2 = 2, \qquad \lambda_3 = 2 + \sqrt{2},$$

$$\mathbf{v}_1 = \begin{pmatrix} 1 \\ \sqrt{2} \\ 1 \end{pmatrix}, \quad \mathbf{v}_2 = \begin{pmatrix} 1 \\ 0 \\ -1 \end{pmatrix}, \quad \mathbf{v}_3 = \begin{pmatrix} 1 \\ -\sqrt{2} \\ 1 \end{pmatrix}.$$

The three normal modes, from slowest to fastest, have frequencies

(a) $\omega_1 = \sqrt{2 - \sqrt{2}}$: all three masses move in tandem, with the middle one moving $\sqrt{2}$ times as far.

(b) $\omega_2 = \sqrt{2}$: the two outer masses move in opposing directions, while the middle mass does not move.

(c) $\omega_3 = \sqrt{2 + \sqrt{2}}$: the two outer masses move in tandem, while the inner mass moves $\sqrt{2}$ times as far in the opposing direction.

The general motion is a quasi-periodic combination of these three normal modes. As such, to the naked eye it can look very complicated. Our mathematical analysis unmasks the innate simplicity, where the complex dynamics are, in fact, entirely governed by just three fundamental modes of vibration. ●

EXERCISES 9.5

9.5.1. A 6 kilogram mass is connected to a spring with stiffness 21 kg/sec^2. Determine the frequency of vibration in Hertz (cycles per second).

9.5.2. The lowest audible frequency is about 20 Hertz = 20 cycles per second. How small a mass would need to be connected to a unit spring to produce a fast enough vibration to be audible? (As always, we assume the spring has negligible mass, which is probably not so reasonable in this situation.)

9.5.3. Graph the following functions. Which are periodic? quasi-periodic? If periodic, what is the (minimal) period?

(a) $\sin 4t + \cos 6t$ (b) $1 + \sin \pi t$

(c) $\cos \frac{1}{2}\pi t + \cos \frac{1}{3}\pi t$ (d) $\cos t + \cos \pi t$

(e) $\sin \frac{1}{4} t + \sin \frac{1}{5} t + \sin \frac{1}{6} t$

(f) $\cos t + \cos \sqrt{2}\, t + \cos 2t$

(g) $\sin t \ \sin 3t$

9.5.4. What is the minimal period of a function of the form

$$\cos \frac{p}{q}\, t + \cos \frac{r}{s}\, t,$$

assuming that each fraction is in lowest terms, i.e., its numerator and denominator have no common factors?

9.5.5. (a) Determine the natural frequencies of the Newtonian system

$$\frac{d^2\mathbf{u}}{dt^2} + \begin{pmatrix} 3 & -2 \\ -2 & 6 \end{pmatrix} \mathbf{u} = \mathbf{0}.$$

(b) What is the dimension of the space of solutions? Explain your answer.

(c) Write out the general solution.

(d) For which initial conditions is the resulting motion

(i) periodic? (ii) quasi-periodic?
(iii) both? (iv) neither?

Justify your answer.

9.5.6. Answer Exercise 9.5.5 for the system

$$\frac{d^2\mathbf{u}}{dt^2} + \begin{pmatrix} 73 & 36 \\ 36 & 52 \end{pmatrix} \mathbf{u} = \mathbf{0}.$$

9.5.7. Find the general solution to the following second order systems:

(a) $\dfrac{d^2u}{dt^2} = -3u + 2v, \quad \dfrac{d^2v}{dt^2} = 2u - 3v$

(b) $\dfrac{d^2u}{dt^2} = -11u - 2v, \quad \dfrac{d^2v}{dt^2} = -2u - 14v$

(c) $\dfrac{d^2\mathbf{u}}{dt^2} + \begin{pmatrix} 1 & 0 & 0 \\ 0 & 4 & 0 \\ 0 & 0 & 9 \end{pmatrix} \mathbf{u} = \mathbf{0}$

(d) $\dfrac{d^2\mathbf{u}}{dt^2} = \begin{pmatrix} -6 & 4 & -1 \\ 4 & -6 & 1 \\ -1 & 1 & -11 \end{pmatrix} \mathbf{u}$

9.5.8. Show that a single mass that is connected to both the top and bottom supports by two springs of stiffnesses c_1, c_2 will vibrate in the same manner as if it were connected to only one support by a spring with the combined stiffness $c = c_1 + c_2$.

9.5.9. Two masses are connected by three springs to top and bottom supports. Can you find a collection of spring constants c_1, c_2, c_3 such that all vibrations are periodic?

♠ **9.5.10.** Suppose the bottom support in the mass-spring chain in Example 9.36 is removed.

(a) Do you predict that the vibration rate will

(i) speed up,
(ii) slow down, or
(iii) stay the same?

(b) Verify your prediction by computing the new vibrational frequencies.

(c) Suppose the middle mass is displaced by a unit amount and then let go. Compute and graph the solutions in both situations. Discuss what you observe.

♠ **9.5.11.** (a) Describe, quantitatively and qualitatively, the normal modes of vibration for a mass–spring chain consisting of 3 unit masses, connected to top and bottom by unit springs.

(b) Answer the same question when the bottom support is removed.

♡ **9.5.12.** Find the vibrational frequencies for a mass–spring chain with n identical masses, connected by $n+1$ identical springs to both top and bottom supports. Is there any sort of limiting behavior as $n \to \infty$? *Hint*: See Exercise 8.2.48.

♣ **9.5.13.** Suppose you are given n different springs. In which order should you connect them to unit masses so that the mass-spring chain vibrates the fastest? Does your answer depend upon the relative sizes of the spring constants? Does it depend upon whether the bottom mass is attached to a support or left hanging free? First try the case of three springs with spring stiffnesses $c_1 = 1, c_2 = 2, c_3 = 3$. Then try varying the stiffnesses. Finally, predict what will happen with 4 or 5 springs, and see if you can make a conjecture in the general case.

♣ **9.5.14.** Suppose the illustrated planar structure has unit masses at the nodes and the bars are all of unit stiffness.

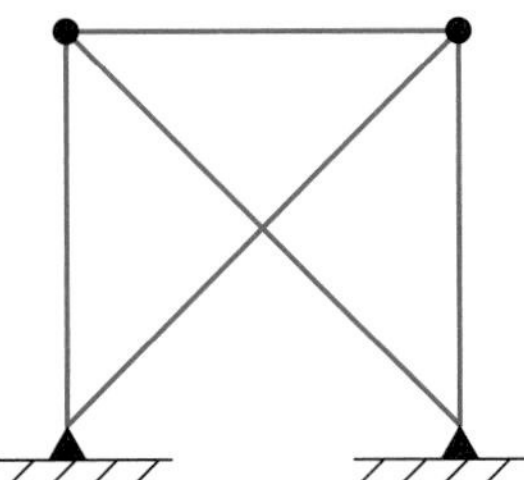

(a) Write down the system of differential equations that describes the dynamical vibrations of the structure.

(b) How many independent modes of vibration are there?

(c) Find numerical values for the vibrational frequencies.

(d) Describe what happens when the structure vibrates in each of the normal modes.

(e) Suppose the left-hand mass is displaced a unit horizontal distance. Determine the subsequent motion.

9.5.15. When does a real first order linear system $\dot{\mathbf{u}} = A\mathbf{u}$ have a quasi-periodic solution? What is the smallest dimension in which this can occur?

Unstable Structures

So far, we have just dealt with the stable case, when the stiffness matrix K is positive definite. Unstable configurations, which can admit rigid motions and/or mechanisms, will provide additional complications. The simplest is a single mass that is not attached to any spring. Since the mass experiences no restraining force, its motion is governed by the elementary second order ordinary differential equation

$$m \frac{d^2u}{dt^2} = 0. \tag{9.73}$$

The general solution is

$$u(t) = c\,t + d. \tag{9.74}$$

If $c = 0$, the mass sits at a fixed position, while when $c \neq 0$ it moves along a straight line with constant velocity.

More generally, suppose that the stiffness matrix K for our structure is only positive semi-definite. Each vector $\mathbf{0} \neq \mathbf{v} \in \ker K$ represents a mode of instability of the system. Since $K\mathbf{v} = \mathbf{0}$, the vector $\mathbf{v}$ is a *null eigenvector* with eigenvalue $\lambda = 0$. Lemma 9.34 provides us with two solutions to the dynamical equations (9.65) of "frequency" $\omega = \sqrt{\lambda} = 0$. The first, $\mathbf{u}(t) = \cos(\omega\,t)\,\mathbf{v} \equiv \mathbf{v}$ is a constant solution, i.e., an equilibrium configuration of the system. Thus, an unstable system does not have a unique equilibrium position, since every null eigenvector $\mathbf{v} \in \ker K$ is a constant solution. On the other hand, the second solution, $\mathbf{u}(t) = \sin(\omega\,t)\,\mathbf{v} \equiv \mathbf{0}$, is trivial, and so doesn't help in constructing the requisite $2n$ linearly independent basis solutions. To find the missing solution(s), let us again argue in analogy with the scalar case (9.74), and try $\mathbf{u}(t) = t\,\mathbf{v}$. Fortunately, this works, since

$$\frac{d\mathbf{u}}{dt} = \mathbf{v}, \qquad \text{so} \qquad \frac{d^2\mathbf{u}}{dt^2} = \mathbf{0}.$$

Also,

$$K\mathbf{u} = t\,K\mathbf{v} = \mathbf{0},$$

and hence $\mathbf{u}(t) = t\,\mathbf{v}$ solves the system

$$\frac{d^2\mathbf{u}}{dt^2} + K\mathbf{u} = \mathbf{0}.$$

Therefore, to each element of the kernel of the stiffness matrix—i.e., each rigid motion and mechanism—there is a two-dimensional family of solutions

$$\mathbf{u}(t) = (c\,t + d)\,\mathbf{v}. \tag{9.75}$$

When $c = 0$, the solution $\mathbf{u}(t) = d\,\mathbf{v}$ reduces to a constant equilibrium; when $c \neq 0$, it is moving off to ∞ with constant velocity in the null direction $\mathbf{v}$, and so represents an unstable mode of the system. The general solution will be a linear superposition of the vibrational modes corresponding to the positive eigenvalues and the unstable linear motions corresponding to the independent null eigenvectors.

REMARK: If the null direction $\mathbf{v} \in \ker K$ represents a rigid translation, then the entire structure will move in that direction. If $\mathbf{v}$ represents an infinitesimal rotation, then, because our model is based on a linear approximation to the true nonlinear motions, the individual masses will move along straight lines, which are the tangent approximations to the circular motion that occurs in the true physical, nonlinear regime. We refer to the earlier discussion in Chapter 6 for details. Finally, if we excite a mechanism, then the masses will again follow straight lines, moving in different directions, whereas in the real world the masses may move along much more complicated curved trajectories. For small motions, the distinction is not so important, while larger displacements, such as occur in the design of robots, [43], will require dealing with the vastly more complicated nonlinear dynamical equations.

EXAMPLE 9.37 Consider a system of three unit masses connected in a line by two unit springs, but not attached to any fixed supports, as illustrated in Figure 9.9. This chain could be viewed as a simplified model of an (unbent) triatomic molecule that is only allowed to move in the vertical direction. The incidence matrix is

$$A = \begin{pmatrix} -1 & 1 & 0 \\ 0 & -1 & 1 \end{pmatrix}$$

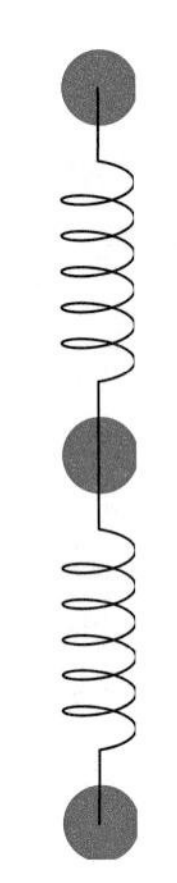

Figure 9.9 A triatomic molecule.

and, since we are dealing with unit springs, the stiffness matrix is

$$K = A^T A = \begin{pmatrix} -1 & 0 \\ 1 & -1 \\ 0 & 1 \end{pmatrix} \begin{pmatrix} -1 & 1 & 0 \\ 0 & -1 & 1 \end{pmatrix} = \begin{pmatrix} 1 & -1 & 0 \\ -1 & 2 & -1 \\ 0 & -1 & 1 \end{pmatrix}.$$

The eigenvalues and eigenvectors of K are easily found:

$$\lambda_1 = 0, \qquad \lambda_2 = 1, \qquad \lambda_3 = 3,$$
$$\mathbf{v}_1 = \begin{pmatrix} 1 \\ 1 \\ 1 \end{pmatrix}, \quad \mathbf{v}_2 = \begin{pmatrix} 1 \\ 0 \\ -1 \end{pmatrix}, \quad \mathbf{v}_3 = \begin{pmatrix} 1 \\ -2 \\ 1 \end{pmatrix}.$$

Each positive eigenvalue provides two trigonometric solutions, while the zero eigenvalue leads to solutions that depend linearly on t. This yields the required six basis solutions:

$$\mathbf{u}_1(t) = \begin{pmatrix} 1 \\ 1 \\ 1 \end{pmatrix}, \quad \mathbf{u}_3(t) = \begin{pmatrix} \cos t \\ 0 \\ -\cos t \end{pmatrix}, \quad \mathbf{u}_5(t) = \begin{pmatrix} \cos \sqrt{3}\, t \\ -2\cos \sqrt{3}\, t \\ \cos \sqrt{3}\, t \end{pmatrix},$$
$$\mathbf{u}_2(t) = \begin{pmatrix} t \\ t \\ t \end{pmatrix}, \quad \mathbf{u}_4(t) = \begin{pmatrix} \sin t \\ 0 \\ -\sin t \end{pmatrix}, \quad \mathbf{u}_6(t) = \begin{pmatrix} \sin \sqrt{3}\, t \\ -2\sin \sqrt{3}\, t \\ \sin \sqrt{3}\, t \end{pmatrix}.$$

The first solution $\mathbf{u}_1(t)$ is a constant, equilibrium mode, where the masses rest at a fixed common distance from their reference positions. The second solution $\mathbf{u}_2(t)$ is the unstable mode, corresponding to a uniform rigid translation of the molecule that does not stretch the interconnecting springs. The final four solutions represent vibrational modes. In the first pair, $\mathbf{u}_3(t), \mathbf{u}_4(t)$, the two outer masses move in opposing directions, while the middle mass remains fixed, while the final pair, $\mathbf{u}_5(t), \mathbf{u}_6(t)$ has the two outer masses moving in tandem, while the inner mass moves twice as far in the opposite direction. The general solution is a linear combination of the six normal modes,

$$\mathbf{u}(t) = c_1\, \mathbf{u}_1(t) + \cdots + c_6\, \mathbf{u}_6(t), \tag{9.76}$$

and corresponds to the entire molecule moving at a fixed velocity while the individual masses perform a quasi-periodic vibration.

Let us see if we can predict the motion of the molecule from its initial conditions

$$\mathbf{u}(0) = \mathbf{a}, \qquad \dot{\mathbf{u}}(0) = \mathbf{b},$$

where $\mathbf{a} = (a_1, a_2, a_3)^T$ indicates the initial displacements of the three atoms, while $\mathbf{b} = (b_1, b_2, b_3)^T$ are their initial velocities. Substituting the solution formula (9.76) leads to the two linear systems

$$c_1\, \mathbf{v}_1 + c_3\, \mathbf{v}_2 + c_5\, \mathbf{v}_3 = \mathbf{a}, \qquad c_2\, \mathbf{v}_1 + c_4\, \mathbf{v}_2 + \sqrt{3}\, c_6\, \mathbf{v}_3 = \mathbf{b},$$

for the coefficients $c_1, \ldots, c_6$. As in (9.72), we can use the orthogonality of the eigenvectors to immediately compute the coefficients:

$$c_1 = \frac{\mathbf{a}\cdot\mathbf{v}_1}{\|\mathbf{v}_1\|^2} = \frac{a_1+a_2+a_3}{3}, \qquad c_4 = \frac{\mathbf{b}\cdot\mathbf{v}_2}{\|\mathbf{v}_2\|^2} = \frac{b_1-b_3}{2},$$
$$c_2 = \frac{\mathbf{b}\cdot\mathbf{v}_1}{\|\mathbf{v}_1\|^2} = \frac{b_1+b_2+b_3}{3}, \qquad c_5 = \frac{\mathbf{a}\cdot\mathbf{v}_3}{\|\mathbf{v}_3\|^2} = \frac{a_1-2a_2+a_3}{6},$$
$$c_3 = \frac{\mathbf{a}\cdot\mathbf{v}_2}{\|\mathbf{v}_2\|^2} = \frac{a_1-a_3}{2}, \qquad c_6 = \frac{\mathbf{b}\cdot\mathbf{v}_3}{\sqrt{3}\,\|\mathbf{v}_3\|^2} = \frac{b_1-2b_2+b_3}{6\sqrt{3}}.$$

In particular, the unstable translational mode is excited if and only if $c_2 \neq 0$, and this occurs if and only if there is a nonzero net initial velocity of the molecule: $b_1+b_2+b_3 \neq 0$. In this case, the vibrating molecule will move off to ∞ at a uniform velocity $c = c_2 = \frac{1}{3}(b_1+b_2+b_3)$ equal to the average of the individual initial velocities. On the other hand, if $b_1+b_2+b_3 = 0$, then the atoms will vibrate quasi-periodically, with frequencies 1 and $\sqrt{3}$, around a fixed location. ●

The observations established in this example hold, in fact, in complete generality. Let us state the result, leaving the details of the proof as an exercise for the reader.

THEOREM 9.38 The solution to an unstable second order linear system with positive semi-definite coefficient matrix K is a combination of a quasi-periodic vibration and a uniform translational motion at a fixed velocity in the direction of a null eigenvector $\mathbf{v} \in \ker K$. In particular, the system will just vibrate around a fixed position if and only if the initial velocity $\dot{\mathbf{u}}(t_0) \in (\ker K)^\perp = \operatorname{rng} K$ lies in the orthogonal complement to the kernel.

As in Chapter 6, the unstable modes $\mathbf{v} \in \ker K$ correspond to either rigid motions or to mechanisms of the structure. Thus, to prevent a structure from exhibiting an unstable motion, one has to ensure that the initial velocity is orthogonal to all of the unstable modes. (The value of the initial position is not an issue.) This is the dynamical counterpart of the requirement that an external force be orthogonal to all unstable modes in order to maintain equilibrium in the structure, cf. Theorem 6.8.

Systems with Differing Masses

When a chain or structure has differing masses at the nodes, the (unforced) Newtonian equations of motion take the more general form

$$M\,\ddot{\mathbf{u}} + K\,\mathbf{u} = \mathbf{0}, \qquad \text{or, equivalently,} \qquad \ddot{\mathbf{u}} = -\,M^{-1}K\,\mathbf{u} = -\,P\,\mathbf{u}. \tag{9.77}$$

The mass matrix M is always positive definite (and, almost always, diagonal, although this is not required by the general theory), while the stiffness matrix $K = A^T C A$ is either positive definite or, in the unstable situation when $\ker A \neq \{\mathbf{0}\}$, positive semi-definite. The coefficient matrix

$$P = M^{-1}K = M^{-1}A^T C\,A \tag{9.78}$$

is *not* in general symmetric, and so we cannot directly apply the preceding constructions. However, P does have the more general self-adjoint form (7.78) based on the weighted inner products

$$\langle \mathbf{u}, \widetilde{\mathbf{u}} \rangle = \mathbf{u}^T M\,\widetilde{\mathbf{u}}, \qquad \langle\!\langle \mathbf{v}, \widetilde{\mathbf{v}} \rangle\!\rangle = \mathbf{v}^T C\,\widetilde{\mathbf{v}}, \tag{9.79}$$

on, respectively, the domain and target spaces for the (reduced) incidence matrix A. Moreover, in the stable case when $\ker A = \{\mathbf{0}\}$, then P is positive definite in the generalized sense of Definition 7.59.

To solve the system of differential equations, we substitute the same trigonometric solution ansatz $\mathbf{u}(t) = \cos(\omega\, t)\,\mathbf{v}$. This results in a *generalized matrix eigenvalue problem*

$$K\,\mathbf{v} = \lambda\,M\,\mathbf{v}, \qquad \text{or, equivalently,} \qquad P\,\mathbf{v} = \lambda\,\mathbf{v}, \qquad \text{with} \quad \lambda = \omega^2. \tag{9.80}$$

The matrix M assumes the role of the identity matrix in the standard eigenvalue equation (8.13), and λ is an eigenvalue if and only if it satisfies the generalized characteristic equation

$$\det(K - \lambda\,M) = 0. \tag{9.81}$$

According to Exercise 8.4.8, if $M > 0$ and $K > 0$, then all the generalized eigenvalues are real and non-negative. Moreover the generalized eigenvectors form an orthogonal basis of $\mathbb{R}^n$, but now with respect to the weighted inner product $\langle \mathbf{u}, \mathbf{v} \rangle = \mathbf{u}^T M\,\mathbf{v}$ governed by the mass matrix. The general solution is a quasi-periodic linear combination of the eigensolutions, of the same form as in (9.70). In the unstable case, when $K \geq 0$ (but M necessarily remains positive definite), one must include enough null eigenvectors to span $\ker K$, each of which leads to an unstable mode of the form (9.75). Further details are relegated to the exercises.

EXERCISES 9.5

9.5.16. Find the general solution to the following systems. Distinguish between the vibrational and unstable modes. What constraints on the initial conditions ensure that the unstable modes are not excited?

(a) $\dfrac{d^2u}{dt^2} = -4u - 2v, \quad \dfrac{d^2v}{dt^2} = -2u - v$

(b) $\dfrac{d^2u}{dt^2} = -u - 3v, \quad \dfrac{d^2v}{dt^2} = -3u - 9v$

(c) $\dfrac{d^2u}{dt^2} = -2u + v - 2w, \quad \dfrac{d^2v}{dt^2} = u - v,$ $\quad \dfrac{d^2w}{dt^2} = -2u - 4w$

(d) $\dfrac{d^2u}{dt^2} = -u + v - 2w, \quad \dfrac{d^2v}{dt^2} = u - v + 2w,$ $\quad \dfrac{d^2w}{dt^2} = -2u + 2v - 4w$

9.5.17. Let $K = \begin{pmatrix} 3 & 0 & -1 \\ 0 & 2 & 0 \\ -1 & 0 & 3 \end{pmatrix}$.

(a) Find an orthogonal matrix Q and a diagonal matrix Λ such that $K = Q\,\Lambda\,Q^T$.

(b) Is K positive definite?

(c) Solve the second order system $\dfrac{d^2\mathbf{u}}{dt^2} = A\,\mathbf{u}$ subject to the initial conditions

$$\mathbf{u}(0) = \begin{pmatrix} 1 \\ 0 \\ 1 \end{pmatrix}, \quad \frac{d\mathbf{u}}{dt}(0) = \begin{pmatrix} 0 \\ 1 \\ 0 \end{pmatrix}.$$

(d) Is your solution periodic? If your answer is "yes", indicate the period.

(e) Is the general solution to the system periodic?

9.5.18. Answer Exercise 9.5.17 when

$$A = \begin{pmatrix} 2 & -1 & 0 \\ -1 & 1 & -1 \\ 0 & -1 & 2 \end{pmatrix}.$$

9.5.19. Compare the solutions to the mass-spring system (9.61) with tiny spring constant $k = \varepsilon \ll 1$ to those of the completely unrestrained system (9.73). Are they close? Discuss.

♠ **9.5.20.** Find the vibrational frequencies and instabilities of the following structures, assuming they have unit masses at all the nodes. Explain in detail how each normal mode moves the structure:

(a) the three bar planar structure in Figure 6.13

(b) its reinforced version in Figure 6.16

(c) the swing set in Figure 6.18

♡ **9.5.21.** Discuss the three-dimensional motions of the triatomic molecule of Example 9.37. Are the vibrational frequencies the same as the one-dimensional model?

♠ **9.5.22.** Assuming unit masses at the nodes, find the vibrational frequencies and describe the normal modes for the following planar structures. What initial conditions will not excite its instabilities (rigid motions and/or mechanisms)?

(a) An equilateral triangle

(b) a square

(c) a regular hexagon

♠ **9.5.23.** Answer Exercise 9.5.22 for the three-dimensional motions of a regular tetrahedron.

♡ **9.5.24.** (a) Show that if a structure contains all unit masses and bars with unit stiffness, $c_i = 1$, then its frequencies of vibration are the nonzero singular values of the reduced incidence matrix.

(b) How would you recognize when a structure is close to being unstable?

9.5.25. Prove that if the initial velocity satisfies $\dot{\mathbf{u}}(t_0) = \mathbf{b} \in \operatorname{corng} A$, then the solution to the initial value problem (9.65, 71) remains bounded.

9.5.26. Find the general solution to the system (9.77) for the following matrix pairs:

(a) $M = \begin{pmatrix} 2 & 0 \\ 0 & 3 \end{pmatrix}, \quad K = \begin{pmatrix} 3 & -1 \\ -1 & 2 \end{pmatrix}$

(b) $M = \begin{pmatrix} 3 & 0 \\ 0 & 5 \end{pmatrix}, \quad K = \begin{pmatrix} 4 & -2 \\ -2 & 3 \end{pmatrix}$

(c) $M = \begin{pmatrix} 2 & 0 \\ 0 & 1 \end{pmatrix}, \quad K = \begin{pmatrix} 2 & -1 \\ -1 & 2 \end{pmatrix},$

(d) $M = \begin{pmatrix} 2 & 0 & 0 \\ 0 & 3 & 0 \\ 0 & 0 & 6 \end{pmatrix}, \quad K = \begin{pmatrix} 5 & -1 & -1 \\ -1 & 6 & 3 \\ -1 & 3 & 9 \end{pmatrix}$

(e) $M = \begin{pmatrix} 2 & 1 \\ 1 & 2 \end{pmatrix}, \quad K = \begin{pmatrix} 3 & -1 \\ -1 & 3 \end{pmatrix}$

(f) $M = \begin{pmatrix} 1 & 1 & 0 \\ 1 & 3 & 1 \\ 0 & 1 & 1 \end{pmatrix}, \quad K = \begin{pmatrix} 1 & 2 & 0 \\ 2 & 8 & 2 \\ 0 & 2 & 1 \end{pmatrix}$

9.5.27. A mass-spring chain of two masses, $m_1 = 1$ and $m_2 = 2$, connected to top and bottom supports by identical springs with unit stiffness. The upper mass is displaced by a unit distance. Find the subsequent motion of the system.

9.5.28. Answer Exercise 9.5.27 when the bottom support is removed.

♠ **9.5.29.** Suppose you have masses $m_1 = 1, m_2 = 2, m_3 = 3$, connected to top and bottom supports by identical unit springs. Does rearranging the order of the masses change the fundamental frequencies? If so, which order produces the fastest vibrations?

♣ **9.5.30.** (a) A water molecule consists of two hydrogen atoms connected at an angle of $105°$ to an oxygen atom whose relative mass is 16 times that of the hydrogen atoms. If the bonds are modeled as linear unit springs, determine the fundamental frequencies and describe the corresponding vibrational modes.

(b) Do the same for a carbon tetrachloride molecule, in which the chlorine atoms, with atomic weight 35, are positioned on the vertices of a regular tetrahedron and the carbon atom, with atomic weight 12, is at the center.

(c) Finally try a benzene molecule, consisting of 6 carbon atoms arranged in a regular hexagon. In this case, every other bond is double strength because two electrons are shared. (Ignore the six extra hydrogen atoms for simplicity.)

9.5.31. So far, our mass-spring chains have only been allowed to move in the vertical direction.

(a) Set up the system governing the planar motions of a mass–spring chain consisting of two unit masses attached to top and bottom supports by unit springs, where the masses are allowed to move in the longitudinal and transverse directions. Compare the resulting vibrational frequencies with the one-dimensional case.

(b) Repeat the analysis when the bottom support is removed.

(c) Can you make any conjectures concerning the planar motions of general mass-spring chains?

9.5.32. Repeat Exercise 9.5.31 for fully 3-dimensional motions of the chain.

◇ **9.5.33.** Suppose M is a nonsingular matrix. Prove that λ is a generalized eigenvalue of the matrix pair K, M if and only if it is an ordinary eigenvalue of the matrix $P = M^{-1} K$. How are the eigenvectors related? How are the characteristic equations related?

9.5.34. Suppose that $\mathbf{u}(t)$ is a solution to (9.77). Let $N = \sqrt{M}$ denote the positive definite square root of the mass matrix M, as defined in Exercise 8.4.27.

(a) Prove that the "weighted" displacement vector $\widetilde{\mathbf{u}}(t) = N\,\mathbf{u}(t)$ solves $d^2\widetilde{\mathbf{u}}/dt^2 = -\widetilde{K}\,\widetilde{\mathbf{u}}$, where $\widetilde{K} = N^{-1} K\, N^{-1}$ is a symmetric, positive semi-definite matrix.

(b) Explain in what sense this can serve as an alternative to the generalized eigenvector solution method.

◇ **9.5.35.** Provide the details of the proof of Theorem 9.38.

Friction and Damping

So far, we have not allowed friction to affect the motion of our dynamical equations. In the standard physical model, the frictional force on a mass in motion is directly proportional to its velocity. In the simplest case of a single mass attached to a spring, one amends the balance of forces in the undamped Newton equation (9.61) to obtain

$$m\,\frac{d^2u}{dt^2} + \beta\,\frac{du}{dt} + k\,u = 0. \tag{9.82}$$

As before, $m > 0$ is the mass, and $k > 0$ the spring stiffness, while $\beta > 0$ measures the effect of a velocity-dependent frictional force—the larger β the greater the frictional damping of the motion.

The solution of this more general second order homogeneous linear ordinary differential equation is found by substituting the usual exponential ansatz $u(t) = e^{\lambda t}$, reducing it to the quadratic characteristic equation

$$m\,\lambda^2 + \beta\,\lambda + k = 0. \tag{9.83}$$

Assuming that $m, \beta, k > 0$, there are three possible cases:

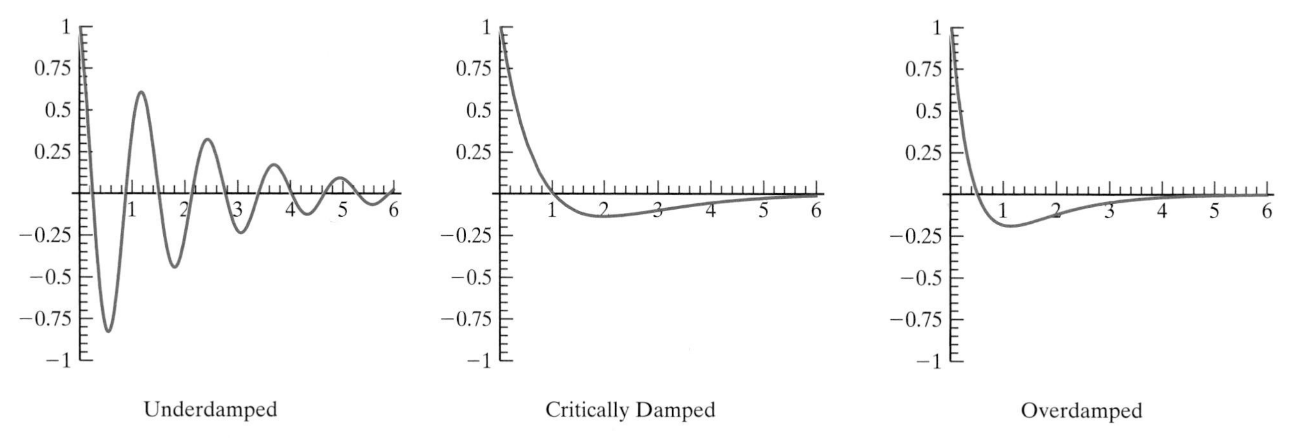

Figure 9.10 Damped vibrations.

Underdamped: If $0 < \beta < 2\sqrt{m\,k}$, then (9.83) has two complex-conjugate roots:

$$\lambda = -\frac{\beta}{2m} \pm \mathrm{i}\,\frac{\sqrt{4m\,k - \beta^2}}{2m} = -\mu \pm \mathrm{i}\,\nu\,. \tag{9.84}$$

The general solution to the differential equation,

$$u(t) = e^{-\mu t}\bigl(c_1 \cos \nu\,t + c_2 \sin \nu\,t\bigr) = r\,e^{-\mu t}\cos(\nu\,t - \delta), \tag{9.85}$$

represents a damped periodic motion. The mass continues to oscillate at a fixed frequency

$$\nu = \frac{\sqrt{4m\,k - \beta^2}}{2m} = \sqrt{\frac{k}{m} - \frac{\beta^2}{4m^2}}\,, \tag{9.86}$$

but the vibrational amplitude $r\,e^{-\mu t}$ decays to zero at an exponential rate as $t \to \infty$. Observe that, in a rigorous mathematical sense, the mass never quite returns to equilibrium, although in the real world, after a sufficiently long time the residual vibrations are not noticeable and equilibrium is physically (but not mathematically) achieved. The rate of decay, $\mu = \beta/(2m)$, is directly proportional to the friction, and inversely proportional to the mass. Thus, greater friction and/or less mass accelerates the return to equilibrium. The friction also has an effect on the vibrational frequency (9.86); the larger β is, the slower the oscillations become and the more rapid the damping effect. As the friction approaches the critical threshold $\beta_\star = 2\sqrt{m\,k}$, the vibrational frequency goes to zero, $\nu \to 0$, and so the period $P = 2\pi/\nu$ becomes longer and longer.

Overdamped: If $\beta > 2\sqrt{m\,k}$, then the characteristic equation (9.83) has two negative real roots

$$\lambda_1 = -\frac{\beta + \sqrt{\beta^2 - 4\,m\,k}}{2\,m} < \lambda_2 = -\frac{\beta - \sqrt{\beta^2 - 4\,m\,k}}{2\,m} < 0,$$

The solution

$$u(t) = c_1\, e^{\lambda_1 t} + c_2\, e^{\lambda_2 t} \tag{9.87}$$

is a linear combination of two decaying exponentials. An overdamped system models the motion of, say, a mass in a vat of molasses. Its "vibration" is so slow that it can pass at most once through the equilibrium position, and then only when its initial velocity is quite large. In the long term, the first exponential in the solution will go to zero faster, and hence, as long as $c_2 \neq 0$, the overall decay rate of the solution is governed by the dominant (least negative) eigenvalue λ_2.

Critically Damped: The borderline case occurs when $\beta = \beta_\star = 2\sqrt{m\,k}$, which means that the characteristic equation (9.83) has only a single negative real root:

$$\lambda_1 = -\frac{\beta}{2\,m}\,.$$

In this case, our ansatz only supplies one exponential solution $e^{\lambda_1 t} = e^{-\beta t/2m}$. In a scalar equation with a repeated root, the second independent solution is obtained by multiplication by t, leading to the general solution

$$u(t) = (c_1\, t + c_2) e^{-\beta t/2m}. \tag{9.88}$$

Even though the formula looks quite different, its qualitative behavior is very similar to the overdamped case. The factor t plays an unimportant role, since the asymptotics of this solution are almost entirely governed by the decaying exponential function. This represents a nonvibrating solution that has the slowest possible decay rate since any further reduction of the frictional coefficient will allow a damped periodic vibration to appear.

In all three cases, the zero solution is globally asymptotically stable. Physically, no matter how small the frictional contribution, all solutions to the unforced system eventually return to equilibrium as friction eventually overwhelms the motion.

This concludes our discussion of the scalar case. Similar considerations apply to mass-spring chains, and to two- and three-dimensional structures. A frictionally damped structure is modeled by a second order system of the form

$$M\,\frac{d^2\mathbf{u}}{dt^2} + B\,\frac{d\mathbf{u}}{dt} + K\,\mathbf{u} = \mathbf{0}, \tag{9.89}$$

where the mass matrix M and the matrix of frictional coefficients B are both diagonal and positive definite, while the stiffness matrix $K = A^T C\, A \geq 0$ is a positive semi-definite Gram matrix constructed from the reduced incidence matrix A. Under these assumptions, it can be proved that the zero equilibrium solution is globally asymptotically stable. However, the mathematical details in this case are sufficiently complicated that we shall leave their analysis as an advanced project for the highly motivated student.

EXERCISES 9.5

9.5.36. Solve the following mass-spring initial value problems, and classify as to

(i) overdamped

(ii) critically damped

(iii) underdamped, or

(iv) undamped

(a) $\ddot{u} + 6\dot{u} + 9u = 0,\ u(0) = 0,\ \dot{u}(0) = 1$

(b) $\ddot{u} + 2\dot{u} + 10u = 0,\ u(0) = 1,\ \dot{u}(0) = 1$

(c) $\ddot{u} + 16u = 0,\ u(1) = 0,\ \dot{u}(1) = 1$

(d) $\ddot{u} + 3\dot{u} + 9u = 0,\ u(0) = 0,\ \dot{u}(0) = 1$

(e) $2\ddot{u} + 3\dot{u} + u = 0,\ u(0) = 2,\ \dot{u}(0) = 0$

(f) $\ddot{u} + 6\dot{u} + 10u = 0,\ u(0) = 3,\ \dot{u}(0) = -2$

9.5.37. Consider the overdamped mass-spring equation $\ddot{u} + 6\dot{u} + 5u = 0$. If the mass starts out a distance 1 away from equilibrium, how large must the initial velocity be in order that it pass through equilibrium once?

9.5.38. (a) A mass weighing 16 pounds stretches a spring 6.4 feet. Assuming no friction, determine the equation of motion and the natural frequency of vibration of the mass-spring system. Use the value $g = 32\ \text{ft}/\text{sec}^2$ for the gravitational acceleration.

(b) The mass-spring system is placed in a jar of oil, whose frictional resistance equals the speed of the mass. Assume the spring is stretched an additional 2 feet from its equilibrium position and let go. Determine the motion of the mass.

(c) Is the system overdamped or underdamped? Are the vibrations more rapid or less rapid than the undamped system?

9.5.39. Suppose you convert the second order equation (9.82) into its phase plane equivalent. What are the phase portraits corresponding to

(a) undamped

(b) underdamped

(c) critically damped, and

(d) overdamped motion?

◇ **9.5.40.** (a) Prove that, for any non-constant solution to an overdamped mass-spring system, there is at most one time where $u(t_\star) = 0$.

(b) Is this statement also valid in the critically damped case?

9.5.41. Discuss the possible behaviors of a mass moving in a frictional medium that is not attached to a spring, i.e., set $k = 0$ in (9.82).

9.6 FORCING AND RESONANCE

So far, our structure has been left free to vibrate on its own. It is now time to see what happens when we shake it. In this section, we will investigate the effects of periodic external forcing on both undamped and damped systems. More general types of forcing can be handled by adapting the variation of parameters method presented in Section 9.4.

The simplest case is that of a single mass connected to a spring that has no frictional damping. We append an external forcing function $f(t)$ to the homogeneous (unforced) equation (9.61), leading to the inhomogeneous second order equation

$$m \frac{d^2u}{dt^2} + k\,u = f(t), \tag{9.90}$$

in which $m > 0$ is the mass and $k > 0$ the spring stiffness. We are particularly interested in the case of periodic forcing

$$f(t) = \alpha \cos \eta\, t \tag{9.91}$$

of frequency $\eta > 0$ and amplitude α. To find a particular solution to (9.90–91), we use the method of undetermined coefficients* which tells us to guess a solution ansatz of the form

$$u^\star(t) = a \cos \eta\, t + b \sin \eta\, t, \tag{9.92}$$

*One can also use variation of parameters, although the intervening calculations are slightly more complicated.

where a, b are constants to be determined. Substituting into the differential equation, we find

$$m \frac{d^2 u^\star}{dt^2} + k\,u^\star = a\,(k - m\,\eta^2)\cos\eta\,t + b\,(k - m\,\eta^2)\sin\eta\,t = \alpha\cos\eta\,t.$$

We can solve for

$$a = \frac{\alpha}{k - m\,\eta^2} = \frac{\alpha}{m(\omega^2 - \eta^2)}, \qquad b = 0, \tag{9.93}$$

where

$$\omega = \sqrt{\frac{k}{m}} \tag{9.94}$$

refers to the natural, unforced vibrational frequency of the system. The solution formula (9.93) is valid, provided that the denominator is nonzero:

$$k - m\,\eta^2 = m(\omega^2 - \eta^2) \neq 0.$$

Therefore, provided the forcing frequency is *not* equal to the system's natural frequency, $\eta \neq \omega$, there exists a particular solution

$$u^\star(t) = a\,\cos\eta\,t = \frac{\alpha}{m(\omega^2 - \eta^2)}\cos\eta\,t \tag{9.95}$$

that vibrates at the same frequency as the forcing function.

The general solution to the inhomogeneous system (9.90) is found, as usual, by adding in an arbitrary solution to the homogeneous equation, as in (9.62), yielding

$$u(t) = \frac{\alpha}{m(\omega^2 - \eta^2)}\cos\eta\,t \;+\; r\,\cos(\omega\,t - \delta), \tag{9.96}$$

where r and δ are determined by the initial conditions. The solution is therefore a quasi-periodic combination of two simple periodic motions—the first, vibrating with frequency ω, represents the internal or natural vibrations of the system, while the second, with frequency η, represents the response to the periodic forcing. Due to the factor $\omega^2 - \eta^2$ in the denominator of (9.96), the closer the forcing frequency is to the natural frequency, the larger the overall amplitude of the response.

Suppose we start the mass at equilibrium, so the initial conditions are

$$u(0) = 0, \qquad \dot u(0) = 0. \tag{9.97}$$

Substituting (9.96) and solving for r, δ, we find that

$$r = -\frac{\alpha}{m(\omega^2 - \eta^2)}, \qquad \delta = 0.$$

Thus, the solution to the initial value problem can be written in the form

$$\begin{aligned} u(t) &= \frac{\alpha}{m(\omega^2 - \eta^2)}\big(\cos\eta\,t - \cos\omega\,t\big) \\ &= \frac{2\,\alpha}{m(\omega^2 - \eta^2)}\,\sin\left(\frac{\omega+\eta}{2}\,t\right)\,\sin\left(\frac{\omega-\eta}{2}\,t\right), \end{aligned} \tag{9.98}$$

where we have employed a standard trigonometric identity, cf. Exercise 3.6.17. The first trigonometric factor, $\sin\frac{1}{2}(\omega+\eta)t$, represents a periodic motion at a frequency equal to the average of the natural and the forcing frequencies. If the forcing frequency η is close to the natural frequency ω, then the second factor, $\sin\frac{1}{2}(\omega-\eta)t$, has a much smaller frequency, and so oscillates on a much longer time scale. As a result, it *modulates* the amplitude of the more rapid vibrations, and is responsible for the phenomenon of *beats*, in which a rapid vibration is subject to a slowly varying amplitude. An everyday illustration of beats is two tuning forks with nearby pitch. When they vibrate near each other, the sound you hear waxes and wanes in intensity. As an example, Figure 9.11 displays the graph of the particular function

$$\cos 14\,t - \cos 15.6\,t = 2\,\sin .8\,t\,\sin 14.8\,t.$$

The slowly varying amplitude $2\,\sin .8\,t$ is clearly visible as the envelope of the relatively rapid vibrations of frequency 14.8.

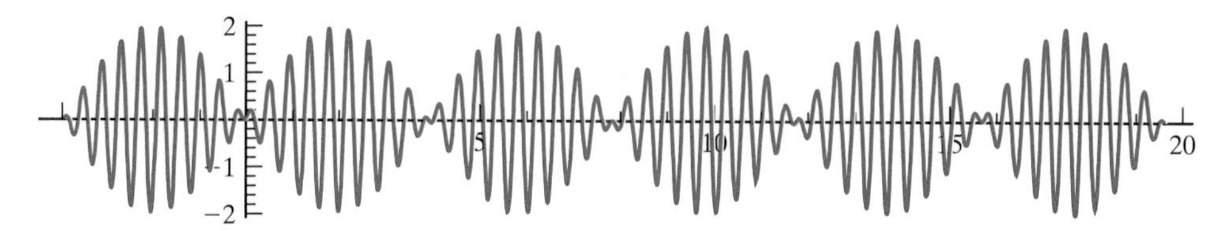

Figure 9.11 Beats in a periodically forced vibration.

When we force the system at exactly the natural frequency $\eta = \omega$, the trigonometric ansatz (9.92) no longer works. This is because both terms are now solutions to the homogeneous equation, and so cannot be combined to form a solution to the inhomogeneous version. In this situation, there is a simple modification to the ansatz, namely multiplication by t, that does the trick. Substituting

$$u^\star(t) = a\,t\,\cos\omega\,t + b\,t\,\sin\omega\,t \tag{9.99}$$

into the differential equation (9.90), we find

$$m\,\frac{d^2u^\star}{dt^2} + k\,u^\star = -\,2\,a\,m\,\omega\,\sin\omega\,t + 2\,b\,m\,\omega\,\cos\omega\,t = \alpha\,\cos\omega\,t.$$

provided

$$a = 0, \qquad b = \frac{\alpha}{2m\,\omega}, \qquad \text{and so} \quad u^\star(t) = \frac{\alpha}{2m\,\omega}\,t\,\sin\omega\,t.$$

Combining the resulting particular solution with the solution to the homogeneous equation leads to the general solution

$$u(t) = \frac{\alpha}{2m\,\omega}\,t\,\sin\omega\,t \;+\; r\,\cos(\omega\,t - \delta). \tag{9.100}$$

Both terms vibrate with frequency ω, but the amplitude of the first grows larger and larger as $t \to \infty$. As illustrated in Figure 9.12, the mass will oscillate more and more wildly. In this situation, the system is said to be in *resonance*, and the increasingly large oscillations are provoked by forcing it at its natural frequency ω. In a physical apparatus, once the amplitude of resonant vibrations stretches the spring beyond its elastic limits, the linear Hooke's law model is no longer applicable, and either the spring breaks or the system enters a nonlinear regime.

Furthermore, if we are very close to resonance, the oscillations induced by the particular solution (9.98) will have extremely large, although bounded, amplitude. The lesson is, never force a system at or close to its natural frequency (or frequencies) of vibration. A classic example was the 1831 collapse of a bridge when a British infantry regiment marched in unison across it, apparently inducing a resonant vibration of the structure. (Learning their lesson, soldiers nowadays no longer march in step across bridges.) An even more dramatic case is the 1940 Tacoma Narrows Bridge disaster, when the vibrations due to a strong wind caused the bridge to oscillate wildly and break apart! The collapse was caught on film, and is very impressive. The traditional explanation was the excitement of the bridge's resonant

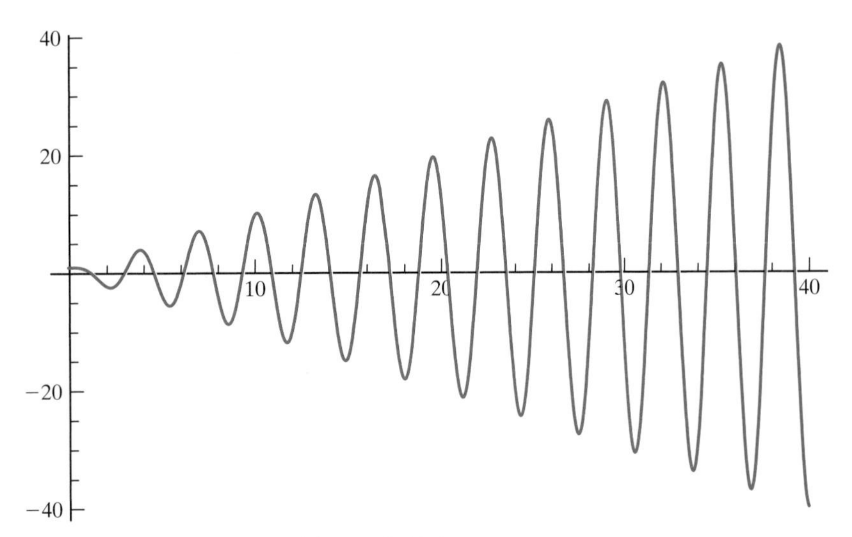

Figure 9.12 Resonance.

frequencies, although later studies revealed a more sophisticated mathematical explanation of the collapse, [19, p. 118]. But resonance is not exclusively harmful. In a microwave oven, the electromagnetic waves are tuned to the resonant frequencies of water molecules so as to excite them into large vibrations and thereby heat up your dinner. Blowing into a clarinet or other wind instrument excites the resonant frequencies in the column of air contained within it, and this produces the musical sound vibrations that we hear.

Frictional effects can partially mollify the extreme behavior near the resonant frequency. The frictionally damped vibrations of a mass on a spring, when subject to periodic forcing, are described by the inhomogeneous differential equation

$$m\,\frac{d^2u}{dt^2} + \beta\,\frac{du}{dt} + k\,u = \alpha\cos\eta\,t. \tag{9.101}$$

Let us assume that the friction is sufficiently small so as to be in the underdamped regime $\beta < 2\sqrt{m\,k}$. Since neither summand solves the homogeneous system, we can use the trigonometric solution ansatz (9.92) to construct the particular solution

$$u^\star(t) = \frac{\alpha}{\sqrt{m^2(\omega^2-\eta^2)^2+\beta^2\eta^2}}\,\cos(\eta\,t-\varepsilon) \quad \text{where} \quad \omega = \sqrt{\frac{k}{m}} \tag{9.102}$$

continues to denote the undamped resonant frequency (9.94), while ε, defined by

$$\tan\varepsilon = \frac{\beta\,\eta}{m(\omega^2-\eta^2)}, \tag{9.103}$$

represents a frictionally induced *phase lag* in the response of the system. Thus, the larger the friction β, the more pronounced the phase lag ε in the response of the system to the external forcing. Speeding up the forcing frequency η increases the overall phase shift, which has the value of $\frac{1}{2}\pi$ at the resonant frequency $\eta = \omega$, so the system lags a quarter period behind the forcing, and reaches a maximum $\varepsilon = \pi$ as $\eta \to \infty$. Thus, the response to a high frequency forcing is almost exactly out of phase—the mass is moving downwards when the force is pulling it upwards, and vice versa!

The general solution is

$$u(t) = \frac{\alpha}{\sqrt{m^2(\omega^2-\eta^2)^2+\beta^2\eta^2}}\,\cos(\eta\,t-\varepsilon) + r\,e^{-\mu t}\cos(\nu\,t-\delta), \tag{9.104}$$

where $\lambda = \mu \pm i\,\nu$ are the roots of the characteristic equation, while r, δ are determined by the initial conditions, cf. (9.84). The second term—the solution to the homogeneous equation—is known as the *transient* since it decays exponentially fast to zero. Thus, at large times, any internal motion of the system that might have been excited by the initial conditions dies out, and only the particular solution (9.102) incited by the forcing persists. The amplitude of the persistent response (9.102) is at a maximum at the resonant frequency $\eta = \omega$, where it takes the value $\alpha/(\beta\,\omega)$. Thus, the smaller the frictional coefficient β (or the slower the resonant frequency ω) the more likely the breakdown of the system due to an overly large response.

EXERCISES 9.6

9.6.1. Graph the following functions. Describe the fast oscillation and beat frequencies:
(a) $\cos 8t - \cos 9t$ (b) $\cos 26t - \cos 24t$
(c) $\cos 10t + \cos 9.5t$ (d) $\cos 5t - \sin 5.2t$

9.6.2. Solve the following initial value problems:
(a) $\ddot{u} + 36u = \cos 3t,\ u(0) = 0,\ \dot{u}(0) = 0$
(b) $\ddot{u} + 6\dot{u} + 9u = \cos t,\ u(0) = 0,\ \dot{u}(0) = 1$
(c) $\ddot{u} + \dot{u} + 4u = \cos 2t,\ u(0) = 1,\ \dot{u}(0) = -1$
(d) $\ddot{u} + 9u = 3\sin 3t,\ u(0) = 1,\ \dot{u}(0) = -1$
(e) $2\ddot{u} + 3\dot{u} + u = \cos \frac{1}{2}t,\ u(0) = 3,\ \dot{u}(0) = -2$
(f) $3\ddot{u} + 4\dot{u} + u = \cos t,\ u(0) = 0,\ \dot{u}(0) = 0$

9.6.3. Solve the following initial value problems. In each case, graph the solution and explain what type of motion is represented.
(a) $\ddot{u} + 25u = 3\cos 4t,\ u(0) = 1,\ \dot{u}(0) = 1$
(b) $\ddot{u} + 4\dot{u} + 40u = 125\cos 5t,\ u(0) = 0,\ \dot{u}(0) = 0$
(c) $\ddot{u} + 6\dot{u} + 5u = 25\sin 5t,\ u(0) = 4,\ \dot{u}(0) = 2$
(d) $\ddot{u} + 16u = \sin 4t,\ u(0) = 0,\ \dot{u}(0) = 0$

9.6.4. A mass $m = 25$ is attached to a unit spring with $k = 1$, and frictional coefficient $\beta = .01$. The spring will break when it moves more than 1 unit. Ignoring the effect of the transient, what is the maximum allowable amplitude α of periodic forcing at frequency $\eta =$
(a) .19 ? (b) .2 ? (c) .21 ?

9.6.5. For what range of frequencies η can you force the mass in Exercise 9.6.4 with amplitude $\alpha = .5$ without breaking the spring?

9.6.6. How large should the friction in Exercise 9.6.4 be so that you can safely force the mass with amplitude $\alpha = .5$ at any frequency?

9.6.7. Suppose the mass-spring-oil system of Exercise 9.5.38b is subject to a periodic external force $2\cos 2t$. Discuss, in as much detail as you can, the long term motion of the mass.

♠ **9.6.8.** (a) Does a function of the form $u(t) = a\cos\eta\, t - b\cos\omega\, t$ still exhibit beats when $\eta \approx \omega$, but $a \neq b$? Use a computer to graph some particular cases and discuss what you observe.
(b) Explain to what extent the conclusions based on (9.98) do not depend upon the choice of initial conditions (9.97).

◇ **9.6.9.** Write down the solution $u(t, \eta)$ to the initial value problem

$$m\,\frac{d^2u}{dt^2} + k\,u = \alpha\cos\eta\, t, \qquad u(0) = \dot{u}(0) = 0,$$

for
(a) a non-resonant forcing function at frequency $\eta \neq \omega$;
(b) a resonant forcing function at frequency $\eta = \omega$.
(c) Show that, as $\eta \to \omega$, the limit of the non-resonant solution equals the resonant solution. Conclude that the solution $u(t, \eta)$ depends continuously on the frequency η even though its mathematical formula changes significantly at resonance.

9.6.10. Justify the solution formulae (9.102) and (9.103).

Electrical Circuits

The Electrical–Mechanical Correspondence outlined in Section 6.2 will continue to operate in the dynamical universe. The equations governing the equilibria of simple electrical circuits and the mechanical systems such as mass/spring chains and structures all have the same underlying mathematical structure. In a similar manner, although they are based on a completely different set of physical principles, circuits with dynamical currents and voltages are modeled by second order linear dynamical systems of the Newtonian form presented earlier.

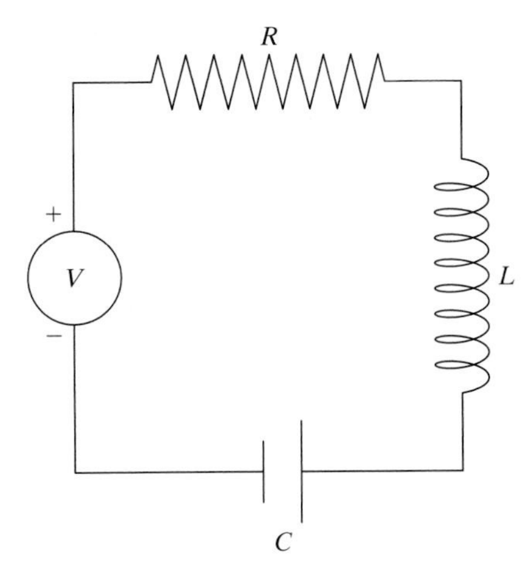

Figure 9.13 The basic $R\,L\,C$ circuit.

In this section, we analyze the very simplest situation: a single loop containing a *resistor* R, an *inductor* L and a *capacitor* C, as illustrated in Figure 9.13. This basic $R\,L\,C$ *circuit* serves as the prototype for more general electrical networks linking various resistors, inductors, capacitors, batteries, voltage sources, etc. (Extending the mathematical analysis to more complicated circuits would make an excellent in-depth student research project.) Let $u(t)$ denote the current in the circuit at time t. We use v_R, v_L, v_C to denote the induced voltages in the three circuit elements; these are prescribed by the fundamental laws of electrical circuitry.

(a) First, as we learned in Section 6.2, the resistance $R \geq 0$ is the proportionality factor between voltage and current, so $v_R = R\,u$.

(b) The voltage passing through an inductor is proportional to the rate of change in the current. Thus, $v_L = L\,\dot{u}$, where $L > 0$ is the *inductance*, and the dot indicates time derivative.

(c) On the other hand, the current passing through a capacitor is proportional to the rate of change in the voltage, and so $u = C\,\dot{v}_C$, where $C > 0$ denotes the *capacitance*. We integrate† this relation to produce the capacitor voltage

$$v_C = \int \frac{u(t)}{C}\,dt.$$

The *Voltage Balance Law* tells us that the total of these individual voltages must equal any externally applied voltage $v_E = F(t)$ coming from, say, a battery or generator. Therefore,

$$v_R + v_L + v_C = v_E.$$

Substituting the preceding formulae, we deduce that the current $u(t)$ in our circuit satisfies the following linear integro-differential equation

$$L\,\frac{du}{dt} + R\,u + \int \frac{u}{C}\,dt = F(t). \tag{9.105}$$

We can convert this into a differential equation by differentiating both sides with respect to t. Assuming, for simplicity, that L, R and C are constant, the result is the

† The integration constant is not important, since we end up differentiating the equation.

linear second order ordinary differential equation

$$L\frac{d^2u}{dt^2} + R\frac{du}{dt} + \frac{1}{C}u = f(t) = F'(t). \tag{9.106}$$

The current will be uniquely specified by the initial conditions $u(t_0) = a$, $\dot{u}(t_0) = b$.

Comparing (9.106) with the equation (9.82) for a mechanically vibrating mass, we see that the correspondence between electrical circuits and mechanical structures developed in Chapter 6 continues to hold in the dynamical regime. The current u corresponds to the displacement. The inductance L plays the role of mass, the resistance R corresponds, as before, to friction, while the reciprocal $1/C$ of capacitance is analogous to the spring stiffness. Thus, all of our analytical conclusions regarding stability of equilibria, qualitative behavior and formulae for solutions, etc., that we established in the mechanical context can, suitably re-interpreted, be immediately applied to electrical circuit theory.

In particular, an RLC circuit is *underdamped* if $R^2 < 4L/C$, and the current $u(t)$ oscillates with frequency

$$\nu = \sqrt{\frac{1}{CL} - \frac{R^2}{4L^2}}\,, \tag{9.107}$$

while slowly dying off to zero. In the overdamped and critically damped cases $R^2 \geq 4L/C$, the resistance in the circuit is so large that the current merely decays to zero at an exponential rate and no longer exhibits any oscillatory behavior. Attaching an alternating current source $F(t) = \alpha\cos\eta t$ to the circuit can induce a catastrophic resonance if there is no resistance and the forcing frequency is equal to the circuit's natural frequency.

EXERCISES 9.6

9.6.11. Classify the following RLC circuits as
(i) underdamped
(ii) critically damped, or
(iii) overdamped
(a) $R = 1, L = 2, C = 4$
(b) $R = 4, L = 3, C = 1$
(c) $R = 2, L = 3, C = 3$
(d) $R = 4, L = 10, C = 2$
(e) $R = 1, L = 1, C = 3$

9.6.12. Find the current in each of the unforced RLC circuits in Exercise 9.6.11 induced by the initial data $u(0) = 1$, $\dot{u}(0) = 0$.

9.6.13. A circuit with $R = 1, L = 2, C = 4$ includes an alternating current source $F(t) = 25\cos 2t$. Find the solution to the initial value problem $u(0) = 1$, $\dot{u}(0) = 0$.

9.6.14. A superconducting LC circuit has no resistance: $R = 0$. Discuss what happens when the circuit is wired to an alternating current source $F(t) = \alpha\cos\eta t$.

9.6.15. A circuit with $R = .002$, $L = 12.5$, and $C = 50$ can carry a maximum current of 250. Ignoring the effect of the transient, what is the maximum allowable amplitude α of an applied periodic current $F(t) = \alpha\cos\eta t$ at frequency $\eta =$
(a) .04 ? (b) .05 ? (c) .1 ?

9.6.16. Given the circuit in Exercise 9.6.15, what range of frequencies η can you supply a unit amplitude periodic current source?

9.6.17. How large should the resistance in the circuit in Exercise 9.6.15 be so that you can safely apply any unit amplitude periodic current?

Forcing and Resonance in Systems

Let us very briefly discuss the effect of periodic forcing on a system of second order ordinary differential equations. Periodically forcing an undamped mass-spring chain or structure, or a resistanceless electrical network, leads to a second order system of the form

$$M \frac{d^2\mathbf{u}}{dt^2} + K\,\mathbf{u} = \cos(\eta\, t)\, \mathbf{a}. \tag{9.108}$$

Here $\mathbf{a}$ is a constant vector representing both a magnitude and a "direction" of the forcing, while η is the frequency of forcing. As always, the solution to the inhomogeneous system is composed of one particular response to the external force combined with the general solution to the homogeneous system which, in the stable case $K > 0$, is a quasi-periodic combination of the normal vibrational modes.

To find a particular solution to the inhomogeneous system, let us try the trigonometric ansatz

$$\mathbf{u}^\star(t) = \cos(\eta\, t)\, \mathbf{w} \tag{9.109}$$

in which $\mathbf{w}$ is a constant vector. Substituting into (9.108) leads to a linear algebraic system

$$(K - \mu\, M)\, \mathbf{w} = \mathbf{a}, \qquad \text{where} \quad \mu = \eta^2. \tag{9.110}$$

If equation (9.110) has a solution, then our ansatz (9.109) is valid, and we have produced a particular vibration of the system that has the same frequency as the forcing vibration. In particular, if $\mu = \eta^2$ is *not* a generalized eigenvalue of the matrix pair K, M, as in (9.80), then the coefficient matrix $K - \mu\, M$ is nonsingular, and so (9.110) can be solved for any right hand side $\mathbf{a}$. The general solution, then, will be a quasi-periodic combination of this particular solution coupled with the normal mode vibrations at the system's natural, unforced frequencies.

The more interesting case is when $\eta^2 = \mu$ is a generalized eigenvalue, and so $K - \mu\, M$ is singular, its kernel being equal to the generalized eigenspace V_μ. In this case, (9.110) will have a solution $\mathbf{w}$ if and only if $\mathbf{a}$ lies in the range of $K - \mu\, M$. According to the Fredholm Alternative Theorem 5.55, the range is the orthogonal complement of the cokernel, which, since the coefficient matrix is symmetric, is the same as the kernel. Therefore, (9.110) will have a solution if and only if $\mathbf{a}$ is orthogonal to V_μ, i.e., $\mathbf{a} \cdot \mathbf{v} = \mathbf{0}$ for every eigenvector $\mathbf{v}$ for the eigenvalue μ. Thus, one can force a system at a natural frequency without inciting resonance, provided that the "direction" of forcing, as determined by the vector $\mathbf{a}$, is orthogonal to the natural directions of motion of the system, as governed by the eigenvectors for that particular frequency.

If the orthogonality condition is not satisfied, then the periodic solution ansatz (9.109) does not apply, and we are in a truly resonant situation. Inspired by the scalar solution, let us try a *resonant solution ansatz*

$$\mathbf{u}^\star(t) = t \sin(\eta t)\, \mathbf{y} + \cos(\eta\, t)\, \mathbf{w}. \tag{9.111}$$

Since

$$\frac{d^2\mathbf{u}^\star}{dt^2} = -\,\eta^2\, t \sin(\eta\, t)\, \mathbf{y} + \cos(\eta\, t)\, (2\,\eta\, \mathbf{y} - \eta^2\, \mathbf{w}),$$

the function (9.111) will solve the differential equation (9.108) provided

$$(K - \mu\, M)\mathbf{y} = \mathbf{0}, \qquad (K - \mu\, M)\mathbf{w} = \mathbf{a} - 2\,\eta\, \mathbf{y}, \quad \mu = \eta^2. \tag{9.112}$$

The first equation requires that $\mathbf{y} \in V_\mu$ be a generalized eigenvector of the matrix pair K, M. The Fredholm Alternative Theorem 5.55 implies that, since the coefficient matrix $K - \mu\, M$ is symmetric, the second equation will be solvable for $\mathbf{w}$ if and only if $\mathbf{a} - 2\,\eta\, \mathbf{y}$ is orthogonal to the generalized eigenspace

$$V_\mu = \operatorname{coker}(K - \mu\, M) = \ker(K - \mu\, M).$$

Thus, the vector $2\eta\mathbf{y}$ is required to be the orthogonal projection of $\mathbf{a}$ onto the eigenspace V_μ. With this choice of $\mathbf{y}$ and $\mathbf{w}$, formula (9.111) defines the resonant solution to the system.

Summarizing, we have shown that, generically, forcing a system at one of its natural frequencies induces resonance.

THEOREM 9.39 An undamped vibrational system will be periodically forced into resonance if and only if the forcing $\mathbf{f} = \cos(\eta\, t)\,\mathbf{a}$ is at a natural frequency of the system and the direction of forcing $\mathbf{a}$ is not orthogonal to the natural direction(s) of motion of the system at that frequency.

EXAMPLE 9.40 Consider the periodically forced system

$$\frac{d^2\mathbf{u}}{dt^2} + \begin{pmatrix} 3 & -2 \\ -2 & 3 \end{pmatrix}\mathbf{u} = \begin{pmatrix} \cos t \\ 0 \end{pmatrix}.$$

The eigenvalues of the coefficient matrix are $\lambda_1 = 5, \lambda_2 = 1$, with corresponding orthogonal eigenvectors

$$\mathbf{v}_1 = \begin{pmatrix} -1 \\ 1 \end{pmatrix}, \qquad \mathbf{v}_2 = \begin{pmatrix} 1 \\ 1 \end{pmatrix}.$$

The resonant frequencies are $\omega_1 = \sqrt{\lambda_1} = \sqrt{5}, \omega_2 = \sqrt{\lambda_2} = 1$, and hence we are forcing at a resonant frequency. To obtain the resonant solution (9.111), we first note that $\mathbf{a} = (1,0)^T$ has orthogonal projection $\mathbf{p} = \left(\frac{1}{2}, \frac{1}{2}\right)^T$ onto the eigenline spanned by $\mathbf{v}_2$, and hence $\mathbf{y} = \frac{1}{2}\mathbf{p} = \left(\frac{1}{4}, \frac{1}{4}\right)^T$. We can then solve

$$(K - \mathrm{I})\mathbf{w} = \begin{pmatrix} 2 & -2 \\ -2 & 2 \end{pmatrix}\mathbf{w} = \mathbf{a} - \mathbf{p} = \begin{pmatrix} \frac{1}{2} \\ -\frac{1}{2} \end{pmatrix} \quad \text{for}^{\ddagger} \quad \mathbf{w} = \begin{pmatrix} \frac{1}{4} \\ 0 \end{pmatrix}.$$

Therefore, the particular resonant solution is

$$\mathbf{u}^\star(t) = t \sin t\, \mathbf{y} + \cos t\, \mathbf{w} = \begin{pmatrix} \frac{1}{4} t \sin t + \frac{1}{4}\cos t \\ \frac{1}{4} t \sin t \end{pmatrix}.$$

The general solution to the system is

$$\mathbf{u}(t) = \begin{pmatrix} \frac{1}{4} t \sin t + \frac{1}{4}\cos t \\ \frac{1}{4} t \sin t \end{pmatrix} + r_1 \cos(\sqrt{5}\, t - \delta_1) \begin{pmatrix} -1 \\ 1 \end{pmatrix} + r_2 \cos(t - \delta_2) \begin{pmatrix} 1 \\ 1 \end{pmatrix},$$

where the amplitudes r_1, r_2 and phase shifts δ_1, δ_2, are fixed by the initial conditions. Eventually the resonant terms involving $t \sin t$ dominate the solution, inducing progressively larger and larger oscillations. ●

‡ We can safely ignore the arbitrary multiple of the eigenvector that can be added to $\mathbf{w}$ as we only need find one particular solution; these will reappear anyway once we assemble the general solution to the system.

EXERCISES 9.6

9.6.18. Find the general solution to the following forced second order systems:

(a) $\dfrac{d^2\mathbf{u}}{dt^2} + \begin{pmatrix} 7 & -2 \\ -2 & 4 \end{pmatrix} \mathbf{u} = \begin{pmatrix} \cos t \\ 0 \end{pmatrix}$

(b) $\dfrac{d^2\mathbf{u}}{dt^2} + \begin{pmatrix} 5 & -2 \\ -2 & 3 \end{pmatrix} \mathbf{u} = \begin{pmatrix} 0 \\ 5\sin 3t \end{pmatrix}$

(c) $\dfrac{d^2\mathbf{u}}{dt^2} + \begin{pmatrix} 13 & -6 \\ -6 & 8 \end{pmatrix} \mathbf{u} = \begin{pmatrix} 5\cos 2t \\ \cos 2t \end{pmatrix}$

(d) $\begin{pmatrix} 2 & 0 \\ 0 & 3 \end{pmatrix} \dfrac{d^2\mathbf{u}}{dt^2} + \begin{pmatrix} 3 & -1 \\ -1 & 2 \end{pmatrix} \mathbf{u} = \begin{pmatrix} \cos \frac{1}{2} t \\ -\cos \frac{1}{2} t \end{pmatrix}$

(e) $\begin{pmatrix} 3 & 0 \\ 0 & 5 \end{pmatrix} \dfrac{d^2\mathbf{u}}{dt^2} + \begin{pmatrix} 4 & -2 \\ -2 & 3 \end{pmatrix} \mathbf{u} = \begin{pmatrix} \cos t \\ 11\sin 2t \end{pmatrix}$

(f) $\dfrac{d^2\mathbf{u}}{dt^2} + \begin{pmatrix} 6 & -4 & 1 \\ -4 & 6 & -1 \\ 1 & -1 & 11 \end{pmatrix} \mathbf{u} = \begin{pmatrix} \cos t \\ 0 \\ \cos t \end{pmatrix}$

(g) $\begin{pmatrix} 2 & 0 & 0 \\ 0 & 3 & 0 \\ 0 & 0 & 6 \end{pmatrix} \dfrac{d^2\mathbf{u}}{dt^2} + \begin{pmatrix} 5 & -1 & -1 \\ -1 & 6 & 3 \\ -1 & 3 & 9 \end{pmatrix} \mathbf{u} = \begin{pmatrix} 0 \\ \cos t \\ \cos t \end{pmatrix}$

9.6.19. (a) Find the resonant frequencies of a mass-spring chain consisting of two masses, $m_1 = 1$ and $m_2 = 2$ connected to top and bottom supports by identical springs with unit stiffness.

(b) Write down an explicit forcing function that will excite the resonance.

9.6.20. Suppose one of the supports is removed from the mass-spring chain of Exercise 9.6.19. Does your forcing function still excite the resonance? Do the internal vibrations of the masses

(i) speed up

(ii) slow down, or

(iii) remain the same?

Does your answer depend upon which of the two supports is removed?

♣ **9.6.21.** Find the resonant frequencies of the following structures, assuming the nodes all have unit mass. Then find a means of forcing the structure at one of the resonant frequencies, and yet not exciting the resonance. Can you also force the structure without exciting any mechanism or rigid motion?

(a) the square truss of Exercise 6.3.5

(b) the joined square truss of Exercise 6.3.6

(c) the house of Exercise 6.3.8

(d) the triangular space station of Example 6.6

(e) the triatomic molecule of Example 9.37

(f) the water molecule of Exercise 9.5.30

CHAPTER 10

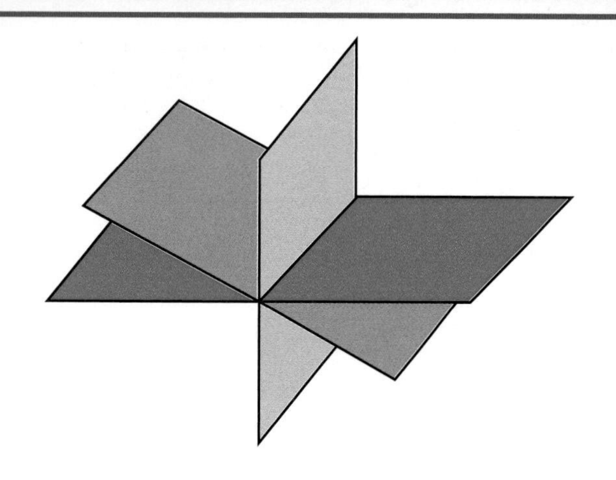

Iteration of Linear Systems

Iteration, meaning the repeated application of a process or function, appears in a surprisingly wide range of applications. Discrete dynamical systems, in which the time variable has been "quantized" into individual units (seconds, days, years, etc.) are modeled by iterative systems. Most numerical solution algorithms, for both linear and nonlinear systems, are based on an iterative procedure. Starting with an initial guess, the successive iterates lead to closer and closer approximations to the true solution. For linear systems of equations, there are several iterative solution schemes that can, in favorable situations, be employed as efficient alternatives to Gaussian Elimination. Iterative methods are particularly effective for solving the very large, sparse systems arising in the numerical solution of both ordinary and partial differential equations. All practical methods for computing eigenvalues and eigenvectors are based on iteration. Probabilistic iterative models known as Markov chains govern basic stochastic processes and appear in a wide variety of application ranging over genetics, population biology, scheduling, internet search engines, financial markets, and many more.

In this book, we will only treat iteration of linear systems. (Nonlinear iteration is of similar importance in applied mathematics and numerical analysis, and we refer the interested reader to [32, 36, 51, 59] for details.) Linear iteration coincides with multiplication by successive powers of a matrix; convergence of the iterates depends on the magnitude of its eigenvalues. We discuss in some detail a variety of convergence criteria based on the spectral radius, on matrix norms, and on eigenvalue estimates provided by the simple, but effective, Gerschgorin Circle Theorem.

We will then turn our attention to the three most important iterative schemes used to accurately approximate the solutions to linear algebraic systems. The classical Jacobi method is the simplest, while an evident serialization leads to the popular Gauss–Seidel method. Completely general convergence criteria are hard to formulate, although convergence is assured for the important class of diagonally dominant matrices that arise in many applications. A simple modification of the Gauss–Seidel scheme, known as Successive Over-Relaxation (SOR), can dramatically speed up the convergence rate, and is the method of choice in many modern applications. Finally, we introduce the method of conjugate gradients, a powerful "semi-direct" iterative scheme that, in contrast to the classical iterative schemes, is guaranteed to eventually produce the exact solution.

In the final section we discuss practical methods for computing eigenvalues and eigenvectors of matrices. Needless to say, we completely avoid trying to solve (or even write down) the characteristic polynomial equation. The very basic power method and its variants, which is based on linear iteration, is used to effectively approximate selected eigenvalues. To determine the complete system of eigenvalues and eigenvectors, the remarkable QR algorithm, which relies on the Gram–Schmidt orthogonalization procedure, is the method of choice, and we shall close with a new proof of its convergence.

10.1 LINEAR ITERATIVE SYSTEMS

We begin with the basic definition of an iterative system of linear equations.

Definition 10.1 A *linear iterative system* takes the form

$$\mathbf{u}^{(k+1)} = T\,\mathbf{u}^{(k)}, \qquad \mathbf{u}^{(0)} = \mathbf{a}. \tag{10.1}$$

The *coefficient matrix* T has size $n \times n$. We will consider both real and complex systems, and so the *iterates** $\mathbf{u}^{(k)}$ are vectors either in $\mathbb{R}^n$ (which assumes that the coefficient matrix T is also real) or in $\mathbb{C}^n$. A linear iterative system can be viewed as a discretized version of a first order system of linear ordinary differential equations, as in (8.9), in which the state of system, as represented by the vector $\mathbf{u}^{(k)}$, changes at discrete time intervals, labeled by the index k. For $k = 1, 2, 3, \ldots$, the solution $\mathbf{u}^{(k)}$ is uniquely determined by the *initial conditions* $\mathbf{u}^{(0)} = \mathbf{a}$.

Scalar Systems

As usual, to study systems one begins with an in-depth analysis of the scalar version. Consider the iterative equation

$$u^{(k+1)} = \lambda\,u^{(k)}, \qquad u^{(0)} = a. \tag{10.2}$$

The general solution to (10.2) is easily found:

$$u^{(1)} = \lambda\,u^{(0)} = \lambda\,a, \qquad u^{(2)} = \lambda\,u^{(1)} = \lambda^2\,a, \qquad u^{(3)} = \lambda\,u^{(2)} = \lambda^3\,a,$$

and, in general,

$$u^{(k)} = \lambda^k\,a. \tag{10.3}$$

If the initial condition is $a = 0$, then the solution $u^{(k)} \equiv 0$ is constant. Therefore, 0 is a *fixed point* or *equilibrium solution* for the iterative system.

EXAMPLE 10.2 Banks add interest to a savings account at discrete time intervals. For example, if the bank offers 5% interest compounded yearly, this means that the account balance will increase by 5% each year. Thus, assuming no deposits or withdrawals, the balance $u^{(k)}$ after k years will satisfy the iterative equation (10.2) with $\lambda = 1 + r$ where $r = .05$ is the interest rate, and the 1 indicates that all the money remains in the account. Thus, after k years, your account balance is

$$u^{(k)} = (1+r)^k a, \qquad \text{where} \quad a = u^{(0)} \tag{10.4}$$

* *Warning*: The superscripts on $\mathbf{u}^{(k)}$ refer to the iterate number, and should not be mistaken for derivatives.

is your initial deposit. For example, if $a = \$1{,}000$, after 1 year your account has $u^{(1)} = \$1{,}050$, after 10 years $u^{(10)} = \$1{,}628.89$, after 50 years $u^{(50)} = \$11{,}467.40$, and after 200 years $u^{(200)} = \$17{,}292{,}580.82$.

When the interest is compounded monthly, the rate is still quoted on a yearly basis, and so you receive $\frac{1}{12}$ of the interest each month. If $\hat{u}^{(k)}$ denotes the balance after k months, then, after n years, the account balance is

$$\hat{u}^{(12n)} = \left(1 + \tfrac{1}{12} r\right)^{12n} a.$$

Thus, when the interest rate of 5% is compounded monthly, your account balance is $\hat{u}^{(12)} = \$1{,}051.16$ after 1 year, $\hat{u}^{(120)} = \$1{,}647.01$ after 10 years, $\hat{u}^{(600)} = \$12{,}119.38$ after 50 years, and $\hat{u}^{(2400)} = \$21{,}573{,}572.66$ dollars after 200 years. So, if you wait sufficiently long, compounding will have a dramatic effect. Similarly, daily compounding replaces 12 by 365.25, the number of days in a year. After 200 years, the balance is $\$22{,}011{,}396.03$. ●

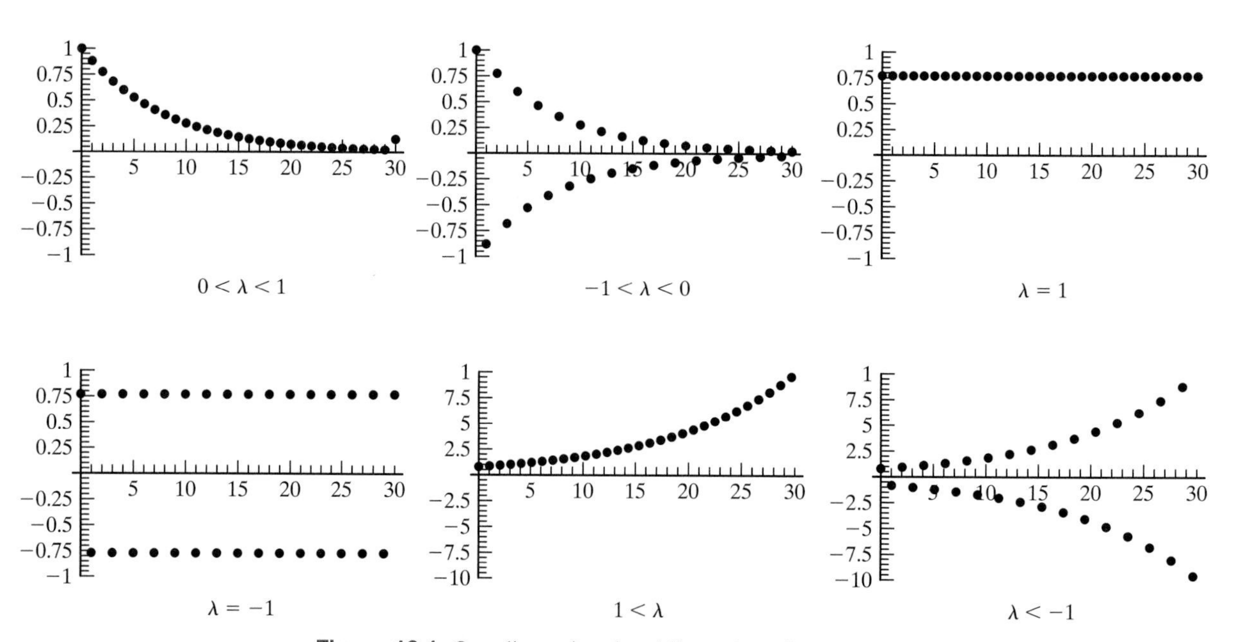

Figure 10.1 One dimensional real linear iterative systems.

Let us analyze the solutions of scalar iterative equations, starting with the case when $\lambda \in \mathbb{R}$ is a real constant. Aside from the equilibrium solution $u^{(k)} \equiv 0$, the iterates exhibit five qualitatively different behaviors, depending on the size of the coefficient λ.

(a) If $\lambda = 0$, the solution immediately becomes zero, and stays there, so $u^{(k)} = 0$ for all $k \geq 1$.

(b) If $0 < \lambda < 1$, then the solution is of one sign, and tends monotonically to zero, so $u^{(k)} \to 0$ as $k \to \infty$.

(c) If $-1 < \lambda < 0$, then the solution tends to zero: $u^{(k)} \to 0$ as $k \to \infty$. Successive iterates have alternating signs.

(d) If $\lambda = 1$, the solution is constant: $u^{(k)} = a$, for all $k \geq 0$.

(e) If $\lambda = -1$, the solution switches back and forth between two values; $u^{(k)} = (-1)^k a$.

(f) If $1 < \lambda < \infty$, then the iterates $u^{(k)}$ become unbounded. If $a > 0$, they go monotonically to $+\infty$; if $a < 0$, to $-\infty$.

(g) If $-\infty < \lambda < -1$, then the iterates $u^{(k)}$ also become unbounded, with alternating signs.

In Figure 10.1 we exhibit representative *scatter plots* for the nontrivial cases (b – g). The horizontal axis indicates the index k and the vertical axis the solution value u. Each dot in the scatter plot represents an iterate $u^{(k)}$.

To describe the different scenarios, we adopt a terminology that already appeared in the continuous realm. In the first three cases, the fixed point $u = 0$ is said to be *globally asymptotically stable* since all solutions tend to 0 as $k \to \infty$. In cases (d) and (e), the zero solution is *stable*, since solutions with nearby initial data, $|a| \ll 1$, remain nearby. In the final two cases, the zero solution is *unstable*; any nonzero initial data $a \neq 0$—no matter how small—will give rise to a solution that eventually goes arbitrarily far away from equilibrium.

Let us also treat the case of a complex scalar iterative system. The coefficient λ and the initial datum a in (10.2) are allowed to be complex numbers. The solution is the same, (10.3), but now we need to know what happens when we raise a complex number λ to a high power. The secret is to write $\lambda = r\, e^{\,i\theta}$ in polar form (3.85), where $r = |\lambda|$ is its modulus and $\theta = \operatorname{ph} \lambda$ its angle or phase. Then $\lambda^k = r^k\, e^{\,i k\theta}$. Since $|e^{\,ik\theta}| = 1$, we have $|\lambda^k| = |\lambda|^k$, and so the solutions (10.3) have modulus $|u^{(k)}| = |\lambda^k a| = |\lambda|^k\, |a|$. As a result, $u^{(k)}$ will remain bounded if and only if $|\lambda| \leq 1$, and will tend to zero as $k \to \infty$ if and only if $|\lambda| < 1$.

We have thus established the basic stability criteria for scalar, linear systems.

THEOREM 10.3 The zero solution to a (real or complex) scalar iterative system $u^{(k+1)} = \lambda\, u^{(k)}$ is

(a) *asymptotically stable* if and only if $|\lambda| < 1$,

(b) *stable* if and only if $|\lambda| \leq 1$,

(c) *unstable* if and only if $|\lambda| > 1$.

EXERCISES 10.1

10.1.1. Suppose $u^{(0)} = 1$. Find $u^{(1)}$, $u^{(10)}$, and $u^{(20)}$ when
(a) $u^{(k+1)} = 2\,u^{(k)}$ (b) $u^{(k+1)} = -.9\,u^{(k)}$
(c) $u^{(k+1)} = i\,u^{(k)}$ (d) $u^{(k+1)} = (1 - 2\,i)\,u^{(k)}$
Is the system
(i) stable? (ii) asymptotically stable?
(iii) unstable?

10.1.2. A bank offers 3.25% interest compounded yearly. Suppose you deposit \$100.
(a) Set up a linear iterative equation to represent your bank balance.
(b) How much money do you have after 10 years?
(c) What if the interest is compounded monthly?

10.1.3. Show that the yearly balances of an account whose interest is compounded monthly satisfy a linear iterative system. How is the effective yearly interest rate determined from the original annual interest rate?

10.1.4. Show that, as the time interval of compounding goes to zero, the bank balance after k years approaches an exponential function $e^{r k}\, a$, where r is the yearly interest rate and a the initial balance.

10.1.5. Let $u(t)$ denote the solution to the linear ordinary differential equation $\dot u = \alpha\, u$, $u(0) = a$. Let $h > 0$ be fixed. Show that the sample values $u^{(k)} = u(k\,h)$ satisfy a linear iterative system. What is the coefficient λ? Compare the stability properties of the differential equation and the corresponding iterative system.

10.1.6. For which values of λ does the scalar iterative system (10.2) have a periodic solution, meaning that $u^{(k+m)} = u^{(k)}$ for some m?

♠ **10.1.7.** Investigate the solutions of the linear iterative equation $u^{(k+1)} = \lambda\, u^{(k)}$ when λ is a complex number with $|\lambda| = 1$, and look for patterns.

10.1.8. Consider the iterative systems $u^{(k+1)} = \lambda\, u^{(k)}$ and $v^{(k+1)} = \mu\, v^{(k)}$, where $|\lambda| > |\mu|$. Prove that, for any nonzero initial data $u^{(0)} = a \neq 0$, $v^{(0)} = b \neq 0$, the solution to the first is eventually larger (in modulus) than that of the second: $|u^{(k)}| > |v^{(k)}|$, for $k \gg 0$.

10.1.9. Let λ, c be fixed. Solve the *affine* (or inhomogeneous linear) iterative equation

$$u^{(k+1)} = \lambda\, u^{(k)} + c, \qquad u^{(0)} = a. \tag{10.5}$$

Discuss the possible behaviors of the solutions. *Hint*: Write the solution in the form $u^{(k)} = u^{\star} + v^{(k)}$, where $u^{\star}$ is the equilibrium solution.

10.1.10. A bank offers 5% interest compounded yearly. Suppose you deposit \$120 in the account each year. Set up an affine iterative equation (10.5) to represent your bank balance. How much money do you have after 10 years? After you retire in 50 years? After 200 years?

10.1.11. Redo Exercise 10.1.10 in the case when the interest is compounded monthly and you deposit \$10 each month.

♡**10.1.12.** Each spring the deer in Minnesota produce offspring at a rate of roughly 1.2 times the total population, while approximately 5% of the population dies as a result of predators and natural causes. In the fall hunters are allowed to shoot 3,600 deer. This winter the Department of Natural Resources (DNR) estimates that there are 20,000 deer. Set up an affine iterative equation (10.5) to represent the deer population each subsequent year. Solve the system and find the population in the next 5 years. How many deer in the long term will there be? Using this information, formulate a reasonable policy of how many deer hunting licenses the DNR should allow each fall, assuming one kill per license.

Powers of Matrices

The solution to the general linear iterative system

$$\mathbf{u}^{(k+1)} = T\,\mathbf{u}^{(k)}, \qquad \mathbf{u}^{(0)} = \mathbf{a}, \tag{10.6}$$

is also, at least at first glance, immediate. Clearly,

$$\mathbf{u}^{(1)} = T\,\mathbf{u}^{(0)} = T\,\mathbf{a}, \quad \mathbf{u}^{(2)} = T\,\mathbf{u}^{(1)} = T^2\mathbf{a}, \quad \mathbf{u}^{(3)} = T\,\mathbf{u}^{(2)} = T^3\mathbf{a},$$

and, in general,

$$\mathbf{u}^{(k)} = T^k\mathbf{a}. \tag{10.7}$$

Thus, the iterates are simply determined by multiplying the initial vector $\mathbf{a}$ by the successive powers of the coefficient matrix T. And so, unlike differential equations, proving the existence and uniqueness of solutions to an iterative system is completely trivial.

However, unlike real or complex scalars, the general formulae and qualitative behavior of the powers of a square matrix are not nearly so immediately apparent. (Before continuing, the reader is urged to experiment with simple 2×2 matrices, trying to detect patterns.) To make progress, recall how we managed to solve linear systems of differential equations by suitably adapting the known exponential solution from the scalar version. In the iterative case, the scalar solution formula (10.3) is written in terms of powers, not exponentials. This motivates us to try the power ansatz

$$\mathbf{u}^{(k)} = \lambda^k\,\mathbf{v}, \tag{10.8}$$

in which λ is a scalar and $\mathbf{v}$ is a fixed vector, as a possible solution to the system. We find

$$\mathbf{u}^{(k+1)} = \lambda^{k+1}\,\mathbf{v}, \quad \text{while} \quad T\,\mathbf{u}^{(k)} = T(\lambda^k\,\mathbf{v}) = \lambda^k\,T\,\mathbf{v}.$$

These two expressions will be equal if and only if

$$T\,\mathbf{v} = \lambda\,\mathbf{v}.$$

Therefore, (10.8) is a nontrivial solution to (10.6) if and only if λ is an *eigenvalue* of the coefficient matrix T and $\mathbf{v} \neq \mathbf{0}$ an associated *eigenvector*.

Thus, to each eigenvector and eigenvalue of the coefficient matrix, we can construct a solution to the iterative system. We can then appeal to linear superposition to combine the basic power solutions to form more general solutions. In particular, if the coefficient matrix is complete, then this method will, as in the case of linear ordinary differential equations, produce the general solution.

THEOREM 10.4 If the coefficient matrix T is complete, then the general solution to the linear iterative system $\mathbf{u}^{(k+1)} = T\,\mathbf{u}^{(k)}$ is given by

$$\mathbf{u}^{(k)} = c_1\,\lambda_1^k\,\mathbf{v}_1 + c_2\,\lambda_2^k\,\mathbf{v}_2 + \cdots + c_n\,\lambda_n^k\,\mathbf{v}_n, \tag{10.9}$$

where $\mathbf{v}_1, \ldots, \mathbf{v}_n$ are the linearly independent eigenvectors and $\lambda_1, \ldots, \lambda_n$ the corresponding eigenvalues of T. The coefficients $c_1, \ldots, c_n$ are arbitrary scalars and are uniquely prescribed by the initial conditions $\mathbf{u}^{(0)} = \mathbf{a}$.

Proof Since we already know that (10.9) is a solution to the system for arbitrary $c_1, \ldots, c_n$, it suffices to show that we can match any prescribed initial conditions. To this end, we need to solve the linear system

$$\mathbf{u}^{(0)} = c_1\,\mathbf{v}_1 + \cdots + c_n\,\mathbf{v}_n = \mathbf{a}. \tag{10.10}$$

Completeness of T implies that its eigenvectors form a basis of $\mathbb{C}^n$, and hence (10.10) always admits a solution. In matrix form, we can rewrite (10.10) as

$$S\,\mathbf{c} = \mathbf{a}, \qquad \text{so that} \qquad \mathbf{c} = S^{-1}\mathbf{a},$$

where $S = (\,\mathbf{v}_1\ \mathbf{v}_2\ \ldots\ \mathbf{v}_n\,)$ is the (nonsingular) matrix whose columns are the eigenvectors. ■

REMARK: Solutions in the incomplete cases rely on the Jordan basis of Section 8.6. As with systems of differential equations, the formulas are more complicated; see Exercise 10.1.36 for details.

EXAMPLE 10.5 Consider the iterative system

$$x^{(k+1)} = \tfrac{3}{5}\,x^{(k)} + \tfrac{1}{5}\,y^{(k)}, \qquad y^{(k+1)} = \tfrac{1}{5}\,x^{(k)} + \tfrac{3}{5}\,y^{(k)}, \tag{10.11}$$

with initial conditions

$$x^{(0)} = a, \qquad y^{(0)} = b. \tag{10.12}$$

The system can be rewritten in our matrix form (10.6), with

$$T = \begin{pmatrix} .6 & .2 \\ .2 & .6 \end{pmatrix}, \qquad \mathbf{u}^{(k)} = \begin{pmatrix} x^{(k)} \\ y^{(k)} \end{pmatrix}, \qquad \mathbf{a} = \begin{pmatrix} a \\ b \end{pmatrix}.$$

Solving the characteristic equation

$$\det(T - \lambda\,\mathrm{I}) = \lambda^2 - 1.2\lambda - .32 = 0$$

produces the eigenvalues $\lambda_1 = .8$, $\lambda_2 = .4$. We then solve the associated linear systems $(T - \lambda_j\,\mathrm{I})\mathbf{v}_j = \mathbf{0}$ for the corresponding eigenvectors:

$$\lambda_1 = .8, \quad \mathbf{v}_1 = \begin{pmatrix} 1 \\ 1 \end{pmatrix}, \qquad \lambda_2 = .4, \quad \mathbf{v}_2 = \begin{pmatrix} -1 \\ 1 \end{pmatrix}.$$

Therefore, the basic power solutions are

$$\mathbf{u}_1^{(k)} = (.8)^k \begin{pmatrix} 1 \\ 1 \end{pmatrix}, \qquad \mathbf{u}_2^{(k)} = (.4)^k \begin{pmatrix} -1 \\ 1 \end{pmatrix}.$$

Theorem 10.4 tells us that the general solution is given as a linear combination,

$$\begin{aligned} \mathbf{u}^{(k)} = c_1\,\mathbf{u}_1^{(k)} + c_2\,\mathbf{u}_2^{(k)} &= c_1\,(.8)^k \begin{pmatrix} 1 \\ 1 \end{pmatrix} + c_2\,(.4)^k \begin{pmatrix} -1 \\ 1 \end{pmatrix} \\ &= \begin{pmatrix} c_1\,(.8)^k - c_2\,(.4)^k \\ c_1\,(.8)^k + c_2\,(.4)^k \end{pmatrix}, \end{aligned}$$

where c_1, c_2 are determined by the initial conditions:

$$\mathbf{u}^{(0)} = \begin{pmatrix} c_1 - c_2 \\ c_1 + c_2 \end{pmatrix} = \begin{pmatrix} a \\ b \end{pmatrix}, \qquad \text{and hence} \quad c_1 = \frac{a+b}{2}, \qquad \mathbf{c}_2 = \frac{b-a}{2}.$$

Therefore, the explicit formula for the solution to the initial value problem (10.11–12) is

$$x^{(k)} = (.8)^k \,\frac{a+b}{2} + (.4)^k \,\frac{a-b}{2}, \qquad y^{(k)} = (.8)^k \,\frac{a+b}{2} + (.4)^k \,\frac{b-a}{2}.$$

In particular, as $k \to \infty$, the iterates $\mathbf{u}^{(k)} \to \mathbf{0}$ converge to zero at a rate governed by the dominant eigenvalue $\lambda_1 = .8$. Thus, (10.11) defines a stable iterative system. Figure 10.2 illustrates the cumulative effect of the iteration. The initial conditions consist of a large number of points on the unit circle $x^2 + y^2 = 1$, which are successively mapped to points on progressively smaller and flatter ellipses, all converging towards the origin. ●

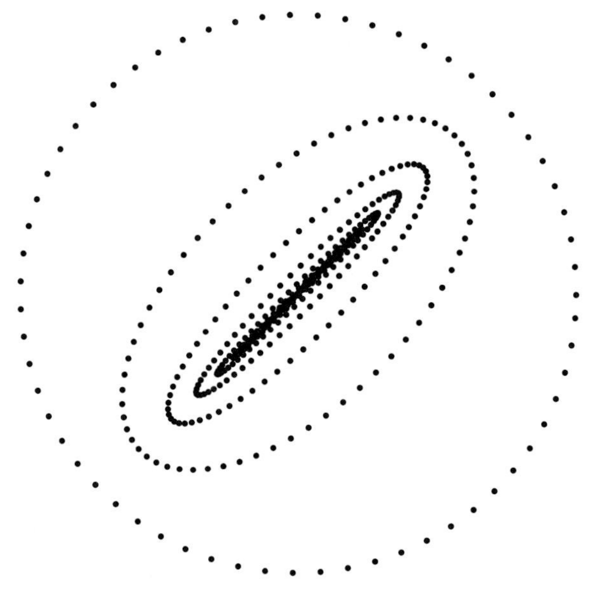

Figure 10.2 Stable iterative system.

EXAMPLE 10.6 The *Fibonacci numbers* are defined by the second order[†] iterative scheme

$$u^{(k+2)} = u^{(k+1)} + u^{(k)}, \tag{10.13}$$

with initial conditions

$$u^{(0)} = a, \qquad u^{(1)} = b. \tag{10.14}$$

In short, to obtain the next Fibonacci number, add the previous two. The classical *Fibonacci integers* start with $a = 0$, $b = 1$; the next few are

$$\begin{aligned} &u^{(0)} = 0, \quad u^{(1)} = 1, \quad u^{(2)} = 1, \quad u^{(3)} = 2, \\ &u^{(4)} = 3, \quad u^{(5)} = 5, \quad u^{(6)} = 8, \quad u^{(7)} = 13, \quad \ldots . \end{aligned}$$

[†] In general, an iterative system $\mathbf{u}^{(k+j)} = T_1 \mathbf{u}^{(k+j-1)} + \cdots + T_j \mathbf{u}^{(k)}$ in which the new iterate depends upon the preceding j values is said to have *order* j.

The Fibonacci integers occur in a surprising variety of natural objects, including leaves, flowers, and fruit, [63]. They were originally introduced by the Italian Renaissance mathematician Fibonacci (Leonardo of Pisa) as a crude model of the growth of a population of rabbits. In Fibonacci's model, the kth Fibonacci number $u^{(k)}$ measures the total number of pairs of rabbits at year k. We start the process with a single juvenile pair[‡] at year 0. Once a year, each pair of rabbits produces a new pair of offspring, but it takes a full year for a rabbit pair to mature enough to produce offspring of their own.

Just as every higher order ordinary differential equation can be replaced by an equivalent first order system, so every higher order iterative equation can be replaced by a first order iterative system. In this particular case, we define the vector

$$\mathbf{u}^{(k)} = \begin{pmatrix} u^{(k)} \\ u^{(k+1)} \end{pmatrix} \in \mathbb{R}^2,$$

and note that (10.13) is equivalent to the matrix system

$$\begin{pmatrix} u^{(k+1)} \\ u^{(k+2)} \end{pmatrix} = \begin{pmatrix} 0 & 1 \\ 1 & 1 \end{pmatrix} \begin{pmatrix} u^{(k)} \\ u^{(k+1)} \end{pmatrix}, \qquad \text{or} \qquad \mathbf{u}^{(k+1)} = T\,\mathbf{u}^{(k)},$$

where

$$T = \begin{pmatrix} 0 & 1 \\ 1 & 1 \end{pmatrix}.$$

To find the explicit formula for the Fibonacci numbers, we must determine the eigenvalues and eigenvectors of the coefficient matrix T. A straightforward computation produces

$$\lambda_1 = \frac{1+\sqrt{5}}{2} = 1.618034\ldots, \qquad \lambda_2 = \frac{1-\sqrt{5}}{2} = -.618034\ldots,$$

$$\mathbf{v}_1 = \begin{pmatrix} \frac{-1+\sqrt{5}}{2} \\ 1 \end{pmatrix}, \qquad \mathbf{v}_2 = \begin{pmatrix} \frac{-1-\sqrt{5}}{2} \\ 1 \end{pmatrix}.$$

Therefore, according to (10.9), the general solution to the Fibonacci system is

$$\begin{aligned} \mathbf{u}^{(k)} &= \begin{pmatrix} u^{(k+1)} \\ u^{(k)} \end{pmatrix} \\ &= c_1 \left(\frac{1+\sqrt{5}}{2}\right)^k \begin{pmatrix} \frac{-1+\sqrt{5}}{2} \\ 1 \end{pmatrix} + c_2 \left(\frac{1-\sqrt{5}}{2}\right)^k \begin{pmatrix} \frac{-1-\sqrt{5}}{2} \\ 1 \end{pmatrix}. \end{aligned} \tag{10.15}$$

The initial data

$$\mathbf{u}^{(0)} = c_1 \begin{pmatrix} \frac{-1+\sqrt{5}}{2} \\ 1 \end{pmatrix} + c_2 \begin{pmatrix} \frac{-1-\sqrt{5}}{2} \\ 1 \end{pmatrix} = \begin{pmatrix} a \\ b \end{pmatrix}$$

uniquely specifies the coefficients

$$c_1 = \frac{2a + (1+\sqrt{5})\,b}{2\sqrt{5}}, \qquad c_2 = -\,\frac{2a + (1-\sqrt{5})\,b}{2\sqrt{5}}.$$

[‡] We ignore important details like the sex of the offspring.

The first entry of the solution vector (10.15) produces the explicit formula

$$u^{(k)} = \frac{(-1+\sqrt{5})\,a + 2b}{2\sqrt{5}} \left(\frac{1+\sqrt{5}}{2}\right)^k + \frac{(1+\sqrt{5})\,a - 2b}{2\sqrt{5}} \left(\frac{1-\sqrt{5}}{2}\right)^k \tag{10.16}$$

for the kth Fibonacci number. For the particular initial conditions $a = 0$, $b = 1$, (10.16) reduces to the classical *Binet formula*

$$u^{(k)} = \frac{1}{\sqrt{5}} \left[\left(\frac{1+\sqrt{5}}{2}\right)^k - \left(\frac{1-\sqrt{5}}{2}\right)^k \right] \tag{10.17}$$

for the kth Fibonacci integer. It is a remarkable fact that, for every value of k, all the $\sqrt{5}$'s cancel out, and the Binet formula does indeed produce the Fibonacci integers listed above. Another useful observation is that, since

$$0 < |\lambda_2| = \frac{\sqrt{5}-1}{2} < 1 < \lambda_1 = \frac{1+\sqrt{5}}{2},$$

the terms involving λ_1^k go to ∞ (and so the zero solution to this iterative system is unstable) while the terms involving λ_2^k go to zero. Therefore, even for k moderately large, the first term in (10.16) is an excellent approximation (and one that gets more and more accurate with increasing k) to the kth Fibonacci number. A plot of the first 4 iterates, starting with the initial data consisting of equally spaced points on the unit circle, can be seen in Figure 10.3. As in the previous example, the circle is mapped to a sequence of progressively more eccentric ellipses; however, their major semi-axes become more and more stretched out, and almost all points end up going off to ∞.

The dominant eigenvalue $\lambda_1 = \frac{1}{2}(1+\sqrt{5}\,) = 1.618034\ldots$ is known as the *golden ratio* and plays an important role in spiral growth in nature, as well as in art, architecture and design, [63]. It describes the overall growth rate of the Fibonacci

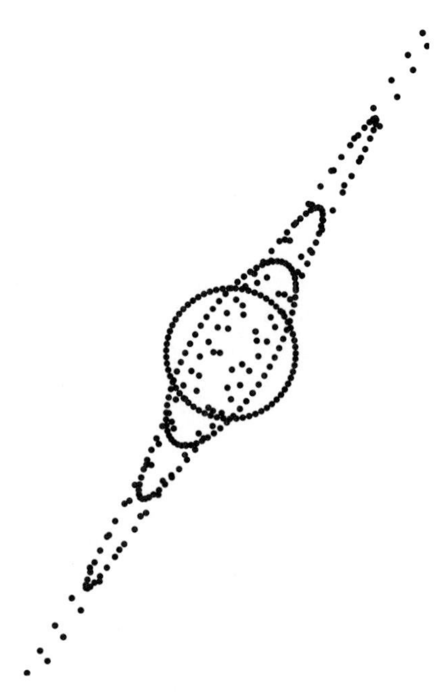

Figure 10.3 Fibonacci iteration.

integers, and, in fact, any sequence of Fibonacci numbers with initial conditions $b \neq \frac{1}{2}(1-\sqrt{5}\,)\,a$. ●

EXAMPLE 10.7 Let

$$T = \begin{pmatrix} -3 & 1 & 6 \\ 1 & -1 & -2 \\ -1 & -1 & 0 \end{pmatrix}$$

be the coefficient matrix for a three-dimensional iterative system $\mathbf{u}^{(k+1)} = T\,\mathbf{u}^{(k)}$. Its eigenvalues and corresponding eigenvectors are

$$\lambda_1 = -2, \qquad \lambda_2 = -1 + \mathrm{i}, \quad \lambda_3 = -1 - \mathrm{i},$$
$$\mathbf{v}_1 = \begin{pmatrix} 4 \\ -2 \\ 1 \end{pmatrix}, \quad \mathbf{v}_2 = \begin{pmatrix} 2-\mathrm{i} \\ -1 \\ 1 \end{pmatrix}, \quad \mathbf{v}_3 = \begin{pmatrix} 2+\mathrm{i} \\ -1 \\ 1 \end{pmatrix}.$$

Therefore, according to (10.9), the general complex solution to the iterative system is

$$\mathbf{u}^{(k)} = b_1\,(-2)^k \begin{pmatrix} 4 \\ -2 \\ 1 \end{pmatrix} + b_2\,(-1+\mathrm{i})^k \begin{pmatrix} 2-\mathrm{i} \\ -1 \\ 1 \end{pmatrix} + b_3\,(-1-\mathrm{i})^k \begin{pmatrix} 2+\mathrm{i} \\ -1 \\ 1 \end{pmatrix},$$

where b_1, b_2, b_3 are arbitrary complex scalars.

If we are only interested in real solutions, we can, as in the case of systems of differential equations, break up any complex solution into its real and imaginary parts, each of which constitutes a real solution. (This is another manifestation of the general Reality Principle of Theorem 7.48, but is not hard to prove directly.) We begin by writing $\lambda_2 = -1 + \mathrm{i} = \sqrt{2}\, e^{3\pi\,\mathrm{i}/4}$, and hence

$$(-1+\mathrm{i})^k = 2^{k/2}\, e^{3k\pi\,\mathrm{i}/4} = 2^{k/2}\left(\cos\tfrac{3}{4}k\pi + \mathrm{i}\,\sin\tfrac{3}{4}k\pi\right).$$

Therefore, the complex solution

$$(-1+\mathrm{i})^k \begin{pmatrix} 2-\mathrm{i} \\ -1 \\ 1 \end{pmatrix} = 2^{k/2} \begin{pmatrix} 2\cos\frac{3}{4}k\pi + \sin\frac{3}{4}k\pi \\ -\cos\frac{3}{4}k\pi \\ \cos\frac{3}{4}k\pi \end{pmatrix} + \mathrm{i}\;2^{k/2} \begin{pmatrix} 2\sin\frac{3}{4}k\pi - \cos\frac{3}{4}k\pi \\ -\sin\frac{3}{4}k\pi \\ \sin\frac{3}{4}k\pi \end{pmatrix}$$

is a combination of two independent real solutions. The complex conjugate eigenvalue $\lambda_3 = -1 - \mathrm{i}$ leads, as before, to the complex conjugate solution—and the same two real solutions. The general real solution $\mathbf{u}^{(k)}$ to the system can be written as a linear combination of the three independent real solutions:

$$c_1\,(-2)^k \begin{pmatrix} 4 \\ -2 \\ 1 \end{pmatrix} + c_2\,2^{k/2} \begin{pmatrix} 2\cos\frac{3}{4}k\pi + \sin\frac{3}{4}k\pi \\ -\cos\frac{3}{4}k\pi \\ \cos\frac{3}{4}k\pi \end{pmatrix} + c_3\,2^{k/2} \begin{pmatrix} 2\sin\frac{3}{4}k\pi - \cos\frac{3}{4}k\pi \\ -\sin\frac{3}{4}k\pi \\ \sin\frac{3}{4}k\pi \end{pmatrix}, \tag{10.18}$$

where c_1, c_2, c_3 are arbitrary real scalars, uniquely prescribed by the initial conditions. ●

Diagonalization and Iteration

An alternative, equally efficient approach to solving iterative systems is based on diagonalization of the coefficient matrix, cf. (8.27). Specifically, assuming the coefficient matrix T is complete, we can factor it as a product

$$T = S\,\Lambda\,S^{-1}, \tag{10.19}$$

in which $\Lambda = \operatorname{diag}(\lambda_1, \lambda_2, \ldots, \lambda_n)$ is the diagonal matrix containing the eigenvalues of T, while the columns of $S = (\mathbf{v}_1, \ldots, \mathbf{v}_n)$ are the corresponding eigenvectors. Consequently, the powers of T are given by

$$T^2 = (S\,\Lambda\,S^{-1})\,(S\,\Lambda\,S^{-1}) = S\,\Lambda^2\,S^{-1},$$
$$T^3 = (S\,\Lambda\,S^{-1})\,(S\,\Lambda\,S^{-1})\,(S\,\Lambda\,S^{-1}) = S\,\Lambda^3\,S^{-1},$$

and, in general

$$T^k = S\,\Lambda^k\,S^{-1}. \tag{10.20}$$

Moreover, since Λ is a diagonal matrix, its powers are trivial to compute:

$$\Lambda^k = \operatorname{diag}(\lambda_1^k, \lambda_2^k, \ldots, \lambda_n^k). \tag{10.21}$$

Thus, by combining (10.20–21), we obtain an explicit formula for the powers of a complete matrix T. Furthermore, the solution to the associated linear iterative system

$$\mathbf{u}^{(k+1)} = T\,\mathbf{u}^{(k)}, \quad \mathbf{u}^{(0)} = \mathbf{a}, \quad \text{is given by} \quad \mathbf{u}^{(k)} = T^k\mathbf{a} = S\,\Lambda^k\,S^{-1}\mathbf{a}. \tag{10.22}$$

You should convince yourself that this gives precisely the same solution as before. Computationally, there is not a significant difference between the two solution methods, and the choice is left to the user.

EXAMPLE 10.8 Suppose

$$T = \begin{pmatrix} 7 & 6 \\ -9 & -8 \end{pmatrix}.$$

Its eigenvalues and eigenvectors are readily computed:

$$\lambda_1 = -2, \quad \mathbf{v}_1 = \begin{pmatrix} -2 \\ 3 \end{pmatrix}, \qquad \lambda_2 = 1, \quad \mathbf{v}_2 = \begin{pmatrix} -1 \\ 1 \end{pmatrix}.$$

We assemble these into the diagonal eigenvalue matrix Λ and the eigenvector matrix S given by

$$\Lambda = \begin{pmatrix} -2 & 0 \\ 0 & 1 \end{pmatrix}, \qquad S = \begin{pmatrix} -2 & -1 \\ 3 & 1 \end{pmatrix},$$

whence

$$\begin{pmatrix} 7 & 6 \\ -9 & -8 \end{pmatrix} = T = S\,\Lambda\,S^{-1} = \begin{pmatrix} -2 & -1 \\ 3 & 1 \end{pmatrix}\begin{pmatrix} -2 & 0 \\ 0 & 1 \end{pmatrix}\begin{pmatrix} 1 & 1 \\ -3 & -2 \end{pmatrix},$$

as you can readily check. Therefore, according to (10.20)

$$T^k = S\,\Lambda^k\,S^{-1} = \begin{pmatrix} -2 & -1 \\ 3 & 1 \end{pmatrix}\begin{pmatrix} (-2)^k & 0 \\ 0 & 1 \end{pmatrix}\begin{pmatrix} 1 & 1 \\ -3 & -2 \end{pmatrix}$$
$$= \begin{pmatrix} 3 - 2\,(-2)^k & 2 - 2\,(-2)^k \\ -3 + 3\,(-2)^k & -2 + 3\,(-2)^k \end{pmatrix}.$$

You may wish to check this formula directly for the first few values of $k = 1, 2, \ldots$. As a result, the solution to the particular iterative system

$$\mathbf{u}^{(k+1)} = \begin{pmatrix} 7 & 6 \\ -9 & -8 \end{pmatrix} \mathbf{u}^{(k)}, \qquad \mathbf{u}^{(0)} = \begin{pmatrix} 1 \\ 1 \end{pmatrix},$$

is

$$\mathbf{u}^{(k)} = T^k \begin{pmatrix} 1 \\ 1 \end{pmatrix} = \begin{pmatrix} 5 - 4(-2)^k \\ -5 + 6(-2)^k \end{pmatrix}.$$

●

EXERCISES 10.1

10.1.13. Find the explicit formula for the solution to the following linear iterative systems:

(a) $u^{(k+1)} = u^{(k)} - 2v^{(k)},\ v^{(k+1)} = -2u^{(k)} + v^{(k)},\ u^{(0)} = 1,\ v^{(0)} = 0$

(b) $u^{(k+1)} = u^{(k)} - \frac{2}{3}v^{(k)},\ v^{(k+1)} = \frac{1}{2}u^{(k)} - \frac{1}{6}v^{(k)},\ u^{(0)} = -2,\ v^{(0)} = 3$

(c) $u^{(k+1)} = u^{(k)} - v^{(k)},\ v^{(k+1)} = -u^{(k)} + 5v^{(k)},\ u^{(0)} = 1,\ v^{(0)} = 0$

(d) $u^{(k+1)} = \frac{1}{2}u^{(k)} + v^{(k)},\ v^{(k+1)} = v^{(k)} - 2w^{(k)},\ w^{(k+1)} = \frac{1}{3}w^{(k)},\ u^{(0)} = 1,\ v^{(0)} = -1,\ w^{(0)} = 1$

(e) $u^{(k+1)} = -u^{(k)} + 2v^{(k)} - w^{(k)},\ v^{(k+1)} = -6u^{(k)} + 7v^{(k)} - 4w^{(k)},\ w^{(k+1)} = -6u^{(k)} + 6v^{(k)} - 4w^{(k)},\ u^{(0)} = 0,\ v^{(0)} = 1,\ w^{(0)} = 3$

10.1.14. Find the explicit formula for the general solution to the linear iterative systems with the following coefficient matrices:

(a) $\begin{pmatrix} -1 & 2 \\ 1 & -1 \end{pmatrix}$ (b) $\begin{pmatrix} -2 & 7 \\ -1 & 3 \end{pmatrix}$

(c) $\begin{pmatrix} -3 & 2 & -2 \\ -6 & 4 & -3 \\ 12 & -6 & -5 \end{pmatrix}$ (d) $\begin{pmatrix} -\frac{5}{6} & \frac{1}{3} & -\frac{1}{6} \\ 0 & -\frac{1}{2} & \frac{1}{3} \\ 1 & -1 & \frac{2}{3} \end{pmatrix}$

10.1.15. The kth *Lucas number* is defined as

$$L^{(k)} = \left(\frac{1+\sqrt{5}}{2}\right)^k + \left(\frac{1-\sqrt{5}}{2}\right)^k.$$

(a) Explain why the Lucas numbers satisfy the Fibonacci iterative equation $L^{(k+2)} = L^{(k+1)} + L^{(k)}$.

(b) Write down the first 7 Lucas numbers.

(c) Prove that every Lucas number is a positive integer.

10.1.16. Prove that all the Fibonacci integers $u^{(k)}$, $k \geq 0$, can be found by just computing the first term in the Binet formula (10.17) and then rounding off to the nearest integer.

10.1.17. What happens to the Fibonacci integers $u^{(k)}$ if we go "backward in time", i.e., for $k < 0$? How is $u^{(-k)}$ related to $u^{(k)}$?

10.1.18. Use formula (10.20) to compute the kth power of the following matrices:

(a) $\begin{pmatrix} 5 & 2 \\ 2 & 2 \end{pmatrix}$ (b) $\begin{pmatrix} 4 & 1 \\ -2 & 1 \end{pmatrix}$

(c) $\begin{pmatrix} 1 & 1 \\ -1 & 1 \end{pmatrix}$ (d) $\begin{pmatrix} 1 & 1 & 2 \\ 1 & 2 & 1 \\ 2 & 1 & 1 \end{pmatrix}$

(e) $\begin{pmatrix} 0 & 1 & 0 \\ 0 & 0 & 1 \\ -1 & 0 & 2 \end{pmatrix}$

10.1.19. Use your answer from Exercise 10.1.18 to solve the following iterative systems:

(a) $u^{(k+1)} = 5u^{(k)} + 2v^{(k)},\ v^{(k+1)} = 2u^{(k)} + 2v^{(k)},\ u^{(0)} = -1,\ v^{(0)} = 0$

(b) $u^{(k+1)} = 4u^{(k)} + v^{(k)},\ v^{(k+1)} = -2u^{(k)} + v^{(k)},\ u^{(0)} = 1,\ v^{(0)} = -3$

(c) $u^{(k+1)} = u^{(k)} + v^{(k)},\ v^{(k+1)} = -u^{(k)} + v^{(k)},\ u^{(0)} = 0,\ v^{(0)} = 2$

(d) $u^{(k+1)} = u^{(k)} + v^{(k)} + 2w^{(k)},\ v^{(k+1)} = u^{(k)} + 2v^{(k)} + w^{(k)},\ w^{(k+1)} = 2u^{(k)} + v^{(k)} + w^{(k)},\ u^{(0)} = 1,\ v^{(0)} = 0,\ w^{(0)} = 1$

(e) $u^{(k+1)} = v^{(k)},\ v^{(k+1)} = w^{(k)},\ w^{(k+1)} = -u^{(k)} + 2w^{(k)},\ u^{(0)} = 1,\ v^{(0)} = 0,\ w^{(0)} = 0$

10.1.20. (a) Given initial data $\mathbf{u}^{(0)} = (1, 1, 1)^T$, explain why the resulting solution $\mathbf{u}^{(k)}$ to the system in Example 10.7 has all integer entries.

(b) Find the coefficients c_1, c_2, c_3 in the explicit solution formula (10.18).

(c) Check the first few iterates to convince yourself that the solution formula does, in spite of appearances, always give an integer value.

10.1.21. (a) Show how to convert the higher order linear iterative equation

$$u^{(k+j)} = c_1 u^{(k+j-1)} + c_2 u^{(k+j-2)} + \cdots + c_j u^{(k)}$$

into a first order system $\mathbf{u}^{(k)} = T\mathbf{u}^{(k)}$. *Hint*: See Example 10.6.

(b) Write down initial conditions that guarantee a unique solution $u^{(k)}$ for all $k \geq 0$.

10.1.22. Apply the method of Exercise 10.1.21 to solve the following iterative equations:

(a) $u^{(k+2)} = -u^{(k+1)} + 2u^{(k)}$, $u^{(0)} = 1$, $u^{(1)} = 2$

(b) $12u^{(k+2)} = u^{(k+1)} + u^{(k)}$, $u^{(0)} = -1$, $u^{(1)} = 2$

(c) $u^{(k+2)} = 4u^{(k+1)} + u^{(k)}$, $u^{(0)} = 1$, $u^{(1)} = -1$

(d) $u^{(k+2)} = 2u^{(k+1)} - 2u^{(k)}$, $u^{(0)} = 1$, $u^{(1)} = 3$

(e) $u^{(k+3)} = 2u^{(k+2)} + u^{(k+1)} - 2u^{(k)}$, $u^{(0)} = 0$, $u^{(1)} = 2$, $u^{(2)} = 3$

(f) $u^{(k+3)} = u^{(k+2)} + 2u^{(k+1)} - 2u^{(k)}$, $u^{(0)} = 0$, $u^{(1)} = 1$, $u^{(2)} = 1$

♣**10.1.23.** Starting with $u^{(0)} = 0$, $u^{(1)} = 0$, $u^{(2)} = 1$, we define the sequence of *tri–Fibonacci integers* $u^{(k)}$ by adding the previous three to get the next one. For instance, $u^{(3)} = u^{(0)} + u^{(1)} + u^{(2)} = 1$.

(a) Write out the next four tri–Fibonacci integers.

(b) Find a third order iterative equation for the kth tri–Fibonacci integer.

(c) Find an explicit formula for the solution, using a computer to approximate the eigenvalues.

(d) Do they grow faster than the usual Fibonacci numbers? What is their overall rate of growth?

♣**10.1.24.** Suppose that Fibonacci's rabbits only live for eight years, [34].

(a) Write out an iterative equation to describe the rabbit population.

(b) Write down the first few terms.

(c) Convert your equation into a first order iterative system, using the method of Exercise 10.1.21.

(d) At what rate does the rabbit population grow?

10.1.25. Find the general solution to the iterative system $u_i^{(k+1)} = u_{i-1}^{(k)} + u_{i+1}^{(k)}$, $i = 1, \ldots, n$, where we set $u_0^{(k)} = u_{n+1}^{(k)} = 0$ for all k. *Hint*: Use Exercise 8.2.47.

10.1.26. Prove that the curves $E_k = \{ T^k\mathbf{x} \mid \|\mathbf{x}\| = 1 \}$, $k = 0, 1, 2, \ldots$, sketched in Figure 10.2 form a family of ellipses with the same principal axes. What are the semi-axes? *Hint*: Use Exercise 8.5.19.

♠**10.1.27.** Plot the ellipses $E_k = \{ T^k\mathbf{x} \mid \|\mathbf{x}\| = 1 \}$ for $k = 1, 2, 3, 4$ for the following matrices T. Then determine their principal axes, semi-axes, and areas. *Hint*: Use Exercise 8.5.19.

(a) $\begin{pmatrix} \frac{2}{3} & -\frac{1}{3} \\ -\frac{1}{3} & \frac{2}{3} \end{pmatrix}$ (b) $\begin{pmatrix} 0 & -1.2 \\ .4 & 0 \end{pmatrix}$

(c) $\begin{pmatrix} \frac{3}{5} & \frac{1}{5} \\ \frac{2}{5} & \frac{4}{5} \end{pmatrix}$

10.1.28. Let T be a positive definite 2×2 matrix. Let $E_n = \{ T^n\mathbf{x} \mid \|\mathbf{x}\| = 1 \}$, $n = 0, 1, 2, \ldots$, be the image of the unit circle under the nth power of T.

(a) Prove that E_n is an ellipse.

True or false:

(b) The ellipses E_n all have the same principal axes.

(c) The semi-axes are given by $r_n = r_1^n$, $s_n = s_1^n$.

(d) The areas are given by $A_n = \pi\alpha^n$ where $\alpha = A_1/\pi$.

10.1.29. Answer Exercise 10.1.28 when T is an arbitrary nonsingular 2×2 matrix. *Hint*: Use Exercise 8.5.19.

10.1.30. Given the general solution (10.9) of the iterative system $\mathbf{u}^{(k+1)} = T\mathbf{u}^{(k)}$, write down the solution to $\mathbf{v}^{(k+1)} = \alpha T\mathbf{v}^{(k)} + \beta\mathbf{v}^{(k)}$, where α, β are fixed scalars.

◇**10.1.31.** Prove directly that if the coefficient matrix of a linear iterative system is real, both the real and imaginary parts of a complex solution are real solutions.

◇**10.1.32.** Explain why the solution $\mathbf{u}^{(k)}$, $k \geq 0$, to the initial value problem (10.6) exists and is uniquely defined. Does this hold if we allow negative $k < 0$?

10.1.33. Prove that if T is a symmetric matrix, then the coefficients in (10.9) are given by the formula $c_j = \mathbf{a}^T\mathbf{v}_j / \mathbf{v}_j^T\mathbf{v}_j$.

10.1.34. Explain why the jth column $\mathbf{c}_j^{(k)}$ of the matrix power T^k satisfies the linear iterative system $\mathbf{c}_j^{(k+1)} = T\mathbf{c}_j^{(k)}$ with initial data $\mathbf{c}_j^{(0)} = \mathbf{e}_j$, the jth standard basis vector.

10.1.35. Let $z^{(k+1)} = \lambda z^{(k)}$ be a complex scalar iterative equation with $\lambda = \mu + i\nu$. Show that its real and imaginary parts $x^{(k)} = \operatorname{Re} z^{(k)}$, $y^{(k)} = \operatorname{Im} z^{(k)}$, satisfy a two-dimensional real linear iterative system. Use the eigenvalue method to solve the real 2×2 system, and verify that your solution coincides with the solution to the original complex equation.

◇**10.1.36.** Let T be an incomplete matrix, and suppose $\mathbf{w}_1, \ldots, \mathbf{w}_j$ is a Jordan chain associated with an incomplete eigenvalue λ.

(a) Prove that, for any $i = 1, \ldots, j$,

$$T^k\mathbf{w}_i = \lambda^k\mathbf{w}_i + k\lambda^{k-1}\mathbf{w}_{i-1} + \binom{k}{2}\lambda^{k-2}\mathbf{w}_{i-2} + \cdots. \tag{10.23}$$

(b) Explain how to use a Jordan basis of T to construct the general solution to the linear iterative system $\mathbf{u}^{(k+1)} = T\,\mathbf{u}^{(k)}$.

10.1.37. Use the method Exercise 10.1.36 to find the general real solution to the following iterative systems:

(a) $u^{(k+1)} = 2\,u^{(k)} + 3\,v^{(k)},\ v^{(k+1)} = 2\,v^{(k)}$

(b) $u^{(k+1)} = u^{(k)} + v^{(k)},\ v^{(k+1)} = -4\,u^{(k)} + 5\,v^{(k)}$

(c) $u^{(k+1)} = -u^{(k)} + v^{(k)} + w^{(k)},$
$v^{(k+1)} = -v^{(k)} + w^{(k)},\ w^{(k+1)} = -w^{(k)}$

(d) $u^{(k+1)} = 3\,u^{(k)} - v^{(k)},$
$v^{(k+1)} = -u^{(k)} + 3\,v^{(k)} + w^{(k)},$
$w^{(k+1)} = -v^{(k)} + 3\,w^{(k)}$

(e) $u^{(k+1)} = u^{(k)} - v^{(k)} - w^{(k)},$
$v^{(k+1)} = 2\,u^{(k)} + 2\,v^{(k)} + 2\,w^{(k)},$
$w^{(k+1)} = -u^{(k)} + v^{(k)} + w^{(k)}$

(f) $u^{(k+1)} = v^{(k)} + z^{(k)},\ v^{(k+1)} = -u^{(k)} + w^{(k)},$
$w^{(k+1)} = z^{(k)},\ z^{(k+1)} = -w^{(k)}$

10.1.38. Find a formula for the kth power of a Jordan block matrix. *Hint*: Use Exercise 10.1.36.

10.1.39. Suppose $\mathbf{u}^{(k)}$ and $\widetilde{\mathbf{u}}^{(k)}$ are two solutions to the same iterative system $\mathbf{u}^{(k+1)} = T\,\mathbf{u}^{(k)}$.

(a) Suppose $\mathbf{u}^{(k_0)} = \widetilde{\mathbf{u}}^{(k_0)}$ for some $k_0 \geq 0$. Can you conclude that these are the same solution: $\mathbf{u}^{(k)} = \widetilde{\mathbf{u}}^{(k)}$ for all k?

(b) What can you say if $\mathbf{u}^{(k_0)} = \widetilde{\mathbf{u}}^{(k_1)}$ for $k_0 \neq k_1$?

♡**10.1.40.** An *affine iterative system* has the form $\mathbf{u}^{(k+1)} = T\,\mathbf{u}^{(k)} + \mathbf{b},\ \mathbf{u}^{(0)} = \mathbf{c}$.

(a) Under what conditions does the system have an equilibrium solution $\mathbf{u}^{(k)} \equiv \mathbf{u}^{\star}$?

(b) In such cases, find a formula for the general solution. *Hint*: Look at $\mathbf{v}^{(k)} = \mathbf{u}^{(k)} - \mathbf{u}^{\star}$.

(c) Solve the following affine iterative systems:

(i) $\mathbf{u}^{(k+1)} = \begin{pmatrix} 6 & 3 \\ -3 & -4 \end{pmatrix}\mathbf{u}^{(k)} + \begin{pmatrix} 1 \\ 2 \end{pmatrix},\quad \mathbf{u}^{(0)} = \begin{pmatrix} 4 \\ -3 \end{pmatrix}$

(ii) $\mathbf{u}^{(k+1)} = \begin{pmatrix} -1 & 2 \\ 1 & -1 \end{pmatrix}\mathbf{u}^{(k)} + \begin{pmatrix} 1 \\ 0 \end{pmatrix},\quad \mathbf{u}^{(0)} = \begin{pmatrix} 0 \\ 1 \end{pmatrix}$

(iii) $\mathbf{u}^{(k+1)} = \begin{pmatrix} -3 & 2 & -2 \\ -6 & 4 & -3 \\ 12 & -6 & -5 \end{pmatrix}\mathbf{u}^{(k)} + \begin{pmatrix} 1 \\ -3 \\ 0 \end{pmatrix},\quad \mathbf{u}^{(0)} = \begin{pmatrix} 1 \\ 0 \\ -1 \end{pmatrix}$

(iv) $\mathbf{u}^{(k+1)} = \begin{pmatrix} -\frac{5}{6} & \frac{1}{3} & -\frac{1}{6} \\ 0 & -\frac{1}{2} & \frac{1}{3} \\ 1 & -1 & \frac{2}{3} \end{pmatrix}\mathbf{u}^{(k)} + \begin{pmatrix} \frac{1}{6} \\ -\frac{1}{3} \\ -\frac{1}{2} \end{pmatrix},\quad \mathbf{u}^{(0)} = \begin{pmatrix} \frac{1}{6} \\ -\frac{2}{3} \\ \frac{1}{3} \end{pmatrix}$

(d) Discuss what happens in cases when there is no fixed point, assuming that T is complete.

♣**10.1.41.** A well-known method of generating a sequence of "pseudo-random" integers $x_0, x_1, \ldots$ in the interval from 0 to n is based on the Fibonacci equation $u^{(k+2)} = u^{(k+1)} + u^{(k)} \bmod n$, with initial values $u^{(0)}$, $u^{(1)}$ chosen from the integers $0, 1, 2, \ldots, n-1$.

(a) Generate the sequence of pseudo-random numbers that result from the choices $n = 10$, $u^{(0)} = 3$, $u^{(1)} = 7$. Keep iterating until the sequence starts repeating.

(b) Experiment with other sequences of pseudo-random numbers generated by the method.

10.2 STABILITY

With the solution formula (10.9) in hand, we are now in a position to understand the qualitative behavior of solutions to (complete) linear iterative systems. The most important case for applications is when all the iterates converge to $\mathbf{0}$.

Definition 10.9 The equilibrium solution $\mathbf{u}^{\star} = \mathbf{0}$ to a linear iterative system (10.1) is called *asymptotically stable* if and only if all solutions $\mathbf{u}^{(k)} \to \mathbf{0}$ as $k \to \infty$.

Asymptotic stability relies on the following property of the coefficient matrix:

Definition 10.10 A matrix T is called *convergent* if its powers converge to the zero matrix, $T^k \to \mathrm{O}$, meaning that the individual entries of T^k all go to 0 as $k \to \infty$.

The equivalence of the convergence condition and stability of the iterative system follows immediately from the solution formula (10.7).

Proposition 10.11 The linear iterative system $\mathbf{u}^{(k+1)} = T\,\mathbf{u}^{(k)}$ has asymptotically stable zero solution if and only if T is a convergent matrix.

Proof If $T^k \to \mathrm{O}$, and $\mathbf{u}^{(k)} = T^k\mathbf{a}$ is any solution, then clearly $\mathbf{u}^{(k)} \to \mathbf{0}$ as $k \to \infty$, proving stability. Conversely, the solution $\mathbf{u}_j^{(k)} = T^k\mathbf{e}_j$ is the same as the jth column of T^k. If the origin is asymptotically stable, then $\mathbf{u}_j^{(k)} \to \mathbf{0}$. Thus, the individual columns of T^k all tend to $\mathbf{0}$, proving that $T^k \to \mathrm{O}$. ■

To facilitate the analysis of convergence, we shall adopt a norm $\|\cdot\|$ on our underlying vector space, $\mathbb{R}^n$ or $\mathbb{C}^n$. The reader may be inclined to choose the Euclidean (or Hermitian) norm, but, in practice, the ∞ norm

$$\|\mathbf{u}\|_\infty = \max\{\,|u_1|,\ldots,|u_n|\,\} \tag{10.24}$$

prescribed by the vector's maximal entry (in modulus) is usually much easier to work with. Convergence of the iterates is equivalent to convergence of their norms:

$$\mathbf{u}^{(k)} \to \mathbf{0} \quad \text{if and only if} \quad \|\mathbf{u}^{(k)}\| \to \mathbf{0} \quad \text{as} \quad k \to \infty.$$

The fundamental stability criterion for linear iterative systems relies on the size of the eigenvalues of the coefficient matrix.

THEOREM 10.12 A linear iterative system (10.1) is asymptotically stable if and only if all its (complex) eigenvalues have modulus strictly less than one: $|\lambda_j| < 1$.

Proof Let us prove this result assuming that the coefficient matrix T is complete. (The proof in the incomplete case relies on the Jordan canonical form, and is outlined in the exercises.) If λ_j is an eigenvalue such that $|\lambda_j| < 1$, then the corresponding basis solution $\mathbf{u}_j^{(k)} = \lambda_j^k\mathbf{v}_j$ tends to zero as $k \to \infty$; indeed,

$$\|\mathbf{u}_j^{(k)}\| = \|\lambda_j^k\mathbf{v}_j\| = |\lambda_j|^k\,\|\mathbf{v}_j\| \longrightarrow 0 \quad \text{since} \quad |\lambda_j| < 1.$$

Therefore, if all eigenvalues are less than 1 in modulus, all terms in the solution formula (10.9) tend to zero, which proves asymptotic stability: $\mathbf{u}^{(k)} \to \mathbf{0}$. Conversely, if any eigenvalue satisfies $|\lambda_j| \geq 1$, then the solution $\mathbf{u}^{(k)} = \lambda_j^k\mathbf{v}_j$ does not tend to $\mathbf{0}$ as $k \to \infty$, and hence $\mathbf{0}$ is not asymptotically stable. ■

Consequently, the necessary and sufficient condition for asymptotic stability of a linear iterative system is that all the eigenvalues of the coefficient matrix lie strictly inside the unit circle* in the complex plane: $|\lambda_j| < 1$.

Definition 10.13 The *spectral radius* of a matrix T is defined as the maximal modulus of all of its real and complex eigenvalues: $\rho(T) = \max\{\,|\lambda_1|,\ldots,|\lambda_k|\,\}$.

We can then restate the Stability Theorem 10.12 as follows:

THEOREM 10.14 The matrix T is convergent if and only if its spectral radius is strictly less than one: $\rho(T) < 1$.

*Note that this is *not* the same as the stability criterion for ordinary differential equations, which requires the eigenvalues of the coefficient matrix to lie in the left half plane, cf. Theorem 9.15.

If T is complete, then we can apply the triangle inequality to (10.9) to estimate

$$\begin{aligned} \|\mathbf{u}^{(k)}\| &= \|c_1 \lambda_1^k \mathbf{v}_1 + \cdots + c_n \lambda_n^k \mathbf{v}_n\| \\ &\le |\lambda_1|^k \, \|c_1 \mathbf{v}_1\| + \cdots + |\lambda_n|^k \, \|c_n \mathbf{v}_n\| \\ &\le \rho(T)^k \big(|c_1| \, \|\mathbf{v}_1\| + \cdots + |c_n| \, \|\mathbf{v}_n\| \big) = C \rho(T)^k, \end{aligned} \tag{10.25}$$

for some constant $C > 0$ that depends only upon the initial conditions. In particular, if $\rho(T) < 1$, then

$$\|\mathbf{u}^{(k)}\| \le C \rho(T)^k \longrightarrow 0 \quad \text{as} \quad k \to \infty, \tag{10.26}$$

in accordance with Theorem 10.14. Thus, the spectral radius prescribes the rate of convergence of the solutions to equilibrium. The smaller the spectral radius, the faster the solutions go to $\mathbf{0}$.

If T has only one largest (simple) eigenvalue, so $|\lambda_1| > |\lambda_j|$ for all $j > 1$, then the first term in the solution formula (10.9) will eventually dominate all the others: $\|\lambda_1^k \mathbf{v}_1\| \gg \|\lambda_j^k \mathbf{v}_j\|$ for $j > 1$ and $k \gg 0$. Therefore, provided that $c_1 \ne 0$, the solution (10.9) has the asymptotic formula

$$\mathbf{u}^{(k)} \approx c_1 \lambda_1^k \mathbf{v}_1, \tag{10.27}$$

and so most solutions end up parallel to $\mathbf{v}_1$. In particular, if $|\lambda_1| = \rho(T) < 1$, such a solution approaches $\mathbf{0}$ along the direction of the dominant eigenvector $\mathbf{v}_1$ at a rate governed by the modulus of the dominant eigenvalue. The exceptional solutions, with $c_1 = 0$, tend to $\mathbf{0}$ at a faster rate, along one of the other eigendirections. In practical computations, one rarely observes the exceptional solutions. Indeed, even if the initial condition does not involve the dominant eigenvector, round-off error during the iteration will almost inevitably introduce a small component in the direction of $\mathbf{v}_1$, which will, if you wait long enough, eventually dominate the computation.

Warning: The inequality (10.25) only applies to complete matrices. In the general case, one can prove that the solution satisfies the slightly weaker inequality

$$\|\mathbf{u}^{(k)}\| \le C \sigma^k \quad \text{for all} \quad k \ge 0, \qquad \text{where} \quad \sigma > \rho(T) \tag{10.28}$$

is any number larger than the spectral radius, while $C > 0$ is a positive constant (whose value may depend on how close σ is to ρ).

EXAMPLE 10.15 According to Example 10.7, the matrix

$$T = \begin{pmatrix} -3 & 1 & 6 \\ 1 & -1 & -2 \\ -1 & -1 & 0 \end{pmatrix} \quad \text{has eigenvalues} \quad \begin{aligned} \lambda_1 &= -2 \\ \lambda_2 &= -1 + \mathrm{i} \\ \lambda_3 &= -1 - \mathrm{i}\,. \end{aligned}$$

Since $|\lambda_1| = 2 > |\lambda_2| = |\lambda_3| = \sqrt{2}$, the spectral radius is $\rho(T) = |\lambda_1| = 2$. We conclude that T is not a convergent matrix. As the reader can check, either directly, or from the solution formula (10.18), the vectors $\mathbf{u}^{(k)} = T^k \mathbf{u}^{(0)}$ obtained by repeatedly multiplying any nonzero initial vector $\mathbf{u}^{(0)}$ by T rapidly go off to ∞, at a rate roughly equal to $\rho(T)^k = 2^k$.

On the other hand, the matrix

$$\widetilde{T} = -\tfrac{1}{3} T = \begin{pmatrix} 1 & -\frac{1}{3} & -2 \\ -\frac{1}{3} & \frac{1}{3} & \frac{2}{3} \\ \frac{1}{3} & \frac{1}{3} & 0 \end{pmatrix} \quad \text{with eigenvalues} \quad \begin{aligned} \lambda_1 &= \tfrac{2}{3} \\ \lambda_2 &= \tfrac{1}{3} - \tfrac{1}{3}\,\mathrm{i} \\ \lambda_3 &= \tfrac{1}{3} + \tfrac{1}{3}\,\mathrm{i} \end{aligned}$$

has spectral radius $\rho(\widetilde{T}) = \frac{2}{3}$, and hence is a convergent matrix. According to (10.27), if we write the initial data $\mathbf{u}^{(0)} = c_1 \mathbf{v}_1 + c_2 \mathbf{v}_2 + c_3 \mathbf{v}_3$ as a linear combination of the eigenvectors, then, provided $c_1 \neq 0$, the iterates have the asymptotic form $\mathbf{u}^{(k)} \approx c_1 \left(-\frac{2}{3}\right)^k \mathbf{v}_1$, where $\mathbf{v}_1 = (4, -2, 1)^T$ is the eigenvector corresponding to the dominant eigenvalue $\lambda_1 = -\frac{2}{3}$. Thus, for most initial vectors, the iterates end up decreasing in length by a factor of almost exactly $\frac{2}{3}$, eventually becoming parallel to the dominant eigenvector $\mathbf{v}_1$. This is borne out by a sample computation: starting with $\mathbf{u}^{(0)} = (1, 1, 1)^T$, the first ten iterates are

$$\begin{pmatrix} -.0936 \\ .0462 \\ -.0231 \end{pmatrix}, \begin{pmatrix} -.0627 \\ .0312 \\ -.0158 \end{pmatrix}, \begin{pmatrix} -.0416 \\ .0208 \\ -.0105 \end{pmatrix}, \begin{pmatrix} -.0275 \\ .0138 \\ -.0069 \end{pmatrix}, \begin{pmatrix} -.0182 \\ .0091 \\ -.0046 \end{pmatrix},$$

$$\begin{pmatrix} -.0121 \\ .0061 \\ -.0030 \end{pmatrix}, \begin{pmatrix} -.0081 \\ .0040 \\ -.0020 \end{pmatrix}, \begin{pmatrix} -.0054 \\ .0027 \\ -.0013 \end{pmatrix}, \begin{pmatrix} -.0036 \\ .0018 \\ -.0009 \end{pmatrix}, \begin{pmatrix} -.0024 \\ .0012 \\ -.0006 \end{pmatrix}. \quad \bullet$$

EXERCISES 10.2

10.2.1. Determine the spectral radius of the following matrices:

(a) $\begin{pmatrix} 1 & 2 \\ 3 & 4 \end{pmatrix}$ (b) $\begin{pmatrix} \frac{1}{3} & -\frac{1}{4} \\ \frac{1}{2} & -\frac{1}{3} \end{pmatrix}$

(c) $\begin{pmatrix} 0 & 1 & 0 \\ 0 & 0 & 1 \\ -2 & 1 & 2 \end{pmatrix}$ (d) $\begin{pmatrix} -1 & 5 & -9 \\ 4 & 0 & -1 \\ 4 & -4 & 3 \end{pmatrix}$

10.2.2. Determine whether or not the following matrices are convergent:

(a) $\begin{pmatrix} 2 & -3 \\ 3 & 2 \end{pmatrix}$ (b) $\begin{pmatrix} .6 & .3 \\ .3 & .7 \end{pmatrix}$

(c) $\frac{1}{5}\begin{pmatrix} 5 & -3 & -2 \\ 1 & -2 & 1 \\ 1 & -5 & 4 \end{pmatrix}$ (d) $\begin{pmatrix} .8 & .3 & .2 \\ .1 & .2 & .6 \\ .1 & .5 & .2 \end{pmatrix}$

10.2.3. Which of the listed coefficient matrices defines a linear iterative system with asymptotically stable zero solution?

(a) $\begin{pmatrix} -3 & 0 \\ -4 & -1 \end{pmatrix}$ (b) $\begin{pmatrix} \frac{1}{2} & \frac{3}{4} \\ \frac{2}{3} & \frac{1}{3} \end{pmatrix}$

(c) $\begin{pmatrix} \frac{1}{2} & \frac{1}{2} \\ -\frac{1}{2} & \frac{1}{2} \end{pmatrix}$ (d) $\begin{pmatrix} -1 & 3 & 0 \\ -1 & 1 & -1 \\ 0 & -1 & -1 \end{pmatrix}$

(e) $\begin{pmatrix} \frac{1}{2} & \frac{1}{4} & -\frac{1}{4} \\ \frac{1}{2} & \frac{3}{4} & -\frac{1}{2} \\ -\frac{1}{4} & -\frac{1}{4} & \frac{1}{2} \end{pmatrix}$

(f) $\begin{pmatrix} 3 & 0 & -1 \\ 0 & 1 & 0 \\ 2 & 0 & 0 \end{pmatrix}$

(g) $\begin{pmatrix} 1 & 0 & -3 & -2 \\ -\frac{1}{2} & \frac{1}{2} & 2 & \frac{3}{2} \\ -\frac{1}{6} & 0 & \frac{3}{2} & \frac{2}{3} \\ \frac{2}{3} & 0 & -3 & -\frac{5}{3} \end{pmatrix}$

10.2.4. (a) Determine the eigenvalues and spectral radius of the matrix

$$T = \begin{pmatrix} 3 & 2 & -2 \\ -2 & 1 & 0 \\ 0 & 2 & 1 \end{pmatrix}.$$

(b) Use this information to find the eigenvalues and spectral radius of

$$\widehat{T} = \begin{pmatrix} \frac{3}{5} & \frac{2}{5} & -\frac{2}{5} \\ -\frac{2}{5} & \frac{1}{5} & 0 \\ 0 & \frac{2}{5} & \frac{1}{5} \end{pmatrix}.$$

(c) Write down an asymptotic formula for the solutions to the iterative system $\mathbf{u}^{(k+1)} = \widehat{T}\, \mathbf{u}^{(k)}$.

10.2.5. (a) Show that the spectral radius of

$$T = \begin{pmatrix} 1 & 1 \\ 0 & 1 \end{pmatrix}$$

is $\rho(T) = 1$.

(b) Show that most iterates $\mathbf{u}^{(k)} = T^k \mathbf{u}^{(0)}$ become unbounded as $k \to \infty$.

(c) Discuss why the inequality $\| \mathbf{u}^{(k)} \| \leq C\, \rho(T)^k$ does not hold when the coefficient matrix is incomplete.

(d) Can you prove that (10.28) holds in this example?

10.2.6. Given a linear iterative system with non-convergent matrix, which solutions, if any, will converge to $\mathbf{0}$?

10.2.7. Prove that if A is any square matrix, then there exists $c \neq 0$ such that the scalar multiple $c\,A$ is a convergent matrix. Find a formula for the largest possible such c.

◇ **10.2.8.** Suppose T is a complete matrix.

(a) Prove that every solution to the corresponding linear iterative system is bounded if and only if $\rho(T) \leq 1$.

(b) Can you generalize this result to incomplete matrices? *Hint*: Look at Exercise 10.1.36.

10.2.9. Suppose a convergent iterative system has a single dominant real eigenvalue λ_1. Discuss how the asymptotic behavior of the real solutions depends on the sign of λ_1.

♡**10.2.10.** Discuss the asymptotic behavior of solutions to an iterative system that has two eigenvalues of largest modulus, e.g., $\lambda_1 = -\lambda_2$, or $\lambda_1 = \overline{\lambda_2}$ are complex conjugate eigenvalues. How can you detect this? How can you determine the eigenvalues and eigenvectors?

10.2.11. Suppose T has spectral radius $\rho(T)$. Can you predict the spectral radius of $c\,T + d\,\mathrm{I}$, where c, d are scalars? If not, what additional information do you need?

10.2.12. Let A have singular values $\sigma_1 \geq \cdots \geq \sigma_n$. Prove that $A^T A$ is a convergent matrix if and only if $\sigma_1 < 1$. (Later we will show that this implies that A itself is convergent.)

♡**10.2.13.** Let M_n be the $n \times n$ tridiagonal matrix with all 1's on the sub- and super-diagonals, and zeros on the main diagonal.

(a) What is the spectral radius of M_n? *Hint*: Use Exercise 8.2.47.

(b) Is M_n convergent?

(c) Find the general solution to the iterative system $\mathbf{u}^{(k+1)} = M_n\,\mathbf{u}^{(k)}$.

♡**10.2.14.** Let α, β be scalars. Let $T_{\alpha,\beta}$ be the $n \times n$ tridiagonal matrix that has all α's on the sub- and super-diagonals, and β's on the main diagonal.

(a) Solve the iterative system $\mathbf{u}^{(k+1)} = T_{\alpha,\beta}\,\mathbf{u}^{(k)}$.

(b) For which values of α, β is the system asymptotically stable? *Hint*: Combine Exercises 10.2.13 and 10.1.30.

10.2.15. (a) Prove that if $|\det T| > 1$ then the iterative system $\mathbf{u}^{(k+1)} = T\,\mathbf{u}^{(k)}$ is unstable.

(b) If $|\det T| < 1$ is the system necessarily asymptotically stable? Prove or give a counterexample.

10.2.16. *True or false*:

(a) $\rho(c\,A) = c\,\rho(A)$

(b) $\rho(S^{-1} A\, S) = \rho(A)$

(c) $\rho(A^2) = \rho(A)^2$

(d) $\rho(A^{-1}) = 1/\rho(A)$

(e) $\rho(A + B) = \rho(A) + \rho(B)$

(f) $\rho(A\,B) = \rho(A)\,\rho(B)$

10.2.17. *True or false*:

(a) If A is convergent, then A^2 is convergent.

(b) If A is convergent, then $A^T A$ is convergent.

10.2.18. *True or false*: If the zero solution of the differential equation $\dot{\mathbf{u}} = A\,\mathbf{u}$ is asymptotically stable, so is the zero solution of the iterative system $\mathbf{u}^{(k+1)} = A\,\mathbf{u}^{(k)}$.

10.2.19. Suppose $T^k \to A$ as $k \to \infty$.

(a) Prove that $A^2 = A$.

(b) Can you characterize all such matrices A?

(c) What are the conditions on the matrix T for this to happen?

10.2.20. Prove that a matrix with all integer entries is convergent if and only if it is nilpotent, i.e., $A^k = \mathrm{O}$ for some k. Give a nonzero example of such a matrix.

♡**10.2.21.** Consider a second order iterative scheme $\mathbf{u}^{(k+2)} = A\,\mathbf{u}^{(k+1)} + B\,\mathbf{u}^{(k)}$. Define a *quadratic eigenvalue* to be a complex number that satisfies

$$\det(\lambda^2\,\mathrm{I} - \lambda\,A - B) = 0.$$

Prove that the system is asymptotically stable if and only if all its quadratic eigenvalues satisfy $|\lambda| < 1$. *Hint*: Look at the equivalent first order system and use Exercise 1.9.23b.

◇**10.2.22.** Let $p(t)$ be a polynomial. Assume $0 < \lambda < \mu$. Prove that there is a positive constant C such that $p(n)\,\lambda^n < C\,\mu^n$ for all $n > 0$.

◇**10.2.23.** Prove the inequality (10.28) when T is incomplete. Use it to complete the proof of Theorem 10.14 in the incomplete case. *Hint*: Use Exercises 10.1.36, 10.2.22.

◇**10.2.24.** Suppose that M is a nonsingular matrix.

(a) Prove that the *implicit iterative scheme* $M\,\mathbf{u}^{(n+1)} = \mathbf{u}^{(n)}$ is asymptotically stable if and only if all the eigenvalues of M are strictly greater than one in magnitude: $|\mu_i| > 1$.

(b) Let K be another matrix. Prove that iterative scheme $M\,\mathbf{u}^{(n+1)} = K\,\mathbf{u}^{(n)}$ is asymptotically stable if and only if all the generalized eigenvalues of the matrix pair K, M, as in Exercise 8.4.8, are strictly less than 1 in magnitude: $|\lambda_i| < 1$.

Fixed Points

The zero vector $\mathbf{0}$ is always a *fixed point* for a linear iterative system $\mathbf{u}^{(k+1)} = T\,\mathbf{u}^{(k)}$, since $\mathbf{0} = T\,\mathbf{0}$, and so $\mathbf{u}^{(k)} \equiv \mathbf{0}$ is an equilibrium solution. Are there any others? The answer is immediate: $\mathbf{u}^\star$ is a fixed point if and only if $\mathbf{u}^\star = T\,\mathbf{u}^\star$, and hence $\mathbf{u}^\star$ satisfies the eigenvector equation for T with eigenvalue $\lambda = 1$. Thus, the system admits a nonzero fixed point if and only if the coefficient matrix T has 1 as an eigenvalue. Since any scalar multiple of the eigenvector $\mathbf{u}^\star$ is also an eigenvector, in such cases the system has infinitely many fixed points, namely all elements of the eigenspace V_1. We are interested in whether the fixed points are *stable* in the sense that solutions having nearby initial conditions remain nearby. More precisely:

Definition 10.16 A fixed point $\mathbf{u}^\star$ of an iterative system $\mathbf{u}^{(k+1)} = T\,\mathbf{u}^{(k)}$ is called *stable* if for every $\varepsilon > 0$ there exists a $\delta > 0$ such that whenever $\|\mathbf{u}^{(0)} - \mathbf{u}^\star\| < \delta$ then the resulting iterates satisfy $\|\mathbf{u}^{(k)} - \mathbf{u}^\star\| < \varepsilon$ for all k.

The stability of the fixed points, at least if the coefficient matrix is complete, is governed by the same solution formula (10.9). If the eigenvalue $\lambda_1 = 1$ is simple, and all other eigenvalues are less than one in modulus, so

$$1 = \lambda_1 > |\lambda_2| \geq \cdots \geq |\lambda_n|,$$

then the solution takes the asymptotic form

$$\mathbf{u}^{(k)} = c_1\,\mathbf{v}_1 + c_2\,\lambda_2^k\,\mathbf{v}_2 + \cdots + c_n\,\lambda_n^k\,\mathbf{v}_n \longrightarrow c_1\,\mathbf{v}_1, \quad \text{as} \quad k \longrightarrow \infty, \tag{10.29}$$

converging to one of the fixed points, i.e., to a multiple of the eigenvector $\mathbf{v}_1$. The coefficient c_1 is prescribed by the initial conditions, cf. (10.10). The rate of convergence of the solution is governed by the modulus $|\lambda_2|$ of the *subdominant eigenvalue*.

THEOREM 10.17 Suppose that T has a simple (or, more generally, complete) eigenvalue $\lambda_1 = 1$, and, moreover, all other eigenvalues satisfy $|\lambda_j| < 1$. Then all solutions to the linear iterative system $\mathbf{u}^{(k+1)} = T\,\mathbf{u}^{(k)}$ converge to a vector $\mathbf{v} \in V_1$ in the $\lambda_1 = 1$ eigenspace. Moreover, all the fixed points $\mathbf{v} \in V_1$ of T are *stable*.

Stability of a fixed point does not imply asymptotic stability, since nearby solutions may converge to a nearby fixed point, i.e., to a slightly different element of the eigenspace V_1. The general necessary and sufficient conditions for stability follows; the proof is left as an exercise for the reader.

THEOREM 10.18 The fixed points of an iterative system $\mathbf{u}^{(k+1)} = T\,\mathbf{u}^{(k)}$ are stable if and only if $\rho(T) \leq 1$ and, moreover, every eigenvalue of modulus $|\lambda| = 1$ is complete.

Thus, when dealing with linear iterative systems, either all fixed points are stable or all are unstable. Keep in mind that the fixed points are the elements of the eigenspace V_1 corresponding to the eigenvalue $\lambda = 1$, if such exists. If 1 is not an eigenvalue of T, then $\mathbf{u}^\star = \mathbf{0}$ is the only fixed point.

EXAMPLE 10.19 Consider the iterative system with coefficient matrix

$$T = \begin{pmatrix} \frac{3}{2} & -\frac{1}{2} & -3 \\ -\frac{1}{2} & \frac{1}{2} & 1 \\ \frac{1}{2} & \frac{1}{2} & 0 \end{pmatrix}.$$

The eigenvalues and corresponding eigenvectors are

$$\lambda_1 = 1, \qquad \lambda_2 = \frac{1+\mathrm{i}}{2}, \qquad \lambda_3 = \frac{1-\mathrm{i}}{2},$$

$$\mathbf{v}_1 = \begin{pmatrix} 4 \\ -2 \\ 1 \end{pmatrix}, \quad \mathbf{v}_2 = \begin{pmatrix} 2-\mathrm{i} \\ -1 \\ 1 \end{pmatrix}, \quad \mathbf{v}_3 = \begin{pmatrix} 2+\mathrm{i} \\ -1 \\ 1 \end{pmatrix}.$$

Since $\lambda_1 = 1$, any scalar multiple of the eigenvector $\mathbf{v}_1$ is a fixed point. The fixed points are stable, since the remaining eigenvalues have modulus $|\lambda_2| = |\lambda_3| = \frac{1}{2}\sqrt{2} \approx .7071 < 1$. Thus, the iterates $\mathbf{u}^{(k)} = T^k \mathbf{a} \longrightarrow c_1 \mathbf{v}_1$ will eventually converge to a multiple of the first eigenvector; in almost all cases the convergence rate is $\frac{1}{2}\sqrt{2}$. For example, starting with $\mathbf{u}^{(0)} = (1,1,1)^T$, leads to the iterates†

$$\mathbf{u}^{(5)} = \begin{pmatrix} -9.5 \\ 4.75 \\ -2.75 \end{pmatrix}, \quad \mathbf{u}^{(10)} = \begin{pmatrix} -7.9062 \\ 3.9062 \\ -1.9062 \end{pmatrix}, \quad \mathbf{u}^{(15)} = \begin{pmatrix} -7.9766 \\ 4.0 \\ -2.0 \end{pmatrix},$$

$$\mathbf{u}^{(20)} = \begin{pmatrix} -8.0088 \\ 4.0029 \\ -2.0029 \end{pmatrix}, \quad \mathbf{u}^{(25)} = \begin{pmatrix} -7.9985 \\ 3.9993 \\ -1.9993 \end{pmatrix}, \quad \mathbf{u}^{(30)} = \begin{pmatrix} -8.0001 \\ 4.0001 \\ -2.0001 \end{pmatrix},$$

which are gradually converging to the particular eigenvector $(-8,4,-2)^T = -2\,\mathbf{v}_1$. This can be predicted in advance by decomposing the initial vector into a linear combination of the eigenvectors:

$$\mathbf{u}^{(0)} = \begin{pmatrix} 1 \\ 1 \\ 1 \end{pmatrix} = -2\begin{pmatrix} 4 \\ -2 \\ 1 \end{pmatrix} + \frac{3+3\,\mathrm{i}}{2}\begin{pmatrix} 2-\mathrm{i} \\ -1 \\ 1 \end{pmatrix} + \frac{3-3\,\mathrm{i}}{2}\begin{pmatrix} 2+\mathrm{i} \\ -1 \\ 1 \end{pmatrix},$$

whence

$$\mathbf{u}^{(k)} = \begin{pmatrix} -8 \\ 4 \\ -2 \end{pmatrix} + \frac{3+3\,\mathrm{i}}{2}\left(\frac{1+\mathrm{i}}{2}\right)^k \begin{pmatrix} 2-\mathrm{i} \\ -1 \\ 1 \end{pmatrix} + \frac{3-3\,\mathrm{i}}{2}\left(\frac{1-\mathrm{i}}{2}\right)^k \begin{pmatrix} 2+\mathrm{i} \\ -1 \\ 1 \end{pmatrix},$$

and so $\mathbf{u}^{(k)} \to (-8,4,-2)^T$ as $k \to \infty$. ●

EXERCISES 10.2

10.2.25. Find all fixed points for the linear iterative systems with the following coefficient matrices:

(a) $\begin{pmatrix} .7 & .3 \\ .2 & .8 \end{pmatrix}$ (b) $\begin{pmatrix} .6 & 1.0 \\ .3 & -.7 \end{pmatrix}$

(c) $\begin{pmatrix} -1 & -1 & -4 \\ -2 & 0 & -4 \\ 1 & -1 & 0 \end{pmatrix}$

(d) $\begin{pmatrix} 2 & 1 & -1 \\ 2 & 3 & -2 \\ -1 & -1 & 2 \end{pmatrix}$

10.2.26. Discuss the stability of each fixed point and the asymptotic behavior(s) of the solutions to the systems in Exercise 10.2.25. Which fixed point, if any, does the solution with initial condition $\mathbf{u}^{(0)} = \mathbf{e}_1$ converge to?

10.2.27. Suppose T is a symmetric matrix that satisfies the hypotheses of Theorem 10.17 with a simple eigenvalue $\lambda_1 = 1$. Prove the solution to the linear iterative system has limiting value

$$\lim_{k\to\infty} \mathbf{u}^{(k)} = \frac{\mathbf{u}^{(0)} \cdot \mathbf{v}_1}{\|\mathbf{v}_1\|^2}\,\mathbf{v}_1.$$

†Since the convergence is slow, we only display every fifth one.

10.2.28. *True or false*: If T has a stable nonzero fixed point, then it is a convergent matrix.

10.2.29. *True or false*: If every point $\mathbf{u} \in \mathbb{R}^n$ is a fixed point, then they are all stable. Characterize such systems.

♡**10.2.30.** (a) Under what conditions does the linear iterative system $\mathbf{u}^{(k+1)} = T\,\mathbf{u}^{(k)}$ have a period 2 solution, i.e., $\mathbf{u}^{(k+2)} = \mathbf{u}^{(k)} \neq \mathbf{u}^{(k+1)}$? Give an example of such a system.
(b) Under what conditions is there a unique period 2 solution?
(c) What about a period m solution?

◇**10.2.31.** Prove Theorem 10.18
(a) assuming T is complete,
(b) for general T. *Hint*: Use Exercise 10.1.36.

10.3 MATRIX NORMS AND THE GERSCHGORIN THEOREM

The convergence of a linear iterative system is governed by the spectral radius or largest eigenvalue (in modulus) of the coefficient matrix. Unfortunately, finding accurate approximations to the eigenvalues of most matrices is a nontrivial computational task. Indeed, as we will learn in Section 10.6, all practical numerical algorithms rely on some form of iteration. But using iteration to determine the spectral radius defeats the purpose, which is to predict the behavior of the iterative system in advance!

In this section, we present two alternative approaches for directly investigating convergence and stability issues. Matrix norms form a natural class of norms on the vector space of $n \times n$ matrices and can, in many instances, be used to establish convergence with a minimal effort. The second approach relies on the relatively straightforward yet powerful Gerschgorin Circle Theorem, which serves to restrict the possible locations of the eigenvalues and hence bound the spectral radius of a matrix. Both are effective weapons in the applied mathematician's arsenal.

Matrix Norms

We work exclusively with real $n \times n$ matrices in this section, although the results straightforwardly extend to complex matrices. We begin by fixing a norm $\|\cdot\|$ on $\mathbb{R}^n$. The norm may or may not come from an inner product—this is irrelevant as far as the construction goes. Each norm on $\mathbb{R}^n$ will naturally induce a norm on the vector space $\mathcal{M}_{n\times n}$ of all $n \times n$ matrices. Roughly speaking, the matrix norm tells us how much a linear transformation stretches vectors relative to the given norm.

THEOREM 10.20 If $\|\cdot\|$ is any norm on $\mathbb{R}^n$, then the quantity

$$\|A\| = \max\{\,\|A\,\mathbf{u}\| \mid \|\mathbf{u}\| = 1\,\} \tag{10.30}$$

defines a norm on $\mathcal{M}_{n\times n}$, known as the *natural matrix norm*.

Proof First note that $\|A\| < \infty$, since the maximum is taken on a closed and bounded subset, namely the unit sphere $S_1 = \{\,\|\mathbf{u}\| = 1\,\}$ for the given norm. To show that (10.30) defines a norm, we need to verify the three basic axioms of Definition 3.12.

Non-negativity, $\|A\| \geq 0$, is immediate. Suppose $\|A\| = 0$. This means that, for every unit vector, $\|A\,\mathbf{u}\| = 0$, and hence $A\,\mathbf{u} = \mathbf{0}$ whenever $\|\mathbf{u}\| = 1$. If $\mathbf{0} \neq \mathbf{v} \in \mathbb{R}^n$ is any nonzero vector, then $\mathbf{u} = \mathbf{v}/r$, where $r = \|\mathbf{v}\|$, is a unit vector, so

$$A\,\mathbf{v} = A(r\,\mathbf{u}) = r\,A\,\mathbf{u} = \mathbf{0}. \tag{10.31}$$

Therefore, $A\,\mathbf{v} = \mathbf{0}$ for every $\mathbf{v} \in \mathbb{R}^n$, which implies $A = \mathrm{O}$ is the zero matrix. This serves to prove the positivity property. As for homogeneity, if $c \in \mathbb{R}$ is any scalar,

$$\|c\,A\| = \max\{\,\|c\,A\,\mathbf{u}\|\,\} = \max\{\,|c|\,\|A\,\mathbf{u}\|\,\} = |c|\,\max\{\,\|A\,\mathbf{u}\|\,\} = |c|\,\|A\|.$$

Finally, to prove the triangle inequality, we use the fact that the maximum of the sum of quantities is bounded by the sum of their individual maxima. Therefore, since the norm on $\mathbb{R}^n$ satisfies the triangle inequality,

$$\begin{aligned}\|A+B\| &= \max\{\,\|A\mathbf{u}+B\mathbf{u}\|\,\} \le \max\{\,\|A\mathbf{u}\|+\|B\mathbf{u}\|\,\} \\ &\le \max\{\,\|A\mathbf{u}\|\,\}+\max\{\,\|B\mathbf{u}\|\,\} = \|A\|+\|B\|.\end{aligned}$$ ■

The property that distinguishes a matrix norm from a generic norm on the space of matrices is the fact that it also obeys a very useful *product inequality*.

THEOREM 10.21 A natural matrix norm satisfies

$$\|A\mathbf{v}\| \le \|A\|\,\|\mathbf{v}\|, \qquad \text{for all} \quad A\in\mathcal{M}_{n\times n}, \quad \mathbf{v}\in\mathbb{R}^n. \tag{10.32}$$

Furthermore,

$$\|A\,B\| \le \|A\|\,\|B\|, \qquad \text{for all} \quad A,B\in\mathcal{M}_{n\times n}. \tag{10.33}$$

Proof Note first that, by definition $\|A\mathbf{u}\| \le \|A\|$ for all unit vectors $\|\mathbf{u}\|=1$. Then, letting $\mathbf{v}=r\,\mathbf{u}$ where $\mathbf{u}$ is a unit vector and $r=\|\mathbf{v}\|$, we have

$$\|A\mathbf{v}\| = \|A(r\,\mathbf{u})\| = r\,\|A\mathbf{u}\| \le r\,\|A\| = \|\mathbf{v}\|\,\|A\|,$$

proving the first inequality. To prove the second, we apply the first to compute

$$\begin{aligned}\|A\,B\| &= \max\{\,\|A\,B\,\mathbf{u}\|\,\} = \max\{\,\|A\,(B\,\mathbf{u})\|\,\} \\ &\le \max\{\,\|A\|\,\|B\,\mathbf{u}\|\,\} = \|A\|\,\max\{\,\|B\,\mathbf{u}\|\,\} = \|A\|\,\|B\|.\end{aligned}$$ ■

REMARK: In general, a norm on the vector space of $n\times n$ matrices is called a *matrix norm* if it also satisfies the multiplicative inequality (10.33). Most, but not all, matrix norms used in applications come from norms on the underlying vector space.

The multiplicative inequality (10.33) implies, in particular, that $\|A^2\| \le \|A\|^2$; equality is not necessarily valid. More generally:

Proposition 10.22 If A is a square matrix, then $\|A^k\| \le \|A\|^k$. In particular, if $\|A\|<1$, then $\|A^k\| \to 0$ as $k\to\infty$, and hence A is a convergent matrix: $A^k \to \mathrm{O}$.

The converse is not quite true; a convergent matrix does not necessarily have matrix norm less than 1, or even ≤ 1—see Example 10.27 below. An alternative proof of Proposition 10.22 can be based on the following useful estimate:

THEOREM 10.23 The spectral radius of a matrix is bounded by its matrix norm:

$$\rho(A) \le \|A\|. \tag{10.34}$$

Proof If λ is a real eigenvalue, and $\mathbf{u}$ a corresponding unit eigenvector, so that $A\mathbf{u}=\lambda\,\mathbf{u}$ with $\|\mathbf{u}\|=1$, then

$$\|A\mathbf{u}\| = \|\lambda\,\mathbf{u}\| = |\lambda|\,\|\mathbf{u}\| = |\lambda|. \tag{10.35}$$

Since $\|A\|$ is the maximum of $\|A\mathbf{u}\|$ over all possible unit vectors, this implies that

$$|\lambda| \le \|A\|. \tag{10.36}$$

If all the eigenvalues of A are real, then the spectral radius is the maximum of their absolute values, and so it too is bounded by $\|A\|$, proving (10.34).

If A has complex eigenvalues, then we need to work a little harder to establish (10.36). (This is because the matrix norm is defined by the effect of A on *real* vectors, and so we cannot directly use the complex eigenvectors to establish the required bound.) Let $\lambda = r\,e^{\,\mathrm{i}\,\theta}$ be a complex eigenvalue with complex eigenvector $\mathbf{z} = \mathbf{x} + \mathrm{i}\,\mathbf{y}$. Define

$$m = \min\left\{\, \|\mathrm{Re}(e^{\,\mathrm{i}\,\varphi}\,\mathbf{z})\| = \|(\cos\varphi)\,\mathbf{x} - (\sin\varphi)\,\mathbf{y}\| \;\middle|\; 0 \le \varphi \le 2\pi \,\right\}. \tag{10.37}$$

Since the indicated subset is a closed curve (in fact, an ellipse) that does not go through the origin,* $m > 0$. Let φ_0 denote the value of the angle that produces the minimum, so

$$m = \|(\cos\varphi_0)\,\mathbf{x} - (\sin\varphi_0)\,\mathbf{y}\| = \|\mathrm{Re}\big(e^{\,\mathrm{i}\,\varphi_0}\,\mathbf{z}\big)\|.$$

Define the real unit vector

$$\mathbf{u} = \frac{\mathrm{Re}\big(e^{\,\mathrm{i}\,\varphi_0}\,\mathbf{z}\big)}{m} = \frac{(\cos\varphi_0)\,\mathbf{x} - (\sin\varphi_0)\,\mathbf{y}}{m}, \qquad \text{so that} \quad \|\mathbf{u}\| = 1.$$

Then

$$A\,\mathbf{u} = \frac{1}{m}\,\mathrm{Re}\big(e^{\,\mathrm{i}\,\varphi_0}\,A\,\mathbf{z}\big) = \frac{1}{m}\,\mathrm{Re}\big(e^{\,\mathrm{i}\,\varphi_0}\,r\,e^{\,\mathrm{i}\,\theta}\,\mathbf{z}\big) = \frac{r}{m}\,\mathrm{Re}\big(e^{\,\mathrm{i}\,(\varphi_0+\theta)}\,\mathbf{z}\big).$$

Therefore, keeping in mind that m is the minimal value in (10.37),

$$\|A\| \ge \|A\,\mathbf{u}\| = \frac{r}{m}\,\|\mathrm{Re}\big(e^{\,\mathrm{i}\,(\varphi_0+\theta)}\,\mathbf{z}\big)\| \ge r = |\lambda|, \tag{10.38}$$

and so (10.36) also holds for complex eigenvalues. ■

Explicit Formulae

Let us now determine the explicit formulae for the matrix norms induced by our most important vector norms on $\mathbb{R}^n$. The simplest to handle is the ∞ norm

$$\|\mathbf{v}\|_\infty = \max\{\,|v_1|, \ldots, |v_n|\,\}.$$

Definition 10.24 The ith *absolute row sum* of a matrix A is the sum of the absolute values of the entries in the ith row:

$$s_i = |a_{i1}| + \cdots + |a_{in}| = \sum_{j=1}^{n} |a_{ij}|. \tag{10.39}$$

Proposition 10.25 The ∞ *matrix norm* of a matrix A is equal to its maximal absolute row sum:

$$\|A\|_\infty = \max\{s_1, \ldots, s_n\} = \max\left\{\, \sum_{j=1}^{n} |a_{ij}| \;\middle|\; 1 \le i \le n \,\right\}. \tag{10.40}$$

Proof Let $s = \max\{s_1, \ldots, s_n\}$ denote the right hand side of (10.40). Given any $\mathbf{v} \in \mathbb{R}^n$, we compute

$$\begin{aligned}\|A\,\mathbf{v}\|_\infty &= \max\left\{\left|\sum_{j=1}^{n} a_{ij}v_j\right|\right\} \le \max\left\{\sum_{j=1}^{n} |a_{ij}v_j|\right\} \\ &\le \max\left\{\sum_{j=1}^{n} |a_{ij}|\right\} \max\left\{\,|v_j|\,\right\} = s\,\|\mathbf{v}\|_\infty.\end{aligned}$$

In particular, by specializing to $\|\mathbf{v}\|_\infty = 1$, we deduce that $\|A\|_\infty \le s$.

*This relies on the fact that $\mathbf{x}, \mathbf{y}$ are linearly independent, which was shown in Exercise 8.3.14.

On the other hand, suppose the maximal absolute row sum occurs at row i, so

$$s_i = \sum_{j=1}^{n} |a_{ij}| = s. \tag{10.41}$$

Let $\mathbf{u} \in \mathbb{R}^n$ be the specific vector that has the following entries: $u_j = +1$ if $a_{ij} > 0$, while $u_j = -1$ if $a_{ij} < 0$. Then $\| \mathbf{u} \|_\infty = 1$. Moreover, since $a_{ij} u_j = |a_{ij}|$, the ith entry of $A\mathbf{u}$ is equal to the ith absolute row sum (10.41). This implies that

$$\| A \|_\infty \geq \| A\mathbf{u} \|_\infty \geq s.$$

■

Combining Propositions 10.22 and 10.25, we have established the following convergence criterion.

Corollary 10.26 If all the absolute row sums of A are strictly less than 1, then $\| A \|_\infty < 1$ and hence A is a convergent matrix.

EXAMPLE 10.27 Consider the symmetric matrix

$$A = \begin{pmatrix} \frac{1}{2} & -\frac{1}{3} \\ -\frac{1}{3} & \frac{1}{4} \end{pmatrix}.$$

Its two absolute row sums are $\left| \frac{1}{2} \right| + \left| -\frac{1}{3} \right| = \frac{5}{6}$, $\left| -\frac{1}{3} \right| + \left| \frac{1}{4} \right| = \frac{7}{12}$ so

$$\| A \|_\infty = \max \left\{ \tfrac{5}{6}, \tfrac{7}{12} \right\} = \tfrac{5}{6} \approx .83333\ldots.$$

Since the norm is less than 1, A is a convergent matrix. Indeed, its eigenvalues are

$$\lambda_1 = \frac{9 + \sqrt{73}}{24} \approx .7310\ldots, \qquad \lambda_2 = \frac{9 - \sqrt{73}}{24} \approx .0190\ldots,$$

and hence the spectral radius is

$$\rho(A) = \frac{9 + \sqrt{73}}{24} \approx .7310\ldots,$$

which is slightly smaller than its ∞ norm.

The row sum test for convergence is not always conclusive. For example, the matrix

$$A = \begin{pmatrix} \frac{1}{2} & -\frac{3}{5} \\ \frac{3}{5} & \frac{1}{4} \end{pmatrix} \quad \text{has matrix norm} \quad \| A \|_\infty = \tfrac{11}{10} > 1.$$

On the other hand, its eigenvalues are $(15 \pm \sqrt{601}\,)/40$, and hence its spectral radius is

$$\rho(A) = \frac{15 + \sqrt{601}}{40} \approx .98788\ldots,$$

which implies that A is (just barely) convergent, even though its maximal row sum is larger than 1. ●

Next, let us investigate the matrix norm associated with the Euclidean norm

$$\| \mathbf{v} \|_2 = \sqrt{v_1^2 + \cdots + v_n^2}\,.$$

The result relies on the singular value decomposition of Theorem 8.34.

Proposition 10.28 The matrix norm corresponding to the Euclidean norm equals the maximal singular value:

$$\| A \|_2 = \sigma_1 = \max \{ \sigma_1, \ldots , \sigma_r \}, \qquad r = \operatorname{rank} A > 0, \qquad \text{while} \quad \| \mathrm{O} \|_2 = 0. \tag{10.42}$$

Proof Let $\mathbf{q}_1, \ldots , \mathbf{q}_n$ be an orthonormal basis of $\mathbb{R}^n$ consisting of the singular vectors $\mathbf{q}_1, \ldots , \mathbf{q}_r$ along with an orthonormal basis $\mathbf{q}_{r+1}, \ldots , \mathbf{q}_n$ of $\ker A$. Thus, by (8.41),

$$A\,\mathbf{q}_i = \begin{cases} \sigma_i\,\mathbf{p}_i, & i = 1, \ldots , r, \\ 0, & i = r+1, \ldots n, \end{cases}$$

where $\mathbf{p}_1, \ldots , \mathbf{p}_r$ form an orthonormal basis for $\operatorname{rng} A$. Suppose $\mathbf{u}$ is any unit vector, so

$$\mathbf{u} = c_1\,\mathbf{q}_1 + \cdots + c_n\,\mathbf{q}_n, \qquad \text{where} \quad \| \mathbf{u} \|_2 = \sqrt{c_1^2 + \cdots + c_n^2} = 1,$$

using the orthonormality of the basis vectors and (5.5). Then

$$A\,\mathbf{u} = c_1 \sigma_1 \mathbf{p}_1 + \cdots + c_r \sigma_r \mathbf{p}_r, \qquad \text{and hence} \quad \| A\,\mathbf{u} \|_2 = \sqrt{c_1^2 \sigma_1^2 + \cdots + c_r^2 \sigma_r^2}\,.$$

Now, since $\sigma_1 \geq \sigma_2 \geq \cdots \geq \sigma_r$,

$$\begin{aligned} \| A\,\mathbf{u} \|_2 &= \sqrt{c_1^2 \sigma_1^2 + \cdots + c_r^2 \sigma_r^2} \\ &\leq \sqrt{c_1^2 \sigma_1^2 + \cdots + c_r^2 \sigma_1^2} \\ &= \sigma_1 \sqrt{c_1^2 + \cdots + c_r^2} \\ &\leq \sigma_1 \sqrt{c_1^2 + \cdots + c_n^2} = \sigma_1. \end{aligned}$$

Moreover, if $c_1 = 1, c_2 = \cdots = c_n = 0$, then $\| A\,\mathbf{u} \|_2 = \sigma_1$. We conclude that

$$\| A \|_2 = \max \{ \, \| A\,\mathbf{u} \|_2 \mid \| \mathbf{u} \|_2 = 1 \, \} = \sigma_1.$$

■

Corollary 10.29 If A is symmetric, its Euclidean matrix norm is equal to its spectral radius.

Proof This follows directly from the fact, proved in Proposition 8.33, that the singular values of a symmetric matrix are just the absolute values of its nonzero eigenvalues. ■

EXAMPLE 10.30 Consider the matrix

$$A = \begin{pmatrix} 0 & -\frac{1}{3} & \frac{1}{3} \\ \frac{1}{4} & 0 & \frac{1}{2} \\ \frac{2}{5} & \frac{1}{5} & 0 \end{pmatrix}.$$

The corresponding Gram matrix

$$A^T A = \begin{pmatrix} .2225 & .0800 & .1250 \\ .0800 & .1511 & -.1111 \\ .1250 & -.1111 & .3611 \end{pmatrix},$$

has eigenvalues $\lambda_1 = .4472$, $\lambda_2 = .2665$, $\lambda_3 = .0210$, and hence the singular values of A are their square roots: $\sigma_1 = .6687$, $\sigma_2 = .5163$, $\sigma_3 = .1448$. The Euclidean matrix norm of A is the largest singular value, and so $\| A \|_2 = .6687$, proving that A is a convergent matrix. Note that, as always, the matrix norm overestimates the spectral radius $\rho(A) = .5$. ●

Unfortunately, as we discovered in Example 10.27, matrix norms are not a foolproof test of convergence. There exist convergent matrices such that $\rho(A) < 1$ and yet have matrix norm $\| A \| \geq 1$. In such cases, the matrix norm is not able to predict convergence of the iterative system, although one should expect the convergence to be quite slow. Although such pathology might show up in the chosen matrix norm, it turns out that one can always rig up some matrix norm for which $\| A \| < 1$. This follows from a more general result, whose proof can be found in [48].

THEOREM 10.31 Let A have spectral radius $\rho(A)$. If $\varepsilon > 0$ is any positive number, then there exists a matrix norm $\| \cdot \|$ such that

$$\rho(A) \leq \| A \| < \rho(A) + \varepsilon. \tag{10.43}$$

Corollary 10.32 If A is a convergent matrix, then there exists a matrix norm such that $\| A \| < 1$.

Proof By definition, A is convergent if and only if $\rho(A) < 1$. Choose $\varepsilon > 0$ such that $\rho(A) + \varepsilon < 1$. Any norm that satisfies (10.43) has the desired property. ■

REMARK: Based on the accumulated evidence, one might be tempted to speculate that the spectral radius itself defines a matrix norm. Unfortunately, this is not the case. For example, the nonzero matrix $A = \begin{pmatrix} 0 & 1 \\ 0 & 0 \end{pmatrix}$ has zero spectral radius, $\rho(A) = 0$, in violation of a basic norm axiom.

EXERCISES 10.3

10.3.1. Compute the ∞ matrix norm of the following matrices. Which are guaranteed to be convergent?

(a) $\begin{pmatrix} \frac{1}{2} & \frac{1}{4} \\ \frac{1}{3} & \frac{1}{6} \end{pmatrix}$ (b) $\begin{pmatrix} \frac{5}{3} & \frac{4}{3} \\ -\frac{7}{6} & -\frac{5}{6} \end{pmatrix}$

(c) $\begin{pmatrix} \frac{2}{7} & -\frac{2}{7} \\ -\frac{2}{7} & \frac{6}{7} \end{pmatrix}$ (d) $\begin{pmatrix} \frac{1}{4} & \frac{3}{2} \\ -\frac{1}{2} & \frac{5}{4} \end{pmatrix}$

(e) $\begin{pmatrix} \frac{2}{7} & \frac{2}{7} & -\frac{4}{7} \\ 0 & \frac{2}{7} & \frac{6}{7} \\ \frac{2}{7} & \frac{4}{7} & \frac{2}{7} \end{pmatrix}$ (f) $\begin{pmatrix} 0 & .1 & .8 \\ -.1 & 0 & .1 \\ -.8 & -.1 & 0 \end{pmatrix}$

(g) $\begin{pmatrix} 1 & -\frac{2}{3} & -\frac{2}{3} \\ 1 & -\frac{1}{3} & -1 \\ \frac{1}{3} & -\frac{2}{3} & 0 \end{pmatrix}$ (h) $\begin{pmatrix} \frac{1}{3} & 0 & 0 \\ -\frac{1}{3} & 0 & \frac{1}{3} \\ 0 & \frac{2}{3} & \frac{1}{3} \end{pmatrix}$

10.3.2. Compute the Euclidean matrix norm of each matrix in Exercise 10.3.1. Have your convergence conclusions changed?

10.3.3. Compute the spectral radii of the matrices in Exercise 10.3.1. Which are convergent? Compare your conclusions with those of Exercises 10.3.1 and 10.3.2.

10.3.4. Let k be an integer and set

$$A_k = \begin{pmatrix} k & -1 \\ k^2 & -k \end{pmatrix}.$$

Compute

(a) $\| A_k \|_\infty$ (b) $\| A_k \|_2$ (c) $\rho(A_k)$

(d) Explain why every A_k is a convergent matrix, even though their matrix norms can be arbitrarily large.

(e) Why does this not contradict Corollary 10.32?

10.3.5. Find a matrix A such that

(a) $\| A^2 \|_\infty \neq \| A \|_\infty^2$; (b) $\| A^2 \|_2 \neq \| A \|_2^2$.

10.3.6. Show that if $|c| < 1/\| A \|$, then $c\,A$ is a convergent matrix.

◇ **10.3.7.** Prove that the spectral radius function does *not* satisfy the triangle inequality by finding matrices A, B such that $\rho(A + B) > \rho(A) + \rho(B)$.

10.3.8. *True or false*: If all singular values of A satisfy $\sigma_i < 1$ then A is convergent.

10.3.9. Find a convergent matrix that has dominant singular value $\sigma_1 > 1$.

10.3.10. *True or false*: If $B = S^{-1} A S$ are similar matrices, then

(a) $\| B \|_\infty = \| A \|_\infty$ (b) $\| B \|_2 = \| A \|_2$

(c) $\rho(B) = \rho(A)$

10.3.11. Prove that the condition number of a nonsingular matrix is given by $\kappa(A) = \| A \|_2 \, \| A^{-1} \|_2$.

◇10.3.12. (i) Find an explicit formula for the 1 matrix norm $\| A \|_1$.

(ii) Compute the 1 matrix norm of the matrices in Exercise 10.3.1, and discuss convergence.

10.3.13. Prove directly from the axioms of Definition 3.12 that (10.40) defines a norm on the space of $n \times n$ matrices.

◇10.3.14. Let $K > 0$ be a positive definite matrix. Characterize the matrix norm induced by the inner product $\langle \mathbf{x}, \mathbf{y} \rangle = \mathbf{x}^T K \mathbf{y}$. *Hint*: Use Exercise 8.4.45.

10.3.15. Let

$$A = \begin{pmatrix} 1 & 1 \\ 1 & -2 \end{pmatrix}.$$

Compute the matrix norm $\| A \|$ using the following norms in $\mathbb{R}^2$:

(a) the weighted ∞ norm

$$\| \mathbf{v} \| = \max\{ 2 \, | v_1 |, 3 \, | v_2 | \}$$

(b) the weighted 1 norm $\| \mathbf{v} \| = 2 \, | v_1 | + 3 \, | v_2 |$;

(c) the weighted inner product norm

$$\| \mathbf{v} \| = \sqrt{2 \, v_1^2 + 3 \, v_2^2}$$

(d) the norm associated with the positive definite matrix

$$K = \begin{pmatrix} 2 & -1 \\ -1 & 2 \end{pmatrix}$$

♡10.3.16. The *Frobenius norm* of an $n \times n$ matrix A is defined as

$$\| A \|_F = \sqrt{ \sum_{i,j=1}^{n} a_{ij}^2 } \, .$$

Prove that this defines a matrix norm by checking the three norm axioms plus the multiplicative inequality (10.33).

10.3.17. Let A be an $n \times n$ matrix with singular value vector $\boldsymbol{\sigma} = (\sigma_1, \ldots, \sigma_r)$. Prove that

(a) $\| \boldsymbol{\sigma} \|_\infty = \| A \|_2$;

(b) $\| \boldsymbol{\sigma} \|_2 = \| A \|_F$, the Frobenius norm of Exercise 10.3.16.

Remark: $\| \boldsymbol{\sigma} \|_1$ also defines a useful matrix norm, known as the *Ky Fan norm*.

10.3.18. Explain why $\| A \| = \max | a_{ij} |$ defines a norm on the space of $n \times n$ matrices. Show by example that this is *not* a matrix norm, i.e., (10.33) is not necessarily valid.

10.3.19. Prove that the closed curve parametrized in (10.37) is an ellipse. What are its semi-axes?

◇10.3.20. (a) Prove that the individual entries a_{ij} of a matrix A are bounded in absolute value by its ∞ matrix norm: $| a_{ij} | \le \| A \|_\infty$.

(b) Prove that if the series

$$\sum_{n=0}^{\infty} \| A_n \|_\infty < \infty$$

converges, then the matrix series

$$\sum_{n=0}^{\infty} A_n = A^\star$$

converges to some matrix $A^\star$.

(c) Prove that the exponential matrix series (9.46) converges.

10.3.21. (a) Use Exercise 10.3.20 to prove that the geometric matrix series

$$\sum_{n=0}^{\infty} A^n$$

converges whenever $\rho(A) < 1$. *Hint*: Apply Corollary 10.32.

(b) Prove that the sum is $(\mathrm{I} - A)^{-1}$. How do you know $\mathrm{I} - A$ is invertible?

The Gerschgorin Circle Theorem

In general, precisely computing the eigenvalues, and hence the spectral radius of a matrix, is not easy, and, in most cases, must be done through a numerical eigenvalue routine. In applications, though, we may not require their exact numerical values, but only approximate locations. The *Gerschgorin Circle Theorem*, due to the early twentieth century Russian mathematician Semen Gerschgorin, serves to restrict the eigenvalues to a certain well-defined region in the complex plane. In favorable situations, this information, which is relatively easy to obtain, will suffice for establishing convergence.

Definition 10.33 Let A be an $n \times n$ matrix, either real or complex. For each $1 \le i \le n$, define the *Gerschgorin disk*

$$D_i = \{\, |z - a_{ii}| \le r_i \mid z \in \mathbb{C} \,\}, \quad \text{where} \quad r_i = \sum_{\substack{j=1 \\ j \ne i}}^{n} |a_{ij}|. \tag{10.44}$$

The *Gerschgorin domain* $D_A = \bigcup_{i=1}^{n} D_i \subset \mathbb{C}$ is the union of the Gerschgorin disks.

Thus, the ith Gerschgorin disk D_i is centered at the ith diagonal entry a_{ii}, and has radius r_i equal to the sum of the absolute values of the off-diagonal entries that are in the ith row of A. We can now state the Gerschgorin Circle Theorem.

THEOREM 10.34 All real and complex eigenvalues of the matrix A lie in its Gerschgorin domain D_A.

EXAMPLE 10.35 The matrix

$$A = \begin{pmatrix} 2 & -1 & 0 \\ 1 & 4 & -1 \\ -1 & -1 & -3 \end{pmatrix}$$

has Gerschgorin disks

$$D_1 = \{\, |z-2| \le 1 \,\}, \qquad D_2 = \{\, |z-4| \le 2 \,\}, \qquad D_3 = \{\, |z+3| \le 2 \,\},$$

which are plotted in Figure 10.4. The eigenvalues of A are

$$\lambda_1 = 3, \qquad \lambda_2 = \sqrt{10} = 3.1623\ldots, \qquad \lambda_3 = -\sqrt{10} = -3.1623\ldots.$$

Observe that λ_1 belongs to both D_1 and D_2, while λ_2 lies in D_2, and λ_3 is in D_3. We thus confirm that all three eigenvalues are in the Gerschgorin domain $D_A = D_1 \cup D_2 \cup D_3$. ●

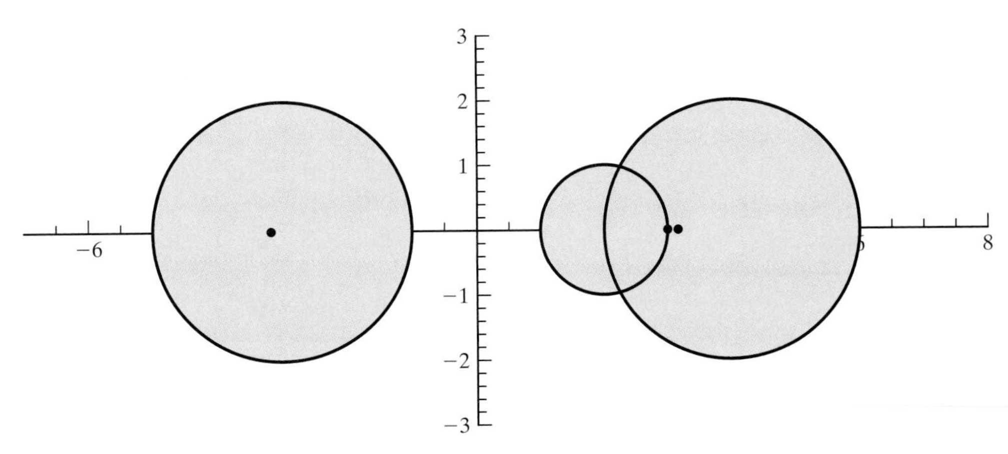

Figure 10.4 Gerschgorin disks and eigenvalues.

Proof of Theorem 10.34 Let $\mathbf{v}$ be an eigenvector of A with eigenvalue λ. Let $\mathbf{u} = \mathbf{v}/\|\mathbf{v}\|_\infty$ be the corresponding unit eigenvector with respect to the ∞ norm, so

$$\|\mathbf{u}\|_\infty = \max\{\, |u_1|, \ldots, |u_n| \,\} = 1.$$

Let u_i be an entry of $\mathbf{u}$ that achieves the maximum: $|u_i| = 1$. Writing out the ith component of the eigenvalue equation $A\mathbf{u} = \lambda\mathbf{u}$, we find

$$\sum_{j=1}^{n} a_{ij}\, u_j = \lambda\, u_i, \qquad \text{which we rewrite as} \qquad \sum_{\substack{j=1 \\ j \neq i}}^{n} a_{ij}\, u_j = (\lambda - a_{ii})\, u_i.$$

Therefore, since all $|u_j| \leq 1$ while $|u_i| = 1$,

$$|\lambda - a_{ii}| = |\lambda - a_{ii}|\,|u_i| = \left| \sum_{j \neq i} a_{ij}\, u_j \right| \leq \sum_{j \neq i} |a_{ij}|\,|u_j| \leq \sum_{j \neq i} |a_{ij}| = r_i.$$

This immediately implies that $\lambda \in D_i \subset D_A$ belongs to the ith Gerschgorin disk. ■

The Gerschgorin Theorem 10.34 can be used to give an alternative proof of Corollary 10.26. If A is any matrix, then the modulus of all points $z \in D_i$ contained in its ith Gerschgorin disk is bounded by the ith absolute row sum,

$$|z| \leq |z - a_{ii}| + |a_{ii}| \leq r_i + |a_{ii}| = s_i,$$

where the final equality follows by comparison of (10.44) and (10.39). Thus, every point $z \in D_A$ in the Gerschgorin set has modulus $|z| \leq \max\{s_1, \ldots, s_n\} = \|A\|_\infty$ bounded by the maximal row sum. Since all eigenvalues λ_j of A are contained in D_A, they too satisfy $|\lambda_j| \leq \|A\|_\infty$, and hence $\rho(A) \leq \|A\|_\infty < 1$, proving convergence.

As a second application, we give a simple direct test that guarantees invertibility of a matrix without requiring Gaussian Elimination or computing determinants. Recall that a matrix A is nonsingular if and only if it does not admit zero as an eigenvalue. Thus, if its Gerschgorin domain does not contain 0, it cannot be an eigenvalue, and hence A is necessarily invertible. The condition $0 \notin D_A$ requires that the matrix have large diagonal entries, as quantified by the following definition.

Definition 10.36 A square matrix A is called *strictly diagonally dominant* if

$$|a_{ii}| > \sum_{\substack{j=1 \\ j \neq i}}^{n} |a_{ij}|, \qquad \text{for all} \quad i = 1, \ldots, n. \tag{10.45}$$

In other words, strict diagonal dominance requires each diagonal entry to be larger, in absolute value, than the sum of the absolute values of *all* the other entries in its row. For example, the matrix

$$\begin{pmatrix} 3 & -1 & 1 \\ 1 & -4 & 2 \\ -2 & -1 & 5 \end{pmatrix}$$

is strictly diagonally dominant since

$$|3| > |-1| + |1|, \qquad |-4| > |1| + |2|, \qquad |5| > |-2| + |-1|.$$

Diagonally dominant matrices appear frequently in numerical solution methods for both ordinary and partial differential equations. As we shall see, they are the most common class of matrices to which iterative solution methods can be successfully applied.

THEOREM 10.37 A strictly diagonally dominant matrix is nonsingular.

Proof The diagonal dominance inequalities (10.45) imply that the radius of the ith Gerschgorin disk is strictly less than the modulus of its center: $r_i < |a_{ii}|$. Thus, the disk cannot contain 0; indeed, if $z \in D_i$, then, by the triangle inequality,

$$r_i > |z - a_{ii}| \geq |a_{ii}| - |z| > r_i - |z|, \qquad \text{and hence} \qquad |z| > 0.$$

Thus, $0 \notin D_A$ does not lie in the Gerschgorin domain and so cannot be an eigenvalue. ■

Warning: The converse to this result is obviously not true; there are plenty of nonsingular matrices that are not diagonally dominant.

EXERCISES 10.3

10.3.22. For each of the following matrices,

(i) find all Gerschgorin disks;

(ii) plot the Gerschgorin domain in the complex plane;

(iii) compute the eigenvalues and confirm the truth of the Circle Theorem 10.34.

(a) $\begin{pmatrix} 1 & -2 \\ -2 & 1 \end{pmatrix}$ (b) $\begin{pmatrix} 1 & -\frac{2}{3} \\ \frac{1}{2} & -\frac{1}{6} \end{pmatrix}$

(c) $\begin{pmatrix} 2 & 3 \\ -1 & 0 \end{pmatrix}$ (d) $\begin{pmatrix} 3 & -1 & 0 \\ -1 & 2 & -1 \\ 0 & -1 & 3 \end{pmatrix}$

(e) $\begin{pmatrix} -1 & 1 & 1 \\ 2 & 2 & -1 \\ 0 & 3 & -4 \end{pmatrix}$ (f) $\begin{pmatrix} \frac{1}{2} & 0 & 0 \\ 0 & 0 & \frac{1}{3} \\ \frac{1}{4} & \frac{1}{6} & 0 \end{pmatrix}$

(g) $\begin{pmatrix} 0 & 1 & 0 \\ 0 & 1 & 1 \\ 0 & -1 & 1 \end{pmatrix}$

(h) $\begin{pmatrix} 3 & 2 & 0 & 0 \\ 1 & 2 & 0 & 0 \\ 0 & 0 & 0 & 1 \\ 0 & 0 & 2 & 1 \end{pmatrix}$

10.3.23. *True or false*: The Gerschgorin domain of the transpose of a matrix A^T is the same as the Gerschgorin domain of the matrix A, that is $D_{A^T} = D_A$.

◇10.3.24. (i) Explain why the eigenvalues of A must lie in its *refined Gerschgorin domain* $D_A^* = D_{A^T} \cap D_A$.

(ii) Find the refined Gerschgorin domains for each of the matrices in Exercise 10.3.22 and confirm the result in part (i).

◇10.3.25. Let A be a square matrix. Prove that $\max\{0, t\} \leq \rho(A) \leq s$, where $s = \max\{s_1, \ldots, s_n\}$ is the maximal absolute row sum of A, as defined in (10.39), and $t = \min\{\, |a_{ii}| - r_i \,\}$, with r_i given by (10.44).

10.3.26. (a) Suppose that every entry of the $n \times n$ matrix A is bounded by

$$|a_{ij}| < \frac{1}{n}.$$

Prove that A is a convergent matrix. *Hint*: Use Exercise 10.3.25.

(b) Produce a matrix of size $n \times n$ with one or more entries satisfying

$$|a_{ij}| = \frac{1}{n}$$

that is not convergent.

10.3.27. Suppose the largest entry (in modulus) of A is $|a_{ij}| = a_\star$. How large can its radius of convergence be?

10.3.28. Write down an example of a diagonally dominant matrix that is also convergent.

10.3.29. *True or false*:

(a) A positive definite matrix is diagonally dominant.

(b) A diagonally dominant matrix is positive definite.

◇10.3.30. Prove that if K is symmetric, diagonally dominant, and each diagonal entry is positive, then K is positive definite.

10.3.31. (a) Write down an invertible matrix A whose Gerschgorin domain contains 0.

(b) Can you find an example which is also diagonally dominant?

10.3.32. Prove that if A is diagonally dominant and each diagonal entry is negative, then the zero equilibrium solution to the linear system of ordinary differential equations $\dot{\mathbf{u}} = A\,\mathbf{u}$ is asymptotically stable.

10.4 MARKOV PROCESSES

A discrete probabilistic process in which the future state of a system depends only upon its current configuration is known as a *Markov chain*, in honor of the pioneering early twentieth studies of the Russian mathematician Andrei Markov. Markov chains are described by linear iterative systems whose coefficient matrices have a special form. They define the simplest examples of stochastic processes, [4, 20], which have many profound physical, biological, economic, and statistical applications. A striking recent application is the immensely popular internet search engine Google Ⓣ. Google's page ranking algorithm is based on modeling the links between web sites as a gigantic Markov process, [50].

To take a very simple (albeit slightly artificial) example, suppose you would like to be able to predict the weather in your city. Consulting local weather records over the past decade, you determine that

(i) If today is sunny, there is a 70% chance that tomorrow will also be sunny,

(ii) But, if today is cloudy, the chances are 80% that tomorrow will also be cloudy.

Question: given that today is sunny, what is the probability that next Saturday's weather will also be sunny?

To mathematically formulate this process, we let $s^{(k)}$ denote the probability that day k is sunny and $c^{(k)}$ the probability that it is cloudy. If we assume that these are the only possibilities, then the individual probabilities must sum to 1, so

$$s^{(k)} + c^{(k)} = 1.$$

According to our data, the probability that the next day is sunny or cloudy is expressed by the equations

$$s^{(k+1)} = .7\, s^{(k)} + .2\, c^{(k)}, \qquad c^{(k+1)} = .3\, s^{(k)} + .8\, c^{(k)}. \tag{10.46}$$

Indeed, day $k+1$ could be sunny either if day k was, with a 70% chance, or, if day k was cloudy, there is still a 20% chance of day $k+1$ being sunny. We rewrite (10.46) in a more convenient matrix form:

$$\mathbf{u}^{(k+1)} = T\,\mathbf{u}^{(k)}, \quad \text{where} \quad T = \begin{pmatrix} .7 & .2 \\ .3 & .8 \end{pmatrix}, \quad \mathbf{u}^{(k)} = \begin{pmatrix} s^{(k)} \\ c^{(k)} \end{pmatrix}. \tag{10.47}$$

In a Markov process, the vector of probabilities $\mathbf{u}^{(k)}$ is known as the kth *state vector* and the matrix T is known as the *transition matrix*, whose entries fix the transition probabilities between the various states.

By assumption, the initial state vector is $\mathbf{u}^{(0)} = (1, 0)^T$, since we know for certain that today is sunny. Rounding off to three decimal places, the subsequent state vectors are

$$\mathbf{u}^{(1)} = \begin{pmatrix} .7 \\ .3 \end{pmatrix}, \quad \mathbf{u}^{(2)} = \begin{pmatrix} .55 \\ .45 \end{pmatrix}, \quad \mathbf{u}^{(3)} = \begin{pmatrix} .475 \\ .525 \end{pmatrix}, \quad \mathbf{u}^{(4)} = \begin{pmatrix} .438 \\ .563 \end{pmatrix},$$

$$\mathbf{u}^{(5)} = \begin{pmatrix} .419 \\ .581 \end{pmatrix}, \quad \mathbf{u}^{(6)} = \begin{pmatrix} .410 \\ .591 \end{pmatrix}, \quad \mathbf{u}^{(7)} = \begin{pmatrix} .405 \\ .595 \end{pmatrix}, \quad \mathbf{u}^{(8)} = \begin{pmatrix} .402 \\ .598 \end{pmatrix}.$$

The iterates converge fairly rapidly to $(.4, .6)^T$, which is, in fact, a fixed point for the iterative system (10.47). Thus, in the long run, 40% of the days will be sunny and 60% will be cloudy. Let us explain why this happens.

Definition 10.38 A vector $\mathbf{u} = (u_1, u_2, \ldots, u_n)^T \in \mathbb{R}^n$ is called a *probability vector* if all its entries lie between 0 and 1, so $0 \leq u_i \leq 1$ for $i = 1, \ldots, n$, and, moreover, their sum is $u_1 + \cdots + u_n = 1$.

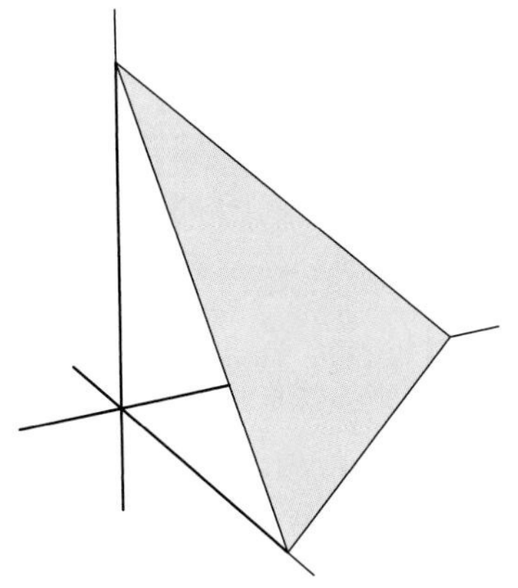

Figure 10.5 Probability vectors in $\mathbb{R}^3$.

We interpret the entry u_i of a probability vector as the probability the system is in state number i. The fact that the entries add up to 1 means that they represent a complete list of probabilities for the possible states of the system. The set of probability vectors defines a *simplex* in $\mathbb{R}^n$. For example, the possible probability vectors $\mathbf{u} \in \mathbb{R}^3$ fill the equilateral triangle plotted in Figure 10.5.

REMARK: Any nonzero vector $\mathbf{0} \neq \mathbf{v} = (v_1, v_2, \ldots, v_n)^T$ with all non-negative entries, $v_i \geq 0$ for $i = 1, \ldots, n$, can be converted into a parallel probability vector by dividing by the sum of its entries:

$$\mathbf{u} = \frac{\mathbf{v}}{v_1 + \cdots + v_n}. \tag{10.48}$$

For example, if $\mathbf{v} = (3, 2, 0, 1)^T$, then $\mathbf{u} = \left(\frac{1}{2}, \frac{1}{3}, 0, \frac{1}{6}\right)^T$ is the corresponding probability vector.

In general, a *Markov chain* is represented by a first order linear iterative system

$$\mathbf{u}^{(k+1)} = T\,\mathbf{u}^{(k)}. \tag{10.49}$$

The entries of the *transition matrix* T must satisfy

$$0 \leq t_{ij} \leq 1, \qquad t_{1j} + \cdots + t_{nj} = 1. \tag{10.50}$$

The entry t_{ij} represents the *transitional probability* that the system will switch from state j to state i. (Note the reversal of indices.) Since this covers all possible transitions, the *column sums* of the transition matrix are all equal to 1, and hence each column of T is a probability vector. In Exercise 10.4.24 you are asked to show that, under these assumptions, if $\mathbf{u}^{(k)}$ is a probability vector, so is $\mathbf{u}^{(k+1)} = T\,\mathbf{u}^{(k)}$. Thus, the solution $\mathbf{u}^{(k)} = T^k\,\mathbf{u}^{(0)}$ to the Markov process represents a sequence or "chain" of probability vectors.

Let us now investigate the convergence of the Markov chain. Not all Markov chains converge—see Exercise 10.4.8 for an example—and so we impose some additional mild restrictions on the transition matrix.

Definition 10.39 A transition matrix (10.50) is *regular* if some power T^k contains no zero entries. In particular, if T itself has no zero entries, then it is regular.

Warning: The term "regular transition matrix" has nothing to do with our earlier term "regular matrix", which was used to describe matrices with an $L\,U$ factorization.

The entries of T^k describe the probabilities of getting from one state to another in k steps. Thus, in words, regularity of the transition matrix means that there is a nonzero probability of getting from any state to any other state in exactly k steps for some $k \geq 1$.

The asymptotic behavior of a regular Markov chain is governed by the following basic result, originally due to Oskar Perron and Georg Frobenius in the early part of the twentieth century. A proof of this theorem can be found at the end of this section.

THEOREM 10.40 If T is a regular transition matrix, then it admits a unique *probability eigenvector* $\mathbf{u}^\star$ with eigenvalue $\lambda_1 = 1$. Moreover, any Markov chain with coefficient matrix T will converge to the probability eigenvector: $\mathbf{u}^{(k)} \to \mathbf{u}^\star$ as $k \to \infty$.

EXAMPLE 10.41 For the weather transition matrix (10.47), the eigenvalues and eigenvectors are

$$\lambda_1 = 1, \quad \mathbf{v}_1 = \begin{pmatrix} \frac{2}{3} \\ 1 \end{pmatrix}, \qquad \lambda_2 = .5, \quad \mathbf{v}_2 = \begin{pmatrix} -1 \\ 1 \end{pmatrix}.$$

The first eigenvector is then converted into a probability vector via (10.48):

$$\mathbf{u}^\star = \mathbf{u}_1 = \frac{1}{1+\frac{2}{3}} \begin{pmatrix} \frac{2}{3} \\ 1 \end{pmatrix} = \begin{pmatrix} \frac{2}{5} \\ \frac{3}{5} \end{pmatrix}.$$

This distinguished probability eigenvector represents the final asymptotic state of the system after many iterations, *no matter what the initial state is*. Thus, our earlier observation that about 40% of the days will be sunny and 60% will be cloudy does not depend upon today's weather. ●

EXAMPLE 10.42 A taxi company in Minnesota serves the cities of Minneapolis and St. Paul, as well as the nearby suburbs. Records indicate that, on average, 10% of the customers taking a taxi in Minneapolis go to St. Paul and 30% go to the suburbs. Customers boarding in St. Paul have a 30% chance of going to Minneapolis and 30% chance of going to the suburbs, while suburban customers choose Minneapolis 40% of the time and St. Paul 30% of the time. The owner of the taxi company is interested in knowing where the taxis will end up, on average.

Let us write this as a Markov process. The entries of the state vector $\mathbf{u}^{(k)} = (u_1^{(k)}, u_2^{(k)}, u_3^{(k)})^T$ tell what proportion of the taxi fleet is, respectively, in Minneapolis, St. Paul and the suburbs, or, equivalently, the probability an individual taxi will end up in one of the three locations. Using the data, we construct the relevant transition matrix

$$T = \begin{pmatrix} .6 & .3 & .4 \\ .1 & .4 & .3 \\ .3 & .3 & .3 \end{pmatrix}.$$

Note that T is regular since it has no zero entries. The probability eigenvector

$$\mathbf{u}^\star \approx (\,.4714,\ .2286,\ .3\,)^T$$

corresponding to the unit eigenvalue $\lambda_1 = 1$ is found by first solving the linear system $(T - \mathrm{I})\mathbf{v}^\star = 0$ and then converting the solution* $\mathbf{v}^\star$ into a valid probability vector $\mathbf{u}^\star$ by use of formula (10.48). According to Theorem 10.40, no matter how the taxis are initially distributed, eventually about 47% of the taxis will be in Minneapolis, 23% in St. Paul, and 30% in the suburbs. This can be confirmed by running numerical experiments on the system. ●

REMARK: The convergence rate of the Markov chain to its steady state is governed by the size of the *subdominant eigenvalue* λ_2. The smaller $|\lambda_2|$ is, the faster the process converges. In the taxi example, $\lambda_2 = .3$ (and $\lambda_3 = 0$), and so the convergence to steady state is fairly rapid.

*Theorem 10.40 guarantees that there is an eigenvector $\mathbf{v}$ with all non-negative entries.

Proof of Theorem 10.40 We begin the proof by replacing T by its transpose[†] $M = T^T$, keeping in mind that every eigenvalue of T is also an eigenvalue of M, cf. Exercise 8.2.35. The conditions (10.50) tell us that the matrix M has entries $0 \le m_{ij} = t_{ji} \le 1$, and, moreover, the *row sums*

$$s_i = \sum_{i=1}^{n} m_{ij} = 1$$

of M, being the same as the corresponding column sums of T, are all equal to 1. Since $M^k = (T^k)^T$, regularity of T implies that some power M^k has all positive entries.

According to Exercise 1.2.29, if $\mathbf{z} = (1, \ldots, 1)^T$ is the column vector all of whose entries are equal to 1, then the entries of $M\,\mathbf{z}$ are the row sums of M. Therefore, $M\,\mathbf{z} = \mathbf{z}$, which implies that $\mathbf{z}$ is an eigenvector of M with eigenvalue $\lambda_1 = 1$. As a consequence, T also has 1 as an eigenvalue. *Warning*: $\mathbf{z}$ is *not* in general an eigenvector of T.

We claim that $\lambda_1 = 1$ is a simple eigenvalue. To this end, we prove that $\mathbf{z}$ spans the one-dimensional eigenspace V_1. In other words, we need to show that if $M\,\mathbf{v} = \mathbf{v}$, then its entries $v_1 = \cdots = v_n = a$ are all equal, and so $\mathbf{v} = a\,\mathbf{z}$ is a scalar multiple of the known eigenvector $\mathbf{z}$. Let us first prove this assuming all of the entries of M are strictly positive, and so $0 < m_{ij} = t_{ji} < 1$ for all i, j. Suppose $\mathbf{v}$ is an eigenvector with not all equal entries. Let v_k be the minimal entry of $\mathbf{v}$, so $v_k \le v_i$ for all $i \ne k$, and at least one inequality is strict, say $v_k < v_j$. Then the kth entry of the eigenvector equation $\mathbf{v} = M\,\mathbf{v}$ is

$$v_k = \sum_{j=1}^{n} m_{kj}\, v_j \; > \; \sum_{j=1}^{n} m_{kj}\, v_k = v_k,$$

where the strict inequality follows from the assumed positivity of the entries of M, and the final equality follows from the fact that M has unit row sums. Thus, we are led to a contradiction, and the claim follows. If M has one or more 0 entries, but M^k has all positive entries, then we apply the previous argument to the equation $M^k\,\mathbf{v} = \mathbf{v}$ which follows from $M\,\mathbf{v} = \mathbf{v}$. If $\lambda_1 = 1$ is a complete eigenvalue, then we are finished. The proof that this is indeed the case is a bit technical, and we refer the reader to [4] for the complete details.

Finally, let us prove that all the other eigenvalues of M are less than 1 in modulus. For this we appeal to the Gerschgorin Circle Theorem 10.34. The Gerschgorin disk D_i is centered at m_{ii} and has radius $r_i = s_i - m_{ii} = 1 - m_{ii}$. Thus the disk lies strictly inside the open unit disk $|z| < 1$ *except* for a single boundary point at $z = 1$; see Figure 10.6. The Circle Theorem 10.34 implies that all eigenvalues except the unit eigenvalue $\lambda_1 = 1$ must lie strictly inside the unit disk, and so $|\lambda_j| < 1$ for $j \ge 2$.

Therefore, the matrix M, and, hence, also T, satisfies the hypotheses of Theorem 10.17. We conclude that the iterates $\mathbf{u}^{(k)} = T^k \mathbf{u}^{(0)} \to \mathbf{u}^\star$ converge to a multiple of the probability eigenvector of T. If the initial condition $\mathbf{u}^{(0)}$ is a probability vector, then so is every subsequent state vector $\mathbf{u}^{(k)}$, and so their limit $\mathbf{u}^\star$ must also be a probability vector. This completes the proof of the theorem. ■

[†] We apologize for the unfortunate clash of notation when writing the transpose of the matrix T.

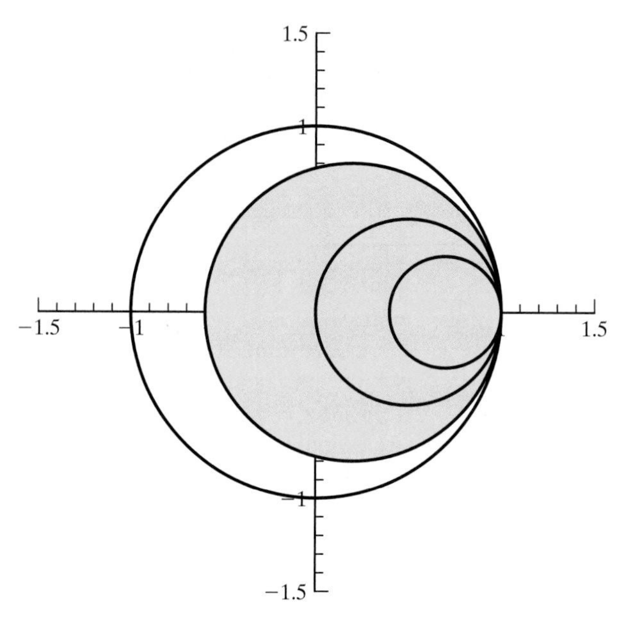

Figure 10.6 Gerschgorin disks for a transition matrix.

EXERCISES 10.4

10.4.1. Determine if the following matrices are regular transition matrices. If so, find the associated probability eigenvector.

(a) $\begin{pmatrix} \frac{1}{2} & \frac{1}{3} \\ \frac{3}{4} & \frac{2}{3} \end{pmatrix}$ (b) $\begin{pmatrix} \frac{1}{4} & \frac{3}{4} \\ \frac{2}{3} & \frac{1}{3} \end{pmatrix}$

(c) $\begin{pmatrix} \frac{1}{4} & \frac{2}{3} \\ \frac{3}{4} & \frac{1}{3} \end{pmatrix}$ (d) $\begin{pmatrix} 0 & \frac{1}{5} \\ 1 & \frac{4}{5} \end{pmatrix}$

(e) $\begin{pmatrix} 0 & 1 & 0 \\ 1 & 0 & 0 \\ 0 & 0 & 1 \end{pmatrix}$ (f) $\begin{pmatrix} .3 & .5 & .2 \\ .3 & .2 & .5 \\ .4 & .3 & .3 \end{pmatrix}$

(g) $\begin{pmatrix} .1 & .5 & 0 \\ .1 & .2 & 1 \\ .8 & .3 & 0 \end{pmatrix}$ (h) $\begin{pmatrix} .1 & .5 & .4 \\ .6 & .1 & .3 \\ .3 & 0 & .7 \end{pmatrix}$

(i) $\begin{pmatrix} \frac{1}{2} & \frac{1}{2} & \frac{1}{3} \\ \frac{1}{2} & 0 & \frac{1}{3} \\ 0 & \frac{1}{2} & \frac{1}{3} \end{pmatrix}$ (j) $\begin{pmatrix} 0 & .2 & 0 & 1 \\ .5 & 0 & .3 & 0 \\ 0 & .8 & 0 & 0 \\ .5 & 0 & .7 & 0 \end{pmatrix}$

(k) $\begin{pmatrix} .1 & .2 & .3 & .4 \\ .2 & .5 & .3 & .1 \\ .3 & .3 & .1 & .3 \\ .4 & .1 & .3 & .2 \end{pmatrix}$

(l) $\begin{pmatrix} 0 & .6 & 0 & .4 \\ .5 & 0 & .3 & .1 \\ 0 & .4 & 0 & .5 \\ .5 & 0 & .7 & 0 \end{pmatrix}$

(m) $\begin{pmatrix} .1 & .3 & .7 & 0 \\ .1 & .2 & 0 & .8 \\ 0 & .5 & 0 & .2 \\ .8 & 0 & .3 & 0 \end{pmatrix}$.

10.4.2. A study has determined that, on average, the occupation of a boy depends on that of his father. If the father is a farmer, there is a 30% chance that the son will be a blue collar laborer, a 30% chance he will be a white collar professional, and a 40% chance he will also be a farmer. If the father is a laborer, there is a 30% chance that the son will also be one, a 60% chance he will be a professional, and a 10% chance he will be a farmer. If the father is a professional, there is a 70% chance that the son will also be one, a 25% chance he will be a laborer, and a 5% chance he will be a farmer.

(a) What is the probability that the grandson of a farmer will also be a farmer?

(b) In the long run, what proportion of the male population will be farmers?

10.4.3. The population of an island is divided into city and country residents. Each year, 5% of the residents of the city move to the country and 15% of the residents of the country move to the city. In 2003, 35,000 people live in the city and 25,000 in the country. Assuming no growth in the population, how many people will live in the city and how many will live in the country between the years 2004 and 2008? What is the eventual population distribution of the island?

10.4.4. A student has the habit that if she doesn't study one night, she is 70% certain of studying the next night. Furthermore, the probability that she studies two nights in a row is 50%. How often does she study in the long run?

10.4.5. A traveling salesman visits the three cities of Atlanta, Boston, and Chicago. The matrix

$$\begin{pmatrix} 0 & .5 & .5 \\ 1 & 0 & .5 \\ 0 & .5 & 0 \end{pmatrix}$$

describes the transition probabilities of his trips. Describe his travels in words, and calculate how often he visits each city on average.

10.4.6. A business executive is managing three branches, labeled A, B and C, of a corporation. She never visits the same branch on consecutive days. If she visits branch A one day, she visits branch B the next day. If she visits either branch B or C that day, then the next day she is twice as likely to visit branch A as to visit branch B or C. Explain why the resulting transition matrix is regular. Which branch does she visit the most often in the long run?

10.4.7. A certain plant species has either red, pink, or white flowers, depending on its genotype. If you cross a pink plant with any other plant, the probability distribution of the offspring are prescribed by the transition matrix

$$T = \begin{pmatrix} .5 & .25 & 0 \\ .5 & .5 & .5 \\ 0 & .25 & .5 \end{pmatrix}.$$

On average, if you continue only crossing with pink plants, what percentage of the three types of flowers would you expect to see in your garden?

10.4.8. Explain why the irregular Markov process with transition matrix

$$T = \begin{pmatrix} 0 & 1 \\ 1 & 0 \end{pmatrix}$$

does not reach a steady state. Use a population model as is Exercise 10.4.3 to interpret what is going on.

10.4.9. A genetic model describing inbreeding, in which mating takes place only between individuals of the same genotype, is given by the Markov process $\mathbf{u}^{(n+1)} = T\,\mathbf{u}^{(n)}$, where

$$T = \begin{pmatrix} 1 & \frac{1}{4} & 0 \\ 0 & \frac{1}{2} & 0 \\ 0 & \frac{1}{4} & 1 \end{pmatrix}$$

is the transition matrix and

$$\mathbf{u}^{(n)} = \begin{pmatrix} p_n \\ q_n \\ r_n \end{pmatrix},$$

whose entries are, respectively, the proportion of populations of genotype AA, Aa, aa in the nth generation. Find the solution to this Markov process and analyze your result.

10.4.10. A bug crawls along the edges of the pictured triangular lattice with six vertices. Upon arriving at a vertex, there is an equal probability of its choosing any edge to leave the vertex. Set up the Markov chain described by the bug's motion, and determine how often, on average, it visits each vertex.

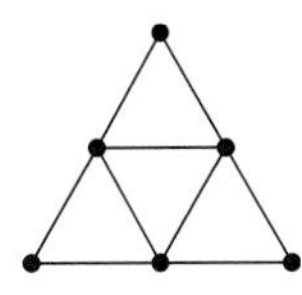

10.4.11. Answer Exercise 10.4.10 for the larger triangular lattice.

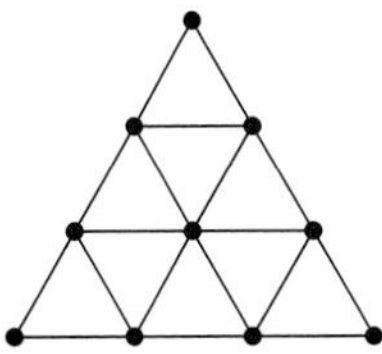

10.4.12. Suppose the bug of Exercise 10.4.10 crawls along the edges of the pictured square lattice. What can you say about its behavior?

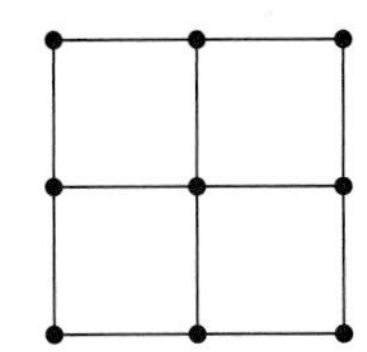

◇**10.4.13.** Let T be a regular transition matrix with probability eigenvector $\mathbf{v}$. Prove that

$$\lim_{k\to\infty} T^k = (\,\mathbf{v}\;\mathbf{v}\;\ldots\;\mathbf{v}\,)$$

is a matrix with every column equal to $\mathbf{v}$.

10.4.14. Find $\lim_{k\to\infty} T^k$ when

$$T = \begin{pmatrix} .8 & .1 & .1 \\ .1 & .8 & .1 \\ .1 & .1 & .8 \end{pmatrix}.$$

10.4.15. Prove that, for all $0 \le p, q \le 1$ with $p + q > 0$, the probability eigenvector of the transition matrix

$$T = \begin{pmatrix} 1-p & q \\ p & 1-q \end{pmatrix}$$

is

$$\mathbf{v} = \left(\frac{q}{p+q},\; \frac{p}{p+q} \right)^T.$$

10.4.16. A transition matrix is called *doubly stochastic* if both its row and column sums are equal to 1. What is the limiting probability state of a Markov chain with doubly stochastic transition matrix?

10.4.17. Describe the final state of a Markov chain with symmetric transition matrix $T = T^T$.

10.4.18. *True or false*: If T and T^T are both transition matrices, then $T = T^T$.

10.4.19. *True or false*: If T is a transition matrix, so is T^{-1}.

10.4.20. *True or false*: The set of all probability vectors forms a subspace of $\mathbb{R}^n$.

10.4.21. *Multiple Choice*: Every probability vector in $\mathbb{R}^n$ lies on the unit sphere for the

(a) 1 norm (b) 2 norm (c) ∞ norm (d) all of the above (e) none of the above

10.4.22. *True or false*: Every probability eigenvector of a regular transition matrix has eigenvalue equal to 1.

10.4.23. Write down an example of

(a) an irregular transition matrix;

(b) a regular transition matrix that has one or more zero entries.

◇**10.4.24.** Let T be a transition matrix. Prove that if $\mathbf{u}$ is a probability vector, so is $\mathbf{v} = T\,\mathbf{u}$.

◇**10.4.25.** (a) Prove that if T and S are transition matrices, so is their product $T\,S$.

(b) Prove that if T is a transition matrix, so is T^k for any $k \geq 0$.

10.5 ITERATIVE SOLUTION OF LINEAR ALGEBRAIC SYSTEMS

In this section, we return to the most basic problem in linear algebra: solving the linear algebraic system

$$A\,\mathbf{u} = \mathbf{b}, \tag{10.51}$$

consisting of n equations in n unknowns. We assume that the coefficient matrix A is nonsingular, and so the solution $\mathbf{u} = A^{-1}\mathbf{b}$ is unique.

We will introduce several popular iterative methods that can be used to approximate the solution for certain classes of coefficient matrices. The resulting algorithms will provide an attractive alternative to Gaussian Elimination, particularly when dealing with the large, sparse systems that arise in the numerical solution to differential equations. One major advantage of an iterative technique is that it (typically) produces progressively more and more accurate approximations to the solution, and hence, by prolonging the iterations, can, at least in principle, compute the solution to any desired order of accuracy. Moreover, even performing just a few iterations may produce a reasonable approximation to the true solution—in stark contrast to Gaussian Elimination, where one must continue the algorithm through to the bitter end before any useful information can be extracted. A partially completed Gaussian Elimination is of scant use! A significant weakness is that iterative schemes are not universally applicable, and their design relies upon the detailed structure of the coefficient matrix.

We shall be attempting to solve the linear system (10.51) by replacing it with an iterative system of the form

$$\mathbf{u}^{(k+1)} = T\,\mathbf{u}^{(k)} + \mathbf{c}, \qquad \mathbf{u}^{(0)} = \mathbf{u}_0, \tag{10.52}$$

in which T is an $n \times n$ matrix and $\mathbf{c}$ a vector. This represents a slight generalization of our earlier iterative system (10.1), in that the right hand side is now an affine function of $\mathbf{u}^{(k)}$. Suppose that the solutions to the affine iterative system converge: $\mathbf{u}^{(k)} \to \mathbf{u}^\star$ as $k \to \infty$. Then, by taking the limit of both sides of (10.52), we discover that the limit point $\mathbf{u}^\star$ solves the *fixed-point equation*

$$\mathbf{u}^\star = T\,\mathbf{u}^\star + \mathbf{c}. \tag{10.53}$$

Thus, we need to design our iterative system so that

(a) the solution to the fixed-point system $\mathbf{u} = T\,\mathbf{u} + \mathbf{c}$ coincides with the solution to the original system $A\,\mathbf{u} = \mathbf{b}$, and

(b) the iterates defined by (10.52) are known to converge to the fixed point.

Before exploring these issues in depth, let us look at a simple example.

EXAMPLE 10.43 Consider the linear system

$$3x + y - z = 3, \quad x - 4y + 2z = -1, \quad -2x - y + 5z = 2, \tag{10.54}$$

which has the vectorial form $A\mathbf{u} = \mathbf{b}$, with

$$A = \begin{pmatrix} 3 & 1 & -1 \\ 1 & -4 & 2 \\ -2 & -1 & 5 \end{pmatrix}, \quad \mathbf{u} = \begin{pmatrix} x \\ y \\ z \end{pmatrix}, \quad \mathbf{b} = \begin{pmatrix} 3 \\ -1 \\ 2 \end{pmatrix}.$$

One easy way to convert a linear system into a fixed-point form is to rewrite it as

$$\mathbf{u} = \mathrm{I}\,\mathbf{u} - A\mathbf{u} + A\mathbf{u} = (\mathrm{I} - A)\mathbf{u} + \mathbf{b} = T\mathbf{u} + \mathbf{c}, \quad \text{where} \quad T = \mathrm{I} - A, \quad \mathbf{c} = \mathbf{b}.$$

In the present case,

$$T = \mathrm{I} - A = \begin{pmatrix} -2 & -1 & 1 \\ -1 & 5 & -2 \\ 2 & 1 & -4 \end{pmatrix}, \quad \mathbf{c} = \mathbf{b} = \begin{pmatrix} 3 \\ -1 \\ 2 \end{pmatrix}.$$

The resulting iterative system $\mathbf{u}^{(k+1)} = T\mathbf{u}^{(k)} + \mathbf{c}$ has the explicit form

$$\begin{aligned} x^{(k+1)} &= -2x^{(k)} - y^{(k)} + z^{(k)} + 3, \\ y^{(k+1)} &= -x^{(k)} + 5y^{(k)} - 2z^{(k)} - 1, \\ z^{(k+1)} &= 2x^{(k)} + y^{(k)} - 4z^{(k)} + 2. \end{aligned} \tag{10.55}$$

Another possibility is to solve the first equation in (10.54) for x, the second for y, and the third for z, so that

$$x = -\tfrac{1}{3}y + \tfrac{1}{3}z + 1, \quad y = \tfrac{1}{4}x + \tfrac{1}{2}z + \tfrac{1}{4}, \quad z = \tfrac{2}{5}x + \tfrac{1}{5}y + \tfrac{2}{5}.$$

The resulting equations have the form of a fixed-point system

$$\mathbf{u} = \widehat{T}\mathbf{u} + \widehat{\mathbf{c}}, \quad \text{in which} \quad \widehat{T} = \begin{pmatrix} 0 & -\frac{1}{3} & \frac{1}{3} \\ \frac{1}{4} & 0 & \frac{1}{2} \\ \frac{2}{5} & \frac{1}{5} & 0 \end{pmatrix}, \quad \widehat{\mathbf{c}} = \begin{pmatrix} 1 \\ \frac{1}{4} \\ \frac{2}{5} \end{pmatrix}.$$

The corresponding iteration $\mathbf{u}^{(k+1)} = \widehat{T}\mathbf{u}^{(k)} + \widehat{\mathbf{c}}$ takes the explicit form

$$\begin{aligned} x^{(k+1)} &= -\tfrac{1}{3}y^{(k)} + \tfrac{1}{3}z^{(k)} + 1, \\ y^{(k+1)} &= \tfrac{1}{4}x^{(k)} + \tfrac{1}{2}z^{(k)} + \tfrac{1}{4}, \\ z^{(k+1)} &= \tfrac{2}{5}x^{(k)} + \tfrac{1}{5}y^{(k)} + \tfrac{2}{5}. \end{aligned} \tag{10.56}$$

Do the resulting iterative schemes converge to the solution $x = y = z = 1$? The results, starting with initial guess $\mathbf{u}^{(0)} = (0, 0, 0)$, are tabulated as follows:

k	$\mathbf{u}^{(k+1)} = T\,\mathbf{u}^{(k)} + \mathbf{b}$			$\mathbf{u}^{(k+1)} = \widehat{T}\,\mathbf{u}^{(k)} + \widehat{\mathbf{c}}$		
0	0	0	0	0	0	0
1	3	−1	2	1	.25	.4
2	0	−13	−1	1.05	.7	.85
3	15	−64	−7	1.05	.9375	.96
4	30	−322	−4	1.0075	.9925	1.0075
5	261	−1633	−244	1.005	1.00562	1.0015
6	870	−7939	−133	.9986	1.002	1.0031
7	6069	−40300	−5665	1.0004	1.0012	.9999
8	22500	−196240	−5500	.9995	1.0000	1.0004
9	145743	−992701	−129238	1.0001	1.0001	.9998
10	571980	−4850773	−184261	.9999	.9999	1.0001
11	3522555	−24457324	−2969767	1.0000	1.0000	1.0000

For the first scheme, the answer is clearly no—the iterates become wilder and wilder. Indeed, this occurs no matter how close the initial guess $\mathbf{u}^{(0)}$ is to the actual solution—unless $\mathbf{u}^{(0)} = \mathbf{u}^\star$ happens to be exactly equal. In the second case, the iterates do converge to the solution, and it does not take too long, even starting from a poor initial guess, to obtain a reasonably accurate approximation. Of course, in such a simple example, it would be silly to use iteration, when Gaussian Elimination can be done by hand and produces the solution almost immediately. However, we use the small examples for illustrative purposes, bringing the full power of iterative schemes to bear on the large linear systems arising in applications. ●

The convergence of solutions to (10.52) to the fixed point $\mathbf{u}^\star$ is based on the behavior of the *error vectors*

$$\mathbf{e}^{(k)} = \mathbf{u}^{(k)} - \mathbf{u}^\star, \tag{10.57}$$

which measure how close the iterates are to the true solution. Let us find out how the successive error vectors are related. We compute

$$\mathbf{e}^{(k+1)} = \mathbf{u}^{(k+1)} - \mathbf{u}^\star = (T\,\mathbf{u}^{(k)} + \mathbf{a}) - (T\,\mathbf{u}^\star + \mathbf{a}) = T(\mathbf{u}^{(k)} - \mathbf{u}^\star) = T\,\mathbf{e}^{(k)},$$

showing that the error vectors satisfy a *linear* iterative system

$$\mathbf{e}^{(k+1)} = T\,\mathbf{e}^{(k)}, \tag{10.58}$$

with the *same* coefficient matrix T. Therefore, they are given by the explicit formula

$$\mathbf{e}^{(k)} = T^k\,\mathbf{e}^{(0)}.$$

Now, the solutions to (10.52) converge to the fixed point, $\mathbf{u}^{(k)} \to \mathbf{u}^\star$, if and only if the error vectors converge to zero: $\mathbf{e}^{(k)} \to \mathbf{0}$ as $k \to \infty$. Our analysis of linear iterative systems, as summarized in Proposition 10.11, establishes the following basic convergence result.

Proposition 10.44 The iterative system (10.52) will converge to the solution to the fixed point equation (10.53) if and only if T is a convergent matrix: $\rho(T) < 1$.

The spectral radius $\rho(T)$ of the coefficient matrix will govern the speed of convergence. Therefore, our main goal is to construct an iterative scheme whose coefficient matrix has as small a spectral radius as possible. At the very least, the spectral radius must be less than 1. For the two iterative schemes presented in Example 10.43, the spectral radii of the coefficient matrices are found to be

$$\rho(T) \approx 4.9675, \qquad \rho(\widehat{T}) = .5.$$

Therefore, T is not a convergent matrix, which explains the wild behavior of its iterates, whereas $\widehat{T}$ is convergent, and one expects the error to roughly decrease by a factor of $\frac{1}{2}$ at each step.

The Jacobi Method

The first general iterative scheme for solving linear systems is based on the same simple idea used in our illustrative Example 10.43. Namely, we solve the ith equation in the system $A\,\mathbf{u} = \mathbf{b}$, which is

$$\sum_{j=1}^{n} a_{ij}\,u_j = b_i,$$

for the ith variable u_i. To do this, we need to assume that all the diagonal entries of A are nonzero: $a_{ii} \neq 0$. The result is

$$u_i = -\frac{1}{a_{ii}} \sum_{\substack{j=1 \\ j\neq i}}^{n} a_{ij}\,u_j + \frac{b_i}{a_{ii}} = \sum_{j=1}^{n} t_{ij}\,u_j + c_i, \tag{10.59}$$

where

$$t_{ij} = \begin{cases} -\dfrac{a_{ij}}{a_{ii}}, & i \neq j, \\ 0, & i = j, \end{cases} \qquad \text{and} \qquad c_i = \frac{b_i}{a_{ii}}. \tag{10.60}$$

The result has the form of a fixed-point system $\mathbf{u} = T\,\mathbf{u} + \mathbf{c}$, and forms the basis of the *Jacobi method*

$$\mathbf{u}^{(k+1)} = T\,\mathbf{u}^{(k)} + \mathbf{c}, \qquad \mathbf{u}^{(0)} = \mathbf{u}_0, \tag{10.61}$$

named after the influential nineteenth century German analyst Carl Jacobi. The explicit form of the Jacobi iterative scheme is

$$u_i^{(k+1)} = -\frac{1}{a_{ii}} \sum_{\substack{j=1 \\ j\neq i}}^{n} a_{ij}\,u_j^{(k)} + \frac{b_i}{a_{ii}}. \tag{10.62}$$

It is instructive to rederive the Jacobi method in a direct matrix form. We begin by decomposing the coefficient matrix

$$A = L + D + U \tag{10.63}$$

into the sum of a strictly lower triangular matrix L, a diagonal matrix D, and a strictly upper triangular matrix U, each of which is uniquely specified. For example, when

$$A = \begin{pmatrix} 3 & 1 & -1 \\ 1 & -4 & 2 \\ -2 & -1 & 5 \end{pmatrix}, \tag{10.64}$$

the decomposition (10.63) yields

$$L = \begin{pmatrix} 0 & 0 & 0 \\ 1 & 0 & 0 \\ -2 & -1 & 0 \end{pmatrix}, \quad D = \begin{pmatrix} 3 & 0 & 0 \\ 0 & -4 & 0 \\ 0 & 0 & 5 \end{pmatrix}, \quad U = \begin{pmatrix} 0 & 1 & -1 \\ 0 & 0 & 2 \\ 0 & 0 & 0 \end{pmatrix}.$$

Warning: The L, D, U in the elementary additive decomposition (10.63) have nothing to do with the L, D, U appearing in factorizations arising from Gaussian Elimination. The latter play no role in the iterative solution methods considered here.

We then rewrite the system

$$A\mathbf{u} = (L + D + U)\,\mathbf{u} = \mathbf{b}$$

in the alternative form

$$D\,\mathbf{u} = -\,(L + U)\,\mathbf{u} + \mathbf{b}.$$

The Jacobi fixed point equations (10.59) amounts to solving for

$$\mathbf{u} = T\,\mathbf{u} + \mathbf{c}, \qquad \text{where} \quad T = -\,D^{-1}(L + U), \quad \mathbf{c} = D^{-1}\mathbf{b}. \tag{10.65}$$

For the example (10.64), we recover the Jacobi iteration matrix as

$$T = -\,D^{-1}(L + U) = \begin{pmatrix} 0 & -\frac{1}{3} & \frac{1}{3} \\ \frac{1}{4} & 0 & \frac{1}{2} \\ \frac{2}{5} & \frac{1}{5} & 0 \end{pmatrix}.$$

Deciding in advance whether or not the Jacobi method will converge is not easy. However, it can be shown that Jacobi iteration *is* guaranteed to converge when the original coefficient matrix has large diagonal entries, in accordance with Definition 10.36.

THEOREM 10.45 If A is strictly diagonally dominant, then the associated Jacobi iteration scheme converges.

Proof We shall prove that $\| T \|_\infty < 1$, and so Corollary 10.26 implies that T is a convergent matrix. The absolute row sums of the Jacobi matrix $T = -\,D^{-1}(L + U)$ are, according to (10.60),

$$s_i = \sum_{j=1}^{n} |t_{ij}| = \frac{1}{|a_{ii}|} \sum_{\substack{j=1 \\ j \neq i}}^{n} |a_{ij}| < 1, \tag{10.66}$$

because A is strictly diagonally dominant. Thus, $\| T \|_\infty = \max\{s_1, \ldots, s_n\} < 1$, and the result follows. ■

EXAMPLE 10.46 Consider the linear system

$$\begin{aligned} 4x + y + w &= 1 \\ x + 4y + z + v &= 2 \\ y + 4z + w &= -1 \\ x + z + 4w + v &= 2 \\ y + w + 4v &= 1. \end{aligned}$$

The Jacobi method solves the respective equations for x, y, z, w, v, leading to the iterative scheme

$$\begin{aligned} x^{(k+1)} &= -\tfrac{1}{4}\,y^{(k)} - \tfrac{1}{4}\,w^{(k)} + \tfrac{1}{4} \\ y^{(k+1)} &= -\tfrac{1}{4}\,x^{(k)} - \tfrac{1}{4}\,z^{(k)} - \tfrac{1}{4}\,v^{(k)} + \tfrac{1}{2} \\ z^{(k+1)} &= -\tfrac{1}{4}\,y^{(k)} - \tfrac{1}{4}\,w^{(k)} - \tfrac{1}{4} \\ w^{(k+1)} &= -\tfrac{1}{4}\,x^{(k)} - \tfrac{1}{4}\,z^{(k)} - \tfrac{1}{4}\,v^{(k)} + \tfrac{1}{2} \\ v^{(k+1)} &= -\tfrac{1}{4}\,y^{(k)} - \tfrac{1}{4}\,w^{(k)} + \tfrac{1}{4}. \end{aligned}$$

The coefficient matrix of the original system

$$A = \begin{pmatrix} 4 & 1 & 0 & 1 & 0 \\ 1 & 4 & 1 & 0 & 1 \\ 0 & 1 & 4 & 1 & 0 \\ 1 & 0 & 1 & 4 & 1 \\ 0 & 1 & 0 & 1 & 4 \end{pmatrix}$$

is diagonally dominant, and so we are guaranteed that the Jacobi iterations will eventually converge to the solution. Indeed, the Jacobi scheme takes the iterative form (10.65), with

$$T = \begin{pmatrix} 0 & -\frac14 & 0 & -\frac14 & 0 \\ -\frac14 & 0 & -\frac14 & 0 & -\frac14 \\ 0 & -\frac14 & 0 & -\frac14 & 0 \\ -\frac14 & 0 & -\frac14 & 0 & -\frac14 \\ 0 & -\frac14 & 0 & -\frac14 & 0 \end{pmatrix}, \qquad \mathbf{c} = \begin{pmatrix} \frac14 \\ \frac12 \\ -\frac14 \\ \frac12 \\ \frac14 \end{pmatrix}.$$

Note that $\| T \|_\infty = \frac34 < 1$, validating convergence of the scheme. Thus, to obtain, say, four decimal place accuracy in the solution, we estimate that it would take less than $\log(.5 \times 10^{-4}) / \log .75 \approx 34$ iterates, assuming a moderate initial error. But the matrix norm always underestimates the true rate of convergence, as prescribed by the spectral radius $\rho(T) = .6124$, which would imply about $\log(.5 \times 10^{-4}) / \log .6124 \approx 20$ iterations to obtain the desired accuracy. Indeed, starting with the initial guess $x^{(0)} = y^{(0)} = z^{(0)} = w^{(0)} = v^{(0)} = 0$, the Jacobi iterates converge to the exact solution

$$x = -.1, \quad y = .7, \quad z = -.6, \quad w = .7, \quad v = -.1,$$

to within four decimal places in exactly 20 iterations. ●

EXERCISES 10.5

10.5.1. (a) Find the spectral radius of the matrix

$$T = \begin{pmatrix} 1 & 1 \\ -1 & -\frac76 \end{pmatrix}.$$

(b) Predict the long term behavior of the iterative system

$$\mathbf{u}^{(k+1)} = T\,\mathbf{u}^{(k)} + \mathbf{b}, \qquad \text{where} \quad \mathbf{b} = \begin{pmatrix} -1 \\ 2 \end{pmatrix},$$

in as much detail as you can.

10.5.2. Answer Exercise 10.5.1 when

(a) $T = \begin{pmatrix} 1 & -\frac12 \\ -1 & \frac32 \end{pmatrix}, \quad \mathbf{b} = \begin{pmatrix} 0 \\ 1 \end{pmatrix}$

(b) $T = \begin{pmatrix} \frac14 & \frac14 & 0 \\ 0 & 0 & \frac14 \\ 1 & 1 & \frac14 \end{pmatrix}, \quad \mathbf{b} = \begin{pmatrix} 1 \\ -1 \\ 3 \end{pmatrix}$

(c) $T = \begin{pmatrix} -.05 & .15 & .15 \\ .35 & .15 & -.35 \\ -.2 & -.2 & .3 \end{pmatrix}, \quad \mathbf{b} = \begin{pmatrix} -1.5 \\ 1.6 \\ 1.7 \end{pmatrix}$

10.5.3. Which of the following systems have a diagonally dominant coefficient matrix?

(a) $\begin{aligned} 5x - y &= 1 \\ -x + 3y &= -1 \end{aligned}$

(b) $\begin{aligned} \tfrac12 x + \tfrac13 y &= 1 \\ \tfrac15 x + \tfrac14 y &= 6 \end{aligned}$

(c) $\begin{aligned} -5x + y &= 3 \\ -3x + 2y &= -2 \end{aligned}$

(d) $\begin{aligned} -2x + y + z &= 1 \\ -x + 2y - z &= -2 \\ x - y + 3z &= 1 \end{aligned}$

(e) $\begin{aligned} -x + \tfrac12 y + \tfrac13 z &= 1 \\ \tfrac13 x + 2y + \tfrac34 z &= -3 \\ \tfrac23 x + \tfrac14 y - \tfrac32 z &= 2 \end{aligned}$

(f) $\begin{aligned} x - 2y + z &= 1 \\ -x + 2y + z &= -1 \\ x + 3y - 2z &= 3 \end{aligned}$

(g) $-4x+2y+z=2$
$-x+3y+z=-1$
$x+4y-6z=3$

♠ **10.5.4.** For the diagonally dominant systems in Exercise 10.5.3, starting with the initial guess $x=y=z=0$, compute the solution to 2 decimal places using the Jacobi method. Check your answer by solving the system directly by Gaussian Elimination.

♠ **10.5.5.** (a) Do any of the non-diagonally dominant systems in Exercise 10.5.3 lead to convergent Jacobi schemes? *Hint*: Check the spectral radius of the Jacobi matrix.

(b) For the convergent systems in Exercise 10.5.3, starting with the initial guess $x=y=z=0$, compute the solution to 2 decimal places by using the Jacobi method, and check your answer by solving the system directly by Gaussian Elimination.

10.5.6. The following linear systems have positive definite coefficient matrices. Use the Jacobi method starting with $\mathbf{u}^{(0)}=\mathbf{0}$ to find the solution to 4 decimal place accuracy.

(a) $\begin{pmatrix} 3 & -1 \\ -1 & 5 \end{pmatrix}\mathbf{u}=\begin{pmatrix} 2 \\ 1 \end{pmatrix}$

(b) $\begin{pmatrix} 2 & 1 \\ 1 & 1 \end{pmatrix}\mathbf{u}=\begin{pmatrix} -3 \\ 1 \end{pmatrix}$

(c) $\begin{pmatrix} 6 & -1 & -3 \\ -1 & 7 & 4 \\ -3 & 4 & 9 \end{pmatrix}\mathbf{u}=\begin{pmatrix} -1 \\ -2 \\ 7 \end{pmatrix}$

(d) $\begin{pmatrix} 3 & -1 & 0 \\ -1 & 2 & 1 \\ 0 & 1 & 5 \end{pmatrix}\mathbf{u}=\begin{pmatrix} 1 \\ -5 \\ 0 \end{pmatrix}$

(e) $\begin{pmatrix} 5 & 1 & 1 & 1 \\ 1 & 5 & 1 & 1 \\ 1 & 1 & 5 & 1 \\ 1 & 1 & 1 & 5 \end{pmatrix}\mathbf{u}=\begin{pmatrix} 4 \\ 0 \\ 0 \\ 0 \end{pmatrix}$

(f) $\begin{pmatrix} 3 & 1 & 0 & -1 \\ 1 & 3 & 1 & 0 \\ 0 & 1 & 3 & 1 \\ -1 & 0 & 1 & 3 \end{pmatrix}\mathbf{u}=\begin{pmatrix} 1 \\ 2 \\ 0 \\ -1 \end{pmatrix}$

♣ **10.5.7.** Let A be the $n \times n$ tridiagonal matrix with all its diagonal entries equal to c and all 1's on the sub- and super-diagonals.

(a) For which values of c is A diagonally dominant?

(b) For which values of c does the Jacobi iteration for $A\mathbf{u}=\mathbf{b}$ converge to the solution? What is the rate of convergence? *Hint*: Use Exercise 8.2.48.

(c) Set $c=2$ and use the Jacobi method to solve the linear systems $K\mathbf{u}=\mathbf{e}_1$, for $n=5, 10$, and 20. Starting with an initial guess of $\mathbf{0}$, how many Jacobi iterations does it take to obtain 3 decimal place accuracy? Does the convergence rate agree with what you computed in part (c)?

10.5.8. Prove that $\mathbf{0} \neq \mathbf{u} \in \ker A$ if and only if $\mathbf{u}$ is a eigenvector of the Jacobi iteration matrix with eigenvalue 1. What does this imply about convergence?

◇ **10.5.9.** Prove that if A is a nonsingular coefficient matrix, then one can always arrange that all its diagonal entries are nonzero by suitably permuting its rows.

10.5.10. Consider the iterative system (10.52) with spectral radius $\rho(T)<1$. Explain why it takes roughly $-1/\log_{10}\rho(T)$ iterations to produce one further decimal digit of accuracy in the solution.

10.5.11. *True or false*: If a system $A\mathbf{u}=\mathbf{b}$ has a diagonally dominant coefficient matrix A, then the equivalent system obtained by applying an elementary row operation to A also has a diagonally dominant coefficient matrix.

The Gauss–Seidel Method

The Gauss–Seidel method relies on a slightly more refined implementation of the Jacobi process. To understand how it works, it will help to write out the Jacobi iteration scheme (10.61) in full detail:

$$\begin{aligned}
u_1^{(k+1)} &= \qquad\qquad t_{12}u_2^{(k)} + t_{13}u_3^{(k)} + \cdots + t_{1,n-1}u_{n-1}^{(k)} + t_{1n}u_n^{(k)} + c_1,\\
u_2^{(k+1)} &= t_{21}u_1^{(k)} \qquad\qquad + t_{23}u_3^{(k)} + \cdots + t_{2,n-1}u_{n-1}^{(k)} + t_{2n}u_n^{(k)} + c_2,\\
u_3^{(k+1)} &= t_{31}u_1^{(k)} + t_{32}u_2^{(k)} \qquad\qquad \cdots + t_{3,n-1}u_{n-1}^{(k)} + t_{3n}u_n^{(k)} + c_3,\\
&\ \ \vdots \\
u_n^{(k+1)} &= t_{n1}u_1^{(k)} + t_{n2}u_2^{(k)} + t_{n3}u_3^{(k)} + \cdots + t_{n,n-1}u_{n-1}^{(k)} \qquad\qquad + c_n,
\end{aligned} \tag{10.67}$$

where we are explicitly noting the fact that the diagonal entries of T vanish. Observe that we are using the entries of $\mathbf{u}^{(k)}$ to compute *all* of the updated values of $\mathbf{u}^{(k+1)}$. Presumably, if the iterates $\mathbf{u}^{(k)}$ are converging to the solution $\mathbf{u}^\star$, then their individual entries are also converging, and so each $u_j^{(k+1)}$ should be a better approximation to $u_j^\star$ than $u_j^{(k)}$ is. Therefore, if we begin the kth Jacobi iteration by computing $u_1^{(k+1)}$ using the first equation, then we are tempted to use this new and improved value to replace $u_1^{(k)}$ in each of the subsequent equations. In particular, we employ the modified equation

$$u_2^{(k+1)} = t_{21}\,u_1^{(k+1)} + t_{23}\,u_3^{(k)} + \cdots + t_{1n}\,u_n^{(k)} + c_2$$

to update the second component of our iterate. This more accurate value should then be used to update $u_3^{(k+1)}$, and so on.

The upshot of these considerations is the *Gauss–Seidel method*

$$\begin{aligned} u_i^{(k+1)} &= t_{i1}\,u_1^{(k+1)} + \cdots + t_{i,i-1}\,u_{i-1}^{(k+1)} \\ &\quad + t_{i,i+1}\,u_{i+1}^{(k)} + \cdots + t_{in}\,u_n^{(k)} + c_i, \end{aligned} \qquad i = 1, \ldots, n, \tag{10.68}$$

named after Gauss (as usual!) and the German astronomer/mathematician Philipp von Seidel. At the kth stage of the iteration, we use (10.68) to compute the revised entries $u_1^{(k+1)}, u_2^{(k+1)}, \ldots, u_n^{(k+1)}$ in their numerical order. Once an entry has been updated, the new value is immediately used in all subsequent computations.

EXAMPLE 10.47 For the linear system

$$3x + y - z = 3, \qquad x - 4y + 2z = -1, \qquad -2x - y + 5z = 2,$$

the Jacobi iteration method was given in (10.56). To construct the corresponding Gauss–Seidel scheme we use updated values of x, y and z as they become available. Explicitly,

$$\begin{aligned} x^{(k+1)} &= -\tfrac{1}{3}\,y^{(k)} + \tfrac{1}{3}\,z^{(k)} + 1 \\ y^{(k+1)} &= \tfrac{1}{4}\,x^{(k+1)} + \tfrac{1}{2}\,z^{(k)} + \tfrac{1}{4} \\ z^{(k+1)} &= \tfrac{2}{5}\,x^{(k+1)} + \tfrac{1}{5}\,y^{(k+1)} + \tfrac{2}{5}. \end{aligned} \tag{10.69}$$

The resulting iterates starting with $\mathbf{u}^{(0)} = \mathbf{0}$ are

$$\mathbf{u}^{(1)} = \begin{pmatrix} 1.0000 \\ .5000 \\ .9000 \end{pmatrix}, \quad \mathbf{u}^{(2)} = \begin{pmatrix} 1.1333 \\ .9833 \\ 1.0500 \end{pmatrix}, \quad \mathbf{u}^{(3)} = \begin{pmatrix} 1.0222 \\ 1.0306 \\ 1.0150 \end{pmatrix}, \quad \mathbf{u}^{(4)} = \begin{pmatrix} .9948 \\ 1.0062 \\ .9992 \end{pmatrix},$$

$$\mathbf{u}^{(5)} = \begin{pmatrix} .9977 \\ .9990 \\ .9989 \end{pmatrix}, \quad \mathbf{u}^{(6)} = \begin{pmatrix} 1.0000 \\ .9994 \\ .9999 \end{pmatrix}, \quad \mathbf{u}^{(7)} = \begin{pmatrix} 1.0001 \\ 1.0000 \\ 1.0001 \end{pmatrix}, \quad \mathbf{u}^{(8)} = \begin{pmatrix} 1.0000 \\ 1.0000 \\ 1.0000 \end{pmatrix},$$

and have converged to the solution, to 4 decimal place accuracy, after only 8 iterations—as opposed to the 11 iterations required by the Jacobi method. ●

The Gauss–Seidel iteration scheme is particularly suited to implementation on a serial computer, since one can immediately replace each component $u_i^{(k)}$ by its updated value $u_i^{(k+1)}$, thereby also saving on storage in the computer's memory. In contrast, the Jacobi scheme requires us to retain all the old values $\mathbf{u}^{(k)}$ until the new approximation $\mathbf{u}^{(k+1)}$ has been computed. Moreover, Gauss–Seidel typically (although not always) converges faster than Jacobi, making it the iterative algorithm of choice for serial processors. On the other hand, with the advent of parallel processing machines, variants of the parallelizable Jacobi scheme have recently been making a comeback.

What is Gauss–Seidel really up to? Let us rewrite the basic iterative equation (10.68) by multiplying by a_{ii} and moving the terms involving $\mathbf{u}^{(k+1)}$ to the left hand side. In view of the formula (10.60) for the entries of T, the resulting equation is

$$a_{i1}\,u_1^{(k+1)} + \cdots + a_{i,i-1}\,u_{i-1}^{(k+1)} + a_{ii}\,u_i^{(k+1)} = -\,a_{i,i+1}\,u_{i+1}^{(k)} - \cdots - a_{in}\,u_n^{(k)} + b_i.$$

In matrix form, taking (10.63) into account, this reads

$$(L + D)\mathbf{u}^{(k+1)} = -\,U\,\mathbf{u}^{(k)} + \mathbf{b}, \tag{10.70}$$

and so can be viewed as a linear system of equations for $\mathbf{u}^{(k+1)}$ with lower triangular coefficient matrix $L + D$. Note that the fixed point of (10.70), namely the solution to

$$(L + D)\,\mathbf{u} = -\,U\,\mathbf{u} + \mathbf{b},$$

coincides with the solution to the original system

$$A\,\mathbf{u} = (L + D + U)\,\mathbf{u} = \mathbf{b}.$$

In other words, the Gauss–Seidel procedure is merely implementing Forward Substitution to solve the lower triangular system (10.70) for the next iterate:

$$\mathbf{u}^{(k+1)} = -\,(L + D)^{-1}U\,\mathbf{u}^{(k)} + (L + D)^{-1}\,\mathbf{b}.$$

The latter is in our more usual iterative form

$$\mathbf{u}^{(k+1)} = \widetilde{T}\,\mathbf{u}^{(k)} + \widetilde{\mathbf{c}}, \qquad \text{where} \quad \widetilde{T} = -(L + D)^{-1}U, \quad \widetilde{\mathbf{c}} = (L + D)^{-1}\,\mathbf{b}. \tag{10.71}$$

Consequently, the convergence of the Gauss–Seidel iterates is governed by the spectral radius of the coefficient matrix $\widetilde{T}$.

Returning to Example 10.47, we have

$$A = \begin{pmatrix} 3 & 1 & -1 \\ 1 & -4 & 2 \\ -2 & -1 & 5 \end{pmatrix}, \quad L+D = \begin{pmatrix} 3 & 0 & 0 \\ 1 & -4 & 0 \\ -2 & -1 & 5 \end{pmatrix}, \quad U = \begin{pmatrix} 0 & 1 & -1 \\ 0 & 0 & 2 \\ 0 & 0 & 0 \end{pmatrix}.$$

Therefore, the Gauss–Seidel matrix is

$$\widetilde{T} = -\,(L + D)^{-1}U = \begin{pmatrix} 0 & -.3333 & .3333 \\ 0 & -.0833 & .5833 \\ 0 & -.1500 & .2500 \end{pmatrix}.$$

Its eigenvalues are 0 and $.0833 \pm .2444\,\mathrm{i}$, and hence its spectral radius is $\rho(\widetilde{T}) \approx .2582$. This is roughly the square of the Jacobi spectral radius of .5, which tell us that the Gauss–Seidel iterations will converge about twice as fast to the solution. This can be verified by more extensive computations. Although examples can be constructed where the Jacobi method converges faster, in many practical situations Gauss–Seidel tends to converge roughly twice as fast as Jacobi.

Completely general conditions guaranteeing convergence of the Gauss–Seidel method are hard to establish. But, like the Jacobi scheme, it is guaranteed to converge when the original coefficient matrix is strictly diagonally dominant.

THEOREM 10.48 If A is strictly diagonally dominant, then the Gauss–Seidel iteration scheme for solving $A\,\mathbf{u} = \mathbf{b}$ converges.

Proof Let $\mathbf{e}^{(k)} = \mathbf{u}^{(k)} - \mathbf{u}^\star$ denote the kth Gauss–Seidel error vector. As in (10.58), the error vectors satisfy the linear iterative system $\mathbf{e}^{(k+1)} = \widetilde{T}\,\mathbf{e}^{(k)}$, but a direct estimate of $\| \widetilde{T} \|_\infty$ is not so easy. Instead, let us write out the linear iterative system in components:

$$e_i^{(k+1)} = t_{i1}\, e_1^{(k+1)} + \cdots + t_{i,i-1}\, e_{i-1}^{(k+1)} + t_{i,i+1}\, e_{i+1}^{(k)} + \cdots + t_{in}\, e_n^{(k)}. \tag{10.72}$$

Let

$$m^{(k)} = \| \mathbf{e}^{(k)} \|_\infty = \max\left\{\, |e_1^{(k)}|, \ldots, |e_n^{(k)}| \,\right\} \tag{10.73}$$

denote the ∞ norm of the kth error vector. To prove convergence, $\mathbf{e}^{(k)} \to \mathbf{0}$, it suffices to show that $m^{(k)} \to 0$ as $k \to \infty$. We claim that diagonal dominance of A implies that

$$m^{(k+1)} \le s\, m^{(k)}, \qquad \text{where} \quad s = \| T \|_\infty < 1 \tag{10.74}$$

denotes the ∞ matrix norm of the *Jacobi* matrix (not the Gauss–Seidel matrix), which, by (10.66), is less than 1. We infer that $m^{(k)} \le s^k\, m^{(0)} \to 0$ as $k \to \infty$, demonstrating the theorem.

To prove (10.74), we use induction on $i = 1, \ldots, n$. Our induction hypothesis is

$$|e_j^{(k+1)}| \le s\, m^{(k)} < m^{(k)} \quad \text{for} \quad j = 1, \ldots, i-1.$$

(When $i = 1$, there is no assumption.) Moreover, by (10.73),

$$|e_j^{(k)}| \le m^{(k)} \quad \text{for all} \quad j = 1, \ldots, n.$$

We use these two inequalities to estimate $|e_i^{(k+1)}|$ from (10.72):

$$\begin{aligned} |e_i^{(k+1)}| &\le |t_{i1}|\, |e_1^{(k+1)}| + \cdots + |t_{i,i-1}|\, |e_{i-1}^{(k+1)}| + |t_{i,i+1}|\, |e_{i+1}^{(k)}| + \cdots + |t_{in}|\, |e_n^{(k)}| \\ &\le \left(\, |t_{i1}| + \cdots + |t_{in}| \,\right) m^{(k)} \le s\, m^{(k)}, \end{aligned}$$

which completes the induction step. As a result, the maximum

$$m^{(k+1)} = \max\left\{\, |e_1^{(k+1)}|, \ldots, |e_n^{(k+1)}| \,\right\} \le s\, m^{(k)}$$

also satisfies the same bound, and hence (10.74) follows. ■

EXAMPLE 10.49 For the linear system considered in Example 10.46, the Gauss–Seidel iterations take the form

$$\begin{aligned} x^{(k+1)} &= -\tfrac14\, y^{(k)} - \tfrac14\, w^{(k)} + \tfrac14 \\ y^{(k+1)} &= -\tfrac14\, x^{(k+1)} - \tfrac14\, z^{(k)} - \tfrac14\, v^{(k)} + \tfrac12 \\ z^{(k+1)} &= -\tfrac14\, y^{(k+1)} - \tfrac14\, w^{(k)} - \tfrac14 \\ w^{(k+1)} &= -\tfrac14\, x^{(k+1)} - \tfrac14\, z^{(k+1)} - \tfrac14\, v^{(k)} + \tfrac12 \\ v^{(k+1)} &= -\tfrac14\, y^{(k+1)} - \tfrac14\, w^{(k+1)} + \tfrac14. \end{aligned}$$

Starting with $x^{(0)} = y^{(0)} = z^{(0)} = w^{(0)} = v^{(0)} = 0$, the Gauss–Seidel iterates converge to the solution $x = -.1$, $y = .7$, $z = -.6$, $w = .7$, $v = -.1$, to four decimal places in 11 iterations, again roughly twice as fast as the Jacobi scheme.

Indeed, the convergence rate is governed by the corresponding Gauss–Seidel matrix $\widetilde{T}$, which is

$$\begin{pmatrix} 4 & 0 & 0 & 0 & 0 \\ 1 & 4 & 0 & 0 & 0 \\ 0 & 1 & 4 & 0 & 0 \\ 1 & 0 & 1 & 4 & 0 \\ 0 & 1 & 0 & 1 & 4 \end{pmatrix}^{-1} \begin{pmatrix} 0 & 1 & 0 & 1 & 0 \\ 0 & 0 & 1 & 0 & 1 \\ 0 & 0 & 0 & 1 & 0 \\ 0 & 0 & 0 & 0 & 1 \\ 0 & 0 & 0 & 0 & 0 \end{pmatrix}$$

$$= \begin{pmatrix} 0 & -.2500 & 0 & -.2500 & 0 \\ 0 & .0625 & -.2500 & .0625 & -.2500 \\ 0 & -.0156 & .0625 & -.2656 & .0625 \\ 0 & .0664 & -.0156 & .1289 & -.2656 \\ 0 & -.0322 & .0664 & -.0479 & .1289 \end{pmatrix}.$$

Its spectral radius is $\rho(\widetilde{T}) = .3936$, which is, as in the previous example, approximately the square of the spectral radius of the Jacobi coefficient matrix, which explains the speed up in convergence. ●

EXERCISES 10.5

♡**10.5.12.** Consider the linear system $A\mathbf{x} = \mathbf{b}$, where

$$A = \begin{pmatrix} 4 & 1 & -2 \\ -1 & 4 & -1 \\ 1 & -1 & 4 \end{pmatrix}, \qquad \mathbf{b} = \begin{pmatrix} 4 \\ 0 \\ 4 \end{pmatrix}.$$

(a) First, solve the equation directly by Gaussian Elimination.

(b) Using the initial approximation $\mathbf{x}^{(0)} = \mathbf{0}$, carry out three iterations of the Jacobi algorithm to compute $\mathbf{x}^{(1)}$, $\mathbf{x}^{(2)}$ and $\mathbf{x}^{(3)}$. How close are you to the exact solution?

(c) Write the Jacobi iteration in the form $\mathbf{x}^{(k+1)} = T\mathbf{x}^{(k)} + \mathbf{c}$. Find the 3×3 matrix T and the vector c explicitly.

(d) Using the initial approximation $\mathbf{x}^{(0)} = \mathbf{0}$, carry out three iterations of the Gauss–Seidel algorithm. Which is a better approximation to the solution—Jacobi or Gauss–Seidel?

(e) Write the Gauss–Seidel iteration in the form $\mathbf{x}^{(k+1)} = \widetilde{T}\mathbf{x}^{(k)} + \mathbf{c}$. Find the 3×3 matrix T and the vector $\mathbf{c}$ explicitly.

(f) Determine the spectral radius of the Jacobi matrix T, and use this to prove that the Jacobi method iteration will converge to the solution of $A\mathbf{x} = \mathbf{b}$ for any choice of the initial approximation $\mathbf{x}^{(0)}$.

(g) Determine the spectral radius of the Gauss–Seidel matrix T. Which method converges faster?

(h) For the faster method, how many iterations would you expect to need to obtain 5 decimal place accuracy?

(i) Test your prediction by computing the solution to the desired accuracy.

♠**10.5.13.** For the diagonally dominant systems in Exercise 10.5.3, starting with the initial guess $x = y = z = 0$, compute the solution to 3 decimal places using the Gauss–Seidel method. Check your answer by solving the system directly by Gaussian Elimination.

10.5.14. Which of the systems in Exercise 10.5.3 lead to convergent Gauss–Seidel schemes? In each case, which converges faster, Jacobi or Gauss–Seidel?

10.5.15. (a) Solve the positive definite linear systems in Exercise 10.5.6 using the Gauss–Seidel scheme to achieve 4 decimal place accuracy.

(b) Compare the convergence rate with the Jacobi method.

♣**10.5.16.** Let $A = \begin{pmatrix} c & 1 & 0 & 0 \\ 1 & c & 1 & 0 \\ 0 & 1 & c & 1 \\ 0 & 0 & 1 & c \end{pmatrix}$.

(a) For what values of c is A diagonally dominant?

(b) Use a computer to find the smallest positive value of $c > 0$ for which Jacobi iteration converges.

(c) Find the smallest positive value of $c > 0$ for which Gauss–Seidel iteration converges. Is your answer the same?

(d) When they both converge, which converges faster—Jacobi or Gauss–Seidel? How much faster? Does your answer depend upon the value of c?

♠**10.5.17.** Consider the linear system

$$\begin{aligned} 2.4x - .8y + .8z &= 1, \\ -.6x + 3.6y - .6z &= 0, \\ 15x + 14.4y - 3.6z &= 0. \end{aligned}$$

Show, by direct computation, that Jacobi iteration converges to the solution, but Gauss–Seidel does not.

♠**10.5.18.** Discuss convergence of Gauss–Seidel iteration for the system

$$\begin{aligned} 5x + 7y + 6z + 5w &= 23 \\ 7x + 10y + 8z + 7w &= 32 \\ 6x + 8y + 10z + 9w &= 33 \\ 5x + 7y + 9z + 10w &= 31. \end{aligned}$$

10.5.19. Let

$$A = \begin{pmatrix} 2 & 4 & -4 \\ 3 & 3 & 3 \\ 2 & 2 & 1 \end{pmatrix}.$$

Find the spectral radius of the Jacobi and Gauss–Seidel iteration matrices, and discuss their convergence.

♠**10.5.20.** Consider the linear system $H_5 \mathbf{u} = \mathbf{e}_1$, where H_5 is the 5×5 Hilbert matrix. Does the Jacobi method converge to the solution? If so, how fast? What about Gauss–Seidel?

◇**10.5.21.** How many arithmetic operations are needed to perform k steps of the Jacobi iteration? What about Gauss–Seidel? Under what conditions is Jacobi or Gauss–Seidel more efficient than Gaussian Elimination?

♣**10.5.22.** Consider the linear system $A\mathbf{x} = \mathbf{e}_1$ based on the 10×10 pentadiagonal matrix

$$A = \begin{pmatrix} z & -1 & 1 & 0 & & & \\ -1 & z & -1 & 1 & 0 & & \\ 1 & -1 & z & -1 & 1 & 0 & \\ 0 & 1 & -1 & z & -1 & 1 & \ddots \\ & 0 & 1 & -1 & z & -1 & \ddots \\ & & 0 & 1 & -1 & z & \ddots \\ & & & \ddots & \ddots & \ddots & \ddots \end{pmatrix}.$$

(a) For what values of z are the Jacobi and Gauss–Seidel methods guaranteed to converge?

(b) Set $z = 4$. How many iterations are required to approximate the solution to 3 decimal places?

(c) How small can $|z|$ be before the methods diverge?

♣**10.5.23.** The *naïve iterative method* for solving $A\mathbf{u} = \mathbf{b}$ is to rewrite it in fixed point form $\mathbf{u} = T\mathbf{u} + \mathbf{c}$, where $T = \mathrm{I} - A$ and $\mathbf{c} = \mathbf{b}$.

(a) What conditions on the eigenvalues of A ensure convergence of the naïve method?

(b) Use the Gerschgorin Theorem 10.34 to prove that the naïve method converges to the solution to

$$\begin{pmatrix} .8 & -.1 & -.1 \\ .2 & 1.5 & -.1 \\ .2 & -.1 & 1.0 \end{pmatrix} \begin{pmatrix} x \\ y \\ z \end{pmatrix} = \begin{pmatrix} 1 \\ -1 \\ 2 \end{pmatrix}.$$

(c) Check by implementing the method.

Successive Over–Relaxation (SOR)

As we know, the smaller the spectral radius (or matrix norm) of the coefficient matrix, the faster the convergence of the iterative scheme. One of the goals of researchers in numerical linear algebra is to design new methods for accelerating the convergence. In his 1950 thesis, the American mathematician David Young discovered a simple modification of the Jacobi and Gauss–Seidel methods that can, in favorable situations, lead to a dramatic speed up in the rate of convergence. The method, known as *successive over-relaxation*, and often abbreviated as SOR, has become the iterative method of choice in many modern applications, [18, 65]. In this subsection, we provide a brief overview.

In practice, finding the optimal iterative algorithm to solve a given linear system is as hard as solving the system itself. Therefore, researchers have relied on a few tried and true techniques for designing iterative schemes that can be used in the more common applications. Consider a linear algebraic system

$$A\mathbf{u} = \mathbf{b}.$$

Every decomposition

$$A = M - N \tag{10.75}$$

of the coefficient matrix into the difference of two matrices leads to an equivalent system of the form

$$M\,\mathbf{u} = N\,\mathbf{u} + \mathbf{b}. \tag{10.76}$$

Provided that M is nonsingular, we can rewrite the system in the fixed point form

$$\mathbf{u} = M^{-1}N\,\mathbf{u} + M^{-1}\mathbf{b} = T\,\mathbf{u} + \mathbf{c}, \qquad \text{where} \qquad T = M^{-1}N, \quad \mathbf{c} = M^{-1}\mathbf{b}.$$

Now, we are free to choose any such M, which then specifies $N = A - M$ uniquely. However, for the resulting iterative scheme $\mathbf{u}^{(k+1)} = T\,\mathbf{u}^{(k)} + \mathbf{c}$ to be practical we must arrange that

(a) $T = M^{-1}N$ is a convergent matrix, and

(b) M can be easily inverted.

The second requirement ensures that the iterative equations

$$M\,\mathbf{u}^{(k+1)} = N\,\mathbf{u}^{(k)} + \mathbf{b} \tag{10.77}$$

can be solved for $\mathbf{u}^{(k+1)}$ with minimal computational effort. Typically, this requires that M be either a diagonal matrix, in which case the inversion is immediate, or upper or lower triangular, in which case one employs Back or Forward Substitution to solve for $\mathbf{u}^{(k+1)}$.

With this in mind, we now introduce the SOR method. It relies on a slight generalization of the Gauss–Seidel decomposition (10.70) of the matrix into lower plus diagonal and upper triangular parts. The starting point is to write

$$A = L + D + U = \bigl[\,L + \alpha\,D\,\bigr] - \bigl[\,(\alpha - 1)\,D - U\,\bigr], \tag{10.78}$$

where $0 \neq \alpha$ is an adjustable scalar parameter. We decompose the system $A\,\mathbf{u} = \mathbf{b}$ as

$$(L + \alpha\,D)\mathbf{u} = \bigl[\,(\alpha - 1)\,D - U\,\bigr]\mathbf{u} + \mathbf{b}. \tag{10.79}$$

It turns out to be slightly more convenient to divide (10.79) through by α and write the resulting iterative system in the form

$$(\omega\,L + D)\mathbf{u}^{(k+1)} = \bigl[\,(1 - \omega)\,D - \omega\,U\,\bigr]\mathbf{u}^{(k)} + \omega\,\mathbf{b}, \tag{10.80}$$

where $\omega = 1/\alpha$ is called the *relaxation parameter*. Assuming, as usual, that all diagonal entries of A are nonzero, the matrix $\omega\,L + D$ is an invertible lower triangular matrix, and so we can use Forward Substitution to solve the iterative system (10.80) to recover $\mathbf{u}^{(k+1)}$. The explicit formula for its ith entry is

$$\begin{aligned} u_i^{(k+1)} &= \omega\,t_{i1}\,u_1^{(k+1)} + \cdots + \omega\,t_{i,i-1}\,u_{i-1}^{(k+1)} + (1 - \omega)\,u_i^{(k)} + \\ &\quad + \omega\,t_{i,i+1}\,u_{i+1}^{(k)} + \cdots + \omega\,t_{in}\,u_n^{(k)} + \omega\,c_i, \end{aligned} \tag{10.81}$$

where t_{ij} and c_i denote the original Jacobi values (10.60). As in the Gauss–Seidel approach, we update the entries $u_i^{(k+1)}$ in numerical order $i = 1, \ldots, n$. Thus, to obtain the SOR scheme (10.81), we merely multiply the right hand side of the Gauss–Seidel scheme (10.68) by the adjustable relaxation parameter ω and append the diagonal term $(1 - \omega)\,u_i^{(k)}$. In particular, if we set $\omega = 1$, then the SOR method reduces to the Gauss–Seidel method. Choosing $\omega < 1$ leads to an *under-relaxed* method, while $\omega > 1$, known as *over-relaxation*, is the choice that works in most practical instances.

To analyze the SOR scheme in detail, we rewrite (10.80) in the fixed point form

$$\mathbf{u}^{(k+1)} = T_\omega \, \mathbf{u}^{(k)} + \mathbf{c}_\omega, \tag{10.82}$$

where

$$\begin{aligned} T_\omega &= (\omega L + D)^{-1}\big[\,(1-\omega)\,D - \omega\,U\,\big], \\ \mathbf{c}_\omega &= (\omega L + D)^{-1}\,\omega\,\mathbf{b}. \end{aligned} \tag{10.83}$$

The rate of convergence is governed by the spectral radius of the matrix T_ω. The goal is to choose the relaxation parameter ω so as to make the spectral radius of T_ω as small as possible. As we will see, a clever choice of ω can result in a dramatic speed up in the convergence rate. Let us look at an elementary example.

EXAMPLE 10.50 Consider the matrix

$$A = \begin{pmatrix} 2 & -1 \\ -1 & 2 \end{pmatrix},$$

which we decompose as $A = L + D + U$, where

$$L = \begin{pmatrix} 0 & 0 \\ -1 & 0 \end{pmatrix}, \qquad D = \begin{pmatrix} 2 & 0 \\ 0 & 2 \end{pmatrix}, \qquad U = \begin{pmatrix} 0 & -1 \\ 0 & 0 \end{pmatrix}.$$

Jacobi iteration is based on the coefficient matrix

$$T = -D^{-1}(L+U) = \begin{pmatrix} 0 & \frac{1}{2} \\ \frac{1}{2} & 0 \end{pmatrix}.$$

Its spectral radius is $\rho(T) = .5$, and hence the Jacobi scheme takes, on average, roughly $3.3 \approx -1/\log_{10} .5$ iterations to produce each new decimal place in the solution.

The SOR scheme (10.80) takes the explicit form

$$\begin{pmatrix} 2 & 0 \\ -\omega & 2 \end{pmatrix} \mathbf{u}^{(k+1)} = \begin{pmatrix} 2(1-\omega) & \omega \\ 0 & 2\,(1-\omega) \end{pmatrix} \mathbf{u}^{(k)} + \omega\,\mathbf{b},$$

where Gauss–Seidel is the particular case $\omega = 1$. The SOR coefficient matrix is

$$T_\omega = \begin{pmatrix} 2 & 0 \\ -\omega & 2 \end{pmatrix}^{-1} \begin{pmatrix} 2\,(1-\omega) & \omega \\ 0 & 2\,(1-\omega) \end{pmatrix} = \begin{pmatrix} 1-\omega & \frac{1}{2}\omega \\ \frac{1}{2}\omega(1-\omega) & \frac{1}{4}(2-\omega)^2 \end{pmatrix}.$$

To compute the eigenvalues of T_ω, we form its characteristic equation

$$\begin{aligned} 0 = \det(T_\omega - \lambda\,\mathrm{I}) &= \lambda^2 - \big(2 - 2\omega + \tfrac{1}{4}\omega^2\big)\lambda + (1-\omega)^2 \\ &= (\lambda + \omega - 1)^2 - \tfrac{1}{4}\lambda\,\omega^2. \end{aligned} \tag{10.84}$$

Our goal is to choose ω so that

(a) both eigenvalues are less than 1 in modulus, so $|\lambda_1|, |\lambda_2| < 1$. This is the minimal requirement for convergence of the method.

(b) the largest eigenvalue (in modulus) is as small as possible. This will give the smallest spectral radius for T_ω and hence the fastest convergence rate.

By (8.25), the product of the two eigenvalues is the determinant,

$$\lambda_1 \lambda_2 = \det T_\omega = (1-\omega)^2.$$

If $\omega \le 0$ or $\omega \ge 2$, then $\det T_\omega \ge 1$, and hence at least one of the eigenvalues would have modulus larger than 1. Thus, in order to ensure convergence, we must require $0 < \omega < 2$. For Gauss–Seidel, at $\omega = 1$, the eigenvalues are $\lambda_1 = \frac{1}{4}$, $\lambda_2 = 0$, and the spectral radius is $\rho(T_1) = .25$. This is exactly the square of the Jacobi spectral radius, and hence the Gauss–Seidel iterates converge twice as fast; so it only takes, on average, about $-1/\log_{10} .25 = 1.66$ Gauss–Seidel iterations to produce each new decimal place of accuracy. It can be shown (Exercise 10.5.34) that as ω increases above 1, the two eigenvalues move along the real axis towards each other. They coincide when

$$\omega = \omega_\star = 8 - 4\sqrt{3} \approx 1.07,$$

at which point

$$\lambda_1 = \lambda_2 = \omega_\star - 1 = .07 = \rho(T_\omega),$$

which is the convergence rate of the optimal SOR scheme. Each iteration produces slightly more than one new decimal place in the solution, which represents a significant improvement over the Gauss–Seidel convergence rate. It takes about twice as many Gauss–Seidel iterations (and four times as many Jacobi iterations) to produce the same accuracy as this optimal SOR method. ●

Of course, in such a simple 2×2 example, it is not so surprising that we can construct the best value for the relaxation parameter by hand. Young was able to find the optimal value of the relaxation parameter for a broad class of matrices that includes most of those arising in the finite difference and finite element numerical solutions to ordinary and partial differential equations. For the matrices in Young's class, the Jacobi eigenvalues occur in signed pairs. If $\pm\mu$ are a pair of eigenvalues for the Jacobi method, then the corresponding eigenvalues of the SOR iteration matrix satisfy the quadratic equation

$$(\lambda + \omega - 1)^2 = \lambda\, \omega^2 \mu^2. \tag{10.85}$$

If $\omega = 1$, so we have standard Gauss–Seidel, then $\lambda^2 = \lambda\,\mu^2$, and so the eigenvalues are $\lambda = 0$, $\lambda = \mu^2$. The Gauss–Seidel spectral radius is therefore the square of the Jacobi spectral radius, and so (at least for matrices in the Young class) its iterates converge twice as fast. The quadratic equation (10.85) has the same properties as in the 2×2 version (10.84) (which corresponds to the case $\mu = \frac{1}{2}$), and hence the optimal value of ω will be the one at which the two roots are equal:

$$\lambda_1 = \lambda_2 = \omega - 1,$$

which occurs when

$$\omega = \frac{2 - 2\sqrt{1 - \mu^2}}{\mu^2} = \frac{2}{1 + \sqrt{1 - \mu^2}}.$$

Therefore, if $\rho_J = \max |\mu|$ denotes the spectral radius of the Jacobi method, then the Gauss–Seidel has spectral radius $\rho_{GS} = \rho_J^2$, while the SOR method with optimal relaxation parameter

$$\omega_\star = \frac{2}{1 + \sqrt{1 - \rho_J^2}}, \qquad \text{has spectral radius} \qquad \rho_\star = \omega_\star - 1. \tag{10.86}$$

For example, if $\rho_J = .99$, which is rather slow convergence (but common for iterative numerical solution schemes for partial differential equations), then $\rho_{GS} = .9801$, which is twice as fast, but still quite slow, while SOR with $\omega_\star = 1.7527$ has $\rho_\star = .7527$, which is dramatically faster*. Indeed, since $\rho_\star \approx (\rho_{GS})^{14} \approx (\rho_J)^{28}$,

*More precisely, since the SOR matrix is not diagonalizable, the overall convergence rate is slightly slower than the spectral radius. However, this technical detail does not affect the overall conclusion.

it takes about 14 Gauss–Seidel (and 28 Jacobi) iterations to produce the same accuracy as one SOR step. It is amazing that such a simple idea can have such a dramatic effect.

EXERCISES 10.5

♡**10.5.24.** Consider the linear system $A\mathbf{u} = \mathbf{b}$, where

$$A = \begin{pmatrix} 2 & 1 \\ 1 & 3 \end{pmatrix}, \qquad \mathbf{b} = \begin{pmatrix} 3 \\ 2 \end{pmatrix}.$$

(a) What is the solution?

(b) Discuss the convergence of the Jacobi iteration method.

(c) Discuss the convergence of the Gauss–Seidel iteration method.

(d) Write down the explicit formulas for the SOR method.

(e) What is the optimal value of the relaxation parameter ω for this system? How much faster is the convergence as compared to the Jacobi and Gauss–Seidel methods?

(f) Suppose your initial guess is $\mathbf{u}^{(0)} = \mathbf{0}$. Give an estimate as to how many steps each iterative method (Jacobi, Gauss–Seidel, SOR) would require in order to approximate the solution to the system to within 5 decimal places.

(g) Verify your answer by direct computation.

♣**10.5.25.** Consider the linear system

$$\begin{aligned} 4x - y - z &= 1 \\ -x + 4y - w &= 2 \\ -x + 4z - w &= 0 \\ -y - z + 4w &= 1. \end{aligned}$$

(a) Find the solution by using Gaussian Elimination and Back Substitution.

(b) Using $\mathbf{0}$ as your initial guess, how many iterations are required to approximate the solution to within five decimal places using

(i) Jacobi iteration?

(ii) Gauss–Seidel iteration?

Can you estimate the spectral radii of the relevant matrices in each case?

(c) Try to find the solution by using the SOR method with the parameter ω taking various values between .5 and 1.5. Which value of ω gives the fastest convergence? What is the spectral radius of the SOR matrix?

♠**10.5.26.** (a) Find the spectral radius of the Jacobi and Gauss–Seidel iteration matrices when

$$A = \begin{pmatrix} 2 & 1 & 0 & 0 \\ 1 & 2 & 1 & 0 \\ 0 & 1 & 2 & 1 \\ 0 & 0 & 1 & 2 \end{pmatrix}.$$

(b) Is A diagonally dominant?

(c) Use (10.86) to fix the optimal value of the SOR parameter. Verify that the spectral radius of the resulting iteration matrix agrees with the second formula in (10.86).

(d) For each iterative scheme, predict how many iterations are needed to solve the linear system $A\mathbf{x} = \mathbf{e}_1$ to 4 decimal places, and then verify your predictions by direct computation.

♠**10.5.27.** Change the matrix in Exercise 10.5.26 to

$$A = \begin{pmatrix} 2 & -1 & 0 & 0 \\ 1 & 2 & -1 & 0 \\ 0 & 1 & 2 & -1 \\ 0 & 0 & 1 & 2 \end{pmatrix},$$

and answer the same questions. Does the SOR method with parameter given by (10.86) speed the iterations up? Why not? Can you find a value of the SOR parameter that does?

♠**10.5.28.** Consider the linear system $A\mathbf{u} = \mathbf{e}_1$ in which A is the 8×8 tridiagonal matrix with all 2's on the main diagonal and all -1's on the sub- and super-diagonal.

(a) Use Exercise 8.2.47 to find the spectral radius of the Jacobi iteration method to solve $A\mathbf{u} = \mathbf{b}$. Does the Jacobi scheme converge?

(b) What is the optimal value of the SOR parameter based on (10.86)? How many Jacobi iterations are needed to match the effect of a single SOR step?

(c) Test out your conclusions by using both Jacobi and SOR to approximate the solution to 3 decimal places.

♠**10.5.29.** In Exercise 10.5.18 you were asked to solve a system by Gauss–Seidel. How much faster can you design an SOR scheme to converge? Experiment with several values of the relaxation parameter ω, and discuss what you find.

♠**10.5.30.** Investigate the three basic iterative techniques—Jacobi, Gauss–Seidel, SOR—for solving the linear system $K^\star \mathbf{u}^\star = \mathbf{f}^\star$ for the cubical circuit in Example 6.4.

♣**10.5.31.** The matrix

$$A = \begin{pmatrix} 4 & -1 & 0 & -1 & 0 & 0 & 0 & 0 & 0 \\ -1 & 4 & -1 & 0 & -1 & 0 & 0 & 0 & 0 \\ 0 & -1 & 4 & 0 & 0 & -1 & 0 & 0 & 0 \\ -1 & 0 & 0 & 4 & -1 & 0 & -1 & 0 & 0 \\ 0 & -1 & 0 & -1 & 4 & -1 & 0 & -1 & 0 \\ 0 & 0 & -1 & 0 & -1 & 4 & 0 & 0 & -1 \\ 0 & 0 & 0 & -1 & 0 & 0 & 4 & -1 & 0 \\ 0 & 0 & 0 & 0 & -1 & 0 & -1 & 4 & -1 \\ 0 & 0 & 0 & 0 & 0 & -1 & 0 & -1 & 4 \end{pmatrix}$$

arises in the finite difference (and finite element) discretization of the Poisson equation on a nine point square grid. Solve the linear system $A\mathbf{u} = \mathbf{e}_5$ using

(a) Gaussian Elimination

(b) Jacobi iteration

(c) Gauss–Seidel iteration

(d) SOR based on the Jacobi spectral radius

♣**10.5.32.** The generalization of Exercise 10.5.31 to an $n \times n$ grid results in an $n^2 \times n^2$ matrix in block tridiagonal form

$$A = \begin{pmatrix} K & -\mathrm{I} & & & \\ -\mathrm{I} & K & -\mathrm{I} & & \\ & -\mathrm{I} & K & -\mathrm{I} & \\ & & \ddots & \ddots & \ddots \end{pmatrix},$$

in which K is the tridiagonal $n \times n$ matrix with 4's on the main diagonal and -1's on the sub- and super-diagonal, while I denotes an $n \times n$ identity matrix. Use the known value of the Jacobi spectral radius

$$\rho_J = \cos \frac{\pi}{n+1},$$

[65], to design an SOR method to solve the linear system $A\mathbf{u} = \mathbf{f}$. Run your method on the cases $n = 5$ and $\mathbf{f} = \mathbf{e}_{13}$ and $n = 25$ and $\mathbf{f} = \mathbf{e}_{313}$ corresponding to a unit force at the center of the grid. How much faster is the convergence rate?

♣**10.5.33.** How much can you speed up the convergence of the iterative solution to the pentadiagonal linear system in Exercise 10.5.22 when $z = 4$ by using SOR? Discuss.

◇**10.5.34.** For the matrix treated in Example 10.50, prove that

(a) as ω increases from 1 to $8 - 4\sqrt{3}$, the two eigenvalues move towards each other, with the larger one decreasing in magnitude;

(b) if $\omega > 8 - 4\sqrt{3}$, the eigenvalues are complex conjugates, with larger modulus than the optimal value.

(c) Can you conclude that $\omega_\star = 8 - 4\sqrt{3}$ is the optimal value for the SOR parameter?

♡**10.5.35.** If $\mathbf{u}^{(k)}$ is an approximation to the solution to $A\mathbf{u} = \mathbf{b}$, then the *residual vector* $\mathbf{r}^{(k)} = \mathbf{b} - A\mathbf{u}^{(k)}$ measures how accurately the approximation solves the system.

(a) Show that the Jacobi iteration can be written in the form $\mathbf{u}^{(k+1)} = \mathbf{u}^{(k)} + D^{-1}\mathbf{r}^{(k)}$.

(b) Show that the Gauss–Seidel iteration has the form $\mathbf{u}^{(k+1)} = \mathbf{u}^{(k)} + (L + D)^{-1}\mathbf{r}^{(k)}$.

(c) Show that the SOR iteration has the form $\mathbf{u}^{(k+1)} = \mathbf{u}^{(k)} + (\omega L + D)^{-1}\mathbf{r}^{(k)}$.

(d) If $\|\mathbf{r}^{(k)}\|$ is small, does this mean that $\mathbf{u}^{(k)}$ is close to the solution? Explain your answer and illustrate with a couple of examples.

10.5.36. Let K be a positive definite $n \times n$ matrix with eigenvalues $\lambda_1 \geq \lambda_2 \geq \cdots \geq \lambda_n > 0$. For what values of ε does the iterative system $\mathbf{u}^{(k+1)} = \mathbf{u}^{(k)} + \varepsilon\, \mathbf{r}^{(k)}$, where $\mathbf{r}^{(k)} = \mathbf{f} - K\mathbf{u}^{(k)}$ is the current *residual vector*, converge to the solution? What is the optimal value of ε, and what is the convergence rate?

Conjugate Gradients

So far, we have established two broad classes of algorithms for solving linear systems. The first, known as *direct methods*, are based on some version of Gaussian Elimination or matrix factorization. Direct methods eventually[†] obtain the exact solution, but must be carried through to completion before any useful information is obtained. The second class contains the *iterative methods* discussed in the present chapter that lead to closer and closer approximations to the solution, but almost never reach the exact value. One might ask whether there are algorithms that com-

[†]This assumes that we are dealing with a fully accurate implementation, i.e., without round-off or other numerical error. In this discussion, numerical instability will be left aside as a separate, albeit ultimately important, concern.

bine the best of both: *semi-direct methods* whose intermediate computations lead to closer and closer approximations, and, moreover, are guaranteed to terminate in a finite number of steps with the exact solution in hand.

For instance, one might seek an algorithm that successively computes each entry of the solution vector $\mathbf{u}_\star$. This seems a rather unlikely scenario, but if we recall that the entries of the solution are merely its coordinates with respect to the standard basis $\mathbf{e}_1, \ldots, \mathbf{e}_n$, then one might try instead to compute the coordinates $t_1, \ldots, t_n$ of $\mathbf{u}_\star = t_1 \mathbf{v}_1 + \cdots + t_n \mathbf{v}_n$ with respect to some basis that is especially adapted to the linear system. Ideally, $\mathbf{v}_1, \ldots, \mathbf{v}_n$ should be an orthogonal basis—but orthogonality with respect to the standard Euclidean dot product may not be so relevant. A better idea is to arrange that the basis be orthogonal with respect to an inner product that arises from the system under consideration. In particular, if the linear system to be solved takes the form

$$K\mathbf{u} = \mathbf{f}, \tag{10.87}$$

in which the coefficient matrix K is *positive definite*—as occurs in many applications — then orthogonality with respect to the induced inner product

$$\langle\!\langle \mathbf{v}, \mathbf{w} \rangle\!\rangle = \mathbf{v}^T K \mathbf{w} \tag{10.88}$$

is very natural. Vectors that are orthogonal with respect to (10.88) are known as *conjugate vectors*, which explain half the name of the conjugate gradient algorithm, first introduced in 1952 by Hestenes and Stiefel, [31].

The term "gradient" stems from the minimization principle. According to Theorem 4.1, the solution $\mathbf{u}_\star$ to the positive definite linear system (10.87) is the unique minimizer of the quadratic function

$$p(\mathbf{u}) = \tfrac{1}{2}\mathbf{u}^T K \mathbf{u} - \mathbf{u}^T \mathbf{f}. \tag{10.89}$$

Thus, one way to solve the system is to minimize $p(\mathbf{u})$. Suppose we find ourselves at a point $\mathbf{u}$ which is not the minimizer. In which direction should we travel to find $\mathbf{u}_\star$? Multivariable calculus tells us that the gradient vector $\nabla p(\mathbf{u})$ of a function points in the direction of its steepest increase at the point, while its negative $-\nabla p(\mathbf{u})$ points in the direction of steepest decrease, [2, 58]. (Our discussion of gradient flow systems (9.18) was based on the same idea.) The gradient of the particular quadratic function (10.89) is easily found:

$$-\nabla p(\mathbf{u}) = \mathbf{f} - K\mathbf{u} = \mathbf{r},$$

where $\mathbf{r}$ is known as the *residual vector* for $\mathbf{u}$. Note that $\mathbf{r} = \mathbf{0}$ if and only if $\mathbf{u} = \mathbf{u}_\star$ is the solution, and so the size of $\mathbf{r}$ measures, in a certain sense, how accurately $\mathbf{u}$ comes to solving the system. Moreover, the residual vector indicates the direction of steepest decrease in the quadratic function, and is thus a good choice of direction to head off in search of the true minimizer.

The initial result is the *gradient descent algorithm*, in which each successive approximation $\mathbf{u}_k$ to the solution is obtained by going a certain distance in the residual direction:

$$\mathbf{u}_{k+1} = \mathbf{u}_k + t_k \mathbf{r}_k, \qquad \text{where} \qquad \mathbf{r}_k = \mathbf{f} - K\mathbf{u}_k. \tag{10.90}$$

The scalar factor t_k can be specified by the requirement that $p(\mathbf{u}_{k+1})$ is as small as possible; in Exercise 10.5.43 you are asked to find this value. A second option is to make the residual vector at $\mathbf{u}_{k+1}$ as small as possible by minimizing, say, its Euclidean norm $\|\mathbf{r}_{k+1}\|$. The initial guess $\mathbf{u}_0$ for the solution can be chosen as desired, with $\mathbf{u}_0 = \mathbf{0}$ the default choice. Gradient descent is a reasonable algorithm, and will lead to the solution in favorable situations. It is also effectively used to find minima of more general nonlinear functions. However, in many circumstances, the iterative method based on gradient descent can take an exceedingly long time to converge to an accurate approximation to the solution, and so is typically not a competitive algorithm.

However, if we supplement the gradient descent idea by the use of conjugate vectors, we are led to a competitive semi-direct solution algorithm. We shall construct the solution $\mathbf{u}_\star$ by successive approximation, with the kth iterate having the form

$$\mathbf{u}_k = t_1 \mathbf{v}_1 + \cdots + t_k \mathbf{v}_k, \qquad \text{so that} \qquad \mathbf{u}_{k+1} = \mathbf{u}_k + t_{k+1} \mathbf{v}_{k+1},$$

where, as advertised, the conjugate vectors $\mathbf{v}_1, \ldots, \mathbf{v}_n$ form a K–orthogonal basis. The secret is not to try to specify the conjugate basis vectors in advance, but rather to successively construct them during the course of the algorithm. We begin, merely for convenience, with an initial guess $\mathbf{u}_0 = \mathbf{0}$ for the solution. The residual vector $\mathbf{r}_0 = \mathbf{f} - K\mathbf{u}_0 = \mathbf{f}$ indicates the direction of steepest decrease of $p(\mathbf{u})$ at $\mathbf{u}_0$, and we update our original guess by moving in this direction, taking $\mathbf{v}_1 = \mathbf{r}_0 = \mathbf{f}$ as our first conjugate direction. The next iterate is $\mathbf{u}_1 = \mathbf{u}_0 + t_1 \mathbf{v}_1 = t_1 \mathbf{v}_1$, and we choose the parameter t_1 so that the corresponding residual vector

$$\mathbf{r}_1 = \mathbf{f} - K\mathbf{u}_1 = \mathbf{r}_0 - t_1 K \mathbf{v}_1 \tag{10.91}$$

is as close to $\mathbf{0}$ (in the Euclidean norm) as possible. This occurs when $\mathbf{r}_1$ is orthogonal to $\mathbf{r}_0$ (why?), and so we require

$$\begin{aligned} 0 = \mathbf{r}_0^T \mathbf{r}_1 &= \|\mathbf{r}_0\|^2 - t_1 \mathbf{r}_0^T K \mathbf{v}_1 \\ &= \|\mathbf{r}_0\|^2 - t_1 \langle\!\langle \mathbf{r}_0, \mathbf{v}_1 \rangle\!\rangle \\ &= \|\mathbf{r}_0\|^2 - t_1 \langle\!\langle \mathbf{v}_1, \mathbf{v}_1 \rangle\!\rangle. \end{aligned} \tag{10.92}$$

Note: We will consistently use $\|\mathbf{v}\|$ to denote the Euclidean norm, and $\langle\!\langle \mathbf{v}, \mathbf{w} \rangle\!\rangle$ the adapted inner product (10.88), which has its own norm: $\sqrt{\langle\!\langle \mathbf{v}, \mathbf{v} \rangle\!\rangle} = \sqrt{\mathbf{v}^T K \mathbf{v}}$.

Therefore, we set

$$t_1 = \frac{\|\mathbf{r}_0\|^2}{\langle\!\langle \mathbf{v}_1, \mathbf{v}_1 \rangle\!\rangle}, \qquad \text{and so} \qquad \mathbf{u}_1 = \mathbf{u}_0 + \frac{\|\mathbf{r}_0\|^2}{\langle\!\langle \mathbf{v}_1, \mathbf{v}_1 \rangle\!\rangle} \mathbf{v}_1 \tag{10.93}$$

is our new approximation to the solution. We can assume that $t_1 \neq 0$, since otherwise the residual $\mathbf{r}_0 = \mathbf{0}$. In this case, $\mathbf{u}_0 = \mathbf{0}$ would be the exact solution of the system, and there would be no reason to continue the procedure.

The gradient descent algorithm would tell us to update $\mathbf{u}_1$ by moving in the residual direction $\mathbf{r}_1$. But in the conjugate gradient algorithm, we choose a direction $\mathbf{v}_2$ which is conjugate, meaning K–orthogonal to the first direction $\mathbf{v}_1 = \mathbf{r}_0$. Thus, as in the Gram–Schmidt process, we slightly modify the residual direction by setting $\mathbf{v}_2 = \mathbf{r}_1 + s_1 \mathbf{v}_1$, where the scalar factor s_1 is determined by the orthogonality requirement

$$0 = \langle\!\langle \mathbf{v}_1, \mathbf{v}_2 \rangle\!\rangle = \langle\!\langle \mathbf{r}_1 + s_1 \mathbf{v}_1, \mathbf{v}_2 \rangle\!\rangle = \langle\!\langle \mathbf{r}_1, \mathbf{v}_1 \rangle\!\rangle + s_1 \langle\!\langle \mathbf{v}_1, \mathbf{v}_1 \rangle\!\rangle,$$

so

$$s_1 = -\frac{\langle\!\langle \mathbf{r}_1, \mathbf{v}_1 \rangle\!\rangle}{\langle\!\langle \mathbf{v}_1, \mathbf{v}_1 \rangle\!\rangle}.$$

Now, in view of (10.91) and the orthogonality of $\mathbf{r}_0$ and $\mathbf{r}_1$,

$$\langle\!\langle \mathbf{r}_1, \mathbf{v}_1 \rangle\!\rangle = \mathbf{r}_1^T K \mathbf{v}_1 = \mathbf{r}_1^T \left(\frac{\mathbf{r}_0 - \mathbf{r}_1}{t_1} \right) = -\frac{1}{t_1} \|\mathbf{r}_1\|^2,$$

while, by (10.93),

$$\langle\!\langle \mathbf{v}_1, \mathbf{v}_1 \rangle\!\rangle = \mathbf{v}_1^T K \mathbf{v}_1 = \frac{1}{t_1} \|\mathbf{r}_0\|^2.$$

Therefore, the second conjugate direction is

$$\mathbf{v}_2 = \mathbf{r}_1 + s_1 \mathbf{v}_1, \qquad \text{where} \quad s_1 = \frac{\|\mathbf{r}_1\|^2}{\|\mathbf{r}_0\|^2}.$$

We then update

$$\mathbf{u}_2 = \mathbf{u}_1 + t_2 \mathbf{v}_2 = \mathbf{u}_0 + t_1 \mathbf{v}_1 + t_2 \mathbf{v}_2 = t_1 \mathbf{v}_1 + t_2 \mathbf{v}_2$$

so as to make the corresponding residual vector

$$\mathbf{r}_2 = \mathbf{f} - K \mathbf{u}_2 = \mathbf{r}_1 - t_2 K \mathbf{v}_2$$

as small as possible, which is accomplished by requiring it to be orthogonal to $\mathbf{r}_1$. Thus, using the K–orthogonality of $\mathbf{v}_1$ and $\mathbf{v}_2$,

$$0 = \mathbf{r}_1^T \mathbf{r}_2 = \|\mathbf{r}_1\|^2 - t_2 \mathbf{r}_1^T K \mathbf{v}_2 = \|\mathbf{r}_1\|^2 - t_2 \langle\!\langle \mathbf{v}_2, \mathbf{v}_2 \rangle\!\rangle,$$

and so

$$t_2 = \frac{\|\mathbf{r}_1\|^2}{\langle\!\langle \mathbf{v}_2, \mathbf{v}_2 \rangle\!\rangle}.$$

Again, we can assume that $t_2 \neq 0$, as otherwise $\mathbf{r}_1 = \mathbf{0}$ and $\mathbf{u}_1$ would be the exact solution, so the algorithm should be terminated.

Continuing in this manner, at the kth stage, we have already constructed the conjugate vectors $\mathbf{v}_1, \ldots, \mathbf{v}_k$, and the solution approximation $\mathbf{u}_k$ as a suitable linear combination of them. The next conjugate direction is given by

$$\mathbf{v}_{k+1} = \mathbf{r}_k + s_k \mathbf{v}_k, \qquad \text{where} \quad s_k = \frac{\|\mathbf{r}_k\|^2}{\|\mathbf{r}_{k-1}\|^2} \tag{10.94}$$

results from the K–orthogonality requirement: $\langle\!\langle \mathbf{v}_i, \mathbf{v}_k \rangle\!\rangle = 0$ for $i < k$. The updated solution approximation

$$\mathbf{u}_{k+1} = \mathbf{u}_k + t_{k+1} \mathbf{v}_{k+1}, \qquad \text{where} \quad t_{k+1} = \frac{\|\mathbf{r}_k\|^2}{\langle\!\langle \mathbf{v}_{k+1}, \mathbf{v}_{k+1} \rangle\!\rangle} \tag{10.95}$$

is then specified so as to make the corresponding residual

$$\mathbf{r}_{k+1} = \mathbf{f} - K \mathbf{u}_{k+1} = \mathbf{r}_k - t_{k+1} K \mathbf{v}_{k+1}$$

as small as possible, by requiring that it be orthogonal to $\mathbf{r}_k$.

Starting with an initial guess $\mathbf{u}_0$, the iterative equations (10.94–95) implement the *Conjugate Gradient Method.* Observe that the algorithm does not require solving any linear systems: apart from multiplication of a matrix times a vector to evaluate $K \mathbf{v}_k$, all other operations are rapidly evaluated Euclidean dot products. Unlike Gaussian Elimination, the method produces a sequence of successive approximations $\mathbf{u}_1, \mathbf{u}_2, \ldots$ to the solution $\mathbf{u}_\star$, and so the iteration can be stopped as soon as a desired solution accuracy is reached—which can be assessed by comparing how close the successive iterates are to each other. On the other hand, unlike purely iterative methods, *the Conjugate Gradient Method does eventually terminate at the exact solution*, because, as remarked at the outset, there are at most n conjugate directions, forming a orthogonal basis of $\mathbb{R}^n$ for the inner product induced by K. Therefore, $\mathbf{u}_n = t_1 \mathbf{v}_1 + \cdots + t_n \mathbf{v}_n = \mathbf{u}_\star$ must be the solution since its residual $\mathbf{r}_n = \mathbf{f} - K \mathbf{u}_n$ is orthogonal to all the conjugate basis vectors $\mathbf{v}_1, \ldots, \mathbf{v}_n$, and hence must be $\mathbf{0}$.

Conjugate Gradient Method for Solving $K\mathbf{u} = \mathbf{f}$

```
start
    choose u_0, e.g. u_0 = 0
    set v_0 = 0
    for k = 0 to m - 1
        set r_k = f - K u_k
        if r_k = 0 print "u_k is the exact solution"; end
        if k = 0 set v_1 = r_0 else set v_{k+1} = r_k + (||r_k||^2 / ||r_{k-1}||^2) v_k
        set u_{k+1} = u_k + (||r_k||^2 / (v_{k+1}^T K v_{k+1})) v_{k+1}
    next k
end
```

A pseudocode program is included; at each stage, $\mathbf{u}_k$ is the current approximation to the solution. The initial guess $\mathbf{u}_0$ can be chosen by the user, with $\mathbf{u}_0 = \mathbf{0}$ the default. The number of iterations $m \le n$ can be chosen by the user in advance; alternatively, one can impose a stopping criterion based on the size of the residual vector, $\|\mathbf{r}_k\|$, or, alternatively, the distance between successive iterates, $\|\mathbf{u}_{k+1} - \mathbf{u}_k\|$. If the process is carried on to the bitter end, i.e., for $m = n$, then, in the absence of round-off errors, the result will be the exact solution to the system.

EXAMPLE 10.51 Consider the linear system $K\mathbf{u} = \mathbf{f}$ with

$$K = \begin{pmatrix} 3 & -1 & 0 \\ -1 & 2 & 1 \\ 0 & 1 & 1 \end{pmatrix}, \qquad \mathbf{f} = \begin{pmatrix} 1 \\ 2 \\ -1 \end{pmatrix}.$$

The exact solution is $\mathbf{u}_\star = (2, 5, -6)^T$. Let us implement the method of conjugate gradients, starting with the initial guess $\mathbf{u}_0 = (0, 0, 0)^T$. The corresponding residual vector is merely $\mathbf{r}_0 = \mathbf{f} - K\mathbf{u}_0 = \mathbf{f} = (1, 2, -1)^T$. The first conjugate direction is $\mathbf{v}_1 = \mathbf{r}_0 = (1, 2, -1)^T$, and we use formula (10.93) to obtain the updated approximation to the solution

$$\mathbf{u}_1 = \mathbf{u}_0 + \frac{\|\mathbf{r}_0\|^2}{\langle\!\langle \mathbf{v}_1, \mathbf{v}_1 \rangle\!\rangle}\, \mathbf{v}_1 = \frac{6}{4} \begin{pmatrix} 1 \\ 2 \\ -1 \end{pmatrix} = \begin{pmatrix} \frac{3}{2} \\ 3 \\ -\frac{3}{2} \end{pmatrix},$$

noting that $\langle\!\langle \mathbf{v}_1, \mathbf{v}_1 \rangle\!\rangle = \mathbf{v}_1^T K \mathbf{v}_1 = 4$. In the next stage of the algorithm, we compute the corresponding residual $\mathbf{r}_1 = \mathbf{f} - K\mathbf{u}_1 = \left(-\frac{1}{2}, -1, -\frac{5}{2}\right)^T$. The conjugate direction is

$$\mathbf{v}_2 = \mathbf{r}_1 + \frac{\|\mathbf{r}_1\|^2}{\|\mathbf{r}_0\|^2}\, \mathbf{v}_1 = \begin{pmatrix} -\frac{1}{2} \\ -1 \\ -\frac{5}{2} \end{pmatrix} + \frac{\frac{15}{2}}{6} \begin{pmatrix} 1 \\ 2 \\ -1 \end{pmatrix} = \begin{pmatrix} \frac{3}{4} \\ \frac{3}{2} \\ -\frac{15}{4} \end{pmatrix},$$

which, as designed, satisfies the conjugacy condition $\langle\!\langle \mathbf{v}_1, \mathbf{v}_2 \rangle\!\rangle = \mathbf{v}_1^T K \mathbf{v}_2 = 0$. Each entry of the ensuing approximation

$$\mathbf{u}_2 = \mathbf{u}_1 + \frac{\|\mathbf{r}_1\|^2}{\langle\langle \mathbf{v}_2, \mathbf{v}_2 \rangle\rangle}\, \mathbf{v}_2 = \begin{pmatrix} \frac{3}{2} \\ 3 \\ -\frac{3}{2} \end{pmatrix} + \frac{\frac{15}{2}}{\frac{27}{4}} \begin{pmatrix} \frac{3}{4} \\ \frac{3}{2} \\ -\frac{15}{4} \end{pmatrix} = \begin{pmatrix} \frac{7}{3} \\ \frac{14}{3} \\ -\frac{17}{3} \end{pmatrix} \approx \begin{pmatrix} 2.3333 \\ 4.6667 \\ -5.6667 \end{pmatrix}$$

is now within a $\frac{1}{3}$ of the exact solution $\mathbf{u}_\star$.

Since we are dealing with a 3×3 system, we will recover the exact solution by one more iteration of the algorithm. The new residual is $\mathbf{r}_2 = \mathbf{f} - K\mathbf{u}_2 = \left(-\frac{4}{3}, \frac{2}{3}, 0\right)^T$. The final conjugate direction is

$$\mathbf{v}_3 = \mathbf{r}_2 + \frac{\|\mathbf{r}_2\|^2}{\|\mathbf{r}_1\|^2}\, \mathbf{v}_2 = \begin{pmatrix} -\frac{4}{3} \\ \frac{2}{3} \\ 0 \end{pmatrix} + \frac{\frac{20}{9}}{\frac{15}{2}} \begin{pmatrix} \frac{3}{4} \\ \frac{3}{2} \\ -\frac{15}{4} \end{pmatrix} = \begin{pmatrix} -\frac{10}{9} \\ \frac{10}{9} \\ -\frac{10}{9} \end{pmatrix},$$

which, as you can check, is conjugate to both $\mathbf{v}_1$ and $\mathbf{v}_2$. The solution is obtained from

$$\mathbf{u}_3 = \mathbf{u}_2 + \frac{\|\mathbf{r}_2\|^2}{\langle\langle \mathbf{v}_3, \mathbf{v}_3 \rangle\rangle}\, \mathbf{v}_3 = \begin{pmatrix} \frac{7}{3} \\ \frac{14}{3} \\ -\frac{17}{3} \end{pmatrix} + \frac{\frac{20}{9}}{\frac{200}{27}} \begin{pmatrix} -\frac{10}{9} \\ \frac{10}{9} \\ -\frac{10}{9} \end{pmatrix} = \begin{pmatrix} 2 \\ 5 \\ -6 \end{pmatrix}. \qquad \bullet$$

Of course, in larger examples, one would not carry through the algorithm to the bitter end since an approximation to the solution is typically obtained with only a few iterations. The result can be a substantial saving in computational time and effort required to produce an approximation to the solution. For further developments and applications, see [18, 51, 67].

EXERCISES 10.5

10.5.37. Solve the following linear systems by the conjugate gradient method, keeping track of the residual vectors and solution approximations as you iterate.

(a) $\begin{pmatrix} 3 & -1 \\ -1 & 5 \end{pmatrix} \mathbf{u} = \begin{pmatrix} 2 \\ 1 \end{pmatrix}$

(b) $\begin{pmatrix} 6 & 2 & 1 \\ 2 & 3 & -1 \\ 1 & -1 & 2 \end{pmatrix} \mathbf{u} = \begin{pmatrix} 1 \\ 0 \\ -2 \end{pmatrix}$

(c) $\begin{pmatrix} 6 & -1 & -3 \\ -1 & 7 & 4 \\ -3 & 4 & 9 \end{pmatrix} \mathbf{u} = \begin{pmatrix} -1 \\ -2 \\ 7 \end{pmatrix}$

(d) $\begin{pmatrix} 6 & -1 & -1 & 5 \\ -1 & 7 & 1 & -1 \\ -1 & 1 & 3 & -3 \\ 5 & -1 & -3 & 6 \end{pmatrix} \mathbf{u} = \begin{pmatrix} 1 \\ 2 \\ 0 \\ -1 \end{pmatrix}$

(e) $\begin{pmatrix} 5 & 1 & 1 & 1 \\ 1 & 5 & 1 & 1 \\ 1 & 1 & 5 & 1 \\ 1 & 1 & 1 & 5 \end{pmatrix} \mathbf{u} = \begin{pmatrix} 4 \\ 0 \\ 0 \\ 0 \end{pmatrix}$

♣**10.5.38.** Use the conjugate gradient method to solve the system in Exercise 10.5.31. How many iterations do you need to obtain the solution that is accurate to 2 decimal places? How does this compare to the Jacobi and SOR methods?

♣**10.5.39.** According to Example 3.33, the $n \times n$ Hilbert matrix H_n is positive definite, and hence we can apply the conjugate gradient method to solve the linear system $H_n \mathbf{u} = \mathbf{f}$. For the values $n = 5, 10, 30$, let $\mathbf{u}^\star \in \mathbb{R}^n$ be the vector with all entries equal to 1.

(a) Compute $\mathbf{f} = H_n \mathbf{u}^\star$.

(b) Use Gaussian Elimination to solve $H_n \mathbf{u} = \mathbf{f}$. How close is your solution to $\mathbf{u}^\star$?

(c) Does pivoting improve the solution in part (b)?

(d) Does the conjugate gradient algorithm do any better?

10.5.40. Try applying the Conjugate Gradient Algorithm to the system $-x + 2y + z = -2$, $y + 2z = 1$, $3x + y - z = 1$. Do you obtain the solution? Why?

10.5.41. *True or false*: If the residual vector satisfies $\| \mathbf{r} \| < .01$, then $\mathbf{u}$ approximates the solution to within two decimal places.

10.5.42. How many arithmetic operations are needed to implement one iteration of the conjugate gradient method? How many iterations can you perform before the method becomes more work that direct Gaussian Elimination? *Remark*: If the matrix is sparse, the number of operations can decrease dramatically.

◇10.5.43. In (10.90), find the value of t_k that minimizes $p(\mathbf{u}_{k+1})$.

10.6 NUMERICAL COMPUTATION OF EIGENVALUES

The importance of the eigenvalues of a square matrix in a broad range of applications has been amply demonstrated in this chapter and its predecessor. However, finding the eigenvalues and associated eigenvectors is not such an easy task. The direct method of constructing the characteristic equation of the matrix through the determinantal formula, then solving the resulting polynomial equation for the eigenvalues, and finally producing the eigenvectors by solving the associated homogeneous linear system, is hopelessly inefficient, and fraught with numerical pitfalls. We are in need of a completely new idea if we have any hopes of designing efficient numerical approximation schemes.

In this section, we develop a few of the most basic numerical schemes for computing eigenvalues and eigenvectors. All are iterative in nature. The most direct are based on the connections between the eigenvalues and the high powers of a matrix. A more sophisticated approach, based on the $Q\,R$ factorization that we learned in Section 5.3, will be presented at the end of the section.

The Power Method

We have already noted the role played by the eigenvalues and eigenvectors in the solution to linear iterative systems. Now we are going to turn the tables, and use the iterative system as a mechanism for approximating the eigenvalues, or, more correctly, selected eigenvalues of the coefficient matrix. The simplest of the resulting computational procedures is known as the *power method.*

We assume, for simplicity, that A is a complete* $n \times n$ matrix. Let $\mathbf{v}_1, \ldots, \mathbf{v}_n$ denote its eigenvector basis, and $\lambda_1, \ldots, \lambda_n$ the corresponding eigenvalues. As we have learned, the solution to the linear iterative system

$$\mathbf{v}^{(k+1)} = A\,\mathbf{v}^{(k)}, \qquad \mathbf{v}^{(0)} = \mathbf{v}, \tag{10.96}$$

is obtained by multiplying the initial vector $\mathbf{v}$ by the successive powers of the coefficient matrix: $\mathbf{v}^{(k)} = A^k\,\mathbf{v}$. If we write the initial vector in terms of the eigenvector basis

$$\mathbf{v} = c_1\,\mathbf{v}_1 + \cdots + c_n\,\mathbf{v}_n, \tag{10.97}$$

then the solution takes the explicit form given in Theorem 10.4, namely

$$\mathbf{v}^{(k)} = A^k\,\mathbf{v} = c_1\,\lambda_1^k\,\mathbf{v}_1 + \cdots + c_n\,\lambda_n^k\,\mathbf{v}_n. \tag{10.98}$$

Suppose further that A has a single dominant *real* eigenvalue, λ_1, that is larger than all others in magnitude, so

$$|\lambda_1| > |\lambda_j| \quad \text{for all} \quad j > 1. \tag{10.99}$$

As its name implies, this eigenvalue will eventually dominate the iteration (10.98). Indeed, since

$$|\lambda_1|^k \gg |\lambda_j|^k \qquad \text{for all} \quad j > 1 \quad \text{and all} \quad k \gg 0,$$

*This is not a very severe restriction. Most matrices are complete. Moreover, perturbations caused by round off and/or numerical inaccuracies will almost inevitably make an incomplete matrix complete.

the first term in the iterative formula (10.98) will eventually be much larger than the rest, and so, provided $c_1 \neq 0$,

$$\mathbf{v}^{(k)} \approx c_1 \lambda_1^k \mathbf{v}_1 \quad \text{for} \quad k \gg 0.$$

Therefore, the solution to the iterative system (10.96) will, almost always, end up being a multiple of the dominant eigenvector of the coefficient matrix.

To compute the corresponding eigenvalue, we note that the ith entry of the iterate $\mathbf{v}^{(k)}$ is approximated by $v_i^{(k)} \approx c_1 \lambda_1^k v_{1,i}$, where $v_{1,i}$ is the ith entry of the eigenvector $\mathbf{v}_1$. Thus, as long as $v_{1,i} \neq 0$, we can recover the dominant eigenvalue by taking a ratio between selected components of successive iterates:

$$\lambda_1 \approx \frac{v_i^{(k)}}{v_i^{(k-1)}}, \quad \text{provided that} \quad v_i^{(k-1)} \neq 0. \tag{10.100}$$

EXAMPLE 10.52 Consider the matrix

$$A = \begin{pmatrix} -1 & 2 & 2 \\ -1 & -4 & -2 \\ -3 & 9 & 7 \end{pmatrix}.$$

As you can check, its eigenvalues and eigenvectors are

$$\lambda_1 = 3, \quad \mathbf{v}_1 = \begin{pmatrix} 1 \\ -1 \\ 3 \end{pmatrix}, \qquad \lambda_2 = -2, \quad \mathbf{v}_2 = \begin{pmatrix} 0 \\ 1 \\ -1 \end{pmatrix},$$

$$\lambda_3 = 1, \quad \mathbf{v}_3 = \begin{pmatrix} -1 \\ 1 \\ -2 \end{pmatrix}.$$

Repeatedly multiplying an initial vector $\mathbf{v} = (1, 0, 0)^T$, say, by A results in the iterates $\mathbf{v}^{(k)} = A^k \mathbf{v}$ listed in the accompanying table. The last column indicates the ratio $\lambda^{(k)} = v_1^{(k)} / v_1^{(k-1)}$ between the first components of successive iterates. (One could equally well use the second or third components.) The ratios are converging

k		$\mathbf{v}^{(k)}$		$\lambda^{(k)}$
0	1	0	0	
1	−1	−1	−3	−1.
2	−7	11	−27	7.
3	−25	17	−69	3.5714
4	−79	95	−255	3.1600
5	−241	209	−693	3.0506
6	−727	791	−2247	3.0166
7	−2185	2057	−6429	3.0055
8	−6559	6815	−19935	3.0018
9	−19681	19169	−58533	3.0006
10	−59047	60071	−178167	3.0002
11	−177145	175097	−529389	3.0001
12	−531439	535535	−1598415	3.0000

to the dominant eigenvalue $\lambda_1 = 3$, while the vectors $\mathbf{v}^{(k)}$ are converging to a very large multiple of the corresponding eigenvector $\mathbf{v}_1 = (1, -1, 3)^T$. ●

The success of the power method lies in the assumption that A has a unique dominant eigenvalue of maximal modulus, which, by definition, equals its spectral radius: $|\lambda_1| = \rho(A)$. The rate of convergence of the method is governed by the ratio $|\lambda_2/\lambda_1|$ between the subdominant and dominant eigenvalues. Thus, the farther the dominant eigenvalue lies away from the rest, the faster the power method converges. We also assumed that the initial vector $\mathbf{v}^{(0)}$ includes a nonzero multiple of the dominant eigenvector, i.e., $c_1 \neq 0$. As we do not know the eigenvectors, it is not so easy to guarantee this in advance, although we must be quite unlucky to make such a poor choice of initial vector. (Of course, the stupid choice $\mathbf{v}^{(0)} = \mathbf{0}$ is not counted.) Moreover, even if c_1 happens to be 0 initially, numerical round-off error will typically come to one's rescue, since it will almost inevitably introduce a tiny component of the eigenvector $\mathbf{v}_1$ into some iterate, and this component will eventually dominate the computation. The trick is to wait long enough for it to show up!

Since the iterates of A are, typically, getting either very large—when $\rho(A) > 1$—or very small—when $\rho(A) < 1$—the iterated vectors will be increasingly subject to numerical over- or under-flow, and the method may break down before a reasonable approximation is achieved. One way to avoid this outcome is to restrict our attention to unit vectors relative to a given norm, e.g., the Euclidean norm or the ∞ norm, since their entries cannot be too large, and so are less likely to cause numerical errors in the computations. As usual, the unit vector $\mathbf{u}^{(k)} = \|\mathbf{v}^{(k)}\|^{-1}\mathbf{v}^{(k)}$ is obtained by dividing the iterate by its norm; it can be computed directly by the modified iterative scheme

$$\mathbf{u}^{(0)} = \frac{\mathbf{v}^{(0)}}{\|\mathbf{v}^{(0)}\|}, \qquad \text{and} \qquad \mathbf{u}^{(k+1)} = \frac{A\,\mathbf{u}^{(k)}}{\|A\,\mathbf{u}^{(k)}\|}. \tag{10.101}$$

If the dominant eigenvalue $\lambda_1 > 0$ is positive, then $\mathbf{u}^{(k)} \to \mathbf{u}_1$ will converge to one of the two dominant unit eigenvectors (the other is $-\mathbf{u}_1$). If $\lambda_1 < 0$, then the iterates will switch back and forth between the two eigenvectors, so $\mathbf{u}^{(k)} \approx \pm\mathbf{u}_1$. In either case, the dominant eigenvalue λ_1 is obtained as a limiting ratio between nonzero entries of $A\,\mathbf{u}^{(k)}$ and $\mathbf{u}^{(k)}$. If some other sort of behavior is observed, it means that one of our assumptions is not valid; either A has more than one dominant eigenvalue of maximum modulus, e.g., it has a complex conjugate pair of eigenvalues of largest modulus, or it is not complete. In such cases, one can apply the more general long-term behavior described in Exercise 10.2.10 to pin down the dominant eigenvalues.

EXAMPLE 10.53 For the matrix considered in Example 10.52, starting the iterative scheme (10.101) with $\mathbf{u}^{(k)} = (1, 0, 0)^T$ by A, the resulting unit vectors are tabulated below. The last column, being the ratio between the first components of $A\,\mathbf{u}^{(k-1)}$ and $\mathbf{u}^{(k-1)}$, again converges to the dominant eigenvalue $\lambda_1 = 3$. ●

Variants of the power method for computing the other eigenvalues of the matrix are explored in the exercises.

k	$\mathbf{u}^{(k)}$			λ
0	1	0	0	
1	−.3015	−.3015	−.9045	−1.0000
2	−.2335	.3669	−.9005	7.0000
3	−.3319	.2257	−.9159	3.5714
4	−.2788	.3353	−.8999	3.1600
5	−.3159	.2740	−.9084	3.0506
6	−.2919	.3176	−.9022	3.0166
7	−.3080	.2899	−.9061	3.0055
8	−.2973	.3089	−.9035	3.0018
9	−.3044	.2965	−.9052	3.0006
10	−.2996	.3048	−.9041	3.0002
11	−.3028	.2993	−.9048	3.0001
12	−.3007	.3030	−.9043	3.0000

EXERCISES 10.6

♠ **10.6.1.** Use the power method to find the dominant eigenvalue and associated eigenvector of the following matrices:

(a) $\begin{pmatrix} -1 & -2 \\ 3 & 4 \end{pmatrix}$ (b) $\begin{pmatrix} -5 & 2 \\ -3 & 0 \end{pmatrix}$

(c) $\begin{pmatrix} 3 & -1 & 0 \\ -1 & 2 & -1 \\ 0 & -1 & 3 \end{pmatrix}$ (d) $\begin{pmatrix} -2 & 0 & 1 \\ -3 & -2 & 0 \\ -2 & 5 & 4 \end{pmatrix}$

(e) $\begin{pmatrix} -1 & -2 & -2 \\ 1 & 2 & 5 \\ -1 & 4 & 0 \end{pmatrix}$ (f) $\begin{pmatrix} 2 & 2 & 1 \\ 1 & 3 & 1 \\ 2 & 2 & 2 \end{pmatrix}$

(g) $\begin{pmatrix} 2 & -1 & 0 & 0 \\ -1 & 2 & -1 & 0 \\ 0 & -1 & 2 & -1 \\ 0 & 0 & -1 & 2 \end{pmatrix}$

(h) $\begin{pmatrix} 4 & 1 & 0 & 1 \\ 1 & 4 & 1 & 0 \\ 0 & 1 & 4 & 1 \\ 1 & 0 & 1 & 4 \end{pmatrix}$

♠ **10.6.2.** Let T_n be the tridiagonal matrix whose diagonal entries are all equal to 2 and whose sub- and super-diagonal entries all equal 1. Use the power method to find the dominant eigenvalue of T_n for $n = 10, 20, 50$. Do your values agree with those in Exercise 8.2.47? How many iterations do you require to obtain 4 decimal place accuracy?

♠ **10.6.3.** Use the power method to find the largest singular value of the following matrices:

(a) $\begin{pmatrix} 1 & 2 \\ -1 & 3 \end{pmatrix}$ (b) $\begin{pmatrix} 2 & 1 & -1 \\ -2 & 3 & 1 \end{pmatrix}$

(c) $\begin{pmatrix} 2 & 2 & 1 & -1 \\ 1 & -2 & 0 & 1 \end{pmatrix}$ (d) $\begin{pmatrix} 3 & 1 & -1 \\ 1 & -2 & 2 \\ 2 & -1 & 1 \end{pmatrix}$

10.6.4. Prove that, for the iterative scheme (10.101), $\| A\,\mathbf{u}^{(k)} \| \to |\lambda_1|$. Assuming λ_1 is real, explain how to deduce its sign.

◇ **10.6.5.** *The Inverse Power Method.* Let A be a nonsingular matrix.

(a) Show that the eigenvalues of A^{-1} are the reciprocals $1/\lambda$ of the eigenvalues of A. How are the eigenvectors related?

(b) Show how to use the power method on A^{-1} to produce the smallest (in modulus) eigenvalue of A.

(c) What is the rate of convergence of the algorithm?

(d) Design a practical iterative algorithm based on the (permuted) $L\,U$ decomposition of A.

♠ **10.6.6.** Apply the inverse power method of Exercise 10.6.7 to the find the smallest eigenvalue of the matrices in Exercise 10.6.1.

◇ **10.6.7.** *The Shifted Inverse Power Method.* Suppose that μ is *not* an eigenvalue of A.

(a) Show that the iterative scheme $\mathbf{u}^{(k+1)} = (A - \mu\, \mathrm{I})^{-1}\mathbf{u}^{(k)}$ converges to the eigenvector of A corresponding to the eigenvalue $\lambda^\star$ that is *closest* to μ. Explain how to find the eigenvalue $\lambda^\star$.

(b) What is the rate of convergence of the algorithm?

(c) What happens if μ is an eigenvalue?

♠ **10.6.8.** Apply the shifted inverse power method of Exercise 10.6.7 to the find the eigenvalue closest to $\mu = .5$ of the matrices in Exercise 10.6.1.

♠ **10.6.9.** (i) Explain how to use the deflation method of Exercise 8.2.52 to find the subdominant eigenvalue of a nonsingular matrix A.

(ii) Apply your method to the matrices in Exercise 10.6.1.

10.6.10. Suppose that $A\,\mathbf{u}^{(k)} = \mathbf{0}$ in the iterative procedure (10.101). What does this indicate?

The *QR* Algorithm

As stated, the power method only produces the dominant (largest in magnitude) eigenvalue of a matrix A. The inverse power method of Exercise 10.6.5 can be used to find the smallest eigenvalue. Additional eigenvalues can be found by using the shifted inverse power method of Exercise 10.6.7, or the deflation method of Exercise 10.6.9. However, if we need to know *all* the eigenvalues, such piecemeal methods are too time-consuming to be of much practical value.

The most popular scheme for simultaneously approximating all the eigenvalues of a matrix A is the remarkable $Q\,R$ algorithm, first proposed in 1961 by Francis, [23], and Kublanovskaya, [39]. The underlying idea is simple, but surprising. The first step is to factor the matrix

$$A = A_0 = Q_0\, R_0$$

into a product of an orthogonal matrix Q_0 and a positive (i.e., with all positive entries along the diagonal) upper triangular matrix R_0 by using the Gram–Schmidt orthogonalization procedure of Theorem 5.24, or, even better, the numerically stable version described in (5.28). Next, multiply the two factors together *in the wrong order*! The result is the new matrix

$$A_1 = R_0\, Q_0.$$

We then repeat these two steps. Thus, we next factor

$$A_1 = Q_1\, R_1$$

using the Gram–Schmidt process, and then multiply the factors in the reverse order to produce

$$A_2 = R_2\, Q_2.$$

The complete algorithm can be written as

$$A = Q_0\, R_0, \qquad A_{k+1} = R_k\, Q_k = Q_{k+1}\, R_{k+1}, \qquad k = 0, 1, 2, \ldots, \tag{10.102}$$

where Q_k, R_k come from the previous step, and the subsequent orthogonal matrix Q_{k+1} and positive upper triangular matrix R_{k+1} are computed by using the numerically stable form of the Gram–Schmidt algorithm.

The astonishing fact is that, for many matrices A, the iterates $A_k \longrightarrow V$ converge to an upper triangular matrix V whose diagonal entries are the eigenvalues of A. Thus, after a sufficient number of iterations, say $k^\star$, the matrix $A_{k^\star}$ will have very small entries below the diagonal, and one can read off a complete system of (approximate) eigenvalues along its diagonal. For each eigenvalue, the computation of the corresponding eigenvector can be done by solving the appropriate homogeneous linear system, or by applying the shifted inverse power method of Exercise 10.6.7.

EXAMPLE 10.54 Consider the matrix

$$A = \begin{pmatrix} 2 & 1 \\ 2 & 3 \end{pmatrix}.$$

The initial Gram–Schmidt factorization $A = Q_0 R_0$ yields

$$Q_0 = \begin{pmatrix} .7071 & .7071 \\ -.7071 & .7071 \end{pmatrix}, \qquad R_0 = \begin{pmatrix} 2.8284 & 2.8284 \\ 0 & 1.4142 \end{pmatrix}.$$

These are multiplied in the reverse order to give

$$A_1 = R_0 Q_0 = \begin{pmatrix} 4 & 0 \\ 1 & 1 \end{pmatrix}.$$

We refactor $A_1 = Q_1 R_1$ via Gram–Schmidt, and then reverse multiply to produce

$$Q_1 = \begin{pmatrix} .9701 & -.2425 \\ .2425 & .9701 \end{pmatrix}, \qquad R_1 = \begin{pmatrix} 4.1231 & .2425 \\ 0 & .9701 \end{pmatrix},$$

$$A_2 = R_1 Q_1 = \begin{pmatrix} 4.0588 & -.7647 \\ .2353 & .9412 \end{pmatrix}.$$

The next iteration yields

$$Q_2 = \begin{pmatrix} .9983 & -.0579 \\ .0579 & .9983 \end{pmatrix}, \qquad R_2 = \begin{pmatrix} 4.0656 & -.7090 \\ 0 & .9839 \end{pmatrix},$$

$$A_3 = R_2 Q_2 = \begin{pmatrix} 4.0178 & -.9431 \\ .0569 & .9822 \end{pmatrix}.$$

Continuing in this manner, after 9 iterations we find, to four decimal places,

$$Q_9 = \begin{pmatrix} 1 & 0 \\ 0 & 1 \end{pmatrix}, \qquad R_9 = \begin{pmatrix} 4 & -1 \\ 0 & 1 \end{pmatrix}, \qquad A_{10} = R_9 Q_9 = \begin{pmatrix} 4 & -1 \\ 0 & 1 \end{pmatrix}.$$

The eigenvalues of A, namely 4 and 1, appear along the diagonal of A_{10}. Additional iterations produce very little further change, although they can be used for increasing the accuracy of the computed eigenvalues. ●

If the original matrix A happens to be symmetric and positive definite, then the limiting matrix $A_k \longrightarrow V = \Lambda$ is, in fact, the diagonal matrix containing the eigenvalues of A. Moreover, if, in this case, we recursively define

$$S_k = S_{k-1} Q_k = Q_0 Q_1 \cdots Q_{k-1} Q_k, \tag{10.103}$$

then $S_k \longrightarrow S$ have, as their limit, an orthogonal matrix whose columns are the orthonormal eigenvector basis of A.

EXAMPLE 10.55 Consider the symmetric matrix

$$A = \begin{pmatrix} 2 & 1 & 0 \\ 1 & 3 & -1 \\ 0 & -1 & 6 \end{pmatrix}.$$

The initial $A = Q_0 R_0$ factorization produces

$$S_0 = Q_0 = \begin{pmatrix} .8944 & -.4082 & -.1826 \\ .4472 & .8165 & .3651 \\ 0 & -.4082 & .9129 \end{pmatrix},$$

$$R_0 = \begin{pmatrix} 2.2361 & 2.2361 & -.4472 \\ 0 & 2.4495 & -3.2660 \\ 0 & 0 & 5.1121 \end{pmatrix},$$

and so

$$A_1 = R_0\, Q_0 = \begin{pmatrix} 3.0000 & 1.0954 & 0 \\ 1.0954 & 3.3333 & -2.0870 \\ 0 & -2.0870 & 4.6667 \end{pmatrix}.$$

We refactor $A_1 = Q_1 R_1$ and reverse multiply to produce

$$Q_1 = \begin{pmatrix} .9393 & -.2734 & -.2071 \\ .3430 & .7488 & .5672 \\ 0 & -.6038 & .7972 \end{pmatrix},$$

$$S_1 = S_0\, Q_1 = \begin{pmatrix} .7001 & -.4400 & -.5623 \\ .7001 & .2686 & .6615 \\ -.1400 & -.8569 & .4962 \end{pmatrix},$$

$$R_1 = \begin{pmatrix} 3.1937 & 2.1723 & -.7158 \\ 0 & 3.4565 & -4.3804 \\ 0 & 0 & 2.5364 \end{pmatrix},$$

$$A_2 = R_1\, Q_1 = \begin{pmatrix} 3.7451 & 1.1856 & 0 \\ 1.1856 & 5.2330 & -1.5314 \\ 0 & -1.5314 & 2.0219 \end{pmatrix}.$$

Continuing in this manner, after 10 iterations we find

$$Q_{10} = \begin{pmatrix} 1.0000 & -.0067 & 0 \\ .0067 & 1.0000 & .0001 \\ 0 & -.0001 & 1.0000 \end{pmatrix},$$

$$S_{10} = \begin{pmatrix} .0753 & -.5667 & -.8205 \\ .3128 & -.7679 & .5591 \\ -.9468 & -.2987 & .1194 \end{pmatrix},$$

$$R_{10} = \begin{pmatrix} 6.3229 & .0647 & 0 \\ 0 & 3.3582 & -.0006 \\ 0 & 0 & 1.3187 \end{pmatrix},$$

$$A_{11} = \begin{pmatrix} 6.3232 & .0224 & 0 \\ .0224 & 3.3581 & -.0002 \\ 0 & -.0002 & 1.3187 \end{pmatrix}.$$

After 20 iterations, the process has completely settled down, and

$$Q_{20} = \begin{pmatrix} 1 & 0 & 0 \\ 0 & 1 & 0 \\ 0 & 0 & 1 \end{pmatrix},$$

$$S_{20} = \begin{pmatrix} .0710 & -.5672 & -.8205 \\ .3069 & -.7702 & .5590 \\ -.9491 & -.2915 & .1194 \end{pmatrix},$$

$$R_{20} = \begin{pmatrix} 6.3234 & .0001 & 0 \\ 0 & 3.3579 & 0 \\ 0 & 0 & 1.3187 \end{pmatrix},$$

$$A_{21} = \begin{pmatrix} 6.3234 & 0 & 0 \\ 0 & 3.3579 & 0 \\ 0 & 0 & 1.3187 \end{pmatrix}.$$

The eigenvalues of A appear along the diagonal of A_{21}, while the columns of S_{20} are the corresponding orthonormal eigenvector basis, listed in the same order as the eigenvalues, both correct to 4 decimal places. ●

We will devote the remainder of this section to a justification of the QR algorithm for a class of matrices. We will assume that A is symmetric, and that its eigenvalues satisfy

$$|\lambda_1| > |\lambda_2| > \cdots > |\lambda_n| > 0. \tag{10.104}$$

According to the Spectral Theorem 8.26, the corresponding unit eigenvectors $\mathbf{u}_1, \ldots, \mathbf{u}_n$ form an orthonormal basis of $\mathbb{R}^n$. The analysis can be adapted to a broader class of matrices, but this will suffice to expose the main ideas without unduly complicating the exposition.

The secret is that the QR algorithm is, in fact, a well-disguised adaptation of the more primitive power method. If we were to use the power method to capture all the eigenvectors and eigenvalues of A, the first thought might be to try to perform it simultaneously on a complete basis $\mathbf{v}_1^{(0)}, \ldots, \mathbf{v}_n^{(0)}$ of $\mathbb{R}^n$ instead of just one individual vector. The problem is that, for almost all vectors, the power iterates $\mathbf{v}_j^{(k)} = A^k \mathbf{v}_j^{(0)}$ all tend to a multiple of the dominant eigenvector $\mathbf{u}_1$. Normalizing the vectors at each step, as in (10.101), is not any better, since then they merely converge to one of the two dominant unit eigenvectors $\pm\mathbf{u}_1$. However, if, inspired by the form of the eigenvector basis, we *orthonormalize* the vectors at each step, then we effectively prevent them from all accumulating at the same dominant unit eigenvector, and so, with some luck, the resulting vectors will converge to the full system of eigenvectors. Since orthonormalizing a basis via the Gram–Schmidt process is equivalent to a QR matrix factorization, the mechanics of the algorithm is not so surprising.

In detail, we start with any orthonormal basis, which, for simplicity, we take to be the standard basis vectors of $\mathbb{R}^n$, and so $\mathbf{u}_1^{(0)} = \mathbf{e}_1, \ldots, \mathbf{u}_n^{(0)} = \mathbf{e}_n$. At the kth stage of the algorithm, we set $\mathbf{u}_1^{(k)}, \ldots, \mathbf{u}_n^{(k)}$ to be the orthonormal vectors that result from applying the Gram–Schmidt algorithm to the power vectors $\mathbf{v}_j^{(k)} = A^k \mathbf{e}_j$. In matrix language, the vectors $\mathbf{v}_1^{(k)}, \ldots, \mathbf{v}_n^{(k)}$ are merely the columns of A^k, and the orthonormal basis $\mathbf{u}_1^{(k)}, \ldots, \mathbf{u}_n^{(k)}$ are the columns of the orthogonal matrix S_k in the QR decomposition of the kth power of A, which we denote by

$$A^k = S_k P_k, \tag{10.105}$$

where P_k is positive upper triangular. Note that, in view of (10.102)

$$\begin{aligned} A &= Q_0 R_0, \\ A^2 &= Q_0 R_0 Q_0 R_0 = Q_0 Q_1 R_1 R_0, \\ A^3 &= Q_0 R_0 Q_0 R_0 Q_0 R_0 = Q_0 Q_1 R_1 Q_1 R_1 R_0 = Q_0 Q_1 Q_2 R_2 R_1 R_0, \end{aligned}$$

and, in general,

$$A^k = \left(Q_0 Q_1 \cdots Q_{k-1} Q_k \right) \left(R_k R_{k-1} \cdots R_1 R_0 \right). \tag{10.106}$$

The product of orthogonal matrices is also orthogonal. The product of positive upper triangular matrices is also positive upper triangular. Therefore, comparing (10.105, 106) and invoking the uniqueness of the QR factorization, we conclude that

$$\begin{aligned} S_k &= Q_0 Q_1 \cdots Q_{k-1} Q_k = S_{k-1} Q_k, \\ P_k &= R_k R_{k-1} \cdots R_1 R_0 = R_k P_{k-1}. \end{aligned} \tag{10.107}$$

Let $S = (\mathbf{u}_1\ \mathbf{u}_2 \ldots \mathbf{u}_n)$ denote an orthogonal matrix whose columns are unit eigenvectors of A. The Spectral Theorem 8.26 tells us that

$$A = S\,\Lambda\,S^T, \qquad \text{where} \qquad \Lambda = \operatorname{diag}(\lambda_1, \ldots, \lambda_n)$$

is the diagonal eigenvalue matrix. Substituting the spectral factorization into (10.105), we find

$$A^k = S\,\Lambda^k\,S^T = S_k\,P_k.$$

We now make one additional assumption on the matrix A by requiring that S^T be a regular matrix. This holds generically, and is the analog of the condition that our initial vector in the power method includes a nonzero component of the dominant eigenvector. Regularity means that we can factor $S^T = L\,U$ into a product of special lower and upper triangular matrices. We can assume that, without loss of generality, the diagonal entries of U — that is, the pivots of S^T—are all positive. Indeed, by Exercise 1.3.30, this can be arranged by multiplying each row of S^T by the sign of its pivot, which amounts to possibly changing the signs of some of the unit eigenvectors $\mathbf{u}_i$, which is allowed since it does not affect their status as an orthonormal eigenvector basis.

Under these two assumptions,

$$A^k = S\,\Lambda^k\,L\,U = S_k\,P_k,$$

and hence

$$S\,\Lambda^k\,L = S_k\,P_k\,U^{-1}.$$

Multiplying on the right by Λ^{-k}, we obtain

$$S\,\Lambda^k\,L\,\Lambda^{-k} = S_k\,T_k, \qquad \text{where} \qquad T_k = P_k\,U^{-1}\,\Lambda^{-k} \tag{10.108}$$

is also a positive upper triangular matrix.

Now consider what happens as $k \to \infty$. The entries of the lower triangular matrix $N = \Lambda^k\,L\,\Lambda^{-k}$ are

$$n_{ij} = \begin{cases} l_{ij}(\lambda_i/\lambda_j)^k, & i > j, \\ l_{ii} = 1, & i = j, \\ 0, & i < j. \end{cases}$$

Since we are assuming $|\lambda_i| < |\lambda_j|$ when $i > j$, we immediately deduce that

$$\Lambda^k\,L\,\Lambda^{-k} \longrightarrow \mathrm{I},$$

and hence

$$S_k\,T_k = S\,\Lambda^k\,L\,\Lambda^{-k} \longrightarrow S \quad \text{as} \quad k \longrightarrow \infty.$$

We now appeal to the following lemma, whose proof will be given after we finish the justification of the $Q\,R$ algorithm.

Lemma 10.56 Let $S_1, S_2, \ldots$ and S be orthogonal matrices and $T_1, T_2, \ldots$ positive upper triangular matrices. Then $S_k\,T_k \to S$ as $k \to \infty$ if and only if $S_k \to S$ and $T_k \to \mathrm{I}$.

Therefore, as claimed, the orthogonal matrices S_k do converge to the orthogonal eigenvector matrix S. Moreover, by (10.107–108),

$$R_k = P_k\,P_{k-1}^{-1} = \left(T_k\,\Lambda^k\,U^{-1}\right)\left(T_{k-1}\,\Lambda^{k-1}\,U^{-1}\right)^{-1} = T_k\,\Lambda\,T_{k-1}^{-1}.$$

Since both T_k and T_{k-1} converge to the identity matrix, in the limit $R_k \to \Lambda$ converges to the diagonal eigenvalue matrix, as claimed. The eigenvalues appear in decreasing order along the diagonal—this is a consequence of our regularity assumption on the transposed eigenvector matrix S^T.

THEOREM 10.57 If A is positive definite, satisfies (10.104), and its transposed eigenvector matrix S^T is regular, then the matrices $S_k \to S$ and $R_k \to \Lambda$ appearing in the QR algorithm applied to A converge to, respectively, the eigenvector matrix S and the diagonal eigenvalue matrix Λ.

The last remaining detail is a proof of Lemma 10.56. We write

$$S = (\mathbf{u}_1 \;\; \mathbf{u}_2 \;\; \ldots \;\; \mathbf{u}_n), \qquad S_k = (\mathbf{u}_1^{(k)}, \ldots, \mathbf{u}_n^{(k)})$$

in columnar form. Let $t_{ij}^{(k)}$ denote the entries of the positive upper triangular matrix T_k. The last column of the limiting equation $S_k\, T_k \to S$ reads

$$t_{nn}^{(k)}\, \mathbf{u}_n^{(k)} \longrightarrow \mathbf{u}_n.$$

Since both $\mathbf{u}_n^{(k)}$ and $\mathbf{u}_n$ are unit vectors, and $t_{nn}^{(k)} > 0$,

$$\| t_{nn}^{(k)}\, \mathbf{u}_n^{(k)} \| = t_{nn}^{(k)} \longrightarrow \| \mathbf{u}_n \| = 1, \qquad \text{and hence the last column} \quad \mathbf{u}_n^{(k)} \longrightarrow \mathbf{u}_n.$$

The next to last column reads

$$t_{n-1,n-1}^{(k)}\, \mathbf{u}_{n-1}^{(k)} + t_{n-1,n}^{(k)}\, \mathbf{u}_n^{(k)} \longrightarrow \mathbf{u}_{n-1}.$$

Taking the inner product with $\mathbf{u}_n^{(k)} \to \mathbf{u}_n$ and using orthonormality, we deduce $t_{n-1,n}^{(k)} \to 0$, and so $t_{n-1,n-1}^{(k)}\, \mathbf{u}_{n-1}^{(k)} \to \mathbf{u}_{n-1}$, which, by the previous reasoning, implies $t_{n-1,n-1}^{(k)} \to 1$ and $\mathbf{u}_{n-1}^{(k)} \to \mathbf{u}_{n-1}$. The proof is completed by working backwards through the remaining columns, using a similar argument at each step. The remaining details are left to the interested reader.

EXERCISES 10.6

10.6.11. Apply the QR algorithm to the following symmetric matrices to find their eigenvalues and eigenvectors to 2 decimal places:

(a) $\begin{pmatrix} 1 & 2 \\ 2 & 6 \end{pmatrix}$ (b) $\begin{pmatrix} 3 & -1 \\ -1 & 5 \end{pmatrix}$

(c) $\begin{pmatrix} 2 & 1 & 0 \\ 1 & 2 & 3 \\ 0 & 3 & 1 \end{pmatrix}$ (d) $\begin{pmatrix} 2 & 5 & 0 \\ 5 & 0 & -3 \\ 0 & -3 & 3 \end{pmatrix}$

(e) $\begin{pmatrix} 3 & -1 & 0 & 0 \\ -1 & 3 & -1 & 0 \\ 0 & -1 & 3 & -1 \\ 0 & 0 & -1 & 3 \end{pmatrix}$

(f) $\begin{pmatrix} 6 & 1 & -1 & 0 \\ 1 & 8 & 1 & -1 \\ -1 & 1 & 4 & 1 \\ 0 & -1 & 1 & 3 \end{pmatrix}$

10.6.12. Show that applying the QR algorithm to the matrix

$$A = \begin{pmatrix} 4 & -1 & 1 \\ -1 & 7 & 2 \\ 1 & 2 & 7 \end{pmatrix}$$

results in a diagonal matrix with the eigenvalues on the diagonal, but not in decreasing order. Explain.

10.6.13. Apply the QR algorithm to the following non-symmetric matrices to find their eigenvalues to 3 decimal places:

(a) $\begin{pmatrix} -1 & -2 \\ 3 & 4 \end{pmatrix}$ (b) $\begin{pmatrix} 2 & 3 \\ 1 & 5 \end{pmatrix}$

(c) $\begin{pmatrix} 2 & 1 & 0 \\ 2 & 0 & -3 \\ 0 & -2 & 1 \end{pmatrix}$ (d) $\begin{pmatrix} 2 & 5 & 1 \\ 2 & -1 & 3 \\ 4 & 5 & 3 \end{pmatrix}$

(e) $\begin{pmatrix} 6 & 1 & 7 & 9 \\ 6 & 8 & 14 & 9 \\ 3 & 1 & 4 & 6 \\ 3 & 2 & 5 & 3 \end{pmatrix}$

10.6.14. The matrix

$$A = \begin{pmatrix} -1 & 2 & 1 \\ -2 & 3 & 1 \\ -2 & 2 & 2 \end{pmatrix}$$

has a double eigenvalue of 1, and so our proof of convergence of the QR algorithm doesn't apply. Does the QR algorithm find its eigenvalues?

10.6.15. Explain why the QR algorithm fails to find the eigenvalues of the matrix

$$A = \begin{pmatrix} 0 & 1 \\ 1 & 0 \end{pmatrix}.$$

◇10.6.16. Prove that all of the matrices A_k defined in (10.102) have the same eigenvalues.

◇10.6.17. (a) Prove that if A is symmetric and tridiagonal, then all matrices A_k appearing in the QR algorithm are also symmetric and tridiagonal. *Hint*: First prove symmetry.

(b) Is the result true if A is not symmetric—only tridiagonal?

Tridiagonalization

In practical implementations, the direct QR algorithm often takes too long to provide reasonable approximations to the eigenvalues of large matrices. Fortunately, the algorithm can be made much more efficient by a simple preprocessing step. The key observation is that the QR algorithm preserves the class of symmetric tridiagonal matrices, and, moreover, like Gaussian Elimination, is much faster when applied to this class of matrices. Moreover, by applying a sequence of Householder reflection matrices (5.35), we can convert any symmetric matrix into tridiagonal form while preserving all the eigenvalues. Thus, by first applying the Householder tridiagonalization process, and then applying the QR method to the resulting tridiagonal matrix, we obtain an efficient and practical algorithm for computing eigenvalues of symmetric matrices. Generalizations to non-symmetric matrices will be briefly considered at the end of the section.

In Householder's approach to the QR factorization, we were able to convert the matrix A to upper triangular form R by a sequence of elementary reflection matrices. Unfortunately, this procedure does not preserve the eigenvalues of the matrix—the diagonal entries of R are *not* the eigenvalues—and so we need to be a bit more clever here. We begin by recalling, from Exercise 8.2.32, that similar matrices have the same eigenvalues.

Lemma 10.58 If $H = \mathrm{I} - 2\mathbf{u}\,\mathbf{u}^T$ is a elementary reflection matrix, with $\mathbf{u}$ a unit vector, then A and $B = H A H$ are similar matrices and hence have the same eigenvalues.

Proof According to (5.36), $H^{-1} = H$, and hence $B = H^{-1} A H$ is similar to A. ■

Given a symmetric $n \times n$ matrix A, our goal is to devise a similar tridiagonal matrix by applying a sequence of Householder reflections. We begin by setting

$$\mathbf{x}_1 = \begin{pmatrix} 0 \\ a_{21} \\ a_{31} \\ \vdots \\ a_{n1} \end{pmatrix}, \qquad \mathbf{y}_1 = \begin{pmatrix} 0 \\ \pm r_1 \\ 0 \\ \vdots \\ 0 \end{pmatrix}, \qquad \text{where} \quad r_1 = \|\mathbf{x}_1\| = \|\mathbf{y}_1\|,$$

so that $\mathbf{x}_1$ contains all the off-diagonal entries of the first column of A. Let

$$H_1 = \mathrm{I} - 2\mathbf{u}_1\,\mathbf{u}_1^T, \qquad \text{where} \quad \mathbf{u}_1 = \frac{\mathbf{x}_1 - \mathbf{y}_1}{\|\mathbf{x}_1 - \mathbf{y}_1\|}$$

be the corresponding elementary reflection matrix that maps $\mathbf{x}_1$ to $\mathbf{y}_1$. Either $\pm$ sign in the formula for $\mathbf{y}_1$ works in the algorithm; a good choice is to set it to be the opposite of the sign of the entry a_{21}, which helps minimize the possible effects of round-off error when computing the unit vector $\mathbf{u}_1$. By direct computation, based on Lemma 5.28 and the fact that the first entry of $\mathbf{u}_1$ is zero,

$$A_2 = H_1\, A\, H_1 = \begin{pmatrix} a_{11} & r_1 & 0 & \dots & 0 \\ r_1 & \widetilde{a}_{22} & \widetilde{a}_{23} & \dots & \widetilde{a}_{2n} \\ 0 & \widetilde{a}_{32} & \widetilde{a}_{33} & \dots & \widetilde{a}_{3n} \\ \vdots & \vdots & \vdots & \ddots & \vdots \\ 0 & \widetilde{a}_{n2} & \widetilde{a}_{n3} & \dots & \widetilde{a}_{nn} \end{pmatrix} \tag{10.109}$$

for certain $\widetilde{a}_{ij}$; the explicit formulae are not needed. Thus, by a single Householder transformation, we convert A into a similar matrix A_2 whose first row and column are in tridiagonal form. We repeat the process on the lower right $(n-1)\times(n-1)$ submatrix of A_2. We set

$$\mathbf{x}_2 = \begin{pmatrix} 0 \\ 0 \\ \widetilde{a}_{32} \\ \widetilde{a}_{42} \\ \vdots \\ \widetilde{a}_{n2} \end{pmatrix}, \qquad \mathbf{y}_1 = \begin{pmatrix} 0 \\ 0 \\ \pm r_2 \\ 0 \\ \vdots \\ 0 \end{pmatrix}, \qquad \text{where} \quad r_2 = \|\mathbf{x}_2\| = \|\mathbf{y}_2\|,$$

and the $\pm$ sign is chosen to be the opposite of that of $\widetilde{a}_{32}$. Setting

$$H_2 = \mathrm{I} - 2\,\mathbf{u}_2\,\mathbf{u}_2^T, \qquad \text{where} \quad \mathbf{u}_2 = \frac{\mathbf{x}_2 - \mathbf{y}_2}{\|\mathbf{x}_2 - \mathbf{y}_2\|},$$

we construct the similar matrix

$$A_3 = H_2\, A_2\, H_2 = \begin{pmatrix} a_{11} & r_1 & 0 & 0 & \dots & 0 \\ r_1 & \widetilde{a}_{22} & r_2 & 0 & \dots & 0 \\ 0 & r_2 & \widehat{a}_{33} & \widehat{a}_{34} & \dots & \widehat{a}_{3n} \\ 0 & 0 & \widehat{a}_{43} & \widehat{a}_{44} & \dots & \widehat{a}_{4n} \\ \vdots & \vdots & \vdots & \vdots & \ddots & \vdots \\ 0 & 0 & \widehat{a}_{n3} & \widehat{a}_{n4} & \dots & \widehat{a}_{nn} \end{pmatrix}.$$

whose first two rows and columns are now in tridiagonal form. The remaining steps in the algorithm should now be clear. Thus, the final result is a tridiagonal matrix $T = A_n$ that has the *same eigenvalues* as the original symmetric matrix A. Let us illustrate the method by an example.

EXAMPLE 10.59 To tridiagonalize

$$A = \begin{pmatrix} 4 & 1 & -1 & 2 \\ 1 & 4 & 1 & -1 \\ -1 & 1 & 4 & 1 \\ 2 & -1 & 1 & 4 \end{pmatrix},$$

we begin with its first column. We set

$$\mathbf{x}_1 = \begin{pmatrix} 0 \\ 1 \\ -1 \\ 2 \end{pmatrix},$$

so that

$$\mathbf{y}_1 = \begin{pmatrix} 0 \\ \sqrt{6} \\ 0 \\ 0 \end{pmatrix} \approx \begin{pmatrix} 0 \\ 2.4495 \\ 0 \\ 0 \end{pmatrix}.$$

Therefore, the unit vector is

$$\mathbf{u}_1 = \frac{\mathbf{x}_1 - \mathbf{y}_1}{\| \mathbf{x}_1 - \mathbf{y}_1 \|} = \begin{pmatrix} 0 \\ .8391 \\ -.2433 \\ .4865 \end{pmatrix},$$

with corresponding Householder matrix

$$H_1 = \mathrm{I} - 2\,\mathbf{u}_1\,\mathbf{u}_1^T = \begin{pmatrix} 1 & 0 & 0 & 0 \\ 0 & -.4082 & .4082 & -.8165 \\ 0 & .4082 & .8816 & .2367 \\ 0 & -.8165 & .2367 & .5266 \end{pmatrix}.$$

Thus,

$$A_2 = H_1\,A\,H_1 = \begin{pmatrix} 4.0000 & -2.4495 & 0 & 0 \\ -2.4495 & 2.3333 & -.3865 & -.8599 \\ 0 & -.3865 & 4.9440 & -.1246 \\ 0 & -.8599 & -.1246 & 4.7227 \end{pmatrix}.$$

In the next phase,

$$\mathbf{x}_2 = \begin{pmatrix} 0 \\ 0 \\ -.3865 \\ -.8599 \end{pmatrix}, \qquad \mathbf{y}_2 = \begin{pmatrix} 0 \\ 0 \\ -.9428 \\ 0 \end{pmatrix},$$

so

$$\mathbf{u}_2 = \begin{pmatrix} 0 \\ 0 \\ -.8396 \\ -.5431 \end{pmatrix},$$

and

$$H_2 = \mathrm{I} - 2\,\mathbf{u}_2\,\mathbf{u}_2^T = \begin{pmatrix} 1 & 0 & 0 & 0 \\ 0 & 1 & 0 & 0 \\ 0 & 0 & -.4100 & -.9121 \\ 0 & 0 & -.9121 & .4100 \end{pmatrix}.$$

The resulting matrix

$$T = A_3 = H_2\,A_2\,H_2 = \begin{pmatrix} 4.0000 & -2.4495 & 0 & 0 \\ -2.4495 & 2.3333 & .9428 & 0 \\ 0 & .9428 & 4.6667 & 0 \\ 0 & 0 & 0 & 5 \end{pmatrix}$$

is now in tridiagonal form.

Since the final tridiagonal matrix T has the same eigenvalues as A, we can apply the QR algorithm to T to approximate the common eigenvalues. (The eigenvectors must then be computed separately, e.g., by the shifted inverse power method.)

According to Exercise 10.6.17, if $A = A_1$ is tridiagonal, so are all the iterates $A_2, A_3, \ldots$. Moreover, far fewer arithmetic operations are required; in Exercise 10.6.25, you are asked to quantify this. For instance, in the preceding example, after we apply 20 iterations of the QR algorithm directly to T, the upper triangular factor has become

$$R_{20} = \begin{pmatrix} 6.0000 & -.0065 & 0 & 0 \\ 0 & 4.5616 & 0 & 0 \\ 0 & 0 & 5.0000 & 0 \\ 0 & 0 & 0 & .4384 \end{pmatrix}.$$

The eigenvalues of T, and hence also of A, appear along the diagonal, and are correct to 4 decimal places. ●

Finally, even if A is not symmetric, one can still apply the same sequence of Householder transformations to simplify it. The final result is no longer tridiagonal, but rather a similar *upper Hessenberg matrix*, which means that all entries below the subdiagonal are zero, but those above the superdiagonal are not necessarily zero. For instance, a 5×5 upper Hessenberg matrix looks like

$$\begin{pmatrix} * & * & * & * & * \\ * & * & * & * & * \\ 0 & * & * & * & * \\ 0 & 0 & * & * & * \\ 0 & 0 & 0 & * & * \end{pmatrix},$$

where the starred entries can be anything. It can be proved that the QR algorithm maintains the upper Hessenberg form, and, while not as efficient as in the tridiagonal case, still yields a significant savings in computational effort required to find the common eigenvalues. Further details and analysis can be found in [18, 51, 67].

EXERCISES 10.6

10.6.18. Use Householder matrices to convert the following matrices into tridiagonal form:

(a) $\begin{pmatrix} 8 & -7 & 2 \\ -7 & 17 & -7 \\ 2 & -7 & 8 \end{pmatrix}$

(b) $\begin{pmatrix} 5 & 1 & -2 & 1 \\ 1 & 5 & 1 & -2 \\ -2 & 1 & 5 & 1 \\ 1 & -2 & 1 & 5 \end{pmatrix}$

(c) $\begin{pmatrix} 4 & 0 & -1 & 1 \\ 0 & 1 & 0 & -1 \\ -1 & 0 & 2 & 0 \\ 1 & -1 & 0 & 3 \end{pmatrix}$

♠**10.6.19.** Find the eigenvalues, to 2 decimal places, of the matrices in Exercise 10.6.18 by applying the QR algorithm to the tridiagonal form.

♠**10.6.20.** Use the tridiagonal QR method to find the singular values of the matrix

$$\begin{pmatrix} 2 & 2 & 1 & -1 \\ 1 & -2 & 0 & 1 \\ 0 & -1 & 2 & 2 \end{pmatrix}.$$

10.6.21. Use Householder matrices to convert the following matrices into upper Hessenberg form:

(a) $\begin{pmatrix} 3 & -1 & 2 \\ 1 & 3 & -4 \\ 2 & -1 & -1 \end{pmatrix}$

(b) $\begin{pmatrix} 3 & 2 & -1 & 1 \\ 2 & 4 & 0 & 1 \\ 0 & 1 & 2 & -6 \\ 1 & 0 & -5 & 1 \end{pmatrix}$

(c) $\begin{pmatrix} 1 & 0 & -1 & 1 \\ 2 & 1 & 1 & -1 \\ -1 & 0 & 1 & 3 \\ 3 & -1 & 1 & 4 \end{pmatrix}$

♠**10.6.22.** Find the eigenvalues, to 2 decimal places, of the matrices in Exercise 10.6.21 by applying the $Q\,R$ algorithm to the upper Hessenberg form.

10.6.23. Prove that the effect of the first Householder reflection is as given in (10.109).

10.6.24. What is the effect of tridiagonalization on the eigenvectors of the matrix?

◇**10.6.25.** (a) How many arithmetic operations (multiplications/divisions and additions/subtractions) are required to place a generic $n \times n$ matrix into tridiagonal form?
(b) How many operations are needed to perform one iteration of the $Q\,R$ algorithm on an $n \times n$ tridiagonal matrix?
(c) How much faster, on average, is the tridiagonal algorithm than the direct $Q\,R$ algorithm for finding the eigenvalues of a matrix?

10.6.26. Write out a pseudocode program to tridiagonalize a matrix. The input should be an $n \times n$ matrix A and the output should be the Householder unit vectors $\mathbf{u}_1, \ldots, \mathbf{u}_{n-1}$ and the tridiagonal matrix R. Does your program produce the upper Hessenberg form when the input matrix is not symmetric?

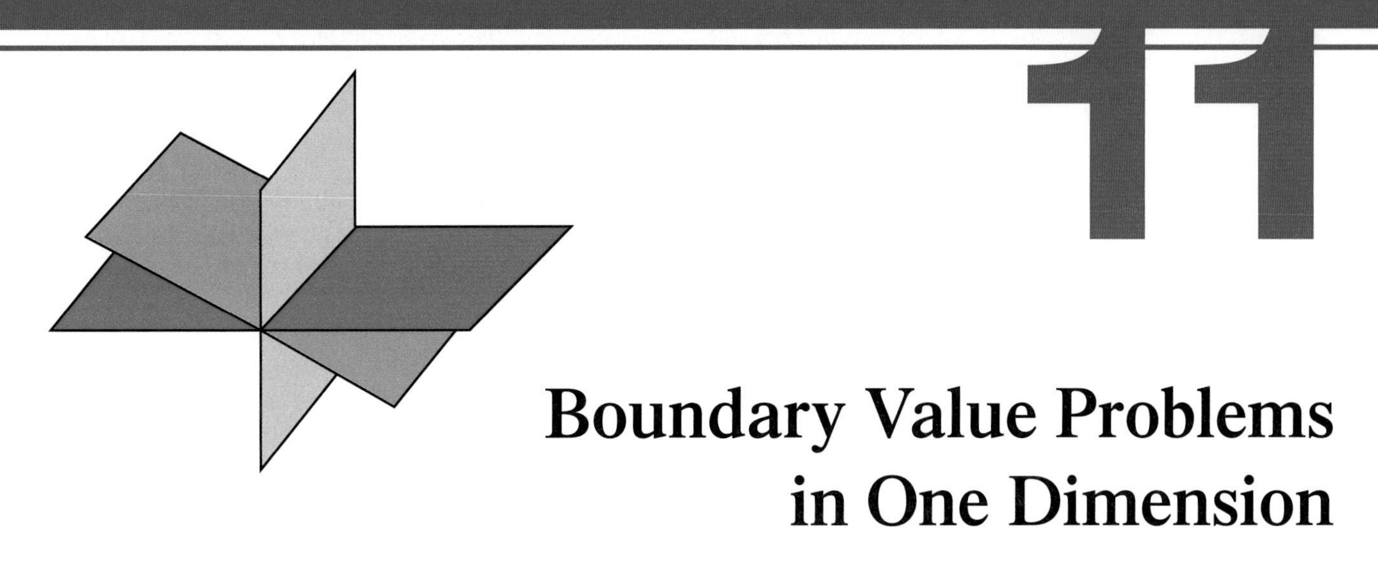

Boundary Value Problems in One Dimension

While its roots are firmly planted in the finite-dimensional world of matrices and vectors, the full scope of linear algebra is much broader. Its historical development and, hence, its structures, concepts, and methods, were strongly influenced by linear analysis—specifically, the need to solve linear differential equations, linear boundary value problems, linear integral equations, and the like. The time has come for us to fully transcend our finite dimensional limitations, and see linear algebra in action in infinite-dimensional function spaces. A complete development of the subject would span the full range of linear applied mathematics, and we refer interested students to Strang's book, [59], or our own forthcoming text, [47], for a much broader perspective and range of applications. But we cannot resist whetting the reader's appetite by looking at the simplest such context— one-dimensional linear boundary value problems.

The equilibrium equation of a one-dimensional continuum—an elastic bar, a bendable beam, and so on—is formulated as a boundary value problem for a scalar ordinary differential equation. The framework introduced for discrete mechanical systems in Chapter 6 will carry over, in its essence, to the infinite-dimensional setting appropriate to such problems. The underlying Euclidean vector space $\mathbb{R}^n$ is replaced by a function space. Vectors become functions, while matrices turn into linear differential operators. Physical boundary value problems are based on self-adjoint boundary value problems, founded on a suitable inner product on the function space. As in the discrete context, the positive definite cases are stable, and the equilibrium solution can be characterized by a minimization principle based on a quadratic energy functional.

Finite-dimensional linear algebra not only provides us with important insights into the underlying mathematical structure, but also motivates basic analytical and numerical solution schemes. In the function space framework, the general superposition principle is reformulated in terms of the effect of a combination of impulse forces concentrated at a single point of the continuum. However, constructing a function that represents a concentrated impulse turns out to be a highly non-trivial mathematical issue. Ordinary functions do not suffice, and we are led to develop a new calculus of generalized functions or distributions, including the remarkable delta function. The response of the system to a unit impulse force is known as the Green's function of the boundary value problem, in honor of the self-taught English mathematician (and miller) George Green. With the Green's function in hand, the general solution to the inhomogeneous system can be reconstructed by superimposing the effects of suitably scaled impulses on the entire domain. We begin with second order boundary value problems describing the equilibria of stretchable bars. We continue on to fourth order boundary value problems that govern the equilibrium of elastic beams, including piecewise cubic spline interpolants that play a key role in modern computer graphics and numerical analysis, and to more general second order boundary value problems of Sturm–Liouville type, which arise in a host of physical applications that involve partial differential equations.

The simplest boundary value problems can be solved by direct integration. However, more complicated systems do not admit explicit formulae for their solutions, and one must rely on numerical approximations. In the final section, we introduce the powerful finite element method. The key idea is to restrict the infinite-dimensional minimization principle characterizing the exact solution to a suitably chosen finite-dimensional subspace of the function space. When properly formulated, the solution to the resulting finite-dimensional minimization problem approximates the true minimizer. As in Chapter 4, the finite-dimensional minimizer is found by solving the induced linear algebraic system, using either direct or iterative methods. On that note, we will, reluctantly, finish our text, leaving the reader on the threshold of the wide range of advanced linear (and nonlinear) analysis that pervades modern science and mathematics.

11.1 ELASTIC BARS

A *bar* is a mathematical idealization of a one-dimensional linearly elastic continuum that can be stretched or contracted in the longitudinal direction, but is not allowed to bend in a transverse direction. (Materials that can bend are called beams, and will be analyzed in Section 11.4.) We will view the bar as the continuum limit of a one-dimensional chain of masses and springs, a system that we already analyzed in Section 11.4. Intuitively, the continuous bar consists of an infinite number of masses connected by infinitely short springs. The individual masses can be thought of as the "atoms" in the bar, although you should not try to read too much into the physics behind this interpretation.

We shall derive the basic equilibrium equations for the bar from first principles. Recall the three basic steps we already used to establish the corresponding equilibrium equations for a discrete mechanical system such as a mass–spring chain:

(i) First, use geometry to relate the displacement of the masses to the elongation in the connecting springs.

(ii) Second, use the constitutive assumptions such as Hooke's Law to relate the strain to the stress or internal force in the system.

(iii) Finally, impose a force balance between external and internal forces.

The remarkable fact, which will, when suitably formulated, carry over to the continuum, is that the force balance law is directly related to the geometrical displacement law by a transpose or, more correctly, adjoint operation.

Consider a bar of length ℓ hanging from a fixed support, with the bottom end left free, as illustrated in Figure 11.1. We use $0 \le x \le \ell$ to refer to the reference or unstressed configuration of the bar, so x measures the distance along the bar from the fixed end $x = 0$ to the free end $x = \ell$. Note that we are adopting the convention

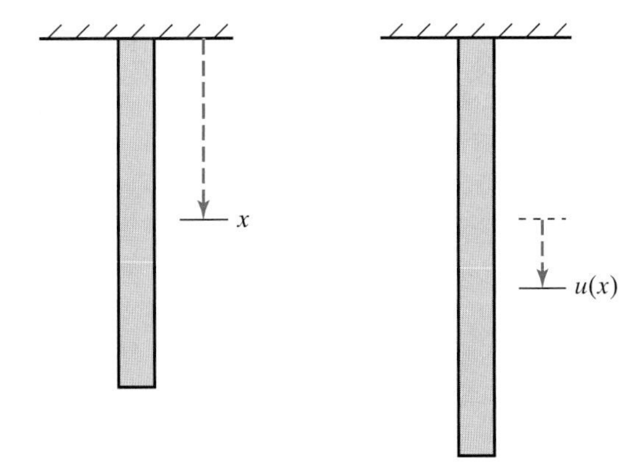

Figure 11.1 Bar with one fixed support.

that the positive x axis points *down*. Let $u(x)$ denote the *displacement* of the bar from its reference configuration. This means that the "atom" that started at position x has moved to position $x + u(x)$. With our convention, $u(x) > 0$ means that the atom has moved down, while if $u(x) < 0$, the atom has moved up. In particular,

$$u(0) = 0, \tag{11.1}$$

because we are assuming that the top end is fixed and cannot move.

The *strain* in the bar measures the relative amount of stretching. Two nearby atoms, at respective positions x and $x + \Delta x$, are moved to positions $x + u(x)$ and $x + \Delta x + u(x + \Delta x)$. The original, unstressed length of this small section of bar was Δx, while in the new configuration the same section has length

$$\big[\, x + \Delta x + u(x + \Delta x) \,\big] - \big[\, x + u(x) \,\big] = \Delta x + \big[\, u(x + \Delta x) - u(x) \,\big].$$

Therefore, this segment has been elongated by an amount $u(x + \Delta x) - u(x)$. The dimensionless strain measures the relative elongation, and so is obtained by dividing by the reference length: $[u(x + \Delta x) - u(x)]/\Delta x$. We now take the continuum limit by letting the interatomic spacing $\Delta x \to 0$. The result is the strain function

$$v(x) = \lim_{\Delta x \to 0} \frac{u(x + \Delta x) - u(x)}{\Delta x} = \frac{du}{dx} \tag{11.2}$$

that measures the local stretch in the bar at position x.

As noted above, we can approximate the bar by a chain of n masses connected by n springs, letting the bottom mass hang free. The mass-spring chain should also have total length ℓ, and so the individual springs have reference length

$$\Delta x = \frac{\ell}{n}.$$

The bar is to be viewed as the *continuum limit*, in which the number of masses $n \to \infty$ and the spring lengths $\Delta x \to 0$. The kth mass starts out at position

$$x_k = k\,\Delta x = \frac{k\,\ell}{n},$$

and, under forcing, experiences a displacement u_k. The relative elongation of the kth spring* is

$$v_k = \frac{e_k}{\Delta x} = \frac{u_{k+1} - u_k}{\Delta x}. \tag{11.3}$$

*We will find it helpful to label the springs from $k = 0$ to $k = n - 1$. This will facilitate comparisons with the bar, which, by convention, starts at position $x_0 = 0$.

In particular, since the fixed end cannot move, the first value $u_0 = 0$ is omitted from the subsequent equations.

The relation (11.3) between displacement and strain takes the familiar matrix form

$$\mathbf{v} = A\,\mathbf{u}, \qquad \mathbf{v} = (v_0, v_1, \ldots, v_{n-1})^T, \qquad \mathbf{u} = (u_1, u_2, \ldots, u_n)^T,$$

where

$$A = \frac{1}{\Delta x}\begin{pmatrix} 1 & & & & & \\ -1 & 1 & & & & \\ & -1 & 1 & & & \\ & & -1 & 1 & & \\ & & & \ddots & \ddots & \\ & & & & -1 & 1 \end{pmatrix} \quad \longrightarrow \quad \frac{d}{dx} \tag{11.4}$$

is the *scaled incidence matrix* of the mass-spring chain. As indicated by the arrow, the derivative operator d/dx that relates displacement to strain in the bar equation (11.2) can be viewed as its continuum limit, as the number of masses $n \to \infty$ and the spring lengths $\Delta x \to 0$. Vice versa, the incidence matrix can be viewed as a discrete, numerical approximation to the derivative operator. Indeed, if we regard the discrete displacements and strains as approximations to the sample values of their continuous counterparts, so

$$u_k \approx u(x_k), \qquad v_k \approx v(x_k),$$

then (11.3) takes the form

$$v(x_k) \;=\; \frac{u(x_{k+1}) - u(x_k)}{\Delta x} \;=\; \frac{u(x_k + \Delta x) - u(x_k)}{\Delta x} \;\approx\; \frac{du}{dx}(x_k).$$

justifying the identification (11.4). The passage back and forth between the discrete and the continuous is the foundation of continuum mechanics—solids, fluids, gases, and plasmas. Discrete models both motivate and provide numerical approximations to continuum systems, which, in turn, simplify and give further insight into the discrete domain.

The next part of the mathematical framework is to use the constitutive relations to relate the strain to the *stress*, or internal force experienced by the bar. To keep matters simple, we shall only consider bars that have a linear relation between stress and strain. For a physical bar, this is a pretty good assumption as long as it is not stretched beyond its elastic limits. Let $w(x)$ denote the stress on the point of the bar that was at reference position x. Hooke's Law implies that

$$w(x) = c(x)\, v(x), \tag{11.5}$$

where $c(x)$ measures the stiffness of the bar at position x. For a homogeneous bar, made out of a uniform material, $c(x) \equiv c$ is a constant function. The constitutive function $c(x)$ can be viewed as the continuum limit of the diagonal matrix

$$C = \begin{pmatrix} c_0 & & & \\ & c_1 & & \\ & & \ddots & \\ & & & c_{n-1} \end{pmatrix}$$

of individual spring constants c_k appearing in the discrete version

$$w_k = c_k\, v_k, \qquad \text{or} \qquad \mathbf{w} = C\,\mathbf{v}, \tag{11.6}$$

that relates stress to strain (internal force to elongation) in the individual springs. Indeed, (11.6) can be identified as the sampled version, $w(x_k) = c(x_k)\, v(x_k)$, of the continuum relation (11.5).

Finally, we need to impose a force balance at each point of the bar. Suppose $f(x)$ is an external force at position x on the bar, where $f(x) > 0$ means the force is acting downwards. Physical examples include mechanical, gravitational, or magnetic forces acting solely in the vertical direction. In equilibrium, the bar will deform so as to balance the external force with its own internal force resulting from stretching. Now, the internal force per unit length on the section of the bar lying between nearby positions x and $x + \Delta x$ is the relative difference in stress at the two ends, namely $[\, w(x + \Delta x) - w(x)\,]/\Delta x$. The force balance law requires that, in the limit,

$$0 = f(x) + \lim_{\Delta x \to 0} \frac{w(x + \Delta x) - w(x)}{\Delta x} = f(x) + \frac{dw}{dx},$$

or

$$f = -\frac{dw}{dx}. \tag{11.7}$$

This can be viewed as the continuum limit of the mass–spring chain force balance equations

$$f_k = \frac{w_{k-1} - w_k}{\Delta x}, \qquad w_n = 0, \tag{11.8}$$

where the final condition ensures the correct formula for the force on the free-hanging bottom mass. (Remember that the springs are numbered from 0 to $n - 1$.) This indicates that we should also impose an analogous boundary condition

$$w(\ell) = 0 \tag{11.9}$$

at the bottom end of the bar, which is hanging freely and so is unable to support any internal stress. The matrix form of the discrete system (11.8) is

$$\mathbf{f} = A^T \mathbf{w},$$

where the transposed scaled incidence matrix

$$A^T = \frac{1}{\Delta x} \begin{pmatrix} 1 & -1 & & & & \\ & 1 & -1 & & & \\ & & 1 & -1 & & \\ & & & 1 & -1 & \\ & & & & \ddots & \ddots \end{pmatrix} \longrightarrow -\frac{d}{dx} \tag{11.10}$$

should approximate the differential operator $-d/dx$ that appears in the continuum force balance law (11.7). Thus, we should somehow interpret $-d/dx$ as the "transpose" or "adjoint" of the differential operator d/dx. This important point will be developed properly in Section 11.3. But before trying to push the theory any further, we will pause to analyze the mathematical equations governing some simple configurations.

But first, let us summarize our progress so far. The three basic equilibrium equations (11.2, 5, 7) are

$$v(x) = \frac{du}{dx}, \qquad w(x) = c(x)\, v(x), \qquad f(x) = -\frac{dw}{dx}. \tag{11.11}$$

Substituting the first into the second, and then the resulting formula into the last equation, leads to the equilibrium equation

$$K[u] = -\frac{d}{dx}\left(c(x)\,\frac{du}{dx}\right) = f(x), \qquad 0 < x < \ell. \tag{11.12}$$

Thus, the displacement $u(x)$ of the bar is obtained as the solution to a second order ordinary differential equation. As such, it will depend on two arbitrary constants, which will be uniquely determined by the boundary conditions† (11.1, 9) at the two ends:

$$u(0) = 0, \qquad w(\ell) = c(\ell)\,u'(\ell) = 0. \tag{11.13}$$

Usually $c(\ell) > 0$, in which case it can be omitted from the second boundary condition, which simply becomes $u'(\ell) = 0$. The resulting boundary value problem is to be viewed as the continuum limit of the linear system

$$K\mathbf{u} = A^T C\,A\,\mathbf{u} = \mathbf{f} \tag{11.14}$$

modeling a mass-spring chain with one free end, cf. (6.11), in which

$$A \longrightarrow \frac{d}{dx}, \qquad C \longrightarrow c(x), \qquad A^T \longrightarrow -\frac{d}{dx},$$
$$\mathbf{u} \longrightarrow u(x), \qquad \mathbf{f} \longrightarrow f(x).$$

And, as we will see, most features of the finite-dimensional linear algebraic system have, when suitably interpreted, direct counterparts in the continuous boundary value problem.

EXAMPLE 11.1 Consider the simplest case of a uniform bar of unit length $\ell = 1$ subjected to a uniform force, e.g., gravity. The equilibrium equation (11.12) is

$$-c\,\frac{d^2u}{dx^2} = f, \tag{11.15}$$

where we are assuming that the force f is constant. This elementary second order ordinary differential equation can be immediately integrated:

$$u(x) = -\tfrac{1}{2}\alpha x^2 + a\,x + b, \qquad \text{where} \quad \alpha = \frac{f}{c} \tag{11.16}$$

is the ratio of the force to the stiffness of the bar. The values of the integration constants a and b are fixed by the boundary conditions (11.13), so

$$u(0) = b = 0, \qquad u'(1) = -\alpha + a = 0.$$

Therefore, there is a unique solution to the boundary value problem, yielding the displacement

$$u(x) = \alpha\left(x - \tfrac{1}{2}x^2\right), \tag{11.17}$$

which is graphed in Figure 11.2 for $\alpha = 1$. Note the parabolic shape, with zero derivative, indicating no strain, at the free end. The displacement reaches its maximum, $u(1) = \frac{1}{2}\alpha$, at the free end of the bar, which is the point which moves downwards the farthest. The stronger the force or the weaker the bar, the farther the overall displacement. ●

†We will sometimes use primes, as in $u' = du/dx$, to denote derivatives with respect to x.

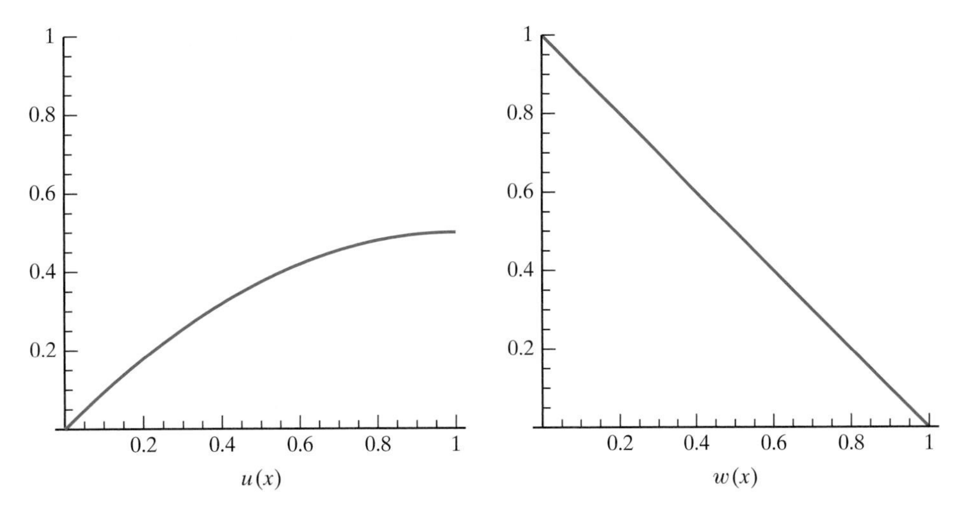

Figure 11.2 Displacement and stress of bar with one fixed end.

REMARK: This example illustrates the simplest way to solve boundary value problems, which is adapted from the usual solution technique for initial value problems. First, solve the differential equation by standard methods (if possible). For a second order equation, the general solution will involve two arbitrary constants. The values of the constants are found by substituting the solution formula into the two boundary conditions. Unlike initial value problems, the existence and/or uniqueness of the solution to a general boundary value problem is not guaranteed, and you may encounter situations where you are unable to complete the solution; see, for instance, Example 7.42. A more sophisticated method, based on the Green's function, will be presented in the following section.

As in the discrete situation, this particular mechanical configuration is *statically determinate*, meaning that we can solve directly for the stress $w(x)$ in terms of the external force $f(x)$ without having to compute the displacement $u(x)$ first. In this particular example, we need to solve the first order boundary value problem

$$-\frac{dw}{dx} = f, \qquad w(1) = 0,$$

arising from the force balance law (11.7). Since f is constant,

$$w(x) = f\,(1-x), \qquad \text{and} \qquad v(x) = \frac{w(x)}{c} = \alpha\,(1-x).$$

Note that the boundary condition uniquely determines the integration constant. We can then find the displacement $u(x)$ by solving another boundary value problem

$$\frac{du}{dx} = v(x) = \alpha\,(1-x), \qquad u(0) = 0,$$

resulting from (11.2), which again leads to (11.17). As before, the appearance of one boundary condition implies that we can find a unique solution to the differential equation.

REMARK: We motivated the boundary value problem for the bar by taking the continuum limit of the mass–spring chain. Let us see to what extent this limiting procedure can be justified. To compare the solutions, we keep the reference length of the chain fixed at $\ell = 1$. So, if we have n identical masses, each spring has length

$\Delta x = 1/n$. The kth mass will start out at reference position $x_k = k/n$. Using static determinacy, we can solve the system (11.8), which reads

$$w_{k-1} = w_k + \frac{f}{n}, \qquad w_n = 0,$$

directly for the stresses:

$$w_k = f\left(1 - \frac{k}{n}\right) = f\,(1 - x_k), \qquad k = 0, \ldots, n-1.$$

Thus, in this particular case, the continuous bar and the discrete chain have equal stresses at the sample points: $w(x_k) = w_k$. The strains are also in agreement:

$$v_k = \frac{1}{c}\, w_k = \alpha\left(1 - \frac{k}{n}\right) = \alpha\,(1 - x_k) = v(x_k),$$

where $\alpha = f/c$, as before. We then obtain the displacements by solving

$$u_{k+1} = u_k + \frac{v_k}{n} = u_k + \frac{\alpha}{n}\left(1 - \frac{k}{n}\right).$$

Since $u_0 = 0$, the solution is

$$\begin{aligned} u_k &= \frac{\alpha}{n}\sum_{i=0}^{k-1}\left(1 - \frac{i}{n}\right) \\ &= \alpha\left(\frac{k}{n} - \frac{k(k-1)}{2n^2}\right) \\ &= \alpha\left(x_k - \tfrac{1}{2}x_k^2\right) + \frac{\alpha\, x_k}{2n} \\ &= u(x_k) + \frac{\alpha\, x_k}{2n}. \end{aligned} \tag{11.18}$$

The sampled displacement $u(x_k)$ is not exactly equal to u_k, but their difference tends to zero as the number of masses $n \to \infty$. In this way, we have completely justified our limiting interpretation.

EXAMPLE 11.2 Consider the same uniform, unit length bar as in the previous example, again subject to a uniform constant force, but now with two fixed ends. We impose inhomogeneous boundary conditions

$$u(0) = 0, \qquad u(1) = d,$$

so the top end is fixed, while the bottom end is displaced an amount d. (Note that $d > 0$ means the bar is stretched, while $d < 0$ means it is compressed.) The general solution to the equilibrium equation (11.15) is, as before, given by (11.16). The values of the arbitrary constants a, b are determined by plugging into the boundary conditions, so

$$u(0) = b = 0, \qquad u(1) = -\tfrac{1}{2}\alpha + d = 0.$$

Thus

$$u(x) = \tfrac{1}{2}\alpha\,(x - x^2) + d\,x \tag{11.19}$$

is the unique solution to the boundary value problem, The displacement is a linear superposition of two functions; the first is induced by the external force f, while the second represents a uniform stretch induced by the boundary condition. In Figure 11.3, the dotted curves represent the two constituents, and the solid graph is their sum, which is the actual displacement.

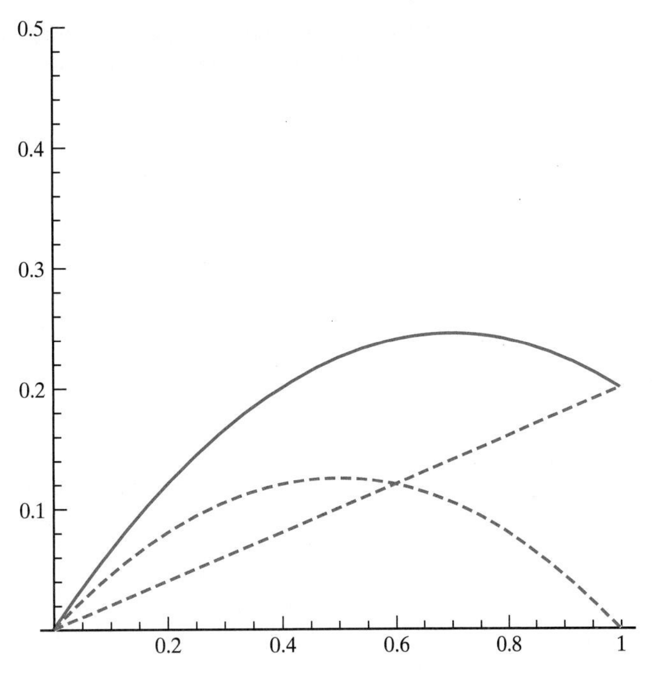

Figure 11.3 Displacements of a bar with two fixed ends.

Unlike a bar with a free end, this configuration is *statically indeterminate*. There is no boundary condition on the force balance equation

$$-\frac{dw}{dx} = f,$$

and so the integration constant a in the stress $w(x) = a - f\,x$ *cannot* be determined without first figuring out the displacement (11.19):

$$w(x) = c\,\frac{du}{dx} = f\left(\tfrac{1}{2} - x\right) + cd. \qquad \bullet$$

EXAMPLE 11.3 Finally, consider the case when both ends of the bar are left free. The boundary value problem

$$-u'' = f(x), \quad u'(0) = 0, \quad u'(\ell) = 0, \tag{11.20}$$

represents the continuum limit of a mass–spring chain with two free ends and corresponds to a bar floating in outer space, subject to a nonconstant external force. Based on our finite-dimensional experience, we expect the solution to manifest an underlying instability of the physical problem. Solving the differential equation, we find that

$$u(x) = a\,x + b - \int_0^x \left(\int_0^y f(z)\,dz \right) dy,$$

where the constants a, b are to be determined by the boundary conditions. Since

$$u'(x) = a - \int_0^x f(z)\,dz,$$

the first boundary condition $u'(0) = 0$ requires $a = 0$. The second boundary condition requires

$$u'(\ell) = \int_0^\ell f(x)\,dx = 0, \tag{11.21}$$

which is not automatically valid! The integral represents the total force per unit length exerted on the bar. As in the case of a mass-spring chain with two free ends, if there is a non-zero net force, the bar cannot remain in equilibrium, but will move off in space and the equilibrium boundary value problem has no solution. On the other hand, if the forcing satisfies the constraint (11.21), then the resulting solution of the boundary value problem has the form

$$u(x) = b - \int_0^x \left(\int_0^y f(z)\, dz \right) dy, \tag{11.22}$$

where the constant b is arbitrary. Thus, when it exists, the solution to the boundary value problem is not unique. The constant b solves the corresponding homogeneous problem, and represents a rigid translation of the entire bar by a distance b.

Physically, the free boundary value problem corresponds to an unstable structure: there is a translational instability in which the bar moves off rigidly in the longitudinal direction. Only balanced forces of mean zero can maintain equilibrium. Furthermore, when it does exist, the equilibrium solution is not unique since there is nothing to tie the bar down to any particular spatial position.

This dichotomy should remind you of our earlier study of linear algebraic systems. An inhomogeneous system $K\mathbf{u} = \mathbf{f}$ consisting of n equations in n unknowns either admits a unique solution for all possible right hand sides $\mathbf{f}$, or, when K is singular, either no solution exists or the solution is not unique. In the latter case, the constraints on the right hand side are prescribed by the Fredholm alternative (5.80), which requires that $\mathbf{f}$ be orthogonal, with respect to the Euclidean inner product, to all elements of $\operatorname{coker} K$. In physical equilibrium systems K is symmetric, and so $\operatorname{coker} K = \ker K$. Thus, the Fredholm alternative requires that the forcing be orthogonal to all the unstable modes. In the function space for the bar, the finite-dimensional dot product is replaced by the L^2 inner product

$$\langle f, g \rangle = \int_0^\ell f(x)\, g(x)\, dx.$$

Since the kernel or solution space to the homogeneous boundary value problem is spanned by the constant function 1, the Fredholm alternative‡ requires that the forcing function be orthogonal to it,

$$\langle f, 1 \rangle = \int_0^\ell f(x)\, dx = 0.$$

This is precisely the condition (11.21) required for existence of a (non-unique) equilibrium solution, and so the analogy between the finite and infinite dimensional categories is complete. ●

REMARK: The boundary value problems that govern the mechanical equilibria of a simple bar arise in many other physical systems. For example, the equation for the thermal equilibrium of a bar under an external heat source is modeled by the same boundary value problem (11.12); in this case, $u(x)$ represents the temperature of the bar, $c(x)$ represents the *diffusivity* or *thermal conductivity* of the material at position x, while $f(x)$ represents an external heat source. A fixed boundary condition $u(\ell) = a$ corresponds to an end that is held at a fixed temperature a, while a free boundary condition $u'(\ell) = 0$ represents an insulated end that does not allow heat energy to enter or leave the bar.

‡In fact, historically, Fredholm first made his discovery while studying such infinite-dimensional systems. Only later was it realized that the same underlying ideas are equally valid in finite-dimensional linear algebraic systems.

EXERCISES 11.1

11.1.1. In the bar with fixed ends discussed in Example 11.2, which point experiences the greatest displacement? Which point experiences the greatest stress? The greatest strain?

11.1.2. Analyze the displacement and stress of a homogeneous bar, $c = 1$, of length $\ell = 1$, when subject to a discontinuous external force

$$f(x) = \begin{cases} 1, & 0 \le x < \frac{1}{2}, \\ -1, & \frac{1}{2} < x \le 1, \end{cases}$$

with both ends are fixed at their reference positions: $u(0) = u(1) = 0$. Plot graphs of both the displacement and the stress.

11.1.3. (a) Find all possible equilibrium configurations in the free bar of Example 11.3 when subject to the external force

$$f(x) = \begin{cases} 1, & 0 \le x < \frac{1}{2}\ell, \\ -1, & \frac{1}{2}\ell < x \le \ell. \end{cases}$$

(b) Find the corresponding stress. Does the stress depend upon which equilibrium configuration is assumed?

(c) Choose one of your solutions and plot the displacement and stress.

11.1.4. A bar of length $\ell = 1$ has stiffness function $c(x) = 1 + x$. Find and graph the displacement, stress and strain when the bar is subjected to a unit gravitational force, assuming there is no displacement at the ends. Which point moves the farthest? Where is the bar most likely to break?

11.1.5. Repeat Exercise 11.1.4 in the case when the bar has one fixed and one free end.

11.1.6. Discuss what happens when both ends of the bar in Exercise 11.1.4 are free.

11.1.7. A bar of length $\ell = 2$ has a constant stiffness function $c = \frac{1}{3}$. Find and graph the displacement and stress when the bar is subjected to a force $f(x) = 1 - x$ and both ends are fixed at their unstressed positions: $u(0) = u(2) = 0$. Which point moves the farthest? Where is the bar most likely to break?

11.1.8. A composite bar is obtained by gluing a uniform bar of length 1 and stiffness $c = 1$ to a second bar of length 1 and stiffness $c = 2$. Using the fact that the stress remains continuous, find and graph the displacement and stress when the bar is subject to a constant external force $f = 1$ with fixed ends $u(0) = u(2) = 0$. Which point is subject to the greatest stress? The greatest strain?

11.1.9. Suppose the composite bar in Exercise 11.1.8 is only fixed at its top end and subject to a uniform gravitational force. Will the free end hang lower if its stiffer half is on top or bottom? Make a prediction, and then confirm it by setting up and solving the two boundary value problems.

11.1.10. A bar of length $\ell = 1$ has one fixed and one free end and stiffness function $c(x) = 1 - x$. Find the displacement when subjected to a unit force. Pay careful attention to the boundary condition at the free end—what does it say about the displacement $u(1)$ and the strain $u'(1)$?

♡**11.1.11.** Analyze the periodic boundary value problem for the angular displacement of a circular ring along the same lines as the free case done in Example 11.3. Let $0 \le x \le 2\pi$ denote the angular coordinate along the ring, so that the boundary conditions are $u(0) = u(2\pi)$ and $u'(0) = u'(2\pi)$. Assume that the ring has constant stiffness $c(x) \equiv 1$. Characterize the forcing functions that maintain equilibrium. Is this in accordance with the Fredholm alternative? Write down a forcing function that maintains equilibrium and determine the corresponding displacement and stress in the ring.

11.1.12. What are the physical units (using metric grams, meters and seconds) for measuring the various quantities appearing in the equilibrium equations for a bar?

11.2 GENERALIZED FUNCTIONS AND THE GREEN'S FUNCTION

The general superposition principle for inhomogeneous linear systems inspires an alternative, powerful approach to the solution of boundary value problems. This method relies on the solution to a particular type of inhomogeneity, namely a concentrated unit impulse. The resulting solutions are collectively known as the Green's function for the boundary value problem. Once the Green's function is known, the response of the system to any other external forcing can be constructed through a continuous superposition of these fundamental solutions. However, rigorously formulating a concentrated impulse force turns out to be a serious mathematical challenge.

To motivate the construction, let us return briefly to the case of a mass–spring chain. Given the equilibrium equations

$$K\mathbf{u} = \mathbf{f}, \tag{11.23}$$

let us decompose the external forcing $\mathbf{f} = (f_1, f_2, \ldots, f_n)^T \in \mathbb{R}^n$ into a linear combination

$$\mathbf{f} = f_1\,\mathbf{e}_1 + f_2\,\mathbf{e}_2 + \cdots + f_n\,\mathbf{e}_n \tag{11.24}$$

of the standard basis vectors of $\mathbb{R}^n$. Suppose we know how to solve each of the individual systems

$$K\mathbf{u}_i = \mathbf{e}_i, \qquad i = 1, \ldots, n. \tag{11.25}$$

The vector $\mathbf{e}_i$ represents a unit force or, more precisely, a *unit impulse*, which is applied solely to the ith mass in the chain; the solution $\mathbf{u}_i$ represents the response of the chain. Since we can decompose any other force vector as a superposition of impulse forces, as in (11.24), the superposition principle tells us that the solution to the inhomogeneous system (11.23) is the self-same linear combination of the individual responses, so

$$\mathbf{u} = f_1\,\mathbf{u}_1 + f_2\,\mathbf{u}_2 + \cdots + f_n\,\mathbf{u}_n. \tag{11.26}$$

REMARK: The alert reader will recognize that $\mathbf{u}_1, \ldots, \mathbf{u}_n$ are the columns of the inverse matrix, K^{-1}, and so we are, in fact, reconstructing the solution to the linear system (11.23) by inverting the coefficient matrix K. Thus, this observation, while noteworthy, does not lead to an efficient solution technique for discrete systems. In contrast, in the case of continuous boundary value problems, this approach leads to one of the most valuable solution paradigms in both practice and theory.

The Delta Function

Our aim is to extend this algebraic technique to boundary value problems. The key question is how to characterize an impulse force that is concentrated on a single atom* of the bar. In general, a *unit impulse* at position $x = y$ will be described by something called the *delta function*, and denoted by $\delta_y(x)$. Since the impulse is supposed to be concentrated solely at $x = y$, we should have

$$\delta_y(x) = 0 \quad \text{for} \quad x \neq y. \tag{11.27}$$

Moreover, since it is a *unit* impulse, we want the total amount of force exerted on the bar to be equal to one. Since we are dealing with a continuum, the total force is represented by an integral over the length of the bar, and so we also require that the delta function to satisfy

$$\int_0^\ell \delta_y(x)\,dx = 1, \qquad \text{provided that} \quad 0 < y < \ell. \tag{11.28}$$

Alas, there is no *bona fide* function that enjoys both of the required properties! Indeed, according to the basic facts of Riemann (or even Lebesgue) integration, two functions which are the same everywhere except at one single point have exactly the same integral, [16, 54]. Thus, since δ_y is zero except at one point, its integral should be 0, not 1. The mathematical conclusion is that the two requirements, (11.27–28) are inconsistent!

*As before, "atom" is used in a figurative sense.

This unfortunate fact stopped mathematicians dead in their tracks. It took the imagination of a British engineer, with the unlikely name Oliver Heaviside, who was not deterred by the lack of rigorous justification, to start utilizing delta functions in practical applications—with remarkable effect. Despite his success, Heaviside was ridiculed by the pure mathematicians of his day, and eventually succumbed to mental illness. But, some thirty years later, the great theoretical physicist Paul Dirac resurrected the delta function for quantum mechanical applications, and this finally made theoreticians sit up and take notice. (Indeed, the term "Dirac delta function" is quite common.) In 1944, the French mathematician Laurent Schwartz finally established a rigorous theory of *distributions* that incorporated such useful, but non-standard objects, [36, 53]. (Thus, to be more accurate, we should really refer to the *delta distribution*; however, we will retain the more common, intuitive designation "delta function" throughout.) It is beyond the scope of this introductory text to develop a fully rigorous theory of distributions. Rather, in the spirit of Heaviside, we shall concentrate on learning, through practice with computations and applications, how to tame these wild mathematical beasts.

There are two distinct ways to introduce the delta function. Both are important and both worth knowing.

Method #1. Limits: The first approach is to regard the delta function $\delta_y(x)$ as a limit of a sequence of ordinary smooth functions[†] $g_n(x)$. These functions will represent more and more concentrated unit forces, which, in the limit, converge to the desired unit impulse concentrated at a single point, $x = y$. Thus, we require

$$\lim_{n\to\infty} g_n(x) = 0, \qquad x \neq y, \tag{11.29}$$

while the total amount of force remains fixed at

$$\int_0^\ell g_n(x)\,dx = 1. \tag{11.30}$$

On a formal level, the limit "function"

$$\delta_y(x) = \lim_{n\to\infty} g_n(x)$$

will satisfy the key properties (11.27–28).

An explicit example of such a sequence is provided by the rational functions

$$g_n(x) = \frac{n}{\pi\left(1 + n^2x^2\right)}\,. \tag{11.31}$$

These functions satisfy

$$\lim_{n\to\infty} g_n(x) = \begin{cases} 0, & x \neq 0, \\ \infty, & x = 0, \end{cases} \tag{11.32}$$

while[‡]

$$\int_{-\infty}^{\infty} g_n(x)\,dx = \frac{1}{\pi}\tan^{-1} n\,x\,\bigg|_{x=-\infty}^{\infty} = 1. \tag{11.33}$$

Therefore, formally, we identify the limiting function

$$\lim_{n\to\infty} g_n(x) = \delta(x) = \delta_0(x),$$

[†]To keep the notation compact, we suppress the dependence of the functions g_n on the point y where the limiting delta function is concentrated.

[‡]For the moment, it will be slightly simpler here to consider the entire real line — corresponding to a bar of infinite length. Exercise 11.2.11 discusses how to modify the construction for a finite interval.

with the unit impulse delta function concentrated at $x = 0$. As sketched in Figure 11.4, as n gets larger and larger, each successive function $g_n(x)$ forms a more and more concentrated spike, while maintaining a unit total area under its graph. The limiting delta function can be thought of as an infinitely tall spike of zero width, entirely concentrated at the origin.

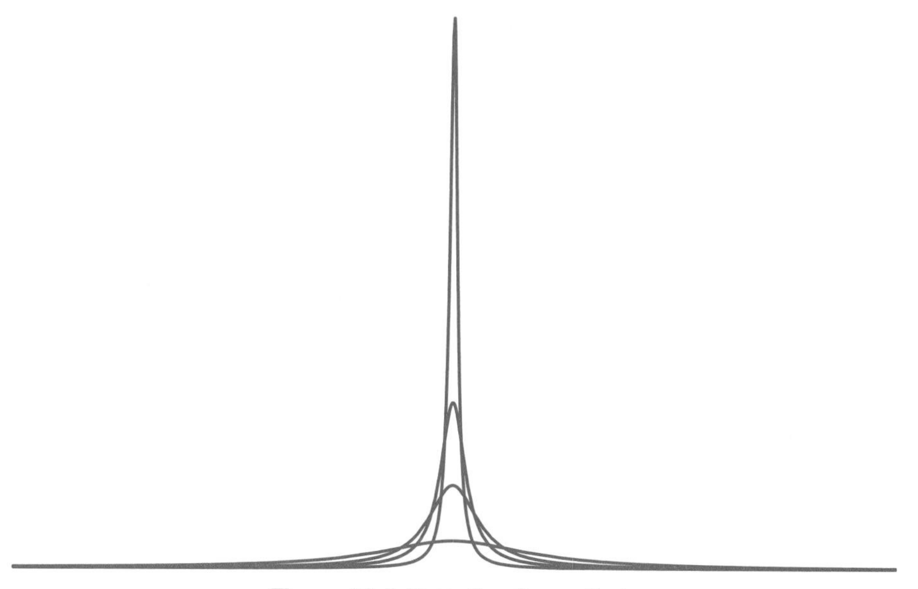

Figure 11.4 Delta function as limit.

REMARK: There are many other possible choices for the limiting functions $g_n(x)$. See Exercise 11.2.10 for another useful example.

REMARK: This construction of the delta function highlights the perils of interchanging limits and integrals without proper justification. In any standard theory of integration, the limit of the functions g_n would be indistinguishable from the zero function, so the limit of their integrals (11.33) would *not* equal the integral of their limit:

$$1 = \lim_{n\to\infty} \int_{-\infty}^{\infty} g_n(x)\,dx \neq \int_{-\infty}^{\infty} \lim_{n\to\infty} g_n(x)\,dx = 0.$$

The delta function is, in a sense, a means of sidestepping this analytic inconvenience. The full ramifications and theoretical constructions underlying such limits must, however, be deferred to a rigorous course in real analysis, [16, 54].

Once we have found the basic delta function $\delta(x) = \delta_0(x)$, which is concentrated at the origin, we can obtain a delta function concentrated at any other position y by a simple translation:

$$\delta_y(x) = \delta(x - y). \tag{11.34}$$

Thus, $\delta_y(x)$ can be realized as the limit of the translated functions

$$\widehat{g}_n(x) = g_n(x - y) = \frac{n}{\pi\left(1 + n^2(x - y)^2\right)}. \tag{11.35}$$

Method #2. Duality: The second approach is a bit more abstract, but much closer to the proper rigorous formulation of the theory of distributions like the delta function. The critical observation is that if $u(x)$ is any continuous function, then

$$\int_0^\ell \delta_y(x)\, u(x)\, dx = u(y), \qquad \text{for} \quad 0 < y < \ell. \tag{11.36}$$

Indeed, since $\delta_y(x) = 0$ for $x \neq y$, the integrand only depends on the value of u at the point $x = y$, and so

$$\int_0^\ell \delta_y(x)\, u(x)\, dx = \int_0^\ell \delta_y(x)\, u(y)\, dx = u(y) \int_0^\ell \delta_y(x)\, dx = u(y).$$

Equation (11.36) serves to define a linear functional[§] $L_y\colon \mathrm{C}^0[0,\ell] \to \mathbb{R}$ that maps a continuous function $u \in \mathrm{C}^0[0,\ell]$ to its value at the point $x = y$:

$$L_y[u] = u(y) \in \mathbb{R}.$$

In the dual approach to generalized functions, the delta function is, in fact, *defined* as this particular linear functional. The function $u(x)$ is sometimes referred to as a *test function*, since it serves to "test" the form of the linear functional L.

REMARK: If the impulse point y lies outside the integration domain, then

$$\int_0^\ell \delta_y(x)\, u(x)\, dx = 0, \qquad \text{when} \quad y < 0 \quad \text{or} \quad y > \ell, \tag{11.37}$$

because the integrand is identically zero on the entire interval. For technical reasons, we will not attempt to define the integral (11.37) if the impulse point $y = 0$ or $y = \ell$ lies on the boundary of the interval of integration.

The interpretation of the linear functional L_y as representing a kind of function $\delta_y(x)$ is based on the following line of thought. According to Theorem 7.10, every scalar-valued linear function $L\colon V \to \mathbb{R}$ on a finite-dimensional inner product space is given by an inner product with a fixed element $\mathbf{a} \in V$, so

$$L[\mathbf{u}] = \langle \mathbf{a}, \mathbf{u} \rangle.$$

In this sense, linear functions on $\mathbb{R}^n$ are the "same" as vectors. (But bear in mind that the identification does depend upon the choice of inner product.) Similarly, on the infinite-dimensional function space $\mathrm{C}^0[0,\ell]$, the L^2 inner product

$$L_g[u] = \langle g, u \rangle = \int_0^\ell g(x)\, u(x)\, dx \tag{11.38}$$

taken with a fixed function $g \in \mathrm{C}^0[0,\ell]$ defines a real-valued linear functional $L_g\colon \mathrm{C}^0[0,\ell] \to \mathbb{R}$. However, unlike the finite-dimensional situation, *not* every real-valued linear functional has this form! In particular, there is no actual function $\delta_y(x)$ such that the identity

$$\langle \delta_y, u \rangle = \int_0^\ell \delta_y(x)\, u(x)\, dx = u(y) \tag{11.39}$$

holds for every continuous function $u(x)$. Every (continuous) function defines a linear functional, but not conversely. Or, stated another way, while the dual space to a finite-dimensional vector space like $\mathbb{R}^n$ can be identified, via an inner product,

[§]Linearity, which requires that $L_y[c\,f + d\,g] = c\,L_y[f] + d\,L_y[g]$ for all functions f, g and all scalars (constants) $c, d \in \mathbb{R}$, is easily established; see also Example 7.7.

with the space itself, this is not the case in infinite-dimensional function space; the dual is an entirely different creature. This disconcerting fact highlights yet another of the profound differences between finite- and infinite-dimensional vector spaces!

But the dual interpretation of generalized functions acts as if this were true. *Generalized functions are real-valued linear functionals on function space, but viewed as a kind of function via the inner product.* Although the identification is not to be taken literally, one can, with a little care, manipulate generalized functions as if they were actual functions, but always keeping in mind that a rigorous justification of such computations must ultimately rely on their true characterization as linear functionals.

The two approaches—limits and duality—are completely compatible. Indeed, with a little extra work, one can justify the dual formula (11.36) as the limit

$$u(y) = \lim_{n \to \infty} \int_0^\ell g_n(x)\, u(x)\, dx = \int_0^\ell \delta_y(x)\, u(x)\, dx \tag{11.40}$$

of the inner products of the function u with the approximating concentrated impulse functions $g_n(x)$ satisfying (11.29–30). In this manner, the linear functional $L[u] = u(y)$ represented by the delta function is the limit, $L_y = \lim_{n \to \infty} L_n$, of the approximating linear functionals

$$L_n[u] = \int_0^\ell g_n(x)\, u(x)\, dx.$$

Thus, the choice of interpretation of the generalized delta function is, on an operational level, a matter of taste. For the novice, the limit interpretation of the delta function is perhaps the easier to digest at first. However, the dual, linear functional interpretation has stronger connections with the rigorous theory and, even in applications, offers some significant advantages.

Although on the surface, the delta function might look a little bizarre, its utility in modern applied mathematics and mathematical physics more than justifies including it in your analytical toolbox. Even though you are probably not yet comfortable with either definition, you are advised to press on and familiarize yourself with its basic properties, to be discussed next. With a little care, you usually won't go far wrong by treating it as if it were a genuine function. After you gain more practical experience, you can, if desired, return to contemplate just exactly what kind of object the delta function really is.

REMARK: If you are familiar with basic measure theory, [54], there is yet a third interpretation of the delta function as a point mass or atomic measure. However, the measure-theoretic approach has definite limitations, and does not cover the full gamut of generalized functions.

Calculus of Generalized Functions

In order to develop a working relationship with the delta function, we need to understand how it behaves under the basic operations of linear algebra and calculus. First, we can take linear combinations of delta functions. For example,

$$f(x) = 2\,\delta(x) + 3\,\delta(x-1)$$

represents a combination of an impulse of magnitude 2 concentrated at $x = 0$ and one of magnitude 3 concentrated at $x = 1$. Since $\delta_y(x) = 0$ for any $x \neq y$, multiplying the delta function by an ordinary function is the same as multiplying by a constant:

$$g(x)\,\delta_y(x) = g(y)\,\delta_y(x), \tag{11.41}$$

provided that $g(x)$ is continuous at $x = y$. For example, $x\,\delta(x) \equiv 0$ is the same as the constant zero function.

Warning: It is *not* permissible to multiply delta functions together, or to use more complicated algebraic operations. Expressions like $\delta(x)^2$, $1/\delta(x)$, $e^{\delta(x)}$, etc., are *not* well defined in the theory of generalized functions. This makes their application to nonlinear systems much more problematic—but this will not concern us.

The integral of the delta function is known as a *step function*. More specifically, the basic formulae (11.36, 37) imply that

$$\int_a^x \delta_y(t)\,dt = \sigma_y(x) = \sigma(x-y) = \begin{cases} 0, & a < x < y, \\ 1, & x > y > a. \end{cases} \tag{11.42}$$

Figure 11.5 shows the graph of $\sigma(x) = \sigma_0(x)$. Unlike the delta function, the step function $\sigma_y(x)$ is an ordinary function. It is continuous—indeed constant—except at $x = y$. The value of the step function at the discontinuity $x = y$ is left unspecified, although a popular choice, motivated by Fourier theory, [16, 47], is to set $\sigma_y(y) = \frac{1}{2}$, the average of its left and right hand limits.

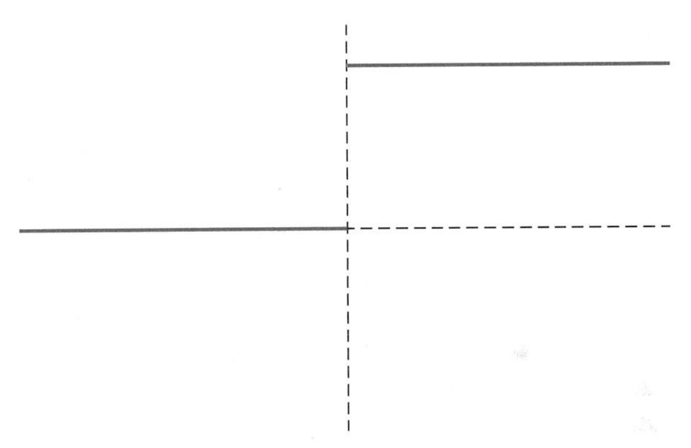

Figure 11.5 Step function.

We note that the integration formula (11.42) is compatible with our characterization of the delta function as the limit of highly concentrated forces. If we integrate the approximating functions (11.31), we obtain

$$f_n(x) = \int_{-\infty}^x g_n(t)\,dt = \frac{1}{\pi}\tan^{-1} n\,x + \frac{1}{2}.$$

Since

$$\lim_{y\to\infty} \tan^{-1} y = \tfrac{1}{2}\pi, \qquad \text{while} \qquad \lim_{y\to-\infty} \tan^{-1} y = -\tfrac{1}{2}\pi,$$

these functions converge to the step function:

$$\lim_{n\to\infty} f_n(x) = \sigma(x) = \begin{cases} 1, & x > 0, \\ \frac{1}{2}, & x = 0, \\ 0, & x < 0. \end{cases} \tag{11.43}$$

A graphical illustration of this limiting process appears in Figure 11.6.

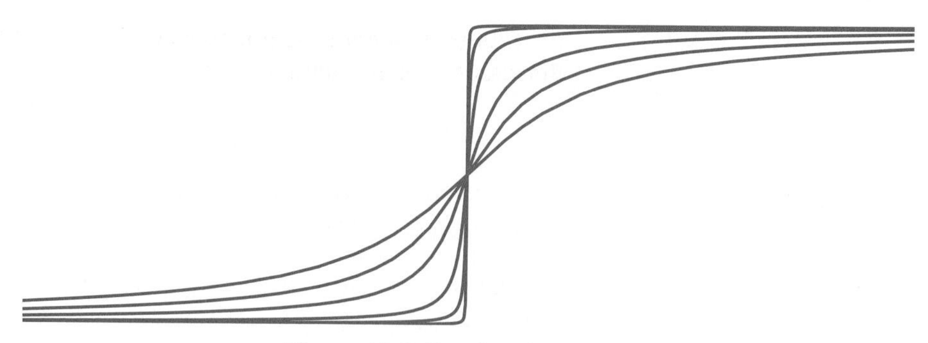

Figure 11.6 Step function as limit.

The integral of the discontinuous step function (11.42) is the continuous *ramp function*

$$\int_a^x \sigma_y(z)\,dz = \rho_y(x) = \rho(x-y) = \begin{cases} 0, & a < x < y, \\ x-y, & x > y > a, \end{cases} \tag{11.44}$$

which is graphed in Figure 11.7.

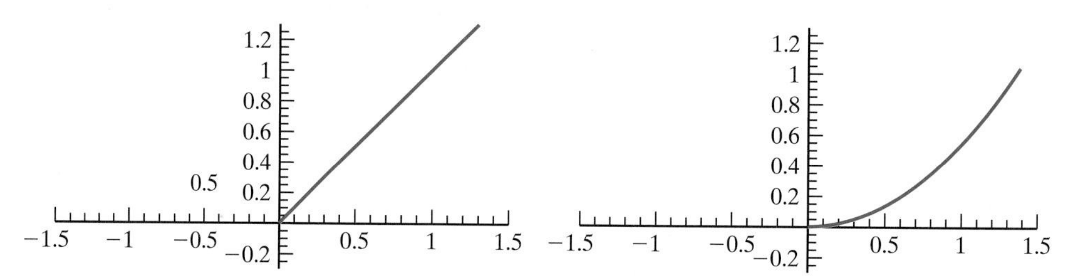

Figure 11.7 First and second order ramp functions.

Note that $\rho(x-y)$ has a corner at $x = y$, and so is not differentiable there; indeed, its derivative

$$\frac{d\rho}{dx} = \sigma$$

has a jump discontinuity, and its second derivative

$$\frac{d^2\rho}{dx^2} = \delta$$

is no longer an ordinary function. We can continue to integrate; the nth integral of the delta function is the *nth order ramp function*

$$\rho_n(x-y) = \begin{cases} \dfrac{(x-y)^n}{n!}, & x > y, \\ 0, & x < y. \end{cases} \tag{11.45}$$

What about differentiation? Motivated by the Fundamental Theorem of Calculus, we shall use formula (11.42) to identify the derivative of the step function with the delta function

$$\frac{d\sigma}{dx} = \delta. \tag{11.46}$$

This fact is highly significant. In basic calculus, one is not allowed to differentiate a discontinuous function. Here, we discover that the derivative can be defined, not as an ordinary function, but rather as a generalized delta function.

This basic identity is a particular instance of a general rule for differentiating functions with discontinuities. We use

$$f(y^-) = \lim_{x \to y^-} f(x), \qquad f(y^+) = \lim_{x \to y^+} f(x), \tag{11.47}$$

to denote, respectively, the left and right sided limits of a function at a point y. The function $f(x)$ is *continuous* at the point y if and only if its one-sided limits exist and are equal to its value: $f(y) = f(y^-) = f(y^+)$. If the one-sided limits are the same, but not equal to $f(y)$, then the function is said to have a *removable discontinuity*, since redefining $f(y) = f(y^-) = f(y^+)$ serves to make f continuous at the point in question. An example is the function $f(x)$ that is equal to 0 for all $x \neq 0$, but has[¶] $f(0) = 1$. Removing the discontinuity by setting $f(0) = 0$ makes $f(x) \equiv 0$ equal to a continuous constant function. Since removable discontinuities play no role in our theory or applications, they will always be removed without penalty.

Warning: Although $\delta(0^+) = 0 = \delta(0^-)$, we will emphatically *not* call 0 a removable discontinuity of the delta function. Only standard functions have removable discontinuities.

Finally, if both the left and right limits exist, but are not equal, then f is said to have a *jump discontinuity* at the point y. The *magnitude* of the jump is the difference

$$\beta = f(y^+) - f(y^-) = \lim_{x \to y^+} f(x) \; - \; \lim_{x \to y^-} f(x) \tag{11.48}$$

between the right and left limits. The magnitude of the jump is positive if the function jumps up, when moving from left to right, and negative if it jumps down. For example, the step function $\sigma(x)$ has a unit, i.e., magnitude 1, jump discontinuity at the origin:

$$\sigma(0^+) - \sigma(0^-) = 1 - 0 = 1,$$

and is continuous everywhere else. Note the value of the function at the point, namely $f(y)$, which may not even be defined, plays no role in the specification of the jump.

In general, the derivative of a function with jump discontinuities is a generalized function that includes delta functions concentrated at each discontinuity. More explicitly, suppose that $f(x)$ is differentiable, in the usual calculus sense, everywhere except at the point y where it has a jump discontinuity of magnitude β. We can re-express the function in the convenient form

$$f(x) = g(x) + \beta\, \sigma(x - y), \tag{11.49}$$

where $g(x)$ is continuous everywhere, with a removable discontinuity at $x = y$, and differentiable except possibly at the jump. Differentiating (11.49), we find that

$$f'(x) = g'(x) + \beta\, \delta(x - y), \tag{11.50}$$

has a delta spike of magnitude β at the discontinuity. Thus, the derivatives of f and g coincide everywhere except at the discontinuity.

EXAMPLE 11.4 Consider the function

$$f(x) = \begin{cases} -x, & x < 1, \\ \frac{1}{5}x^2, & x > 1, \end{cases} \tag{11.51}$$

[¶]This function is *not* a version of the delta function. It is an ordinary function, and its integral is 0, not 1.

which we graph in Figure 11.8. We note that f has a single jump discontinuity of magnitude $\frac{6}{5}$ at $x = 1$. This means that

$$f(x) = g(x) + \tfrac{6}{5}\,\sigma(x-1), \qquad \text{where} \qquad g(x) = \begin{cases} -x, & x < 1, \\ \frac{1}{5}x^2 - \frac{6}{5}, & x > 1, \end{cases}$$

is continuous everywhere, since its right and left hand limits at the original discontinuity are equal: $g(1^+) = g(1^-) = -1$. Therefore,

$$f'(x) = g'(x) + \tfrac{6}{5}\,\delta(x-1), \qquad \text{where} \qquad g'(x) = \begin{cases} -1, & x < 1, \\ \frac{2}{5}x, & x > 1, \end{cases}$$

while $g'(1)$, and hence $f'(1)$, is not defined. In Figure 11.8, the delta spike in the derivative of f is symbolized by a vertical line—although this pictorial device fails to indicate its magnitude of $\frac{6}{5}$.

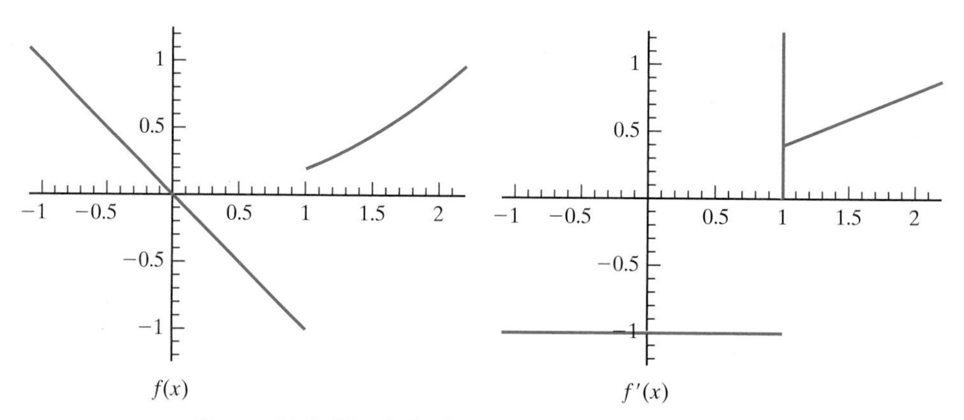

Figure 11.8 The derivative of a discontinuous function.

Since $g'(x)$ can be found by directly differentiating the formula for $f(x)$, once we determine the magnitude and location of the jump discontinuities of $f(x)$, we can compute its derivative directly without introducing to the auxiliary function $g(x)$. ●

EXAMPLE 11.5 As a second, more streamlined example, consider the function

$$f(x) = \begin{cases} -x, & x < 0, \\ x^2 - 1, & 0 < x < 1, \\ 2e^{-x}, & x > 1, \end{cases}$$

which is plotted in Figure 11.9. This function has jump discontinuities of magnitude -1 at $x = 0$, and of magnitude $2/e$ at $x = 1$. Therefore, in light of the preceding remark,

$$f'(x) = -\,\delta(x) + \frac{2}{e}\,\delta(x-1) + \begin{cases} -1, & x < 0, \\ 2x, & 0 < x < 1, \\ -2e^{-x}, & x > 1, \end{cases}$$

where the final terms are obtained by directly differentiating $f(x)$. ●

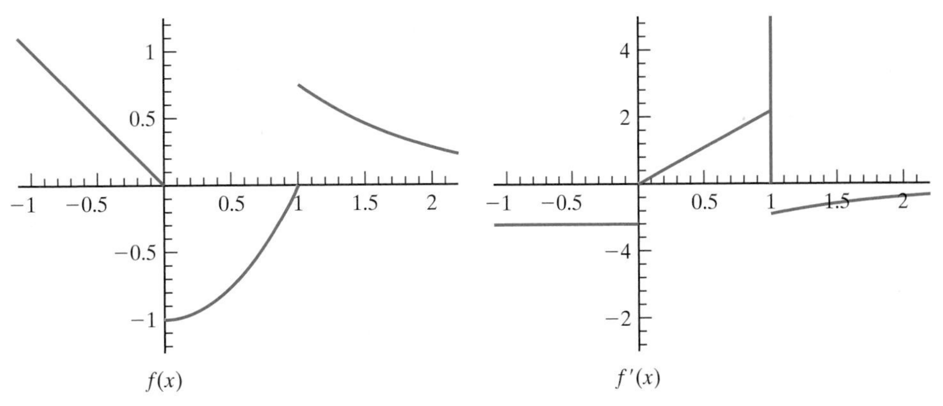

Figure 11.9 The derivative of a discontinuous function.

EXAMPLE 11.6 The derivative of the absolute value function

$$a(x) = |x| = \begin{cases} x, & x > 0, \\ -x, & x < 0, \end{cases}$$

is the *sign function*

$$s(x) = a'(x) = \begin{cases} +1, & x > 0, \\ -1, & x < 0. \end{cases} \tag{11.52}$$

Note that there is no delta function in $a'(x)$ because $a(x)$ is continuous everywhere. Since $s(x)$ has a jump of magnitude 2 at the origin and is otherwise constant, its derivative $s'(x) = a''(x) = 2\,\delta(x)$ is twice the delta function. ●

We are even allowed to differentiate the delta function. Its first derivative

$$\delta_y'(x) = \delta'(x - y)$$

can be interpreted in two ways. First, we may view $\delta'(x)$ as the limit of the derivatives of the approximating functions (11.31):

$$\frac{d\delta}{dx} = \lim_{n\to\infty} \frac{dg_n}{dx} = \lim_{n\to\infty} \frac{-2n^3 x}{\pi(1 + n^2x^2)^2}. \tag{11.53}$$

The graphs of these rational functions take the form of more and more concentrated spiked "doublets", as illustrated in Figure 11.10. To determine the effect of the derivative on a test function $u(x)$, we compute the limiting integral

$$\begin{aligned} \langle \delta', u \rangle &= \int_{-\infty}^{\infty} \delta'(x)\,u(x)\,dx = \lim_{n\to\infty} \int_{-\infty}^{\infty} g_n'(x)\,u(x)\,dx \\ &= -\lim_{n\to\infty} \int_{-\infty}^{\infty} g_n(x)\,u'(x)\,dx = -\int_{-\infty}^{\infty} \delta(x)\,u'(x)\,dx = -u'(0). \end{aligned} \tag{11.54}$$

In the middle step, we used an integration by parts, noting that the boundary terms at $\pm\infty$ vanish, provided that $u(x)$ is continuously differentiable and bounded as $|x| \to \infty$. Pay attention to the minus sign in the final answer.

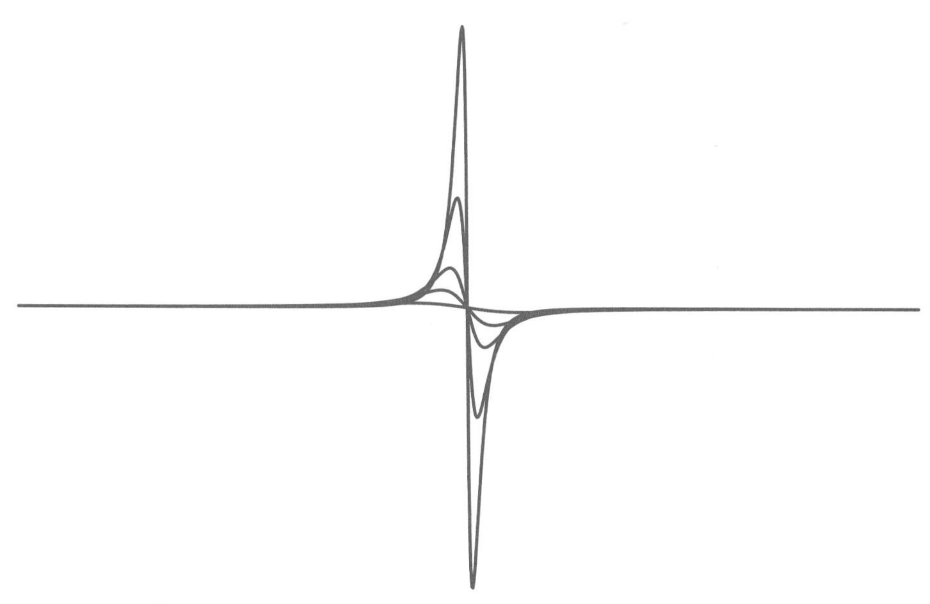

Figure 11.10 Derivative of delta function as limit of doublets.

In the dual interpretation, the generalized function $\delta_y'(x)$ corresponds to the linear functional

$$L_y'[u] = -u'(y) = \langle \delta_y', u \rangle = \int_0^\ell \delta_y'(x)\, u(x)\, dx, \qquad \text{where} \quad 0 < y < \ell, \tag{11.55}$$

that maps a continuously differentiable function $u(x)$ to *minus* its derivative at the point y. We note that (11.55) is compatible with a formal integration by parts

$$\int_0^\ell \delta'(x-y)\, u(x)\, dx = \delta(x-y)\, u(x) \Big|_{x=0}^{\ell} - \int_0^\ell \delta(x-y)\, u'(x)\, dx$$
$$= -u'(y).$$

The boundary terms at $x = 0$ and $x = \ell$ automatically vanish since $\delta(x-y) = 0$ for $x \neq y$.

Warning: The functions $\widetilde{g}_n(x) = g_n(x) + g_n'(x)$ satisfy

$$\lim_{n\to\infty} \widetilde{g}_n(x) = 0$$

for all $x \neq y$, while

$$\int_{-\infty}^{\infty} \widetilde{g}_n(x)\, dx = 1.$$

However,

$$\lim_{n\to\infty} \widetilde{g}_n = \lim_{n\to\infty} g_n + \lim_{n\to\infty} g_n' = \delta + \delta'.$$

Thus, our original conditions (11.29–30) are *not* in fact sufficient to characterize whether a sequence of functions has the delta function as a limit. To be absolutely sure, one must, in fact, verify the more comprehensive limiting formula (11.40).

EXERCISES 11.2

11.2.1. Evaluate the following integrals:

(a) $\displaystyle\int_{-\pi}^{\pi} \delta(x)\cos x\,dx$ (b) $\displaystyle\int_{1}^{2} \delta(x)\,(x-2)\,dx$

(c) $\displaystyle\int_{0}^{3} \delta_1(x)\,e^x\,dx$ (d) $\displaystyle\int_{1}^{e} \delta(x-2)\,\log x\,dx$

(e) $\displaystyle\int_{0}^{1} \delta\bigl(x-\tfrac{1}{3}\bigr)\,x^2\,dx$ (f) $\displaystyle\int_{-1}^{1} \frac{\delta(x+2)\,dx}{1+x^2}$

11.2.2. Simplify the following generalized functions; then write out how they act on a suitable test function $u(x)$:

(a) $e^x\,\delta(x)$ (b) $x\,\delta(x-1)$

(c) $3\,\delta_1(x) - 3x\,\delta_{-1}(x)$ (d) $\dfrac{\delta(x-1)}{x+1}$

(e) $(\cos x)\bigl[\,\delta(x)+\delta(x-\pi)+\delta(x+\pi)\,\bigr]$

(f) $\dfrac{\delta_1(x)-\delta_2(x)}{x^2+1}$

11.2.3. Justify the formula $x\,\delta(x) = 0$ by using

(a) limits (b) duality

11.2.4. Define the generalized function $\varphi(x) = \delta(x) - 3\,\delta(x-1)$

(a) as a limit of ordinary functions;

(b) by using duality.

◇ **11.2.5.** (a) Justify the formula $\delta(2x) = \frac{1}{2}\,\delta(x)$ by

(i) limits (ii) duality

(b) Find a similar formula for $\delta(ax)$ when $a > 0$.

11.2.6. Find and sketch a graph of the derivative (in the context of generalized functions) of the following functions:

(a) $f(x) = \begin{cases} x^2, & 0 < x < 3, \\ x, & -1 < x < 0, \\ 0, & \text{otherwise} \end{cases}$

(b) $g(x) = \begin{cases} \sin|x|, & |x| < \frac{1}{2}\pi, \\ 0, & \text{otherwise} \end{cases}$

(c) $h(x) = \begin{cases} \sin \pi x, & x > 1, \\ 1-x^2, & -1 < x < 1, \\ e^x, & x < -1 \end{cases}$

(d) $k(x) = \begin{cases} \sin x, & x < -\pi, \\ x^2-\pi^2, & -\pi < x < 0, \\ e^{-x}, & x > 0 \end{cases}$

11.2.7. Find the first and second derivatives of the functions

(a) $f(x) = \begin{cases} x+1, & -1 < x < 0, \\ 1-x, & 0 < x < 1, \\ 0, & \text{otherwise} \end{cases}$

(b) $k(x) = \begin{cases} |x|, & -2 < x < 2, \\ 0, & \text{otherwise} \end{cases}$

(c) $s(x) = \begin{cases} 1+\cos \pi x, & -1 < x < 1, \\ 0, & \text{otherwise} \end{cases}$

11.2.8. Find the first and second derivatives of

(a) $e^{-|x|}$ (b) $2\,|x| - |x-1|$

(c) $|x^2+x|$ (d) $x\ \mathrm{sgn}(x^2-4)$

(e) $\sin|x|$ (f) $|\sin x|$ (g) $\mathrm{sgn}(\sin x)$

11.2.9. (a) Prove that $\sigma(\lambda x) = \sigma(x)$ for any $\lambda > 0$.

(b) What about if $\lambda < 0$?

(c) Use parts (a), (b) to deduce that

$$\delta(\lambda x) = \frac{1}{|\lambda|}\,\delta(x)$$

for any $\lambda \neq 0$.

◇**11.2.10.** Explain why the Gaussian functions

$$g_n(x) = \frac{n}{\sqrt{\pi}}\,e^{-n^2x^2}$$

have the delta function $\delta(x)$ as their limit as $n \to \infty$. *Hint*: Use the known formula

$$\int_{-\infty}^{\infty} e^{-x^2}\,dx = \sqrt{\pi}.$$

◇**11.2.11.** In this exercise, we realize the delta function $\delta_y(x)$ as a limit of functions on a finite interval $[0,\ell]$. Let $0 < y < \ell$.

(a) Prove that the functions

$$\widetilde{g}_n(x) = \frac{g_n(x-y)}{m_n},$$

where $g_n(x)$ is given by (11.31), and

$$m_n = \int_0^\ell g_n(x-y)\,dx$$

satisfy (11.29–30), and hence

$$\lim_{n\to\infty} \widetilde{g}_n(x) = \delta_y(x).$$

(b) One can, alternatively, relax the second condition (11.30) to

$$\lim_{n\to\infty} \int_0^\ell g_n(x-y)\,dx = 1.$$

Show that, under this relaxed definition,

$$\lim_{n\to\infty} g_n(x-y) = \delta_y(x)$$

on a finite interval $[0,\ell]$.

♡**11.2.12.** For each positive integer n, let

$$g_n(x) = \begin{cases} \frac{1}{2}n, & |x| < 1/n, \\ 0, & \text{otherwise.} \end{cases}$$

(a) Sketch a graph of $g_n(x)$.
(b) Show that $\lim_{n\to\infty} g_n(x) = \delta(x)$.
(c) Evaluate

$$f_n(x) = \int_{-\infty}^{x} g_n(y)\,dy$$

and sketch a graph. Does the sequence $f_n(x)$ converge to the step function $\sigma(x)$ as $n \to \infty$?
(d) Find the derivative $h_n(x) = g_n'(x)$.
(e) Does the sequence $h_n(x)$ converge to $\delta'(x)$ as $n \to \infty$?

♡**11.2.13.** Answer Exercise 11.2.12 for the *hat functions*

$$g_n(x) = \begin{cases} n - n^2 |x|, & |x| < 1/n, \\ 0, & \text{otherwise.} \end{cases}$$

11.2.14. Justify the formula

$$\lim_{n\to\infty} n\left[\,\delta\left(x - \tfrac{1}{n}\right) - \delta\left(x + \tfrac{1}{n}\right)\right] = -2\,\delta'(x).$$

11.2.15. Let $y < a$. Sketch the graphs of
(a) $s(x) = \int_a^x \delta_y(z)\,dz$
(b) $r(x) = \int_a^x \sigma_y(z)\,dz$

11.2.16. Let

$$\langle f, g\rangle = \int_0^{\ell} f(x)\,g(x)\,dx$$

denote the L^2 inner product on the interval $[0, \ell]$. Suppose $u(x)$ satisfies $u(0) = u(\ell) = 0$. Write out the integration by parts formula $\langle \delta_y, u\rangle = \langle \sigma_y', u\rangle = -\langle \sigma_y, u'\rangle$, and then justify it by direct analysis of the resulting integrals.

11.2.17. Let $\delta_y^{(k)}(x)$ denote the kth derivative of the delta function $\delta_y(x)$. Justify the formula $\langle \delta_y^{(k)}, u\rangle = (-1)^k u^{(k)}(y)$ whenever u is k times continuously differentiable.

11.2.18. According to (11.41), $x\,\delta(x) = 0$. On the other hand, by Leibniz' rule, $(x\,\delta(x))' = \delta(x) + x\,\delta'(x)$ is apparently not zero. Can you explain this paradox?

11.2.19. If $f \in C^1$, should $(f\,\delta)' = f\,\delta'$ or $f'\,\delta + f\,\delta'$?

◇**11.2.20.** (a) Use duality to justify the formula $f(x)\,\delta'(x) = f(0)\,\delta'(x) - f'(0)\,\delta(x)$ when $f \in C^1$.
(b) Find a similar formula for $f(x)\,\delta^{(n)}(x)$ for the product of a sufficiently smooth function and the nth derivative of the delta function.

11.2.21. Use Exercise 11.2.20 to simplify the following generalized functions; then write out how they act on a suitable test function $u(x)$:
(a) $\varphi(x) = (x-2)\,\delta'(x)$
(b) $\psi(x) = (1 + \sin x)\left[\,\delta(x) + \delta'(x)\,\right]$
(c) $\chi(x) = x^2\left[\,\delta(x-1) - \delta'(x-2)\,\right]$
(d) $\gamma(x) = e^x\,\delta''(x+1)$

◇**11.2.22.** Prove that if $f(x)$ is a continuous function, and

$$\int_a^b f(x)\,dx = 0$$

for every interval $[a, b]$, then $f(x) \equiv 0$ everywhere.

◇**11.2.23.** Explain in detail why there is no continuous function $\delta_y(x)$ such that the inner product identity (11.39) holds for every continuous function $u(x)$.

◇**11.2.24.** Explain in detail why the sequence (11.43) converges non-uniformly.

The Green's Function

To further cement our new-found friendship, we now put the delta function to work to solve inhomogeneous boundary value problems. Consider a bar of length ℓ subject to a unit impulse force $\delta_y(x) = \delta(x-y)$ concentrated at position $0 < y < \ell$. The underlying differential equation (11.12) takes the special form

$$-\frac{d}{dx}\left(c(x)\,\frac{du}{dx}\right) = \delta(x-y), \qquad 0 < x < \ell, \tag{11.56}$$

which we supplement with homogeneous boundary conditions that lead to a unique solution. The solution is known as the *Green's function* for the boundary value problem, and will be denoted by $G_y(x) = G(x, y)$.

EXAMPLE 11.7 Let us look at the simple case of a homogeneous bar, of unit length $\ell = 1$, with constant stiffness c, and fixed at both ends. The boundary value problem for the

Green's function $G(x, y)$ takes the form

$$-c\,u'' = \delta(x-y), \qquad u(0) = 0 = u(1), \tag{11.57}$$

where $0 < y < 1$ indicates the point at which we apply the impulse force. The solution to the differential equation is obtained by direct integration. First, by (11.42),

$$u'(x) = -\frac{\sigma(x-y)}{c} + a,$$

where a is a constant of integration. A second integration leads to

$$u(x) = -\frac{\rho(x-y)}{c} + a\,x + b, \tag{11.58}$$

where ρ is the ramp function (11.44). The integration constants a, b are fixed by the boundary conditions; since $0 < y < 1$, we have

$$u(0) = b = 0, \qquad u(1) = -\frac{1-y}{c} + a + b = 0, \qquad \text{and so} \qquad a = \frac{1-y}{c}.$$

Therefore, the Green's function for the problem is

$$G(x, y) = -\rho(x-y) + (1-y)x = \begin{cases} x(1-y)/c, & x \le y, \\ y(1-x)/c, & x \ge y, \end{cases} \tag{11.59}$$

Figure 11.11 sketches a graph of $G(x, y)$ when $c = 1$. Note that, for each fixed y, it is a continuous and piecewise affine function of x—meaning that its graph consists of connected straight line segments, with a corner where the unit impulse force is being applied. ●

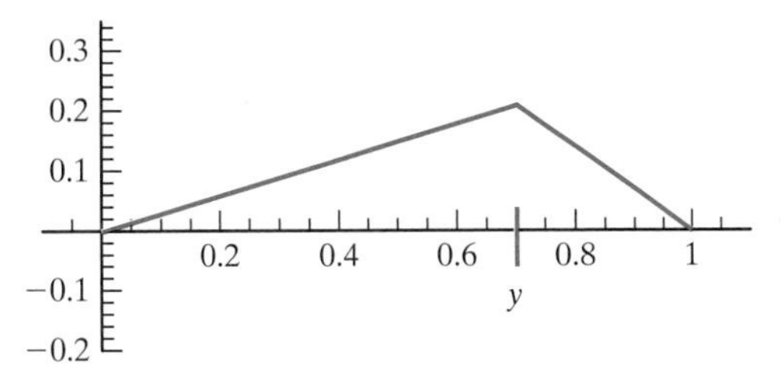

Figure 11.11 Green's function for a bar with fixed ends.

Once we have determined the Green's function, we are able to solve the general inhomogeneous boundary value problem

$$-u'' = f(x), \qquad u(0) = 0 = u(1), \tag{11.60}$$

The solution formula is a consequence of linear superposition. We first express the forcing function $f(x)$ as a linear combination of impulses concentrated at points along the bar. Since there is a continuum of possible positions $0 \le y \le 1$ at which impulse forces may be applied, we will use an integral to sum them up, thereby writing the external force as

$$f(x) = \int_0^1 f(y)\,\delta(x-y)\,dy. \tag{11.61}$$

In the continuous context, sums are replaced by integrals, and we will interpret (11.61) as the (continuous) superposition of an infinite collection of impulses $f(y)\,\delta(x-y)$, of magnitude $f(y)$ and concentrated at position y.

The superposition principle states that, for linear systems, linear combinations of inhomogeneities produce linear combinations of solutions. Again, we adapt this principle to the continuum by replacing the sums by integrals. Thus, we claim that the solution to the boundary value problem is the self-same linear superposition

$$u(x) = \int_0^1 f(y)\, G(x,y)\, dy \tag{11.62}$$

of the Green's function solutions to the individual unit impulse problems.

For the particular boundary value problem (11.60), we use the explicit formula (11.59) for the Green's function. Breaking the integral (11.62) into two parts, for $y < x$ and $y > x$, we arrive at the explicit solution formula

$$u(x) = \frac{1}{c}\int_0^x (1-x)\, y\, f(y)\, dy \; + \; \frac{1}{c}\int_x^1 x\,(1-y)\, f(y)\, dy. \tag{11.63}$$

For example, under a constant unit force f, (11.63) reduces to

$$\begin{aligned} u(x) &= \frac{f}{c}\int_0^x (1-x)\, y\, dy \; + \; \frac{f}{c}\int_x^1 x\,(1-y)\, dy \\ &= \frac{f}{2c}(1-x)\,x^2 + \frac{f}{2c}\,x\,(1-x)^2 = \frac{f}{2c}(x - x^2), \end{aligned}$$

in agreement with our earlier solution (11.19) in the special case $d = 0$. Although this relatively simple problem was perhaps easier to solve directly, the Green's function approach helps crystallize our understanding, and provides a unified framework that covers the full range of linear boundary value problems arising in applications, including those governed by partial differential equations, [36, 59, 68].

Let us, finally, convince ourselves that the superposition formula (11.63) does indeed give the correct answer. First,

$$\begin{aligned} \frac{du}{dx} &= (1-x)\,x\, f(x) + \int_0^x [\,-y\, f(y)\,]\, dy \\ &\qquad - x\,(1-x)\, f(x) + \int_x^1 (1-y)\, f(y)\, dy \\ &= -\int_0^1 y\, f(y)\, dy + \int_x^1 f(y)\, dy. \end{aligned}$$

Differentiating again, we conclude that $\dfrac{d^2u}{dx^2} = -f(x)$, as claimed.

Remark: In computing the derivatives of u, we made use of the calculus formula

$$\frac{d}{dx}\int_{\alpha(x)}^{\beta(x)} F(x,y)\, dy = F(x,\beta(x))\,\frac{d\beta}{dx} - F(x,\alpha(x))\,\frac{d\alpha}{dx} + \int_{\alpha(x)}^{\beta(x)} \frac{\partial F}{\partial x}(x,y)\, dy \tag{11.64}$$

for the derivative of an integral with variable limits, which is a straightforward consequence of the Fundamental Theorem of Calculus and the chain rule, [2, 58]. As with all limiting processes, one must always be careful when interchanging the order of differentiation and integration.

We note the following fundamental properties, that serve to uniquely characterize the Green's function. First, since the delta forcing vanishes except at the point $x = y$, the Green's function satisfies the homogeneous differential equation[‖]

$$\frac{\partial^2 G}{\partial x^2}(x, y) = 0 \quad \text{for all} \quad x \neq y. \tag{11.65}$$

Secondly, by construction, it must satisfy the boundary conditions,

$$G(0, y) = 0 = G(1, y).$$

Thirdly, for each fixed y, $G(x, y)$ is a continuous function of x, but its derivative $\partial G / \partial x$ has a jump discontinuity of magnitude $-1/c$ at the impulse point $x = y$. The second derivative $\partial^2 G / \partial x^2$ has a delta function discontinuity there, and thereby solves the original impulse boundary value problem (11.57).

Finally, we cannot help but notice that the Green's function is a symmetric function of its two arguments: $G(x, y) = G(y, x)$. Symmetry has the interesting physical consequence that the displacement of the bar at position x due to an impulse force concentrated at position y is exactly the same as the displacement of the bar at y due to an impulse of the same magnitude being applied at x. This turns out to be a rather general, although perhaps unanticipated phenomenon. (For the finite-dimensional counterpart for mass-spring chains, circuits, and structures see Exercises 6.1.6, 6.2.18, and 6.3.17.) Symmetry is a consequence of the underlying symmetry or "self-adjointness" of the boundary value problem, to be developed properly in the following section.

REMARK: The Green's function $G(x, y)$ should be be viewed as the continuum limit of the inverse of the stiffness matrix, $G = K^{-1}$, appearing in the discrete equilibrium equations $K\mathbf{u} = \mathbf{f}$. Indeed, the entries G_{ij} of the inverse matrix are approximations to the sampled values $G(x_i, x_j)$. In particular, symmetry of the Green's function, whereby $G(x_i, x_j) = G(x_j, x_i)$, corresponds to symmetry, $G_{ij} = G_{ji}$, of the inverse of the symmetric stiffness matrix. In Exercise 11.2.33, you are asked to study this limiting procedure in some detail.

Let us summarize the fundamental properties that serve to characterize the Green's function, in a form that applies to general second order boundary value problems.

Basic Properties of the Green's Function

(i) Solves the homogeneous differential equation:

$$-\frac{\partial}{\partial x}\left(c(x)\, \frac{\partial}{\partial x}\, G(x, y) \right) = 0, \qquad \text{for all} \quad x \neq y. \tag{11.66}$$

(ii) Satisfies the homogeneous boundary conditions.

(iii) Is a continuous function of its arguments.

(iv) As a function of x, its derivative $\dfrac{\partial G}{\partial x}$ has a jump discontinuity of magnitude $-1/c(y)$ at $x = y$.

(v) Is a symmetric function of its arguments:

$$G(x, y) = G(y, x). \tag{11.67}$$

[‖]Since $G(x, y)$ is a function of two variables, we switch to partial derivative notation to indicate its derivatives.

(vi) Generates a superposition principle for the solution under general forcing functions:

$$u(x) = \int_0^\ell G(x,y)\, f(y)\, dy. \tag{11.68}$$

EXAMPLE 11.8 Consider a uniform bar of length $\ell = 1$ with one fixed and one free end, subject to an external force. The displacement $u(x)$ satisfies the boundary value problem

$$-c u'' = f(x), \quad u(0) = 0, \quad u'(1) = 0, \tag{11.69}$$

where c is the elastic constant of the bar. To determine the Green's function, we appeal to its characterizing properties, although one could equally well use direct integration as in the preceding Example 11.7.

First, since, as a function of x, it must satisfy the homogeneous differential equation $-c\,u'' = 0$ for all $x \neq y$, the Green's function must be of the form

$$G(x,y) = \begin{cases} p\,x + q, & x \le y, \\ r\,x + s, & x \ge y, \end{cases}$$

for certain constants p, q, r, s. Second, the boundary conditions require

$$q = G(0,y) = 0, \qquad r = \frac{\partial G}{\partial x}(1,y) = 0.$$

Continuity of the Green's function at $x = y$ imposes the further constraint

$$p\,y = G(y^-, y) = G(y^+, y) = s.$$

Finally, the derivative $\partial G/\partial x$ must have a jump discontinuity of magnitude $-1/c$ at $x = y$, and so

$$-\frac{1}{c} = \frac{\partial G}{\partial x}(y^+, y) - \frac{\partial G}{\partial x}(y^-, y) = 0 - p, \qquad \text{and so} \quad p = s = \frac{1}{c}.$$

We conclude that the Green's function for this problem is

$$G(x,y) = \begin{cases} x/c, & x \le y, \\ y/c, & x \ge y, \end{cases} \tag{11.70}$$

which, for $c = 1$, is graphed in Figure 11.12. Note that $G(x,y) = G(y,x)$ is indeed symmetric, which helps check the correctness of our computation. Finally, the superposition principle (11.68) implies that the solution to the boundary value problem (11.69) can be written as a single integral, namely

$$u(x) = \int_0^1 G(x,y) f(y)\, dy = \frac{1}{c}\int_0^x y\, f(y)\, dy + \frac{1}{c}\int_x^1 x\, f(y)\, dy. \tag{11.71}$$

The reader may wish to verify this directly, as we did in the previous example. ●

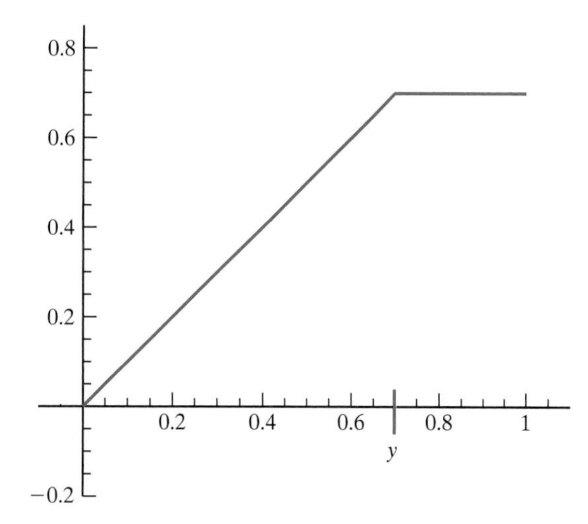

Figure 11.12 Green's function for bar with one fixed and one free end.

EXERCISES 11.2

11.2.25. A point 2 cm along a 10 cm bar experiences a displacement of 1 mm under a force of 2 Newtons applied at the midpoint of the bar. How far does the midpoint deflect when a force of 1 Newton is applied at the point 2 cm along the bar?

11.2.26. Derive the Green's function for the boundary value problem (11.69) by direct integration. Make sure that you obtain the same answer (11.70).

11.2.27. A uniform bar of unit length $\ell = 4$ has constant stiffness $c = 2$. Find the Green's function for the case when

(a) both ends are fixed

(b) one end is fixed and the other is free

(c) Why is there no Green's function when both ends are free?

11.2.28. A bar of length $\ell = 1$ and stiffness function $c(x) = 1 + x$ has both ends clamped: $u(0) = u(1) = 0$.

(a) Construct the Green's function for this boundary value problem.

(b) Use the Green's function to construct an integral formula for the solution when the bar is subject to a general external force $f(x)$.

(c) Evaluate the integrals and compare with your solution in Exercise 11.1.4 in the case of a unit gravitational force $f(x) \equiv 1$.

♡**11.2.29.** An elastic bar has constitutive function

$$c(x) = \frac{1}{1+x^2} \quad \text{for} \quad 0 \le x \le 1.$$

(a) Find the displacement and internal force when the bar is subjected to a constant external force, $f \equiv 1$, and the ends are held fixed: $u(0) = u(1) = 0$.

(b) Find the Green's function for the boundary value problem describing the displacements of the bar.

(c) Use the resulting superposition formula to check your solution to part (a).

(d) Which point y on the bar is the "weakest", i.e., the bar experiences the largest displacement under a unit impulse concentrated at that point?

11.2.30. Consider the boundary value problem $-u'' = f(x)$, $u(0) = 0$, $u(1) = 2u'(1)$.

(a) Find the Green's function.

(b) Which of the fundamental properties does your Green's function satisfy?

(c) Write down an explicit integral formula for the solution to the boundary value problem, and prove its validity by a direct computation.

(d) Explain why the related boundary value problem $-u'' = f$, $u(0) = 0$, $u(1) = u'(1)$, does not have a Green's function.

♡**11.2.31.** When n is a positive integer, set

$$f_n(x) = \begin{cases} \frac{n}{2}, & |x - y| < \frac{1}{n}, \\ 0 & \text{otherwise.} \end{cases}$$

(a) Find the solution $u_n(x)$ to the boundary value problem $-u'' = f_n(x)$, $u(0) = u(1) = 0$, assuming $0 < y - \frac{1}{n} < y + \frac{1}{n} < 1$.

(b) Prove that

$$\lim_{n \to \infty} u_n(x) = G(x, y)$$

converges to the Green's function (11.59). Why should this be the case?

(c) Reconfirm the result in part (b) by graphing $u_5(x), u_{15}(x), u_{25}(x)$ along with $G(x,y)$ for $y = .3$.

11.2.32. Prove directly that formula (11.71) gives the solution to the boundary value problem (11.69).

♠**11.2.33.** A system of n masses is connected to a top support by n unit springs of length $\Delta x = \frac{1}{n}$, so that the overall length of the mass-spring chain is $\ell = 1$. Let $K = A^T A$, where A is the rescaled incidence matrix (11.4), and let $G = K^{-1}/\Delta x$.

(a) Explain why the matrix entries $g_{ij} \approx G(x_i, x_j)$ should approximate the sampled values of the Green's function (11.70) for the continuum limit boundary value problem with $c = 1$.

(b) Test this expectation for $n = 5, 10, 20$. How close is the matrix G to the sampled Green's function?

♠**11.2.34.** Reformulate Exercise 11.2.33 for a mass-spring chain connected to both top and bottom supports. Do the same conclusions hold?

◇**11.2.35.** Prove the differentiation formula (11.64).

11.3 ADJOINTS AND MINIMUM PRINCIPLES

One of the profound messages of this text is that the linear algebraic structures that were initially* designed for finite-dimensional problems all have direct counterparts in the infinite-dimensional function spaces. To further develop this theme, let us now discuss how the boundary value problems for continuous elastic bars fit into our general equilibrium framework of positive (semi-)definite linear systems. As we will see, the associated energy minimization principle not only leads to a new mathematical characterization of the equilibrium solution, it also, through the finite element method, underlies the most important class of numerical approximation algorithms for such boundary value problems.

Adjoints of Differential Operators

In discrete systems, a key step was the recognition that the matrix appearing in the force balance law is the transpose of the incidence matrix relating displacements and elongations. In the continuum limit, the discrete incidence matrix has turned into a differential operator. But how do you take the "transpose" of a differential operator? The abstract answer to this quandary can be found in Section 7.5. The transpose of a matrix is a particular instance of the general notion of the *adjoint* of a linear function, which relies on the specification of inner products on its domain and target spaces. In the case of the matrix transpose, the adjoint is taken with respect to the standard dot product on Euclidean space. Thus, the correct interpretation of the "transpose" of a differential operator is as the adjoint linear operator with respect to suitable inner products on function space.

For bars and similar one-dimensional media, the role of the incidence matrix is played by the derivative $v = D[u] = du/dx$, which defines a linear operator $D\colon U \to V$ from the vector space of possible displacements $u(x)$, denoted by U, to the vector space of possible strains $v(x)$, denoted by V. In order to compute its adjoint, we need to impose inner products on both the displacement space U and the strain space V. The simplest is to adopt the same standard L^2 inner product

$$\langle u, \widetilde{u} \rangle = \int_0^\ell u(x)\,\widetilde{u}(x)\,dx, \qquad \langle\!\langle v, \widetilde{v} \rangle\!\rangle = \int_0^\ell v(x)\,\widetilde{v}(x)\,dx, \tag{11.72}$$

on both vector spaces. These are the continuum analogs of the Euclidean dot product, and, as we shall see, will be appropriate when dealing with homogeneous bars. According to the defining equation (7.74), the adjoint D^* of the derivative operator

*Sometimes, the order is reversed, and, at least historically, basic linear algebra concepts make their first appearance in function space. Examples include the Cauchy–Schwarz inequality, the Fredholm alternative, and the Fourier transform.

must satisfy the inner product identity

$$\langle\!\langle D[u], v \rangle\!\rangle = \langle u, D^*[v] \rangle \quad \text{for all} \quad u \in U, \quad v \in V. \tag{11.73}$$

First, we compute the left hand side:

$$\langle\!\langle D[u], v \rangle\!\rangle = \left\langle\!\!\left\langle \frac{du}{dx}, v \right\rangle\!\!\right\rangle = \int_0^\ell \frac{du}{dx}\, v\, dx. \tag{11.74}$$

On the other hand, the right hand side should equal

$$\langle u, D^*[v] \rangle = \int_0^\ell u\, D^*[v]\, dx. \tag{11.75}$$

Now, in the latter integral, we see u multiplying the result of applying the linear operator D^* to v. To identify this integrand with that in the previous integral (11.74), we need to somehow remove the derivative from u. The secret is integration by parts, which allows us to rewrite the first integral in the form

$$\int_0^\ell \frac{du}{dx}\, v\, dx = \big[\, u(\ell)\, v(\ell) - u(0)\, v(0) \,\big] - \int_0^\ell u\, \frac{dv}{dx}\, dx. \tag{11.76}$$

Ignoring the two boundary terms for a moment, the remaining integral has the form of an inner product

$$-\int_0^\ell u\, \frac{dv}{dx}\, dx = \int_0^\ell u \left[-\frac{dv}{dx} \right] dx = \left\langle u, -\frac{dv}{dx} \right\rangle = \langle u, -D[v] \rangle. \tag{11.77}$$

Equating (11.74) and (11.77), we deduce that

$$\langle\!\langle D[u], v \rangle\!\rangle = \left\langle\!\!\left\langle \frac{du}{dx}, v \right\rangle\!\!\right\rangle = \left\langle u, -\frac{dv}{dx} \right\rangle = \langle u, -D[v] \rangle.$$

Thus, to satisfy the adjoint equation (11.73), we must have

$$\langle u, D^*[v] \rangle = \langle u, -D[v] \rangle \quad \text{for all} \quad u \in U, \quad v \in V,$$

and so

$$\left(\frac{d}{dx} \right)^* = D^* = -D = -\frac{d}{dx}. \tag{11.78}$$

The final equation confirms our earlier identification of the derivative operator D as the continuum limit of the incidence matrix A, and its negative $-D = D^*$ as the limit of the transposed (or adjoint) incidence matrix $A^T = A^*$.

However, the preceding argument is valid *only* if the boundary terms in the integration by parts formula (11.76) vanish:

$$u(\ell)\, v(\ell) - u(0)\, v(0) = 0, \tag{11.79}$$

which necessitates imposing suitable boundary conditions on the functions u and v. For example, in the case of a bar with both ends fixed, the boundary conditions

$$u(0) = 0, \qquad u(\ell) = 0, \tag{11.80}$$

will ensure that (11.79) holds, and therefore validate (11.78). The homogeneous boundary conditions serve to define the vector space

$$U = \big\{\, u(x) \in \mathrm{C}^2[0, \ell] \ \big|\ u(0) = u(\ell) = 0 \,\big\}$$

of allowable displacements, consisting of all twice continuously differentiable functions that vanish at the ends of the bar.

The fixed boundary conditions (11.80) are not the only possibilities that ensure the vanishing of the boundary terms (11.79). An evident alternative is to require that the strain vanish at both endpoints, $v(0) = v(\ell) = 0$. In this case, the strain space

$$V = \left\{ \, v(x) \in \mathrm{C}^1[0, \ell\,] \;\middle|\; v(0) = v(\ell) = 0 \, \right\}$$

consists of all functions that vanish at the endpoints. Since the derivative $D\colon U \to V$ must map a displacement $u(x)$ to an *allowable* strain $v(x)$, the vector space of possible displacements takes the form

$$U = \left\{ \, u(x) \in \mathrm{C}^2[0, \ell\,] \;\middle|\; u'(0) = u'(\ell) = 0 \, \right\}.$$

Thus, this case corresponds to the free boundary conditions of Example 11.3. Again, restricting $D\colon U \to V$ to these particular vector spaces ensures that the boundary terms (11.79) vanish, and so (11.78) holds in this situation too.

Let us list the most important combinations of boundary conditions that imply the vanishing of the boundary terms (11.79), and so ensure the validity of the adjoint equation (11.78).

Self-Adjoint Boundary Conditions for a Bar

(a) Both ends fixed: $u(0) = u(\ell) = 0$.

(b) One free and one fixed end: $u(0) = u'(\ell) = 0$ or $u'(0) = u(\ell) = 0$.

(c) Both ends free: $u'(0) = u'(\ell) = 0$.

(d) Periodic bar or ring: $u(0) = u(\ell), \quad u'(0) = u'(\ell)$.

In all cases, the boundary conditions impose restrictions on the displacement space U and, in cases cases (b–d) when identifying $v(x) = u'(x)$, the strain space V also.

In mathematics, a fixed boundary condition, $u(a) = 0$, is commonly referred to as a *Dirichlet boundary condition*, to honor the nineteenth century French analyst Lejeune Dirichlet. A free boundary condition, $u'(a) = 0$, is known as a *Neumann boundary condition*, after his German contemporary Carl Gottfried Neumann. The *Dirichlet boundary value problem* (a) has both ends fixed, while the *Neumann boundary value problem* (c) has both ends free. The intermediate case (b) is known as a *mixed boundary value problem*. The periodic boundary conditions (d) represent a bar that has its ends joined together to form a circular† elastic ring, and represents the continuum limit of the periodic mass–spring chain discussed in Exercise 6.3.11.

Summarizing, for a homogeneous bar with unit stiffness $c(x) \equiv 1$, the displacement, strain, and external force are related by the adjoint formulae

$$v = D[\,u\,] = u', \qquad f = D^*[\,v\,] = -\,v',$$

provided that we impose a suitable pair of homogeneous boundary conditions. The equilibrium equation has the self-adjoint form

$$K[\,u\,] = f, \qquad \text{where} \qquad K = D^* \circ D = -\,D^2. \tag{11.81}$$

We note that

$$K^* = (D^* \circ D)^* = D^* \circ (D^*)^* = D^* \circ D = K, \tag{11.82}$$

which proves self-adjointness of the differential operator. In gory detail,

$$\langle K[\,u\,], \widetilde{u} \rangle = \int_0^\ell \big[-u''(x)\, \widetilde{u}(x) \,\big]\, dx = \int_0^\ell \big[-u(x)\, \widetilde{u}''(x) \,\big]\, dx = \langle u, K[\,\widetilde{u}\,] \rangle \tag{11.83}$$

† The circle is sufficiently large so that we can safely ignore any curvature effects.

for all displacements $u, \widetilde{u} \in U$. A direct verification of this formula relies on two integration by parts, employing the selected boundary conditions to eliminate the boundary contributions.

To deal with nonuniform materials, we must modify the inner products. Let us retain the ordinary L^2 inner product

$$\langle u, \widetilde{u} \rangle = \int_0^\ell u(x)\, \widetilde{u}(x)\, dx, \qquad u, \widetilde{u} \in U, \tag{11.84}$$

on the vector space of possible displacements, but adopt a weighted inner product

$$\langle\!\langle v, \widetilde{v} \rangle\!\rangle = \int_0^\ell v(x)\, \widetilde{v}(x)\, c(x)\, dx, \qquad v, \widetilde{v} \in V, \tag{11.85}$$

on the space of strain functions. The weight function $c(x) > 0$ coincides with the stiffness of the bar; its positivity, which is required for (11.85) to define a *bona fide* inner product, is in accordance with the underlying physical assumptions.

Let us recompute the adjoint of the derivative operator $D\colon U \to V$, this time with respect to the inner products (11.84–85). Now we need to compare

$$\langle\!\langle D[u], v \rangle\!\rangle = \int_0^\ell \frac{du}{dx}\, v(x)\, c(x)\, dx,$$

with

$$\langle u, D^*[v] \rangle = \int_0^\ell u(x)\, D^*[v]\, dx.$$

Integrating the first expression by parts, we find

$$\begin{aligned} \int_0^\ell \frac{du}{dx}\, c\, v\, dx &= \big[\, u(\ell)\, c(\ell)\, v(\ell) - u(0)\, c(0)\, v(0) \,\big] - \int_0^\ell u\, \frac{d(c\, v)}{dx}\, dx \\ &= \int_0^\ell u \left[-\frac{d(c\, v)}{dx} \right] dx, \end{aligned} \tag{11.86}$$

provided that we choose our boundary conditions so that

$$u(\ell)\, c(\ell)\, v(\ell) - u(0)\, c(0)\, v(0) = 0. \tag{11.87}$$

As you can check, this follows from any of the listed boundary conditions: Dirichlet, Neumann, or mixed, as well as the periodic case, assuming $c(0) = c(\ell)$. Therefore, in such situations, the weighted adjoint of the derivative operator is

$$D^*[v] = -\frac{d(c\, v)}{dx} = -c\, \frac{dv}{dx} - c'\, v. \tag{11.88}$$

The self-adjoint combination $K = D^* \circ D$ is

$$K[u] = -\frac{d}{dx}\left(c(x)\, \frac{du}{dx} \right), \tag{11.89}$$

and hence we have formulated the original equation (11.12) for a nonuniform bar in the same abstract self-adjoint form.

As an application, let us show how the self-adjoint formulation leads directly to the symmetry of the Green's function $G(x, y)$. As a function of x, the Green's function satisfies

$$K[G(x, y)] = \delta(x - y).$$

Thus, by the definition of the delta function and the self-adjointness identity (11.83),

$$\begin{aligned} G(z,y) &= \int_0^\ell G(x,y)\,\delta(x-z)\,dx = \langle G(x,y), \delta(x-z)\rangle \\ &= \langle G(x,y), K[G(x,z)]\rangle = \langle K[G(x,y)], G(x,z)\rangle \\ &= \langle \delta(x-y), G(x,z)\rangle = \int_0^\ell G(x,z)\,\delta(x-y)\,dx = G(y,z), \end{aligned} \tag{11.90}$$

for any $0 < y, z < \ell$, which validates[‡] the symmetry equation (11.67).

Positivity and Minimum Principles

We are now able to characterize the solution to a stable self-adjoint boundary value problem by a quadratic minimization principle. Again, the development shadows the finite-dimensional case presented in Chapter 6. So the first step is to understand how a differential operator defining a boundary value problem can be positive definite.

According to the abstract Definition 7.59, a linear operator $K: U \to U$ on an inner product space U is *positive definite*, provided that it is

(a) self-adjoint, so $K^* = K$, and

(b) satisfies the positivity criterion $\langle K[u], u\rangle > 0$ for all $0 \neq u \in U$.

Self-adjointness of the product operator $K = D^* \circ D$ was established in (11.82). Furthermore, Theorem 7.62 tells us that K is positive definite if and only if $\ker D = \{0\}$. Indeed, by the definition of the adjoint,

$$\langle K[u], u\rangle = \langle D^*[D[u]], u\rangle = \langle\!\langle D[u], D[u]\rangle\!\rangle = \| D[u]\|^2 \geq 0, \tag{11.91}$$

so $K = D^* \circ D$ is automatically positive semi-definite. Moreover, $\langle K[u], u\rangle = 0$ if and only if $D[u] = 0$, i.e., $u \in \ker K$. Thus $\ker D = \{0\}$ is both necessary and sufficient for the positivity criterion to hold.

Now, in the absence of constraints, the kernel of the derivative operator D is *not* trivial. Indeed, $D[u] = u' = 0$ if and only if $u(x) \equiv c$ is constant, and hence $\ker D$ is the one-dimensional subspace of $\mathrm{C}^1[0,\ell]$ consisting of all constant functions. However, we are viewing D as a linear operator on the vector space U of allowable displacements, and so the elements of $\ker D \subset U$ must also be allowable, meaning that they must satisfy the boundary conditions. Thus, positivity reduces, in the present situation, to the question of whether or not there are any nontrivial constant functions that satisfy the prescribed homogeneous boundary conditions.

Clearly, the only constant function that satisfies a homogeneous Dirichlet boundary condition is the zero function. Therefore, when restricted to the Dirichlet displacement space $U = \{u(0) = u(\ell) = 0\}$, the derivative operator has trivial kernel, $\ker D = \{0\}$, so $K = D^* \circ D$ defines a positive definite linear operator on U. A similar argument applies to the mixed boundary value problem, which is also positive definite. On the other hand, any constant function satisfies the homogeneous Neumann boundary conditions, and so $\ker D \subset \widetilde{U} = \{u'(0) = u'(\ell) = 0\}$ is a one-dimensional subspace. Therefore, the Neumann boundary value problem is only positive semi-definite. A similar argument shows that the periodic problem is also positive semi-definite. Observe that, just as in the finite-dimensional version, the positive definite cases are stable, and the boundary value problem admits a unique equilibrium solution under arbitrary external forcing, whereas the semi-definite cases are unstable, and have either no solution or infinitely many equilibrium solutions, depending on the nature of the external forcing.

[‡]Symmetry at the endpoints follows from continuity.

In the positive definite, stable cases, we can characterize the equilibrium solution as the unique function $u \in U$ that minimizes the quadratic functional

$$\mathcal{P}[u] = \tfrac{1}{2}\,\|D[u]\|^2 - \langle u, f\rangle = \int_0^\ell \left[\,\tfrac{1}{2}\,c(x)\,u'(x)^2 - f(x)\,u(x)\,\right] dx. \tag{11.92}$$

A proof of this general fact appears following Theorem 7.61. Pay attention: the norm in (11.92) refers to the strain space V, and so is associated with the weighted inner product (11.85), whereas the inner product term refers to the displacement space U, which has been given the L^2 inner product. Physically, the first term measures the internal energy due to the stress in the bar, while the second term is the potential energy induced by the external forcing. Thus, as always, the equilibrium solution seeks to minimize the total energy in the system.

EXAMPLE 11.9 Consider the homogeneous Dirichlet boundary value problem

$$-u'' = f(x), \qquad u(0) = 0, \qquad u(\ell) = 0. \tag{11.93}$$

for a uniform bar with two fixed ends. This is a stable case, and so the underlying differential operator $K = D^* \circ D = -D^2$, when acting on the space of displacements satisfying the boundary conditions, is positive definite. Explicitly, positive definiteness requires

$$\langle K[u], u\rangle = \int_0^\ell [\,-u''(x)\,u(x)\,]\,dx = \int_0^\ell [\,u'(x)\,]^2\,dx > 0 \tag{11.94}$$

for all nonzero $u(x) \not\equiv 0$ with $u(0) = u(\ell) = 0$. Notice how we used an integration by parts, invoking the boundary conditions to eliminate the boundary contributions, to expose the positivity of the integral. The corresponding energy functional is

$$\mathcal{P}[u] = \tfrac{1}{2}\|u'\|^2 - \langle u, f\rangle = \int_0^\ell \left[\,\tfrac{1}{2}\,u'(x)^2 - f(x)\,u(x)\,\right] dx.$$

Its minimum value, taken over all possible displacement functions that satisfy the boundary conditions, occurs precisely when $u = u_\star$ is the solution to the boundary value problem.

A direct verification of the latter fact may be instructive. As in our derivation of the adjoint operator, it relies on an integration by parts. Since $-u_\star'' = f$,

$$\begin{aligned}
\mathcal{P}[u] &= \int_0^\ell \left[\,\tfrac{1}{2}\,(u')^2 + u_\star''\,u\,\right] dx \\
&= u_\star'(\ell)\,u(\ell) - u_\star'(0)\,u(0) + \int_0^\ell \left[\,\tfrac{1}{2}(u')^2 - u_\star'\,u'\,\right] dx \\
&= \int_0^\ell \tfrac{1}{2}(u' - u_\star')^2\,dx \;-\; \int_0^\ell \tfrac{1}{2}(u_\star')^2\,dx,
\end{aligned} \tag{11.95}$$

where the boundary terms vanish owing to the boundary conditions on $u_\star$ and u. In the final expression for $\mathcal{P}[u]$, the first integral is always ≥ 0, and is actually equal to 0 if and only if $u'(x) = u_\star'(x)$ for all $0 \leq x \leq \ell$. On the other hand, the second integral does not depend upon u at all. Thus, for $\mathcal{P}[u]$ to achieve a minimum, $u(x) = u_\star(x) + c$ for some constant c. But the boundary conditions force $c = 0$, and hence the energy functional will assume its minimum value if and only if $u = u_\star$. ●

EXERCISES 11.3

11.3.1. Let $\mathcal{P}[u] = \int_0^1 \left[\frac{1}{2}(u')^2 - 5u \right] dx$.

(a) Find the function $u_\star(x)$ that minimizes $\mathcal{P}[u]$ among all C^2 functions that satisfy $u(0) = u(1) = 0$.

(b) Test your answer by computing $\mathcal{P}[u_\star]$ and then comparing with the value of $\mathcal{P}[u]$ when $u(x) =$

(i) $x - x^2$ (ii) $\frac{3}{2}x - \frac{3}{2}x^3$

(iii) $\frac{2}{3}\sin \pi x$ (iv) $x^2 - x^4$

11.3.2. Consider the boundary value problem $-u'' = x$, $u(0) = u(1) = 0$.

(i) Find the solution.

(ii) Write down a minimization principle that characterizes the solution.

(iii) What is the value of the quadratic energy functional on the solution?

(iv) Write down at least two other functions that satisfy the boundary conditions and check that they produce larger values for the energy. (Use numerical integration to evaluate the integrals if necessary.)

11.3.3. Answer Exercise 11.3.2 for the boundary value problems

(a) $\dfrac{d}{dx}\left(\dfrac{1}{1+x^2}\dfrac{du}{dx}\right) = x^2$, $u(-1) = u(1) = 0$

(b) $-(e^x u')' = e^{-x}$, $u(0) = u'(1) = 0$

(c) $x^2 u'' + 2x u' = 3x^2$, $u'(1) = u(2) = 0$

(d) $x u'' + 3u' = 1$, $u(-2) = u(-1) = 0$

11.3.4. For each of the following functionals and associated boundary conditions,

(i) write down a boundary value problem satisfied by the minimizing function, and

(ii) find the minimizing function $u(x)$:

(a) $\displaystyle\int_0^1 \left[\tfrac{1}{2}(u')^2 - 3u \right] dx$, $u(0) = u(1) = 0$

(b) $\displaystyle\int_0^1 \left[\tfrac{1}{2}(x+1)(u')^2 - 5u \right] dx$, $u(0) = u(1) = 0$

(c) $\displaystyle\int_1^3 \left[x(u')^2 + 2u \right] dx$, $u(1) = u(3) = 0$

(d) $\displaystyle\int_0^1 \left[\tfrac{1}{2}e^x(u')^2 - (1+e^x)u \right] dx$, $u(0) = u(1) = 0$

(e) $\displaystyle\int_{-1}^1 \frac{(x^2+1)(u')^2 + xu}{(x^2+1)^2}\, dx$, $u(-1) = u(1) = 0$

11.3.5. Which of the following quadratic functionals possess a unique minimizer among all functions satisfying the indicated boundary conditions? Find the minimizer if it exists.

(a) $\displaystyle\int_1^2 \left[\tfrac{1}{2}x(u')^2 + 2(x-1)u \right] dx$, $u(1) = u(2) = 0$

(b) $\displaystyle\int_{-\pi}^{\pi} \left[\tfrac{1}{2}x(u')^2 - u\cos x \right] dx$, $u(-\pi) = u(\pi) = 0$

(c) $\displaystyle\int_{-1}^1 \left[(u')^2\cos x - u\sin x \right] dx$, $u(-1) = 0,\ u'(1) = 0$

(d) $\displaystyle\int_{-2}^2 \left[(1-x^2)(u')^2 - u \right] dx$, $u(-2) = 0,\ u(2) = 0$

(e) $\displaystyle\int_0^1 \left[(x+1)(u')^2 - u \right] dx$, $u'(0) = u'(1) = 0$

11.3.6. Does the quadratic functional

$$\mathcal{P}[u] = \int_0^1 \left[\tfrac{1}{2}(u')^2 - \left(x - \tfrac{1}{2}\right)u \right] dx$$

have a minimum value when $u(x)$ is subject to the homogeneous Neumann boundary value conditions $u'(0) = u'(1) = 0$? If so, find all functions that minimize $\mathcal{P}[u]$. *Hint*: What are the solutions to the associated boundary value problem?

11.3.7. Let $c(x) > 0$ for $0 \le x \le 1$. Prove that, when subject to the Neumann boundary conditions $u'(0) = u'(1) = 0$, the quadratic functional

$$\mathcal{P}[u] = \int_0^1 \left[\tfrac{1}{2}c(x)(u')^2 - f(x)u \right] dx$$

achieves a minimum value if and only if $f(x)$ has mean zero.

◇ **11.3.8.** Show that the internal energy in a bar (the first term in (11.92)) is one half the unweighted L^2 inner product between stress and strain: $\frac{1}{2}\|D[u]\|^2 = \frac{1}{2}\langle v, w\rangle$.

11.3.9. Let $u_\star(x)$ be the equilibrium solution to the Dirichlet boundary value problem for a bar: $K[u_\star] = f$, $u_\star(0) = u_\star(\ell) = 0$. Prove that if $f(x) \not\equiv 0$, then the total energy at equilibrium is strictly negative: $\mathcal{P}[u_\star] < 0$.

11.3.10. Is the conclusion of Exercise 11.3.9 valid for the mixed boundary value problem $u(0) = u'(\ell) = 0$?

11.3.11. Find a function $u(x)$ such that

$$\int_0^1 u''(x)\, u(x)\, dx > 0.$$

How do you reconcile this with the claimed positivity in (11.94)?

11.3.12. Does the inequality (11.94) hold when $u(x) \not\equiv 0$ is subject to the Neumann boundary conditions $u'(0) = u'(\ell) = 0$?

11.3.13. Let $U = C^0[0,1]$. Find the adjoint I^* of the identity operator $I : U \to U$ under the weighted inner products

$$\langle u, \widetilde{u} \rangle = \int_0^\ell u(x)\, \widetilde{u}(x)\, \rho(x)\, dx, \qquad \langle\!\langle v, \widetilde{v} \rangle\!\rangle = \int_0^\ell v(x)\, \widetilde{v}(x)\, c(x)\, dx, \tag{11.96}$$

where $\rho(x) > 0$ and $c(x) > 0$ are positive weight functions, on, respectively, the domain and target copies of U.

11.3.14. Let $c(x) \in C^0[a,b]$ be a continuous function. Prove that the linear multiplication operator $K[u] = c(x)\, u(x)$ is self-adjoint with respect to the L^2 inner product. What sort of boundary conditions need to be imposed?

11.3.15. Compute the adjoint of the derivative operator $v = D[u] = u'$ under the weighted inner products (11.96) on, respectively, the displacement and strain spaces. Verify that all four types of boundary conditions are allowed. Choose one set of boundary conditions and write out the self-adjoint boundary value problem $D^* \circ D[u] = f$. *Remark*: In physics, the weight function $\rho(x)$ can be identified with the density of the bar.

♡**11.3.16.** (a) Determine the adjoint of the differential operator $v = L[u] = u' + 2x\,u$ with respect to the L^2 inner products on $[0,1]$ when subject to the fixed boundary conditions $u(0) = u(1) = 0$.
(b) Is the self-adjoint operator $K = L^* \circ L$ is positive definite? Explain your answer.
(c) Write out the boundary value problem represented by $K[u] = f$.
(d) Find the solution to the boundary value problem when $f(x) = e^{x^2}$. *Hint*: To integrate the differential equation, work with the factored form of the differential operator.
(e) Why can't you impose the free boundary conditions $u'(0) = u'(1) = 0$ in this situation?

♡**11.3.17.** (a) Show that a differential equation of the form $a(x)\, u'' + b(x)\, u' = f(x)$ is in self-adjoint form (11.12) if and only if $b(x) = a'(x)$.
(b) If $b(x) \ne a'(x)$ and $a(x) \ne 0$ everywhere, show that you can multiply the differential equation by a suitable *integrating factor* $\rho(x)$ so that the resulting ordinary differential equation $\rho(x)\, a(x)\, u'' + \rho(x)\, b(x)\, u' = \rho(x)\, f(x)$ is in self-adjoint form.
(c) Place the following differential equations in self-adjoint form, using an *integrating factor* if required:
(i) $-x^2 u'' - 2x\,u' = x - 1$
(ii) $e^x u'' + e^x u' = e^{2x}$
(iii) $u'' + 2u' = 1$
(iv) $-x\,u'' - 3u' = x$
(v) $\cos x\; u'' + \sin x\; u' = \cos x$

11.3.18. Consider the linear operator

$$L[u] = \begin{pmatrix} u' \\ u \end{pmatrix}$$

that maps $u(x) \in C^1$ to the vector-valued function whose components consist of the function and its first derivative.
(a) Compute the adjoint L^* with respect to the L^2 inner products on both the domain and target spaces, subject to the boundary conditions $u(0) = u(1)$.
(b) Write down and solve the self-adjoint boundary value problem $L^* \circ L[u] = x - 1$.

11.3.19. Prove that the complex differential operator

$$L[u] = \mathrm{i}\, \frac{du}{dx}$$

is self-adjoint with respect to the L^2 Hermitian inner product

$$\langle u, v \rangle = \int_0^{2\pi} u(x)\, \overline{v(x)}\, dx$$

on the vector space of continuously differentiable, complex-valued, 2π periodic functions: $u(x + 2\pi) = u(x)$.

11.3.20. In Exercise 7.5.6, you determined the adjoint of the derivative operator D when acting on the space of quadratic polynomials with respect to the L^2 inner product

$$\langle p, q \rangle = \int_0^1 p(x)\, q(x)\, dx.$$

Can you explain why $D^* \ne -D$ in this situation?

Inhomogeneous Boundary Conditions

So far, we have restricted our attention to homogeneous boundary value problems. Inhomogeneous boundary conditions are a little trickier, since the spaces of allowable displacements and allowable strains are no longer vector spaces, and so the abstract theory, as developed in Chapter 7, is not directly applicable.

One way to circumvent this difficulty is to slightly modify the displacement function so as to satisfy homogeneous boundary conditions. Consider, for example, the inhomogeneous Dirichlet boundary value problem

$$K[u] = -\frac{d}{dx}\left(c(x)\,\frac{du}{dx}\right) = f(x), \qquad u(0) = \alpha, \qquad u(\ell) = \beta. \tag{11.97}$$

We shall choose a function $h(x)$ that satisfies the boundary conditions:

$$h(0) = \alpha, \qquad h(\ell) = \beta.$$

Note that we are *not* requiring h to satisfy the differential equation, and so one, but by no means the only, possible choice is the linear interpolating polynomial

$$h(x) = \alpha + \frac{\beta - \alpha}{\ell}\, x. \tag{11.98}$$

Since u and h have the same boundary values, their difference

$$\widetilde{u}(x) = u(x) - h(x) \tag{11.99}$$

satisfies the homogeneous Dirichlet boundary conditions

$$\widetilde{u}(0) = \widetilde{u}(\ell) = 0. \tag{11.100}$$

Moreover, by linearity, $\widetilde{u}$ satisfies the modified equation

$$K[\widetilde{u}\,] = K[u - h] = K[u] - K[h] = f - K[h] \equiv \widetilde{f},$$

or, explicitly,

$$-\frac{d}{dx}\left(c(x)\,\frac{d\widetilde{u}}{dx}\right) = \widetilde{f}(x), \qquad \text{where} \quad \widetilde{f}(x) = f(x) + \frac{d}{dx}\left(c(x)\,\frac{dh}{dx}\right). \tag{11.101}$$

For the particular choice (11.98),

$$\widetilde{f}(x) = f(x) + \frac{\beta - \alpha}{\ell}\, c'(x).$$

Thus, we have managed to convert the original inhomogeneous problem for u into a homogeneous boundary value problem for $\widetilde{u}$. Once we have solved the latter, the solution to the original is simply reconstructed from the formula

$$u(x) = \widetilde{u}(x) + h(x). \tag{11.102}$$

We know that the homogeneous Dirichlet boundary value problem (11.100–101) is positive definite, and so we can characterize its solution by a minimum principle, namely as the minimizer of the quadratic energy functional

$$\mathcal{P}[\widetilde{u}\,] = \tfrac{1}{2}\,\|\widetilde{u}\,'\|^2 - \langle \widetilde{u}, \widetilde{f}\,\rangle = \int_0^\ell \left[\,\tfrac{1}{2}\,c(x)\,\widetilde{u}\,'(x)^2 - \widetilde{f}(x)\,\widetilde{u}(x)\,\right] dx. \tag{11.103}$$

Let us rewrite the minimization principle in terms of the original displacement function $u(x)$. Replacing $\widetilde{u}$ and $\widetilde{f}$ by their formulae (11.99, 101) yields

$$\begin{aligned}\mathcal{P}[\widetilde{u}] &= \tfrac{1}{2}\|u' - h'\|^2 - \langle u - h, f - K[h]\rangle \\ &= \left[\tfrac{1}{2}\|u'\|^2 - \langle u, f\rangle\right] - \left[\langle\!\langle u', h'\rangle\!\rangle - \langle u, K[h]\rangle\right] \\ &\quad + \left[\tfrac{1}{2}\|h'\|^2 + \langle h, f - K[h]\rangle\right] \\ &= \mathcal{P}[u] - \left[\langle\!\langle u', h'\rangle\!\rangle - \langle u, K[h]\rangle\right] + C_0. \end{aligned} \tag{11.104}$$

In the middle expression, the last pair of terms depend only on the initial choice of $h(x)$, and not on $u(x)$; thus, once h has been selected, they can be regarded as a fixed constant, here denoted by C_0. The first pair of terms reproduces the quadratic energy functional (11.92) for the actual displacement $u(x)$. The middle terms can be explicitly evaluated:

$$\begin{aligned}\langle\!\langle u', h'\rangle\!\rangle - \langle u, K[h]\rangle &= \int_0^\ell \left[c(x)\,h'(x)\,u'(x) + \big(c(x)\,h'(x)\big)'\,u(x)\right] dx \\ &= \int_0^\ell \frac{d}{dx}\left[c(x)\,h'(x)\,u(x)\right] dx \\ &= c(\ell)\,h'(\ell)\,u(\ell) - c(0)\,h'(0)\,u(0). \end{aligned} \tag{11.105}$$

In particular, if $u(x)$ satisfies the inhomogeneous Dirichlet boundary conditions $u(0) = \alpha$, $u(\ell) = \beta$, then

$$\langle\!\langle u', h'\rangle\!\rangle - \langle u, K[h]\rangle = c(\ell)\,h'(\ell)\,\beta - c(0)\,h'(0)\,\alpha = C_1$$

also depends only on the interpolating function h and not on u. Therefore,

$$\mathcal{P}[\widetilde{u}] = \mathcal{P}[u] - C_1 + C_0$$

differ by a constant. We conclude that, if the function $\widetilde{u}$ minimizes $\mathcal{P}[\widetilde{u}]$, then $u = \widetilde{u} + h$ necessarily minimizes $\mathcal{P}[u]$. In this manner, we have characterized the solution to the inhomogeneous Dirichlet boundary value problem by the *same* minimization principle.

THEOREM 11.10 The solution $u_\star(x)$ to the Dirichlet boundary value problem

$$-\frac{d}{dx}\left(c(x)\,\frac{du}{dx}\right) = f(x), \qquad u(0) = \alpha, \qquad u(\ell) = \beta,$$

is the unique C^2 function that satisfies the indicated boundary conditions and minimizes the energy functional

$$\mathcal{P}[u] = \int_0^\ell \left[\tfrac{1}{2}\,c(x)\,u'(x)^2 - f(x)\,u(x)\right] dx.$$

Warning: The inhomogeneous mixed boundary value problem is trickier, since the extra terms (11.105) *will* depend upon the value of $u(x)$. The details are worked out in Exercise 11.3.26.

EXERCISES 11.3

11.3.21. Find the function $u_\star(x)$ that minimizes the integral

$$\mathcal{P}[u] = \int_1^2 \left[\frac{x}{2} \left(\frac{du}{dx} \right)^2 + x^2 u \right] dx$$

subject to the boundary conditions $u(1) = 0$, $u(2) = 1$.

11.3.22. For each of the following functionals and associated boundary conditions,

(i) write down a boundary value problem satisfied by the minimizing function, and

(ii) find the minimizing function $u_\star(x)$:

(a) $\displaystyle\int_0^1 \left[\tfrac{1}{2}(u')^2 + u \right] dx, \ u(0) = 2, \ u(1) = 3$

(b) $\displaystyle\int_1^e \left[x\,(u')^2 - 2\,x\,u \right] dx, \ u(1) = 1, \ u(e) = 1$

(c) $\displaystyle\int_{-1}^1 \frac{(u')^2}{x^2+1}\, dx, \ u(-1) = 1, \ u(1) = -1$

(d) $\displaystyle\int_0^1 \left[\tfrac{1}{2} e^{-x} (u')^2 - u \right] dx,$

$u(0) = -1, \ u(1) = 0$

11.3.23. For each the following boundary value problems,

(i) write down a minimization principle, carefully specifying the space of functions, and

(ii) find the solution:

(a) $-u'' = \cos x, \ u(0) = 1, \ u(\pi) = -2$

(b) $\displaystyle -\frac{d}{dx}\left(\frac{1}{1+x^2} \frac{du}{dx} \right) = x^2, \ u(-1) = -1,$ $u(1) = 1$

(c) $e^x (u'' + u') = 1, \ u(0) = 1, \ u(1) = 0$

(d) $x\,u'' + 2u' = 1 - x, \ u(1) = -1, \ u(3) = 2$

11.3.24. Explain how to solve the inhomogeneous boundary value problem $-u'' = f(x)$, $u(0) = \alpha$, $u(1) = \beta$, by using the Green's function (11.59).

11.3.25. A bar 1 meter long has stiffness $c(x) = 1+x$ at position $0 \le x \le 1$. It is subject to an external force $f(x) = 1-x$. The left end of the bar is fixed, while the right end is extended 1 cm.

(a) Write out and solve the boundary value problem governing the displacement of the bar.

(b) Write down the potential energy functional as the corresponding minimization principle.

(c) What is the energy of your solution?

(d) Construct two other possible displacement functions that satisfy the boundary conditions, and show, by direct computation, that they have larger potential energy.

◇11.3.26. Prove that the solution to the mixed boundary value problem

$$-\frac{d}{dx}\left(c(x) \frac{du}{dx} \right) = f(x), \quad u(0) = \alpha, \quad u'(\ell) = \beta,$$

is the unique C^2 function that minimizes the modified energy functional

$$\widetilde{\mathcal{P}}[u] = -\beta\, c(\ell)\, u(\ell) + \int_0^\ell \left[\tfrac{1}{2} c(x)\, u'(x)^2 - f(x)\, u(x) \right] dx \qquad (11.106)$$

when subject to the inhomogeneous boundary conditions. *Hint*: Mimic the derivation of Theorem 11.10.

Remark: Physically, the inhomogeneous Neumann boundary condition $u'(\ell) = \beta$ represents an applied strain at the free end, and contributes an additional term to the total energy of the mechanical system.

11.3.27. Find the function $u(x)$ that minimizes the integral

$$\mathcal{P}[u] = \int_1^2 \left[x(u')^2 + x^2 u \right] dx$$

subject to the boundary conditions $u(1) = 1$, $u'(2) = 0$. *Hint*: Use Exercise 11.3.26.

11.3.28. Prove that the functional

$$\mathcal{P}[u] = \int_0^1 (u')^2\, dx$$

subject to the mixed boundary conditions $u(0) = 0$, $u'(1) = 1$ has no minimizer! Thus, omitting the extra boundary term in (11.106) is a fatal mistake.

◇11.3.29. Suppose

$$\int_0^1 f(x)\, dx = 0.$$

Prove that all solutions to the inhomogeneous Neumann boundary value problem

$$-u'' = f, \quad u'(0) = \alpha, \quad u'(1) = \beta,$$

are minimizers of the modified energy functional

$$\widehat{\mathcal{P}}[u] = \alpha\, u(0) - \beta\, u(b) + \int_a^b \left[\tfrac{1}{2} u'(x)^2 - f(x)\, u(x) \right] dx.$$

11.4 BEAMS AND SPLINES

Unlike a bar, which can only stretch longitudinally, a *beam* is allowed to bend. To keep the geometry simple, we treat the case in which the beam is restricted to the $x\,y$ plane, as sketched in Figure 11.13. Let $0 \le x \le \ell$ represent the reference position along a horizontal beam of length ℓ. To further simplify the physics, we shall ignore stretching, and assume that the "atoms" in the beam can only move in the transverse direction, with $y = u(x)$ representing the vertical displacement of the "atom" that starts out at position x.

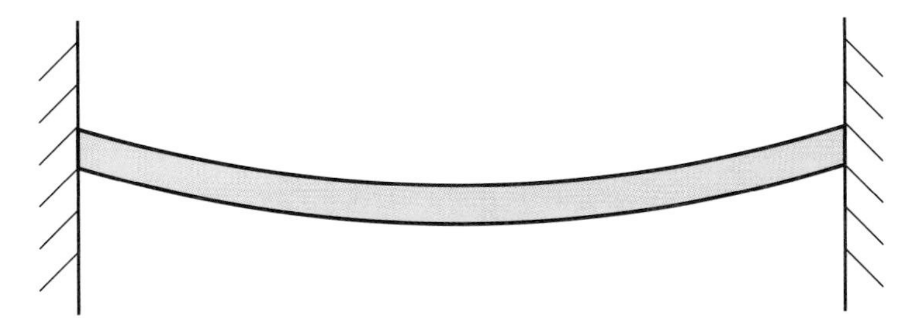

Figure 11.13 Bending of a beam.

The *strain* in a beam depends on how much it is bent. Mathematically, bending is equal to the *curvature** of the graph of the displacement function $u(x)$, and is computed by the usual calculus formula

$$\kappa = \frac{u''}{(1 + (u')^2)^{3/2}} \,. \tag{11.107}$$

Thus, for beams, the strain is a *nonlinear* function of displacement. Since we are only willing to deal with linear systems in this text, we shall suppress the nonlinearity by assuming that the beam is not bent too far; more specifically, we assume that the derivative $u'(x) \ll 1$ is small and so the tangent line is nearly horizontal. Under this assumption, the curvature function (11.107) is replaced by its linear approximation

$$\kappa \approx u''. \tag{11.108}$$

From now on, we will identify $v = D^2[\,u\,] = u''$ as the *strain* in a bending beam. The second derivative operator $L = D^2$ that maps displacement u to strain $v = L[\,u\,]$ thereby describes the beam's intrinsic (linearized) geometry.

The next step is to formulate a constitutive relation between stress and strain. Physically, the *stress* $w(x)$ represents the bending moment of the beam, defined as the product of internal force and angular deflection. Our small bending assumption implies an elastic Hooke's law relation

$$w(x) = c(x)\,v(x) = c(x)\,\frac{d^2u}{dx^2}\,, \tag{11.109}$$

where the proportionality factor $c(x) > 0$ measures the *stiffness* of the beam at the point x. In particular, a uniform beam has constant stiffness, $c(x) \equiv c$.

Finally, the differential equation governing the equilibrium configuration of the beam will follow from a balance of the internal and external forces. To compute the internal force, we appeal to our general equilibrium framework, which tells us

*By definition, [2, 58], the *curvature* of a curve at a point is equal to the reciprocal, $\kappa = 1/r$ of the radius of the osculating circle.

to apply the adjoint of the incidence operator $L = D^2$ to the strain, leading to the force balance law

$$L^*[v] = L^* \circ L[u] = f. \tag{11.110}$$

Let us compute the adjoint. We use the ordinary L^2 inner product on the space of displacements $u(x)$, and adopt a weighted inner product, based on the stiffness function $c(x)$, between strain functions:

$$\langle u, \widetilde{u} \rangle = \int_a^b u(x)\, \widetilde{u}(x)\, dx, \qquad \langle\!\langle v, \widetilde{v} \rangle\!\rangle = \int_a^b v(x)\, \widetilde{v}(x)\, c(x)\, dx. \tag{11.111}$$

According to the general adjoint equation (7.74), we need to equate

$$\int_0^\ell L[u]\, v\, c\, dx = \langle\!\langle L[u], v \rangle\!\rangle = \langle u, L^*[v] \rangle = \int_0^\ell u\, L^*[v]\, dx. \tag{11.112}$$

As before, the computation relies on (in this case two) integrations by parts:

$$\begin{aligned} \langle\!\langle L[u], v \rangle\!\rangle &= \int_0^\ell \frac{d^2u}{dx^2}\, c\, v\, dx \\ &= \left[\frac{du}{dx}\, c\, v \right] \Bigg|_{x=0}^{\ell} - \int_0^\ell \frac{du}{dx}\, \frac{d(c\,v)}{dx}\, dx \\ &= \left[\frac{du}{dx}\, c\, v - u\, \frac{d(c\,v)}{dx} \right] \Bigg|_{x=0}^{\ell} + \int_0^\ell u\, \frac{d^2(c\,v)}{dx^2}\, dx. \end{aligned}$$

Comparing with (11.112), we conclude that $L^*[v] = D^2(c\,v)$ provided the boundary terms vanish:

$$\begin{aligned} \left[\frac{du}{dx}\, c\, v - u\, \frac{d(c\,v)}{dx} \right] \Bigg|_{x=0}^{\ell} &= \left[\frac{du}{dx}\, w - u\, \frac{dw}{dx} \right] \Bigg|_{x=0}^{\ell} \\ &= \big[u'(\ell)\, w(\ell) - u(\ell)\, w'(\ell) \big] - \big[u'(0)\, w(0) - u(0)\, w'(0) \big] = 0. \end{aligned} \tag{11.113}$$

Thus, under suitable boundary conditions, the force balance equations are

$$L^*[v] = \frac{d^2(c\,v)}{dx^2} = f(x). \tag{11.114}$$

A justification of (11.114) based on physical principles can be found in [64]. Combining (11.109, 114), we conclude that the equilibrium configuration of the beam is characterized as a solution to the fourth order ordinary differential equation

$$\frac{d^2}{dx^2} \left(c(x)\, \frac{d^2u}{dx^2} \right) = f(x). \tag{11.115}$$

As such, the general solution will depend upon 4 arbitrary constants, and so we need to impose a total of four boundary conditions—two at each end—in order to uniquely specify the equilibrium displacement. The (homogeneous) boundary conditions should be chosen so as to make the boundary terms in our integration by parts computation vanish, cf. (11.113). There are a variety of ways in which this can be arranged, and the most important possibilities are the following:

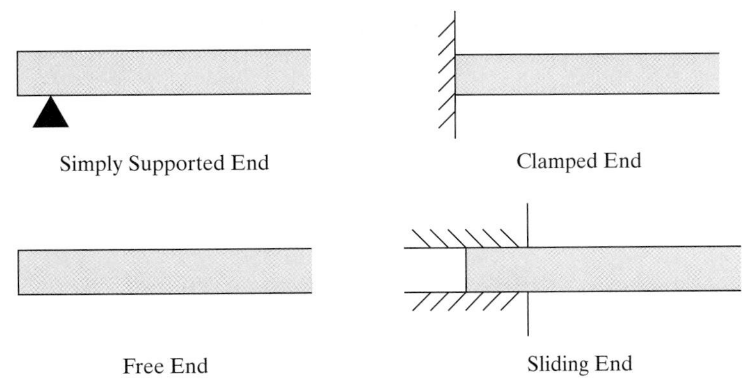

Figure 11.14 Boundary conditions for a beam.

Self-Adjoint Boundary Conditions for a Beam

(a) Simply supported end: $u(0) = w(0) = 0$
(b) Fixed (clamped) end: $u(0) = u'(0) = 0$
(c) Free end: $w(0) = w'(0) = 0$
(d) Sliding end: $u'(0) = w'(0) = 0$

In these conditions, $w(x) = c(x)\, v(x) = c(x)\, u''(x)$ is the stress resulting from the displacement $u(x)$.

A second pair of boundary conditions must be imposed at the other end $x = \ell$. You can mix or match these conditions in any combination—for example, a pair of simply supported ends, or one free end and one fixed end, and so on. Inhomogeneous boundary conditions are also allowed and used to model applied displacements or applied forces at the ends. Yet another option is to consider a bendable circular ring, which is subject to *periodic boundary conditions*

$$u(0) = u(\ell), \quad u'(0) = u'(\ell), \quad w(0) = w(\ell), \quad w'(0) = w'(\ell),$$

indicating that the ends of the beam have been welded together.

Let us concentrate our efforts on the uniform beam, of unit length $\ell = 1$, choosing units so that its stiffness $c(x) \equiv 1$. In the absence of external forcing, the differential equation (11.115) reduces to the elementary fourth order ordinary differential equation

$$\frac{d^4 u}{dx^4} = 0. \tag{11.116}$$

The general solution is an arbitrary cubic polynomial,

$$u = a\, x^3 + b\, x^2 + c\, x + d. \tag{11.117}$$

Let us use this formula to solve a couple of representative boundary value problems.

First, suppose we clamp both ends of the beam, imposing the boundary conditions

$$u(0) = 0, \quad u'(0) = \beta, \quad u(1) = 0, \quad u'(1) = 0, \tag{11.118}$$

so that the left end is tilted by a (small) angle $\tan^{-1} \beta$. We substitute the solution formula (11.117) into the boundary conditions (11.118) and solve for

$$a = \beta, \quad b = -2\,\beta, \quad c = \beta, \quad d = 0.$$

The resulting solution

$$u(x) = \beta\,(x^3 - 2x^2 + x) = \beta\,x(1-x)^2 \tag{11.119}$$

is known as a *Hermite cubic spline*† and is graphed in Figure 11.15.

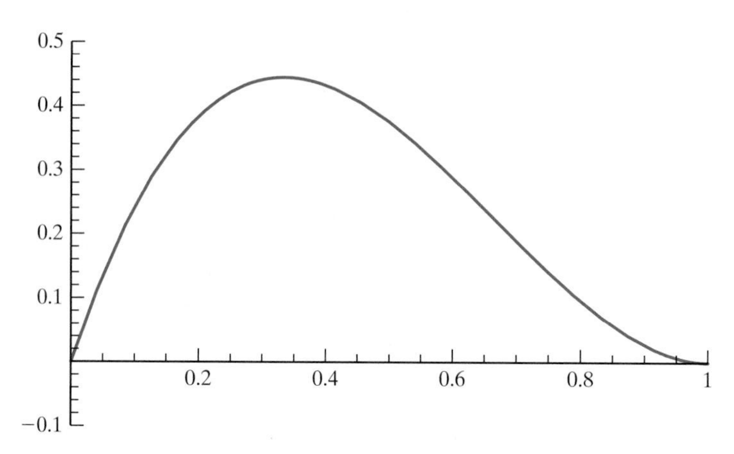

Figure 11.15 Hermite cubic spline.

As a second example, suppose that we raise the left hand end of the beam without tilting, which corresponds to the boundary conditions

$$u(0) = \alpha, \quad u'(0) = 0, \quad u(1) = 0, \quad u'(1) = 0. \tag{11.120}$$

Substituting (11.117) and solving for a, b, c, d, we find that the solution is

$$u(x) = \alpha\,(1-x)^2\,(2x+1). \tag{11.121}$$

Observe that if we simultaneously raise and tilt the left end, so $u(0) = \alpha$, $u'(0) = \beta$, then we can simply use superposition to write the solution as the sum of (11.119) and (11.121):

$$u(x) = \alpha\,(1-x)^2\,(2x+1) + \beta\,x(1-x)^2.$$

To analyze a forced beam, we can adapt the Green's function approach. As we know, the Green's function will depend on the choice of (homogeneous) boundary conditions. Let us treat the case when the beam has two fixed ends, and so

$$u(0) = 0, \quad u'(0) = 0, \quad u(1) = 0, \quad u'(1) = 0. \tag{11.122}$$

To construct the Green's function, we must solve the forced differential equation

$$\frac{d^4u}{dx^4} = \delta(x-y) \tag{11.123}$$

corresponding to a concentrated unit impulse applied at position y along the beam. Integrating (11.123) four times, using (11.45) with $n = 4$, we produce the general solution

$$u(x) = a\,x^3 + b\,x^2 + c\,x + d + \begin{cases} \frac{1}{6}\,(x-y)^3, & x > y, \\ 0, & x < y, \end{cases}$$

to the differential equation (11.123). The boundary conditions (11.122) require

$$u(0) = d = 0, \qquad u(1) = a + b + \tfrac{1}{6}\,(1-y)^3 = 0,$$

$$u'(0) = c = 0, \qquad u'(1) = 3a + 2b + \tfrac{1}{2}\,(1-y)^2 = 0,$$

† We first met Charles Hermite in Section 3.6, and the term "spline" will be explained shortly.

and hence

$$a = \tfrac{1}{3}(1-y)^3 - \tfrac{1}{2}(1-y)^2, \quad b = -\tfrac{1}{2}(1-y)^3 + \tfrac{1}{2}(1-y)^2.$$

Therefore, the Green's function is

$$G(x,y) = \begin{cases} \tfrac{1}{6}x^2(1-y)^2(3y - x - 2xy), & x < y, \\ \tfrac{1}{6}y^2(1-x)^2(3x - y - 2xy), & x > y. \end{cases} \tag{11.124}$$

Observe that, as with the second order bar system, the Green's function is symmetric, $G(x,y) = G(y,x)$, which is a manifestation of the self-adjointness of the underlying boundary value problem, cf. (11.90). Symmetry implies that the deflection of the beam at position x due to a concentrated impulse force applied at position y is the same as the deflection at y due to an impulse force of the same magnitude applied at x.

As a function of x, the Green's function $G(x,y)$ satisfies the homogeneous differential equation (11.116) for all $x \neq y$. Its first and second derivatives $\partial G/\partial x$, $\partial^2 G/\partial x^2$ are continuous, while $\partial^3 G/\partial x^3$ has a unit jump discontinuity at $x = y$, which then produces the required delta function impulse in $\partial^4 G/\partial x^4$. The Green's function (11.124) is graphed in Figure 11.16, and appears to be quite smooth. Evidently, the human eye cannot easily discern discontinuities in third order derivatives!

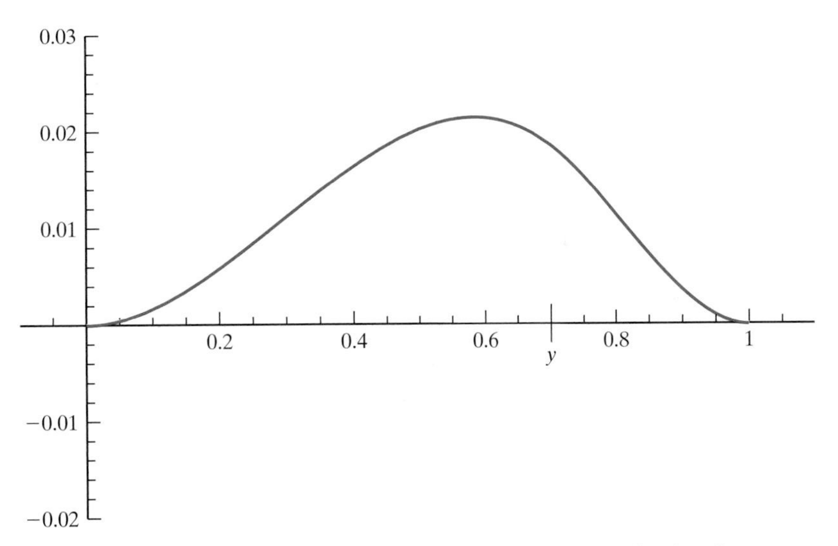

Figure 11.16 Green's function for a beam with two fixed ends.

The solution to the forced boundary value problem

$$\frac{d^4 u}{dx^4} = f(x), \quad u(0) = u'(0) = u(1) = u'(1) = 0, \tag{11.125}$$

for a beam with fixed ends is then obtained by invoking the superposition principle. We view the forcing function as a linear superposition

$$f(x) = \int_0^\ell f(y)\,\delta(x-y)\,dx$$

of impulse delta forces. The solution is the self-same linear superposition of Green's function responses:

$$\begin{aligned} u(x) &= \int_0^1 G(x,y)\,f(y)\,dy \\ &= \frac{1}{6}\int_0^x y^2\,(1-x)^2\,(3x-y-2xy)\,f(y)\,dy \\ &\quad + \frac{1}{6}\int_x^1 x^2\,(1-y)^2\,(3y-x-2xy)\,f(y)\,dy. \end{aligned} \tag{11.126}$$

For example, under a constant unit downwards force $f(x) \equiv 1$, e.g., gravity, the deflection of the beam is given by

$$u(x) = \tfrac{1}{24}x^4 - \tfrac{1}{12}x^3 + \tfrac{1}{24}x^2 = \tfrac{1}{24}x^2(1-x)^2,$$

and graphed in Figure 11.17. Although we could, of course, obtain $u(x)$ by integrating the original differential equation (11.125) directly, writing the solution formula (11.126) as a single integral has evident advantages.

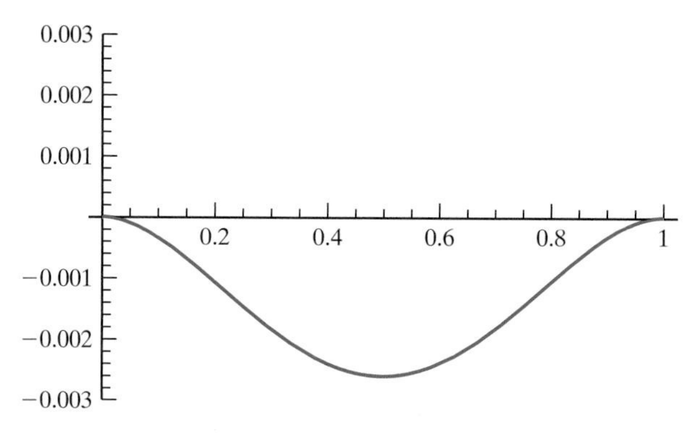

Figure 11.17 Deflection of a uniform beam under gravity.

Since the beam operator $K = L^* \circ L$ assumes the standard self-adjoint, positive semi-definite form, the boundary value problem will be positive definite and hence stable if and only if $\ker L = \ker D^2 = \{\mathbf{0}\}$ when restricted to the space of allowable displacement functions. Since the second derivative D^2 annihilates all linear polynomials

$$u(x) = \alpha + \beta\,x,$$

positive definiteness requires that no non-zero linear polynomials satisfy all four homogeneous boundary conditions. For example, any beam with one fixed end is stable since $u(x) \equiv 0$ is the only linear polynomial that satisfies $u(0) = u'(0) = 0$. On the other hand, a beam with two free ends is unstable since every linear polynomial displacement has zero stress $w(x) = u''(x) \equiv 0$, and so satisfies the boundary conditions $w(0) = w'(0) = w(\ell) = w'(\ell) = 0$. Similarly, a beam with a simply supported plus a free end is not positive definite since $u(x) = \beta\,x$ satisfies the four boundary conditions $u(0) = u'(0) = 0$, $w(\ell) = w'(\ell) = 0$. In the stable cases, the equilibrium solution can be characterized as the unique minimizer of the quadratic energy functional[‡]

$$\mathcal{P}[u] = \tfrac{1}{2}\,\|L[u]\|^2 - \langle u, f\rangle = \int_a^b \left[\,\tfrac{1}{2}c(x)\,u''(x)^2 - f(x)\,u(x)\,\right]dx \tag{11.127}$$

[‡]Keep in mind that the norm on the strain functions $v = L[u] = u''$ is based on the weighted inner product $\langle\!\langle v, \widetilde{v}\rangle\!\rangle$ in (11.111).

among all C^4 functions satisfying the homogeneous boundary conditions. Inhomogeneous boundary conditions require some extra analysis, since the required integration by parts may introduce additional boundary contributions.

EXERCISES 11.4

11.4.1. When possible, find and graph the displacement $u(x)$ and bending moment $w(x)$ of a beam of unit length $\ell = 1$ with constant bending stiffness $c(x) \equiv 1$ subject to a unit gravitational force when it has

(a) two simply supported ends,

(b) one simply supported end and one free end,

(c) one simply supported end and one sliding end,

(d) one fixed end and one free end,

(e) one fixed end and one sliding end.

♡ **11.4.2.** Which of the boundary conditions in Exercise 11.4.1 has resulted in the furthest displacement of the beam? Which leads to the most stress?

11.4.3. Write down a minimization principle, when possible, for each of the problems in Exercise 11.4.1.

11.4.4. For each of the problems in Exercise 11.4.1,

(i) find the Green's function $G(x, y)$, when possible;

(ii) graph $G\left(x, \frac{1}{2}\right)$;

(iii) write down an integral formula for the displacement of the beam under a general external force;

(iv) find the point along the beam that experiences the most displacement under a concentrated unit force.

♡ **11.4.5.** Suppose an elastic bar and an elastic beam have the same length and the same elastic constant $c(x) \equiv 1$; both have fixed ends, and are both subjected to a constant external force of the same magnitude. Which has the greater displacement—the bar or the beam?

11.4.6. An elastic beam of unit length has bending stiffness $c(x) = 1 + x^2$ for $0 \le x \le 1$. Find the strain in the beam under a constant unit force if both ends are simply supported. Would you call this problem statically determinate or indeterminate?

11.4.7. Consider a beam of unit length $\ell = 1$ of constant bending stiffness $c(x) \equiv 1$ with two free ends.

(a) Determine, by direct integration two constraints that must be imposed on the external force $f(x)$ in order that there be an equilibrium solution.

(b) Can you relate your conditions to the Fredholm alternative?

(c) Write down a forcing function that satisfies both constraints, and then find all corresponding solutions to the beam equation.

11.4.8. Answer Exercise 11.4.7 for a beam with one sliding end and one simply supported end.

11.4.9. *True or false*: The Green's function for a beam is symmetric, $G(x, y) = G(y, x)$ if and only if the same boundary conditions are imposed at both ends.

11.4.10. Verify the solution formula (11.126) by direct substitution into the differential equation and boundary conditions (11.125).

♡**11.4.11.** (a) List all possible combinations of the listed self-adjoint homogeneous boundary conditions that lead to a unique equilibrium solution to the boundary value problem when subject to any external force.

(b) List all possible combinations of the self-adjoint homogeneous boundary conditions that lead to a positive definite boundary value problem.

(c) List all possible combinations of the self-adjoint homogeneous boundary conditions for which a Green's function can be constructed.

11.4.12. A beam has unit length $\ell = 1$ and constant bending stiffness $c(x) \equiv 1$. Explain the physical configuration that each of the following inhomogeneous boundary conditions represents, and then, if possible find the equilibrium solution, assuming no external forcing:

(a) $u(0) = 0,\ w(0) = 0,\ u(1) = 1,\ w(1) = 0$

(b) $u(0) = -1,\ u'(0) = 0,\ u(1) = 0,\ u'(1) = 2$

(c) $u(0) = 0,\ u'(0) = 1,\ w(1) = 1,\ w'(1) = 0$

(d) $u(0) = 2,\ w(0) = 0,\ u'(1) = -2,\ w'(1) = 0$

11.4.13. Write down a minimization principle that characterizes the solution to the inhomogeneous boundary value problems in Exercise 11.4.12. *Hint*: Adapt the argument at the end of Section 11.3; see also Exercise 11.3.26 for further hints.

Splines

In pre–CAD (computer aided design) draftsmanship, a *spline* was a long, thin, flexible strip of wood that was used to draw a smooth curve through prescribed points. The points were marked by small pegs, and the spline rested on the pegs. The mathematical theory of splines was first developed in the 1940's by the Romanian mathematician Isaac Schoenberg as an attractive alternative to polynomial interpolation and approximation. Splines have since become ubiquitous in numerical analysis, in geometric modeling, in design and manufacturing, in computer graphics and animation, and in many other applications.

We suppose that the spline coincides with the graph of a function $y = u(x)$. The pegs are fixed at the prescribed data points $(x_0, y_0), \ldots, (x_n, y_n)$, and this requires $u(x)$ to satisfy the interpolation conditions

$$u(x_j) = y_j, \qquad j = 0, \ldots, n. \tag{11.128}$$

The *mesh points* $x_0 < x_1 < x_2 < \cdots < x_n$ are distinct and labeled in increasing order. The spline is modeled as an elastic beam, and so satisfies the homogeneous beam equation (11.116). Therefore,

$$\begin{gathered} u(x) = a_j + b_j (x - x_j) + c_j (x - x_j)^2 + d_j (x - x_j)^3, \\ x_j \le x \le x_{j+1}, \qquad j = 0, \ldots, n-1, \end{gathered} \tag{11.129}$$

is a piecewise cubic function—meaning that, between successive mesh points, it is a cubic polynomial, but not necessarily the same cubic on each subinterval. The fact that we write the formula (11.129) in terms of $x - x_j$ is merely for computational convenience.

Our problem is to determine the coefficients

$$a_j, \quad b_j, \quad c_j, \quad d_j, \qquad j = 0, \ldots, n-1.$$

Since there are n subintervals, there are a total of $4n$ coefficients, and so we require $4n$ equations to uniquely prescribe them. First, we need the spline to satisfy the interpolation conditions (11.128). Since it is defined by a different formula on each side of the mesh point, this results in a total of $2n$ conditions:

$$\begin{aligned} u(x_j^+) &= a_j = y_j, \\ u(x_{j+1}^-) &= a_j + b_j h_j + c_j h_j^2 + d_j h_j^3 = y_{j+1}, \end{aligned} \qquad j = 0, \ldots, n-1, \tag{11.130}$$

where we abbreviate the length of the jth subinterval by

$$h_j = x_{j+1} - x_j.$$

The next step is to require that the spline be as smooth as possible. The interpolation conditions (11.130) guarantee that $u(x)$ is continuous. The condition $u(x) \in \mathrm{C}^1$ be continuously differentiable requires that $u'(x)$ be continuous at the interior mesh points $x_1, \ldots, x_{n-1}$, which imposes the $n - 1$ additional conditions

$$b_j + 2c_j h_j + 3d_j h_j^2 = u'(x_{j+1}^-) = u'(x_{j+1}^+) = b_{j+1}, \qquad j = 0, \ldots, n-2. \tag{11.131}$$

To make $u \in \mathrm{C}^2$, we impose $n - 1$ further conditions

$$2c_j + 6d_j h_j = u''(x_{j+1}^-) = u''(x_{j+1}^+) = 2c_{j+1}, \qquad j = 0, \ldots, n-2, \tag{11.132}$$

to ensure that u'' is continuous at the mesh points. We have now imposed a total of $4n - 2$ conditions, namely (11.130–132), on the $4n$ coefficients. The two missing constraints will come from boundary conditions at the two endpoints, namely x_0 and x_n. There are three common types:

(i) *Natural boundary conditions*: $u''(x_0) = u''(x_n) = 0$, whereby

$$c_0 = 0, \qquad c_{n-1} + 3 d_{n-1} h_{n-1} = 0. \tag{11.133}$$

Physically, this models a simply supported spline that rests freely on the first and last pegs.

(ii) *Clamped boundary conditions*: $u'(x_0) = \alpha$, $u'(x_n) = \beta$, where α, β, which could be 0, are fixed by the user. This requires

$$b_0 = \alpha, \qquad b_{n-1} + 2 c_{n-1} h_{n-1} + 3 d_{n-1} h_{n-1}^2 = \beta. \tag{11.134}$$

This corresponds to clamping the spline at prescribed angles at each end.

(iii) *Periodic boundary conditions*: $u'(x_0) = u'(x_n)$, $u''(x_0) = u''(x_n)$, so that

$$b_0 = b_{n-1} + 2 c_{n-1} h_{n-1} + 3 d_{n-1} h_{n-1}^2, \qquad c_0 = c_{n-1} + 3 d_{n-1} h_{n-1}. \tag{11.135}$$

The periodic case is used to draw smooth closed curves; see below.

THEOREM 11.11 Suppose we are given mesh points $a = x_0 < x_1 < \cdots < x_n = b$, and corresponding data values $y_0, y_1, \ldots, y_n$, along with one of the three kinds of boundary conditions (11.133), (11.134), or (11.135). Then there exists a unique piecewise cubic spline function $u(x) \in \mathrm{C}^2[a, b]$ that interpolates the data, $u(x_0) = y_0, \ldots, u(x_n) = y_n$, and satisfies the boundary conditions.

Proof We first discuss the natural case. The clamped case is left as an exercise for the reader, while the slightly harder periodic case will be treated at the end of the section. The first set of equations in (11.130) says that

$$a_j = y_j, \qquad j = 0, \ldots, n-1. \tag{11.136}$$

Next, (11.132–133) imply that

$$d_j = \frac{c_{j+1} - c_j}{3 h_j}. \tag{11.137}$$

This equation also holds for $j = n - 1$, provided that we make the convention that[§]

$$c_n = 0.$$

We now substitute (11.136–137) into the second set of equations in (11.130), and then solve the resulting equation for

$$b_j = \frac{y_{j+1} - y_j}{h_j} - \frac{(2 c_j + c_{j+1}) h_j}{3}. \tag{11.138}$$

Substituting this result and (11.137) back into (11.131), and simplifying, we find

$$\begin{aligned} h_j c_j + 2(h_j + h_{j+1}) c_{j+1} + h_{j+1} c_{j+2} &= 3 \left[\frac{y_{j+2} - y_{j+1}}{h_{j+1}} - \frac{y_{j+1} - y_j}{h_j} \right] \\ &= z_{j+1}, \end{aligned} \tag{11.139}$$

where we introduce z_{j+1} as a shorthand for the quantity on the right hand side.

In the case of natural boundary conditions, we have

$$c_0 = 0, \qquad c_n = 0,$$

[§]This is merely for convenience; there is no c_n used in the formula for the spline.

and so (11.139) constitutes a tridiagonal linear system

$$A\,\mathbf{c} = \mathbf{z}, \tag{11.140}$$

for the unknown coefficients $\mathbf{c} = (c_1, c_2, \ldots, c_{n-1})^T$, with coefficient matrix

$$A = \begin{pmatrix} 2(h_0+h_1) & h_1 & & & & \\ h_1 & 2(h_1+h_2) & h_2 & & & \\ & h_2 & 2(h_2+h_3) & h_3 & & \\ & & \ddots & \ddots & \ddots & \\ & & & h_{n-3} & 2(h_{n-3}+h_{n-2}) & h_{n-2} \\ & & & & h_{n-2} & 2(h_{n-2}+h_{n-1}) \end{pmatrix} \tag{11.141}$$

and right hand side $\mathbf{z} = (z_1, z_2, \ldots, z_{n-1})^T$. Once (11.141) has been solved, we will then use (11.136–138) to reconstruct the other spline coefficients a_j, b_j, d_j.

The key observation is that the coefficient matrix A is *strictly diagonally dominant*, cf. Definition 10.36, because all the $h_j > 0$, and so

$$2(h_{j-1} + h_j) > h_{j-1} + h_j.$$

Theorem 10.37 implies that A is nonsingular, and hence the tridiagonal linear system has a unique solution $\mathbf{c}$. This suffices to prove the theorem in the case of natural boundary conditions. ■

To actually solve the linear system (11.140), we can apply our tridiagonal solution algorithm (1.66). Let us specialize to the most important case, when the mesh points are equally spaced in the interval $[a, b]$, so that

$$x_j = a + j\,h, \quad \text{where} \quad h = h_j = \frac{b-a}{n}, \qquad j = 0, \ldots, n-1.$$

In this case, the coefficient matrix $A = h\,B$ is equal to h times the tridiagonal matrix

$$B = \begin{pmatrix} 4 & 1 & & & & & \\ 1 & 4 & 1 & & & & \\ & 1 & 4 & 1 & & & \\ & & 1 & 4 & 1 & & \\ & & & 1 & 4 & 1 & \\ & & & & \ddots & \ddots & \ddots \end{pmatrix}$$

that first appeared in Example 1.37. Its $L\,U$ factorization takes on an especially simple form, since most of the entries of L and U are essentially the same decimal numbers. This makes the implementation of the Forward and Back Substitution procedures almost trivial.

Figure 11.18 shows a particular example —a natural spline passing through the data points $(0,0)$, $(1,2)$, $(2,-1)$, $(3,1)$, $(4,0)$. As with the Green's function for the beam, the human eye is unable to discern the discontinuities in its third derivatives, and so the graph appears completely smooth, even though it is, in fact, only C^2.

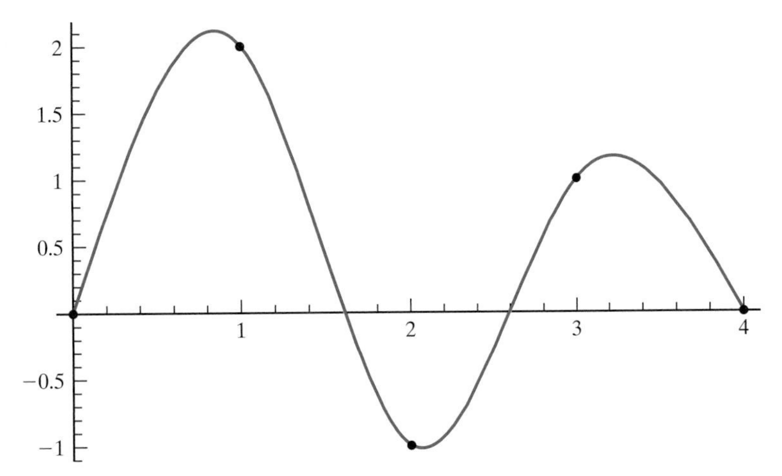

Figure 11.18 A cubic spline.

In the periodic case, we set

$$a_{n+k} = a_n, \qquad b_{n+k} = b_n, \qquad c_{n+k} = c_n, \qquad d_{n+k} = d_n, \qquad z_{n+k} = z_n.$$

With this convention, the basic equations (11.136–139) are the same. In this case, the coefficient matrix for the linear system

$$A\,\mathbf{c} = \mathbf{z}, \qquad \text{with} \quad \mathbf{c} = (c_0, c_1, \ldots, c_{n-1})^T, \quad \mathbf{z} = (z_0, z_1, \ldots, z_{n-1})^T,$$

is of *circulant tridiagonal* form:

$$A = \begin{pmatrix} 2(h_{n-1}+h_0) & h_0 & & & & h_{n-1} \\ h_0 & 2(h_0+h_1) & h_1 & & & \\ & h_1 & 2(h_1+h_2) & h_2 & & \\ & & \ddots & \ddots & \ddots & \\ & & & h_{n-3} & 2(h_{n-3}+h_{n-2}) & h_{n-2} \\ h_{n-1} & & & & h_{n-2} & 2(h_{n-2}+h_{n-1}) \end{pmatrix}. \tag{11.142}$$

Again A is strictly diagonally dominant, and so there is a unique solution $\mathbf{c}$, from which one reconstructs the spline, proving Theorem 11.11 in the periodic case. The $L\,U$ factorization of tridiagonal circulant matrices was discussed in Exercise 1.7.14.

One immediate application of splines is curve fitting in computer aided design and graphics. The basic problem is to draw a smooth parametrized curve $\mathbf{u}(t) = (u(t), v(t))^T$ that passes through a set of prescribed data points $\mathbf{x}_k = (x_k, y_k)^T$ in the plane. We have the freedom to choose the parameter value $t = t_k$ when the curve passes through the kth point; the simplest and most common choice is to set $t_k = k$. We then construct the functions $x = u(t)$ and $y = v(t)$ as cubic splines interpolating the x and y coordinates of the data points, so $u(t_k) = x_k$, $v(t_k) = y_k$. For smooth closed curves, we require that both splines be periodic; for curves with ends, either natural or clamped boundary conditions are used.

Most computer graphics packages include one or more implementations of parametrized spline curves. The same idea also underlies modern font design for laser printing and typography (including the fonts used in this book). The great advantage of spline fonts over their bitmapped counterparts is that they can be readily scaled. Some sample letter shapes parametrized by periodic splines passing through

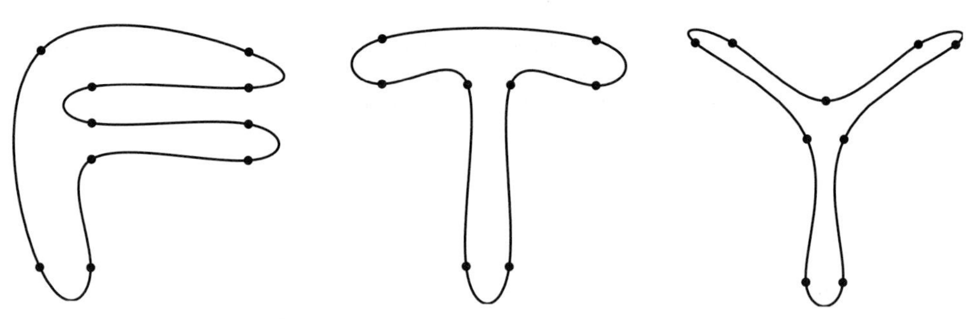

Figure 11.19 Three sample spline letters.

the indicated data points are plotted in Figure 11.19. Better fits can be easily obtained by increasing the number of data points. Various extensions of the basic spline algorithms to space curves and surfaces are an essential component of modern computer graphics, design, and animation, [21, 57].

EXERCISES 11.4

11.4.14. Find and graph the natural cubic spline interpolant for the following data:

(a)

x	-1	0	1
y	-2	1	-1

(b)

x	0	1	2	3
y	1	2	0	1

(c)

x	1	2	4
y	3	0	2

(d)

x	-2	-1	0	1	2
y	5	2	3	-1	1

11.4.15. Repeat Exercise 11.4.14 when the spline has homogeneous clamped boundary conditions.

11.4.16. Find and graph the periodic cubic spline that interpolates the following data:

(a)

x	0	1	2	3
y	1	0	0	1

(b)

x	0	1	2	3
y	1	2	0	1

(c)

x	0	1	2	3	4
y	1	0	0	0	1

(d)

x	-2	-1	0	1	2
y	1	2	-2	-1	1

♠**11.4.17.** (a) Given the known values of $\sin x$ at $x = 0^\circ$, 30°, 45°, 60°, construct the natural cubic spline interpolant.

(b) Compare the accuracy of the spline with the least squares and interpolating polynomials you found in Exercise 4.4.19.

♣**11.4.18.** (a) Using the exact values for $\sqrt{x}$ at $x = 0, \frac{1}{4}, \frac{9}{16}, 1$, construct the natural cubic spline interpolant.

(b) What is the maximal error of the spline on the interval $[0, 1]$?

(c) Compare the error with that of the interpolating cubic polynomial you found in Exercise 4.4.20.

(d) Which is the better approximation?

(e) Answer (d) using the cubic least squares approximant based on the L^2 norm on $[0, 1]$.

♠**11.4.19.** According to Figure 4.10, the interpolating polynomials for the function $1/(1+x^2)$ on the interval $[-3, 3]$ based on equally spaced mesh points are very inaccurate near the ends of the interval. Does the natural spline interpolant based on the same 3, 5, and 11 data points exhibit the same inaccuracy?

♣**11.4.20.** (a) Draw outlines of the block capital letters I, C, H, and S on a sheet of graph paper. Fix several points on the graphs and measure their x and y coordinates.

(b) Use periodic cubic splines $x = u(t)$, $y = v(t)$ to interpolate the coordinates of the data points using equally spaced nodes for the parameter values t_k. Graph the resulting spline letters, and discuss how the method could be used in font design. To get nicer results, you may wish to experiment with different numbers and locations for the points.

♣**11.4.21.** Repeat Exercise 11.4.20, using the Lagrange interpolating polynomials instead of splines to parametrize the curves. Compare the two methods and discuss advantages and disadvantages.

♡**11.4.22.** Let $x_0 < x_1 < \cdots < x_n$. For each $j = 0, \ldots, n$, the jth *cardinal spline* $C_j(x)$ is defined to be the natural cubic spline interpolating the *Lagrange data*

$$y_0 = 0, \quad y_1 = 0, \quad \ldots \quad y_{j-1} = 0, \quad y_j = 1, \quad y_{j+1} = 0, \quad \ldots \quad y_n = 0.$$

(a) Construct and graph the natural cardinal splines corresponding to the nodes $x_0 = 0$, $x_1 = 1$, $x_2 = 2$, and $x_3 = 3$.

(b) Prove that the natural spline that interpolates the data $y_0, \ldots, y_n$ can be uniquely written as a linear combination $u(x) = y_0 C_0(x) + y_1 C_1(x) + \cdots + y_n C_n(x)$ of the cardinal splines.

(c) Explain why the space of natural splines on $n+1$ nodes is a vector space of dimension $n+1$.

(d) Discuss briefly what modifications are required to adapt this method to periodic and to clamped splines.

♡**11.4.23.** A *bell-shaped* or *B–spline* $u = \beta(x)$ interpolates the data

$$\beta(-2) = 0, \quad \beta(-1) = 1, \quad \beta(0) = 4, \quad \beta(1) = 1, \quad \beta(2) = 0.$$

(a) Find the explicit formula for the natural B–spline and plot its graph.

(b) Show that $\beta(x)$ also satisfies the homogeneous clamped boundary conditions $u'(-2) = u'(2) = 0$.

(c) Show that $\beta(x)$ also satisfies the periodic boundary conditions. Thus, for this particular interpolation problem, the natural, clamped, and periodic splines happen to coincide.

(d) Prove that

$$\beta^\star(x) = \begin{cases} \beta(x), & -2 \le x \le 2 \\ 0, & \text{otherwise,} \end{cases}$$

defines a C^2 spline on any interval $[-k, k]$.

♡**11.4.24.** Let $\beta(x)$ denote the B–spline function of Exercise 11.4.23. Assuming $n \ge 4$, let P_n denote the vector space of periodic cubic splines based on the integer nodes $x_j = j$ for $j = 0, \ldots, n$.

(a) Prove that the *B–splines*

$$B_j(x) = \beta\big((x - j - m) \bmod n + m \big), \qquad j = 0, \ldots, n-1,$$

where m denotes the integer part of $n/2$, form a basis for P_n.

(b) Graph the basis periodic B-splines in the case $n = 5$.

(c) Let $u(x)$ denote the periodic spline interpolant for the data values $y_0, \ldots, y_{n-1}$. Explain how to write $u(x) = \alpha_0 B_0(x) + \cdots + \alpha_{n-1} B_{n-1}(x)$ in terms of the B splines by solving a linear system for the coefficients $\alpha_0, \ldots, \alpha_{n-1}$.

(d) Write the periodic spline with $y_0 = y_5 = 0$, $y_1 = 2$, $y_2 = 1$, $y_3 = -1$, $y_4 = -2$, as a linear combination of the periodic basis B–splines $B_0(x), \ldots, B_4(x)$. Plot the resulting periodic spline function.

11.5 STURM–LIOUVILLE BOUNDARY VALUE PROBLEMS

The systems that govern the equilibrium configurations of bars are particular instances of a very general class of second order boundary value problems that was first systematically investigated by the nineteenth century French mathematicians Jacques Sturm and Joseph Liouville. Sturm–Liouville boundary value problems appear in a very wide range of applications, particularly in the analysis of partial differential equations by the method of separation of variables. A partial list of applications includes

(a) heat conduction in non-uniform bars;

(b) vibrations of non-uniform bars and strings;

(c) quantum mechanics and scattering theory;

(d) oscillations of circular membranes (vibrations of drums), cylinders, and spheres; and

(e) thermodynamics of cylindrical and spherical bodies.

In this section, we will show how the class of Sturm–Liouville boundary value problems fits into our general equilibrium framework. Unfortunately, we do not have the analytical tools to delve into any substantial examples in this text; further developments can be found in [6, 47, 68].

The general *Sturm–Liouville boundary value problem* is based on a second order ordinary differential equation of the form

$$-\frac{d}{dx}\left(p(x)\,\frac{du}{dx}\right)+q(x)\,u=-\,p(x)\,\frac{d^2u}{dx^2}-p'(x)\,\frac{du}{dx}+q(x)\,u=f(x), \tag{11.143}$$

which is supplemented by Dirichlet, Neumann, mixed, or periodic boundary conditions. To be specific, let us concentrate on the case of homogeneous Dirichlet boundary conditions

$$u(a)=0, \qquad u(b)=0. \tag{11.144}$$

To avoid singular points of the differential equation, we assume that $p(x)>0$ for all $a\le x\le b$. To ensure positive definiteness of the Sturm–Liouville differential operator, we also assume $q(x)\ge 0$. These assumptions suffice to guarantee existence and uniqueness of the solution to the boundary value problem. A proof of the following theorem can be found in [37].

THEOREM 11.12 Let $p(x)>0$ and $q(x)\ge 0$ for $a\le x\le b$. Then the Sturm–Liouville boundary value problem (11.143–144) admits a unique solution.

Most Sturm–Liouville problems cannot be solved in terms of elementary functions. Indeed, most of the important special functions appearing in mathematical physics, including Bessel functions, Legendre functions, hypergeometric functions, and so on, first arise as solutions to particular Sturm–Liouville equations, [45].

EXAMPLE 11.13 Consider the constant coefficient Sturm–Liouville boundary value problem

$$-\,u''+\omega^2 u=f(x), \qquad u(0)=u(1)=0. \tag{11.145}$$

The functions $p(x)\equiv 1$ and $q(x)\equiv\omega^2>0$ are both constant. We will solve this problem by constructing the Green's function. Thus, we first consider the effect of a delta function inhomogeneity

$$-\,u''+\omega^2 u=\delta(x-y), \qquad u(0)=u(1)=0. \tag{11.146}$$

Rather than try to integrate this differential equation directly, let us appeal to the defining properties of the Green's function. The general solution to the homogeneous equation is a linear combination of the two basic exponentials $e^{\omega x}$ and $e^{-\omega x}$, or better, the hyperbolic functions

$$\cosh\omega x=\frac{e^{\omega x}+e^{-\omega x}}{2}, \qquad \sinh\omega x=\frac{e^{\omega x}-e^{-\omega x}}{2}. \tag{11.147}$$

The solutions satisfying the first boundary condition are multiples of $\sinh\omega x$, while those satisfying the second boundary condition are multiples of $\sinh\omega(1-x)$. Therefore, the solution to (11.146) has the form

$$G(x,y)=\begin{cases} a\,\sinh\omega x, & x<y,\\ b\,\sinh\omega(1-x), & x>y.\end{cases} \tag{11.148}$$

Continuity of $G(x,y)$ at $x=y$ requires

$$a\,\sinh\omega y=b\,\sinh\omega(1-y). \tag{11.149}$$

At $x = y$, the derivative $\partial G/\partial x$ must have a jump discontinuity of magnitude -1 in order that the second derivative term in (11.146) match the delta function. Since

$$\frac{\partial G}{\partial x}(x, y) = \begin{cases} a\,\omega \cosh \omega x, & x < y, \\ -b\,\omega \cosh \omega(1-x), & x > y, \end{cases}$$

the jump condition requires

$$a\,\omega \cosh \omega\, y - 1 = -b\,\omega \cosh \omega\,(1-y). \tag{11.150}$$

If we multiply (11.149) by $\omega \cosh \omega\,(1-y)$ and (11.150) by $\sinh \omega\,(1-y)$ and then add the results together, we find

$$\begin{aligned} \sinh \omega\,(1-y) &= a\,\omega \big[\sinh \omega\, y \cosh \omega\,(1-y) + \cosh \omega\, y \sinh \omega\,(1-y) \big] \\ &= a\,\omega \sinh \omega, \end{aligned}$$

where we used the addition formula for the hyperbolic sine:

$$\sinh(\alpha + \beta) = \sinh \alpha \cosh \beta + \cosh \alpha \sinh \beta. \tag{11.151}$$

Therefore,

$$a = \frac{\sinh \omega\,(1-y)}{\omega \sinh \omega}, \qquad b = \frac{\sinh \omega\, y}{\omega \sinh \omega},$$

and the Green's function is

$$G(x, y) = \begin{cases} \dfrac{\sinh \omega x \sinh \omega\,(1-y)}{\omega \sinh \omega}, & x < y, \\[2ex] \dfrac{\sinh \omega\,(1-x) \sinh \omega\, y}{\omega \sinh \omega}, & x > y. \end{cases} \tag{11.152}$$

Note that $G(x, y) = G(y, x)$ is symmetric, in accordance with the self-adjoint nature of the boundary value problem. A graph appears in Figure 11.20; note that the corner, indicating a discontinuity in the first derivative, appears at the point $x = y$ where the impulse force is applied.

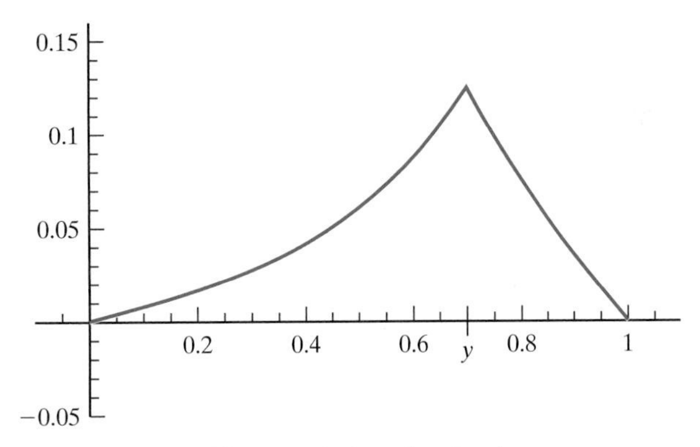

Figure 11.20 Green's function for the constant coefficient Sturm–Liouville problem.

The general solution to the inhomogeneous boundary value problem (11.145) is given by the basic superposition formula (11.62), which becomes

$$\begin{aligned} u(x) &= \int_0^1 G(x, y) f(y)\, dx \\ &= \int_0^x \frac{\sinh \omega\,(1-x) \sinh \omega\, y}{\omega \sinh \omega} f(y)\, dy \\ &\quad + \int_x^1 \frac{\sinh \omega x \sinh \omega\,(1-y)}{\omega \sinh \omega} f(y)\, dy. \end{aligned}$$

For example, under a constant unit force $f(x) \equiv 1$, the solution is

$$\begin{aligned} u(x) &= \int_0^x \frac{\sinh \omega (1-x) \sinh \omega y}{\omega \sinh \omega}\, dy + \int_x^1 \frac{\sinh \omega x \, \sinh \omega (1-y)}{\omega \sinh \omega}\, dy \\ &= \frac{\sinh \omega (1-x) \big(\cosh \omega x - 1 \big)}{\omega^2 \sinh \omega} + \frac{\sinh \omega x \big(\cosh \omega (1-x) - 1 \big)}{\omega^2 \sinh \omega} \\ &= \frac{1}{\omega^2} - \frac{\sinh \omega x + \sinh \omega (1-x)}{\omega^2 \sinh \omega}. \end{aligned} \tag{11.153}$$

For comparative purposes, the reader may wish to rederive this particular solution by a direct calculation, without appealing to the Green's function. ●

To place a Sturm–Liouville boundary value problem in our self-adjoint framework, we proceed as follows. (Exercise 11.5.8 serves to motivate the construction.) Consider the linear operator

$$L[u] = \begin{pmatrix} u' \\ u \end{pmatrix}$$

that maps $u(x)$ to the vector-valued function whose components are the function and its first derivative. For the homogeneous Dirichlet boundary conditions (11.144), the domain of L will be the vector space

$$U = \big\{ \, u(x) \in \mathrm{C}^2[a,b] \;\big|\; u(a) = u(b) = 0 \, \big\}$$

consisting of all twice continuously differentiable functions that vanish at the endpoints. The target space of $L\colon U \to V$ consists of continuously differentiable vector-valued functions $\mathbf{v}(x) = (v_1(x), v_2(x))^T$; we denote this vector space as $V = \mathrm{C}^1([a,b], \mathbb{R}^2)$.

To proceed, we must compute the adjoint of $L\colon U \to V$. To recover the Sturm–Liouville problem, we use the standard L^2 inner product (11.84) on U, but adopt a weighted inner product

$$\langle\!\langle \mathbf{v}, \mathbf{w} \rangle\!\rangle = \int_a^b \big[\, p(x)\, v_1(x)\, w_1(x) + q(x)\, v_2(x)\, w_2(x) \,\big]\, dx, \qquad \mathbf{v} = \begin{pmatrix} v_1 \\ v_2 \end{pmatrix}, \quad \mathbf{w} = \begin{pmatrix} w_1 \\ w_2 \end{pmatrix}, \tag{11.154}$$

on V. The positivity assumptions on the weight functions p, q ensure that this is a *bona fide* inner product. According to the defining equation (7.74), the adjoint $L^*\colon V \to U$ is required to satisfy

$$\langle\!\langle L[u], \mathbf{v} \rangle\!\rangle = \langle u, L^*[\mathbf{v}] \rangle.$$

As usual, the adjoint computation relies on integration by parts. Here, we only need to manipulate the first summand:

$$\begin{aligned} \langle\!\langle L[u], \mathbf{v} \rangle\!\rangle &= \int_a^b \big[\, p\, u'\, v_1 + q\, u\, v_2 \,\big]\, dx \\ &= p(b)\, u(b)\, v_1(b) - p(a)\, u(a)\, v_1(a) + \int_a^b u \, \big[-(p\, v_1)' + q\, v_2 \,\big]\, dx. \end{aligned}$$

The Dirichlet conditions (11.144) ensure that the boundary terms vanish, and therefore,

$$\langle\!\langle L[u], \mathbf{v} \rangle\!\rangle = \int_a^b u \, \big[-(p\, v_1)' + q\, v_2 \,\big]\, dx = \langle u, L^*[\mathbf{v}] \rangle.$$

We conclude that the adjoint operator is given by

$$L^*[\mathbf{v}] = -\frac{d(p\,v_1)}{dx} + q\,v_2.$$

The canonical self-adjoint combination

$$K[u] = L^* \circ L[u] = L^* \begin{pmatrix} u' \\ u \end{pmatrix} = -\frac{d}{dx}\left(p\,\frac{du}{dx}\right) + q\,u \tag{11.155}$$

reproduces the Sturm–Liouville differential operator. Moreover, since $\ker L = \{0\}$ is trivial (why?), the boundary value problem is positive definite. Theorem 7.62 implies that the solution can be characterized as the unique minimizer of the quadratic functional

$$\begin{aligned} \mathcal{P}[u] &= \tfrac{1}{2}\|L[u]\|^2 - \langle u, f\rangle \\ &= \int_a^b \left[\, \tfrac{1}{2}\,p(x)\,u'(x)^2 + \tfrac{1}{2}\,q(x)\,u(x)^2 - f(x)\,u(x)\,\right] dx \end{aligned} \tag{11.156}$$

among all C^2 functions satisfying the prescribed boundary conditions. For example, the solution to the constant coefficient Sturm–Liouville problem (11.145) can be characterized as minimizing the quadratic functional

$$\mathcal{P}[u] = \int_0^1 \left[\, \tfrac{1}{2}\,u'^2 + \tfrac{1}{2}\,\omega^2 u^2 - f\,u\,\right] dx$$

among all C^2 functions satisfying $u(0) = u(1) = 0$.

EXERCISES 11.5

11.5.1. Solve the Sturm–Liouville boundary value problem $-4u'' + 9u = 0$, $u(0) = 0$, $u(2) = 1$. Is your solution unique?

11.5.2. *True or false*: The Neumann boundary value problem $-u'' + u = 1$, $u'(0) = u'(1) = 0$, has a unique solution.

11.5.3. Use the Green's function (11.152) to solve the Sturm–Liouville boundary value problem (11.145) when the forcing function is

$$f(x) = \begin{cases} 1, & 0 \le x < \frac{1}{2}, \\ -1, & \frac{1}{2} < x \le 1. \end{cases}$$

11.5.4. (a) Find the Green's function for the mixed boundary value problem $-u'' + \omega^2 u = f(x)$, $u(0) = 0$, $u'(1) = 0$.

(b) Use your Green's function to find the solution when

$$f(x) = \begin{cases} 1, & 0 \le x < \frac{1}{2}, \\ -1, & \frac{1}{2} < x \le 1. \end{cases}$$

11.5.5. Answer Exercise 11.5.4 for the Neumann boundary conditions $u'(0) = u'(1) = 0$. Explain why, in contrast to the boundary value problem for a bar, this problem does have a Green's function and a unique solution.

◇ **11.5.6.** Explain why the homogeneous Neumann boundary value problem for a Sturm–Liouville operator with $q(x) > 0$ *is* positive definite. What is the minimization principle?

♡ **11.5.7.** (a) For which values of λ does the boundary value problem $u'' + \lambda\,u = h(x)$, $u(0) = 0$, $u(1) = 0$ have a unique solution?

(b) Construct the Green's function for all such λ.

(c) In the non-unique cases, use the Fredholm alternative to find conditions on the forcing function $h(x)$ that guarantee the existence of a solution.

◇ **11.5.8.** Show that each summand in the differential equation for a Sturm–Liouville boundary value problem is self-adjoint. Then use the method of Exercise 7.5.21 to recover the self-adjoint form (11.155).

◇ **11.5.9.** (a) Show that *any* regular second order linear ordinary differential equation

$$a(x)\,u'' + b(x)\,u' + c(x)\,u = f(x),$$

with $a(x) \ne 0$, can be placed in Sturm–Liouville form (11.143) by multiplying the equation by a suitable *integrating factor* $\mu(x)$.

(b) Use this method to place the following differential equations in Sturm–Liouville form:

(i) $-u'' - 2u' + u = e^x$,
(ii) $-x^2 u'' + 2xu' + 3u = 1$,
(iii) $xu'' + (1-x)u' + u = 0$.

(c) In each case, write down a minimization principle that characterizes the solutions to the Dirichlet boundary value problem on the interval $[1, 2]$.

11.5.10. Find all values of $\lambda > 0$ such that the boundary value problem $y'' + 2y' + (\lambda + 1)y = 0$, $y(0) = 0$, $y(2) = 0$, has a nonzero solution $y(x)$.

◇11.5.11. Let $\varepsilon > 0$.

(a) Find the solution $u(x, \varepsilon)$ to the boundary value problem $-u'' + \varepsilon^2 u = 1$, $u(0) = u(1) = 0$.

(b) Show that as $\varepsilon \to 0^+$, the solution $u(x, \varepsilon) \to u_\star(x)$ converges uniformly to the solution to $-u_\star'' = 1$, $u_\star(0) = u_\star(1) = 0$. This is a special case of a general fact about solutions to a *regular perturbation* of a boundary value problem, [36].

◇11.5.12. Let $\varepsilon > 0$.

(a) Find the solution $u(x, \varepsilon)$ to the boundary value problem $-\varepsilon^2 u'' + u = 1$, $u(0) = u(1) = 0$.

(b) Show that as $\varepsilon \to 0^+$, the solution $u(x, \varepsilon) \to u_\star(x)$ converges to the solution to $u_\star = 1$. What happened to the boundary conditions?

(c) Explain why the convergence of this *singular perturbation* is non-uniform.
Remark: The regions close to the end-points are *boundary layers*, and require a separate analysis. Boundary layers appear in fluid flows near solid boundaries, e.g., the flow of air past an airplane wing.

11.5.13. Prove that (11.154) defines an inner product.

◇11.5.14. (a) Prove the addition formula (11.151) for the hyperbolic sine function.

(b) Find a corresponding addition formula for the hyperbolic cosine.

11.6 FINITE ELEMENTS

The characterization of the solution to a positive definite boundary value problem via a minimization principle inspires a very powerful and widely used numerical solution scheme, known as the *finite element method*. In this final section, we give a brief introduction to the finite element method in the context of one-dimensional boundary value problems involving ordinary differential equations.

The underlying idea is strikingly simple. We are trying to find the solution to a boundary value problem by minimizing a quadratic functional $\mathcal{P}[u]$ on an infinite-dimensional vector space U. The solution $u_\star \in U$ to this minimization problem is found by solving a differential equation subject to specified boundary conditions. However, as we learned in Chapter 4.1, minimizing the functional on a *finite-dimensional subspace* $W \subset U$ is a problem in linear algebra, and, moreover, one that we already know how to solve! Of course, restricting the functional $\mathcal{P}[u]$ to the subspace W will not, barring luck, lead to the exact minimizer. Nevertheless, if we choose W to be a sufficiently "large" subspace, the resulting minimizer $w_\star \in W$ may very well provide a reasonable approximation to the actual solution $u_\star \in U$. A rigorous justification of this process, under appropriate hypotheses, requires a full analysis of the finite element method, and we refer the interested reader to [61, 71]. Here we shall concentrate on trying to understand how to apply the method in practice.

To be a bit more explicit, consider the minimization principle

$$\mathcal{P}[u] = \tfrac{1}{2}\, \| L[u] \|^2 - \langle f, u \rangle \tag{11.157}$$

for the linear system

$$K[u] = f, \quad \text{where} \quad K = L^* \circ L,$$

representing our boundary value problem. The norm in (11.157) is typically based on some form of weighted inner product $\langle\langle v, \widetilde{v} \rangle\rangle$ on the space of strains $v = L[u] \in V$, while the inner product term $\langle f, u \rangle$ is typically (although not necessarily) unweighted on the space of displacements $u \in U$. The linear operator takes the self-adjoint form $K = L^* \circ L$, and must be positive definite—which requires $\ker L = \{0\}$. Without the positivity assumption, the boundary value problem has either no solutions, or infinitely many; in either event, the basic finite element method will not apply.

Rather than try to minimize $\mathcal{P}[u]$ on the entire function space U, we now seek to minimize it on a suitably chosen finite-dimensional subspace $W \subset U$. We begin by selecting a basis* $\varphi_1, \ldots, \varphi_n$ of the subspace W. The general element of W is a (uniquely determined) linear combination

$$\varphi(x) = c_1 \varphi_1(x) + \cdots + c_n \varphi_n(x) \tag{11.158}$$

of the basis functions. Our goal, then, is to determine the coefficients $c_1, \ldots, c_n$ such that $\varphi(x)$ minimizes $\mathcal{P}[\varphi]$ among all such functions. Substituting (11.158) into (11.157) and expanding we find

$$\mathcal{P}[\varphi] = \frac{1}{2} \sum_{i,j=1}^{n} m_{ij} c_i c_j - \sum_{i=1}^{n} b_i c_i = \tfrac{1}{2} \mathbf{c}^T M \mathbf{c} - \mathbf{c}^T \mathbf{b}, \tag{11.159}$$

where

(a) $\mathbf{c} = (c_1, c_2, \ldots, c_n)^T$ is the vector of unknown coefficients in (11.158),

(b) $M = (m_{ij})$ is the symmetric $n \times n$ matrix with entries

$$m_{ij} = \langle\!\langle L[\varphi_i], L[\varphi_j] \rangle\!\rangle, \quad i, j = 1, \ldots, n, \tag{11.160}$$

(c) $\mathbf{b} = (b_1, b_2, \ldots, b_n)^T$ is the vector with entries

$$b_i = \langle f, \varphi_i \rangle, \quad i = 1, \ldots, n. \tag{11.161}$$

Observe that, once we specify the basis functions φ_i, the coefficients m_{ij} and b_i are all known quantities. Therefore, we have reduced our original problem to a finite-dimensional problem of minimizing the quadratic function (11.159) over all possible vectors $\mathbf{c} \in \mathbb{R}^n$. The coefficient matrix M is, in fact, positive definite, since, by the preceding computation,

$$\begin{aligned} \mathbf{c}^T M \mathbf{c} = \sum_{i,j=1}^{n} m_{ij} c_i c_j &= \| L[c_1 \varphi_1(x) + \cdots + c_n \varphi_n] \|^2 \\ &= \| L[\varphi] \|^2 > 0 \end{aligned} \tag{11.162}$$

as long as $L[\varphi] \neq 0$. Moreover, our positivity assumption implies that $L[\varphi] = 0$ if and only if $\varphi \equiv 0$, and hence (11.162) is indeed positive for all $\mathbf{c} \neq \mathbf{0}$. We can now invoke the original finite-dimensional minimization Theorem 4.1 to conclude that the unique minimizer to (11.159) is obtained by solving the associated linear system

$$M \mathbf{c} = \mathbf{b}. \tag{11.163}$$

Solving (11.163) relies on some form of Gaussian Elimination, or, alternatively, an iterative linear system solver, e.g., Gauss–Seidel or SOR.

This constitutes the basic abstract setting for the finite element method. The main issue, then, is how to effectively choose the finite-dimensional subspace W. Two candidates that might spring to mind are the space $\mathcal{P}^{(n)}$ of polynomials of degree $\leq n$, or the space $\mathcal{T}^{(n)}$ of trigonometric polynomials of degree $\leq n$. However, for a variety of reasons, neither is well suited to the finite element method. One criterion is that the functions in W must satisfy the relevant boundary conditions—otherwise

---

*In this case, an orthonormal basis is not of any particular help.

W would not be a subspace of U. More importantly, in order to obtain sufficient accuracy, the linear algebraic system (11.163) will typically be rather large, and so the coefficient matrix M should be as sparse as possible, i.e., have lots of zero entries. Otherwise, computing the solution will be too time-consuming to be of much practical value. Such considerations prove to be of absolutely crucial importance when applying the method to solve boundary value problems for partial differential equations in higher dimensions.

The really innovative contribution of the finite element method is to first (paradoxically) *enlarge* the space U of allowable functions upon which to minimize the quadratic functional $\mathcal{P}[u]$. The governing differential equation requires its solutions to have a certain degree of smoothness, whereas the associated minimization principle typically requires only half as many derivatives. Thus, for second order boundary value problems, including bars, (11.92), and general Sturm–Liouville problems, (11.156), $\mathcal{P}[u]$ only involves first order derivatives. It can be rigorously shown that the functional has the *same* minimizing solution, even if one allows (reasonable) functions that fail to have enough derivatives to satisfy the differential equation. Thus, one can try minimizing over subspaces containing fairly "rough" functions. Again, the justification of this method requires some deeper analysis, which lies beyond the scope of this introductory treatment.

For second order boundary value problems, a popular and effective choice of the finite-dimensional subspace is to use continuous, piecewise affine functions. Recall that a function is affine, $f(x) = a\,x + b$, if and only if its graph is a straight line. The function is *piecewise affine* if its graph consists of a finite number of straight line segments; a typical example is plotted in Figure 11.21. Continuity requires that the individual line segments be connected together end to end.

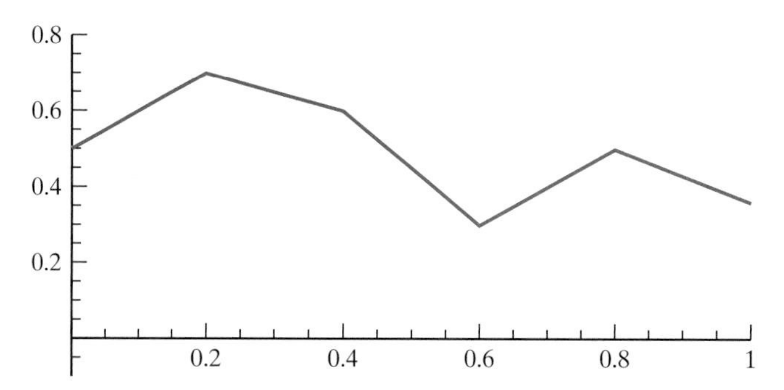

Figure 11.21 A continuous piecewise affine function

Given a boundary value problem on a bounded interval $[a, b]$, let us fix a finite collection of *mesh points*

$$a = x_0 < x_1 < x_2 < \cdots < x_{n-1} < x_n = b.$$

The formulas simplify if one uses equally spaced mesh points, but this is not necessary for the method to apply. Let W denote the vector space consisting of all continuous, piecewise affine functions, with corners at the nodes, that satisfy the homogeneous boundary conditions. To be specific, let us treat the case of Dirichlet (fixed) boundary conditions

$$\varphi(a) = \varphi(b) = 0. \tag{11.164}$$

Thus, on each subinterval

$$\varphi(x) = c_j + b_j(x - x_j), \quad \text{for} \quad x_j \le x \le x_{j+1}, \qquad j = 0, \ldots, n-1.$$

Continuity of $\varphi(x)$ requires

$$c_j = \varphi(x_j^+) = \varphi(x_j^-) = c_{j-1} + b_{j-1} h_{j-1}, \qquad j = 1, \ldots, n-1, \tag{11.165}$$

where $h_{j-1} = x_j - x_{j-1}$ denotes the length of the jth subinterval. The boundary conditions (11.164) require

$$\varphi(a) = c_0 = 0, \qquad \varphi(b) = c_{n-1} + h_{n-1} b_{n-1} = 0. \tag{11.166}$$

The function $\varphi(x)$ involves a total of $2n$ unspecified coefficients $c_0, \ldots, c_{n-1}, b_0, \ldots, b_{n-1}$. The continuity conditions (11.165) and the second boundary condition (11.166) uniquely determine the b_j. The first boundary condition specifies c_0, while the remaining $n-1$ coefficients $c_1 = \varphi(x_1), \ldots, c_{n-1} = \varphi(x_{n-1})$ are arbitrary. We conclude that the finite element subspace W has dimension $n-1$, which is the number of interior mesh points.

Remark: Every function $\varphi(x)$ in our subspace has piecewise constant first derivative $w'(x)$. However, the jump discontinuities in $\varphi'(x)$ imply that its second derivative $\varphi''(x)$ has a delta function impulse at each mesh point, and is therefore far from being a solution to the differential equation. Nevertheless, the finite element minimizer $\varphi_\star(x)$ will, in practice, provide a reasonable approximation to the actual solution $u_\star(x)$.

The most convenient basis for W consists of the *hat functions*, which are continuous, piecewise affine functions that interpolate the same basis data as the Lagrange polynomials (4.47), namely

$$\varphi_j(x_k) = \begin{cases} 1, & j = k, \\ 0, & j \neq k, \end{cases} \qquad \text{for} \quad j = 1, \ldots, n-1, \qquad k = 0, \ldots, n.$$

The graph of a typical hat function appears in Figure 11.22. The explicit formula is easily established:

$$\varphi_j(x) = \begin{cases} \dfrac{x - x_{j-1}}{x_j - x_{j-1}}, & x_{j-1} \le x \le x_j, \\[2ex] \dfrac{x_{j+1} - x}{x_{j+1} - x_j}, & x_j \le x \le x_{j+1}, \\[2ex] 0, & x \le x_{j-1} \ \text{ or } \ x \ge x_{j+1}, \end{cases} \qquad j = 1, \ldots, n-1. \tag{11.167}$$

An advantage of using these basis elements is that the resulting coefficient matrix (11.160) turns out to be tridiagonal. Therefore, the tridiagonal Gaussian Elimination algorithm in (1.66) will rapidly produce the solution to the linear system (11.163). Since the accuracy of the finite element solution increases with the number of mesh points, this solution scheme allows us to easily compute very accurate numerical approximations.

EXAMPLE 11.14 Consider the equilibrium equations

$$K[u] = -\frac{d}{dx}\left(c(x) \frac{du}{dx} \right) = f(x), \qquad 0 < x < \ell,$$

for a non-uniform bar subject to homogeneous Dirichlet boundary conditions. In order to formulate a finite element approximation scheme, we begin with the minimization principle (11.92) based on the quadratic functional

$$\mathcal{P}[u] = \tfrac{1}{2} \, \| u' \|^2 - \langle f, u \rangle = \int_0^\ell \left[\tfrac{1}{2} c(x) u'(x)^2 - f(x) u(x) \right] dx.$$

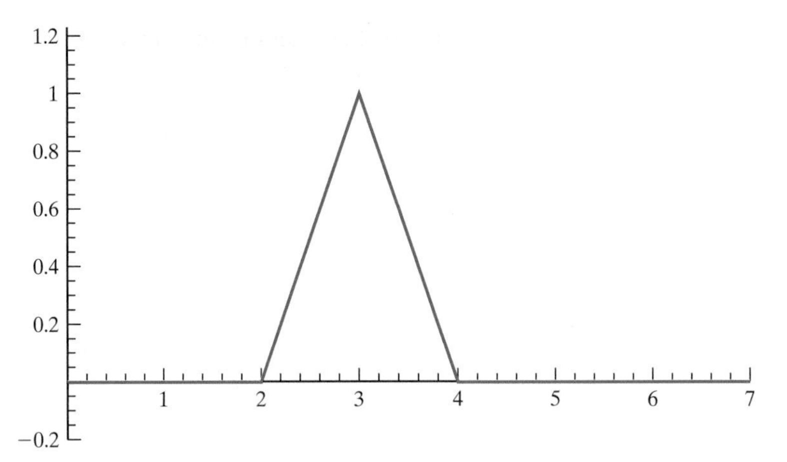

Figure 11.22 A hat function.

We divide the interval $[0, \ell]$ into n equal subintervals, each of length $h = \ell/n$. The resulting uniform mesh has

$$x_j = j\,h = \frac{j\,\ell}{n}, \qquad j = 0, \ldots, n.$$

The corresponding finite element basis hat functions are explicitly given by

$$\varphi_j(x) = \begin{cases} (x - x_{j-1})/h, & x_{j-1} \le x \le x_j, \\ (x_{j+1} - x)/h, & x_j \le x \le x_{j+1}, \\ 0, & \text{otherwise}, \end{cases} \qquad j = 1, \ldots, n-1. \tag{11.168}$$

The associated linear system (11.163) has coefficient matrix entries

$$m_{ij} = \langle\!\langle \varphi_i', \varphi_j' \rangle\!\rangle = \int_0^\ell \varphi_i'(x)\,\varphi_j'(x)\,c(x)\,dx, \qquad i, j = 1, \ldots, n-1.$$

Since the function $\varphi_i(x)$ vanishes except on the interval $x_{i-1} < x < x_{i+1}$, while $\varphi_j(x)$ vanishes outside $x_{j-1} < x < x_{j+1}$, the integral will vanish unless $i = j$ or $i = j \pm 1$. Moreover,

$$\varphi_j'(x) = \begin{cases} 1/h, & x_{j-1} \le x \le x_j, \\ -1/h, & x_j \le x \le x_{j+1}, \\ 0, & \text{otherwise}, \end{cases} \qquad j = 1, \ldots, n-1.$$

Therefore, the coefficient matrix has the tridiagonal form

$$M = \frac{1}{h^2} \begin{pmatrix} s_0 + s_1 & -s_1 & & & & \\ -s_1 & s_1 + s_2 & -s_2 & & & \\ & -s_2 & s_2 + s_3 & -s_3 & & \\ & & \ddots & \ddots & \ddots & \\ & & & -s_{n-3} & s_{n-3} + s_{n-2} & -s_{n-2} \\ & & & & -s_{n-2} & s_{n-2} + s_{n-1} \end{pmatrix}, \tag{11.169}$$

where

$$s_j = \int_{x_j}^{x_{j+1}} c(x)\,dx \tag{11.170}$$

is the total stiffness of the jth subinterval. For example, in the homogeneous case $c(x) \equiv 1$, the coefficient matrix (11.169) reduces to the very special form

$$M = \frac{1}{h} \begin{pmatrix} 2 & -1 & & & & & \\ -1 & 2 & -1 & & & & \\ & -1 & 2 & -1 & & & \\ & & \ddots & \ddots & \ddots & & \\ & & & -1 & 2 & -1 \\ & & & & -1 & 2 \end{pmatrix}. \tag{11.171}$$

The corresponding right hand side has entries

$$\begin{aligned} b_j = \langle f, \varphi_j \rangle &= \int_0^\ell f(x)\,\varphi_j(x)\,dx \\ &= \frac{1}{h}\left[\int_{x_{j-1}}^{x_j} (x - x_{j-1}) f(x)\,dx + \int_{x_j}^{x_{j+1}} (x_{j+1} - x) f(x)\,dx \right]. \end{aligned} \tag{11.172}$$

In this manner, we have assembled the basic ingredients for determining the finite element approximation to the solution. ●

In practice, we do not have to explicitly evaluate the integrals (11.170, 172), but may replace them by a suitably close numerical approximation. When $h \ll 1$ is small, then the integrals are taken over small intervals, and we can use the trapezoid rule[†], [10, 58], to approximate them:

$$s_j \approx \frac{h}{2}\left[\, c(x_j) + c(x_{j+1}) \,\right], \qquad b_j \approx h\, f(x_j). \tag{11.173}$$

EXAMPLE 11.15 Consider the boundary value problem

$$-\frac{d}{dx}(x+1)\frac{du}{dx} = 1, \qquad u(0) = 0, \qquad u(1) = 0. \tag{11.174}$$

The explicit solution is easily found by direct integration:

$$u(x) = -x + \frac{\log(x+1)}{\log 2}. \tag{11.175}$$

It minimizes the associated quadratic functional

$$\mathcal{P}[u] = \int_0^\ell \left[\, \tfrac{1}{2}(x+1)\,u'(x)^2 - u(x) \,\right] dx \tag{11.176}$$

over all possible functions $u \in \mathrm{C}^1$ that satisfy the given boundary conditions. The finite element system (11.163) has coefficient matrix given by (11.169) and right hand side (11.172), where

$$s_j = \int_{x_j}^{x_{j+1}} (1+x)\,dx = h\,(1 + x_j) + \tfrac{1}{2}h^2 = h + h^2\left(j + \frac{1}{2}\right),$$

$$b_j = \int_{x_j}^{x_{j+1}} 1\,dx = h.$$

The resulting solution is plotted in Figure 11.23. The first three graphs contain, respectively, 5, 10, 20 points in the mesh, so that $h = .2, .1, .05$, while the last plots

[†]One might be tempted use more accurate numerical integration procedures, but the improvement in accuracy of the final answer is not very significant, particularly if the step size h is small.

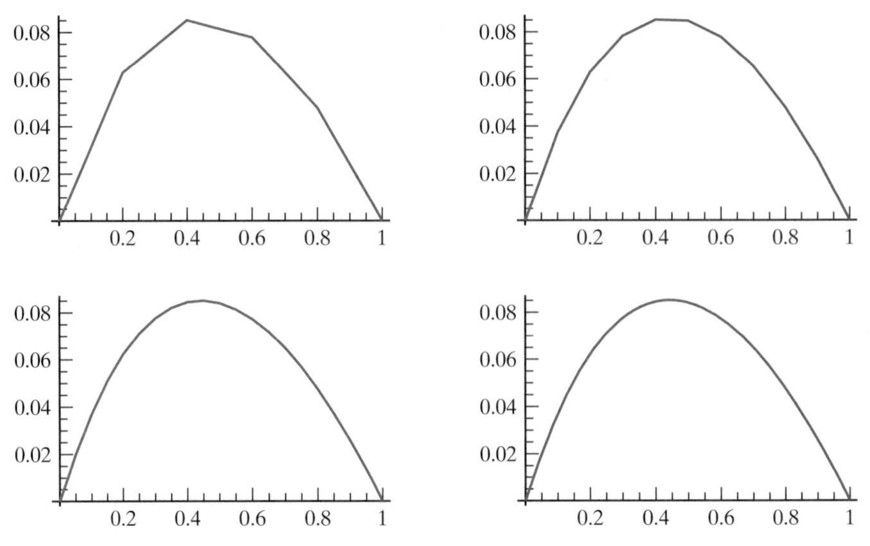

Figure 11.23 Finite element solution to (11.174).

the exact solution (11.175). Even when computed on rather coarse meshes, the finite element approximation is quite respectable. ●

EXAMPLE 11.16 Consider the Sturm–Liouville boundary value problem

$$-u'' + (x+1)u = x\,e^x, \qquad u(0) = 0, \qquad u(1) = 0. \tag{11.177}$$

The solution minimizes the quadratic functional (11.156), which in this particular case is

$$\mathcal{P}[u] = \int_0^1 \left[\, \tfrac{1}{2}\, u'(x)^2 + \tfrac{1}{2}\,(x+1)\,u(x)^2 - e^x\, u(x)\,\right] dx, \tag{11.178}$$

over all functions $u(x)$ that satisfy the boundary conditions. We lay out a uniform mesh of step size $h = 1/n$. The corresponding finite element basis hat functions as in (11.168). The matrix entries are given by[‡]

$$m_{ij} = \int_0^1 \left[\, \varphi_i'(x)\,\varphi_j'(x) + (x+1)\,\varphi_i(x)\,\varphi_j(x)\,\right] dx$$
$$\approx \begin{cases} \dfrac{2}{h} + \dfrac{2h}{3}\,(x_i+1), & i = j, \\[2mm] -\dfrac{1}{h} + \dfrac{h}{6}\,(x_i+1), & |i-j| = 1, \\[2mm] 0, & \text{otherwise}, \end{cases}$$

while

$$b_i = \langle x\,e^x, \varphi_i \rangle = \int_0^1 x\,e^x\,\varphi_i(x)\,dx \;\approx\; x_i\,e^{x_i}\,h.$$

The resulting solution is plotted in Figure 11.24. As in the previous figure, the first three graphs contain, respectively, 5, 10, 20 points in the mesh, while the last plots the exact solution, which can be expressed in terms of Airy functions, cf. [45]. ●

[‡]The integration is made easier by noting that the integrand is zero except on a small subinterval. Since the function $x+1$ (but not φ_i or φ_j) does not vary significantly on this subinterval, it can be approximated by its value $1+x_i$ at a mesh point. A similar simplification is used in the ensuing integral for b_i.

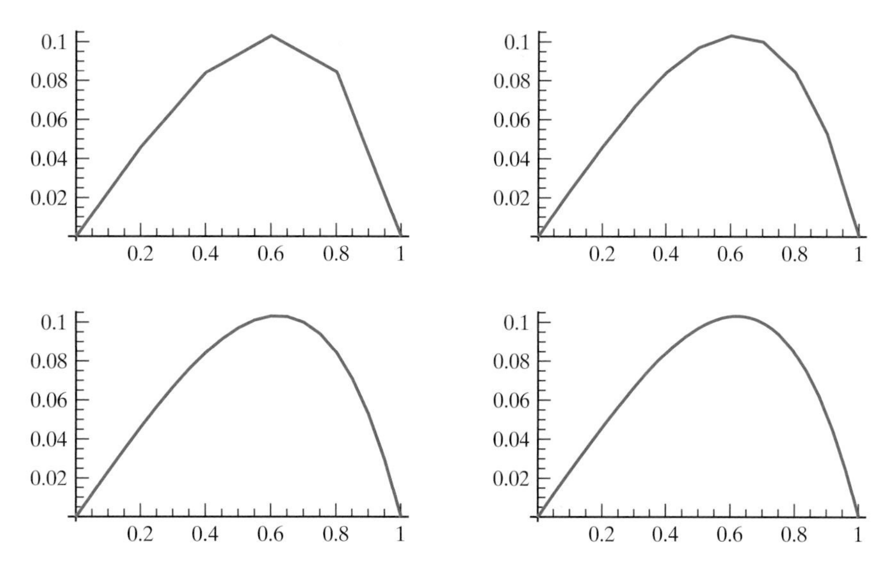

Figure 11.24 Finite element solution to (11.177).

So far, we have only treated homogeneous boundary conditions. An inhomogeneous boundary value problem does not immediately fit into our framework since the set of functions satisfying the boundary conditions does *not* form a vector space. As discussed at the end of Section 11.3, one way to get around this problem is to replace $u(x)$ by $\widetilde{u}(x) = u(x) - h(x)$, where $h(x)$ is any convenient function that satisfies the boundary conditions. For example, for the inhomogeneous Dirichlet conditions

$$u(a) = \alpha, \qquad u(b) = \beta,$$

we can subtract off the affine function

$$h(x) = \frac{(\beta - \alpha)x + \alpha\, b - \beta\, a}{b - a}.$$

Another option is to choose an appropriate combination of elements at the endpoints:

$$h(x) = \alpha\, \varphi_0(x) + \beta\, \varphi_n(x).$$

Linearity implies that the difference $\widetilde{u}(x) = u(x) - h(x)$ satisfies the amended differential equation

$$K[\,\widetilde{u}\,] = \widetilde{f}, \qquad \text{where} \qquad \widetilde{f} = f - K[\,h\,],$$

now supplemented by homogeneous boundary conditions. The modified boundary value problem can then be solved by the standard finite element method. Further details are left as a project for the motivated student.

Finally, one can employ other functions beyond the piecewise affine hat functions (11.167) to span finite element subspace. Another popular choice, which is essential for higher order boundary value problems such as beams, is to use splines. Thus, once we have chosen our mesh points, we can let $\varphi_j(x)$ be the basis B–splines discussed in Exercise 11.4.24. Since $\varphi_j(x) = 0$ for $x \le x_{j-2}$ or $x \ge x_{j+2}$, the resulting coefficient matrix (11.160) is *pentadiagonal*, which means $m_{ij} = 0$ whenever $|\,i - j\,| > 2$. Pentadiagonal matrices are not quite as pleasant as their tridiagonal cousins, but are still rather sparse. Positive definiteness of M implies that an iterative solution technique, e.g., SOR, can effectively and rapidly solve the linear system, and thereby produce the finite element spline approximation to the boundary value problem.

EXERCISES 11.6

♣ **11.6.1.** Use the finite element method to approximate the solution to the boundary value problem

$$-\frac{d}{dx}\left(e^{-x}\frac{du}{dx}\right)=1, \qquad u(0)=u(2)=0.$$

Carefully explain how you are setting up the problem. Plot the resulting solutions and compare your answer with the exact solution. You should use an equally spaced mesh, but try at least three different mesh spacings and compare your results. By inspecting the errors in your various approximations, can you predict how many mesh points would be required for 6 digit accuracy?

♠ **11.6.2.** For each of the following boundary value problems,

(i) Solve the problem exactly.

(ii) Approximate the solution using the finite element method based on 10 equally spaced nodes.

(iii) Compare the graphs of the exact solution and its piecewise affine finite element approximation.

(iv) What is the maximal error?

(a) $-u''=\sigma(x-1),\ u(0)=u(2)=0$

(b) $-\dfrac{d}{dx}\left((1+x)\dfrac{du}{dx}\right)=1,\ u(0)=u(1)=0$

(c) $-\dfrac{d}{dx}\left(x^2\dfrac{du}{dx}\right)=-x,\ u(1)=u(3)=0$

(d) $-\dfrac{d}{dx}\left(e^x\dfrac{du}{dx}\right)=e^x,\ u(-1)=u(1)=0$

♣ **11.6.3.** The exact solution to the boundary value problem $-u''=3x,\ u(0)=u(1)=0$, is $u(x)=\frac{1}{2}x-\frac{1}{2}x^3$.

(a) Use the finite element method based on 5 equally spaced nodes to approximate the solution.

(b) Compare the graphs of the exact solution and its piecewise affine finite element approximation. What is the maximal error

(i) at the mesh points?

(ii) on the entire interval?

(c) Suppose you employ a natural cubic spline to interpolate the sample values of the finite element solution at the nodes. How does that affect the errors in part (c)?

♣ **11.6.4.** (a) Devise a finite element method for solving the mixed boundary value problem

$$-\frac{d}{dx}\left(c(x)\frac{du}{dx}\right)=f(x), \qquad a<x<b,$$
$$u(a)=0, \qquad u'(b)=0.$$

(b) Apply your method to approximate the solution to

$$-\frac{d}{dx}\left((1+x)\frac{du}{dx}\right)=1, \qquad 0<x<1$$
$$u(0)=0, \qquad u'(1)=0,$$

using 10 equally spaced mesh points. Compare your numerical approximation with the exact solution.

♣ **11.6.5.** Consider the periodic boundary value problem $-u''+u=x,\ u(0)=u(2\pi),\ u'(0)=u'(2\pi)$.

(a) Write down the analytic solution.

(b) Write down a minimization principle.

(c) Divide the interval $[0,2\pi]$ into $n=5$ equal subintervals, and let W_n denote the subspace consisting of all piecewise affine functions that satisfy the boundary conditions. What is the dimension of W_n? Write down a basis.

(d) Construct the finite element approximation to the solution to the boundary value problem by minimizing the functional in part (b) on the subspace W_n. Graph the result and compare with the exact solution. What is the maximal error on the interval?

(e) Repeat part (d) for $n=10, 20$, and 40 subintervals, and discuss the convergence of your solutions.

♣ **11.6.6.** Answer Exercise 11.6.5 when the finite element subspace W_n consists of all periodic piecewise affine functions of period 1, so $\varphi(x+1)=\varphi(x)$. Which version is better?

♣ **11.6.7.** Use the method of Exercise 11.6.6 to approximate the solution to the periodic *Mathieu boundary value problem* $-u''+(1+\cos x)u=1$, $u(0)=u(2\pi),\ u'(0)=u'(2\pi)$. *Note*: In this case, the exact solution cannot be written in terms of elementary functions, [45].

11.6.8. Give a direct proof that the finite element matrix (11.171) is positive definite, by either

(a) determining its $L\,D\,L^T$ decomposition, or

(b) finding its eigenvalues, using Exercise 8.2.48.

11.6.9. (a) Is the finite element matrix (11.171) diagonally dominant?

(b) Does the Jacobi iteration scheme converge to the solution to $M\mathbf{c}=\mathbf{b}$? How fast? *Hint*: Use Exercise 8.2.48.

♣**11.6.10.** Consider the boundary value problem solved in Example 11.15. Let W_n be the subspace consisting of all polynomials $u(x)$ of degree $\le n$ satisfying the boundary conditions $u(0) = u(1) = 0$. In this project, we will try to approximate the exact solution to the boundary value problem by minimizing the functional (11.176) on the polynomial subspace W_n.

(a) For each $n = 5$, 10 and 20:

(b) First, determine a basis for W_n.

(c) Set up the minimization problem as a system of linear equations for the coefficients of the polynomial minimizer relative to your basis.

(d) Solve the polynomial minimization problem and graph the solution.

(e) Compare your "polynomial finite element" solution with the exact solution and the piecewise affine finite element solution graphed in Figure 11.23. Discuss what you observe.

♣**11.6.11.** Consider the boundary value problem $-u'' + \lambda u = x$, for $0 < x < \pi$, with $u(0) = 0$, $u(1) = 0$.

(a) For what values of λ does the system have a unique solution?

(b) For which values of λ can you find a minimization principle that characterizes the solution? Is the solution unique for all such values of λ?

(c) Using n equally spaced mesh points, write down the finite element equations for approximating the solution to the boundary value problem. *Note*: Although the finite element construction is only supposed to work when there is a minimization principle, we will consider the resulting linear algebraic system for any value of λ.

(d) For which values of λ does the finite element system have a unique solution? *Hint*: Use Exercise 8.2.48. How do these values compare to those in part (a)?

(e) Select a value of λ for which the solution can be characterized by a minimization principle and verify that the finite element approximation with $n = 10$ approximates the exact solution.

(f) Experiment with other values of λ. Does your finite element solution give a good approximation to the exact solution, when it exists? What happens at values of λ for which the solution does not exist or is not unique?

♡**11.6.12.** Given data points $(x_0, u_0), \ldots, (x_n, u_n)$ with $x_i \neq x_j$, $i \neq j$, prove that there is a unique continuous piecewise affine interpolant $u = f(x)$, so $u_i = f(x_i)$, $i = 0, \ldots, n$.

(a) Prove that the piecewise affine interpolant can be written in the form

$$f(x) = \sum_{i=0}^{n} u_i \, \varphi_i(x),$$

where φ_i denotes the hat function (11.167).

(b) Write down and graph the piecewise affine interpolant for the following data:

x_i	0	1	2	3
u_i	2	3	6	11

◇**11.6.13.** Prove that every continuous piecewise affine function based on the mesh points $x_0 < x_1 < \cdots < x_n$ can be written in the form

$$f(x) = a\,x + b + \sum_{i=0}^{n} c_i \, |x - x_i|, \qquad (11.179)$$

where

$$c_j = \tfrac{1}{2}\big[\, f'(x_j^+) - f'(x_j^-) \,\big] = \frac{f(x_{j+1}) - f(x_j)}{2\,(x_{j+1} - x_j)} - \frac{f(x_j) - f(x_{j-1})}{2\,(x_j - x_{j-1})}.$$

(a) What are the values of a and b?

(b) Write the hat function (11.168) in the form (11.179).

(c) Write the piecewise affine interpolant constructed in Exercise 11.6.12(c), in the form (11.179).
Remark: The multidimensional version of (11.179), in which the absolute value is replaced by the Euclidean norm $\|\mathbf{x}\|$, is the simplest of a powerful new class of multivariate interpolation schemes known as *radial basis functions*, [9].

References

[1] Abraham, R., Marsden, J.E., Ratiu, T., *Manifolds, Tensor Analysis, and Applications*, Springer–Verlag, New York, 1988

[2] Apostol, T.M., *Calculus*, Blaisdell Publishing Co., Waltham, Mass., 1967–69

[3] Baker, G.A., Jr., Graves–Morris, P., *Padé Approximants*, Encyclopedia of Mathematics and Its Applications, v. 59, Cambridge Univ. Press, Cambridge, 1996

[4] Behrends, E., *Introduction to Markov Chains*, Vieweg, Braunschweig/Wiesbaden, Germany, 2000

[5] Bollobás, B., *Graph Theory: an Introductory Course*, Graduate Texts in Mathematics, vol. 63, Springer–Verlag, New York, 1993

[6] Boyce, W.E., DiPrima, R.C., *Elementary Differential Equations and Boundary Value Problems*, 7th ed., John Wiley & Sons, Inc., New York, 2001

[7] Brigham, E.O., *The Fast Fourier Transform*, Prentice–Hall, Inc., Englewood Cliffs, N.J., 1974

[8] Briggs, W.L., Henson, V.E., *The DFT. An Owner's Manual for the Discrete Fourier Transform*, SIAM, Philadelphia, PA, 1995

[9] Buhmann, M.D., *Radial Basis Functions: Theory and Implementations*, Cambridge University Press, Cambridge, 2003

[10] Burden, R.L., Faires, J.D., *Numerical Analysis*, Seventh Edition, Brooks/Cole, Pacific Grove, CA, 2001

[11] Bürgisser, P., Clausen, M., Shokrollahi, M.A., *Algebraic Complexity Theory*, Springer–Verlag, New York, 1997

[12] Buss, S.A., *3D Computer Graphics*, Cambridge University Press, Cambridge, 2003

[13] Cantwell, B.J., *Introduction to Symmetry Analysis*, Cambridge University Press, Cambridge, 2003

[14] Cooley, J.W., Tukey, J.W., An algorithm for the machine computation of complex Fourier series, Math. Comp., 19 (1965) 297–301

[15] Crowe, M.J., *A History of Vector Analysis*, Dover Publ., New York, 1985

[16] Davidson, K.R., Donsig, A.P., *Real Analysis with Real Applications*, Prentice–Hall, Inc., Upper Saddle River, N.J., 2002

[17] DeGroot, M.H., Schervish, M.J., *Probability and Statistics*, 3rd ed., Addison–Wesley, Boston, 2002

[18] Demmel, J.W., *Applied Numerical Linear Algebra*, SIAM, Philadelphia, PA, 1997

[19] Diacu, F., *An Introduction to Differential Equations*, W.H. Freeman, New York, 2000

[20] Durrett, R., *Essentials of Stochastic Processes*, Springer–Verlag, New York, 1999

[21] Farin, G.E., *Curves and Surfaces for CAGD: A Practical Guide*, Academic Press, London, 2002

[22] Fine, B., Rosenberger, G., *The Fundamental Theorem of Algebra*, Undergraduate Texts in Mathematics, Springer–Verlag, New York, 1997

[23] Francis, J.G.F., The QR transformation I, II, Comput. J., 4 (1961–2) 265–271, 332–345

[24] Graver, J.E., Counting on Frameworks: Mathematics to Aid the Design of Rigid Structures, Dolciani Math. Expo. No. 25, Mathematical Association of America, Washington, DC, 2001

[25] Gohberg, I., Koltracht, I., Triangular factors of Cauchy and Vandermonde matrices, Integral Eq. Operator Theory, 26 (1996) 46–59

[26] Goldstein, H., *Classical Mechanics*, Second Edition, Addison–Wesley, Reading, MA, 1980

[27] Golub, G.H, Van Loan, C.F., *Matrix Computations*, Johns Hopkins Univ. Press, Baltimore, 1989

[28] Hale, J.K., *Ordinary Differential Equations*, Second Edition, R. E. Krieger Pub. Co., Huntington, N.Y., 1980

[29] Herrlich, H., Strecker, G.E., *Category Theory, an Introduction*, Allyn and Bacon, Boston, 1973

[30] Herstein, I.N., *Abstract Algebra*, John Wiley & Sons, Inc., New York, 1999

[31] Hestenes, M.R., Stiefel, E., Methods of conjugate gradients for solving linear systems, J. Res. Nat. Bur. Standards, 49 (1952) 409–436

[32] Higham, N.J., *Accuracy and Stability of Numerical Algorithms*, Second Edition, SIAM, Philadelphia, 2002

[33] Hirsch, M.W., Smale, S., *Differential Equations, Dynamical Systems, and Linear Algebra*, Academic Press, New York, 1974

[34] Hoggatt, V.E., Jr., Lind, D.A., The dying rabbit problem, Fib. Quart., 7 (1969) 482–487

[35] Jolliffe, I.T., *Principal Component Analysis*, 2nd ed., Springer–Verlag, New York, 2002

[36] Keener, J.P., *Principles of Applied Mathematics. Transformation and Approximation*, Addison–Wesley Publ. Co., New York, 1988

[37] Keller, H.B., *Numerical Methods for Two-Point Boundary-Value Problems*, Blaisdell, Waltham, MA, 1968

[38] Krall, A.M., *Applied Analysis*, D. Reidel Publishing Co., Boston, 1986

[39] Kublanovskaya, V.N., On some algorithms for the solution of the complete eigenvalue problem, USSR Comput. Math. Math. Phys., 3 (1961) 637–657

[40] Messiah, A., *Quantum Mechanics*, John Wiley & Sons, New York, 1976

[41] Misner, C.W., Thorne, K.S., Wheeler, J.A., *Gravitation*, W.H. Freeman, San Francisco, 1973

[42] Moon, F.C., *Chaotic Vibrations*, John Wiley & Sons, New York, 1987

[43] Murray, R.N., Li, Z.X., Sastry, S.S., *A Mathematical Introduction to Robotic Manipulation*, CRC Press, Boca Raton, FL, 1994

[44] Nilsson, J.W., Riedel, S., *Electric Circuits*, 7th ed., Prentice–Hall, Inc., Upper Saddle River, N.J., 2005

[45] Olver, F.W.J., *Asymptotics and Special Functions*, Academic Press, New York, 1974

[46] Olver, P.J., *Applications of Lie Groups to Differential Equations*, 2nd ed., Graduate Texts in Mathematics, vol. 107, Springer–Verlag, New York, 1993

[47] Olver, P.J., Shakiban, C., *Applied Mathematics*, Prentice–Hall, Inc., Upper Saddle River, N.J., to appear

[48] Ortega, J.M., *Numerical Analysis, a Second Course*, Academic Press, New York, 1972

[49] Oruç, H., Phillips, G. M., Explicit factorization of the Vandermonde matrix, Linear Algebra Appl., 315 (2000) 113–123

[50] Page, L., Brin, S., Motwani, R., Winograd, T., The PageRank citation ranking: bringing order to the web, preprint, Stanford University, 1998

[51] Press, W.H., Teukolsky, S.A., Vetterling, W.T., Flannery, B.P., *Numerical Recipes in C: The Art of Scientific Computing*, 2nd ed., Cambridge University Press, Cambridge, 1995

[52] Reed, M., Simon, B., *Methods of Modern Mathematical Physics*, Academic Press, New York, 1972

[53] Richards, I., Youn, H., *Theory of Distributions: a Non-Technical Introduction*, Cambridge University Press, Cambridge, 1990

[54] Royden, H.L., *Real Analysis*, Macmillan Co., New York, 1988

[55] Salomon, D., *Computer Graphics and Geometric Modeling*, Springer–Verlag, New York, 1999

[56] Sapiro, G., *Geometric Partial Differential Equations and Image Analysis*, Cambridge Univ. Press, Cambridge, 2001

[57] Schumaker, L.L., *Spline Functions: Basic Theory*, John Wiley & Sons, New York, 1981

[58] Stewart, J., *Calculus: Early Transcendentals*, 5th ed., Thomson Brooks Cole, Belmont, CA, 2003

[59] Strang, G., *Introduction to Applied Mathematics*, Wellesley Cambridge Press, Wellesley, Mass., 1986

[60] Strang, G., *Linear Algebra and its Applications*, Third Ed., Harcourt, Brace, Jovanovich, San Diego, 1988

[61] Strang, G., Fix, G.J., *An Analysis of the Finite Element Method*, Prentice–Hall, Inc., Englewood Cliffs, N.J., 1973

[62] Strassen, V., Gaussian elimination is not optimal, Numer. Math., 13 (1969) 354–356

[63] Tannenbaum, P., *Excursions in Modern Mathematics*, 5th ed., Prentice–Hall, Inc., Upper Saddle River, N.J, 2004

[64] Ugural, A.C., Fenster, S.K., *Advanced Strength and Applied Elasticity*, 4th ed., Prentice–Hall, Inc., Upper Saddle River, N.J., 2003

[65] Varga, R.S., *Matrix Iterative Analysis*, 2nd ed., Springer–Verlag, New York, 2000

[66] Walter, G.G., Shen, X., *Wavelets and Other Orthogonal Systems*, 2nd ed., Chapman & Hall/CRC, Boca Raton, Fl, 2001

[67] Watkins, D.S., *Fundamentals of Matrix Computations*, Wiley–Interscience, New York, 2002

[68] Weinberger, H.F., *A First Course in Partial Differential Equations*, Ginn and Co., Waltham, Mass., 1965

[69] Yaglom, I.M., *Felix Klein and Sophus Lie*, Birkhäuser, Boston, 1988

[70] Yale, P.B., *Geometry and Symmetry*, Holden–Day, San Francisco, 1968

[71] Zienkiewicz, O.C., Taylor, R.L., *The Finite Element Method*, 4th ed., McGraw–Hill, New York, 1989

Answers to Selected Exercises

Note: Some answers are left incomplete. For True–False problems, a reason should also be supplied with the answer.

Chapter 1

1.1.1 (b) $u = 1,\ v = -1$;
(d) $u = \frac{1}{3},\ v = -\frac{4}{3},\ w = 0$;
(f) $x = 2,\ y = 2,\ z = -3,\ w = 1$.

1.1.3 (a) With Forward Substitution, we just start with the top equation and work down: $x = -3,\ y = -3,\ z = -22$.

1.2.1 (a) 3×4, (c) 6, (e) $\begin{pmatrix} 0 \\ 2 \\ -6 \end{pmatrix}$.

1.2.2 (a) $\begin{pmatrix} 1 & 2 & 3 \\ 4 & 5 & 6 \\ 7 & 8 & 9 \end{pmatrix}$, (c) $\begin{pmatrix} 1 & 2 & 3 & 4 \\ 4 & 5 & 6 & 7 \\ 7 & 8 & 9 & 3 \end{pmatrix}$,
(e) $\begin{pmatrix} 1 \\ 2 \\ 3 \end{pmatrix}$.

1.2.4 (b) $A = \begin{pmatrix} 6 & 1 \\ 3 & -2 \end{pmatrix}$, $\mathbf{x} = \begin{pmatrix} u \\ v \end{pmatrix}$, $\mathbf{b} = \begin{pmatrix} 5 \\ 5 \end{pmatrix}$;
(d) $A = \begin{pmatrix} 2 & 1 & 2 \\ -1 & 3 & 3 \\ 4 & -3 & 0 \end{pmatrix}$, $\mathbf{x} = \begin{pmatrix} u \\ v \\ w \end{pmatrix}$,
$\mathbf{b} = \begin{pmatrix} 3 \\ -2 \\ 7 \end{pmatrix}$;
(f) $A = \begin{pmatrix} 1 & 0 & 1 & -2 \\ 2 & -1 & 2 & -1 \\ 0 & -6 & -4 & 2 \\ 1 & 3 & 2 & -1 \end{pmatrix}$, $\mathbf{x} = \begin{pmatrix} x \\ y \\ z \\ w \end{pmatrix}$,
$\mathbf{b} = \begin{pmatrix} -3 \\ 3 \\ 2 \\ 1 \end{pmatrix}$.

1.2.5 (a) $x - y = -1,\ 2x + 3y = -3$;
solution: $x = -\frac{6}{5},\ y = -\frac{1}{5}$.
(c)
$3x_1 - x_3 = 1,\ -2x_1 - x_2 = 0,\ x_1 + x_2 - 3x_3 = 1$;
solution: $x_1 = \frac{1}{5},\ x_2 = -\frac{2}{5},\ x_3 = -\frac{2}{5}$.

1.2.7 (b) undefined, (d) undefined,
(f) $\begin{pmatrix} 1 & 11 & 9 \\ 3 & -12 & -12 \\ 7 & 8 & 8 \end{pmatrix}$, (h) $\begin{pmatrix} 9 & -2 & 14 \\ -8 & 6 & -17 \\ 12 & -3 & 28 \end{pmatrix}$.

1.2.9 1, 6, 11, 16.

1.2.11 (a) True

1.2.14 $B = \begin{pmatrix} x & y \\ 0 & x \end{pmatrix}$ where x, y are arbitrary.

1.2.15 (a) $(A + B)^2 = (A + B)(A + B) = AA + AB + BA + BB = A^2 + 2AB + B^2$, since $AB = BA$.
$A = \begin{pmatrix} 1 & 2 \\ 0 & 1 \end{pmatrix}$, $B = \begin{pmatrix} 0 & 0 \\ 1 & 0 \end{pmatrix}$.

1.2.17 $A\,\mathrm{O}_{n\times p} = \mathrm{O}_{m\times p}$, $\mathrm{O}_{l\times m}\,A = \mathrm{O}_{l\times n}$.

1.2.19 False.

1.2.22 (a) A must be a square matrix.
(b) By associativity, $A\,A^2 = A\,A\,A = A^2\,A = A^3$.

1.2.23 $A = \begin{pmatrix} 0 & 1 \\ 0 & 0 \end{pmatrix}$.

1.2.25 (a) $X = \begin{pmatrix} -1 & 1 \\ 3 & -2 \end{pmatrix}$.

1.2.27 (a) $X = \mathrm{O}$.
(b) Example: $A = \begin{pmatrix} 1 & 2 \\ 0 & 1 \end{pmatrix}$, $B = \begin{pmatrix} 3 & 2 \\ -2 & -1 \end{pmatrix}$,
$X = \begin{pmatrix} 1 & 0 \\ 1 & 1 \end{pmatrix}$.

1.2.31 (b) $A\,B - B\,A = \mathrm{O}$ if and only if $A\,B = B\,A$;
(c) (i) $\begin{pmatrix} -1 & 2 \\ 6 & 1 \end{pmatrix}$;
(d) (i) $[\,c\,A + d\,B, C\,] = (c\,A + d\,B)C - C(c\,A + d\,B) = c(A\,C - C\,A) + d(B\,C - C\,B) = c\,[\,A, B\,] + d\,[\,B, C\,]$;
(ii) $[\,A, B\,] = A\,B - B\,A = -\,(B\,A - A\,B) = -\,[\,B, A\,]$..

1.2.36 (a) $S = \begin{pmatrix} 2 & 0 \\ 0 & 2 \end{pmatrix}$, or, more generally, 2 times any of the matrices in part (c).
(c) Any of the matrices $\begin{pmatrix} \pm 1 & 0 \\ 0 & \pm 1 \end{pmatrix}$, $\begin{pmatrix} a & b \\ c & -a \end{pmatrix}$, where a is arbitrary and $bc = 1 - a^2$.

1.3.1 (a) $\left(\begin{array}{cc|c} 1 & 7 & 4 \\ -2 & -9 & 2 \end{array}\right)$; $x_1 = -10,\ x_2 = 2$.
(c) $\left(\begin{array}{ccc|c} 1 & -2 & 1 & 0 \\ 0 & 2 & -8 & 8 \\ -4 & 5 & 9 & -9 \end{array}\right)$;
$x = 29,\ y = 16,\ z = 3$.
(e) $\left(\begin{array}{cccc|c} 1 & 0 & -2 & 0 & -1 \\ 0 & 1 & 0 & -1 & 2 \\ 0 & -3 & 2 & 0 & 0 \\ -4 & 0 & 0 & 7 & -5 \end{array}\right)$;
$x_1 = -4,\ x_2 = -1,\ x_3 = -\frac{3}{2},\ x_4 = -3$.

1.3.2 (a) $3x + 2y = 2,\ -4x - 3y = -1$;
solution: $x = 4,\ y = -5$,

(c) $3x - y + 2z = -3$, $-2y - 5z = -1$, $6x - 2y + z = -3$; solution: $x = \frac{2}{3}$, $y = 3$, $z = -1$.

1.3.3 (b) $u = 1$, $v = -1$;
(d) $x_1 = \frac{11}{3}$, $x_2 = -\frac{10}{3}$, $x_3 = -\frac{2}{3}$;
(f) $a = \frac{1}{3}$, $b = 0$, $c = \frac{4}{3}$, $d = -\frac{2}{3}$.

1.3.5 (a) regular, (d) not regular.

1.3.6 (a) $x = 1 - 2\,\mathrm{i}$, $y = 1$. (c) $x = \frac{1}{2}$, $y = -\frac{1}{4} + \frac{3}{4}\,\mathrm{i}$.

1.3.12 (a) Set $l_{ij} = \begin{cases} a_{ij}, & i < j, \\ 0, & i \ge j, \end{cases}$ $u_{ij} = \begin{cases} a_{ij}, & i > j, \\ 0, & i \le j, \end{cases}$

$d_{ij} = \begin{cases} a_{ij}, & i = j, \\ 0, & i \ne j. \end{cases}$

(b) $L = \begin{pmatrix} 0 & 0 & 0 \\ 1 & 0 & 0 \\ -2 & 0 & 0 \end{pmatrix}$, $D = \begin{pmatrix} 3 & 0 & 0 \\ 0 & -4 & 0 \\ 0 & 0 & 5 \end{pmatrix}$,

$U = \begin{pmatrix} 0 & 1 & -1 \\ 0 & 0 & 2 \\ 0 & 0 & 0 \end{pmatrix}$.

1.3.14 (a) Add -2 times the second row to the first row of a $2 \times n$ matrix.
(c) Add -5 times the third row to the second row of a $3 \times n$ matrix.

1.3.15 (a) $\begin{pmatrix} 1 & 0 & 0 & 0 \\ 0 & 1 & 0 & 0 \\ 0 & 0 & 1 & 0 \\ 0 & 0 & 1 & 1 \end{pmatrix}$, (c) $\begin{pmatrix} 1 & 0 & 0 & 3 \\ 0 & 1 & 0 & 0 \\ 0 & 0 & 1 & 0 \\ 0 & 0 & 0 & 1 \end{pmatrix}$.

1.3.17 $E_3 E_2 E_1 = \begin{pmatrix} 1 & 0 & 0 \\ -2 & 1 & 0 \\ -2 & \frac{1}{2} & 1 \end{pmatrix}$,

$E_1 E_2 E_3 = \begin{pmatrix} 1 & 0 & 0 \\ -2 & 1 & 0 \\ -1 & \frac{1}{2} & 1 \end{pmatrix}$.

The second is easier to predict since its entries are the same as the corresponding entries of the E_i.

1.3.19 (a) upper triangular, (d) special lower triangular.

1.3.20 (a) $a_{ij} = 0$ for all $i \ne j$.
(c) $a_{ij} = 0$ for all $i < j$ and $a_{ii} = 1$ for all i.

1.3.22 (a) $L = \begin{pmatrix} 1 & 0 \\ -1 & 1 \end{pmatrix}$, $U = \begin{pmatrix} 1 & 3 \\ 0 & 3 \end{pmatrix}$,

(c) $L = \begin{pmatrix} 1 & 0 & 0 \\ -1 & 1 & 0 \\ 1 & 0 & 1 \end{pmatrix}$, $U = \begin{pmatrix} -1 & 1 & -1 \\ 0 & 2 & 0 \\ 0 & 0 & 3 \end{pmatrix}$,

(e) $L = \begin{pmatrix} 1 & 0 & 0 \\ -2 & 1 & 0 \\ -1 & -1 & 1 \end{pmatrix}$,

$U = \begin{pmatrix} -1 & 0 & 0 \\ 0 & -3 & 0 \\ 0 & 0 & 2 \end{pmatrix}$,

(g) $L = \begin{pmatrix} 1 & 0 & 0 & 0 \\ 0 & 1 & 0 & 0 \\ -1 & \frac{3}{2} & 1 & 0 \\ 0 & -\frac{1}{2} & 3 & 1 \end{pmatrix}$,

$U = \begin{pmatrix} 1 & 0 & -1 & 0 \\ 0 & 2 & -1 & -1 \\ 0 & 0 & \frac{1}{2} & \frac{7}{2} \\ 0 & 0 & 0 & -10 \end{pmatrix}$.

1.3.24 (a) $\begin{pmatrix} 1 & 0 & 0 & 0 \\ 2 & 1 & 0 & 0 \\ 3 & 4 & 1 & 0 \\ 5 & 6 & 7 & 1 \end{pmatrix}$

(b) (1) Add -2 times first row to second row.
(2) Add -3 times first row to third row.
(3) Add -5 times first row to fourth row.
(4) Add -4 times second row to third row.
(5) Add -6 times second row to fourth row.
(6) Add -7 times third row to fourth row.

1.3.26 False.

1.3.29 $\begin{pmatrix} 0 & 1 \\ 1 & 0 \end{pmatrix} = \begin{pmatrix} 1 & 0 \\ a & 1 \end{pmatrix} \begin{pmatrix} x & y \\ 0 & z \end{pmatrix} = \begin{pmatrix} x & y \\ a x & a y + z \end{pmatrix}$ implies $x = 0$ and $a x = 1$, which is impossible.

1.3.31 (b) $\mathbf{x} = \begin{pmatrix} \frac{1}{4} \\ \frac{1}{4} \end{pmatrix}$, (d) $\mathbf{x} = \begin{pmatrix} -\frac{4}{7} \\ \frac{2}{7} \\ \frac{5}{7} \end{pmatrix}$,

(f) $\mathbf{x} = \begin{pmatrix} 0 \\ 1 \\ -1 \end{pmatrix}$, (h) $\mathbf{x} = \begin{pmatrix} -\frac{37}{12} \\ -\frac{17}{12} \\ \frac{1}{4} \\ 2 \end{pmatrix}$.

1.3.32 (a) $L = \begin{pmatrix} 1 & 0 \\ -3 & 1 \end{pmatrix}$, $U = \begin{pmatrix} -1 & 3 \\ 0 & 11 \end{pmatrix}$;

$\mathbf{x}_1 = \begin{pmatrix} -\frac{5}{11} \\ \frac{2}{11} \end{pmatrix}$, $\mathbf{x}_2 = \begin{pmatrix} 1 \\ 1 \end{pmatrix}$, $\mathbf{x}_3 = \begin{pmatrix} \frac{9}{11} \\ \frac{3}{11} \end{pmatrix}$;

(c) $L = \begin{pmatrix} 1 & 0 & 0 \\ -\frac{2}{3} & 1 & 0 \\ \frac{2}{9} & \frac{5}{3} & 1 \end{pmatrix}$, $U = \begin{pmatrix} 9 & -2 & -1 \\ 0 & -\frac{1}{3} & \frac{1}{3} \\ 0 & 0 & -\frac{1}{3} \end{pmatrix}$;

$\mathbf{x}_1 = \begin{pmatrix} 1 \\ 2 \\ 3 \end{pmatrix}$, $\mathbf{x}_2 = \begin{pmatrix} -2 \\ -9 \\ -1 \end{pmatrix}$;

(e) $L = \begin{pmatrix} 1 & 0 & 0 & 0 \\ 0 & 1 & 0 & 0 \\ -1 & \frac{3}{2} & 1 & 0 \\ 0 & -\frac{1}{2} & -1 & 1 \end{pmatrix}$,

$U = \begin{pmatrix} 1 & 0 & -1 & 0 \\ 0 & 2 & 3 & -1 \\ 0 & 0 & -\frac{7}{2} & \frac{7}{2} \\ 0 & 0 & 0 & 4 \end{pmatrix}$;

$\mathbf{x}_1 = \begin{pmatrix} \frac{5}{4} \\ -\frac{1}{4} \\ \frac{1}{4} \\ \frac{1}{4} \end{pmatrix}, \mathbf{x}_2 = \begin{pmatrix} \frac{1}{14} \\ -\frac{5}{14} \\ \frac{1}{14} \\ \frac{1}{2} \end{pmatrix}$.

1.4.1 (a) nonsingular, (c) nonsingular, (e) singular, (g) singular.

1.4.2 (a) regular and nonsingular, (c) nonsingular.

1.4.3 (b) $x_1 = 0, x_2 = -1, x_3 = 2$;
(d) $x = -\frac{13}{2}, y = -\frac{9}{2}, z = -1, w = -3$.

1.4.6 True.

1.4.9 (a) $P_1 = \begin{pmatrix} 1 & 0 & 0 & 0 \\ 0 & 0 & 0 & 1 \\ 0 & 0 & 1 & 0 \\ 0 & 1 & 0 & 0 \end{pmatrix}$,

(b) $P_2 = \begin{pmatrix} 0 & 0 & 0 & 1 \\ 0 & 1 & 0 & 0 \\ 0 & 0 & 1 & 0 \\ 1 & 0 & 0 & 0 \end{pmatrix}$,

(c) No. (d) $P_1 P_2$ permutes the rows in the order 4, 1, 3, 2, while $P_2 P_1$ leaves them in the order 2, 4, 3, 1

1.4.10 (a) $\begin{pmatrix} 0 & 1 & 0 \\ 0 & 0 & 1 \\ 1 & 0 & 0 \end{pmatrix}$, (c) $\begin{pmatrix} 0 & 1 & 0 & 0 \\ 1 & 0 & 0 & 0 \\ 0 & 0 & 0 & 1 \\ 0 & 0 & 1 & 0 \end{pmatrix}$.

1.4.12 (b) $\begin{pmatrix} 0 & 1 & 0 & 0 \\ 1 & 0 & 0 & 0 \\ 0 & 0 & 0 & 1 \\ 0 & 0 & 1 & 0 \end{pmatrix}, \begin{pmatrix} 1 & 0 & 0 & 0 \\ 0 & 0 & 0 & 1 \\ 0 & 1 & 0 & 0 \\ 0 & 0 & 1 & 0 \end{pmatrix}$,
$\begin{pmatrix} 0 & 0 & 0 & 1 \\ 0 & 1 & 0 & 0 \\ 1 & 0 & 0 & 0 \\ 0 & 0 & 1 & 0 \end{pmatrix}, \begin{pmatrix} 1 & 0 & 0 & 0 \\ 0 & 1 & 0 & 0 \\ 0 & 0 & 0 & 1 \\ 0 & 0 & 1 & 0 \end{pmatrix}$,
$\begin{pmatrix} 0 & 0 & 0 & 1 \\ 1 & 0 & 0 & 0 \\ 0 & 1 & 0 & 0 \\ 0 & 0 & 1 & 0 \end{pmatrix}, \begin{pmatrix} 0 & 1 & 0 & 0 \\ 0 & 0 & 0 & 1 \\ 1 & 0 & 0 & 0 \\ 0 & 0 & 1 & 0 \end{pmatrix}$.

1.4.13 (a) True. (c) False.

1.4.15 (a) $\begin{pmatrix} 1 & 0 & 0 \\ 0 & 0 & 1 \\ 0 & 1 & 0 \end{pmatrix}$. (b) True.

1.4.16 (b) (i) $\begin{pmatrix} 0 & 1 & 0 \\ 1 & 0 & 0 \\ 0 & 0 & 1 \end{pmatrix}$—elementary matrix;

(iii) $\begin{pmatrix} 1 & 0 & 0 & 0 \\ 0 & 0 & 1 & 0 \\ 0 & 0 & 0 & 1 \\ 0 & 1 & 0 & 0 \end{pmatrix}$—not elementary.

(c) (i) $\begin{pmatrix} 1 & 2 & 3 \\ 2 & 3 & 1 \end{pmatrix}$, (iii) $\begin{pmatrix} 1 & 2 & 3 & 4 \\ 4 & 1 & 2 & 3 \end{pmatrix}$.

1.4.19 (a) $\begin{pmatrix} 0 & 1 \\ 1 & 0 \end{pmatrix}\begin{pmatrix} 0 & 1 \\ 2 & -1 \end{pmatrix} = \begin{pmatrix} 1 & 0 \\ 0 & 1 \end{pmatrix}\begin{pmatrix} 2 & -1 \\ 0 & 1 \end{pmatrix}$;
$\mathbf{x} = \begin{pmatrix} \frac{5}{2} \\ 3 \end{pmatrix}$.

(c) $\begin{pmatrix} 0 & 0 & 1 \\ 1 & 0 & 0 \\ 0 & 1 & 0 \end{pmatrix}\begin{pmatrix} 0 & 1 & -3 \\ 0 & 2 & 3 \\ 1 & 0 & 2 \end{pmatrix}$
$= \begin{pmatrix} 1 & 0 & 0 \\ 0 & 1 & 0 \\ 0 & 2 & 1 \end{pmatrix}\begin{pmatrix} 1 & 0 & 2 \\ 0 & 1 & -3 \\ 0 & 0 & 9 \end{pmatrix}; \mathbf{x} = \begin{pmatrix} -1 \\ 1 \\ 0 \end{pmatrix}$.

(e) $\begin{pmatrix} 0 & 0 & 1 & 0 \\ 1 & 0 & 0 & 0 \\ 0 & 1 & 0 & 0 \\ 0 & 0 & 0 & 1 \end{pmatrix}\begin{pmatrix} 0 & 1 & 0 & 0 \\ 2 & 3 & 1 & 0 \\ 1 & 4 & -1 & 2 \\ 7 & -1 & 2 & 3 \end{pmatrix}$
$= \begin{pmatrix} 1 & 0 & 0 & 0 \\ 0 & 1 & 0 & 0 \\ 2 & -5 & 1 & 0 \\ 7 & -29 & 3 & 1 \end{pmatrix}\begin{pmatrix} 1 & 4 & -1 & 2 \\ 0 & 1 & 0 & 0 \\ 0 & 0 & 3 & -4 \\ 0 & 0 & 0 & 1 \end{pmatrix}$;
$\mathbf{x} = \begin{pmatrix} -1 \\ -1 \\ 1 \\ 3 \end{pmatrix}$.

1.4.20 (b) $\begin{pmatrix} 0 & 0 & 1 & 0 \\ 0 & 1 & 0 & 0 \\ 1 & 0 & 0 & 0 \\ 0 & 0 & 0 & 1 \end{pmatrix}\begin{pmatrix} 0 & 1 & -1 & 1 \\ 0 & 1 & 1 & 0 \\ 1 & -1 & 1 & -3 \\ 1 & 2 & -1 & 1 \end{pmatrix}$
$= \begin{pmatrix} 1 & 0 & 0 & 0 \\ 0 & 1 & 0 & 0 \\ 0 & 1 & 1 & 0 \\ 1 & 3 & \frac{5}{2} & 1 \end{pmatrix}\begin{pmatrix} 1 & -1 & 1 & -3 \\ 0 & 1 & 1 & 0 \\ 0 & 0 & -2 & 1 \\ 0 & 0 & 0 & \frac{3}{2} \end{pmatrix}$;
$x = 4, y = 0, z = 1, w = 1$.

1.4.22 $\begin{pmatrix} 0 & 1 & 0 \\ 1 & 0 & 0 \\ 0 & 0 & 1 \end{pmatrix}\begin{pmatrix} 0 & 1 & 2 \\ 1 & 0 & -1 \\ 1 & 1 & 3 \end{pmatrix}$
$= \begin{pmatrix} 1 & 0 & 0 \\ 0 & 1 & 0 \\ 1 & 1 & 1 \end{pmatrix}\begin{pmatrix} 1 & 0 & -1 \\ 0 & 1 & 2 \\ 0 & 0 & 2 \end{pmatrix}$,

$$\begin{pmatrix}0&1&0\\0&0&1\\1&0&0\end{pmatrix}\begin{pmatrix}0&1&2\\1&0&-1\\1&1&3\end{pmatrix} = \begin{pmatrix}1&0&0\\1&1&0\\0&1&1\end{pmatrix}\begin{pmatrix}1&0&-1\\0&1&4\\0&0&-2\end{pmatrix},$$

$$\begin{pmatrix}0&0&1\\0&1&0\\1&0&0\end{pmatrix}\begin{pmatrix}0&1&2\\1&0&-1\\1&1&3\end{pmatrix} = \begin{pmatrix}1&0&0\\1&1&0\\0&-1&1\end{pmatrix}\begin{pmatrix}1&1&3\\0&-1&-4\\0&0&-2\end{pmatrix},$$

$$\begin{pmatrix}0&0&1\\1&0&0\\0&1&0\end{pmatrix}\begin{pmatrix}0&1&2\\1&0&-1\\1&1&3\end{pmatrix} = \begin{pmatrix}1&0&0\\0&1&0\\1&-1&1\end{pmatrix}\begin{pmatrix}1&1&3\\0&1&2\\0&0&-2\end{pmatrix}.$$

The other two permutation matrices are not regular.

1.4.24 False.

1.5.1 (b) $$\begin{pmatrix}2&1&1\\3&2&1\\2&1&2\end{pmatrix}\begin{pmatrix}3&-1&-1\\-4&2&1\\-1&0&1\end{pmatrix} = \begin{pmatrix}3&-1&-1\\-4&2&1\\-1&0&1\end{pmatrix}\begin{pmatrix}2&1&1\\3&2&1\\2&1&2\end{pmatrix} = \begin{pmatrix}1&0&0\\0&1&0\\0&0&1\end{pmatrix}.$$

1.5.2 $X = \begin{pmatrix}-5&16&6\\3&-8&-3\\-1&3&1\end{pmatrix}.$

1.5.3 (a) $\begin{pmatrix}0&1\\1&0\end{pmatrix}$, (c) $\begin{pmatrix}1&2\\0&1\end{pmatrix}$,

(e) $\begin{pmatrix}1&0&0&0\\0&1&0&0\\0&-6&1&0\\0&0&0&1\end{pmatrix}.$

1.5.6 (a) $A^{-1} = \begin{pmatrix}-1&1\\2&-1\end{pmatrix}$, $B^{-1} = \begin{pmatrix}\frac{2}{3}&\frac{1}{3}\\-\frac{1}{3}&\frac{1}{3}\end{pmatrix}.$

(b) $C = \begin{pmatrix}2&1\\3&0\end{pmatrix}$, $C^{-1} = B^{-1}A^{-1} = \begin{pmatrix}0&\frac{1}{3}\\1&-\frac{2}{3}\end{pmatrix}.$

1.5.8 (a) Set $P_1 = \begin{pmatrix}1&0&0\\0&1&0\\0&0&1\end{pmatrix}$, $P_2 = \begin{pmatrix}0&1&0\\0&0&1\\1&0&0\end{pmatrix}$,

$P_3 = \begin{pmatrix}0&0&1\\1&0&0\\0&1&0\end{pmatrix}$, $P_4 = \begin{pmatrix}0&1&0\\1&0&0\\0&0&1\end{pmatrix}$,

$P_5 = \begin{pmatrix}0&0&1\\0&1&0\\1&0&0\end{pmatrix}$, $P_6 = \begin{pmatrix}1&0&0\\0&0&1\\0&1&0\end{pmatrix}$.

Then $P_1^{-1} = P_1$, $P_2^{-1} = P_3$, $P_3^{-1} = P_2$, $P_4^{-1} = P_4$, $P_5^{-1} = P_5$, $P_6^{-1} = P_6$.

1.5.9 (a) $\begin{pmatrix}0&0&0&1\\0&0&1&0\\0&1&0&0\\1&0&0&0\end{pmatrix}$, (c) $\begin{pmatrix}1&0&0&0\\0&0&1&0\\0&0&0&1\\0&1&0&0\end{pmatrix}$.

1.5.12 This is true if and only if $A^2 = \mathrm{I}$; see Exercise 1.2.36.

1.5.14 $\left(\frac{1}{c}A^{-1}\right)(c\,A) = \frac{1}{c}\,c\,A^{-1}A = \mathrm{I}$.

1.5.18 (a) $A = \mathrm{I}^{-1}A\,\mathrm{I}$.
(b) If $B = S^{-1}AS$, then $A = S\,B\,S^{-1} = T^{-1}B\,T$, where $T = S^{-1}$.

1.5.20 (a) $B\,A = \begin{pmatrix}1&0\\0&1\end{pmatrix}$.
(b) $A\,X = \mathrm{I}$ does not have a solution.

1.5.22 (a) No. The only solutions are complex: $a = \left(-\frac{1}{2} \pm \mathrm{i}\sqrt{\frac{2}{3}}\right)b$.
(b) Yes. One example is $A = \begin{pmatrix}-1&1\\-1&0\end{pmatrix}$, $B = \begin{pmatrix}1&0\\0&1\end{pmatrix}$.

1.5.24 (b) $\begin{pmatrix}-\frac{1}{8}&\frac{3}{8}\\\frac{3}{8}&-\frac{1}{8}\end{pmatrix}$, (d) No inverse,

(f) $\begin{pmatrix}-\frac{5}{8}&\frac{1}{8}&\frac{5}{8}\\-\frac{1}{2}&\frac{1}{2}&-\frac{1}{2}\\\frac{7}{8}&-\frac{3}{8}&\frac{1}{8}\end{pmatrix}$,

(h) $\begin{pmatrix}0&2&1&1\\1&-6&-2&-3\\0&-5&0&-3\\0&2&0&1\end{pmatrix}$.

1.5.25 (b) $\begin{pmatrix}1&0\\3&1\end{pmatrix}\begin{pmatrix}1&0\\0&-8\end{pmatrix}\begin{pmatrix}1&3\\0&1\end{pmatrix} = \begin{pmatrix}1&3\\3&1\end{pmatrix}$;
(d) Not possible;

(f) $$\begin{pmatrix}1&0&0\\3&1&0\\0&0&1\end{pmatrix}\begin{pmatrix}1&0&0\\0&1&0\\2&0&1\end{pmatrix}\begin{pmatrix}1&0&0\\0&1&0\\0&3&1\end{pmatrix}\begin{pmatrix}1&0&0\\0&-1&0\\0&0&1\end{pmatrix}\begin{pmatrix}1&0&0\\0&1&0\\0&0&8\end{pmatrix}\begin{pmatrix}1&0&3\\0&1&0\\0&0&1\end{pmatrix}\begin{pmatrix}1&0&0\\0&1&4\\0&0&1\end{pmatrix}\begin{pmatrix}1&2&0\\0&1&0\\0&0&1\end{pmatrix} = \begin{pmatrix}1&2&3\\3&5&5\\2&1&2\end{pmatrix};$$

(h) $\begin{pmatrix} 1 & 0 & 0 & 0 \\ 0 & 0 & 1 & 0 \\ 0 & 1 & 0 & 0 \\ 0 & 0 & 0 & 1 \end{pmatrix} \begin{pmatrix} 1 & 0 & 0 & 0 \\ \frac{1}{2} & 1 & 0 & 0 \\ 0 & 0 & 1 & 0 \\ 0 & 0 & 0 & 1 \end{pmatrix}$

$$\begin{pmatrix} 1 & 0 & 0 & 0 \\ 0 & 1 & 0 & 0 \\ 0 & 0 & 1 & 0 \\ 0 & 0 & -2 & 1 \end{pmatrix} \begin{pmatrix} 2 & 0 & 0 & 0 \\ 0 & 1 & 0 & 0 \\ 0 & 0 & 1 & 0 \\ 0 & 0 & 0 & 1 \end{pmatrix}$$

$$\begin{pmatrix} 1 & 0 & 0 & 0 \\ 0 & -\frac{1}{2} & 0 & 0 \\ 0 & 0 & 1 & 0 \\ 0 & 0 & 0 & 1 \end{pmatrix} \begin{pmatrix} 1 & 0 & 0 & \frac{1}{2} \\ 0 & 1 & 0 & 0 \\ 0 & 0 & 1 & 0 \\ 0 & 0 & 0 & 1 \end{pmatrix}$$

$$\begin{pmatrix} 1 & 0 & 0 & 0 \\ 0 & 1 & 0 & 3 \\ 0 & 0 & 1 & 0 \\ 0 & 0 & 0 & 1 \end{pmatrix} \begin{pmatrix} 1 & 0 & 0 & 0 \\ 0 & 1 & 0 & 0 \\ 0 & 0 & 1 & 3 \\ 0 & 0 & 0 & 1 \end{pmatrix}$$

$$\begin{pmatrix} 1 & \frac{1}{2} & 0 & 0 \\ 0 & 1 & 0 & 0 \\ 0 & 0 & 1 & 0 \\ 0 & 0 & 0 & 1 \end{pmatrix} = \begin{pmatrix} 2 & 1 & 0 & 1 \\ 0 & 0 & 1 & 3 \\ 1 & 0 & 0 & -1 \\ 0 & 0 & -2 & -5 \end{pmatrix}.$$

1.5.27 (a) $\begin{pmatrix} -\frac{i}{2} & \frac{1}{2} \\ \frac{1}{2} & -\frac{i}{2} \end{pmatrix}$, (c) $\begin{pmatrix} i & 0 & -1 \\ 1-i & -i & 1 \\ -1 & -1 & -i \end{pmatrix}$.

1.5.30 (b) $\begin{pmatrix} \frac{5}{17} & \frac{2}{17} \\ -\frac{1}{17} & \frac{3}{17} \end{pmatrix} \begin{pmatrix} 2 \\ 12 \end{pmatrix} = \begin{pmatrix} 2 \\ 2 \end{pmatrix}$;

(d) $\begin{pmatrix} 9 & -15 & -8 \\ 6 & -10 & -5 \\ -1 & 2 & 1 \end{pmatrix} \begin{pmatrix} 3 \\ -1 \\ 5 \end{pmatrix} = \begin{pmatrix} 2 \\ 3 \\ 0 \end{pmatrix}$;

(f) $\begin{pmatrix} 1 & 0 & 1 & 1 \\ 0 & 0 & -1 & -1 \\ 2 & -1 & -1 & 0 \\ 2 & -1 & -1 & -1 \end{pmatrix} \begin{pmatrix} 4 \\ 11 \\ -7 \\ 6 \end{pmatrix} = \begin{pmatrix} 3 \\ 1 \\ 4 \\ -2 \end{pmatrix}$.

1.5.31 (b) $\begin{pmatrix} \frac{1}{4} \\ \frac{1}{4} \end{pmatrix}$, (d) Singular,

(f) $\begin{pmatrix} \frac{1}{8} \\ -\frac{1}{2} \\ \frac{5}{8} \end{pmatrix}$, (h) $\begin{pmatrix} 4 \\ -10 \\ -8 \\ 3 \end{pmatrix}$.

1.5.32 (b) $\begin{pmatrix} 0 & 1 \\ 1 & 0 \end{pmatrix} \begin{pmatrix} 0 & 4 \\ -7 & 2 \end{pmatrix}$

$= \begin{pmatrix} 1 & 0 \\ 0 & 1 \end{pmatrix} \begin{pmatrix} -7 & 0 \\ 0 & 4 \end{pmatrix} \begin{pmatrix} 1 & -\frac{2}{7} \\ 0 & 1 \end{pmatrix}$

(d) $\begin{pmatrix} 1 & 0 & 0 \\ 0 & 0 & 1 \\ 0 & 1 & 0 \end{pmatrix} \begin{pmatrix} 1 & 1 & 5 \\ 1 & 1 & -2 \\ 2 & -1 & 3 \end{pmatrix}$

$= \begin{pmatrix} 1 & 0 & 0 \\ 2 & 1 & 0 \\ 1 & 0 & 1 \end{pmatrix} \begin{pmatrix} 1 & 0 & 0 \\ 0 & -3 & 0 \\ 0 & 0 & -7 \end{pmatrix} \begin{pmatrix} 1 & 1 & 5 \\ 0 & 1 & \frac{7}{3} \\ 0 & 0 & 1 \end{pmatrix}$;

(f) $\begin{pmatrix} 1 & -1 & 1 & 2 \\ 1 & -4 & 1 & 5 \\ 1 & 2 & -1 & -1 \\ 3 & 1 & 1 & 6 \end{pmatrix}$

$= \begin{pmatrix} 1 & 0 & 0 & 0 \\ 1 & 1 & 0 & 0 \\ 1 & -1 & 1 & 0 \\ 3 & -\frac{4}{3} & 1 & 1 \end{pmatrix} \begin{pmatrix} 1 & 0 & 0 & 0 \\ 0 & -3 & 0 & 0 \\ 0 & 0 & -2 & 0 \\ 0 & 0 & 0 & 4 \end{pmatrix}$

$\begin{pmatrix} 1 & -1 & 1 & 2 \\ 0 & 1 & 0 & -1 \\ 0 & 0 & 1 & 0 \\ 0 & 0 & 0 & 1 \end{pmatrix}$.

1.5.33 (b) $\begin{pmatrix} -8 \\ 3 \end{pmatrix}$, (d) $\begin{pmatrix} 1 \\ -2 \\ 0 \end{pmatrix}$, (f) $\begin{pmatrix} \frac{7}{3} \\ 2 \\ 5 \\ -\frac{5}{3} \end{pmatrix}$.

1.6.1 (b) $\begin{pmatrix} 1 & 0 \\ 1 & 2 \end{pmatrix}$, (d) $\begin{pmatrix} 1 & 2 \\ 2 & 0 \\ -1 & 2 \end{pmatrix}$, (f) $\begin{pmatrix} 1 & 3 & 5 \\ 2 & 4 & 6 \end{pmatrix}$.

1.6.2 $A^T = \begin{pmatrix} 3 & 1 \\ -1 & 2 \\ -1 & 1 \end{pmatrix}$, $B^T = \begin{pmatrix} -1 & 2 & -3 \\ 2 & 0 & 4 \end{pmatrix}$,

$(A\,B)^T = B^T A^T = \begin{pmatrix} -2 & 0 \\ 2 & 6 \end{pmatrix}$,

$(B\,A)^T = A^T B^T = \begin{pmatrix} -1 & 6 & -5 \\ 5 & -2 & 11 \\ 3 & -2 & 7 \end{pmatrix}$.

1.6.5 $(A\,B\,C)^T = C^T B^T A^T$

1.6.6 False.

1.6.10 No; for example, $\begin{pmatrix} 1 \\ 2 \end{pmatrix} \begin{pmatrix} 3 & 4 \end{pmatrix} = \begin{pmatrix} 3 & 4 \\ 6 & 8 \end{pmatrix}$ while $\begin{pmatrix} 3 \\ 4 \end{pmatrix} \begin{pmatrix} 1 & 2 \end{pmatrix} = \begin{pmatrix} 3 & 6 \\ 4 & 8 \end{pmatrix}$.

1.6.13 (b) $A = \begin{pmatrix} 1 & 2 \\ 0 & 1 \end{pmatrix}$, $B = \begin{pmatrix} 1 & 1 \\ 1 & 1 \end{pmatrix}$.

1.6.17 (b) $a = -1, b = 2, c = 3$.

1.6.18 (a) $\begin{pmatrix} 1 & 0 & 0 \\ 0 & 1 & 0 \\ 0 & 0 & 1 \end{pmatrix}$, $\begin{pmatrix} 0 & 1 & 0 \\ 1 & 0 & 0 \\ 0 & 0 & 1 \end{pmatrix}$, $\begin{pmatrix} 0 & 0 & 1 \\ 0 & 1 & 0 \\ 1 & 0 & 0 \end{pmatrix}$, $\begin{pmatrix} 1 & 0 & 0 \\ 0 & 0 & 1 \\ 0 & 1 & 0 \end{pmatrix}$.

1.6.20 True.

1.6.25 (a) $\begin{pmatrix} 1 & 1 \\ 1 & 4 \end{pmatrix} = \begin{pmatrix} 1 & 0 \\ 1 & 1 \end{pmatrix} \begin{pmatrix} 1 & 0 \\ 0 & 3 \end{pmatrix} \begin{pmatrix} 1 & 1 \\ 0 & 1 \end{pmatrix}$,

(c) $\begin{pmatrix} 1 & -1 & -1 \\ -1 & 3 & 2 \\ -1 & 2 & 0 \end{pmatrix} =$

$$\begin{pmatrix} 1 & 0 & 0 \\ -1 & 1 & 0 \\ -1 & \frac{1}{2} & 1 \end{pmatrix} \begin{pmatrix} 1 & 0 & 0 \\ 0 & 2 & 0 \\ 0 & 0 & -\frac{3}{2} \end{pmatrix} \begin{pmatrix} 1 & -1 & -1 \\ 0 & 1 & \frac{1}{2} \\ 0 & 0 & 1 \end{pmatrix}.$$

1.6.27 The matrix is not regular, since after the first set of row operations the (2, 2) entry is 0.

1.7.1 (b) $x = -4$, $y = -5$, $z = -1$.
(i) 17 multiplications and 11 additions;
(ii) 20 multiplications and 11 additions;
(iii) computing A^{-1} takes 27 multiplications and 12 additions, while multiplying $A^{-1}\mathbf{b} = \mathbf{x}$ takes another 9 multiplications and 6 additions.

1.7.3 Back Substitution is about twice as fast.

1.7.6 $\frac{1}{3}n^3 + \frac{3}{2}n^2 - \frac{5}{6}n$ multiplications and $\frac{1}{3}n^3 + \frac{1}{2}n^2 - \frac{5}{6}n$ additions.

1.7.9 (b) $$\begin{pmatrix} 1 & -1 & 0 & 0 \\ -1 & 2 & 1 & 0 \\ 0 & -1 & 4 & 1 \\ 0 & 0 & -1 & 6 \end{pmatrix} = \begin{pmatrix} 1 & 0 & 0 & 0 \\ -1 & 1 & 0 & 0 \\ 0 & -1 & 1 & 0 \\ 0 & 0 & -1 & 1 \end{pmatrix} \begin{pmatrix} 1 & -1 & 0 & 0 \\ 0 & 1 & 1 & 0 \\ 0 & 0 & 5 & 1 \\ 0 & 0 & 0 & 7 \end{pmatrix},$$
$$\mathbf{x} = \begin{pmatrix} 1 \\ 0 \\ 1 \\ 2 \end{pmatrix}.$$

1.7.10 (a) For $n = 4$:
$$\begin{pmatrix} 2 & -1 & 0 & 0 \\ -1 & 2 & -1 & 0 \\ 0 & -1 & 2 & -1 \\ 0 & 0 & -1 & 2 \end{pmatrix} = \begin{pmatrix} 1 & 0 & 0 & 0 \\ -\frac{1}{2} & 1 & 0 & 0 \\ 0 & -\frac{2}{3} & 1 & 0 \\ 0 & 0 & -\frac{3}{4} & 1 \end{pmatrix} \begin{pmatrix} 2 & -1 & 0 & 0 \\ 0 & \frac{3}{2} & -1 & 0 \\ 0 & 0 & \frac{4}{3} & -1 \\ 0 & 0 & 0 & \frac{5}{4} \end{pmatrix};$$
(b) $(2, 3, 3, 2)^T$.

1.7.12 (a) False.

1.7.13 $$\begin{pmatrix} 4 & 1 & 0 & 1 \\ 1 & 4 & 1 & 0 \\ 0 & 1 & 4 & 1 \\ 1 & 0 & 1 & 4 \end{pmatrix} = \begin{pmatrix} 1 & 0 & 0 & 0 \\ \frac{1}{4} & 1 & 0 & 0 \\ 0 & \frac{4}{15} & 1 & 0 \\ \frac{1}{4} & -\frac{1}{15} & \frac{2}{7} & 1 \end{pmatrix} \begin{pmatrix} 4 & 1 & 0 & 1 \\ 0 & \frac{15}{4} & 1 & -\frac{1}{4} \\ 0 & 0 & \frac{56}{15} & \frac{16}{15} \\ 0 & 0 & 0 & \frac{24}{7} \end{pmatrix}.$$

1.7.16 (a) $(-8, 4)^T$, (b) $(-10, -4.1)^T$, (c) $(-8.1, -4.1)^T$.

1.7.18 (a) $x = -2$, $y = 2$, $z = 3$,
(b) $x = -7.3$, $y = 3.3$, $z = 2.9$,
(c) $x = -1.9$, $y = 2.$, $z = 2.9$.

1.7.20 (a) $\left(\frac{6}{5}, -\frac{13}{5}, -\frac{9}{5}\right)^T$, (c) $(0, 1, 1, 0)^T$.

1.7.21 (a) $\left(-\frac{1}{13}, \frac{8}{13}\right)^T = (-.0769, .6154)^T$,
(c) $\left(\frac{2}{121}, \frac{38}{121}, \frac{59}{242}, -\frac{56}{121}\right)^T = (.0165, .3141, .2438, -.4628)^T$.

1.8.1 (a) Unique solution: $(-\frac{1}{2}, -\frac{3}{4})^T$;
(c) no solutions;
(e) infinitely many solutions: $(5 - 2z, 1, z, 0)^T$, where z is arbitrary.

1.8.2 (b) Incompatible
(d) $(1 + 3x_2 - 2x_3, x_2, x_3)^T$, where x_2 and x_3 are arbitrary;
(f) $(-5 - 3x_4, 19 - 4x_4, -6 - 2x_4, x_4)^T$, where x_4 is arbitrary.

1.8.4 (i) $a \neq b$ and $b \neq 0$; $b = 0$, (ii) $a \neq -2$;
(iii) $a = b \neq 0$, or $a = -2$, $b = 0$.

1.8.6 (a) $\left(1 + \mathrm{i} - \frac{1}{2}(1 + \mathrm{i})y, y, -\mathrm{i}\right)^T$, where y is arbitrary;;
(c) $(3 + 2\,\mathrm{i}, -1 + 2\,\mathrm{i}, 3\,\mathrm{i})^T$.

1.8.7 (b) 1, (d) 3, (f) 1, (h) 2.

1.8.8 (b) $$\begin{pmatrix} 2 & 1 & 3 \\ -2 & -1 & -3 \end{pmatrix} = \begin{pmatrix} 1 & 0 \\ -1 & 1 \end{pmatrix} \begin{pmatrix} 2 & 1 & 3 \\ 0 & 0 & 0 \end{pmatrix},$$
(d) $$\begin{pmatrix} 1 & 0 & 0 \\ 0 & 0 & 1 \\ 0 & 1 & 0 \end{pmatrix} \begin{pmatrix} 2 & -1 & 0 \\ 1 & 1 & -1 \\ 2 & -1 & 1 \end{pmatrix} = \begin{pmatrix} 1 & 0 & 0 \\ \frac{1}{2} & 1 & 0 \\ 1 & 0 & 1 \end{pmatrix} \begin{pmatrix} 2 & -1 & 0 \\ 0 & \frac{3}{2} & -1 \\ 0 & 0 & 1 \end{pmatrix},$$
(f) $\begin{pmatrix} 0 & -1 & 2 & 5 \end{pmatrix} = (\,1\,)\begin{pmatrix} 0 & -1 & 2 & 5 \end{pmatrix}$,
(h) $$\begin{pmatrix} 1 & -1 & 2 & 1 \\ 2 & 1 & -1 & 0 \\ 1 & 2 & -3 & -1 \\ 4 & -1 & 3 & 2 \\ 0 & 3 & -5 & -2 \end{pmatrix} = \begin{pmatrix} 1 & 0 & 0 & 0 & 0 \\ 2 & 1 & 0 & 0 & 0 \\ 1 & 1 & 1 & 0 & 0 \\ 4 & 1 & 0 & 1 & 0 \\ 0 & 1 & 0 & 0 & 1 \end{pmatrix} \begin{pmatrix} 1 & -1 & 2 & 1 \\ 0 & 3 & -5 & -2 \\ 0 & 0 & 0 & 0 \\ 0 & 0 & 0 & 0 \\ 0 & 0 & 0 & 0 \end{pmatrix}.$$

1.8.10 (a) $\begin{pmatrix} 1 & 0 & 0 \\ 0 & 1 & 0 \end{pmatrix}$, (c) $\begin{pmatrix} 1 & 0 \\ 0 & 1 \\ 0 & 0 \end{pmatrix}$.

1.8.11 (a) $x^2 + y^2 = 1$, $x^2 - y^2 = 2$,
(c) $y = x^3$, $x - y = 0$.

1.8.13 True.

1.8.16 1.

1.8.18 If $A = \begin{pmatrix} 1 & 0 \\ 0 & 0 \end{pmatrix}$, $B = \begin{pmatrix} 0 & 1 \\ 0 & 0 \end{pmatrix}$, then AB has rank 1 but BA has rank 0.

1.8.22 (a) $x = z$, $y = z$, where z is arbitrary; (c) $x = y = z = 0$; (e) $x = 13z$, $y = 5z$, $w = 0$, where z is arbitrary.

1.8.23 (a) $\left(\frac{1}{3}y, y\right)^T$, where y is arbitrary; (c) $\left(-\frac{11}{5}z + \frac{3}{5}w, \frac{2}{5}z - \frac{6}{5}w, z, w\right)^T$, where z and w are arbitrary; (e) $(-4z, 2z, z)^T$, where z is arbitrary; (g) $(3z, 3z, z, 0)^T$, where z is arbitrary.

1.8.27 (b) $k = 0$ or $k = \frac{1}{2}$.

1.9.1 (b) 0, (d) 6, (f) 40.

1.9.2 $\det A = -2$, $\det B = -11$, $\det AB = 22$.

1.9.4 $\det A = 0$ or 1

1.9.5 (a) True. (c) False. (e) True. (g) True.

1.9.7 $\det B = \det(S^{-1} A S) = \det S^{-1} \det A \det S = \dfrac{1}{\det S} \det A \det S = \det A$.

1.9.13 $\det A \; \det A^{-1} = \det(A A^{-1}) = \det \mathrm{I} = 1$.

1.9.15 $a_{11}a_{22}a_{33}a_{44} - a_{11}a_{22}a_{34}a_{43} - a_{11}a_{23}a_{32}a_{44} + a_{11}a_{23}a_{34}a_{42} - a_{11}a_{24}a_{33}a_{42} + a_{11}a_{24}a_{32}a_{43} - a_{12}a_{21}a_{33}a_{44} + a_{12}a_{21}a_{34}a_{43} + a_{12}a_{23}a_{31}a_{44} - a_{12}a_{23}a_{34}a_{41} + a_{12}a_{24}a_{33}a_{41} - a_{12}a_{24}a_{31}a_{43} + a_{13}a_{21}a_{32}a_{44} - a_{13}a_{21}a_{34}a_{42} - a_{13}a_{22}a_{31}a_{44} + a_{13}a_{22}a_{34}a_{41} - a_{13}a_{24}a_{32}a_{41} + a_{13}a_{24}a_{31}a_{42} - a_{14}a_{21}a_{32}a_{43} + a_{14}a_{21}a_{33}a_{42} + a_{14}a_{22}a_{31}a_{43} - a_{14}a_{22}a_{33}a_{41} + a_{14}a_{23}a_{32}a_{41} - a_{14}a_{23}a_{31}a_{42}$.

1.9.21 (b) $\det \begin{pmatrix} 1 & 1 & 1 \\ t_1 & t_2 & t_3 \\ t_1^2 & t_2^2 & t_3^2 \end{pmatrix} = (t_2 - t_1)(t_3 - t_1)(t_3 - t_2)$.

1.9.22 (b) (i) $x = -2.6$, $y = 5.2$. (d) (i) $x = -\frac{1}{9}$, $y = \frac{7}{9}$, $z = \frac{8}{9}$.

1.9.23 (c) (i) 129, (iii) 15.

Chapter 2

2.1.1 *Commutativity of Addition*: $(x + \mathrm{i}\,y) + (u + \mathrm{i}\,v) = (x + u) + \mathrm{i}\,(y + v) = (u + \mathrm{i}\,v) + (x + \mathrm{i}\,y)$.
Associativity of Addition: $(x + \mathrm{i}\,y) + \big[(u + \mathrm{i}\,v) + (p + \mathrm{i}\,q)\big] = (x + \mathrm{i}\,y) + \big[(u + p) + \mathrm{i}\,(v + q)\big] = (x + u + p) + \mathrm{i}\,(y + v + q) = \big[(x + u) + \mathrm{i}\,(y + v)\big] + (p + \mathrm{i}\,q) = \big[(x + \mathrm{i}\,y) + (u + \mathrm{i}\,v)\big] + (p + \mathrm{i}\,q)$.
Additive Identity: $\mathbf{0} = 0 = 0 + \mathrm{i}\,0$ and $(x + \mathrm{i}\,y) + 0 = x + \mathrm{i}\,y = 0 + (x + \mathrm{i}\,y)$.
Additive Inverse: $-(x + \mathrm{i}\,y) = (-x) + \mathrm{i}\,(-y)$ and $(x + \mathrm{i}\,y) + \big[(-x) + \mathrm{i}\,(-y)\big] = 0 = \big[(-x) + \mathrm{i}\,(-y)\big] + (x + \mathrm{i}\,y)$.
Distributivity: $(c + d)(x + \mathrm{i}\,y) = (c + d)x + \mathrm{i}\,(c + d)y = (cx + dx) + \mathrm{i}\,(cy + dy) = c\,(x + \mathrm{i}\,y) + d\,(x + \mathrm{i}\,y)$, $c[(x + \mathrm{i}\,y) + (u + \mathrm{i}\,v)] = c\,(x + u) + \mathrm{i}\,c\,(y + v) = (cx + cu) + \mathrm{i}\,(cy + cv) = c\,(x + \mathrm{i}\,y) + c\,(u + \mathrm{i}\,v)$.
Associativity of Scalar Multiplication: $c\,[d\,(x + \mathrm{i}\,y)] = c\,[(dx) + \mathrm{i}\,(dy)] = (cdx) + \mathrm{i}\,(cdy) = (cd)(x + \mathrm{i}\,y)$.
Unit for Scalar Multiplication: $1\,(x + \mathrm{i}\,y) = (1x) + \mathrm{i}\,(1y) = x + \mathrm{i}\,y$.

2.1.4 (a) $(1, 1, 1, 1)^T$, $(1, -1, 1, -1)^T$, $(1, 1, 1, 1)^T$, $(1, -1, 1, -1)^T$.

2.1.6 (a) $f(x) = -4x + 3$.

2.1.7 (a) $(x - y, xy)^T$, $(e^x, \cos y)^T$, $(1, 3)^T$; (b) sum: $(x - y + e^x + 1, xy + \cos y + 3)^T$. Multiplied by -5: $(-5x + 5y - 5e^x - 5, -5xy - 5\cos y - 15)^T$.

2.1.11 (j) Let $\mathbf{z} = c\mathbf{0}$. Then $\mathbf{z} + \mathbf{z} = c(\mathbf{0} + \mathbf{0}) = c\mathbf{0} = \mathbf{z}$, and so, as in the proof of (h), $\mathbf{z} = \mathbf{0}$.

2.2.2 (a) Not a subspace, (c) subspace, (e) not a subspace, (g) subspace, (i) subspace.

2.2.3 (a) Subspace, (c) subspace, (e) not a subspace.

2.2.5 False.

2.2.7 (b) Not a subspace, (d) subspace.

2.2.9 If L and M are strictly lower triangular, so is $cL + dM$ for any $c, d \in \mathbb{R}$.

2.2.12 (a) Not a subspace, (c) not a subspace, (e) not a subspace, (g) subspace.

2.2.13 (a) Yes. (c) Yes. (e) No. (g) Yes.

2.2.15 (b) Subspace, (d) subspace, (f) subspace.

2.2.16 (a) Subspace, (c) not a subspace, (e) subspace, (g) subspace.

2.2.18 The zero function $u(x) \equiv 0$ is not a solution.

2.2.20 $\nabla \cdot (c\mathbf{v} + d\mathbf{w}) = c\,\nabla \cdot \mathbf{v} + d\,\nabla \cdot \mathbf{w} = 0$ whenever $\nabla \cdot \mathbf{v} = \nabla \cdot \mathbf{w} = 0$ and $c, d, \in \mathbb{R}$.

2.2.22 (b) If $\mathbf{w} + \mathbf{z}$, $\widetilde{\mathbf{w}} + \widetilde{\mathbf{z}} \in W + Z$ and $c, d, \in \mathbb{R}$, then $c\,(\mathbf{w} + \mathbf{z}) + d\,(\widetilde{\mathbf{w}} + \widetilde{\mathbf{z}}) = (c\,\mathbf{w} + d\,\widetilde{\mathbf{w}}) + (c\,\mathbf{z} + d\,\widetilde{\mathbf{z}}) \in W + Z$.

2.2.24 (b) The lines only intersect at the origin; moreover, every $\mathbf{v} = (x, y)^T = (a, a)^T + (3b, b)^T$, for $a = -\frac{1}{2}x + \frac{3}{2}y$, $b = \frac{1}{2}x - \frac{1}{2}y$.

2.2.29 The Taylor series $1/(1 + x^2) = 1 - x^2 + x^4 - x^6 + \cdots$ converges when $|x| < 1$.

2.3.1 $(-1, 2, 3)^T = 2\,(2, -1, 2)^T - (5, -4, 1)^T$.

2.3.3 (b) Yes, since $(1, -2, -1)^T = \frac{3}{10}(1, 2, 2)^T + \frac{7}{10}(1, -2, 0)^T - \frac{4}{10}(0, 3, 4)^T$.

2.3.4 (a) No (c) Yes (e) Yes.

2.3.5 (b) The plane $z = -\frac{3}{5}x - \frac{6}{5}y$.

2.3.7 (a) Every symmetric matrix has the form $\begin{pmatrix} a & b \\ b & c \end{pmatrix} = a\begin{pmatrix} 1 & 0 \\ 0 & 0 \end{pmatrix} + c\begin{pmatrix} 0 & 0 \\ 0 & 1 \end{pmatrix} + b\begin{pmatrix} 0 & 1 \\ 1 & 0 \end{pmatrix}$.

2.3.8 (b) Yes.

2.3.9 (a) Yes (c) No (e) No.

2.3.10 (a) $\sin 3x = \cos\left(3x - \frac{1}{2}\pi\right)$;
(c) $3\cos 2x + 4\sin 2x = 5\cos\left(2x - \tan^{-1}\frac{4}{3}\right)$.

2.3.13 (a) e^{2x}; (c) $e^{3x}, 1$;
(e) $e^{-x/2}\cos\frac{\sqrt{3}}{2}x,\ e^{-x/2}\sin\frac{\sqrt{3}}{2}x$.

2.3.15 (a) $\begin{pmatrix}2\\1\end{pmatrix} = 2\,\mathbf{f}_1(x) + \mathbf{f}_2(x) - \mathbf{f}_3(x)$;
(d) not in the span.

2.3.17 False.

2.3.21 (b) Linearly dependent; (d) linearly independent; (f) linearly dependent; (h) linearly independent.

2.3.22 (a) The only solution to $c_1(1,0,2,1)^T + c_2(-2,3,-1,1)^T + c_3(2,-2,1,-1)^T = \mathbf{0}$ is $c_1 = c_2 = c_3 = 0$.
(b) All but the second lie in the span.
(c) $a - c + d = 0$.

2.3.24 (a) Linearly dependent; (c) linearly independent; (e) linearly dependent.

2.3.25 False.

2.3.29 False.

2.3.32 (a) Linearly dependent.

2.3.33 (b) Linearly independent; (d) linearly independent; (f) linearly dependent; (h) linearly independent.

2.3.35 (c) Linearly dependent.

2.4.1 (a) Basis. (d) Not a basis.

2.4.2 (a) Not a basis. (c) Not a basis.

2.4.3 (b) $\left(\frac{3}{4},1,0\right)^T,\ \left(\frac{1}{4},0,1\right)^T$.

2.4.4 (a) They do not span $\mathbb{R}^3$; (b) linearly dependent; (c) no; (d) 2.

2.4.6 (b) $\begin{pmatrix}2\\-1\\1\end{pmatrix} = (-1)\begin{pmatrix}2\\1\\0\end{pmatrix} + \begin{pmatrix}4\\0\\1\end{pmatrix}$,
$\begin{pmatrix}0\\2\\-1\end{pmatrix} = 2\begin{pmatrix}2\\1\\0\end{pmatrix} - \begin{pmatrix}4\\0\\1\end{pmatrix}$.

2.4.8 (b) Basis: $-x^2 + x,\ -x^2 + 1$, so $\dim = 2$.

2.4.9 (b) Basis: $(2,0,1)^T,\ (0,-1,3)^T$, so $\dim = 2$.

2.4.11 (b) $1 + t^3 = 2 - 3(1-t) + 3(1-t)^2 - (1-t)^3$.

2.4.14 (a) Basis: $E_{11} = \begin{pmatrix}1&0\\0&0\end{pmatrix}, E_{12} = \begin{pmatrix}0&1\\0&0\end{pmatrix}$,
$E_{21} = \begin{pmatrix}0&0\\1&0\end{pmatrix},\ E_{22} = \begin{pmatrix}0&0\\0&1\end{pmatrix}$.

2.4.17 (a) $E_{11} = \begin{pmatrix}1&0\\0&0\end{pmatrix},\ E_{12} = \begin{pmatrix}0&1\\0&0\end{pmatrix}$,
$E_{22} = \begin{pmatrix}0&0\\0&1\end{pmatrix}$; dimension $= 3$.

2.4.20 If $\mathbf{v}_1 = (1,0)^T,\ \mathbf{v}_2 = (0,1)^T,\ \mathbf{v}_3 = (1,1)^T$, then $(2,1)^T = 2\mathbf{v}_1 + \mathbf{v}_2 = \mathbf{v}_1 + \mathbf{v}_3$. In fact, there are infinitely many different ways of writing this vector as a linear combination of $\mathbf{v}_1, \mathbf{v}_2, \mathbf{v}_3$.

2.4.24 (c) (i) Example: $\left(1,1,\frac{1}{2}\right)^T,\ (1,0,0)^T,\ (0,1,0)^T$.

2.5.1 (a) Range: $\frac{3}{4}b_1 + b_2 = 0$; kernel spanned by $\left(\frac{1}{2},1\right)^T$.
(c) Range: $-2b_1 + b_2 + b_3 = 0$; kernel spanned by $\left(-\frac{5}{4},-\frac{7}{8},1\right)^T$.

2.5.2 (a) $\left(-\frac{5}{2},0,1\right)^T,\ \left(\frac{1}{2},1,0\right)^T$: plane;
(c) $(2,0,1)^T,\ (-3,1,0)^T$: plane;
(e) $(0,0,0)^T$: point.

2.5.4 (a) $\mathbf{b} = (-1,2,-1)^T$;
(b) the general solution is $\mathbf{x} = (1+t, 2+t, 3+t)^T$ where t is arbitrary.

2.5.5 In each case, the solution is $\mathbf{x} = \mathbf{x}^\star + \mathbf{z}$, where $\mathbf{x}^\star$ is the particular solution and $\mathbf{z}$ belongs to the kernel:
(b) $\mathbf{x}^\star = (1,-1,0)^T,\ \mathbf{z} = z\left(-\frac{2}{7},\frac{1}{7},1\right)^T$;
(d) $\mathbf{x}^\star = \left(\frac{5}{6},1,-\frac{2}{3}\right)^T,\ \mathbf{z} = \mathbf{0}$;
(f) $\mathbf{x}^\star = \left(\frac{11}{2},\frac{1}{2},0,0\right)^T$,
$\mathbf{z} = r\left(-\frac{13}{2},-\frac{3}{2},1,0\right)^T + s\left(-\frac{3}{2},-\frac{1}{2},0,1\right)^T$.

2.5.7 The kernel has dimension $n-1$, with basis $-r^{k-1}\mathbf{e}_1 + \mathbf{e}_k = \left(-r^{k-1},0,\ldots,0,1,0,\ldots,0\right)^T$ for $k = 2,\ldots n$. The range has dimension 1, with basis $(1, r^n, r^{2n} \ldots, r^{(n-1)n})^T$.

2.5.9 False.

2.5.12 $\mathbf{x}_1^\star = \left(-2,\frac{3}{2}\right)^T,\ \mathbf{x}_2^\star = \left(-1,\frac{1}{2}\right)^T$;
$\mathbf{x} = \mathbf{x}_1^\star + 4\mathbf{x}_2^\star = \left(-6,\frac{7}{2}\right)^T$.

2.5.14 (b) $\mathbf{x} = \mathbf{x}_1^\star + t\,(\mathbf{x}_2^\star - \mathbf{x}_1^\star)$
$= (1,1,0)^T + t\,(-4,2,-2)^T$.

2.5.16 6 horizontal and -6 vertical units.

2.5.18 False.

2.5.21 (a) range: $\begin{pmatrix}1\\2\end{pmatrix}$; corange: $\begin{pmatrix}1\\-3\end{pmatrix}$; kernel: $\begin{pmatrix}3\\1\end{pmatrix}$;
cokernel: $\begin{pmatrix}-2\\1\end{pmatrix}$.

(c) range: $\begin{pmatrix}1\\1\\2\end{pmatrix}, \begin{pmatrix}1\\0\\3\end{pmatrix}$; corange: $\begin{pmatrix}1\\1\\2\\1\end{pmatrix}, \begin{pmatrix}0\\-1\\-3\\2\end{pmatrix}$;
kernel: $\begin{pmatrix}1\\-3\\1\\0\end{pmatrix}, \begin{pmatrix}-3\\2\\0\\1\end{pmatrix}$; cokernel: $\begin{pmatrix}-3\\1\\1\end{pmatrix}$.

2.5.23 range: $\begin{pmatrix}1\\2\\-3\end{pmatrix}, \begin{pmatrix}0\\4\\1\end{pmatrix}$; corange: $\begin{pmatrix}1\\-3\\0\end{pmatrix}, \begin{pmatrix}0\\0\\4\end{pmatrix}$;
second column: $\begin{pmatrix}-3\\-6\\9\end{pmatrix} = -3\begin{pmatrix}1\\2\\-3\end{pmatrix}$; second and

third rows: $\begin{pmatrix} 2 \\ -6 \\ 4 \end{pmatrix} = 2 \begin{pmatrix} 1 \\ -3 \\ 0 \end{pmatrix} + \begin{pmatrix} 0 \\ 0 \\ 4 \end{pmatrix}$,

$$\begin{pmatrix} -3 \\ 9 \\ 1 \end{pmatrix} = -3 \begin{pmatrix} 1 \\ -3 \\ 0 \end{pmatrix} + \tfrac{1}{4} \begin{pmatrix} 0 \\ 0 \\ 4 \end{pmatrix}.$$

2.5.24 (i) rank = 1; $\dim \operatorname{rng} A = \dim \operatorname{corng} A = 1$, $\dim \ker A = \dim \operatorname{coker} A = 1$; kernel basis: $(-2, 1)^T$; cokernel basis: $(2, 1)^T$; compatibility conditions: $2 b_1 + b_2 = 0$; example: $\mathbf{b} = (1, -2)^T$, with solution $\mathbf{x} = (1, 0)^T + z\,(-2, 1)^T$.
(iii) rank = 2; $\dim \operatorname{rng} A = \dim \operatorname{corng} A = 2$, $\dim \ker A = 0$, $\dim \operatorname{coker} A = 1$; kernel: $\{\mathbf{0}\}$; cokernel basis: $\left(-\frac{20}{13}, \frac{3}{13}, 1\right)^T$; compatibility conditions: $-\frac{20}{13} b_1 + \frac{3}{13} b_2 + b_3 = 0$; example: $\mathbf{b} = (1, -2, 2)^T$, with solution $\mathbf{x} = (1, 0, 0)^T$.
(v) rank = 2; $\dim \operatorname{rng} A = \dim \operatorname{corng} A = 2$, $\dim \ker A = 1$, $\dim \operatorname{coker} A = 2$; kernel basis: $(-1, -1, 1)^T$; cokernel basis: $\left(-\frac{9}{4}, \frac{1}{4}, 1, 0\right)^T$, $\left(\frac{1}{4}, -\frac{1}{4}, 0, 1\right)^T$; compatibility conditions: $-\frac{9}{4} b_1 + \frac{1}{4} b_2 + b_3 = 0$, $\frac{1}{4} b_1 - \frac{1}{4} b_2 + b_4 = 0$; example: $\mathbf{b} = (2, 6, 3, 1)^T$, with solution $\mathbf{x} = (1, 0, 0)^T + z\,(-1, -1, 1)^T$.

2.5.25 (b) $\dim = 1$; basis: $(1, 1, -1)^T$;
(d) $\dim = 3$; basis: $(1, 0, -3, 2)^T$, $(0, 1, 2, -3)^T$, $(1, -3, -8, 7)^T$.

2.5.27 (b) $(1, 1, 1, 0)^T$, $(0, -1, 0, 1)^T$.

2.5.31 (a) Yes. (c) Yes.

2.5.33 Any symmetric matrix.

2.5.36 This is false.

2.5.38 If $\mathbf{v} \in \ker A$ then $A\mathbf{v} = \mathbf{0}$ and so $B A \mathbf{v} = B \mathbf{0} = \mathbf{0}$, so $\mathbf{v} \in \ker(B A)$. The first statement follows from setting $B = A$.

2.5.42 True if the matrices have the same size, but false in general.

2.6.1 (a) (c)

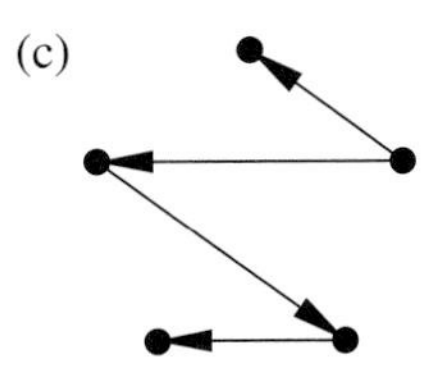

2.6.3 (a) $\begin{pmatrix} -1 & 0 & 1 & 0 \\ 0 & -1 & 1 & 0 \\ 0 & 1 & 0 & -1 \\ 0 & 0 & 1 & -1 \end{pmatrix}$;

(c) $\begin{pmatrix} -1 & 0 & 1 & 0 & 0 \\ -1 & 1 & 0 & 0 & 0 \\ 0 & -1 & 0 & 1 & 0 \\ 0 & -1 & 0 & 0 & 1 \\ 0 & 0 & 1 & -1 & 0 \\ 0 & 0 & 0 & 1 & -1 \end{pmatrix}$;

(e) $\begin{pmatrix} -1 & 0 & 0 & 1 & 0 & 0 \\ 1 & 0 & 0 & 0 & -1 & 0 \\ 0 & 1 & -1 & 0 & 0 & 0 \\ 0 & -1 & 0 & 0 & 0 & 1 \\ 0 & 0 & 1 & 0 & 0 & -1 \\ 0 & 0 & 0 & -1 & 1 & 0 \end{pmatrix}$.

2.6.4 (a) 1 circuit: $\begin{pmatrix} 0 \\ -1 \\ -1 \\ 1 \end{pmatrix}$;

(c) 2 circuits: $\begin{pmatrix} -1 \\ 1 \\ 1 \\ 0 \\ 1 \\ 0 \end{pmatrix}, \begin{pmatrix} 0 \\ 0 \\ -1 \\ 1 \\ 0 \\ 1 \end{pmatrix}$;

(e) 2 circuits: $\begin{pmatrix} 0 \\ 0 \\ 1 \\ 1 \\ 1 \\ 0 \end{pmatrix}, \begin{pmatrix} 1 \\ 1 \\ 0 \\ 0 \\ 0 \\ 1 \end{pmatrix}$.

2.6.6 (a) $\begin{pmatrix} 1 & -1 & 0 & 0 & 0 & 0 & 0 & 0 \\ 1 & 0 & -1 & 0 & 0 & 0 & 0 & 0 \\ 1 & 0 & 0 & -1 & 0 & 0 & 0 & 0 \\ 0 & 1 & 0 & 0 & -1 & 0 & 0 & 0 \\ 0 & 1 & 0 & 0 & 0 & -1 & 0 & 0 \\ 0 & 0 & 1 & 0 & -1 & 0 & 0 & 0 \\ 0 & 0 & 1 & 0 & 0 & 0 & -1 & 0 \\ 0 & 0 & 0 & 1 & 0 & -1 & 0 & 0 \\ 0 & 0 & 0 & 1 & 0 & 0 & -1 & 0 \\ 0 & 0 & 0 & 0 & 1 & 0 & 0 & -1 \\ 0 & 0 & 0 & 0 & 0 & 1 & 0 & -1 \\ 0 & 0 & 0 & 0 & 0 & 0 & 1 & -1 \end{pmatrix}$

Cokernel basis:

$$\begin{pmatrix} -1 \\ 1 \\ 0 \\ -1 \\ 0 \\ 1 \\ 0 \\ 0 \\ 0 \\ 0 \\ 0 \\ 0 \end{pmatrix}, \begin{pmatrix} -1 \\ 0 \\ 1 \\ 0 \\ -1 \\ 0 \\ 0 \\ 1 \\ 0 \\ 0 \\ 0 \\ 0 \end{pmatrix}, \begin{pmatrix} 0 \\ -1 \\ 1 \\ 0 \\ 0 \\ 0 \\ -1 \\ 0 \\ 1 \\ 0 \\ 0 \\ 0 \end{pmatrix}, \begin{pmatrix} 0 \\ 0 \\ 0 \\ -1 \\ 1 \\ 0 \\ 0 \\ 0 \\ 0 \\ -1 \\ 1 \\ 0 \end{pmatrix}, \begin{pmatrix} 0 \\ 0 \\ 0 \\ 0 \\ 0 \\ -1 \\ 1 \\ 0 \\ 0 \\ -1 \\ 0 \\ 1 \end{pmatrix}.$$

2.6.8 (a) (i) $\begin{pmatrix} -1 & 1 & 0 & 0 \\ 0 & 1 & -1 & 0 \\ 0 & 1 & 0 & -1 \end{pmatrix}$,

(iii) $\begin{pmatrix} -1 & 1 & 0 & 0 & 0 & 0 \\ 0 & 1 & -1 & 0 & 0 & 0 \\ 0 & 0 & 1 & -1 & 0 & 0 \\ 0 & 1 & 0 & 0 & -1 & 0 \\ 0 & 1 & 0 & 0 & 0 & -1 \end{pmatrix}$.

(b)

$$\begin{pmatrix} -1 & 1 & 0 & 0 & 0 \\ 0 & -1 & 1 & 0 & 0 \\ 0 & 0 & -1 & 1 & 0 \\ 0 & 0 & 0 & -1 & 1 \end{pmatrix},$$

$$\begin{pmatrix} -1 & 1 & 0 & 0 & 0 \\ 0 & -1 & 1 & 0 & 0 \\ 0 & 0 & -1 & 1 & 0 \\ 0 & 1 & 0 & 0 & -1 \end{pmatrix},$$

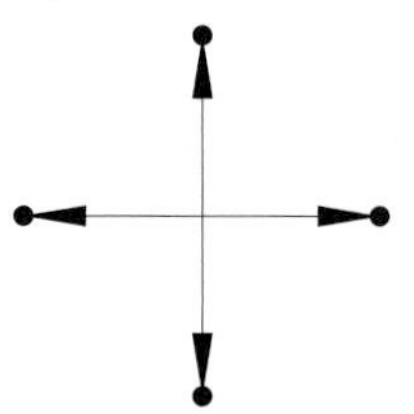

$$\begin{pmatrix} -1 & 1 & 0 & 0 & 0 \\ 0 & 1 & -1 & 0 & 0 \\ 0 & 1 & 0 & -1 & 0 \\ 0 & 1 & 0 & 0 & -1 \end{pmatrix}.$$

2.6.9 K_4:

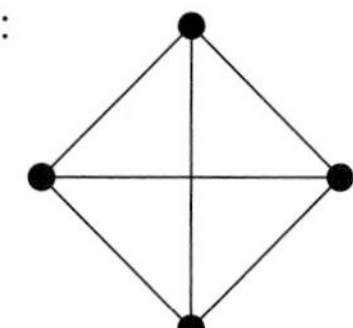

$$\begin{pmatrix} 1 & -1 & 0 & 0 \\ 1 & 0 & -1 & 0 \\ 1 & 0 & 0 & -1 \\ 0 & 1 & -1 & 0 \\ 0 & 1 & 0 & -1 \\ 0 & 0 & 1 & -1 \end{pmatrix}.$$

2.6.10 $K_{2,4}$:

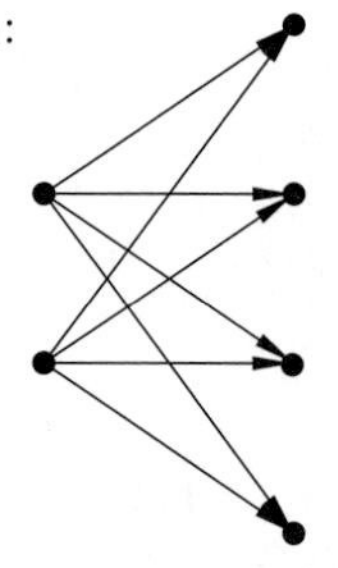

$$\begin{pmatrix} 1 & 0 & -1 & 0 & 0 & 0 \\ 1 & 0 & 0 & -1 & 0 & 0 \\ 1 & 0 & 0 & 0 & -1 & 0 \\ 1 & 0 & 0 & 0 & 0 & -1 \\ 0 & 1 & -1 & 0 & 0 & 0 \\ 0 & 1 & 0 & -1 & 0 & 0 \\ 0 & 1 & 0 & 0 & -1 & 0 \\ 0 & 1 & 0 & 0 & 0 & -1 \end{pmatrix}.$$

2.6.15 False.

Chapter 3

3.1.2 (b) No. (d) No. (f) Yes.

3.1.4 (b) Bilinearity: $\langle c\,\mathbf{u} + d\,\mathbf{v}, \mathbf{w}\rangle = 4(c u_1 + d v_1) w_1 + 2(c u_1 + d v_1) w_2 + 2(c u_2 + d v_2) w_1 + 4(c u_2 + d v_2) w_2 + (c u_3 + d v_3) w_3 = c(4u_1 w_1 + 2u_1 w_2 + 2u_2 w_1 + 4u_2 w_2 + u_3 w_3) + d(4v_1 w_1 + 2v_1 w_2 + 2v_2 w_1 + 4v_2 w_2 + v_3 w_3) = c\,\langle \mathbf{u}, \mathbf{w}\rangle + d\,\langle \mathbf{v}, \mathbf{w}\rangle$,
$\langle \mathbf{u}, c\,\mathbf{v} + d\,\mathbf{w}\rangle = 4u_1(c v_1 + d w_1) + 2u_1(c v_2 + d w_2) + 2u_2(c v_1 + d w_1) + 4u_2(c v_2 + d w_2) + u_3(c v_3 + d w_3) = c(4u_1 v_1 + 2u_1 v_2 + 2u_2 v_1 + 4u_2 v_2 + u_3 v_3) + d(4u_1 w_1 + 2u_1 w_2 + 2u_2 w_1 + 4u_2 w_2 + u_3 w_3) = c\,\langle \mathbf{u}, \mathbf{v}\rangle + d\,\langle \mathbf{u}, \mathbf{w}\rangle$.

Symmetry:
$\langle \mathbf{v}, \mathbf{w}\rangle = 4v_1 w_1 + 2v_1 w_2 + 2v_2 w_1 + 4v_2 w_2 + v_3 w_3 = 4w_1 v_1 + 2w_1 v_2 + 2w_2 v_1 + 4w_2 v_2 + w_3 v_3 = \langle \mathbf{w}, \mathbf{v}\rangle$.

Positivity: $\langle \mathbf{v}, \mathbf{v}\rangle = 4v_1^2 + 4v_1 v_2 + 4v_2^2 + v_3^2 = (2v_1 + v_2)^2 + 3v_2^2 + v_3^2 > 0$ for all $\mathbf{v} = (v_1, v_2, v_3)^T \neq \mathbf{0}$.

3.1.5 (b) $2v_1^2 + 5v_2^2 = 1$; ellipse with semi-axes $\frac{1}{\sqrt{2}}, \frac{1}{\sqrt{5}}$.

3.1.7 $\|c\,\mathbf{v}\| = \sqrt{\langle c\,\mathbf{v}, c\,\mathbf{v}\rangle} = \sqrt{c^2\,\langle \mathbf{v}, \mathbf{v}\rangle} = |c|\,\|\mathbf{v}\|$.

3.1.11 (a) $\|\mathbf{u} + \mathbf{v}\|^2 - \|\mathbf{u} - \mathbf{v}\|^2 = \langle \mathbf{u} + \mathbf{v}, \mathbf{u} + \mathbf{v}\rangle - \langle \mathbf{u} - \mathbf{v}, \mathbf{u} - \mathbf{v}\rangle = \big(\langle \mathbf{u}, \mathbf{u}\rangle + 2\,\langle \mathbf{u}, \mathbf{v}\rangle + \langle \mathbf{v}, \mathbf{v}\rangle\big) - \big(\langle \mathbf{u}, \mathbf{u}\rangle - 2\,\langle \mathbf{u}, \mathbf{v}\rangle + \langle \mathbf{v}, \mathbf{v}\rangle\big) = 4\,\langle \mathbf{u}, \mathbf{v}\rangle$.
(b) $\langle \mathbf{v}, \mathbf{w}\rangle = \frac{1}{4}\big[(v_1 + w_1)^2 - 3(v_1 + w_1)(v_2 + w_2) + 5(v_2 + w_2)^2\big] - \frac{1}{4}\big[(v_1 - w_1)^2 - 3(v_1 - w_1)(v_2 - w_2) + 5(v_2 - w_2)^2\big] = v_1 w_1 - \frac{3}{2} v_1 w_2 - \frac{3}{2} v_2 w_1 + 5\,v_2 w_2$.

3.1.14 $\mathbf{v} \cdot (A\,\mathbf{w}) = \mathbf{v}^T A\,\mathbf{w} = (A^T \mathbf{v})^T\,\mathbf{w} = (A^T \mathbf{v}) \cdot \mathbf{w}$.

3.1.17 Bilinearity: $\langle\!\langle\!\langle c\,\mathbf{u} + d\,\mathbf{v}, \mathbf{w}\rangle\!\rangle\!\rangle = \langle c\,\mathbf{u} + d\,\mathbf{v}, \mathbf{w}\rangle + \langle\!\langle c\,\mathbf{u} + d\,\mathbf{v}, \mathbf{w}\rangle\!\rangle = c\,\langle \mathbf{u}, \mathbf{w}\rangle + d\,\langle \mathbf{v}, \mathbf{w}\rangle + c\,\langle\!\langle \mathbf{u}, \mathbf{w}\rangle\!\rangle + d\,\langle\!\langle \mathbf{v}, \mathbf{w}\rangle\!\rangle = c\,\langle\!\langle\!\langle \mathbf{u}, \mathbf{w}\rangle\!\rangle\!\rangle + d\,\langle\!\langle\!\langle \mathbf{v}, \mathbf{w}\rangle\!\rangle\!\rangle$,

$\langle\!\langle\!\langle \mathbf{u}, c\,\mathbf{v} + d\,\mathbf{w}\rangle\!\rangle\!\rangle = \langle \mathbf{u}, c\,\mathbf{v} + d\,\mathbf{w}\rangle + \langle\!\langle \mathbf{u}, c\,\mathbf{v} + d\,\mathbf{w}\rangle\!\rangle = c\,\langle \mathbf{u}, \mathbf{v}\rangle + d\,\langle \mathbf{u}, \mathbf{w}\rangle + c\,\langle\!\langle \mathbf{u}, \mathbf{v}\rangle\!\rangle + d\,\langle\!\langle \mathbf{u}, \mathbf{w}\rangle\!\rangle = c\,\langle\!\langle\!\langle \mathbf{u}, \mathbf{v}\rangle\!\rangle\!\rangle + d\,\langle\!\langle\!\langle \mathbf{u}, \mathbf{w}\rangle\!\rangle\!\rangle$.

Symmetry: $\langle\!\langle\!\langle \mathbf{v}, \mathbf{w}\rangle\!\rangle\!\rangle = \langle \mathbf{v}, \mathbf{w}\rangle + \langle\!\langle \mathbf{v}, \mathbf{w}\rangle\!\rangle = \langle \mathbf{w}, \mathbf{v}\rangle + \langle\!\langle \mathbf{w}, \mathbf{v}\rangle\!\rangle = \langle\!\langle\!\langle \mathbf{w}, \mathbf{v}\rangle\!\rangle\!\rangle$.

Positivity: $\langle\!\langle\!\langle \mathbf{v}, \mathbf{v}\rangle\!\rangle\!\rangle = \langle \mathbf{v}, \mathbf{v}\rangle + \langle\!\langle \mathbf{v}, \mathbf{v}\rangle\!\rangle > 0$ for all $\mathbf{v} \neq \mathbf{0}$ since both terms are positive.

3.1.19 (a) $\langle 1, x\rangle = \frac{1}{2}$, $\|1\| = 1$, $\|x\| = \frac{1}{\sqrt{3}}$.
(c) $\langle x, e^x\rangle = 1$, $\|x\| = \frac{1}{\sqrt{3}}$, $\|e^x\| = \sqrt{\frac{1}{2}(e^2-1)}$.

3.1.20 (b) $\langle f, g\rangle = 0$, $\|f\| = \sqrt{\frac{2}{3}}$, $\|g\| = \sqrt{\frac{56}{15}}$.

3.1.21 (a) Yes. (c) Yes.

3.1.24 No.

3.1.26 (a) No. (b) Yes.

3.1.28 (a) Bilinearity:
$\langle c f + d g, h\rangle =$
$\int_a^b \left[c f(x) + d g(x) \right] h(x)\, w(x)\, dx =$
$c \int_a^b f(x)\, h(x)\, w(x)\, dx + d \int_a^b g(x)\, h(x)\, w(x)\, dx =$
$c\langle f, h\rangle + d\langle g, h\rangle$. The second identity is similar.

Symmetry: $\langle f, g\rangle = \int_a^b f(x)\, g(x)\, w(x)\, dx =$
$\int_a^b g(x)\, f(x)\, w(x)\, dx = \langle g, f\rangle$.

Positivity: $\langle f, f\rangle = \int_a^b f(x)^2\, w(x)\, dx \ge 0$.
Moreover, since $w(x) > 0$ and the integrand is continuous, $\langle f, f\rangle = 0$ if and only if $f(x)^2\, w(x) \equiv 0$ for all x, and so $f(x) \equiv 0$.

3.1.30 (a) $\langle f, g\rangle = \frac{2}{3}$, $\|f\| = 1$, $\|g\| = \sqrt{\frac{28}{45}}$.

3.2.1 (a) $|\mathbf{v}_1 \cdot \mathbf{v}_2| = 3 \le 5 = \sqrt{5}\,\sqrt{5} = \|\mathbf{v}_1\|\,\|\mathbf{v}_2\|$; angle: $\cos^{-1}\frac{3}{5} \approx .9273$;
(c) $|\mathbf{v}_1 \cdot \mathbf{v}_2| = 0 \le 2\sqrt{6} = \sqrt{2}\,\sqrt{12} = \|\mathbf{v}_1\|\,\|\mathbf{v}_2\|$; angle: $\frac{1}{2}\pi \approx 1.5708$;
(e) $|\mathbf{v}_1 \cdot \mathbf{v}_2| = 4 \le 2\sqrt{15} = \sqrt{10}\,\sqrt{6} = \|\mathbf{v}_1\|\,\|\mathbf{v}_2\|$; angle: $\cos^{-1}\left(-\frac{2}{\sqrt{15}}\right) \approx 2.1134$.

3.2.3 $\|(1,1,0) - (0,0,0)\| = \|(1,1,0) - (1,0,1)\| = \|(1,1,0) - (0,1,1)\| = \cdots = \sqrt{2}$. Edge angle: $\frac{1}{3}\pi = 60^\circ$. Center angle: $\cos\theta = -\frac{1}{3}$, so $\theta = 1.9106 = 109.4712^\circ$.

3.2.4 (a) $|\mathbf{v} \cdot \mathbf{w}| = 5 \le 7.0711 = \sqrt{5}\,\sqrt{10} = \|\mathbf{v}\|\,\|\mathbf{w}\|$.
(c) $|\mathbf{v} \cdot \mathbf{w}| = 22 \le 23.6432 = \sqrt{13}\,\sqrt{43} = \|\mathbf{v}\|\,\|\mathbf{w}\|$.

3.2.5 (b) $|\langle \mathbf{v}, \mathbf{w}\rangle| = 11 \le 11.7473 = \sqrt{23}\,\sqrt{6} = \|\mathbf{v}\|\,\|\mathbf{w}\|$.

3.2.7 Set $\mathbf{v} = (a_1, \ldots, a_n)^T$, $\mathbf{w} = (1, 1, \ldots, 1)^T$, so that Cauchy–Schwarz gives
$|\mathbf{v} \cdot \mathbf{w}| = |a_1 + a_2 + \cdots + a_n| \le$
$\sqrt{n}\,\sqrt{a_1^2 + a_2^2 + \cdots + a_n^2} = \|\mathbf{v}\|\,\|\mathbf{w}\|$.

3.2.9 Since $a \le |a|$ for any real number a, so $\langle \mathbf{v}, \mathbf{w}\rangle \le |\langle \mathbf{v}, \mathbf{w}\rangle| \le \|\mathbf{v}\|\,\|\mathbf{w}\|$.

3.2.12 (b) $|\langle f, g\rangle| = 2/e = .7358 \le 1.555$
$= \sqrt{\frac{2}{3}}\,\sqrt{\frac{1}{2}(e^2 - e^{-2})} = \|f\|\,\|g\|$.

3.2.13 (a) $\frac{1}{2}\pi$.

3.2.14 (a) $|\langle f, g\rangle| = \frac{2}{3} \le \sqrt{\frac{28}{45}} = \|f\|\,\|g\|$.

3.2.16 All scalar multiples of $\left(\frac{1}{2}, -\frac{7}{4}, 1\right)^T$.

3.2.18 $a\,(1, -2, 1, 0)^T + b\,(2, -3, 0, 1)^T$.

3.2.21 (a) All solutions to $a + b = 1$.

3.2.23 Choose $\mathbf{v} = \mathbf{w}$; then $0 = \langle \mathbf{w}, \mathbf{w}\rangle = \|\mathbf{w}\|^2$ and hence $\mathbf{w} = \mathbf{0}$.

3.2.26 (a) $\langle p_1, p_2\rangle = \int_0^1 \left(x - \frac{1}{2}\right) dx = 0$,
$\langle p_1, p_3\rangle = \int_0^1 \left(x^2 - x + \frac{1}{6}\right) dx = 0$,
$\langle p_2, p_3\rangle = \int_0^1 \left(x - \frac{1}{2}\right)\left(x^2 - x + \frac{1}{6}\right) dx = 0$.

3.2.28 $p(x) = a\,((e-1)\,x - 1) + b\,(x^2 - (e-2)\,x)$ for any $a, b \in \mathbb{R}$.

3.2.30 Example: 1 and $x - \frac{2}{3}$.

3.2.32 (b) $\|\mathbf{v}_1 + \mathbf{v}_2\| = \sqrt{2} \le 2\sqrt{2} = \|\mathbf{v}_1\| + \|\mathbf{v}_2\|$;
(d) $\|\mathbf{v}_1 + \mathbf{v}_2\| = \sqrt{3} \le \sqrt{3} + \sqrt{6} = \|\mathbf{v}_1\| + \|\mathbf{v}_2\|$.

3.2.33 (b) $\|\mathbf{v}_1 + \mathbf{v}_2\| = \sqrt{6} \le 3 + \sqrt{19} = \|\mathbf{v}_1\| + \|\mathbf{v}_2\|$.

3.2.34 (b) $\|f + g\| = \sqrt{\frac{2}{3} + \frac{1}{2}e^2 + 4e^{-1} - \frac{1}{2}e^{-2}} =$
$2.40105 \le 2.72093 = \sqrt{\frac{2}{3}} + \sqrt{\frac{1}{2}(e^2 - e^{-2})} =$
$\|f\| + \|g\|$.

3.2.35 (a) $\|f + g\| = \sqrt{\frac{133}{45}} = 1.71917 \le 2.71917 =$
$1 + \sqrt{\frac{28}{45}} = \|f\| + \|g\|$.

3.2.37 (a) $\left| \int_0^1 f(x)\, g(x)\, e^x\, dx \right| \le$
$\sqrt{\int_0^1 f(x)^2\, e^x\, dx}\,\sqrt{\int_0^1 g(x)^2\, e^x\, dx}$;
$\sqrt{\int_0^1 \left[f(x) + g(x) \right]^2 e^x\, dx} \le$
$\sqrt{\int_0^1 f(x)^2\, e^x\, dx} + \sqrt{\int_0^1 g(x)^2\, e^x\, dx}$.
(b) $\langle f, g\rangle = \frac{1}{2}(e^2 - 1) = 3.1945 \le 3.3063 =$
$\sqrt{e-1}\,\sqrt{\frac{1}{3}(e^3 - 1)} = \|f\|\,\|g\|$;
$\|f + g\| = \sqrt{\frac{1}{3}e^3 + e^2 + e - \frac{7}{3}} = 3.8038 \le$
$3.8331 = \sqrt{e-1} + \sqrt{\frac{1}{3}(e^3 - 1)} = \|f\| + \|g\|$.
(c) $\cos\theta = \frac{\sqrt{3}}{2}\,\frac{e^2-1}{\sqrt{(e-1)(e^3-1)}} = .9662$, so $\theta = .2607$.

3.2.40 True.

3.3.1 $\|\mathbf{v} + \mathbf{w}\|_1 = 2 \le 2 = 1 + 1 = \|\mathbf{v}\|_1 + \|\mathbf{w}\|_1$;
$\|\mathbf{v} + \mathbf{w}\|_2 = \sqrt{2} \le 2 = 1 + 1 = \|\mathbf{v}\|_2 + \|\mathbf{w}\|_2$;
$\|\mathbf{v} + \mathbf{w}\|_3 = \sqrt[3]{2} \le 2 = 1 + 1 = \|\mathbf{v}\|_3 + \|\mathbf{w}\|_3$;
$\|\mathbf{v} + \mathbf{w}\|_\infty = 1 \le 2 = 1 + 1 = \|\mathbf{v}\|_\infty + \|\mathbf{w}\|_\infty$.

3.3.2 (b) $\|\mathbf{v} + \mathbf{w}\|_1 = 2 \le 4 = 2 + 2 = \|\mathbf{v}\|_1 + \|\mathbf{w}\|_1$;
$\|\mathbf{v} + \mathbf{w}\|_2 = \sqrt{2} \le 2\sqrt{2} = \sqrt{2} + \sqrt{2} = \|\mathbf{v}\|_2 + \|\mathbf{w}\|_2$;
$\|\mathbf{v} + \mathbf{w}\|_3 = \sqrt[3]{2} \le 2\sqrt[3]{2} = \sqrt[3]{2} + \sqrt[3]{2} = \|\mathbf{v}\|_3 + \|\mathbf{w}\|_3$;
$\|\mathbf{v} + \mathbf{w}\|_\infty = 1 \le 2 = 1 + 1 = \|\mathbf{v}\|_\infty + \|\mathbf{w}\|_\infty$.

3.3.3 (b) $\|\mathbf{u} - \mathbf{v}\|_2 = \sqrt{13}$, $\|\mathbf{u} - \mathbf{w}\|_2 = \sqrt{12}$, $\|\mathbf{v} - \mathbf{w}\|_2 = \sqrt{21}$, so $\mathbf{u}, \mathbf{w}$ are closest.

3.3.4 (a) $\|f\|_\infty = \frac{2}{3}$, $\|g\|_\infty = \frac{1}{4}$;
(b) $\|f + g\|_\infty = \frac{2}{3} \le \frac{2}{3} + \frac{1}{4} = \|f\|_\infty + \|g\|_\infty$.

3.3.6 (b) $\|f - g\|_2 = \sqrt{\frac{1}{3}} = .57735$,
$\|f - h\|_2 = \sqrt{\frac{3}{2} - \frac{4}{\pi}} = .47619$,
$\|g - h\|_2 = \sqrt{\frac{5}{6} - \frac{2}{\pi}} = .44352$, so g, h are closest.

3.3.7 (a) $\|f+g\|_1 = \frac{3}{4} = .75 \le 1.3125 = 1 + \frac{5}{16} = \|f\|_1 + \|g\|_1$;
(b) $\|f+g\|_2 = \sqrt{\frac{31}{48}} = .8036 \le 1.3819 = 1 + \sqrt{\frac{7}{48}} = \|f\|_2 + \|g\|_2$;
(c) $\|f+g\|_3 = \frac{\sqrt[3]{39}}{4} = .8478 \le 1.4310 = 1 + \frac{\sqrt[3]{41}}{8} = \|f\|_3 + \|g\|_3$;
(d) $\|f+g\|_\infty = \frac{5}{4} = 1.25 \le 1.75 = 1 + \frac{3}{4} = \|f\|_\infty + \|g\|_\infty$.

3.3.10 (a) Comes from weighted inner product $\langle \mathbf{v}, \mathbf{w} \rangle = 2 v_1 w_1 + 3 v_2 w_2$.
(c) Clearly positive; $\|c\,v\| = 2|c\,v_1| + |c\,v_2| = |c|\,(2|v_1| + |v_2|) = |c|\,\|\mathbf{v}\|$;
$\|\mathbf{v}+\mathbf{w}\| = 2|v_1 + w_1| + |v_2 + w_2| \le 2|v_1| + |v_2| + 2|w_1| + |w_2| = \|\mathbf{v}\| + \|\mathbf{w}\|$.
(e) Clearly non-negative and equals zero if and only if $v_1 - v_2 = 0 = v_1 + v_2$, so $\mathbf{v} = \mathbf{0}$;
$\|c\,\mathbf{v}\| = \max\{\,|c\,v_1 - c\,v_2|, |c\,v_1 + c\,v_2|\,\} = |c| \max\{\,|v_1 - v_2|, |v_1 + v_2|\,\} = |c|\,\|\mathbf{v}\|$;
$\|\mathbf{v}+\mathbf{w}\| = \max\{\,|v_1 + w_1 - v_2 - w_2|, |v_1 + w_1 + v_2 + w_2|\,\} \le \max\{\,|v_1 - v_2| + |w_1 - w_2|, |v_1 + v_2| + |w_1 + w_2|\,\} \le \max\{\,|v_1 - v_2|, |v_1 + v_2|\,\} + \max\{\,|w_1 - w_2|, |w_1 + w_2|\,\} = \|\mathbf{v}\| + \|\mathbf{w}\|$.

3.3.11 (a) Yes. (d) No.

3.3.13 True for an inner product norm, but false in general.

3.3.14 If $\mathbf{x} = (1,0)^T$, $\mathbf{y} = (0,1)^T$, say, then $\|\mathbf{x}+\mathbf{y}\|_\infty^2 + \|\mathbf{x}-\mathbf{y}\|_\infty^2 = 1 + 1 = 2 \ne 4 = 2\,(\,\|\mathbf{x}\|_\infty^2 + \|\mathbf{y}\|_\infty^2\,)$, which contradicts the identity in Exercise 3.1.12.

3.3.17 (a) $\|f+g\|_1 = \int_a^b |f(x) + g(x)|\,dx \le \int_a^b \big[\,|f(x)| + |g(x)|\,\big]\,dx = \int_a^b |f(x)|\,dx + \int_a^b |g(x)|\,dx = \|f\|_1 + \|g\|_1$.

3.3.19 (a) Clearly positive; $\|c\,\mathbf{v}\| = \max\{\,\|c\,\mathbf{v}\|_1, \|c\,\mathbf{v}\|_2\,\} = |c| \max\{\,\|\mathbf{v}\|_1, \|\mathbf{v}\|_2\,\} = |c|\,\|\mathbf{v}\|$;
$\|\mathbf{v}+\mathbf{w}\| = \max\{\,\|\mathbf{v}+\mathbf{w}\|_1, \|\mathbf{v}+\mathbf{w}\|_2\,\} \le \max\{\,\|\mathbf{v}\|_1 + \|\mathbf{w}\|_1, \|\mathbf{v}\|_2 + \|\mathbf{w}\|_2\,\} \le \max\{\,\|\mathbf{v}\|_1, \|\mathbf{v}\|_2\,\} + \max\{\,\|\mathbf{w}\|_1, \|\mathbf{w}\|_2\,\} = \|\mathbf{v}\| + \|\mathbf{w}\|$.
(c) Yes.

3.3.20 (b) $\left(\frac{1}{3}, \frac{2}{3}, -1\right)^T$; (d) $\left(\frac{1}{3}, \frac{2}{3}, -1\right)^T$.

3.3.21 (b) $\|\mathbf{v}\|^2 = \frac{1}{2}(\cos^2\theta + \sin^2\theta + \cos^2\phi + \sin^2\phi) = 1$.

3.3.22 2, namely $\mathbf{u} = \mathbf{v}/\|\mathbf{v}\|$ and $-\mathbf{u}$.

3.3.24 (a) (c)

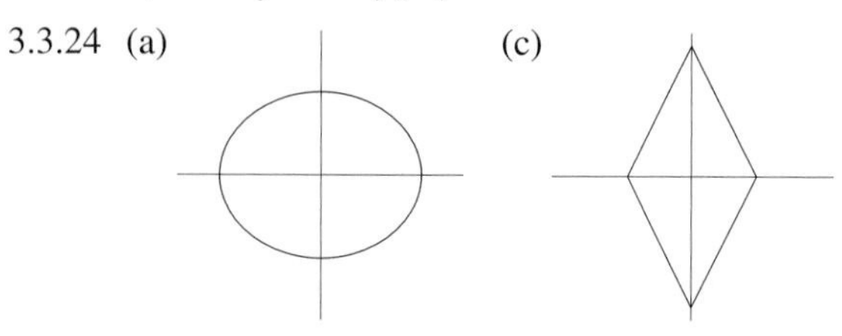

(e)

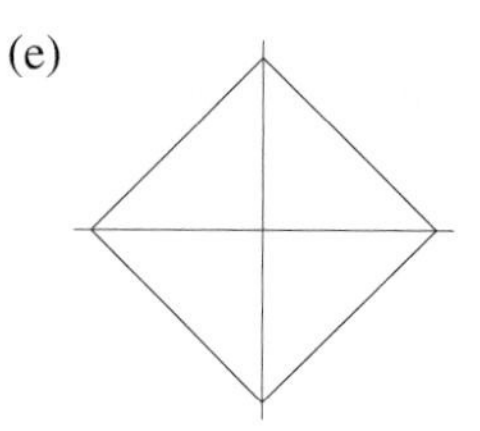

3.3.27 True.

3.3.28 (a) $\frac{18}{5}x - \frac{6}{5}$; (c) $\frac{3}{2}x - \frac{1}{2}$; (e) $\frac{3}{2\sqrt{2}}x - \frac{1}{2\sqrt{2}}$.

3.3.29 (a) Yes. (c) Yes. (e) No.
(g) No — its norm is not defined.

3.3.31 (a) $\|\mathbf{v}\|_2 = \sqrt{2}$, $\|\mathbf{v}\|_\infty = 1$, $\frac{1}{\sqrt{2}}\sqrt{2} \le 1 \le \sqrt{2}$;
(c) $\|\mathbf{v}\|_2 = 2$, $\|\mathbf{v}\|_\infty = 1$, $\frac{1}{2}\,2 \le 1 \le 2$.

3.3.32 (a) $\mathbf{v} = (a,0)^T$ or $(0,a)^T$;
(c) $\mathbf{v} = (a,0)^T$ or $(0,a)^T$.

3.3.35 (i) $\|\mathbf{v}\|_1^2 = \left(\sum_{i=1}^n |v_i|\right)^2 = \sum_{i=1}^n |v_i|^2 + 2\sum_{i<j} |v_i|\,|v_j| \ge \sum_{i=1}^n |v_i|^2 = \|\mathbf{v}\|_2^2$.
On the other hand, since $2xy \le x^2 + y^2$, $\|\mathbf{v}\|_1^2 = \sum_{i=1}^n |v_i|^2 + 2\sum_{i<j} |v_i|\,|v_j| \le n\sum_{i=1}^n |v_i|^2 = n\,\|\mathbf{v}\|_2^2$.
(ii) (a) $\|\mathbf{v}\|_2 = \sqrt{2}$, $\|\mathbf{v}\|_1 = 2$, and $\sqrt{2} \le 2 \le \sqrt{2}\,\sqrt{2}$;
(c) $\|\mathbf{v}\|_2 = 2$, $\|\mathbf{v}\|_1 = 4$, and $2 \le 4 \le 2 \cdot 2$.
(iii) (a) $\mathbf{v} = c\,\mathbf{e}_j$ for some $j = 1, \ldots, n$.

3.3.37 (a) $c^\star = \sqrt{2}$, $C^\star = \sqrt{3}$.

3.3.39 The norms are constant multiples of each other.

3.4.1 (a) positive definite; $\langle \mathbf{v}, \mathbf{w} \rangle = v_1 w_1 + 2 v_2 w_2$.
(c) not positive definite.
(e) positive definite; $\langle \mathbf{v}, \mathbf{w} \rangle = v_1 w_1 - v_1 w_2 - v_2 w_1 + 3 v_2 w_2$.

3.4.4 (a) $k_{ii} = \mathbf{e}_i^T K\,\mathbf{e}_i > 0$.
(b) $\begin{pmatrix} 1 & 2 \\ 2 & 1 \end{pmatrix}$, (c) $\begin{pmatrix} 1 & 0 \\ 0 & 0 \end{pmatrix}$.

3.4.6 First, $(c\,K)^T = c\,K^T = c\,K$ is symmetric. Second, $\mathbf{x}^T(c\,K)\,\mathbf{x} = c\,\mathbf{x}^T K\,\mathbf{x} > 0$ for any $\mathbf{x} \ne \mathbf{0}$ since $c > 0$ and $K > 0$.

3.4.8 $\begin{pmatrix} 3 & 1 \\ 1 & 1 \end{pmatrix}$, $\begin{pmatrix} 1 & 1 \\ 1 & 4 \end{pmatrix}$.

3.4.15 $\mathbf{x}^T K\,\mathbf{x} = (\,1 \quad 1\,)\begin{pmatrix} -1 \\ 1 \end{pmatrix} = \mathbf{0}$, but $K\,\mathbf{x} = \begin{pmatrix} -1 \\ 1 \end{pmatrix} \ne \mathbf{0}$.

3.4.17 (a) False.

3.4.20 Since $q(\mathbf{x})$ is a scalar $q(\mathbf{x}) = \mathbf{x}^T A\,\mathbf{x} = (\mathbf{x}^T A\,\mathbf{x})^T = \mathbf{x}^T A^T \mathbf{x}$, and hence $q(\mathbf{x}) = \frac{1}{2}(\mathbf{x}^T A\,\mathbf{x} + \mathbf{x}^T A^T \mathbf{x}) = \mathbf{x}^T K\,\mathbf{x}$.

3.4.22 (i) $\begin{pmatrix} 10 & 6 \\ 6 & 4 \end{pmatrix}$; positive definite.
(iii) $\begin{pmatrix} 6 & -8 \\ -8 & 13 \end{pmatrix}$; positive definite.

(v) $\begin{pmatrix} 9 & 6 & 3 \\ 6 & 6 & 0 \\ 3 & 0 & 3 \end{pmatrix}$; positive semi-definite;

null vectors: $\begin{pmatrix} -t \\ t \\ t \end{pmatrix}$.

(vii) $\begin{pmatrix} 30 & 0 & -6 \\ 0 & 30 & 3 \\ -6 & 3 & 15 \end{pmatrix}$; positive definite.

3.4.23 (iv) $\begin{pmatrix} 3 & 1 & 2 \\ 1 & 4 & 3 \\ 2 & 3 & 5 \end{pmatrix}$; positive definite.

3.4.24 (vii) $\begin{pmatrix} 10 & -2 & -1 \\ -2 & \frac{145}{12} & \frac{10}{3} \\ -1 & \frac{10}{3} & \frac{41}{6} \end{pmatrix}$.

3.4.25 $K = \begin{pmatrix} 1 & e-1 & \frac{1}{2}(e^2-1) \\ e-1 & \frac{1}{2}(e^2-1) & \frac{1}{3}(e^3-1) \\ \frac{1}{2}(e^2-1) & \frac{1}{3}(e^3-1) & \frac{1}{4}(e^4-1) \end{pmatrix}$ is positive definite.

3.4.32 $0 = \mathbf{x}^T K \mathbf{x} = \mathbf{x}^T A^T C A \mathbf{x} = \mathbf{y}^T C \mathbf{y}$, where $\mathbf{y} = A\mathbf{x}$. Since $C > 0$, this implies $\mathbf{y} = \mathbf{0}$, and hence $\mathbf{x} \in \ker A = \ker K$.

3.5.1 (a) Positive definite; (c) not positive definite; (e) positive definite.

3.5.2 (a) $\begin{pmatrix} 1 & 2 \\ 2 & 3 \end{pmatrix} = \begin{pmatrix} 1 & 0 \\ 2 & 1 \end{pmatrix} \begin{pmatrix} 1 & 0 \\ 0 & -1 \end{pmatrix} \begin{pmatrix} 1 & 2 \\ 0 & 1 \end{pmatrix}$; not positive definite;

(c) $\begin{pmatrix} 3 & -1 & 3 \\ -1 & 5 & 1 \\ 3 & 1 & 5 \end{pmatrix} =$

$\begin{pmatrix} 1 & 0 & 0 \\ -\frac{1}{3} & 1 & 0 \\ 1 & \frac{3}{7} & 1 \end{pmatrix} \begin{pmatrix} 3 & 0 & 0 \\ 0 & \frac{14}{3} & 0 \\ 0 & 0 & \frac{8}{7} \end{pmatrix} \begin{pmatrix} 1 & -\frac{1}{3} & 1 \\ 0 & 1 & \frac{3}{7} \\ 0 & 0 & 1 \end{pmatrix}$;

positive definite;

(e) $\begin{pmatrix} 2 & 1 & -2 \\ 1 & 1 & -3 \\ -2 & -3 & 11 \end{pmatrix} =$

$\begin{pmatrix} 1 & 0 & 0 \\ \frac{1}{2} & 1 & 0 \\ -1 & -4 & 1 \end{pmatrix} \begin{pmatrix} 2 & 0 & 0 \\ 0 & \frac{1}{2} & 0 \\ 0 & 0 & 1 \end{pmatrix} \begin{pmatrix} 1 & \frac{1}{2} & -1 \\ 0 & 1 & -4 \\ 0 & 0 & 1 \end{pmatrix}$;

positive definite;

(g) $\begin{pmatrix} 3 & 2 & 1 & 0 \\ 2 & 3 & 0 & 1 \\ 1 & 0 & 3 & 2 \\ 0 & 1 & 2 & 4 \end{pmatrix} =$

$\begin{pmatrix} 1 & 0 & 0 & 0 \\ \frac{2}{3} & 1 & 0 & 0 \\ \frac{1}{3} & -\frac{2}{5} & 1 & 0 \\ 0 & \frac{3}{5} & 1 & 1 \end{pmatrix} \begin{pmatrix} 3 & 0 & 0 & 0 \\ 0 & \frac{5}{3} & 0 & 0 \\ 0 & 0 & \frac{12}{5} & 0 \\ 0 & 0 & 0 & 1 \end{pmatrix}$

$\begin{pmatrix} 1 & \frac{2}{3} & \frac{1}{3} & 0 \\ 0 & 1 & -\frac{2}{5} & \frac{3}{5} \\ 0 & 0 & 1 & 1 \\ 0 & 0 & 0 & 1 \end{pmatrix}$;

positive definite.

3.5.4 $K = \begin{pmatrix} 1 & \frac{1}{2} & -\frac{1}{2} \\ \frac{1}{2} & 2 & 0 \\ -\frac{1}{2} & 0 & 3 \end{pmatrix}$; positive definite.

3.5.5 (a) $(x + 4y)^2 - 15y^2$; not positive definite; (b) $(x - 2y)^2 + 3y^2$; positive definite.

3.5.6 (b) $\left(x + \frac{3}{2}y - z\right)^2 + \frac{3}{4}(y + 2z)^2 + 4z^2$.

3.5.7 (a) $(x \;\; y \;\; z)^T \begin{pmatrix} 1 & 0 & 2 \\ 0 & 2 & 4 \\ 2 & 4 & 12 \end{pmatrix} \begin{pmatrix} x \\ y \\ z \end{pmatrix}$; not positive definite;

(c) $(x \;\; y \;\; z)^T \begin{pmatrix} 1 & 1 & -2 \\ 1 & 2 & -3 \\ -2 & -3 & 6 \end{pmatrix} \begin{pmatrix} x \\ y \\ z \end{pmatrix}$; positive definite.

3.5.13 For $c \gg 0$ sufficiently large, write $S = (S + c\,\mathrm{I}) + (-c\,\mathrm{I}) = K + N$.

3.5.19 (b) $\begin{pmatrix} 4 & -12 \\ -12 & 45 \end{pmatrix} = \begin{pmatrix} 2 & 0 \\ -6 & 3 \end{pmatrix} \begin{pmatrix} 2 & -6 \\ 0 & 3 \end{pmatrix}$;

(d) $\begin{pmatrix} 2 & 1 & 1 \\ 1 & 2 & 1 \\ 1 & 1 & 2 \end{pmatrix}$

$= \begin{pmatrix} \sqrt{2} & 0 & 0 \\ \frac{1}{\sqrt{2}} & \frac{\sqrt{3}}{\sqrt{2}} & 0 \\ \frac{1}{\sqrt{2}} & \frac{1}{\sqrt{6}} & \frac{2}{\sqrt{3}} \end{pmatrix} \begin{pmatrix} \sqrt{2} & \frac{1}{\sqrt{2}} & \frac{1}{\sqrt{2}} \\ 0 & \frac{\sqrt{3}}{\sqrt{2}} & \frac{1}{\sqrt{6}} \\ 0 & 0 & \frac{2}{\sqrt{3}} \end{pmatrix}$.

3.5.20 (a) $\begin{pmatrix} 4 & -2 \\ -2 & 4 \end{pmatrix} = \begin{pmatrix} 2 & 0 \\ -1 & \sqrt{3} \end{pmatrix} \begin{pmatrix} 2 & -1 \\ 0 & \sqrt{3} \end{pmatrix}$;
(c) no factorization.

3.5.21 (b) $z_1^2 + z_2^2$, where $z_1 = x_1 - x_2$, $z_2 = \sqrt{3}\,x_2$;
(d) $z_1^2 + z_2^2 + z_3^2$, where $z_1 = \sqrt{3}\,x_1 - \frac{1}{\sqrt{3}}x_2 - \frac{1}{\sqrt{3}}x_3$, $z_2 = \sqrt{\frac{5}{3}}\,x_2 - \frac{1}{\sqrt{15}}x_3$, $z_3 = \sqrt{\frac{28}{5}}\,x_3$;
(f) $z_1^2 + z_2^2 + z_3^2$, where $z_1 = 2x_1 - \frac{1}{2}x_2 - x_3$, $z_2 = \frac{1}{2}x_2 - 2x_3$, $z_3 = x_3$.

3.6.2 $e^{k\pi\,\mathrm{i}} = \cos k\pi + \mathrm{i}\,\sin k\pi = (-1)^k$.

3.6.5 (a) $\mathrm{i} = e^{\pi\,\mathrm{i}/2}$;
(b) $\sqrt{\mathrm{i}} = e^{\pi\,\mathrm{i}/4} = \frac{1}{\sqrt{2}} + \frac{\mathrm{i}}{\sqrt{2}}$ and $e^{5\pi\,\mathrm{i}/4} = -\frac{1}{\sqrt{2}} - \frac{\mathrm{i}}{\sqrt{2}}$.

3.6.7 (b) $\overline{z}$ moves in a clockwise direction around a circle of radius r.

3.6.9 Write $z = r\,e^{\mathrm{i}\,\theta}$ so $\theta = \mathrm{ph}\,z$. Then $\mathrm{Re}\,e^{\mathrm{i}\,\varphi}\,z = \mathrm{Re}(r\,e^{\mathrm{i}\,(\varphi+\theta)}) = r\cos(\varphi+\theta) \le r = |z|$, with equality if and only if $\varphi + \theta$ is an integer multiple of 2π.

3.6.11 If $z = r\,e^{\,\mathrm{i}\,\theta}$, $w = s\,e^{\,\mathrm{i}\,\varphi} \neq 0$, then $z/w = (r/s)\,e^{\,\mathrm{i}\,(\theta-\varphi)}$ has phase $\mathrm{ph}\,(z/w) = \theta - \varphi = \mathrm{ph}\,z - \mathrm{ph}\,w$, while $z\,\overline{w} = r\,s\,e^{\,\mathrm{i}\,(\theta-\varphi)}$ also has phase $\mathrm{ph}\,(z\,\overline{w}) = \theta - \varphi = \mathrm{ph}\,z - \mathrm{ph}\,w$.

3.6.13 Set $z = x + \mathrm{i}\,y$, $w = u + \mathrm{i}\,v$, then $z\,\overline{w} = (x+\mathrm{i}\,y)\cdot(u-\mathrm{i}\,v) = (x\,u+y\,v)+\mathrm{i}\,(y\,u-x\,v)$ has real part $\mathrm{Re}(z\,\overline{w}) = x\,u + y\,v$ which is the dot product between $(x, y)^T$ and $(u, v)^T$.

3.6.16 (b) $\cos 3\theta = \cos^3\theta - 3\cos\theta\,\sin^2\theta$, $\sin 3\theta = 3\cos\theta\,\sin^2\theta - \sin^3\theta$.

3.6.18 $e^z = e^x\cos y + \mathrm{i}\,e^x\sin y = r\cos\theta + \mathrm{i}\,r\sin\theta$ implies $r = |e^z| = e^x$ and $\theta = \mathrm{ph}\,e^z = y$.

3.6.20 (a) $\cosh(x + \mathrm{i}\,y) = \cosh x\,\cos y - \mathrm{i}\,\sinh x\,\sin y$, (b) $\cos\mathrm{i}\,z = \cosh z$.

3.6.21 (c) (ii) $\cos\theta\sin\theta = -\frac{1}{4}\,\mathrm{i}\,e^{2\mathrm{i}\theta} + \frac{1}{4}\,\mathrm{i}\,e^{-2\mathrm{i}\theta} = \frac{1}{2}\sin 2\theta$, (iv) $\sin^4\theta = \frac{1}{16}e^{4\mathrm{i}\theta} - \frac{1}{4}e^{2\mathrm{i}\theta} + \frac{3}{8} - \frac{1}{4}e^{-2\mathrm{i}\theta} + \frac{1}{16}e^{-4\mathrm{i}\theta} = \frac{3}{8} - \frac{1}{2}\cos 2\theta + \frac{1}{8}\cos 4\theta$.

3.6.24 (a) $\dfrac{d}{dx}\,e^{\lambda x} = \dfrac{d}{dx}\left(e^{\mu x}\cos\nu x + \mathrm{i}\,e^{\mu x}\sin\nu x\right) = \mu\,e^{\mu x}\cos\nu x - \nu\,e^{\mu x}\sin\nu x + \mathrm{i}\,\mu\,e^{\mu x}\sin\nu x + \mathrm{i}\,\nu\,e^{\mu x}\cos\nu x = (\mu + \mathrm{i}\,\nu)\left(e^{\mu x}\cos\nu x + \mathrm{i}\,e^{\mu x}\sin\nu x\right) = \lambda\,e^{\lambda x}$.

3.6.28 (b) Linearly dependent; (d) linearly dependent; (f) linearly independent.

3.6.30 (b) Dimension = 2; basis: $(\mathrm{i} - 1, 0, 1)^T$, $(-\mathrm{i}, 1, 0)^T$; (d) dimension = 1; basis: $\left(-\frac{14}{5} - \frac{8}{5}\mathrm{i}, \frac{13}{5} - \frac{4}{5}\mathrm{i}, 1\right)^T$.

3.6.32 (b) Range: $(2, -4)^T$, $(-1 + \mathrm{i}, 3)^T$; corange: $(2, -1 + \mathrm{i}, 1 - 2\mathrm{i})^T$, $(0, 1 + \mathrm{i}, 3 - 3\mathrm{i})^T$; kernel: $\left(1 + \frac{5}{2}\mathrm{i}, 3\mathrm{i}, 1\right)^T$; cokernel: $\{\mathbf{0}\}$.

3.6.35 (b) Subspace; (d) not a subspace.

3.6.37 (a) $\sin x = -\frac{1}{2}\,\mathrm{i}\,e^{\mathrm{i}x} + \frac{1}{2}\,\mathrm{i}\,e^{-\mathrm{i}x}$; (c) not an element.

3.6.39 (b) No. (d) Yes.

3.6.42 (a) $\|\mathbf{z} + \mathbf{w}\|^2 = \langle \mathbf{z} + \mathbf{w}, \mathbf{z} + \mathbf{w}\rangle = \|\mathbf{z}\|^2 + \langle \mathbf{z}, \mathbf{w}\rangle + \langle \mathbf{w}, \mathbf{z}\rangle + \|\mathbf{w}\|^2 = \|\mathbf{z}\|^2 + \langle \mathbf{z}, \mathbf{w}\rangle + \overline{\langle \mathbf{z}, \mathbf{w}\rangle} + \|\mathbf{w}\|^2 = \|\mathbf{z}\|^2 + 2\,\mathrm{Re}\langle \mathbf{z}, \mathbf{w}\rangle + \|\mathbf{w}\|^2$.

3.6.47 Infinitely many.

3.6.51 (b) (i) $\langle x + \mathrm{i}, x - \mathrm{i}\rangle = -\frac{2}{3} + \mathrm{i}$; $\|x + \mathrm{i}\| = \|x - \mathrm{i}\| = \frac{2}{\sqrt{3}}$; (ii) $|\langle x + \mathrm{i}, x - \mathrm{i}\rangle| = \frac{\sqrt{13}}{3} \leq \frac{4}{3} = \|x + \mathrm{i}\|\,\|x - \mathrm{i}\|$; $\|(x + \mathrm{i}) + (x - \mathrm{i})\| = \|2x\| = \frac{2}{\sqrt{3}} \leq \frac{4}{\sqrt{3}} = \|x + \mathrm{i}\| + \|x - \mathrm{i}\|$.

Chapter 4

4.1.3 (b) $(0, 2)^T$; (d) $\left(-\frac{3}{2}, \frac{3}{2}\right)^T$.

4.1.4 (i) (b) all points on the line segment $(0, y)^T$ for $1 \leq y \leq 3$; (d) $\left(-\frac{3}{2}, \frac{3}{2}\right)^T$. (ii) (b) $(0, 2)^T$; (d) all points on the line segment $(t, -t)^T$ for $-2 \leq t \leq -1$.

4.1.7 When $\|\mathbf{a}\| = \|\mathbf{b}\| = 1$, the distance is $|\sin\theta|$ where θ is the angle between $\mathbf{a}$, $\mathbf{b}$.

4.1.9 (b) $\frac{1}{\sqrt{14}}$.

4.2.1 $x = \frac{1}{2}$, $y = \frac{1}{2}$, $z = -2$ with $f(x, y, z) = -\frac{3}{2}$.

4.2.3 (b) Minimizer: $x = \frac{2}{9}$, $y = \frac{2}{9}$; minimum value: $\frac{32}{9}$. (d) Minimizer: $x = -\frac{1}{2}$, $y = -1$, $z = 1$; minimum value: $-\frac{5}{4}$. (f) No minimum.

4.2.5 (a) $p(\mathbf{x}) = 4x^2 - 24\,x\,y + 45\,y^2 + x - 4\,y + 3$; minimizer: $\mathbf{x}^\star = \left(\frac{1}{24}, \frac{1}{18}\right)^T = (.0417, .0556)^T$; minimum value: $p(\mathbf{x}^\star) = \frac{419}{144} = 2.9097$. (d) $p(\mathbf{x}) = x^2 + 2\,x\,y + 2\,x\,z + 2\,y^2 - 2\,y\,z + z^2 + 6x + 2\,y - 4z + 1$; no minimizer since K is not positive definite.

4.2.6 $n = 3$: minimum value $-\frac{2}{7}$; minimizer $\mathbf{x}^\star = \left(-\frac{5}{28}, -\frac{3}{14}, -\frac{5}{28}\right)^T$.

4.2.9 Let $\mathbf{x}^\star = K^{-1}\mathbf{f}$ be the minimizer. When $c = 0$, according to the third expression in (**??**), $p(\mathbf{x}^\star) = -(\mathbf{x}^\star)^T K\,\mathbf{x}^\star < 0$ because K is positive definite. The minimum value is 0 if and only if $\mathbf{x}^\star = \mathbf{0}$ if and only if $\mathbf{f} = \mathbf{0}$.

4.2.11 If and only if $\mathbf{f} = \mathbf{0}$ and the function is constant, in which case every $\mathbf{x}$ is a minimizer.

4.3.1 Closest point: $\left(\frac{6}{7}, \frac{38}{35}, \frac{36}{35}\right)^T = (.85714, 1.08571, 1.02857)^T$. Distance: $\frac{1}{\sqrt{35}} = .16903$.

4.3.2 (a) Closest point: $(.8343, 1.0497, 1.0221)^T$. Distance: .1744.

4.3.4 (a) Closest point: $\left(\frac{7}{4}, \frac{7}{4}, \frac{7}{4}, \frac{7}{4}\right)^T$. Distance: $\sqrt{\frac{11}{4}}$. (c) Closest point: $(3, 1, 2, 0)^T$. Distance: 1.

4.3.6 (i) 4.3.4 (a) Closest point: $\left(\frac{3}{2}, \frac{3}{2}, \frac{3}{2}, \frac{3}{2}\right)^T$. Distance: $\sqrt{\frac{7}{4}}$. (c) Closest point: $(3, 1, 2, 0)^T$. Distance: 1. (ii) 4.3.4 (a) Closest point: $\left(\frac{25}{14}, \frac{25}{14}, \frac{25}{14}, \frac{25}{14}\right)^T = (1.78571, 1.78571, 1.78571, 1.78571)^T$. Distance: $\sqrt{\frac{215}{14}} = 3.91882$. (c) Closest point: $\left(\frac{28}{9}, \frac{11}{9}, \frac{16}{9}, 0\right)^T = (3.11111, 1.22222, 1.77778, 0)^T$. Distance: $\sqrt{\frac{32}{9}} = 1.88562$.

4.3.8 (a) $\sqrt{\frac{8}{3}}$; (b) $\frac{7}{\sqrt{6}}$.

4.3.10 (a) $\mathbf{x} = \frac{1}{n}\sum_{i=1}^{n}\mathbf{a}_i$ — the center of mass. (b) (i) $\mathbf{x} = \left(-\frac{1}{2}, 4\right)$.

4.3.14 (b) $x = -\frac{1}{25}$, $y = -\frac{8}{21}$;
(d) $x = \frac{1}{3}$, $y = 2$, $z = \frac{3}{4}$.

4.3.15 (b) $\left(\frac{8}{5}, \frac{28}{65}\right)^T = (1.6, .4308)^T$,
(d) $\left(\frac{227}{941}, \frac{304}{941}\right)^T = (.2412, .3231)^T$.

4.3.17 $\mathbf{x}^\star = (-1, 2, 3)^T$. The least squares error is 0.

4.4.1 (b) $y = 1.9 - 1.1\,t$.

4.4.3 (a) $y = 3.9227\,t - 7717.7$;
(b) \$147,359 and \$166,973.

4.4.5 (a) $y = e^{4.6051 - .1903\,t}$; at $t = 10$, $y = 14.9059$;
(b) $t = 48.3897$ days.

4.4.7 170.62 minutes.

4.4.9 (a) $z = 6.9667 - .8\,x - .9333\,y$.

4.4.12 (b) $p(t) = \frac{1}{3}(t-1)(t-3) - \frac{1}{4}t(t-3) + \frac{1}{24}t(t-1)$
$= 1 - \frac{5}{8}t + \frac{1}{8}t^2$,
(d) $p(t) = \frac{1}{2}t(t-2)(t-3) - 2t(t-1)(t-3) + \frac{3}{2}t(t-1)(t-2) = t^2$.

4.4.13 (b) $y = t^2 + 3t + 6$

(d) $y = -t^3 + 2t^2 - 1$

4.4.16 42.1038 seconds.

4.4.18 (a) $p_2(t) = 1 + t + \frac{1}{2}t^2$, whose maximal error over $[0, 1]$ is .218282.

4.4.19 *Note*: In this solution t is measured in degrees!
(a) $p(t) = .0146352\,t + .0243439$.
Maximum error $\approx .0373$.
(c) $p(t) = \frac{\pi}{180}\,t$. Maximum error $\approx .181$.
(e) $p(t) = \frac{\pi}{180}\,t - \frac{1}{6}\left(\frac{\pi}{180}\,t\right)^3$.
Maximum error $\approx .0102$.

4.4.21 $p(t) = .9409\,t + .4566\,t^2 - .7732\,t^3 + .9330\,t^4$.

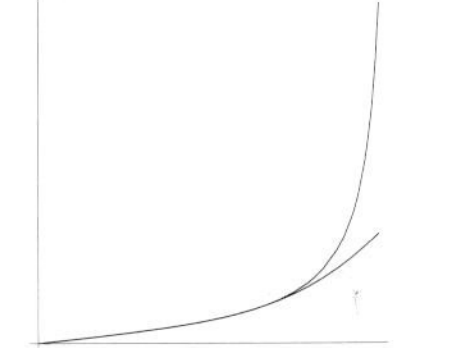

4.4.27 (a) For the data $(0, 0), (1, 1), (2, 2)$, the interpolating polynomial is the straight line $y = t$.
(b) The Lagrange interpolating polynomials are zero at n of the sample points. But the only polynomial of degree $< n$ than vanishes at n points is the zero polynomial, which does not interpolate the final nonzero data value.

4.4.30 $K = V^T V$ is a Gram matrix, where $V = V(x_1, \ldots, x_n)$ is the nonsingular $n \times n$ Vandermonde matrix.

4.4.31 (a) $f'(x) \approx \dfrac{f(x+h) - f(x-h)}{2h}$;
(c) $f'(x) \approx \dfrac{-f(x+2h) + 4f(x+h) - 3f(x)}{2h}$.

4.4.32 (b) Simpson's Rule:
$\int_a^b f(x)\,dx \approx \frac{1}{6}(b-a)\left[f(x_0) + 4f(x_1) + f(x_2)\right]$.
(d) Midpoint Rule: $\int_a^b f(x)\,dx \approx (b-a)\,f(x_0)$.

4.4.33 $g(t) = \frac{3}{8}\cos \pi t + \frac{1}{2}\sin \pi t$.

4.4.35 (a) $g(t) = .538642\,e^t - .004497\,e^{2t}$;
(b) .735894;
(c) The maximal error is .745159 at $t = 3.66351$.

4.4.36 (a) 5 points: $g(t) = -4.4530\cos t + 3.4146\sin t = 5.6115\cos(t - 2.4874)$;
9 points: $g(t) = -4.2284\cos t + 3.6560\sin t = 5.5898\cos(t - 2.4287)$.

4.4.37 (a) $n = 1, k = 4$: $p(t) = .4172 + .4540\cos t$;
maximal error is .1722;

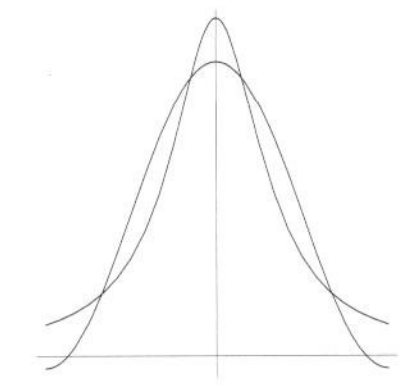

(c) $n = 2, k = 16$:
$p(t) = .4017 + .389329\cos t + .1278\cos 2t$;
maximal error is .0812;

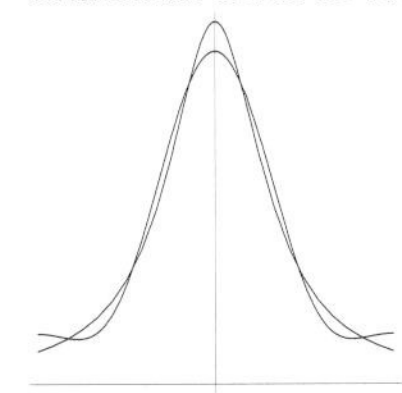

4.4.40 (a) $.29 + .36t$, (c) $-1.2308 + 1.9444\,t$.

4.4.43 (b) $\frac{9}{28} + \frac{9}{7}t - \frac{9}{14}t^2$.

4.4.45 Linear: $p_1(t) = .11477 + .66444\,t$.

4.4.47 (a) $1.875x^2 - .875x$,
(c) $1.7857x^2 - 1.0714x + .1071$.

4.4.50 (ii) $\frac{2}{7} - \frac{25}{14}t + \frac{50}{21}t^2$; maximal error: $\frac{2}{7}$ at $t = 0$.

4.4.52 (b) $z = \frac{9}{10}(x - y)$.

Chapter 5

5.1.1 (a) orthogonal basis, (c) not a basis,
(e) orthogonal basis.

5.1.2 (b) orthonormal basis.

5.1.3 (a) basis, (c) not a basis,
(e) orthonormal basis.

5.1.5 (b) $a = \pm\sqrt{\frac{2}{3}}$.

5.1.6 $a = 2b > 0$.

5.1.9 False.

5.1.13 (c) $A = \begin{pmatrix} 1 & 2 \\ 1 & 3 \end{pmatrix}$, $K = \begin{pmatrix} 10 & -7 \\ -7 & 5 \end{pmatrix}$, with $\langle \mathbf{v}, \mathbf{w} \rangle = 10\, v_1 w_1 - 7\, v_1 w_2 - 7\, v_2 w_1 + 5\, v_2 w_2$.

5.1.14 (a) $\mathbf{v}_1 = \left(\cos\theta,\ \frac{1}{\sqrt{2}}\sin\theta\right)^T$, $\mathbf{v}_2 = \left(\pm\sin\theta,\ \frac{1}{\sqrt{2}}\cos\theta\right)^T$, for any $0 \le \theta < 2\pi$.

5.1.19 $\cos x$ and $\sin x$ span the solution space, and $\int_{-\pi}^{\pi} \sin x\, \cos x\, dx = 0$.

5.1.21 (a) We compute $\langle \mathbf{v}_1, \mathbf{v}_2 \rangle = \langle \mathbf{v}_1, \mathbf{v}_3 \rangle = \langle \mathbf{v}_2, \mathbf{v}_3 \rangle = 0$ and $\| \mathbf{v}_1 \| = \| \mathbf{v}_2 \| = \| \mathbf{v}_3 \| = 1$.
(b) $\langle \mathbf{v}, \mathbf{v}_1 \rangle = \frac{7}{5}$, $\langle \mathbf{v}, \mathbf{v}_2 \rangle = \frac{11}{13}$, and $\langle \mathbf{v}, \mathbf{v}_3 \rangle = -\frac{37}{65}$, and so $(1,1,1)^T = \frac{7}{5}\mathbf{v}_1 + \frac{11}{13}\mathbf{v}_2 - \frac{37}{65}\mathbf{v}_3$.
$\left(\frac{7}{5}\right)^2 + \left(\frac{11}{13}\right)^2 + \left(-\frac{37}{65}\right)^2 = 3 = \| \mathbf{v} \|^2$.

5.1.25 $x = \frac{1}{2} p_1(x) + p_2(x)$.

5.1.26 (b) $\frac{1}{\sqrt{2}},\ \sqrt{\frac{3}{2}}\, t,\ \sqrt{\frac{5}{2}}\left(\frac{3}{2}t^2 - \frac{1}{2}\right),\ \sqrt{\frac{7}{2}}\left(\frac{5}{2}t^3 - \frac{3}{2}t\right)$,
(c) $t^3 = \frac{3}{5} P_1(t) + P_3(t)$.

5.1.28 (b) $\cos x \sin x = \frac{1}{2}\sin 2x$,
(d) $\cos^2 x \sin^3 x = \frac{1}{8}\sin x + \frac{1}{16}\sin 3x - \frac{1}{16}\sin 5x$.

5.1.30 $\langle e^{ikx}, e^{ilx} \rangle = \dfrac{1}{2\pi}\displaystyle\int_{-\pi}^{\pi} e^{ikx}\,\overline{e^{ikx}}\, dx = \dfrac{1}{2\pi}\displaystyle\int_{-\pi}^{\pi} e^{i(k-l)x}\, dx = \begin{cases} 1, & k = l, \\ 0, & k \ne l. \end{cases}$

5.2.1 (b) $\frac{1}{\sqrt{2}}(1,1,0)^T$, $\frac{1}{\sqrt{6}}(-1,1,-2)^T$, $\frac{1}{\sqrt{3}}(1,-1,-1)^T$.

5.2.2 (a) $\left(\frac{1}{\sqrt{2}}, 0, \frac{1}{\sqrt{2}}, 0\right)^T$, $\left(0, \frac{1}{\sqrt{2}}, 0, -\frac{1}{\sqrt{2}}\right)^T$, $\left(\frac{1}{2}, \frac{1}{2}, -\frac{1}{2}, \frac{1}{2}\right)^T$, $\left(-\frac{1}{2}, \frac{1}{2}, \frac{1}{2}, \frac{1}{2}\right)^T$.

5.2.4 (b) $\left(\frac{1}{\sqrt{5}}, \frac{2}{\sqrt{5}}, 0\right)^T$, $\left(-\frac{4}{3\sqrt{5}}, \frac{2}{3\sqrt{5}}, \frac{5}{3\sqrt{5}}\right)^T$.

5.2.6 (a) $\frac{1}{\sqrt{3}}(1,1,-1,0)^T$, $\frac{1}{\sqrt{15}}(-1,2,1,3)^T$, $\frac{1}{\sqrt{15}}(3,-1,2,1)^T$;
(c) $\frac{1}{\sqrt{6}}(2,1,0,-1)^T$, $\frac{1}{\sqrt{6}}(0,1,-2,1)^T$;
(e) $\frac{1}{\sqrt{14}}(2,-1,3,0)^T$, $\frac{1}{9\sqrt{42}}(-34,31,33,14)^T$.

5.2.7 (a) Range: $\frac{1}{\sqrt{10}}\begin{pmatrix} 1 \\ -3 \end{pmatrix}$; kernel: $\frac{1}{\sqrt{2}}\begin{pmatrix} 1 \\ 1 \end{pmatrix}$; corange: $\frac{1}{\sqrt{2}}\begin{pmatrix} 1 \\ -1 \end{pmatrix}$; cokernel: $\frac{1}{\sqrt{10}}\begin{pmatrix} 3 \\ 1 \end{pmatrix}$.
(c) Range: $\frac{1}{\sqrt{3}}\begin{pmatrix} 1 \\ 1 \\ -1 \end{pmatrix}$, $\frac{1}{\sqrt{42}}\begin{pmatrix} 1 \\ 4 \\ 5 \end{pmatrix}$, $\frac{1}{\sqrt{14}}\begin{pmatrix} 3 \\ -2 \\ 1 \end{pmatrix}$;
kernel: $\frac{1}{2}\begin{pmatrix} -1 \\ -1 \\ 1 \\ 1 \end{pmatrix}$;
corange: $\frac{1}{\sqrt{2}}\begin{pmatrix} 1 \\ 0 \\ 1 \\ 0 \end{pmatrix}$, $\frac{1}{\sqrt{2}}\begin{pmatrix} 0 \\ 1 \\ 0 \\ 1 \end{pmatrix}$, $\frac{1}{2}\begin{pmatrix} -1 \\ 1 \\ 1 \\ -1 \end{pmatrix}$;
cokernel: $\{\mathbf{0}\}$.

5.2.8 (i) (b) $\frac{1}{\sqrt{5}}(1,1,0)^T$, $\frac{1}{\sqrt{55}}(-2,3,-5)^T$, $\frac{1}{\sqrt{66}}(2,-3,-6)^T$.
(ii) (b) $\left(\frac{1}{\sqrt{2}}, \frac{1}{\sqrt{2}}, 0\right)^T$, $\left(-\frac{1}{2}, 0, -\frac{1}{2}\right)^T$, $\left(0, -\frac{1}{\sqrt{2}}, -\frac{1}{\sqrt{2}}\right)^T$.

5.2.9 (b) $\left(\frac{1}{2}, 0\right)^T$, $\left(\frac{1}{2\sqrt{3}}, \frac{2}{\sqrt{3}}\right)^T$.

5.2.10 (a) $\left(\frac{1}{2}, 0, 0\right)^T$, $\left(\frac{1}{2\sqrt{2}}, \frac{1}{\sqrt{2}}, 0\right)^T$, $\left(\frac{1}{2\sqrt{6}}, \frac{1}{\sqrt{6}}, \frac{\sqrt{2}}{\sqrt{3}}\right)^T$.

5.2.13 (a) $\left(\frac{1+i}{2}, \frac{1-i}{2}\right)^T$, $\left(\frac{3-i}{2\sqrt{5}}, \frac{1+3i}{2\sqrt{5}}\right)^T$.

5.2.14 (b) $\frac{1}{3}(-1-2i, 2, 0)^T$, $\frac{1}{3\sqrt{19}}(6+2i, 5-5i, 9)^T$.

5.2.15 False.

5.2.17 (a) $(0., .7071, .7071)^T$, $(.8165, -.4082, .4082)^T$, $(.57735, .57735, -.57735)^T$;
(c) $(.5164, .2582, .7746, .2582, 0.)^T$, $(-.21895, -.5200, .4926, -.5200, .4105)^T$, $(.2529, .5454, -.2380, -.3372, .6843)^T$.

5.3.1 (a) Neither, (c) orthogonal, (f) proper orthogonal.

5.3.3 (a) True. (b) False.

5.3.5 (a) $Q^T Q = (y_1^2 + y_2^2 + y_3^2 + y_4^2)^2\, I$ hence $\det Q = (y_1^2 + y_2^2 + y_3^2 + y_4^2)^3 = 1$.
(b) $$Q^{-1} = Q^T = \begin{pmatrix} y_1^2 + y_2^2 - y_3^2 - y_4^2 & 2(y_2 y_3 - y_1 y_4) & 2(y_2 y_4 + y_1 y_3) \\ 2(y_2 y_3 + y_1 y_4) & y_1^2 - y_2^2 + y_3^2 - y_4^2 & 2(y_3 y_4 - y_1 y_2) \\ 2(y_2 y_4 - y_1 y_3) & 2(y_3 y_4 + y_1 y_2) & y_1^2 - y_2^2 - y_3^2 + y_4^2 \end{pmatrix}.$$

5.3.7 $\det(Q_1 Q_2) = \det Q_1 \det Q_2$. If both determinants are $+1$, so is their product.

5.3.9 $(Q^{-1})^T = (Q^T)^T = Q = (Q^{-1})^{-1}$.

5.3.11 All diagonal matrices whose diagonal entries are ± 1.

5.3.14 False; it's true only for row interchanges or multiplication of a row by -1.

5.3.16 (a) $\| Q\mathbf{x} \|^2 = (Q\mathbf{x})^T Q\mathbf{x} = \mathbf{x}^T Q^T Q\mathbf{x} = \mathbf{x}^T\mathbf{x} = \| \mathbf{x} \|^2$.

5.3.19 If $S = (\mathbf{v}_1\ \mathbf{v}_2 \ldots \mathbf{v}_n)$, then $S^{-1} = S^T D$, where $D = \operatorname{diag}(1/\| \mathbf{v}_1 \|^2, \ldots, 1/\| \mathbf{v}_n \|^2)$.

5.3.23 If $S = (\mathbf{v}_1\ \mathbf{v}_2 \ldots \mathbf{v}_n)$, then the (i,j) entry of $S^T K S$ is $\mathbf{v}_i^T K \mathbf{v}_j = \langle \mathbf{v}_i, \mathbf{v}_j \rangle$, so $S^T K S = I$ if and only if $\mathbf{v}_1, \ldots, \mathbf{v}_n$ forms an orthonormal basis for the inner product defined by K.

5.3.25 (b) (ii) $U^{-1} = U^\dagger$

$$= \begin{pmatrix} \frac{1}{\sqrt{3}} & \frac{1}{\sqrt{3}} & \frac{1}{\sqrt{3}} \\ \frac{1}{\sqrt{3}} & -\frac{1}{2\sqrt{3}} - \frac{i}{2} & -\frac{1}{2\sqrt{3}} + \frac{i}{2} \\ \frac{1}{\sqrt{3}} & -\frac{1}{2\sqrt{3}} + \frac{i}{2} & -\frac{1}{2\sqrt{3}} - \frac{i}{2} \end{pmatrix}.$$

(c) (ii) yes.

5.3.27 (a) $\begin{pmatrix} 1 & -3 \\ 2 & 1 \end{pmatrix} = \begin{pmatrix} \frac{1}{\sqrt{5}} & -\frac{2}{\sqrt{5}} \\ \frac{2}{\sqrt{5}} & \frac{1}{\sqrt{5}} \end{pmatrix} \begin{pmatrix} \sqrt{5} & -\frac{1}{\sqrt{5}} \\ 0 & \frac{7}{\sqrt{5}} \end{pmatrix}$;

(c) $\begin{pmatrix} 2 & 1 & -1 \\ 0 & 1 & 3 \\ -1 & -1 & 1 \end{pmatrix} =$

$$\begin{pmatrix} \frac{2}{\sqrt{5}} & -\frac{1}{\sqrt{30}} & \frac{1}{\sqrt{6}} \\ 0 & \sqrt{\frac{5}{6}} & \frac{1}{\sqrt{6}} \\ -\frac{1}{\sqrt{5}} & -\sqrt{\frac{2}{15}} & \sqrt{\frac{2}{3}} \end{pmatrix} \begin{pmatrix} \sqrt{5} & \frac{3}{\sqrt{5}} & -\frac{3}{\sqrt{5}} \\ 0 & \sqrt{\frac{6}{5}} & 7\sqrt{\frac{2}{15}} \\ 0 & 0 & 2\sqrt{\frac{2}{3}} \end{pmatrix};$$

(e) $\begin{pmatrix} 0 & 0 & 2 \\ 0 & 4 & 1 \\ -1 & 0 & 1 \end{pmatrix}$

$$= \begin{pmatrix} 0 & 0 & 1 \\ 0 & 1 & 0 \\ -1 & 0 & 0 \end{pmatrix} \begin{pmatrix} 1 & 0 & -1 \\ 0 & 4 & 1 \\ 0 & 0 & 2 \end{pmatrix}.$$

5.3.28 (ii) (a) $\begin{pmatrix} 2 & 1 & -1 \\ 1 & 0 & 2 \\ 2 & -1 & 3 \end{pmatrix} =$

$$\begin{pmatrix} \frac{2}{3} & \frac{1}{\sqrt{2}} & -\frac{1}{3\sqrt{2}} \\ \frac{1}{3} & 0 & \frac{2\sqrt{2}}{3} \\ \frac{2}{3} & -\frac{1}{\sqrt{2}} & -\frac{1}{3\sqrt{2}} \end{pmatrix} \begin{pmatrix} 3 & 0 & 2 \\ 0 & \sqrt{2} & -2\sqrt{2} \\ 0 & 0 & \sqrt{2} \end{pmatrix};$$

(b) $\begin{pmatrix} x \\ y \\ z \end{pmatrix} = \begin{pmatrix} 1 \\ -1 \\ -1 \end{pmatrix}$.

5.3.29 $\begin{pmatrix} 4 & 1 & 0 & 0 \\ 1 & 4 & 1 & 0 \\ 0 & 1 & 4 & 1 \\ 0 & 0 & 1 & 4 \end{pmatrix}$

$$= \begin{pmatrix} .9701 & -.2339 & .0619 & -.0172 \\ .2425 & .9354 & -.2477 & .0688 \\ 0 & .2650 & .9291 & -.2581 \\ 0 & 0 & .2677 & .9635 \end{pmatrix} \begin{pmatrix} 4.1231 & 1.9403 & .2425 & 0 \\ 0 & 3.773 & 1.9956 & .2650 \\ 0 & 0 & 3.7361 & 1.9997 \\ 0 & 0 & 0 & 3.596 \end{pmatrix}.$$

5.3.30 Exercise 5.3.27(a) $\widehat{\mathbf{v}}_1 = \begin{pmatrix} -1.2361 \\ 2.0000 \end{pmatrix}$,

$H_1 = \begin{pmatrix} .4472 & .8944 \\ .8944 & -.4472 \end{pmatrix}$,

$Q = \begin{pmatrix} .4472 & .8944 \\ .8944 & -.4472 \end{pmatrix}$,

$R = \begin{pmatrix} 2.2361 & -.4472 \\ 0 & -3.1305 \end{pmatrix}$;

(c) $\widehat{\mathbf{v}}_1 = \begin{pmatrix} -.2361 \\ 0 \\ -1 \end{pmatrix}$,

$H_1 = \begin{pmatrix} .8944 & 0 & -.4472 \\ 0 & 1 & 0 \\ -.4472 & 0 & -.8944 \end{pmatrix}$, $\widehat{\mathbf{v}}_2 = \begin{pmatrix} 0 \\ -.0954 \\ .4472 \end{pmatrix}$,

$H_2 = \begin{pmatrix} 1 & 0 & 0 \\ 0 & .9129 & .4082 \\ 0 & .4082 & -.9129 \end{pmatrix}$,

$Q = \begin{pmatrix} .8944 & -.1826 & .4082 \\ 0 & .9129 & .4082 \\ -.4472 & -.3651 & .8165 \end{pmatrix}$,

$R = \begin{pmatrix} 2.2361 & 1.3416 & -1.3416 \\ 0 & 1.0954 & 2.556 \\ 0 & 0 & 1.633 \end{pmatrix}$, ;

(e) $\widehat{\mathbf{v}}_1 = \begin{pmatrix} -1 \\ 0 \\ -1 \end{pmatrix}$, $H_1 = \begin{pmatrix} 0 & 0 & -1 \\ 0 & 1 & 0 \\ -1 & 0 & 0 \end{pmatrix}$,

$\widehat{\mathbf{v}}_2 = \begin{pmatrix} 0 \\ 0 \\ 0 \end{pmatrix}$, $H_2 = \begin{pmatrix} 1 & 0 & 0 \\ 0 & 1 & 0 \\ 0 & 0 & 1 \end{pmatrix}$,

$Q = \begin{pmatrix} 0 & 0 & -1 \\ 0 & 1 & 0 \\ -1 & 0 & 0 \end{pmatrix}$, $R = \begin{pmatrix} 1 & 0 & -1 \\ 0 & 4 & 1 \\ 0 & 0 & -2 \end{pmatrix}$;

5.3.29: 4 × 4 case: $\widehat{\mathbf{v}}_1 = \begin{pmatrix} -.1231 \\ 1 \\ 0 \\ 0 \end{pmatrix}$,

$H_1 = \begin{pmatrix} .9701 & .2425 & 0 & 0 \\ .2425 & -.9701 & 0 & 0 \\ 0 & 0 & 1 & 0 \\ 0 & 0 & 0 & 1 \end{pmatrix}$,

$\widehat{\mathbf{v}}_2 = \begin{pmatrix} 0 \\ -7.411 \\ 1 \\ 0 \end{pmatrix}$,

$H_2 = \begin{pmatrix} 1 & 0 & 0 & 0 \\ 0 & -.9642 & .2650 & 0 \\ 0 & .2650 & .9642 & 0 \\ 0 & 0 & 0 & 1 \end{pmatrix}$,

$\widehat{\mathbf{v}}_3 = \begin{pmatrix} 0 \\ 0 \\ -.1363 \\ 1 \end{pmatrix}$,

$$H_3 = \begin{pmatrix} 1 & 0 & 0 & 0 \\ 0 & 1 & 0 & 0 \\ 0 & 0 & .9635 & .2677 \\ 0 & 0 & .2677 & -.9635 \end{pmatrix},$$

$$Q = \begin{pmatrix} .9701 & -.2339 & .0619 & .0172 \\ .2425 & .9354 & -.2477 & -.0688 \\ 0 & .2650 & .9291 & .2581 \\ 0 & 0 & .2677 & -.9635 \end{pmatrix},$$

$$R = \begin{pmatrix} 4.1231 & 1.9403 & .2425 & 0 \\ 0 & 3.773 & 1.9956 & .2650 \\ 0 & 0 & 3.7361 & 1.9997 \\ 0 & 0 & 0 & -3.596 \end{pmatrix};$$

5.3.33 (c) (i) $\begin{pmatrix} 1 & -1 \\ 2 & 3 \\ 0 & 2 \end{pmatrix} = \begin{pmatrix} \frac{1}{\sqrt5} & -\frac23 \\ \frac{2}{\sqrt5} & \frac13 \\ 0 & \frac23 \end{pmatrix} \begin{pmatrix} \sqrt5 & \sqrt5 \\ 0 & 3 \end{pmatrix};$

(iii) $\begin{pmatrix} -1 & 1 \\ 1 & -2 \\ -1 & 2 \\ 1 & -1 \end{pmatrix} = \begin{pmatrix} -\frac12 & -\frac12 \\ \frac12 & -\frac12 \\ -\frac12 & \frac12 \\ \frac12 & \frac12 \end{pmatrix} \begin{pmatrix} 2 & -3 \\ 0 & 1 \end{pmatrix}.$

5.3.34 (b) (i) $\begin{pmatrix} \mathrm{i} & 1 \\ -1 & 2\,\mathrm{i} \end{pmatrix}$
$= \begin{pmatrix} \frac{\mathrm{i}}{\sqrt2} & -\frac{1}{\sqrt2} \\ -\frac{1}{\sqrt2} & \frac{\mathrm{i}}{\sqrt2} \end{pmatrix} \begin{pmatrix} \sqrt2 & -\frac{3\,\mathrm{i}}{\sqrt2} \\ 0 & \frac{1}{\sqrt2} \end{pmatrix};$

(iii) $\begin{pmatrix} \mathrm{i} & 1 & 0 \\ 1 & \mathrm{i} & 1 \\ 0 & 1 & \mathrm{i} \end{pmatrix} =$

$$\begin{pmatrix} \frac{\mathrm{i}}{\sqrt2} & \frac{1}{\sqrt3} & -\frac{\mathrm{i}}{\sqrt6} \\ \frac{1}{\sqrt2} & \frac{\mathrm{i}}{\sqrt3} & \frac{1}{\sqrt6} \\ 0 & \frac{1}{\sqrt3} & \mathrm{i}\sqrt{\frac23} \end{pmatrix} \begin{pmatrix} \sqrt2 & 0 & \frac{1}{\sqrt2} \\ 0 & \sqrt3 & 0 \\ 0 & 0 & \sqrt{\frac32} \end{pmatrix}.$$

5.4.1 (b) $t^4 + t^2 = q_4(t) + \frac{13}{7}\, q_2(t) + \frac{8}{15}\, q_0(t)$.

5.4.2 (a) $q_5(t) = t^5 - \frac{10}{9}\, t^3 + \frac{5}{21}\, t$;
(b) $t^5 = q_5(t) + \frac{10}{9}\, q_3(t) + \frac37\, q_1(t)$.

5.4.6 $q_k(t) = \dfrac{k\,!}{(2k)\,!}\, \dfrac{d^k}{dt^k}\,(t^2-1)^k$,
$\|q_k\| = \dfrac{2^k\,(k\,!)^2}{(2k)\,!} \sqrt{\dfrac{2}{2k+1}}$.

5.4.8 Write $P_k(t) = \dfrac{1}{2^k\, k\,!}\, \dfrac{d^k}{dt^k}\,(t-1)^k\,(t+1)^k$. Differentiating using Leibniz' Rule, the only term that does not contain a factor of $t-1$ is when the derivatives are applied to $(t-1)^k$ every time. Thus, $P_k(t) = 2^{-k}\,(t+1)^k + (t-1)\, S_k(t)$ for some polynomial $S_k(t)$, and so $P_k(1) = 1$.

5.4.11 (a) $P_0(t) = 1$, $P_1(t) = t - \frac32$,
$P_2(t) = t^2 - 3t + \frac{13}{6}$,
$P_3(t) = t^3 - \frac92\, t^2 + \frac{33}{5}\, t - \frac{63}{20}$;
(c) $P_0(t) = 1$, $P_1(t) = t$, $P_2(t) = t^2 - \frac35$,
$P_3(t) = t^3 - \frac57\, t$.

5.4.13 $1,\ \frac12\, t,\ \frac38\, t^2 - \frac12,\ \frac{5}{16}\, t^3 - \frac34\, t,\ \frac{35}{128}\, t^4 - \frac{15}{16}\, t^2 + \frac38,$
$\frac{63}{256}\, t^5 - \frac{35}{32}\, t^3 + \frac{15}{16}\, t$.

5.4.15 $p_0(t) = 1$, $p_1(t) = t - \frac12$, $p_2(t) = t^2 - t + \frac16$,
$p_3(t) = t^3 - \frac32\, t^2 + \frac{33}{65}\, t - \frac{1}{260}$.

5.4.17 $L_4(t) = t^4 - 16t^3 + 72t^2 - 96t + 24$, $\|L_4\| = 24$.

5.4.19 $p_0(t) = 1$, $p_1(t) = t$, $p_2(t) = t^2 - \frac12$,
$p_3(t) = t^3 - \frac32\, t$, $p_4(t) = t^4 - 3t^2 + \frac34$.

5.4.22 $e^{2x},\ \dfrac{2\,(e^3-1)}{3\,(e^2-1)}\, e^x$.

5.5.1 (a) $\mathbf{v}_2, \mathbf{v}_4$, (c) $\mathbf{v}_2$, (e) $\mathbf{v}_1$.

5.5.2 (a) $\left(-\frac13, \frac13, \frac13\right)^T$; (c) $\left(\frac79, \frac{11}{9}, \frac19\right)^T$.

5.5.5 (a) $\left(\frac{11}{21}, \frac{10}{21}, -\frac27, -\frac{10}{21}\right)^T$; (c) $\left(\frac23, \frac73, -1, \frac53\right)^T$.

5.5.6 (i) (a) $\left(-\frac15, \frac15, \frac15\right)^T$; (c) $\left(\frac{15}{17}, \frac{19}{17}, \frac{1}{17}\right)^T$.
(ii) (a) $(0,0,0)^T$; (c) $\left(\frac{13}{22}, \frac{9}{22}, -\frac{1}{22}\right)^T$.

5.5.8 (b) (ii) $\begin{pmatrix} \frac49 & -\frac49 & \frac29 \\ -\frac49 & \frac49 & -\frac29 \\ \frac29 & -\frac29 & \frac19 \end{pmatrix}$,

(iv) $\begin{pmatrix} \frac19 & -\frac29 & \frac29 & 0 \\ -\frac29 & \frac89 & 0 & -\frac29 \\ \frac29 & 0 & \frac89 & -\frac29 \\ 0 & -\frac29 & -\frac29 & \frac19 \end{pmatrix}$.

5.5.10 $\left(\frac12, -\frac12, 2\right)^T$.

5.5.11 (b) $\left(\frac{9}{14}, \frac{4}{31}\right)^T$.

5.5.13 orthogonal basis: $(1,0,2,1)^T$, $(1,1,0,-1)^T$, $\left(\frac12, -1, 0, -\frac12\right)^T$; closest point: $\left(-\frac23, 2, \frac23, \frac43\right)^T$.

5.5.15 (b) $p_1(t) = .285714 + 1.01429t$,
$p_2(t) = .285714 + 1.01429\,t - .0190476\,(t^2-4)$,
$p_3(t) = .285714 + 1.01429\,t - .0190476\,(t^2-4) - .008333\,(t^3 - 7t)$.

5.5.17 (b) $p_4(t) = .2857 + 1.0143\,t - .019048\,(t^2-4) - .008333\,(t^3-7t) + .011742\left(t^4 - \frac{67}{7}\,t^2 + \frac{72}{7}\right)$.

5.5.20 $q_0(t) = 1$, $q_1(t) = t - \overline{t}$,
$q_2(t) = t^2 - \dfrac{\overline{t^3} - \overline{t}\,\overline{t^2}}{\overline{t^2} - \overline{t}^2}\,(t - \overline{t}) - \overline{t^2}$,
$\mathbf{q}_0 = \mathbf{t}_0$, $\mathbf{q}_1 = \mathbf{t}_1 - \overline{t}\,\mathbf{t}_0$,
$\mathbf{q}_2 = \mathbf{t}_2 - \dfrac{\overline{t^3} - \overline{t}\,\overline{t^2}}{\overline{t^2} - \overline{t}^2}\,(\mathbf{t}_1 - \overline{t}) - \overline{t^2}\,\mathbf{t}_0$,
$\|\mathbf{q}_0\|^2 = 1$, $\|\mathbf{q}_1\|^2 = \overline{t^2} - \overline{t}^2$,
$\|\mathbf{q}_2\|^2 = \overline{t^4} - \left(\overline{t^2}\right)^2 - \dfrac{\left(\overline{t^3} - \overline{t}\,\overline{t^2}\right)^2}{\overline{t^2} - \overline{t}^2}$.

5.5.23 (b) $Q = \begin{pmatrix} .8 & -.43644 \\ .4 & .65465 \\ .2 & -.43644 \\ .4 & .43644 \end{pmatrix}$,

$R = \begin{pmatrix} 5 & 0 \\ 0 & 4.58258 \end{pmatrix}$, $\mathbf{x} = \begin{pmatrix} -.04000 \\ -.38095 \end{pmatrix}$;

(d) $Q = \begin{pmatrix} .18257 & .36515 & .12910 \\ .36515 & -.18257 & .90370 \\ 0 & .91287 & .12910 \\ -.91287 & 0 & .38730 \end{pmatrix}$,

$R = \begin{pmatrix} 5.47723 & -2.19089 & 0 \\ 0 & 1.09545 & -3.65148 \\ 0 & 0 & 2.58199 \end{pmatrix}$,

$\mathbf{x} = \begin{pmatrix} .33333 \\ 2.00000 \\ .75000 \end{pmatrix}$.

5.5.25 (a) If $A = Q$ has orthonormal columns, then $\| Q\mathbf{x}^\star - \mathbf{b} \|^2 = \|\mathbf{b}\|^2 - \| Q^T \mathbf{b} \|^2 = \sum_{i=1}^m b_i^2 - \sum_{i=1}^n (\mathbf{u}_i \cdot \mathbf{b})^2$.

5.5.27 (a) $P = \begin{pmatrix} .25 & -.25 & -.35 & .05 \\ -.25 & .25 & .35 & -.05 \\ -.35 & .35 & .49 & -.07 \\ .05 & -.05 & -.07 & .01 \end{pmatrix}$,

$P\mathbf{v} = \begin{pmatrix} .25 \\ -.25 \\ -.35 \\ .05 \end{pmatrix}$;

(c) $P = \begin{pmatrix} .28 & -.4 & .2 & .04 \\ -.4 & .6 & -.2 & -.2 \\ .2 & -.2 & .4 & -.4 \\ .04 & -.2 & -.4 & .72 \end{pmatrix}$,

$P\mathbf{v} = \begin{pmatrix} .28 \\ -.4 \\ .2 \\ .04 \end{pmatrix}$.

5.5.28 (a) $\frac{1}{5} + \frac{4}{7}\left(-\frac{1}{2} + \frac{3}{2}t^2\right) = -\frac{3}{35} + \frac{6}{7}t^2$.

5.5.30 $1.718282 + .845155\,(2t - 1) + .139864\,(6t^2 - 6t + 1) + .0139313\,(20t^3 - 30t^2 + 12t - 1) = .99906 + 1.0183\,t + .421246\,t^2 + .278625\,t^3$.

5.5.31 Quadratic: $\frac{1}{4} + \frac{9}{20}(2t - 1) + \frac{1}{4}(6t^2 - 6t + 1) = \frac{1}{20} - \frac{3}{5}t + \frac{3}{2}t^2$; value: $\frac{1}{2800} = .0003571$.

5.5.33 $.459698 + .427919\,(2t - 1) - .0392436\,(6t^2 - 6t + 1) - .00721219\,(20t^3 - 30t^2 + 12t - 1) = -.000252739 + 1.00475\,t - .0190961\,t^2 - .144244\,t^3$.

5.5.35 (a) $\frac{3}{2} - \frac{10}{3}\left(t - \frac{3}{4}\right) + \frac{35}{4}\left(t^2 - \frac{4}{3}t + \frac{2}{5}\right) = \frac{15}{2} - 15t + \frac{35}{4}t^2$.

5.6.1 (a) span of $\left(\frac{1}{3}, 1, 0\right)^T, \left(-\frac{1}{3}, 0, 1\right)^T$, $\dim W^\perp = 2$; (d) span of $(2, 1, 1)^T$, $\dim W^\perp = 1$.

5.6.2 (a) $(3, 4, -5)^T$; (c) $(-1, -1, 1)^T$.

5.6.3 (a) $(-1, 3, 2, 1)^T$; (c) $\left(-1, \frac{2}{7}, 1, 0\right)^T, \left(0, \frac{4}{7}, 0, 1\right)^T$.

5.6.4 (a) $\mathbf{w} = \left(\frac{3}{10}, -\frac{1}{10}\right)^T$, $\mathbf{z} = \left(\frac{7}{10}, \frac{21}{10}\right)^T$; (c) $\mathbf{w} = \left(\frac{1}{3}, -\frac{1}{3}, -\frac{1}{3}\right)^T$, $\mathbf{z} = \left(\frac{2}{3}, \frac{1}{3}, \frac{1}{3}\right)^T$.

5.6.5 (a) span of $\left(\frac{2}{3}, 1, 0\right)^T, (-1, 0, 1)^T$, $\dim W^\perp = 2$; (d) span of $\left(6, \frac{3}{2}, 1\right)^T$, $\dim W^\perp = 1$.

5.6.7 (b) Basis: $t^3 - \frac{3}{5}t,\ t^4 - \frac{6}{7}t^2 + \frac{3}{35}$; $\dim W^\perp = 2$.

5.6.9 (a) If $\mathbf{w} \in W \cap W^\perp$ then $\mathbf{w} \in W^\perp$ must be orthogonal to every vector in W and so $\mathbf{w} \in W$ is orthogonal to itself, which implies $\mathbf{w} = \mathbf{0}$.

5.6.17 (a) (i) Range: $\begin{pmatrix} 1 \\ 2 \end{pmatrix}$; cokernel: $\begin{pmatrix} -2 \\ 1 \end{pmatrix}$; corange: $\begin{pmatrix} 1 \\ -2 \end{pmatrix}$; kernel: $\begin{pmatrix} 2 \\ 1 \end{pmatrix}$;

(ii) $\begin{pmatrix} 1 \\ 2 \end{pmatrix} \cdot \begin{pmatrix} -2 \\ 1 \end{pmatrix} = 0$;

(iii) $\begin{pmatrix} 1 \\ -2 \end{pmatrix} \cdot \begin{pmatrix} 2 \\ 1 \end{pmatrix} = 0$.

(c) (i) Range: $\begin{pmatrix} 0 \\ -1 \\ -2 \end{pmatrix}, \begin{pmatrix} 1 \\ 0 \\ 3 \end{pmatrix}$; cokernel: $\begin{pmatrix} -3 \\ -2 \\ 1 \end{pmatrix}$;

corange: $\begin{pmatrix} -1 \\ 0 \\ -3 \end{pmatrix}, \begin{pmatrix} 0 \\ 1 \\ 2 \end{pmatrix}$; kernel: $\begin{pmatrix} -3 \\ -2 \\ 1 \end{pmatrix}$;

(ii) $\begin{pmatrix} 0 \\ -1 \\ -2 \end{pmatrix} \cdot \begin{pmatrix} -3 \\ -2 \\ 1 \end{pmatrix} = \begin{pmatrix} 1 \\ 0 \\ 3 \end{pmatrix} \cdot \begin{pmatrix} -3 \\ -2 \\ 1 \end{pmatrix} = 0$;

(iii) $\begin{pmatrix} -1 \\ 0 \\ -3 \end{pmatrix} \cdot \begin{pmatrix} -3 \\ -2 \\ 1 \end{pmatrix} = \begin{pmatrix} 0 \\ 1 \\ 2 \end{pmatrix} \cdot \begin{pmatrix} -3 \\ -2 \\ 1 \end{pmatrix} = 0$.

(e) (i) Range: $\begin{pmatrix} 3 \\ 1 \\ 5 \end{pmatrix}, \begin{pmatrix} 1 \\ 1 \\ 2 \end{pmatrix}$; cokernel: $\begin{pmatrix} -3 \\ -1 \\ 2 \end{pmatrix}$;

corange: $\begin{pmatrix} 3 \\ 1 \\ 4 \\ 2 \\ 7 \end{pmatrix}, \begin{pmatrix} 0 \\ 1 \\ 1 \\ -1 \\ 1 \end{pmatrix}$;

kernel: $\begin{pmatrix} -1 \\ -1 \\ 1 \\ 0 \\ 0 \end{pmatrix}, \begin{pmatrix} -1 \\ 1 \\ 0 \\ 1 \\ 0 \end{pmatrix}, \begin{pmatrix} -2 \\ -1 \\ 0 \\ 0 \\ 1 \end{pmatrix}$;

(ii) $\begin{pmatrix} 3 \\ 1 \\ 5 \end{pmatrix} \cdot \begin{pmatrix} -3 \\ -1 \\ 2 \end{pmatrix} = \begin{pmatrix} 1 \\ 1 \\ 2 \end{pmatrix} \cdot \begin{pmatrix} -3 \\ -1 \\ 2 \end{pmatrix} = 0$;

(iii) $\begin{pmatrix}3\\1\\4\\2\\7\end{pmatrix}\cdot\begin{pmatrix}-1\\-1\\1\\0\\0\end{pmatrix}=\begin{pmatrix}3\\1\\4\\2\\7\end{pmatrix}\cdot\begin{pmatrix}-1\\1\\0\\1\\0\end{pmatrix}=$
$\begin{pmatrix}3\\1\\4\\2\\7\end{pmatrix}\cdot\begin{pmatrix}-2\\-1\\0\\0\\1\end{pmatrix}=\begin{pmatrix}0\\1\\1\\-1\\1\end{pmatrix}\cdot\begin{pmatrix}-1\\-1\\1\\0\\0\end{pmatrix}=$
$\begin{pmatrix}0\\1\\1\\-1\\1\end{pmatrix}\cdot\begin{pmatrix}-1\\1\\0\\1\\0\end{pmatrix}=\begin{pmatrix}0\\1\\1\\-1\\1\end{pmatrix}\cdot\begin{pmatrix}-2\\-1\\0\\0\\1\end{pmatrix}=0.$

5.6.18 (a) The compatibility condition is $\frac{2}{3}b_1+b_2=0$ and so the basis is $\left(\frac{2}{3},1\right)^T$.
(c) There are no compatibility conditions, and so the cokernel is $\{\mathbf{0}\}$.

5.6.20 (a) $2a-b+c=0$;
(c) $-3b_1+b_2+b_3=2b_1-5b_2+b_4=0$.

5.6.21 (a) $\mathbf{z}=\begin{pmatrix}\frac{1}{2}\\0\\-\frac{1}{2}\end{pmatrix}$,
$\mathbf{w}=\begin{pmatrix}\frac{1}{2}\\0\\\frac{1}{2}\end{pmatrix}=-\frac{3}{2}\begin{pmatrix}1\\-2\\1\end{pmatrix}+\begin{pmatrix}2\\-3\\2\end{pmatrix}$;
(c) $\mathbf{z}=\begin{pmatrix}\frac{14}{17}\\-\frac{1}{17}\\-\frac{4}{17}\\-\frac{5}{17}\end{pmatrix}$,
$\mathbf{w}=\begin{pmatrix}\frac{3}{17}\\\frac{1}{17}\\\frac{4}{17}\\\frac{5}{17}\end{pmatrix}=\frac{1}{51}\begin{pmatrix}1\\-1\\0\\3\end{pmatrix}+\frac{4}{51}\begin{pmatrix}2\\1\\3\\3\end{pmatrix}$.

5.6.22 (a) (i) Fredholm requires that the cokernel basis $\left(\frac{1}{2},1\right)^T$ be orthogonal to the right hand side $(-6,3)^T$;
(ii) the general solution is $x=-3+2y$ with y free;
(iii) the minimum norm solution is $x=-\frac{3}{5}$, $y=\frac{6}{5}$.
(c) (i) Fredholm requires that the cokernel basis $(-1,3)^T$ be orthogonal to the right hand side $(12,4)^T$
(ii) the general solution is $x=2+\frac{1}{2}y-\frac{3}{2}z$ with y,z free;
(iii) the minimum norm solution is $x=\frac{4}{7}$, $y=-\frac{2}{7}$, $z=\frac{6}{7}$.
(e) (i) Fredholm requires that the cokernel basis $(-10,-9,7,0)^T$, $(6,4,0,7)^T$ be orthogonal to the right hand side $(-8,5,-5,4)^T$;
(ii) the general solution is $x_1=1-t$, $x_2=3+2t$, $x_3=t$ with t free;
(iii) the minimum norm solution is $x_1=\frac{11}{6}$, $x_2=\frac{4}{3}$, $x_3=-\frac{5}{6}$.

5.6.27 False.

5.7.1 (a) (i) $c_0=0$, $c_1=-\frac{1}{2}\,\mathrm{i}$, $c_2=c_{-2}=0$, $c_3=c_{-1}=\frac{1}{2}\,\mathrm{i}$,
(ii) $\frac{1}{2}\,\mathrm{i}\,e^{-\mathrm{i}x}-\frac{1}{2}\,\mathrm{i}\,e^{\mathrm{i}x}=\sin x$;
(c) (i) $c_0=\frac{1}{3}$, $c_1=\frac{3-\sqrt{3}\,\mathrm{i}}{12}$, $c_2=\frac{1-\sqrt{3}\,\mathrm{i}}{12}$, $c_3=c_{-3}=0$, $c_4=c_{-2}=\frac{1+\sqrt{3}\,\mathrm{i}}{12}$, $c_5=c_{-1}=\frac{3+\sqrt{3}\,\mathrm{i}}{12}$,
(ii) $\frac{1+\sqrt{3}\,\mathrm{i}}{12}e^{-2\mathrm{i}x}+\frac{3+\sqrt{3}\,\mathrm{i}}{12}e^{-\mathrm{i}x}+\frac{1}{3}+\frac{3-\sqrt{3}\,\mathrm{i}}{12}e^{\mathrm{i}x}+\frac{1-\sqrt{3}\,\mathrm{i}}{12}e^{2\mathrm{i}x}=\frac{1}{3}+\frac{1}{2}\cos x+\frac{1}{2\sqrt{3}}\sin x+\frac{1}{6}\cos 2x+\frac{1}{2\sqrt{3}}\sin 2x$.

5.7.2 (a) (i) $f_0=f_3=2$, $f_1=-1$, $f_2=-1$.
(ii) $e^{-\mathrm{i}x}+e^{\mathrm{i}x}=2\cos x$;
(c) (i) $f_0=f_5=6$,
$f_1=2+2e^{2\pi\mathrm{i}/5}+2e^{-4\pi\mathrm{i}/5}=1+.7265\,\mathrm{i}$,
$f_2=2+2e^{2\pi\mathrm{i}/5}+2e^{4\pi\mathrm{i}/5}=1+3.0777\,\mathrm{i}$,
$f_3=2+2e^{-2\pi\mathrm{i}/5}+2e^{-4\pi\mathrm{i}/5}=1-3.0777\,\mathrm{i}$,
$f_4=2+2e^{-2\pi\mathrm{i}/5}+2e^{4\pi\mathrm{i}/5}=1-.7265\,\mathrm{i}$,
(ii) $2e^{-2\mathrm{i}x}+2+2e^{\mathrm{i}x}=2+2\cos x+2\mathrm{i}\sin x+2\cos 2x-2\mathrm{i}\sin 2x$;

5.7.4 (a)

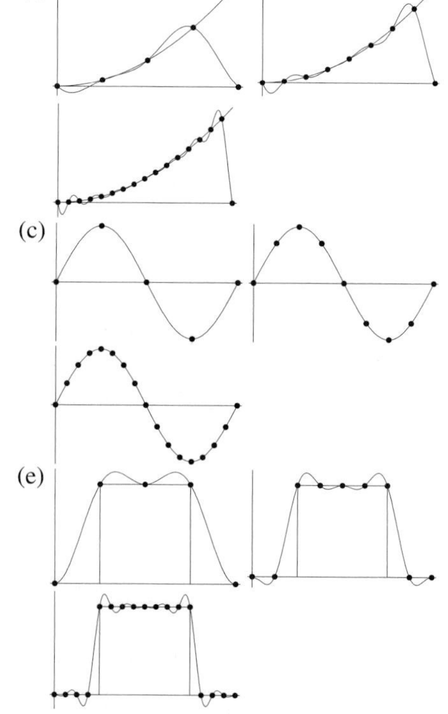

5.7.6 (a) The roots all have modulus $|\zeta^k|=1$ and phase ph $\zeta^k = 2\pi k/n$; the angle between successive roots is $2\pi/n$. The sides meet at an angle of $\pi - 2\pi/n$.

5.7.7 (a) (ii) $e^{2\pi k\,\mathrm{i}/5}$ with $k = 1, 2, 3, 4$.

5.7.10 Original function:

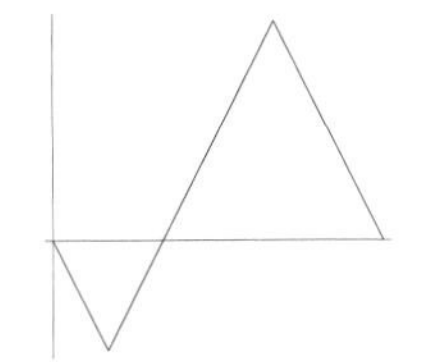

11 mode compression:

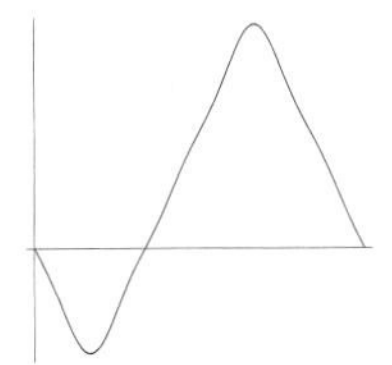

21 mode compression:

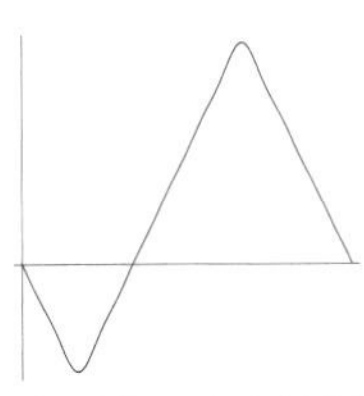

The maximal errors are .08956 and .04836.

5.7.11 (b) Original function:

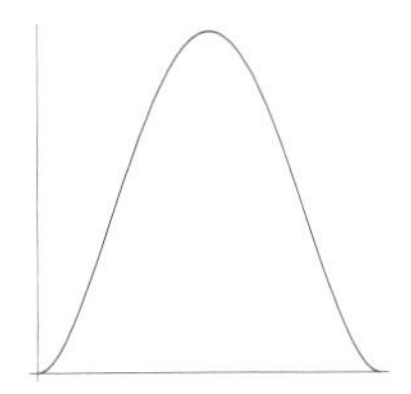

11 mode compression:

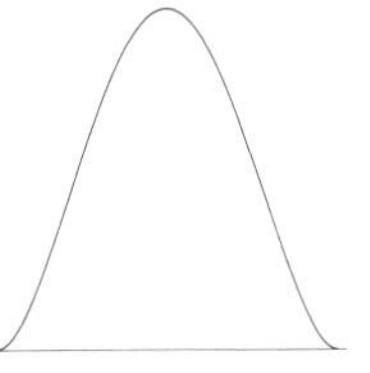

21 mode compression:

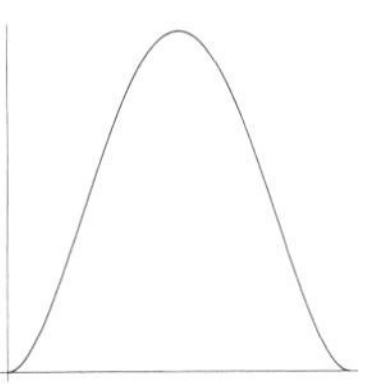

The maximal errors are .09462 and .013755.

5.7.13 Very few are needed. In fact, if you take too many modes, you do worse! For example, if $\varepsilon = .1$,

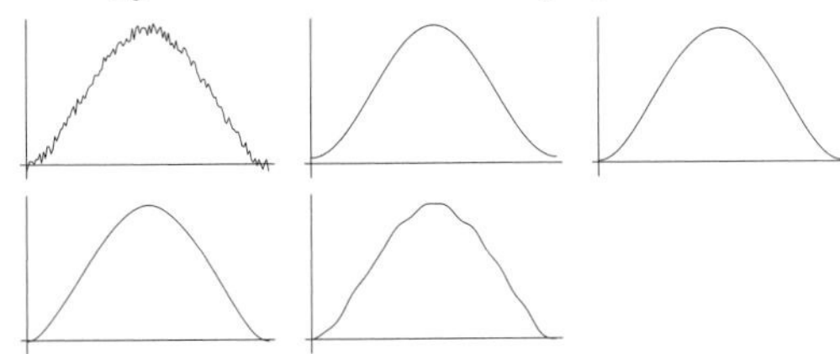

plots the noisy signal and the effect of retaining $2l + 1 = 3, 5, 11, 21$ modes. Only the first three give reasonable results. When $\varepsilon = .5$ the effect is even more pronounced:

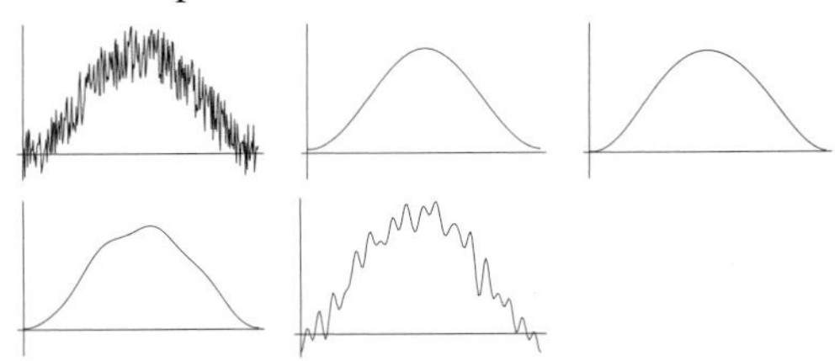

5.7.16 True for the odd case (**??**), but false for the even case (**??**).

5.7.17 (a) $\mathbf{f} = \begin{pmatrix} 0 \\ \frac12 \\ 1 \\ \frac32 \end{pmatrix}, \quad c^{(0)} = \begin{pmatrix} 0 \\ 1 \\ \frac12 \\ \frac32 \end{pmatrix}, \quad c^{(1)} = \begin{pmatrix} \frac12 \\ -\frac12 \\ 1 \\ -\frac12 \end{pmatrix},$

$$c = c^{(2)} = \begin{pmatrix} \frac34 \\ -\frac14 + \frac14\,\mathrm{i} \\ -\frac14 \\ -\frac14 + \frac14\,\mathrm{i} \end{pmatrix};$$

(c) $\mathbf{f} = \begin{pmatrix} \pi \\ \frac34\pi \\ \frac12\pi \\ \frac14\pi \\ 0 \\ \frac14\pi \\ \frac12\pi \\ \frac34\pi \end{pmatrix}, \quad c^{(0)} = \begin{pmatrix} \pi \\ 0 \\ \frac12\pi \\ \frac12\pi \\ \frac34\pi \\ \frac14\pi \\ \frac14\pi \\ \frac34\pi \end{pmatrix}, \quad c^{(1)} = \begin{pmatrix} \frac12\pi \\ \frac12\pi \\ \frac12\pi \\ 0 \\ \frac12\pi \\ \frac14\pi \\ \frac12\pi \\ -\frac14\pi \end{pmatrix},$

$$c^{(2)} = \begin{pmatrix} \frac12\pi \\ \frac14\pi \\ 0 \\ \frac14\pi \\ \frac12\pi \\ \frac{1+\mathrm{i}}{8}\pi \\ 0 \\ \frac{1-\mathrm{i}}{8}\pi \end{pmatrix}, \quad c = c^{(3)} = \begin{pmatrix} \frac12\pi \\ \frac{\sqrt2+1}{8\sqrt2}\pi \\ 0 \\ \frac{\sqrt2-1}{8\sqrt2}\pi \\ 0 \\ \frac{\sqrt2-1}{8\sqrt2}\pi \\ 0 \\ \frac{\sqrt2+1}{8\sqrt2}\pi \end{pmatrix}.$$

5.7.18 (a) $\mathbf{c} = \begin{pmatrix} 1 \\ -1 \\ 1 \\ -1 \end{pmatrix}, \quad \mathbf{f}^{(0)} = \begin{pmatrix} 1 \\ 1 \\ -1 \\ -1 \end{pmatrix}, \quad \mathbf{f}^{(1)} = \begin{pmatrix} 2 \\ 0 \\ -2 \\ 0 \end{pmatrix},$

$$\mathbf{f} = \mathbf{f}^{(2)} = \begin{pmatrix} 0 \\ 0 \\ 4 \\ 0 \end{pmatrix}.$$

Chapter 6

6.1.1 (a) $K = \begin{pmatrix} 3 & -2 \\ -2 & 3 \end{pmatrix}$; (b) $\mathbf{u} = \begin{pmatrix} 3.6 \\ 3.4 \end{pmatrix}$;
(c) the first mass has moved the farthest;
(d) $\mathbf{e} = (3.6, -.2, -3.4)^T$, so the first spring has stretched the most, while the third spring experiences the most compression.

6.1.3 Exercise 6.1.1: (a) $K = \begin{pmatrix} 3 & -2 \\ -2 & 2 \end{pmatrix}$;
(b) $\mathbf{u} = \begin{pmatrix} 7.0 \\ 8.5 \end{pmatrix}$;
(c) the second mass has moved the farthest;
(d) $\mathbf{e} = (7., 1.5)^T$, so the first spring has stretched the most.

6.1.8 (a) For maximum displacement of the bottom mass, the springs should be arranged from weakest at the top to strongest at the bottom, so $c_1 = c = 1$, $c_2 = c' = 2$, $c_3 = c'' = 3$.

6.1.10 The sub-diagonal entries of L are $l_{i,i-1} = -1/i$, while the diagonal entries of D are $d_{ii} = (i+1)/i$.

6.1.13 (b) $p(\mathbf{u}) = u_1^2 - u_1 u_2 + u_2^2 - u_2 u_3 + u_3^2 - u_3 u_4 + u_4^2 - g\,(u_1 + u_2 + u_3 + u_4)$.

6.1.14 (a) $p(\mathbf{u}) = \frac{3}{2} u_1^2 - 2 u_1 u_2 + \frac{3}{2} u_2^2 - 4 u_1 - 3 u_2$, so $p(\mathbf{u}^\star) = -12.3$.
(b) For instance, $p(1,0) = -2.5$, $p(0,1) = -1.5$, $p(3,3) = -12$.

6.1.16 (a) Two masses, both ends fixed, $c_1 = 2$, $c_2 = 4$, $c_3 = 2$, $\mathbf{f} = (-1, 3)^T$; equilibrium: $\mathbf{u}^\star = (.3, .7)^T$.
(c) Three masses, top end fixed, $c_1 = 1$, $c_2 = 3$, $c_3 = 5$, $\mathbf{f} = (1, 1, -1)^T$; equilibrium: $\mathbf{u}^\star = \left(-\frac{1}{2}, -\frac{5}{6}\right)^T$.

6.2.1 (b) (d)

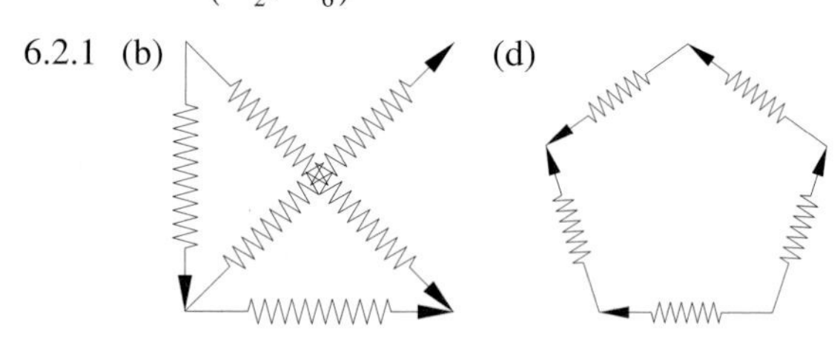

6.2.2 (a) $A = \begin{pmatrix} 1 & -1 & 0 & 0 \\ 1 & 0 & -1 & 0 \\ 1 & 0 & 0 & -1 \\ 0 & 1 & -1 & 0 \\ 0 & 1 & 0 & -1 \end{pmatrix}$;
(b) $\begin{pmatrix} 3 & -1 & -1 \\ -1 & 3 & -1 \\ -1 & -1 & 2 \end{pmatrix} \begin{pmatrix} u_1 \\ u_2 \\ u_3 \end{pmatrix} = \begin{pmatrix} 3 \\ 0 \\ 0 \end{pmatrix}$;
(c) $\mathbf{u} = \begin{pmatrix} \frac{15}{8} \\ \frac{9}{8} \\ \frac{3}{2} \end{pmatrix}$; (d) $\mathbf{v} = \left(\frac{3}{4}, \frac{3}{8}, \frac{15}{8}, -\frac{3}{8}, \frac{9}{8}\right)^T$;
(e) wire 3.

6.2.6 None.

6.2.8 (a) The potentials remain the same, but the currents are all twice as large.

6.2.10 (a) $n = 3$: potentials:
$$\begin{pmatrix} .0288 & .0577 & .0769 & .0577 & .0288 \\ .0577 & .1250 & .1923 & .1250 & .0577 \\ .0769 & .1923 & .4423 & .1923 & .0769 \\ .0577 & .1250 & .1923 & .1250 & .0577 \\ .0288 & .0577 & .0769 & .0577 & .0288 \end{pmatrix};$$
currents along the horizontal wires:
$$\begin{pmatrix} -.0288 & -.0288 & -.0192 & .0192 & .0288 & .0288 \\ -.0577 & -.0673 & -.0673 & .0673 & .0673 & .0577 \\ -.0769 & -.1153 & -.2500 & .2500 & .1153 & .0769 \\ -.0577 & -.0673 & -.0673 & .0673 & .0673 & .0577 \\ -.0288 & -.0288 & -.0192 & .0192 & .0288 & .0288 \end{pmatrix},$$
where all wires are oriented from left to right, so the currents are all going away from the center. The currents in the vertical wires are given by the transpose of the matrix.

6.2.12 (a) (i) $\mathbf{u} = (2, 1, 1, 0)^T$, $\mathbf{e} = (1, 0, 1)^T$;
(iii) $\mathbf{u} = (3, 2, 1, 1, 1, 0)^T$, $\mathbf{e} = (1, 1, 0, 0, 1)^T$.

6.2.13 (i) $\mathbf{u} = \left(\frac{3}{2}, \frac{1}{2}, 0, 0\right)^T$, $\mathbf{e} = \left(1, \frac{1}{2}, \frac{1}{2}\right)^T$;
(iii) $\mathbf{u} = \left(\frac{7}{3}, \frac{4}{3}, \frac{1}{3}, 0, 0, 0\right)^T$, $\mathbf{e} = \left(1, 1, \frac{1}{3}, \frac{1}{3}, \frac{1}{3}\right)^T$.

6.2.15 (a) True.

6.2.17 (a) $A^T C\, A\, \mathbf{u} = \mathbf{f} - A^T C\, \mathbf{b}$.

6.3.1 8 cm

6.3.3 (a) $\mathbf{u} = (1.5, -.5, 2.5, 2.5)^T$, $\mathbf{e} = (.7071, 1, 0, -1.5811)^T$.

6.3.5 (a) $A = \begin{pmatrix} 0 & 1 & 0 & 0 \\ -1 & 0 & 1 & 0 \\ 0 & 0 & 0 & 1 \\ 0 & 0 & \frac{1}{\sqrt{2}} & \frac{1}{\sqrt{2}} \\ -\frac{1}{\sqrt{2}} & \frac{1}{\sqrt{2}} & 0 & 0 \end{pmatrix}$;
(b)
$$\begin{aligned} \tfrac{3}{2} u_1 - \tfrac{1}{2} v_1 - u_2 &= 0, \\ -\tfrac{1}{2} u_1 + \tfrac{3}{2} v_1 &= 0, \\ -u_1 + \tfrac{3}{2} u_2 + \tfrac{1}{2} v_2 &= 0, \\ \tfrac{1}{2} u_2 + \tfrac{3}{2} v_2 &= 0. \end{aligned}$$
(c) Stable, statically indeterminate.
(d) $\mathbf{f}_1 = \left(\frac{3}{2}, -\frac{1}{2}\right)^T$, $\mathbf{f}_2 = (-1, 0)^T$; the horizontal bar.

6.3.8 (a) $A = \begin{pmatrix} 0 & 1 & 0 & 0 & 0 & 0 \\ -\frac{3}{\sqrt{10}} & -\frac{1}{\sqrt{10}} & \frac{3}{\sqrt{10}} & \frac{1}{\sqrt{10}} & 0 & 0 \\ -1 & 0 & 0 & 0 & 1 & 0 \\ 0 & 0 & -\frac{3}{\sqrt{10}} & \frac{1}{\sqrt{10}} & \frac{3}{\sqrt{10}} & -\frac{1}{\sqrt{10}} \\ 0 & 0 & 0 & 0 & 0 & 1 \end{pmatrix}.$

(b) One instability: simultaneous horizontal motion of the three nodes.

(c) $f_1 + f_2 + f_3 = 0$. For example: $\mathbf{f}_1 = \mathbf{f}_2 = \mathbf{f}_3 = (0, 1)^T$, $\mathbf{e} = \left(\frac{3}{2}, \sqrt{\frac{5}{2}}, -\frac{3}{2}, \sqrt{\frac{5}{2}}, \frac{3}{2}\right)^T$.

(d) To stabilize, add in one more bar starting at one of the fixed nodes and going to one of the two movable nodes not already connected to it.

(e) In every case, $\mathbf{e} = \left(\frac{3}{2}, \sqrt{\frac{5}{2}}, -\frac{3}{2}, \sqrt{\frac{5}{2}}, \frac{3}{2}, 0\right)^T$, so the stresses on the previous bars are all the same, while the reinforcing bar experiences no stress.

6.3.12 (a) $A = \begin{pmatrix} -1 & 0 & 0 & 1 & 0 & 0 & 0 & 0 & 0 & 0 & 0 & 0 \\ 0 & -1 & 0 & 0 & 0 & 0 & 0 & 1 & 0 & 0 & 0 & 0 \\ 0 & 0 & -1 & 0 & 0 & 0 & 0 & \frac{1}{\sqrt{2}} & 0 & 0 & 0 & 0 \\ 0 & 0 & 0 & \frac{1}{\sqrt{2}} & -\frac{1}{\sqrt{2}} & 0 & -\frac{1}{\sqrt{2}} & 0 & 0 & 0 & 0 & 1 \\ 0 & 0 & 0 & \frac{1}{\sqrt{2}} & 0 & -\frac{1}{\sqrt{2}} & 0 & 0 & 0 & -\frac{1}{\sqrt{2}} & 0 & \frac{1}{\sqrt{2}} \\ 0 & 0 & 0 & 0 & 0 & 0 & 0 & \frac{1}{\sqrt{2}} & -\frac{1}{\sqrt{2}} & 0 & -\frac{1}{\sqrt{2}} & \frac{1}{\sqrt{2}} \end{pmatrix};$

(b) $\mathbf{v}_1 = \begin{pmatrix} 1 \\ 0 \\ 0 \\ 1 \\ 0 \\ 0 \\ 1 \\ 0 \\ 0 \\ 1 \\ 0 \\ 0 \end{pmatrix}, \mathbf{v}_2 = \begin{pmatrix} 0 \\ 1 \\ 0 \\ 0 \\ 1 \\ 0 \\ 0 \\ 1 \\ 0 \\ 0 \\ 1 \\ 0 \end{pmatrix}, \mathbf{v}_3 = \begin{pmatrix} 0 \\ 0 \\ 1 \\ 0 \\ 0 \\ 1 \\ 0 \\ 0 \\ 1 \\ 0 \\ 0 \\ 1 \end{pmatrix}, \mathbf{v}_4 = \begin{pmatrix} 0 \\ 0 \\ 0 \\ 0 \\ 0 \\ 0 \\ 0 \\ 0 \\ -1 \\ 0 \\ 1 \\ 0 \end{pmatrix}, \mathbf{v}_5 = \begin{pmatrix} 0 \\ 0 \\ 0 \\ 0 \\ 0 \\ -1 \\ 0 \\ 0 \\ 0 \\ 1 \\ 0 \\ 0 \end{pmatrix}, \mathbf{v}_6 = \begin{pmatrix} 0 \\ 0 \\ 0 \\ 0 \\ -1 \\ 0 \\ 1 \\ 0 \\ 0 \\ 0 \\ 0 \\ 0 \end{pmatrix};$

(c) $\mathbf{v}_1, \mathbf{v}_2, \mathbf{v}_3$; (d) $\mathbf{v}_4, \mathbf{v}_5, \mathbf{v}_6$;

(e) $K = \begin{pmatrix} 1 & 0 & 0 & -1 & 0 & 0 & 0 & 0 & 0 & 0 & 0 & 0 \\ 0 & 1 & 0 & 0 & 0 & 0 & 0 & -1 & 0 & 0 & 0 & 0 \\ 0 & 0 & 1 & 0 & 0 & 0 & 0 & 0 & 0 & 0 & 0 & -1 \\ -1 & 0 & 0 & 2 & -\frac{1}{2} & -\frac{1}{2} & -\frac{1}{2} & \frac{1}{2} & 0 & -\frac{1}{2} & 0 & \frac{1}{2} \\ 0 & 0 & 0 & -\frac{1}{2} & \frac{1}{2} & 0 & \frac{1}{2} & -\frac{1}{2} & 0 & 0 & 0 & 0 \\ 0 & 0 & 0 & -\frac{1}{2} & 0 & \frac{1}{2} & 0 & 0 & 0 & \frac{1}{2} & 0 & -\frac{1}{2} \\ 0 & 0 & 0 & -\frac{1}{2} & \frac{1}{2} & 0 & \frac{1}{2} & -\frac{1}{2} & 0 & 0 & 0 & 0 \\ 0 & -1 & 0 & \frac{1}{2} & -\frac{1}{2} & 0 & -\frac{1}{2} & 2 & -\frac{1}{2} & 0 & -\frac{1}{2} & \frac{1}{2} \\ 0 & 0 & 0 & 0 & 0 & 0 & 0 & -\frac{1}{2} & \frac{1}{2} & 0 & \frac{1}{2} & -\frac{1}{2} \\ 0 & 0 & 0 & -\frac{1}{2} & 0 & \frac{1}{2} & 0 & 0 & 0 & \frac{1}{2} & 0 & -\frac{1}{2} \\ 0 & 0 & 0 & 0 & 0 & 0 & 0 & -\frac{1}{2} & \frac{1}{2} & 0 & \frac{1}{2} & -\frac{1}{2} \\ 0 & 0 & -1 & \frac{1}{2} & 0 & -\frac{1}{2} & 0 & \frac{1}{2} & -\frac{1}{2} & -\frac{1}{2} & -\frac{1}{2} & 2 \end{pmatrix};$

(f) $f_1 + f_2 + f_3 + f_4 = 0$, $g_1 + g_2 + g_3 + g_4 = 0$, $h_1 + h_2 + h_3 + h_4 = 0$, $h_3 = g_4$, $h_2 = f_4$, $g_2 = f_3$; (g) you need to fix three nodes; (h) only the vertical bar experiences compression of magnitude 1.

6.3.14 True.

6.3.15 (a) True.

6.3.18 False.

6.3.22 (a) $A^\star = \begin{pmatrix} \frac{1}{\sqrt{2}} & \frac{1}{\sqrt{2}} & 0 & 0 & 0 \\ -1 & 0 & 1 & 0 & 0 \\ 0 & 0 & -\frac{1}{\sqrt{2}} & \frac{1}{\sqrt{2}} & \frac{1}{\sqrt{2}} \end{pmatrix}$; $K^\star \mathbf{u} = \mathbf{f}^\star$

where $K^\star = \begin{pmatrix} \frac{3}{2} & \frac{1}{2} & -1 & 0 & 0 \\ \frac{1}{2} & \frac{1}{2} & 0 & 0 & 0 \\ -1 & 0 & \frac{3}{2} & -\frac{1}{2} & -\frac{1}{2} \\ 0 & 0 & -\frac{1}{2} & \frac{1}{2} & \frac{1}{2} \\ 0 & 0 & -\frac{1}{2} & \frac{1}{2} & \frac{1}{2} \end{pmatrix}$.

(b) Unstable: 0 rigid motions; 2 mechanisms.

Chapter 7

7.1.1 (b) Not linear, (d) linear, (f) not linear.

7.1.2 (a) Linear, (c) not linear, (e) not linear.

7.1.3 (a) $F(0,0) = (2,0)^T \neq \mathbf{0}$.
(c) $F(-x,-y) = F(x,y) \neq -F(x,y)$.

7.1.5 (b) $\begin{pmatrix} 1 & 0 & 0 \\ 0 & \frac{1}{2} & -\frac{\sqrt{3}}{2} \\ 0 & \frac{\sqrt{3}}{2} & \frac{1}{2} \end{pmatrix}$, (d) $\begin{pmatrix} 0 & 0 & 1 \\ 1 & 0 & 0 \\ 0 & 1 & 0 \end{pmatrix}$,

(f) $\begin{pmatrix} 1 & 0 & 0 \\ 0 & 1 & 0 \\ 0 & 0 & 0 \end{pmatrix}$.

7.1.6 $L\begin{pmatrix} x \\ y \end{pmatrix} = \frac{5}{2}x - \frac{1}{2}y$. Yes, because $\begin{pmatrix} 1 \\ 1 \end{pmatrix}, \begin{pmatrix} 1 \\ -1 \end{pmatrix}$ form a basis.

7.1.9 No, because linearity would require

$$L\begin{pmatrix} 0 \\ 1 \\ -1 \end{pmatrix} = L\left[\begin{pmatrix} 1 \\ -1 \\ 0 \end{pmatrix} - \begin{pmatrix} 1 \\ -1 \\ 0 \end{pmatrix}\right] = L\begin{pmatrix} 1 \\ 0 \\ -1 \end{pmatrix} - L\begin{pmatrix} 1 \\ -1 \\ 0 \end{pmatrix} = 3 \neq -2.$$

7.1.11 No, since $N(-\mathbf{v}) = N(\mathbf{v}) \neq -N(\mathbf{v})$.

7.1.14 (a) $L[cX + dY] = A(cX + dY) = cAX + dAY = cL[X] + dL[Y]$;

$\begin{pmatrix} a & 0 & b & 0 \\ 0 & a & 0 & b \\ c & 0 & d & 0 \\ 0 & c & 0 & d \end{pmatrix}$.

7.1.15 (b) Not linear; target space = $\mathcal{M}_{n\times n}$.
(d) Not linear; target space = $\mathcal{M}_{n\times n}$.
(f) Linear; target space = $\mathbb{R}$.
(h) Linear; target space = $\mathbb{R}^n$.

7.1.19 (b) Not linear; target space = $\mathbb{R}$.
(d) Linear; target space = $\mathbb{R}$.
(f) Linear; target space = $\mathrm{C}^1(\mathbb{R})$.
(h) Linear; target space = $\mathrm{C}^0(\mathbb{R})$.
(j) Linear; target space = $\mathrm{C}^0(\mathbb{R})$.
(l) Linear; target space = $\mathbb{R}$.
(n) Linear; target space = $\mathrm{C}^2(\mathbb{R})$.
(p) Not linear; target space = $\mathrm{C}^1(\mathbb{R})$.
(r) Linear; target space = $\mathbb{R}$.

7.1.20 True.

7.1.22 $I_w[cf + dg] = \int_a^b \left[c f(x) + d g(x) \right] w(x)\,dx = c\int_a^b f(x)\,w(x)\,dx + d\int_a^b g(x)\,w(x)\,dx = c\,I_w[f] + d\,I_w[g]$.

7.1.24 $\Delta[cf + dg] = \dfrac{\partial^2}{\partial x^2}\left[c f(x,y) + d g(x,y) \right] + \dfrac{\partial^2}{\partial y^2}\left[c f(x,y) + d g(x,y) \right] = c\left(\dfrac{\partial^2 f}{\partial x^2} + \dfrac{\partial^2 f}{\partial y^2} \right) + d\left(\dfrac{\partial^2 g}{\partial x^2} + \dfrac{\partial^2 g}{\partial y^2} \right) = c\,\Delta[f] + d\,\Delta[g]$.

7.1.26 (b) Curl: $\nabla \times (c\mathbf{f} + d\mathbf{g}) = c\nabla\times\mathbf{f} + d\nabla\times\mathbf{g}$; domain is space of continuously differentiable vector fields; target is space of continuous vector fields.

7.1.27 (b) dimension = 4; basis:
$\begin{pmatrix} 1 & 0 \\ 0 & 0 \end{pmatrix}, \begin{pmatrix} 0 & 1 \\ 0 & 0 \end{pmatrix}, \begin{pmatrix} 0 & 0 \\ 1 & 0 \end{pmatrix}, \begin{pmatrix} 0 & 0 \\ 0 & 1 \end{pmatrix}$.
(d) dimension = 4; basis given by L_0, L_1, L_2, L_3, where $L_i[a_3x^3 + a_2x^2 + a_1x + a_0] = a_i$.

7.1.29 False.

7.1.30 (a) $\mathbf{a} = (3, -1, 2)^T$, (c) $\mathbf{a} = \left(\frac{5}{4}, -\frac{1}{2}, \frac{5}{4}\right)^T$.

7.1.31 (b) (ii) $\mathbf{a} = \left(\frac{2}{3}, -1\right)^T$.

7.1.33 (b) $\mathbf{v}_1 = \left(\frac{1}{7}, \frac{3}{7}\right), \mathbf{v}_2 = \left(\frac{2}{7}, -\frac{1}{7}\right)$,
(d) $\mathbf{v}_1 = (8, 1, 3), \mathbf{v}_2 = (10, 1, 4), \mathbf{v}_3 = (7, 1, 3)$.

7.1.34 (a) $9 - 36x + 30x^2$, (c) 1.

7.1.37 (a) $S \circ T = T \circ S =$ clockwise rotation by $60°$ = counterclockwise rotation by $300°$;
(c) $S \circ T = T \circ S =$ rotation by $180°$;
(e) $S \circ T = T \circ S = \mathrm{O}$;
(g) $S \circ T$ maps $(x,y)^T$ to $(y,0)^T$; $T \circ S$ maps $(x,y)^T$ to $(0,x)^T$.

7.1.39 (a) $R = \begin{pmatrix} 1 & 0 & 0 \\ 0 & 0 & -1 \\ 0 & 1 & 0 \end{pmatrix}$, $S = \begin{pmatrix} 0 & -1 & 0 \\ 1 & 0 & 0 \\ 0 & 0 & 1 \end{pmatrix}$;

(b) $R \circ S = \begin{pmatrix} 0 & -1 & 0 \\ 0 & 0 & -1 \\ 1 & 0 & 0 \end{pmatrix}$,

$S \circ R = \begin{pmatrix} 0 & 0 & 1 \\ 1 & 0 & 0 \\ 0 & 1 & 0 \end{pmatrix}$; Under $R \circ S$, $\mathbf{e}_1, \mathbf{e}_2, \mathbf{e}_3$ go to $\mathbf{e}_3, -\mathbf{e}_1, -\mathbf{e}_2$, respectively; under $S \circ R$, they go to $\mathbf{e}_2, \mathbf{e}_3, \mathbf{e}_1$.

7.1.41 (a) $L = E \circ D$ where $D[f(x)] = f'(x)$, $E[g(x)] = g(0)$. They do not commute — $D \circ E$ is not even defined since the target of E, namely $\mathbb{R}$, is

not the domain of D, the space of differentiable functions.
(b) $e = 0$.

7.1.43 (a) According to Lemma ??, $M_a \circ D$ is linear, and hence, for the same reason, $L = D \circ (M_a \circ D)$ is also linear.
(b) $L = a(x)\, D^2 + a'(x)\, D$.

7.1.47 (b) (ii) $\begin{pmatrix} 0 & -2 \\ -2 & 0 \end{pmatrix}$.
(d) $\bigl[\,[\,L, M\,], N\,\bigr] = \begin{pmatrix} -3 & 2 \\ 2 & 3 \end{pmatrix}$,
$\bigl[\,[\,N, L\,], M\,\bigr] = \begin{pmatrix} 0 & 0 \\ -2 & 0 \end{pmatrix}$,
$\bigl[\,[\,M, N\,], L\,\bigr] = \begin{pmatrix} 3 & -2 \\ 0 & -3 \end{pmatrix}$, which add up to O.

7.1.48 (a) $[\,P, Q\,][\,f\,] = (x\,f)' - x\,f' = f$.

7.1.51 (b) Yes: Counterclockwise rotation by 45°.
(d) No.

7.1.52 (b) Function: $\begin{pmatrix} \frac{1}{\sqrt2} & \frac{1}{\sqrt2} \\ -\frac{1}{\sqrt2} & \frac{1}{\sqrt2} \end{pmatrix}$;
inverse: $\begin{pmatrix} \frac{1}{\sqrt2} & -\frac{1}{\sqrt2} \\ \frac{1}{\sqrt2} & \frac{1}{\sqrt2} \end{pmatrix}$.
(d) Function: $\begin{pmatrix} \frac12 & \frac12 \\ \frac12 & \frac12 \end{pmatrix}$; no inverse.

7.1.53 $L^{-1}[\mathbf{e}_1] = (-2, 1)^T$, $L^{-1}[\mathbf{e}_2] = (-3, 1)^T$.

7.1.56 (d) (ii) $L = \begin{pmatrix} \frac13 & \frac13 \\ 1 & -1 \end{pmatrix}$, $L^{-1} = \begin{pmatrix} \frac32 & \frac12 \\ \frac32 & -\frac12 \end{pmatrix}$.

7.1.60 (a) $L[a\,x^2 + b\,x + c] = a\,x^2 + (b + 2a)\,x + (c + b)$;
$L^{-1}[a\,x^2 + b\,x + c] =$
$a\,x^2 + (b - 2a)\,x + (c - b + 2a) = e^{-x} \int_{-\infty}^{x} e^{y}\, p(y)\, dy$.

7.2.1 (a) $\begin{pmatrix} \frac{1}{\sqrt2} & -\frac{1}{\sqrt2} \\ \frac{1}{\sqrt2} & \frac{1}{\sqrt2} \end{pmatrix}$. (i) The line $y = x$;
(ii) the rotated square $0 \le x + y,\ x - y \le \sqrt2$;
(iii) the unit disk.
(c) $\begin{pmatrix} -\frac35 & \frac45 \\ \frac45 & \frac35 \end{pmatrix}$. (i) The line $4x + 3y = 0$;
(ii) the rotated square with vertices $(0, 0)^T$, $\left(\frac{1}{\sqrt2}, \frac{1}{\sqrt2}\right)^T$, $\left(0, \sqrt2\right)^T$, $\left(-\frac{1}{\sqrt2}, \frac{1}{\sqrt2}\right)^T$;
(iii) the unit disk.
(e) $\begin{pmatrix} -\frac12 & \frac32 \\ -\frac32 & \frac52 \end{pmatrix}$. (i) The line $y = 3x$;
(ii) the parallelogram with vertices $(0, 0)^T$, $\left(-\frac12, -\frac32\right)^T$, $(1, 1)^T$, $\left(\frac32, \frac52\right)^T$;
(iii) the ellipse $\frac{17}{2}x^2 - 9xy + \frac52 y^2 \le 1$.

7.2.2

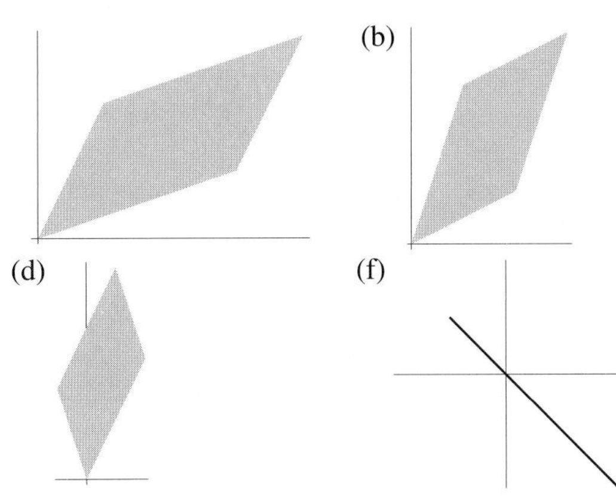

7.2.4 $\begin{pmatrix} 0 & 1 \\ 1 & 0 \end{pmatrix}^2 = \begin{pmatrix} 1 & 0 \\ 0 & 1 \end{pmatrix}$; reflecting twice through the line $y = x$ brings you back where you started.

7.2.6 The line that goes through $(-1, 2)^T$, $(-4, -1)^T$.

7.2.8 Example: $\begin{pmatrix} 1 & 0 & 0 \\ 0 & 2 & 0 \\ 0 & 0 & 4 \end{pmatrix}$.

7.2.9 (b) True. (d) False.

7.2.11 (b) $R = 2\,\dfrac{\mathbf{v}\,\mathbf{v}^T}{\|\mathbf{v}\|^2} - \mathrm{I}$.
(c) (i) $\begin{pmatrix} 1 & 0 \\ 0 & -1 \end{pmatrix}$; (iii) $\begin{pmatrix} 0 & 1 \\ 1 & 0 \end{pmatrix}$.

7.2.12 (b) $\begin{pmatrix} 1 & 1 \\ -1 & 1 \end{pmatrix} = \begin{pmatrix} 1 & 0 \\ -1 & 1 \end{pmatrix}\begin{pmatrix} 1 & 0 \\ 0 & 2 \end{pmatrix}\begin{pmatrix} 1 & 1 \\ 0 & 1 \end{pmatrix}$:
a shear of magnitude -1 along the x axis, followed by a scaling in the y direction a factor of 2, followed by a shear of magnitude -1 along the y axis.
(d) $\begin{pmatrix} 1 & 1 & 0 \\ 1 & 0 & 1 \\ 0 & 1 & 1 \end{pmatrix} = \begin{pmatrix} 1 & 0 & 0 \\ 1 & 1 & 0 \\ 0 & 0 & 1 \end{pmatrix}\begin{pmatrix} 1 & 0 & 0 \\ 0 & 1 & 0 \\ 0 & -1 & 1 \end{pmatrix}$
$\begin{pmatrix} 1 & 0 & 0 \\ 0 & 1 & 0 \\ 0 & 0 & 2 \end{pmatrix}\begin{pmatrix} 1 & 0 & 0 \\ 0 & -1 & 0 \\ 0 & 0 & 1 \end{pmatrix}\begin{pmatrix} 1 & 0 & 0 \\ 0 & 1 & -1 \\ 0 & 0 & 1 \end{pmatrix}$
$\begin{pmatrix} 1 & 1 & 0 \\ 0 & 1 & 0 \\ 0 & 0 & 1 \end{pmatrix}$: a shear of magnitude 1 along the x axis that fixes the xz plane, followed a shear of magnitude -1 along the y axis that fixes the xy plane, followed by a reflection in the xz plane, followed by a scaling in the z direction by a factor of 2, followed a shear of magnitude -1 along the z axis that fixes the xz plane, followed a shear of magnitude 1 along the y axis that fixes the yz plane.

7.2.14 (a) $\begin{pmatrix} 1 & 0 & 0 \\ 0 & \frac12 & -\frac{\sqrt3}{2} \\ 0 & \frac{\sqrt3}{2} & \frac12 \end{pmatrix}$

(b) $\begin{pmatrix} 1 & 0 & 0 \\ 0 & 1 & 0 \\ 0 & \sqrt{3} & 1 \end{pmatrix}\begin{pmatrix} 1 & 0 & 0 \\ 0 & \frac{1}{2} & 0 \\ 0 & 0 & 1 \end{pmatrix}\begin{pmatrix} 1 & 0 & 0 \\ 0 & 1 & 0 \\ 0 & 0 & 2 \end{pmatrix}$
$\begin{pmatrix} 1 & 0 & 0 \\ 0 & 1 & -\sqrt{3} \\ 0 & 0 & 1 \end{pmatrix}$: a shear of magnitude $-\sqrt{3}$ along the y axis that fixes the xy plane, following by a scaling in the z direction by a factor of 2, following by a scaling in the y direction by a factor of $\frac{1}{2}$, following by a shear of magnitude $\sqrt{3}$ along the z axis that fixes the xz plane.

7.2.15 (b) $\begin{pmatrix} \frac{1}{2} & \frac{1}{2} \\ \frac{1}{2} & \frac{1}{2} \end{pmatrix}$.

7.2.17 $\begin{pmatrix} 1 & 0 & 0 \\ 0 & 1 & 0 \\ 0 & 0 & 1 \end{pmatrix}$ is the identity transformation;
$\begin{pmatrix} 0 & 0 & 1 \\ 0 & 1 & 0 \\ 1 & 0 & 0 \end{pmatrix}$ is a reflection in the xz plane;
$\begin{pmatrix} 0 & 1 & 0 \\ 0 & 0 & 1 \\ 1 & 0 & 0 \end{pmatrix}$ is rotation by 120° around the line $x = y = z$.

7.2.19 $\det \mathrm{I}_2 = +1$, representing a 180° rotation, while $\det \mathrm{I}_3 = -1$, and so is a reflection.

7.2.21 (c) (i) $\begin{pmatrix} \frac{7}{25} & 0 & -\frac{24}{25} \\ 0 & 1 & 0 \\ -\frac{24}{25} & 0 & -\frac{7}{25} \end{pmatrix}$.

7.2.24 (b) $\begin{pmatrix} 1 & -6 \\ -\frac{4}{3} & 3 \end{pmatrix}$, (d) $\begin{pmatrix} -1 & 0 \\ 0 & 5 \end{pmatrix}$..

7.2.25 (b) $\begin{pmatrix} -1 & 0 & 0 \\ 0 & -2 & 0 \\ 0 & 0 & 1 \end{pmatrix}$.

7.2.26 (b) $\begin{pmatrix} 1 \\ 0 \\ 0 \end{pmatrix}, \begin{pmatrix} -4 \\ 0 \\ 1 \end{pmatrix}, \begin{pmatrix} 0 \\ 4 \\ 3 \end{pmatrix}$, and $\begin{pmatrix} 1 \\ -2 \end{pmatrix}, \begin{pmatrix} 2 \\ 1 \end{pmatrix}$;
canonical form: $\begin{pmatrix} 1 & 0 & 0 \\ 0 & 0 & 0 \end{pmatrix}$.
(d) $\begin{pmatrix} 1 \\ 0 \\ 0 \end{pmatrix}, \begin{pmatrix} 0 \\ 1 \\ 0 \end{pmatrix}, \begin{pmatrix} 1 \\ -2 \\ 3 \end{pmatrix}$, and $\begin{pmatrix} 1 \\ 1 \\ 2 \end{pmatrix}, \begin{pmatrix} 2 \\ -1 \\ 1 \end{pmatrix}$,
$\begin{pmatrix} -1 \\ -1 \\ 1 \end{pmatrix}$; canonical form: $\begin{pmatrix} 1 & 0 & 0 \\ 0 & 1 & 0 \\ 0 & 0 & 0 \end{pmatrix}$.

7.3.1 (a) (i) The horizontal line $y = -1$;
(ii) the disk $(x-2)^2 + (y+1)^2 \le 1$;
(iii) the square $\{2 \le x \le 3, -1 \le y \le 0\}$.
(c) (i) The horizontal line $y = 2$;
(ii) the ellipse $x^2 - 4xy + 5y^2 + 6x - 16y + 12 \le 0$;
(iii) the parallelogram with vertices $(1,2)^T, (2,2)^T, (4,3)^T, (3,3)^T$.
(e) (i) The line $4x + 3y + 6 = 0$;
(ii) the disk $(x+3)^2 + (y-2)^2 \le 1$;
(iii) the rotated square with corners $(-3,2)$, $(-2.4, 1.2)$, $(-1.6, 1.8)$, $(-2.2, 2.6)$.
(g) (i) The line $x + y + 1 = 0$;
(ii) the disk $(x-2)^2 + (y+3)^2 = 2$ of radius $\sqrt{2}$ centered at $(2,-3)^T$;
(iii) the rotated square with corners $(2,-3)$, $(3,-4)$, $(4,-3)$, $(3,-2)$.

7.3.2 (a) $T_3 \circ T_4[\mathbf{x}] = \begin{pmatrix} -2 & 1 \\ -1 & 0 \end{pmatrix}\mathbf{x} + \begin{pmatrix} 2 \\ 2 \end{pmatrix}$, with
$\begin{pmatrix} -2 & 1 \\ -1 & 0 \end{pmatrix} = \begin{pmatrix} 1 & 2 \\ 0 & 1 \end{pmatrix}\begin{pmatrix} 0 & 1 \\ -1 & 0 \end{pmatrix}$,
$\begin{pmatrix} 2 \\ 2 \end{pmatrix} = \begin{pmatrix} 1 & 2 \\ 0 & 1 \end{pmatrix}\begin{pmatrix} 1 \\ 0 \end{pmatrix} + \begin{pmatrix} 1 \\ 2 \end{pmatrix}$;
(c) $T_3 \circ T_6[\mathbf{x}] = \begin{pmatrix} \frac{3}{2} & \frac{3}{2} \\ \frac{1}{2} & \frac{1}{2} \end{pmatrix}\mathbf{x} + \begin{pmatrix} 2 \\ 2 \end{pmatrix}$, with
$\begin{pmatrix} \frac{3}{2} & \frac{3}{2} \\ \frac{1}{2} & \frac{1}{2} \end{pmatrix} = \begin{pmatrix} 1 & 2 \\ 0 & 1 \end{pmatrix}\begin{pmatrix} \frac{1}{2} & \frac{1}{2} \\ \frac{1}{2} & \frac{1}{2} \end{pmatrix}$,
$\begin{pmatrix} 2 \\ 2 \end{pmatrix} = \begin{pmatrix} 1 & 2 \\ 0 & 1 \end{pmatrix}\begin{pmatrix} 1 \\ 0 \end{pmatrix} + \begin{pmatrix} 1 \\ 2 \end{pmatrix}$;
(e) $T_7 \circ T_8[\mathbf{x}] = \begin{pmatrix} 0 & 0 \\ -4 & -2 \end{pmatrix}\mathbf{x} + \begin{pmatrix} 2 \\ 2 \end{pmatrix}$, with
$\begin{pmatrix} 0 & 0 \\ -4 & -2 \end{pmatrix} = \begin{pmatrix} 1 & 1 \\ -1 & 1 \end{pmatrix}\begin{pmatrix} 2 & 1 \\ -2 & -1 \end{pmatrix}$,
$\begin{pmatrix} 4 \\ -3 \end{pmatrix} = \begin{pmatrix} 1 & 1 \\ -1 & 1 \end{pmatrix}\begin{pmatrix} 1 \\ 1 \end{pmatrix} + \begin{pmatrix} 2 \\ -3 \end{pmatrix}$.

7.3.3 (b) True. (d) False.

7.3.6 (c) $T_2^{-1}\begin{pmatrix} x \\ y \end{pmatrix} = \begin{pmatrix} \frac{1}{3} & 0 \\ 0 & \frac{1}{2} \end{pmatrix}\begin{pmatrix} x \\ y \end{pmatrix} + \begin{pmatrix} \frac{1}{3} \\ 0 \end{pmatrix}$,
$T_4^{-1}\begin{pmatrix} x \\ y \end{pmatrix} = \begin{pmatrix} 0 & -1 \\ 1 & 0 \end{pmatrix}\begin{pmatrix} x \\ y \end{pmatrix} + \begin{pmatrix} 0 \\ -1 \end{pmatrix}$,
T_6 has no inverse.

7.3.8 The dimension is $n^2 + n$.

7.3.10 (b) Isometry, (d) not an isometry.

7.3.14 (b) $F[\mathbf{x}] = \begin{pmatrix} \frac{\sqrt{3}}{2} & -\frac{1}{2} \\ \frac{1}{2} & \frac{\sqrt{3}}{2} \end{pmatrix}\left[\mathbf{x} - \begin{pmatrix} 1 \\ 1 \end{pmatrix}\right] + \begin{pmatrix} 1 \\ 1 \end{pmatrix}$
$= \begin{pmatrix} \frac{\sqrt{3}}{2} & -\frac{1}{2} \\ \frac{1}{2} & \frac{\sqrt{3}}{2} \end{pmatrix}\mathbf{x} + \begin{pmatrix} \frac{3-\sqrt{3}}{2} \\ \frac{1-\sqrt{3}}{2} \end{pmatrix}$;
$G[\mathbf{x}] = \begin{pmatrix} 0 & -1 \\ 1 & 0 \end{pmatrix}\left[\mathbf{x} - \begin{pmatrix} -2 \\ 1 \end{pmatrix}\right] + \begin{pmatrix} -2 \\ 1 \end{pmatrix}$
$= \begin{pmatrix} 0 & -1 \\ 1 & 0 \end{pmatrix}\mathbf{x} + \begin{pmatrix} -1 \\ 3 \end{pmatrix}$;
$F \circ G[\mathbf{x}] = \begin{pmatrix} -\frac{1}{2} & -\frac{\sqrt{3}}{2} \\ \frac{\sqrt{3}}{2} & -\frac{1}{2} \end{pmatrix}\mathbf{x} + \begin{pmatrix} -\sqrt{3} \\ \sqrt{3} \end{pmatrix}$

$= \begin{pmatrix} -\frac{1}{2} & -\frac{\sqrt{3}}{2} \\ \frac{\sqrt{3}}{2} & -\frac{1}{2} \end{pmatrix} \left[\mathbf{x} - \begin{pmatrix} \frac{-1-\sqrt{3}}{2} \\ \frac{-1+\sqrt{3}}{2} \end{pmatrix} \right] + \begin{pmatrix} \frac{-1-\sqrt{3}}{2} \\ \frac{-1+\sqrt{3}}{2} \end{pmatrix}$

counterclockwise rotation around $\left(\frac{-1-\sqrt{3}}{2}, \frac{-1+\sqrt{3}}{2} \right)^T$ by $120°$;

$G \circ F[\mathbf{x}] = \begin{pmatrix} -\frac{1}{2} & -\frac{\sqrt{3}}{2} \\ \frac{\sqrt{3}}{2} & -\frac{1}{2} \end{pmatrix} \mathbf{x} + \begin{pmatrix} \frac{-3+\sqrt{3}}{2} \\ \frac{9-\sqrt{3}}{2} \end{pmatrix}$

$= \begin{pmatrix} -\frac{1}{2} & -\frac{\sqrt{3}}{2} \\ \frac{\sqrt{3}}{2} & -\frac{1}{2} \end{pmatrix} \left[\mathbf{x} - \begin{pmatrix} \frac{-1-\sqrt{3}}{2} \\ \frac{5-\sqrt{3}}{2} \end{pmatrix} \right] + \begin{pmatrix} \frac{-1-\sqrt{3}}{2} \\ \frac{5-\sqrt{3}}{2} \end{pmatrix}$

counterclockwise rotation around $\left(\frac{-1-\sqrt{3}}{2}, \frac{5-\sqrt{3}}{2} \right)^T$ by $120°$.

7.3.16 (b) $\left(y + \frac{3}{\sqrt{2}}, x + \frac{3}{\sqrt{2}} \right)^T$.

7.3.21 (a) $F[\mathbf{x}] = Q\,\mathbf{x} + \mathbf{b}$, where $\mathbf{b}$ is arbitrary and Q is a rotation by 0, 90, 180 or 270 degrees, or a reflection in the x axis, the y axis, the line $y = x$ or the line $y = -x$.

7.3.23 (a) $q(H\,\mathbf{x}) = (x \cosh\alpha + y \sinh\alpha)^2 - (x \sinh\alpha + y \cosh\alpha)^2 = x^2 - y^2 = q(\mathbf{x})$.

7.3.24 (a) $\mathbf{q} = (\,x/(1-y),\ 0\,)^T$.

7.4.1 (a) $L(x) = 3x$; domain $\mathbb{R}$; target $\mathbb{R}$; right hand side -5; inhomogeneous.

(c) $L(u,v,w) = \begin{pmatrix} u - 2v \\ v - w \end{pmatrix}$; domain $\mathbb{R}^3$; target $\mathbb{R}^2$; right hand side $\begin{pmatrix} -3 \\ -1 \end{pmatrix}$; inhomogeneous.

(e) $L[u] = u'(x) + 3x\,u(x)$; domain $C^1(\mathbb{R})$; target $C^0(\mathbb{R})$; right hand side 0; homogeneous.

(g) $L[u] = \begin{pmatrix} u'(x) - u(x) \\ u(0) \end{pmatrix}$; domain $C^1(\mathbb{R})$; target $C^0(\mathbb{R}) \times \mathbb{R}$; right hand side $\begin{pmatrix} 0 \\ 1 \end{pmatrix}$; inhomogeneous.

(i) $L[u] = \begin{pmatrix} u''(x) + x^2 u(x) \\ u(0) \\ u'(0) \end{pmatrix}$; domain $C^2(\mathbb{R})$; target $C^0(\mathbb{R}) \times \mathbb{R}^2$; right hand side $\begin{pmatrix} 3x \\ 1 \\ 0 \end{pmatrix}$; inhomogeneous.

(k) $L[u,v] = \begin{pmatrix} u''(x) - v''(x) - 2u(x) + v(x) \\ u(0) - v(0) \\ u(1) - v(1) \end{pmatrix}$; domain $C^2(\mathbb{R}) \times C^2(\mathbb{R})$; target $C^0(\mathbb{R}) \times \mathbb{R}^2$; right hand side $\begin{pmatrix} 0 \\ 0 \\ 0 \end{pmatrix}$; homogeneous.

(m) $L[u] = \int_0^\infty u(t)\,e^{-st}\,dt$; domain $C^0(\mathbb{R})$; target $C^0(\mathbb{R})$; right hand side $1 + s^2$; inhomogeneous.

(o) $L[u,v] = \int_0^1 u(y)\,dy - \int_0^1 y\,v(y)\,dy$; domain $C^0(\mathbb{R}) \times C^0(\mathbb{R})$; target $\mathbb{R}$; right hand side 0; homogeneous.

(q) $L[u] = \begin{pmatrix} \partial u/\partial x - \partial v/\partial y \\ \partial u/\partial y + \partial v/\partial x \end{pmatrix}$; domain $C^1(\mathbb{R}^2) \times C^1(\mathbb{R}^2)$; target $C^0(\mathbb{R}^2)$; right hand side $\mathbf{0}$; homogeneous.

7.4.4 (b) (ii) $u(t) = e^{t^2-1}$.

7.4.6 (a) $u(x) = c_1 e^{2x} + c_2 e^{-2x}$, $\dim = 2$;
(c) $u(x) = c_1 + c_2 e^{3x} + c_3 e^{-3x}$, $\dim = 3$.

7.4.7 (a) $L[c\,y + d\,z] = (c\,y + d\,z)'' + (c\,y + d\,z) = c\,(y'' + y) + d\,(z'' + z) = c\,L[y] + d\,L[z]$.
(b) $\ker L$ is the span of the basic solutions $\cos x$, $\sin x$.

7.4.9 (a) $p(D) = D^3 + 5D^2 + 3D - 9$.
(b) $e^x, e^{-3x}, x\,e^{-3x}$. The general solution is $y(x) = c_1 e^x + c_2 e^{-3x} + c_2 x\,e^{-3x}$.

7.4.10 (a) minimal order 2: $u'' + u' - 6u = 0$.
(c) minimal order 2: $u'' - 2u' + u = 0$.

7.4.11 (a) $u = c_1 x + c_2 x^{-5}$,
(c) $u = c_1 |x|^{(1+\sqrt{5})/2} + c_2 |x|^{(1-\sqrt{5})/2}$,
(e) $u = c_1 x^3 + c_2 x^{-1/3}$.

7.4.13 (ii) (a) $v'' + 4v' - 5v = 0$, $v(t) = c_1 e^t + c_2 e^{-5t}$.
(c) $v'' - v' - v = 0$, $v(t) = c_1 e^{(1+\sqrt{5})t/2} + c_2 e^{(1-\sqrt{5})t/2}$.
(e) $3v'' - 8v' - 3v = 0$, $v(t) = c_1 e^{3t} + c_2 e^{-t/3}$.

7.4.14 (b) (i) $u(x) = c_1 x + c_2 x \log|x|$.

7.4.16 True if S is a connected interval. False otherwise.

7.4.18 $u = c_1 + c_2 \log r$. The solutions form a two-dimensional vector space.

7.4.21 (a) Basis: $1, x, y, z, x^2 - y^2, x^2 - z^2, x\,y, x\,z, y\,z$; dimension = 9.

7.4.24 (a) Not in the range.
(d) $\mathbf{x} = \begin{pmatrix} -2 \\ 0 \\ 2 \\ 0 \end{pmatrix} + \left[y \begin{pmatrix} 3 \\ 1 \\ 0 \\ 0 \end{pmatrix} + w \begin{pmatrix} 2 \\ 0 \\ -3 \\ 1 \end{pmatrix} \right]$.

7.4.25 (b) $x = -\frac{1}{7} + \frac{3}{7}z$, $y = \frac{4}{7} + \frac{2}{7}z$, not unique;
(d) $u = 2$, $v = -1$, $w = 0$, unique.

7.4.26 (b) $u(x) = \frac{1}{6} e^x \sin x + c_1 e^{2x/5} \cos \frac{4}{5} x + c_2 e^{2x/5} \sin \frac{4}{5} x$.

7.4.27 (b) $u(x) = \frac{1}{4} - \frac{1}{4} \cos 2x$,
(d) $u(x) = -\frac{1}{10} \cos x + \frac{1}{5} \sin x + \frac{11}{10} e^{-x} \cos 2x + \frac{9}{10} e^{-x} \sin 2x$.

7.4.28 (a) Unique solution: $u(x) = x - \pi \dfrac{\sin \sqrt{2}\,x}{\sin \sqrt{2}\,\pi}$.
(d) Infinitely many solutions: $u(x) = \frac{1}{2} + c\,e^{-x} \sin x$.
(f) No solution.

7.4.29 (b) $u(x) = \frac{1}{2} \log x + \frac{3}{4} + c_1 x + c_2 x^2$.

7.4.30 (a) If $\mathbf{b} \in \text{rng}(L \circ M)$, then $\mathbf{b} = L \circ M[\mathbf{x}]$ for some $\mathbf{x}$, and so $\mathbf{b} = L[M[\mathbf{x}]] = L[\mathbf{y}]$ belongs to rng L.

7.4.33 (b) $\{\mathbf{0}\}$, the x axis, $\mathbb{R}^2$.

7.4.35 (b) $u(x) = -\frac{1}{9}x - \frac{1}{10}\sin x + c_1 e^{3x} + c_2 e^{-3x}$.
(d) $u(x) = \frac{1}{6}x e^x - \frac{1}{18}e^x + \frac{1}{4}e^{-x} + c_1 e^x + c_2 e^{-2x}$.

7.4.37 $u(x) = -7\cos\sqrt{x} - 3\sin\sqrt{x}$.

7.4.39 (a) $u(x) = \frac{1}{9}x + \cos 3x + \frac{1}{27}\sin 3x$.
(c) $u(x) = 3\cos 2x + \frac{3}{10}\sin 2x - \frac{1}{5}\sin 3x$.

7.4.41 (b) (i) $u(x) = c_1 e^x + c_2 x e^x$,
(iii) $u(x) = c_1 e^{-x^2} + c_2 x e^{-x^2}$.

7.4.43 Example: $A = \begin{pmatrix} 1 & 2 \\ \mathrm{i} & 2\,\mathrm{i} \end{pmatrix}$.

7.4.44 (b) $u(x) = c_1 e^{-3x}\cos x + c_2 e^{-3x}\sin x$,
(d) $u(x) = c_1 e^{x/\sqrt{2}}\cos\frac{1}{\sqrt{2}}x + c_2 e^{-x/\sqrt{2}}\cos\frac{1}{\sqrt{2}}x + c_3 e^{x/\sqrt{2}}\sin\frac{1}{\sqrt{2}}x + c_4 e^{-x/\sqrt{2}}\sin\frac{1}{\sqrt{2}}x$,
(f) $u(x) = c_1 x\cos(\sqrt{2}\log|x|) + c_2 x\sin(\sqrt{2}\log|x|)$.

7.4.45 (a) minimal order 2: $u'' + 2u' + 10u = 0$.
(c) minimal order 5: $u^{(v)} + 4u^{(iv)} + 14u''' + 20u'' + 25u' = 0$.

7.4.46 (b) $u(x) = c_1 e^x + c_2 e^{(\mathrm{i}-1)x} = \left(c_1 e^x + c_2 e^{-x}\cos x\right) + \mathrm{i}\, e^{-x}\sin x$.

7.4.47 (a) $x^4 - 6x^2y^2 + y^4,\ 4x^3y - 4xy^3$.

7.4.48 (a) $\dfrac{\partial u}{\partial t} = -k^2 e^{-k^2 t + \mathrm{i}kx} = \dfrac{\partial^2 u}{\partial x^2}$;
(c) $e^{-k^2 t}\cos kx,\ e^{-k^2 t}\sin kx$.

7.4.50 (a) Conjugated, (d) not conjugated.

7.4.55 (b) $\left(\left(-\frac{3}{2} + \frac{1}{2}\,\mathrm{i}\right)y + \left(-\frac{1}{2} - \frac{1}{2}\,\mathrm{i}\right)z,\ y,\ z\right)^T$ where $y, z \in \mathbb{C}$.

7.5.1 (a) $\begin{pmatrix} 1 & -1 \\ 2 & 3 \end{pmatrix}$, (c) $\begin{pmatrix} \frac{13}{7} & -\frac{10}{7} \\ \frac{5}{7} & \frac{15}{7} \end{pmatrix}$.

7.5.2 Domain (a), target (b): $\begin{pmatrix} 2 & -3 \\ 4 & 9 \end{pmatrix}$;
domain (b), target (c): $\begin{pmatrix} \frac{3}{2} & -\frac{5}{2} \\ \frac{1}{3} & \frac{10}{3} \end{pmatrix}$;

7.5.3 (b) $\begin{pmatrix} 1 & -2 & 0 \\ \frac{1}{2} & 0 & -\frac{3}{2} \\ 0 & \frac{2}{3} & 2 \end{pmatrix}$.

7.5.4 Domain (a), target (b): $\begin{pmatrix} 1 & -2 & 0 \\ 1 & 0 & -3 \\ 0 & 2 & 6 \end{pmatrix}$;
domain (b), target (c): $\begin{pmatrix} 1 & -1 & -1 \\ 1 & 0 & -1 \\ \frac{1}{3} & \frac{4}{3} & \frac{5}{3} \end{pmatrix}$.

7.5.5 Domain (a), target (a): $\begin{pmatrix} 1 & 0 & -1 \\ 3 & 2 & 1 \end{pmatrix}$;
domain (a), target (c): $\begin{pmatrix} 2 & 0 & -2 \\ 8 & 8 & 4 \end{pmatrix}$;
domain (b), target (c): $\begin{pmatrix} 1 & 0 & -1 \\ \frac{8}{3} & \frac{8}{3} & \frac{4}{3} \end{pmatrix}$;
domain (c), target (a): $\begin{pmatrix} 1 & \frac{2}{7} & -\frac{3}{7} \\ 1 & \frac{4}{7} & \frac{1}{7} \end{pmatrix}$.

7.5.8 (b) $\langle \mathbf{u}, (cL)^*[\mathbf{v}] \rangle = \langle\langle (cL)[\mathbf{u}], \mathbf{v} \rangle\rangle = c\,\langle\langle L[\mathbf{u}], \mathbf{v} \rangle\rangle = c\,\langle \mathbf{u}, L^*[\mathbf{v}] \rangle = \langle \mathbf{u}, c\,L^*[\mathbf{v}] \rangle$.

7.5.9 (a) $A = \begin{pmatrix} -1 & 0 \\ 0 & -1 \end{pmatrix} = A^T$;
(c) $A = \begin{pmatrix} 3 & 0 \\ 0 & 3 \end{pmatrix} = A^T$.

7.5.11 $MA = \begin{pmatrix} 2 & 0 \\ 0 & 3 \end{pmatrix}\begin{pmatrix} 6 & 3 \\ 2 & 4 \end{pmatrix} = \begin{pmatrix} 12 & 6 \\ 6 & 12 \end{pmatrix}$ is symmetric.

7.5.12 (a) $a_{12} = \frac{1}{2}a_{21},\ a_{13} = \frac{1}{3}a_{31},\ \frac{1}{2}a_{23} = \frac{1}{3}a_{32}$.

7.5.15 False.

7.5.19 (a) $\langle M_a[u], v \rangle = \int_a^b M_a[u(x)]\,v(x)\,dx = \int_a^b a(x)\,u(x)\,v(x)\,dx = \int_a^b u(x)\,M_a[v(x)]\,dx = \langle u, M_a[v] \rangle$.

7.5.22 Minimizer: $\left(\frac{1}{5}, -\frac{1}{5}\right)^T$; minimum value: $-\frac{1}{5}$.

7.5.24 Minimizer: $\left(\frac{2}{3}, \frac{1}{3}\right)^T$; minimum value: -2.

7.5.25 (b) Minimizer: $\left(\frac{5}{3}, \frac{4}{3}\right)^T$; minimum value: -5.

7.5.26 (a) Minimizer: $\left(\frac{7}{13}, \frac{2}{13}\right)^T$; minimum value: $-\frac{7}{26}$.
(c) Minimizer: $\left(\frac{12}{13}, \frac{5}{26}\right)^T$; minimum value: $-\frac{43}{52}$.

7.5.27 (b) $\frac{6}{11}$.

Chapter 8

8.1.1 (b) $u(t) = 3\,e^{2(t-1)}$.

8.1.2 $\gamma = \log 2/100 \approx .0069$; after 10 years: 93.3033 grams.

8.1.5 $u(t) = u(0)\,e^{1.3t}$. To double, $t = \log 2/1.3 = .5332$.

8.1.6 19.6234 years.

8.1.8 (a) $u(t) = \frac{1}{2} + \frac{1}{2}e^{2t}$, (c) $u(t) = 2 - 3\,e^{-3(t-2)}$.

8.1.10 (b) 24.6094 years.

8.1.12 (a) $u(t) = \frac{1}{3}e^{2t/7}$.
(b) One unit: $t = 36.0813$; 1000 units: $t = 60.2585$.

8.2.1 (a) Eigenvalues: $3, -1$; eigenvectors: $\begin{pmatrix} -1 \\ 1 \end{pmatrix}, \begin{pmatrix} 1 \\ 1 \end{pmatrix}$.
(c) Eigenvalue: 2; eigenvector: $\begin{pmatrix} -1 \\ 1 \end{pmatrix}$.
(e) Eigenvalues: 4, 3, 1;
eigenvectors: $\begin{pmatrix} 1 \\ -1 \\ 1 \end{pmatrix}, \begin{pmatrix} -1 \\ 0 \\ 1 \end{pmatrix}, \begin{pmatrix} 1 \\ 2 \\ 1 \end{pmatrix}$.
(g) Eigenvalues: $0,\ 1 + \mathrm{i},\ 1 - \mathrm{i}$;

eigenvectors: $\begin{pmatrix} 3 \\ 1 \\ 0 \end{pmatrix}, \begin{pmatrix} 3-2\,\mathrm{i} \\ 3-\mathrm{i} \\ 1 \end{pmatrix}, \begin{pmatrix} 3+2\,\mathrm{i} \\ 3+\mathrm{i} \\ 1 \end{pmatrix}$.

(i) -1: simple eigenvalue, eigenvector $\begin{pmatrix} 2 \\ -1 \\ 1 \end{pmatrix}$;

2: double eigenvalue, eigenvectors $\begin{pmatrix} \frac{1}{3} \\ 0 \\ 1 \end{pmatrix}, \begin{pmatrix} -\frac{2}{3} \\ 1 \\ 0 \end{pmatrix}$.

8.2.2 (a) Eigenvalues: $e^{\pm \mathrm{i}\theta} = \cos\theta \pm \mathrm{i}\,\sin\theta$; eigenvectors: $(1, \mp \mathrm{i})^T$, which are real only for $\theta = 0$ and π.
(b) Because $R_\theta - a\,\mathrm{I}$ has an inverse if and only if a is not an eigenvalue.

8.2.4 (a) O is a trivial example.

8.2.7 (a) Eigenvalues: $\mathrm{i}, -1+\mathrm{i}$; eigenvectors: $\begin{pmatrix} 1 \\ 0 \end{pmatrix}$, $\begin{pmatrix} -1 \\ 1 \end{pmatrix}$.
(c) Eigenvalues: $-3, 2\,\mathrm{i}$; eigenvectors: $\begin{pmatrix} -1 \\ 1 \end{pmatrix}$, $\begin{pmatrix} \frac{3}{5}+\frac{1}{5}\,\mathrm{i} \\ 1 \end{pmatrix}$.

8.2.9 For $n = 2$, the eigenvalues are 0, 2, and the eigenvectors are $\begin{pmatrix} -1 \\ 1 \end{pmatrix}, \begin{pmatrix} 1 \\ 1 \end{pmatrix}$.

8.2.11 True.

8.2.15 (a) $\operatorname{tr} A = 2 = 3 + (-1)$; $\det A = -3 = 3 \cdot (-1)$.
(c) $\operatorname{tr} A = 4 = 2 + 2$; $\det A = 4 = 2 \cdot 2$.
(e) $\operatorname{tr} A = 8 = 4 + 3 + 1$; $\det A = 12 = 4 \cdot 3 \cdot 1$.
(g) $\operatorname{tr} A = 2 = 0 + (1+\mathrm{i}) + (1-\mathrm{i})$; $\det A = 0 = 0 \cdot (1+\mathrm{i}\sqrt{2}) \cdot (1-\mathrm{i}\sqrt{2})$.
(i) $\operatorname{tr} A = 3 = (-1) + 2 + 2$; $\det A = -4 = (-1) \cdot 2 \cdot 2$.

8.2.19 (b) If $A\mathbf{v} = \lambda\,\mathbf{v}$ then $B\mathbf{v} = (A + d\,\mathrm{I})\mathbf{v} = (\lambda + d)\,\mathbf{v}$.

8.2.21 (a) False.

8.2.22 False.

8.2.24 (a) Starting with $A\mathbf{v} = \lambda\,\mathbf{v}$, multiply both sides by A^{-1} and divide by λ to obtain $A^{-1}\mathbf{v} = (1/\lambda)\,\mathbf{v}$. Therefore, $\mathbf{v}$ is an eigenvector of A^{-1} with eigenvalue $1/\lambda$.

8.2.29 (b) False.

8.2.31 (a) (ii) $Q = \begin{pmatrix} \frac{7}{25} & -\frac{24}{25} \\ -\frac{24}{25} & -\frac{7}{25} \end{pmatrix}$. Eigenvalues $-1, 1$; eigenvectors $\begin{pmatrix} \frac{3}{5} \\ \frac{4}{5} \end{pmatrix}, \begin{pmatrix} \frac{4}{5} \\ -\frac{3}{5} \end{pmatrix}$.
(iii) $Q = \begin{pmatrix} 1 & 0 & 0 \\ 0 & -1 & 0 \\ 0 & 0 & 1 \end{pmatrix}$. Eigenvalue -1 has eigenvector: $\begin{pmatrix} 0 \\ 1 \\ 0 \end{pmatrix}$; eigenvalue 1 has eigenvectors: $\begin{pmatrix} 1 \\ 0 \\ 0 \end{pmatrix}, \begin{pmatrix} 0 \\ 0 \\ 1 \end{pmatrix}$.

8.2.33 (a) $p_{A^{-1}}(\lambda) = \det(A^{-1} - \lambda\,\mathrm{I}) = \det\left[\lambda\,A^{-1}\left(\frac{1}{\lambda}\,\mathrm{I} - A\right)\right] = \dfrac{(-\lambda)^n}{\det A}\,p_A\left(\dfrac{1}{\lambda}\right)$.
(b) (i) $A^{-1} = \begin{pmatrix} -2 & 1 \\ \frac{3}{2} & -\frac{1}{2} \end{pmatrix}$. Then $p_A(\lambda) = \lambda^2 - 5\lambda - 2$, while $p_{A^{-1}}(\lambda) = \lambda^2 + \frac{5}{2}\lambda - \frac{1}{2} = \frac{\lambda^2}{2}\left(-\frac{2}{\lambda^2} - \frac{5}{\lambda} + 1\right)$.

8.2.34 (a) If $A\mathbf{v} = \lambda\,\mathbf{v}$ then $\mathbf{0} = A^k\mathbf{v} = \lambda^k\,\mathbf{v}$ and hence $\lambda^k = 0$.

8.2.39 False.

8.2.41 (a) If $Q\mathbf{v} = \lambda\,\mathbf{v}$, then $Q^T\mathbf{v} = Q^{-1}\mathbf{v} = \lambda^{-1}\mathbf{v}$ and so λ^{-1} is an eigenvalue of Q^T, and so, by Exercise 8.2.35, also of Q.
(b) If $Q\mathbf{v} = \lambda\,\mathbf{v}$, then, by Exercise 5.3.16, $\|\mathbf{v}\| = \|Q\mathbf{v}\| = |\lambda|\,\|\mathbf{v}\|$, and hence $|\lambda| = 1$.

8.2.43 (b) Axis: $(2, -5, 1)^T$; angle: $\cos^{-1}\frac{7}{13} \approx 1.00219$.

8.2.45 (a) $(Q - \mathrm{I})^T(Q - \mathrm{I}) = Q^TQ - Q - Q^T + \mathrm{I} = 2\,\mathrm{I} - Q - Q^T = K$ is a Gram matrix, which is positive semi-definite by Theorem **??**.

8.2.47 (a) For M_2: eigenvalues $1, -1$; eigenvectors $(1, 1)^T, (-1, 1)^T$.

8.3.1 (a) Complete. dim $= 1$ with basis $(1, 1)^T$.
(c) Complete. dim $= 1$ with basis $(0, 1, 0)^T$.
(e) Complete. dim $= 2$ with basis $(1, 0, 0)^T, (0, -1, 1)^T$.
(g) Not an eigenvalue.

8.3.2 (a) Eigenvalue: 2; eigenvector: $(2, 1)^T$; not complete.
(c) Eigenvalues: $1 \pm 2\,\mathrm{i}$; eigenvectors: $(1 \pm \mathrm{i}, 2)^T$; complete.
(e) Eigenvalue 3 has eigenvectors $(1, 1, 0)^T$, $(1, 0, 1)^T$; not complete.
(g) Eigenvalue 3 has eigenvector $(0, 1, 1)^T$; eigenvalue -2 has eigenvector $(-1, 1, 1)^T$; not complete.

8.3.3 (a) Eigenvalues: $-2, 4$; the eigenvectors $(-1, 1)^T$, $(1, 1)^T$ form a basis for $\mathbb{R}^2$.
(c) Eigenvalue: 1; the eigenvector $(1, 0)^T$ spans a one-dimensional subspace of $\mathbb{R}^2$.
(e) The eigenvalue 1 has eigenvector $(1, 0, 0)^T$, while -1 has eigenvector $(0, 0, 1)^T$. The eigenvectors span a two-dimensional subspace of $\mathbb{R}^3$.
(g) The eigenvalues are $\mathrm{i}, -\mathrm{i}, 1$. The eigenvectors are $(-\mathrm{i}, 0, 1)^T$, $(\mathrm{i}, 0, 1)^T$ and $(0, 1, 0)^T$. The real eigenvectors span only a one-dimensional subspace of $\mathbb{R}^3$.

8.3.4 (a) Complex eigenvector basis;
(c) no eigenvector basis;
(e) no eigenvector basis;
(g) complex eigenvector basis.

8.3.6 (a) True.

8.3.8 (a) Every eigenvector of A is an eigenvector of A^2 with eigenvalue λ^2, and hence if A has a basis of eigenvectors, so does A^2.

8.3.15 (a) $S = \begin{pmatrix} 3 & 3 \\ 1 & 2 \end{pmatrix}$, $D = \begin{pmatrix} 0 & 0 \\ 0 & -3 \end{pmatrix}$.

(c) $S = \begin{pmatrix} -\frac{3}{5} + \frac{1}{5}\,\mathrm{i} & -\frac{3}{5} - \frac{1}{5}\,\mathrm{i} \\ 1 & 1 \end{pmatrix}$,

$D = \begin{pmatrix} -1 + \mathrm{i} & 0 \\ 0 & -1 - \mathrm{i} \end{pmatrix}$.

(e) $S = \begin{pmatrix} 0 & 21 & 1 \\ 1 & -10 & 6 \\ 0 & 7 & 3 \end{pmatrix}$, $D = \begin{pmatrix} 0 & 0 & 0 \\ 0 & 7 & 0 \\ 0 & 0 & -1 \end{pmatrix}$.

(g) $S = \begin{pmatrix} 1 & 0 & -1 \\ 0 & -1 & 0 \\ 0 & 1 & 1 \end{pmatrix}$, $D = \begin{pmatrix} 2 & 0 & 0 \\ 0 & 2 & 0 \\ 0 & 0 & -3 \end{pmatrix}$.

(i) $S = \begin{pmatrix} 0 & -1 & 0 & 1 \\ -1 & 0 & 1 & 0 \\ 0 & 1 & 0 & 1 \\ 1 & 0 & 1 & 0 \end{pmatrix}$,

$D = \begin{pmatrix} -1 & 0 & 0 & 0 \\ 0 & -1 & 0 & 0 \\ 0 & 0 & 1 & 0 \\ 0 & 0 & 0 & 1 \end{pmatrix}$.

8.3.18 (a) $\begin{pmatrix} 0 & -1 & 0 \\ 1 & 0 & 0 \\ 0 & 0 & 1 \end{pmatrix} =$

$$\begin{pmatrix} \mathrm{i} & -\mathrm{i} & 0 \\ 1 & 1 & 0 \\ 0 & 0 & 1 \end{pmatrix} \begin{pmatrix} \mathrm{i} & 0 & 0 \\ 0 & -\mathrm{i} & 0 \\ 0 & 0 & 1 \end{pmatrix} \begin{pmatrix} -\frac{\mathrm{i}}{2} & \frac{1}{2} & 0 \\ \frac{\mathrm{i}}{2} & \frac{1}{2} & 0 \\ 0 & 0 & \frac{1}{2} \end{pmatrix}.$$

8.3.19 (a) Yes. (c) No. (e) Yes.

8.3.20 (a) $S = \begin{pmatrix} 1 & -1 \\ 1 & 1 \end{pmatrix}$, $D = \begin{pmatrix} 1 + \mathrm{i} & 0 \\ 0 & -1 + \mathrm{i} \end{pmatrix}$.

(c) $S = \begin{pmatrix} 1 & -\frac{1}{2} - \frac{1}{2}\,\mathrm{i} \\ 1 & 1 \end{pmatrix}$, $D = \begin{pmatrix} 4 & 0 \\ 0 & -1 \end{pmatrix}$.

8.3.21 (a) $\begin{pmatrix} 7 & 4 \\ -8 & -5 \end{pmatrix}$, (c) $\begin{pmatrix} 3 & 0 \\ 0 & 3 \end{pmatrix}$,

(e) example: $\begin{pmatrix} 0 & 0 & 4 \\ 0 & 1 & 0 \\ 0 & 0 & -2 \end{pmatrix}$.

8.3.22 (b) $\begin{pmatrix} -1 & 0 \\ 0 & 2 \end{pmatrix}$.

8.3.25 True.

8.3.27 All eigenvalues are ± 1.

8.4.1 (b) Eigenvalues: 7, 3; eigenvectors: $\frac{1}{\sqrt{2}}(-1, 1)^T$, $\frac{1}{\sqrt{2}}(1, 1)^T$.
(d) eigenvalues: 6, 1, −4; eigenvectors:
$\left(\frac{4}{5\sqrt{2}}, \frac{3}{5\sqrt{2}}, \frac{1}{\sqrt{2}}\right)^T, \left(-\frac{3}{5}, \frac{4}{5}, 0\right)^T, \left(-\frac{4}{5\sqrt{2}}, -\frac{3}{5\sqrt{2}}, \frac{1}{\sqrt{2}}\right)^T$.

8.4.2 (a) eigenvalues $\frac{5}{2} \pm \frac{1}{2}\sqrt{17}$; positive definite.
(c) eigenvalues 0, 1, 3; positive semi-definite.

8.4.4 If all eigenvalues are distinct, there are 2^n different bases, governed by the choice of sign in the unit eigenvectors $\pm \mathbf{u}_k$. If the eigenvalues are repeated, there are infinitely many.

8.4.6 (a) If $A\mathbf{v} = \lambda \mathbf{v}$ and $\mathbf{v} \neq \mathbf{0}$ is real, then $\lambda \|\mathbf{v}\|^2 = (A\mathbf{v}) \cdot \mathbf{v} = (A\mathbf{v})^T \mathbf{v} = \mathbf{v}^T A^T \mathbf{v} = -\mathbf{v}^T A \mathbf{v} = -\mathbf{v} \cdot (A\mathbf{v}) = -\lambda \|\mathbf{v}\|^2$, and hence $\lambda = 0$.
(c) Since $\det A = 0$, cf. Exercise 1.9.10, at least one of the eigenvalues of A must be 0.
(e) The eigenvalues are: (i) $\pm 2\,\mathrm{i}$, (iii) $0, \pm\sqrt{3}\,\mathrm{i}$.

8.4.7 (c) (ii) eigenvalues 4, −2; eigenvectors: $(2 - \mathrm{i}, 1)^T, (-2 + \mathrm{i}, 5)^T$.

8.4.9 (a) eigenvalues: $\frac{5}{3}, \frac{1}{2}$; eigenvectors: $(-3, 1)^T$, $\left(\frac{1}{2}, 1\right)^T$;
(c) eigenvalues: 7, 1; eigenvectors: $\left(\frac{1}{2}, 1\right)^T, (1, 0)^T$;
(e) eigenvalues: 3, 1, 0; eigenvectors: $(1, -2, 1)^T$, $(-1, 0, 1)^T, \left(1, -\frac{1}{2}, 1\right)^T$.

8.4.13 (d) (i) Eigenvalues 3, −1; eigenvectors $(1, 1)^T$, $(1, -1)^T$.
(iii) Eigenvalues $0, 2 - 2\,\mathrm{i}, 0, 2 + 2\,\mathrm{i}$; eigenvectors $(1, 1, 1, 1)^T, (1, \mathrm{i}, -1, -\mathrm{i})^T, (1, -1, 1, -1)^T$, $(1, -\mathrm{i}, -1, \mathrm{i})^T$.
(e) The eigenvalues are (i) 6, 3, 3;
(iii) $6, \frac{7+\sqrt{5}}{2}, \frac{7+\sqrt{5}}{2}, \frac{7-\sqrt{5}}{2}, \frac{7-\sqrt{5}}{2}$.

8.4.14 (a) $\begin{pmatrix} -3 & 4 \\ 4 & 3 \end{pmatrix} =$

$$\begin{pmatrix} \frac{1}{\sqrt{5}} & -\frac{2}{\sqrt{5}} \\ \frac{2}{\sqrt{5}} & \frac{1}{\sqrt{5}} \end{pmatrix} \begin{pmatrix} 5 & 0 \\ 0 & -5 \end{pmatrix} \begin{pmatrix} \frac{1}{\sqrt{5}} & \frac{2}{\sqrt{5}} \\ -\frac{2}{\sqrt{5}} & \frac{1}{\sqrt{5}} \end{pmatrix}.$$

(c) $\begin{pmatrix} 1 & 1 & 0 \\ 1 & 2 & 1 \\ 0 & 1 & 1 \end{pmatrix} = \begin{pmatrix} \frac{1}{\sqrt{6}} & -\frac{1}{\sqrt{2}} & \frac{1}{\sqrt{3}} \\ \frac{2}{\sqrt{6}} & 0 & -\frac{1}{\sqrt{3}} \\ \frac{1}{\sqrt{6}} & \frac{1}{\sqrt{2}} & \frac{1}{\sqrt{3}} \end{pmatrix}$

$$\begin{pmatrix} 3 & 0 & 0 \\ 0 & 1 & 0 \\ 0 & 0 & 0 \end{pmatrix} \begin{pmatrix} \frac{1}{\sqrt{6}} & \frac{2}{\sqrt{6}} & \frac{1}{\sqrt{6}} \\ -\frac{1}{\sqrt{2}} & 0 & \frac{1}{\sqrt{2}} \\ \frac{1}{\sqrt{3}} & -\frac{1}{\sqrt{3}} & \frac{1}{\sqrt{3}} \end{pmatrix}.$$

8.4.15 (b) $\begin{pmatrix} 5 & -2 \\ -2 & 5 \end{pmatrix} = \begin{pmatrix} -\frac{1}{\sqrt{2}} & \frac{1}{\sqrt{2}} \\ \frac{1}{\sqrt{2}} & \frac{1}{\sqrt{2}} \end{pmatrix}$

$$\begin{pmatrix} 5 & 0 \\ 0 & -10 \end{pmatrix} \begin{pmatrix} -\frac{1}{\sqrt{2}} & \frac{1}{\sqrt{2}} \\ \frac{1}{\sqrt{2}} & \frac{1}{\sqrt{2}} \end{pmatrix}.$$

(d) $\begin{pmatrix}1&0&4\\0&1&3\\4&3&1\end{pmatrix} = \begin{pmatrix}\frac{4}{5\sqrt2}&-\frac35&-\frac{4}{5\sqrt2}\\ \frac{3}{5\sqrt2}&\frac45&\frac{3}{5\sqrt2}\\ \frac1{\sqrt2}&0&\frac1{\sqrt2}\end{pmatrix}$
$\begin{pmatrix}6&0&0\\0&1&0\\0&0&-4\end{pmatrix}\begin{pmatrix}\frac{4}{5\sqrt2}&\frac{3}{5\sqrt2}&\frac1{\sqrt2}\\-\frac35&\frac45&0\\-\frac{4}{5\sqrt2}&-\frac{3}{5\sqrt2}&\frac1{\sqrt2}\end{pmatrix}.$

8.4.16 (a) $\begin{pmatrix}\frac{57}{25}&-\frac{24}{25}\\-\frac{24}{25}&\frac{43}{25}\end{pmatrix}$;
(c) none, since eigenvectors are not orthogonal.

8.4.17 (b) $7\left(\frac1{\sqrt5}x+\frac2{\sqrt5}y\right)^2+\frac{11}{2}\left(-\frac2{\sqrt5}x+\frac1{\sqrt5}y\right)^2 =$
$\frac75(x+2y)^2+\frac{2}{5}(-2x+y)^2$,
(d) $\frac12\left(\frac1{\sqrt3}x+\frac1{\sqrt3}y+\frac1{\sqrt3}z\right)^2+$
$\left(-\frac1{\sqrt2}y+\frac1{\sqrt2}z\right)^2+2\left(-\frac2{\sqrt6}x+\frac1{\sqrt6}y+\frac1{\sqrt6}z\right)^2$
$=\frac16(x+y+z)^2+\frac12(-y+z)^2+\frac13(-2x+y+z)^2.$

8.4.20 True, assuming that the eigenvector basis is real. (For complex eigenvector bases, the result is false, even for real matrices.)

8.4.22 Principal stretches: $4+\sqrt3,\ 4-\sqrt3,\ 1$; directions:
$\begin{pmatrix}1\\-1+\sqrt3\\1\end{pmatrix},\ \begin{pmatrix}1\\-1-\sqrt3\\1\end{pmatrix},\ \begin{pmatrix}-1\\0\\1\end{pmatrix}.$

8.4.24 (b) (ii)

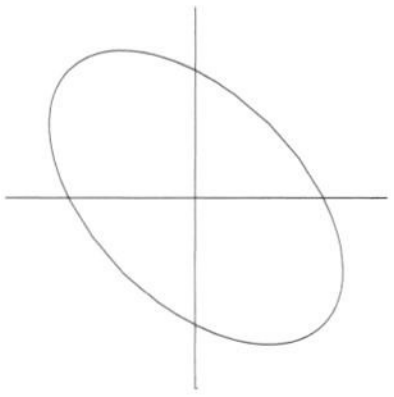

ellipse, with semi-axes $\sqrt2,\ \sqrt{\frac23}$, and principal axes $\begin{pmatrix}-1\\1\end{pmatrix},\ \begin{pmatrix}1\\1\end{pmatrix}$.

8.4.27 (b) (i) $\frac12\begin{pmatrix}\sqrt3+1&\sqrt3-1\\\sqrt3-1&\sqrt3+1\end{pmatrix}$; $\begin{pmatrix}\sqrt2&0&0\\0&\sqrt5&0\\0&0&3\end{pmatrix}$.

8.4.30 (b) $\begin{pmatrix}2&-3\\1&6\end{pmatrix}=\begin{pmatrix}\frac2{\sqrt5}&-\frac1{\sqrt5}\\\frac1{\sqrt5}&\frac2{\sqrt5}\end{pmatrix}\begin{pmatrix}\sqrt5&0\\0&3\sqrt5\end{pmatrix}$,
(d) $\begin{pmatrix}0&-3&8\\1&0&0\\0&4&6\end{pmatrix}$
$=\begin{pmatrix}0&-\frac35&\frac45\\1&0&0\\0&\frac45&\frac35\end{pmatrix}\begin{pmatrix}1&0&0\\0&5&0\\0&0&10\end{pmatrix}.$

8.4.31 (ii) (a) $\begin{pmatrix}-3&4\\4&3\end{pmatrix}=5\begin{pmatrix}\frac15&\frac25\\\frac25&\frac45\end{pmatrix}-5\begin{pmatrix}\frac45&-\frac25\\-\frac25&\frac15\end{pmatrix}.$

(c) $\begin{pmatrix}1&1&0\\1&2&1\\0&1&1\end{pmatrix}=$
$3\begin{pmatrix}\frac16&\frac13&\frac16\\\frac13&\frac23&\frac13\\\frac16&\frac13&\frac16\end{pmatrix}+\begin{pmatrix}\frac12&0&-\frac12\\0&0&0\\-\frac12&0&\frac12\end{pmatrix}.$

8.4.33 (b) $\begin{pmatrix}6&1-2i\\1+2i&2\end{pmatrix}=$
$\begin{pmatrix}\frac{1-2i}{\sqrt6}&\frac{-1+2i}{\sqrt{30}}\\\frac1{\sqrt6}&\frac{\sqrt5}{\sqrt6}\end{pmatrix}\begin{pmatrix}7&0\\0&1\end{pmatrix}\begin{pmatrix}\frac{1+2i}{\sqrt6}&\frac1{\sqrt6}\\-\frac{1+2i}{\sqrt{30}}&\frac{\sqrt5}{\sqrt6}\end{pmatrix}.$

8.4.34 Maximum: 7; minimum: 3.

8.4.36 (a) $\frac{5+\sqrt5}{2}=\max\{\,2x^2-2xy+3y^2 \mid x^2+y^2=1\,\}$,
$\frac{5-\sqrt5}{2}=\min\{\,2x^2-2xy+3y^2 \mid x^2+y^2=1\,\}$;
(c) $12=\max\{\,6x^2-8xy+2xz+6y^2-2yz+11z^2 \mid x^2+y^2+z^2=1\,\}$,
$2=\min\{\,6x^2-8xy+2xz+6y^2-2yz+11z^2 \mid x^2+y^2+z^2=1\,\}$.

8.4.37 (c) $9=\max\{\,6x^2-8xy+2xz+6y^2-2yz+11z^2 \mid x^2+y^2+z^2=1,\ x-y+2z=0\,\}$.

8.4.38 (a) Maximum: 3; minimum: -2.
(c) Maximum: $\frac{8+\sqrt5}{2}$; minimum: $\frac{8-\sqrt5}{2}$.

8.4.40 Maximum: $r^2\lambda_1$; minimum $r^2\lambda_n$, where λ_1,λ_n are, respectively, the largest and smallest eigenvalues of K.

8.4.46 (a) $\max=\frac34$, $\min=\frac25$; (c) $\max=2$, $\min=\frac12$.

8.5.1 (a) $3\pm\sqrt5$, (c) $5\sqrt2$; (e) $\sqrt7,\ \sqrt2$.

8.5.2 (a) $\begin{pmatrix}1&1\\0&2\end{pmatrix}=\begin{pmatrix}\frac{-1+\sqrt5}{\sqrt{10-2\sqrt5}}&\frac{-1-\sqrt5}{\sqrt{10+2\sqrt5}}\\\frac{2}{\sqrt{10-2\sqrt5}}&\frac{2}{\sqrt{10+2\sqrt5}}\end{pmatrix}$
$\begin{pmatrix}\sqrt{3+\sqrt5}&0\\0&\sqrt{3-\sqrt5}\end{pmatrix}\begin{pmatrix}\frac{-2+\sqrt5}{\sqrt{10-4\sqrt5}}&\frac{1}{\sqrt{10-4\sqrt5}}\\\frac{-2-\sqrt5}{\sqrt{10+4\sqrt5}}&\frac{1}{\sqrt{10+4\sqrt5}}\end{pmatrix}$,
(c) $\begin{pmatrix}1&-2\\-3&6\end{pmatrix}=\begin{pmatrix}-\frac1{\sqrt{10}}\\\frac3{\sqrt{10}}\end{pmatrix}\left(5\sqrt2\right)\begin{pmatrix}-\frac1{\sqrt5}&\frac2{\sqrt5}\end{pmatrix}$,
(e) $\begin{pmatrix}2&1&0&-1\\0&-1&1&1\end{pmatrix}=\begin{pmatrix}-\frac2{\sqrt5}&\frac1{\sqrt5}\\\frac1{\sqrt5}&\frac2{\sqrt5}\end{pmatrix}$
$\begin{pmatrix}\sqrt7&0\\0&\sqrt2\end{pmatrix}\begin{pmatrix}-\frac4{\sqrt{35}}&-\frac3{\sqrt{35}}&\frac1{\sqrt{35}}&\frac3{\sqrt{35}}\\\frac2{\sqrt{10}}&-\frac1{\sqrt{10}}&\frac2{\sqrt{10}}&\frac1{\sqrt{10}}\end{pmatrix}.$

8.5.4 (a) 14.9330, (c) 4348.17, (e) 181594.

8.5.5 (b) Exact solution: $x=-1$, $y=-109$, $z=231$; three digit rounding: $x=-2.06$, $y=-75.7$, $z=162$; condition number: 1.17×10^5.

8.5.7 Let $A=\mathbf{v}\in\mathbb{R}^n$ be the matrix (column vector) in question.
(a) One singular value: $\|\mathbf{v}\|$;

(b) $P = \mathbf{v}/\|\mathbf{v}\|$, $\Sigma = \big(\, \|\mathbf{v}\| \,\big)$, $Q = (1)$;
(c) $\mathbf{v}^+ = \mathbf{v}^T/\|\mathbf{v}\|^2$.

8.5.9 True, with one exception — the zero matrix.

8.5.11 True.

8.5.14 False.

8.5.16 False.

8.5.18 False.

8.5.20 $109\,u^2 + 72\,u\,v + 13\,v^2 = 121$; semi-axes: 11, 1; principal axes: $\begin{pmatrix} -1 \\ 3 \end{pmatrix}, \begin{pmatrix} 3 \\ 1 \end{pmatrix}$; area: $11\,\pi$.

8.5.22 (b) True if rank $A = n$, but false otherwise.

8.5.25 (b) $\begin{pmatrix} \frac{1}{5} & \frac{2}{5} \\ -\frac{2}{5} & \frac{1}{5} \end{pmatrix}$, (d) $\begin{pmatrix} 0 & 0 & 0 \\ 0 & -1 & 0 \\ 1 & 0 & 0 \end{pmatrix}$,
(f) $\begin{pmatrix} \frac{1}{140} & \frac{1}{70} & \frac{3}{140} \\ \frac{3}{140} & \frac{3}{70} & \frac{9}{140} \end{pmatrix}$.

8.5.26 (b) $A = \begin{pmatrix} 1 & -3 \\ 2 & 1 \\ 1 & 1 \end{pmatrix}$, $A^+ = \begin{pmatrix} \frac{1}{6} & \frac{1}{3} & \frac{1}{6} \\ -\frac{3}{11} & \frac{1}{11} & \frac{1}{11} \end{pmatrix}$,
$\mathbf{x}^\star = A^+ \begin{pmatrix} 2 \\ -1 \\ 0 \end{pmatrix} = \begin{pmatrix} 0 \\ -\frac{7}{11} \end{pmatrix}$.

8.5.27 (a) Since $A^+ = Q\,\Sigma^{-1} P^T$ is the singular value decomposition of A^+, we find $(A^+)^+ = P\,(\Sigma^{-1})^{-1} Q^T = P\,\Sigma\,Q^T = A$.
(c) $A^+ A\, A^+ = (Q\,\Sigma^{-1} P^T)(P\,\Sigma\,Q^T)(Q\,\Sigma^{-1} P^T) = Q\,\Sigma^{-1}\Sigma\,\Sigma^{-1} P^T = Q\,\Sigma^{-1} P^T = A^+$.

8.6.1 (a) $U = \begin{pmatrix} \frac{1}{\sqrt{2}} & \frac{1}{\sqrt{2}} \\ -\frac{1}{\sqrt{2}} & \frac{1}{\sqrt{2}} \end{pmatrix}$, $\Delta = \begin{pmatrix} 2 & -2 \\ 0 & 2 \end{pmatrix}$;
(c) $U = \begin{pmatrix} \frac{3}{\sqrt{13}} & \frac{2}{\sqrt{13}} \\ -\frac{2}{\sqrt{13}} & \frac{3}{\sqrt{13}} \end{pmatrix}$, $\Delta = \begin{pmatrix} 2 & 15 \\ 0 & -1 \end{pmatrix}$;
(e) $U = \begin{pmatrix} -\frac{1}{\sqrt{5}} & \frac{4}{3\sqrt{5}} & \frac{2}{3} \\ 0 & \frac{\sqrt{5}}{3} & -\frac{2}{3} \\ \frac{2}{\sqrt{5}} & \frac{2}{3\sqrt{5}} & \frac{1}{3} \end{pmatrix}$,
$\Delta = \begin{pmatrix} -2 & -1 & \frac{22}{\sqrt{5}} \\ 0 & 1 & -\frac{9}{\sqrt{5}} \\ 0 & 0 & 1 \end{pmatrix}$.

8.6.3 If $U_1^\dagger U_1 = \mathrm{I} = U_2^\dagger U_2$, then $(U_1 U_2)^\dagger (U_1 U_2) = U_2^\dagger U_1^\dagger U_1 U_2 = U_2^\dagger U_2 = \mathrm{I}$, and so $U_1 U_2$ is also orthogonal.

8.6.6 (b) Two 1×1 Jordan blocks; eigenvalues $-3, 6$; eigenvectors $\mathbf{e}_1, \mathbf{e}_2$.
(d) One 3×3 Jordan block; eigenvalue 0; eigenvector $\mathbf{e}_1$.

8.6.8 $\begin{pmatrix} 2 & 0 & 0 \\ 0 & 2 & 0 \\ 0 & 0 & 5 \end{pmatrix}, \begin{pmatrix} 2 & 0 & 0 \\ 0 & 5 & 0 \\ 0 & 0 & 2 \end{pmatrix}, \begin{pmatrix} 5 & 0 & 0 \\ 0 & 2 & 0 \\ 0 & 0 & 2 \end{pmatrix},$
$\begin{pmatrix} 2 & 0 & 0 \\ 0 & 5 & 0 \\ 0 & 0 & 5 \end{pmatrix}, \begin{pmatrix} 5 & 0 & 0 \\ 0 & 2 & 0 \\ 0 & 0 & 5 \end{pmatrix}, \begin{pmatrix} 5 & 0 & 0 \\ 0 & 5 & 0 \\ 0 & 0 & 2 \end{pmatrix},$
$\begin{pmatrix} 2 & 1 & 0 \\ 0 & 2 & 0 \\ 0 & 0 & 5 \end{pmatrix}, \begin{pmatrix} 5 & 0 & 0 \\ 0 & 2 & 1 \\ 0 & 0 & 2 \end{pmatrix}, \begin{pmatrix} 2 & 0 & 0 \\ 0 & 5 & 1 \\ 0 & 0 & 5 \end{pmatrix},$
$\begin{pmatrix} 5 & 1 & 0 \\ 0 & 5 & 0 \\ 0 & 0 & 2 \end{pmatrix}$.

8.6.9 (a) Eigenvalue: 2. Jordan basis: $\mathbf{v}_1 = \begin{pmatrix} 1 \\ 0 \end{pmatrix}, \mathbf{v}_2 = \begin{pmatrix} 0 \\ \frac{1}{3} \end{pmatrix}$.
Jordan canonical form: $\begin{pmatrix} 2 & 1 \\ 0 & 2 \end{pmatrix}$.
(c) Eigenvalue: 1. Jordan basis:
$\mathbf{v}_1 = \begin{pmatrix} 1 \\ 0 \\ 0 \end{pmatrix}, \mathbf{v}_2 = \begin{pmatrix} 0 \\ 1 \\ 0 \end{pmatrix}, \mathbf{v}_3 = \begin{pmatrix} 0 \\ -1 \\ 1 \end{pmatrix}$.
Jordan canonical form: $\begin{pmatrix} 1 & 1 & 0 \\ 0 & 1 & 1 \\ 0 & 0 & 1 \end{pmatrix}$.
(e) Eigenvalues: $-2, 0$. Jordan basis:
$\mathbf{v}_1 = \begin{pmatrix} -1 \\ 0 \\ 1 \end{pmatrix}, \mathbf{v}_2 = \begin{pmatrix} 0 \\ -1 \\ 0 \end{pmatrix}, \mathbf{v}_3 = \begin{pmatrix} 0 \\ -1 \\ 1 \end{pmatrix}$.
Jordan canonical form: $\begin{pmatrix} -2 & 1 & 0 \\ 0 & -2 & 0 \\ 0 & 0 & 0 \end{pmatrix}$.

8.6.11 True.

8.6.13 True.

8.6.15 (a) $A = \begin{pmatrix} 0 & 1 \\ 0 & 0 \end{pmatrix}$.

8.6.16 Not necessarily.

Chapter 9

9.1.1 (ii) (a) $u(t) = c_1 e^{-2t} + c_2 e^{2t}$.
(b) $\dfrac{d\mathbf{u}}{dt} = \begin{pmatrix} 0 & 1 \\ 4 & 0 \end{pmatrix} \mathbf{u}$.
(c) $\mathbf{u}(t) = \begin{pmatrix} c_1 e^{-2t} + c_2 e^{2t} \\ -2 c_1 e^{-2t} + 2 c_2 e^{2t} \end{pmatrix}$.
(d) (e)
(iv) (a) $u(t) = c_1 e^{-t} + c_2 e^{-3t}$.
(b) $\dfrac{d\mathbf{u}}{dt} = \begin{pmatrix} 0 & 1 \\ -3 & -4 \end{pmatrix} \mathbf{u}$.

(c) $\mathbf{u}(t)=\begin{pmatrix} c_1 e^{-t}+c_2 e^{-3t} \\ -c_1 e^{-t}-3c_2 e^{-3t}\end{pmatrix}$.

(d) (e)

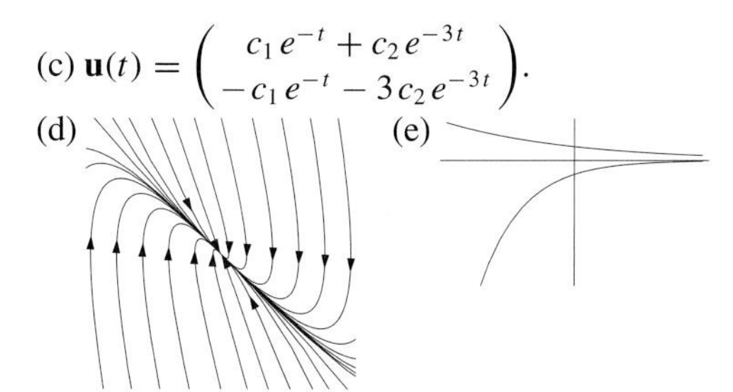

9.1.4 False.

9.1.5 (c) (i) $\dfrac{d\mathbf{v}}{dt}=\begin{pmatrix} 0 & -1 \\ 4 & 0\end{pmatrix}\mathbf{v}$;

$\mathbf{v}(t)=\begin{pmatrix} c_1\cos 2t - c_2\sin 2t \\ 2c_1\sin 2t+2c_2\cos 2t\end{pmatrix}$.

(iii) $\dfrac{d\mathbf{v}}{dt}=\begin{pmatrix} 0 & -1 \\ 1 & 2\end{pmatrix}\mathbf{v}$;

$\mathbf{v}(t)=\begin{pmatrix} c_1 e^t - c_2 t e^t \\ (c_2-c_1)e^t + c_2 t e^t\end{pmatrix}$.

9.1.8 False.

9.1.10 (a) $\ddot u-(a+d)\dot u+(ad-bc)u=0$.
(c) (i) $\ddot u+u=0$, hence $u(t)=c_1\cos t+c_2\sin t$, $v(t)=-c_1\sin t+c_2\cos t$.
(iii) $\ddot u-\dot u-6u=0$, hence $u(t)=c_1 e^{3t}+c_2 e^{-2t}$, $v(t)=c_1 e^{3t}+6c_2 e^{-2t}$.

9.1.12 (b) $x_1(t)=-c_1 e^{-5t}+3c_2 e^{5t}$, $x_2(t)=3c_1 e^{-5t}+c_2 e^{5t}$.
(d) $y_1(t)=-c_1 e^{-t}-c_2 e^{t}-\frac{2}{3}c_3$, $y_2(t)=c_1 e^{-t}-c_2 e^{t}$, $y_3(t)=c_1 e^{-t}+c_2 e^{t}+c_3$.

9.1.13 (a) $\mathbf{u}(t)=$ $\left(\frac12 e^{2-2t}+\frac12 e^{-2+2t},\,-\frac12 e^{2-2t}+\frac12 e^{-2+2t}\right)^T$;
(c) $\mathbf{u}(t)=\left(e^t\cos\sqrt2\,t,\,-\frac{1}{\sqrt2}e^t\sin\sqrt2\,t\right)^T$;
(e) $\mathbf{u}(t)=(-4-6\cos t-9\sin t,\; 2+3\cos t+6\sin t,\, -1-3\sin t)$.

9.1.15 $(x(1),y(1))^T=(-1.10719,\,.34303)^T$.

9.1.17 $\mathbf{u}(t)=\frac32 e^{-5t}\begin{pmatrix}1\\1\end{pmatrix}+\frac12 e^{-7t}\begin{pmatrix}-1\\1\end{pmatrix}$.

9.1.20 (a) Linearly independent; (c) linearly independent; (e) linearly independent; (g) linearly dependent.

9.1.25 (ii) (a) $\mathbf{u}(t)=\begin{pmatrix}-1 & 1\\ 1 & 1\end{pmatrix}\begin{pmatrix}c_1 e^{-2t}\\ c_2 e^{2t}\end{pmatrix}$;

(c) $\mathbf{u}(t)=\begin{pmatrix}-\sqrt2\,\mathrm{i} & \sqrt2\,\mathrm{i}\\ 1 & 1\end{pmatrix}\begin{pmatrix}c_1 e^{(1+\mathrm{i}\sqrt2)t}\\ c_2 e^{(1-\mathrm{i}\sqrt2)t}\end{pmatrix}$;

(e) $\mathbf{u}(t)=\begin{pmatrix}4 & 3+2\,\mathrm{i} & 3-2\,\mathrm{i}\\ -2 & -2-\mathrm{i} & -2+\mathrm{i}\\ 1 & 1 & 1\end{pmatrix}\begin{pmatrix}c_1\\ c_2 e^{\mathrm{i}t}\\ c_3 e^{-\mathrm{i}t}\end{pmatrix}$.

9.1.26 (a) $\begin{pmatrix}c_1 e^{2t}+c_2 t e^{2t}\\ c_2 e^{2t}\end{pmatrix}$

(c) $\begin{pmatrix}c_1 e^{-3t}+c_2\left(\frac12+t\right)e^{-3t}\\ 2c_1 e^{-3t}+2c_2 t e^{-3t}\end{pmatrix}$

(e) $\big(c_1 e^{-3t}+c_2 t e^{-3t}+c_3\left(1+\frac12 t^2\right)e^{-3t},$ $c_2 e^{-3t}+c_3 t e^{-3t},\, c_1 e^{-3t}+c_2 t e^{-3t}+\frac12 c_3 t^2 e^{-3t}\big)^T$

(g) $\big(c_1 e^{3t}+c_2 t e^{3t}-\frac14 c_3 e^{-t}-\frac14 c_4(t+1)e^{-t},$ $c_3 e^{-t}+c_4 t e^{-t},\, c_2 e^{3t}-\frac14 c_4 e^{-t},\, c_4 e^{-t}\big)^T$

9.1.27 (a) $\dfrac{d\mathbf{u}}{dt}=\begin{pmatrix}2 & -\frac12\\ 0 & 1\end{pmatrix}\mathbf{u}$, (c) $\dfrac{d\mathbf{u}}{dt}=\begin{pmatrix}0 & 0\\ 1 & 0\end{pmatrix}\mathbf{u}$,

(e) $\dfrac{d\mathbf{u}}{dt}=\begin{pmatrix}2 & 0 & 0\\ 0 & -3 & 0\\ 2 & 3 & 0\end{pmatrix}\mathbf{u}$,

(g) $\dfrac{d\mathbf{u}}{dt}=\begin{pmatrix}0 & \frac12 & \frac12\\ -2 & 0 & 0\\ -2 & 0 & 0\end{pmatrix}\mathbf{u}$.

9.1.28 (a) No. (c) No.
(e) Yes: $\dot{\mathbf{u}}=\begin{pmatrix}2 & 3\\ -3 & 2\end{pmatrix}\mathbf{u}$.
(g) Yes: $\dot u=\begin{pmatrix}0 & 0 & 0\\ 1 & 0 & 0\\ -1 & 0 & 0\end{pmatrix}\mathbf{u}$.

9.1.30 $u(t)=c_1 e^{-t}+c_2+c_3 e^{t}+c_4 e^{2t}$, $v(t)=-c_1 e^{-t}+c_2+c_3 e^{t}-c_4 e^{2t}$.

9.1.33 (b) (i) $u(t)=-3c_1 e^{2t}+c_2 e^{-2t}-\frac14$, $v(t)=c_1 e^{2t}+c_2 e^{-2t}+\frac14$.

9.2.1 (a) asymptotically stable; (d) stable; (f) unstable.

9.2.3 (a) $\dot u=-2u$, $\dot v=-2v$; $u(t)=c_1 e^{-2t}$, $v(t)=c_2 e^{-2t}$.
(c) $\dot u=-8u+2v$, $\dot v=2u-2v$;
$u(t)=-c_1\frac{\sqrt{13}+3}{2}e^{-(5+\sqrt{13})t}+c_2\frac{\sqrt{13}-3}{2}e^{-(5-\sqrt{13})t}$,
$v(t)=c_1 e^{-(5+\sqrt{13})t}+c_2 e^{-(5-\sqrt{13})t}$.

9.2.4 (a) $\dot u=2v$, $\dot v=-2u$; $u(t)=c_1\cos 2t+c_2\sin 2t$, $v(t)=-c_1\sin 2t+c_2\cos 2t$; stable.

9.2.5 (b) Neither. Unstable.
(d) Hamiltonian flow. Stable.

9.2.7 (b) $\mathbf{u}(t)=$
$$\begin{pmatrix}c_1\cos t+c_2\sin t+c_3 t\cos t+c_4 t\cos t\\ -c_1\sin t+c_2\cos t-c_3 t\sin t+c_4 t\cos t\\ c_3\cos t+c_4\sin t\\ -c_3\sin t+c_4\cos t\end{pmatrix}.$$

9.2.11 True.

9.2.14 True if A is nonsingular; false if A is singular.

9.2.15 (a) True.

9.2.18 False.

9.3.1 (ii) $A=\begin{pmatrix}-2 & 3\\ -1 & 1\end{pmatrix}$;

$\lambda_1=\frac12+\mathrm{i}\frac{\sqrt3}{2}$, $\mathbf{v}_1=\begin{pmatrix}\frac32+\mathrm{i}\frac{\sqrt3}{2}\\ 1\end{pmatrix}$,

$\lambda_2=\frac12-\mathrm{i}\frac{\sqrt3}{2}$, $\mathbf{v}_2=\begin{pmatrix}\frac32-\mathrm{i}\frac{\sqrt3}{2}\\ 1\end{pmatrix}$,

$u_1(t)=e^{-t/2}\left[\left(\frac32 c_1-\frac{\sqrt3}{2}c_2\right)\cos\frac{\sqrt3}{2}t\right.$

$+\left(\frac{\sqrt{3}}{2}c_1+\frac{3}{2}c_2\right)\sin\frac{\sqrt{3}}{2}t\Big]$,
$u_2(t)=e^{-t/2}\left[c_1\cos\frac{\sqrt{3}}{2}t+c_2\sin\frac{\sqrt{3}}{2}t\right]$;
stable focus.

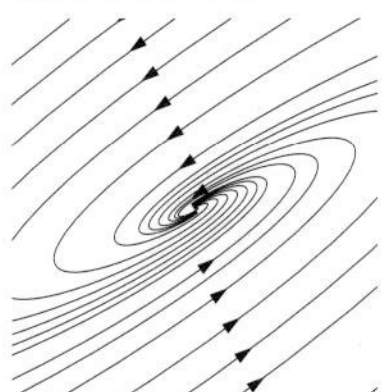

9.3.2 (ii) $\mathbf{u}(t)=c_1e^{-t}\begin{pmatrix}2\cos t-\sin t\\ 5\cos t\end{pmatrix}$
$+c_2e^{-t}\begin{pmatrix}2\sin t+\cos t\\ 5\sin t\end{pmatrix}$; stable focus.

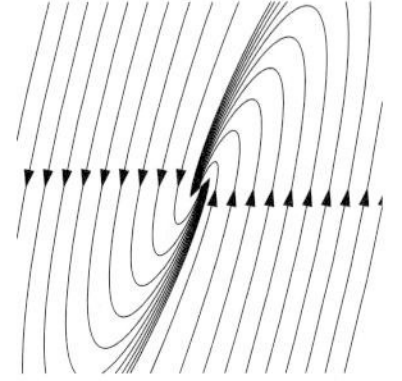

9.3.3 (a) Unstable saddle point (c) unstable node

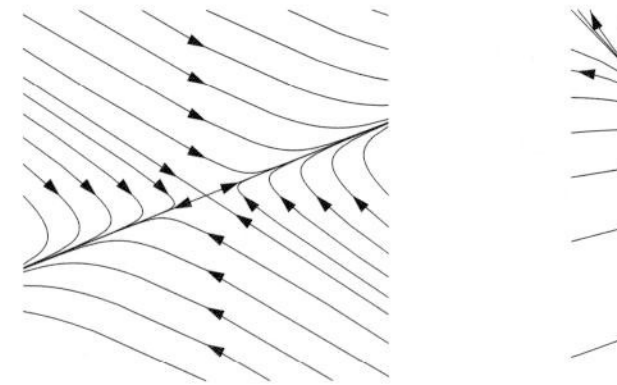

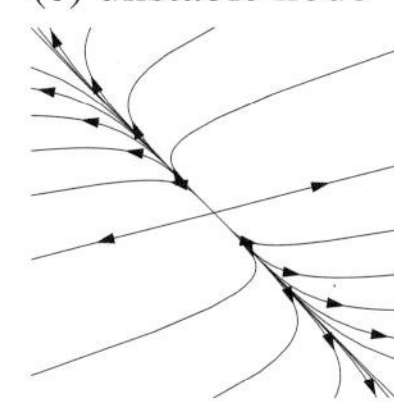

9.3.4 (b) (d)

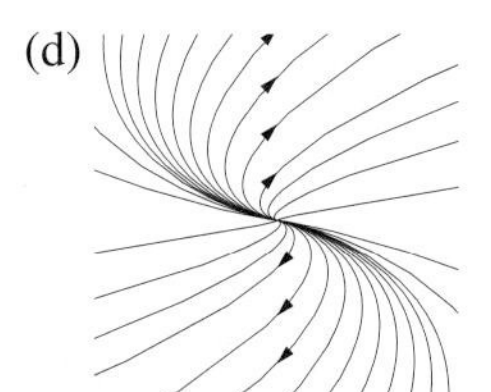

9.4.1 (a) $\begin{pmatrix}\frac{4}{3}e^t-\frac{1}{3}e^{-2t} & -\frac{1}{3}e^t+\frac{1}{3}e^{-2t}\\ \frac{4}{3}e^t-\frac{4}{3}e^{-2t} & -\frac{1}{3}e^t+\frac{4}{3}e^{-2t}\end{pmatrix}$;
(c) $\begin{pmatrix}\cos t & -\sin t\\ \sin t & \cos t\end{pmatrix}$;
(e) $\begin{pmatrix}e^{2t}\cos t-3e^{2t}\sin t & 2e^{2t}\sin t\\ -5e^{2t}\sin t & e^{2t}\cos t+3e^{2t}\sin t\end{pmatrix}$.

9.4.2 (a) $\begin{pmatrix}1 & 0 & 0\\ 2\sin t & \cos t & \sin t\\ 2\cos t-2 & -\sin t & \cos t\end{pmatrix}$;
(c) $\begin{pmatrix}e^{-2t}+te^{-2t} & te^{-2t} & te^{-2t}\\ -1+e^{-2t} & e^{-2t} & -1+e^{-2t}\\ 1-e^{-2t}-te^{-2t} & -te^{-2t} & 1-te^{-2t}\end{pmatrix}$.

9.4.3 Exercise 9.4.1: (a) $\det e^{tA}=e^{-t}=e^{t\,\mathrm{tr}\,A}$,
(c) $\det e^{tA}=1=e^{t\,\mathrm{tr}\,A}$,
(e) $\det e^{tA}=e^{4t}=e^{t\,\mathrm{tr}\,A}$.
Exercise 9.4.2: (a) $\det e^{tA}=1=e^{t\,\mathrm{tr}\,A}$,
(c) $\det e^{tA}=e^{-4t}=e^{t\,\mathrm{tr}\,A}$.

9.4.4 (b) $\begin{pmatrix}e\cos\sqrt{2} & -\sqrt{2}\,e\sin\sqrt{2}\\ \frac{1}{\sqrt{2}}e\sin\sqrt{2} & e\cos\sqrt{2}\end{pmatrix}$,
(d) $\begin{pmatrix}e & 0 & 0\\ 0 & e^{-2} & 0\\ 0 & 0 & e^{-5}\end{pmatrix}$.

9.4.5 (b) $\mathbf{u}(t)=\left(-6e^{-t}+5e^{-3t},\,-4e^{-t}+5e^{-3t}\right)^T$.

9.4.11 (a) False.

9.4.15 (a) Let $V(t)=(e^{tA})^T$. Then $dV/dt=$ $\left(\frac{d}{dt}e^{tA}\right)^T=(e^{tA}A)^T=A^T(e^{tA})^T=A^TV$, and $V(0)=I$. Therefore, by the the definition of matrix exponential, $V(t)=e^{tA^T}$.

9.4.17 (c) 9.4.1: (a) $\begin{pmatrix}1 & 1\\ 1 & 4\end{pmatrix}\begin{pmatrix}e^t & 0\\ 0 & e^{-2t}\end{pmatrix}\begin{pmatrix}1 & 1\\ 1 & 4\end{pmatrix}^{-1}=$
$\begin{pmatrix}\frac{4}{3}e^t-\frac{1}{3}e^{-2t} & -\frac{1}{3}e^t+\frac{1}{3}e^{-2t}\\ \frac{4}{3}e^t-\frac{4}{3}e^{-2t} & -\frac{1}{3}e^t+\frac{4}{3}e^{-2t}\end{pmatrix}$;
(c) $\begin{pmatrix}\mathrm{i} & -\mathrm{i}\\ 1 & 1\end{pmatrix}\begin{pmatrix}e^{\mathrm{i}t} & 0\\ 0 & e^{-\mathrm{i}t}\end{pmatrix}\begin{pmatrix}\mathrm{i} & -\mathrm{i}\\ 1 & 1\end{pmatrix}^{-1}=$
$\begin{pmatrix}\cos t & -\sin t\\ \sin t & \cos t\end{pmatrix}$.
9.4.2: (a) $\begin{pmatrix}-1 & 0 & 0\\ 0 & -\mathrm{i} & \mathrm{i}\\ 2 & 1 & 1\end{pmatrix}\begin{pmatrix}1 & 0 & 0\\ 0 & e^{\mathrm{i}t} & 0\\ 0 & 0 & e^{-\mathrm{i}t}\end{pmatrix}$
$\begin{pmatrix}-1 & 0 & 0\\ 0 & -\mathrm{i} & \mathrm{i}\\ 2 & 1 & 1\end{pmatrix}^{-1}=$
$\begin{pmatrix}1 & 0 & 0\\ 2\sin t & \cos t & \sin t\\ 2\cos t-2 & -\sin t & \cos t\end{pmatrix}$;
(c) not diagonalizable.

9.4.20 Lemma 9.28 implies $\det e^{tA}=e^{t\,\mathrm{tr}\,A}=1$ for all t if and only if $\mathrm{tr}\,A=0$.

9.4.25 (a) If $U(t)=Ce^{tB}$, then $\frac{dU}{dt}=Ce^{tB}B=UB$, and so U satisfies the differential equation. Moreover, $C=U(0)$. Thus, $U(t)$ is the unique solution to the initial value problem $\dot{U}=UB$, $U(0)=C$.

9.4.32 (b) $u_1(t)=e^{t-1}-e^t+t\,e^t$, $u_2(t)=e^{t-1}-e^t+t\,e^t$;
(d) $u(t)=\frac{13}{16}e^{4t}+\frac{3}{16}-\frac{1}{4}t$, $v(t)=\frac{13}{16}e^{4t}-\frac{29}{16}+\frac{3}{4}t$.

9.4.33 (a) $u_1(t)=\frac{1}{2}\cos 2t+\frac{1}{4}\sin 2t+\frac{1}{2}-\frac{1}{2}t$,
$u_2(t)=2e^{-t}-\frac{1}{2}\cos 2t-\frac{1}{4}\sin 2t-\frac{3}{2}+\frac{3}{2}t$,
$u_3(t)=2e^{-t}-\frac{1}{4}\cos 2t-\frac{3}{4}\sin 2t-\frac{7}{4}+\frac{3}{2}t$.

9.4.36 (b) $\begin{pmatrix}1 & 0\\ t & 1\end{pmatrix}$, (d) $\begin{pmatrix}\cos 2t & -\sin 2t\\ 2\sin 2t & 2\cos 2t\end{pmatrix}$.

9.4.37 (b) $\begin{pmatrix}1 & 0 & t\\ 0 & 1 & 0\\ 0 & 0 & 1\end{pmatrix}$, (d) $\begin{pmatrix}\cos t & \sin t & 0\\ -\sin t & \cos t & 0\\ 0 & 0 & e^t\end{pmatrix}$.

9.4.42 $\left[\begin{pmatrix} 2 & 0 \\ 0 & 0 \end{pmatrix}, \begin{pmatrix} 0 & 0 \\ 1 & 0 \end{pmatrix}\right] = \begin{pmatrix} 0 & 0 \\ -2 & 0 \end{pmatrix}$,
$\left[\begin{pmatrix} 2 & 0 \\ 0 & 0 \end{pmatrix}, \begin{pmatrix} 0 & 3 \\ -3 & 0 \end{pmatrix}\right] = \begin{pmatrix} 0 & 6 \\ 6 & 0 \end{pmatrix}$,
$\left[\begin{pmatrix} 2 & 0 \\ 0 & 0 \end{pmatrix}, \begin{pmatrix} 0 & -1 \\ 4 & 0 \end{pmatrix}\right] = \begin{pmatrix} 0 & -2 \\ -8 & 0 \end{pmatrix}$,
$\left[\begin{pmatrix} 2 & 0 \\ 0 & 0 \end{pmatrix}, \begin{pmatrix} 0 & 1 \\ 1 & 0 \end{pmatrix}\right] = \begin{pmatrix} 0 & 2 \\ -2 & 0 \end{pmatrix}$,
$\left[\begin{pmatrix} 2 & 0 & 0 \\ 0 & 1 & 0 \\ 0 & 0 & 0 \end{pmatrix}, \begin{pmatrix} 0 & 0 & 1 \\ 0 & 0 & 0 \\ 0 & 0 & 0 \end{pmatrix}\right] = \begin{pmatrix} 0 & 0 & 2 \\ 0 & 0 & 0 \\ 0 & 0 & 0 \end{pmatrix}$,
$\left[\begin{pmatrix} 2 & 0 & 0 \\ 0 & 1 & 0 \\ 0 & 0 & 0 \end{pmatrix}, \begin{pmatrix} 0 & 0 & -2 \\ 0 & 0 & 0 \\ 2 & 0 & 0 \end{pmatrix}\right] =$
$\begin{pmatrix} 0 & 0 & -4 \\ 0 & 0 & 0 \\ -4 & 0 & 0 \end{pmatrix}$,
$\left[\begin{pmatrix} 2 & 0 & 0 \\ 0 & 1 & 0 \\ 0 & 0 & 0 \end{pmatrix}, \begin{pmatrix} 0 & 1 & 0 \\ -1 & 0 & 0 \\ 0 & 0 & 1 \end{pmatrix}\right] =$
$\begin{pmatrix} 0 & 1 & 0 \\ 1 & 0 & 0 \\ 0 & 0 & 0 \end{pmatrix}$,
$\left[\begin{pmatrix} 2 & 0 & 0 \\ 0 & 1 & 0 \\ 0 & 0 & 0 \end{pmatrix}, \begin{pmatrix} 0 & 0 & 1 \\ 0 & 0 & 0 \\ 1 & 0 & 0 \end{pmatrix}\right] =$
$\begin{pmatrix} 0 & 0 & 2 \\ 0 & 0 & 0 \\ -2 & 0 & 0 \end{pmatrix}$.

9.4.44 $[[A,B],C] = (AB - BA)C - C(AB - BA) = ABC - BAC - CAB + CBA$,
$[[C,A],B] = (CA - AC)B - B(CA - AB) = CAB - ACB - BCA + BAC$,
$[[B,C],A] = (BC - CB)A - A(BC - CB) = BCA - CBA - ABC + ACB$,
The sum is zero.

9.4.47 (a) $(x+t, y)^T$: translations in x direction.
(c) $((x+1)\cos t - y\sin t - 1,$
$(x+1)\sin t + y\cos t)^T$: rotations around the point $(-1,0)^T$.

9.5.1 $\sqrt{21/6}/(2\pi) \approx .297752$ Hertz.

9.5.3 (a) Periodic of period π;

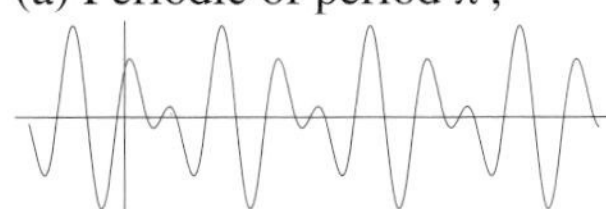

(c) Periodic of period 12;

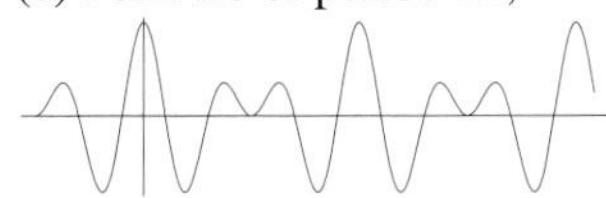

(f) Quasi-periodic;

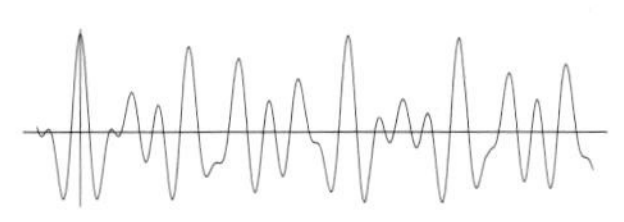

9.5.5 (a) $\sqrt{2}, \sqrt{7}$; (b) 4;
(c) $\mathbf{u}(t) = r_1 \cos(\sqrt{2}\,t - \delta_1)\begin{pmatrix} 2 \\ 1 \end{pmatrix}$
$+ r_2\cos(\sqrt{7}\,t - \delta_2)\begin{pmatrix} -1 \\ 2 \end{pmatrix}$;
(d) The solution is periodic if only one frequency is excited, i.e., $r_1 = 0$ or $r_2 = 0$; all other solutions are quasiperiodic.

9.5.7 (a) $u(t) = r_1\cos(t - \delta_1) + r_2\cos(\sqrt{5}\,t - \delta_2)$,
$v(t) = r_1\cos(t - \delta_1) - r_2\cos(\sqrt{5}\,t - \delta_2)$;
(c) $\mathbf{u}(t) = (r_1\cos(t - \delta_1), r_2\cos(2t - \delta_2),$
$r_3\cos(3t - \delta_1)^T$.

9.5.9 Yes. For example, $c_1 = 16$, $c_2 = 36$, $c_3 = 37$.

9.5.11 (a) Vibrational frequencies:
$\omega_1 = \sqrt{2 - \sqrt{2}} = .7654$, $\omega_2 = \sqrt{2} = 1.4142$,
$\omega_3 = \sqrt{2 + \sqrt{2}} = 1.847$;
eigenvectors: $\mathbf{v}_1 = \left(1, \sqrt{2}, 1\right)^T$, $\mathbf{v}_2 = (-1, 0, 1)^T$,
$\mathbf{v}_3 = \left(1, -\sqrt{2}, 1\right)^T$.

9.5.15 The smallest dimension is 4.

9.5.16 (a) $u(t) = at + b + 2r\cos(\sqrt{5}\,t - \delta)$,
$v(t) = -2at - 2b + r\cos(\sqrt{5}\,t - \delta)$.
The unstable mode consists of the terms with a in them; it will not be excited if the initial conditions satisfy $\dot{u}(t_0) - 2\dot{v}(t_0) = 0$.
(c) $u(t) =$
$-2at - 2b - \frac{1-\sqrt{13}}{4} r_1 \cos\left(\sqrt{\frac{7+\sqrt{13}}{2}}\, t - \delta_1\right) -$
$\frac{1+\sqrt{13}}{4} r_2 \cos\left(\sqrt{\frac{7-\sqrt{13}}{2}}\, t - \delta_2\right)$,
$v(t) =$
$-2at - 2b + \frac{3-\sqrt{13}}{4} r_1 \cos\left(\sqrt{\frac{7+\sqrt{13}}{2}}\, t - \delta_1\right) +$
$\frac{3+\sqrt{13}}{4} r_2 \cos\left(\sqrt{\frac{7-\sqrt{13}}{2}}\, t - \delta_2\right)$,
$w(t) = at + b + r_1\cos\left(\sqrt{\frac{7+\sqrt{13}}{2}}\, t - \delta_1\right) +$
$r_2\cos\left(\sqrt{\frac{7-\sqrt{13}}{2}}\, t - \delta_2\right)$.
The unstable mode is the term containing a; it will not be excited if the initial conditions satisfy $-2\dot{u}(t_0) - 2\dot{v}(t_0) + \dot{w}(t_0) = 0$.

9.5.17 (a) $Q = \begin{pmatrix} -\frac{1}{\sqrt{2}} & \frac{1}{\sqrt{2}} & 0 \\ 0 & 0 & 1 \\ \frac{1}{\sqrt{2}} & \frac{1}{\sqrt{2}} & 0 \end{pmatrix}$, $\Lambda = \begin{pmatrix} 4 & 0 & 0 \\ 0 & 2 & 0 \\ 0 & 0 & 2 \end{pmatrix}$;
(c) $\mathbf{u}(t) = \begin{pmatrix} \cos\sqrt{2}\,t \\ \frac{1}{\sqrt{2}}\sin\sqrt{2}\,t \\ \cos\sqrt{2}\,t \end{pmatrix}$.

9.5.20 (a) Frequencies: $\omega_1 = \sqrt{\frac{3}{2} - \frac{1}{2}\sqrt{5}} = .61803$, $\omega_2 = 1$, $\omega_3 = \sqrt{\frac{3}{2} + \frac{1}{2}\sqrt{5}} = 1.618034$; stable eigenvectors: $\mathbf{v}_1 = \left(-2-\sqrt{5}, -1, -2+\sqrt{5}, 1\right)^T$, $\mathbf{v}_2 = (-1,-1,-1,1)^T$, $\mathbf{v}_3 = \left(2+\sqrt{5}, 1, -2-\sqrt{5}, 1\right)^T$; unstable eigenvector: $\mathbf{v}_4 = (1,-1,1,1)^T$.

9.5.22 (b) 4 normal modes of vibration, all of frequency $\sqrt{2}$; 4 unstable modes: 3 rigid motions and one mechanism.

9.5.26 (a) $\mathbf{u}(t) = r_1 \cos\left(\frac{1}{\sqrt{2}}t - \delta_1\right)\begin{pmatrix}1\\2\end{pmatrix} + r_2\cos\left(\sqrt{\frac{5}{3}}\,t - \delta_2\right)\begin{pmatrix}-3\\1\end{pmatrix}$;

(c) $\mathbf{u}(t) = r_1\cos\left(\sqrt{\frac{3-\sqrt{3}}{2}}\,t - \delta_1\right)\begin{pmatrix}\frac{1+\sqrt{3}}{2}\\1\end{pmatrix} + r_2\cos\left(\sqrt{\frac{3+\sqrt{3}}{2}}\,t - \delta_2\right)\begin{pmatrix}\frac{1-\sqrt{3}}{2}\\1\end{pmatrix}$;

(e) $\mathbf{u}(t) = r_1\cos\left(\sqrt{\frac{2}{3}}\,t - \delta_1\right)\begin{pmatrix}1\\1\end{pmatrix} + r_2\cos(2t - \delta_2)\begin{pmatrix}-1\\1\end{pmatrix}$.

9.5.27 $u_1(t) = \frac{\sqrt{3}-1}{2\sqrt{3}}\cos\sqrt{\frac{3-\sqrt{3}}{2}}\,t + \frac{\sqrt{3}+1}{2\sqrt{3}}\cos\sqrt{\frac{3+\sqrt{3}}{2}}\,t$, $u_2(t) = \frac{1}{2\sqrt{3}}\cos\sqrt{\frac{3-\sqrt{3}}{2}}\,t - \frac{1}{2\sqrt{3}}\cos\sqrt{\frac{3+\sqrt{3}}{2}}\,t$.

9.5.31 (a) $\frac{d^2}{dt^2}\begin{pmatrix}x_1\\y_1\\x_2\\y_2\end{pmatrix} + \begin{pmatrix}2&0&-1&0\\0&0&0&0\\-1&0&2&0\\0&0&0&0\end{pmatrix}\begin{pmatrix}x_1\\y_1\\x_2\\y_2\end{pmatrix} = \begin{pmatrix}0\\0\\0\\0\end{pmatrix}$; vibrational frequencies: $\omega_1 = 1$, $\omega_2 = \sqrt{3}$; four unstable modes.

9.5.36 (a) $u(t) = t\,e^{-3t}$. Critically damped.
(c) $u(t) = \frac{1}{4}\sin 4(t-1)$. Undamped.
(e) $u(t) = 4e^{-t/2} - 2e^{-t}$. Overdamped.

9.5.38 (a) $.5\,\ddot{u} + 2.5\,u = 0$; natural frequency $\omega = \sqrt{5}$.
(b) $u(t) = e^{-t}(2\cos 2t + \sin 2t)$.
(c) underdamped, with less rapid vibrations.

9.6.1 (a) $\cos 8t - \cos 9t = 2\sin\frac{1}{2}t\,\sin\frac{17}{2}t$. Fast frequency: $\frac{17}{2}$, beat frequency: $\frac{1}{2}$;

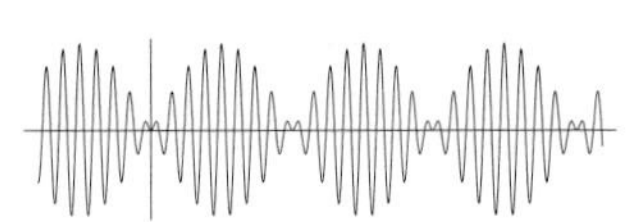

(c) $\cos 10t + \cos 9.5t = 2\sin .25t\,\sin 9.75t$. Fast frequency: 9.75, beat frequency: .25;

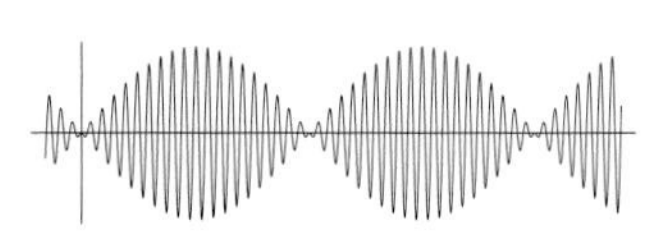

9.6.2 (a) $u(t) = \frac{1}{27}\cos 3t - \frac{1}{27}\cos 6t$;
(c) $u(t) = \frac{1}{2}\sin 2t + e^{-t/2}\left(\cos\frac{\sqrt{15}}{2}t - \frac{\sqrt{15}}{5}\sin\frac{\sqrt{15}}{2}t\right)$;
(e) $u(t) = \frac{1}{5}\cos\frac{1}{2}t + \frac{3}{5}\sin\frac{1}{2}t + \frac{9}{5}e^{-t} + e^{-t/2}$.

9.6.3 (a) $u(t) = \frac{1}{3}\cos 4t + \frac{2}{3}\cos 5t + \frac{1}{5}\sin 5t$; undamped periodic motion with fast frequency 4.5 and beat frequency .5:

(c) $u(t) = -\frac{60}{29}\cos 2t + \frac{5}{29}\sin 2t - \frac{56}{29}e^{-5t} + 8e^{-t}$; the transient is overdamped; the persistent motion is periodic:

9.6.4 (b) .002.

9.6.7 $u(t) = \frac{4}{17}\cos 2t + \frac{16}{17}\sin 2t + e^{-t}\left(\frac{30}{17}\cos 2t - \frac{1}{17}\sin 2t\right)$

9.6.9 (a) $u(t) = \dfrac{\alpha(\cos\eta t - \cos\omega t)}{m(\omega^2 - \eta^2)}$;
(b) $u(t) = \dfrac{\alpha t}{2m\omega}\sin\omega t$.

9.6.11 (b) Overdamped, (d) underdamped.

9.6.12 (b) $u(t) = \frac{3}{2}e^{-t/3} - \frac{1}{2}e^{-t}$,
(d) $u(t) = e^{-t/5}\cos\frac{1}{10}t + 2e^{-t/5}\sin\frac{1}{10}t$.

9.6.15 (b) 2.8126.

9.6.17 $R \geq .10051$.

9.6.18 (b) $\mathbf{u}(t) = \sin 3t\begin{pmatrix}\frac{1}{2}\\-1\end{pmatrix} + r_1\cos(\sqrt{4+\sqrt{5}}\,t - \delta_1)\begin{pmatrix}-1-\sqrt{5}\\2\end{pmatrix} + r_2\cos(4-\sqrt{5}\,t - \delta_2)\begin{pmatrix}-1+\sqrt{5}\\2\end{pmatrix}$.

(d) $\mathbf{u}(t) = \cos\frac{1}{2}t\begin{pmatrix}\frac{2}{17}\\-\frac{12}{17}\end{pmatrix} + r_1\cos(\sqrt{\frac{5}{3}}\,t - \delta_1)\begin{pmatrix}-3\\1\end{pmatrix} + r_2\cos(\frac{1}{\sqrt{2}}\,t - \delta_2)\begin{pmatrix}1\\2\end{pmatrix}$.

(f) $\mathbf{u}(t) = \cos t\begin{pmatrix}\frac{6}{11}\\\frac{5}{11}\\\frac{1}{11}\end{pmatrix} + r_1\cos(\sqrt{12}\,t - \delta_1)\begin{pmatrix}1\\-1\\2\end{pmatrix} + r_2\cos(3t - \delta_2)\begin{pmatrix}-1\\1\\1\end{pmatrix} + r_3\cos(\sqrt{2}\,t - \delta_3)\begin{pmatrix}1\\1\\0\end{pmatrix}$.

9.6.19 (a) $\sqrt{\frac{3-\sqrt{3}}{2}} = .796225$, $\sqrt{\frac{3+\sqrt{3}}{2}} = 1.53819$.
(b) For example, a forcing function of the form $\cos\left(\sqrt{\frac{3+\sqrt{3}}{2}}\,t\right)\begin{pmatrix} w_1 \\ w_2 \end{pmatrix}$ where $w_2 \neq (1+\sqrt{3})w_1$, will excite resonance.

9.6.21 (a) Resonant frequencies: $\omega_1 = .5412$, $\omega_2 = 1.1371$, $\omega_3 = 1.3066$, $\omega_4 = 1.6453$.
Eigenvectors: $\mathbf{v}_1 = \begin{pmatrix} .6533 \\ .2706 \\ .6533 \\ -.2706 \end{pmatrix}$, $\mathbf{v}_2 = \begin{pmatrix} .2706 \\ .6533 \\ -.2706 \\ .6533 \end{pmatrix}$, $\mathbf{v}_3 = \begin{pmatrix} .2706 \\ -.6533 \\ .2706 \\ .6533 \end{pmatrix}$, $\mathbf{v}_4 = \begin{pmatrix} -.6533 \\ .2706 \\ .6533 \\ .2706 \end{pmatrix}$. No unstable modes.
(c) Resonant frequencies: $\omega_1 = .3542$, $\omega_2 = .9727$, $\omega_3 = 1.0279$, $\omega_4 = 1.6894$, $\omega_5 = 1.7372$.
Eigenvectors: $\mathbf{v}_1 = \begin{pmatrix} -.0989 \\ -.0706 \\ 0 \\ -.9851 \\ .0989 \\ -.0706 \end{pmatrix}$, $\mathbf{v}_2 = \begin{pmatrix} -.1160 \\ .6780 \\ .2319 \\ 0 \\ -.1160 \\ -.6780 \end{pmatrix}$, $\mathbf{v}_3 = \begin{pmatrix} .1251 \\ -.6940 \\ 0 \\ .0744 \\ -.1251 \\ -.6940 \end{pmatrix}$, $\mathbf{v}_4 = \begin{pmatrix} .3914 \\ .2009 \\ -.7829 \\ 0 \\ .3914 \\ -.2009 \end{pmatrix}$,
$\mathbf{v}_5 = \begin{pmatrix} .6889 \\ .1158 \\ 0 \\ -.1549 \\ -.6889 \\ .1158 \end{pmatrix}$. Unstable mode: $\mathbf{z} = (1,0,1,0,1,0)^T$.
(e) Resonant frequencies: $\omega_1 = 1$, $\omega_2 = \sqrt{3} = 1.73205$. Eigenvectors: $\mathbf{v}_1 = \begin{pmatrix} 1 \\ 0 \\ -1 \end{pmatrix}$, $\mathbf{v}_2 = \begin{pmatrix} 1 \\ -2 \\ 1 \end{pmatrix}$. Unstable mode: $\mathbf{z} = \begin{pmatrix} 1 \\ 1 \\ 1 \end{pmatrix}$.

Chapter 10

10.1.1 (a) $u^{(1)} = 2$, $u^{(10)} = 1024$ and $u^{(20)} = 1048576$; unstable.
(c) $u^{(1)} = \mathrm{i}$, $u^{(10)} = -1$ and $u^{(20)} = 1$; stable.

10.1.2 (a) $u^{(k+1)} = 1.0325\,u^{(k)}$, $u^{(0)} = 100$, where $u^{(k)}$ represents the balance after k years.
(b) $u^{(10)} = 1.0325^{10} \times 100 = 137.69$ dollars.

10.1.5 Since $u(t) = a\,e^{\alpha t}$ we have $u^{(k+1)} = u((k+1)h) = a\,e^{\alpha(k+1)h} = e^{\alpha h}\left(a\,e^{\alpha k h}\right) = e^{\alpha h}\,u^{(k)}$, and so $\lambda = e^{\alpha h}$. The stability properties are the same: $|\alpha| < 1$ for asymptotic stability; $|\alpha| \le 1$ for stability, $|\alpha| > 1$ for an unstable system.

10.1.10 $u^{(k+1)} = 1.05\,u^{(k)} + 120$, $u^{(0)} = 0$; $u^{(10)} = \$1,509.35$.

10.1.13 (a) $u^{(k)} = \dfrac{3^k + (-1)^k}{2}$, $v^{(k)} = \dfrac{-3^k + (-1)^k}{2}$
(c) $u^{(k)} = \dfrac{(\sqrt{5}+2)(3-\sqrt{5})^k + (\sqrt{5}-2)(3+\sqrt{5})^k}{2\sqrt{5}}$
$v^{(k)} = \dfrac{(3-\sqrt{5})^k - (3+\sqrt{5})^k}{2\sqrt{5}}$

10.1.14 (a) $\mathbf{u}^{(k)} = c_1(-1-\sqrt{2})^k\begin{pmatrix} -\sqrt{2} \\ 1 \end{pmatrix} + c_2(-1+\sqrt{2})^k\begin{pmatrix} \sqrt{2} \\ 1 \end{pmatrix}$;
(c) $\mathbf{u}^{(k)} = c_1\begin{pmatrix} 1 \\ 2 \\ 0 \end{pmatrix} + c_2(-2)^k\begin{pmatrix} 2 \\ 3 \\ 2 \end{pmatrix} + c_3(-3)^k\begin{pmatrix} 2 \\ 3 \\ 3 \end{pmatrix}$.

10.1.15 (b) 2, 1, 3, 4, 7, 11, 18.

10.1.18 (b) $\begin{pmatrix} 4 & 1 \\ -2 & 1 \end{pmatrix}^k = \begin{pmatrix} -1 & -1 \\ 1 & 2 \end{pmatrix}\begin{pmatrix} 3^k & 0 \\ 0 & 2^k \end{pmatrix}\begin{pmatrix} -2 & -1 \\ 1 & 1 \end{pmatrix}$,
(d) $\begin{pmatrix} 1 & 1 & 2 \\ 1 & 2 & 1 \\ 2 & 1 & 1 \end{pmatrix}^k = \begin{pmatrix} 1 & 1 & -1 \\ 1 & -2 & 0 \\ 1 & 1 & 1 \end{pmatrix}\begin{pmatrix} 4^k & 0 & 0 \\ 0 & 1 & 0 \\ 0 & 0 & (-1)^k \end{pmatrix}\begin{pmatrix} \frac{1}{3} & \frac{1}{3} & \frac{1}{3} \\ \frac{1}{6} & -\frac{1}{3} & \frac{1}{6} \\ -\frac{1}{2} & 0 & \frac{1}{2} \end{pmatrix}$.

10.1.19 (b) $\begin{pmatrix} u^{(k)} \\ v^{(k)} \end{pmatrix} = \begin{pmatrix} -1 & -1 \\ 1 & 2 \end{pmatrix}\begin{pmatrix} 3^k \\ -2^{k+1} \end{pmatrix}$,
(d) $\begin{pmatrix} u^{(k)} \\ v^{(k)} \\ w^{(k)} \end{pmatrix} = \begin{pmatrix} 1 & 1 & -1 \\ 1 & -2 & 0 \\ 1 & 1 & 1 \end{pmatrix}\begin{pmatrix} \frac{2}{3}4^k \\ \frac{1}{3} \\ 0 \end{pmatrix}$,

10.1.22 (a) $u^{(k)} = \frac{4}{3} - \frac{1}{3}(-2)^k$,
(c) $u^{(k)} = \dfrac{(5-3\sqrt{5})(2+\sqrt{5})^k + (5+3\sqrt{5})(2-\sqrt{5})^k}{10}$,
(e) $u^{(k)} = -\frac{1}{2} - \frac{1}{2}(-1)^k + 2^k$.

10.1.24 (a) $u^{(k)} = u^{(k-1)} + u^{(k-2)} - u^{(k-8)}$.
(b) 0, 1, 1, 2, 3, 5, 8, 13, 21, 33, 53, ...
(d) 1.59^k.

10.1.27 (a)

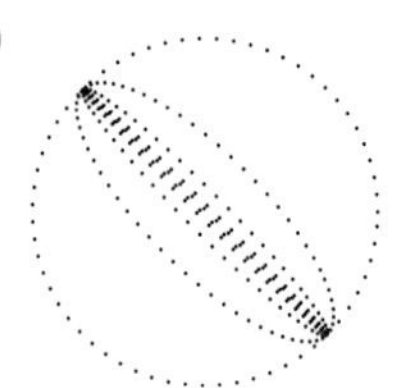

E_1: principal axes: $\begin{pmatrix} -1 \\ 1 \end{pmatrix}, \begin{pmatrix} 1 \\ 1 \end{pmatrix}$, semi-axes: $1, \frac{1}{3}$, area: $\frac{1}{3}\pi$.

E_2: principal axes: $\begin{pmatrix} -1 \\ 1 \end{pmatrix}, \begin{pmatrix} 1 \\ 1 \end{pmatrix}$, semi-axes: $1, \frac{1}{9}$, area: $\frac{1}{9}\pi$.

E_3: principal axes: $\begin{pmatrix} -1 \\ 1 \end{pmatrix}, \begin{pmatrix} 1 \\ 1 \end{pmatrix}$, semi-axes: $1, \frac{1}{27}$, area: $\frac{1}{27}\pi$.

E_4: principal axes: $\begin{pmatrix} -1 \\ 1 \end{pmatrix}, \begin{pmatrix} 1 \\ 1 \end{pmatrix}$, semi-axes: $1, \frac{1}{81}$, area: $\frac{1}{81}\pi$.

10.1.28 (b) True. (d) True.

10.1.30 $\mathbf{v}^{(k)} = c_1 (\alpha\lambda_1 + \beta)^k \mathbf{v}_1 + \cdots + c_n (\alpha\lambda_n + \beta)^k \mathbf{v}_n$.

10.1.37 (a) $u^{(k)} = 2^k\big(c_1 + \frac{1}{2}k c_2\big)$, $v^{(k)} = \frac{1}{3}2^k c_2$;
(c) $u^{(k)} = (-1)^k\big(c_1 - k c_2 + \frac{1}{2}k(k-1)c_3\big)$, $v^{(k)} = (-1)^k\big(c_2 - (k+1)c_3\big)$, $w^{(k)} = (-1)^k c_3$;
(e) $u^{(0)} = -c_2$, $v^{(0)} = -c_1 + c_3$, $w^{(0)} = c_1 + c_2$, while, for $k > 0$, $u^{(k)} = -2^k\big(c_2 + \frac{1}{2}k c_3\big)$, $v^{(k)} = 2^k c_3$, $w^{(k)} = 2^k\big(c_2 + \frac{1}{2}k c_3\big)$.

10.1.40 (c) (i) $\mathbf{u}^{(k)} = \begin{pmatrix} \frac{2}{3} \\ -1 \end{pmatrix} - 5^k\begin{pmatrix} -3 \\ 1 \end{pmatrix} - (-3)^k\begin{pmatrix} -\frac{1}{3} \\ 1 \end{pmatrix}$;

(iii) $\mathbf{u}^{(k)} = \begin{pmatrix} -1 \\ -\frac{3}{2} \\ -1 \end{pmatrix} - 3\begin{pmatrix} 1 \\ 2 \\ 0 \end{pmatrix} + \frac{15}{2}(-2)^k\begin{pmatrix} 2 \\ 3 \\ 2 \end{pmatrix} - 5(-3)^k\begin{pmatrix} 2 \\ 3 \\ 3 \end{pmatrix}$.

10.2.1 (a) $\frac{5+\sqrt{33}}{2} \approx 5.3723$, (c) 2.

10.2.2 (a) Not convergent; (c) convergent.

10.2.3 (b) Unstable; (d) stable; (f) unstable.

10.2.5 (b) If $\mathbf{u}^{(0)} = \begin{pmatrix} a \\ b \end{pmatrix}$, then $\mathbf{u}^{(k)} = \begin{pmatrix} a + k b \\ b \end{pmatrix} \to \infty$ provided $b \neq 0$.

10.2.7 cA is convergent if and only if $|c| < 1/\rho(A)$.

10.2.12 The eigenvalues of $A^T A$ are $\lambda_i = \sigma_i^2$, and so the spectral radius of $A^T A$ is equal to $\rho(A^T A) = \max\{\sigma_1^2, \ldots, \sigma_n^2\}$. Thus $\rho(A^T A) = \lambda_1 < 1$ if and only if $\sigma_1 = \sqrt{\lambda_1} < 1$.

10.2.13 (a) $\rho(M_n) = 2\cos\frac{\pi}{n+1}$. (b) No.
(c) The entries of $\mathbf{u}^{(k)}$ are $u_i^{(k)} = \sum_{j=1}^n c_j \left(2\cos\frac{j\pi}{n+1}\right)^k \sin\frac{i j \pi}{n+1}$, $i = 1, \ldots, n$.

10.2.16 (a) False. (c) True. (e) False.

10.2.18 (a) False.

10.2.20 If $\mathbf{v}$ has integer entries, so does $A^k\mathbf{v}$ for any k, and so the only way in which $A^k\mathbf{v} \to \mathbf{0}$ is if $A^k\mathbf{v} = \mathbf{0}$ for some k. Let k_i be such that $A^{k_i}\mathbf{e}_i = \mathbf{0}$, and $k = \max\{k_1, \ldots, k_n\}$, so $A^k\mathbf{e}_i = \mathbf{0}$ for all $i = 1, \ldots, n$. Then $A^k \mathrm{I} = A^k = \mathrm{O}$, and hence A is nilpotent. The simplest example is $\begin{pmatrix} 0 & 1 \\ 0 & 0 \end{pmatrix}$.

10.2.25 (a) All scalar multiples of $(1, 1)^T$;
(c) all scalar multiples of $(-1, -2, 1)^T$.

10.2.26 (a) The fixed points are stable, while all other solutions go to a unique fixed point at rate $\left(\frac{1}{2}\right)^k$. When $\mathbf{u}^{(0)} = (1, 0)^T$, then $\mathbf{u}^{(k)} \to \left(\frac{3}{5}, \frac{3}{5}\right)^T$.
(c) The fixed points are unstable. Most solutions, specifically those with a nonzero component in the dominant eigenvector direction, become unbounded. However, when $\mathbf{u}^{(0)} = (1, 0, 0)^T$, then $\mathbf{u}^{(k)} = (-1, -2, 1)^T$ for $k \geq 1$, and the solution stays at a fixed point.

10.2.28 False.

10.3.1 (a) $\frac{3}{4}$, convergent; (c) $\frac{8}{7}$, inconclusive;
(e) $\frac{8}{7}$, inconclusive; (g) $\frac{7}{3}$, inconclusive.

10.3.2 (a) .6719, convergent; (c) .9755, convergent;
(e) 1.1066, inconclusive; (g) 2.0343, inconclusive.

10.3.3 (a) $\frac{2}{3}$, convergent; (c) .9755, convergent;
(e) .9437, convergent; (g) $\frac{2}{3}$, convergent.

10.3.5 (a) $A = \begin{pmatrix} 1 & 1 \\ 0 & 1 \end{pmatrix}$.

10.3.8 True.

10.3.10 (a) False. (c) True.

10.3.12 (ii) (a) $\frac{5}{6}$, convergent; (c) $\frac{8}{7}$, inconclusive;
(e) $\frac{12}{7}$, inconclusive; (g) $\frac{7}{3}$, inconclusive.

10.3.15 (a) $\frac{7}{2}$; (c) 2.3544.

10.3.22 (a) Gerschgorin disk: $|z - 1| \leq 2$; eigenvalues: $3, -1$;

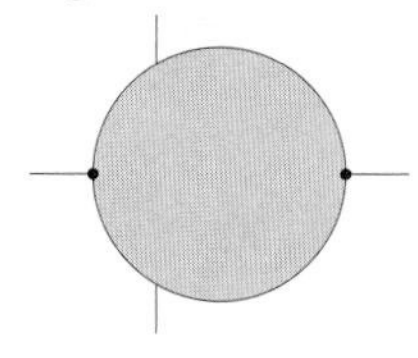

(c) Gerschgorin disks: $|z - 2| \leq 3$, $|z| \leq 1$; eigenvalues: $1 \pm \mathrm{i}\sqrt{2}$;

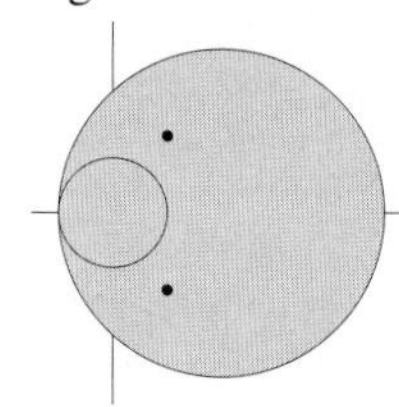

(e) Gerschgorin disks: $|z+1| \le 2$, $|z-2| \le 3$, $|z+4| \le 3$;
eigenvalues: $-2.69805 \pm .806289$, 2.3961;

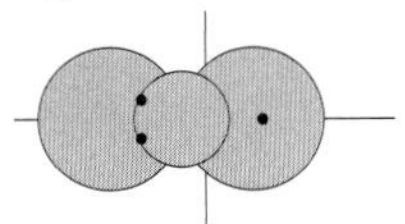

(g) Gerschgorin disks: $|z| \le 1$, $|z-1| \le 1$;
eigenvalues: $0,\ 1 \pm \mathrm{i}$;

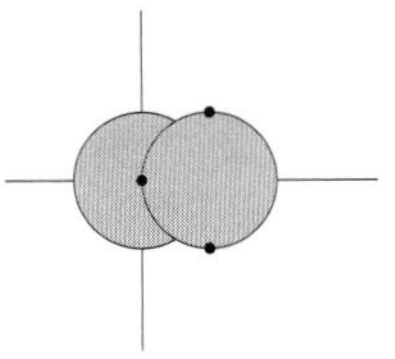

10.3.24 (ii) (a) Gerschgorin disk: $|z-1| \le 2$;
eigenvalues: $3, -1$;

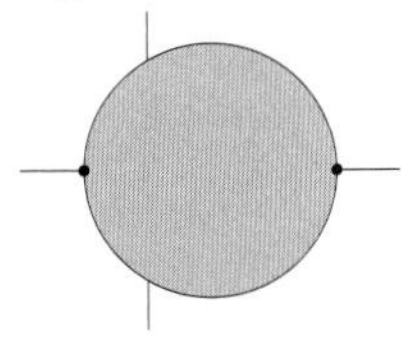

(c) Gerschgorin disks: $|z-2| \le 1$, $|z| \le 3$;
eigenvalues: $1 \pm \mathrm{i}\sqrt{2}$;

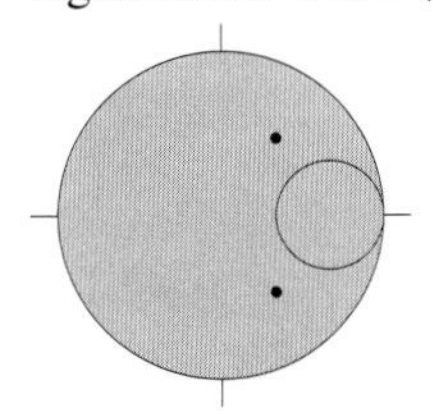

(e) Gerschgorin disks: $|z+1| \le 2$, $|z-2| \le 4$, $|z+4| \le 2$; eigenvalues:
$-2.69805 \pm .806289$, 2.3961;

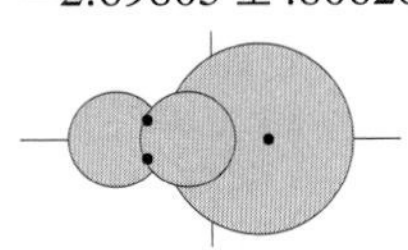

(g) Gerschgorin disks: $z = 0$, $|z-1| \le 2$, $|z-1| \le 1$;
eigenvalues: $0,\ 1 \pm \mathrm{i}$;

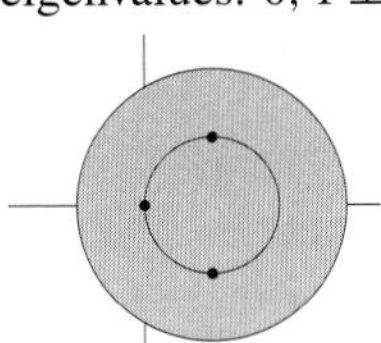

10.3.26 (b) $\begin{pmatrix} \frac12 & \frac12 \\ \frac12 & \frac12 \end{pmatrix}$.

10.3.28 $\begin{pmatrix} \frac12 & 0 \\ 0 & \frac12 \end{pmatrix}$.

10.3.29 (a) False.

10.3.31 (a) $A = \begin{pmatrix} 0 & 1 \\ 1 & 0 \end{pmatrix}$.

10.4.1 (b) Not a transition matrix;
(d) regular transition matrix: $\left(\frac16, \frac56\right)^T$;
(f) regular transition matrix: $\left(\frac13, \frac13, \frac13\right)^T$;
(h) not a transition matrix;
(j) not a regular transition matrix;
(l) regular transition matrix:
$(.2507, .2248, .2348, .2897)^T$.

10.4.3 (a) 20.5%;
(b) 9.76% farmers, 26.83% laborers, 63.41% professionals

10.4.5 58.33% of the nights.

10.4.7 The transition matrix is $T = \begin{pmatrix} 0 & \frac23 & \frac23 \\ 1 & 0 & \frac13 \\ 0 & \frac13 & 0 \end{pmatrix}$, and

$T^4 = \begin{pmatrix} \frac{14}{27} & \frac{26}{81} & \frac{26}{81} \\ \frac29 & \frac{49}{81} & \frac{16}{27} \\ \frac{7}{27} & \frac{2}{27} & \frac{7}{81} \end{pmatrix}$ has all positive entries. She visits branch A 40% of the time, branch B 45% and branch C: 15%.

10.4.10 Numbering the vertices from top to bottom and left to right, the transition matrix is

$$T = \begin{pmatrix} 0 & \frac14 & \frac14 & 0 & 0 & 0 \\ \frac12 & 0 & \frac14 & \frac12 & \frac14 & 0 \\ \frac12 & \frac14 & 0 & 0 & \frac14 & \frac12 \\ 0 & \frac14 & 0 & 0 & \frac14 & 0 \\ 0 & \frac14 & \frac14 & \frac12 & 0 & \frac12 \\ 0 & 0 & \frac14 & 0 & \frac14 & 0 \end{pmatrix}.$$ The probability eigenvector is $\begin{pmatrix} \frac19 \\ \frac29 \\ \frac29 \\ \frac19 \\ \frac29 \\ \frac19 \end{pmatrix}$.

10.4.14 $\begin{pmatrix} \frac13 & \frac13 & \frac13 \\ \frac13 & \frac13 & \frac13 \\ \frac13 & \frac13 & \frac13 \end{pmatrix}$.

10.4.16 All equal probabilities: $\mathbf{z} = (1/n, \ldots, 1/n)^T$.

10.4.18 False.

10.4.20 False.

10.4.22 True.

10.4.23 (a) $\begin{pmatrix} 0 & 1 \\ 1 & 0 \end{pmatrix}$; (b) $\begin{pmatrix} 0 & \frac{1}{2} \\ 1 & \frac{1}{2} \end{pmatrix}$.

10.5.1 (a) $\rho(T) = \frac{1}{2}$.
(b) The iterates will converge to the fixed point $\left(-\frac{1}{6}, 1\right)^T$ at rate $\frac{1}{2}$, asymptotically along the direction of the dominant eigenvector $(-3, 2)^T$.

10.5.2 (b) $\rho(T) = \frac{3}{4}$. The iterates converge to the fixed point $(1.6, .8, 7.2)^T$ at a rate $\frac{3}{4}$, along the dominant eigenvector direction $(1, 2, 6)^T$.

10.5.3 (a) Diagonally dominant;
(c) not diagonally dominant;
(e) diagonally dominant.

10.5.4 (a) $x = .142857, y = -.285714$;
(e) $x = -1.9172$, $y = -.339703$, $z = -2.24204$.

10.5.5 (c) Jacobi spectral radius = .547723, so Jacobi converges to the solution $x = \frac{8}{7} = 1.142857$, $y = \frac{19}{7} = 2.71429$.

10.5.6 (a) $(.7857, .3571)^T$,
(c) $(.3333, -1.0000, 1.3333)^T$,
(e) $(.8750, -.1250, -.1250, -.1250)^T$.

10.5.11 False for elementary row operations of types 1 & 2, but true for those of type 3.

10.5.13 (a) $x = .142857$, $y = -.285714$;
(e) $x = -1.9172$, $y = -.339703$, $z = -2.24204$.

10.5.14 (a) $\rho_J = .2582$, $\rho_{GS} = .0667$; Gauss–Seidel converges faster.
(c) $\rho_J = .5477$, $\rho_{GS} = .3$; Gauss–Seidel converges faster.
(e) $\rho_J = .4541$, $\rho_{GS} = .2887$; Gauss–Seidel converges faster.

10.5.15 (a) Solution: $\mathbf{u} = (.7857, .3571)^T$; spectral radii: $\rho_J = \frac{1}{\sqrt{15}} = .2582$, $\rho_{GS} = \frac{1}{15} = .06667$, so Gauss–Seidel converges twice as fast;
(c) Solution: $\mathbf{u} = (.3333, -1.0000, 1.3333)^T$; spectral radii: $\rho_J = .7291$, $\rho_{GS} = .3104$, so Gauss–Seidel converges 3.7019 times as fast;
(e) Solution: $\mathbf{u} = (.8750, -.1250, -.1250, -.1250)^T$; spectral radii: $\rho_J = .6$, $\rho_{GS} = .1416$, so Gauss–Seidel converges 3.8272 times as fast.

10.5.17 $x = .083799$, $y = .21648$, $z = 1.21508$; $\rho_J = .8166$, $\rho_{GS} = 1.0994$.

10.5.19 $\rho_J = 0$, $\rho_{GS} = 2$.

10.5.21 (a) $|z| > 4$;
(b) The solution is $\mathbf{u} = (.0115385, -.0294314, -.0755853, .0536789, .31505, .0541806, -.0767559, -.032107, .0140468, .0115385)^T$; it takes 41 Jacobi iterations and 6 Gauss–Seidel iterations to compute the first three decimal places.

10.5.25 (a) $x = .5$, $y = .75$, $z = .25$, $w = .5$.
(b) (i) 14 iterations, $\rho_J = .5$;
(ii) 8 iterations; $\rho_{GS} = .25$.
(c) $\omega = 1.0718$ requires 6 iterations; $\rho_{SOR} = .0718$.

10.5.26 (a) $\rho_J = \frac{1+\sqrt{5}}{4} = .809017$, $\rho_{GS} = \frac{3+\sqrt{5}}{8} = .654508$;
(b) no;
(c) $\omega_\star = 1.25962$ and $\rho_\star = .25962$;
(d) $\mathbf{x} = (.8, -.6, .4, -.2)^T$. Jacobi: prediction 44 iterations; actual 45 iterations. Gauss-Seidel: prediction 22 iterations; actual 22 iterations. Optimal SOR: prediction 7 iterations; actual 9 iterations.

10.5.31 (a) $\mathbf{u} = (.0625, .125, .0625, .125, .375, .125, .0625, .125, .0625)^T$.

10.5.33 The Jacobi spectral radius is $\rho_J = .909657$. Using (10.86) to fix the SOR parameter $\omega = 1.41307$ slows down the convergence since $\rho_{SOR} = .509584$ while $\rho_{GS} = .32373$. Computing the spectral radius directly, the optimal SOR parameter is $\omega_\star = 1.17157$ with $\rho_\star = .290435$.

10.5.37 (b) $\mathbf{r}_0 = \begin{pmatrix} 1 \\ 0 \\ -2 \end{pmatrix}$, $\mathbf{u}_1 = \begin{pmatrix} .5 \\ 0 \\ -1 \end{pmatrix}$, $\mathbf{r}_1 = \begin{pmatrix} -1 \\ -2 \\ -.5 \end{pmatrix}$,
$\mathbf{u}_2 = \begin{pmatrix} .51814 \\ -.72539 \\ -1.94301 \end{pmatrix}$, $\mathbf{r}_2 = \begin{pmatrix} 1.28497 \\ -.80311 \\ .64249 \end{pmatrix}$,
$\mathbf{u}_3 = \begin{pmatrix} 1. \\ -1.4 \\ -2.2 \end{pmatrix}$.

(d) $\mathbf{r}_0 = \begin{pmatrix} 1 \\ 2 \\ 0 \\ -1 \end{pmatrix}$, $\mathbf{u}_1 = \begin{pmatrix} .2 \\ .4 \\ 0 \\ -.2 \end{pmatrix}$, $\mathbf{r}_1 = \begin{pmatrix} 1.2 \\ -.8 \\ -.8 \\ -.4 \end{pmatrix}$,
$\mathbf{u}_2 = \begin{pmatrix} .90654 \\ .46729 \\ -.33645 \\ -.57009 \end{pmatrix}$, $\mathbf{r}_2 = \begin{pmatrix} -1.45794 \\ -.59813 \\ -.26168 \\ -2.65421 \end{pmatrix}$,
$\mathbf{u}_3 = \begin{pmatrix} 4.56612 \\ .40985 \\ -2.92409 \\ -5.50820 \end{pmatrix}$, $\mathbf{r}_3 = \begin{pmatrix} -1.36993 \\ 1.11307 \\ -3.59606 \\ .85621 \end{pmatrix}$,
$\mathbf{u}_4 = \begin{pmatrix} 9.50 \\ 1.25 \\ -10.25 \\ -13.00 \end{pmatrix}$.

10.5.41 False.

10.6.1 (a) After 17 iterations, $\lambda = 2.00002$, $\mathbf{u} = (-.55470, .83205)^T$;
(c) after 38 iterations, $\lambda = 3.99996$, $\mathbf{u} = (.57737, -.57735, .57734)^T$;
(e) after 36 iterations, $\lambda = 5.54911$, $\mathbf{u} = (-.39488, .71005, .58300)^T$;

(g) after 36 iterations, $\lambda = 3.61800$, $\mathbf{u} = (.37176, -.60151, .60150, -.37174)^T$.

10.6.3 (a) $\sigma_1 = 3.6180$; (c) $\sigma_1 = 3.4067$.

10.6.6 (a) After 15 iterations, $\lambda = .99998$, $\mathbf{u} = (.70711, -.70710)^T$; (c) after 12 iterations, $\lambda = 1.00001$, $\mathbf{u} = (.40825, .81650, .40825)^T$; (e) after 7 iterations, $\lambda = -.88536$, $\mathbf{u} = (-.88751, -.29939, .35027)^T$; (g) after 11 iterations, $\lambda = .38197$, $\mathbf{u} = (.37175, .60150, .60150, .37175)^T$.

10.6.8 (a) After 11 iterations, $\lambda^\star = 1.0000$, $\mathbf{u} = (.70711, -.70710)^T$; (c) after 10 iterations, $\lambda = 1.00000$, $\mathbf{u} = (.40825, .81650, .40825)^T$; (e) after 8 iterations, $\lambda = -.88537$, $\mathbf{u} = (.88753, .29937, -.35024)^T$; (g) after 9 iterations, $\lambda = .38197$, $\mathbf{u} = (-.37175, -.60150, -.60150, -.37175)^T$.

10.6.11 (a) eigenvalues: 6.7016, .2984; eigenvectors: $\begin{pmatrix} .3310 \\ .9436 \end{pmatrix}, \begin{pmatrix} .9436 \\ -.3310 \end{pmatrix}$.
(c) eigenvalues: 4.7577, 1.9009, −1.6586; eigenvectors: $\begin{pmatrix} .2726 \\ .7519 \\ .6003 \end{pmatrix}, \begin{pmatrix} .9454 \\ -.0937 \\ -.3120 \end{pmatrix}, \begin{pmatrix} -.1784 \\ .6526 \\ -.7364 \end{pmatrix}$.
(e) eigenvalues: 4.6180, 3.6180, 2.3820, 1.3820; eigenvectors: $\begin{pmatrix} -.3717 \\ .6015 \\ -.6015 \\ .3717 \end{pmatrix}, \begin{pmatrix} -.6015 \\ .3717 \\ .3717 \\ -.6015 \end{pmatrix}, \begin{pmatrix} -.6015 \\ -.3717 \\ .3717 \\ .6015 \end{pmatrix}, \begin{pmatrix} .3717 \\ .6015 \\ .6015 \\ .3717 \end{pmatrix}$.

10.6.13 (a) eigenvalues: 2, 1; eigenvectors: $\begin{pmatrix} -2 \\ 3 \end{pmatrix}, \begin{pmatrix} -1 \\ 1 \end{pmatrix}$.
(c) eigenvalues: 3.5842, −2.2899, 1.7057; eigenvectors: $\begin{pmatrix} -.4466 \\ -.7076 \\ .5476 \end{pmatrix}, \begin{pmatrix} .1953 \\ -.8380 \\ -.5094 \end{pmatrix}, \begin{pmatrix} .7491 \\ -.2204 \\ .6247 \end{pmatrix}$.
(e) eigenvalues: 18.3344, 4.2737, 0, −1.6081; eigenvectors: $\begin{pmatrix} .4136 \\ .8289 \\ .2588 \\ .2734 \end{pmatrix}, \begin{pmatrix} -.4183 \\ .9016 \\ -.0957 \\ .0545 \end{pmatrix}, \begin{pmatrix} -.5774 \\ -.5774 \\ .5774 \\ 0 \end{pmatrix}, \begin{pmatrix} -.2057 \\ .4632 \\ -.6168 \\ .6022 \end{pmatrix}$.

10.6.18 (b) $H_1 = \begin{pmatrix} 1 & 0 & 0 & 0 \\ 0 & -.4082 & .8165 & -.4082 \\ 0 & .8165 & .5266 & .2367 \\ 0 & -.4082 & .2367 & .8816 \end{pmatrix}$,
$T_1 = H_1 A H_1 = \begin{pmatrix} 5.0000 & -2.4495 & 0 & 0 \\ -2.4495 & 3.8333 & 1.3865 & .9397 \\ 0 & 1.3865 & 6.2801 & -.9566 \\ 0 & .9397 & -.9566 & 6.8865 \end{pmatrix}$,
$H_2 = \begin{pmatrix} 1 & 0 & 0 & 0 \\ 0 & 1 & 0 & 0 \\ 0 & 0 & -.8278 & -.5610 \\ 0 & 0 & -.5610 & .8278 \end{pmatrix}$,
$T = H_2 T_1 H_2 = \begin{pmatrix} 5.0000 & -2.4495 & 0 & 0 \\ -2.4495 & 3.8333 & -1.6750 & 0 \\ 0 & -1.6750 & 5.5825 & .0728 \\ 0 & 0 & .0728 & 7.5842 \end{pmatrix}$.

10.6.19 (b) eigenvalues: 7.6180, 7.5414, 5.3820, 1.4586.

10.6.21 (b) $H_1 = \begin{pmatrix} 1 & 0 & 0 & 0 \\ 0 & -.8944 & 0 & -.4472 \\ 0 & 0 & 1 & 0 \\ 0 & -.4472 & 0 & .8944 \end{pmatrix}$, $A_1 = \begin{pmatrix} 3.0000 & -2.2361 & -1.0000 & 0 \\ -2.2361 & 3.8000 & 2.2361 & .4000 \\ 0 & 1.7889 & 2.0000 & -5.8138 \\ 0 & 1.4000 & -4.4721 & 1.2000 \end{pmatrix}$,
$H_2 = \begin{pmatrix} 1 & 0 & 0 & 0 \\ 0 & 1 & 0 & 0 \\ 0 & 0 & -.7875 & -.6163 \\ 0 & 0 & -.6163 & .7875 \end{pmatrix}$, $A_2 = \begin{pmatrix} 3.0000 & -2.2361 & .7875 & .6163 \\ -2.2361 & 3.8000 & -2.0074 & -1.0631 \\ 0 & -2.2716 & -3.2961 & 2.2950 \\ 0 & 0 & .9534 & 6.4961 \end{pmatrix}$.

Chapter 11

11.1.2 $u(x) = \begin{cases} \frac14 x - \frac12 x^2, & 0 \le x \le \frac12, \\ \frac14 - \frac34 x + \frac12 x^2, & \frac12 \le x \le 1, \end{cases}$
$v(x) = \begin{cases} \frac14 - x, & 0 \le x \le \frac12, \\ x - \frac34, & \frac12 \le x \le 1. \end{cases}$

11.1.4 $u(x) = \dfrac{\log(x+1)}{\log 2} - x$,
$v(x) = u'(x) = \dfrac{1}{\log 2\,(x+1)} - 1$,
$w(x) = (1+x)v(x) = \dfrac{1}{\log 2} - 1 - x$. Maximum displacement at $x = 1/\log 2 - 1 \approx .4427$. The bar will break at the point of maximum stress, $x = 0$.

11.1.7 $u(x) = x^3 - \frac32 x^2 - x$, $v(x) = 3x^2 - 3x - 1$. Maximum (absolute) displacement at

$x = 1 - \frac{1}{3}\sqrt{3},\ 1 + \frac{1}{3}\sqrt{3}$; maximal stress at $x = 0$ and 2, so either end is most likely to break.

11.1.10 $-((1-x)u')' = 1,\ u(0) = 0,$ $w(1) = \lim_{x\to 1}(1-x)u'(x) = 0$; the solution is $u = x$.

11.2.1 (a) 1, (c) e, (e) $\frac{1}{9}$.

11.2.2 (a) $\varphi(x) = \delta(x)$; $\int_a^b \varphi(x)\,u(x)\,dx = u(0)$ for $a < 0 < b$.
(c) $\varphi(x) = 3\,\delta(x-1) + 3\,\delta(x+1)$;
$\int_a^b \varphi(x)\,u(x)\,dx = 3u(1) + 3u(-1)$ for $a < -1 < 1 < b$.
(e) $\varphi(x) = \delta(x) - \delta(x-\pi) - \delta(x+\pi)$;
$\int_a^b \varphi(x)\,u(x)\,dx = u(0) - u(\pi) - u(-\pi)$ for $a < -\pi < \pi < b$.

11.2.4 (a) $\varphi(x)$
$$= \lim_{n\to\infty}\left[\frac{n}{\pi\left(1+n^2x^2\right)} - \frac{3n}{\pi\left(1+n^2(x-1)^2\right)}\right];$$
(b) $\int_a^b \varphi(x)\,u(x)\,dx = u(0) - 3u(1)$, for $a < 0 < 1 < b$.

11.2.6 (a) $f'(x) = -\delta(x+1) - 9\,\delta(x-3)$
$$+\begin{cases} 2x, & 0 < x < 3,\\ 1, & -1 < x < 0,\\ 0, & \text{otherwise.}\end{cases}$$
(c) $h'(x) = -e^{-1}\,\delta(x+1)$
$$+\begin{cases} \pi\cos\pi x, & x > 1,\\ -2x, & -1 < x < 1,\\ e^x, & x < -1.\end{cases}$$

11.2.7 (b) $k'(x) = 2\,\delta(x+2) - 2\,\delta(x-2)$
$$+\begin{cases} -1, & -2 < x < 0,\\ 1, & 0 < x < 2,\\ 0, & \text{otherwise,}\end{cases}$$
$= 2\,\delta(x+2) - 2\,\delta(x-2) - \sigma(x+2) + 2\,\sigma(x) - \sigma(x-2)$,
$k''(x) = 2\,\delta'(x+2) - 2\,\delta'(x-2)$
$-\,\delta(x+2) + 2\,\delta(x) - \delta(x-2)$.

11.2.8 (b) $f'(x) = \begin{cases} -1 & x < 0,\\ 3, & 0 < x < 1,\\ 1, & x > 1,\end{cases}$
$= -1 + 4\,\sigma(x) - 2\,\sigma(x-1)$,
$f''(x) = 4\,\delta(x) - 2\,\delta(x-1)$.
(d) $f'(x) = 4\,\delta(x+2) - 4\,\delta(x-2)$
$$+\begin{cases} 1, & |x| > 2,\\ -1, & |x| < 2,\end{cases}$$
$= 4\,\delta(x+2) - 4\,\delta(x-2) + 1 - 2\,\sigma(x+2) + 2\,\sigma(x-2)$,
$f''(x) = 4\,\delta'(x+2) - 4\,\delta'(x-2)$
$-\,2\,\delta(x+2) + 2\,\delta(x-2)$.
(f) $f'(x) = \operatorname{sgn}(\sin x)\,\cos x$,
$f''(x) = -\,|\sin x| + 2\sum_{n=-\infty}^{\infty}\delta(x - n\pi)$.

11.2.10 $\displaystyle \lim_{n\to\infty}\frac{n}{\sqrt{\pi}}\,e^{-n^2x^2} = \begin{cases} 0, & x \neq 0,\\ \infty, & x = 0\end{cases},$
$$\int_{-\infty}^{\infty}\frac{n}{\sqrt{\pi}}\,e^{-n^2x^2}\,dx = \int_{-\infty}^{\infty}\frac{1}{\sqrt{\pi}}\,e^{-y^2}\,dy = 1.$$

11.2.12 (b) $\lim_{n\to\infty} g_n(x) = 0$ for $x \neq 0$ since $g_n(x) = 0$ whenever $n > 1/|x|$. Moreover, $\int_{-\infty}^{\infty} g_n(x)\,dx = 1$, hence $\lim_{n\to\infty} g_n(x) = \delta(x)$.
(d) $h_n(x) = \frac{1}{2}n\,\delta\left(x + \frac{1}{n}\right) - \frac{1}{2}n\,\delta\left(x - \frac{1}{n}\right)$.

11.2.14 By duality, $\lim_{n\to\infty}\left\langle n\left[\delta\left(x - \frac{1}{n}\right) - \delta\left(x + \frac{1}{n}\right)\right], f\right\rangle =$ $\lim_{n\to\infty} n\left[f\left(\frac{1}{n}\right) - f\left(-\frac{1}{n}\right)\right] = 2\,f'(0) =$ $-2\,\langle\delta', f\rangle$, for any $f \in \mathrm{C}^1$.

11.2.18 $x\,\delta'(x) = -\,\delta(x)$ because they both yield the same value on a test function: $\langle x\,\delta', u\rangle =$ $\int_{-\infty}^{\infty} x\,\delta'(x)\,u(x)\,dx = -\left[x\,u(x)\right]'\Big|_{x=0} = u(0) =$ $\int_{-\infty}^{\infty}\delta(x)\,u(x)\,dx = \langle u, \delta\rangle$.

11.2.21 (a) $\varphi(x) = -\,2\,\delta'(x) - \delta(x)$,
$\int_{-\infty}^{\infty}\varphi(x)\,u(x)\,dx = 2u'(0) - u(0)$;
(c) $\chi(x) = \delta(x-1) - 4\,\delta'(x-2) + 4\,\delta(x-2)$,
$\int_{-\infty}^{\infty}\chi(x)\,u(x)\,dx = u(1) + 4u'(2) + 4u(2)$.

11.2.27 (a) $G(x,y) = \begin{cases} \frac{1}{8}x(4-y), & x \le y,\\ \frac{1}{8}y(4-x), & x \ge y.\end{cases}$

11.2.29 (a) $u(x) = \frac{9}{16}x - \frac{1}{2}x^2 + \frac{3}{16}x^3 - \frac{1}{4}x^4$,
$w(x) = \frac{9}{16} - x$;
(b) $G(x,y) =$
$$\begin{cases} \left(1 - \frac{3}{4}y - \frac{1}{4}y^3\right)\left(x + \frac{1}{3}x^3\right), & x < y,\\ \left(1 - \frac{3}{4}x - \frac{1}{4}x^3\right)\left(x + \frac{1}{3}x^3\right), & x > y.\end{cases}$$

11.2.31 (a) $u_n(x) =$
$$\begin{cases} x(y-1), & 0 \le x \le y - \frac{1}{n},\\ \frac{1}{4n} - \frac{1}{2}x + \frac{1}{4}nx^2 - \frac{1}{2}y \\ \quad +\left(1 - \frac{1}{2}n\right)xy + \frac{1}{4}ny^2, & |x - y| \le \frac{1}{n},\\ x(1-y), & y + \frac{1}{n} \le x \le 1.\end{cases}$$

11.3.1 (a) $u_\star(x) = \frac{5}{2}x - \frac{5}{2}x^2$.
(b) $\mathcal{P}[u_\star] = -\frac{25}{24} = -1.04167$;
(i) $\mathcal{P}[x - x^2] = -\frac{2}{3} = -.66667$;
(iii) $\mathcal{P}[\frac{2}{3}\sin\pi x] = -\frac{20}{3\pi} + \frac{1}{9}\pi^2 = -1.02544$.

11.3.3 (a) (i) $u_\star(x) = \frac{1}{18}x^6 + \frac{1}{12}x^4 - \frac{5}{36}$;
(ii) $\mathcal{P}[u] = \int_{-1}^{1}\left[\frac{(u')^2}{2(x^2+1)} + x^2u\right]dx$,
$u(-1) = u(1) = 0$;
(iii) $\mathcal{P}[u_\star] = -.0282187$;
(iv) $\mathcal{P}[-\frac{1}{5}(1 - x^2)] = -.018997$;
$\mathcal{P}[-\frac{1}{5}\cos\frac{1}{2}\pi x] = -.0150593$.
(c) (i) $u_\star(x) = \frac{5}{2} - x^{-1} - \frac{1}{2}x^2$;
(ii) $\mathcal{P}[u] = \int_1^2\left[\frac{1}{2}x^2(u')^2 - 3x^2u\right]dx\,\frac{5}{6} - \frac{1}{x} - \frac{1}{2}x^2$,
$u'(1) = u(2) = 0$.
(iii) $\mathcal{P}[u_\star] = -\frac{37}{20} = -1.85$;
(iv) $\mathcal{P}[2x - x^2] = \frac{11}{6} = -1.83333$;
$\mathcal{P}[\cos\frac{1}{2}\pi(x-1)] = -1.84534$.

11.3.4 (b) $-\left((x+1)u'\right)' = 5,\ u(0) = u(1) = 0$;
$u_\star(x) = \frac{5}{\log 2}\log(x+1) - 5x$.

(d) $-(e^x u')' = 1 + e^x$, $u(0) = u(1) = 0$;
$u_\star(x) = (x-1)e^{-x} - x + 1$.

11.3.5 (a) Unique minimizer:
$u(x) = \frac{1}{2}x^2 - 2x + \frac{3}{2} + \dfrac{\log x}{2\log 2}$.
(d) No minimizer since $1 - x^2$ is not positive for all $-2 < x < 2$.

11.3.12 No.

11.3.15 $K[u] = -\dfrac{1}{\rho(x)}\dfrac{d}{dx}\,c(x)\,\dfrac{du}{dx} = f(x)$, along with the selected boundary conditions.

11.3.17 (c) (ii) no integrating factor needed:
$-(e^x u')' = -e^{2x}$;
integrating factor $\rho(x) = x^2$, so $-(x^3 u')' = x^3$.

11.3.18 (a) $L^*[\mathbf{v}] = -v_1' + v_2$.
(b) $-u'' + u = x - 1$, $u(0) = u(1) = 0$, with solution $u(x) = x - 1 + \dfrac{e^{2-x} - e^x}{e^2 - 1}$.

11.3.22 (a) (i) $-u'' = -1$, $u(0) = 2$, $u(1) = 3$;
(ii) $u_\star(x) = \frac{1}{2}x^2 + \frac{1}{2}x + 2$.
(c) (i) $-\dfrac{d}{dx}\left(\dfrac{1}{1+x^2}\dfrac{du}{dx}\right) = 0$, $u(-1) = 1$, $u(1) = -1$; $u_\star(x) = -\frac{3}{4}x - \frac{1}{4}x^3$.

11.3.23 (a) (i) $\int_0^\pi \left[\frac{1}{2}(u')^2 - (\cos x)\,u\right]dx$, $u(0) = 1$, $u(\pi) = -2$.
(ii) $u_\star(x) = \cos x - x/\pi$;
(c) (i) $\int_0^1 \left[\frac{1}{2}e^x (u')^2 + u\right]dx$, $u(0) = 1$, $u(1) = 0$.
(ii) $u_\star(x) = e^{-x}(1-x)$.

11.3.25 (a) $-d/dx\,[(1+x)du/dx] = 1 - x$, $u(0) = 0$, $u(1) = .01$;
$u_\star(x) = .25x^2 - 1.5x + 1.8178\log(1+x)$;
(b) $\mathcal{P}[u]$
$= \displaystyle\int_0^1 \left(\frac{1}{2}(1+x)\left(\frac{du}{dx}\right)^2 - (1-x)\,u\right)dx$;
$\mathcal{P}[u_\star] = -.0102$.

11.4.1 (a) $u(x) = \frac{1}{24}x^4 - \frac{1}{12}x^3 + \frac{1}{24}x$,
$w(x) = u''(x) = \frac{1}{2}x^2 - \frac{1}{2}x$.
(c) $u(x) = \frac{1}{24}x^4 - \frac{1}{6}x^3 + \frac{1}{3}x$,
$w(x) = u''(x) = \frac{1}{2}x^2 - x$.

11.4.2 (a) Maximal displacement: $u\left(\frac{1}{2}\right) = .01302$; maximal stress: $w\left(\frac{1}{2}\right) = -.125$;
(c) Maximal displacement: $u(1) = .2083$; maximal stress: $w(1) = -.5$.

11.4.3 (a) Minimize $\mathcal{P}[u] = \int_0^1 \left[\frac{1}{2}u''(x)^2 - u(x)\right]dx$ for $u(0) = u''(0) = u(1) = u''(1) = 0$.
(c) Minimize $\mathcal{P}[u] = \int_0^1 \left[\frac{1}{2}u''(x)^2 - u(x)\right]dx$ for $u(0) = u''(0) = u'(1) = u'''(1) = 0$.

11.4.4 (a) (i) $G(x,y) =$
$$\begin{cases} \frac{1}{3}xy - \frac{1}{6}x^3 - \frac{1}{2}xy^2 + \frac{1}{6}x^3y + \frac{1}{6}xy^3, & x < y, \\ \frac{1}{3}xy - \frac{1}{2}x^2y - \frac{1}{6}y^3 + \frac{1}{6}x^3y + \frac{1}{6}xy^3, & x > y; \end{cases}$$
(iii) $u(x) = \int_0^x \left(\frac{1}{3}xy - \frac{1}{2}x^2y - \frac{1}{6}y^3 + \frac{1}{6}x^3y + \frac{1}{6}xy^3\right)f(y)\,dy$
$+ \int_x^1 \left(\frac{1}{3}xy - \frac{1}{6}x^3 - \frac{1}{2}xy^2 + \frac{1}{6}x^3y + \frac{1}{6}xy^3\right)f(y)\,dy$;
(iv) $x = \frac{1}{2}$.
(c) (i) $G(x,y) = \begin{cases} xy - \frac{1}{6}x^3 - \frac{1}{2}xy^2, & x < y, \\ xy - \frac{1}{2}x^2y - \frac{1}{6}y^3, & x > y; \end{cases}$
(iii) $u(x) = \int_0^x (xy - \frac{1}{6}x^3 - \frac{1}{2}xy^2)f(y)\,dy + \int_x^1 (xy - \frac{1}{6}x^3 - \frac{1}{2}xy^2)f(y)\,dy$;
(iv) $x = 1$.

11.4.6 $v(x) = \dfrac{x^2 - x}{2(1+x^2)}$; statically determinate.

11.4.7 (a) $\int_0^1 f(y)\,dy = 0$, $\int_0^1\left(\int_0^z f(y)\,dy\right)dz = 0$.
(b) $\langle f, 1\rangle = \int_0^1 f(x)\,dx = 0$,
$\langle f, x\rangle = \int_0^1 x f(x)\,dx = 0$.
(c) $f(x) = x^2 - x + \frac{1}{6}$ satisfies the constraints; the corresponding solution is
$u(x) = \frac{1}{360}x^6 - \frac{1}{120}x^5 + \frac{1}{144}x^4 + cx + d$, where c, d are arbitrary constants.

11.4.9 False.

11.4.11 (a) Fixed plus any other boundary condition, or simply supported plus any other except free.

11.4.12 (a) Simply supported end with right end raised 1 unit; $u(x) = x$.
(c) Left end is clamped at a 45° angle, right end is free with an induced stress; $u(x) = x + \frac{1}{2}x^2$.

11.4.13 (a) $\mathcal{P}[u] = \int_0^1 \frac{1}{2}u''(x)^2\,dx$,
(c) $\mathcal{P}[u] = u'(1) + \int_0^1 \frac{1}{2}u''(x)^2\,dx$.

11.4.14 (a) $u(x) = \begin{cases} -1.25(x+1)^3 + 4.25(x+1) - 2, & -1 \le x \le 0, \\ 1.25x^3 - 3.75x^2 + .5x - 1, & 0 \le x \le 1. \end{cases}$

(c) $u(x) = \begin{cases} \frac{2}{3}(x-1)^3 - \frac{11}{3}(x-1) + 3, & 1 \le x \le 2, \\ -\frac{1}{3}(x-2)^3 + 2(x-2)^2 - \frac{5}{3}(x-2), & 2 \le x \le 4. \end{cases}$

11.4.16 (a) $u(x) = \begin{cases} x^3 - 2x^2 + 1, & 0 \le x \le 1, \\ (x-1)^2 - (x-1), & 1 \le x \le 2, \\ -(x-2)^3 + (x-2)^2 + (x-2), & 2 \le x \le 3. \end{cases}$

(c) $u(x) = \begin{cases} \frac{5}{4}x^3 - \frac{9}{4}x^2 + 1, & 0 \le x \le 1, \\ -\frac{3}{4}(x-1)^3 + \frac{3}{2}(x-1)^2 - \frac{3}{4}(x-1), & 1 \le x \le 2, \\ \frac{3}{4}(x-2)^3 - \frac{3}{4}(x-2)^2, & 2 \le x \le 3, \\ -\frac{5}{4}(x-3)^3 + \frac{3}{2}(x-3)^2 + \frac{3}{4}(x-3), & 3 \le x \le 4. \end{cases}$

11.4.18 (a) $u(x) = \begin{cases} 2.2718x - 4.3490x^3, & 0 \le x \le 1, \\ .5 + 1.4564\left(x - \frac{1}{4}\right) - 3.2618\left(x - \frac{1}{4}\right)^2 + 3.7164\left(x - \frac{1}{4}\right)^3, & \frac{1}{4} \le x \le \frac{9}{16}, \\ .75 + .5066\left(x - \frac{9}{16}\right) + .2224\left(x - \frac{9}{16}\right)^2 - .1694\left(x - \frac{9}{16}\right)^3, & \frac{9}{16} \le x \le 1. \end{cases}$

11.4.22 (a) $C_0(x) = \begin{cases} 1 - \frac{19}{15}x + \frac{4}{15}x^3, & 0 \le x \le 1, \\ -\frac{7}{15}(x-1) + \frac{4}{5}(x-1)^2 - \frac{1}{3}(x-1)^3, & 1 \le x \le 2, \\ \frac{2}{15}(x-2) - \frac{1}{5}(x-2)^2 + \frac{1}{15}(x-2)^3, & 2 \le x \le 3 \end{cases}$

$C_1(x) = \begin{cases} \frac{8}{5}x - \frac{3}{5}x^3, & 0 \le x \le 1, \\ 1 - \frac{1}{5}(x-1) - \frac{9}{5}(x-1)^2 + (x-1)^3, & 1 \le x \le 2, \\ -\frac{4}{5}(x-2) + \frac{6}{5}(x-2)^2 - \frac{2}{5}(x-2)^3, & 2 \le x \le 3 \end{cases}$

11.5.2 True.

11.5.4 (a) $G(x,y) = \begin{cases} \dfrac{\sinh \omega x \cosh \omega(1-y)}{\omega \cosh \omega}, & x < y, \\ \dfrac{\cosh \omega(1-x) \sinh \omega y}{\omega \cosh \omega}, & x > y. \end{cases}$

(b) $u(x) =$

$\begin{cases} \dfrac{1}{\omega^2} - \dfrac{(e^{\omega/2} - e^{-\omega/2} + e^{-\omega})e^{\omega x} + (e^{\omega} - e^{\omega/2} + e^{-\omega/2})e^{-\omega x}}{\omega^2(e^{\omega} + e^{-\omega})}, & x \le \frac{1}{2}, \\ -\dfrac{1}{\omega^2} + \dfrac{(e^{-\omega/2} - e^{-\omega} + e^{-3\omega/2})e^{\omega x} + (e^{3\omega/2} - e^{\omega} + e^{\omega/2})e^{-\omega x}}{\omega^2(e^{\omega} + e^{-\omega})}, & x \ge \frac{1}{2}. \end{cases}$

11.5.7 (b) $\lambda = -\omega^2 < 0$,

$G(x,y) = \begin{cases} -\dfrac{\sinh \omega(1-y) \sinh \omega x}{\omega \sinh \omega}, & x < y, \\ -\dfrac{\sinh \omega(1-x) \sinh \omega y}{\omega \sinh \omega}, & x > y. \end{cases}$

$\lambda = 0$, $G(x,y) = \begin{cases} x(y-1), & x < y, \\ y(x-1), & x > y. \end{cases}$

$\lambda = \omega^2 \ne n^2\pi^2 > 0$,

$G(x,y) = \begin{cases} \dfrac{\sin \omega(1-y) \sin \omega x}{\omega \sin \omega}, & x < y, \\ \dfrac{\sin \omega(1-x) \sin \omega y}{\omega \sin \omega}, & x > y. \end{cases}$

11.5.9 (b) (ii) $\mu(x) = x^{-4}$ yields $-\dfrac{d}{dx}\left(\dfrac{1}{x^2}\dfrac{du}{dx}\right) + \dfrac{3}{x^4}u = \dfrac{1}{x^4}$.

(c) (ii) Minimize $\mathcal{P}[u] = \displaystyle\int_a^b \left[\frac{u'(x)^2}{2x^2} + \frac{3u(x)^2}{2x^4} - \frac{u(x)}{x^4}\right] dx$ subject to $u(1) = u(2) = 0$.

11.5.11 $u(x,\varepsilon) = \dfrac{1}{\varepsilon^2} - \dfrac{(1 - e^{-\varepsilon})e^{\varepsilon x} + (e^{\varepsilon} - 1)e^{-\varepsilon x}}{\varepsilon^2(e^{\varepsilon} - e^{-\varepsilon})}$,

$\lim_{\varepsilon \to 0^+} u(x,\varepsilon) = \frac{1}{2}x - \frac{1}{2}x^2 = u_\star(x)$.

11.5.14 (a) $\sinh\alpha \cosh\beta + \cosh\alpha \sinh\beta = \frac{1}{4}(e^{\alpha} - e^{-\alpha})(e^{\beta} + e^{-\beta}) + \frac{1}{4}(e^{\alpha} + e^{-\alpha})(e^{\beta} - e^{-\beta}) = \frac{1}{2}(e^{\alpha+\beta} - e^{-\alpha-\beta}) = \sinh(\alpha + \beta)$.

11.6.2 (a) Solution: $u(x) = \frac{1}{4}x - \rho_2(x-1)$

$= \begin{cases} \frac{1}{4}x, & 0 \le x \le 1, \\ \frac{1}{4}x - \frac{1}{2}(x-1)^2, & 1 \le x \le 2; \end{cases}$

finite element sample values: (0., .06, .12, .18, .24, .3, .32, .3, .24, .14, 0.); maximal error at sample points .05; maximal overall error: .05.

(c) Solution: $u(x) = \frac{1}{2}x - 2 + \frac{3}{2}x^{-1}$;

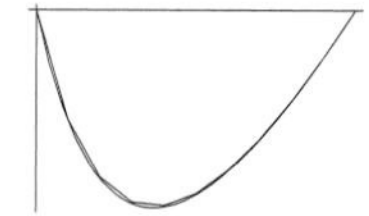

finite element sample values: (0., −.1482, −.2264, −.2604, −.2648, −.2485, −.2170, −.1742, −.1225, −.0640, 0.); maximal error at sample points .002175; maximal overall error: .01224.

11.6.5 (a) $u(x) = x + \dfrac{\pi e^x}{1 - e^{2\pi}} + \dfrac{\pi e^{-x}}{1 - e^{-2\pi}}$.

(b) Minimize $\mathcal{P}[u] = \int_0^{2\pi}\left[\frac{1}{2}u'(x)^2 + \frac{1}{2}u(x)^2 - x\,u(x)\right]dx$ for $u \in C^2$ with $u(0) = u(2\pi)$, $u'(0) = u'(2\pi)$.

(c) $\dim W_5 = 4$

(d) $n = 5$: maximal error: .9435;

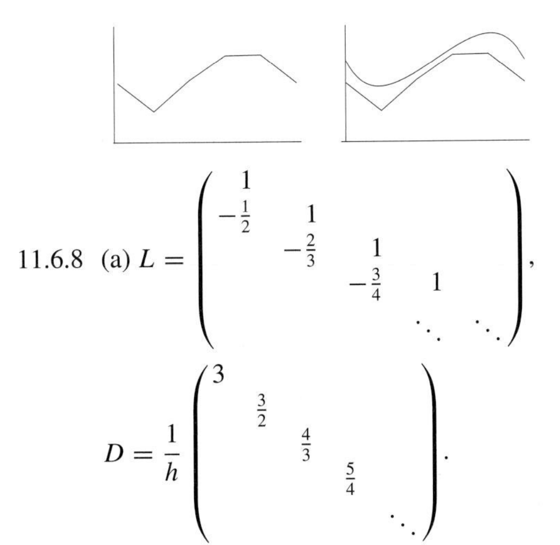

11.6.8 (a) $L = \begin{pmatrix} 1 & & & & \\ -\frac{1}{2} & 1 & & & \\ & -\frac{2}{3} & 1 & & \\ & & -\frac{3}{4} & 1 & \\ & & & \ddots & \ddots \end{pmatrix}$,

$$D = \frac{1}{h}\begin{pmatrix} 3 & & & & \\ & \frac{3}{2} & & & \\ & & \frac{4}{3} & & \\ & & & \frac{5}{4} & \\ & & & & \ddots \end{pmatrix}.$$

Since all pivots are positive, the matrix is positive definite.

11.6.12 (a) We define $f(x) = u_i + \dfrac{u_{i+1} - u_i}{x_{i+1} - x_i}(x - x_i)$ for $x_i \le x \le x_{i+1}$.

(c) $u(x) = 2\varphi_0(x) + 3\varphi_1(x) + 6\varphi_2(x) + 11\varphi_3(x)$

$$= \begin{cases} x + 2, & 0 \le x \le 1, \\ 3x, & 1 \le x \le 2, \\ 5x - 4, & 2 \le x \le 3. \end{cases}$$

Index

J

K

L

M

N

O

P

Q

R

S

T